GENETICS

GENETICS

Third Edition

Peter J. Russell

REED COLLEGE

HarperCollins*Publishers*

Sponsoring Editor: Glyn Davies
Development Editor: Kathleen Dolan
Project Coordination, Text and Cover Design: Proof Positive/Farrowlyne Associates, Inc.
Cover Art: Computer-generated image of natural DNA. Lawrence Berkeley
Laboratories, Courtesy University of California.
Photo Researcher: Cheryl Kucharzak
Production Manager: Michael Weinstein
Compositor: York Graphic Services, Inc.
Printer and Binder: Von Hoffmann Press, Inc.
Cover Printer: The Lehigh Press, Inc.

Genetics, Third Edition

Library of Congress Cataloging-in-Publication Data

Russell, Peter J.
  Genetics/Peter J. Russell.—3rd. ed.
     p.   cm.
  Includes bibliographical references and index.
  ISBN 0-673-52143-5 (student edition)
  ISBN 0-673-52202-4 (teacher edition)
  1. Genetics.   I. Title.
  [DNLM: 1. Genetics.   QH 430 R966g]
QH430.R87   1992
575. 1—dc20
DNLM/DLC
for Library of Congress   91-35366
                                                              CIP

     92   93   94   9   8   7   6   5   4   3

CONTENTS

5

LINKAGE, CROSSING-OVER, AND GENE MAPPING IN EUKARYOTES 126

6

ADVANCED GENETIC MAPPING IN EUKARYOTES 162

7

GENETIC RECOMBINATION IN BACTERIA AND BACTERIOPHAGES 194

8

THE BEGINNINGS OF MOLECULAR GENETICS: GENE FUNCTION 233

9

THE STRUCTURE OF GENETIC MATERIAL 266

10

THE ORGANIZATION OF DNA IN CHROMOSOMES 288

11

DNA REPLICATION AND RECOMBINATION   325

12

TRANSCRIPTION   361

13

RNA MOLECULES AND RNA PROCESSING   381

14

THE GENETIC CODE AND THE TRANSLATION OF THE GENETIC MESSAGE   409

15

RECOMBINANT DNA TECHNOLOGY AND THE MANIPULATION OF DNA   430

PREFACE

Genetics has long been one of the key areas in biology, and in the years since the first unravelings of the DNA mysteries, we have seen an unparalleled explosion of knowledge and understanding about genetics and its related disciplines. Not only is knowledge about genetics accumulating rapidly, but its many applications now affect our daily lives and benefit humanity. In writing *Genetics*, my goal is to provide today's students with a succinct and logical synthesis of this vital field.

The third edition of *Genetics* retains the overall approach, logical progression of ideas, organization, and pedagogical features (such as "Analytical Approaches for Solving Genetics Problems" and "Keynotes") that made the first and second edition readily accessible to students. All chapters have been fine-tuned, and parts have been rewritten to clarify genetic processes and to provide new examples. In addition, the third edition includes the following significant changes and improvements:

1. All the molecular aspects of genetics are updated so that the book continues to reflect our current understanding of genes at the molecular level.
2. The coverage of recombinant DNA technology has been updated and expanded to include new topics, such as polymerase chain reaction (PCR), DNA fingerprinting, dideoxy DNA sequencing, cloning genes by complementation, and constructing a restriction map.
3. The chapter on eukaryotic gene regulation is significantly rewritten and expanded. In particular, the section on genetic control of *Drosophila* development is expanded considerably, and there are new sections on immunogenetics and genetic regulation of development in *C. elegans*.
4. Mendelian genetics in humans is now introduced in Chapter 2, "Mendelian Genetics," rather than in the subsequent chapter, to establish the general applicability of Mendelian principles to humans.
5. The art program is completely revised. All figures are now in full color in order to illustrate facts and concepts with greater clarity, making genetics more accessible to students.
6. Close to 100 new questions and problems of varying difficulty have been added to the set of problems at the end of each chapter. Independent reviewers have checked all answers for accuracy. In addition, "Analytical Approaches for Solving Genetics Problems," a very successful feature in earlier editions, is now used more consistently throughout the text, especially in the more descriptive molecular chapters.
7. Careful editing of each chapter has eliminated excessive detail. As a result, the material is presented at a consistent level that ensures student comprehension of basic concepts.
8. There are several new pedagogical features designed to aid student comprehension and study, including chapter outlines, "Principal Points," and chapter summaries. A Student Problem-Solving Guide and Solutions Manual is also available. Complete descriptions of these features are found in the section "Pedagogical Features."

ORGANIZATION AND COVERAGE

The first seven chapters of *Genetics* deal with transmission of the genetic material. In Chapter 1 we review the structure of viruses and of prokaryotic and eukaryotic cells and the processes of asexual and sexual reproduction. Mitosis and meiosis are discussed in the context of both animal and plant life cycles. Chapter 2 is focused on Mendel's contributions to our understanding of the principles of heredity, and Chapters 2 and 3 both present the basic principles of genetics in relation to Mendel's laws. Mendelian genetics in humans is now introduced in Chapter 2 rather than Chapter 3, as in the second edition. Chapter 3 covers experimental evidence for the relationship between genes and chromosomes, methods of sex determination, and sex linkage. Also in this chapter is a discussion of Mendelian genetics in humans with respect to sex-linked genes.

The exceptions to and extensions of Mendelian analysis (such as the existence of multiple alleles, gene interactions and modified Mendelian ratios, and lethal alleles) are described in Chapter 4. In Chapter 5, we describe how to determine the order of and distance between genes on eukaryotic chromosomes in genetic experiments set up to quantify crossovers occurring during meiosis.

In the next two chapters we consider advanced mapping analysis in eukaryotes and genetic mapping in prokaryotes. Chapter 6 is focused on tetrad analysis, primarily in fungal systems; on mapping eukaryotic genes through mitotic analysis; and using somatic cell hybrids to map genes to human chromosomes. In Chapter 7 we discuss the ways to map genes in bacteriophages and in bacteria, taking advantage of the processes of transformation, conjugation, and transduction. Fine structure analysis of bacteriophage genes concludes this chapter, consolidating the discussion of bacteriophage genes in one chapter. (In the second edition, the material was found in Chapter 8.)

Chapters 8 through 15 comprise the "molecular core" of *Genetics,* detailing the current level of our knowledge about the molecular aspects of genetics. In Chapter 8 we examine genetic fine structure and some aspects of gene function, such as genetic control of the structure and function of proteins and enzymes, and the role of particular sets of genes in directing and controlling biochemical pathways. This chapter also covers the molecular structure of proteins. We have revised the sequence of topics in order to represent more accurately the historical development of the concept of a gene, which was first understood at a functional level, then at a molecular level. Chapter 8's discussion of gene function also enables students to understand immediately that genes specify proteins and enzymes. This concept, in turn, provides a basis for understanding the significance of the molecular material in the following chapters.

In Chapter 9 we cover the structure of DNA, presenting the classical experiments that revealed DNA and RNA to be the genetic material and that established the double helix model as the structure of DNA. The details of DNA structure and organization in prokaryotic chromosomes are set out in Chapter 10. We cover DNA replication in prokaryotes and eukaryotes in Chapter 11.

After thoroughly explaining the nature of the gene and its relationship to chromosome structure, we discuss the first step in the expression of a gene transcription. In Chapter 12 we describe the general process of transcription. Then, in Chapter 13 we present the most current information about the details of the structure, transcription, and processing of messenger RNA, transfer RNA, and ribosomal RNA molecules in both prokaryotes and eukaryotes. Chapter 14 includes discussions of the evidence for the nature of the genetic code, and of our current knowledge about translation in both

prokaryotes and eukaryotes. In Chapter 15 we discuss recombinant DNA technology, the essential tool of modern genetics. There are descriptions of using recombinant DNA technology to clone and characterize genes and to manipulate DNA, followed by an expanded discussion of commercial applications of recombinant DNA technology.

The next two chapters focus on regulation of gene expression in prokaryotes (Chapter 16) and eukaryotes (Chapter 17). In Chapter 16 we discuss the operon as a unit of gene regulation, the current molecular details in the regulation of gene expression in bacterial operons, and regulation of genes in bacteriophages. In Chapter 17, we explain how eukaryotic gene expression is regulated, stressing molecular changes that accompany gene regulation, short-term gene regulation in lower and higher eukaryotes, gene regulation in development and differentiation, genetic defects in human development, immunogenetics, and genetic defects in human development.

The next three chapters concentrate on the ways in which genetic material can change or be changed, or both. Chapter 18 covers the processes of gene mutation, then the procedures that screen for potential mutagens and carcinogens (the Ames test) and that select for particular classes of mutants from a heterogeneous population. This chapter also now includes a new section on site-directed mutagenesis. Chromosomal aberrations—that is, changes in normal chromosome structure or chromosome number—are described in Chapter 19. Chapter 20 presents the structures and movements of transposable genetic elements in prokaryotes and eukaryotes, retroviruses, and oncogenes and their relationship to cancer.

In Chapter 21, we address the organization and genetics of extranuclear genomes of mitochondria and chloroplasts. We cover the classical gene experiments that established that a gene is extranuclear. In this chapter we also discuss current molecular information about the gene organization within the extranuclear genomes.

In Chapters 22 and 23 we describe quantitative genetics and the genetics of populations, respectively. In Chapter 22, "Quantitative Genetics," we develop the concept that some heritable traits—the quantitative traits—show continuous variation over a range of phenotypes. In this chapter we also discuss heritability: the relative extent to which a characteristic is determined by genes or by the environment. In Chapter 23, "Genetics of Populations," we present the basic principles in population genetics, extending our treatment of the gene from the individual organism to a population of organisms. Both Chapters 22 and 23 include discussions of the application of molecular tools to these areas of genetics. Chapter 23, for example, includes a new section on measuring genetic variation with RFLPs and DNA sequencing. In addition, the coverage of molecular evolution has been updated and expanded.

PEDAGOGICAL FEATURES

Because the field of genetics is so complex, making the study of it so difficult, I have incorporated a number of special pedagogical features designed to help students and to enhance their understanding and appreciation of genetic principles. Some of these features are familiar, but this edition also includes new pedagogical tools:

1. With the exception of Chapter 1, all chapters have a section on "Analytical Approaches for Solving Genetics Problems." Genetics principles have always been best taught with a problem-solving approach. However, beginning students often do not acquire the necessary experience with basic concepts that would enable them to attack assigned problems methodically. In the "Analytical Approaches" sections, typical genetic problems are "talked through," step-by-step, to help students understand how to tackle genetics problems by applying fundamental principles.

2. In a feature new to this edition, each chapter opens with an outline of its contents and a section called "Principal Points." Principal Points are short summaries that alert students to the key concepts they will encounter in the material to come.

3. In the problem sets that close the chapters are close to 500 questions and problems designed to give students further practice in solving genetics problems. The problems for each chapter represent a range of topics and difficulty. The answers to questions indicated by an asterisk can be found at the back of the book, and answers to all questions are available in a separate supplement.

4. Throughout each chapter, strategically placed "Keynote" summaries emphasize important ideas and critical points.

5. Chapter summaries, another new feature, now close each chapter, further reinforcing the major points that have been discussed.

6. Important terms and concepts are clearly defined where they are introduced in the text. For easy reference, they are also compiled in a Glossary at the back of the book. Special care has been taken to provide the most useful Index—extensive, accurate, and well cross-referenced.

7. Comprehensive and up-to-date suggested readings for each chapter are listed at the back of the book.

8. Some chapters include boxes covering special topics related to chapter coverage. Some of these boxed topics are: *What Organisms Are Suitable for Use in Genetic Experiments* (Chapter 2); *Equilibrium Density Gradient Centrifugation* (Chapter 10); and *Radioactive Labeling of DNA* (Chapter 18).

SUPPLEMENTS

A new Student Problem-Solving Guide and Solutions Manual has been prepared by Gail Patt of Boston University. In addition to detailed solutions for all the problems in the text, the Guide for each chapter contains these features: a review of important terms and concepts; "Thinking Analytically" sections, which provide guidance and tips on solving problems and avoiding common pitfalls; and additional questions for practice and review.

A set of 125 full-color transparencies and 75 transparency masters is available to adopters of the text.

ACKNOWLEDGMENTS

I would like to extend thanks to Ben Pierce (Baylor University) for his continued rewriting and updating of Chapters 22 and 23, "Quantitative Genetics" and "Genetics of Populations." His work on these chapters has made a significant contribution to the overall quality of this book.

Also making an invaluable contribution to this text are John and Bette Woolsey of J/B Woolsey Associates, Inc., who developed and executed the new full-color art program.

I would also like to thank all the reviewers involved in this edition:

John Belote, *Syracuse University*
Anna W. Berkovitz, *Purdue University*
Allan Bornbusch, *Smith College*
Nathan Chu, *Barnard College*
Max P. Dunford, *New Mexico State University*
Joseph O. Falkinham, III, *Virginia Polytechnic Institute*
Irving Finger, *Haverford College*
Elliot Goldstein, *Arizona State University*
Jeffrey C. Hall, *Brandeis University*
Stanley Hattman, *University of Rochester*
William W. Johnson, *University of New Mexico*
Kenneth C. Jones, *California State University, Northridge*
Peter Kuempel, *University of Colorado*
Clint Magill, *Texas A&M University*
Margaret Mathies, *Claremont College*
Gail Patt, *Boston University*
Ruth B. Phillips, *University of Wisconsin-Milwaukee*
Frank Rice, *Fairfield University*

I also thank Professors Berkovitz, Belote, and Rice, as well as Professor D. A. Whited, North Dakota State University, and especially Muriel Nesbitt, University of California, San Diego, for contributing new problems to the chapters. I am grateful to the Literary Executor of the late Sir Ronald A. Fisher, F.R.S., to Dr. Frank Yates, F.R.S., and to Longman Group Ltd. London, for permission to reprint Table IV from their book *Statistical Tables for Biological, Agricultural, and Medical Research* (6th edition, 1974).

Finally, I wish to thank those who helped make *Genetics* a physical reality. I thank Cheryl Kucharzak and Karen Kobilik of HarperCollins for photo research. Barbara Littlewood produced an excellent index, and Anita Bennett's work in reviewing the art and checking the answers to the genetics problems was painstaking and invaluable. I gratefully acknowledge the hard work and effort of Ellen Pettengell, designer and project manager, and Gail Savage, copy and production editor, of Proof Positive/Farrowlyne Associates in producing the book.

I especially thank Glyn Davies, the sponsoring editor, and Kathleen Dolan, the developmental editor, of HarperCollins for believing in this project and for their help and support in bringing it to fruition. Producing a book as complex as this means juggling many things simultaneously, and Kathleen, in particular, did an outstanding job in this regard. It was a pleasure working with her.

Peter J. Russell

GENETICS

1 Viruses, Cells, and Cellular Reproduction

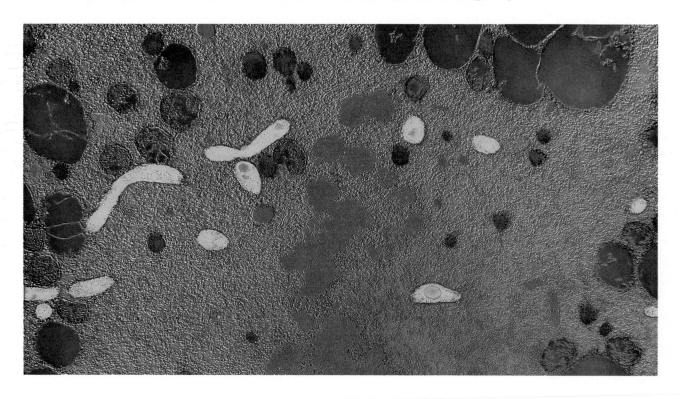

PRINCIPAL POINTS

~ The cell is the smallest building unit of a multicellular organism and can only be produced from another cell by cell division.

~ Prokaryotes are the simplest cellular organisms. Exemplified by the bacteria, prokaryotes lack a membrane-bound nucleus and divide by binary fission. The genetic material of prokaryotes is found in a single, circular DNA molecule that is associated with few proteins.

~ Eukaryotes are organisms that have cells in which the genetic material is located in a membrane-bound nucleus. The genetic material is distributed among several linear chromosomes, each of which consists of a complex of DNA and many proteins.

~ Diploid eukaryotic cells have two sets of chromosomes, one set coming from each parent. The members of a pair of chromosomes are called homologous chromosomes. Haploid eukaryotic cells have only one set of chromosomes.

~ Mitosis is the process of nuclear division in haploid or diploid eukaryotic cells. It results in the production of daughter nuclei that contain identical chromosome numbers and are genetically identical to one another and to the parent nucleus from which they arose.

~ Meiosis occurs in all sexually reproducing eukaryotes. It is a process in which a diploid cell (or cell nucleus) is transformed through one round of DNA replication and two rounds of nuclear division into four haploid cells (or four nuclei).

~ Meiosis generates genetic variability through the various ways in which maternal and paternal chromosomes are combined in progeny nuclei and by crossing-over between members of a homologous pair of chromosomes.

*G*enetics is the science of heredity and involves studying the structure and function of genes, and the way genes are passed from one generation to the next. The differences between organisms are the result of differences in the genes they carry, differences which have resulted from the evolutionary processes of mutation (a heritable change in the genetic material), recombination (exchange of genetic material between chromosomes), and selection (the favoring of particular combinations of genes in a given environment). The principles of heredity were recognized by Gregor Mendel in the 1860s, the development of the subject began about 1900, and the key discoveries have been made in the past four or five decades.

The study of genetics has examined three aspects of the genetic material of living organisms: (1) how physical traits are coded for and expressed in the organism (Figure 1.1a); (2) how these traits are inherited (copied and passed on) from one generation to the next (Figure 1.1b); and (3) how changes in the genetic material have led to past and present biological diversity.

At present, we have a good understanding of the nature of genetic material, how it is replicated and passed on from generation to generation, how it is expressed in the cell, and how that expression is regulated. Clearly, gene activity is of central importance to cell growth and function and to the development and differentiation of organisms. In these respects the study

of genetics is the foundation of molecular and cell biology, embryology, and developmental biology. And, with respect to the study of biological diversity, genetics interfaces with specialties like ecology, evolutionary biology, and morphology.

Beyond the laboratory, genetics has received increasing public attention in recent years. For example, in medicine, a number of diseases have been shown to be caused by genetic defects, and strides are being made in understanding the molecular bases of some of those diseases. In a health-conscious society, such as ours, the public has become more educated about genetics as hopes for the eradication of genetic diseases are raised. Presentations on television and in newspapers and magazines offer excellent opportunities for the layperson to learn about genetics and its relevance to society.

In the past 15 years or so, public interest in genetics has been heightened also because of the development and application of particular molecular techniques known collectively as *recombinant DNA technology*. Research in this area has led to the development of industries dealing in what is called *biotechnology*, or *genetic engineering*. While most of this research was undertaken to study fundamental genetic processes, without regard to practical applications, there has been, and will continue to be, a lot of applied work done that is of direct benefit to humankind. In the area of plant breeding, genetic engineering is being applied

~ FIGURE 1.1

The two roles of genetic material: (a) Coding for the physical traits (phenotype of an organism) through the processes of transcription, translation, and protein manufacture; (b) Ensuring, through replication and division, that the organism's genetic code is accurately transmitted from one generation to the next.

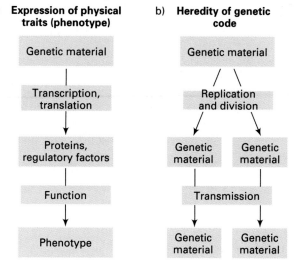

the genetic material may be double stranded or single stranded. Every type of cellular organism can be infected by viruses. Thus there are plant viruses, animal viruses, and bacterial viruses. The latter are also called bacteriophages ("bacteria eaters"), or simply phages.

Viruses vary considerably in size, shape, and composition (Figure 1.2a–g; p. 4) and have evolved specific mechanisms to infect a host cell or cells. Once the genetic material of the virus is within the host cell, it can direct the cellular machinery to produce progeny viruses.

CELLS—THE UNITS OF LIFE

After many years of experiments the **cell theory** was put forward in the early nineteenth century. There are three basic tenets of the cell theory:

1. The **cell** is the smallest building unit of a multicellular organism and, as a unit, is itself an elementary organism.
2. Each cell in a multicellular organism has a specific role.
3. A cell can only be produced from another cell by cell division.

Beyond these three basic ideas our knowledge and understanding of the structure of organisms and cells has been advanced through the use of microscopes, both light microscopes and electron microscopes. With the microscope we can see that while there is a great diversity of cell types, they all have some features in common. We will start our study of the genetics of various life forms by examining cell structure.

Prokaryotes

Prokaryotes (meaning "prenuclear") are the simplest cellular organisms. Included in this group are all the bacteria.

Bacteria are spherical, rod-shaped, or spiral-shaped single-celled or multicellular, filamentous organisms (Figure 1.3; p. 5). The bacteria are divided into two distantly related groups, the *eubacteria* and the *archaebacteria*. Bacteria of the eubacteria are the common varieties found in living organisms (naturally or by infection), soil, and water. Bacteria of the archaebacteria are found in much more inhospitable conditions, such as hot springs, salt marshes, methane-rich marshes, or in the ocean depths.

Bacteria vary in size from about 100 nm (1 nm = 10^{-9} m) in diameter, to 10 μm (1 μm = 10^{-6} m) in diameter and 60 μm long. The genetic material of bacteria is found in a single, circular DNA molecule that is associated with few proteins. In bacteria this genetic

to improve crop yields and to develop plants resistant to pests and plant diseases. In the area of animal breeding, genetic engineering is being employed in the beef, dairy cattle, and poultry industries to improve yields and to develop better strains. In medicine the results are equally impressive. For example, recombinant DNA technology is proving very important in the production of antibiotics, hormones, and other medically important agents and in the diagnosis and treatment of certain human genetic diseases (such as Huntington's disease and Tay-Sachs disease). In short, the science of genetics is currently in a dramatic growth phase. By understanding the basic principles of genetics, as described in this text, you will have a greater capacity to understand, foster, and perhaps contribute to the new and exciting applications of this subject.

In this chapter, some fundamental principles of biology relevant to genetics are reviewed. The first part of the chapter focuses on the organization of viruses, and of prokaryotic and eukaryotic cells. The remainder of the chapter concentrates on cellular reproduction in eukaryotes, with emphasis on mitosis and meiosis.

VIRUSES

A **virus** is a noncellular organism that can reproduce only within a host cell. It contains genetic material, either **DNA (deoxyribonucleic acid)** or **RNA (ribonucleic acid)**, within a protein coat. Depending on the virus,

~ FIGURE 1.2

Electron micrographs of representative virus particles. (a) Icosahedral virus: adenovirus; (b) Helical virus: a segment of tobacco mosaic virus; (c) Icosahedral virus surrounded by an envelope: herpesvirus; (d) Helical virus surrounded by an envelope: influenza virus (a myxovirus); (e) Complex virus: vaccinia virus (a poxvirus); (f) Bacterial virus; lambda (λ); (g) Human immunodeficiency virus (HIV), the disease agent of AIDS.

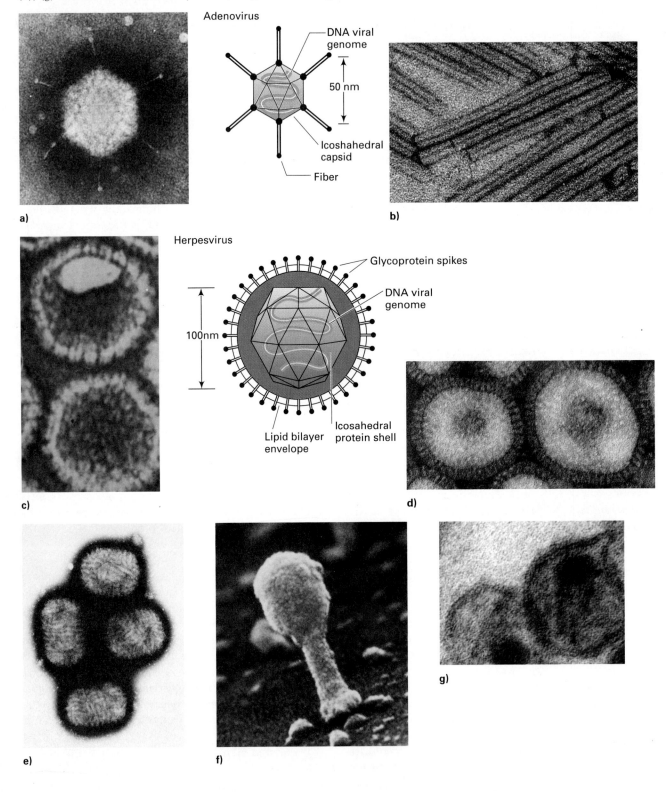

~ FIGURE 1.3

Four prokaryotes: (a) *Escherichia coli*, a rod-shaped bacterium common in human intestines; (b) *Spirillum volutans*, a helically coiled bacterium; (c) *Streptococcus lactis*, a spherical bacterium common in milk; (d) *Actinomycete*, a prokaryote with branching, multicellular filaments.

 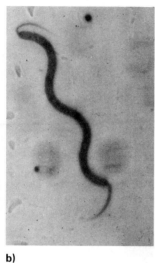 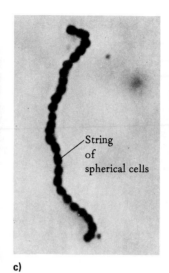 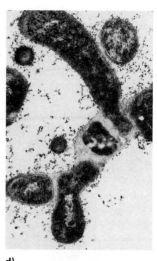

a) b) c) d)

material is not separated from the rest of the cytoplasm by a nuclear membrane such as is found in eukaryotic ("true nuclear") organisms, and this is a major distinguishing feature of prokaryotes. In their cytoplasm bacteria also contain a large number of ribosomes, the organelles on which protein synthesis takes place. The shape of the bacterium is maintained by a rigid cell wall located outside the cell membrane.

In most cases, the bacteria studied in genetics are eubacteria. The most intensely studied bacterium in genetics is *Escherichia coli* (Figure 1.3a), a rod-shaped bacterium common in human intestines. Studies of this bacterium have resulted in significant advances in our understanding of the regulation of gene expression and in the development of the whole field of molecular biology. Today *E. coli* is used extensively in recombinant DNA experiments.

Bacteria reproduce by **binary fission;** that is, after their genetic material replicates, each cell simply divides in two (Figure 1.4).

Eukaryotes

Eukaryotes (meaning "true nucleus") are organisms that have cells in which the genetic material is located in the **nucleus,** a discrete structure within the cell that is bounded by a nuclear membrane. Eukaryotes can be unicellular or multicellular and are often grouped by the terms *lower* and *higher*. Lower eukaryotes have relatively simple cellular organization, and while their

DNA is significantly more complex than that of prokaryotes, it is less complex than that of the higher eukaryotes. Higher eukaryotes include genetically more complex multicellular organisms.

The following eukaryotic organisms are commonly used in genetic research, and you will see a number of them frequently throughout this text: *Saccharomyces cerevisiae* (yeast), *Neurospora crassa* (pink bread mold), *Chlamydomonas reinhardi* (a green alga), *Pisum sativum* (garden pea), *Zea mays* (corn), *Drosophila melanogaster* (fruit fly), *Caenorhabditis elegans* (nematode), *Mus musculus* (mouse), and *Homo sapiens* (humans) (Figure 1.5; p. 6). Of these, *Chlamydomonas* and *Saccharomyces* are unicellular organisms, and the

~ FIGURE 1.4

Binary fission of a bacterium, *E. coli*.

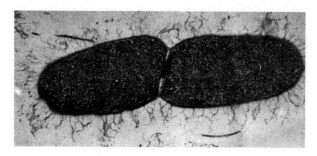

~ FIGURE 1.5

Eukaryotic organisms commonly used in genetics research: (a) *Saccharomyces cerevisiae* (a budding yeast); (b) *Neurospora crassa* (orange bread mold); (c) *Chlamydomonas reinhardi* (a green alga); (d) *Pisum sativum* (a garden pea); (e) *Zea mays* (corn); (f) *Drosophila melanogaster* (fruit fly); (g) *Caenorhabditis elegans* (a nematode); (h) *Mus musculus* (mouse); (i) *Homo sapiens* (humans).

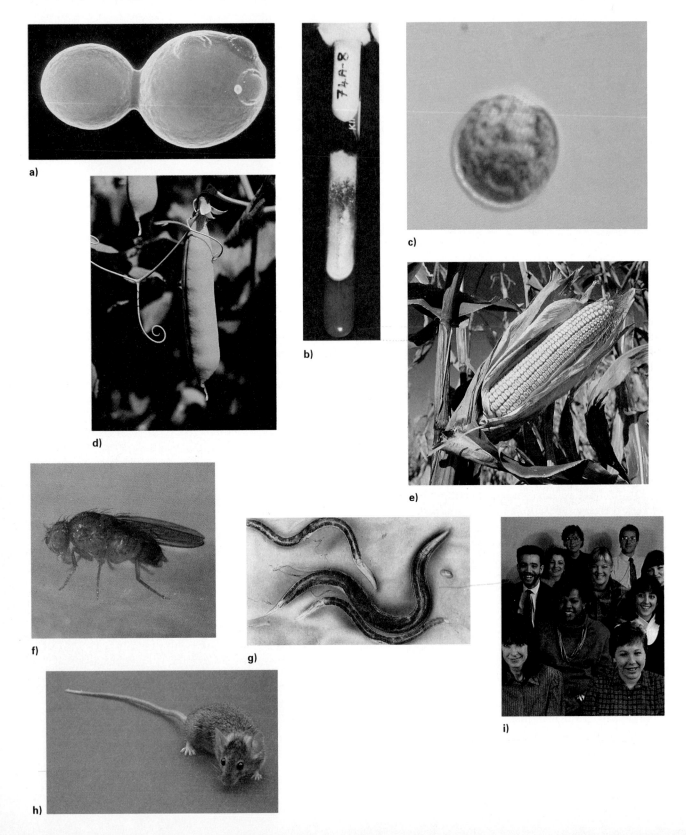

rest are multicellular. *Neurospora, Saccharomyces,* and *Chlamydomonas* are lower eukaryotes; those remaining are higher eukaryotes.

CELL STRUCTURE. The structure of the eukaryotic cell is very different from that of a bacterial cell. Figure 1.6a is an electron micrograph of a cross section through an animal cell, Figure 1.6b is a diagram of a thin section through a generalized eukaryotic (animal) cell, Figure 1.6c (p. 8) is an electron micrograph of a cross section through a higher plant cell, and Figure 1.6d (p. 8) is a diagram of a thin section through a generalized higher plant cell. Surrounding the cytoplasm of both animal and plant cells is a membrane called the **plasma membrane.** Plant cells, but not animal cells, have a rigid cell wall outside of the plasma membrane.

NUCLEUS. The nucleus is separated from the rest of the cell, the cytoplasm and associated organelles, by the double membrane of the nuclear envelope. Figure 1.7a

(p. 8) presents an electron micrograph of a cross-section through the nucleus of an animal cell; Figure 1.7b presents a diagram of an animal cell nucleus. The membrane is selectively permeable and has pores (*nuclear pores*). These two characteristics make it possible for materials to move between the nucleus and the cytoplasm. The pores are about 20–80 nm in diameter, and are each surrounded by a thickened, electron-dense ring, or annulus.

The nucleus contains most of the genetic material of the cell. The genetic material is complexed with protein and is organized into a number of linear structures called **chromosomes.** *Chromosome* means "colored body" and is so named because these threadlike structures are visible under the light microscope only after they are stained with dyes. However, in a nondividing cell such as the one shown in Figure 1.6, the chromosomes are in an extended state and are relatively difficult to see even when stained. Also within the nucleus is the *nucleolus,* a structure within which the ribosomes are assembled.

~ **FIGURE 1.6 a, b**

Eukaryotic cell: (a) Electron micrograph of a thin section through an eukaryotic animal cell. Key: LD: Lipid droplet; Ly: lysosome; M: mitochondrion; MV: microvilli; N: nucleus; RER, SER: rough and smooth endoplasmic reticulum, respectively; (b) Drawing of a thin section through an animal cell, showing the main organizational features and the principal organelles.

a)

b)

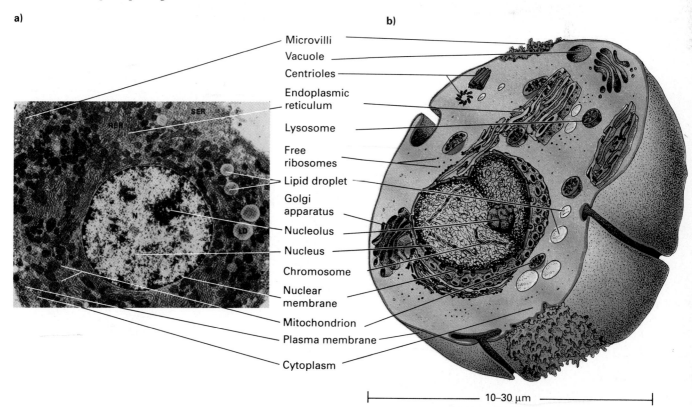

Microvilli
Vacuole
Centrioles
Endoplasmic reticulum
Lysosome
Free ribosomes
Lipid droplet
Golgi apparatus
Nucleolus
Nucleus
Chromosome
Nuclear membrane
Mitochondrion
Plasma membrane
Cytoplasm

10–30 μm

~ FIGURE 1.6 c, d

Eukaryotic cell: (c) Electron micrograph of a thin section of a root tip cell from *Elodea canadensis*. Easily seen are the cell wall, nucleus, vacuoles, mitochondria, endoplasmic reticulum, Golgi apparatus, and ribosomes; (d) Drawing of a thin section through a plant cell. Note that not all cell parts in the diagrams are readily visible in the electron micrographs.

c)   d)

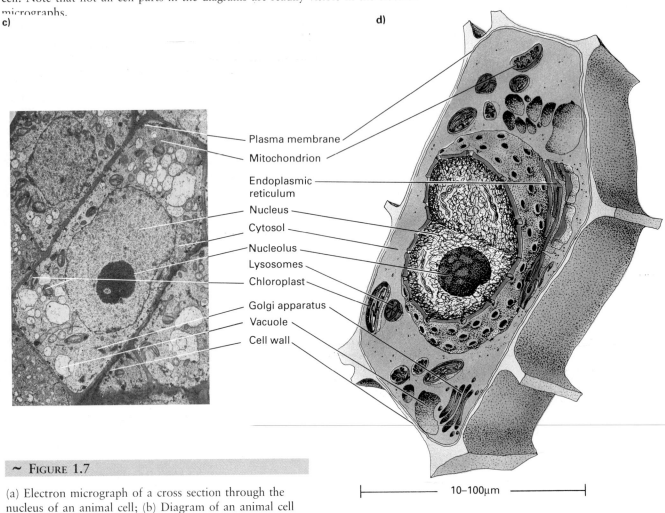

Plasma membrane
Mitochondrion
Endoplasmic reticulum
Nucleus
Cytosol
Nucleolus
Lysosomes
Chloroplast
Golgi apparatus
Vacuole
Cell wall

⊢——— 10–100μm ———⊣

~ FIGURE 1.7

(a) Electron micrograph of a cross section through the nucleus of an animal cell; (b) Diagram of an animal cell nucleus.

a)   b)

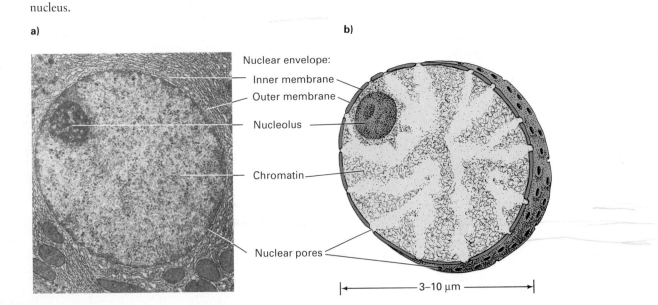

Nuclear envelope:
Inner membrane
Outer membrane
Nucleolus
Chromatin
Nuclear pores

⊢——— 3–10 μm ———⊣

~ FIGURE 1.8

Schematic of the centriole: (left) end view; (right) side view.

CYTOPLASM. The **cytoplasm** contains a vast amount of different materials and organelles. Extending throughout the cytoplasm is a complex network of protein filaments called the **cytoskeleton.** It is the cytoskeleton that is responsible for the different shapes of eukaryotic cells, and for their abilities to move in directed and coordinated ways. Movement of organelles within the cell is also a function of the cytoskeleton. Of special interest for geneticists are the *centrioles, endoplasmic reticulum* (ER), *ribosomes, Golgi apparatus, mitochondria,* and *chloroplasts.*

Centrioles (Figure 1.8) (also called *basal bodies*) are found in nearly all animal cells, but not usually in plant cells (except in cells of lower plants). A centriole is a small, cylindrical organelle about 0.2 μm wide and 0.4 μm long. It consists of a ring of nine groups of three fused microtubules (a microtubule is a specialized protein filament). In animal cells, a pair of centrioles functions as the focus for the *centrosome,* which organizes the spindle fibers that function in mitosis and meiosis. Both mitosis and meiosis are important cell replication processes and will be discussed in detail later.

The endoplasmic reticulum is a double-membrane system that runs through the cell (Figure 1.9). In Figure 1.10 (p. 10), a photograph and diagram illustrates the relationship between the ER and the nuclear membranes. Under the electron microscope we can see two types of ER: rough (Figure 1.9a) and smooth (Figure 1.9b). The former has ribosomes attached to it, giving it a rough appearance, and the latter does not. Ribosomes bound to the endoplasmic reticulum synthesize proteins to be secreted by the cell or to be localized in the cell membrane or vacuolar organelles. These proteins are moved into the space between the two membranes

~ FIGURE 1.9

(a) Rough and (b) smooth endoplasmic reticulum.

a) Rough endoplasmic reticulum

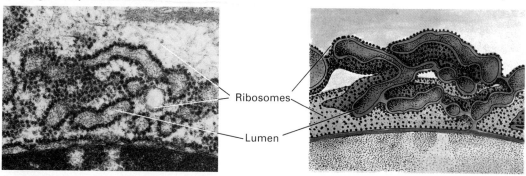

b) Smooth endoplasmic reticulum

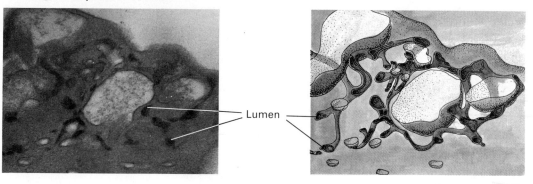

(lumen) and are then translocated to the Golgi apparatus (Figure 1.11), where they are packaged into vesicles. The vesicles are thought to carry material between the Golgi apparatus and different compartments of the cell. In secretory cells, some of the vesicles migrate to the cell surface and, by fusion with the outer membrane, release their contents to the outside of the cell. The synthesis of all other kinds of proteins, such as enzymes and cellular structural proteins, is performed by ribosomes that are free in the cytoplasm.

Eukaryotic cells contain mitochondria (singular: mitochondrion), illustrated in Figure 1.12. These large organelles are surrounded by a double membrane, the inner one of which is highly convoluted. Mitochondria play an extremely important role in energy processing for the cell. They also contain genetic material in the form of a circular molecule of DNA. As in bacteria, probably no structural proteins are associated with the genetic material of mitochondria.

Finally, many plant cells contain chloroplasts, the large, double-membraned, chlorophyll-containing organelles involved in photosynthesis (Figure 1.13). Chloroplasts also contain genetic material, and as in mitochondria, its form is a circular DNA molecule.

Table 1.1 (p. 12) summarizes the key similarities and differences between prokaryotic and eukaryotic cells.

~ **FIGURE 1.11**

(a) Diagram of the Golgi apparatus; (b) Electron micrograph of the Golgi apparatus. The cis region, which takes in vesicles formed in the ER, is positioned toward the ER; the trans region, which releases vesicles into the cytoplasm, faces toward the plasma membrane.

~ **FIGURE 1.10**

(a) Electron micrograph of the endoplasmic reticulum; (b) Diagram illustrating the relationship between the nuclear membrane and the endoplasmic reticulum.

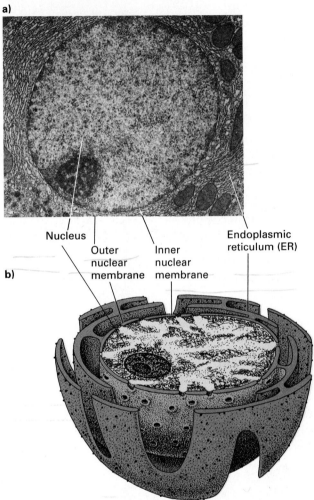

a)

b)

Nucleus Outer nuclear membrane Inner nuclear membrane Endoplasmic reticulum (ER)

a)

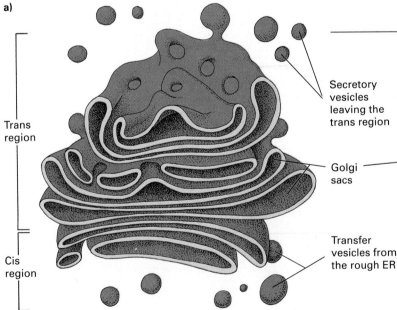

Trans region

Cis region

b)

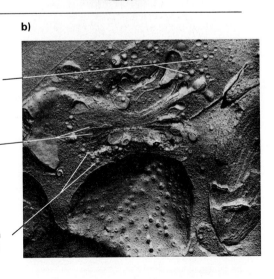

Secretory vesicles leaving the trans region

Golgi sacs

Transfer vesicles from the rough ER

~ FIGURE 1.12

(a) Electron micrograph of a mitochondrion; (b) Diagram of a mitochondrion. Energy-processing mechanisms involve interrelationships between the organelle's intermembranal space, the inner membrane, and the matrix.

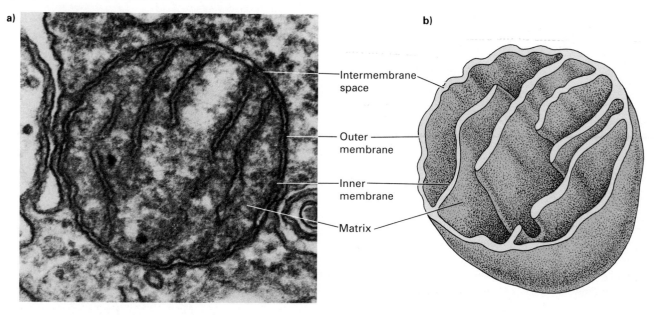

a)

b)

Intermembrane space

Outer membrane

Inner membrane

Matrix

~ FIGURE 1.13

(a) Electron micrograph of a section through a plant leaf, showing chloroplasts;
(b) Diagram of a chloroplast. The organelle's energy-harvesting mechanisms involve interrelationships between the intermembranal space, the stroma, the thylakoids stacked in grana, and the area within each thylakoid.

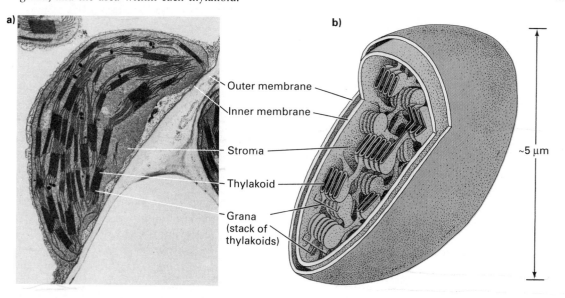

a)

b)

Outer membrane

Inner membrane

Stroma

Thylakoid

Grana
(stack of
thylakoids)

~5 μm

~ TABLE 1.1

Key Differences Among the Cells of Prokaryotes, Higher Plants, and Animals

FEATURE	PROKARYOTIC CELL	EUKARYOTES	
		HIGHER PLANT CELL	ANIMAL CELL
Cell membrane	External only (two separated by a space)	External and internal	External and internal
Supporting structure	Polysaccharide cell wall	Polysaccharide cell wall (cellulose)	Protein cytoskeleton
Nuclear membrane	Absent	Present	Present
Chromosomes	Single, circular DNA only	Multiple, linear, complexed with protein	Multiple, linear, complexed with protein
Membrane-bound organelles	Absent	Many, including mitochondria, large vacuoles, and chloroplasts	Many, including mitochondria, lysosomes
Endoplasmic reticulum	Absent	Present	Present
Ribosomes	Smaller, free	Larger, some membrane-bound	Large, some membrane-bound
Cell division	Fission	Mitosis	Mitosis
Sexual recombination	Only by plasmid-mediated transfer	Meiosis and fertilization	Meiosis and fertilization
Centrioles	None	Usually absent	Usually present

CELLULAR REPRODUCTION IN EUKARYOTES

If all living things are composed of cells, and if all cells arise from other cells, then what processes give rise to accurate copies of a cell? To answer this question we now turn to the subject of cell division in eukaryotes.

Chromosome Complement of Eukaryotes

The genetic material of eukaryotes is distributed among multiple chromosomes; the number of chromosomes, with rare possible exceptions, is characteristic of the species. For example, the nuclei of human cells have 46 chromosomes, those of the fruit fly *Drosophila melanogaster* have 8, and those of the bread mold *Neurospora crassa* have 7. Many eukaryotes have two copies of each type of chromosome in their nuclei, so their chromosome complement is said to be **diploid, or 2N**. Diploid eukaryotes are produced by the fusion of two **gametes**, one from the female parent and one from the male parent. The fusion produces a diploid **zygote**,

which then divides. Each gamete has only one set of chromosomes and is said to be **haploid** (N). The complete haploid complement of genetic information is called the **genome.** Of the three organisms just listed, humans and fruit flys are **diploid** (2N) and the bread mold is haploid (N). Thus humans have 23 pairs of chromosomes, *Drosophila melanogaster* has 4 pairs, and *Neurospora crassa* has 7 single chromosomes.

In diploids, the members of a chromosome pair are called **homologous chromosomes**; each individual member of a pair is called a **homolog**. Homologous chromosomes are usually identical with respect to the arrangement of genes they contain and with respect to their visible structure. Chromosomes from different pairs are called **nonhomologous chromosomes.** Figure 1.14 illustrates the chromosomal organization of haploid and diploid organisms.

In animals and in some plants there are differences in the chromosome complement of male and female cells. One sex has a matched pair of **sex chromosomes,** chromosomes related to the sex of the organism; the other sex has an unmatched pair or an unpaired chromosome. For example, human females have two X chromosomes (XX), while human males have one X and one Y (XY). Chromosomes other than sex chromosomes are called **autosomes.**

KEYNOTE

Diploid eukaryotic cells have a double set of chromosomes, one set coming from each parent. The members of a pair of chromosomes, one from each parent, are called homologous chromosomes. Haploid eukaryotic cells have only one set of chromosomes.

Under the microscope we see that chromosomes differ in their size and morphology within and between species. Each chromosome has a specialized region somewhere along its length that is often seen as a constriction under the microscope. This constriction, called a **centromere** (or **kinetochore**), is important in the activities of the chromosomes during cellular division and can be located in one of four general positions in the chromosome (Figure 1.15). A **metacentric chromosome** has the centromere in approximately the center of the chromosome so that it appears to have two approximately equal arms. **Submetacentric chromosomes** have one arm longer than the other, **acrocentric chromosomes** have one arm plus a stalk and satellite, and **telocentric chromosomes** have only one arm, since the centromere is at the end. Chromosomes also vary in relative size. Chromosomes of mice, for example, are all similar in length, whereas those of humans show a wide range of relative lengths.

~ FIGURE 1.14

Chromosomal organization of haploid and diploid organisms.

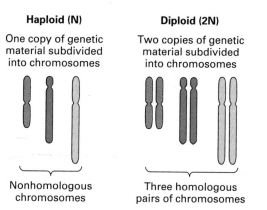

Haploid (N)	Diploid (2N)
One copy of genetic material subdivided into chromosomes	Two copies of genetic material subdivided into chromosomes
Nonhomologous chromosomes	Three homologous pairs of chromosomes

~ FIGURE 1.15

General classification of eukaryotic chromosomes into metacentric, submetacentric, acrocentric, and telocentric types, based on the position of the centromere.

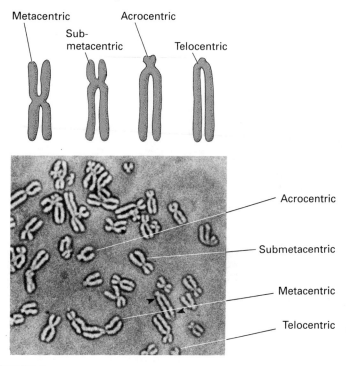

Asexual and Sexual Reproduction

Eukaryotes can reproduce by asexual or sexual reproduction. In **asexual reproduction** a new individual develops from either a single cell or from a group of cells (vegetative reproduction) in the absence of any sexual process. Asexual reproduction is found in both unicellular and multicellular eukaryotes. Single-celled eukaryotes (such as yeast) grow, double their genetic material, and generate two progeny cells, each of which contains an exact copy of the genetic material found in the parental cell. This process repeats as long as there are sufficient nutrients in the growth medium. In multicellular organisms asexual reproduction is sometimes referred to as **vegetative reproduction.** Multicellular fungi, for example, can be propagated vegetatively by taking a small piece of the growth mass and transferring it to new medium. Many higher plants, such as roses and fruit trees, are commercially maintained by grafting, which is an asexual means of propagation.

Sexual reproduction is the fusion of two haploid gametes (sex cells) to produce a single diploid zygote cell. From this zygote a new multicellular individual develops by mitotic division under programmed control from genetic material. Sexual reproduction in-

~ FIGURE 1.16

Cycle of growth and reproduction in a sexually-reproducing higher eukaryote. Sexual reproduction involves the alternation of diploid (somatic) and haploid (gametic) phases.

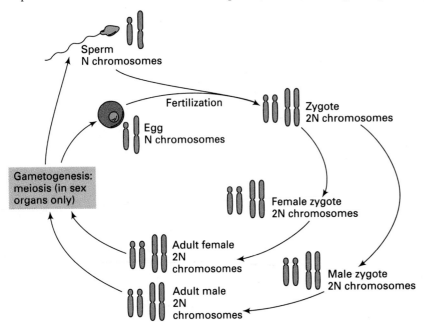

volves the alternation of diploid and haploid phases. The main biological significance of sexual reproduction is that it achieves genetic recombination; that is, it generates gene combinations in the offspring that are distinct from those in the parents. With the exception of self-fertilizing organisms (such as many plants), the two gametes are from different parents, and it is during the production of the gametes that genetic recombination takes place. Figure 1.16 shows the cycle of growth and sexual reproduction in a higher eukaryote and also shows how the number of chromosomes characteristic of an organism is kept constant from generation to generation.

Sexually reproducing animals have two sorts of cells: somatic (body) cells and germ (sex) cells. Somatic cells are haploid or diploid, depending on the eukaryote; lower eukaryotes such as yeast are often haploid, while all higher eukaryotes are diploid. Somatic cells reproduce by a process called mitosis. Germ cells (or gametes), which are always haploid, are produced by a process called meiosis. Figure 1.17 illustrates the differences between asexual and sexual reproduction.

Mitosis: Nuclear Division

In both unicellular and multicellular eukaryotes, cellular reproduction is a cyclical process of growth, **mitosis** (*nuclear division* or *karyokinesis*) and (usually) **cell** division (*cytokinesis*). This process is called the **cell cycle**. In proliferating somatic cells the cell cycle consists of two phases (Figure 1.18): the mitotic (or dividing) phase (M) and an interphase between divisions. Interphase consists of three stages, G_1 (gap 1), S, and G_2 (gap 2). During G_1 (presynthesis stage), the cell prepares for DNA and chromosome replication, which take place in the S stage, during which DNA is replicated. In G_2 (the postsynthesis stage), the cell prepares for cell division, or the M phase. Most of the cell cycle is spent in the G_1 stage, although the relative time spent in each of the four stages varies greatly among cell types.

During interphase of the cell division cycle, the individual chromosomes are extended and are difficult to see under the light microscope. The DNA of each chromosome replicates, and the DNA of the centromere also replicates, although only one centromere structure is seen under the microscope. The product of chromosome duplication is two exact copies, called **sister chromatids,** which are held together by the replicated but unseparated centromere. More precisely, a **chromatid** is one of the two visibly distinct, longitudinal subunits of all replicated chromosomes, which become visible between early prophase and metaphase of mitosis (and between prophase I and the second metaphase of meiosis, discussed later). After mitotic anaphase, they become known as **daughter chromosomes.**

~ FIGURE 1.17

The differences between asexual and sexual reproduction.

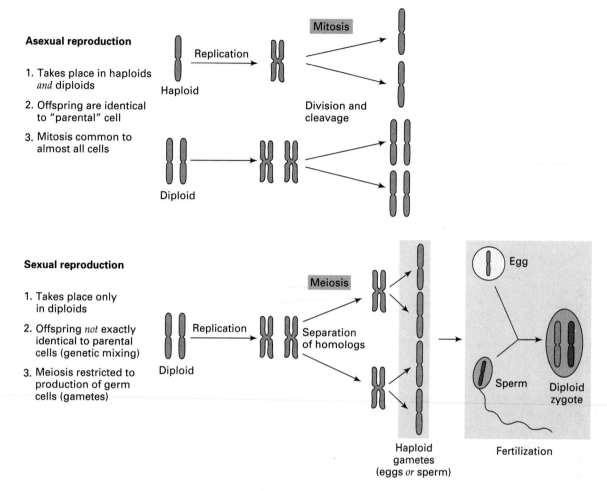

Asexual reproduction

1. Takes place in haploids *and* diploids

2. Offspring are identical to "parental" cell

3. Mitosis common to almost all cells

Sexual reproduction

1. Takes place only in diploids

2. Offspring *not* exactly identical to parental cells (genetic mixing)

3. Meiosis restricted to production of germ cells (gametes)

Cell division in eukaryotic cells involves two processes that may or may not occur together: mitosis (division of the nucleus—karyokinesis), which follows the precise replication of chromosomes and results in the distribution of a complete chromosome set to each progeny nucleus; and cytokinesis, the division of the cytoplasm to form two cells.

Mitosis occurs in both haploid and diploid cells. It is a continuous process, but for purposes of discussion it is usually divided into four cytologically distinguishable stages called *prophase, metaphase, anaphase,* and *telophase.* In Figure 1.19 (p. 16) photographs show the typical chromosome morphology in interphase and in the four stages of mitosis in plant (onion root tip) and animal (newt lung epithelium) cells. Figure 1.20 (p. 17) shows the four stages in simplified diagrams.

PROPHASE. At the beginning of **prophase** (Figure 1.19b; Figure 1.20) the chromatids are very elongated and cannot be seen under the light microscope. In preparation for mitosis they begin to coil tightly so that they

~ FIGURE 1.18

Eukaryotic cell cycle. This cycle assumes a period of 24 hours in culture, although there is great variation between cell types and organisms *in vivo.*

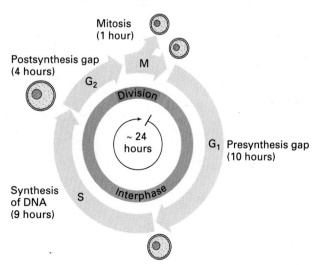

~ FIGURE 1.19

Stages of mitosis in onion root tip (left) and newt lung epithelium (right): (a) Interphase; (b) Prophase; (c) Metaphase; (d) Anaphase; (e) Telophase.

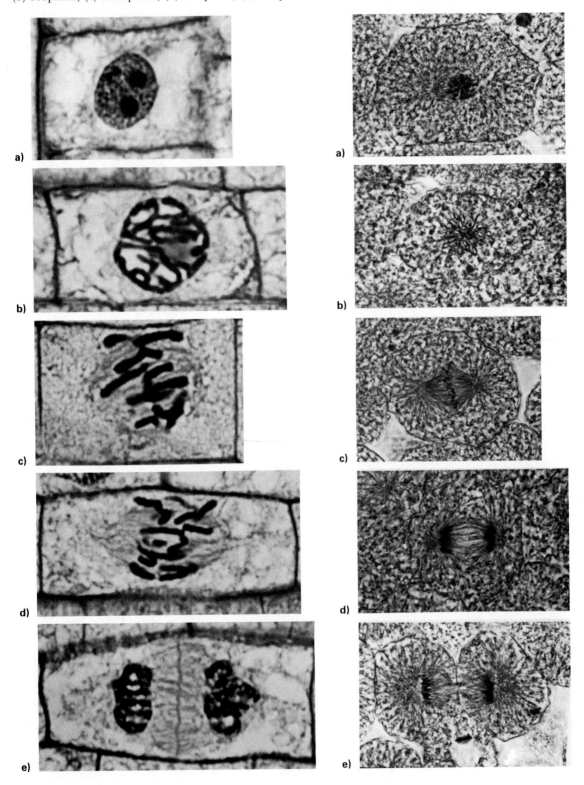

~ FIGURE 1.20

Mitosis in an animal cell.

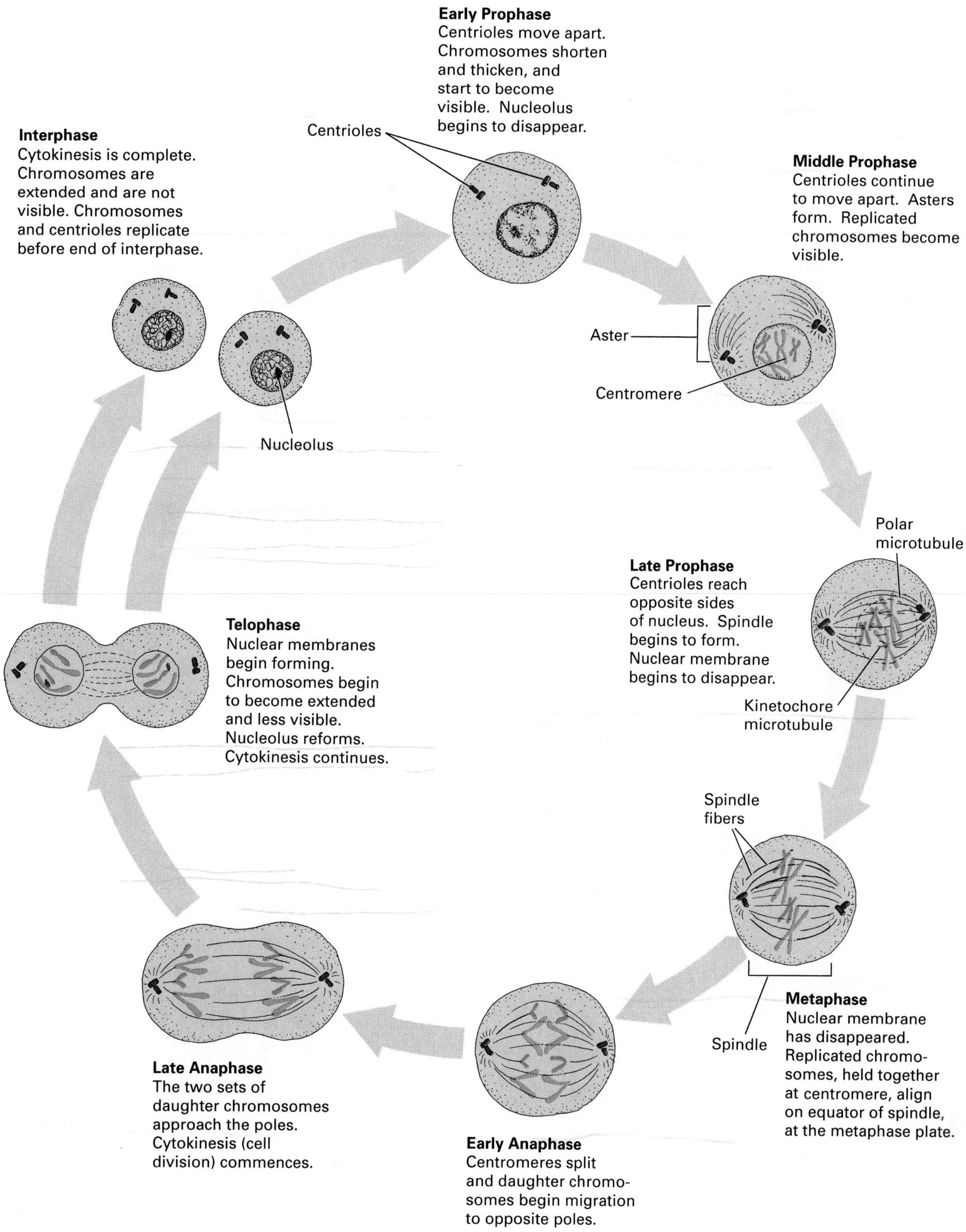

Early Prophase
Centrioles move apart.
Chromosomes shorten
and thicken, and
start to become
visible. Nucleolus
begins to disappear.

Centrioles

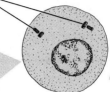

Middle Prophase
Centrioles continue
to move apart. Asters
form. Replicated
chromosomes become
visible.

Interphase
Cytokinesis is complete.
Chromosomes are
extended and are not
visible. Chromosomes
and centrioles replicate
before end of interphase.

Nucleolus

Aster

Centromere

Late Prophase
Centrioles reach
opposite sides
of nucleus. Spindle
begins to form.
Nuclear membrane
begins to disappear.

Polar
microtubule

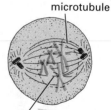

Kinetochore
microtubule

Telophase
Nuclear membranes
begin forming.
Chromosomes begin
to become extended
and less visible.
Nucleolus reforms.
Cytokinesis continues.

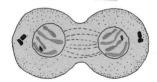

Spindle
fibers

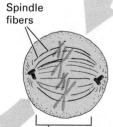

Late Anaphase
The two sets of
daughter chromosomes
approach the poles.
Cytokinesis (cell
division) commences.

Metaphase
Nuclear membrane
has disappeared.
Replicated chromo-
somes, held together
at centromere, align
on equator of spindle,
at the metaphase plate.

Spindle

Early Anaphase
Centromeres split
and daughter chromo-
somes begin migration
to opposite poles.

appear shorter and fatter under the microscope. By late prophase each chromosome, which was duplicated during the preceding S phase, is seen to consist of two sister chromatids.

During prophase, the mitotic spindle (spindle apparatus) assembles outside the nucleus. Each of the spindle fibers in the bipolar mitotic spindle is approximately 25 nm in diameter and consists of microtubules made of special proteins called *tubulins*. In most animal cells, the foci for spindle assembly are the centrioles (see Figure 1.8). (Higher-plant cells usually do not have centrioles, but they do have a mitotic spindle.) Prior to the S phase, the cell's pair of centrioles have replicated and each new centriole pair becomes the focus for a radial array of microtubules called the *aster*. Early in prophase the two asters are adjacent to one another close to the nuclear membrane. By late prophase the two asters are far apart along the outside of the nucleus and are spanned by the microtubular spindle fibers.

Near the end of prophase the nuclear membrane breaks down and the nucleolus or nucleoli disappear, allowing the spindle to enter the nuclear area. Specialized structures called *kinetochores* form on either face of the centromere of each chromosome and become attached to special microtubules called *kinetochore microtubules* (Figure 1.21). These microtubules radiate in opposite directions from each side of each chromosome and interact with the spindle microtubules.

~ **FIGURE 1.21**

Kinetochores and kinetochore microtubules.

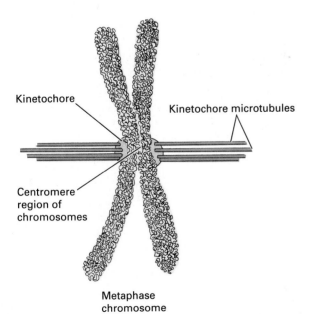

Kinetochore

Kinetochore microtubules

Centromere region of chromosomes

Metaphase chromosome

METAPHASE. Metaphase (Figure 1.19c; Figure 1.20) begins when the nuclear membrane has completely disappeared. During metaphase the chromosomes become arranged so that their centromeres become aligned in one plane halfway between the two spindle poles and with the long axes of the chromosomes at 90 degrees to the spindle axis. The kinetochore microtubules are responsible for this chromosome alignment event. The plane where the chromosomes become aligned is called the **metaphase plate**. Figure 1.22 shows electron micrographs of human chromosomes at this stage of the cell cycle. Note the highly condensed state of the sister chromatids. The electron micrograph (EM) in Figure 1.22c shows a human chromosome from which much of the protein has been removed. In the center is a dense framework of protein called a *scaffold*, which retains the form of the chromosome. The scaffold is surrounded by a halo of DNA filaments that have uncoiled and spread outward.

ANAPHASE. Anaphase (Figure 1.19d; Figure 1.20) is initiated when sister chromatids break apart at the centromere, splitting the chromosome. Once the paired kinetochores on each chromosome separate, the sister chromatid pairs undergo disjunction (separation), and the daughter chromosomes (as each sister chromatid is now called) move toward the poles. In anaphase the two centromeres of the daughter chromosomes migrate toward the opposite poles of the cell. Now the chromosomes assume the characteristic shapes related to the location of the centromere along the chromosome's length. For example, a metacentric chromosome will appear as a V as the two roughly equal-length chromosome arms trail the centromere in its migration toward the pole. Similarly, a submetacentric chromosome will appear as a J. Figure 1.23 gives an interpretation of the mitotic apparatus in anaphase in an animal cell.

TELOPHASE. During telophase (Figure 1.19e; Figure 1.20), the migration of daughter chromosomes to the two poles is completed. The two sets of progeny chromosomes are assembled into two groups at opposite ends of the cell. The chromosomes begin to uncoil and assume the extended state characteristic of interphase. A nuclear membrane forms around each chromosome group, the spindle microtubules disappear, and the nucleolus or nucleoli re-form. At this point, the cell has two nuclei.

CYTOKINESIS. Cytokinesis refers to division of the cytoplasm. In most cases the telophase stage of mitosis is accompanied by cytokinesis. Cytokinesis compartmentalizes the two new nuclei into separate daughter cells, completing the mitotic cell division process. In

~ FIGURE 1.22

~ FIGURE 1.22

Human metaphase chromosome. (a) Transmission electron micrograph of an intact
chromosome; (b) Scanning electron micrograph of an intact chromosome;
(c) Transmission electron micrograph of a chromosome from which much of the protein
has been removed; the DNA filaments have uncoiled.

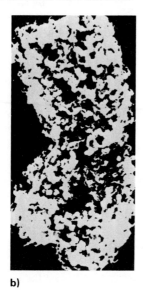
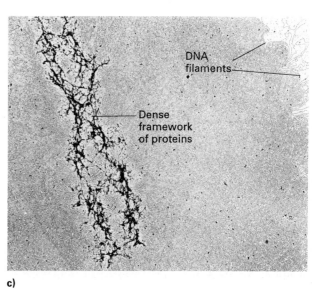

a)   b)   c)

~ FIGURE 1.23

Mitotic apparatus at anaphase in an animal cell. A centriole pair is located at each pole,
with spindle microtubules spanning the cell. The separation of the sister chromatids
depends on spindle microtubules attached to the centromere of each chromatid.

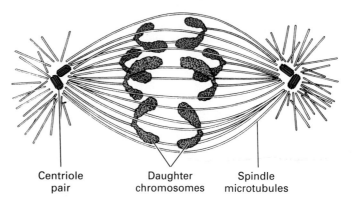

Centriole   Daughter   Spindle
pair   chromosomes   microtubules

~ FIGURE 1.24

Cytokinesis (cell division). (a) Diagram of cytokinesis in an animal cell; (b) Diagram of cytokinesis in a higher plant cell; (c) Scanning electron micrograph of an animal cell in tissue culture shortly after its division.

a) Animal cell   **b) Plant cell**

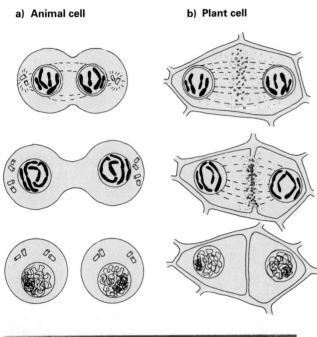

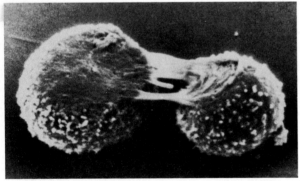

c) Scanning electron micrograph

animal cells cytokinesis occurs by the formation of a constriction in the middle of the cell until two daughter cells are produced (Figure 1.24a). However, most plant cells do not divide by the formation of a constriction. Instead, a new cell membrane and cell wall are assembled between the two new nuclei to form a *cell plate* (Figure 1.24b). Cell wall material coats each side of the cell plate, and the result is two progeny cells. Figure 1.24c is a scanning electron micrograph of an animal cell in tissue culture shortly after its division.

GENETIC SIGNIFICANCE OF MITOSIS. Mitosis maintains the genetic content of a cell from generation to generation. Mitosis occurs in haploid or diploid cells after DNA and chromosome duplication has taken place. It is a highly ordered process in which a duplicated chromosome set is partitioned equally into the two daughter cells. Thus, for a haploid (N) cell, chromosome duplication produces a cell with two sets of chromosomes. The ensuing mitosis results in two progeny haploid cells, each with one set of chromosomes. For a diploid cell, which has two sets of chromosomes, chromosome duplication produces a cell in which each chromosome set has doubled its content. The ensuing mitosis results in two progeny diploid cells, each with two sets of chromosomes.

*K*EYNOTE

Mitosis is the process of nuclear division in eukaryotes. It is one part of the cell cycle (i.e., G_1, S, G_2, and M) and results in the production of daughter nuclei that contain identical chromosome numbers and are genetically identical to one another and to the parent nucleus from which they arose. Prior to mitosis, DNA synthesis occurs to double the amount of DNA concomitant with chromosome duplication. Mitosis is usually followed by cytokinesis. Both haploid and diploid cells can undergo mitosis.

Meiosis

Meiosis is the term applied to the two successive divisions of a diploid nucleus following only one DNA replication cycle. That results in the formation of haploid gametes (**gametogenesis**) or of meiospores (in sporogenesis). During meiosis, homologous chromosomes replicate, then pair, then undergo two divisions. As a consequence, each of the four cells resulting from the two meiotic divisions receives one chromosome of each chromosome set. The two nuclear divisions of a normal meiosis are called **meiosis I** and **meiosis II**. The first meiotic division results in reduction in the number of chromosomes from diploid to haploid (reductional division), and the second division results in separation of the chromatids (equational division). In most cases the divisions are accompanied by cytokinesis, so the result of the meiosis of a single diploid cell is four haploid cells.

MEIOSIS I: THE FIRST MEIOTIC DIVISION.

Meiosis I, in which the chromosome number is reduced from diploid to haploid, consists of four cytologically distinguishable stages: *prophase I, metaphase I, anaphase I,* and *telophase I.* Figure 1.25 (pp. 22–23) presents photographs and diagrams of the meiosis I stages.

Prophase I. In brief, in prophase I the chromosomes become shorter and thicker, crossing-over occurs, the spindle apparatus forms, and the nuclear membrane and nucleolus/nucleoli disappear (Figure 1.25). Prophase I is divided into a number of stages. As it begins, the chromosomes have already replicated. In **leptonema** (early prophase; = the leptotene stage) the chromosomes have begun to coil and are now visible. The threadlike chromosomes seen at this time are the sister chromatids held together at the centromere. The number of chromosomes present at this stage is the same as the number in the diploid cell. Once a cell enters leptonema, it's committed to meiotic process.

In **zygonema** (early/mid-prophase; = the zygotene stage) homologous chromosomes begin to pair in a highly specific way (like a zipper). This chromosome pairing is called **synapsis.** Because of the replication that occurred earlier, each synapsed set of homologous chromosomes consists of four chromatids and is referred to as a **bivalent** or *tetrad.*

Following zygonema is **pachynema** (mid-prophase; = the pachytene stage). At this stage the chromosomes become much shorter and thicker and are now visible as four-strand structures. They are also more intimately synapsed. During pachynema a most significant event takes place: **crossing-over,** the reciprocal exchange of chromosome segments at corresponding positions along pairs of homologous chromosomes. If there are genetic differences between the homologs, crossing-over can produce new gene combinations in a chromatid. There is usually no loss or addition of genetic material to either chromosome, since crossing-over involves reciprocal exchanges.

The process of crossing-over is facilitated by the tight synapse (alignment side by side) of the four chromatids and involves the formation of a zipperlike structure along the length of the chromatids called the **synaptinemal complex** (Figure 1.26; p. 24). Although we know that the structure consists mainly of protein with some RNA attached, we do not know its role in crossing-over. In some organisms, such as male *Drosophila,* the synaptinemal complex is not formed and no crossing-over occurs in that case.

A chromosome that emerges from meiosis with a combination of genes that differs from the combination with which it started is called a **recombinant chromosome.** Therefore crossing-over is a mechanism that can give rise to **genetic recombination,** a concept we will examine more fully in later chapters.

The next stage in prophase I is **diplonema** (mid/late prophase; = the diplotene stage). Here the chromosomes begin to move apart. The process of crossing-over becomes visible during diplonema as a cross-shaped structure called a **chiasma** (plural, chiasmata: Figure 1.27a; p. 24). Since all four chromatids may be involved in crossing-over events along the length of the homologs, the chiasma pattern at this stage may be very complex.

Diplonema is followed rapidly in most organisms by the remaining stages of meiosis. However, in many animals the oocytes (egg cells) can remain in diplonema for very long periods. For example, in the human female, oocytes go through meiosis up to diplonema by the seventh month of fetal development and then remain arrested in this stage for many years. At the onset of puberty and until menopause, one oocyte per menstrual cycle matures into a haploid ovum (egg) and is released from the ovary.

Diakinesis (late prophase) follows diplonema. During this stage, the four chromatids of each bivalent (tetrad) are even more condensed, and the chiasmata often *terminalize;* that is, they move down the chromatids to the ends (Figure 1.27b). As a result, the chromatids at this stage appear to be attached near the tips. That is, the chromosomes are sliding past each other so that, in a sense, the chiasmata delay the separation of the chromatids. In addition, the nucleoli disappear, and the nuclear membrane begins to break down. The chromosomes can be most easily counted at this stage of meiosis.

Metaphase I. By the beginning of metaphase I (see Figure 1.25), the nuclear membrane has completely broken down, and the bivalents (tetrads) become aligned on the equatorial plane of the cell. The spindle apparatus is completely formed now, and the microtubules are attached to the centromeres of the homologs. Note particularly that it is the synapsed pairs of homologs (the bivalents) that are found at the metaphase plate. In contrast, in mitosis of most (but not all) organisms, replicated homologous chromosomes (sister chromatid pairs) align independently at the metaphase plate. *In other words, a key difference between mitosis and meiosis is that sister centromeres remain joined in meiosis I, whereas in mitosis they separate.*

Anaphase I. In anaphase I (see Figure 1.25) the chromosomes in each tetrad separate, so homologous pairs (the bivalents) disjoin and migrate toward opposite

~ FIGURE 1.25

Meiosis. The diagrams show the stages of meiosis in an animal cell; the photographs show the same meiotic stages in a plant (here, chives).

Early prophase I:

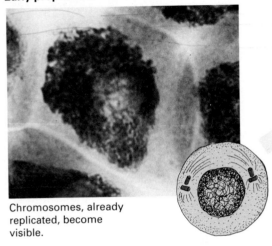

Chromosomes, already replicated, become visible.

Middle prophase I:

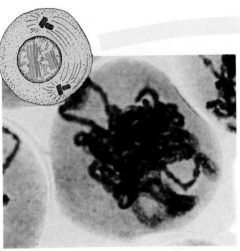

Homologous chromosomes shorten and thicken. The chromosomes synapse and crossing over occurs.

4 gametes

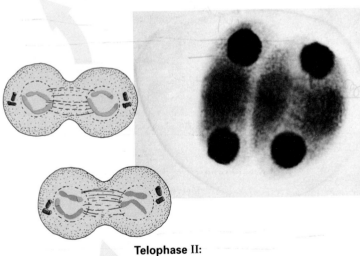

Telophase II:

Anaphase II:

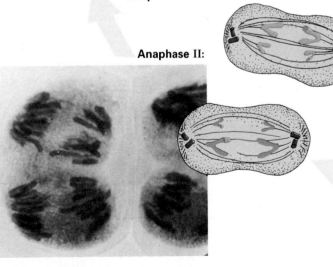

Metaphase II:

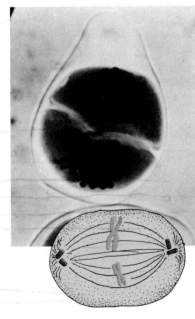

Metaphase I:

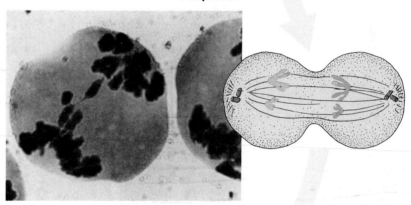

Assembly of spindle is completed. Each chromosome pair (bivalent) aligns across the equatorial plane of the spindle.

Late prophase I:

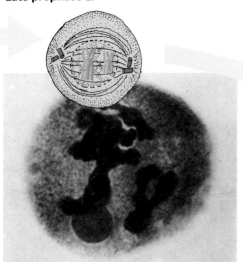

Results of crossing over become visible as chiasmata. Nuclear membrane begins to disappear. Spindle apparatus begins to form.

Anaphase I:

Homologous chromosome pairs (dyads) separate and migrate toward opposite poles.

Telophase I:

Cytokinesis: In most species, cytokinesis occurs to produce two cells. Chromosomes do not replicate before meiosis II

Chromosomes (each with two sister chromatids) complete migration to the poles and new nuclear membranes may form.

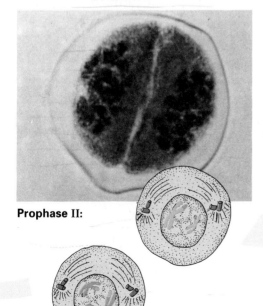

Prophase II:

~ FIGURE 1.26

Electron micrograph of the synaptinemal complex separating two homologous chromatid pairs during the zygonema stage of prophase I.

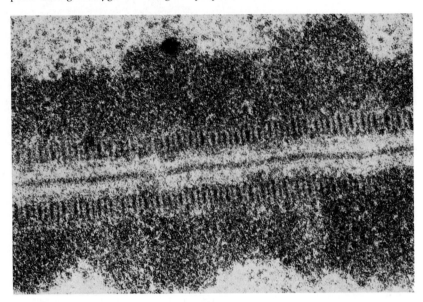

poles, the areas in which new nuclei will form. (At this stage, each pair is called a *dyad*.) This migration gives rise to two things: (1) Maternally-derived and paternally-derived homologues are segregated randomly at each pole (except for the parts of chromosomes exchanged during the crossing-over process); and (2) At each pole there is a haploid complement of replicated chromosomes. It is important to remember that at this time the segregated sister chromatid pairs remain attached at their respective centromeres.

Telophase I. The length of telophase I (see Figure 1.25) varies considerably among species. Nonetheless, in telophase I the dyads complete their migration to opposite poles of the cell, and (in most cases) new nuclear membranes form around each haploid grouping. In most species, cytokinesis follows, producing two haploid cells. Thus, meiosis I, which begins with a diploid cell that contains one maternally-derived and one paternally-derived set of chromosomes, ends with two nuclei, each of which contains one mixed-parental set

~ FIGURE 1.27

(a) Appearance of chiasmata, the visible evidence of crossing-over, in diplonema;
(b) Terminalization of a chiasmata during diakinesis.

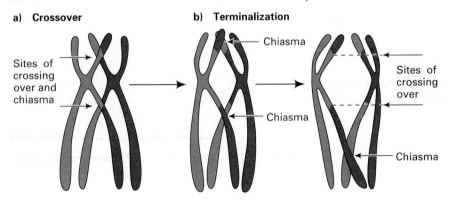

of replicated chromosomes, all still joined at the centromeres. After cytokinesis, each of the two progeny cells has a nucleus with a haploid set of replicated chromosomes.

MEIOSIS II: THE SECOND MEIOTIC DIVISION. The second meiotic division is very similar to a mitotic division. Figure 1.25 presents photographs and diagrams of prophase II, metaphase II, anaphase II, and telophase II of meiosis II.

Prophase II (Figure 1.25) is a stage of chromosome contraction.

In **metaphase II** (see Figure 1.25) each of the two daughter cells organizes a spindle apparatus that attaches to the now-divided centromeres. The centromeres line up on the equator of the second-division spindles.

During **anaphase II** (see Figure 1.25) the centromeres (and therefore the chromatids) are pulled to the opposite poles of the spindle: One sister chromatid of each pair goes to one pole, while the other goes to the opposite pole. The separated chromatids are now referred to as chromosomes in their own right.

In the last stage, **telophase II** (see Figure 1.25), a nuclear membrane forms around each set of chromosomes, and cytokinesis takes place. After telophase II, the chromosomes become more extended and again are invisible under the light microscope.

Since there is *no chromosome duplication between meiosis I and meiosis II, the end* products of the two meiotic divisions are four haploid cells from one original diploid cell. Each of the four progeny cells has one chromosome from each homologous pair of chromosomes. Remember, however, that these chromosomes are not exact copies of the original chromosomes because of the crossing-over that occurs between chromosomes during pachynema of meiosis I. Figure 1.28 (p. 26) compares mitosis and meiosis.

Genetic Significance of Meiosis

Meiosis has three significant results:

1. Meiosis generates cells with half the number of chromosomes found in the diploid cell that entered the process because two division cycles follow only one cycle of DNA replication (S period). Fusion of the haploid nuclei (called fertilization, or syngamy) restores the diploid number. Therefore through a cycle of meiosis and fertilization, the chromosome number is maintained in sexually reproducing organisms.
2. In metaphase I of meiosis, each maternally derived and paternally derived chromosome has an equal chance of aligning on one or the other side of the equatorial metaphase plate. As a result, each nucleus generated by meiosis will have a combination of maternal and paternal chromosomes.

The number of possible chromosome combinations is large, especially when the number of chromosomes in an organism is large. Consider a hypothetical organism with two pairs of chromosomes in a diploid cell entering meiosis. Figure 1.29 (p. 26) shows the two possible combinations of maternal and paternal chromosomes that can occur at the metaphase plate.

The general formula states that the number of possible chromosome arrangements is 2^{n-1}, where n is the number of chromosome pairs. In *Drosophila*, which has four pairs of chromosomes, the number of possible arrangements is 2^3, or 8; in humans, which have 23 chromosome pairs, over 4 million metaphase arrangements are possible. Therefore since there are many gene differences between the maternally derived and paternally derived chromosomes, the nuclei produced by meiosis will be genetically quite different from the parental cell and from each other.

3. The crossing-over between maternal and paternal chromatid pairs during meiosis I results in still more variation in the final combinations. Crossing-over occurs during every meiosis, and because the sites of crossing-over vary from one meiosis to another, the number of different kinds of progeny nuclei produced by the process is extremely large. Given the genetic features of meiosis, this process is of critical importance for understanding the behavior of genes, as will be seen in the following chapters.

𝒦EYNOTE

Meiosis occurs in all sexually reproducing eukaryotes. It is a process by which a diploid (2N) cell or cell nucleus is transformed, through one round of chromosome replication and two rounds of nuclear division, into four haploid (N) cells or nuclei. In the first of two divisions, pairing synapsis of homologous chromosomes occurs. The meiotic process results in the conservation of the number of chromosomes from generation to generation. It also generates genetic variability through the various ways in which maternal and paternal chromosomes are combined in the progeny nuclei and by crossing-over (the physical exchange of genes between maternally or paternally derived homologs). ▬▬▬▬▬

~ FIGURE 1.28

Comparison of mitosis and meiosis in a diploid cell.

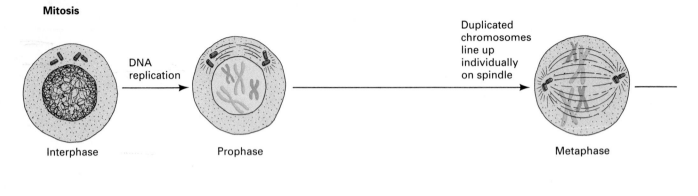

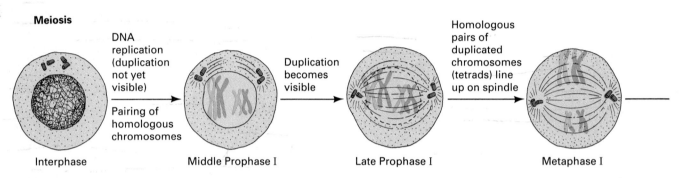

~ FIGURE 1.29

Two possible arrangements of two pairs of homologous chromosomes on the metaphase plate of the first meiotic division. Paternal chromosomes are shown in yellow and green; maternal chromosomes in purple and tan.

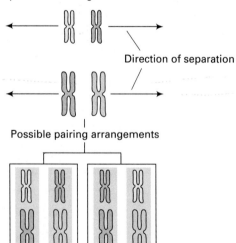

Locations of Meiosis in the Life Cycle

MEIOSIS IN ANIMALS. Most multicellular animals are diploid through most of their life cycle. In such animals meiosis produces haploid gametes, which produce a diploid zygote when their nuclei fuse in the fertilization process. The zygote then divides mitotically to produce the new diploid organism. Thus the gametes are the only haploid stages of the life cycle. Gametes are only formed in specialized cells. In the male the gamete is the sperm, produced through a process called **spermatogenesis**. The female gamete is the egg, produced by **oogenesis**. Spermatogenesis and oogenesis are illustrated in Figure 1.30 (p. 28).

In male animals the **sperm cells** (also called **spermatozoa**) are produced by the testes. The testes contain the primordial germ cells (*primary spermatogonia*), which, through mitotic division, produce *secondary spermatogonia*. Spermatogonia transform into *primary spermatocyctes (meiocytes)*, each of which undergoes meiosis I and gives rise to two *secondary spermatocytes*. Each spermatocyte undergoes meiosis II. As a result of

~ **FIGURE 1.28 continued**

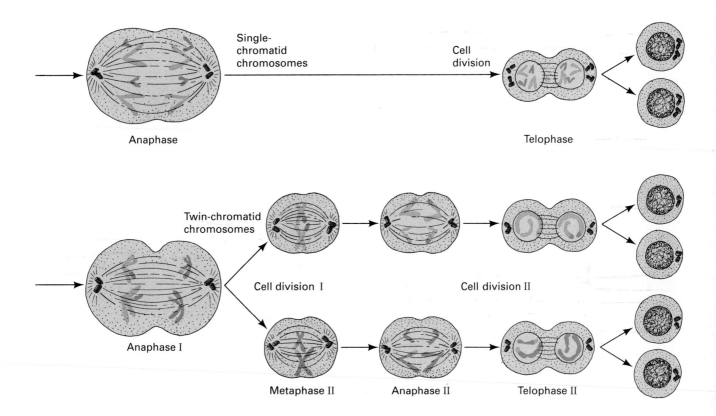

these two divisions, four haploid *spermatids* arise, which eventually differentiate into the mature male gametes, the spermatozoa.

In female animals the ovary contains the primordial germ cells (*primary oogonia*), which, by mitosis, give rise to *secondary oogonia*. These cells transform into *primary oocytes*, which grow until the end of oogenesis. The diploid, primary oocyte goes through meiosis I and unequal cytokinesis to give two cells: a large one called the **secondary oocyte** and a very small one called the *first polar body*. In the second meiotic division the secondary oocyte produces two haploid cells. One is a very small cell and is called a *second polar body;* the other is a large cell that rapidly matures into the mature egg cell, or **ovum**. The first polar body may or may not divide. Only the ovum is a viable gamete. Thus in the female animal only one mature gamete (the ovum) is produced by meiosis of a diploid cell.

MEIOSIS IN PLANTS. The life cycle of sexually reproducing plants typically has two phases, the **gameto-**phyte or haploid stage in which gametes are produced, and the **sporophyte** or diploid stage in which haploid spores are produced by meiosis.

In angiosperms, the flowering plants, the flower is the structure in which sexual reproduction occurs. Figure 1.31 (p. 28) shows a generalized flower containing both male and female reproductive organs, the **stamens** and **pistils**, respectively. Each stamen consists of a single stalk, the filament, on the top of which is an anther. From the anther are released the pollen grains, which are immature male gametophytes. The pistil contains the female gametophytes and typically consists of the stigma, a sticky surface specialized to receive the pollen; the style, a thin stalk down which a pollen tube grows; and at the base of the structure, the ovary, within which are the ovules. Each ovule encloses a female gametophyte with a single egg cell. When the egg cell is fertilized, the ovule develops into a seed.

Meiosis occurs in specific ways in the female and male parts of the flower. Meiosis in the female part of the flower occurs as follows. Within the ovary of the flower, each ovule is a single large 2N cell called the

~ FIGURE 1.30

Spermatogenesis and oogenesis in an animal cell.

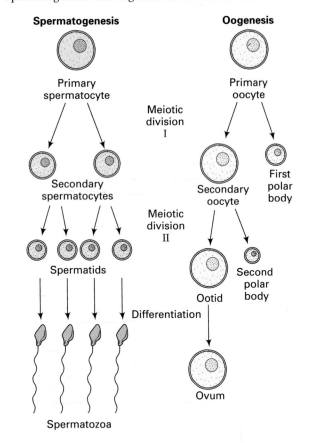

~ FIGURE 1.31

Generalized structure of a flower.

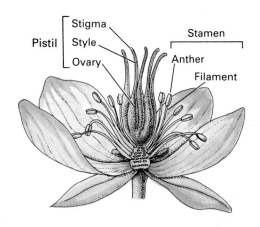

megaspore mother cell. This cell undergoes meiosis, producing four haploid products, only one of which remains viable—the megaspore, or haploid female gametophyte. The production of the megaspore is called **megasporogenesis** (Figure 1.32; p. 29).

Meiosis in the megaspore mother cell involves two divisions as usual, resulting in four haploid cells. Three of the cells disintegrate leaving a single large cell with one haploid nucleus. Three successive nuclear mitotic divisions produce a megagametophyte cell with eight identical haploid nuclei. The eight nuclei migrate in two sets of four to the two ends of the cell, and then two of the nuclei—one from each end—migrate back to the cell center. Cell walls are produced, resulting in unequal division of the cytoplasm. The larger cell forms around the two central nuclei: this binucleate cell, when it is fertilized, gives rise to the **endosperm** of the seed; hence it is called the *endosperm mother cell*. Each of the other six haploid nuclei is now enclosed in its own cell. The entire seven-celled megagametophyte structure is called the *embryo sac*. One of the six haploid cells is the egg cell, which will become the single female gamete of the ovule as the result of random fertilization by a sperm cell.

In the anther, meiosis occurs by similar, although less complex, events to produce the pollen grains, or male gametophytes (Figure 1.33; p. 30). This series of events is called **microsporogenesis.** Anthers contain four pollen sacs within which are many diploid microspore mother cells, each of which undergoes meiosis to produce four haploid microspores. The haploid nucleus of each microspore divides once mitotically, producing a haploid generative nucleus and a haploid tube nucleus. Each microspore, with its two nuclei, then generates a tough coat to become a pollen grain.

During pollination, the pollen grains are deposited on the stigma and each "germinates," one producing a pollen tube (Figure 1.34; p. 30). The pollen tube grows down through the stigma and into the style, producing enzymes to digest the tissues ahead of its growth. As the pollen tube grows down the style, the nuclei stay close to the growing tip: the haploid generative nucleus produces two sperm nuclei (the gametes) by mitosis, while the tube nucleus does not divide at all. The pollen tube eventually penetrates the tissues of the ovule at a very small pore called the *micropyle*, allowing the two sperm nuclei to enter the embryo sac. One sperm nucleus fertilizes the haploid egg cell, producing the diploid zygote. The other sperm nucleus fuses with the two haploid nuclei in the endosperm mother cell, producing a triploid nucleus, thereby completing fertilization. This **double fertilization** is found only in flowering plants.

After fertilization, the diploid zygote, the triploid endosperm cell, and the other tissues begin the cell divisions necessary to produce the plant embryo. In this structure, the embryos derive from the zygote, and the stored food within the seed derives from the triploid endosperm cell.

Among living organisms only plants produce gametes from special bodies called gametophytes. Thus

~ FIGURE 1.32

Megasporogenesis: Meiosis in the female part of the flower and the production of the embryo sac.

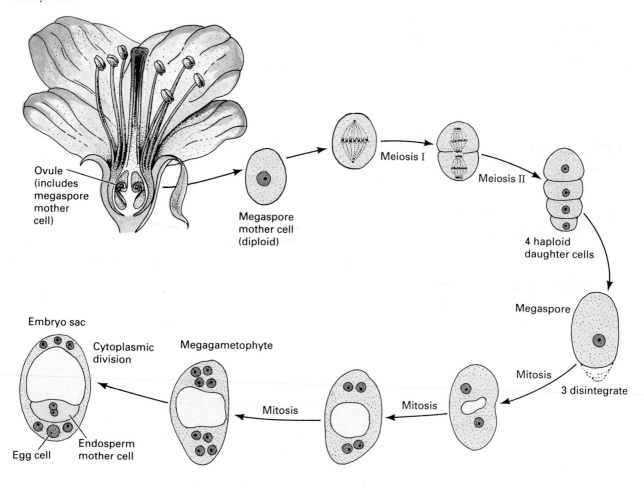

plants have two distinct reproductive phases, called the **alternation of generations** (Figure 1.35; p. 31). Meiosis and fertilization constitute the transition between these stages. The haploid gametophyte generation begins with spores that are produced by meiosis. In flowering plants the equivalents to spores are the cells that ultimately become pollen and ovule. Fertilization initiates the diploid sporophyte generation. The diploid sporophyte generation produces specialized haploid cells called spores, which, in turn, produce gametophytes. Thus the sporophyte is the second alternate generation.

Figure 1.35 also indicates the relationship of each phase of the alternation of generations to meiosis. The products of meiosis in plants are the spores. The spores of ferns and mosses, the cells in gymnosperms (cone-bearing plants) and angiosperms that produce male gametophytes, and the cells in the ovules of flowers that produce the female gametophyte are all produced by meiosis. Thus in all green plants the alternation of generations involves an alternation between stages of hap-

loid cells and diploid cells: The gametophyte cells are haploid, and the sporophyte cells are diploid.

SUMMARY

In this chapter we reviewed basic organizational features of viruses and cells, and discussed cellular reproduction in eukaryotes, with particular emphasis on mitosis and meiosis.

As the simplest cellular organisms, prokaryotes lack a membrane-bound nucleus and divide by binary fission. The genetic material of prokaryotes is, nonetheless, localized within the cell in an area called the nucleoid region. Prokaryotic genetic material consists of a single, circular DNA molecule that is associated with few proteins. Eukaryotic organisms may be single-celled or multicellular and have a membrane-bound nucleus. The genetic material of eukaryotes consists of DNA distributed among several linear chromosomes.

~ FIGURE 1.33

Microsporogenesis: Meiosis in the male part of the flower and the production of a pollen grain.

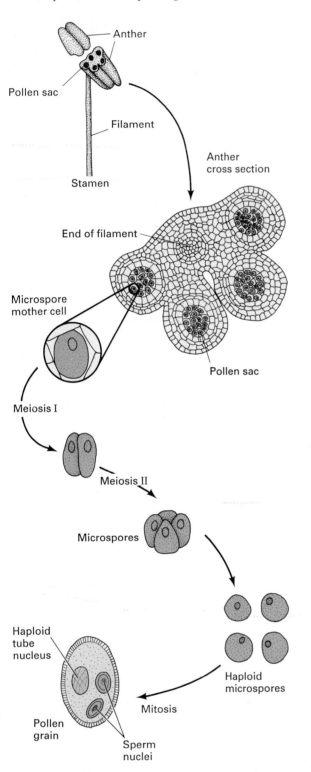

~ FIGURE 1.34

Pollination and fertilization in a flowering plant.

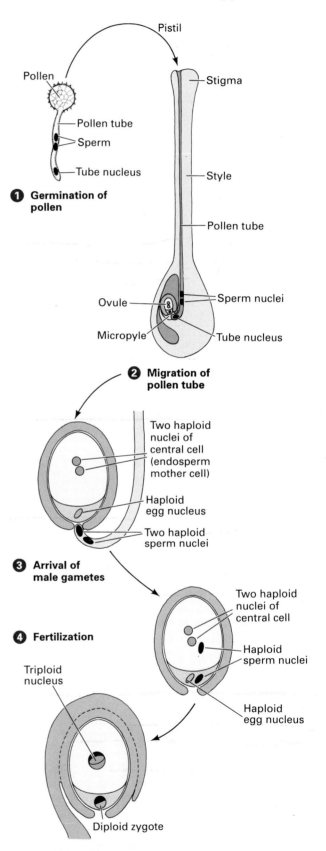

~ FIGURE 1.35

Alternation of gametophyte and sporophyte generations in green plants.

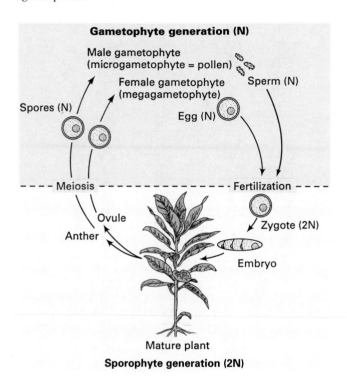

In these chromosomes, the DNA is complexed with many proteins. Typically, eukaryotic cells contain either one or two sets of chromosomes: the former are haploid cells and the latter are diploid cells. Nuclear division in both haploid and diploid cells occurs by mitosis. Mitosis involves one round of DNA replication followed by one round of nuclear division, often accompanied by cell division. Thus, mitosis results in the production of daughter nuclei that contain identical chromosome numbers and that are genetically identical to one another and to the parent nucleus from which they arose.

In all sexually reproducing organisms, meiosis occurs at a particular stage in the life cycle. Meiosis is the process by which a diploid cell (never a haploid cell) or cell nucleus undergoes one round of DNA replication and two rounds of nuclear division to produce four haploid cells or nuclei. In many eukaryotes the products of meiosis are the gametes. Unlike mitosis, meiosis generates genetic variability in two main ways: (1) through the various ways in which maternal and paternal chromosomes are combined in progeny nuclei; and (2) through crossing-over between maternally derived and paternally derived homologs to produce recombinant chromosomes with some maternal and some paternal genes.

ANALYTICAL APPROACHES FOR SOLVING GENETICS PROBLEMS

Because this chapter is primarily descriptive, there are no quantitative problems for this material.

QUESTIONS AND PROBLEMS

In 1.1 through 1.3, select the correct answer.

*1.1 Interphase is a period corresponding to the cell cycle phases of
a. mitosis.   c. $G_1 + S + G_2$.
b. S.   d. $G_1 + S + G_2 + M$.

1.2 Chromatids joined together by a centromere are called
a. sister chromatids.   c. alleles.
b. homologs.   d. bivalents (tetrads).

*1.3 Mitosis and meiosis always differ in regard to the presence of
a. chromatids.   d. centromeres.
b. homologs.   e. spindles.
c. bivalents

*Solutions to Questions and Problems marked with an asterisk are found on pages S-1–S-16.

1.4 State whether each of the following statements is true or false. Explain your choice.
a. The chromosomes in a somatic cell of any organism are all morphologically alike.
b. During mitosis the chromosomes divide and the resulting sister chromatids separate at anaphase, ending up in two nuclei, each of which has the same number of chromosomes as the parental cell.
c. At zygonema, any chromosome may synapse with any other chromosome in the same cell.

*1.5 Decide whether the answer to these statements is *yes* or *no*. Then explain the reasons for your decision.
a. Can meiosis occur in haploid species?
b. Can meiosis occur in a haploid individual?

In 1.6 through 1.8, select the correct answer.

1.6 The general life cycle of a eukaryotic organism has the sequence
a. 1N → meiosis → 2N → fertilization → 1N.
b. 2N → meiosis → 1N → fertilization → 2N.
c. 1N → mitosis → 2N → fertilization → 1N.
d. 2N → mitosis → 1N → fertilization → 2N.

*1.7 Which statement is true?
a. Gametes are 2N; zygotes are 1N.
b. Gametes and zygotes are 2N.
c. The number of chromosomes can be the same in gamete cells and in somatic cells.
d. The zygotic and the somatic chromosome numbers cannot be the same.
e. Haploid organisms have haploid zygotes.

1.8 All of the following happen in prophase I of meiosis *except*
a. chromosome condensation.
b. pairing of homologs.
c. chiasma formation.
d. terminalization.
e. segregation.

*1.9 Give the name of the stages of mitosis or meiosis at which the following events occur:
a. Chromosomes are located in a plane at the center of the spindle.
b. The chromosomes move away from the spindle equator to the poles.

1.10 Given the diploid, meiotic mother cell shown below, diagram the chromosomes as they would appear (a) in late pachynema; (b) in a nucleus at prophase of the second meiotic division; (c) in the first polar body resulting from oogenesis in an animal.

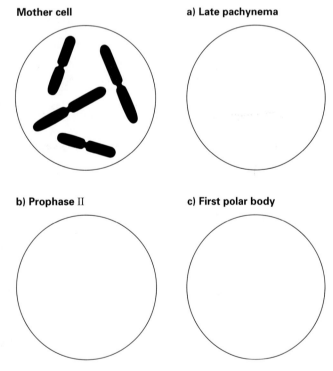

Mother cell **a) Late pachynema**

b) Prophase II **c) First polar body**

1.11 The cells in the figure below were all taken from the same individual (a mammal). Identify the cell division events happening in each cell, and explain your reasoning. What is the sex of the individual? What is the diploid chromosome number?

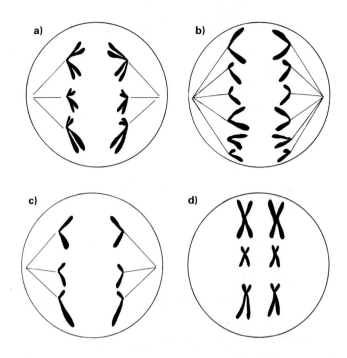

a)   b)

c)   d)

1.12 Does mitosis or meiosis have greater significance in genetics (i.e., the study of heredity)? Explain your choice.

*1.13 Consider a diploid organism that has three pairs of chromosomes. Assume that the organism receives chromosomes A, B, and C from the female parent and A′, B′, and C′ from the male parent. To answer the following questions, assume that no crossing-over occurs.
a. What proportion of the gametes of this organism would be expected to contain all the chromosomes of maternal origin?
b. What proportion of the gametes would be expected to contain some chromosomes from both maternal and paternal origin?

*1.14 Normal diploid cells of a theoretical mammal are examined cytologically at the mitotic metaphase stage for their chromosome complement. One short chromosome, two medium-length chromosomes and three long chromosomes are present. Explain how the cells might have such a set of chromosomes.

1.15 Is the following statement true or false? Explain your decision.
 "Meiotic chromosomes may be seen after appropriate staining in nuclei from rapidly-dividing skin cells."

*1.16 Is the following statement true or false? Explain your decision.
 "All of the sperm from one human male are genetically identical."

1.17 The horse has a diploid set of 64 chromosomes, and a donkey has a diploid set of 62 chromosomes. Mules (viable but usually sterile progeny) are produced when a male donkey is mated to a female horse. How many chromosomes will a mule cell contain?

2 MENDELIAN GENETICS

PRINCIPAL POINTS

~ The genotype is the genetic constitution of an organism, while the phenotype is the physical manifestation of genetic traits.

~ The genes give the potential for the development of characteristics; this potential is affected by interactions with other genes and by the environment.

~ Mendel's principle of segregation states that the two members of a gene pair (alleles) segregate from each other in the formation of gametes.

~ A testcross is a cross of an individual of unknown genotype, usually expressing the dominant phenotype, with a homozygous recessive in order to determine the genotype of the unknown individual.

~ Mendel's principle of independent assortment states that genes on different chromosomes are distributed independently of one another during the production of gametes.

~ Mendelian principles apply to humans as well as to peas and all other eukaryotes. The study of the inheritance of genetic traits in humans is complicated by the fact that no controlled crosses can be done. Instead, human geneticists analyze genetic traits by pedigree analysis; that is, by examining the occurrences of the trait in family trees of individuals who clearly exhibit the trait.

By simple observation it is evident that there is a lot of variation among individuals of a given species. Among dogs, for example, are many breeds, including Bernese mountain dogs, Dalmatians, pointers, dachshunds, Yorkshire terriers, and so on (Figure 2.1; see also Figure 22.15). These breeds are clearly distinguishable by their size, shape, and color, yet they are all representatives of the same species: *Canis familiaris*. Similarly, differences among individual humans include eye color, height, and hair color, even though they all belong to the species *Homo sapiens*. Within a species each generation perpetuates individual differences, yet the species remains clearly identifiable. The differences among individuals within and between species are the result of differences in the DNA sequences that constitute their genes. It is genes that determine the structure, function, and development of the cell or organism. Thus the genetic information coded in DNA is responsible for species and individual variation.

For many centuries breeding experiments on plants and animals have been carried out in an attempt to understand how characteristics are transmitted from one generation to the next. The understanding of how genes are transmitted from parent to offspring began with the work of Gregor Johann Mendel (1822–1884), an Austrian monk. In this chapter we will learn the basic principles of transmission genetics by examining Mendel's experiments and their results. Throughout this chapter we must remember that the segregation of genes is directly related to the behavior of chromosomes. However, while Mendel analyzed the patterns of segregation of hereditary traits that he called characters, he did not know that genes control the characters, or that genes are located in chromosomes. Indeed, he did not know that chromosomes existed.

GENOTYPE AND PHENOTYPE

Before we begin our study of Mendel's work, we must distinguish between the nature of the genetic material each individual organism possesses and the physical characteristics expressed by this genetic material.

The characteristics of an individual that are transmitted from one generation to another are sometimes called **hereditary traits** (Mendel called them **characters**). These traits are under the control of DNA segments called **genes**. The genetic constitution of an organism is called its **genotype**, while the physical manifestation of a genetic trait is called a **phenotype**. Different forms of the gene that controls eye color in humans, for example, produce a number of phenotypes: brown eyes, blue eyes, green eyes, and so on.

The genes that every individual carries are essentially like a blueprint that can be interpreted in several ways. Genes give only the potential for the development of a particular phenotypic characteristic; the extent to which that potential is realized depends not only on interactions with other genes and their products but also on environmental influences (Figure 2.2). A person's height, for example, is controlled by many genes, the expression of which, in turn, can be significantly

~ FIGURE 2.1

Variation among dogs: (a) Bernese mountain dog; (b) Dalmatian; (c) Pointer; (d) Dachshund; (e) Yorkshire terrier.

a)

b)

c)

d)

e)

~ FIGURE 2.2

Influences on the physical manifestation (phenotype) of the genetic blueprint (genotype): interactions with other genes and their products (such as hormones) and with the environment (such as nutrition).

Genotype
(genetic
constitution)

Actions of
other genes and
their products

Environmental
influences

Phenotype
(expression of
physical trait)

affected by internal and external environmental influences. Notable among these influences are nutrition and the effects of hormones during puberty. While we can say that a child is "tall like her father," there is no simple genetic explanation for that statement. She may be tall because excellent nutritional habits are interacting with the genetic potential for height.

Because of the effects of environment, then, individuals may have identical genotypes (as in identical twins) but different phenotypes. Studies have shown, for instance, that identical twins raised apart and exposed to different environmental effects tend to vary more in their appearance than identical twins raised together in the same environment (Figure 2.3; p. 36). Similarly, individuals may have virtually identical phenotypes but very different genotypes. These situations reinforce these points: (1) genes are a starting position for determining the structure and function of an organism, and (2) the road to the mature phenotypic state is highly complex and involves a myriad of interacting biochemical pathways.

It is important to understand, for our future discussions, however, that while the phenotype is the product

~ FIGURE 2.3

Identical twins (a) reared apart, and (b) together.

a)

b)

of interaction between gene(s) and environment, the contribution of each is variable. In some cases, the environmental influence is great, but in others, the environmental contribution is nonexistent. We will develop the relationship between genotype and phenotype in more detail as the text proceeds, but for our current task of examining Mendel's experiments, the important aspects mentioned above will suffice.

𝒦EYNOTE

The genotype is the genetic constitution of an organism. The phenotype is the physical manifestation of the genetic traits. The genes give the potential for the development of characteristics; this potential is often affected by interactions with other genes and with the environment. Thus individuals with the same genotype can have different phenotypes, and individuals with the same phenotypes may have different genotypes.

MENDEL'S EXPERIMENTS

Mendel's Experimental Design

The first person to gain some understanding of the principles of heredity was Gregor Johann Mendel (Figure 2.4), whose work is considered to be the foundation of modern genetics. Mendel was born in 1822. In 1843 he was admitted to the Augustinian Monastery in Brno (now Brünn, Austria), where he later carried out his experiments. From 1851 to 1853 Mendel

attended the University of Vienna where he studied a variety of sciences and mathematics. This education evidently gave him an excellent background for designing experiments and analyzing experimental data.

Mendel returned to the monastery in 1853 and began teaching natural science and physics. In 1854 he began a series of breeding experiments with the garden pea *Pisum sativum* in an attempt to learn something about the mechanisms of heredity.

From the results of crossbreeding pea plants that exhibited differences in characteristics such as height,

~ FIGURE 2.4

Gregor Johann Mendel, the father of the science of genetics.

BOX 2.1
What Organisms Are Suitable for Use in Genetic Experiments?

There are several qualities that make an organism well suited for genetic experimentation.

(1) If an experimenter wants to test a genetic hypothesis, the genetic history of the organism involved must be well known. Therefore the genetic background of the parents used in the experimental crosses must be known.

(2) The organism must have a relatively short life cycle so that a large number of generations occur within a relatively short time. In this way data over many generations can be rapidly obtained.

(3) A large number of offspring must be produced from a mating, since much genetic information can be obtained if there are numerous progeny to study.

(4) The organism should be easy to handle. Hundreds of fruit flies can easily be kept in half-pint milk bottles for experimental purposes, but hundreds of elephants would certainly be more difficult and much more expensive to maintain.

(5) Most importantly, individuals in the population must differ in a number of ways. If there are no discernible differences among the individual organisms under study, it is impossible to study the inheritance of traits. The more marked the differences, the easier the genetic analysis.

flower color, and seed shape, Mendel developed a simple theory to explain the transmission of hereditary characteristics or traits from generation to generation. (Note that Mendel had no knowledge of mitosis and meiosis. Now, of course, we know that the segregation of genes conforms to chromosome behavior.) Figure 2.5 outlines the general procedure of a genetic cross. Although Mendel reported his conclusions in 1865, it wasn't until after his death (1884), until the late 1800s and early 1900s that their significance was fully realized. Today Mendel's methods of analysis continue to be used in genetic studies. In view of his significant contributions, Mendel is regarded as the father of genetics.

Mendel's experimental approach was effective because he developed a simple interpretation of the ratios he obtained and then carried out direct and convincing experiments to test this hypothesis. In his initial breeding experiments he reasoned that he should study one characteristic at a time to see how it was inherited. (This is how you should work genetics problems.) He made carefully controlled matings (crosses) between strains of peas that exhibited differences in heritable characteristics, and most importantly, he kept very careful records of the outcome of the crosses and the number of each type of pea produced. The numerical data he obtained enabled him to do a rigorous analysis of the hereditary transmission of characteristics.

Mendel performed all his significant genetic experiments with the garden pea, a good choice since the pea fits many of the criteria that make organisms suitable for use in genetic experiments (Box 2.1). Notably, the

~ FIGURE 2.5

A genetic cross in animals and its outcome.

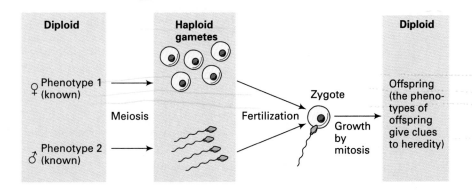

~ FIGURE 2.6

Procedure for crossing pea plants.

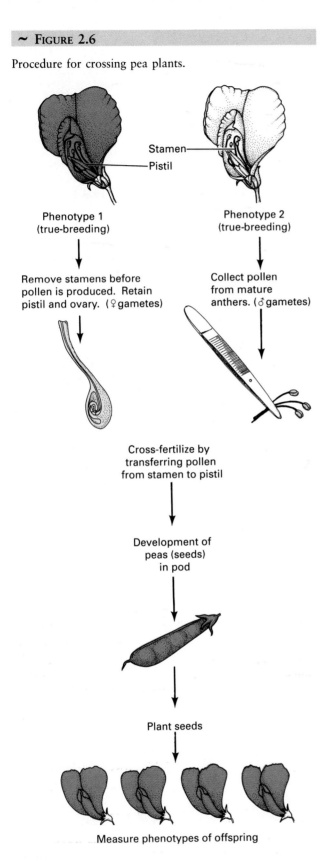

Stamen
Pistil

Phenotype 1
(true-breeding)

Phenotype 2
(true-breeding)

Remove stamens before
pollen is produced. Retain
pistil and ovary. (♀gametes)

Collect pollen
from mature
anthers. (♂gametes)

Cross-fertilize by
transferring pollen
from stamen to pistil

Development of
peas (seeds)
in pod

Plant seeds

Measure phenotypes of offspring

pea plant is easy to grow, it bears flowers and fruit in the same year a seed is planted, and it produces a large number of seeds. When the seeds are planted, they germinate to produce new pea plants.

Figure 2.6 presents a cross section of a flower of the garden pea, showing the stamens and the pistils (the male and the female reproductive organs, respectively). The pea normally reproduces by **self-fertilization;** that is, pollen (the male [♂] gametophytes) produced from the stamens lands on the pistil (containing the female [♀] gametophytes) within the same flower and fertilizes the plant. This process is also called **selfing.** Fortunately, it is a relatively simple procedure to prevent self-fertilization of the pea by removing the stamens from a developing flower bud before they produce any mature pollen. The pistils of that flower can then be pollinated with pollen taken from the stamens of another flower and dusted onto the stigma of the pistil of the emasculated one.

Cross-fertilization, or more simply a **cross,** is the term used for the fusion of male gametes (pollen) from one individual and female gametes (eggs) from another. Once cross-fertilization has occurred, the zygote develops into seeds (peas), which are then planted. The phenotypes of the plants that grow from the seeds are then analyzed.

Mendel knew enough about the reproductive cycle of the garden pea to plan the crosses carefully. For his experiments he obtained 34 strains of pea plants that differed in a number of traits. He allowed each strain to self-fertilize for many generations to ensure that the traits he wanted to study were inherited and to remove from consideration those strains that produced progeny with traits different from the parental type. This preliminary work ensured that he only worked with pea strains in which the trait under investigation remained unchanged from parent to offspring for many generations. Such strains are called **true-breeding** or **pure-breeding strains.**

Next, Mendel selected seven traits to study in breeding experiments. Each trait had two easily distinguishable, alternative appearances (phenotypes), as shown in Figure 2.7. These traits affect the appearance of most parts of the pea plant, including: (1) flower and seed coat color (grey versus white seed coats, and purple versus white flowers [Note: A single gene controls these particular color properties of both seed coats and flowers.]); (2) seed color (yellow versus green); (3) seed shape (smooth versus wrinkled); (4) pod color (green versus yellow); (5) pod shape (inflated versus pinched); (6) stem height (tall versus short); and (7) flower position (axial versus terminal).

And, finally, Mendel looked at only two traits at a time and *kept careful records of his observations,* an important responsibility with all experiments.

~ FIGURE 2.7

Seven character pairs in the garden pea that Mendel studied in his breeding experiments.

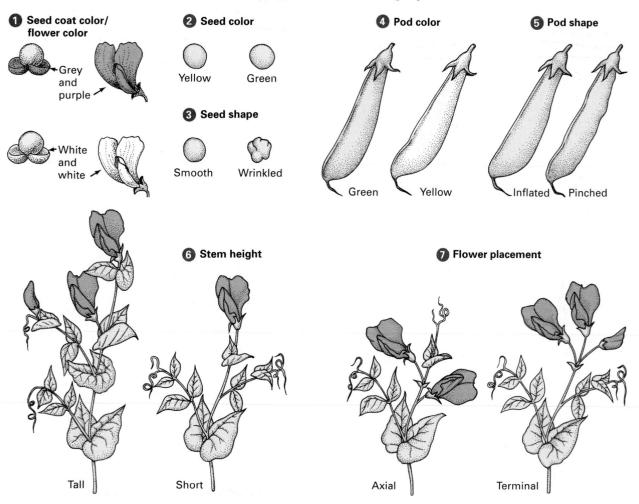

1 Seed coat color/ flower color

← Grey and purple

← White and white

2 Seed color

Yellow   Green

3 Seed shape

Smooth   Wrinkled

4 Pod color

Green   Yellow

5 Pod shape

Inflated   Pinched

6 Stem height

Tall   Short

7 Flower placement

Axial   Terminal

MONOHYBRID CROSSES AND MENDEL'S PRINCIPLE OF SEGREGATION

Before we discuss Mendel's experiment, we will clarify the terminology encountered in breeding experiments. The parental generation is called the **P generation**. The progeny of the P mating is called the **first filial generation**, or **F₁**. The subsequent generation produced by breeding together the F₁ offspring is termed the **F₂ generation**. Interbreeding the offspring of each generation results in F₃, F₄, F₅ generations, and so on.

To begin testing his hypothesis about the nature of heredity, Mendel first performed crosses between true-breeding strains of peas that differed in a single trait. Such crosses are called **monohybrid crosses**. For example, he pollinated pea plants that gave rise only to smooth seeds with pollen from a true-breeding variety

that produced wrinkled seeds.* As Figure 2.8 (p. 40) shows, the outcome of this monohybrid cross was all smooth seeds. The same result was obtained when the parental types were reversed; that is, when the pollen from a smooth-seeded plant was used to pollinate a pea plant that gave wrinkled seeds. [Matings that are done both ways, smooth female (♀) × wrinkled male (♂) and wrinkled female × smooth male, are called **reciprocal crosses**. Conventionally, in many organisms, including plants, the female is given first in crosses.]

The significant point of this cross was that all the F₁ progeny seeds of the smooth × wrinkled reciprocal crosses were smooth; that is, they exactly resembled only one of the parents in this character.

---

* Seeds are the diploid progeny of sexual reproduction. If a phenotype concerns the seed itself, the results of the cross can be directly seen by inspecting of the seeds. If a phenotype concerns a part of the mature plant, such as stem size or flower color, the seeds must be germinated before that phenotype can be observed.

~ FIGURE 2.8

Results of one of Mendel's breeding crosses. In the parental generation he crossed a true-breeding pea strain that produced smooth seeds with one that produced wrinkled seeds. All the F_1 progeny seeds were smooth.

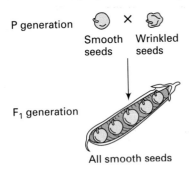

P generation

Smooth seeds   Wrinkled seeds

F_1 generation

All smooth seeds

Next, Mendel planted the seeds and allowed the F_1 plants to self-fertilize and produce the F_2 seed. Both smooth and wrinkled seeds appeared in the F_2 generation, and both types could be found within the same pod. Typical of his analytical approach to the experiments, Mendel counted the number of each type. He found that 5474 were smooth and 1850 were wrinkled (Figure 2.9). The calculated ratio of smooth to wrinkled seeds was 2.96:1, which is very close to a 3:1 ratio.

Using the same quantitative approach, Mendel analyzed the behavior of the six other pairs of traits. Qualitatively and quantitatively, the same results were obtained (Table 2.1). From the seven sets of crosses he made the following generalizations about his data:

1. The results of reciprocal crosses were always the same.
2. All F_1 progeny resembled one of the parental strains.
3. In the F_2 generation the parental trait that had disappeared in the F_1 generation reappeared. Further, the trait seen in the F_1 was always found in the F_2 at about three times the frequency of the other trait.

The key question for Mendel (and for other geneticists before and after Mendel) was how a trait present in the P generation could disappear in the F_1 and then reappear with full expression in the F_2. Mendel observed that, while the F_1 progeny resembled only one of the parents in their phenotype, they did not breed true, a fact that distinguishes the F_1 from the parent they resembled. The F_1 progeny also possessed the potential to produce F_2 progeny that expressed the parental phenotypic characteristic that had disappeared in the F_1 generation. Mendel concluded that the alter-

~ FIGURE 2.9

The F_2 progeny of the cross shown in Figure 2.8. When the plants grown from the F_1 seeds were self-pollinated, both smooth and wrinkled F_2 progeny seeds were produced. Commonly, both seed types were found in the same pod. In Mendel's actual experiments he counted 5474 smooth and 1850 wrinkled F_2 progeny seeds for a ratio of 2.96:1.

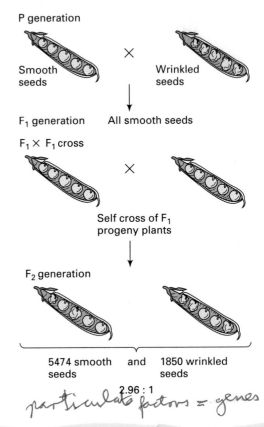

P generation

Smooth seeds   ×   Wrinkled seeds

F_1 generation   All smooth seeds

$F_1 \times F_1$ cross

×

Self cross of F_1 progeny plants

F_2 generation

5474 smooth seeds   and   1850 wrinkled seeds

2.96 : 1

particulate factors = genes

native traits in the crosses—for example, smoothness or wrinkledness of the seeds—were determined by **particulate factors**. He reasoned that these factors, which were transmitted from parents to progeny through the gametes, carried hereditary information. We now know these factors by another name, *genes*.

Since Mendel was examining pairs of traits (e.g., wrinkled/smooth), each factor was considered to exist in alternative forms (which we now call **alleles**), each of which specified one of the traits. For the gene that controls the shape of the pea seed, for example, there is one form, or allele, that results in the production of a smooth seed and another allele that results in a wrinkled seed.

Mendel reasoned further that for pure-breeding strains of the peas, both egg and pollen must have carried or transmitted identical forms of the factor. Since both traits were seen in the F_2, whereas only one appeared in the F_1, then each F_1 individual must have

~ **TABLE 2.1**

Mendel's Results in Crosses Between Plants Differing in One of Seven Characters

CHARACTER[a]	F$_1$	F$_2$ (NUMBER)			F$_2$ (PERCENT)	
		DOMINANT	RECESSIVE	TOTAL	DOMINANT	RECESSIVE
Seeds: smooth versus wrinkled	All smooth	5,474	1850	7,324	74.7	25.3
Seeds: yellow versus green	All yellow	6,022	2001	8,023	75.1	24.9
[b]Seed coats: grey versus white	All grey	705	224	929	75.9	24.1
Flowers: purple versus white	All purple					
Flowers: axial versus terminal	All axial	651	207	858	75.9	24.1
Pods: inflated versus pinched	All inflated	882	299	1,181	74.7	25.3
Pods: green versus yellow	All green	428	152	580	73.8	26.2
Stem: tall versus short	All tall	787	277	1,064	74.0	26.0
Total or average		14,949	5010	19,959	74.9	25.1

[a]The dominant trait is always written first.

[b]A single gene controls both the seed coat and the flower color trait.

contained both factors, one for each of the alternative traits. Furthermore, since only one of the characters was seen in the F$_1$s, the expression of the "missing" trait must somehow have been masked by the visible trait, a feature called *dominance*. For the smooth × wrinkled example the F$_1$ seeds were all smooth. Thus the allele for smoothness is **dominant** to the allele for wrinkledness. Conversely, wrinkled is said to be **recessive** to smooth (Figure 2.10). Similar conclusions can be made for the other six pairs of traits. The dominant and the recessive forms for each pair of traits are indicated in Table 2.1.

We can see the crosses more easily if we use some symbols for the alleles. For the smooth × wrinkled cross we can give the symbol *S* to the factor for smoothness and the symbol *s* to the factor for wrinkledness. The convention is that the dominant allele is given the uppercase (capital) letter and the recessive allele the lowercase (small) letter. Using these symbols, we can diagram the cross as shown in Figure 2.11 (p. 42): the production of the F$_1$ generation is shown in Figure 2.11a, and the F$_2$ generation in Figure 2.11b. (In both Figure 2.10 and 2.11 the genes are located on chromosomes. Keep in mind, gene segregation follows the behavior of chromosomes, although the existence of chromosomes was unknown to Mendel.) Since each parent is true breeding, each must contain two copies of the same allele. Thus the genotype of the parental plant grown from the smooth seeds is *SS* and that of the wrinkled parent is *ss*. True-breeding individuals that contain only one specific allele are **homozygous.**

When plants produce gametes by meiosis, each gamete contains only one copy of the gene (i.e., one allele); the plants from smooth seeds produce *S*-bearing gametes, and the plants from wrinkled seeds produce

~ **FIGURE 2.10**

Dominant and recessive alleles of a gene for seed shape in peas.

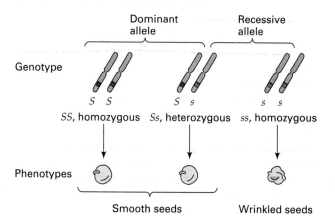

s-bearing gametes. When the gametes fuse during the fertilization process, the resulting zygote has one *S* and one *s* factor, a genotype of *Ss*. Plants that have two different alleles of a specific gene are called **heterozygous.** Because of the dominance of the smooth *S* allele, only smooth seeds develop from the F$_1$ zygotes.

The plants derived from the F$_1$ seeds differ from the smooth parent in that they produce two types of gametes in equal numbers: *S*-bearing and *s*-bearing. All the possible F$_1$ gametic fusions are represented in the matrix in Figure 2.11, called a **Punnett square** after its originator, R. Punnett. These fusions give rise to the zygotes that produce the F$_2$ generation.

In the F$_2$ generation three types of genotypes are produced: *SS, Ss,* and *ss.* As a result of the random fusing of gametes, the relative proportion of these zygotes is 1:2:1, respectively. However, since the *S* factor is dominant to the *s* factor, both the *SS* and *Ss* seeds are smooth, and the F$_2$ generation seeds exhibit a phenotypic ratio of 3 smooth:1 wrinkled seeds. The results are the same for crosses involving the other six character pairs.

The Principle of Segregation

From the sort of data we have discussed, Mendel proposed his **first law,** the **principle of segregation,** which states that *the two members of a gene pair (alleles) segregate (separate) from each other in the formation of gametes.* As a result, half the gametes carry one allele

~ FIGURE 2.11

The same cross as in Figures 2.8 and 2.9. Here, genetic symbols illustrate the principle of segregation of Mendelian factors. (a) Production of F$_1$ generation; (b) Production of F$_2$ generation.

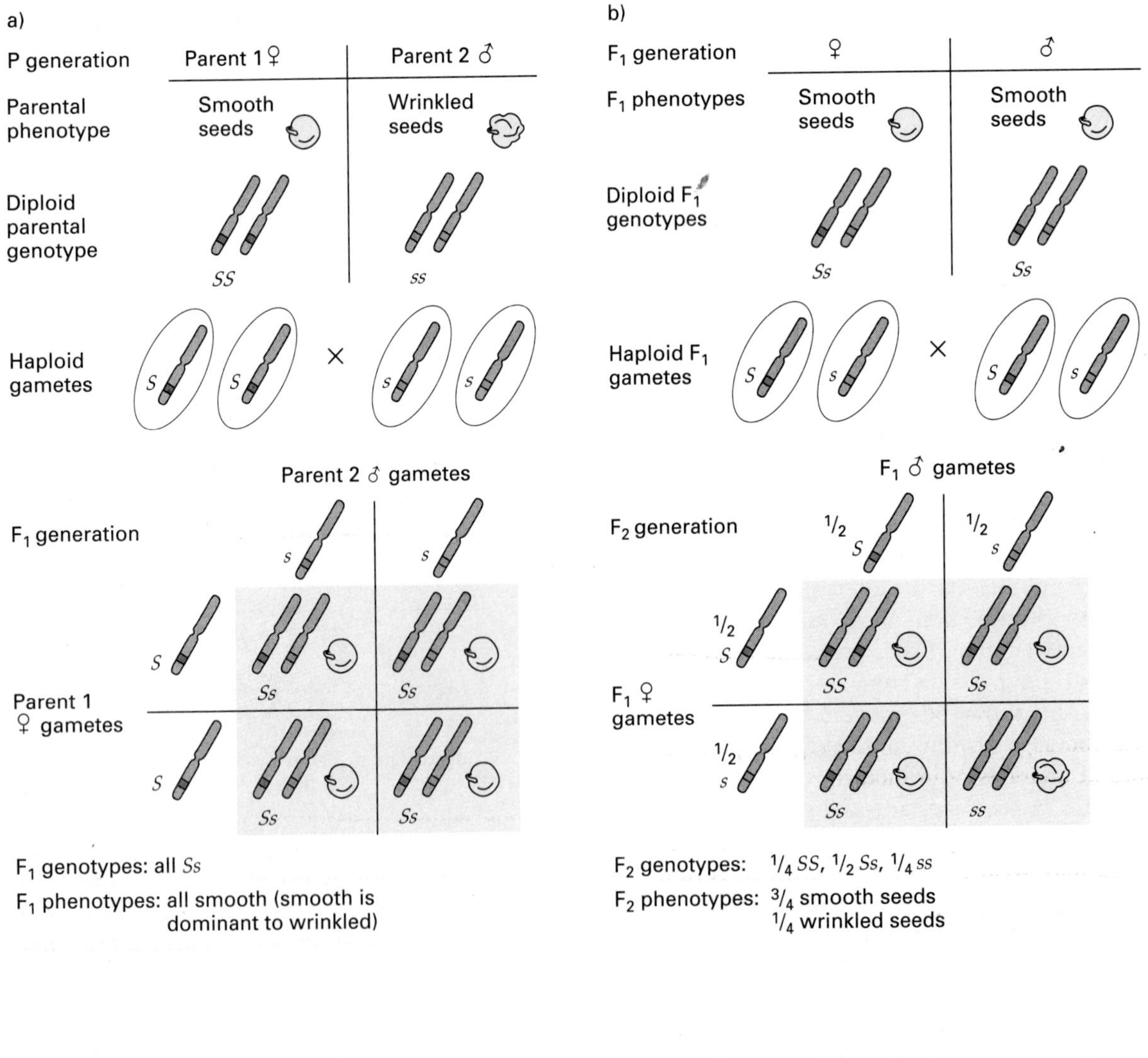

Box 2.2
Genetic Terminology

Cross: a mating between two individuals, leading to the fusion of gametes.

Zygote: the cell produced by the fusion of male and female gametes.

Gene (Mendelian factor): the determinant of a characteristic of an organism.

Locus (gene locus): the specific place on a chromosome where a gene is located.

Alleles: alternative forms of a gene. For example, *S* and *s* alleles represent the smoothness and wrinkledness of the pea seed. The rationale for the symbols used in our discussion of Mendel's work is still applicable in plant genetics: The dominant allele is assigned the uppercase letter and the recessive is assigned the lowercase letter. The actual letter used is based on the dominant phenotype, for instance, *S* for smooth, *s* for wrinkled, *Y* for yellow, and *y* for green. Later we will see a different way of assigning gene symbols, one that is more generally used in genetics today.

Genotype: the genetic constitution of an organism. An organism that has the same allele for a given gene locus on both members of a homologous pair of chromosomes in a diploid organism is said to be **homozygous** for that allele. Homozygotes produce only one gametic type. Thus true-breeding smooth individuals have the genotype *SS*, and true-breeding wrinkled individuals have the genotype *ss*; both are homozygous. The smooth parent is **homozygous dominant;** the wrinkled parent is **homozygous recessive.**

Diploid organisms that have two different alleles at a specific gene locus on homologous chromosomes are said to be **heterozygous.** So F_1 hybrid plants from the cross of *SS* and *ss* parents have one *S* allele and one *s* allele. Individuals heterozygous for two allelic forms of a gene produce two kinds of gametes (*S* and *s*).

Phenotype: the physical manifestation of a genetic trait that results from a specific genotype and its interaction with the environment. In our example the *S* allele was dominant to the *s* allele, so in the heterozygous condition the seed is smooth. Therefore both the homozygous dominant *SS* and the heterozygous *Ss* seeds have the same phenotype (smooth), even though they differ in genotype.

---

and the other half carry the other allele. In other words, each gamete carries only one allele at each gene locus. The progeny are produced by the random combination of gametes from the two parents.

In proposing the principle of segregation, Mendel had clearly differentiated between the factors (genes) that determined the traits (the genotype) and the traits themselves (the phenotype). From a modern perspective this law means that at the gene level the members of a pair of alleles of a gene on a chromosome segregate during meiosis so that any offspring receives only one member of a pair from each parent. Thus gene segregation parallels the separation of homologous pairs of chromosomes at anaphase I in meiosis.

Since the genetics concepts and terms we have learned so far in this chapter are often difficult to assimilate, you may wish to take a moment to review them (see Box 2.2). A thorough understanding of these terms is essential to your study of genetics.

KEYNOTE

Mendel's first law, the principle of segregation, states that the two members of a gene pair (alleles) segregate (separate) from each other in the formation of gametes; half the gametes carry one allele, and the other half carry the other allele. ⎯⎯⎯⎯⎯

Representing Crosses with a Branch Diagram

The use of a Punnett square to consider the pairing of all possible gamete types from the two parents in a monohybrid cross (shown in Figure 2.11) is a relatively simple way to learn how to predict the relative frequencies of genotypes and phenotypes in the next generation. In this section, however, we will describe an alter-

Box 2.3
Elementary Principles of Probability

A **probability** is the ratio of the number of times a particular event occurs to the number of trials during which the event could have happened. For example, the probability of picking a heart (from 13 hearts) from a deck of cards (52 cards) is $p(heart) = 13/52 = 1/4$. That is, we would expect, on the average, to pick a heart from a deck of cards once in every four trials, assuming we put the card back after each trial.

Probabilities and the *laws of chance* are involved in the transmission of genes. As a simple example, let's consider a couple and the chance that their child will be a boy or a girl. Assume that an equal number of boys and girls are born (which is not actually true, but we can assume it to be so for the sake of discussion). The probability that the child will be a boy is 1/2, or 0.5. Similarly, the probability that the child will be a girl is 1/2.

Now a rule of probability can be introduced: the **product rule**. This rule states that the probability of two independent events occurring simultaneously is the product of each of their probabilities. Thus the probability that a family will have two girls in a row is 1/4. That is, the probability of the first child being a girl is 1/2, the probability of the second being a girl is also 1/2, and by the product rule the probability of the first and second being girls is $1/2 \times 1/2$, or 1/4. Similarly, the probability of having three boys in a row is $1/2 \times 1/2 \times 1/2 = 1/8$.

Another rule of probability is the **sum rule**, which states that the probability that two mutually exclusive events will occur is the sum of their individual probabilities. For example, if two dice are thrown, what is the probability of getting two sixes or two ones? The probability of getting two sixes is found by using the product rule. The probability of getting one six, $p(one\ six)$, is 1/6, since there are six faces to a die. Therefore the probability of getting two sixes, $p(two\ sixes)$, when two dice are thrown is $1/6 \times 1/6 = 1/36$. Similarly, $p(two\ ones) = 1/36$. To roll two sixes or two ones involves mutually exclusive events, so the sum rule is used to find the probability. The answer is $1/36 + 1/36 = 2/36 = 1/18$.

~ **FIGURE 2.12**

Calculating the ratios of phenotypes in the F_2 generation of the Figure 2.11 cross by using the branch diagram approach.

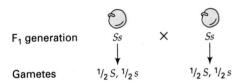

F_1 generation Ss $\times$ Ss

Gametes   1/2 S, 1/2 s 1/2 S, 1/2 s

Random combination of gametes results in:

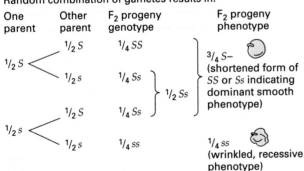

One parent	Other parent	F_2 progeny genotype	F_2 progeny phenotype

1/2 S — 1/2 S — 1/4 SS

1/2 S — 1/2 s — 1/4 Ss

1/2 Ss → 3/4 S– (shortened form of SS or Ss indicating dominant smooth phenotype)

1/2 s — 1/2 S — 1/4 Ss

1/2 s — 1/2 s — 1/4 ss → 1/4 ss (wrinkled, recessive phenotype)

native method, one which you are encouraged to master: the branch diagram. (Box 2.3 discusses some elementary principles of probability that will help you understand this approach.) Figure 2.12 illustrates the application of the branch diagram approach for the analysis of the $F_1 \times F_1$ self of the smooth $\times$ wrinkled cross diagrammed in Figure 2.11.

The F_1 seeds from the cross have the genotype Ss. Both eggs and pollen are produced in the flowers of the plants grown from these seeds. In the meiotic process an equal number of S and s gametes are expected to be produced, so we can say that half the gametes are S and the other half are s. Thus 1/2 is the predicted frequency of these two types. But just as tossing a coin many times does not always give exactly half heads and half tails, the gamete frequency may not be exactly realized. However, the more chances (e.g., tosses), the more likely you will obtain the true frequency.

From the rules of probability, the relative expected frequencies of the three possible genotypes in the F_2 generation can be predicted. To produce an SS plant, an S egg must pair with an S pollen grain. The frequency of S eggs in the population of eggs is 1/2, and the frequency of S pollen grains in the pollen population is also 1/2. Therefore the relative expected proportion of SS plants in the F_2 is $1/2 \times 1/2 = 1/4$. A similar proportion is predicted for the wrinkled, or ss, progeny.

What about the Ss progeny? Again, the frequency of S in one gametic type is 1/2, and the frequency of s in the other gametic type is also 1/2. However, there are two ways in which Ss progeny can be obtained. The first involves the fusion of an S egg with s pollen, and the second is a fusion of an s egg with S pollen. Using the product rule (Box 2.3), the probability of each of these events occurring is $1/2 \times 1/2 = 1/4$. So, using the sum rule (Box 2.3), the probability of one or the other occurring is the sum of the individual probabilities, or $1/4 + 1/4 = 1/2$.

Given the rules of probability, the prediction is that a fourth of the F_2 progeny will be SS, half will be Ss, and a fourth will be ss, exactly as was found with the method shown in Figure 2.11. Either method, then, is applicable to any cross.

Confirming the Principle of Segregation: The Use of Testcrosses

When formulating his principle of segregation, Mendel did a number of tests to ensure the validity of his results. He continued the self-fertilizations to the F_6 generation and found that in every generation both the dominant and recessive characters appeared. He concluded that the principle of segregation was valid no matter how many generations were carried out.

Another important test concerned the F_2 plants. As shown in Figure 2.11, a ratio of $1:2:1$ occurs for the genotypes SS, Ss, and ss for the smooth $\times$ wrinkled example. Phenotypically, the ratio of smooth to wrinkled is $3:1$. At the time of his experiments Mendel could only hypothesize the presence of segregating factors that were responsible for the smooth and wrinkled phenotypes. To be sure of his factor hypothesis, Mendel tested the F_2 plants by allowing them to self-pollinate. As he expected, the plants produced from wrinkled seeds bred true, supporting his contention that they were pure for the s factor (gene).

Selfing the plants derived from the F_2 smooth seeds produced two different types of progeny. One-third produced only smooth seeds, whereas the other two-thirds produced both smooth and wrinkled seeds in each pod (Figure 2.13). In fact, the ratio of the two types of seeds obtained from the second group of plants was the same as that of all the F_2 progeny. These results completely support the principle of segregation of genes. The random combination of gametes that form the zygotes of the F_2 produces two genotypes that give rise to the smooth phenotype (e.g., Figures 2.11 and 2.12). Moreover, the relative proportion of the two

~ **FIGURE 2.13**

Determining the genotypes of the F_2 smooth progeny of Figure 2.11 by selfing the plants grown from the smooth seeds.

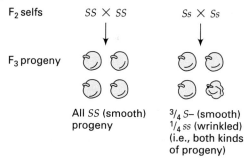

F_2 selfs $SS \times SS$ $Ss \times Ss$

F_3 progeny

All SS (smooth) progeny

$^3/_4$ $S-$ (smooth) $^1/_4$ ss (wrinkled) (i.e., both kinds of progeny)

genotypes SS and Ss is $1:2$. The SS seeds give rise to true-breeding plants, whereas the Ss seeds give rise to plants that behave exactly like the F_1 plants when they are self-pollinated. *Mendel explained these results by proposing that each plant had two factors, while each gamete had only one. He also proposed that the random combination of the gametes generated the progeny in the proportions he found. Mendel obtained the same results in all seven sets of crosses.* The segregation pattern seen results from the behavior of chromosomes.

The self-fertilization test of the F_2s proved to be a useful way of confirming the genotype of a plant with a given phenotype. A more common test to find out the genotype of an organism is to perform a **testcross**, a cross of an individual of unknown genotype, usually expressing the dominant phenotype, with a homozygous recessive individual in order to determine the genotype of the individual.

Considering the cross shown in Figure 2.11, we can predict the outcome of a testcross of the F_2s showing the dominant, smooth seed phenotype. If the F_2s are homozygous SS, the result of a testcross with an ss plant will be all smooth seeds. As Figure 2.14a (p. 46) shows, the Parent 1 smooth SSs produce only S gametes. Parent 2 is homozygous recessive wrinkled, ss, so it produces only s gametes. Therefore, all zygotes are Ss, and all the resulting seeds are smooth in phenotype. In actual practice, therefore, if one performs a testcross with a recessive plant and a plant showing the dominant trait, and the progeny exhibit only dominant phenotypes, the results indicate that the F_1 seed used in the testcross is homozygous.

On the other hand, if heterozygous Ss F_2 plants are testcrossed with a homozygous ss plant, a $1:1$ ratio of dominant to recessive phenotypes is expected. As Figure 2.14b shows, the Parent 1 smooth Ss produces

~ FIGURE 2.14

Determining the genotypes of the F$_2$ generation smooth seeds (Parent 1) of Figure 2.11 by testcrossing plants grown from the seed with a homozygous recessive wrinkled (*ss*) strain (Parent 2). (a) If Parent 1 is *SS*, all progeny seeds are smooth in phenotype; (b) If Parent 1 is *Ss*, 1/2 of the progeny seeds are smooth and 1/2 are wrinkled.

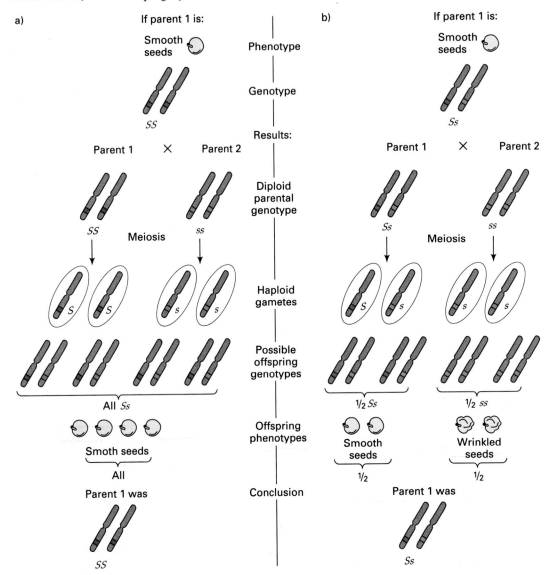

both *S* and *s* gametes in equal proportion, while the homozygous *ss* parent produces only *s* gametes. As a result, half the progeny of the testcross are *Ss* heterozygotes and have a smooth phenotype because of the dominance of the *S* allele, and the other half are *ss* homozygotes and have a wrinkled phenotype. In actual practice, therefore, if one performs a testcross with a recessive plant and a plant showing the dominant trait, and the progeny exhibit a 1:1 ratio of dominant to recessive phenotypes, the results indicate that the F$_1$ seed used in the testcross is heterozygous.

In summary, testcrosses of Mendel's plants grown from F$_2$ seeds with the dominant phenotype would give these results: A third of the plants would give rise only to progeny seeds that express the dominant phenotype. These F$_2$ seeds would be homozygous for the dominant gene. The other two-thirds would give a 1:1 ratio of heterozygous and homozygous recessive seeds, indicating that the F$_2$ seeds must be heterozygous. By doing a testcross, one can determine that the F$_1$ smooth seeds are *Ss* heterozygous, since a 1:1 ratio of smooth: wrinkled will be obtained, just as in Figure 2.14b.

$\mathcal{K}$EYNOTE

A testcross is a cross of an individual of unknown genotype, usually expressing the dominant phenotype, with a homozygous recessive in order to determine the genotype of the unknown individual. The phenotypes of the progeny of the testcross indicate the genotype of the individual tested. If the progeny all show the dominant phenotype, the individual was homozygous dominant. If there is an approximately 1:1 ratio of progeny with dominant and recessive phenotypes, the individual was heterozygous.

DIHYBRID CROSSES AND THE MENDELIAN PRINCIPLE OF INDEPENDENT ASSORTMENT

The next logical extension of Mendel's experiments was to determine what happens when two pairs of characters were simultaneously involved in the cross. Mendel did a number of these crosses, and in each case he obtained the same results. From these experiments he proposed his **second law**, the **principle of independent assortment**, which states that the factors (genes) for different traits assort independently of one another. In modern terms, this means that genes on different chromosomes behave independently in the production of gametes. There are also many examples of genes that are far enough apart on the same chromosome that they assort independently. This principle was unknown to Mendel and will be considered when linkage and crossing-over are discussed in Chapter 5.

Consider an example involving smooth (S)/wrinkled (s) and yellow (Y)/green (y) seed traits (yellow is dominant to green). Mendel made crosses between true-breeding smooth-yellow plants ($SS\ YY$) and wrinkled-green plants ($ss\ yy$), with the results shown in Figure 2.15. All the F$_1$ seeds from this cross were smooth and yellow, as predicted from the results of the monohybrid crosses. As Figure 2.15a shows, the smooth-yellow parent produces only $S\ Y$ gametes, which give rise to $Ss\ Yy$ zygotes upon fusion with the $s\ y$ gametes from the wrinkled-green parent. Because of the dominance of the smooth and the yellow traits, all F$_1$ seeds are smooth and yellow.

The seeds of the F$_1$ generation will grow into plants that are heterozygous for two pairs of alleles at two different loci. Such plants are called dihybrid plants, and a cross between two dihybrids of the same type is called a **dihybrid cross**.

When Mendel self-pollinated the F$_1$ plants to give rise to the F$_2$ generation (Figure 2.15b; p. 48), he considered two possible outcomes. One was that the genes for the traits from the original parents would be transmitted together to the progeny. In this case a phenotypic ratio of 3:1 smooth-yellow:wrinkled-green would be predicted.

The other possibility was that the traits would be inherited independently of one another. The dihybrid plants grown from the F$_1$ seeds produce four types of gametes: $S\ Y$, $S\ y$, $s\ Y$, and $s\ y$. Because of the independence of the two pairs of genes, there is an equal frequency of each gametic type. In the F$_1 \times$ F$_1$ selfing, the four types of gametes fuse randomly in all possible combinations to give rise to the zygotes and, hence, the progeny seeds. All the possible gametic fusions are represented in the Punnett square in Figure 2.15. In a dihybrid cross there are 16 possible gametic fusions. The result is 9 different genotypes but, because of dominance, only 4 phenotypes:

~ FIGURE 2.15a

The principle of independent assortment in a dihybrid cross. This cross, actually done by Mendel, involves the smooth/wrinkled and yellow/green character pairs of the garden pea. (a) Production of the F$_1$ generation.

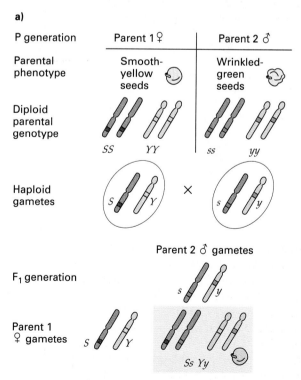

a)

F$_1$ genotypes: all $Ss\ Yy$
F$_1$ phenotypes: all smooth-yellow seeds

~ FIGURE 2.15b

(b) The F$_2$ genotypes and 9:3:3:1 phenotypic ratio of smooth-yellow:smooth-green: wrinkled-yellow:wrinkled-green are derived by using the Punnett square. (Note that, compared with previous figures of this kind, only one progeny box is shown in the F$_1$, instead of four. This is because only one class of gametes exists for Parent 2 and only one class for Parent 1. Previously we showed two gametes from each parent, even though those gametes were identical.)

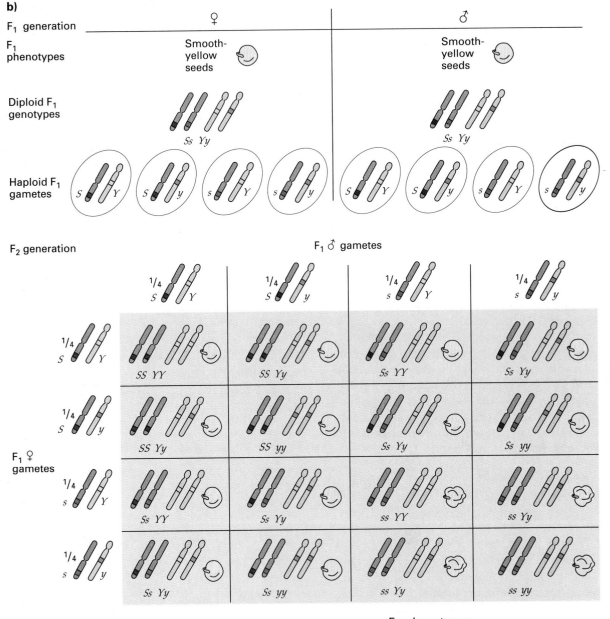

F$_2$ genotypes:
$\frac{1}{16}$ (*SS YY*) + $\frac{2}{16}$ (*Ss YY*) + $\frac{2}{16}$ (*SS Yy*) + $\frac{4}{16}$ (*Ss Yy*) = $\frac{9}{16}$ smooth-yellow seeds
$\frac{1}{16}$ (*SS yy*) + $\frac{2}{16}$ (*Ss yy*) = $\frac{3}{16}$ smooth-green seeds
$\frac{1}{16}$ (*ss YY*) + $\frac{2}{16}$ (*ss Yy*) = $\frac{3}{16}$ wrinkled-yellow seeds
$\frac{1}{16}$ (*ss yy*) = $\frac{1}{16}$ wrinkled-green seeds

1 *SS YY*, 2 *Ss YY*,
2 *SS Yy*, 4 *Ss Yy* = 9 smooth-yellow

1 *SS yy*, 2 *Ss yy* = 3 smooth-green

1 *ss YY*, 2 *ss Yy* = 3 wrinkled-yellow

1 *ss yy* = 1 wrinkled-green

According to the rules of probability, if pairs of characters are inherited independently in a dihybrid cross, then the F_2 from an $F_1 \times F_1$ selfing will give a 9:3:3:1 ratio of the four possible phenotypic classes. This ratio is the result of the independent assortment of the two gene pairs into the gametes and of the random fusion of those gametes. The 9:3:3:1 ratio is the product of two 3:1 ratios; that is, $(3:1)^2 = 9:3:3:1$.

This prediction was met in all dihybrid crosses that Mendel performed. In every case the F_2 ratio was close to 9:3:3:1. For our example he counted 315 smooth-yellow, 108 smooth-green, 101 wrinkled-yellow, and 32 wrinkled-green seeds—very close to the predicted ratio. To Mendel this result meant that the factors (genes) determining the specific, different character pairs he was analyzing were transmitted independently. We now know that this statement is true because genes are located on chromosomes, which are distributed to gametes by the process of meiosis.

$\mathcal{K}$EYNOTE

Mendel's second law, the principle of independent assortment, states that genes on different chromosomes behave independently in the production of gametes. ━━━━━━━━━━

Branch Diagram of Dihybrid Crosses

As perhaps is already apparent, it is quite tedious and slow to construct a Punnett square of gamete combinations and then to count up the numbers of each phenotypic class from all the genotypes produced. This chore is not too difficult for a dihybrid cross, but for more than two gene pairs it becomes rather complex. *It is easier to get into the habit of calculating the expected ratios of phenotypic classes by using a branch diagram.*

Using the same example, in which the two gene pairs assort independently into the gametes, we will consider each gene pair in turn. Earlier we showed that an F_1 self of an *Ss* heterozygote gave rise to progeny of which three-fourths were smooth and a fourth were wrinkled. Genotypically, the former class had at least

~ **FIGURE 2.16**

Using the branch diagram approach to calculate the F_2 phenotypic ratio of the Figure 2.15 cross.

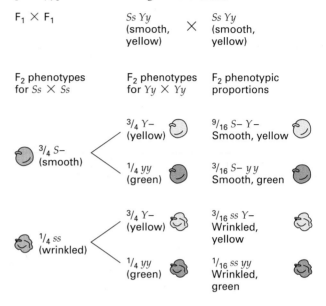

one dominant *S* allele; that is, they were *SS* or *Ss*. A convenient way to signify this situation is to use a dash to indicate an allele that has no effect on the phenotype. Thus *S—* means that phenotypically the seeds are smooth and genotypically they are either *SS* or *Ss*.

Now consider the F_2 produced from a selfing of *Yy* heterozygotes; again, a 3:1 ratio is seen, with three-fourths of the seeds being yellow and one fourth being green. Since this segregation occurs independently of the segregation of the smooth/wrinkled pair, we can consider all possible combinations of the phenotypic classes in the dihybrid cross. For example, the expected proportion of F_2 seeds that are smooth and yellow is the product of the probability that an F_2 seed will be smooth and the probability that it will be yellow, or $3/4 \times 3/4 = 9/16$. Similarly, the expected proportion of F_2s that are wrinkled and yellow is $3/4 \times 1/4 = 3/16$. Extending this calculation to all possible phenotypes, as shown in Figure 2.16, we obtain the 9:3:3:1 ratio.

The testcross can be used to check the genotypes of the F_1s and the F_2s from a dihybrid cross. In the example the F_1 is a double heterozygote, *Ss Yy*. This F_1 produces four types of gametes in equal proportions, as was the case in Figure 2.15: *S Y*, *S y*, *s Y*, and *s y*. In a testcross with a doubly homozygous recessive plant, in this case *ss yy*, the phenotypic ratio of the progeny is a direct reflection of the ratio of gametic types produced by the F_1 parent. In a testcross like this one, then, there will be a 1:1:1:1 ratio in the offspring of *Ss Yy*: *Ss yy*: *ss Yy*: *ss yy* genotypes, which means a 1:1:1:1 ratio of smooth-yellow : smooth-green : wrinkled-yellow :

wrinkled-green phenotypes. This 1:1:1:1 phenotypic ratio is diagnostic of testcrosses in which the "unknown" parent is a double heterozygote.

In the F_2 of a dihybrid cross there were nine different genotypic classes but only four phenotypic classes. The genotypes can be ascertained by testcrossing as we've shown. Table 2.2 lists the expected ratios of progeny phenotypes from such testcrosses. No two patterns are the same, so here the testcross is truly a diagnostic approach to confirm genotypic type.

Trihybrid Crosses

Mendel also confirmed his laws for three characters segregating in other garden pea crosses. Such crosses are called **trihybrid crosses**. Here the proportions of F_2 genotypes and phenotypes are predicted with precisely the same logic used before, considering each character pair independently. Although we will not spell out the results in detail, Figure 2.17 shows a branch diagram derivation of the F_2 phenotypic classes for a trihybrid cross. The independently assorting character pairs in the cross are smooth versus wrinkled seeds, yellow versus green seeds, and purple versus white flowers. There are 64 combinations of 8 maternal and 8 paternal gametes. Combination of these gametes gives rise to 27 different genotypes and 8 different phenotypes.

Now that enough examples have been considered, we can make some generalizations about phenotypic and genotypic classes. In each of the previous examples, the F_1 is heterozygous for each gene involved in the

~ Table 2.2

Proportions of Phenotypic Classes Expected from Testcrosses of Strains with Various Genotypes for Two Gene Pairs

	PROPORTION OF PHENOTYPIC CLASSES			
TESTCROSSES	$A- B-$	$A- bb$	$aa B-$	$aa bb$
$AA\ BB \times aa\ bb$	1	0	0	0
$Aa\ BB \times aa\ bb$	1/2	0	1/2	0
$AA\ Bb \times aa\ bb$	1/2	1/2	0	0
$Aa\ Bb \times aa\ bb$	1/4	1/4	1/4	1/4
$AA\ bb \times aa\ bb$	0	1	0	0
$Aa\ bb \times aa\ bb$	0	1/2	0	1/2
$aa\ BB \times aa\ bb$	0	0	1	0
$aa\ Bb \times aa\ bb$	0	0	1/2	1/2
$aa\ bb \times aa\ bb$	0	0	0	1

~ Table 2.3

Number of Phenotypic and Genotypic Classes Expected from Self-Crosses of Heterozygotes in Which All Genes Show Complete Dominance

NUMBER OF SEGREGATING GENE PAIRS	NUMBER OF PHENOTYPIC CLASSES	NUMBER OF GENOTYPIC CLASSES
1^a	2	3
2	4	9
3	8	27
4	16	81
n	2^n	3^n

[a]For example, from $Aa \times Aa$, two phenotypic classes are expected, with genotypic classes of AA, Aa, and aa.

cross, and the F_2 is generated by selfing where that is possible or, in organisms where it is not possible (e.g., humans, mice, and *Drosophila*), by allowing the F_1s to interbreed. In monohybrid crosses there were 2 phenotypic classes in the F_2, in dihybrid crosses there were 4, and in trihybrid crosses there were 8. The general rule is that there will be 2^n phenotypic classes in the F_2, where n is the number of independently assorting, heterozygous gene pairs (Table 2.3). (This rule holds *only* when a dominant-recessive relation holds for each of the gene pairs.) Furthermore, we saw that there were 3 genotypic classes in the F_2 of monohybrid crosses, 9 in dihybrid crosses, and 27 in trihybrid crosses. A simple rule is that the number of genotypic classes will be 3^n, where n is the number of heterozygous gene pairs.

Incidentally, the phenotypic rule (2^n) can also be used to predict the number of classes that will come from a multiple heterozygous F_1 used in a testcross. Here the number of genotypes in the next generation will be the same as the number of phenotypes.

"Rediscovery" of Mendel's Principles

Mendel published his treatise on heredity in 1865 in *Verhandlungen des Naturforschenden Vereines in Brünn*, but it received little attention from the scientific community until the "rediscovery" of his principles in the early 1900s. In 1985 Iris Sandler and Laurence Sandler proposed a possible reason. They contend that it may have been impossible for the scientific community from 1865 to 1900 to understand the significance

~ FIGURE 2.17

Branch diagram derivation of the relative frequencies of the eight phenotypic classes in the F_2 of a trihybrid cross.

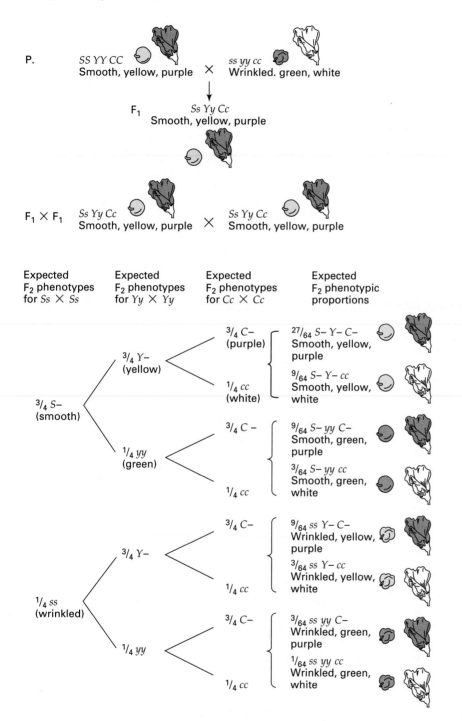

of Mendel's work because it did not fit into that community's conception of the relationship of heredity to other sciences. To Mendel's contemporaries, heredity included not only those ideas that are today understood as genetic, but also those that are considered developmental. In other words, their concept of heredity included what we now know as genetics *and* embryol-

ogy. More pertinently, they also viewed heredity as simply a particular moment in development, and not as a distinct process requiring special analysis. By 1900 conceptions had changed enough that the significance of Mendel's work was more apparent.

At about the turn of the century three researchers working independently on breeding experiments came to the same conclusions as had Mendel. The three men were Carl Correns, Hugo de Vries, and Erich von Tschermak. Correns concentrated mostly on maize (corn) and peas, de Vries worked with a number of different plant species, and von Tschermak studied peas.

The first demonstration that Mendelism applied to animals came in 1902 from the work of William Bateson, who experimented with fowl. Bateson also coined the terms *genetics*, *zygote*, F_1, F_2, and **allelomorph** (literally, "alternative form," meaning one of an array of different forms of a gene). The last was shortened by others to *allele*. The term *gene* as a replacement for Mendelian factor was introduced by W. L. Johannsen in 1909. In the years after Bateson's work, a number of investigators showed the general applicability of Mendelian principles to all eukaryotic organisms.

MENDELIAN GENETICS IN HUMANS

After the rediscovery of Mendel's laws in 1900, geneticists found that the inheritance of genes follows the same principles in all sexually reproducing eukaryotes, including humans. W. Farabee in 1905 was the first to document a Mendelian trait in humans, brachydactyly (abnormally broad and short fingers: Figure 2.18). By analyzing the trait in human families, Farabee learned that brachydactyly is inherited. The pattern of transmission of the abnormality over several generations led to the conclusion that the trait is a simple dominant trait. In this section we will introduce some of the methods used to determine the mechanism of hereditary transmission, and we will learn about some inherited human traits.

Pedigree Analysis

The study of human genetics is complicated because controlled matings of humans cannot be made. This does not mean that genetic analysis of humans is impossible, but it does mean that geneticists must study crosses that happen by *chance* rather than by control. The inheritance patterns of human traits are usually established by examining the way the trait occurs in the family trees of individuals who clearly exhibit the trait. The family tree investigation is called **pedigree analysis**, and it involves the careful compilation of phenotypic records of the family over several generations. The individual through whom the pedigree is discovered is called the **propositus**. The more information there is, the more likely the investigator will be able to make some conclusions about the mechanism of inheritance of the gene (or genes) responsible for the trait being studied.

~ **FIGURE 2.18**

Photograph of (a) a normal hand alongside (b) a human hand with brachydactyly.

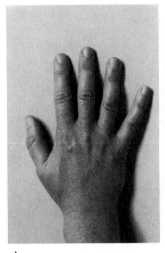
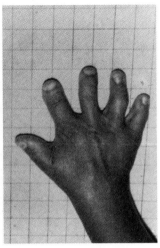

a)   b)

𝒦EYNOTE

In humans, the inheritance patterns of traits are usually established by examining the occurrences of the traits in the family trees of individuals who clearly exhibit the traits. The family tree investigation is called pedigree analysis. _____

One of the modern applications of pedigree analysis is called **genetic counseling**: A geneticist makes predictions about the probabilities of particular traits (deleterious or not) occurring among a couple's children. In most cases the couple comes to the counselor because there is some possibility of a heritable trait in one or both families. As we might expect, pedigree analysis is most useful for traits that are the result of a single gene difference.

Because it is difficult to assign genotypes to family members with respect to a trait, it is not possible to diagram crosses the way we did earlier in this chapter. Therefore, pedigree analysis has its own set of symbols. Figure 2.19 summarizes the basic symbols that will be used here and elsewhere in the text. (The terms *autosomal* and *sex-linked* used in the figure are explained in Chapter 3; they are included here for completeness.) Figure 2.20 presents a hypothetical pedigree to show how the symbols are assigned to the family tree.

The trait presented in Figure 2.20 is determined by a recessive allele *a*. Gene symbols are included here to show the deductive reasoning possible with such pedigrees; normally, such symbols would not be present.

~ **FIGURE 2.19**

Symbols used in human pedigree analysis.

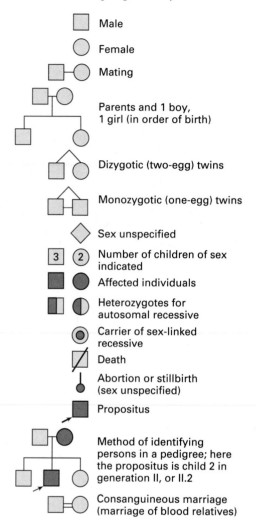

~ **FIGURE 2.20**

A human pedigree, illustrating the use of pedigree symbols.

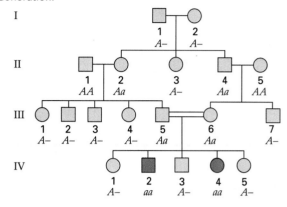

~ **FIGURE 2.21**

(a) Human albino (left); (b) A pedigree showing the transmission of the autosomal recessive trait of albinism.

a)

b) Pedigree

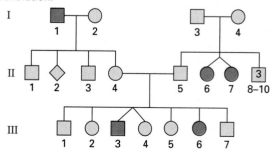

Generations are numbered with roman numerals, while individuals are numbered with arabic numerals, facilitating references to particular people in the pedigree. The trait in this pedigree results from homozygosity for the rare allele, brought about by cousins marrying. Since cousins share a fair proportion of their genes, it can be expected that a number of genes will become homozygous, and in this case one of the genes resulted in an identifiable genetic trait. Since a number of genes in the homozygous condition give rise to serious diseases, marriage between cousins is often prohibited by law.

Examples of Human Genetic Traits

RECESSIVE TRAITS. A large number of human traits are known to be caused by recessive genes. Individuals expressing the recessive trait of albinism (deficient pigmentation) are shown in Figure 2.21a and a pedigree for this trait is shown in Figure 2.21b (p. 53).

For recessive traits to be expressed, the allele must be homozygous (i.e. *aa*). Recessive mutant alleles are commonly associated with serious abnormalities or diseases. Individuals with albinism, for instance, do not produce the melanin pigment that protects the skin from harmful ultraviolet radiation. As a consequence, albinos have considerable skin and eye sensitivity to sunlight. Frequencies of recessive mutant alleles are usually higher than frequencies of dominant mutant alleles because heterozygotes are not at a significant selective disadvantage. Nonetheless, most recessive traits are rare. In the United States approximately 1 in 38,000 of the white population and 1 in 10,000 of the black population are albinos.

It is sometimes difficult to prove that a trait is caused by a recessive mutant allele, especially if the trait is a relatively rare one. For example, environmental factors could lead to some families having more than one affected sibling, thereby giving the false impression that the trait might be inherited. The following lists some general characteristics of recessive inheritance for a relatively rare trait (see Figure 2.21):

1. Most affected individuals have two normal parents, both of whom are heterozygous, so the trait tends to skip generations. (This does not mean that the trait *will* appear in alternating generations, however, since the appearance of the trait depends upon the genotypes of the parents.) The trait appears in the F_1 since a quarter of the progeny are expected to be homozygous for the recessive allele. If the trait is rare or relatively rare, an individual expressing the trait is likely to mate with a homozygous normal individual; thus the next generation is represented by heterozygotes who do not express the trait.
2. Matings between normal heterozygotes should yield both normal progeny and progeny exhibiting the trait, in a ratio of about 3 : 1. The problem here is that it is difficult to obtain enough numbers to make the data statistically significant, especially if a biochemical test is necessary to confirm the presence of the trait. In such cases only the living members of a family can be surveyed. For recessive genes that have less deleterious effects, the allele can reach significant frequencies in the population. An example of such a recessive trait is attached ear lobes. There are significant numbers of heterozygotes and homozygous recessives for this trait in the population. As a result, there is a strong possibility of $Aa \times aa$ matings, and half the progeny will have the trait.
3. Particularly with rare traits, the recessive allele carried by each parent has typically originated with a common ancestor, and the heterozygous, phenotypically normal parents of an individual expressing the trait may be related to each other.
4. When both parents are affected, all their progeny will exhibit the trait.

DOMINANT TRAITS. There are many known dominant human traits. Figure 2.22a illustrates one such trait called woolly hair, in which an individual's hair is very tightly kinked, is very brittle, and breaks before it can grow very long. The best examples of pedigrees for this trait come from Norwegian families; one of these pedigrees is presented in Figure 2.22b. Since it is a fairly rare trait and since not all children of an affected parent show the trait, it can be assumed that most woolly haired individuals are heterozygous for the dominant allele involved.

Several generalizations can be made about dominant traits. Dominant mutant alleles are expressed in a heterozygote when they are in combination with what is usually called the **wild-type allele**—the allele that is designated as the standard ("normal") for the organism. Because many dominant mutant alleles that give rise to recognizable traits are rare, it is extremely unusual to find individuals homozygous for the dominant allele. Thus an affected person in a pedigree is likely to be a heterozygote, and most marriages that involve the mutant allele are between a heterozygote and a homozygous recessive (wild type). Most dominant mutant genes that are clinically significant (that is, they cause medical problems) fall into this category.

Also, most known dominant traits produce less severe clinical effects than do recessive traits. If the dominant trait has serious effects, natural selection tends to decrease the frequency of that allele or to remove that allele from the population. That is, mutant individuals may be unable to reproduce or may not survive.

Another very common feature of dominant traits is that their expression can vary quite widely from one individual to another. **Expressivity** refers to the degree to which an individual manifests a phenotype corresponding to its genotype. Variations in expressivity presumably reflect the biochemical effects of the dominant mutant allele, the activity of other genes whose products interact with the gene or gene product being studied, or the sensitivity of the individual to environmental conditions. Dominant mutant genes in humans also often have variable or low **penetrance**. Penetrance is the frequency with which a dominant gene or a homozygous recessive gene expresses itself in the phenotype of an individual (see Chapter 4). Thus, variable or low penetrance is a major problem for genetic diagnosis and counseling when rare traits are involved (e.g., polydactyly, a trait involving extra fingers and toes).

~ **FIGURE 2.22**

(a) Members of a Norwegian family, some exhibiting the trait of woolly hair;
(b) Part of a pedigree showing the transmission of the autosomal dominant trait of
woolly hair.

a)

b) Generation:

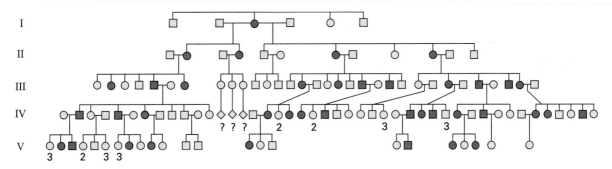

The following are some general characteristics of a dominant trait (refer to Figure 2.22b):

1. Every affected person in the pedigree must have at least one affected parent.
2. Each generation of the pedigree must have individuals who express the mutant gene. In other words, the trait must not skip generations.
3. An affected heterozygous individual must transmit the mutant gene to half his or her progeny. Suppose the dominant mutant allele is designated A, and its wild-type allele is a. Then most crosses will be $Aa \times aa$. From basic Mendelian principles half the progeny will be aa (wild type) and the other half will be Aa, showing the trait.

$\mathcal{K}$EYNOTE

Mendelian principles apply to humans as well as to peas and all other eukaryotes. The study of the human inheritance of genetic traits is complicated by the fact that no controlled crosses can be done. Instead, human geneticists analyze genetic traits by pedigree analysis, that is, by examining the occurrences of the trait in family trees of individuals who clearly exhibit the trait. Many recessively inherited and dominantly inherited genetic traits have been identified as a result of pedigree analysis.

Summary

In this chapter fundamental principles of gene function and gene segregation were discussed. Genes in this chapter are defined as DNA segments that control the biological characteristics that are transmitted from one generation to another; that is, the hereditary traits. An organism's genetic constitution is called its genotype, while the physical manifestation of a genetic trait is called the phenotype. An organism's genes only give the potential for the development of that organism's characteristics. That potential is influenced during development by interactions with other genes as well as by the environment. Thus, individuals with the same genotype can have different phenotypes, and individuals with the same phenotype may have different genotypes.

The first person to obtain some understanding of the principles of heredity—that is, the inheritance of certain traits—was Gregor Mendel. From his breeding experiments with garden peas in the mid-nineteenth century, Mendel proposed two basic principles of genetics. In modern terms, the principle of segregation states that the two members of a single gene pair (the alleles) segregate from each other in the formation of gametes. For each gene with two alleles, half of the gametes carry one allele, and the other half carry the other allele. The principle of independent assortment, proposed on the basis of experiments involving more than one gene, states that genes on different chromosomes behave independently in the production of gametes. Both principles are recognized by characteristic phenotypic ratios—called Mendelian ratios—in particular crosses. For the principle of segregation, in a monohybrid cross between two true-breeding parents, one exhibiting a dominant phenotype and the other a recessive phenotype, the F_2 phenotypic ratio will be $3:1$ for the dominant:recessive phenotypes. For the principle of independent assortment, in a dihybrid cross, the F_2 phenotypic ratio will be $9:3:3:1$ for the four phenotypic classes. The gene segregation patterns can be studied more definitively by determining the genotypes for each phenotypic class. This is done using a testcross, in which an individual of unknown genotype is crossed with a homozygous recessive to determine the genotype of the unknown individual. For example, in a $3:1$ phenotypic ratio, the dominant class can be shown to consist of 1 homozygous dominant: 2 heterozygotes by using a testcross.

After the rediscovery of Mendel's laws in 1900, geneticists found that Mendelian principles of gene segregation apply to humans as well as to peas and all other eukaryotes. The study of the inheritance of genetic traits in humans is complicated by the fact that no controlled crosses can be done. Instead, human geneticists analyze genetic traits by pedigree analysis, that is, by examining the occurrences of the trait in family trees of individuals who clearly exhibit the trait. Many recessively inherited and dominantly inherited genetic traits have been identified as a result of pedigree analysis.

Analytical Approaches for Solving Genetics Problems

The most practical way to reinforce Mendelian principles is to solve genetics problems. These problems generally present familiar and unfamiliar examples and pose questions designed to get problem solvers to work the material forward and backward so that they become totally familiar with it. In this and all following chapters we will discuss how to approach genetics problems by presenting examples of such problems and discussing their answers.

Q.1 A purple-flowered pea plant is crossed with a white-flowered pea plant. All the F_1 plants produced purple flowers. When the F_1 plants are allowed to self-pollinate, 401 of the F_2s have purple flowers and 131 have white flowers. What are the genotypes of the parental and F_1 generation plants?

A.1 The ratio of plant phenotypes in the F_2 is very close to the $3:1$ ratio expected of a monohybrid cross. More specifically, this ratio is expected to result from an $F_1 \times F_1$ cross in which both are identically heterozygous for a specific gene pair. In addition, since only one phenotypic class appeared in the F_1, we can deduce that both parental plants were true breeding. Further, since the F_1 phenotype exactly resembled one of the parental phenotypes, we can say that purple is dominant to white flowers. Assigning the symbol P to the gene that determines purpleness of flowers and the symbol p to the alternative form of the gene that determines whiteness, we can write the genotypes:

P generation:	PP, for the purple-flowered plant
	pp, for the white-flowered plant
F_1 generation:	Pp, which, because of dominance,
	is purple-flowered

We could further deduce that the F$_2$ plants have an approximately $1:2:1$ ratio of $PP:Pp:pp$.

Q.2 Consider three gene pairs Aa, Bb, and Cc, each of which affects a different character. In each case the upper-case letter signifies the dominant allele and the lowercase letter the recessive allele. These three gene pairs assort independently of each other. Calculate the probability of obtaining:

(a) an *Aa BB Cc* zygote from a cross of individuals that are *Aa Bb Cc × Aa Bb Cc*;

(b) an *Aa BB cc* zygote from a cross of individuals that are *aa BB cc × AA bb CC*;

(c) an *A B C* phenotype from a cross of individuals that are *Aa Bb CC × Aa Bb cc*;

(d) an *a b c* phenotype from a cross of individuals that are *Aa Bb Cc × aa Bb cc*.

A.2 Again, we must reduce the question into simple parts in order to apply basic Mendelian principles. The key is that the genes assort independently, so we must multiply the probabilities of the individual occurrences to obtain the answers.

a. First, we must consider the *Aa* gene pair. The cross is *Aa × Aa*, so the probability of the zygote being *Aa* is 2/4 since the expected distribution of genotypes is 1 *AA* : 2 *Aa* : 1 *aa*, as we have discussed. Then the probability of *BB* from *Bb × Bb* is 1/4, and that of *Cc* from *Cc × Cc* is 2/4, following the same sort of logic. The probability of an *Aa BB Cc* zygote is $1/2 \times 1/4 \times 1/2 = 1/16$.

b. Similar logic is needed here, although we must be sure of the genotypes of the parental types since they differ from one gene pair to another. For the *Aa* pair the probability of getting *Aa* from *AA × aa* has to be 1. Next, the probability of getting *BB* from *BB × bb* is 0, so on these grounds alone we cannot get the zygote asked for from the cross given.

c. This question and the next ask for the probability of getting a particular phenotype, so we must start thinking of dominance. Again, we break up the question and consider each character pair in turn. The probability of an *A* phenotype from *Aa × Aa* is 3/4, from basic Mendelian principles. Similarly, the probability of a *B* phenotype from *Bb × Bb* is 3/4. Lastly, the probability of a *C* phenotype from *CC × cc* is 1. Overall, the probability of an *A B C* phenotype is $3/4 \times 3/4 \times 1 = 9/16$.

d. An *a b c* phenotype from *Aa Bb Cc × aa Bb cc* is $1/2 \times 1/4 \times 1/2 = 1/16$.

Q.3 In chickens the white plumage of the leghorn breed is dominant over colored plumage, feathered shanks are dominant over clean shanks, and pea comb is dominant over single comb. Each of the gene pairs segregates independently. If a homozygous white, feathered, pea-combed chicken is crossed with a homozygous colored, clean, single-comb chicken, and the F$_1$s are allowed to inter-breed, what proportion of the birds in the F$_2$ will produce only white, feathered, pea-combed progeny if mated to colored, clean-shanked, single-combed birds?

A.3 This example is typical of a question that presents the unfamiliar in an attempt to get at the familiar. The best approach to such questions is to reduce them to their simple parts and, wherever possible, to assign gene symbols for each character. We are told which character is dominant for each of the three gene pairs, so we can use *W* for white and *w* for colored, *F* for feathered and *f* for clean shanks, and *P* for pea comb and *p* for single comb. The cross involves true-breeding strains and can be written as follows:

P generation: *WW FF PP × ww ff pp*
F$_1$ generation: *Ww Ff Pp*

Now the question asks the proportion of the birds in the F$_2$ that will produce only white, feathered, pea-combed progeny if mated to colored, clean-shanked, single-combed birds. The latter are homozygous recessive for all three genes: that is, *ww ff pp*, as in the parental generation. For the result asked for, the F$_2$ birds must be white, feathered, and pea-combed, and they must be homozygous for the dominant alleles of the respective genes. What we are seeking, then, is the proportion of the F$_2$ chickens that are *WW FF PP* in genotype. We know that each gene pair segregates independently, so the answer can be calculated by using simple probability rules. We consider each gene pair in turn. For the white/colored case the F$_1$ × F$_1$ is *Ww × Ww*, and we know from Mendelian principles that the relative proportion of F$_2$ genotypes will be 1 *WW* : 2 *Ww* : 1 *ww*. Therefore the proportion of the F$_2$s that will be *WW* is 1/4. The same relationship holds for the other two pairs of genes. Since the segregation of the three gene pairs is independent, we must multiply the probabilities of each occurrence to calculate the probability for *WW FF PP* individuals. The answer is $1/4 \times 1/4 \times 1/4 = 1/64$.

QUESTIONS AND PROBLEMS

*2.1 In tomatoes, red fruit color is dominant to yellow. Suppose a tomato plant homozygous for red is crossed with one homozygous for yellow. Determine the appearance of (a) the F$_1$; (b) the F$_2$; (c) the offspring of a cross of the F$_1$ back to the red parent; (d) the offspring of a cross of the F$_1$ back to the yellow parent.

2.2 A red-fruited tomato plant, when crossed with a yellow-fruited one, produces progeny about half of which are red-fruited and half of which are yellow-fruited. What are the genotypes of the parents?

2.3 In maize, a dominant gene A is necessary for seed color as opposed to colorless (a). A recessive gene wx results in waxy starch as opposed to normal starch (Wx). The two genes segregate independently. Give phenotypes and relative frequencies for offspring resulting when a plant of genetic constitution $Aa\ WxWx$ is testcrossed.

*2.4 F_2 plants segregate 3/4 colored : 1/4 colorless. If a colored plant is picked at random and selfed, what is the probability that more than one type will segregate among a large number of its progeny?

*2.5 In guinea pigs rough coat (R) is dominant over smooth coat (r). A rough-coated guinea pig is bred to a smooth one, giving eight rough and seven smooth progeny in the F_1.
a. What are the genotypes of the parents and their offspring?
b. If one of the rough F_1 animals is mated to its rough parent, what progeny would you expect?

2.6 In cattle the polled (hornless) condition (P) is dominant over the horned (p) phenotype. A particular polled bull is bred to three cows. With cow A, which is horned, a horned calf is produced; with a polled cow B a horned calf is produced; and with horned cow C a polled calf is produced. What are the genotypes of the bull and the three cows, and what phenotypic ratios do you expect in the offspring of these three matings?

*2.7 In the Jimsonweed purple flowers are dominant to white. When a particular purple-flowered Jimsonweed is self-fertilized, there are 28 purple-flowered and 10 white-flowered progeny. What proportion of the purple-flowered progeny will breed true?

*2.8 Two black female mice are crossed with the same brown male. In a number of litters female X produced 9 blacks and 7 browns and female Y produced 14 blacks. What is the mechanism of inheritance of black and brown coat color in mice? What are the genotypes of the parents?

2.9 Bean plants may differ in their symptoms when infected with a virus. Some show local lesions that do not seriously harm the plant. Other plants show general systemic infection. The following genetic analysis was made:

P	local lesions × systemic lesions
F_1	all local lesions
F_2	785 local lesions : 269 systemic lesions

What is probably the genetic basis of this difference in beans? Assign gene symbols to all the genotypes occurring in the above experiment. Design a testcross to verify your assumptions.

*2.10 Fur color in babbits, a furry little animal and popular pet, is determined by a pair of alleles B and b. BB and Bb babbits are black, and bb babbits are white.

A farmer wants to breed babbits for sale. Pure (bb) female babbits breed poorly. The farmer purchases a pair of black babbits, and these mate and produce six black and two white babbits. The farmer immediately sells his white babbits, then he comes to consult you for a breeding strategy to produce more white babbits.
a. If he performed random crosses between pairs of F_1 babbits, what proportion of the F_2 would be white?
b. If he crossed an F_1 male to the parental female, what is the probability that this cross will produce white progeny?
c. What would be the farmer's best strategy to maximize the production of white babbits?

2.11 In Jimsonweed purple flower (P) is dominant to white (p), and spiny pods (S) are dominant to smooth (s). In a cross between a Jimsonweed homozygous for white flowers and spiny pods and one homozygous for purple flowers and smooth pods, determine the phenotype of (a) the F_1; (b) the F_2; (c) the progeny of a cross of the F_1 back to the white, spiny parent; (d) the progeny of a cross of the F_1 back to the purple, smooth parent.

2.12 What progeny would you expect from the following Jimsonweed crosses? You are encouraged to use the branch diagram method of analysis.
a. $PP\ ss × pp\ SS$ **d.** $Pp\ Ss × Pp\ Ss$
b. $Pp\ SS × pp\ ss$ **e.** $Pp\ Ss × Pp\ ss$
c. $Pp\ Ss × Pp\ SS$ **f.** $Pp\ Ss × pp\ ss$

*2.13 In summer squash white fruit (W) is dominant over yellow (w), and disk-shaped fruit (D) is dominant over sphere-shaped fruit (d). In the following problems the appearances of the parents and their progeny are given. Determine the genotypes of the parents in each case.
a. White, disk × yellow, sphere gives 1/2 white, disk and 1/2 white, sphere.
b. White, sphere × white, sphere gives 3/4 white, sphere and 1/4 yellow, sphere.
c. Yellow, disk × white, sphere gives all white, disk progeny.
d. White, disk × yellow, sphere gives 1/4 white, disk, 1/4 white, sphere, 1/4 yellow, disk, and 1/4 yellow, sphere.
e. White, disk × white, sphere gives 3/8 white, disk, 3/8 white, sphere, 1/8 yellow, disk, and 1/8 yellow, sphere.

*2.14 Genes a, b, and c assort independently and are recessive to their respective alleles A, B, and C. Two triply heterozygous ($Aa\ Bb\ Cc$) individuals are crossed.
a. What is the probability that a given offspring will be phenotypically ABC, that is, exhibit all three dominant traits?

b. What is the probability that a given offspring will be genotypically homozygous for all three dominant alleles?

2.15 In garden peas tall stem (T) is dominant over short stem (t), green pods (G) are dominant over yellow pods (g), and smooth seeds (S) are dominant over wrinkled seeds (s). Suppose a homozygous short, green, wrinkled pea plant is crossed with a homozygous tall, yellow, smooth one.

a. What will be the appearance of the F_1?

b. What will be the appearance of the F_2?

c. What will be the appearance of the offspring of a cross of the F_1 back to its short, green, wrinkled parent?

d. What will be the appearance of the offspring of a cross of the F_1 back to its tall, yellow, smooth parent?

2.16 Two homozygous strains of corn are hybridized. They are distinguished by six different pairs of genes, all of which assort independently and produce an independent phenotypic effect. The F_1 hybrid is selfed to give an F_2.

a. What is the number of possible genotypes in the F_2?

b. How many of these genotypes will be homozygous at all six gene loci?

c. If all gene pairs act in a dominant-recessive fashion, what proportion of the F_2 will be homozygous for all dominants?

d. What proportion of the F_2 will show all dominant phenotypes?

*2.17 The coat color of mice is controlled by several genes. The agouti pattern, characterized by a yellow band of pigment near the tip of the hairs, is produced by the dominant allele A; homozygous aa mice do not have the band and are nonagouti. The dominant allele B determines black hairs, and the recessive allele b determines brown. Homozygous $c^h c^h$ individuals allow pigments to be deposited only at the extremities (e.g., feet, nose, and ears) in a pattern called Himalayan. The genotype $C-$ allows pigment to be distributed over the entire body.

a. If a true-breeding black mouse is crossed with a true-breeding brown, agouti, Himalayan mouse, what will be the phenotypes of the F_1 and F_2?

b. What proportion of the black agouti F_2 will be of genotype-$Aa\ BB\ Cc^h$?

c. What proportion of the Himalayan mice in the F_2 are expected to show brown pigment?

d. What proportion of all agoutis in the F_2 are expected to show black pigment?

2.18 In cocker spaniels, solid coat color is dominant over spotted coat. Suppose a true-breeding solid-colored dog is crossed with a spotted dog, and the F_1 dogs are interbred.

a. What is the probability that the first puppy born will have a spotted coat?

b. What is the probability that if four puppies are born, all of them will have a solid coat?

3 Chromosomal Basis of Inheritance, Sex Determination, and Sex Linkage

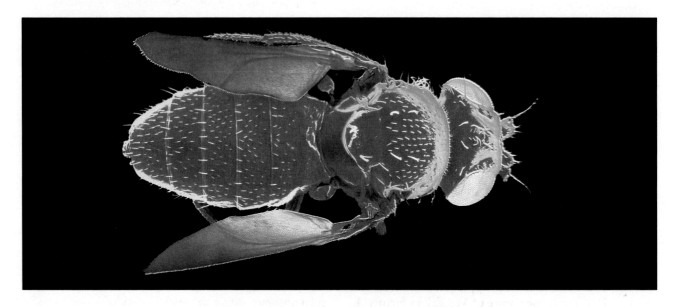

PRINCIPAL POINTS

~ The chromosome theory of heredity states that the chromosomes are the carriers of the genes.

~ A sex chromosome is a chromosome in eukaryotic organisms that is represented differently in the two sexes. In most organisms with sex chromosomes, the female has two X chromosomes (i.e. she is XX) while the male has one X and one Y chromosome (he is XY). The Y chromosome is structurally and functionally different from the X chromosome.

~ Sex linkage is the association of genes with the sex-determining chromosomes of eukaryotes. Such genes are referred to as sex-linked genes.

~ In many eukaryotic organisms, sex determination is related to the sex chromosomes. In humans and other mammals, for example, the presence of a Y chromosome specifies maleness, while its absence results in femaleness. Several other sex-determination mechanisms are known in eukaryotes.

~ In humans (and other eukaryotes with sex chromosomes), the gene responsible for a trait can be inherited in one of five different ways: autosomal recessive, autosomal dominant, sex-linked recessive, sex-linked dominant, or Y-linked.

Soon after the beginning of the twentieth century, scientists realized that Mendelian laws applied to a large variety of organisms and that the principles of Mendelian analysis could be used to predict the outcome of crosses in these organisms. On Mendel's foundation early geneticists began to build genetic hypotheses that could be tested by appropriate crosses and began to investigate the nature of Mendelian factors. In this chapter we will examine the evidence showing that Mendelian factors are what we now know as genes and that genes are located in chromosomes. We will learn about various mechanisms of sex determination, some of which involve special chromosomes (sex chromosomes) that are related to the sex of the organism. We will also learn about the genetic segregation patterns of genes located on the sex chromosomes and in the other chromosomes (the autosomes) of the cell.

CHROMOSOME THEORY OF INHERITANCE

By the end of the nineteenth century, several important discoveries had been made in the field of cytology, the scientific study of cells. Through the research of biologists, notably K. Nageli, M. J. Schleiden, and T. Schwann, many details of cell structure and cell division were known. R. Brown named and described the cell nucleus in 1833, and in the latter part of the century O. Hertwig showed that the nucleus was involved directly in sea urchin fertilization. In plants a similar fusion of sex cell nuclei was described by E. Strasburger. Strasburger also coined the terms **nucleoplasm** and **cytoplasm** for the semisolid matrix of cell material in the nucleus and in the surrounding cell body, respectively.

Toward the end of the century W. C. Schneider and W. Flemming, working independently, showed that chromosomes within the nucleus divided longitudinally during cell division. E. van Beneden subsequently showed that the divided chromosomes were distributed in equal numbers to the two daughter cells.

Cytologists also firmly established that, with a few exceptions, the total number of chromosomes remains constant in all cells of an organism, except during gamete formation, when the chromosome number is reduced by half. They showed that the 2N chromosome number is reestablished in the next generation when gametes fuse during fertilization to produce the zygote cell from which the mature organism develops. While the number of chromosomes in a cell is constant in individuals of the same species (with few exceptions), the chromosome number varies considerably between species (Table 3.1; p. 62).

In 1900 Mendel's work on the nature of hereditary factors was rediscovered. Within a short time Mendel's experiments became known and appreciated throughout the scientific world. In 1903 Walter Sutton and Theodor Boveri independently recognized that the transmission of chromosomes from one generation to the next closely paralleled the pattern of transmission of genes from one generation to the next. To explain this correlation, they put forward the hypothesis that genes are on chromosomes, a proposal known as the

~ TABLE 3.1

Chromosome Number in Various Organisms[a]

ORGANISM	TOTAL CHROMOSOME NUMBER	ORGANISM	TOTAL CHROMOSOME NUMBER
Human	46	Australian ant (*Myrecia pilosula*)	♂ 1, ♀ 2
Chimpanzee	48	Nematode	11 ♂, 12 ♀
Rhesus monkey	42	*Neurospora* (haploid)	7
Dog	78	Sphagnum moss (haploid)	23
Cat	72	Field horsetail	216
Horse	64	Giant sequoia	22
Mallard	80	Tobacco	48
Chicken	78	Cotton	52
Alligator	32	Kidney bean	22
Cobra	38	Broad bean	12
Bullfrog	26	Onion	16
Toad	36	Potato	48
Goldfish	94	Tomato	24
Starfish	36	Bread wheat	42
Fruit fly	8	Rice	24
Housefly	12	Baker's yeast	34
Mosquito	6		

[a]Except as noted, all chromosome numbers are for diploid cells.

chromosome theory of heredity. In the subsequent ten years many experiments by cytologists and geneticists provided irrefutable evidence for the relationship between genes and chromosomes.

𝒦EYNOTE

The chromosome theory of heredity states that the chromosomes are the carriers of the genes. The first clear formulation of the chromosome theory was made in 1903 by Sutton and by Boveri, who independently recognized that the transmission of chromosomes from one generation to the next closely paralleled the pattern of transmission of genes from one generation to the next. ━━━━

Sex Chromosomes

The proof of the chromosome theory of heredity came from experiments that related the hereditary behavior of particular genes to the transmission of the **sex chromosome**, the chromosome in eukaryotic organisms that is represented differently in the two sexes. The discovery of sex chromosomes will be reviewed here as a prelude to discussing the experimental proof of the Sutton–Boveri hypothesis.

In the early 1900s, Clarence E. McClung, Nettie Stevens, and Edmund B. Wilson, all working with insects, independently obtained evidence that particular chromosomes determined the sex of an organism. Working with grasshoppers and other members of the order Orthoptera, McClung proposed in 1901 that a particular chromosome, which he called the *accessory chromosome,* was involved in determining each insect's sex at fertilization. In 1905 Stevens and Wilson provided more information about McClung's proposal. They found that in some Orthoptera the female has an even number of chromosomes while the male has an odd number. There are two copies of one of the chromosomes in the female but only one copy in the male. It was this extra chromosome in the female that McClung had hypothesized as the accessory chromosome. Stevens called the extra chromosome an **X chromosome.** Since this chromosome is directly related to the sex of the organism, the X chromosome is an example of a sex chromosome.

When female grasshoppers form eggs by meiosis, each egg receives one of each chromosome, including an X chromosome. Half the male's sperm cells produced by meiosis receive an X chromosome and the other half do not. Consequently, the sex of the progeny grasshoppers is determined by the chromosome composition of the sperm that fertilizes the egg. If the sperm carries an X chromosome, then the resulting fertilized egg will have a pair of Xs, and the individual that develops from the zygote will be female. If the sperm does not have an X, the fertilized egg will have an unpaired X and the individual will be a male.

Not all insects have an unpaired sex chromosome as do grasshoppers. For example, Stevens found that in the common mealworm, *Tenebrio molitor,* the male has a partner chromosome for the X chromosome. That partner is much smaller and clearly distinguishable from the X chromosome. Stevens called the partner chromosome the **Y chromosome,** and like the X chromosome, it is a sex chromosome. The sperm cells of the mealworm contain either an X or a Y chromosome, and the sex of the offspring is determined by the sperm type that fertilizes the X chromosome-bearing egg: XX mealworms are female, and XY mealworms are male.

Stevens and Wilson observed similar X−Y sex chromosome complements in other organisms. To geneticists perhaps the most important of these organisms is the fruit fly, *Drosophila melanogaster.* Figure 3.1 shows the appearance of male and female *Drosophila,* Figure 3.2 shows the chromosome complements of the two sexes, and Figure 3.3 (p. 64) shows the life cycle. Because the male produces two kinds of gametes with respect to sex chromosome content (X or Y), and

~ **FIGURE 3.1**

Female (left) and male (right) *Drosophila melanogaster* (fruit fly), an organism used extensively in genetics experiments: (a) Adult flies; (b) Drawings of ventral abdominal surface to show differences in genitalia.

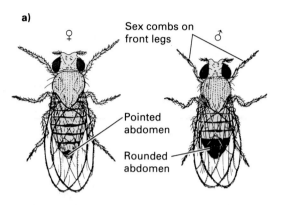

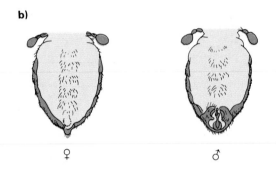

~ **FIGURE 3.2**

Chromosomes of *D. melanogaster* diagrammed to show their morphological differences. A female has four pairs of chromosomes in her somatic cells, including a pair of X chromosomes. The only difference in the male is an XY pair of sex chromosomes instead of two Xs.

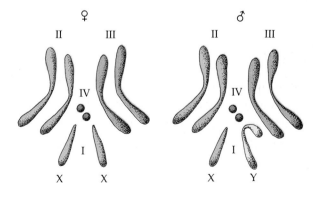

~ FIGURE 3.3

Life cycle of *D. melanogaster*.

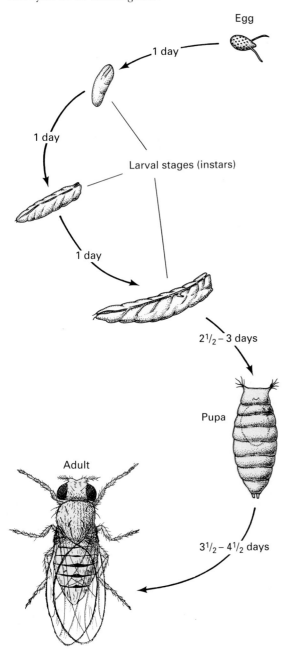

Egg

1 day

1 day

Larval stages (instars)

1 day

$2^{1}/_{2}$ – 3 days

Pupa

Adult

$3^{1}/_{2}$ – $4^{1}/_{2}$ days

As shown in Figure 3.4, the pattern of transmission of X and the Y chromosomes from generation to generation is very straightforward. The production of the F_1 generation is shown in Figure 3.4a, and the F_2 generation in Figure 3.4b. In this figure the X is represented by a straight structure much like a slash mark, and the Y by a similar structure topped by a hook to the right.

KEYNOTE

A sex chromosome is a chromosome in eukaryotic organisms that is represented differently in the two sexes. In many of the organisms encountered in genetic studies, one sex possesses a pair of identical chromosomes (the X chromosomes). The opposite sex possesses a pair of visibly different chromosomes: one is an X chromosome, and the other, structurally and functionally different, is called the Y chromosome. Commonly, the XX sex is female, and the XY sex is male. The XX and XY sexes are called the homogametic and heterogametic sexes, respectively. _____

Sex Linkage

The conclusive evidence to support the chromosome theory of heredity was obtained by using *Drosphila* (fruit flies) in genetic experiments. In 1910 Thomas Hunt Morgan reported his results of crossing experiments with the fruit fly.

Morgan found a male fly in one of his true-breeding stocks that had white eyes instead of the bright red eyes that are characteristic of the **wild type.** The term *wild type* refers to a strain, organism, or gene that is designated as the standard for the organism with respect to genotype and phenotype. For example, a *Drosophila* strain with all wild-type genes will have bright red eyes. Variants of a wild-type strain arise from mutational changes of the wild-type genes that produce **mutant alleles;** the result is strains with mutant characteristics. Mutant alleles may be recessive or dominant to the wild-type allele; for example, the mutant allele that causes white eyes in *Drosophila* is recessive to the wild-type (red-eye) allele.

When Morgan crossed the white-eyed male with a red-eyed female from the same stock, he found that all the F_1 flies were red-eyed and concluded that the white-eyed trait was recessive. Next, he allowed the F_1 progeny to interbreed and counted the phenotypic classes in

because the female produces only one gametic type (X), the male is called the **heterogametic sex** and the female is called the **homogametic sex.** In *Drosophila* the X and Y chromosomes are similar in size but their shapes are different. (Note: In some organisms the male is homogametic and the female is heterogametic. This reversal will be discussed later in the chapter.)

~ FIGURE 3.4

Inheritance pattern of X and Y chromosomes in organisms where the female is XX and the male is XY: (a) Production of the F_1 generation; (b) production of the F_2 generation.

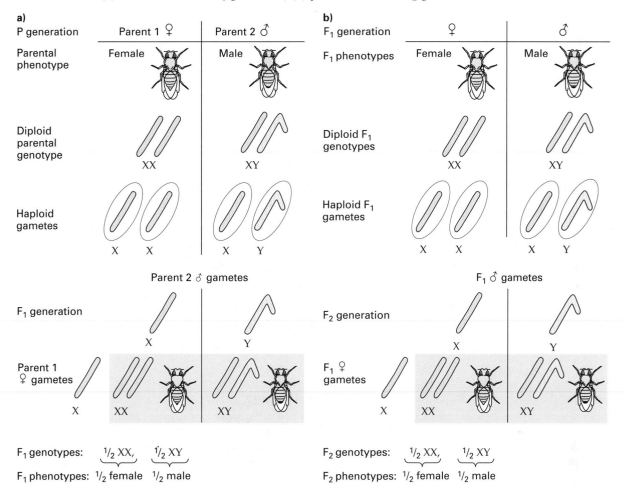

the F_2 generation. There were 3470 red-eyed and 782 white-eyed flies. The number of individuals with the recessive phenotype was too small to fit the Mendelian 3:1 ratio. In addition, *Morgan noticed that all the white-eyed flies were male.* (Later, he determined that there was another reason for the lower-than-expected number of flies with the recessive phenotype: flies with white eyes are not as viable as those with red eyes.)

Next, Morgan crossed a white-eyed male fly with one of its red-eyed daughters in order to find out more about the inheritance of the white trait. The result was 129 red-eyed females, 132 red-eyed males, 88 white-eyed females, and 86 white-eyed males. The closest ratio to the observed results is 1:1:1:1, with the deviation from this ratio again stemming from the decreased

viability of white-eyed flies. Morgan's results showed that females could have white eyes and that approximately equal numbers of red- and white-eyed flies occur in both sexes.

Figure 3.5 (p. 68) diagrams the first set of crosses, which produced the 3:1 ratio of red-eyed:white-eyed flies in the F_2. The *Drosophila* gene symbolism used is described in Box 3.1 (pp. 66–67) and should be understood before proceeding with this discussion. Note as the discussion proceeds that the mother-son inheritance pattern presented in Figure 3.5 is the result of the segregation of genes located on a sex chromosome.

In interpreting the data from his two crosses, Morgan proposed that the gene for the eye color variant is located on the X chromosome and that there is no

BOX 3.1
Genetic Symbols Revisited

Unfortunately, there is no one system of gene symbols used by geneticists. In this chapter we have seen that the gene symbols used for *Drosophila* are different from those used for peas in Chapter 2. This *Drosophila* symbolism is commonly, but not exclusively, used in genetics today. In this system the symbol + indicates a wild-type allele of a gene; that is, the allele most frequently found in natural populations of the organism. A lowercase letter designates mutant alleles of a gene that are recessive to the wild-type allele, and an uppercase letter is used for alleles that are dominant to the wild-type allele. *The letters are chosen on the basis of the phenotype of the organism expressing the mutant allele.* For example, a variant strain of *Drosophila* has bright orange eyes instead of the usual bright red. The mutant allele involved is recessive to the wild-type bright red allele, and because the eye color is close to vermilion in tint, the allele is designated v and is called the vermilion allele. The wild-type allele of v is v^+, but when there is no chance of confusing it with other genes in the cross, it is often shortened to +. In the "Mendelian" terminology used up to now, the recessive mutant allele would also be v, and its wild-type allele would be V.

A conventional way to represent the chromosomes (instead of the way we have been using in the figures) is to use the slash (/). Thus v^+/v or $+/v$ indicates that there are *two* homologous chromosomes, one with the wild-type allele (v^+ or +) and the other with the recessive allele (v). The Y chromosome is usually symbolized as a Y or a bent slash (/). Thus Morgan's cross of a true-breeding red-eyed female fly with a white-eyed male could be written $w^+/w^+ \times w/Y$ or $+/+ \times w/\ell$.

The same rules apply when the alleles involved are dominant to the wild-type allele. For instance, some *Drosophila*, called *Bar* mutants, have bar-shaped eyes instead of the normal round eyes (Box Figure 1). The shape of the eye is related to the number of facets, and the *Bar* mutation results in fewer facets than found in the wild type giving the

~ BOX FIGURE 3.1

Eye of *Drosophila melanogaster*, a typical insect compound eye made up of a large number of facets: (a) Round eye of wild-type flies; (b) Bar-eye in homozygous (B/B) female; (c) Bar-eye in hemizygous (B/Y) male; (d) Intermediate wide Bar-eye phenotype in heterozygous (B/B^+) females.

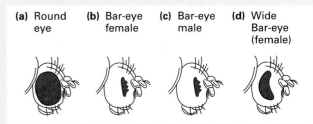

(a) Round eye   **(b)** Bar-eye female   **(c)** Bar-eye male   **(d)** Wide Bar-eye (female)

eye its bar shape. In the heterozygote the allele for bar-shaped eye is dominant although not completely so; as a result, hemizygous males and females homozygous for *Bar* have much smaller eyes than females heterozygous for *Bar*.

The symbol for this allele is B, and the wild-type allele is B^+, or + in the shorthand version. The *Bar* gene is located on the X chromosome, and there is no allele on the Y. Box Figure 2 illustrates the inheritance of the bar-shaped eye characteristic in reciprocal crosses. These crosses show that the bar–eye phenotype is determined by an X-linked gene and that no equivalent allele is present in the Y chromosome.

In the rest of the book, both the A/a ("Mendelian") and a^+/a (*Drosophila*) symbolisms will be used, so it is important that you, the student, be able to work with both. Since it is easier to verbalize the "Mendelian" symbols (e.g. big A, small a), many of our examples will follow that symbolism, even though the *Drosophila* symbolism in many ways is more informative. That is, with the *Drosophila* system, the wild-type and mutant alleles are readily apparent, while that is not the case with the "Mendelian" system. The "Mendelian" system is commonly used in animal and plant breeding.

~ **BOX FIGURE 3.2**
Reciprocal crosses of Bar-eye flies with normal (wild-type) flies. (a) Wild-type B^+/B^+ ♀ × bar-eyed B/Y ♂; (b) Bar-eyed B/B ♀ × wild-type B^+/Y ♂.

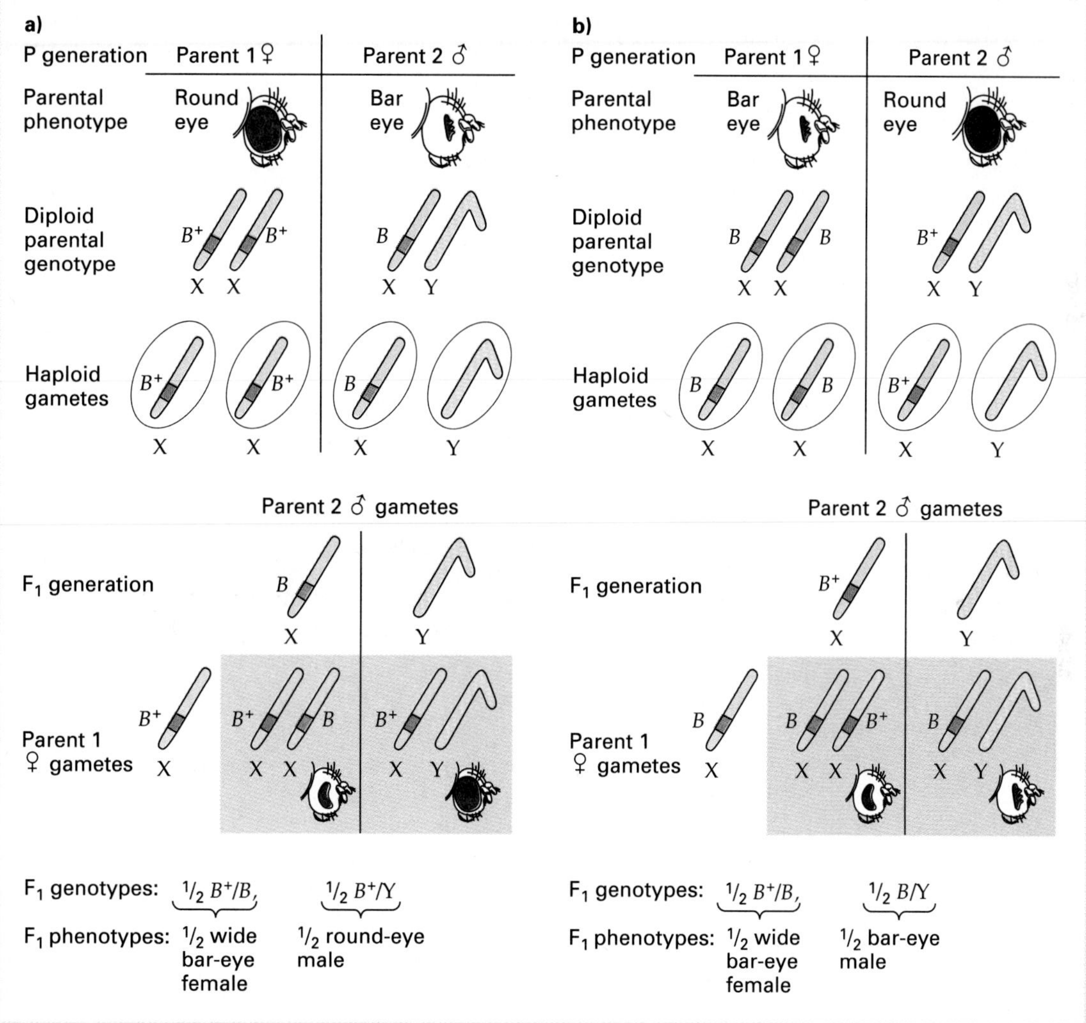
a)
P generation
Parent 1 ♀
Parent 2 ♂
Parental phenotype
Round eye
Bar eye
Diploid parental genotype
B+ B+
X X
B
X Y
Haploid gametes
B+ X
B+ X
B X
Y
Parent 2 ♂ gametes
F1 generation
B X
Y
Parent 1 ♀ gametes
B+ X
B+ B X X
B+ X Y
F1 genotypes: 1/2 B+/B, 1/2 B+/Y
F1 phenotypes: 1/2 wide bar-eye female
1/2 round-eye male
b)
P generation
Parent 1 ♀
Parent 2 ♂
Parental phenotype
Bar eye
Round eye
Diploid parental genotype
B B
X X
B+
X Y
Haploid gametes
B X
B X
B+ X
Y
Parent 2 ♂ gametes
F1 generation
B+ X
Y
Parent 1 ♀ gametes
B X
B B+ X X
B X Y
F1 genotypes: 1/2 B+/B, 1/2 B/Y
F1 phenotypes: 1/2 wide bar-eye female
1/2 bar-eye male

~ FIGURE 3.5

The X-linked inheritance of red eyes and white eyes in *D. melanogaster*. The symbols w and w^+ indicate the white- and red-eyed alleles, respectively. (a) A red-eyed female is crossed with a white-eyed male; (b) The F_1 flies are interbred to produce the F_2s.

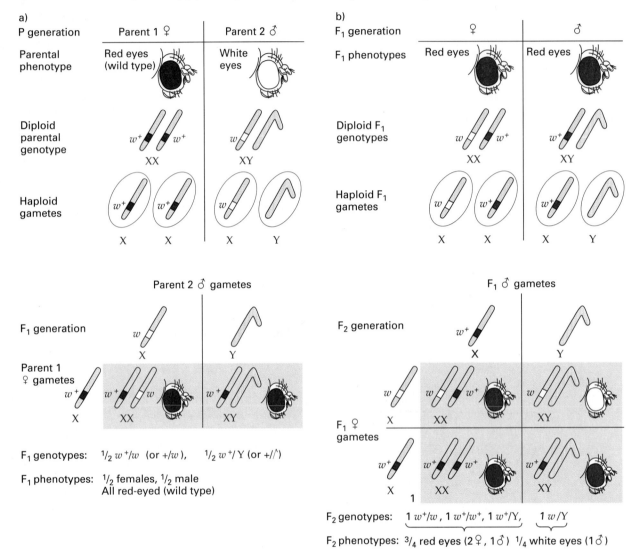

equivalent gene on the Y chromosome. The condition of X-linked genes in males is said to be **hemizygous** since they have no corresponding allele on the Y. For example, the white-eyed *Drosophila* males have an X chromosome with a white allele and no other allele of that gene in their gene complement: These males are said to be hemizygous for the white allele. Since the white allele of the gene is recessive, the original white-eyed male must have had the recessive allele for white eyes (designated w; see Box 3.1) on his X chromosome. The red-eyed female came from a true-breeding stock,

so both of her X chromosomes must have carried the dominant allele for red eyes, w^+.

The F_1s are produced in the following way (Figure 3.5): The males receive their only X chromosome from their mother and hence have the w^+ allele and are red-eyed. The females receive a dominant w^+ allele from their mother and a recessive w allele from their father. As a result of this inheritance pattern, the F_1 females are also red-eyed.

To produce the F_2 flies, Morgan interbred F_1 red-eyed females and red-eyed males (Figure 3.5b). In

~ FIGURE 3.6

Testcross of an F_1 female from the cross in Figure 3.5 with a white-eyed male. The result is a 1:1 ratio of red-eyed:white-eyed flies in each sex.

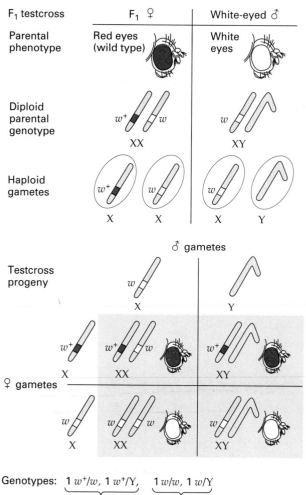

Genotypes: 1 w^+/w, 1 w^+/Y,   1 w/w, 1 w/Y
Phenotypes: ½ red eyes   ½ white eyes
           (1♀, 1♂)   (1♀, 1♂)

the F_2 the males that received an X chromosome with the w allele from their mother are white-eyed; those that received an X chromosome with the w^+ allele are red-eyed.

From these results we can interpret Morgan's second cross, the cross of a white-eyed male with its red-eyed daughter. That cross, shown in Figure 3.6, produced a ratio of approximately 1 red-eyed female:1 red-eyed male:1 white-eyed female:1 white-eyed male. The white-eyed male produces sperm that bears either a Y chromosome or an X chromosome with the recessive

w allele. As a result, the eye phenotype of the progeny is determined by the genetic makeup of the egg. This gene transmission from a male parent to a female offspring ("child") to a male grandchild is called **crisscross inheritance.**

Morgan did a third cross in this series, crossing a red-eyed male with a true breeding white-eyed female (Figure 3.7; p. 70). (This cross is the *reciprocal cross* of Morgan's first cross performed—white male × red female—shown in Figure 3.5.) The parental male here is hemizygous for the w^+ allele, and the parental female is homozygous for the w allele. All the F_1 females receive a w^+-bearing X from their father and a w-bearing X from their mother (Figure 3.7a). Consequently, they are heterozygous w^+/w and have red eyes. All the F_1 males receive a w-bearing X from their mother and a Y from their father, and so they have white eyes (Figure 3.7a). This result is distinct from that of the cross in Figure 3.5. Note also that these results are different from the normal results of a reciprocal cross because of the sex linkage.

Interbreeding of the F_1s (Figure 3.7b) follows the same pattern as the testcross diagramed in Figure 3.6, since the male is w/Y and the female is w^+/w. Thus an approximately equal number of male and female red- and white-eyed flies appear in the F_2. Again, this ratio differs from the results obtained in the first cross, where an approximately 3:1 ratio of red-eyed:white-eyed flies was obtained and where none of the females and approximately half the males exhibited the white-eyed phenotype. The difference in phenotypic ratios in the two sets of crosses reflects the transmission patterns of sex chromosomes and the genes they contain.

Morgan's crosses of *Drosophila* involved eye color characteristics that we now know are coded for by genes found on the X chromosome. These characteristics and the genes that give rise to them are referred to as **sex-linked,** or, more correctly, **X-linked.** *Sex-linked inheritance* is the term used for the pattern of hereditary transmission of sex-linked genes. When the results of reciprocal crosses are not the same, sex-linked characteristics may well be involved. By comparison, the results of reciprocal crosses are the same when they involve genes located on the **autosomes** (chromosomes other than the sex chromosomes). Most importantly, Morgan's results strongly supported the hypothesis that genes were located on chromosomes, although his data only showed a strong correlation between the behavior of genes and chromosomes and did not prove the Sutton–Boveri hypothesis. Morgan found many other examples of genes on the X chromosome in *Drosophila* and in other organisms, thereby showing that his observations were not confined to a single species.

~ FIGURE 3.7

Reciprocal cross of that shown in Figure 3.5. (a) A homozygous white-eyed female is crossed with a red-eyed (wild type) male; (b) The F_1 flies are interbred to produce the F_2s. The results of this cross differ from those in Figure 3.5 because of the way sex chromosomes segregate in crosses.

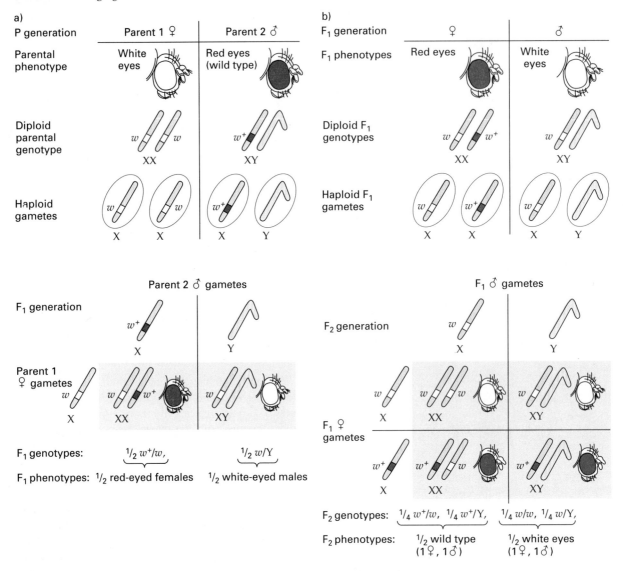

KEYNOTE

Sex linkage is the association of genes with the sex-determining chromosomes of eukaryotes. Such genes, as well as the phenotypic characteristics these genes control, are referred to as sex-linked. Morgan's pioneering work with the inheritance of sex-linked genes of *Drosophila* strongly supported, but did not prove, the chromosome theory of heredity. _____

Nondisjunction of X Chromosomes

The actual proof for the chromosome theory of inheritance came from the work of Morgan's student, Calvin Bridges, who was carrying out basic genetic and cytological studies with *Drosophila*. Morgan's work showed that from a cross of a white-eyed female (w/w ♀) with a red-eyed male (w^+/Y ♂), all the F_1 males should be white-eyed, and all the females should be red-eyed. Bridges found that there are rare excep-

~ FIGURE 3.8

Nondisjunction involving the X chromosome. Nondisjunction of autosomal chromosomes and chromosomes in mitosis occurs in the same way. (a) Normal X chromosome segregation; (b) Nondisjunction of X chromosomes in meiosis I; (c) Nondisjunction of X chromosomes in meiosis II.

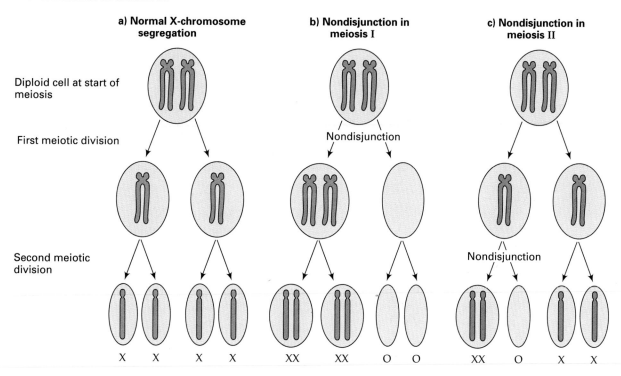

tions to this result: About 1 in 2000 of the F_1 flies from such a cross are either white-eyed females or red-eyed males.

To explain these data, Bridges hypothesized that a problem had occurred with chromosome segregation in meiosis. That is, homologous chromosomes (in meiosis I) or sister chromatids (in meiosis II or mitosis) must move to opposite poles at anaphase. When this fails to take place, chromosome **nondisjunction** results. For the crosses being analyzed, occasionally the two X chromosomes failed to separate, so eggs were produced either with two X chromosomes or with no X chromosomes instead of the usual one. This particular example of nondisjunction is called **X chromosome nondisjunction** (Figure 3.8). When it occurs in an individual with a normal set of chromosomes, it is called **primary nondisjunction.** Normal disjunction of the X chromosomes is illustrated in Figure 3.8a, and nondisjunction of the X chromosomes in meiosis I and meiosis II is shown in Figures 3.8b and 3.8c, respectively.

Primary nondisjunction of the X chromosomes can explain the exceptional flies in Bridges' first experimental cross. When primary nondisjunction occurs in the w/w female (Figure 3.9; p. 72), two classes of exceptional eggs result with equal frequency: those with two X chromosomes and those with no X chromosomes. The male is w^+/Y, producing equal numbers of w^+- and Y-bearing sperm. When these eggs are fertilized by the two types of sperm, the result is four types of zygotes. Two types usually do not survive: the YO class, which has a Y chromosome but no X chromosome, and the triplo-X (XXX) class, which has three X chromosomes and, as shown in Figure 3.9, the genotype $w/w/w^+$. The former die because they lack the X chromosome and its genes that code for essential cell functions. The latter often die because the flies apparently cannot function with three doses of each of the X chromosome genes, at least under standard culture conditions.

The surviving classes are the red-eyed XO males (in *Drosophila* the XO pattern produces a sterile male), with no Y chromosome and a w^+ allele on the X, and the white-eyed XXY females, with a w allele on each X. The males are red-eyed because they received their X chromosome from their fathers, and the females have white eyes because their two X chromosomes came from their mothers. This result is unusual, since sons

~ FIGURE 3.9

Primary nondisjunction during meiosis in a white-eyed female *D. melanogaster,* and results of a cross with a normal red-eyed male. Note the decreased viability of XXX and YO progeny.

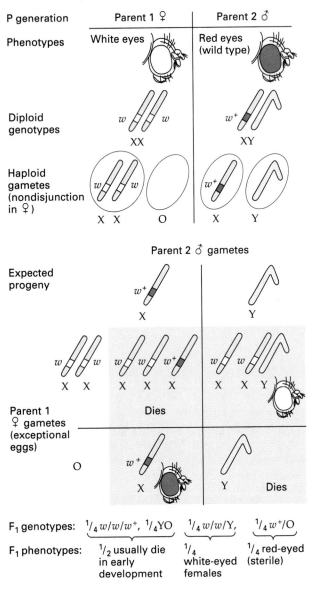

F₁ genotypes: $\frac{1}{4}w/w/w^+$, $\frac{1}{4}$YO $\frac{1}{4}w/w/Y$, $\frac{1}{4}w^+/O$

F₁ phenotypes: $\frac{1}{2}$ usually die in early development $\frac{1}{4}$ white-eyed females $\frac{1}{4}$ red-eyed (sterile)

normal set of chromosomes either are missing, or are present in more than the usual number of copies.)

Bridges found that XXY *Drosophila* females are normal and fertile, and to test his hypothesis further, he crossed the exceptional white-eyed XXY females with normal red-eyed XY males (Figure 3.10). The XXY female is homozygous for the *w* allele on her two Xs. The male has the *w*⁺ allele on his X. Both parents have no equivalent eye color allele on their Y chromosomes. The expectation was that the two X chromosomes of the XXY parent would segregate into different gametes: one gamete would be X and the other XY. Fusion with an X-bearing sperm from the male would give XX and XXY progeny, both of which would be heterozygous *w*⁺/*w* and therefore would have red eyes. Again, flies with unexpected phenotypes resulted from this cross: A low percentage of the male progeny had red eyes and a similarly low percentage of the female progeny had white eyes.

To account for these unusual phenotypes, Bridges hypothesized that segregation of chromosomes in the meiosis of an XXY female can occur in two ways. In normal disjunction, the two X chromosomes separate and migrate to opposite poles, with one of them accompanied by the Y, to produce equal numbers of X- and XY-bearing eggs. This pattern is the one that the Xs should follow during meiosis, and, in fact, it occurs 92 percent of the time (Figure 3.10a).

In the second pattern, which takes place the remaining 8 percent of the time, there is nondisjunction of the Xs (Figure 3.10b). Bridges called this segregation **secondary nondisjunction** because it occurred in the progenies of females produced by primary nondisjunction. Secondary nondisjunction results in the two Xs migrating together to one pole and the Y migrating to the other; the eggs are XX and Y. When these eggs are fertilized by the two classes of sperm (X and Y), the two surviving classes are exceptional red-eyed (XO) males and white-eyed (XXY) females. The other two classes, the XXX and YO, die, usually early in development. Thus, of the 4 percent fertilized cells produced through secondary nondisjunction, only half survive to adulthood. Bridges verified his secondary nondisjunction hypothesis by microscopic examination of the chromosomes of the flies collected from the cross.

Thus, Bridges' experiments showed that the odd pattern of inheritance always went hand-in-hand with the specific aneuploid types (XO and XXY). Such a correlation could not be due to some accidental parallelism, proving without any doubt that a specific gene was associated with a specific chromosome. Bridges' results were taken as proof of the chromosome theory of heredity.

In sum, there is a parallel behavior of Mendelian genes and chromosomes, in that gene segregation pat-

normally get their X from their mothers, and daughters get one X from each parent.

Bridges' hypothesis was very plausible and readily testable by examining the chromosome composition of the exceptional flies. This cytological examination showed that, indeed, the white-eyed females were XXY and the red-eyed males were XO. (**Aneuploidy** is the term used for the abnormal condition—as is the case here—in which one or more whole chromosomes of a

~ FIGURE 3.10

Results of a cross between the exceptional white-eyed XXY female of Figure 3.9 with a normal red-eyed XY male. Again, XXX and YY progeny usually die. (a) Normal disjunction of the X chromosomes in the XXY female. (b) Secondary nondisjunction of the homologous X chromosomes in meiosis I of the XXY female.

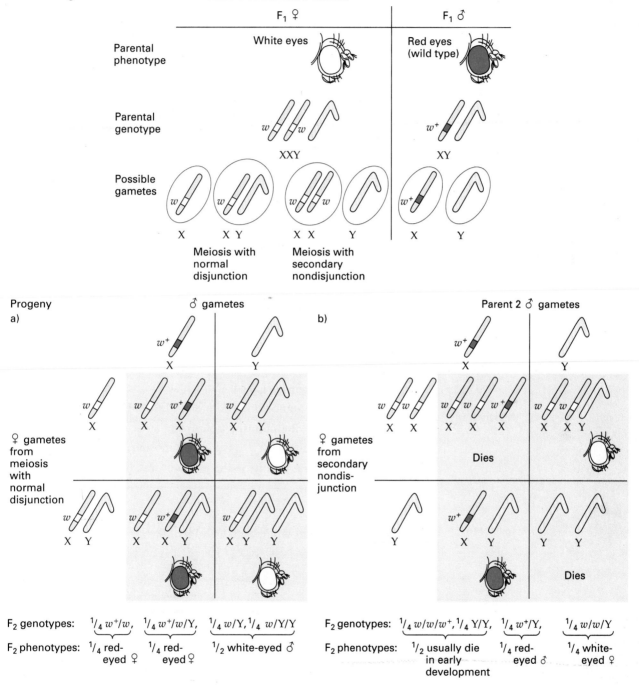

~ FIGURE 3.11

Illustration of the parallel behavior between Mendelian genes and chromosomes in meiosis. In the hypothetical *Aa Bb* diploid cell there is a homologous pair of metacentric chromosomes, which carry the *A/a* gene pair, and a homologous pair of telocentric chromosomes, which carry the *B/b* gene pair. The independent alignment of the two homologous pairs of chromosomes at metaphase I results in equal frequencies of the four meiotic products, *AB*, *ab*, *Ab*, and *aB*, illustrating Mendel's principle of independent assortment.

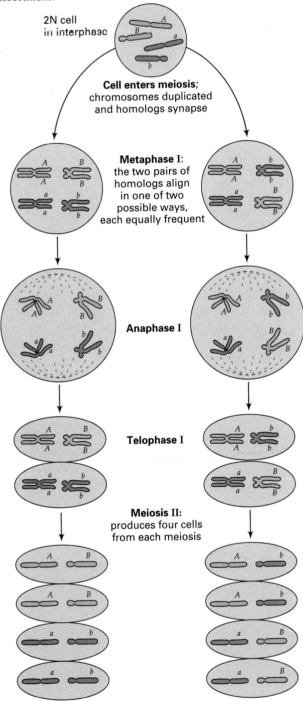

terns follow the patterns of chromosome behavior in meiosis. In Figure 3.11 this parallel is illustrated for a diploid cell with a homologous pair of metacentric chromosomes and a homologous pair of telocentric chromosomes. The cell is genotypically *Aa Bb*, with the *A/a* gene pair on the metacentric homologs and the *B/b* gene pair on the telocentric homologs. As the figure shows, the two homologous pairs of chromosomes align on the metaphase plate independently, giving rise to two different segregation patterns for the two gene pairs. Since each of the two alignments, and hence segregation patterns, is equally likely, the result of meiosis is cells that exhibit equal frequencies of the genotypes *AB*, *ab*, *Ab*, and *aB*. Genotypes *AB* and *ab* result from one chromosome alignment, and genotypes *Ab* and *aB* result from the other alignment. In terms of Mendel's laws, the principle of segregation applies to the segregation pattern of one homologous pair of chromosomes and the associated gene pair in Figure 3.11, while the principle of independent assortment applies to the segregation pattern of both homologous pairs of chromosomes and the associated two gene pairs in Figure 3.11.

𝒦EYNOTE

Bridges observed an unexpected inheritance pattern of an X-linked mutant gene in *Drosophila*. He correlated this pattern directly with a rare event during meiosis, called nondisjunction, in which members of a homologous pair of chromosomes do not segregate to the opposite poles. The association of a specific gene with a specific chromosome was taken as definite proof for the chromosome theory of heredity.

SEX DETERMINATION AND SEX LINKAGE IN EUKARYOTIC SYSTEMS

In this section some of the mechanisms for sex determination will be discussed. In *genotypic sex determination systems*, sex is governed by the genotype of the zygote or spores; in *phenotypic* (or *environmental*) *sex determination systems*, sex is governed by internal and external environmental conditions.

Genotypic Sex Determination Systems

Genotypic sex determination, in which the sex chromosomes play a decisive role in the inheritance and determination of sex, may occur in one of two ways. In the

X chromosome-autosome balance system (seen in *Drosophila* and the nematode, *Caenorhabditis elegans*), the main factor in sex determination is the ratio between the number of X chromosomes and the number of sets of autosomes. Sex is determined at the time of fertilization, and sex differences are assumed to arise during development, due to the action of two sets of genes located on the X chromosomes and on the autosomes. In this system the Y chromosome has no effect on sex determination, but is required for male fertility.

In the other system (seen in humans), the Y chromosome of the heterogametic sex is active in determining the sex of an individual. Individuals carrying the Y chromosome are male, irrespective of their autosomal genotype. In short, the embryo's sex is determined at the time of fertilization by the sex chromosomes contributed by the sperm and the ovum.

SEX DETERMINATION IN DROSOPHILA.

Drosophila melanogaster has four pairs of chromosomes: three pairs of autosomes and one pair of sex chromosomes. In this organism the homogametic sex is the female (XX) and the heterogametic sex is the male (XY). However, the sex of the fly is not the consequence of the presence or absence of a Y chromosome: An XXY fly is female and an XO fly is male. In fact, the sex of the fly, as many studies have shown, is determined by the ratio of the number of X chromosomes (X) to the number of sets of autosomes (A). *Drosophila* is the prototype of the *X chromosome-autosome balance system of sex determination.*

Table 3.2 presents some chromosome complements and the sex of the resulting flies. In a normal female there are two Xs and two sets of autosomes; hence the X:A ratio is 1.00. A normal male has a ratio of 0.50. If the X:A ratio is greater than or equal to 1.00, the fly will be female; if the X:A ratio is less than or equal to 0.50, the fly will be male. If the ratio is between 0.50 and 1.00, the fly is neither male nor female; it is an intersex. Intersex flies are variable in appearance,

~ **TABLE 3.2**

Sex Balance Theory of Sex Determination in *Drosophila melanogaster*

SEX CHROMOSOME COMPLEMENT	AUTOSOME COMPLEMENT (A)	X:A RATIO[a]	SEX OF FLIES
XX	AA	1.00	♀
XY	AA	0.50	♂
XXX	AA	1.50	Metafemale (sterile)
XXY	AA	1.00	♀
XXX	AAAA	0.75	Intersex (sterile)
XXXX	AAA	1.33	Metafemale (sterile)
XX	AAA	0.67	Intersex (sterile)
X	AA	0.50	♂ (sterile)

[a]If the X chromosome:autosome ratio is greater than or equal to 1.00 (X:A ≥ 1.00), the fly will be a female. If the X chromosome:autosome ratio is less than or equal to 0.50 (X:A ≤ 0.50), the fly will be male. Between these two ratios, the fly will be an intersex.

having, in general, complex mixtures of male and female attributes for the internal sex organs and external genitalia. Such flies are sterile.

SEX DETERMINATION IN THE NEMATODE CAENORHABDITIS ELEGANS.

The nematode *Caenorhabditis elegans* (Figure 3.12 and Figure 3.13 [p. 76]) has become a popular tool for developmental geneticists because an adult has only a few thousand cells and the lineages of every one of those cells has been carefully defined from egg to adult. Moreover, development from an egg to an adult takes only a little over two days.

~ **FIGURE 3.12**

Caenorhabditis elegans; (a) Photograph and (b) drawing of a hermaphrodite (XX).

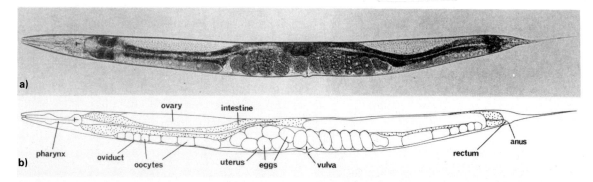

~ FIGURE 3.13

Diagrams of the hermaphrodite and the male *Caenorhabditis elegans;* (a) Photograph and (b) diagram of male (XO).

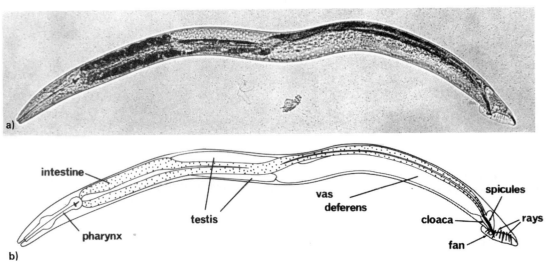

Caenorhabditis elegans has two sexual types: hermaphrodites (Figure 13.12) and males (Figure 3.13). Most individuals are hermaphrodites; that is, they have both ovaries and testes. They make sperm when they are larvae and store that sperm as development continues. In the adult, the ovary produces eggs that are fertilized by the stored sperm as the eggs migrate to the uterus. Self-fertilization in this way almost always produces more hermaphrodites. However, in 0.2 percent of the time males are produced from self-fertilization. These males can fertilize hermaphrodites if the two mate, and such matings result in about equal numbers of hermaphrodite and male progeny because their sperm has a competitive advantage over the sperm stored in the hermaphrodite.

In terms of sex chromosomes hermaphrodites are XX and males are XO. As in *Drosophila*, sex in this diploid organism is determined by the ratio of X chromosomes to autosomes: a ratio of 1.00 results in hermaphrodites and a ratio of 0.50 results in males.

SEX DETERMINATION IN MAMMALS. In contrast, in humans and other mammals the presence of the Y chromosome determines maleness, and its absence determines femaleness. This mode of sex determination is called the *chromosomal mechanism of sex determination*. If there is no Y chromosome present, the gonadal primordia develop into ovaries. If a Y chromosome is present, it regulates the production of *testis-determining factor* (encoded by the *Tdf* gene), which causes the gonadal primordia to develop into a testis instead of an ovary. Evidence for the chromosomal basis of sex determination mechanism came from studies of humans and other mammals in which meiotic nondis-

junction produces an abnormal sex chromosome complement. While sex determination in these cases follows directly from the sex chromosomes present, those individuals who have unusual chromosome complements display many unusual characteristics.

Nondisjunction can produce, for example, exceptional XO individuals that have an X but no Y chromosome. In *Drosophila*, XO flies are male if there is a normal set of autosomes. In humans, XO individuals with the normal two sets of autosomes are female and sterile, and they exhibit **Turner syndrome.** A Turner syndrome individual is shown in Figure 3.14a, and her set of metaphase chromosomes is shown in 3.14b. (A complete set of metaphase chromosomes in a cell is called its **karyotype;** this is described more fully in Chapter 10.) Note that there is only one sex chromosome, an X chromosome. These aneuploid females have a genomic complement of 45X, indicating that they have a total of 45 chromosomes, in contrast to the normal 46, and that the sex chromosome complement consists of one X chromosome. They have few noticeable major defects until puberty, when they fail to develop secondary sexual characteristics. They tend to be shorter than average, and they have weblike necks, poorly developed breasts, and immature internal sexual organs. They exhibit mental deficiencies and are usually infertile.

Nondisjunction can also result in XXY humans, which are male and have **Klinefelter syndrome.** Figure 3.15a shows a Klinefelter syndrome individual, and Figure 3.15b shows a karyotype for such an individual. Note the presence of three sex chromosomes; two X chromosomes and one Y chromosome. These 47,XXY males have underdeveloped testes, and are often taller than the average male. Some have lower-than-average

~ FIGURE 3.14

Turner syndrome (XO): (a) individual; (b) karyotype.

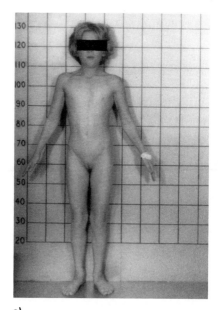

a)

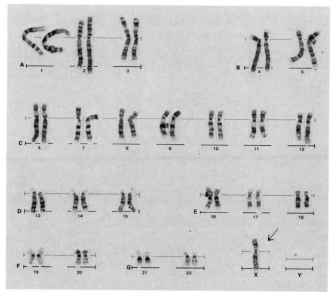

b)

~ FIGURE 3.15

Klinefelter syndrome (XXY): (a) individual; (b) karyotype.

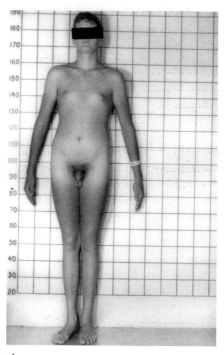

a)

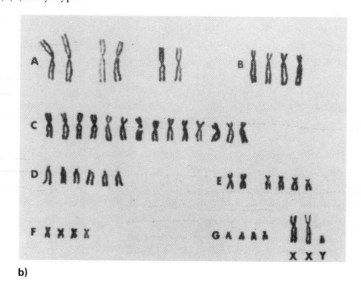

b)

~ TABLE 3.3

Consequences of Various Numbers of X- and Y-Chromosome Abnormalities in Humans, Showing Role of the Y in Sex Determination

CHROMOSOME CONSTITUTION[a]	DESIGNATION OF INDIVIDUAL	NUMBER OF BARR BODIES
46,XX	Normal ♀	1
46,XY	Normal ♂	0
45,X	Turner syndrome ♀	0
47,XXX	Triplo-X ♀	2
48,XXXX	Tetra-X ♀	3
47,XXY	Klinefelter syndrome ♂	1
48,XXXY	Klinefelter syndrome ♂	2
48,XXYY	Klinefelter syndrome ♂	1
47,XYY	XYY ♂	0

[a]The first number indicates the total number of chromosomes in the nucleus, and the Xs and Ys indicate the sex chromosome complement.

IQs. Individuals with similar phenotypes are also found with higher numbers of X and/or Y chromosomes, e.g. 48,XXXY and 48,XXYY. Some individuals have one X and two Y chromosomes. These 47,XYY individuals are male because of the presence of the Y; they result from nondisjunction of the Y chromosome in meiosis. They tend to be taller than average and occasionally there are adverse effects on fertility. Some females have three X chromosomes instead of the normal two. These 47,XXX (triplo-X) females are sometimes mentally deficient and infertile.

Table 3.3 summarizes the consequences of exceptional X and Y chromosome complements in humans. In each case two sets of autosomes are associated with the sex chromosomes. With rare exceptions mammals cannot tolerate any variation in the number of autosomes; mammals with an unusual number of autosomes usually die. Furthermore, a mechanism in mammals compensates for X chromosomes in excess of the normal complement but not for extra autosomes. This process is called **dosage compensation.** As we know, normal human males are XY and normal females are XX. The nuclei of normal females have a highly condensed mass of chromatin not found in the nuclei of normal male cells. This mass, which represents a cytologically condensed and inactivated X chromosome, was first discovered by Murray Barr, and it has been named the **Barr body.** That is, XX individuals normally have two X chromosomes and one Barr body in their somatic cells, and XY individuals have one X chromosome, one Y chromosome, and no Barr bodies (Figure 3.16).

In the somatic cells of individuals with a greater-than-normal number of X chromosomes, all but one of the X chromosomes become inactivated and produce Barr bodies. (A general formula for the number of Barr bodies is the number of X chromosomes minus one.) In 1961 this concept was expanded by Mary Lyon in what is called the *Lyon hypothesis.* Lyon proposed that the Barr body is an inactive (or mostly inactive) X chromosome, and her hypothesis explained the survival of individuals with X chromosome aberrations. The components of the Lyon hypothesis are as follows:

1. The Barr body is a genetically inactive X chromosome (it has become "lyonized"; the process is called **lyonization**);
2. The inactivation occurs at about the sixteenth day following fertilization;
3. The X chromosome to be inactivated is randomly chosen from the maternal and paternal X chromosomes in a process that is independent from cell to cell. (Once a maternal or paternal X chromosome is inactivated in a cell, all descendants of that cell inherit the inactivation pattern.)

Lyonization, then, is a mechanism that, when it operates in cells with extra X chromosomes, serves to minimize the effects of multiple copies of X-linked genes. In normal individuals, it results in adults, both sexes of which have the same number of active X chromosomes.

The number of Barr bodies associated with the normal and abnormal human X chromosome constitutions we have discussed are given in Table 3.3. Normal

~ FIGURE 3.16

Barr bodies: (a) Nucleus of normal human female cells (XX), showing Barr bodies; (b) Nucleus of normal human male cells (XY), showing no Barr bodies.

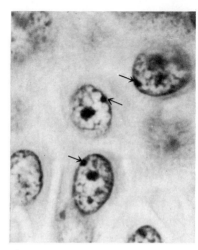
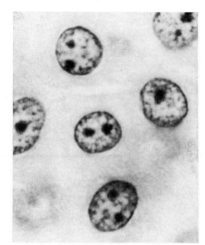

a)   b)

46,XX females have one Barr body, while normal 46,XY males have no Barr bodies. Turner syndrome females (45,X) have no Barr bodies; 47,XXY Klinefelter syndrome males have one Barr body, triplo-X (47,XXX) females have two Barr bodies, and 47,XYY males have no Barr bodies.

In 1955 proof that the Y chromosome plays an active role in sex determination in humans and other mammals was obtained when E. J. Eichwald and C. R. Silmser set out to see whether they could detect differences between males and females that could be related to sex chromosomes. The background for their experiments is as follows: Individuals that are genetically very similar can exchange skin grafts or other organ grafts without having antibodies formed to counter, or reject, the grafts. (An **antibody** is a protein molecule that recognizes and binds to a foreign substance introduced into the organism. Any large molecule that stimulates the production of specific antibodies or that binds specifically to an antibody is called an **antigen.**) The grafts are not recognized as foreign since the cell surfaces of the graft have the same antigens as the cells of the recipient organism.

Eichwald and Silmser knew that strains of experimental organisms have been produced by inbreeding them (making brother-sister matings) for so many generations that the individuals have essentially the same genotype. These strains are called **isogenic.** Eichwald and Silmser reasoned that the only genetic differences between isogenic brothers and sisters would result from sex chromosome differences between the two sexes.

Using individuals from an inbred mouse strain, they transplanted skin from females to male mice, and vice versa. In the former there was no rejection of the transplant, but in the latter the skin transplanted to the female mice was eventually rejected. They concluded that there must have been a functional gene (or genes) in males that produced an antigen unique to males and to which the female made antibodies. Since the only genetic difference in males and females of the strain was the presence of the Y chromosome in the males, they reasoned that the antigen was coded for by Y-linked genes.

The acceptance of tissue grafts is called **histocompatibility,** and the Y-linked gene is said to have a histocompatibility locus, named the *H−Y locus.* The product of this locus is the **H−Y antigen.** Years after Eichwald and Silmser did their experiments, E. Goldberg discovered that the H−Y antigen is conserved in evolution and that it is present at the earliest stages of embryo development, even in an eight-cell mouse embryo, for example. In 1975, S. Ohno proposed the H−Y hypothesis, that the H−Y antigen is the diffusible substance that results in testes formation and Wölffian duct development in mammals.

Subsequently, in 1984, Anne McLaren and her colleagues disproved Ohno's hypothesis. Her group bred mice in which the gene controlling the H−Y antigen segregated separately from maleness (i.e. testis development). That is, these families of mice included females which had the H−Y antigen and males which lacked H−Y antigen. Since the males lacking H−Y antigen had a defect in spermatogenesis, McLaren concluded that the H−Y antigen may have a role in sperm formation.

More recently David Page looked for individuals in which the chromosomes and the gonads did not match or in which there was an abnormal Y chromosome. By studying patients who were attending infertility clinics he found human equivalents of McLaren's mice. Among the patients he analyzed were 6 females who were XY instead of the normal XX and 20 normal-looking males who had two X chromosomes instead of the normal XY. All of the unusual XY females were found to have deletions of small segments near the tip of the short arm of the Y chromosome (because of its centromere position, the Y chromosome is divided into two uneven portions, a short arm and a long arm). The unusual XX males all had a segment of the short arm of the Y chromosome—including the segment deleted in the XY females—attached to one of the X chromosomes.

The conclusion was that the Y segment identified in the XX males and the XY females contained the gene for a factor important for male sexual development, the testis-determining factor (*Tdf*). In 1987 Page's group showed that the *Tdf* gene and the H−Y antigen gene map to different regions of the Y chromosome. The *Tdf* gene is near the tip of the short arm, while the H−Y gene is near the centromere. The *Tdf* gene has yet to be isolated.

SEX CHROMOSOMES IN OTHER ORGANISMS. Not all organisms have an X−Y sex chromosome makeup like that found in mammals and in *Drosophila*. In birds, butterflies, moths, and some fish, the sex chromosome composition is the opposite of that in mammals: The male in these organisms is the homogametic sex and the female is the heterogametic sex. To prevent confusion with the X and Y chromosome convention, we designate the sex chromosomes in these organisms as Z and W. Thus the males are ZZ and the females are ZW. Genes on the Z chromosome behave just like X-linked genes in the earlier examples except that hemizygosity is found only in females. All the daughters of a male homozygous for a Z-linked recessive gene express the recessive trait, and so on. Figure 3.17 presents an example, the inheritance of the sex-linked (Z-linked), dominant barred plumage (*B*) in poultry.

Higher plants exhibit quite a variety of sexual situations. Some species (the ginko, for example) have plants of separate sexes, with male plants producing flowers that contain only stamens and female plants producing flowers that contain only pistils. These species are called **dioecious.** Other species have both male and female sex organs on the same plant. If the sex organs are in the same flower, the plant is said to be **hermaphroditic** (e.g., the rose and the buttercup), and the flower is said to be a *perfect flower*. If the sex organs are in different flowers on the same plant, it is said to be **monoecious** (e.g., corn), and the flower is said to be an *imperfect flower*.

Some dioecious plants have sex chromosomes that differ between the sexes, and a large proportion of these plants have an X−Y system. Only in a few instances has sex determination in such plants been studied with anything like the rigor seen in *Drosophila* studies. Plant research has shown that in dioecious X−Y plants an X chromosome-autosome balance system like that in *Drosophila* determines the sex of the individual. However, many other sex determination systems are seen in dioecious plants. For example, some of these plants have visibly different (heteromorphic) chromosomes in the two sexes, but the constitution of the different sexes is more complex than that of the *Drosophila* X−Y type. Still other dioecious plants have no visibly heteromorphic pair of chromosomes; in these instances the mechanism of sex determination is not known.

Many species, particularly eukaryotic microorganisms, do not have sex chromosomes but instead rely on a *genic system* for the determination of sex, that is, a system in which the sexes are specified by simple allelic differences at a small number of gene loci. For example, the orange bread mold *Neurospora crassa* is a haploid fungus that has two sexes referred to as **mating types.** The sexes are morphologically indistinguishable and are determined by the *A* and *a* alleles of a single gene locus in chromosome I of this organism. If two strains of *Neurospora* are placed together on the appropriate medium, they will not mate if each carries the same allele. However, if one is mating type *A* and the other is mating type *a*, the two strains will fuse, a diploid nucleus will be produced by fusion of two haploid nuclei, and meiosis will be initiated. Similarly, in the yeast *Saccharomyces cerevisiae*, there are two mating types: *a* and *α*. These mating types are controlled by the *MATa* and *MATα* alleles, respectively, of a single gene.

Phenotypic Sex Determination Systems

The environment plays a major role in systems of **phenotypic sex determination.** In the marine worm *Bonellia* (Figure 3.18; p. 82), for example, the free-swimming larval forms are sexually undifferentiated. If an individual settles down alone, it becomes female. If a larva attaches to the body of an adult female, the larva will differentiate into a male. Thus sex differentiation in *Bonellia* is not determined at fertilization by some genetic component but is probably directed by environmental factors related either to the association or lack of association with other members of the species.

~ FIGURE 3.17

Sex-linked inheritance in chickens. The barred-feather (*B*) phenotype is caused by a gene that is dominant to the allele for nonbarred (*b*) plumage. (a) A barred-feather female is crossed with a nonbarred male; (b) An F$_1$ nonbarred female is crossed with an F$_1$ barred male.

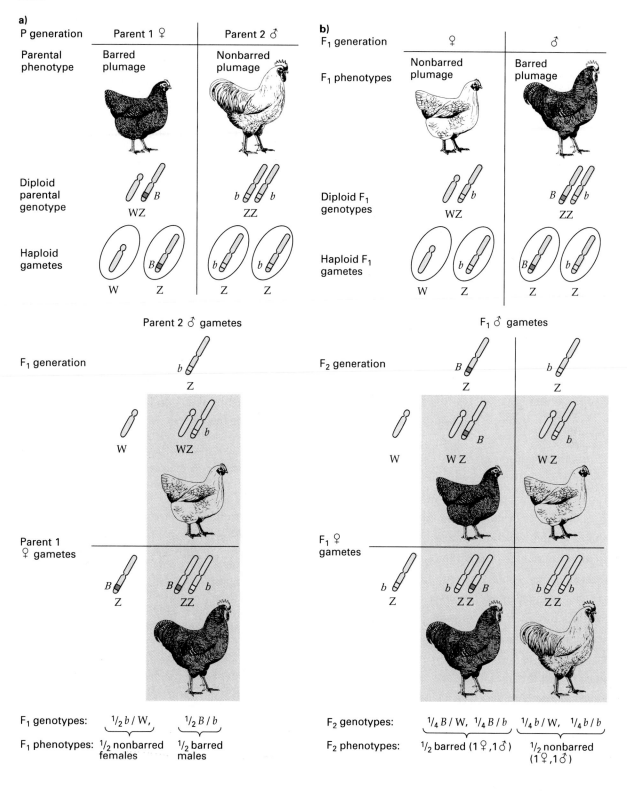

~ FIGURE 3.18

Drawing of a female and male *Bonellia*, a marine worm.

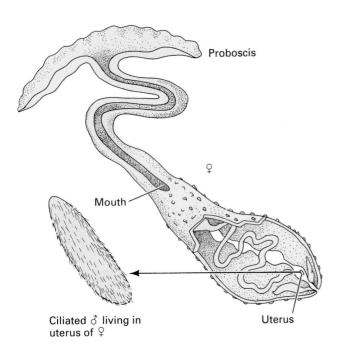

Proboscis

♀

Mouth

Ciliated ♂ living in
uterus of ♀

Uterus

𝒦EYNOTE

Many eukaryotic organisms have sex chromosomes that are represented differentially in the two sexes; in humans and many other eukaryotes the male is XY and the female is XX. In many cases sex determination is related to the sex chromosomes. For humans and many other mammals, for instance, the presence of the Y chromosome confers maleness, and its absence results in femaleness. *Drosophila* and *Caenorhabditis* has an X chromosome–autosome balance system of sex determination: The sex of the individual is related to the ratio of the number of X chromosomes to the number of sets of autosomes. Several other sex-determining systems are known in the eukaryotes, including genic systems, found particularly in the lower eukaryotes, and phenotypic (environmental) systems. ⎯⎯⎯⎯⎯⎯⎯

Sex Linkage in Humans

The use of pedigree analysis to analyze the inheritance of human genetic traits was introduced in Chapter 2. In this chapter we have learned about the relationship between genes and chromosomes, and about the existence of sex chromosomes and autosomes. In humans the gene responsible for a trait can be inherited in one of five different ways: autosomal recessive, autosomal dominant, X-linked recessive, X-linked dominant, and Y-linked. These five mechanisms of inheritance are also found in all eukaryotes with X and Y chromosomes. The recessive and dominant human traits we discussed in Chapter 2 were all autosomal traits. In this section, we will discuss examples of X-linked and Y-linked traits.

X-LINKED RECESSIVE INHERITANCE. A trait due to a recessive mutant gene carried on the X chromosome is called an **X-linked recessive trait**. At least 100 human traits are known whose inheritance has been traced to the X chromosome. Most of the traits involve X-linked recessive genes. The most well known X-linked recessive pedigree is that of hemophilia A in Queen Victoria's family (Figure 3.19). Hemophilia (the bleeder's disease) is a serious ailment in which the blood lacks a clotting factor. A cut or even a serious bruise can be fatal to a hemophiliac. In Queen Victoria's pedigree the first instance of hemophilia was in one of her sons, so she was a carrier for this trait.

In X-linked recessive traits the female must be homozygous for the recessive allele in order to express the mutant trait. The trait is expressed in the male who possesses but one copy of the mutant allele on the X chromosome (except in the rare instance of a homologous gene on the Y chromosome). Therefore affected males normally transmit the mutant gene to all their daughters but to none of their sons. The instance of a father-to-son inheritance of a trait in a pedigree would tend to rule out X-linked recessive inheritance. As mentioned earlier in this chapter, the passage of sex-linked traits from mother to son and from father to daughter is called crisscross inheritance.

Other characteristics of X-linked recessive inheritance are as follows (refer to Figure 3.19):

1. For rare X-linked recessive genes many more males than females should exhibit the trait, owing to the different number of X chromosomes in the two sexes. Most females will be heterozygous under these conditions.
2. The sons of heterozygous (carrier) mothers should show an approximately 1:1 ratio of normal indi-

~ FIGURE 3.19

(a) Photograph of Queen Victoria as a young woman; (b) Pedigree of Queen Victoria (III.2) and her descendants, showing the inheritance of hemophilia. Either Queen Victoria was heterozygous for the sex-linked recessive hemophilia allele, or the trait arose as a mutation in her germ cells (the cells that give rise to the gametes). In many cases the marriage partner has been omitted as a shorthand way to save space.

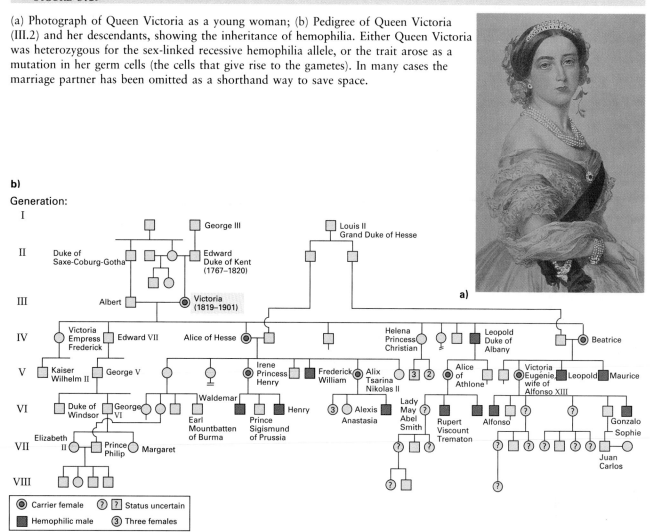

viduals to individuals expressing the trait. That is, $a^+/a \times a^+/Y$ gives half a^+/Y and half a/Y sons. If a small family produces only unaffected sons, the trait will skip a generation.

3. From a mating of a carrier female with a normal male all daughters will be normal, but half will be carriers. That is, $a^+/a \times a^+/Y$ gives half a^+/a^+ and half a^+/a females. In turn, half the sons of the F_1 carrier females will exhibit the trait.

4. A male expressing the trait, when mated with a homozygous normal female, will produce all normal children, but all the female progeny will be carriers. That is, $a^+/a^+ \times a/Y$ gives a^+/a females and a^+/Y (normal) males.

X-LINKED DOMINANT INHERITANCE. A trait due to a dominant mutant gene carried on the X chromosome is called an **X-linked dominant trait**. Only a few dominant X-linked traits have been identified.

An example of an X-linked dominant trait that causes faulty tooth enamel and dental discoloration is shown in Figure 3.20a (p. 84) and a pedigree for this trait is shown in Figure 3.20b. Note that all the daughters and none of the sons of an affected father (III.1) are affected, and that heterozygous mothers (IV.3) transmit the trait to half their sons and half their daughters. Webbing to the tips of the toes in an Irish family in South Dakota (studied in the 1930s) has also been ascribed to an X-linked dominant mutant allele, as has

~ FIGURE 3.20

(a) A person with the X-linked dominant trait of faulty enamel; (b) A pedigree showing the transmission of the faulty enamel trait. This pedigree illustrates a shorthand convention which omits parents who do not exhibit the trait. Thus it is a given that the female in generation I paired with a male who did not exhibit the trait.

a)

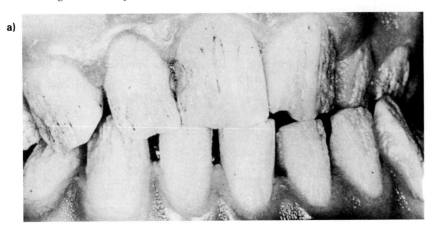

b) Pedigree

Generation:

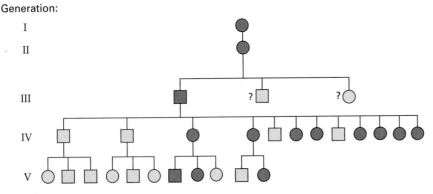

a severe bleeding anomaly called constitutional throm-bopathy, studied in two Finnish families on an archipel-ago in the Baltic Sea. In the latter (also studied in the 1930s), bleeding is not due to the absence of a clotting factor (as in hemophilia) but to interference in the formation of blood platelets, which are needed for blood clotting.

The X-linked dominant traits follow the same sort of inheritance rules as the X-linked recessives, except that heterozygous females express the trait. In general, X-linked dominant traits tend to be milder in the female than in the male. Also, since females have twice the number of X chromosomes as males, X-linked dominant traits are more frequent in females than in males. If the trait is a rare one, we expect two-thirds of the affected persons to be female. For rare traits most females in a pedigree would be heterozygous. These females should pass on the trait to half their progeny, regardless of their sex. As in X-linked recessive traits, males who have an X-linked dominant trait transmit the allele to their daughters but not to their sons. As a

result, we see crisscross inheritance (father-to-daughter) of the trait.

Y-LINKED INHERITANCE. A trait due to a mutant gene carried on the Y chromosome but with no coun-terpart on the X is called a **Y-linked,** or **holandric** ("wholly male"), **trait.** Such traits should be readily recognizable, since every son of an affected male should have the trait and no females should ever express it. Several traits with Y-linked inheritance have been sug-gested. In most cases the genetic evidence for such inheritance is poor or nonexistent. In fact, there is no clear evidence for genetic loci on the Y chromosome other than those involved with sex determination in the male.

A possible example of Y-linked inheritance is the hairy ears, or hairy pinnae, trait in which bristly hairs of atypical length grow from the ears (Figure 3.21). This trait is common in parts of India, although some white populations also exhibit it. While this trait shows father-to-son inheritance, there is no doubt that it is a

~ FIGURE 3.21

Individuals exhibiting the possibly Y-linked trait of hairy ears.

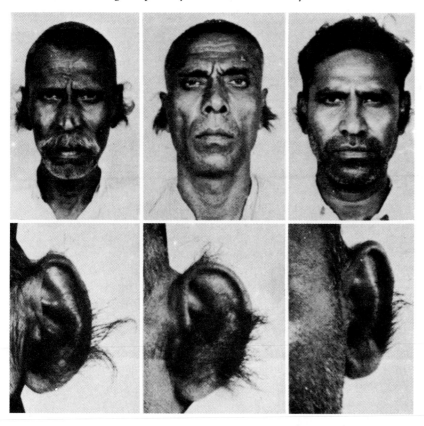

complex phenotype, and many of the collected pedigrees can be interpreted in other ways, such as autosomal inheritance. The trait could also be the result of the interaction of a gene with the male hormone testosterone, similar to the appearance of hair on the face and chest. Another putative Y-linked trait is bilateral radio-ulnar synostosis. In this trait the radius and ulna bones of both arms are fused.

𝒦EYNOTE

Controlled matings cannot be made in humans, so analysis of the inheritance of genes in humans must rely on pedigree analysis. This technique involves the careful study of the phenotypic records of the family extending over several generations. The data obtained from pedigree analysis enable geneticists to make judgments, with varying degrees of confidence, about whether a mutant gene is inherited as an autosomal recessive, an autosomal dominant, an X-linked recessive, an X-linked dominant, or a Y-linked allele.

SUMMARY

In Chapter 2, genes were considered as abstract entities that control hereditary characteristics. Prior to the beginning of this century, genes were called Mendelian factors. As a result of the efforts of cytologists working in the nineteenth century, information was being accumulated about cell structure and cell division. The fields of genetics and cytology came together in 1903 when Sutton and Boveri independently hypothesized that genes are on chromosomes, "bodies" within the cell nucleus. The chromosome theory of heredity states that the chromosome transmission from one generation to the next closely parallels the patterns of gene transmission from one generation to the next.

Proof of the chromosome theory of heredity came from experiments that related the hereditary behavior of particular genes to the transmission of the sex chromosome. The sex chromosome in eukaryotic organisms is the chromosome that is represented differently in the two sexes. In most organisms with sex chromosomes, the female has two X chromosomes, while the male has one X and one Y chromosome. The Y chromosome is structurally and genetically different from the X chromosome. The association of genes with the sex determining chromosomes of eukaryotes is called sex linkage. Such genes, and the phenotypes they control, are called sex-linked.

In this chapter, a number of sex determination mechanisms were discussed. In many cases sex determination is related to the sex chromosomes. In humans, for example, the presence of a Y chromosome specifies maleness, while its absence results in femaleness. Several other sex-determination systems are known in the eukaryotes, including X chromosome-autosome balance systems (in which sex is determined by the ratio of the number of X chromosomes to the number of sets of autosomes), genic systems (in which a simple allelic difference determines the sex of an individual), and phenotypic systems (in which sex is determined by environmental cues).

Also discussed in this chapter were sex-linked traits in humans. In Chapter 2 we discussed the fact that the inheritance patterns of traits in humans are usually studied by charting the family trees of individuals exhibiting the trait, a method called pedigree analysis.

Five different mechanisms of inheritance of the gene responsible for a trait are possible: autosomal recessive, autosomal dominant, X-linked recessive, X-linked dominant, or Y-linked. In this chapter we considered examples of X- and Y-linked human traits to illustrate the features of those mechanisms of inheritance in pedigrees. Autosomal traits were discussed in Chapter 2. It is important to realize that collecting reliable human pedigree data is a difficult task. In many cases the accuracy of record keeping within the families involved is open to question. Also, particularly with small families, there may not be enough affected people to allow an unambiguous determination of the inheritance mechanism involved, especially when rare traits are involved. Moreover, the degree to which a trait is expressed may vary, so that some individuals may be erroneously classified as normal. It is also possible for a mutant phenotype to be produced by alleles of different genes, and therefore different pedigrees may, quite correctly, indicate different mechanisms of inheritance of the "same" trait.

ANALYTICAL APPROACHES FOR SOLVING GENETICS PROBLEMS

The concepts introduced in this chapter may be reinforced by solving genetics problems. The types of problems are similar to those introduced in Chapter 2. When sex linkage is involved, remember that one sex has two kinds of sex chromosomes, whereas the other sex has only one; this feature alters the inheritance patterns slightly. Most sample problems presented in this section center on interpreting data and predicting the outcome of particular crosses.

Q.1 A female from a pure-breeding strain of *Drosophila* with vermilion-colored eyes is crossed with a male from a pure-breeding wild-type, red-eyed strain. All the F_1 males have vermilion-colored eyes, and all the females have wild-type red eyes. What conclusions can you draw about the mode of inheritance of the vermilion trait, and how could you test them?

A.1 The observation is the classic one that suggests a sex-linked trait is involved. Since none of the F_1 daughters have the trait and all the F_1 males do, the trait is presumably X-linked recessive. The results fit this hypothesis, because the F_1 males receive the X chromosome from their homozygous v/v mother, with the v gene on the X chromosome. Furthermore, the F_1 females are v^+/v since they receive a v^+-bearing X chromosome from the wild-type

male parent and a v-bearing X chromosome from the female parent. If the trait were autosomal recessive, all the F_1 flies would have had wild-type eyes. If it were autosomal dominant, both the F_1 males and females would have had vermilion-colored eyes. If the trait were X-linked dominant, all the F_1 flies would have had vermilion eyes.

The easiest way to verify this hypothesis is to let the F_1 flies interbreed. This cross is v^+/v ♀ $\times$ v/Y ♂, and the expectation is that there will be a 1:1 ratio of wild-type:vermilion eyes in both sexes in the F_2. That is, half the females are v^+/v and half are v/v; half the males are v^+/Y and half are v/Y. This ratio is certainly not the 3:1 ratio that would result from an $F_1 \times F_1$ cross for an autosomal gene.

Q.2 In humans, hemophilia A or B is caused by an X-linked recessive gene. A woman who is a nonbleeder had a father who was a hemophiliac. (Actually, hemophiliac fathers are relatively rare, but we can at least consider one for this question.) She marries a nonbleeder, and they plan to have children. Calculate the probability of hemophilia in the female and male offspring.

A.2 Since hemophilia is an X-linked trait, and since her father was a hemophiliac, the woman must be heterozygous for this recessive gene. If we assign the symbol h to this recessive mutation and h^+ to the wild-type (non-bleeder) allele, she must be h^+/h. The man she marries is normal with regard to blood clotting and hence must be hemizygous for h^+, that is, h^+/Y. All their daughters receive an X chromosome from the father, and so each must have an h^+ gene. In fact, half the daughters are h^+/h^+ and the other half are h^+/h. Since the wild-type allele is dominant, none of the daughters are hemophiliacs. However, all the sons of the marriage receive their X chromosome from their mother. Therefore they have a probability of 1/2 that they will receive the chromosome carrying the h allele, which means they will be hemophiliacs. Thus the probability of hemophilia among daughters of this marriage is 0, and among sons it is 1/2.

Q.3 Tribbles are hypothetical animals that have an X−Y sex determination mechanism like that of humans. The trait bald (b) is X-linked and recessive to furry (b^+), and the trait long leg (l) is autosomal and recessive to short leg (l^+). You make reciprocal crosses between true-breeding bald, long-legged tribbles and true-breeding furry, short-legged tribbles. Do you expect a 9:3:3:1 ratio in the F_2 of either or both of these crosses? Explain your answer.

A.3 This question focused on the fundamentals of X chromosome and autosome segregation during a genetic cross and tests whether or not you have grasped the principles involved in gene segregation. Figure 3.A diagrams the two crosses involved, and we can discuss the answer by referring to it.

Let us first consider the cross of a wild-type female tribble (b^+/b^+, l^+/l^+) with a male double-mutant tribble (b/Y, l/l). Part (a) of Figure 3.A diagrams this cross. These F_1s are all normal—that is, with furry bodies and short legs—because for the autosomal character both sexes are heterozygous, and for the X-linked character the female is heterozygous and the male is hemizygous for the b^+ allele donated by the normal mother. With the production of the F_2 progeny the best approach is to treat the X-linked and autosomal traits separately. For the X-linked trait, random combination of the gametes produced gives a 1:1:1:1 genotypic ratio of b^+/b^+ (furry female):b^+/b (furry female):b^+/Y (furry male):b/Y (bald male) progeny. Collecting by phenotypes, we see that all the females are furry; half the males are furry and half are bald. For the autosomal leg trait the $F_1 \times F_1$ is a cross of two hetero-

FIGURE 3.A

a) Furry, short-legged (wild-type) ♀ ×
bald, long-legged ♂

b) Bald, long-legged ♀ ×
furry, short-legged (wild-type) ♂

P generation

b^+/b^+ l^+/l^+ ♀
(furry, short)   × b/Y l/l ♂
(bald, long)

P generation

b/b l/l ♀
(bald, long)   × b^+/Y l^+/l^+ ♂
(furry, short)

F_1 generation

b^+/b l^+/l ♀
(furry, short)   × b^+/Y l^+/l ♂
(furry, short)

F_1 generation

b^+/b l^+/l ♀
(furry, short)   × b/Y l^+/l ♂
(bald, short)

F_2 generation

Sex-linked phenotypes and genotypes	Autosomal phenotypes and genotypes

Genotypic results:

$\frac{1}{2}$ b^+ ($\frac{1}{2}$ b^+/b^+, $\frac{1}{2}$ b^+/b; furry) ♀ < $\frac{3}{4}$ l^+ (l^+/l^+ and l^+/l; short) / $\frac{1}{4}$ l (l/l; long)

$\frac{1}{4}$ b^+ (b^+/Y; furry) ♂ < $\frac{3}{4}$ l^+ (short) / $\frac{1}{4}$ l (long)

$\frac{1}{4}$ b (b/Y; bald) ♂ < $\frac{3}{4}$ l^+ (short) / $\frac{1}{4}$ l (long)

F_2 generation

Sex-linked phenotypes and genotypes	Autosomal phenotypes and genotypes

Genotypic results:

$\frac{1}{4}$ b^+ (b^+/b^+; furry) ♀ < $\frac{3}{4}$ l^+ (l^+/l^+ and l^+/l; short) / $\frac{1}{4}$ l (l/l; long)

$\frac{1}{4}$ b (b/b; bald) ♀ < $\frac{3}{4}$ l^+ (short) / $\frac{1}{4}$ l (long)

$\frac{1}{4}$ b^+ (b^+/Y; furry) ♂ < $\frac{3}{4}$ l^+ (short) / $\frac{1}{4}$ l (long)

$\frac{1}{4}$ b (b/Y; bald) ♂ < $\frac{3}{4}$ l^+ (short) / $\frac{1}{4}$ l (long)

Phenotypic ratios:

	Furry-short		Furry-long		Bald-short		Bald-long
	b^+ l^+		b^+ l		b l^+		b l
♀	6	:	2	:	0	:	0
♂	3	:	1	:	3	:	1
Total	9	:	3	:	3	:	1

Phenotypic ratios:

	Furry-short		Furry-long		Bald-short		Bald-long
	b^+ l^+		b^+ l		b l^+		b l
♀	3	:	1	:	3	:	1
♂	3	:	1	:	3	:	1
Total	6	:	2	:	6	:	2

zygotes, so we expect a 3:1 phenotypic ratio of short-legged:long-legged tribbles in the F_2. Since autosome segregation is independent of the inheritance of the X chromosome, we can multiply the probabilities of the occurrence of the X-linked and autosomal traits to calculate their relative frequencies. The calculations are presented in part (a) of Figure 3.A, from which we see that the ratio of the four possible phenotypic classes differs in females and males.

The first cross, then, has a 9:3:3:1 ratio of the four possible phenotypes in the F_2. However, note that the ratio in each sex is not 9:3:3:1, owing to the inheritance pattern of the X chromosome. This result contrasts markedly with the pattern of two autosomal genes segregating independently, where the 9:3:3:1 ratio is found for both sexes.

The second cross (a reciprocal cross) is diagrammed in part (b) of Figure 3.A. Since the parental female in this cross is homozygous for the sex-linked trait, all the F_1 males are bald. Genotypically, the F_1 males and females differ from those in the first cross with respect to the sex chromosome but are just the same with respect to the autosome. Again, considering the X chromosome first as we go to the F_2, we find a 1:1:1:1 genotypic ratio of furry females:bald females:furry males:bald males. In this case, then, half of both males and females are furry and half are bald, in contrast to the results of the first cross in which no bald females were produced in the F_2. For the autosomal trait we expect a 3:1 ratio of short:long in the F_2, as before. Putting the two traits together, we get the calculations presented in part (b) of the figure. (*Note:* We use the total 6:2:6:2 here rather than 3:1:3:1 because the numbers add to 16, as does 9 + 3 + 3 + 1.) So in this case we do not get a 9:3:3:1 ratio; moreover, the ratio is the same in both sexes.

This question has forced us to think through the segregation of two types of chromosomes and has shown that we must be careful about predicting the outcomes of crosses in which sex chromosomes are involved. Nonetheless, the basic principles for the analysis were no different from those used before: Reduce the questions to their basic parts and then put the puzzle together step by step.

QUESTIONS AND PROBLEMS

*3.1 At the time of synapsis, preceding the reduction division in meiosis, the homologous chromosomes align in pairs and one member of each pair passes to each of the daughter nuclei (see Chapter 1). In an animal with five pairs of chromosomes, assume that chromosomes 1, 2, 3, 4, and 5 have come from the father, and 1′, 2′, 3′, 4′, and 5′ have come from the mother. In what proportion of the germ cells of this animal will all the paternal chromosomes be present together?

*3.2 In a male *Homo sapiens*, from which grandparent could each sex chromosome have been derived? (Indicate "Yes" or "No" for each option.)

	MOTHER'S		FATHER'S	
	MOTHER	FATHER	MOTHER	FATHER
X chromosome	_____	_____	_____	_____
Y chromosome	_____	_____	_____	_____

3.3 In *Drosophila*, white eyes are a sex-linked character. The mutant allele for white eyes (w) is recessive to the wild-type allele for bright red eye color (w^+).
a. A white-eyed female is crossed with a red-eyed male. An F_1 female from this cross is mated with her father, and an F_1 male is mated with his mother. What will be the eye color of the offspring of these last two crosses?
b. A white-eyed female is crossed with a red-eyed male, and the F_2 from this cross is interbred. What will be the eye color of the F_3?

*3.4 One form of color blindness (c) in humans is caused by a sex-linked recessive mutant gene. A woman with normal color vision (c^+) and whose father was color-blind marries a man of normal vision whose father was also color-blind. What proportion of their offspring will be color-blind? (Give your answer separately for males and females.)

3.5 In humans, red-green color blindness is due to an X-linked recessive gene. A color-blind daughter is born to a mother with normal color vision and a father who is color-blind. What is the mother's genotype with respect to the alleles concerned?

*3.6 In humans, red-green color blindness is recessive and X-linked, while albinism is recessive and autosomal. What types of children can be produced as the result of marriages between two homozygous parents, a normal-visioned albino woman and a color-blind, normally pigmented man?

*3.7 In *Drosophila*, vestigial (partially formed) wings (vg) are recessive to normal long wings (vg^+), and the gene for this trait is autosomal. The gene for the white eye trait is on the X chromosome. Suppose a homozygous white-eyed, long-winged female fly is crossed with a homozygous red-eyed, vestigial-winged male.
a. What will be the appearance of the F_1?
b. What will be the appearance of the F_2?
c. What will be the appearance of the offspring of a cross of the F_1 back to each parent?

3.8 In *Drosophila*, two red-eyed, long-winged flies are bred together and produce the offspring given in the following table:

	FEMALES	MALES
Red-eyed, long-winged	3/4	3/8
Red-eyed, vestigial-winged	1/4	1/8
White-eyed, long-winged	—	3/8
White-eyed, vestigial-winged	—	3/8

What are the genotypes of the parents?

3.9 In poultry a dominant sex-linked gene (B) produces barred feathers, and the recessive allele (b), when homozygous, produces nonbarred feathers. Suppose a nonbarred cock is crossed with a barred hen.
a. What will be the appearance of the F_1 birds?
b. If an F_1 female is mated with her father, what will be the appearance of the offspring?
c. If an F_1 male is mated with his mother, what will be the appearance of the offspring?

*__3.10__ A man (A) suffering from defective tooth enamel, which results in brown-colored teeth, marries a normal woman. All their daughters have brown teeth, but the sons are normal. The sons of man A marry normal women, and all their children are normal. The daughters of man A marry normal men, and 50 percent of their children have brown teeth. Explain these facts.

3.11 In humans, differences in the ability to taste phenylthiourea are due to a pair of autosomal alleles. Inability to taste is recessive to ability to taste. A child who is a nontaster is born to a couple who can both taste the substance. What is the probability that their next child will be a taster?

*__3.12__ Cystic fibrosis is inherited as an autosomal recessive. Two noncystic fibrosis parents have two children with cystic fibrosis and three children who do not have cystic fibrosis. They come to you for genetic counseling.
a. What is the numerical probability that their next child will have cystic fibrosis?
b. Their non-affected children are concerned about being heterozygous. What is the numerical probability that a given non-affected child in the family is heterozygous?

3.13 Huntington's disease is a human disease inherited as a Mendelian autosomal dominant. The disease results in choreic (uncontrolled) movements, progressive mental deterioration, and eventually death. The disease affects the carriers of the trait anytime between 15 and 65 years of age. The American folksinger Woody Guthrie died of Huntington's disease, as did one of his parents. Marjorie Mazia, Woody's wife, had no history of this disease in her family. The Guthries had three children. What is the probability that a particular Guthrie child will die of Huntington's disease?

3.14 Suppose gene A is on the X chromosome, and genes B, C, and D are on three different autosomes. Thus $A-$ signifies the dominant phenotype in the male or female. An equivalent situation holds for $B-$, $C-$, and $D-$. The cross $AA\ BB\ CC\ DD$ females $\times aY\ bb\ cc\ dd$ males is made.
a. What is the probability of obtaining an $A-$ individual in the F_1?
b. What is the probability of obtaining an a male in the F_1?
c. What is the probability of obtaining an $A-\ B-\ C-\ D-$ female in the F_1?
d. How many different F_2 genotypes will there be?
e. What proportion of F_2s will be heterozygous for the four genes?
f. Determine the probabilities of obtaining each of the following types in the F_2: (1) $A-\ bb\ CC\ dd$ (female); (2) $aY\ BB\ Cc\ Dd$ (male); (3) $AY\ bb\ CC\ dd$ (male); (4) $aa\ bb\ Cc\ Dd$ (female).

*__3.15__ As a famous mad scientist, you have cleverly devised a method to isolate *Drosophila* ova that have undergone primary nondisjunction of the sex chromosomes. In one experiment you used females homozygous for the sex-linked recessive mutation causing white eyes (w) as your source of nondisjunction ova. The ova were collected and fertilized with sperm from red-eyed males. The progeny of this "engineered" cross were then backcrossed separately to the two parental strains (this is called "backcrossing"). What classes of progeny (genotype and phenotype) would you expect to result from these backcrosses? (The genotype of the original parents may be denoted as ww for the females and w^+Y for the males.)

3.16 In *Drosophila* the bobbed gene (bb^+) is located on the X chromosome. Unlike most X-linked genes, however, the Y chromosome also carries a bobbed gene. The mutant allele bb is recessive to bb^+. If a wild-type F_1 female that resulted from primary nondisjunction in oogenesis in a cross of a bobbed female with a wild-type male is mated to a bobbed male, what will be the phenotypes and their frequencies in the offspring? List males and females separately in your answer. (*Hint:* Refer to the chapter for information about the frequency of nondisjunction in *Drosophila*.)

3.17 A Turner syndrome individual would be expected to have the following number of Barr bodies in the majority of cells:
a. 0   **b.** 1   **c.** 2   **d.** 3

3.18 An XXY Klinefelter syndrome individual would be expected to have the following number of Barr bodies in the majority of cells:
a. 0   **b.** 1   **c.** 2   **d.** 3

*__3.19__ In human genetics the pedigree is used for analysis of inheritance patterns. The female is represented by a circle and the male by a square. The following figure presents three, two-generation family pedigrees for a trait in humans. Normal individuals are represented by unshaded symbols and people with the trait by shaded symbols. For each pedigree (A, B, and C), state, by answering yes or no in the appropriate blank space, whether transmission of the trait can be accounted for on the basis of each of the listed simple modes of inheritance:

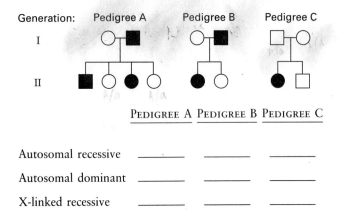

PEDIGREE A   PEDIGREE B   PEDIGREE C

	PEDIGREE A	PEDIGREE B	PEDIGREE C
Autosomal recessive	_____	_____	_____
Autosomal dominant	_____	_____	_____
X-linked recessive	_____	_____	_____
X-linked dominant	_____	_____	_____

3.20 Looking at the following pedigree, in which shaded symbols represent a "trait," which of the progeny (as designated by numbers) eliminate X-linked recessiveness as a mode of inheritance for the trait?

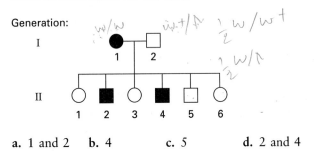

a. 1 and 2   **b.** 4   **c.** 5   **d.** 2 and 4

*3.21 When constructing human pedigrees, geneticists often refer to particular persons by number. The generations are labeled by roman numerals and the individuals in each generation by arabic numerals. For example, in the pedigree in the figure below, the female with the asterisk would be I.2. Use this means to designate specific individuals in the pedigree. Determine the probable inheritance mode for the trait shown in the affected individuals (the shaded symbols) by answering the following questions. Assume the condition is caused by a single gene.

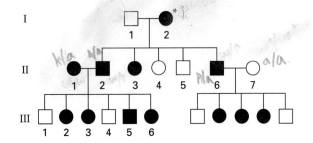

a. Y-linked inheritance can be excluded at a glance. What two other mechanisms of inheritance can be definitely excluded? Why can these be excluded?

b. Of the remaining mechanisms of inheritance which is the most likely? Why?

3.22 Phenylketonuria (PKU) is an inborn error in the metabolism of the amino acid phenylalanine. The characteristic feature of PKU is severe mental retardation; many untreated patients with this trait have IQs below 20. The three-generation pedigree shown in the following figure is of an affected family.

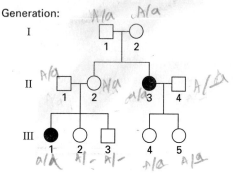

a. What is the mechanism of inheritance of PKU?

b. Which persons in the pedigree are known to be heterozygous for PKU?

c. What is the probability that III.2 is a carrier (heterozygous)?

d. If III.3 and III.4 marry, what is the probability that their first child will have PKU?

3.23 For the more complex pedigrees shown in Figure 3.B, indicate the probable mechanism of inheritance: autosomal recessive, autosomal dominant, X-linked recessive, X-linked dominant, Y-linked.

*3.24 If a rare genetic disease is inherited on the basis of an X-linked dominant gene, one would expect to find the following:

a. Affected fathers have 100 percent affected sons.

b. Affected mothers have 100 percent affected daughters.

c. Affected fathers have 100 percent affected daughters.

d. Affected mothers have 100 percent affected sons.

3.25 If a genetic disease is inherited on the basis of an autosomal dominant gene, one would expect to find the following:

a. Affected fathers have only affected children.

b. Affected mothers never have affected sons.

c. If both parents are affected, all of their offspring have the disease.

d. If a child has the disease, one of his or her grandparents also had it.

*3.26 If a genetic disease is inherited as an autosomal recessive, one would expect to find the following:

a. Two affected individuals never have an unaffected child.

FIGURE 3.B

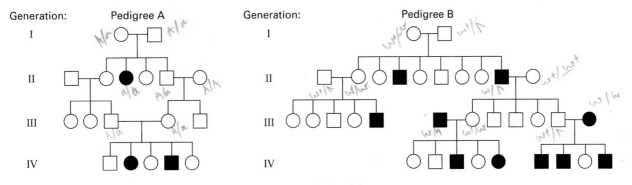

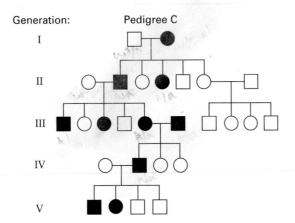

b. Two affected individuals have affected male offspring but no affected female children.

c. If a child has the disease, one of his or her grandparents will have had it.

d. In a marriage between an affected individual and an unaffected one, all the children are unaffected.

3.27 Which of the following statements is *not* true for a disease that is inherited as a rare X-linked dominant?

a. All daughters of an affected male will inherit the disease.

b. Sons will inherit the disease only if their mothers have the disease.

c. Both affected males and affected females will pass the trait to half the children.

d. Daughters will inherit the disease only if their fathers have the disease.

*3.28 Women who were known to be carriers of the X-linked, recessive, hemophilia gene were studied in order to determine the amount of time required for the blood clotting reaction. It was found that the time required for clotting was extremely variable from individual to individual. The values obtained ranged from normal clotting time at one extreme, all the way to clinical hemophilia at the other extreme. What is the most probable explanation for these findings?

*3.29 In a species of amphibian, sex can be reversed by placing hormones in the water where fertilized eggs are developing. Eggs exposed to estrogens develop as phenotypic females, whatever their genetic sex, while eggs exposed to androgens all develop as phenotypic males.

In one experiment several phenotypic females raised in estrogen-containing water were mated to normal males. About half of these females produced only male offspring, while the other half produced males and females in the expected 1:1 ratio.

In another experiment several phenotypic males raised in androgen-containing water were mated to normal females. About half of these males produced the normal 1:1 ratio of daughters and sons, but the other half produced about two females for every male.

a. How is sex determined in these amphibians? Explain your reasoning.

b. If the offspring from the mating above that yielded a ratio of 2 females to 1 male were crossed to normal mates, how many kinds of matings would there be? What proportion of the two sexes would be seen in the offspring of each kind of mating?

4 Extensions of Mendelian Genetic Analysis

PRINCIPAL POINTS

~ Many allelic forms of a gene can exist. This phenomenon is called multiple allelism. Any given diploid individual can possess only two different alleles.

~ In complete dominance the same phenotype results whether an allele is heterozygous or homozygous. In incomplete dominance the phenotype of the heterozygote is intermediate between those of the two homozygotes, whereas in codominance the heterozygote exhibits the phenotypes of both homozygotes.

~ In many cases, nonallelic genes do not function independently in determining phenotypic characteristics. In epistasis, for example, modified Mendelian ratios occur because of interactions of nonallelic genes: the phenotypic expression of one gene depends upon the genotype of another gene locus.

~ Alleles of certain genes, when expressed, may be fatal to the individual. The existence of such lethal alleles of a gene indicates that the product usually produced by the gene is essential for the function of the organism.

~ The zygote's genetic constitution only specifies the organism's potential to develop and function. As the organism develops and differentiates, many things can influence gene expression. One such influence is the organism's environment, both internal and external. For the former, examples may include age and sex of the individual. For the latter, factors include nutrition, light, chemicals, temperature, and infectious agents.

It was originally thought that Mendel's principles apply to all eukaryotic organisms and that they form the foundation for predicting the outcome of crosses in which segregation and independent assortment might be occurring. As more and more geneticists did experiments, though, they found that there are exceptions and extensions to Mendel's principles. Several of these cases will be discussed in this chapter. We will examine examples of genes that have many different alleles rather than just two, cases in which one allele is not completely dominant to another allele at the gene locus, and situations in which products of different genes interact to produce modified Mendelian ratios. We will also look at lethal genes, and at the effects of the environment on gene expression. This discussion will give us a broader knowledge of genetic analysis, particularly in terms of how genes relate to the phenotypes of an organism.

MULTIPLE ALLELES

So far in our genetic analyses we have considered only pairs of alleles controlling characters, such as smooth versus wrinkled seeds in peas, red versus white eyes in *Drosophila*, and unattached versus attached ear lobes in humans. The allele found in the standard laboratory strain of the organism is the **wild-type allele,** and the alternative allele is the variant or **mutant allele.** In a population of individuals, however, there can be many alleles of a given gene, not just two. Such genes are said to have **multiple alleles,** and the alleles are said to constitute a *multiple allelic series* (Figure 4.1). Although a gene may have multiple alleles in a given population of individuals, a *single diploid individual can possess only two of these alleles.*

Drosophila Eye Color

Alleles are alternative forms of a gene, each of which may affect a character differently. The w^+ and w alleles in *Drosophila*, for example, are involved in the devel-

~ **FIGURE 4.1**

Allelic forms of a gene.

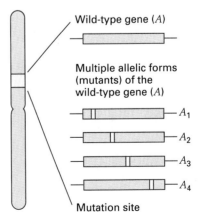

opment of eye color. The w^+ allele results in red eyes, and the w allele, when homozygous or hemizygous, results in white eyes.

It was Morgan's work with a white-eyed variant of *Drosophila* that indicated the presence of genes on the X chromosome. Soon after that discovery he found evidence for other distinct genes on the *Drosophila* X chromosome (such as the bar-eye shape and the vermilion eye color genes). Morgan experimented with strains that had different eye color genes. When he crossed a strain homozygous for the recessive white-eye trait $(w/w)^*$ with a strain homozygous for the recessive vermilion-eye trait (v/v), the results were unexpected: the white-eye and the vermilion genes are nonallelic. The F_1 females were all red-eyed. The simplest explanation is that the white and vermilion eye color traits are specified by *nonallelic genes*. Both X chromosomes of the white-eyed parental strain have the genetic makeup $w \, v^+$—carrying the mutant w allele and the wild-type vermilion allele, v^+. While the vermilion-eyed parental strain has X chromosomes of genetic type $w^+ \, v$—carrying the mutant v allele and the wild-type allele of white, w^+. The F_1 females from this pairing are doubly heterozygous $w \, v^+/w^+ \, v$ and therefore show the dominant effect resulting from the presence of the wild-type allele for each allelic pair; namely, the brick red eyes characteristic of wild-type flies. (*Note:* In the absence of any other mutant genes, a w^+ or v^+ allele alone will give wild-type bright red eyes.)

In 1912 Morgan obtained data for X-linked eye color genes that could not be explained so simply. The eye color variants were white and eosin (reddish orange). Like white, eosin is recessive to the wild-type eye color. However, when a female from an eosin-eyed strain is crossed with a male from a white-eyed strain, all F_1 females have eosin eyes. In 1913, Alfred Sturtevant concluded that (1) red (wild-type) eye color is dominant to eosin and to white, and (2) eosin is recessive to wild type but dominant to white. These conclusions were based on an important assumption: eosin and white are both mutant alleles of a single gene. In other words, there are multiple alleles of the white gene.

Sturtevant's concept becomes clearer if we assign symbols to the alleles. A lowercase letter (or letters) designates the gene, and superscripts designate the different alleles, as in the following example: w^+ is the wild-type allele of the white-eye gene, w is the recessive white allele, and w^e is the eosin allele. Figure 4.2a uses this notation to track the cross of an eosin-eyed female with a white-eyed male. The F_1 females are w^e/w and have eosin eyes because w^e is dominant over w. If these

F_1 females are crossed with red-eyed males (Figure 4.2b), all the female progeny are heterozygous and red-eyed since they contain the w^+ allele; they are either w^+/w^e or w^+/w. Half the male progeny are eosin-eyed (w^e/Y), and the other half are white (w/Y).

Multiple allelic series exist for all types of genes, not just X-linked ones. But let us expand on the example of the white-eyed gene of *Drosophila* a little. There are many known alleles of the white gene that are distinguishable because they produce different eye colors when the alleles are homozygous or hemizygous. Eye color phenotype is related to the amount of pigment deposited in the eye cells. It is possible to extract those pigments from the eyes and to quantify the amount of pigment present in wild type and in strains with different mutant alleles of the white locus (Table 4.1). As expected, the original mutant allele w has the least amount of pigment. Between white and wild, there is a whole range of pigment amounts. The phenotypic expressions of the same gene reflect the varying extent to which the biological activity of the white gene product has been altered.

The number of allelic forms of a gene is not restricted to three, and indeed many hundreds of alleles are known for some genes. The number of possible genotypes in a multiple allelic series depends on the number of alleles involved (Table 4.2). With one allele,

---

~ TABLE 4.1

Eye Pigment Quantification for *Drosophila white* Alleles

GENOTYPES	RELATIVE AMOUNT OF TOTAL PIGMENT
w^+/w^+ (wild type)	1.0000
w/w (white)	0.0044
w^t/w^t (tinged)	0.0062
w^a/w^a (apricot)	0.0197
w^{bl}/w^{bl} (blood)	0.0310
w^e/w^e (eosin)	0.0324
w^{ch}/w^{ch} (cherry)	0.0410
w^{a3}/w^{a3} (apricot-3)	0.0632
w^w/w^w (wine)	0.0650
w^{co}/w^{co} (coral)	0.0798
w^{sat}/w^{sat} (satsuma)	0.1404
w^{col}/w^{col} (colored)	0.1636

Source: From spectrophotometric data of D. J. Nolte, 1959. The eye-pigmentary system of *Drosophila. Heredity* 13:219–281.

---

* As a reminder, the / represents the homologous chromosomes on which the alleles are found (see Box 3.1).

~ FIGURE 4.2

Results of crosses of *Drosophila melanogaster* involving two mutant alleles of the same locus, white (w) and white-eosin (w^e). (a) white-eosin-eyed (w^e/w^e) ♀ × white eyed (w/Y) ♂. (b) F$_1$ (w^e/w) ♀ × red-eyed (wild type) (w^+/Y) ♂.

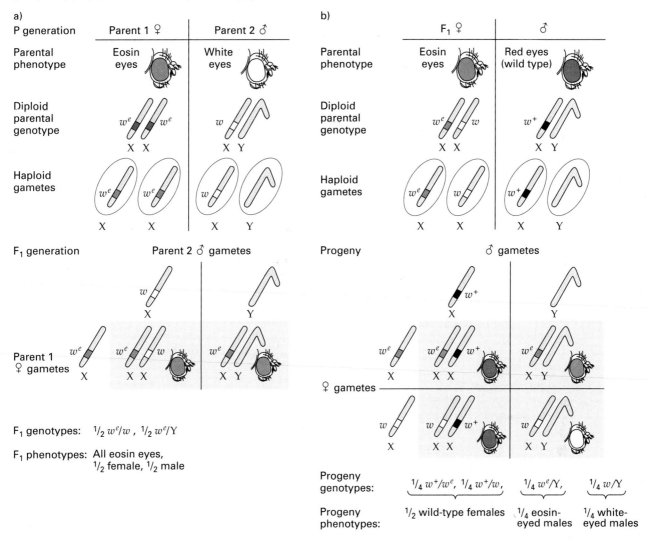

only one genotype is possible. With two alleles, A^1 and A^2, three genotypes are possible: A^1/A^1 and A^2/A^2 homozygotes and the A^1/A^2 heterozygote. When there are n alleles, $n(n + 1)/2$ genotypes are possible, of which n are homozygous and $n(n - 1)/2$ are heterozygotes.

ABO Blood Groups

Another example of multiple alleles of a gene is found in the human ABO blood group series, which was discovered by Karl Landsteiner, in the early 1900s. Since certain ABO blood groups are incompatible, these alleles are of particular importance when blood transfusions are contemplated. (There are many blood group

~ TABLE 4.2

Genotype Number in Multiple Alleles

NUMBER OF ALLELES	KINDS OF GENOTYPES	KINDS OF HOMO-ZYGOTES	KINDS OF HETERO-ZYGOTES
1	1	1	0
2	3	2	1
3	6	3	3
4	10	4	6
5	15	5	10
n	$\dfrac{n(n + 1)}{2}$	n	$\dfrac{n(n - 1)}{2}$

~ TABLE 4.3

ABO Blood Groups in Humans, Determined by the Alleles I^A, I^B, and I^O

PHENOTYPE (BLOOD GROUP)	GENOTYPE
O	I^O/I^O
A	I^A/I^A or I^A/I^O
B	I^B/I^B or I^B/I^O
AB	I^A/I^B

series other than ABO; these also can cause problems in blood transfusions and need to be checked through the process of cross-matching.)

Four blood group phenotypes occur in the ABO system; O, A, B, and AB; Table 4.3 gives their possible genotypes. The six genotypes that give rise to the four phenotypes represent various combinations of three ABO blood group alleles, I^A, I^B, and I^O. People homozygous for the recessive I^O allele are of blood group O. Both I^A and I^B are dominant to I^O. Thus you will express blood group A if you are either I^A/I^A or I^A/I^O, and you will express blood group B if you are either I^B/I^B or I^B/I^O. Since heterozygous I^A/I^B individuals express blood group AB, I^A and I^B alleles exhibit codominance.

The genetics of this system follows basic Mendelian principles. An individual who expresses blood group O, for example, must be I^O/I^O in genotype. The parents of this person could both be O ($I^O/I^O \times I^O/I^O$), they could both be A ($I^A/I^O \times I^A/I^O$, to produce one-fourth I^O/I^O progeny), they could both be B ($I^B/I^O \times I^B/I^O$), or one could be A and one could be B ($I^A/I^O \times I^B/I^O$). Simply put, each parent would have to be either homozygous I^O or heterozygous, with I^O as one of the two alleles.

Blood-typing (the determination of an individual's blood group) and the analysis of the blood group inheritance are sometimes used in legal medicine in cases of disputed paternity or maternity or in cases of an inadvertent baby switch in a hospital. In such cases genetic data cannot prove that he or she is, indeed, the parent. Genetic analysis on the basis of blood group can only be used to show that an individual is not the parent of a particular child; for example, a child of phenotype AB (genotype I^A/I^B) could not be the child of a parent of phenotype O (genotype I^O/I^O). (Note: Blood-type data alone are usually not sufficient for a legal decision in many cases, according to the laws in most states.)

When giving blood transfusions, doctors must carefully match the blood groups of donors and recipients because the blood group alleles specify molecular groups, called *cellular antigens*, that attach to the outside of the red blood cells. The I^A allele specifies the A antigen, and in people with blood group A (I^A/I^A or I^A/I^O) the blood serum contains naturally occurring antibodies for the B antigen (called anti-B antibodies). Conversely, people of the B blood group have serum containing naturally occurring anti-A antibodies. For the AB blood type, both A and B antigens are found on the blood cells, so neither anti-A nor anti-B antibodies occur in the serum of people with this blood type. In people with blood type O the blood cells have neither A nor B antigen, and therefore the serum contains both anti-A and anti-B antibodies. These antigen-antibody relationships are shown in Figure 4.3.

Figure 4.3 also shows the reaction that occurs when blood cells from the four groups are added to serum from each of the four groups. In a number of instances the blood cells clump, or agglutinate. Since clumped cells cannot move through the fine capillaries, agglutination may lead to organ failure and possibly death. Note that, for the ABO blood group series, individuals with the AB blood group can receive blood of any group in transfusions because their serum contains no antibodies—they are universal acceptors. Similarly, individuals with O blood group have no antigens on their blood cells, so they can give blood to any group; they are universal donors.

Because of their medical significance, the ABO blood groups have been studied in a wide variety of human populations. The data indicate a large variation in the relative distribution of O, A, B, and AB blood groups in different populations (Table 4.4) and have served for the study of the genetics of populations (Chapter 23).

$\mathcal{K}$EYNOTE

Many allelic forms of a gene can exist in a population. When they do, the gene is said to show multiple allelism, and the alleles involved constitute a multiple allelic series. Any given diploid individual, however, can possess only two different alleles. Multiple alleles obey the same rule of transmission as alleles of which there are only two kinds, although the dominance relationships among multiple alleles vary from one group to another.

~ **FIGURE 4.3**

Antigen-antibody reactions that characterize the human ABO blood groups. Blood serum from each of the four blood groups was mixed with blood cells from the four types, in all possible combinations. In some cases, such as a mix of B serum with A cells, the cells become clumped, because the antigens on the blood cell surface react with the antibodies in the serum.

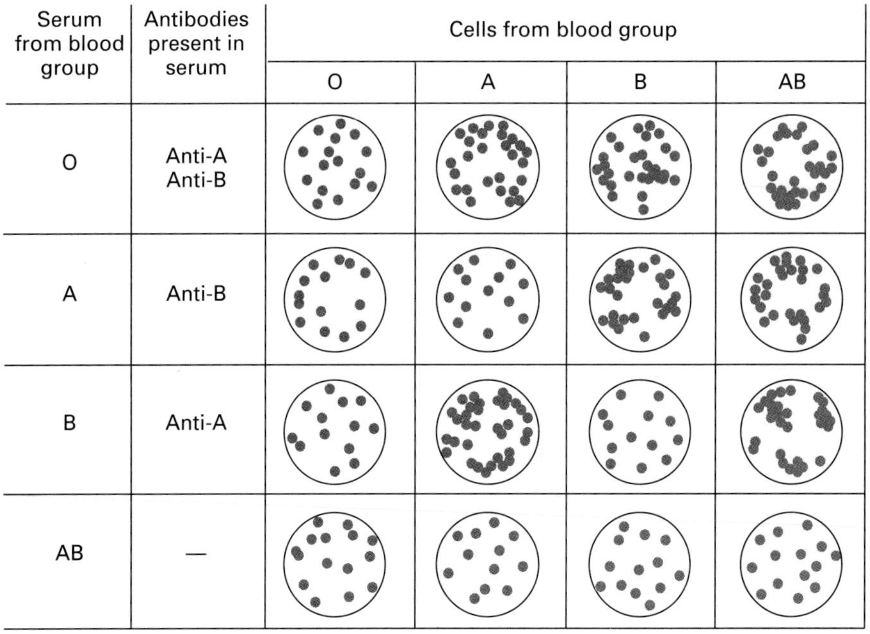

Serum from blood group	Antibodies present in serum	Cells from blood group			
		O	A	B	AB
O	Anti-A Anti-B				
A	Anti-B				
B	Anti-A				
AB	—				

~ **TABLE 4.4**

Relative Frequencies of ABO Blood Groups in Some Human Populations

	BLOOD GROUP			
POPULATION	O	A	B	AB
Armenians	.289	.499	.132	.080
Austrians	.427	.391	.115	.066
Bolivian Indians	.931	.053	.016	.001
Chinese	.439	.270	.233	.058
Danes	.423	.434	.101	.042
Eskimos (Greenland)	.472	.452	.059	.017
French	.417	.453	.091	.039
Irish	.542	.323	.106	.029
Nigerians	.515	.214	.232	.039
U.S. whites (St. Louis)	.453	.413	.099	.035
U.S. blacks (Iowa)	.491	.265	.201	.043

Source: From A. E. Mourant, A. C. Kopec, and K. Domaniewska-Sobczak, 1976. *The Distribution of the Human Blood Groups.* 2nd ed., London, UK: Oxford University Press.

MODIFICATIONS OF DOMINANCE RELATIONSHIPS

In most of the genetic examples discussed so far, one allele is dominant to the other, so that at the phenotypic level the heterozygote is essentially indistinguishable from the homozygous dominant. This phenomenon is called **complete dominance.** In **complete recessiveness,** the allele is phenotypically expressed only in a homozygous organism. Complete dominance and complete recessiveness are the two extremes of a range of dominance relationships. Such dominance/recessiveness relationships characterized the allelic pairs controlling all the character pairs that Mendel studied. Many allelic pairs, however, do not exhibit this dominance relationship.

Incomplete Dominance

In many cases one allele is not completely dominant to another allele, a phenomenon called **incomplete,** or **partial, dominance.** In incomplete dominance the heterozygote's phenotype is between that of individuals homozygous for either individual allele involved.

Plumage color in chickens presents a good example of incomplete dominance. Crosses between a true-

~ FIGURE 4.4

Incomplete dominance in Andalusian fowls. (a) A cross between a white and a black bird produces F_1s of intermediate grey color, called Andalusian blues. (b) The F_2 generation shows the 1:2:1 phenotype ratio characteristic of incomplete dominance.

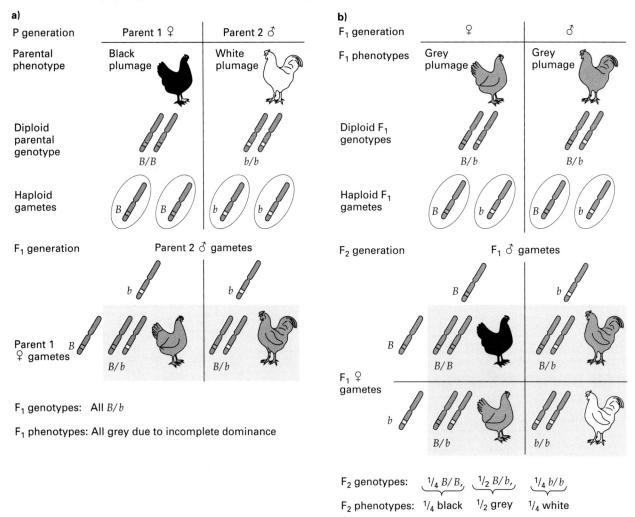

breeding black strain and a true-breeding white strain give F_1 birds with grey plumage (Figure 4.4a), called Andalusian blues by chicken breeders. An Andalusian blue cannot be true breeding because it is heterozygous, so in Andalusian × Andalusian crosses (Figure 4.4b) the two alleles will segregate in the offspring and produce black, Andalusian blue, and white fowl in a ratio of 1:2:1. The most efficient way to produce Andalusian blues is to cross black × white, since all progeny of this mating are Andalusian blues.

Another example of incomplete dominance is the Palomino horse, which has a golden-yellow body color and a mane and tail that are almost white (Figure 4.5a). Palominos do not breed true. When they are interbred, one-quarter of the progeny, called cremellos, are extremely light in color (Figure 4.5b), one-half are

Palomino, and one-quarter are light chestnuts (Figure 4.5c). The 1:2:1 ratio resulting from interbreeding characterizes incomplete dominance. Here is why the phenotype arises. Basically, a Palomino is a light chestnut horse, which has also inherited an incompletely dominant gene, *Dilute* (*D*). In one dose, *D* dilutes the normal reddish-brown of light chestnuts to yellow or cream. When homozygous, *D* produces an extreme dilution of the phenotype: a cremello. When the *Dilute* gene alleles interact with the combination of genes that would otherwise cause the light chestnut color, three different phenotypes result: *D/D*, Cremello; *D/d*, Palomino; and *d/d* Light chestnut (also called sorrel).

In the plant kingdom there are many examples of incomplete dominance, such as flower color in the snapdragon (Figure 4.6). A cross of a red-flowered vari-

~ FIGURE 4.5

Incomplete dominance in horses. (a) Palomino horse, genotype *D/d* for the *Dilute* gene; (b) Cremello horse, genotype *D/D*; and (c) Light chestnut (sorrel) horse, genotype *d/d*.

a)

b)

c)

~ FIGURE 4.6

Incomplete dominance in snapdragons. (a) A cross of true-breeding red-flowered variety with a white-flowered variety produces F$_1$s with pink flowers. (b) A 1:2:1 ratio of red:pink:white is seen in the F$_2$.

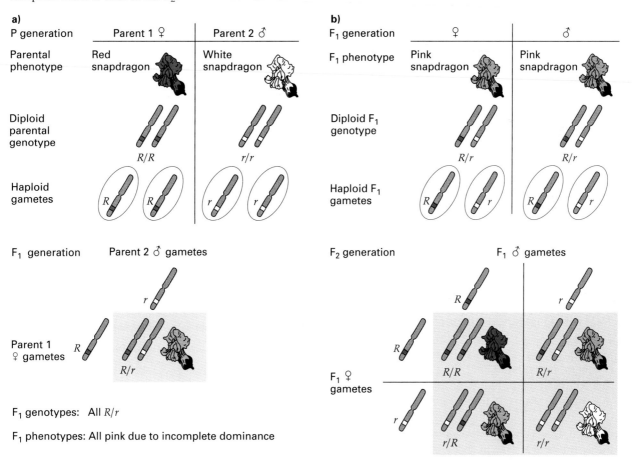

a)

	Parent 1 ♀	Parent 2 ♂
P generation		
Parental phenotype	Red snapdragon	White snapdragon
Diploid parental genotype	*R/R*	*r/r*
Haploid gametes	*R* *R*	*r* *r*

F$_1$ generation — Parent 2 ♂ gametes

Parent 1 ♀ gametes

R/r

F$_1$ genotypes: All *R/r*

F$_1$ phenotypes: All pink due to incomplete dominance

b)

	♀	♂
F$_1$ generation		
F$_1$ phenotype	Pink snapdragon	Pink snapdragon
Diploid F$_1$ genotype	*R/r*	*R/r*
Haploid F$_1$ gametes	*R* *r*	*R* *r*

F$_2$ generation — F$_1$ ♂ gametes

F$_1$ ♀ gametes

	R	*r*
R	*R/R*	*R/r*
r	*r/R*	*r/r*

F$_2$ genotypes: ¼ *R/R*, ½ *R/r*, ¼ *r/r*

F$_2$ phenotypes: ¼ red ½ pink ¼ white

~ TABLE 4.5

Inheritance Patterns for the M−N Blood Types

PARENTAL PHENOTYPES	PARENTAL GENOTYPES	PROGENY PHENOTYPES	PROGENY GENOTYPES
M × M	$L^M/L^M \times L^M/L^M$	All M	All L^M/L^M
N × N	$L^N/L^N \times L^N/L^N$	All N	All L^N/L^N
N × M	$L^N/L^N \times L^M/L^M$	All MN	All L^M/L^N
MN × M	$L^M/L^N \times L^M/L^M$	1/2 MN : 1/2 M	1/2 L^M/L^N : 1/2 L^M/L^M
MN × N	$L^M/L^N \times L^N/L^N$	1/2 MN : 1/2 N	1/2 L^M/L^N : 1/2 L^N/L^N
MN × MN	$L^M/L^N \times L^M/L^N$	1/4 MM : 1/2 MN : 1/4 NN	1/4 L^M/L^M : 1/2 L^M/L^N : 1/4 L^N/L^N

ety with a white-flowered variety produces F_1 plants with pink flowers (Figure 4.6a). The F_2 shows a 1:2:1 ratio of red:pink:white (Figure 4.6b). The red and white flowers are determined by homozygosity for the respective color alleles, the pink flowers by heterozygosity for the two alleles.

In Chapter 3, we discussed another example of incomplete dominance, although not specifically in those terms: the dominant X-linked bar eyes in *Drosophila*. Recall that in females the heterozygous B/B^+ flies* had a wide bar, a less extreme intermediate bar eye, than B/B females. Wide bar results from incomplete dominance of the mutant B allele.

Codominance

Codominance is another modification of the dominance relationship. In codominance the heterozygote exhibits the phenotypes of *both* homozygotes. It differs from incomplete dominance, in which the heterozygote exhibits a phenotype intermediate between the two homozygotes.

The ABO blood series discussed earlier in this chapter is a good example of codominance. Heterozygous I^A/I^B individuals are blood group AB because both the A antigen (product of the I^A allele) and the B antigen (product of the I^B allele) are produced. Thus, the I^A and I^B alleles are codominant.

The M−N human blood group system, with a locus on chromosome 4, is another example of codominance. In terms of transfusion compatibility, this system is of less clinical importance than the ABO system. In the M−N system three blood types occur:

M, N, and MN. They are determined by the genotypes L^M/L^M, L^N/L^N, and L^M/L^N, respectively. As in the ABO system, the M−N alleles result in the formation of antigens on the red blood cell surface. The heterozygote in this case has both the M and the N antigens and shows the phenotypes of both homozygotes. Inheritance patterns for the M−N blood groups are illustrated in Table 4.5.

There is no distinct dividing line between codominance and incomplete dominance, so in many situations it is not easy to distinguish which is involved. To make such a distinction requires the ability to detect, in the heterozygote, the distinct substances specified by each member of an allelic pair. Thus, the molecular basis of a gene's expression must be understood for accurate description of the basis for each phenotype.

$\mathcal{K}$EYNOTE

In complete dominance the same phenotype results whether an allele is heterozygous or homozygous. In complete recessiveness the allele is phenotypically expressed only when it is homozygous; the recessive allele has no effect on the phenotype of the heterozygote. Complete dominance and complete recessiveness are two extremes between which all transitional degrees of dominance are possible. In incomplete dominance, for example, the phenotype of the heterozygote is intermediate between those of the two homozygotes, while in codominance the heterozygote exhibits the phenotypes of both homozygotes (essentially, no dominance is involved). ▬▬▬▬

---

* As a reminder, in *Drosphila* gene symbolism, a dominant mutation such as *Bar* is given an upper-case letter (*B*), and its wild-type allele is the same symbol with an added superscript "+" (B^+) (see Box 3.1).

GENE INTERACTIONS AND MODIFIED MENDELIAN RATIOS

No gene acts by itself in determining an individual's phenotype; the phenotype is the result of highly complex and integrated patterns of molecular reactions that are under direct gene control. All the genetic examples we have discussed and will discuss have discrete biochemical bases; in a number of cases complex interactions between genes can be detected by genetic analysis. Some examples will be discussed in this section.

In Chapter 2 we discussed Mendel's principle of independent assortment, which states that genes on different chromosomes behave independently in the production of gametes. Let's assume that there are two independently assorting gene pairs, each with two alleles; *A* and *a*, and *B* and *b*. The outcome of a cross between individuals, each of whom is doubly heterozygous (*A/a B/b* × *A/a B/b*), will be nine genotypes in the following proportions: 1/16 *A/A B/B* : 2/16 *A/A B/b* : 1/16 *A/A b/b* : 2/16 *A/a B/B* : 4/16 *A/a B/b* : 2/16 *A/a b/b* : 1/16 *a/a B/B* : 2/16 *a/a B/b* : 1/16 *a/a b/b*. If the phenotypes determined by the two allelic pairs are distinct—for example, smooth versus wrinkled peas, long versus short stems—then we get the familiar dihybrid phenotypic ratio of 9:3:3:1. That is, 9/16 of the progeny show both dominant phenotypes, 3/16 show one dominant phenotype and the other gene pair's recessive phenotype, 3/16 show the first gene pair's recessive phenotype and the other pair's dominant phenotype, and 1/16 show both recessive phenotypes. Any alteration in this standard 9:3:3:1 ratio indicates that the phenotype is the product of the interaction of two or more genes. As we discuss each modified Mendelian ratio, we will refer to this genotypic distribution.

For the 9:3:3:1 phenotypic ratio the genotypes for these phenotypes can be represented in a shorthand way as, respectively, *A/− B/−*, *A/− b/b*, *a/a B/−*, *a/a b/b*. The dash indicates that the phenotype is the same, whether the gene is homozygous dominant or heterozygous (e.g., *A/−* means *A/A* and/or *A/a*). This system cannot be used when incomplete dominance or codominance is involved.

The following sections discuss the main processes that result in modified Mendelian ratios. The first section describes examples of interactions between nonallelic genes that control the same general phenotypic attribute. In the second section, we will discuss examples of interactions of nonallelic genes in which the phenotypic expression of one gene depends upon the genotype of another gene locus. This second type of interaction is called *epistasis*. In both cases the discussions are confined to dihybrid crosses in which the two pairs of alleles assort independently. In the "real world" there are many more complex examples of gene interactions, involving more than two pairs of alleles and/or genes that do not assort independently.

Gene Interactions That Produce New Phenotypes

In all the previous examples of dihybrid crosses, the two character pairs have acted independently in terms of phenotype. For example, the allelic pair for round/wrinkled peas had no effect on the allelic pair for long/short stem. If, however, the two allelic pairs affect the same phenotypic characteristic, there is a chance for gene product interaction to give novel phenotypes, and the result can be modified phenotypic ratios, depending on the type and the extent of interaction between the products of the nonallelic genes.

COMB SHAPE IN CHICKENS. A classic example of such gene interactions is comb shape in chickens. New comb shape phenotypes are a consequence of interactions between two allelic pairs. Figure 4.7 shows the four comb phenotypes that result from the interaction of the alleles of two gene loci. Each of these types can be bred true.

~ FIGURE 4.7

Four distinct comb shape phenotypes in chickens, resulting from all possible combinations of a dominant and a recessive allele at each of two gene loci: (a) rose comb (*R/− p/p*); (b) walnut comb (*R/− P/−*); (c) pea comb (*r/r P/−*); (d) single comb (*r/r p/p*).

a) Rose comb

b) Walnut comb

c) Pea comb

d) Single comb

~ FIGURE 4.8

Results of genetic crosses, showing the interaction of genes for comb shape in fowl. (a) The cross of a true-breeding rose-combed bird with a true-breeding pea-combed bird gives all walnut-combed offspring in the F_1. (b) When the F_1 birds are interbred, a 9:3:3:1 ratio of walnut:rose:pea:single occurs in the F_2.

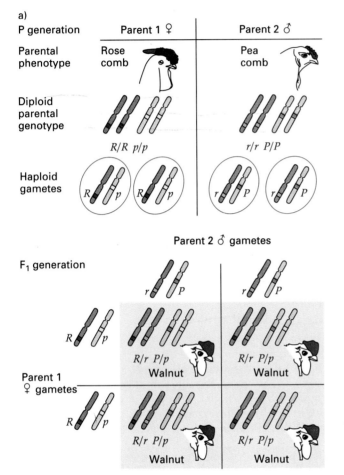

a)
P generation

| | Parent 1 ♀ | Parent 2 ♂ |

Parental phenotype: Rose comb / Pea comb

Diploid parental genotype: $R/R\ p/p$ / $r/r\ P/P$

Haploid gametes: $R\ p$ $R\ p$ / $r\ P$ $r\ P$

Parent 2 ♂ gametes

F₁ generation

Parent 1 ♀ gametes

	$r\ P$	$r\ P$
$R\ p$	$R/r\ P/p$ Walnut	$R/r\ P/p$ Walnut
$R\ p$	$R/r\ P/p$ Walnut	$R/r\ P/p$ Walnut

F₁ genotypes: All $R/r\ P/p$

F₁ phenotypes: All walnut comb

b)
F₂ generation

F₂ phenotypic ratio for $R/r \times R/r$	F₂ phenotypic ratio for $P/p \times P/p$	Combined F₂ ratios	Expected F₂ phenotypic proportions
$3/4\ R/-$	$3/4\ P/-$	$9/16\ R/-\ P/-$	$9/16$ walnut
	$1/4\ p/p$	$3/16\ R/-\ p/p$	$3/16$ rose
$1/4\ r/r$	$3/4\ P/-$	$3/16\ r/r\ P/-$	$3/16$ pea
	$1/4\ p/p$	$1/16\ r/r\ p/p$	$1/16$ single

Crosses made between true-breeding rose-combed and single-combed varieties showed that rose was dominant over single. When the F_1 rose-combed birds were bred together, there was a clear segregation into 3 rose:1 single in the F_2. Similarly, pea comb was found to be dominant over single, with a 3 pea:1 single ratio in the F_2. When true-breeding rose and pea varieties were crossed, however, the result was new and interesting (Figure 4.8). Instead of showing either rose or pea combs, all birds in the F_1 showed a new comb form, different from either the rose or the pea comb (Figure 4.8a). This new form was called walnut comb, since it resembles half a walnut meat.

When the F_1 walnut-combed birds were bred together, another fascinating result was observed in the F_2 (Figure 4.8b): Not only did walnut-, rose-, and pea-combed birds appear, but so did single-combed birds. These four comb types occurred in a ratio of 9 walnut:3 rose:3 pea:1 single. Such a ratio is characteristic in F_2 progeny from a cross of parents differing in two genes. The doubly dominant class in the F_2 was walnut, while the proportion of the singles indicated that this class contained both recessive alleles.

The overall explanation of the results is as follows: The walnut comb depends on the presence of two dominant alleles, R and P, both located at two independently assorting gene loci. In the presence of at least one R allele, and with homozygosity for the recessive p allele, a rose comb results. Birds with at least one P allele, and homozygous for the recessive r allele, have a pea comb. Doubly homozygous recessive $r/r\ p/p$ birds have a single comb. These interpretations are reflected in the branch diagram of Figure 4.8b.

Thus it is the interaction of both dominant alleles, each of which individually produces a different phenotype, that produces a new phenotype. The biochemical basis for the four comb types is not known. At a very general level we can propose that the single-comb phenotype results from the activities of a number of genes other than the R and P genes. In other words, $r/r\ p/p$ birds do not produce any functional gene product that influences the comb phenotype beyond the basic single appearance. The dominant R allele might produce a gene product that interacts with the products of genes controlling the single-comb phenotype to produce a rose-shaped comb. Similarly, the dominant *P* allele might produce a gene product that interacts with the products of the single-comb genes to produce a pea-shaped comb. When the products of both the *R* and *P* alleles are present, they interact to produce another comb variation, the walnut comb.

FRUIT SHAPE IN SUMMER SQUASH (THE 9:6:1 RATIO).

In the comb shape example, each dominant allele alone produces a different phenotype, but both dominant alleles together produce a new phenotype. And when either recessive allele is homozygous, yet another comb phenotype results. Similarly, fruit shape in summer squashes shows complete dominance at both gene pairs, and interaction between both dominants results in a new phenotype. In this case, though, each dominant allele alone produces the same phenotype, so this example differs from the comb shape example discussed above.

Two of the many varieties of summer squash have spherical fruit and long fruit (Figure 4.9). The long-fruit varieties are always true breeding. However, in some crosses between different varieties of true-breed-

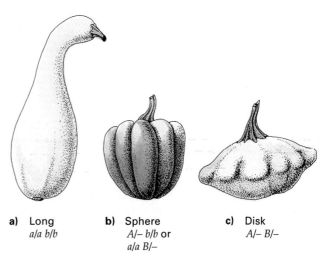

~ FIGURE 4.9

Three true-breeding fruit shapes of summer squash; (a) long; (b) sphere; (c) disk.

a) Long	**b)** Sphere	**c)** Disk
a/a b/b	*A/– b/b* or *a/a B/–*	*A/– B/–*

ing sphere-shaped plants, the F_1 fruit is disk-shaped (Figure 4.9). In such instances, the F_2 fruit show approximately 9/16 disk-shaped, 6/16 sphere-shaped, and 1/16 long-shaped, as shown in Figure 4.10. The

~ FIGURE 4.10

Generation of an F_2 9:6:1 ratio for fruit shape in summer squash.

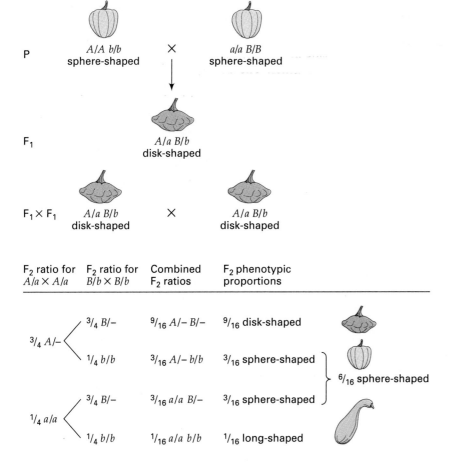

explanation is as follows: Either dominant allele alone (*A/− b/b* or *a/a B/−*) specifies spherical fruit, while the two nonallelic dominant alleles (*A/− B/−*) interact together to produce a new phenotype, namely, disk-shaped fruit. The doubly homozygous recessive (*a/a b/b*) gives a long fruit shape. Thus in the cross above the sphere-shaped parentals are *A/A b/b* and *a/a B/B*. The F₁s are disk-shaped and doubly heterozygous *A/a B/b*. The F₂ disk-shaped fruits are *A/− B/−*, the spherical fruits are *A/− b/b* or *a/a B/−*, and the long-shaped fruits are *a/a b/b*, giving the 9:6:1 ratio.

Again, the precise biochemical bases for the different shapes of squash fruit are not known. In the *a/a b/b* squash, in the absence of *A* and *B* allele products, the functions of other genes determine the long fruit shape. If either the *A* or the *B* allele product, *but not both*, is present, the basic long fruit shape is modified into a sphere shape. The disk fruit shape presumably occurs through a modification of the sphere shape due to the interaction of *A* and *B* products.

Epistasis

Epistasis is the interaction of nonallelic genes, in which the phenotypic expression of one gene depends upon the genotype of another. Genes whose expression is masked by nonallelic genes are said to be *hypostatic*. If we think about the F₂ genotypes presented at the beginning of this section, epistasis may be caused by the presence of homozygous recessives of one gene pair, so that *a/a* masks the effect of the *B* gene. Or epistasis may result from the presence of one dominant allele in a gene pair. For example, *A* might mask the effect of the *B* gene. Moreover, the epistatic effect need not be just in one direction, as in the examples we just mentioned. Epistasis can occur in both directions between two gene pairs. All these possibilities can produce quite a number of modifications of the 9:3:3:1 ratio.

COAT COLOR IN RODENTS (THE 9:3:4 RATIO). In one type of epistasis, *a/a B/−* and *a/a b/b* individuals have the same phenotype, so the overall phenotypic ratio is 9:3:4. An example is coat color in rodents. The ancestral coat color of mice is the greyish color seen in ordinary wild mice, due to the presence of two pigments in the fur. Individual hairs are mostly black with narrow yellow bands near the tip (Figure 4.11). This coloration, the agouti pattern, has a camouflage function and is found in many wild rodents, including the wild rabbit, the guinea pig, the grey squirrel, and wild mice.

Several variations in the agouti coat coloration have been preserved under domestication. The most

familiar example is the albino, in which the complete absence of pigment in the fur and in the irises of the eyes causes a white coat and pink eyes. Albinos are true breeding, and this variation behaves as a complete recessive to any other color. Another variant has black

Pigment patterns in rodent fur: (a) agouti; (b) albino; and (c) black.

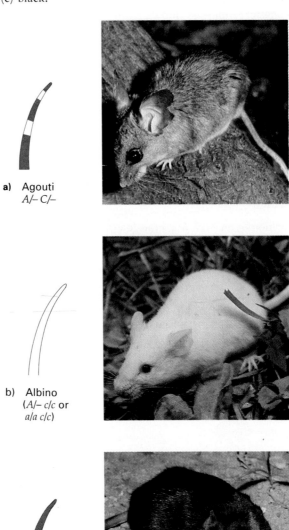

a) Agouti
A/− C/−

b) Albino
(*A/− c/c* or
a/a c/c)

c) Black
a/a C/−

coat color, presumably the result of the absence of the yellow pigment found in the agouti pattern. Black also breeds true and is recessive to agouti.

When black mice are crossed with albinos, the F_1 progeny are all agouti. When these F_1 agoutis are interbred, the F_2 progeny consist of approximately 9/16 agouti animals, 3/16 black, and 4/16 albino as shown in Figure 4.12. This pattern occurs because the parents differ in a gene necessary for the development of any color, which the black mice have but the albinos do not, and in a gene for the agouti pattern, which results in a banding of the black hairs with yellow. Genotypically, $A/-$ $C/-$ are agouti, a/a $C/-$ are black, and $A/-$ c/c and a/a c/c are albino, giving a 9:3:4 phenotypic ratio of agouti:black:albino.

It is now known that three gene loci are involved in the phenotypes of rodent coat color. At one locus the dominant C allele specifies a product that is necessary for the production of any pigment in the coat; the recessive c allele, when homozygous, prevents pigment formation and hence the mice are albino. At a second locus the dominant allele A specifies a product that determines the agouti factor. Its recessive allele a is present in the homozygous state in all nonagouti mice, such as blacks. The dominant allele B of the third locus specifies a product that governs the synthesis of black pigment. The recessive allele b, when homozygous, results in brown pigment. This latter locus is relevant to the example only insofar as the basic hair color is black. All the mice involved must have at least one B allele.

~ FIGURE 4.12

Generation of an F_2 9:3:4 ratio for coat color in rodents.

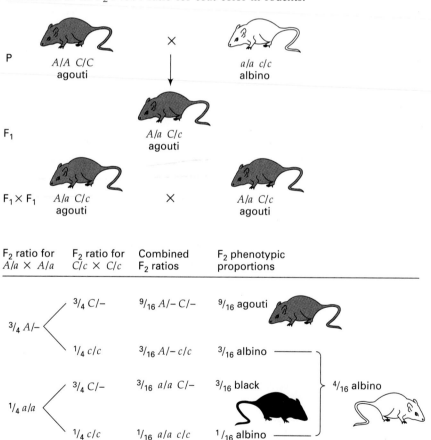

FLOWER COLOR IN SWEET PEAS (THE 9:7 RATIO). The sweet pea occurs in a number of true-breeding varieties, most of which have descended from a purple-flowered wild sweet pea of Sicily. Purple flower color is dominant to white and gives a typical 3:1 ratio in the F_2. The white-flowered varieties breed true, and crosses between different white varieties usually produce white-flowered progeny. In some cases, however, crosses of two true-breeding white varieties give only purple-flowered F_1 plants. When these F_1 hybrids are self-fertilized, they produce an F_2 generation consisting of about 9/16 purple-flowered sweet peas and 7/16 white-flowered, as shown in Figure 4.13. All the F_2 white-flowered plants breed true when self-fertilized. A ninth of the purple-flowered F_2 plants—the C/C P/P genotypes—breed true.

These results may be explained by considering the interaction of two nonallelic genes. In contrast to the example of comb shape in chickens, no new traits appear in the F_1 or F_2. The occurrence of purple flowers in about 9/16 of the F_2 plants suggests that colored flowers appear only when two independent dominant factors are present together and that the color purple results from some interaction between them. White flower color would then be due to the absence of either or both of these factors. Thus we propose that gene pair C/c specifies whether or not the flower can be colored, and gene pair P/p specifies whether or not purple flower color will result.

Figure 4.14 presents a hypothetical pathway for the production of purple pigment. In this pathway a colorless precursor compound is converted through several steps (via compounds 1, 2, and 3) to a purple end product. Each step is controlled by a functional gene product. To explain the F_2 ratio in the sweet pea example, a hypothesis can be put forward that genes C and P control different steps early in the pathway; C controls the conversion of white compound 1 to white compound 2, and P controls the conversion of compound 2 to compound 3 (Figure 4.14a). Therefore, homozygosity for the recessive allele of either or both of the C and P genes will result in a block in the pathway. As a result, only white pigment rather than purple pigment will accumulate. That is, $C/- p/p$, $c/c P/-$, and $c/c p/p$

~ **FIGURE 4.13**

Generation of an F_2 9:7 ratio for flower color in sweet peas.

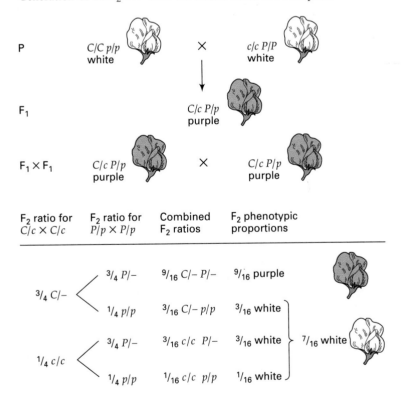

~ FIGURE 4.14

Hypothetical pathway for the production of (a) purple pigment and (b) white pigment in sweet peas to explain the 9 purple:7 white ratio in the F_2 of a dihybrid self.

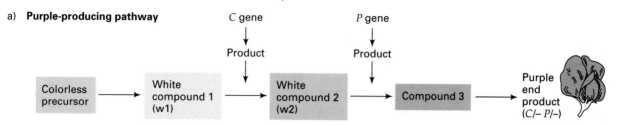

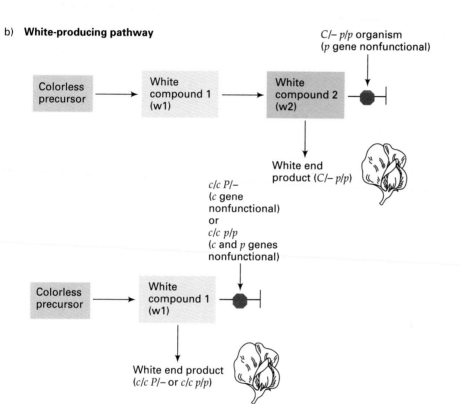

a) **Purple-producing pathway**

b) **White-producing pathway**

genotypes will all be white (Figure 4.14b). The only plants that produce purple flowers will be those in which the two steps are completed so that the rest of the pathway leading to the colored pigment can be carried out. This situation occurs only in $C/- P/-$ plants. (The hypothetical pathway proposed for this sweet pea example is applicable to other 9:7 ratios, with modification, of course, for the particular phenotypes encountered.)

In the cross described, the two white parentals were $C/C\ p/p$ and $c/c\ P/P$ and the F_1 plants were purple and doubly heterozygous. Interbreeding the F_1 gives a 9:7 ratio of purple:white in the F_2. The true-breeding purple F_2 plants were $C/C\ P/P$. In this case epistasis occurs in both directions between two gene pairs. The consequence of this gene interaction is that the same phenotype (white) is exhibited whenever one or the other gene pair is homozygous recessive.

FRUIT COLOR IN SUMMER SQUASH (THE 12:3:1 RATIO). Summer squash has three common fruit colors: white, yellow, and green. In crosses of white and yellow and of white and green, white is always expressed. In crosses of yellow and green, yellow is expressed. Yellow thus is recessive to white but dominant to green. The simplest explanation is an allele for white is epistatic to those for yellow and green.

Consider two gene pairs, W/w and Y/y. In squashes that are $W/-$ in genotype, the fruit is white no matter what genotype is at the other locus. In w/w plants, (1) the fruit will be yellow if a dominant allele of the other locus is present, and (2) green if it is absent. In other words, $W/- Y/-$ and $W/- y/y$ plants have white fruits, $w/w Y/-$ plants have yellow fruits, and $w/w y/y$ plants have green fruits. The F_2 progeny of an F_1 self of doubly heterozygous individuals shows a 12:3:1 ratio of white:yellow:green fruits in the plants (Figure 4.15). This example shows complete dominance at both gene pairs, with one gene (the white one here), when dominant, epistatic to the other gene.

A hypothetical biochemical pathway to explain the 12:3:1 ratio of squash color is shown in Figure 4.16. The assumption is that a white substance is converted to a yellow end product via a green intermediate. Both steps are controlled by genes, the green to yellow step being controlled by the Y gene. The F_2 phenotypic ratio will be generated if the dominant Y allele is needed for the conversion of the green substance to yellow, and if the dominant W allele specifies a product that inhibits the white-to-green conversion step. Thus all plants that have at least one W allele will be white-fruited, no matter which alleles are present at the Y locus, since the W gene product prevents the making of green substance (Figure 4.16a). As a result, 12/16 of the F_2 squashes are white-fruited and have the genotypes $W/- Y/-$ and $W/- y/y$. Also, 3/16 of the F_2s are yellow-fruited and are the $w/w Y/-$ individuals (Figure 4.16b). In this case green substance is made since there is no inhibition of the white-to-green step. Further, functional Y allele product is present and catalyzes the conversion of green substance to yellow substance. Lastly, 1/16 of the F_2s are green-fruited and are the $w/w y/y$ individuals (Figure 4.16c). Again, green substance is produced because there is no inhibition of the white-to-green step, but in the absence of a dominant Y allele, the green substance cannot be converted to yellow substance.

~ FIGURE 4.15

Generation of an F_2 12:3:1 ratio for fruit color in summer squash.

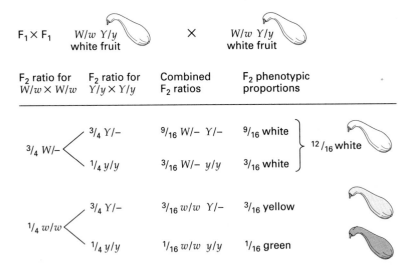

~ FIGURE 4.16

Hypothetical pathway to explain the F_2 ratio of 12 white:3 yellow:1 green color in squashes. (a) Production of white color; (b) Production of yellow color; (c) Production of green color.

a) White fruit-producing pathway

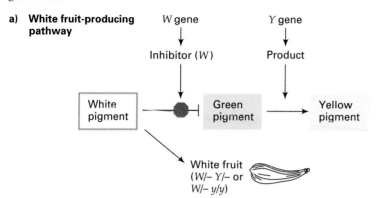

b) Yellow fruit-producing pathway

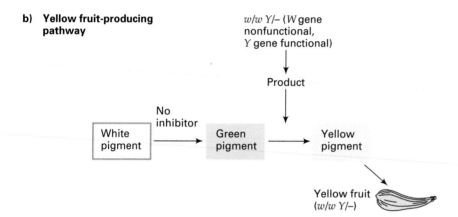

c) Green fruit-producing pathway

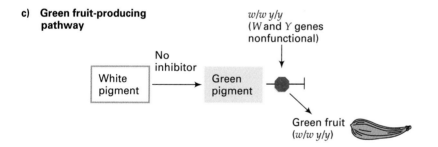

Many types of interaction are possible between the products of gene pairs (Table 4.6; p. 110). Geneticists detect such interactions when they observe modifications of the expected phenotypic ratios in crosses. We have discussed some examples in which two nonallelic genes assort independently and in which complete dominance is exhibited in each gene pair. The ratios we discussed would necessarily be modified further if the genes did not assort independently or if incomplete dominance or codominance prevailed.

𝒦EYNOTE

In many instances in plants and animals, nonallelic genes do not function independently in determining phenotypic characteristics. In some cases interaction between gene products results in new phenotypes without modification of typical Mendelian ratios. In another type of gene interaction, called epistasis, interaction between gene products causes modifica-

~ **TABLE 4.6**

Summary of Dominance Modifications

		A/A B/B	A/A B/b	A/a B/B	A/a B/b	A/A b/b	A/a b/b	a/a B/B	a/a B/b	a/a b/b
More than four phenotypic classes	A and B both incompletely dominant	1	2	2	4	1	2	1	2	1
	A incompletely dominant; B completely dominant	3		6		1	2	3		1
Four phenotypic classes	A and B both completely dominant (classic ratio)	9				3		3		1
Fewer than four phenotypic classes	a/a epistatic to B and b; recessive epistasis	9				3		4		
	A epistatic to B and b; dominant epistasis	12						3		1
	A epistatic to B and b; b/b epistatic to A and a; dominant and recessive epistasis	13[a]							3	
	a/a epistatic to B and b; b/b epistatic to A and a; duplicate recessive epistasis	9						7		
	A epistatic to B and b; B epistatic to A and a; duplicate dominant epistasis	15								1
	Duplicate interaction	9						6		1

[a]The 13 is composed of the 12 classes immediately above plus the one a/a b/b from the last column.

tions of Mendelian ratios because one gene product interferes with the phenotypic expression of another nonallelic gene (or genes). The phenotype is controlled largely by the former gene and not the latter when both genes occur together in the genotype. Epistasis is complicated further when one or both gene pairs involve incomplete dominance or codominance, or when gene pairs do not assort independently. ————

ESSENTIAL GENES AND LETHAL ALLELES

For a few years after the rediscovery of Mendelism, geneticists believed that mutations only changed the appearance of a living organism, but then they discovered that a mutant allele could cause the death of an organism. In a sense this mutation is still a change in phenotype, with the new phenotype being lethality. An allele that results in the death of an organism is called a **lethal allele,** and the gene involved is called an *essential* *gene.* **Essential genes** are genes which, when they are mutated, can result in a lethal phenotype. If the mutation is due to a **dominant lethal allele,** the lethal phenotype will be exhibited by the heterozygotes. If the mutation is due to a **recessive lethal allele,** only homozygotes for that allele will exhibit the lethal phenotype. In 1905 Lucien Cuenot reported the results of his studies of a mouse body color gene. Wild-type mice have a dark body color, and Cuenot was interested in the inheritance of a gene that causes yellow body color. The yellow variety resembles Andalusian fowl, in that it never breeds true. Matings between two yellows produce progeny of which about 2/3 are yellow and 1/3 are of another color (black, brown, or grey). When yellows were bred to nonyellows, about half the young were yellow and half were nonyellow. This ratio is what is expected from the mating of a heterozygote with a recessive, which suggests that yellow mice are heterozygous. When heterozygotes are interbred, however, we would expect 1/4 to be pure yellow, 1/2 to be heterozygous yellow, and 1/4 to be nonyellow.

Cuenot's data were correctly interpreted in 1910 by W. Castle and C. Little, who proposed that the yellow homozygotes were aborted in utero. In other words, the yellow allele has a dominant effect with regard to coat color, but acts as a recessive allele with respect to the lethality phenotype, since only homozygotes die. The correctness of the Castle-Little hypothesis was provided by examination of pregnant mice. The yellow × cross is shown in Figure 4.17. With the symbol Y assigned to the yellow allele, and Y^+ assigned to the normal, wild-type allele, the cross is $Y/Y^+ \times Y/Y^+$. We expect a genotypic ratio of 1 Y/Y : 2 Y/Y^+ : 1 Y^+/Y^+ among the progeny. The 1/4 Y/Ys die before birth, leaving 2/4 Y/Y^+ and 1/4 Y^+/Y^+, which gives a 2:1 phenotypic ratio of yellow Y/Y^+: nonyellow Y^+/Y^+ survivors. Since the Y allele causes lethality in the homozygous state, it is called a *recessive lethal*. Characteristically, when two heterozygotes are crossed, recessive lethal alleles are recognized by a 2:1 ratio of progeny types.

Numerous examples of lethal genes are found in all diploid organisms. For example, the recessive lethal Dexter gene in cattle has properties very similar to those of yellow gene in the mouse. Heterozygous Dexter cattle are viable but have short legs. (In fact, someone once suggested that Dexter cattle would be a useful strain to raise since they would be less likely to escape from fenced pastures.) In the homozygous state the Dexter gene produces calves with extremely short legs and a bulldoglike appearance. These calves are usually aborted.

In humans there are also many known recessive lethal alleles. One particular autosomal recessive

~ **FIGURE 4.17**

~ **FIGURE 4.17**

Inheritance of a lethal gene Y in mice. A mating of yellow by yellow gives 1/4 nonyellow (black) mice, 1/2 yellow mice, and 1/4 dead embryos. The viable yellow mice are heterozygous Y/Y^+, and the dead individuals are homozygous Y/Y.

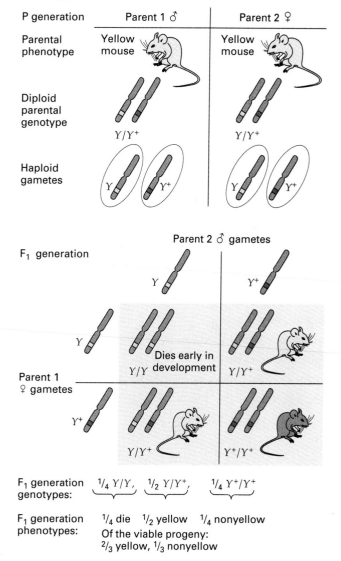

mutant allele, when homozygous, causes *Tay-Sachs disease.* Homozygotes appear normal at birth, but before about one year of age they begin to show symptoms of central nervous system deterioration. Progressive mental retardation, blindness, and loss of neuromuscular control follow. The afflicted children usually die at three to four years of age. The genetic defect in Tay-Sachs results in a deficiency in an enzyme called *hexosaminidase A* (hex A), a substance required for the metabolism of complex chemical structures called *sphingolipids,* necessary for proper nerve function.

There are sex-linked lethal mutations as well as autosomal lethal mutations, and dominant lethal mutations as well as recessive lethals. By definition, dominant lethals exert their effect in heterozygotes, so, in general, the organism dies at conception or at quite a young age. Dominant lethals cannot be studied genetically unless death occurs after the organism has reached reproductive age. One example is the autosomal dominant trait Huntington's disease. This disease results in involuntary movements, progressive central nervous system degeneration, and eventually death. The onset of the symptoms may not occur until the early thirties or so, and death occurs usually when the afflicted persons are in their forties or fifties. As a result, the individuals may have passed on the gene to their offspring before knowing whether they are afflicted with the disease. As mentioned earlier, the American folksinger Woody Guthrie died from Huntington's disease.

KEYNOTE

An allele that is fatal to the individual is called a lethal allele. Recessive lethal and dominant lethal alleles exist, and they can be sex-linked or autosomal. The existence of lethal alleles of a gene indicates that the gene's normal product is essential for the function of the organism. _____

THE ENVIRONMENT AND GENE EXPRESSION

The genetic constitution of the zygote only specifies the organism's potential to develop and function. Many things can influence gene expression as the organism develops and differentiates; one such influence is the organism's environment.

A gene is a segment of DNA, and an organism's set of genes (its genome) is found in its chromosomes. The *development* of an organism from a zygote is a process of regulated *growth* and *differentiation* that results from the interactions of the genome with the internal cellular environment and with the external environment. Development is a programmed series of phenotypic changes that are under temporal, spatial, and quantitative control. Development is essentially irreversible under normal environmental conditions. Four

major processes interact with one another to constitute the complex process of development: (1) replication of the genetic material, (2) growth, (3) differentiation of the various cell types, and (4) the aggregation of differentiated cells into defined tissues and organs.

We must think of development as a series of intertwined, complex biochemical pathways whose steps are under gene control. Any of these pathways are susceptible to environmental influences if the products of the genes controlling the pathways are affected by the internal or external environmental parameters. Though genes influence development, no gene by itself totally determines a particular phenotype. In other words, a phenotype is the result of closely interwoven interactions among many factors during development. Therefore, even though individual organisms have the same genetic constitution, that does not necessarily mean the same phenotypes will result. This phenomenon is most readily studied in experimental organisms where the genotype is unequivocally known. The extent to which the gene manifests its effects under varying environmental conditions can then be seen.

In some cases not all individuals who are known to have a particular gene show the phenotype specified by that gene. The frequency with which a dominant or homozygous recessive gene manifests itself in individuals in a population is called the **penetrance** of the gene. Penetrance depends on both the genotype (e.g., the presence of epistatic or other genes) and the environment. Figure 4.18 illustrates the concept of penetrance. Penetrance is complete (100 percent) when all the homozygous recessives show one phenotype, when all the homozygous dominants show another phenotype, and when all the heterozygotes are alike. For example, if all individuals carrying a dominant mutant gene show the mutant phenotype, the gene is exhibiting complete penetrance. Many genes show complete penetrance: All the seven gene pairs in Mendel's experiments and the alleles in the human ABO blood group system are two examples. If less than 100 percent of the carriers of a particular genotype exhibit the phenotype expected, penetrance is incomplete. For example, an organism may be genotypically *A/–* or *a/a* but may not display the phenotype typically associated with that genotype. If 80 percent, say, of the individuals carrying a particular gene show the corresponding phenotype, we say that there is 80 percent penetrance.

An example of incomplete penetrance is retinoblastoma (tumor of the eye) disease in humans. This disease is inherited as an autosomal dominant, and the mutant gene shows incomplete penetrance, since not all the individuals who have the gene for retinoblastoma express that trait. Another human gene that shows

~ FIGURE 4.18

Complete and incomplete penetrance.

Known genotype

Expected phenotype produced

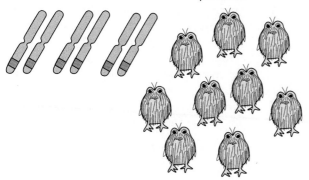

Complete penetrance

Identical genotypes and environment

Expected phenotype produced (100%)

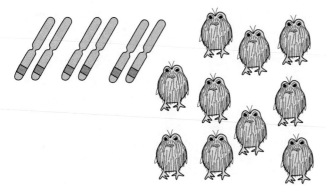

Incomplete penetrance

Identical genotypes and environment

Expected phenotype produced (<100%)

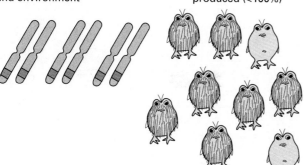

incomplete penetrance is the autosomal dominant gene that determines polydactyly (extra fingers or toes).

At a different level we can also determine the degree to which a gene influences a phenotype. **Expres-**

sivity refers to the degree to which a penetrant gene or genotype is phenotypically expressed. Expressivity may be described in either qualitative or quantitative terms (Figure 4.19); for example, expressivity may be referred to as severe, intermediate, or slight. Like penetrance, expressivity depends on both the genotype and the external environment, and it may be constant or variable. An example of variation in expression is found in the human condition osteogenesis imperfecta. The three main features of this disease are blue sclerae (the whites of the eyes), very fragile bones, and deafness. This condition is inherited as an autosomal dominant with almost 100 percent penetrance; that is, 100 percent of individuals with the genotype have the condition. However, the trait shows variable expressivity: A person with the gene may have any one or any combination of the three traits. Moreover, the fragility of the

~ FIGURE 4.19

Expressivity of a gene: severe, intermediate, and slight.

Known genotype

Expected phenotype produced

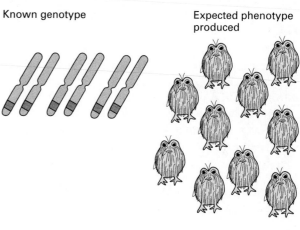

Identical, known penetrant genotype and environment

A range of phenotypes produced

Severe   Intermediate   Slight

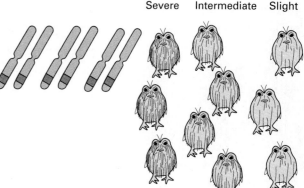

~ FIGURE 4.20

Wild-type *Drosophila* eye (top) and eye of a fly that is homozygous for the recessive mutant gene *eyeless (ey)* (bottom).

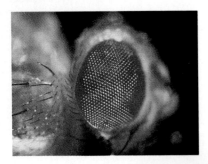

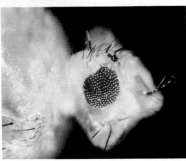

bones for those who exhibit this condition is also very variable. Therefore, we see that in medical genetics it is important to recognize that a gene may vary widely in its expression, a qualification that makes the task of genetic counseling that much more difficult. Another example of expressivity occurs in *Drosophila* homozygous for the recessive mutant gene, *eyeless*. Such flies may have phenotypes varying from no eyes to completely normal eyes, although the common condition is an eye significantly smaller than normal (Figure 4.20).

ℛEYNOTE

Penetrance is the frequency with which a dominant or homozygous recessive gene manifests itself in the phenotype of an individual. Expressivity is the kind or degree of phenotypic manifestation of a penetrant gene or genotype.

Effects of the Internal Environment

Complex biochemical responses occur in cells and in the organism as a result of certain stimuli from the internal environment. Not enough is known about cell biochemistry as it relates to development and adult

function, though, to describe these environmental effects at the molecular level.

AGE. The age of the organism reflects a relevant internal environmental changes that can affect gene function. All genes do not continually function; instead over time, as the organism develops and functions, there is programmed activation and deactivation of genes. Again, the Huntington's disease gene is one whose effects are not manifested until later in the organism's existence. Somehow, the gene responds to the age of the individual, and the phenotypic effects, devastating as they are, commence. Numerous other age-dependent genetic diseases occur in humans; pattern baldness (i.e., the bald spot creeping forward from the crown of the head) appears in males between 20 and 30 years, and Duchenne severe muscular dystrophy appears in children between 2 and 5 years. While such age-related genetic effects can be described, and while in some cases we know the defective gene product involved, in most cases the nature of the age dependency is not understood.

SEX. The expression of particular genes may be influenced by the sex of the individual. In the case of sex-linked genes, as mentioned earlier, differences in the phenotypes of the two sexes are related to different complements of genes on the sex chromosomes. However, in some cases genes that are not located on the sex chromosomes affect a particular character that appears in one sex but not the other. Traits of this kind are called **sex-limited traits.**

One such trait is the feathering pattern of chickens (Figure 4.21). A common difference between male and female fowls is in the structure of many of the feathers. Cocks often have long, narrow, pointed feathers on their hackles and saddle, pointed feathers on the cape, back, and wing, and long, curving, pointed sickle feathers on the tail. This pattern is called cock feathering. Hens usually show the contrasting feather characteristics of short, broad, blunt, and straight feathers. This pattern is called hen feathering. If F_1 heterozygous (h^+/h) hen-feathered males and females are interbred, all F_2 females are hen-feathered, whereas F_2 males exhibit a 3:1 ratio of hen-feathered:cock-feathered.

The interpretation of this pattern is that the trait is controlled by an autosomal gene, with hen feathering (h^+) dominant to cock feathering (h). Assigning gene symbols, we can conclude that all three possible F_2 genotypes in the females, h^+/h^+, h^+/h, and h/h, give rise to hen feathering of the tail, but in the F_2 males the h^+/h^+ and h^+/h chickens are hen-feathered while the h/h chickens are cock-feathered. In other words, the fact that in the homozygous condition (h/h) the cock-

~ FIGURE 4.21

Sex-limited inheritance of feathering in chickens. From a cross between heterozygous F_1 parents the F_2 progenies show a 3:1 ratio of hen-feathered:cock-feathered male birds, while all females are hen-feathered. The gene controlling the phenotype is autosomal.

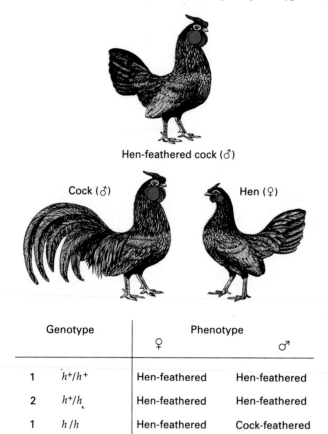

Hen-feathered cock (♂)

Cock (♂)   Hen (♀)

Genotype		Phenotype	
		♀	♂
1	h^+/h^+	Hen-feathered	Hen-feathered
2	h^+/h	Hen-feathered	Hen-feathered
1	h/h	Hen-feathered	Cock-feathered

feathered trait is male-limited shows that sex as an internal environmental parameter can affect gene expression. Several experiments have indicated that biochemical differences in sex hormones between the two sexes in association with the genetic constitution of the skin (in which the feathers originate) determine the state of feathering in the fowl.

Other examples of sex-limited traits are cryptorchidism in dogs, in which one or both testes fail to descend through the inguinal canal into the scrotum, and hereditary inguinal hernia condition in swine.

A slightly different situation is found in **sex-influenced traits.** Such traits appear in both sexes, but either the frequency of occurrence in the two sexes is different or the relationship between genotype and phenotype is different.

An example of a sex-influenced trait is pattern baldness in humans (Figure 4.22; p. 116). One pair of alleles of an autosomal gene is involved in this trait, b^+ and b. The b/b genotype specifies pattern baldness in both males and females, and the b^+/b^+ genotype gives a nonbald phenotype. The difference lies in the heterozygote: In males it leads to the bald phenotype, and in females it leads to the nonbald phenotype. In other words, the b allele acts as a dominant in males but as a recessive in females; its expression is influenced by the sex hormones of the individual. This pattern of inheritance and gene expression explains why pattern baldness is far more frequent among men than among women. From a large sampling of the progeny of matings between two heterozygotes, 3/4 of the daughters are nonbald and 1/4 are bald, and 3/4 of the sons are bald and 1/4 are nonbald.

Other examples of sex-influenced traits are horn type in Dorset Horn sheep, in which the ram's horns are heavier and more coiled than those of the ewe

~ **Figure 4.22**

Sex-influenced inheritance of pattern baldness in humans. The allele is recessive in one sex and dominant in the other, so a cross between two heterozygotes produces a 3:1 ratio of bald:nonbald in males and 1:3 ratio of bald:nonbald in females. (a) Production of F_1 generation; (b) Production of F_2 generation.

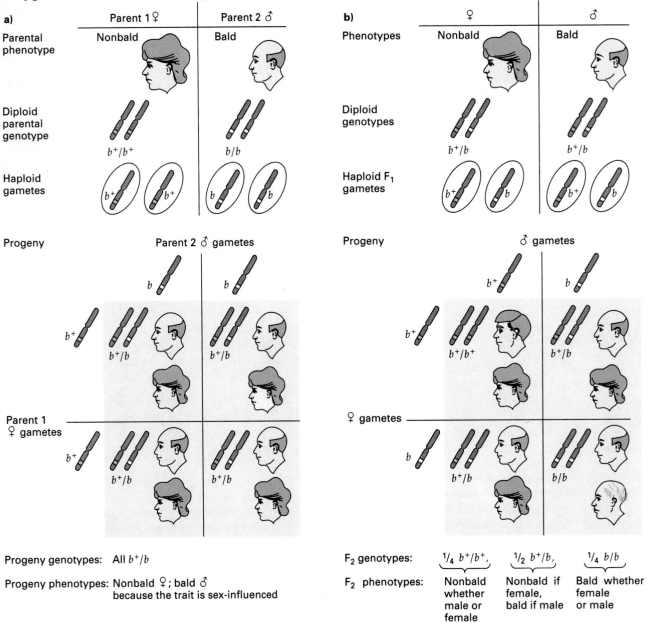

(Figure 4.23), and the crooked breast bone syndrome of chickens and turkeys, a common defect that has both genetic and environmental influences and that is more common in males than females.

Effects of the External Environment

Although many factors in the external environment influence gene expression (temperature, nutrition, light,

chemicals, infectious agents, etc.), we will only discuss two factors here, temperature and chemicals.

TEMPERATURE. The rate at which a chemical reaction occurs depends in large part on temperature. The same is also true, within limits, for biochemical reactions taking place within organisms. Therefore it is reasonable to expect that temperature might affect gene expression, having significant effects on development.

~ FIGURE 4.23

Sex-influenced horns in Dorset Horn sheep. (a) The ram's horns are much heavier and more coiled than (b) the ewe's.

a)   b)

A good example is fur color in Himalayan rabbits (Figure 4.24). Certain genotypes of this white rabbit cause dark fur to develop at the extremities (ears, nose, and paws), where the local surface temperature is lower. Since all body cells develop from a single zygote, this distinct fur pattern cannot be the result of a genotypic difference of the cells in those areas. It must result from external environmental influence.

This hypothesis can be tested by rearing Himalayan rabbits under different temperature conditions (Figure 4.25). When a rabbit is reared at a temperature above 30°C, all its fur, including that of the ears, nose, and paws, is white (Figure 4.25a). If a rabbit is raised at a temperature of approximately 25°C, the typical Himalayan phenotype results (Figure 4.25b). Lastly, if a rabbit is raised at 25°C while its left rear flank is artificially cooled to a temperature below 25°C, the rabbit develops the Himalayan coat phenotype, exhibiting additional dark fur on the cooled flank area (Figure 4.25c).

CHEMICALS.

Phenylketonuria. Ultimately, all the interactions we have discussed occur at the chemical level. The human

~ FIGURE 4.24

Himalayan rabbit.

~ FIGURE 4.25

Phenotypes of Himalayan rabbits raised under different temperature conditions.

a) White extremities, reared at >30°C

b) Normal Himalayan pattern, reared at 25°C

c) Himalayan pattern with dark patch on flank, reared at 25°C, flank cooled to below 25°C

disease phenylketonuria (PKU), for example, is a disorder controlled by an autosomal recessive gene. In individuals homozygous for the recessive allele, a variety of symptoms appear, most notably mental retardation at an early age. The cause of the phenotypes is a defective gene product in a biochemical pathway that is used for the metabolism of phenylalanine, an amino acid. (An amino acid is a protein building block.) The degree to which the symptoms of PKU manifest themselves depends upon diet. Problem foods include protein containing phenylalanine, such as the protein of mother's milk, which may be considered to be an external environment effect. Phenylketonuria can be diagnosed soon after birth, for example, by a simple blood test and then treated by restricting the amount of phenylalanine in the diet.

Phenocopies induced by chemicals. Changes in the environment can influence the expression of one or more genes. The most sensitive time period is early development, since relatively small changes at that time can result in great changes later. When administered to a developing embryo, certain drugs, chemicals, and viruses produce effects that mimic those of known specific mutant alleles. The abnormal individual resulting from this treatment is called a **phenocopy** (phenotypic copies). A phenocopy is defined as a nonhereditary, phenotypic modification (caused by special environmental conditions) that mimics a similar phenotype caused by a gene mutation. The agent that produces a phenocopy is called a *phenocopying agent*. There are many examples of phenocopies, and, in some instances,

studies of phenocopies have provided useful information about the actual molecular defects caused by the mutant counterpart. Remember that the abnormalities seen in these situations occur in a genotypically *normal* individual without any alteration of the genetic constitution.

Phenocopies occur in all sorts of organisms, from the simple to the complex. In the bread mold *Neurospora crassa*, for example, the normal-growth phenotype is a many-branched, web-like mass in which individual filaments extend fairly rapidly. The result is an orange-colored, fluffy growth similar to the grey or green mold that sometimes grows on bread. In some mutants of *Neurospora* branching is much more frequent and growth is slower, resulting in what is called a colonial-like growth habit, since it extends only slowly from the point of inoculation and looks like a button on a solid medium.

Figure 4.26 shows these two growth phenotypes. Colonial growth is a result of a defect in cell wall structure, which makes the cell wall weaker. Consequently, the internal pressure of the cytoplasm in the filaments induces many more branches than normal. The colonial phenotype can be phenocopied by adding to the growth medium chemicals that affect normal cell wall development. In wild-type *Neurospora* adding the unusual sugar sorbose brings about colonial growth. If a piece of the colonial culture is now placed onto a normal medium, the wild-type growth pattern is reestablished. At the biochemical level, cell wall abnormalities similar to those found in some colonial mutants can be detected in the sorbose-induced phenocopies.

~ FIGURE 4.26

The fungus *Neurospora crassa: (left)* the wild-type-growth phenotype and *(right)* a colonial-growth phenocopy. The phenocopy is produced by growing the wild-type strain on a medium containing sorbose. The colonial morphology is very similar to that seen in genetic mutants of this fungus.

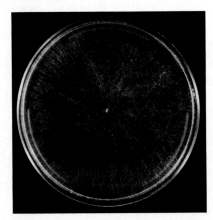

There are also many examples of phenocopies in humans. Cataracts, deafness, and heart defects are sometimes produced when an individual is homozygous for rare recessive alleles; these disorders may also result if the mother is infected with rubella (German measles) virus during the first 12 weeks of pregnancy. Another human trait for which there is a phenocopy is *phocomelia,* a suppression of the development of the long bones of the limbs, which is caused by a rare dominant allele with variable expressivity. Between 1959 and 1961 similar phenotypes were produced by the sedative thalidomide when taken by expectant mothers during the 35th to 50th day of gestation. The drug was removed from the market when its devastating effects were discovered.

In conclusion, at least for the more complex organisms, phenocopies are produced when an organism is exposed to a phenocopying agent during a sharply defined period of the organism's development. The same effect can often be produced by very different agents.

Twin Studies

We are left with the nature/nurture question: How do we decide to what extent a character is controlled by genes and to what extent it is controlled by the environment? One solution is to study genetically identical individuals in varying environments. In the case of laboratory-reared animals, these studies remove the uncertainty of the variation in genotype on experimental results. Typically, the experimental subjects are lines of mice or rats that have been inbred (brother × sister matings) over a large number of generations. As a consequence of this inbreeding, most genes are homozygous and therefore researchers have, as nearly as is reasonably possible, a genetically homogeneous population of organisms. Such inbred (isogenic) lines are used routinely in studies of the effects of unknown agents, such as new pharmaceutical agents, food additives, food colorings, and dyes, where the animal is used to assess the potential risk to humans.

Although we cannot produce inbred lines of humans for such studies, we can study twins. Two types of twins are possible: **fraternal (dizygotic) twins** and **identical (monozygotic) twins.** Fraternal twins result from the fertilization of two eggs released by the ovaries at the same time. Two eggs mean two distinct meioses, and since meiosis is the time when genetic exchange and chromosome segregation take place, the two eggs are likely to be very different genetically because most humans are heterozygous for many of their genes. Note that we expect a roughly 1:2:1 ratio of boy-boy:boy-girl:girl-girl pairs among the two-egg twins. Identical twins, on the other hand, result from the fertilization of a single egg. After the first mitotic division of the fertilized egg, the two cells separate, and each can give rise to a separate individual. These two individuals are genetically identical, since they come from the same fertilized egg: No boy-girl identical twins are expected here.

In human populations, then, it is possible to study whether a trait is inherited by examining its occurrence in twins, since twins, if they are raised together, are exposed to essentially the same environmental influences. In brief, an experimenter collects data on the frequency with which both members of a set of twins possess a trait. If both have the trait or if both do not have the trait, they are said to be **concordant.** Conversely, if one twin has the trait and the other does not, they are **discordant.** By determining the concordance-discordance frequencies for a particular trait, the experimenter can make some conclusions about the relative contributions of the genetic and nongenetic environments to that trait.

More specifically, the results for identical and fraternal twins can be compared; the conclusions depend on the results. If there is a high degree of concordance for identical twins but a great discordance for fraternal twins, the trait in question can be said to have a strong hereditary component. On the other hand, nearly equal concordance and discordance ratios for identical and fraternal twins would suggest that the trait was determined more by environmental factors than by hereditary ones.

Let us briefly consider some examples, with the understanding that these examples involve complex genotypic environmental components. Lots of children catch measles; therefore there is a high concordance frequency in both monozygotic and dizygotic twins. Thus whether an individual gets measles is largely determined by environmental influences. There is also about equal concordance for handedness (i.e., whether you are left- or right-handed), for the age at which a baby sits up, and for coffee drinking. However, marked differences show up for such things as eye color and hair color, as we might expect, since these characters are known to involve genes. There is also a marked difference in concordance in the two types of twins, with a high concordance in monozygotic twins for traits such as pulse rate, blood pressure and schizophrenia. It would do little good to catalog those traits that are or are not controlled mainly by genes. Instead, we can conclude that by studying twins, we can obtain some evidence for the involvement or noninvolvement of genes in various traits. Such studies are difficult, but, if done carefully, can lead to interesting results. Nonetheless we must interpret the results of such a study very carefully, since the variation in penetrance can complicate the results.

KEYNOTE

The phenotypic expression of a gene depends on several factors, including its dominance relations, the genetic constitution of the rest of the genome (e.g., the presence of epistatic or modifier genes), and the influences of the internal and external environments. In some cases special environmental conditions can cause a phenocopy, a nonhereditary and phenotypic modification that mimics a similar phenotype caused by a gene mutation. ————————

SUMMARY

A variety of exceptions and extensions to Mendel's principles were discussed in this chapter. These included:

1. Multiple alleles: Rather than having only two allelic forms (wild type and mutant), a gene may have many allelic forms, and these alleles are called multiple alleles of a gene. Any given diploid individual can have only two different alleles of the multiple allelic series.
2. Modified dominance relationships: In complete dominance, the same phenotype results whether an allele is heterozygous or homozygous. All examples in earlier chapters involved complete dominance. In incomplete dominance, the phenotype of the heterozygote is intermediate between those of the two homozygotes. In codominance, the heterozygote exhibits the phenotypes of both homozygotes.
3. Gene interactions and modified Mendelian ratios: In many cases, nonallelic genes do not function

independently in determining phenotypic characteristics. In epistasis, modified Mendelian ratios occur because of interactions of nonallelic genes: the phenotypic expression of one gene depends upon the genotype of another gene locus.

4. Essential genes and lethal alleles: Alleles of certain genes, when expressed, are fatal. Such lethal alleles may be recessive or dominant. The existence of lethal alleles of a gene indicates that the normal product of the gene is essential for the organism to function.
5. It is not always the case that a gene's phenotypes are simply wild type and mutant. The phenomenon of penetrance describes the condition in which not all individuals who are known to have a particular gene show the phenotype specified by that gene. That is, penetrance is the frequency with which a dominant or homozygous recessive gene manifests itself in individuals in the population. The related phenomenon of expressivity describes the degree to which a gene influences a phenotype. That is, expressivity is the degree to which a penetrant gene or genotype is phenotypically expressed. Both penetrances and expressivity depend on both the genotype and the external environment.
6. An organism's potential to develop and function is specified by the zygote's genetic constitution. As an organism develops and differentiates, gene expression is influenced by a number of factors, including the internal and external environment. Examples of the former include age and sex of the individual, and examples of the latter include nutrition, chemicals, and temperature.

ANALYTICAL APPROACHES FOR SOLVING GENETICS PROBLEMS

Q.1 In snapdragons red flower color (R) is incompletely dominant to white flower color (r); the heterozygote has pink flowers. Also, normal broad leaves (B) are incompletely dominant to narrow, grasslike leaves (b); the heterozygote has an intermediate leaf breadth. If a red-flowered, narrow-leaved snapdragon is crossed with a white-flowered, broad-leaved one, what will be the phenotypes of the F_1 and F_2 generations, and what will be the frequencies of the different classes?

A.1 This basic question on gene segregation includes the issue of incomplete dominance. In the case of incomplete

dominance, remember that the genotype can be directly determined from the phenotype. Therefore we do not need to ask whether or not a strain is true breeding because all phenotypes have a different (and, therefore, known) genotype.

The best approach here is to assign genotypes to the parental snapdragons. Let $R/R\ b/b$ represent the red, narrow plant, and $r/r\ B/b$ represent the white, broad plant. The F_1 plants from this cross will all be double heterozygotes, $R/r\ B/b$. Owing to the incomplete dominance, these plants are pink-flowered and have leaves of intermediate breadth. Interbreeding the F_1s gives the F_2 generation, but

FIGURE 4.A

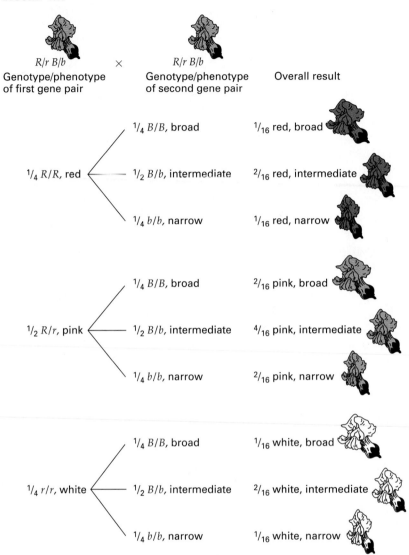

it does not have the usual 9:3:3:1 ratio. Instead, there is a different phenotype for each genotype. These genotypes and phenotypes and their relative frequencies are shown in the Figure 4.A.

Q.2 In snapdragons, red flower color is incompletely dominant to white, with the heterozygote being pink; normal flowers are completely dominant to peloric-shaped ones; and tallness is completely dominant to dwarfness. The three gene pairs segregate independently. If a homozygous red, tall, normal-flowered plant is crossed with a homozygous white, dwarf, peloric-flowered one, what proportion of the F_2 will resemble the F_1 in appearance?

A.2 Let us assign symbols: R = red and r = white; N = normal flowers and n = peloric; T = tall and t = dwarf. Then the initial cross becomes $R/R\ T/T\ N/N \times r/r\ t/t\ n/n$. From this cross we see that all the F_1 plants are triple heterozygotes with the genotype $R/r\ T/t\ N/n$ and the phenotype pink, tall, normal-flowered. Interbreeding the F_1 generation will produce 27 different genotypes in the F_2; this answer follows from the rule that the number of genotypes is 3^n, where n is the number of heterozygous gene pairs involved in the $F_1 \times F_1$ cross.

Here we are asked specifically for the proportion of F_2 progeny that resemble the F_1 in appearance. We can calculate this proportion in a direct way, without needing to display all the possible genotypes and collect those classes with the appropriate phenotype. First, we calculate the frequency of pink-flowered plants in the F_2; then we determine the proportion of these plants that have the other two attributes. From an $R/r \times R/r$ cross we calculate that half of the progeny will be heterozygous R/r and therefore pink. Next, we determine the proportion of F_2 plants that are phenotypically like the F_1 with respect to height (i.e., tall). Either T/T or T/t plants will be tall, and so 3/4 of the F_2 will be tall. Similarly, 3/4 of the F_2 plants will be normal-flowered like the F_1s. To obtain the probability of occurrence of all three of these phenotypes together (i.e., pink, tall, normal), we must multiply the individual probabilities since the gene pairs segregate independently. The answer is $1/2 \times 3/4 \times 3/4$, or 9/32.

Q.3 a. An $F_1 \times F_1$ self gives a 9:7 phenotypic ratio in the F_2. What phenotypic ratio would you expect if you testcrossed the F_1?

b. Answer the same question for an $F_1 \times F_1$ cross that gives a 9:3:4 ratio.

c. Answer the same question for a 15:1 ratio.

A.3 This question deals with epistatic effects. In answering the question, we must consider the interaction between the different genotypes in order to proceed with the testcross. Let us set up the general genotypes that we will deal with throughout. The simplest are allelic pairs a^+ and a, and b^+ and b, where the wild-type alleles are completely dominant to the other member of the pair.

a. A 9:7 ratio in the F_2 implies that both members of the F_1 are double heterozygotes and that epistasis is involved. Essentially, any genotype with a homozygous recessive condition has the same phenotype, so the 3, 3, and 1 parts of a 9:3:3:1 ratio are phenotypically combined into one class. In terms of genotype, 9/16 are $a^+/-\ b^+/-$ types, and the other 7/16 are $a^+/-\ b/b$, $a/a\ b^+/-$, and $a/a\ b/b$. (As always, the use of the—after a wild-type allele signifies that the same phenotype results, whether the missing allele is a wild type or a mutant.) Now the testcross asked for is $a^+/a\ b^+/b \times a/a\ b/b$. Following the same logic used in questions like this one, we can predict a 1:1:1:1 ratio of $a^+/a\ b/b:a^+/a\ b/b:a/a\ b^+/b:a/a\ b/b$. The first genotype will have the same phenotype as the 9/16 class of the F_2, but because of epistasis, the other three genotypes will have the same phenotype as the 7/16 class of the F_2. In sum, the answer is a phenotypic ratio of 1:3 in the progeny of a testcross of the F_1.

b. We are asked to answer the same question for a 9:3:4 ratio in the F_2. Again, this question involves a modified dihybrid ratio, where two classes of the 9:3:3:1 have the same phenotype. Complete dominance for each of the two gene pairs occurs here also, so the F_1 individu-

als are $a^+/a\ b^+/b$. Perhaps both the $a^+/-\ b^+/b$ and $a/a\ b/b$ classes in the F_2 will have the same phenotype, while the $a^+/-\ b^+/-$ and $a/a\ b^+/-$ classes will have phenotypes distinct from each other and from the interaction class. The genotypic ratio of a testcross of the F_1 is the same as in part a. Considering them in the same order as we did in part a, the second and fourth classes would have the same phenotype, owing to epistasis. So there are only three possible phenotypic classes instead of the four found in the testcross of a dihybrid F_1, where there is complete dominance and no interaction. The phenotypic ratio here is 1:1:2, where these phenotypes are listed in the same relative order as in the 9:3:4.

c. This question is yet another example of epistasis. Since $15 + 1$ is 16, this number gives the outcome of an F_1 self of a dihybrid where there is complete dominance for each gene pair and interaction between the dominant alleles. In this case the $a^+/-\ b^+/-$, $a^+/-\ b/b$, and $a/a\ b^+/-$ classes have one phenotype and include 15/16 of the F_2 progeny, and the $a/a\ b/b$ class has the other phenotype and 1/16 of the F_2. The genotypic results of a testcross of the F_1 are the same as in parts a and b; that is, the F_2 progeny exhibit a 1:1:1:1 ratio of $a^+/a\ b^+/b:a^+/a\ b/b:a/a\ b^+/b:a/a\ b/b$. The first three classes have the same phenotype, which is the same as that of the 15/16 of the F_2s, and the last class has the other phenotype. The answer, then, is a 3:1 phenotypic ratio.

QUESTIONS AND PROBLEMS

4.1 In rabbits, C = agouti coat color, c^{ch} = chinchilla, c^h = Himalayan, and c = albino. The four alleles constitute a multiple allelic series. The agouti C is dominant to the three other alleles, c is recessive to all three other alleles, and chinchilla is dominant to Himalayan. Determine the phenotypes of progeny from the following crosses:

a. $C/C \times c/c$ d. $C/c^h \times c^h/c$ g. $c^h/c \times c/c$

b. $C/c^{ch} \times C/c$ e. $C/c^h \times c/c$ h. $C/c^h \times C/c$

c. $C/c \times C/c$ f. $c^{ch}/c^h \times c^h/c$ i. $C/c^h \times C/c^{ch}$

*4.2 If a given population of diploid organisms contains three, and only three, alleles of a particular gene (say w, $w1$, and $w2$), how many different diploid genotypes are possible in the populations? List all possible genotypes of diploids (consider only these three alleles).

4.3 The genetic basis of the ABO blood types seems most likely to be

a. multiple alleles.

b. polyexpressive hemizygotes.

c. allelically excluded alternates.

d. three independently assorting genes.

4.4 In humans the three alleles I^A, I^B, and I^O constitute a multiple allelic series that determine the ABO blood group system, as we described in this chapter. For the following problems, state whether the child mentioned can actually be produced from the marriage. Explain your answer.
a. An O child from the marriage of two A individuals.
b. An O child from the marriage of an A to a B.
c. An AB child from the marriage of an A to an O.
d. An O child from the marriage of an AB to an A.
e. An A child from the marriage of an AB to a B.

4.5 A man is blood type O,M. A woman is blood type A,M and her child is type A,MN. The aforesaid man cannot be the father of the child because:
a. O men cannot have type A children.
b. O men cannot have MN children.
c. An O man and an A woman cannot have an A child.
d. An M man and an M woman cannot have an MN child.

*4.6 A woman of blood group AB marries a man of blood group A whose father was group O. What is the probability that
a. their two children will both be group A?
b. one child will be group B and the other group O?
c. the first child will be a son of group AB, and their second child a son of group B?

4.7 If a mother and her child belong to blood group O, what blood group could the father *not* belong to?

*4.8 A man of what blood group could not be a father to a child of blood type AB?

4.9 In snapdragons, red flower color (R) is incompletely dominant to white (r); the R/r heterozygotes are pink. A red-flowered snapdragon is crossed with a white-flowered one. Determine the flower color of (a) the F_1; (b) the F_2; (c) the progeny of a cross of the F_1 to the red parent; (d) the progeny of a cross of the F_1 to the white parent.

*4.10 In Shorthorn cattle the heterozygous condition of the alleles for red coat color (R) and white coat color (r) is roan coat color. If two roan cattle are mated, what proportion of the progeny will resemble their parents in coat color?

4.11 What progeny will a roan Shorthorn have if bred to (a) red; (b) roan; (c) white?

*4.12 In peaches, fuzzy skin (F) is completely dominant to smooth (nectarine) skin (f), and the heterozygous conditions of oval glands at the base of the leaves (O) and no glands (o) give round glands. A homozygous fuzzy, no-gland peach variety is bred to a smooth, oval-gland variety.
a. What will be the appearance of the F_1?
b. What will be the appearance of the F_2?

c. What will be the appearance of the offspring of a cross of the F_1 back to the smooth, oval-glanded parent?

4.13 In guinea pigs, short hair (L) is dominant to long hair (l), and the heterozygous conditions of yellow coat (W) and white coat (w) give cream coat. A short-haired, cream guinea pig is bred to a long-haired, white guinea pig, and a long-haired, cream baby guinea pig is produced. When the baby grows up, it is bred back to the short-haired, cream parent. What phenotypic classes and in what proportions are expected among the offspring?

4.14 The shape of radishes may be long (L/L), oval (L/l), or round (l/l), and the color of radishes may be red (R/R), purple (R/r), or white (r/r). If a long, red radish plant is crossed with a round, white plant, what will be the appearance of the F_1 and the F_2?

4.15 In poultry the genes for rose comb (R) and pea comb (P), if present together, give walnut comb. The recessive alleles of each gene, when present together in a homozygous state, give single comb. What will be the comb characters of the offspring of the following crosses?
a. $R/R\ P/p \times r/r\ P/p$
b. $r/r\ P/P \times R/r\ P/p$
c. $R/r\ p/p \times r/r\ P/p$
d. $R/r\ P/p \times R/r\ P/p$
e. $R/r\ p/p \times R/r\ p/p$

4.16 For the following crosses involving the comb character in poultry, determine the genotypes of the two parents:
a. A walnut crossed with a single produces offspring 1/4 of which are walnut, 1/4 rose, 1/4 pea, and 1/4 single.
b. A rose crossed with a walnut produces offspring 3/8 of which are walnut, 3/8 rose, 1/8 pea, and 1/8 single.
c. A rose crossed with a pea produces five walnut and six rose offspring.
d. A walnut crossed with a walnut produces one rose, two walnut, and one single offspring.

4.17 In poultry feathered shanks (F) are dominant to clean (f), and white plumage of white leghorns (I) is dominant to black (i).
a. A feathered-shanked, white, rose-combed bird crossed with a clean-shanked, white, walnut-combed bird produces these offspring: two feathered, white, rose; four clean, white, walnut; three feathered, black, pea; one clean, black, single; one feathered white, single; two clean, white, rose. What are the genotypes of the parents?
b. A feathered-shanked, white, walnut-combed bird crossed with a clean-shanked, white, pea-combed bird produces a single offspring, which is clean-shanked, black, and single-combed. In further offspring from this cross, what proportion may be expected to resemble each parent, respectively?

*4.18 F_2 plants segregate 9/16 colored : 7/16 colorless. If a colored plant from the F_2 is chosen at random and selfed, what is the probability that there will be no segregation of the two phenotypes among its progeny?

4.19 The gene *l* in *Drosophila* is recessive, sex-linked, and lethal when homozygous or hemizygous (the condition in the male). If a female of genotype *L/l* is crossed with a normal male, what is the probability that the first two surviving progeny to be observed will be males?

*4.20 A locus in mice is involved with pigment production; when parents heterozygous at this locus are mated, 3/4 of the progeny are colored and 1/4 are albino. Another phenotype concerns the coat color produced in the mice; when two yellow mice are mated, 2/3 of the progeny are yellow and 1/3 are agouti. The albino mice cannot express whatever alleles they may have at the independently assorting agouti locus.

a. When yellow mice are crossed with albino, they produce an F_1 consisting of 1/2 albino, 1/3 yellow, and 1/6 agouti. What are the probable genotypes of the parents?

b. If yellow F_1 mice are crossed among themselves, what phenotypic ratio would you expect among the progeny? What proportion of the yellow progeny produced here would be expected to be true breeding?

4.21 In *Drosophila melanogaster*, a recessive autosomal gene, ebony (*e*), produces a black body color when homozygous, and an independently assorting autosomal gene (*bl*) also produces a black body color when homozygous. Flies with genotypes *e/e bl+/−*, *e+/− bl/bl*, and *e/e bl/bl* are phenotypically identical with respect to body color. If true-breeding *e/e bl+/bl+* ebony flies are crossed with true-breeding *e+/e+ bl/bl* flies,

a. what will be the phenotype of the F_1s?

b. what phenotypes and what proportions would occur in the F_2 generation?

c. what phenotypic ratios would you expect to find in the progeny of these backcrosses: (1) F_1 × true-breeding ebony and (2) F_1 × true-breeding black?

*4.22 In four-o'clocks two genes, *Y* and *R*, affect flower color. Neither is completely dominant, and the two interact on each other to produce seven different flower colors:

Y/Y R/R = crimson	*Y/y R/R* = magenta
Y/Y R/r = orange-red	*Y/y R/r* = magenta-rose
Y/Y r/r = yellow	*Y/y r/r* = pale yellow
y/y R/R, y/y R/r, and *y/y r/r* = white	

a. In a cross of a crimson-flowered plant with a white one (*y/y r/r*), what will be the appearances of the F_1, the F_2, and the offspring of the F_1 backcrossed to the crimson parent?

b. What will be the flower colors in the offspring of a cross of orange-red × pale yellow?

c. What will be the flower colors in the offspring of a cross of a yellow with a *y/y R/r* white?

4.23 Two four-o'clock plants were crossed and gave the following offspring: 1/8 crimson, 1/8 orange-red, 1/4 magenta, 1/4 magenta-rose, and 1/4 white. Unfortunately, the person who made the crosses was color-blind and could not record the flower colors of the parents. From the results of the cross, deduce the genotypes and flower colors of the two parents.

*4.24 Genes *A*, *B*, and *C* are independently assorting and control production of a black pigment.

a. Assume that *A*, *B*, and *C* act in a pathway as follows:

$$\text{colorless} \xrightarrow{A} \xrightarrow{B} \xrightarrow{C} \text{black}$$

The alternative alleles that give abnormal functioning of these genes are designated *a*, *b*, and *c*, respectively. A black *A/A B/B C/C* is crossed by a colorless *a/a b/b c/c* to give a black F_1. The F_1 is selfed. What proportion of the F_2 is colorless? (Assume that the products of each step except the last are colorless, so only colorless and black phenotypes are observed.)

b. Assume that *C* produces an inhibitor that prevents the formation of black by destroying the ability of *B* to carry out its function, as follows:

$$\text{colorless} \xrightarrow{A} \xrightarrow{B} \text{black}$$
$$\uparrow$$
$$C \text{ (inhibitor)}$$

A colorless *A/A B/B C/C* individual is crossed with a colorless *a/a b/b c/c*, giving a colorless F_1. The F_1 is selfed to give an F_2. What is the ratio of colorless to black in the F_2? (Only colorless and black phenotypes are observed, as in Part a.)

4.25 In *Drosophila* a mutant strain has plum-colored eyes. A cross between a plum-eyed male and a plum-eyed female gives 2/3 plum-eyed and 1/3 red-eyed (wild-type) progeny flies. A second mutant strain of *Drosophila*, called stubble, has short bristles instead of the normal long bristles. A cross between a stubble female and a stubble male gives 2/3 stubble and 1/3 normal-bristled flies in the offspring. Assuming that the plum gene assorts independently from the stubble gene, what will be the phenotypes and their relative proportions in the progeny of a cross between two plum-eyed, stubble-bristled flies? (Both genes are autosomal.)

*4.26 In sheep, white fleece (*W*) is dominant over black (*w*), and horned (*H*) is dominant over hornless (*h*) in males but recessive in females. If a homozygous horned white ram is bred to a homozygous hornless black ewe, what will be the appearance of the F_1 and the F_2?

4.27 A horned black ram bred to a hornless white ewe has the following offspring: Of the males 1/4 are horned, white; 1/4 are horned, black; 1/4 are hornless, white; and 1/4 are hornless, black. Of the females 1/2 are hornless and black, and 1/2 are hornless and white. What are the genotypes of the parents?

*__4.28__ A horned white ram is bred to the following four ewes and has one offspring by the first three and two by the fourth: Ewe A is hornless and black; the offspring is a horned white female. Ewe B is hornless and white; the offspring is a hornless black female. Ewe C is horned and black; the offspring is a horned white female. Ewe D is hornless and white; the offspring are one hornless black male and one horned white female. What are the genotypes of the five parents?

*__4.29__ Common pattern baldness is more frequent in males than in females. This appreciable difference in frequency is assumed to be due to
a. Y-linkage of this trait.
b. X-linked recessive mode of inheritance for the trait in question.
c. sex-influenced autosomal inheritance.
d. excessive beer-drinking in males, consumption of gin being approximately equal between the sexes.

5 Linkage, Crossing-Over, and Gene Mapping in Eukaryotes

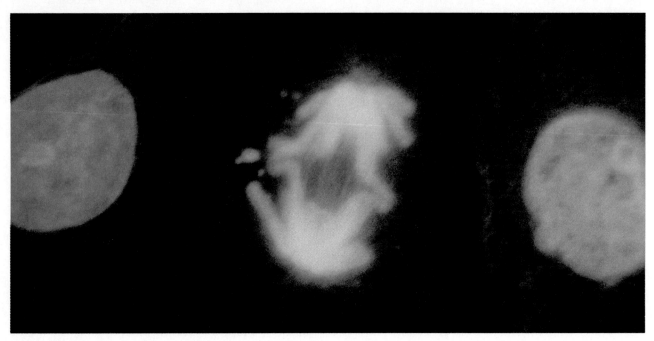

Principal Points

~ The production of genetic recombinants results from physical exchanges between homologous chromosomes in meiosis. A chiasma is the site of crossing-over, and crossing-over is the reciprocal exchange of chromosome parts at corresponding positions along homologous chromosomes by symmetrical breakage and crosswise rejoining.

~ The proof that genetic recombination occurs when crossing-over takes place in meiosis came from experiments in which recombinants for genetic markers occurred only when exchanges of cytological markers occurred.

~ Crossing-over is a reciprocal event that, in eukaryotes, occurs at the four-strand stage in prophase I of meiosis.

~ The map distance between two genes is based on the frequency of recombination between the two genes. The recombination frequency is an approximation of the frequency of crossovers between the two genes. As distance between genes increases, the incidence of multiple crossovers causes the recombination frequency to be an underestimate of the crossover frequency and hence of the map distance. Mapping functions may be used to correct for this problem and, hence, to give a more accurate estimate of map distance.

~ The occurrence of a chiasma between two chromatids may physically impede the occurrence of a second chiasma nearby, a phenomenon called chiasma interference.

---

*I*n the examples throughout Chapters 1–4, genes assorted independently during meiosis as a result of their location on nonhomologous chromosomes. In many instances, however, certain genes (and hence the phenotypic characters they control) are inherited together because they are located on the same chromosome. Genes on the same chromosome are called *linked genes* and are said to belong to a *linkage group* (Figure 5.1). The number of linkage groups in an organism equals the haploid number of chromosomes. Through testcrosses we can determine which genes are linked to each other and can then construct a *linkage map,* or *genetic map,* of each chromosome.

Gene mapping and gene loci are fundamental to the entire concept of the gene. Genetic mapping provides information that is useful in many aspects of genetic analysis. Mapping can tell us, for example, whether genes that work together to produce a given phenotype are located together on the chromosome. Also, knowing the locations of genes on chromosomes has been useful in recombinant DNA research and in experiments directed toward understanding the DNA sequences in and around genes and those sequences at parts of the chromosomes (such as the centromeres) that are of particular functional interest. In this chapter, we will discuss the discovery of genetic linkage, the evidence that genetic recombination involves a physical exchange of chromosome parts, and the methods of gene mapping analysis in eukaryotes.

~ Figure 5.1

Linked genes and linkage groups. Genes located on the same chromosome belong to the same linkage group.

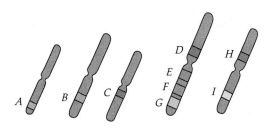

A, B, C:	Unlinked genes
D–E–F–G:	Genes on the same linkage group
H–I:	Genes on the same linkage group
[A], [B], [C], [D–E–F–G], [H–I]:	Five linkage groups

Discovery of Genetic Linkage

Partial Linkage in the Sweet Pea

In 1905, W. Bateson, E. R. Saunders, and R. C. Punnett, in their studies of heredity in the sweet pea, *Lathyrus odoratus L*, discovered the first exception to the law of independent assortment of gene pairs. They crossed two true-breeding strains, one that produced purple flowers and long pollen and another that produced red flowers and round pollen. All F_1 plants had purple flowers and long pollen, indicating that purple flower color was dominant to red flower color and that long pollen was dominant to round pollen. The 381 plants of the F$_2$ consisted of 305 plants with purple petals and 76 plants with red petals. (This is not a good fit to the 3:1 ratio predicted by the law of segregation for a cross of two heterozygotes, so presumably there was a difference in viability for the two phenotypic classes.) For the pollen shape trait the F$_2$ plants segregated into 305 with long pollen and 76 with round pollen. These results indicated that both the flower color trait and the pollen shape trait are controlled by a single pair of alleles at their respective genes.

A natural extension of the data analysis was to see whether the two character pairs assorted independently in the cross. From the F$_1$ × F$_1$ cross done by Punnett in 1917 the F$_2$ progeny displayed the following phenotypes:

284 (74.6 percent) purple flowers, long pollen
 21 (5.5 percent) purple flowers, round pollen
 21 (5.5 percent) red flowers, long pollen
 55 (14.4 percent) red flowers, round pollen

More precise frequencies determined by Punnett in 1917 and based on 6952 F$_2$ plants derived from the same type of purple and long × red and round cross gave these results:

4831 (69.5 percent) purple flowers, long pollen
 390 (5.6 percent) purple flowers, round pollen
 393 (5.6 percent) red flowers, long pollen
1338 (19.3 percent) red flowers, round pollen

If the genes assorted independently, the expected proportions would be 9/16:3/16:3/16:1/16 for the four phenotypes in the order given, or in percentages, 56.25 percent:18.75 percent:18.75 percent:6.25 percent. However, the observed values differ significantly from those expected according to the independent assortment hypothesis. (What constitutes a "significant" difference between observed results and the results expected on the basis of a particular hypothesis may be determined by a *chi-square test*, a statistical procedure described on pp. 137–39.)

Bateson, Saunders, and Punnett recognized that the dominant purple and long characters occurred together more often than predicted by Mendel's second law but could not explain why the discrepancy occurred. They tried to explain their exceptional results in terms of modified Mendelian ratios but without success. Today we know that the genes they were studying are located on the same chromosome. Genes located on the same chromosome are said to exhibit **linkage** and are considered to be **linked genes**. Thus the results can be ex-

~ **FIGURE 5.2**

Expected results of F$_1$ × F$_1$ cross if genes for flower color and pollen size assort independently.

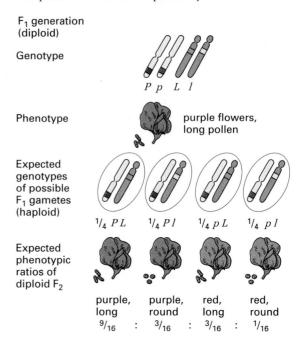

F$_1$ generation (diploid)

Genotype

$P\ p\quad L\ l$

Phenotype

purple flowers, long pollen

Expected genotypes of possible F$_1$ gametes (haploid)

¼ $P\,L$ ¼ $P\,l$ ¼ $p\,L$ ¼ $p\,l$

Expected phenotypic ratios of diploid F$_2$

purple, long : purple, round : red, long : red, round
9/16 : 3/16 : 3/16 : 1/16

plained as follows: If we symbolize the purple allele as *P*, the red allele as *p*, the long pollen allele as *L*, and the round pollen allele as *l*, the cross is *PP LL* × *pp ll*. The F$_1$ double heterozygote is *Pp Ll*. If independent assortment were occurring (Figure 5.2), equal proportions of *P L, P l, p L*, and *p l* gametes would be produced by the F$_1$ plants, and the resulting F$_2$ generation would exhibit the expected 9:3:3:1 ratio.

Instead, the observed ratio of F$_2$ phenotypes can best be explained if 44 percent of the gametes are *P L*, 44 percent are *p l*, and 6 percent each are *P l* and *p L* (Figure 5.3). At the time of pollen and egg formation in the F$_1$ sweet peas, there was a tendency for the parental *P* and *L* alleles to stay together and for the parental *p* and *l* alleles to stay together. Since two *nonparental* combinations of phenotypes were produced (purple and round, and red and long), albeit at low frequency, *p L* and *P l* gametes must also have been generated. Thus the linkage between the two alleles in these crosses was not complete. Such **partial linkage** occurs when homologous chromosomes exchange corresponding parts during meiosis through the process called *crossing-over* (see Chapter 1 and later discussion in this chapter). Figure 5.4 shows what the observed results would be if the linkage between the two alleles was complete. In the F$_1$ all are purple, long, and in the F$_2$ a 3:1 ratio is observed of purple, long:red, round.

~ FIGURE 5.3

Observed results indicating partial linkage of genes: (a) Under assumption that *P L* and *p l* gametes (pollen and eggs) each occur with a frequency of 44 percent and the *P l* and *p L* gametes each occur with a frequency of 6 percent; (b) Production of gametes from F$_1$ double heterozygote; (c) Combination of gametes to produce the F$_2$ classes.

a)

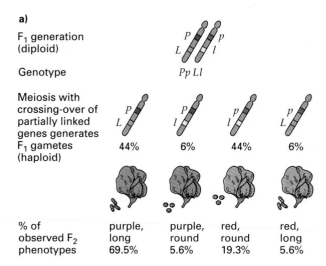

F$_1$ generation
(diploid)

Genotype   *Pp Ll*

Meiosis with
crossing-over of
partially linked
genes generates
F$_1$ gametes
(haploid)   44%   6%   44%   6%

| % of observed F$_2$ phenotypes | purple, long 69.5% | purple, round 5.6% | red, round 19.3% | red, long 5.6% |

b) Random union of F$_1$ gametes to produce F$_2$

Gametes from F$_1$ heterozygote (%)	Pollen			
	P L 44	*P l* 6	*p l* 44	*p L* 6
	Genotype frequencies in F$_2$ (%)			
Eggs *P L* 44	19.36	2.64	19.36	2.64
P l 6	2.64	0.36	2.64	0.36
p l 44	19.36	2.64	19.36	2.64
p L 6	2.64	0.36	2.64	0.36

c) Phenotype frequencies in F$_2$ (%)

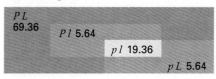

F$_2$ phenotype
totals (%)

P L 69.36
P l 5.64
p l 19.36
p L 5.64

Morgan's Linkage Experiments with *Drosophila*

Our modern understanding of genetic linkage stems from T. H. Morgan and his colleagues' work with linkage in *Drosophila melanogaster*, done around 1911. By 1911 Morgan had accumulated a number of spontaneously arising sex-linked mutants. On the basis of an accumulation of evidence from a number of experiments (see Chapter 4), he concluded that two recessive

~ FIGURE 5.4

Expected results of F$_1$ × F$_1$ cross if genes are completely linked.

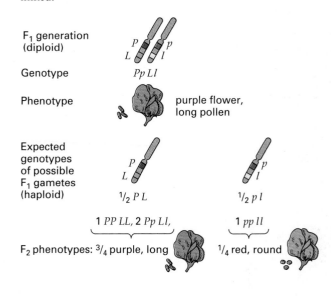

F$_1$ generation
(diploid)

Genotype   *Pp Ll*

Phenotype   purple flower,
                            long pollen

Expected
genotypes
of possible
F$_1$ gametes
(haploid) $^1/_2$ *P L* $^1/_2$ *p l*

1 *PP LL*, 2 *Pp Ll*,   1 *pp ll*

F$_2$ phenotypes: $^3/_4$ purple, long $^1/_4$ red, round

genes *w* (white eye) and *m* (miniature wing) were X-linked. Since both were on the X chromosome, then, the two genes could be *linked* to each other. Strains carrying these mutant genes were used in experiments to see whether or not genetic exchange could occur between the linked genes on the chromosomes.

Morgan crossed a female white miniature (*w m/ w m*) with a wild-type male (*w$^+$ m$^+$/Y*) (Figure 5.5; p. 130). As expected, all the F$_1$ males were white-eyed and had miniature wings, while all females were wild type for both the eye color and wing size traits. The F$_1$ flies were interbred and the 2441 F$_2$ flies were analyzed. (Note that in crosses of linked X-linked genes set up as in Figure 5.5, the F$_1$ × F$_1$ is equivalent to doing a testcross.) In the F$_2$, the most frequent phenotypic classes in both sexes were the grandparental phenotypes of mutant white eyes plus miniature wings or wild-type red eyes plus large wings. Conventionally, we refer to the original genotypes of the two chromosomes as **parental genotypes, parental classes,** or, more simply, **parentals.** The term is also used to describe phenotypes, so the original white miniature females and wild-type males in these particular crosses are defined as the parentals.

Morgan observed a significant number of flies with the nonparental phenotypic combinations of white eyes plus normal wings and red eyes plus miniature wings. Nonparental combinations of linked genes are called **recombinants.** In all, 900 of 2441 F$_2$ flies, or 36.9 percent, exhibited recombinant phenotypes. This percent-

~ FIGURE 5.5

Morgan's experimental crosses of white-eye and miniature-wing variants of
D. melanogaster, showing evidence of linkage and recombination in the X chromosome.

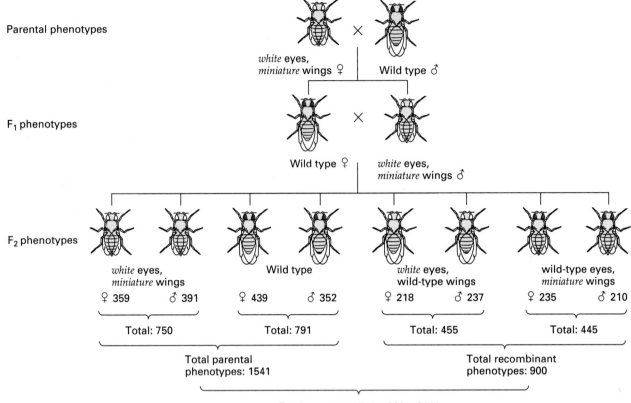

Parental phenotypes

white eyes,
miniature wings ♀   ×   Wild type ♂

F_1 phenotypes

Wild type ♀   ×   *white* eyes,
miniature wings ♂

F_2 phenotypes

white eyes,
miniature wings

♀ 359   ♂ 391

Wild type

♀ 439   ♂ 352

white eyes,
wild-type wings

♀ 218   ♂ 237

wild-type eyes,
miniature wings

♀ 235   ♂ 210

Total: 750   Total: 791   Total: 455   Total: 445

Total parental
phenotypes: 1541

Total recombinant
phenotypes: 900

Total progeny: 1541 + 900 = 2441

Percent recombinants: $^{900}/_{2441} \times 100 = 36.9$

age is significantly less than 50 percent, the figure that would have meant non-linkage (independent assortment). To explain the recombinants, Morgan proposed that in 36.9 percent of the meioses an exchange of genes had occurred between the two X chromosomes of the F_1 females. (No such exchange needs to be postulated for sperm production since the males are hemizygous and no genetic exchange occurs between the non-homologous X and Y chromosomes.)

Morgan extended the work to include other sex-linked characters. He crossed a true-breeding white-eyed, yellow-bodied mutant female with a wild-type red-eyed, grey-bodied male. (Red eyes and a grey body are dominant traits.) As expected, the F_1 progeny consisted of wild-type females and white-eyed, yellow-bodied males; that is, the two parental phenotypic classes. Among the 2205 F_2 flies obtained, the original grandparental combinations of white eyes plus yellow body and red eyes plus grey body were the most fre-

quent phenotypic classes, while only 1.3 percent of the flies (29 of 2205) showed recombinant phenotypes. Morgan concluded that the genes that controlled eye color and body color traits were more closely linked than were the genes for eye color and wing size.

Morgan's group conducted a large number of other crosses of this type, and the conclusions were always the same. *In each case the parental phenotypic classes were the most frequent while the recombinant classes occurred much less frequently.* Approximately equal numbers of each of the two parental classes were obtained, and similar results were obtained for the recombinant classes. Morgan's general conclusion was that *during segregation of alleles at meiosis, certain ones tend to remain together because they lie near each other on the same chromosome.* To take this conclusion one step further, characters remain together to a greater extent the closer the corresponding genes are on the chromosome.

Morgan was also able to relate the production of the phenotypic classes in the F_2 progeny of the cross to chromsomal segregation. Earlier, in 1909, F. Janssens had described *chiasmata* (singular = chiasma; cytologically observable reciprocal exchange between homologous chromosomes) during prophase I of meiosis in the salamander. Janssens thought (but could not prove) that the chiasmata might be sites of physical exchange between maternal and paternal homologues (see Chapter 1). Morgan hypothesized that the partial linkage occurred when two genes on the same chromosome were physically separated from one another by chiasma during meiosis. In 1912 Morgan and E. Cattell introduced the term **crossing-over** to describe the process of chromosomal interchange by which new combinations of linked genes (recombinants) arise.

The terminology here can be confusing. To clarify:

1. A chiasma is the place on a homologous pair of chromosomes at which a physical exchange is occurring; that is, it is the site of crossing-over.
2. Crossing-over is the actual process of reciprocal exchange of chromosome segments at corresponding positions along homologous chromosomes; the process involves symmetrical breakage and crosswise rejoining.
3. Crossing-over is also defined as the events leading to genetic recombination between linked genes in both prokaryotes and eukaryotes.

Figure 5.6 presents a simplified diagram of this highly complex process.

KEYNOTE

The production of genetic recombinants results from physical exchanges between homologous chromosomes that have become tightly aligned during meiotic prophase I. A chiasma is the site of crossing-over. Crossing-over is the reciprocal exchange of chromosome parts at corresponding positions along homologous chromosomes by symmetrical breakage and crosswise rejoining. Crossing-over is also used to describe the events leading to genetic recombination between linked genes. ————

GENE RECOMBINATION AND THE ROLE OF CHROMOSOMAL EXCHANGE

Morgan's hypothesis that during meiosis physical exchange between chromosomes leads to genetic recombination is now universally accepted. At the time, however, it was only a hypothesis, and convincing evidence that the appearance of recombinants was associated with crossing-over was not obtained until the 1930s.

Corn Experiments

The first evidence for the association of gene recombination with chromosomal exchange came in 1931, from the work of Harriet B. Creighton and Barbara

~ FIGURE 5.6

Mechanism of crossing-over: a highly simplified diagram of a crossover between two nonsister chromatids during meiotic prophase, giving rise to recombinant (nonparental) combinations of linked genes.

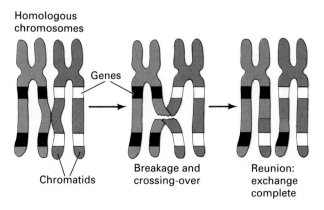

McClintock with corn, *Zea mays*. They performed an experiment in which the two chromosomes under study differed cytologically on both sides of the chromosomal segment in which the crossing-over event occurs.

Crosses were made of a strain of corn that was heterozygous for two genes on chromosome 9 (Figure 5.7a). One of the genes determines colored (*C*) or colorless (*c*) seeds. The other gene determines the forms of starch synthesized by the plants; standard-type plants (*Wx*) produced two forms of starch, amylose and amylopectin, while waxy plants (*wx*) produce only amylopectin. One of the chromosomes had a normal appearance and had the genotype *c Wx* genes. Its homolog had the genotype *C wx*, possessed a large, darkly staining knob at the end nearer *C*, and was longer than the *c Wx* chromosome because a piece of chromosome 8 had broken off and become attached to the end nearer *wx*. (The process of a chromosome segment breaking off from one chromosome and reattaching to another is called *translocation*—see Chapter 19.) Cytologically distinguishable features such as these are called **cytological markers**; the genes involved are called **genetic markers** or **gene markers**.

During meiosis, crossing-over occurs between the two loci (Figure 5.7b). When the two classes of recombinants, *c wx* and *C Wx*, that occurred in the progeny were examined, Creighton and McClintock found that whenever the genes had recombined, the cytological features (the knob and the extra piece) had also recom-

bined (Figure 5.7c). No such exchange was evident in the parental (nonrecombinant) classes of progeny. These data provided strong evidence that genetic recombination is associated with physical exchange of parts between homologous chromosomes.

Drosophila Experiments

Within a few weeks of the publication of Creighton and McClintock's results, Curt Stern reported identical conclusions for experiments done with *Drosophila melanogaster*, indicating that the results from corn were not peculiar to that organism. In these fly experiments, the approach was similar: genetic markers and cytological markers were followed through a genetic cross (Figure 5.8).

In Stern's work, the male parent carried normal X and Y chromosomes with the recessive allele *car* (carnation-colored eye) and the wild-type allele of the *B* (bareye) gene. The *car* gene is located closer to the end of the X chromosome than the *B* gene. Since the male parent is hemizygous, these flies had wild-type-shaped, carnation-colored eyes. The female parent had two abnormal and cytologically distinct X chromosomes. One X chromosome, on which were the wild-type alleles of both the *car* and the *B* genes, had a portion of the Y chromosome attached to it. The other X chromosome had the recessive *car* allele and the dominant *B* allele. In addition, the latter X chromosome was distinctly shorter than normal X since part of it had broken off and was attached to the small chromosome 4 (see the unattached piece in Figure 5.8). Phenotypically, the parental females had a wide-bar eye since they were *B/+*, and the eye was wild-type in color since they were heterozygous *+/car*.

In gamete formation only two classes of sperm were produced: the Y-bearing and the X-bearing. The Y sperm had no genetic markers of relevance here, and the X sperm carried the *car* and *B*+ alleles. However, four classes of eggs were produced. Two—the parentals—resulted from meioses in which no crossing-over had occurred between the chromosomes, and the other two were recombinant gametes produced by crossing-over. As with the corn experiments, every case in which genetic recombination occurred was accompanied by an exchange of identifiable chromosome segments.

If no recombination occurred, the two phenotypic classes of progeny were wild-type and carnation-colored plus bar. No exchanges of chromosome parts were evident among these nonrecombinants. There were also two classes of recombinants: carnation-colored round eyes, and wild-type colored bar eyes. The carnation flies had a complete X chromosome, and the bar flies had a shorter-than-normal X chromosome

Evidence for association of gene recombination with chromosomal exchange in corn. (a) Physical and genetic constitutions of the two chromosomes in the heterozygote; (b) Crossover occurs; (c) Physical and genetic constitutions of the recombinant chromosomes.

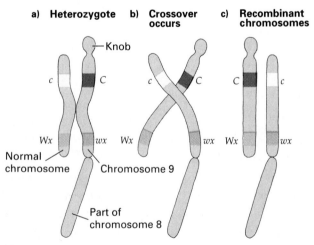

a) **Heterozygote** b) **Crossover occurs** c) **Recombinant chromosomes**

~ FIGURE 5.8

Stern's experiment to demonstrate the relationship between genetic recombination and chromosomal exchange in *D. melanogaster*.

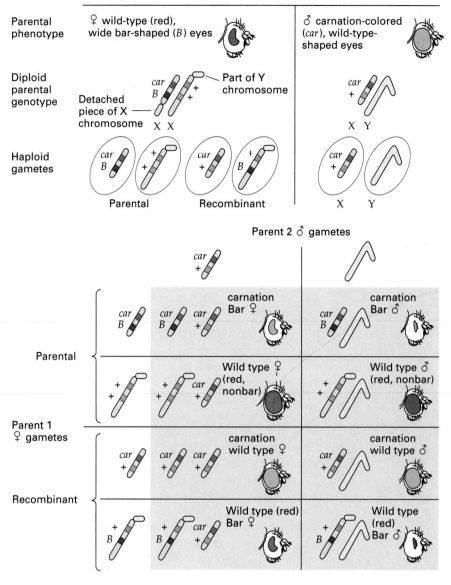

F₁ genotypes: *car B/car +; car B/Y*
         + +/car +; ++/Y
        car +/car +; car +/Y
        + B/car +; +B/Y

F₁ phenotypes: carnation, Bar; 1♀, 1♂
        Wild type (red), nonbar; 1♀, 1♂
        carnation, wild type; 1♀, 1♂
        Wild type (red), Bar; 1♀, 1♂

to which a piece of the Y chromosome was attached, while the rest of the X chromosome was attached to chromosome 4. This chromosomal makeup could only have resulted from physical exchanges of homologous chromosome parts.

There is no doubt, therefore, *that genetic recombination results from physical crossing-over between chromosomes.* (Note that as with Bridges's work on nondisjunction, the application of a combined genetic and cytological approach generated an important and fundamental piece of information.)

ᛕEYNOTE

The proof that genetic recombination occurs when crossing over takes place during meiosis came from breeding experiments in which the parents differed with respect to both genetic and cytological markers. Results of these experiments showed that when recombinant phenotypes occurred, the cytological markers indicated that crossing-over had occurred.

Crossing-Over at the Tetrad (Four-Chromatid) Stage of Meiosis

In Chapter 1 we found that crossing-over occurs in prophase I of meiosis; further studies have shown that crossing-over occurs at the tetrad (four-chromatid) stage in prophase I. The organism *Neurospora crassa* (the orange bread mold) is used here to show how evidence was obtained to support this conclusion and, therefore, to disprove the alternative hypothesis that crossing-over occurred before meiosis in interphase.

LIFE CYCLE OF *NEUROSPORA CRASSA*. To understand the proof for crossing-over at the tetrad stage, we must understand the life cycle of *Neurospora crassa* (Figure 5.9). Then we can appreciate how suitable this organism is for the experiment.

Neurospora crassa is a mycelial-form fungus, meaning that it spreads over its growth medium in a weblike pattern (see Figure 4.26). It produces a furry mycelial mat when it grows on bread, and its designation as an orange bread mold comes from the color of the asexual spores (called conidia) it produces when it

~ **FIGURE 5.9**

Life cycle of the haploid, mycelial-form fungus *Neurospora crassa*. (Parts not to scale)

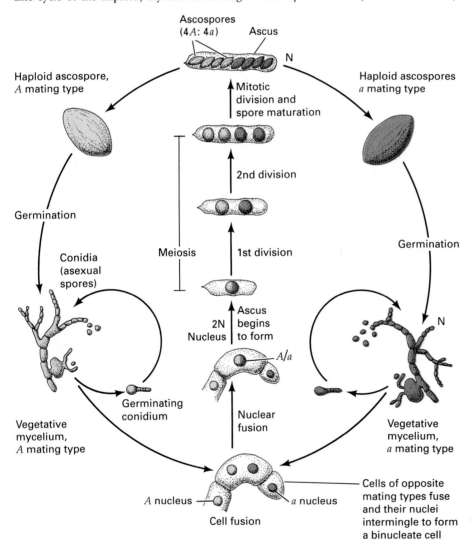

grows. *Neurospora* has several properties that make it useful for genetic and biochemical studies: It is a haploid (N) organism, so the effects of mutations may be seen directly, and its short life cycle facilitates studying the segregation of genetic defects.

Neurospora can be propagated vegetatively (asexually) to produce unlimited populations of strains with particular genotypes for study. To propagate *Neurospora* asexually, pieces of the mycelial growth or the asexual spores (conidia) can be used to inoculate a suitable growth medium to give rise to a new mycelium. This growth involves only mitotic division of the nuclei and the production of new cell mass.

Neurospora crassa can also reproduce by sexual means. There are two mating types (sexes, in a loose sense), called *A* and *a*. Both mating types look identical and can only be distinguished by the fact that strains of the *A* mating type do not mate with other *A* strains, and *a* strains do not mate with other *a* strains. Under normal conditions *N. crassa* reproduces asexually. However, if nutrients (particularly nitrogen sources) become seriously depleted in the growth medium and if both *A* and *a* mating strains are present, the sexual cycle is initiated. Cells of the two mating types fuse, followed by fusion of two haploid nuclei to produce an *A/a* diploid nucleus, which is the only diploid stage of the life cycle. The diploid nucleus immediately undergoes meiosis and produces four haploid nuclei (two *A* and two *a*) within an elongating sac called an *ascus*. A subsequent mitotic division results in a linear arrangement of eight haploid nuclei around which spore walls form to produce eight sexual ascospores (four *A* and four *a*). Each ascus, then, contains all the products of the initial, single meiosis. When the ascus is ripe, the ascospores (sexual spores) are shot out of the ascus and out of the fruiting body that encloses the ascus to be dispersed by wind currents. A particularly important feature of sexual reproduction in *N. crassa* is that all the products of a single meiosis are found together in the ascus and can be isolated and cultured for analysis. Moreover, the order of the four spore pairs within an ascus reflects exactly the orientation of the four chromatids of each tetrad at the metaphase plate in the first meiotic division: These are called **ordered tetrads**.

RECOMBINATION AT THE TETRAD STAGE IN *NEUROSPORA*. To determine whether crossing-over occurs at the tetrad stage of prophase I, or prior to meiosis at the two-chromatid phase in interphase, crosses are made between haploid strains of *Neurospora* that differ genetically. A theoretical example is shown in Figure 5.10 (p. 136). The two strains have the following genetic constitution: One of them, of mating

type *A*, is unable to synthesize the amino acid methionine, so that compound must be provided in the growth medium in order for the strain to grow. This Met⁻ strain (here the symbol refers to the phenotype) is wild-type for all other genes. The other strain, of mating type *a*, can synthesize methionine but is unable to synthesize another amino acid, histidine. The latter strain is referred to as the His⁻ strain. Genotypically, the Met⁻ strain is *met his⁺*, while the latter is *met⁺ his*. Both genes are in the same linkage group; that is, on the same chromosome.

The diploid nucleus produced from the mating is heterozygous for the two genes. We can now make some theoretical predictions. In simplified terms, in meiosis, crossing-over will occur between the two genes at a frequency that is a function of the distance between them on the chromosome. Recombinant progeny will be produced with the haploid genotypes of *met⁺ his⁺* and *met his*. The former is wild-type for both genes, and the latter is a double mutant strain that requires both methionine and histidine in the growth medium in order to grow. There is a relatively easy way to test the progeny strains for their phenotypes (which are equivalent to their genotypes since the organism is haploid). It is only necessary to determine whether they can grow on an unsupplemented medium, a medium with methionine, one with histidine, or one with both methionine and histidine.

If no crossing-over occurs, the resulting ascus will contain four spores of each parental type. Figure 5.10a shows the consequences of crossovers occurring at the two-strand stage, that is, prior to meiosis in interphase: The resulting asci will contain only recombinant spores (four of each type). If, however, crossing-over occurs at the four-strand stage, the results would be as diagramed in Figure 5.10b. Of the four chromatids, crossing-over occurs between just two, one from each parental strain. Those chromatids become recombinant for the gene markers and give rise to four recombinant spores in the ascus. The other two chromatids, not involved in the crossover, give rise to four parental spores in the same ascus. The overall result is an ascus with four parental and four recombinant progeny spores, representing all four possible phenotypic classes. Asci like this are found routinely in such experiments.

These results provide unequivocal proof that crossing-over occurs at the tetrad stage of prophase I and involves only two of the four strands for a given crossover event. This has to be the case because chromosomes enter meiosis already replicated into two chromatids; only later recombination events could produce such results. Furthermore, the results showed that crossing over is a reciprocal process.

~ FIGURE 5.10

Experiment showing that crossing-over occurs at the four-chromatid stage of meiosis; two haploid strains with different genetic markers (*met* and *his*) are crossed. (a) If a single crossover occurred at the two-chromatid stage (i.e., in interphase, before chromosome duplication), the resulting ascus would contain only the two types of recombinant ascospores. This result differs from experimental findings that asci resulting from a single crossover contain both types of parental ascospores and both types of recombinant ascospores. (b) If a single crossover occurred at the four-chromatid (tetrad) stage, an ascus would be produced that contains both types of parental ascospores and both types of recombinant ascospores. This outcome matches observed results.

(a) **Production of progeny spores if crossing-over occurs at two-chromatid stage prior to meiosis**

(b) **Production of progeny spores if crossing-over occurs at four-chromatid stage of meiosis**

KEYNOTE

Crossing-over occurs at the four-chromatid (tetrad) stage in prophase I in meiosis. The evidence for this principle came from the analysis of genetic experiments with the haploid, mycelial fungus *Neurospora crassa,* in which all the products of a single meiotic event are contained within a structure called the ascus. The crucial result was that a single ascus contained two spores with the parental genotypes and two spores with the recombinant genotypes.

LOCATING GENES ON CHROMOSOMES: MAPPING TECHNIQUES

So far we have learned two significant concepts. The first, derived principally from Morgan's experiments with *Drosophila,* is that genetic recombinants result from crossing-over between homologous chromosomes. More specifically, the number of genetic recombinants produced is characteristic of the two linked genes involved. The second concept, derived from genetic studies with the fungus *Neurospora crassa,* is that crossing-over takes place at the tetrad stage in prophase I of meiosis and involves only two of the four chromatids for a given crossing-over event. We will now examine how genetic experiments can be used to determine the relative position of genes on chromosomes in eukaryotic organisms. This process is called **genetic mapping.**

Detecting Linkage Through Testcrosses

Before beginning any experiments to construct a **genetic map** (also called a **linkage map**), geneticists must show that the genes under consideration are linked. If they are not linked, then the genes are assorting independently. Therefore, the simplest way to test for linkage is to analyze the results of crosses to see whether the data deviate significantly from those expected by independent assortment.

In Chapter 2 we showed that in diploid organisms, if two allelic pairs assort independently, and if there is complete dominance of one allele of each pair, then the progeny from a self will show a $9:3:3:1$ phenotypic ratio. However, this cross is not the best one to use to check for linkage. The testcross (a cross of an individual with an unknown genotype with an individual homozygous recessive for all genes involved) is better because the distribution of phenotypes is the result of

segregation events in only one of the parents; the other parent contributes only recessive alleles to the progeny and those alleles are not phenotypically expressed. We saw in Chapter 2 that a testcross between $a^+/a\ b^+/b$ and $a/a\ b/b$, where genes a and b are unlinked, gives a $1:1:1:1$ ratio of the four possible phenotypic classes $a^+\ b^+:a^+\ b:a\ b^+:a\ b$. Since any *significant* deviation from this ratio in the direction of too many parental types and too few recombinant types would suggest that the two genes are linked, it is important to know how large a deviation must be in order to be considered "significant." The **chi-square test** can be used to make such a decision.

THE CHI-SQUARE TEST. The observed phenotypic ratios among progeny of a cross rarely match the expected or predicted ratios even when the hypothesis on which the expected ratios are based is correct. For example, none of the F_2s of Mendel's monohybrid crosses exhibited an exact $3:1$ ratio, although the observed ratios were close enough so that he was not bothered by the discrepancies. Observed results may not match the expected results for several reasons, such as inadequate sample size, sampling error, or decreased viability of some of the genotypes involved.

On the other hand, if the deviation between the observed and the expected result is large enough the data may indicate that the hypothesis might be wrong. A common way to decide is to use a statistical test called the **chi-square** (χ^2) **test**, which, as used in the analysis of genetic data, is, essentially, a *goodness-of-fit test*.

To illustrate the use of the chi-square test, we will analyze the progeny data from a testcross involving fruit flies (see Question 5.7). In *Drosophila,* b is a recessive autosomal mutation which, when homozygous, results in black body color, and vg is a recessive autosomal mutation which, when homozygous, results in flies with vestigial (short, crumpled) wings. Wild-type flies have grey bodies and long, uncrumpled (normal) wings. True breeding black, normal ($b/b\ vg^+/vg^+$) flies were crossed with true-breeding grey, vestigial ($b^+/b^+\ vg/vg$) flies. The F_1 grey, normal ($b^+/b\ vg^+/vg$) flies were testcrossed to black, vestigial ($b/b\ vg/vg$) flies. The progeny data were

283	grey, normal
1294	grey, vestigial
1418	black, normal
241	black, vestigial
Total flies	3236

We hypothesize that the two genes are unlinked and use the chi-square test to test the hypothesis, as shown in Table 5.1, p. 138.

~ TABLE 5.1

Chi-Square Test Example

(1) PHENOTYPES	(2) OBSERVED NUMBER (o)	(3) EXPECTED NUMBER (e)	(4) d	(5) d^2	(6) d^2/e
grey, normal	283	809	−526	276,676	342.00
grey, vestigial	1294	809	485	235,225	290.76
black, normal	1418	809	609	370,881	458.44
black, vestigial	241	809	−568	322,624	398.79
Total	3236	3236			1,490.00

(7) $\chi^2 = 1,490.00$ (8) df = 3

If the two genes are unlinked, then a testcross should result in a 1:1:1:1 ratio of the four phenotypic classes. First, in column 1 we list the four phenotypes expected in the progeny of the cross. Then we list the observed (o) numbers for each phenotype, using actual numbers and not percentages or proportions (column 2). Next, we calculate the expected (e) number for each phenotypic class, given the total number of progeny (3236) and the hypothesis under evaluation (in this case 1:1:1:1 (column 3). Thus we list $1/4 \times 3236 = 809$, and so on. Now we subtract the expected number (e) from the observed number (o) for each class to find differences, called the deviation value (d). The sum of the d values is always zero (column 4).

In column 5 the deviation squared (d^2) is computed by multiplying each deviation value in column 4 by itself. In column 6 the deviation squared is then divided by the expected number (e). The chi-square value, χ^2 (item 7 in the table), is the total of the four values in column 6. In our example χ^2 is 1490.00. The general formula is

$$\chi^2 = \Sigma \frac{d^2}{e}$$

The last value in the table, item 8, is the degrees of freedom (df) for the set of data. The degrees of freedom in a test involving n classes are usually equal to $n - 1$. That is, if the total number of progeny (3236 in our example) is divided among n classes (four phenotypic classes in the example), then once the expected numbers have been computed for $n - 1$ classes (three in our example), the expected number of the last class is set. Thus in our example there are only three degrees of freedom in the analysis.

The χ^2 value and the degrees of freedom are next used with a table to determine the probability (P) that the deviation of the observed values from the expected values is due to chance. Generally, if the probability of obtaining the observed χ^2 values is greater than 5 in 100 ($P > 0.05$), the deviation is considered not statistically significant and could have occurred by chance alone. Thus the hypothesis being tested could apply to the data being analyzed. If $P \leq 0.05$, we consider the deviation from the expected values to be statistically significant and not due to chance alone; the hypothesis may well be invalid. If $P \leq 0.01$, the deviation is highly significant and some nonchance factor is involved; the hypothesis is almost certainly invalid.

The P value for a set of data is obtained from tables of χ^2 values for various degrees of freedom. Table 5.2 presents part of a table of chi-square probabilities. As the number of degrees of freedom increases, a higher χ^2 value is acceptable in terms of corresponding to a good fit (that is, high) P value. For our example, $\chi^2 = 1,490.00$ with three degrees of freedom, the P value is much lower than 0.001; in fact it is not on the table. This is interpreted to mean that, if the hypothesis was correct, much lower than 0.001 times out of 100, we could expect χ^2 values of this magnitude. Thus, we must consider the independent assortment hypothesis to be invalid, and we must think of an alternative hypothesis, the simplest of which is that the genes are linked.

Let us consider that, in another chi-square analysis of a different set of data, we obtained $\chi^2 = 0.4700$ with three degrees of freedom. By looking up the value in Table 5.2, we see that the P value is greater than 0.90 and less than 0.95 ($0.90 < P < 0.95$). Thus from 90 to 95 times out of 100 we could expect χ^2 values of this

~ **TABLE 5.2**

Chi-Square Probabilities

					PROBABILITIES					
df	*0.95*	*0.90*	*0.70*	*0.50*	*0.30*	*0.20*	*0.10*	*0.05*	*0.01*	*0.001*
1	0.004	0.016	0.15	0.46	1.07	1.64	2.71	3.84	6.64	10.83
2	0.10	0.21	0.71	1.39	2.41	3.22	4.61	5.99	9.21	13.82
3	0.35	0.58	1.42	2.37	3.67	4.64	6.25	7.82	11.35	16.27
4	0.71	1.06	2.20	3.36	4.88	5.99	7.78	9.49	13.28	18.47
5	1.15	1.61	3.00	4.35	6.06	7.29	9.24	11.07	15.09	20.52
6	1.64	2.20	3.83	5.35	7.23	8.56	10.65	12.59	16.81	22.46
7	2.17	2.83	4.67	6.35	8.38	9.80	12.02	14.07	18.48	24.32
8	2.73	3.49	5.53	7.34	9.52	11.03	13.36	15.51	20.09	26.13
9	3.33	4.17	6.39	8.34	10.66	12.24	14.68	16.92	21.67	27.88
10	3.94	4.87	7.27	9.34	11.78	13.44	15.99	18.31	23.21	29.59
11	4.58	5.58	8.15	10.34	12.90	14.63	17.28	19.68	24.73	31.26
12	5.23	6.30	9.03	11.34	14.01	15.81	18.55	21.03	26.22	32.91
13	5.89	7.04	9.93	12.34	15.12	16.99	19.81	22.36	27.69	34.53
14	6.57	7.79	10.82	13.34	16.22	18.15	21.06	23.69	29.14	36.12
15	7.26	8.55	11.72	14.34	17.32	19.31	22.31	25.00	30.58	37.70
20	10.85	12.44	16.27	19.34	22.78	25.04	28.41	31.41	37.57	45.32
25	14.61	16.47	20.87	24.34	28.17	30.68	34.38	37.65	44.31	52.62
30	18.49	20.60	25.51	29.34	33.53	36.25	40.26	43.77	50.89	59.70
50	34.76	37.69	44.31	49.34	54.72	58.16	63.17	67.51	76.15	86.66

<--------------------|-------------------->
                 Accept | Reject
                at 0.05 level

SOURCE: Taken from Table IV of Fisher and Yates, *Statistical Tables for Biological, Agricultural and Medical Research*, published by Oliver & Boyd, Edinburgh, and by permission of the authors and publishers.

magnitude or greater. This *P* value, being greater than 0.05, indicates that the results are consistent with the hypothesis being tested.

THE CONCEPT OF A GENETIC MAP. The data obtained by Morgan from *Drosophila* crosses indicated that the frequency of crossing-over (and hence of recombinants) for linked genes is characteristic of the gene pairs involved: For the X-linked genes for white (*w*) and miniature (*m*) the frequency of crossing-over is 36.9 percent, and for the white and yellow (*y*) genes it is 1.3 percent. Moreover, the frequency of recombinants

for two linked genes is approximately constant, regardless of how the alleles involved are arranged on the chromosomes. In an individual doubly heterozygous for the *w* and *m* alleles, for example, the alleles can be arranged in two ways:

$$\frac{w^+ \ m^+}{w \ m} \quad \text{or} \quad \frac{w^+ \ m}{w \ m^+}$$

In the arrangement on the left, the two wild-type alleles are on one homolog and the two recessive mutant alleles are on the other homolog, an arrangement called

coupling (or the *cis* configuration). Crossing-over between the two loci produces $w^+ m$ and $w m^+$ recombinants. (**Loci** is the plural of **locus,** the position of a gene on a genetic map.) In the arrangement on the right, each homolog carries the wild-type allele of one gene and the mutant allele of the other gene, an arrangement called **repulsion** (or the *trans* configuration). Crossing-over between the two genes produces $w^+ m^+$ and $w m$ recombinants. While the actual phenotypes of the recombinant classes are different for the two arrangements, the percentage of recombinants among the total progeny will be 36.9 percent in each case.

Morgan thought that the characteristic crossover frequencies for linked genes might be related to the physical distances separating the genes on the chromosome. In 1913 a student of Morgan's, Alfred Sturtevant, devised the testcross method to analyze the linkage relationships between genes. He suggested that the percentage of recombinants (produced by crossovers in gamete production) could be used as a quantitative measure of the genetic distance between two gene pairs on a genetic map (Figure 5.11). This distance is measured in **map units** (mu). A crossover frequency of 1 percent between two genes equals 1 map unit. That is, one map unit is the distance between gene pairs for which 1 product out of 100 is recombinant. The map unit is sometimes called a **centi-Morgan** (cM) in honor of Morgan.

Sturtevant's concept of genetic map distance led him to the important discovery that on the map the genetic distances between a series of linked genes are additive. Thus the genes on a chromosome can be represented by a one-dimensional, genetic map that shows in linear order the genes belonging to the chromosome. Crossover and recombination values delineate a linear order of the genes on a chromosome and to provide information about the genetic distance between any two genes. Genetically speaking, the greater the crossover frequency between two genes, the farther apart they are. Thus, in Figure 5.12, the probability of recombination occurring between genes A and B is much less than between genes B and C, because A and B are closer together than are B and C.

The first genetic map ever constructed was based on recombination frequencies from *Drosophila* crosses involving the sex-linked genes w, m, and y discussed earlier, where w gives white eyes, m gives miniature wings, and y gives yellow body. From these mapping experiments, the recombination frequencies for the $w \times m$, $w \times y$, and $m \times y$ crosses were established as 32.6, 1.3, and 33.9 percent, respectively. The percentages are quantitative measures of the distance between the genes involved.

~ **FIGURE 5.11**

Percentage of recombinants as a quantitative measure of the genetic distance (map unit) between two genes pairs on a genetic map.

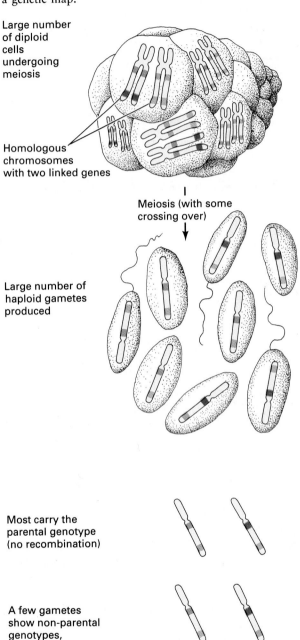

Large number of diploid cells undergoing meiosis

Homologous chromosomes with two linked genes

Meiosis (with some crossing over)

Large number of haploid gametes produced

Most carry the parental genotype (no recombination)

A few gametes show non-parental genotypes, indicating recombination

The logic used in constructing a genetic map of these three genes is as follows: The genes must be arranged at different points on a line that best accommodate the data. The recombination frequencies show that w and y are closely linked and that m is quite a distance away from the other two genes. Since the $w-m$ genetic distance is less than the $y-m$ distance (as

~ FIGURE 5.12

Recombination between linked genes and map distance. The farther apart two genes are, the greater the number of possible sites for recombination. Thus, the probability of recombination occurring between genes A and B, for example, is much less than that between genes B and C. The percentage of recombinants can provide information about the relative genetic distance between two linked genes.

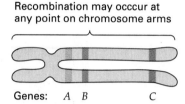

Recombination may occcur at any point on chromosome arms

Genes: A B   C

shown by the smaller percentage of the recombinants in the $w \times m$ cross), the order of genes must be $y-w-m$. Thus the three genes are ordered and spaced as follows, with the y gene assigned an arbitrary value of 0 and the other map units indicating genetic distance from y:

0   1.3   33.9   map units

y   w   m

Recombination frequencies are used to construct genetic maps in all gene-mapping experiments, whether the organism is a eukaryote or a prokaryote.

Gene Mapping by Using Two-Point Testcrosses

From Sturtevant's work we have seen that the percentage of recombinants resulting from crossing-over is used as a measurement of the genetic distance between two linked genes. Quite simply, by carrying out two-point testcrosses such as those shown in Figure 5.13 (p. 142), we can determine the relative numbers of parental and recombinant classes in the progeny. For autosomal recessives a double heterozygote is crossed with a doubly homozygous recessive mutant strain (Figure 5.13a). When the double heterozygous $a^+ b^+/ a\, b$ F_1 progeny from a cross of $a^+ b^+/a^+ b^+$ with $a\, b/a\, b$ are testcrossed with $a\, b/a\, b$, four phenotypic classes are found among the F_2 progeny. Two of these classes have the parental phenotypes $a^+ b^+$ and $a\, b$ and derive from zygotes in which no crossover had occurred. Since both classes were the result of the same no-crossover event, approximately equal numbers of these two types are expected.

The other two F_2 phenotypic classes have recombinant phenotypes $a^+ b$ and $a\, b^+$, which derive from zygotes in which a single crossover occurred between the chromosomes. Again, we expect approximately equal numbers of these two recombinant classes. Because a single crossover event occurs more rarely than no crossing over, an excess of parental phenotypes over recombinant phenotypes in the progeny of a testcross indicates linkage between the genes involved. For autosomal dominants a double heterozygote is crossed with a wild-type strain, since the recessive allele of a dominant mutant is the wild-type allele (Figure 5.13b). When the doubly heterozygous $A\, B/A^+ B^+$ F_1 progeny from a cross of $A\, B/A\, B$ with $A^+ B^+/A^+ B^+$ are testcrossed with $A^+ B^+/A^+ B^+$, four phenotypic classes are found among the F_2 progeny. Two of these classes have the parental phenotypes $A\, B$ and $A^+ B^+$ and derive from zygotes in which no crossover had occurred. The other two F_2 phenotypic classes have recombinant phenotypes $A\, B^+$ and $A^+ B$, and these derive from zygotes in which a single crossover occurred between the chromosomes. As for the autosomal recessives, an excess of parental phenotypes over recombinant phenotypes is expected.

Two-point testcrosses are essentially the same when sex-linked genes are involved. For X-linked recessives a double heterozygous female is crossed with a hemizygous male carrying the recessive alleles

$$\frac{+\ +}{a\ b} \times \xrightarrow{a\ b}$$

(Note that "+ +" here is a shorthand for the linked wild-type alleles $a^+ b^+$; cf. Box 3.1.) For X-linked dominants a doubly heterozygous female is crossed with a male carrying wild-type alleles on the X chromosome:

$$\frac{A\ B}{+\ +} \times \xrightarrow{+\ +}$$

(Here, since A and B are dominant mutations, "+ +" is a shorthand for the linked wild-type alleles $A^+ B^+$.) For both X-linked cases, the females can be crossed with males of any genotype. If only male progeny are analyzed, the contribution of the male's X chromosome to the progeny is ignored.

In all cases a two-point testcross should yield a pair of parental types that occur with about equal frequencies, and a pair of recombinant types that also occur with equal frequencies. The phenotypes will, of course, depend on the configuration of the two allelic pairs in the homologous chromosomes, that is, whether they are in coupling or repulsion. To get a count of the representatives of each progeny class (this will give the

~ Figure 5.13

Testcross to show that two genes are linked: (a) Genes a and b are recessive mutant alleles linked on the same autosome. A homozygous $a^+ b^+/a^+ b^+$ individual is crossed with a homozygous recessive $a b/a b$ individual, and the doubly heterozygous F$_1$ progeny ($a^+ b^+/a b$) are testcrossed with homozygous $a b/a b$ individuals; (b) Genes A and B are dominant mutant alleles linked on the same autosome. A homozygous dominant individual ($A B/A B$) is crossed with a wild-type $A^+ B^+/A^+ B^+$ (homozygous recessive) individual, and doubly heterozygous F$_1$ progeny $A B/A^+ B^+$ are testcrossed with the wild-type $A^+ B^+/A^+ B^+$.

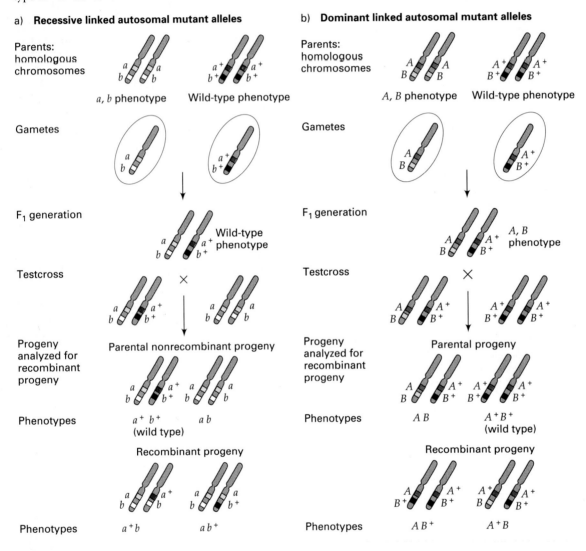

The value for the percentage of recombinants is usually directly converted into map units, so, for example, 20 percent recombination between genes a and b indicates that the two genes are 20 mu apart.

The two-point method of mapping is most accurate when the two genes examined are close together; when genes are far apart, other factors can influence the percentage of recombinant types), the following formula is used:

$$\frac{\text{number of recombinants}}{\text{total number of testcross progeny}} \times 100 = \text{percent recombinants}$$

production of recombinant progeny. When we are drawing a map of the genes on a chromosome, we must take these factors into consideration. Large numbers of progeny must also be counted (scored) to ensure a high degree of accuracy. From Sturtevant's work we have already seen the logic behind constructing an actual linkage map from recombination frequencies involving all possible pairwise crosses of the genes under study. From mapping experiments carried out in all types of organisms, we know that genes are linearly arranged in linkage groups. There is a one-to-one correspondence of linkage groups and chromosomes, so the sequence of genes on the linkage group reflects the sequence of genes on the chromosome.

Generating a Genetic Map

A genetic map is generated from estimating the number of times a crossover event occurred in a particular segment of the chromosome out of all meioses examined (in the form of progeny organisms). Since in many cases the probability of a crossing-over event is not uniform along a chromosome, we must be cautious about how far we extrapolate the genetic map (which derives from data produced by genetic crosses) to the chromosomal map (which derives from cytogenetic determinations of the locations of genes along the chromosome itself). Nonetheless, a simple working hypothesis is to consider crossovers as being randomly distributed along the chromosome.

The recombination frequencies observed between genes may also be used to predict the outcome of genetic crosses. For example, a recombination frequency of 20 percent between genes indicates that the probability of a crossover occurring in that region of the chromosome in a meiosis is 0.1. This is because each crossover event produces two recombinants (a reciprocal pair). In simple terms, if we assume that crossing over occurs randomly along a chromosome, more than one crossover may occur in a given region in a meiosis. The probabilities of these so-called **multiple crossovers** can readily be computed using the product rule (Chapter 2): the probability of two independent events occurring simultaneously is equal to the product of the individual probabilities of two single events. That is, the probability of two crossovers (called a **double crossover**) is

$$0.2 \times 0.2 = 0.04$$

The probability of three crossovers (a triple crossover) is

$$0.2 \times 0.2 \times 0.2 = 0.008$$

and so on.

For any testcross, the percentage of recombinants in the progeny cannot exceed 50 percent. That is, if the genes are assorting independently, an equal number of recombinants and parentals is expected in the progeny, so the frequency of recombinants is 50 percent. With such a value in hand, we state that the two genes are unlinked. Genes may be unlinked (that is, show 50 percent recombination) in two ways. First, the genes may be on different chromosomes, a case we discussed before. Second, *the genes may be on the same chromosome but lie so far apart that many different crossovers occur between them.*

The second case can be explained by referring to Figure 5.14 (p. 144), which shows the consequences of single- and double-crossovers on the production of parental and recombinant chromatids. The loci are far apart on the same chromosome; as a result, in a given meiosis at least a single (and usually more) crossover always occurs between the two loci. For single crossovers occurring between any pair of non-sister chromatids, the result is two parental and two recombinant chromatids; that is, for two loci 50 percent of the products are recombinant (Figure 5.14a).

Double crossovers can involve two, three, or all four of the chromatids (Figure 5.14b). For double crossovers involving the same two non-sister chromatids (called a *two-strand double crossover*), all four resulting chromatids are parental. For *three-strand double crossovers* (double crossovers involving three of the four chromatids), two parental and two recombinant chromatids result. For a *four-strand double crossover,* all four resulting chromatids are recombinant. Considering all of the possible double crossover patterns together, 50 percent of the products are recombinant for the two loci. Similarly, for any multiple number of crossovers between loci that are far apart, examination of a large number of meioses will show 50 percent of the resulting chromatids are recombinant. This is the reason for the recombination frequency limit of 50 percent exhibited by unlinked genes on the same chromosome.

When genes are closer together, however, there will be no crossovers between the two loci in some meioses, resulting in four parental chromatid products. In mapping linked genes, therefore, the map distance depends on the ratio of the meioses with no crossovers to meioses with any number of crossovers between the loci.

The point is, if two genes are unlinked, and therefore show 50 percent recombination, it does not necessarily mean that they are on different chromosomes. More data would be needed to determine whether the genes are on the same chromosome or different chro-

~ FIGURE 5.14

Demonstration that the recombination frequency between two genes located far apart on the same chromosome cannot exceed 50 percent. (a) Single crossovers produce one-half parental and one-half recombinant chromatids; (b) Double crossovers (two-strand, three-strand, and four-strand) collectively produce one-half parental and one-half recombinant chromatids.

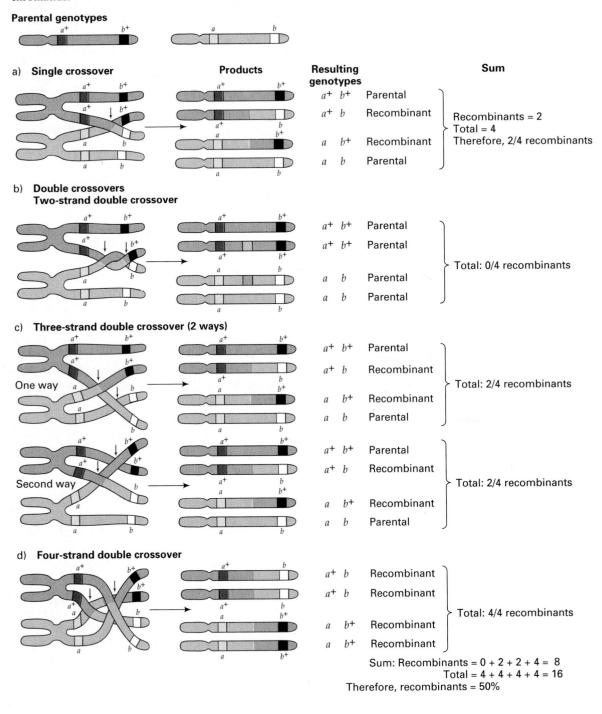

~ TABLE 5.3

Data for the F_2 Progeny of a Cross Between a Yellow-bodied, White-eyed, Miniature-winged
($y\ w\ m/y\ w\ m$) Mutant Female and a Wild-Type ($+\ +\ +/Y$) Male *Drosophila*

PHENOTYPES				CROSSOVERS		
BODY COLOR	EYE COLOR	WING SIZE	NUMBER	BODY COLOR AND EYE COLOR	EYE COLOR AND WING SIZE	BODY COLOR AND WING SIZE
normal	red (normal)	long (normal)	758	—	—	—
yellow	white	miniature	700	—	—	—
normal	red	miniature	401	—	401	401
yellow	white	long	317	—	317	317
normal	white	miniature	16	16	—	16
yellow	red	long	12	12	—	12
normal	white	long	1	1	1	—
yellow	red	miniature	0	0	0	—
		Total	2205	29	719	746
		Percentage	100	1.3	32.6	33.8

mosomes. One way to find out is to map a number of other genes in the linkage group. For example, if a and m show 50 percent recombination, perhaps we will find that a shows 27 percent recombination with e, and e shows 26 percent recombination with m. This result would indicate that a and m are in the same linkage group.

Double Crossovers

Data for the F_2 progeny of a cross between a yellow-bodied, white-eyed, miniature-winged female and a normal male *Drosophila* are shown in Table 5.3. We have already shown that the gene order is $y-w-m$. The data show 29 exchanges between the genes for body color and eye color and 719 exchanges between genes for eye color and wing size, giving a total of 748. However, only 746 exchanges are apparent between the body color and wing size genes. The discrepancy is due to the single normal-bodied, white-eyed, normal-winged fly. In the production of the egg from which this fly developed, two exchanges (a double crossover) must have occurred between the X chromosomes, with the result that the body color and wing size genes (normal

body color and normal-length wings) appeared together, as in one grandparent, although the intervening gene (for white eyes) was derived from the other grandparent. This example shows that large crossover frequencies give an inaccurate map. Unless intervening genes are available, there is no way of detecting double crossovers (or even-numbered multiple crossovers).

To see the significance of the mechanics of double crossovers in genetic-mapping experiments, consider a hypothetical case of two allelic pairs (a^+/a and b^+/b) in coupling and separated by quite a distance on the same chromosome. Figure 5.15a, top (p. 146) shows that a single crossover results in recombination of the two allelic pairs. Figure 5.15b (p. 146) shows that a two-strand crossover involving two of the four chromatids does not result in recombination of the allelic pairs, so only parental progeny result. Exactly the same would occur if there were no crossing-over between the two chromatids in this region. However, the percentage of crossing-over between genes *is* a measure of the distance between them. Therefore, since the double cross-over in Figure 5.15b, top, did not generate recombinants, two crossover events will be uncounted, and the estimate of map distance between genes a and b will be

~ FIGURE 5.15

Progeny of single and double crossovers. (In this figure, only the two chromatids involved in crossing over are shown.) (a) A single crossover between linked genes generates recombinant gametes. (b) Top: A double crossover between linked genes gives parental gametes. Thus inaccurate map distances between genes result, since not all crossovers can be accounted for. Bottom: A possible solution to the double-crossover problem. The presence of a third allelic pair between the two allelic pairs of Figure 5.15a enables us to detect the double-crossover event. In a double crossover, the middle gene will change positions relative to the outside genes.

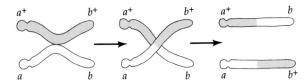

a) Single crossover

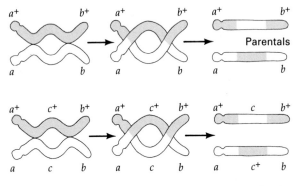

b) Double crossover

Parentals

Gametes are recombinant for the c^+/c gene pair relative to the a^+/a and b^+/b gene pairs

low, depending on how many double crossovers occur per meiosis in the region being examined.

In sum, genetic map distance is derived from the average frequency of crossing over occurring between linked genes, whereas recombination frequency is a measure of the crossovers that result in the detectable exchange of the gene markers. If no multiple crossovers occurred between genes, there would be a direct linear relationship between genetic map distance and recombination frequency. In practice, such a relationship is seen only when genetic map distances are small; that is, when genes are closely linked. As genes become farther apart, the chances of multiple crossovers between them increases, and there is no longer a linear relationship between map distance and recombination frequency. As a result, it is difficult to obtain an accurate measure of map distance when multiple crossovers are involved.

In many systems, it has been possible to derive mathematical formulas called **mapping functions**, which are used to correct the observed recombination values for the incidence of multiple crossovers. Mapping functions all require one to make some basic assumptions. Hence the usefulness of applying the mapping functions depends upon the validity of the assumptions. In operation, mapping functions convert experimentally-observed recombination frequencies into genetic map distances that are additive by correcting for the effects of multiple crossovers, which can otherwise cause one to underestimate genetic distance.

Three Point Cross

How can we avoid this pitfall? Double crossing-over occurs quite rarely within distances of 5 mu or less, and it occurs only sometimes in greater physical distances for certain chromosomal regions. Consequently, one way to get accurate map distances is to study linked genes closely. Another efficient way to obtain such data is to use a **three-point testcross**, which involves three genes within a relatively short section of a chromosome. (The evidence for the closely linked nature of three genes would derive from separate two-point testcrosses and the drawing of a preliminary genetic map.)

The potential advantage of the three-point testcross can be seen in the theoretical case if we have a third allelic pair c^+/c between the a^+/a and b^+/b allelic pairs of Figure 5.15a, as diagramed in Figure 5.15b, bottom. A two-strand double-crossover event between a and b might be detected by the recombination of the c^+/c allelic pair in relation to the other two allelic pairs. Of course, we can now ask: What if the double crossover occurred between genes a and c? Nonetheless, the principle holds: Accurate mapping data can be obtained when the genes are relatively close together. In fact, three-point testcrosses are routinely used as an efficient way of mapping genes both for their order in the chromosome and for the distances between them, as we will see in the next section.

$\mathcal{K}$EYNOTE

The map distance between two genes is based on the frequency of recombination between the two genes (an approximation of the frequency of crossovers between the two genes). By using appropriate genetic markers in crosses, geneticists are able to compute the recombination frequency between genes in chromosomes as the percentage of progeny showing the reciprocal recombinant phenotypes. The more closely the recombination frequency parallels the crossover frequency, the closer the genes are. But when the genetic distance increases between genes, the incidence of multiple crossovers, and particularly double crossovers, causes the recombination frequency to be an underestimate of the crossover frequency and hence of the map distance. Mapping functions may be used to correct for the effects of multiple crossovers and hence to give a more accurate estimate of map distance. _____

Mapping Chromosomes by Using Three-Point Testcrosses

A three-point testcross is preferable to a two-point testcross for mapping purposes because it aids in the detection of double crossovers and because the relative order of the genes on the chromosome can be established from a single cross. In diploid organisms the three-point testcross is a cross of a triple heterozygote with a triply homozygous recessive. If the mutant genes in the cross are all recessive, a typical three-point testcross might be

$$\frac{+\ +\ +}{a\ b\ c} \times \frac{a\ b\ c}{a\ b\ c}$$

If any of the mutant genes are dominant, the homozygous parent in the testcross will carry the wild-type allele of these genes. For example, a testcross of a strain heterozygous for two recessive mutations (a and c) and one dominant mutation (B) might be

$$\frac{+\ +\ +}{a\ B\ c} \times \frac{a\ +\ c}{a\ +\ c}$$

In a testcross involving sex-linked genes, the female is the heterozygous strain (assuming that the female is the homogametic sex) and the male is hemizygous for the recessive markers.

Three-point mapping is also done in haploid eukaryotic organisms, but in this case a testcross is not needed. Thus in *Neurospora crassa* we might cross two haploid strains to generate the triply heterozygous diploid cell. A cross of an $a\ b\ c$ strain with a $+ + +$ strain, for instance, would give a $+ + +/a\ b\ c$ diploid cell. That cell goes through meiosis to generate haploid spores from which progeny *Neurospora* cultures grow. Since the cells are haploid, the phenotypes of the progeny are determined directly by the genotypes.

Let us suppose we have a (hypothetical) plant in which there are three linked genes, all of which control fruit phenotypes. A recessive allele p of the first gene determines purple fruit color, in contrast to the white color of the wild type. A recessive allele r of the second gene results in a round fruit shape, as compared with an elongated fruit in the wild type. A recessive allele j of the third gene gives a juicy fruit instead of the wild-type dry fruit. The task before us is to determine the order of the genes on the chromosome and to determine the map distance between the genes. To do so, we must carry out the appropriate testcross of a triple heterozygote ($+ + +/p\ r\ j$) with a triply homozygous recessive (*p r j/p r j*) and then count the different phenotypic classes in the progeny (Figure 5.16).

Three genes are heterozygous in the cross. For each, two different phenotypes occur in the testcross progeny. Therefore, for the three genes a total of $(2)^3 = 8$ phenotypic classes will appear in the progeny, representing all possible combinations of phenotypes. In an actual experiment not all the phenotypic classes may be generated. Nevertheless, the absence of a phenotypic class is important information, and the experimenter should enter a 0 in the class for which no progeny are found.

ESTABLISHING THE ORDER OF THE GENES. The first step in finding the genetic map distances of the three genes is to determine the order in which the genes are located on the chromosome; that is, which gene is in the middle? Each progeny type is generated from a

~ FIGURE 5.16

Theoretical analysis of a three-point mapping, showing the testcross used and the resultant progeny.

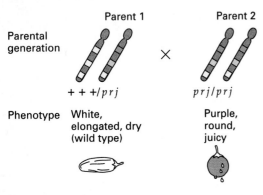

	Class	Phenotype		Number	Genotype of gamete from heterozygous parent
Parental phenotypes	1	Wild type (white, elongated, dry)		179	+ + +
	2	purple, round, juicy		173	*p r j*
Recombinant phenotypes	3	purple, elongated, dry		52	*p* + +
	4	white, round, juicy		46	+ *r j*
	5	purple, elongated, juicy		22	*p* + *j*
	6	white, round, dry		22	+ *r* +
	7	white, elongated, juicy		4	+ + *j*
	8	purple, round, dry		2	*p r* +

Total = 500

zygote formed by the fusion of the gametes from each parent. One parent carries the recessive alleles for all three genes; the other is heterozygous for all three genes. Therefore, the phenotype of the resulting individual is determined by the alleles in the gamete from the triply heterozygous parent; the gamete from the other parent will carry only recessive alleles. We know from the genotypes of the original parents that all three genes are in coupling. Since the heterozygous parent in the testcross was $+ + +/p\ r\ j$, classes 1 and 2 in Figure 5.16 are parental progeny: Class 1 is produced by the fusion of a $+ + +$ gamete with a $p\ r\ j$ gamete from the triply homozygous recessive parent. Class 2 is produced by the fusion of a $p\ r\ j$ gamete from the triply heterozygous parent and a $p\ r\ j$ gamete from the triply homozygous recessive parent. These classes are generated from meiosis in which no crossing-over occurs on the region of the chromosome in which the three genes are located.

The progeny classes deriving from a double crossover involving the same two chromatids between the loci can often be found by inspecting the numbers of each phenotypic class. Since the frequency of a double crossover in a region is expected to be lower than the frequency of a single crossover, *double-crossover gametes are the least frequent to be produced.* Thus to identify the double-crossover progeny, we can examine the progeny to find the pair of classes that have the lowest number of representatives. In Figure 5.16 classes 7 and 8 are such a pair. The genotypes of the gametes from the heterozygous parent that give rise to these phenotypes are $+ + j$ and $p\ r\ +$.

Referring now to Figure 5.15b, bottom, we see that a double crossover involving the same two chromatids changes the orientation of the gene in the center of the sequence (here, c^+/c) with respect to the two flanking allelic pairs. Therefore genes p, r, and j must be arranged in such a way that the center gene switches to give classes 7 and 8. To determine the arrangement, we must check the relative organization of the genes in the parental heterozygote so that it is clear which genes are in coupling and which are in repulsion. In this example the parental (noncrossover) gametes are $+ + +$ and $p\ r\ j$, so all are in coupling. The double-crossover gametes are $+ + j$ and $p\ r\ +$, so the only possible gene order that is compatible with the data is $p\ j\ r$, with the genotype of the heterozygous parent being $+ + +/p\ j\ r$. Figure 5.17 illustrates the generation of the double-crossover gametes from that parent.

CALCULATING THE MAP DISTANCES BETWEEN GENES.

Now the data can be rewritten as shown in Figure 5.18 to reflect the newly determined gene order.

~ FIGURE 5.17

Rearrangement of the three genes of Figure 5.16 to $p\ j\ r$. The evidence is that a double crossover involving the same two chromatids (as it is shown in this figure) generates the least frequent pair of recombinant phenotypes (in this case class 7 with six progeny and class 8 with zero progeny).

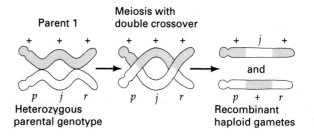

~ FIGURE 5.18

Rewritten form of the testcross and testcross progeny of Figure 5.16, based on the actual gene order $p\ j\ r$.

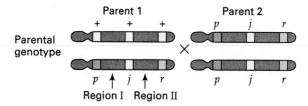

Testcross progeny

Class	Genotype of gamete from heterozygous parent	Number	Origin
1	$+ + +$	179	Parentals, no crossover
2	$p\ j\ r$	173	
3	$p\ + +$	52	Recombinants, single crossover region I
4	$+ j\ r$	46	
5	$p\ j\ +$	22	Recombinants, single crossover region II
6	$+ + r$	22	
7	$+ j\ +$	4	Recombinants, double crossover
8	$p\ + r$	2	

Total = 500

For convenience in this analysis, the region between genes p and j is called region I and that between genes j and r is called region II.

Map distances can now be calculated as described previously. The frequency of crossing-over (genetic recombination) is computed between two genes at a time. For the $p-j$ distance all the crossovers that occurred in region I must be added together. Thus we must consider the recombinant progeny resulting from a single crossover in that region (classes 3 and 4) and the recombinant progeny produced by a double crossover (classes 7 and 8). The latter must be included since each double crossover includes a crossover in region I and therefore is a recombination event between genes p and j. From Figure 5.18 there are 98 recombinant progeny in classes 3 and 4, and 6 in classes 7 and 8, giving a total of 104 progeny that result from a recombination event in region I. Since there are a total of 500 progeny, the percentage of progeny generated by crossing-over in region I is 20.8 percent, determined as follows:

$$\frac{\text{single crossovers in region I } (p-j) + \text{double crossovers}}{\text{total progeny}} \times 100$$

$$= \frac{98+6}{500} \times 100$$

$$= \frac{104}{500} \times 100$$

$$= 20.8\%$$

In other words, the distance between genes p and j is 20.8 mu. This map distance, which is quite large, is chosen mainly for illustration. If this cross were an actual cross, we would have to be cautious about that map distance value, since it is in the range where double crossovers might occur at a significant frequency and might go undetected. In an actual cross the value of 20.8 mu would probably underestimate the true distance.

The same method is used to obtain the map distance between genes j and r. That is, we calculate the frequency of crossovers in the cross that gave rise to progeny recombinant for genes j and r and directly relate that frequency to map distance. In this case all the crossovers that occurred in region II (see Figure 5.18) must be added (classes 5, 6, 7, and 8). The percentage of crossovers is calculated in the following manner:

$$\frac{\text{single crossovers in region II } (j-r) + \text{double crossovers}}{\text{total progeny}} \times 100$$

$$= \frac{44+6}{500} \times 100$$

$$= \frac{50}{500} \times 100$$

$$= 10.0\%$$

Thus the map distance between genes j and r is 10.0 map units.

In summary, a genetic map of the three genes in the example has been generated (Figure 5.19). The example has illustrated that the three-point testcross is an effective way of establishing the order of genes and of calculating map distances. In calculating map distances from three-point testcross data, the double-crossover figure must be added to each of the single-crossover figures, since in each case a double crossover represents single-crossover events occurring simultaneously in regions I and II in the same meiosis.

To compute the map distance between the two outside genes, we simply add the two map distances. Thus in the example the $p-r$ distance is $20.8 + 10.0 = 30.8$ mu. This map distance can be computed directly from the data by combining the two formulas discussed previously. So the $p-r$ distance is calculated in the following manner:

$$\text{distance} = \frac{\begin{array}{c}(\text{single crossovers in region I} \\ + \\ \text{double crossovers}) \\ + \\ (\text{single crossovers in region II} \\ + \\ \text{double crossovers})\end{array}}{\text{total progeny}} \times 100$$

$$= \frac{\begin{array}{c}(\text{single crossovers in region I} \\ + \\ \text{single crossovers in region II}) \\ + \\ 2(\text{double crossovers})\end{array}}{\text{total progeny}} \times 100$$

$$= \frac{98 + 44 + 2(6)}{500} \times 100$$

$$= 30.8 \text{ map units}$$

~ FIGURE 5.19

Genetic linkage map of the *p-j-r* region of the chromosome is computed from the recombination data of Figure 5.18.

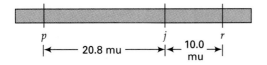

$\mathcal{K}$EYNOTE

The map distance between genes is calculated from the results of testcrosses between strains carrying appropriate genetic markers. The most accurate testcross is the three-point testcross using a diploid organism in which a triple heterozygote is crossed with a homozygous recessive for all three genes. The most accurate map distances are obtained when the three genes are reasonably close to one another, with, say 5–10 map units between each locus. The unit of genetic distance is the map unit (mu), which is the distance between gene pairs for which 1 product out of 100 is recombinant, or, in other words, a recombinant frequency of 1 percent is 1 mu. For illustration, the genetic map for *Drosophila melanogaster* is shown in Figure 5.20 (pp. 152–53).

INTERFERENCE AND COINCIDENCE. The map distances determined by three-point mapping are primarily useful in elaborating the overall organization of genes on a chromosome. In addition, the map distances obtained are useful in telling us a little about the recombination mechanisms themselves. For example, the map distance of 20.8 mu between gene *p* and *j* means that 20.8 percent of the gametes should result from crossing-over between the two gene loci. Similarly, the map distance of 10.0 mu between *j* and *r* indicates that 10.0 percent of the gametes should result from crossing-over between these two gene loci.

In the example of Figure 5.18, if crossing-over in region I is independent of crossing-over in region II, then the probability of simultaneous crossing-over (i.e., a double crossover) in the two regions is equal to the product of the probabilities of the two events occurring

separately. Thus if crossing-over in the two regions is independent, we have

$$0.208 \times 0.100 = 0.0208$$

or 2.08 percent double crossovers are expected to occur. However, only $6/500 = 1.2\%$ double crossovers occurred in this cross (in classes 7 and 8).

This discrepancy is not just a simple experimental error due to small sample size, misscored progeny, and so on. It is characteristic of such mapping crosses that double-crossover progeny typically do not appear as often as the map distances between the genes lead us to expect. Thus, once a crossing-over event has occurred in one part of the meiotic tetrad, the probability of another crossing-over event occurring nearby is reduced, most probably by physical interference caused by the breaking and rejoining chromatids. This phenomenon is called **chiasma interference** (also **chromosomal interference**).

In most organisms, there is no way to predict the extent of interference, since it tends to vary from chromosome to chromosome and even between segments of the same chromosome. But for the gene maps to be useful in predicting the outcome of crosses, we should know the magnitude of the interference throughout the map. The usual way to express the extent of interference is as a **coefficient of coincidence**, which is the ratio

$$\frac{\text{observed double-crossover frequency}}{\text{expected double-crossover frequency}}$$

And

$$\text{interference} = 1 - \text{coefficient of coincidence}$$

For the portion of the map in our example the coefficient of coincidence is

$$0.012/0.0208 = 0.577$$

The coefficient of coincidence typically varies between zero and one and may be interpreted as follows: A coincidence of one means that in a given region all double crossovers occurred that were expected on the basis of two independent events; there is no interference, so the interference value is zero. However, if the coefficient of coincidence is zero, none of the expected double crossovers occur. Here there is total interference, with one crossover completely preventing a second crossover in the region under examination. The interference value is one. These examples show that coincidence values and interference values are

~ FIGURE 5.20

Part of the genetic map of *Drosophila melanogaster,* with illustrations of several commonly used genetic markers.

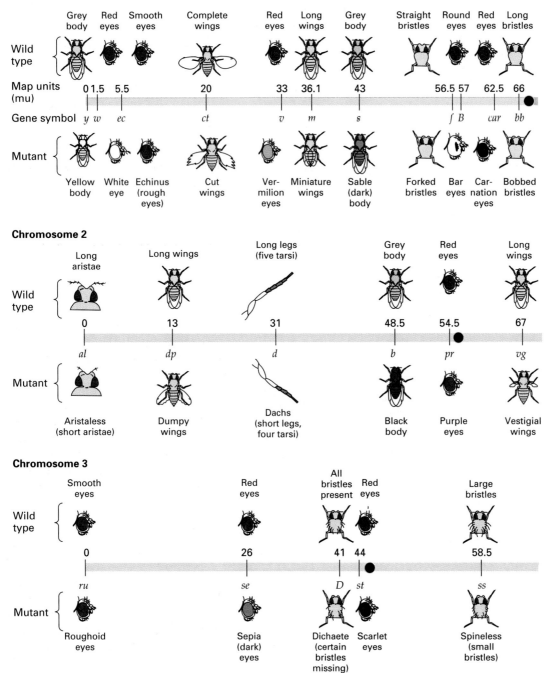

~ FIGURE 5.20 continued

Chromosome 4

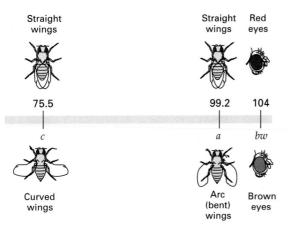

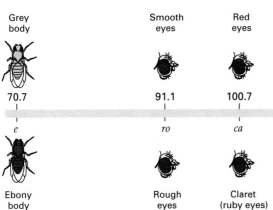

inversely related. In the example the coefficient of coincidence of 0.577 means that the interference value is 0.423. Only 57.7 percent of the expected double crossovers took place in the cross.

KEYNOTE

The occurrence of a chiasma between two chromatids may physically impede the occurrence of a second chiasma nearby, a phenomenon called chiasma interference. The extent of interference is expressed by the coefficient of coincidence, which is calculated by dividing the number of observed double crossovers by the number of expected double crossovers. The coefficient of coincidence ranges from zero to one, and the extent of interference is measured as 1 − coefficient of coincidence.

SUMMARY

In this chapter, linkage, crossing-over, and gene mapping in eukaryotes were discussed. The production of genetic recombinants results from physical exchanges between homologous chromosomes in meiosis. The site of crossing-over is called a chiasma, and the exchange of parts of chromatids is called crossing-over. Proof that genetic recombination occurs when crossing-over takes place in meiosis came from classical experiments in which recombinants for genetic markers occurred only when exchanges of cytological markers occurred. In most cases crossing-over is a reciprocal event'that, in eukaryotes, occurs at the four-strand stage in prophase I of meiosis.

Genetic mapping is the process of locating the position of genes in relation to one another on chromosomes. To map genes in this way, it is first necessary to show that genes are linked, a characteristic indicated by the fact that they do not assort independently in crosses. The map distance between two linked genes is then calculated based on the frequency of recombination between the two genes. The recombination frequency is an approximation of the frequency of crossovers between the two genes. The most accurate map distances are determined for genes that are closely linked because, as map distance increases, the incidence of multiple crossovers causes the recombination frequency to be an underestimate of the crossover frequency, and hence of the map distance.

ANALYTICAL APPROACHES FOR SOLVING GENETICS PROBLEMS

Q.1 In corn the gene for colored (C) seeds is completely dominant to the gene for colorless (c) seeds. Similarly, for the character of the endosperm (the part of the seed that contains the food stored for the embryo), a single gene pair controls whether the endosperm is full or shrunken. Full (Sh) is dominant to shrunken (sh). A true-breeding colored, full-seeded plant was crossed with a colorless, shrunken-seeded one. The F₁ colored, full plants were testcrossed to the doubly recessive type, that is, colorless and shrunken. The result was as follows:

colored, full	4032
colored, shrunken	149
colorless, full	152
colorless, shrunken	4035
Total	8368

Is there evidence that the gene for color and the gene for endosperm shape are linked? If so, what is the map distance between the two loci?

A.1 The best approach is to diagram the cross using gene symbols:

$$P: \quad \text{colored and full} \times \text{colorless and shrunken}$$
$$CC\ SS \qquad\qquad cc\ ss$$
$$\downarrow$$
$$F_1: \qquad \text{colored and full}$$
$$Cc\ Ss$$
$$\text{Testcross:} \quad \text{colored and full} \times \text{colorless and shrunken}$$
$$Cc\ Ss \qquad\qquad cc\ ss$$

If the genes were unlinked, a 1:1:1:1 ratio of colored and full:colored and shrunken:colorless and full:colorless and shrunken would result in the progeny of this testcross. By inspection, we can see that the actual progeny deviate a great deal from this ratio, showing a 27:1:1:27 ratio. If we did a chi-square test (using the actual numbers, not the percentages or ratios), we would see immediately that the hypothesis that the genes are unlinked is invalid, and we must consider the two genes to be linked in coupling. More specifically, the parental combinations (colored plus full and colorless plus shrunken) are more numerous than expected, while the recombinant types (colorless plus full and colored plus shrunken) are correspondingly less numerous than expected. This result

comes directly from the inequality of the four gamete types produced by meiosis in the colored and full F₁ parent.

To calculate the map distance between the two genes, we need to compute the frequency of crossovers in that region of the chromosome during meiosis. We cannot do that directly, but we can compute the percentage of recombinant progeny that must have resulted from such crossovers:

Parental types:	colored, full	4032
	colorless, shrunken	4035
		8067
Recombinant types:	colored, shrunken	149
	colorless, full	152
		301

This calculation gives about 3.6 percent recombinant types (i.e. $\frac{301}{8368} \times 100$) and about 96.4 percent parental types (i.e. $\frac{8067}{8368} \times 100$). Since the recombination frequency can be used directly as an indication of map distance, especially when the distance is small, we can conclude that the distance between the two genes is 3.6 mu (3.6 cM).

We would get approximately the same result if the two genes were in repulsion rather than in coupling. That is, the crossovers are occurring between homologous chromosomes, regardless of whether or not there are genetic differences in the two homologs that we, as experimenters, use as markers in genetic crosses. This same cross in repulsion would be as follows:

$$P: \text{colorless and full} \times \text{colored and shrunken}$$
$$\frac{c\ S}{c\ S} \qquad\qquad \frac{C\ s}{C\ s}$$
$$\downarrow$$
$$F_1: \qquad \text{colored and full}$$
$$\frac{C\ s}{c\ S}$$

The testcross of the F₁ with colorless and shrunken (cc ss) gives 638 colored and full (recombinant):21,379 colored and shrunken (parental):21,906 colorless and full (parental):672 colorless and shrunken (recombinant) with a total of 44,595 progeny. Thus 2.94 percent are recombinants, for a map distance between the two genes of 2.94 mu, a figure reasonably close to the results of the cross made in coupling.

Q.2 In the Chinese primrose, slate-colored flower (s) is recessive to blue flower (S); red stigma (r) is recessive to green stigma (R); and long style (l) is recessive to short style (L). The F_1 of a cross between two true-breeding strains, when testcrossed, gave the following progeny:

PHENOTYPE	NUMBER OF PROGENY
slate flower, green stigma, short style	27
slate flower, red stigma, short style	85
blue flower, red stigma, short style	402
slate flower, red stigma, long style	977
slate flower, green stigma, long style	427
blue flower, green stigma, long style	95
blue flower, green stigma, short style	960
blue flower, red stigma, long style	27
Total	3000

a. What were the genotypes of the parents in the cross of the two true-breeding strains?
b. Make a map of these genes, showing gene order, and distance between them.
c. Derive the coefficient of coincidence for interference between these genes.

A.2 a. With three gene pairs, eight phenotypic classes are expected, and eight are observed. The reciprocal pairs of classes with the most representatives are those resulting from no crossovers, and these pairs can tell us the genotypes of the original parents. The two classes are slate, red, long and blue, green, short. Thus the F_1 triply heterozygous parent of this generation must have been $S R L/s r l$, so the true-breeding parents were $S R L/S R L$ (blue, green, short) and $s r l/s r l$ (slate, red, long).

b. The order of the genes can be determined by inspecting the reciprocal pairs of phenotypic classes that represent the results of double crossing-over. These classes have the least numerous representatives, so the double-crossover classes are slate plus green plus short ($s R L$) and blue plus red plus long ($S r l$). The gene pair that has changed its position relative to the other two pairs of alleles is the central gene, S/s in this case.

Therefore the order of genes is $R S L$ (or $L S R$). We can diagram the F_1 testcross as follows:

$$\frac{R S L}{r s l} \times \frac{r s l}{r s l}$$

A single crossover between the R and S genes gives the green plus slate plus long ($R s l$) and red plus blue plus short ($r S L$) classes, which have 427 and 402 members, respectively, for a total of 829. The double-crossover classes have already been defined, and they yield 54 progeny. The map distance between R and S is given by the crossover frequency in that region, which is the sum of the single crossovers and double crossovers divided by the total number of progeny, then multiplied by 100 percent. Thus

$$\frac{829 + 54}{3000} \times 100\% = \frac{883}{3000} \times 100\%$$

$$= 29.43\% \text{ or } 29.43 \text{ map units}$$

With similar logic the distance between S and L is given by the crossover frequency in that region, which is the sum of the single-crossover and double-crossover progeny classes. The single-crossover progeny classes are green plus blue plus long ($R S l$) and red plus slate plus short ($r s L$), which have 95 and 85 members, respectively, for a total of 180. The map distance is given by

$$\frac{180 + 54}{3000} + 100\% = \frac{234}{3000} \times 100\%$$

$$= 7.8\% = 7.8 \text{ map units}$$

The data we have derived give us the following map:

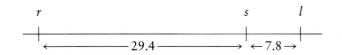

c. The coefficient of coincidence is given by

$$\frac{\text{frequency of observed double crossovers}}{\text{frequency of expected double crossovers}}$$

The frequency of observed double crossovers is $(54/3000) \times 100\% = 1.8\%$. The expected frequency of double crossovers is the product of the map distances between r and s and between s and l, that is, $29.4 \times 7.8\% = 2.3\%$. The coefficient of coincidence, therefore, is $1.8/2.3 = 0.78$. In other words, 78% of the expected double crossovers did indeed take place; there was only 22% interference.

QUESTIONS AND PROBLEMS

5.1 A cross $a^+a^+ \, b^+b^+ \times aa \, bb$ results in an F_1 of phenotype $a^+ \, b^+$; the following numbers are obtained in the F_2 (phenotypes):

$a^+ \, b^+$	110
$a^+ \, b$	16
$a \, b^+$	19
$a \, b$	15
Total	160

Are genes at the a and b loci linked or independent? What F_2 numbers would otherwise be expected?

5.2 In the F_2 of his cross of red-flowered × white-flowered *Pisum*, Mendel obtained 705 plants with red flowers and 224 with white.
a. Is this result consistent with his hypothesis of factor segregation, from which a 3:1 ratio would be predicted?
b. In how many similar experiments would a deviation as great as or greater than this one be expected? (Calculate χ^2 and obtain the approximate value of P from the table.)

*5.3 In corn a dihybrid for the recessive a and b is testcrossed. The distribution of the phenotypes was as follows:

$A \, B$	122
$A \, b$	118
$a \, B$	81
$a \, b$	79

Are the genes assorting independently? Test the hypothesis with a χ^2 test. Explain tentatively any deviation from expectation, and tell how you would test your explanation.

5.4 Why are crosses used for chromosome mapping in *Drosophila* set up with females heterozygous and males homozygous recessive?

5.5 The F_1 from a cross of $A \, B/A \, B \times a \, b/a \, b$ is testcrossed, resulting in the following phenotypic ratios:

$A \, B$	308
$A \, b$	190
$a \, b$	292
$a \, B$	210

What is the frequency of recombination between genes a and b?

*5.6 In rabbits the English type of coat (white-spotted) is dominant over non-English (unspotted), and short hair is dominant over long hair (Angora). When homozygous English, short-haired rabbits were crossed with non-English Angoras and the F_1 crossed back to non-English Angoras, the following offspring were obtained: 72 English and short-haired; 69 non-English and Angora; 11 English and Angora; and 6 non-English and short-haired. What is the map distance between the genes for coat color and hair length?

5.7 In *Drosophila* the mutant black (b) has a black body, and the wild type has a grey body; the mutant vestigial (vg) has wings that are much shorter and very crumpled when compared to the long wings of the wild type. In the following cross, the true-breeding parents are given together with the counts of offspring of F_1 females × black and vestigial males:

P black and normal × grey and vestigial
F_1 females × black and vestigial males

Progeny:		
grey, normal	283	recom
grey, vestigial	1294	P
black, normal	1418	P
black, vestigial	241	recom

From these data, calculate the map distance between the black and vestigial genes.

*5.8 A gene controlling wing size is located on chromosome 2 in *Drosophila*. The recessive allele vg results in vestigial wings when homozygous; the vg^+ allele determines long wings. A new eye mutation, which we will call "maroon-like," is isolated. Homozygous maroon-like (m/m) results in maroon colored eyes; the m^+ allele is bright red. The location of the m gene is unknown, and you are asked to design an experiment to determine whether m is located on chromosome 2.

You cross true-breeding virgin maroon females to true-breeding vg/vg males and obtain all wild-type F_1 progeny. Then you allow the F_1 to interbreed. As soon as the F_2 start to hatch, you begin to classify the flies, and among the first six newly-hatched flies, you find four wild type; one vestigial-winged and red-eyed fly; and one vestigial-winged, maroon-eyed fly. You immediately draw the conclusions that (1) maroon-like is not X-linked and (2) maroon-like is not linked to vestigial. Based on this small sample, how could you tell? On what chromosome is *m* located? (Hint: There is no crossing-over in *Drosophila* males.)

*5.9 Use the following two-point recombination data to map the genes concerned. Show the order and the length of the shortest intervals.

GENE LOCI	% RECOM-BINATION	GENE LOCI	% RECOM-BINATION
a,b	50	*b,d*	13
a,c	15	*b,e*	50
a,d	38	*c,d*	50
a,e	8	*c,e*	7
b,c	50	*d,e*	45

5.10 Use the following two-point recombination data to map the genes concerned. Show the order and the length of the shortest intervals.

LOCI	% RECOM-BINATION	LOCI	% RECOM-BINATION
a,b	50	*c,d*	50
a,c	17	*c,e*	50
a,d	50	*c,f*	7
a,e	50	*c,g*	19
a,f	12	*d,e*	7
a,g	3	*d,f*	50
b,c	50	*d,g*	50
b,d	2	*e,f*	50
b,e	5	*e,g*	50
b,f	50	*f,g*	15
b,g	50		

5.11 The following data are from Bridges and Morgan's work on the recombination between the genes black, curved, purple, speck, and vestigial in chromosome 2 of *Drosophila*. On the basis of the data, map the chromosome for these five genes as accurately as possible. Remember that determinations for short distances are more accurate than those for long ones.

GENES IN CROSS	TOTAL PROGENY	NUMBER OF RECOMBINANTS
black, curved	62,679	14,237
black, purple	48,931	3,026
black, speck	685	326
black, vestigial	20,153	3,578
curved, purple	51,136	10,205
curved, speck	10,042	3,037
curved, vestigial	1,720	141
purple, speck	11,985	5,474
purple, vestigial	13,601	1,609
speck, vestigial	2,054	738

*5.12 Genes *a* and *b* are linked, with 10 percent recombination. What would be the phenotypes, and the probability of each, among progeny of the following cross?

$$\frac{a\ b^+}{a^+\ b} \times \frac{a\ b}{a\ b}$$

*5.13 Genes *a* and *b* are sex-linked and are located 7 mu apart in the X chromosome of *Drosophila*. A female of genotype $a^+\ b/a\ b^+$ is mated with a wild type ($a^+\ b^+$) male.

a. What is the probability that one of her sons will be either $a^+\ b^+$ or $a\ b^+$ in phenotype?

b. What is the probability that one of her daughters will be $a^+\ b^+$ in phenotype?

5.14 *a* and *b* are linked autosomal genes whose recombination frequency in females is 5 percent; *c* and *d* are X-linked genes, located 10 map units apart. A homozygous dominant female is mated to a recessive male, and the daughters are testcrossed. Which of the following would you expect to observe in the testcross progeny?

a. Different ratios in males and females.

b. Nearly equal frequency of $a^+\ b\ c^+\ d^+$, $a^+\ b\ c\ d$, $a\ b^+\ c^+\ d^+$, and $a\ b^+\ c\ d$ classes.

c. Independent segregation of some genes with respect to others involved in the cross.

d. Double-crossover classes less frequent than expected because of interference between the two marked regions.

5.15 In maize the dominant genes *A* and *C* are both necessary for colored seeds. Homozygous recessive plants give colorless seed, regardless of the genes at the second locus. Genes *A* and *C* show independent segregation, while the recessive mutant gene waxy endosperm (*wx*) is linked with *C* (20 percent recombination). The dominant *Wx* allele results in starchy endosperm.

a. What phenotypic ratios would be expected when a plant of constitution *c Wx/C wx A/A* is testcrossed?

b. What phenotypic ratios would be expected when a plant of constitution *c Wx/C wx A/a* is testcrossed?

*5.16 Assume that genes *a* and *b* are linked and show 20 percent crossing-over.

a. If a homozygous *A B/A B* individual is crossed with an *a b/a b* individual, what will be the genotype of the F₁? What gametes will the F₁ produce and in what proportions? If the F₁ is testcrossed with a doubly homozygous recessive individual, what will be the proportions and genotypes of the offspring?

b. If, instead, the original cross is *A b/A b × a B/a B*, what will be the genotype of the F₁? What gametes will the F₁ produce and in what proportions? If the F₁ is testcrossed with a doubly homozygous recessive, what will be the proportions and genotypes of the offspring?

5.17 In tomatoes, tall vine is dominant over dwarf, and spherical fruit shape is dominant over pear shape. Vine height and fruit shape are linked, with a recombinant percentage of 20. A certain tall, spherical-fruited tomato plant is crossed with a dwarf, pear-fruited plant. The progeny are 81 tall, spherical; 79 dwarf, pear; 22 tall, pear; and 17 dwarf, spherical. Another tall and spherical plant crossed with a dwarf and pear plant produces 21 tall, pear; 18 dwarf, spherical; 5 tall, spherical; and 4 dwarf, pear. What are the genotypes of the two tall and spherical plants? If they were crossed, what would their offspring be?

*5.18 Genes a and b are in one chromosome, 20 mu apart; c and d are in another chromosome, 10 mu apart. Genes e and f are in yet another chromosome and are 30 mu apart. Cross a homozygous $A B C D E F$ individual with an $a b c d e f$ one, and cross the F_1 back to an $a b c d e f$ individual. What are the chances of getting individuals of the following phenotypes in the progeny?

a. $A B C D E F$
b. $A B C d e f$
c. $A b c D E f$
d. $a B C d e f$
e. $a b c D e F$

*5.19 Genes d and p occupy loci 5 map units apart in the same autosomal linkage group. Gene h is a separate autosomal linkage group and therefore segregates independently of the other two. What types of offspring are expected, and what is the probability of each, when individuals of the following genotypes are testcrossed:

a. $\dfrac{D\,P}{d\,p}\ \dfrac{h}{h}$

b. $\dfrac{d\,P}{D\,p}\ \dfrac{H}{h}$

5.20 A hairy-winged (h) *Drosophila* female is mated with a yellow-bodied (y), white-eyed (w) male. The F_1 are all normal. The F_1 progeny are then crossed, and the F_2 that emerge are as follows:

Females:	wild type	757
	hairy	243
Males:	wild type	390
	hairy	130
	yellow	4
	white	3
	hairy, yellow	1
	hairy, white	2
	yellow, white	360
	hairy, yellow, white	110

Give genotypes of the parents and the F_1, and note the linkage relations and distances, where appropriate.

5.21 In the Maltese bippy amiable (A) is dominant to nasty (a), benign (B) is dominant to active (b), and crazy (C) is dominant to sane (c). A true-breeding amiable, active, crazy bippy was mated, with some difficulty, to a true-breeding nasty, benign, sane bippy. An F_1 individual from this cross was then used in a testcross (to a nasty, active, sane bippy) and produced, in typical prolific bippy fashion, 4000 offspring. From an ancient manuscript entitled *The Genetics of the Bippy, Maltese and Other*, you discover that all three genes are autosomal, a is linked to b but not to c, and the map distance between a and b is 20 mu.

a. Predict all the expected phenotypes and the numbers of each type from this cross.
b. Which phenotypic classes would be missing had a and b shown complete linkage?
c. Which phenotypic classes would be missing if a and b were unlinked?
d. Again, assuming a and b to be unlinked, predict all the expected phenotypes of nasty bippies and the frequencies of each type resulting from a self-cross of the F_1.

5.22 How many possible linear orders are there for three linked genes? How many are there if the left versus the right end of the chromosome is taken into consideration?

5.23 Fill in the blanks. Continuous bars indicate linkage, and the order of linked genes is correct as shown. If all types of gametes are equally probable, write "none" in the right column headed "Least frequent classes." In the right column, show two gamete genotypes, unless all types are equally frequent, in which case write "none."

PARENT GENOTYPES	NUMBER OF DIFFERENT POSSIBLE GAMETE GENOTYPES	LEAST FREQUENT CLASSES	
$\dfrac{A\ \ b\ \ C}{a\ \ B\ \ c}$	_____	____	____
$\dfrac{A\ b\ C}{a\ B\ c}$	_____	____	____
$\dfrac{A\ b\ C\ D}{a\ B\ c\ d}$	_____	____	____
$\dfrac{A\ b\ C\ D\ e\ f}{a\ B\ C\ d\ e\ f}$	_____	____	____
$\dfrac{b\ D}{B\ d}$	_____	____	____

5.24 For each of the following tabulations of testcross progeny phenotypes and numbers, state which locus is in the middle, and reconstruct the genotype of the tested triple heterozygotes.

a.	A B C	191		b.	C D E	9
	a b c	180			c d e	11
	A b c	5			C d e	35
	a B C	5			c D E	27
	A B c	21			C D e	78
	a b C	31			c d E	81
	A b C	104			C d E	275
	a B c	109			c D e	256

c.	F G H	110
	f g h	114
	F g h	37
	f G H	33
	F G h	202
	f g H	185
	F g H	4
	f G h	0

*5.25 Genes at loci *f*, *m*, and *w* are linked, but their order is unknown. The F_1 heterozygotes from a cross of *FF MM WW* × *ff mm ww* are testcrossed. The most frequent phenotypes in testcross progeny will be *FMW* and *fmw*, regardless of what the gene order turns out to be.
a. What classes of testcross progeny (phenotypes) would be least frequent if locus *m* is in the middle?
b. What classes would be least frequent if locus *f* is in the middle?
c. What classes would be least frequent if locus *w* is in the middle?

5.26 The following numbers were obtained for testcross progeny in *Drosophila* (phenotypes):

+	m	+	218
w	+	f	236
+	+	f	168
w	m	+	178
+	m	f	95
w	+	+	101
+	+	+	3
w	m	f	1
		Total	1000

Construct a genetic map.

*5.27 Three of the many recessive mutations in *Drosophila melanogaster* that affect body color, wing shape, or bristle morphology are black (*b*) body versus grey in the wild type, dumpy (*dp*), obliquely truncated wings versus long wings in the wild type, and hooked (*hk*) bristles at the tip versus not hooked in the wild type. From a cross of a dumpy female with a black and hooked male, all the F_1 were wild type for all three characters. The testcross of an F_1 female with a dumpy, black, hooked male gave the following results:

wild type	169
black	19
black, hooked	301
dumpy, hooked	21
hooked	8
hooked, dumpy, black	172
dumpy, black	6
dumpy	305
Total	1000

a. Construct a genetic map of the linkage group (or groups) these genes occupy. If applicable, show the order and give the map distances between the genes.
b. (1) Determine the coefficient of coincidence for the portion of the chromosome involved in the cross. (2) How much interference is there?

5.28 In corn, colorless aleurone (*c*) is recessive to colored (*C*), shrunken endosperm (*sh*) is recessive to full (*Sh*), and waxy endosperm (*wx*) is recessive to starchy (*Wx*). The F_1 plants from the cross of true-breeding colored, shrunken, and starchy × true-breeding colorless, full, and waxy were crossed with colorless, shrunken, and waxy plants, and the following progeny were generated:

colored, shrunken, starchy	2538
colorless, full, waxy	2708
colored, full, waxy	116
colorless, shrunken, starchy	113
colored, shrunken, waxy	601
colorless, full, starchy	626
colored, full, starchy	4
colorless, shrunken, waxy	2

Map the positions of the *c*, *sh*, and *wx* genes in the chromosome.

5.29 In Chinese primroses long style (*l*) is recessive to short (*L*), red flower (*r*) is recessive to magenta (*R*), and red stigma (*rs*) is recessive to green (*Rs*). From a cross of homozygous short, magenta flower, and green stigma with long, red flower, and red stigma, the F_1 was crossed back to long, red flower, and red stigma. The following offspring were obtained:

STYLE	FLOWER	STIGMA	NUMBER
short	magenta	green	1063
long	red	red	1032
short	magenta	red	634
long	red	green	526
short	red	red	156
long	magenta	green	180
short	red	green	39
long	magenta	red	54

Map the genes involved.

5.30 The frequencies of gametes of different genotypes, determined by testcrossing a triple heterozygote, are as follows:

GAMETE GENOTYPE	%
+ + +	12.9
a b c	13.5
+ + c	6.9
a b +	6.5
+ b c	26.4
a + +	27.2
a + c	3.1
+ b +	3.5
Total	100.0

a. Which gametes are known to have been involved in double crossovers?
b. Which gamete types have not been involved in any exchanges?
c. The order shown is not necessarily correct. Which gene locus is in the middle?

*5.31 Genes a, b, and c are recessive. Females heterozygous at these three loci are crossed to phenotypically wild-type males. The progeny are phenotypically as follows:

Daughters:	all + + +	
Sons:	+ + +	23
	a b c	26
	+ + c	45
	a b +	54
	+ b c	427
	a + +	424
	a + c	1
	+ b +	0
	Total	1000

a. What is known of the genotype of the females' parents with respect to these three loci? Give gene order and the arrangement in the homologs.
b. What is known of the genotype of the male parents?
c. Map the three genes.

5.32 Two normal-looking *Drosophila* are crossed and yield the following phenotypes among the progeny:

Females:	+ + +	2000
Males:	+ + +	3
	a b c	1
	+ b c	839
	a + +	825
	a b +	86
	+ + c	90
	a + c	81
	+ b +	75
	Total	4000

Give parental genotypes, gene arrangement in the female parent, map distance, and the coefficient of coincidence.

5.33 Three different semidominant mutations affect the tail of mice. They are linked genes, and all three are lethal in the embryo when homozygous. Fused-tail (Fu) and kinky-tail (Ki) mice have kinky-appearing tails, while brachyury (T) mice have short tails. A fourth gene, histocompatibility-2 ($H-2$), is linked to the three tail genes and is concerned with tissue transplantation. Mice that are $H-2/+$ will accept tissue grafts, whereas $+/+$ mice will not. In the following crosses the normal allele is represented by a +. The phenotypes of the progeny are given for four crosses.

(1) $\dfrac{Fu\ +}{+\ Ki} \times \dfrac{+\ +}{+\ +}$

$\begin{cases} \text{Fused tail} & 106 \\ \text{Kinky tail} & 92 \\ \text{Normal tail} & 1 \\ \text{Fused-kinky tail} & 1 \end{cases}$

(2) $\dfrac{Fu\ H-2}{+\ \ +} \times \dfrac{+\ +}{+\ +}$

$\begin{cases} \text{Fused tail, accepts graft} & 88 \\ \text{Normal tail, rejects graft} & 104 \\ \text{Normal tail, accepts graft} & 5 \\ \text{Fused tail, rejects graft} & 3 \end{cases}$

(3) $\dfrac{T\ H-2}{+\ \ +} \times \dfrac{+\ +}{+\ +}$

$\begin{cases} \text{Brachy tail, accepts graft} & 1048 \\ \text{Normal tail, rejects graft} & 1152 \\ \text{Brachy tail, rejects graft} & 138 \\ \text{Normal tail, accepts graft} & 162 \end{cases}$

(4) $\dfrac{Fu\ +}{+\ T} \times \dfrac{+\ +}{+\ +}$

$\begin{cases} \text{Fused tail} & 146 \\ \text{Brachy tail} & 130 \\ \text{Normal tail} & 14 \\ \text{Fused-brachy tail} & 10 \end{cases}$

Make a map of the four genes involved in these crosses, giving gene order and map distances between the genes.

*5.34 The cross in *Drosophila* of

$$\frac{a^+\ b^+\ c\ d\ e}{a\ b\ c^+\ d^+\ e^+} \times \frac{a\ b\ c\ d\ e}{a\ b\ c\ d\ e}$$

gave 1000 progeny of the following 16 phenotypes:

GENOTYPE	NUMBER
(1) $a^+\ b^+\ c\ d\ e$	220
(2) $a^+\ b^+\ c\ d\ e^+$	230
(3) $a\ b\ c^+\ d^+\ e$	210
(4) $a\ b\ c^+\ d^+\ e^+$	215
(5) $a\ b^+\ c^+\ d^+\ e$	12
(6) $a\ b^+\ c^+\ d^+\ e^+$	13
(7) $a^+\ b\ c\ d\ e^+$	16
(8) $a^+\ b\ c\ d\ e$	14
(9) $a\ b^+\ c^+\ d\ e^+$	14
(10) $a\ b^+\ c^+\ d\ e$	13
(11) $a^+\ b\ c\ d^+\ e^+$	8
(12) $a^+\ b\ c\ d^+\ e$	8
(13) $a^+\ b^+\ c^+\ d\ e^+$	7
(14) $a^+\ b^+\ c^+\ d\ e$	7
(15) $a\ b\ c\ d^+\ e^+$	6
(16) $a\ b\ c\ d^+\ e$	7

a. Draw a genetic map of the chromosome, indicating the linkage of the five genes and the number of map units separating each.

b. From the single-crossover frequencies, what would be the expected frequency of $a^+ b^+ c^+ d^+ e^+$ flies?

5.35 Complete all diagrams in Figure 5.A by correctly showing the centromeres, chromosome strands, and alleles on each strand.

FIGURE 5.A

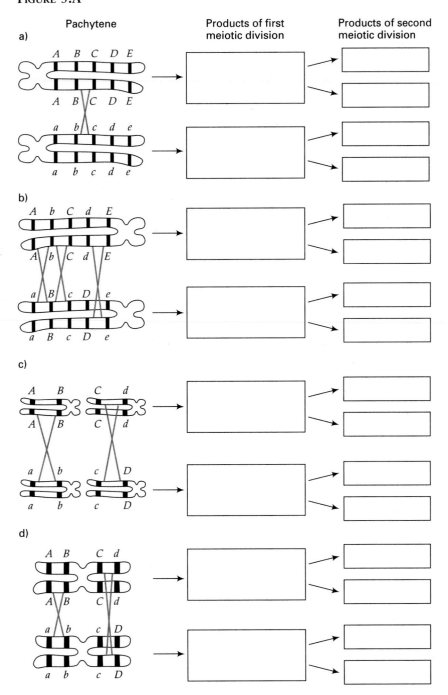

6 ADVANCED GENETIC MAPPING IN EUKARYOTES

TETRAD ANALYSIS
Life Cycle of Yeast
Life Cycle of *Chlamydomonas reinhardi*
Using Random-Spore Analysis to Map Genes in Haploid Eukaryotes
Using Tetrad Analysis to Map Two Linked Genes
Calculating Gene-Centromere Distance in Organisms with Linear Tetrads

MITOTIC RECOMBINATION
Discovery of Mitotic Recombination
Mechanism of Mitotic Crossing-over
Mitotic Recombination in Fungal Systems

MAPPING GENES IN HUMAN CHROMOSOMES
Mapping Human Genes by Recombination Analysis
Mapping Human Genes by Somatic Cell Hybridization Techniques

PRINCIPAL POINTS

~ Tetrad analysis is a mapping technique that can only be used to map the genes of certain haploid eukaryotic organisms in which the products of a single meiosis, the meiotic tetrad, are contained within a single structure.

~ In organisms in which tetrads can be analyzed, the analysis of the relative proportion of tetrad types provides another way of computing the map distance between genes.

~ In organisms in which the meiotic tetrads are arranged in a linear fashion, it is possible to map the distance of a gene from its centromere.

~ Crossing-over can occur during mitosis (rarely) as well as during meiosis.

~ In certain organisms, parasexual genetic analysis is used to map genes. In these organisms, genetic recombination is achieved by means other than meiosis and fertilization, for example by crossing-over during mitosis.

~ Mapping genes to human chromosomes can be accomplished using somatic cell hybridization techniques, for example, by fusing human and mouse cells and analyzing the descendants of the hybrid cells.

---

*I*n the previous chapter we considered the classical experiments involving gene linkage in eukaryotes. We saw that the outcome of crosses can be used to determine map distances and that the map distances obtained are useful in predicting the outcome of crosses. In this chapter we will see how the basic mapping principles are applied and extended in determining map distance by tetrad analysis in certain appropriate haploid organisms, by nonmeiotic means, and in the mapping of human chromosomes.

TETRAD ANALYSIS

Tetrad analysis is a mapping technique that can be used to map the genes of those eukaryotic organisms in which the products of a single meiosis, the meiotic tetrad, are contained within a single structure. The eukaryotic organisms in which this phenomenon occurs are either fungi or single-celled algae, all of which are haploid. *Neurospora crassa* and yeast (both fungi) and *Chlamydomonas reinhardi* (a single-celled alga) are frequently used in tetrad analysis.

By analyzing the phenotypes of the meiotic tetrads, geneticists can directly infer the genotypes of each member of the tetrad. Haploid organisms exhibit no genetic dominance since there is only one copy of each gene; hence the genotype is expressed directly in the phenotype. Much valuable information about how and when genetic recombination (crossing-over) occurs and about the process of gene segregation has been discovered through tetrad analysis of various organisms, and we will outline this technique in the following sections.

Since tetrad analysis is done primarily on the three microorganisms mentioned above, it will be helpful to begin by outlining the life cycles of yeast and *Chlamydomonas reinhardi*. The life cycle of *Neurospora crassa* has already been described in Chapter 5.

Life Cycle of Yeast

Figure 6.1 (p. 164) diagrams the life cycle of baker's yeast, *Saccharomyces cerevisiae* (the yeast responsible for making bread rise). Two mating types occur in yeast, *a* and *α*. The haploid vegetative cells of this organism reproduce mitotically, with the new cell arising from the parental cell by budding. Fusion of haploid *a* and *α* cells produces a diploid cell that is stable and that also reproduces by budding.

As we saw in *Neurospora crassa*, under conditions of nitrogen starvation diploid *a*/*α* cells sporulate; that is, they go through meiosis. The four haploid meiotic products of a diploid cell, the ascospores, are contained within a roughly spherical ascus. Two of these ascospores are of mating type *a*, and two are of mating type *α*. When the ascus is ripe, the ascospores are released, and they germinate to produce haploid vegetative cells. On a solid medium each ascospore develops into a discrete colony. In yeast the four ascospores are arranged randomly within the ascus so that only unordered tetrads can be isolated from this organism (in contrast to *Neurospora crassa*, where the ascospores are linearly arranged in the ascus in a way that reflects the orientation of the four chromatids of the meiotic tetrad at metaphase I).

~ FIGURE 6.1

(a) Life cycle of the yeast *Saccharomyces cerevisiae;* (b) Scanning electron micrograph of *Saccharomyces cerevisiae,* many of which have buds.

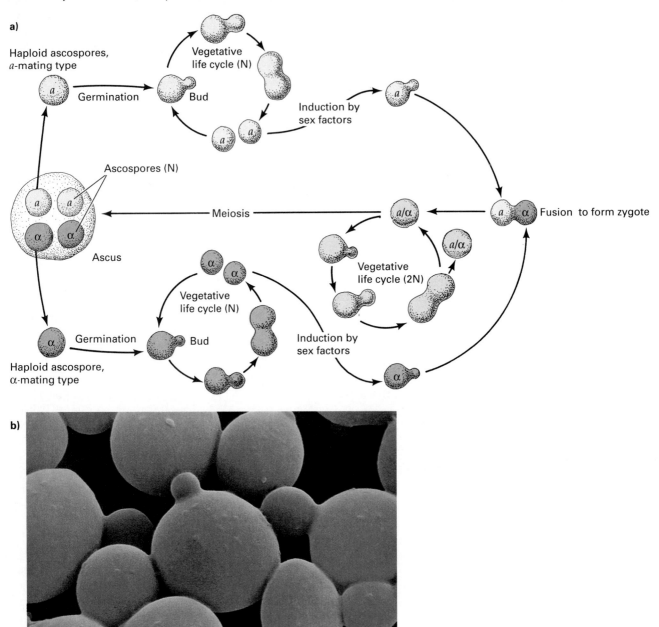

Life Cycle of *Chlamydomonas reinhardi*

Figure 6.2 diagrams the life cycle of *Chlamydomonas reinhardi.* Like yeast, *Chlamydomonas reinhardi* has haploid vegetative cells. Each individual is a single green algal cell that can swim freely as a result of the motion of its two flagella. When nitrogen is limited, the cells change morphologically to become gametes so that mating is possible. There are two mating types, designated plus (+) and minus (−). Only gametes of opposite mating types (in this case, + and −) can fuse to produce diploid zygotes. No fusion occurs between gametes of like mating types. After a maturation process, the zygote enters meiosis. The four haploid mei-

~ FIGURE 6.2

(a) Life cycle of the unicellular green alga *Chlamydomonas reinhardi;* (b) Light micrograph of an individual *Chlamydomonas;* (c) Light micrograph of + and − *Chlamydomonas* fusing.

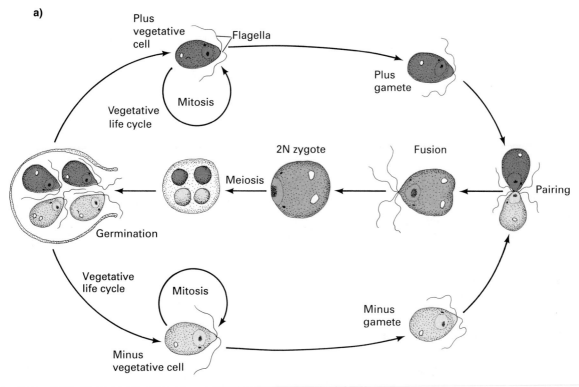

a)

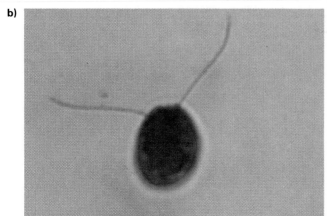

b)

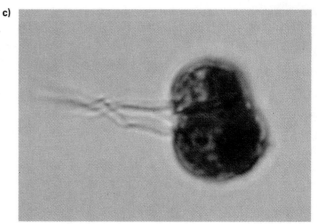

c)

otic products are contained within a sac as an unordered tetrad, and there are two + and two − cells. When these cells are released, they are free-swimming, and by mitosis they give rise to clones of those meiotic products. Again, any one of these haploid vegetative cells can become differentiated into a gamete under nitrogen-poor conditions.

Using Random-Spore Analysis to Map Genes in Haploid Eukaryotes

In the three haploid eukaryotes we have discussed, the meiotic products can be collected after they have been released from the ascus or sac. In the case of the fungi (*Neurospora crassa* and yeast), spores can be induced to germinate and the resulting cultures can be analyzed. The free-swimming meiotic products of *Chlamydomonas* can be analyzed directly. In fact, the haploid nature of the mature stages of all three organisms simplifies the analysis, since this stage is exactly equivalent to that of the gametes produced after meiosis in a diploid eukaryote. Thus we can make three-point crosses to map genes on a chromosome, using essentially the same approach as the one we used for a diploid eukaryote.

Figure 6.3 shows a three-point-mapping cross in a haploid eukaryote to illustrate the similarities with three-point testcrosses in diploids. Here a wild-type (haploid) strain is crossed with a strain that carries three mutant genes in the same chromosome. The result is a diploid zygote that is triply heterozygous. When this zygote undergoes meiosis to produce the haploid progeny organisms, there is the potential to form eight genotypic (and hence phenotypic) classes in a fashion identical to that in three-point testcrosses involving diploid eukaryotes. The only difference is that the parents and the progeny are haploid, not diploid. Thus, random-spore analysis can be used with haploid eukaryotes to obtain data for drawing a genetic map of their chromosomes. This information is important for understanding the structural and functional organization of the genome in such organisms.

Using Tetrad Analysis to Map Two Linked Genes

Formally, random-spore analysis to map genes in *N. crassa*, yeast, and *C. reinhardi* is little different from analysis of the progeny of testcrosses in diploid eukaryotes. Such analysis produces data for drawing genetic maps. Let us consider how we can determine the map distance between two linked genes by analyzing a number of complete meiotic tetrads. The analysis to be described requires only unordered tetrads, and hence is applicable to *N. crassa*, yeast, and *C. reinhardi*.

By making an appropriate cross, a diploid zygote is constructed that is heterozygous for both genes, and after meiosis, the resulting tetrads are analyzed. For a cross of *a b* × + +, three possible tetrad types result, called **parental-ditype (PD)**, **tetratype (T)**, and **nonparental-ditype (NPD)** tetrads (Figure 6.4).

The PD tetrads contain only two types of meiotic products, both of which are of the parental type *a b* and + + (hence the name parental ditype). The T tetrads contain all four possible types of meiotic products, that is, the two parental types and the two recombinant types *a* + and + *b*. The NPD tetrads contain two types of meiotic products, both of which are of the nonparental (recombinant) types *a* + and + *b*.

~ FIGURE 6.3

Typical genetic cross for mapping three genes in a haploid organism such as yeast or *Neurospora crassa*.

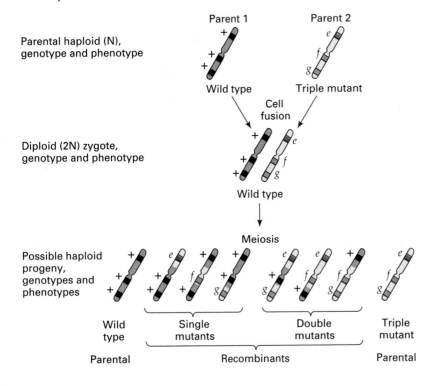

~ FIGURE 6.4

Three types of tetrads produced from a cross of $a\ b \times + +$: parental-ditype (PD), tetratype (T), and nonparental-ditype (NPD) tetrads.

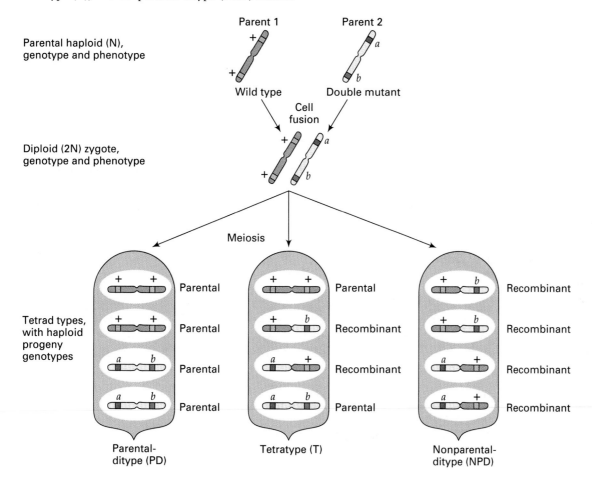

In organisms in which the products of a meiosis (the meiotic tetrad) are contained within a single structure, three types of tetrads are possible when two genes are segregating in a cross. The parental-ditype (PD) tetrad contains four nuclei, all of parental genotypes: two of one parent and two of the other parent. The nonparental-ditype (NPD) tetrad contains four nuclei, all of recombinant (nonparental) genotypes, that is, two of each possible type. The tetratype (T) tetrad contains two parental and two recombinant nuclei (one of each parental type and one of each recombinant type). ⎯⎯⎯⎯⎯

Tetrad analysis helps to determine whether or not two genes are linked. This determination is based on the ways each tetrad type is produced when genes are unlinked and when genes are linked.

Figure 6.5 (p. 168) shows how PD, NPD, and T tetrads are produced when two genes are on different chromosomes. In this case the PD and NPD tetrads result from events in which no crossovers are involved; the relative metaphase plate orientation of the four chromatids for the two chromosomes determines whether a PD or an NPD tetrad results. Since the two sets of four chromatids align at the metaphase plate independently, the PD and NPD orientations shown occur with approximately equal frequency. Thus if two genes are unlinked, the frequency of PD tetrads will equal the frequency of NPD tetrads. For linked genes the results are different: T tetrads are produced when there is, for example, a single crossover between one of the genes and the centromere on that chromosome. The frequency of such T tetrads would depend on the distance between the genes and their respective centromeres.

~ FIGURE 6.5

Origin of tetrad types for a cross $a\ b \times +\ +$, in which the two genes are located on different, independently assorting chromosomes.

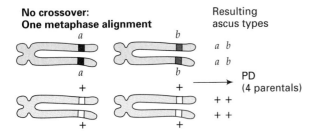

No crossover:
One metaphase alignment

Resulting ascus types

PD (4 parentals)

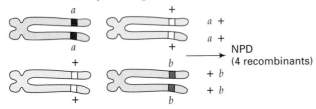

Alternative metaphase alignment

NPD (4 recombinants)

Sums to 50% parentals
50% recombinants

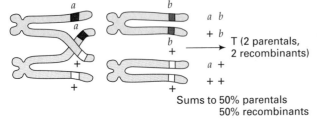

Single crossover:
Single crossover between *a* gene and its centromere

T (2 parentals, 2 recombinants)

Sums to 50% parentals
50% recombinants

Figure 6.6 shows the origins of each tetrad type when two genes are linked on the same chromosome. If no crossing-over (Figure 6.6a) occurs between the genes, then a PD ascus results. A single crossover (Figure 6.6b) produces two parental and two recombinant chromatids and hence a T ascus. As we learned in Chapter 5, in double crossovers we must take into account the chromatid strands involved. In a two-strand double crossover (Figure 6.6c) the two crossover events between the two genes take place between the same two chromatids. This crossover results in a PD ascus, since no recombinant progeny are produced. Three-strand double crossovers (Figure 6.6d) include three of the four chromatids, and there are two possible ways in which this event can happen. In either case two recombinant and two parental progeny types are produced in each ascus, which is a T ascus. Lastly, in four-

strand double crossovers (Figure 6.6e) each crossover event involves two distinct chromatids, so all four chromatids of the tetrad are involved. This crossover is the only way in which NPD tetrads are produced. Therefore since PD tetrads are produced either when there is no crossing-over or when there is a two-strand double crossover, and since NPD tetrads are only produced by four-strand double crossovers (which are 1/4 of all the possible double crossovers and, hence, are rare), we can state that two genes are linked if the frequency of PD tetrads is far greater than the frequency of NPD tetrads (i.e., PD ≫ NPD).

Once we know that two genes are linked, and once we have data on the relative numbers of each type of meiotic tetrad, the distance between the two genes can be computed by using the basic mapping formula:

$$\frac{\text{number of recombinants}}{\text{total number of progeny}} \times 100$$

In tetrad analysis, however, we analyze types of tetrads, rather than individual progeny. By examining the nature of the four products in each tetrad, we see that a PD tetrad has all parental meiotic products, a T tetrad has two parental and two recombinant meiotic products, and an NPD tetrad has four recombinant meiotic products. To convert the basic mapping formula into tetrad terms, the recombination frequency between genes *a* and *b* becomes

$$\frac{1/2\ T + NPD}{\text{total tetrads}} \times 100$$

In this formula the 1/2 T and the NPD represent the only recombinants; the other 1/2 T and the PD represent the nonrecombinants (the parentals). Thus the formula does indeed compute the percentage of recombinants. For instance, if there are 200 asci with 140 PD, 48 T, and 12 NPD, the recombination frequency between the genes is

$$\frac{1/2(48) + 12}{200} \times 100 = 18\%$$

This conversion produces a formula for calculating the map distance between two linked genes by using the frequencies of the three possible types of tetrads rather than by analyzing individual progeny. If more than two genes are linked in a cross, the data may best be analyzed by considering two genes at a time and by classifying each tetrad into PD, NPD, and T for each pair. Box 6.1 (p. 170) presents a modified formula, one that more accurately calculates map distance from tetrad data.

~ FIGURE 6.6

Origin of tetrad types for a cross $a\ b \times +\ +$, in which both genes are located on the same chromosome. The arrows indicate the sites at which crossovers occurred. (a) No crossover; (b) Single crossover; (c-e) Three types of double crossovers.

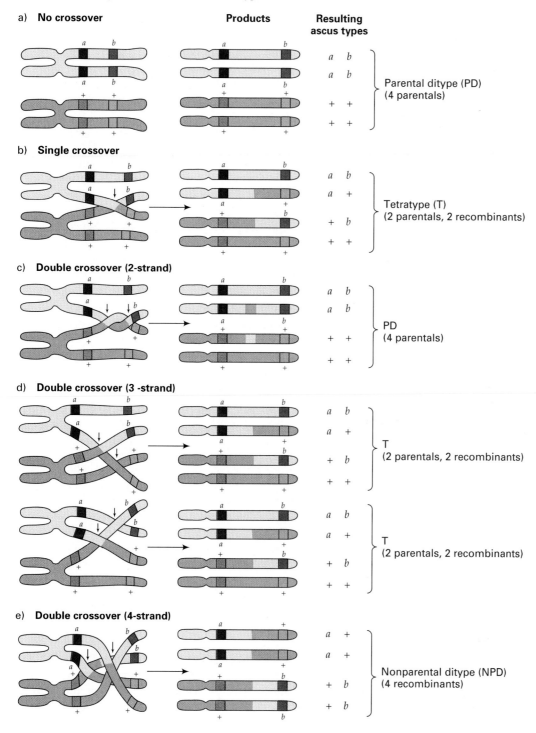

Box 6.1
Calculating Map Distance from Tetrad Data on the Basis of Crossover Frequency

The formula just derived in the text calculates map distance on the basis of recombination frequency. As such, the formula is formally analogous to those formulas used in Chapter 5 to calculate map distance in map units. However, map distance between genes is more accurately calculated on the basis of crossover frequency. How can we modify the tetrad analysis formula to be more accurate? In Figure 6.6 we saw that there were three types of double crossovers: two-strand, three-strand, and four-strand double crossovers. Theoretically, if crossovers are occurring randomly, these three types of double crossovers should occur in a ratio of 1:2:1, respectively. The following discussion follows from that assumption. In actuality, the stated ratio may not be the case; any deviation will erode the accuracy of the new formula.

The four-strand double crossover frequency is directly determined from the number of NPD tetrads. Two-strand double crossovers result in PD tetrads but, even though crossovers have occurred between the two loci, recombinants do not result (see Figure 6.6c). But in order to account for all crossover activity, the two crossovers still need to be counted. From the theoretical ratio of the three different types of double crossovers, the number of PD tetrads that result from double crossovers should equal the number of NPD tetrads.

There are two types of three-strand double crossovers and each gives rise to a T tetrad, as does a single crossover. Thus, for each three-strand double crossover, the original formula only considers one of the two crossovers. Between the two three-strand double crossover types, then, the calculations do not include the equivalent of two crossovers or one double crossover. Again, the NPD frequency theoretically is equivalent to the frequency of a double crossover. Therefore, a more accurate formula for calculating the distance between two genes from tetrad data is:

$$
\begin{aligned}
\text{Recombinants} = \quad & 1/2\ T + NPD \\
& + NPD \text{ (to add 2-strand double crossover, which gives rise to PD tetrads with a} \\
& \qquad \text{frequency equal to NPD frequency)} \\
& + NPD \text{ (to add the equivalent of a double crossover not included in the two 3-strand} \\
& \qquad \underline{\text{double crossovers—again the frequency equals NPD frequency)}} \\
\text{Total} = \quad & \overline{1/2\ T + 3NPD}
\end{aligned}
$$

∴ Modified formula to determine map distance between two genes is:

$$
\frac{1/2\ T + 3NPD}{\text{Total}} \times 100
$$

The reader should note that, while the newly-derived formula gives a more accurate estimate of map distance, the text formula rather than this Box formula should be used in working the end-of-chapter problems.

$\mathcal{K}$EYNOTE

In organisms in which all products of meiosis are contained within a single structure, the analysis of the relative proportion of tetrad types provides another way to compute the map distance between genes. The general formula when two linked genes are being mapped is

$$
\frac{1/2\ T + NPD}{\text{total tetrads}} \times 100
$$

Calculating Gene-Centromere Distance in Organisms with Linear Tetrads

In *Neurospora crassa* and some other haploid microorganisms, the meiotic products are arranged linearly within the ascus in what is called an ordered tetrad. By contrast, in organisms such as yeast and *Chlamydomonas reinhardi*, the meiotic products are arranged randomly within the ascus or sac in what is called an unordered tetrad. The tetrad analysis we have discussed so far is applicable to both types of tetrad arrangement.

Neurospora crassa actually has eight spores, since the four meiotic products undergo one more mitotic division before the ascospores are produced. Since mitotic division produces, in essence, clones, for the purposes of our genetic discussions the eight spores can be considered as four pairs and we will ignore the last mitotic division. The most interesting thing about them is that their genetic content directly reflects the orientation of the four chromatids of each chromosome pair in the diploid zygote nucleus at metaphase I. *This fact allows us to map the distance between genes and their centromeres,* which is not possible with unordered tetrads. Locating the centromeres on the genetic maps of chromosomes makes the maps more complete. In higher organisms centromeres are located primarily through cytological studies (which are usually not possible in lower eukaryotes because their chromosomes are too small).

In this example we will map the position of the mating-type locus of *N. crassa* in relation to its centromere. Mating type is a function of which allele, *A* or *a*, is present at a locus in linkage group I. If an *N. crassa* strain of mating type *A* is crossed with one of mating type *a*, a diploid zygote of genotype *A/a* results. Figure 6.7 (p. 172) shows the various ways in which this zygote can give rise to the four meiotic products, the ascospores. For the purposes of illustration, a ● will be used to indicate the centromeres of the *A* parent and a ○ will be used to indicate the centromere of the *a* parent. (In reality there is no difference between the two.) At the zygote stage of the life cycle the chromosomes have divided but the centromeres have not.

If no crossing-over occurs between the mating-type locus and the centromere, the resulting ascospores have the genotypes shown in Figure 6.7b. An important point here is that the centromeres do not separate until just before the second meiotic division. Therefore the spores in the top half of the ascus always have the centromere from one parent (the ● centromere, in this case), and the spores in the other half of the ascus always have the centromere from the other parent (○, here).

Since the two types of centromeres segregate to different nuclear areas after the first meiotic division, we say that they show *first-division segregation.* And since no crossing-over occurs between the mating-type locus and the centromere, each allelic pair also shows first-division segregation. Note particularly that all four spores in an ascus showing first-division segregation of alleles are parental types: The *A* allele is on the chromosome with a ● centromere and the *a* allele is on the chromosome with the ○ centromere. Furthermore, since it is equally likely that the four chromatids in the diploid zygote will be rotated 180°, we expect equal numbers of first-division segregation asci in which the ● *A* spores are in the bottom half and the ○ *a* spores are in the top half.

To determine the map distance between a gene and its centromere, we must measure the crossover frequency between the two chromosomal sites. Thus we need to predict the consequences of a single crossover between the mating-type locus and the centromere. Figure 6.7b shows that four ascus types are produced. These four types are generated in equal frequencies, since they reflect the four possible orientations of the four chromatids in the diploid zygote at metaphase I. Each has first-division segregation of the centromere; that is, segregation occurs during meiosis I. By contrast, *A* and *a* do not segregate into separate nuclei until the second division. This situation is called *second-division segregation* for the gene; that is, segregation occurs during meiosis II. Here, the pattern of gene segregation depends upon which chromatids are involved in the crossover event. The 1:1:1:1 (*A a A a* and *a A a A*) and 1:2:1 (*A a, a A,* and *a A, A a*) second-division segregation patterns are readily distinguishable from the usual 2:2 (*A A a a* and *a a A A*) first-division segregation pattern.

By analyzing ordered tetrads, we can count the number of asci that show second-division segregation for a particular gene marker. For the mating-type locus, about 14 percent of the asci show second-division segregation. This value must be converted to a map distance, and in this case it is not a direct conversion. So the diagrammatic distinction of the two centromere types comes in useful here. If we consider the centromere to be a gene marker (and in a sense it is), then the parental types are ● *A* and ○ *a*, and the recombinant types are ● *a* and ○ *A*. In a first-division segregation ascus (Figure 6.7a) all the spores are parentals, while in a second-division segregation ascus (Figure 6.7b) half are parentals and half are recombinants. Therefore to convert the tetrad data to crossover or recombination data, we divide the percentage of second-division segregation asci (14 percent) by 2. Thus the mating-type gene is 7 mu from the centromere of linkage group I.

In essence, the computation of gene-centromere distance is a special case of mapping the distance between two genes. If ordered tetrads are isolated, then not only can genes be mapped one to another, but each can be mapped to its centromere.

KEYNOTE

In some microorganisms the products of a meiosis are arranged in a specialized structure in a way that reflects the orientation of the four chromatids of each homologous pair of chromosomes at metaphase I. The ordered tetrads allow us to map the distance of a gene from its centromere. Where no crossover occurs between the gene and the centro-

~ FIGURE 6.7

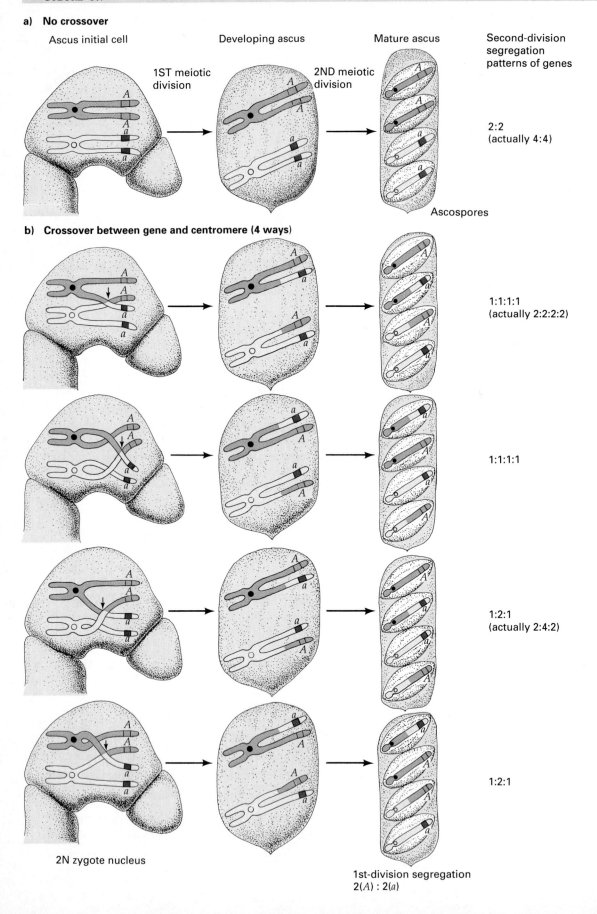

a) **No crossover**

Ascus initial cell

Developing ascus

Mature ascus

Second-division
segregation
patterns of genes

1ST meiotic
division

2ND meiotic
division

2:2
(actually 4:4)

Ascospores

b) **Crossover between gene and centromere (4 ways)**

1:1:1:1
(actually 2:2:2:2)

1:1:1:1

1:2:1
(actually 2:4:2)

1:2:1

2N zygote nucleus

1st-division segregation
2(*A*) : 2(*a*)

~ FIGURE 6.7 continued

Gene-centromere distance of the mating-type locus in *N. crassa*. (a) Production of an ascus from a diploid zygote in which no crossing-over occurred between the centromere and the mating-type locus, and first-division segregation for the mating-type alleles. (In *N. crassa* a mitotic division after the second meiotic division produces eight spores, thereby doubling each progeny type: i.e., the 2:2 ratio is actually 4:4. The mitotic division is not considered here for simplicity's sake.) (b) Production of asci after a single crossover occurs between the mating-type locus and its centromere. The arrows show the points at which the crossovers occurred. The asci show second-division segregation for the mating-type locus, and the four types of asci are produced in equal proportions.

mere, the result is a first-division segregation tetrad, in which one parental type is found in half the ordered tetrad and the other parental type is found in the other half, giving a 2:2 segregation pattern. When a single crossover occurs between the gene and its centromere, several different tetrad segregation patterns are found, and each exemplifies second-division segregation. The gene-centromere map distance is computed as the percentage of second-division tetrads divided by 2. ─────

MITOTIC RECOMBINATION

Discovery of Mitotic Recombination

Experimental evidence shows that in some organisms crossing-over can occur during mitosis as well as during meiosis. The first demonstration of **mitotic crossing-over** was observed by Curt Stern in 1936, in crosses involving *Drosophila* strains carrying recessive sex-linked mutations that cause yellow body color (y) instead of the normal grey body color and short, twisty bristles (singed, sn) instead of the normal long, curved bristles. In flies with a wild-type grey body color, all bristles are black; in yellow-bodied (mutant) flies the bristles are yellow.

From a cross of homozygous $y^+ sn/y^+ sn$ females (grey bodies, singed bristles) with $y sn^+//$ males (yellow bodies, normal bristles), Stern found, as expected, that the female progeny were mostly wild type in appearance: They had grey bodies and normal bristles. However, instead of being all one color, some females had

sectors of yellow and/or singed bristles. The origin of these flies could be explained by chromosome nondisjunction or by chromosomal loss. (These changes are somatic rather than germline changes.)

Other females had **twin spots,** two adjacent regions of bristles, one showing the yellow phenotype and the other showing the singed phenotype, that is, a *mosaic* phenotype. Surrounding the twin spot, and constituting essentially the rest of the bristles, the phenotype was wild type. Figure 6.8 (p. 174) diagrams this twin spot and contrasts it with the single-phenotype spots. Stern reasoned that since the two parts of a twin spot were always adjacent, the twin spots must be the reciprocal products of the same genetic event. The best explanation was that they were generated by a mitotic crossing-over event, an event that has been found in a number of organisms, although it occurs rarely. This explanation was deemed most likely because Stern knew that the tissue involved is produced by mitotic events after the *Drosophila* embryo has differentiated.

The production of twin spots by mitotic crossing-over is shown in Figure 6.9 (p. 174).

The starting point is a wild-type diploid cell that is heterozygous for both the y (yellow) and sn (singed) genes. The mutant alleles are on different homologs. Crossing-over can occur as a rare mitotic event, either between the centromere and the sn locus (Figure 6.9, left), or between the sn and the y locus (Figure 6.9, right). (Mitotic crossing-over will be described in detail in the next section.) In the former case, if chromatids 1 and 3 segregate to one progeny nucleus, and chromatids 2 and 4 segregate to the other, one progeny cell is homozygous y/y and the other is homozygous sn/sn. When these divide and produce two tissue patches, one of which is yellow in phenotype and the other of which

~ FIGURE 6.8

Body surface phenotype segregation in a *Drosophila* strain
$y^+ sn/y sn^+$. The *sn* allele causes short, twisted (singed)
bristles, and the *y* allele results in a yellow body
coloration: (a) single yellow spot in normal-body color
background; (b) twin spot of yellow color and singed
bristles; (c) single singed-bristle spot in normal-bristle
phenotype background.

a) **Single   b) Twin spot   c) Single**
    yellow spot   singed spot

~ FIGURE 6.9

Production of the twin spot and of the single yellow spot shown in Figure 6.8 by mitotic
crossing-over.

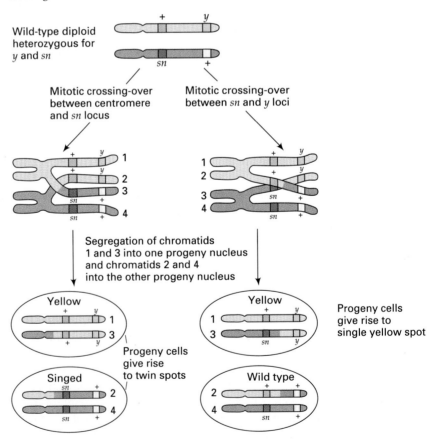

is singed, a twin spot is produced. The surrounding tissue, not involved in any mitotic crossing-over, will be wild type in phenotype. In the case of crossing-over between *sn* and *y*, and of chromatid segregation as above, the result is one progeny cell that is yellow in phenotype (*y/y*) and the other that is wild type in phenotype, producing a single yellow spot.

Mechanism of Mitotic Crossing-over

Mitotic crossing-over (mitotic recombination) can only be studied in diploid cells. It is a process during mitosis that produces a progeny cell with a combination of genes that differs from that of the diploid parental cell that entered the mitotic cycle. *Mitotic crossing-over occurs at a stage similar to the four-strand stage of meiosis.* Recall that during mitosis each pair of homologous chromosomes replicates and that the two pairs of chromatids then align independently at the metaphase plate. When the chromatids separate, each progeny cell receives one copy of each parental homolog so that each cell has the same genetic constitution as the parental cell. Figure 6.10 shows the pattern of gene segregation for a theoretical cell heterozygous for all the genes on its one pair of homologous chromosomes. Both the parental and progeny cells are wild type in phenotype.

Under rare circumstances, after each chromosome has replicated and prior to metaphase, the two pairs of maternally derived and paternally derived chromatids come together to form a tetrad that is analogous to the four-strand stage in meiosis. It is during this stage that crossing-over can occur. Figure 6.11 (p. 176) diagrams the consequences of mitotic crossing-over between genes *c* and *d* in a cell of the same genotype as the cell in Figure 6.10. After the crossover has occurred, the two pairs of chromatids separate and migrate independently to the metaphase plate, resulting in two possible orientations (a and b in Figure 6.11).

When type a completes mitosis, the two diploid cells 1 and 2 are produced. Progeny cell 1 is homozygous $d^+ e^+/d^+ e^+$ and therefore has a wild-type phenotype indistinguishable from that of the parental cell. Progeny cell 2 is homozygous *d e/d e* and hence expresses the phenotypes associated with these two mutant alleles. For both progeny cell types, all other genes are heterozygous, and thus the phenotypes expressed are those of the dominant, wild-type alleles of those genes: Progeny cell 1 is phenotypically like the parental cell. In general, then, mitotic crossing-over can make all those genes distal to the crossover point homozygous if the chromatid pairs align appropriately at the metaphase plate. This phenomenon applies only to those genes on the same chromosome arm as the crossover; that is, to those genes from the centromere

~ FIGURE 6.10

Normal mitotic segregation of genes in a theoretical diploid cell with one homologous pair of chromosomes. The cell is heterozygous for all five genes on the chromosome. During mitosis each homolog replicates, and the two pairs of chromatids align independently on the metaphase plate. Each progeny cell receives one chromatid of each pair and hence is heterozygous for all the five genes. The progeny are both genetically identical to the parental cell and phenotypically are wild type.

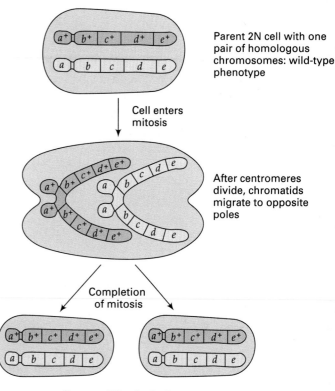

Parent 2N cell with one pair of homologous chromosomes: wild-type phenotype

Cell enters mitosis

After centromeres divide, chromatids migrate to opposite poles

Completion of mitosis

Progeny 2N cells, both with wild-type phenotype

outward. Therefore the crossover between *c* and *d* has no effect on gene *a*, which is on the other arm of this chromosome.

When type b completes mitosis, the two diploid cells 3 and 4 are produced. In both progeny cells, as a result of the orientation of the chromatid pairs at metaphase, all genes are heterozygous, and the cells are phenotypically wild type. Nonetheless, inspection of Figure 6.11 shows that genetic recombination has occurred, since all the genes in 3 are not in coupling, as they were in the parental cell. Since genetic-mapping studies depend on the detection and counting of progeny with phenotypes that are recombined from those found in the parents, the progeny cells from type a metaphase are the ones to analyze in mitotic recombination studies.

~ **FIGURE 6.11**

Result of a mitosis of the same cell type as the cell in Figure 6.10 but in which a rare mitotic crossing-over event occurs.

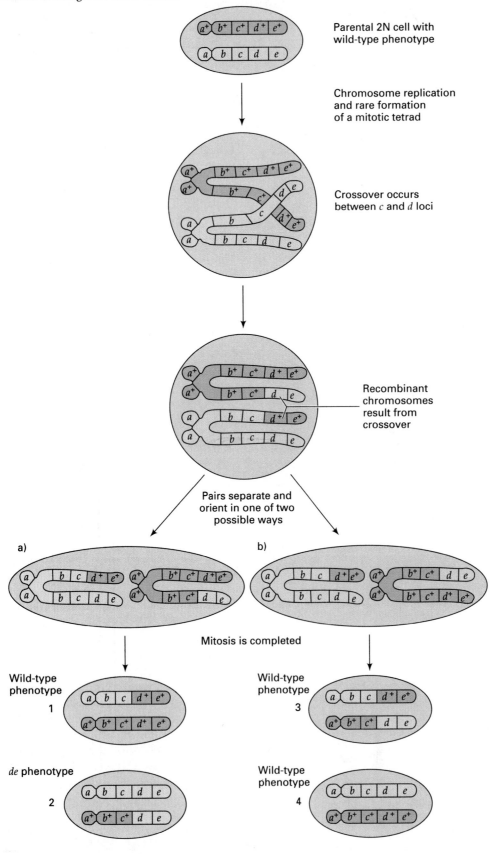

KEYNOTE

Crossing-over can occur during mitosis as well as during meiosis, although it occurs much more rarely during mitosis. As in meiosis, mitotic crossing-over occurs at a four-strand stage. Single crossovers during mitosis can be detected in a heterozygote because loci distal to the crossover and on the same chromosome arm may become homozygous.

Mitotic Recombination in Fungal Systems

Mitotic recombination has been studied most extensively in fungi. Here we will look at some classic experiments done since the 1950s with the fungus *Aspergillus nidulans*, which showed that mitotic recombination analysis can be used to construct genetic maps.

LIFE CYCLE OF *ASPERGILLUS NIDULANS*. The fungus *Aspergillus nidulans* is a mycelial-form fungus like *Neurospora*, and its colonies are greenish. Its use in genetic studies is summarized in Figure 6.12. *Aspergil-*

lus is a suitable organism for genetic studies using mitotic recombination since (1) its asexual spores (cf. the conidia of *Neurospora*) have only a single nucleus—that is, they are uninucleate; (2) the phenotype of each asexual spore is controlled by the genotype of the nucleus it carries; and (3) two haploid strains can be fused by mixing them together. The result of fusion is a mycelium (the typical growth habit of this organism) in which the two nuclear types from the two original haploid strains (distinguished by shading in Figure 6.12) coexist and divide mitotically within the same cytoplasm. A cell—or a collection of cells, as in a mycelium—that possesses any number of genetically different nuclei in a common cytoplasm is called a **heterokaryon** (literally, "different nuclei"). As we shall see, the formation of heterokaryons in *Aspergillus* enables geneticists to construct heterozygous strains that can be used in mitotic recombination studies.

When a heterokaryon of *Aspergillus* produces asexual spores, each spore contains only one of the two haploid nuclear types since the spores are uninucleate (see Figure 6.12). Rarely, two haploid nuclei in a heterokaryon will fuse to produce a diploid nucleus, in a process called *diploidization*. Each uninucleate, asexual spore produced by a diploid strain of *Aspergillus* will have a diploid nucleus. Diploid cells can undergo mitotic crossing-over, thus producing mitotic genetic recombinants. These diploid nuclei are unstable and eventually divide by mitosis to produce haploid progeny nuclei (haploid segregants), which may contain combinations of alleles that differ from those of the two parental haploid nuclei. The process in which haploid nuclei are formed from a diploid nucleus is called **haploidization**.

In haploidization the two homologs of each chromosome independently assort. As with the independent assortment during meiosis, which homolog ends up in which haploid segregant is random. Thus the result of haploidization is similar to that of independent assortment during meiosis—the independent segregation of blocks of genes during haploidization will indicate on which chromosomes the genes belong.

Systems that achieve genetic recombination by means other than the regular alternation of meiosis and fertilization are called **parasexual systems**. In fungi (such as *Aspergillus*) the parasexual cycle consists of the following sequence of events: (1) the formation of a heterokaryon in a multinucleate mycelium, (2) the rare fusion of haploid nuclei differing in genotype within the heterokaryon, (3) mitotic crossing-over within diploid fusion nuclei that multiply side by side with the haploid nuclei, and (4) the subsequent haploidization of the diploid fusion nuclei without a meiotic process. The parasexual cycle has the same effect as a regular meiotic sexual cycle in that it results in genetic recombination.

~ **FIGURE 6.12**

Parasexual reproduction: heterokaryons and diploids in *Aspergillus nidulans*.

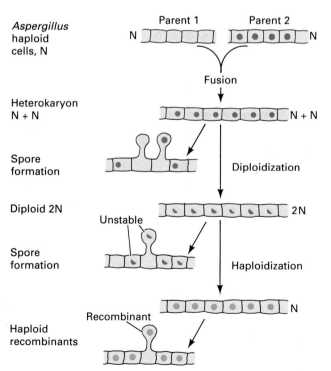

Mitotic recombination analysis in *Asper-gillus nidulans*. As in meiotic recombination analysis, in mitotic recombination analysis it is necessary to construct a strain that is heterozygous for the genes that are to be studied. In *Aspergillus* a heterokaryon is constructed between two strains that differ genetically; then the rare diploid nuclei that form are selected. Usually, the strains to be fused have genotypes that are conducive to the formation of heterokaryons, as we shall see.

For example, consider the following two haploid strains:

Strain 1: $w \quad ad^+ \, pro \quad paba^+ \, y^+ \, bi$
Strain 2: $w^+ \, ad \quad pro^+ \, paba \quad y \quad bi^+$

The alleles *ad, pro, paba,* and *bi* are recessive to their respective wild-type alleles, and they specify that adenine, proline, para-aminobenzoic acid, and biotin, respectively, must be added to the growth medium in order for the strain carrying those mutant alleles to survive. Thus either strain alone cannot grow without the appropriate growth supplements. However, a heterokaryon resulting from the fusion of the two strains requires no growth supplements since all four genes are then heterozygous; this procedure is the typical way of directing the formation of, or forcing, a heterokaryon for experimental analysis.

The *w* and *y* alleles control the color of the asexual spores and hence the overall color of the colony (Figure 6.13). The *w* allele is recessive to the wild-type green allele w^+ and results in white spores. The *y* allele is also recessive to the wild-type green allele y^+ and results in yellow spores. A strain with genotype $w^+ \, y^+$ is green, a $w \, y^+$ strain is white, a $w^+ \, y$ strain is yellow, and a $w \, y$ strain is white because of epistatic effects (see Chapter 4). The heterokaryon of strain 1 and strain 2, however, is not green, as we would expect from the heterozygosity of both genes. Since the spores are uninucleate, their color (and hence the colony coloration) is controlled by the genotype of the nucleus contained in each spore. Therefore the heterokaryon has a mixture of mostly yellow and white spores and has a mottled appearance.

However, some of the spores on the heterokaryotic colony will be green as a result of the rare diploidization process. With respect to the spore coloration the diploid nucleus genotype is $w^+ \, y/w \, y^+$. The diploid spores (detectably larger than haploid spores), which makes it possible to isolate and culture them for study. The diploid cultures derived from these spores require

Stylized diagram of a green *Aspergillus* colony with the starting diploid genotype $w^+ \, y^+/w \, y$. Through haploidization and mitotic crossovers, new haploid and diploid genotypes are produced and are detected as colored sectors: green (not detectable in the green colony background), yellow, and white.

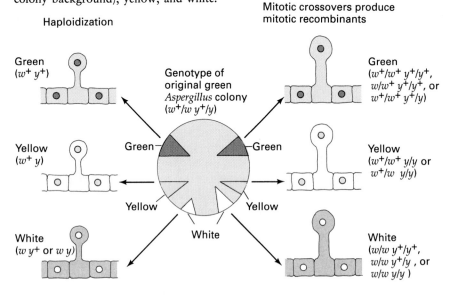

no growth supplements, since wild-type alleles for all the nutritional-requirement genes are present: $w^+ ad^+ pro\ paba^+ y^+ bi/w\ ad\ pro^+ paba\ y\ bi^+$.

When a diploid *Aspergillus* spore with the genotype above is used to inoculate a solid growth medium, a colony will be produced that radiates from the point of inoculation. As long as the spores produced are diploid and have the same spore coloration genotype as the inoculating spore (i.e., $w^+ y/w\ y^+$), the colony will be green. During the growth of the green colony, either haploidization, mitotic crossing over, or both may occur. When spore coloration genes are involved, the cells and their offspring give rise to sectors of different colors as the colony continues to grow (see Figure 6.13). Three types of colored sectors are produced, with genotypes that differ from the diploid spore used to inoculate the medium: green (indistinguishable from the parental colony phenotype), yellow, and white. With respect to the spore coloration genes, the haploid segregants and diploid recombinant genotypes for the three types of sectors are as shown in Table 6.1.

When diploid *Aspergillus* cells reproduce to yield colonies, the colonies are predominantly green. Both haploid and diploid white and yellow sectors may also be produced, although their incidence is rare (see Figure 6.13). Whether the sector is haploid or diploid can be determined by examining the size of the spores since diploid spores are the larger of the two. All haploid sectors are produced by haploidization. As an example, let us consider the haploid white sectors. About half the white haploid sectors have the genotype $w\ ad^+ pro\ paba^+ y^+ bi$, and half have the genotype $w\ ad\ pro^+ paba\ y\ bi^+$. With the exception of the common white allele, these two genotypes are reciprocals.

~ TABLE 6.1

Haploid Segregants and Diploid Recombinant Genotypes for *Aspergillus* Experiment

PHENOTYPE	GENOTYPE	
	HAPLOID	DIPLOID
green	$w^+ y^+$	$w^+/w^+\ y^+/y^+$
		$w^+/w\ \ y^+/y^+$
		$w^+/w^+\ y^+/y$
yellow	$w^+ y$	$w^+/w\ \ y\ /y$
		$w^+/w^+\ y\ /y$
white	$w\ \ y^+$	$w\ /w\ \ y^+/y^+$
	$w\ \ y$	$w\ /w\ \ y^+/y$
		$w\ /w\ \ y\ /y$

The 50:50 segregation of the two sets of five alleles indicates that they are located on a different chromosome from that carrying the w gene. Thus, in some haploid white sectors a chromosome v $ad\ pro^+ paba\ y\ bi^+$ has segregated, while in others its homolog with $ad^+ pro\ paba^+ y^+ bi$ alleles is segregated. The interpretation of the haploid white sector data is that the six gene loci are located on two nonhomologous chromosomes; the white gene is on one chromosome, and the other five genes are on the other. It is not possible to determine gene order by this analysis, although the correct gene order is given in the following diagram:

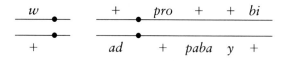

To put the above analysis in context, remember that the study of haploid segregants provides essential information about which genes are linked on which chromosomes. Once that information is known, the next stage of the mitotic analysis is to establish gene order and determine map distances between genes on the same chromosome.

To determine gene order and establish map distance, we must study a second type of segregant, the diploid segregants. From the diploid cell we constructed, the easily identifiable diploid segregants are those having white or yellow sectors. That such sectors are diploid would be established by examining spore size. The selected diploid segregants may then be analyzed phenotypically to determine the genotypes for the other genes on the chromosomes.

The diploid segregants are produced by mitotic crossing-over. Since this event is very rare, for all intents and purposes, only single crossovers need to be considered in each chromosome arm. As was discussed earlier, a crossing-over event in mitosis makes all those genes distal to the crossing-over point (on the same chromosome arm) homozygous. Thus any recessive alleles that are heterozygous in the diploid cell may become homozygous recessive as a result of the crossover, and a new phenotype, the recessive one, will be manifested.

One way in which a diploid yellow sector can arise is diagramed in Figure 6.14 (p. 180). A mitotic crossover between the *pro* and *paba* genes has produced a segregant that is homozygous for the y allele and hence is yellow. The same crossover produces a twin spot that is homozygous y^+/y^+, but since it is green, it is not detected in the overall green color of the colony. The crossover diagramed has made all genes distal to that

~ Figure 6.14

Between the *pro* and *paba* loci, a possible mitotic crossing-over event that can give rise to a diploid yellow sector in the green diploid *Aspergillus* strain of Figure 6.12.

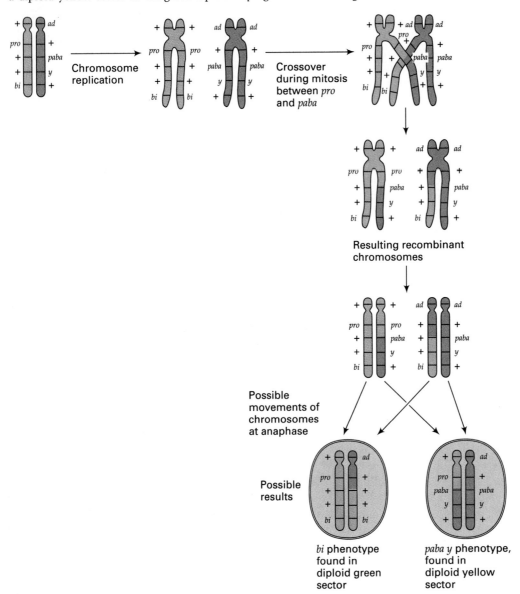

point homozygous. Therefore the yellow segregant is also *paba/paba* and requires para-amino-benzoic acid in order to grow. The homozygosity for the *bi⁺* allele goes undetected. One other crossover that could produce a diploid yellow sector is a crossover between the *paba* and *y* loci (Figure 6.15). In this case the yellow sector is still heterozygous for *paba*, since mitotic crossing-over produces homozygosity only for genes distal to the crossover point.

These facts give us a way to determine the gene order for each chromosome arm. A gene marker that is far away from the centromere is chosen, and then diploid segregants homozygous for that gene as a result of mitotic recombination are isolated. With a number of genes marking the chromosome from the centromere out to the distal marker, it is then a simple matter to see which sets of genes become homozygous in the various segregants. In the example, *y* and *paba y* genotypes were found in the various yellow sectors. From the mechanics of mitotic recombination the order of genes is centromere-*paba-y*. We cannot assign a position for those genes that become homozygous wild type, so

~ **FIGURE 6.15**

Production of diploid yellow sector in a green diploid *Aspergillus* strain by mitotic crossing-over between the *paba* and *y* loci.

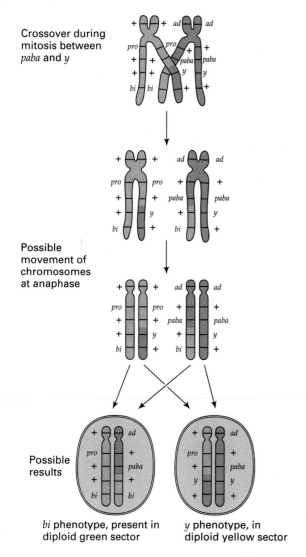

Crossover during mitosis between *paba* and *y*

Possible movement of chromosomes at anaphase

Possible results

bi phenotype, present in diploid green sector

y phenotype, in diploid yellow sector

KEYNOTE

The parasexual cycle describes genetic systems that achieve genetic recombination by means other than the regular alternation of meiosis and fertilization. The parasexual cycle in fungi such as *Aspergillus* consists of (1) the formation of a heterokaryon by mycelial fusion and then fusion of the two haploid nuclei to give a diploid nucleus; (2) mitotic crossing-over within the diploid nucleus; (3) haploidization of the diploid nucleui without meiosis, a process that produces haploid nuclei into which one or the other of each parental chromosome has segregated randomly. ─────────

MAPPING GENES IN HUMAN CHROMOSOMES

For practical reasons, with humans it is not possible to do genetic mapping experiments of the kind discussed for other organisms. Nonetheless, we have a strong interest in mapping genes in human chromosomes, since there are so many known diseases and traits that have a genetic basis. In Chapters 2 and 3 we saw that pedigree analysis could be used to determine the mode by which a particular genetic trait is inherited. In this way many genes have been localized to the X chromosome. However, pedigree analysis cannot show on which chromosome a particular autosomal gene is located.

Although modern statistical and computer analysis has made it possible to obtain some linkage data for autosomal genes, many of the more than 1200 known autosomal genes have not been located on one of the 23 autosomes. Moreover, probably between 10,000 and 50,000 genes have not even been identified yet, much less located on all the chromosomes. In this section, we will discuss some traditional ways to map human genes. One traditional method involves recombination analysis. Another method entails the fusion of human and rodent cultured cells in a process called **somatic cell hybridization**. In later chapters we will introduce some of the modern approaches that are now being used.

Mapping Human Genes by Recombination Analysis

As we have seen, it is not possible to set up appropriate testcrosses for human genetic mapping by recombination analysis. In a very few cases certain pedigrees have included individuals with appropriate genotypes to

their relative positions must be determined from other mitotic recombination experiments.

After diploid segregants, which are produced by mitotic recombination, are obtained, they can be counted. Such quantitative data can be used to compute map distance between the genes; that is, the mitotic recombination frequency between genes can be computed by using the same formula used in meiotic recombination studies. The map distance between the *paba* and *y* genes, for example, is given by the percentage of yellow segregants that result from crossing-over in the *paba-y* region. Those particular segregants are still heterozygous for *paba* and *pro*.

permit analysis of linkage between autosomal genes. Recombination analysis in humans has been carried out more for X-linked genes, however, because the hemizygosity of the X chromosome in males has provided a rich source of useful genotypic pairings in pedigrees.

Consider the following theoretical example (Figure 6.16). A male with two rare X-linked recessive alleles a and b marries a woman who expresses neither of the traits involved. Since the traits are rare, it is likely that the woman is homozygous for the wild-type allele of each gene; that is, she is $a^+ b^+/a^+ b^+$. A female offspring from these parents would be doubly heterozygous $a^+ b^+/a b$. Recombinant gametes from this female would be produced by crossing-over between the two genes at a frequency related to the genetic distance that separates them.

If this female pairs with a normal $a^+ b^+/Y$ male, all female progeny will be $a^+ b^+$ in phenotype because of the $a^+ b^+$ chromosome transmitted from the father. The male progeny, however, will express all four possible phenotype classes because of the hemizygosity of the X chromosome; that is, the parental $a^+ b^+$ and $a b$, and the recombinant $a^+ b$ and $a b^+$. Thus, analysis of the male progeny from pairings such as this ($a^+ b^+/a b \times a^+ b^+/Y$) in a large number of pedigrees will produce a value for the frequency of recombination between the two loci involved, and an estimate of genetic map distance can be obtained.

Using this approach a number of genes have been mapped along the human X chromosome. For example, it was found that the distance between the green weakness gene, g (the recessive allele responsible for a form of color blindness) and the hemophilia A gene, h, is 8 map units.

Mapping Human Genes by Somatic Cell Hybridization Techniques

FORMATION OF SOMATIC CELL HYBRIDS. Somatic cells can be isolated from a multicellular eukaryote and under appropriate conditions can be cultured in vitro to give rise to what is known as a *cell culture line*. If the cell used to initiate the line is taken from mature somatic tissue, the cell line tends to have a finite life time, during which there is no change in the chromosomal constitution of progeny cells from that found in the parental cell. More useful, though, are established cell lines. The cells for such lines have been selected to grow and divide essentially indefinitely in cell culture. Often the cell lines are derived from malignant tissues, and often the chromosomal constitution differs from that characteristic of the wild-type parental organism.

Figure 6.17 diagrams the formation of a somatic cell hybrid of a human cell and an established cell line from the mouse. To cause the cells to fuse, geneticists mix suspensions of the two cell types with polyethylene glycol (PEG) or inactivated sendai which causes the membranes to fuse. Once the cells have fused, the nuclei of the now binucleate heterokaryon often fuse to form a single nucleus enclosing both sets of chromosomes. Each cell of this uninucleate cell line is called a **synkaryon.** The cell line reproduces by mitosis, so that each original synkaryotic cell gives rise to a colony of synkaryotic cells that can then be studied.

USING THE HAT TECHNIQUE TO LOCATE HYBRID CELLS. One of the problems in producing a hybrid cell is finding it among the parental cells. Fortunately, a convenient selective procedure prevents parental cells from growing but allows hybrid cells to reproduce. This procedure is the *HAT technique* (hypoxanthine-aminopterin-thymidine). The HAT technique takes advantage of conditions under which only hybrid cells are able to synthesize DNA and thus divide and increase in number. The nonhybrid cells are unable to synthesize DNA and hence they cannot divide.

To understand how the HAT technique can work, we must understand a little about the biochemistry of DNA synthesis. In the cell, DNA synthesis requires the presence of molecules called purine and pyrimidine triphosphates, which are made from their respective

~ **FIGURE 6.16**

Calculation of recombination frequency for two X-linked human genes by analyzing the male progeny of a woman doubly heterozygous for the two genes.

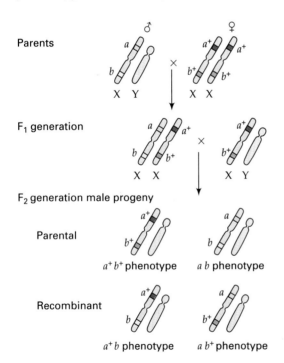

monophosphates. (The structure and synthesis of DNA will be developed more fully in later chapters.) The purine and pyrimidine monophosphates can be generated either by their synthesis from simpler precursor molecules or by what are called *salvage pathways,* in

which enzymes recycle purines and pyrimidines produced by degradation of DNA or RNA. If the drug aminopterin is present in the culture medium (as it is in the HAT technique), the synthesis of new purines and pyrimidines from precursors is inhibited. As a consequence, the cells become completely dependent on the salvage pathways for the production of purine and pyrimidine triphosphates needed for the synthesis of DNA.

Figure 6.18 illustrates the use of the HAT technique to produce a human-mouse hybrid cell. In this technique the two fused cells have different genetic defects in the purine and pyrimidine salvage pathways. In general, aminopterin will inhibit the new synthesis of

~ **FIGURE 6.17**

Technique for producing a human-mouse hybrid somatic cell. Cell fusion results in cells that contain all the mouse chromosomes and a set of human chromosomes. (A fibroblast cell is a somatic cell from fibrous connective tissue.)

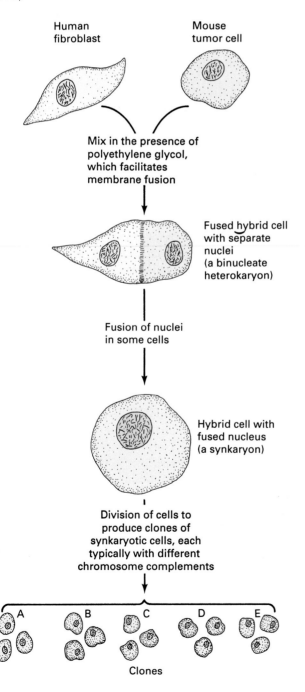

~ **FIGURE 6.18**

HAT technique for selecting fused, hybrid mouse-human cells.

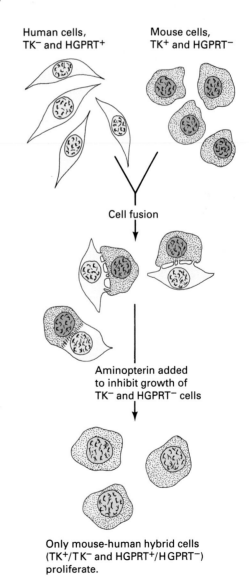

purines and pyrimidines from precursors so that the cells have to rely on the salvage pathways for purine and pyrimidine production. The genetic defects facilitate the isolation of hybrid cells: One cell line will lack an enzyme needed in the purine salvage pathway, and the other cell line will lack an enzyme needed in the pyrimidine salvage pathway. In the example, the mouse cell line is defective in the activity of hypoxanthine phosphoribosyl transferase (HGPRT), an enzyme needed for the purine salvage pathway. These cells do have the enzyme thymidine kinase (TK), which is needed for the pyrimidine salvage pathway. On the other hand, the human cell line has normal HGPRT activity but is deficient in TK activity. In the presence of aminopterin neither human nor mouse cells can grow, since the human cells cannot make pyrimidine monophosphates, and the mouse cells cannot make the purine monophosphates. However, any hybrid cells formed by fusion of the two cell types will grow, because the human chromosomes contribute a normal HGPRT gene and the mouse chromosomes contribute a normal TK gene. Furthermore, the HAT medium contains hypoxanthine (which is a precursor of purines) and thymidine (which is a precursor of pyrimidines). Therefore both enzymes can be made, both salvage pathways can operate, and DNA synthesis can occur. The HAT technique is an excellent example of using gene defects in a powerful selection scheme.

KEYNOTE

Geneticists can make a somatic cell hybrid between a human cell and a mouse cell line. The two cell types are fused by the presence of a suitable agent such as polyethylene glycol. Once the cells have fused, the nuclei often fuse to form a uninucleate cell line called a synkaryon.

USING HYBRID SOMATIC CELLS TO LOCATE GENES. Two interesting properties of human/rodent hybrid cells make them particularly useful for somatic cell genetics. First, the human and rodent chromosomes are readily distinguishable under the microscope, particularly with the use of fluorescent dyes, which produce characteristic banding patterns on the chromosomes. The 46 human chromosomes vary extensively in size, while the 40 mouse chromosomes are quite small and uniform. And, since all mouse cells are telocentric, distinguishing between them and human chromosomes is simplified even more. Second, as the hybrid cell

reproduces, the number of chromosomes decreases because some chromosomes are discarded; for unknown reasons human chromosomes are preferentially lost. On the selective medium with aminopterin just described, the eventual surviving, stable hybrid cell line is usually one that contains a complete set of mouse chromosomes plus a small number of human chromosomes, which vary in number and type from cell line to cell line. If a line is to survive, it must contain the human chromosome that supplies a function that is deficient in the rodent genome.

To map human genes by using hybrid somatic cell lines, we start with a human cell that carries one or more genetic markers. Generally, we use markers that can readily be detected by analyzing tissue culture cells. Examples of markers include genes that control resistance to antibiotics and to other drugs, that code for enzymes, that determine nutritional requirements, and that specify cell surface antigens detectable by the use of fluorescence-labeled antibodies.

Once the human genetic markers are chosen, various stable hybrid cell lines (i.e., cultures no longer losing chromosomes) are analyzed for the presence or absence of such markers. These data are correlated for a number of somatic cell lines with the presence or absence of particular chromosomes. Given enough cell lines, we can show that a particular marker is present only when one particular chromosome is present, and the marker is absent when that chromosome is absent. For example, if a drug resistance phenotype is detected in several different cell lines that have in common human chromosome 14, then we would conclude that the gene controlling resistance to the drug is on chromosome 14. In this way genes amenable to analysis can be localized to individual chromosomes. Genes localized to a particular chromosome through this experimental approach are called **syntenic** ("together thread," i.e., another term for *linked*).

With this experimental technique many genes have been localized to individual chromosomes in the human genome. In some instances, further localization of genes to particular regions of chromosomes has also been possible. In these experiments the cells chosen have chromosomal abnormalities. For example, in some cells a part of a chromosome may spontaneously break off and be lost (**deleted**) or become attached to another chromosome (**translocated**). If either of these events occurs, we can reinvestigate the location of genes known to be on the affected chromosome or chromosomes by forming hybrid somatic cell lines between the human cell that contains the chromosome and a deletion or a translocated chromosome. The loss or addition of various pieces of a particular chromosome can

~ FIGURE 6.19

Using somatic cell hybridization for determining that a human gene responsible for production of an enzyme is located in the short arm of chromosome 6. (a) In cell line A the two human chromosomes remaining are 6 and 12 and the enzyme can be detected. (b) In cell line B both chromosomes 6 and 12 are present, but chromosome 6 has suffered a deletion. Since the enzyme is not detectable, the gene for the enzyme must have been on the deleted segment of chromosome 6.

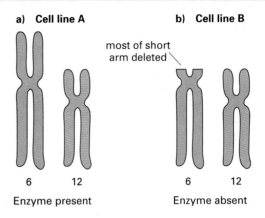

then be correlated with the presence or absence of specific genetic markers.

This concept, involving a deletion of part of a chromosome, is depicted in Figure 6.19. Here, the enzyme is detectable in cell line A which contains both chromosomes 6 and 12 (Figure 6.19a), thereby indicating that the gene for the enzyme is on either chromosome 6 or 12. Obviously, a cell line with only one of the two chromosomes would indicate which chromosome carried the gene by the presence or absence of the enzyme activity. In cell line B (Figure 6.19b) both chromosomes are present, but chromosome 6 has suffered a deletion of most of the short arm of that chromosome. The absence of enzyme activity in this cell line indicates that the gene for the enzyme must be on the short arm of chromosome 6. While this method gives fairly specific information about the location of genes on human chromosomes, it is still not as precise a method as conventional genetic-mapping methods used with laboratory organisms.

As a summary, Figure 6.20 (p. 186) shows a genetic map of human chromosomes, indicating some of the details of the banding patterns. Table 6.2 (pp. 187–189) gives the key to the symbols used in the figure. Despite the wealth of information in the figure and the table, hundreds of known human genes still remain to be localized.

𝒦EYNOTE

Hybrid somatic cell lines help in localizing human genes to particular chromosomes. Within the nucleus the chromosomes of the two cell types are easily distinguished. Suitable genes for mapping are those that can be detected by analyzing tissue culture cells, such as genes that affect biochemical requirements or enzyme activities. As a somatic cell hybrid reproduces, a preferential and random loss of human chromosomes occurs. Eventually, stable hybrid lines are established with various subsets of the human chromosomes. With enough cell lines of this kind, a particular trait can be identified with a particular human chromosome.

SUMMARY

In this chapter we have learned how to map genes in certain microorganisms by using tetrad analysis. In a subset of those microorganisms, the meiotic tetrads are ordered in a way that reflects the orientation of the four chromatids of each homologous pair of chromosomes at metaphase I. Ordered tetrads make it possible to map a gene's location relative to its centromere. In this chapter we also discovered how to map genes by using nonmeiotic means. The meiotic gene-mapping methods described in the previous chapter cannot be used on certain eukaryotes, so the discovery of nonmeiotic mapping procedures has been beneficial in the sense that it allows us to map genes in those eukaryotes. For example, genes have been located on human chromosomes by using somatic cell hybrids between human and rodent cells; the hybrids preferentially and randomly discard human chromosomes.

~ FIGURE 6.20

Genetic map of human chromosomes. The key to the symbols is found in Table 6.2.

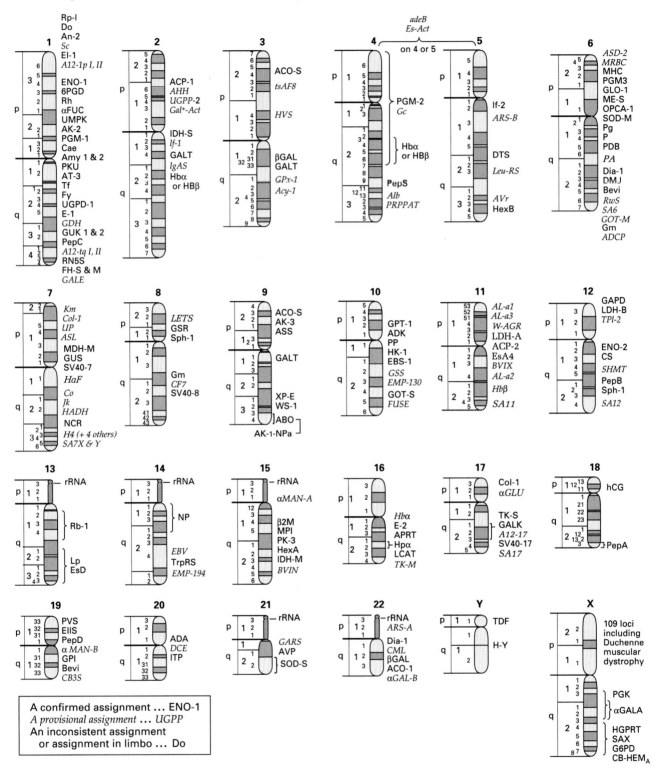

A confirmed assignment ... ENO-1
A provisional assignment ... UGPP
An inconsistent assignment
or assignment in limbo ... Do

~ TABLE 6.2

Key to Gene Symbols in Figure 6.20

ABO	= ABO blood group (chr. 9)
ACO–M	= Aconitase, mitochondrial (chr. 22)
ACO–S	= Aconitase, soluble (chr. 9)
ACP–1	= Acid phosphatase-1 (chr. 2)
#ACP–2	= Acid phosphatase-2 (chr. 11)
Acy–1	= Aminoacylase-1 (chr. 3)
adeB	= Formylglycinamide ribotide amidotransferase (chr. 4 or 5)
#ADA	= Adenosine deaminase (chr. 20)
#ADCP	= Adenosine deaminase complexing protein (chr. 6)
ADK	= Adenosine kinase (chr. 10)
A12–1pI,II	= Adenovirus-12 chromosome modification site-1p I and II (chr. 1 p)
A12–1qI,II	= Adenovirus-12 chromosome modification site-1q I and II (chr. 1 q)
A12–17	= Adenovirus-12 chromosome modification site-17 (chr. 17)
#AH–3	= Adrenal hyperplasia III (21-hydroxylase deficiency) (chr. 6)
AHH	= Aryl hydrocarbon hydroxylase (chr. 2)
AK–1	= Adenylate kinase-1 (chr. 9)
AK–2	= Adenylate kinase-2 (chr. 1)
AK–3	= Adenylate kinase-3 (chr. 9)
AL	= Lethal antigen: 3 loci (a1, a2, a3) (chr. 11)
#Alb	= Albumin (chr. 4)
Amy–1	= Amylase, salivary (chr. 1)
Amy–2	= Amylase, pancreatic (chr. 1)
#An–2	= Aniridia, type II Baltimore (chr. 1)
#ARS–A	= Arylsulfatase A (chr. 22)
#ARS–B	= Arylsulfatase B (chr. 5)
#ARPT	= Adenine phosphoribosyltransferase (chr. 16)
#ASD–2	= Atrial septal defect, secundum type (chr. 6)
#ASL	= Argininosuccinate lyase (chr. 7)
#ASS	= Argininosuccinate synthetase (chr. 9)
#AT–3	= Antithrombin III (chr. 1)
AVP	= Antiviral protein (chr. 21)
AVr	= Antiviral state regulator (chr. 5)
Bevi	= Baboon M7 virus infection (chr. 6 or 19)
Bf	= Properdin factor B (chr. 6)
$\beta 2M$	= $\beta 2$–microglobulin (chr. 15)
BVIN	= BALB virus induction, N-tropic (chr. 15)
BVIX	= BALB virus induction xenotropic (chr. 11)
#C2	= Complement component-2 (chr. 6)
C4F	= Complement component-4 fast (chr. 6)
C4S	= Complement component-4 slow (chr. 6)
C6	= Complement component-6 (chr. 6)
#C8	= Complement component-8 (chr. 6)
#Cae	= Cataract, zonular pulverulent (chr. 1)

CB	= Color blindness (deutan and protan) (X chr.)
CB3S	= Coxsackie B3 virus susceptibility (chr. 19)
#CF7	= Clotting factor VII (chr. 8)
Ch	= Chido blood group (chr. 6)—same as C4S
#CML	= Chronic myeloid leukemia (chr. 22)
Co	= Colton blood group (chr. 7)
#Col–1	= Collagen I (α—1 and α—2) (chr. 7 and 17)
CS	= Citrate synthase, mitochondrial (chr. 12)
#Dia–1	= NADH-diaphorase (chr. 6 or 22)
#DMJ	= Juvenile diabetes mellitus (chr. 6)
Do	= Dombrock blood group (chr. 1)
DCE	= Desmosterol-to-cholesterol enzyme (chr. 20)
#DTS	= Diphtheria toxin sensitivity (chr. 5)
#EBS–1	= Epidermolysis bullosa, Ogna type (chr. 10)
#EBV	= Epstein-Barr virus integration site (chr. 14)
#E–1	= Pseudocholinesterase-1 (chr. 1)
#E–2	= Pseudocholinesterase-2 (chr. 16)
E11–S	= Echo 11 sensitivity (chr. 19)
E1–1	= Elliptocytosis-1 (chr. 1)
EMP–130	= External membrane protein-130 (chr. 10)
EMP–195	= External membrane protein-195 (chr. 14)
ENO–1	= Enolase-1 (chr. 1)
ENO–2	= Enolase-2 (chr. 12)
Es–Act	= Esterase activator (chr. 4 or 5)
EsA4	= Esterase-A4 (chr. 11)
EsD	= Esterase D (chr. 13)
FH–M	= Fumarate hydratase, mitochondrial (chr. 1)
FH–S	= Fumarate hydratase, soluble (chr. 1)
#αFUC	= Alpha-L-fucosidase (chr. 1)
FUSE	= Polykaryocytosis inducer (chr. 10)
#Fy	= Duffy blood group (chr. 1)
Gal+-Act	= Galactose + activator (chr. 2)
#αGAL A	= α-galactosidase A (Fabry disease) (X chr.)
αGAL B	= α-galactosidase B (chr. 22)
#βGAL	= β-galactosidase (chr. 3) (see also chr. 12 and 22)
#GALK	= Galactokinase (chr. 17)
#GALT	= Galactose-1-phosphate uridyltransferase (chr. 2, 3, or 9)
#GALE	= Galactose-4-epimerase (chr. 1)
#αGLU	= α-glucosidase (chr. 17)
GAPD	= Glyceraldehyde-3-phosphate dehydrogenase (chr. 12)
GAPS	= Phosphoribosyl glycinamide synthetase (chr. 21)

~ TABLE 6.2 continued

Key to Gene Symbols in Figure 6.20

Gc	= Group-specific component (chr. 4)
GDH	= Glucose dehydrogenase (chr. 1)
GLO–1	= Glyoxylase I (chr. 6)
Gm	= Immunoglobulin heavy chain (chr. 8)
GOT–M	= Glutamate oxaloacetate transaminase, mitochondrial (chr. 6)
GOT–S	= Glutamate oxaloacetate transaminase, soluble (chr. 10)
#GPI	= Glucosephosphate isomerase (chr. 19)
GPT–1	= Glutamate pyruvate transaminase, soluble (chr. 10)
#GPx–1	= Glutathione peroxidase-1 (chr. 3)
#G6PD	= Glucose-6-phosphate dehydrogenase (X chr.)
#GSR	= Glutathione reductase (chr. 8)
GSS	= Glutamate-gamma-semialdehyde synthetase (chr. 10)
GARS	= Glycinamide ribonucleotide synthetase (chr. 21)
GUK–1 & 2	= Guanylate kinase-1 & 2 (chr. 1)
#GUS	= Beta-glucuronidase (chr. 7)

H4	= Histone H4 and 4 other histone genes (chr. 7)
HADH	= Hydroxyacyl-CoA dehydrogenase (chr. 7)
HaF	= Hageman factor (chr. 7)
#Hbα	= Hemoglobin alpha chain (chr. 2, 4, 5, or 16)
#Hbβ	= Hemoglobin beta chain (chr. 2, 4, 5, or 11)
hCG	= Human chorionic gonadotropin (chr. 18)
#Hch	= Hemochromatosis (chr. 6)
#HEM–A	= Classic hemophilia (X chr.)
#HexA	= Hexosaminidase A (chr. 15)
#HexB	= Hexosaminidase B (chr. 5)
#HGPRT	= Hypoxanthine-guanine phosphoribosyltransferase (X chr.)
HK–1	= Hexokinase-1 (chr. 10)
HLA (A–D)	= Human leukocyte antigens (chr. 6)
HLA–DR	= Human leukocyte antigen, D-related (chr. 6)
Hpα	= Haptoglobin, alpha (chr. 16)
#HVS	= Herpes virus sensitivity (chr. 6)
H–Y	= Y histocompatibility antigen (Y chr.)

IgAS	= Immunoglobulin heavy chains attachment site (chr. 2)
If–1	= Interferon-1 (chr. 2)
If–2	= Interferon-2 (chr. 5)
IDH–M	= Isocitrate dehydrogenase, mitochondrial (chr. 15)
IDH–S	= Isocitrate dehydrogenase, soluble (chr. 2)
#ITP	= Inosine triphosphatase (chr. 20)

Jk	= Kidd blood group (chr. 7)
Km	= Kappa immunoglobulin light chains, Inv (chr. 7)

#LAP	= Laryngeal adductor paralysis (chr. 6)
#LCAT	= Lecithin-cholesterol acyltransferase (chr. 16)
LDH–A	= Lactate dehydrogenase A (chr. 11)
LDH–B	= Lactate dehydrogenase B (chr. 12)
LETS	= Large, external, transformation-sensitive protein (chr. 8)
LeuRS	= Leucyl-tRNA synthetase (chr. 5)
Lp	= Lipoprotein β–Lp (chr. 13)

αMAN–A	= Cytoplasmic α–D-mannosidase (chr. 15)
#αMAN–B	= Lysosomal α–D-mannosidase (chr. 19)
#MDH–M	= Malate dehydrogenase, mitochondrial (chr. 7)
MDH–S	= Malate dehydrogenase, soluble (chr. 2)
ME–S	= Malic enzyme, soluble (chr. 6)
MHC	= Major histocompatibility complex (chr. 6)
MLC–W	= Mixed lymphocyte culture, weak (chr. 6)
MPI	= Mannosephosphate isomerase (chr. 15)
MRBC	= Monkey red blood cell receptor (chr. 6)
MTR	= 5-Methyltetrahydrofolate: L-homocysteine S–S methyltransferase (chr. 1)

NCR	= Neutrophil chemotactic response (chr. 7)
NDF	= Neutrophil differentiation factor (chr. 6)
#NP	= Nucleoside phosphorylase (chr. 14)
#NPa	= Nail-patella syndrome (chr. 9)

#OPCA–1	= Olivopontocerebellar atrophy I (chr. 6)

P	= P blood group (chr. 6)
PA	= Plasminogen activator (chr. 6)
#PDB	= Paget disease of bone (chr. 6)
PepA	= Peptidase A (chr. 18)
PepB	= Peptidase B (chr. 12)
PepC	= Peptidase C (chr. 1)
PepD	= Peptidase D (chr. 19)
PepS	= Peptidase S (chr. 4)
Pg	= Pepsinogen (chr. 6)
PGK	= Phosphoglycerate kinase (X chr.)
PGM–1	= Phosphoglucomutase-1 (chr. 1)
PGM–2	= Phosphoglucomutase-2 (chr. 4)
PGM–3	= Phosphoglucomutase-3 (chr. 6)
6PGD	= 6-phosphogluconate dehydrogenase (chr. 1)
PRPPAT	= Phosphoribosylpyrophosphate amidotransferase (chr. 4)
PK–3	= Pyruvate kinase-3 (chr. 15)
#PKU	= Phenylketonuria (chr. 1)
PP	= Inorganic pyrophosphatase (chr. 10)
#PVS	= Polio sensitivity (chr. 19)

#RB–1	= Retinoblastoma-1 (chr. 13)
rC3b	= Receptor for C3b (chr. 6)
rC3d	= Receptor for C3d (chr. 6)
Rg	= Rodgers blood group (chr. 6)—same as C4F

~ TABLE 6.2 continued

Key to Gene Symbols in Figure 6.20

#Rh	= Rhesus blood group (chr. 1)
RN5S	= 5S RNA gene(s) (chr. 1)
#RP–1	= Retinitis pigmentosa-1 (chr. 1)
rRNA	= Ribosomal RNA (chr. 13, 14, 15, 21, 22)
#RwS	= Ragweed sensitivity (chr. 6)

SA6	= Surface antigen 6 (chr. 6)
SA7	= Surface antigen 7 (chr. 7)
SA11	= Surface antigen 11 (chr. 11)
SA12	= Surface antigen 12 (chr. 12)
SA17	= Surface antigen 17 (chr. 17)
SAX	= X-linked species (or surface) antigen (X chr.)
Sc	= Scianna blood group (chr. 1)
SHMT	= Serine hydroxymethyltransferase (chr. 12)
SOD–M	= Superoxide dismutase, mito-chondrial (chr. 6)
SOD–S	= Superoxide dismutase, soluble (chr. 21)
Sph–1	= Sherocytosis, Denver type (chr. 8 or 12)
SV40–7	= SV40-integration site-7 (chr. 7)
SV40–8	= SV40-integration site-8 (chr. 8)
SV40–17	= SV40-integration site-17 (chr. 17)

TDF	= Testis-determining factor (Y chr.)—probably same as H−Y
Tf	= Transferrin (chr. 1)
TK–M	= Thymidine kinase, mitochondrial (chr. 16)
TK–S	= Thymidine kinase, soluble (chr. 17)
#TPI–1 & 2	= Triosephosphate isomerase-1 and −2 (chr. 12)
TrpRS	= Tryptophanyl-tRNA synthetase (chr. 14)
tsAF8	= Temperature-sensitive (AF8) complement (chr. 3)

UGPP–1	= Uridyl diphosphate glucose pyrophosphorylase-1 (chr. 1)
UGPP–2	= Uridyl diphosphate glucose pyrophosphorylase-2 (chr. 2)
UMPK	= Uridine monophosphate kinase (chr. 1)
UP	= Uridine phosphorylase (chr. 7)

#WS–1	= Waardenburg syndrome-1 (chr. 9)
#W–AGR	= Wilms tumor—aniridia/ambiguous genitalia/mental retardation (chr. 11)

#XP–E	= Xeroderma pigmentosum, Egyptian (chr. 9)

ANALYTICAL APPROACHES FOR SOLVING GENETICS PROBLEMS

Q.1 A *Neurospora* strain that required both adenine (*ad*) and tryptophan (*trp*) for growth was mated to a wild-type strain (*ad*$^+$ *trp*$^+$), and this cross produced the following ordered tetrads:

Spore pair 1:	*ad trp*	*ad* +	*ad trp*	*ad trp*
Spore pair 2:	*ad trp*	*ad* +	*ad* +	+ *trp*
Spore pair 3:	+ +	+ *trp*	+ *trp*	*ad* +
Spore pair 4:	+ +	+ *trp*	+ +	+ +
	(1) 63	(2) 3	(3) 15	(4) 9

Spore pair 1:	*ad trp*	*ad* +	+ *trp*
Spore pair 2:	+ +	+ *trp*	+ +
Spore pair 3:	*ad trp*	*ad* +	*ad trp*
Spore pair 4:	+ +	+ *trp*	*ad* +
	(5) 3	(6) 1	(7) 6

a. Determine the gene-centromere distance for the two genes.

b. From the data given, calculate the map distance between the two genes.

A.1 a. The gene-centromere distance is given by the formula

$$\frac{\text{percent second-division tetrads}}{2} = \chi \text{ mu from centromere}$$

For the *ad* gene tetrads 4, 5, and 6 show second-division segregation; the total number of such tetrads is 13. There are 100 tetrads, so the *ad* gene is (13/2)% = 6.5 mu from its centromere. For the *trp* gene tetrads, 3, 5, 6, and 7 show second-division segregation, and the total number of such tetrads is 25, indicating the *trp* gene is 12.5 mu from its centromere.

b. The linkage relationship between the genes can be determined by analyzing the relative number of parental-ditype (PD), nonparental-ditype (NPD), and tetratype (T) tetrads. Tetrads 1 and 5 are PD, 2 and 6 are NPD, and 3, 4, and 7 are T. If two genes are unlinked, the frequency of PD tetrads will approximately equal the frequency of NPD tetrads. If two genes are linked, the frequency of PD tetrads will greatly exceed that of the NPD tetrads. Here the latter case prevails, so the two genes must be linked. The map distance between two genes is given by the general formula

$$\frac{1/2 \text{ T} + \text{NPD}}{\text{total}} \times 100$$

For this example the number of T tetrads is 30, and the number of NPD tetrads is 4. Thus the map distance between *ad* and *trp* is

$$\frac{(1/2 \times 30) + 4}{100} \times 100 = 19 \text{ mu}$$

Thus we have the following map, with ● indicating the centromere:

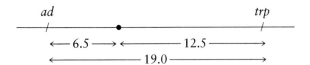

Q.2 In *Aspergillus*, forced diploids were constructed between a wild-type strain and a strain containing the mutant genes *y* (yellow), *w* (white), *pro* (proline requirement), *met* (methionine requirement), and *ad* (adenine requirement). All these genes are known to be in a single chromosome.

Homozygous yellow and homozygous white segregants were isolated and analyzed for the presence of the other gene markers. The following phenotypic results were obtained:

y/y segregants:	w^+ pro^+	met^+	ad^+	15
	w^+ pro	met^+	ad^+	28
w/w segregants:	y^+ pro^+	met	ad	6
	y^+ pro^+	met^+	ad	12

Draw a map of the chromosome, giving the order of the genes and the position of the centromere.

A.2 The segregants are all diploid. In mitotic recombination a single crossover renders all gene loci distal to that point homozygous. In this regard the crossover events in one chromosome arm are independent of those in the other chromosome arm. Therefore we must inspect the data with these concepts in mind.

There are two classes of *y/y* segregants: wild-type segregants and proline-requiring segregants, which are homozygous for the *pro* gene. Thus of the four loci other than yellow, only *pro* is in the same chromosome arm as *y*. Furthermore, since not all the *y/y* segregants are *pro* in phenotype, the *pro* locus must be closer to the centromere than the *y* locus, as shown in the following map:

```
          y            pro
   ──────┼───────────┼────────●
```

A single crossover between *pro* and *y* will give *y/y* segregants that are wild-type for all other genes, whereas a single crossover between the centromere and *pro* will give homozygosity for both *pro* and *y*—hence, the *pro* requirement.

Similar logic can be applied to the *w/w* segregants. Again, there are two classes. Both classes are also phenotypically *ad*, indicating that the *ad* locus is further from the centromere than the *w* locus. Hence every time *w* becomes homozygous, so does *ad*. The remaining gene to be located is *met*. Some of the *w/w* segregants are met^+ and some are *met*, so the *met* gene is closer to the centromere than the *w* gene. The reasoning here is analogous to that for the placement of the *pro* gene in the other arm. Taking all the conclusions together, we have the following gene order:

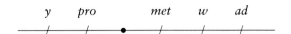

QUESTIONS AND PROBLEMS

6.1 In *Saccharomyces, Neurospora,* and *Chlamydomonas,* what meiotic events give rise to PD, NPD, and T tetrads?

6.2 What important item of information regarding crossing-over can be obtained from tetrad analysis (as in *Neurospora*) but not from single-strand analysis (as in *Drosophila*)?

6.3 A cross was made between a pantothenate-requiring (*pan*) strain and a lysine-requiring (*lys*) strain of *Neurospora crassa,* and 750 random ascospores were analyzed for their ability to grow on a minimal medium (a medium lacking pantothenate and lysine). Thirty colonies subsequently grew. Map the *pan* and *lys* loci.

*6.4 In *Neurospora* the following crosses yielded the progeny as shown:

$$a^+ b \times a b^+ \rightarrow$$

981	a^+ b
1000	a b^+
10	a^+ b^+
9	a b
2000	

$$a^+ c \times a c^+ \rightarrow$$

850	a^+ c
833	a c^+
169	a^+ c^+
148	a c
2000	

$$b^+ c \times b c^+ \rightarrow$$

850	b^+ c
850	b c^+
140	b^+ c^+
160	b c
2000	

What is the probable gene order, and what are the approximate map distances between adjacent genes?

6.5 Four different albino strains of *Neurospora* were each crossed to the wild type. All crosses resulted in half wild-type and half albino progeny. Crosses were made between the first strain and the other three with the following results:

1 × 2:	975 albino, 25 wild type
1 × 3:	1000 albino
1 × 4:	750 albino, 250 wild type

Which mutations represent different genes, and which genes are linked? How did you arrive at your conclusions?

*6.6 Genes *met* and *thi* are linked in *Neurospora crassa*; we wish to locate *arg* with respect to *met* and *thi*. From the cross *arg + + × + thi met*, the following random ascospore isolates were obtained. Map these three genes.

arg thi met	26		*arg + +*	51
arg thi +	17		*+ thi +*	4
arg + met	3		*+ + met*	14
+ thi met	56		*+ + +*	29

6.7 Given a *Neurospora* zygote of the constitution shown in the figure below, diagram the significant events producing an ascus where the *A* alleles segregate at the first division and the *B* alleles segregate at the second division.

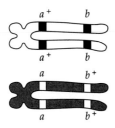

6.8 Double exchanges between two loci can be of several types, called two-strand, three-strand, and four-strand doubles.

a. Four recombination gametes would be produced from a tetrad in which the first of two exchanges is depicted in the figure below. Draw in the second exchange.

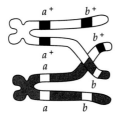

b. In the following figure, draw in the second exchange so that four nonrecombination gametes would result.

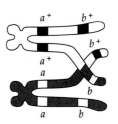

c. Other possible double-crossover types would result in two recombination and two parental gametes. If all possible multiple-crossover types occur at random, the frequency of recombination between genes at two loci will never exceed a certain percentage, regardless of how far apart they are on the chromosome. What is that percentage? Explain the reason why the percentage cannot theoretically exceed that value.

*6.9 A cross between a pink (p^-) yeast strain of mating type mt^+ and a grey strain (p^+) of mating type mt^- produced the following tetrads:

NUMBER OF TETRADS	KIND OF TETRAD
18	$p^+ \, mt^+, p^+ \, mt^+, p^- \, mt^-, p^- \, mt^-$
8	$p^+ \, mt^+, p^- \, mt^+, p^+ \, mt^-, p^- \, mt^-$
20	$p^+ \, mt^-, p^+ \, mt^-, p^- \, mt^+, p^- \, mt^+$

On the basis of these results, are the *p* and *mt* genes on separate chromosomes?

6.10 In *Neurospora* the peach gene (*pe*) is on one chromosome and the colonial gene (*col*) is on another. Disregarding the occurrence of chiasmata, what kinds of tetrads (asci) would you expect and in what proportions if these two strains were crossed?

6.11 The following unordered asci were obtained from the cross *leu + × + rib* in yeast. Draw the linkage map and determine the map distance.

110	45
leu +	*leu rib*
+ rib	*leu +*
leu +	*+ +*
+ rib	*+ rib*

6	39
+ +	*leu +*
leu rib	*+ rib*
leu rib	*+ +*
+ +	*leu rib*

*6.12 The genes *a*, *b*, and *c* are on the same chromosome arm in *Neurospora crassa*. The following ordered asci were obtained from the cross *a b + × + + c*:

45	5	146	1
a b +	*a b +*	*a b +*	*a b +*
+ b c	*a + +*	*a b +*	*+ + +*
a + +	*+ b c*	*+ + c*	*a b c*
+ + c	*+ + c*	*+ + c*	*+ + c*

10	20	15	58
a b +	*a b +*	*a b +*	*a b +*
a + c	*+ + c*	*a b c*	*+ b +*
+ b +	*a b +*	*+ + +*	*a + c*
+ + c	*+ + c*	*+ + c*	*+ + c*

a. Determine the correct gene order.
b. Determine all gene-gene and gene-centromere distances.

*6.13 If 10 percent of the asci analyzed in a particular two-point cross of *Neurospora crassa* show that an exchange has occurred between two loci (i.e., exhibit second-division segregation), what is the map distance between the two loci?

6.14 A diploid strain of *Aspergillus nidulans* (forced between wild type and a multiple mutant) that was heterozygous for the recessive mutations *y* (yellow), *w* (white), *ad* (adenine), *sm* (small), *phe* (phenylalanine), and *pu* (putrescine) produced haploid segregants. Forty-one haploid white and yellow segregants were tested and found to have the following genotypes and numbers:

$$\text{white} \begin{cases} y\ w\ pu\ ad\ sm\ phe & 7 \\ y\ w\ pu\ ad\ +\ \ + & 11 \end{cases}$$

$$\text{yellow} \begin{cases} y\ +\ +\ \ +\ sm\ phe & 16 \\ y\ +\ +\ \ +\ +\ \ + & 7 \end{cases}$$

What are the linkage relationships of these genes?

6.15 A heterokaryon was established in the fungus *Aspergillus nidulans* between a *met⁻, trp⁻* auxotroph and a *leu⁻, nic⁻* auxotroph. A diploid strain was selected from this heterokaryon. From this diploid strain, the following eight haploid strains were obtained from conidial isolates, via the parasexual cycle, in approximately equal frequencies:

1. *nic⁺ leu⁺ met⁻ trp⁻*   5. *leu⁺ nic⁻ met⁺trp⁺*
2. *met⁺ leu⁺ nic⁺ trp⁺*   6. *met⁻nic⁻ leu⁻ trp⁻*
3. *trp⁻ met⁻ leu⁺ nic⁻*   7. *trp⁺ leu⁻ met⁺nic⁺*
4. *leu⁻ nic⁻ trp⁺ met⁺*   8. *nic⁺ met⁻trp⁻ leu⁻*

Which, if any, of these four marker genes are linked, and which are unlinked?

*6.16 A (green) diploid of *Aspergillus nidulans* is heterozygous for *each* of the following recessive mutant genes: *sm*, *pu*, *phe*, *bi*, *w* (white), *y* (yellow), and *ad*. Analysis of white and yellow haploid segregants from this diploid indicated several classes with the following genotypes:

GENOTYPE

sm	*pu*	*phe*	*bi*	*w*	*y*	*ad*
sm	*pu*	*phe*	*+*	*w*	*y*	*ad*
+	*pu*	*+*	*+*	*w*	*y*	*ad*
+	*pu*	*+*	*bi*	*w*	*+*	*ad*
sm	*+*	*phe*	*+*	*+*	*y*	*+*
+	*+*	*+*	*+*	*+*	*y*	*+*
sm	*pu*	*phe*	*bi*	*w*	*+*	*ad*

How many linkage groups are involved, and which genes are on which linkage group?

6.17 A (green) diploid of *Aspergillus nidulans* is homozygous for the recessive mutant gene *ad* and heterozygous for the following recessive mutant genes: *paba*, *ribo*, *y* (yellow), *an*, *bi*, *pro*, and *su-ad*. Those recessive alleles that are on the same chromosome are in coupling. The *su-ad* allele is a recessive suppressor of the *ad* allele: The *+/su-ad* genotype does not suppress the adenine requirement of the *ad/ad* diploid, whereas the *su-ad/su-ad* genotype does suppress that requirement. Therefore the parental diploid requires adenine for growth. From this diploid two classes of segregants were selected: yellow and adenine-independent. The accompanying table lists the types of segregants obtained.

SEGREGANT TYPE SELECTED	PHENOTYPE
Adenine-independent	*+*
	ribo
	ribo an
	ribo an pro paba y bi
Yellow	*y ad bi*
	paba y ad bi
	pro paba y ad bi
	ribo an pro paba y bi

Analyze these results as completely as possible to determine the location of the centromere and the relative locations of the genes.

6.18 The accompanying table shows the only human chromosomes present in stable human-mouse somatic cell hybrid lines.

HUMAN CHROMOSOMES

	2	4	10	19
A	−	+	+	−
B	+	−	+	+
C	−	+	+	+
D	+	+	−	−

The presence of four enzymes, I, II, III, and IV, was investigated: I was present in A, B, and C but absent in D; II was in B and D but absent in A and C; III was in A, C, and D but not in B; and IV was in B and C but not in A and D. On what chromosomes are the genes for the four enzymes?

7 Genetic Recombination in Bacteria and Bacteriophages

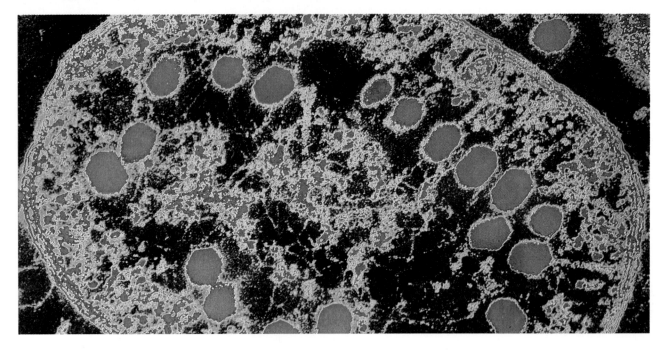

PRINCIPAL POINTS

~ Transformation is the transfer of genetic material between organisms by means of extracellular pieces of DNA. By a genetic recombination event, part of the transforming DNA molecule can exchange with part of the recipient's chromosomal DNA. Transformation can be used to determine gene order and map distance between genes.

~ Conjugation is a process in which there is a unidirectional transfer of genetic information through direct cellular contact between a donor and a recipient bacterial cell. The donor state is conferred by the presence of a plasmid called an *F* factor. Conjugation results in the transfer of the *F* factor from donor to recipient.

~ The *F* factor can integrate into the bacterial chromosome. Strains in which this has occurred—*Hfr* strains—can conjugate with recipient strains and transfer of the bacterial chromosome ensues. Using conjugation, bacterial genes can be mapped by determining the relative times at which donor genes enter the recipient and become evident as recombinants.

~ Transduction is a process whereby bacteriophages (phages) mediate the transfer of bacterial DNA from one bacterium (the donor) to another (the recipient). Transduction permits fine-structure mapping of small chromosome segments.

~ The same principles used to map eukaryotic genes are used to map phage genes.

~ The same general principles of recombinational mapping can be applied to mapping the distance between mutational sites in different genes (intergenic mapping) and to mapping mutational sites within the same gene (intragenic mapping).

~ Intragenic mapping of a bacteriophage region resulted in a fine structure map of the region and demonstrated that the unit of mutation and recombination is the base pair in DNA. This definition replaced the classical view that genes were indivisible by mutation and recombination.

~ The number of units of function (genes) defined by a set of mutations expressing the same mutant phenotypes is determined by the complementation, or *cistrans*, test. If two mutants, each carrying a mutation in a different gene, are combined, the mutations will complement and a wild-type function will result. If two mutants, each with a mutation in the same gene, are combined, the mutations will not complement and the mutant phenotype will still be seen.

---

*I*n the previous two chapters we considered the principles of genetic mapping in eukaryotic organisms. To map genes in bacteria and bacteriophages, geneticists use essentially the same experimental strategies. Crosses are made between strains that differ in genetic markers, and recombinants—the products of the exchange of genetic material—are detected and counted. The principle applied in analyzing the data obtained from such prokaryotic crosses is also the same; that is, the frequency with which exchanges occur between two sets of genes gives us an idea of the map distances between the genes' loci. In this chapter we describe genetic recombination in bacteria and bacteriophages and some approaches for mapping genes.

In addition, we describe a series of classical genetic experiments from the 1950s and early 1960s, experiments that explored the nature of the gene at a highly-detailed level. These experiments investigated the fine structure of the gene: that is, the detailed molecular organization of the gene as it relates to the mutational, recombinational, and functional events in which the

gene is involved. These experiments indicated that the gene was not like a bead on a string but, rather, was a linear sequence of nucleotides in DNA, a concept that had not been proved by any other set of genetic experiments. In the past few years, of course, the nucleotide sequences of many genes have been obtained from DNA-sequencing procedures, and the sequence changes for many mutations have been defined.

MAPPING GENES IN BACTERIA

An extensively-used bacterium for genetic and molecular analysis is *Escherichia coli* (*E. coli*), which is found most commonly in the large intestines of most animals (including humans). This bacterium is a good subject for study since it can be grown on a simple, defined medium and can be handled with simple microbiological techniques.

As we learned in Chapter 1, bacterial cells are relatively small compared with eukaryotic cells. *E. coli* is a

~ FIGURE 7.1

Scanning electron micrograph of several *Escherichia coli* bacteria.

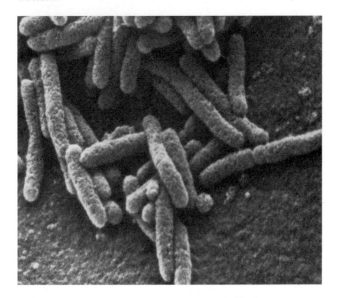

~ FIGURE 7.2

Bacterial colonies growing on a nutrient medium in a Petri dish.

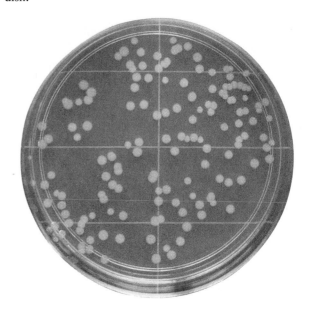

cylindrical organism about 1 μm long and 0.5 μm in diameter (Figure 7.1). Its cytoplasm is full of ribosomes (the organelles responsible for protein synthesis), while the central nucleoid region contains the genetic material, a single circular chromosome consisting of DNA. As in all prokaryotes, there is no membrane between the bacterium's nucleoid region and the rest of the cell.

Like other bacteria, *E. coli* can be grown both in liquid medium and on the surface of growth medium solidified with agar. Genetic analysis of bacteria typically is done by spreading ("plating") on agar-solidified media. Wherever a single bacterium lands on the agar surface it will grow and divide repeatedly, resulting in the formation of a visible cluster of genetically identical cells called a *colony* (Figure 7.2). The concentration of bacterial cells in a suspension (the *titer*) can be determined by spreading known volumes of the suspension or of a known dilution of the suspension on the agar surface, incubating the plates, and then counting the number of resulting colonies. By varying the nutrient composition of the media it is possible to detect and quantify the various genotypic classes in a genetic analysis.

Genetic material can be transferred between bacteria by three main processes: transformation, conjugation, and transduction. In each case: (1) transfer is unidirectional; (2) there is no true diploid zygote; and (3) only genes included in the circular chromosome will be stably inherited. Thus, these genetic transfer processes are different from those in eukaryotes.

1. **Transformation** is a process in which there is a transfer of genetic information by means of extracellular pieces of DNA. The DNA fragments are derived from donor bacteria and may be taken up by another living bacterium, the recipient. If the donor and the recipient bacteria differ genetically, then genetic recombinants are produced by crossover events involving the DNA fragments and the recipient's chromosome. Cells that have been transformed are called **transformants.**

2. **Conjugation** is a process in which there is a unidirectional transfer of genetic information through direct cellular contact between a donor and a recipient bacterial cell. The contact is followed by the formation of a cellular bridge physically connecting both cells and by fusion of the membranes of the two cells where they are joined. Then a segment (rarely all) of the donor's chromosome may be transferred into the recipient and may undergo genetic recombination with a homologous chromosome segment of the recipient cell. The recipients receiving donor DNA are called **transconjugants.**

3. **Transduction** is a process by which bacteriophages (bacterial viruses) function as intermediaries in the transfer of bacterial genetic information from one bacterium (the donor) to another (the recipient). Since the amount of DNA a phage can carry is limited, the amount of genetic material that can be transferred is usually less than 1 percent of that in the bacterial chromosome. Once the donor genetic material has been introduced into the recipient, it

may undergo genetic recombination with a homologous chromosome region. The recipients inheriting donor DNA in this way are called **transductants**.

It is possible to map bacterial genes by using any one of these methods. However, not all methods can be used for each species of bacterium, and the size of the region that one can map varies according to method. For example, conjugation allows mapping of large chromosome segments, while transduction permits fine structure mapping of small chromosome segments.

BACTERIAL TRANSFORMATION

Transformation of bacteria was first observed by Frederick Griffith in 1928, and in 1944 Oswald Avery and his colleagues showed that DNA was responsible for the genetic change that was observed (see Chapter 8). In transformation, fragments of free DNA are taken up by a living bacterium, bringing about a stable genetic change in that cell. This process is generally used to map genes in bacterial species in which conjugation or transduction does not occur. In transformation mapping experiments, DNA from a donor strain is extracted and purified. The extraction process breaks the DNA into small linear double-stranded fragments, and this genetic material is then added to a suspension of recipient bacteria with a different genotype. The donor DNA that is taken up by the recipient cell may undergo genetic recombination with the homologous parts of the recipient's chromosome to produce a recombinant chromosome. Those recipients whose phenotypes are changed by a recombination event are called *transformants*. In transformation (in principle) any bacterial strain can serve as a donor strain, and any strain can serve as a recipient. Only if there are genetic differences between donors and recipients will transformants be obtained.

Most bacterial species can probably undergo recombination through the transformational process. However, there is a great deal of variability in their ability to take up DNA. Some bacteria, such as *Bacillus subtilis,* can be easily transformed in the test tube, so mapping has been done with this bacterium by using this technique. Wild-type *E. coli,* however, is not readily transformable because the bacterium produces enzymes that rapidly degrade incoming DNA. Over the years, though, scientists have found ways to increase the efficiency of cells of a number of bacterial species in taking up DNA. Typically, cells are treated chemically to make the membrane more permeable to DNA. Such cells are called *competent cells.* As a result of such experimentation, it is now routine to transform *E. coli* in the test tube, and that ability has been extremely important in advancing our knowledge of gene function in both prokaryotes and eukaryotes through the use of recombinant DNA techniques (see Chapter 15). In terms of mechanisms, then, there are two types of bacterial transformation: *natural transformation,* in which bacteria are naturally able to take up DNA and be genetically transformed by it; and *engineered transformation,* in which bacteria have been genetically altered to enable them to take up and be genetically transformed by added DNA. *Bacillus subtilis* exemplifies bacteria in which natural transformation occurs; *E. coli* exemplifies bacteria in which engineered transformation occurs.

No matter which type of bacteria is involved in a transformation experiment, only a small proportion of the cells will actually take up DNA. Once a donor DNA fragment has been taken up by a recipient bacterium with a different genetic constitution, we can detect the transformants produced by recombination between the transforming DNA pieces and the recipient cell's chromosome. Let us consider an example of natural transformation of *Bacillus subtilis* (Figure 7.3; p. 198). (Other systems may differ in the details of the process.) The donor double-stranded DNA fragment is wild type for a mutant allele *a* in the recipient cell (Figure 7.3a). During DNA uptake, one of the two DNA strands is degraded so that only one intact linear DNA strand ultimately is found within the cell (Figure 7.3b). This single, linear strand pairs (synapses) with the homologous DNA of the recipient cell's circular chromosome, forming a triple-stranded structure over the length of the synapsed region (Figure 7.3c). Recombination can then occur by a double crossover event involving the donor single-stranded DNA strand and the recipient double-stranded DNA (Figure 7.3d). The result is a recombinant recipient chromosome: in the region between the two crossovers, one DNA strand has the donor a^+ DNA segment, and the other strand has the recipient *a* DNA segment. In other words, in that region, *the two DNA strands are part donor–part recipient for the genetic information.* (The other product of the double crossover event is a single-stranded piece of DNA carrying an *a* DNA segment; that DNA fragment is degraded.) After one round of replication of the recipient chromosome, one progeny chromosome has donor genetic information on both DNA strands and is an a^+ transformant. The other progeny chromosome has recipient genetic information on both DNA strands and is an *a* nontransformant. Equal numbers of a^+ transformants and *a* nontransformants are expected. Given highly competent recipient cells and an excess of DNA fragments, transformation of most genes will occur at a frequency of about 10^{-3}, or in about one cell in every 10^3 cells.

~ FIGURE 7.3

Transformation in bacteria: (a) bacterium with *a* recipient DNA and linear donor *a⁺* DNA fragment; (b) one strand of donor DNA enters cell; (c) donor transforming DNA fragment synapses with homologous region on the bacterial chromosome; (d) double crossover results in exchange of homologous DNA segments and, after replication, in a viable *a⁺* transformant. The other product of crossing over, a linear DNA fragment, is degraded.

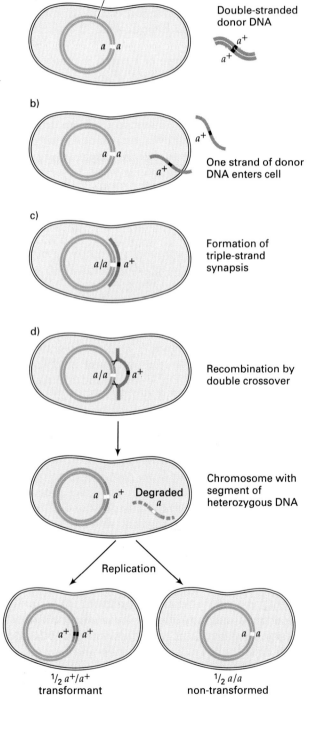

Using transformation it is possible to determine gene linkage, gene order, and map distance. The principles are as follows. If two genes, x^+ and y^+, are far apart on the donor chromosome, they will always be found on different DNA fragments. Thus, given an $x^+ y^+$ donor and an $x y$ recipient, the probability of simultaneous transformation (*cotransformation*) of the recipient to $x^+ y^+$ (from the product rule) is the product of the probability of transformation with each gene alone. If transformation occurred at a frequency of 1 in 10^3 cells per gene (see above), $x^+ y^+$ transformants would be expected to appear at a frequency of 1 in 10^6 recipient cells ($10^{-3} \times 10^{-3}$). Therefore, if two genes are close enough that they often are carried on the same DNA fragment, the cotransformation frequency would be close to the frequency of transformation by a single gene. In practical terms, this argument is turned around: if cotransformation occurs at a frequency that is substantially higher than the product of the two single-gene transformations, the two genes must be close together.

Gene order can also be determined from cotransformation data. For example, if genes p and q are often cotransformed, and genes q and o are often cotransformed, but genes o and p are never cotransformed, the gene order must be p-q-o (Figure 7.4).

Since geneticists can control the size of DNA fragments used in transformation experiments, the probability of cotransformation of two genes can be related

~ FIGURE 7.4

Demonstration of determining gene order by cotransformation.

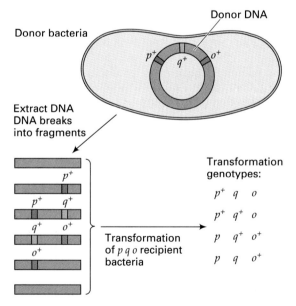

to the average molecular size of the transforming DNA. By relating cotransformation frequency to average size of the transforming DNA, one can derive a physical map of the genes (i.e., a map of the relative physical locations of genes along the DNA).

$\mathcal{K}$EYNOTE

Transformation is the transfer of genetic material between organisms by means of extracellular pieces of DNA. In transformation, DNA is extracted from a donor strain and added to recipient cells. A DNA fragment taken up by the recipient cell may associate with the homologous region of the recipient's chromosome. By a genetic recombination event, part of the transforming DNA molecule can exchange with part of the recipient's chromosomal DNA. Frequent cotransformation of donor genes indicates close physical linkage of those genes. Analysis of cotransformants can be used to determine gene order and map distance between genes. Transformation has been used to construct genetic maps for bacterial species for which conjugation or transduction analyses are not possible.

CONJUGATION IN BACTERIA

Discovery of Conjugation in *E. coli*

In 1946 another method of genetic transfer between bacteria, conjugation, was discovered by Joshua Lederberg and Edward Tatum. They studied two *E. coli* strains that differed in their nutritional requirements. Strain A had the genotype *met bio thr⁺ leu⁺ thi⁺*, and strain B had the genotype *met⁺ bio⁺ thr leu thi*. Strains carrying any of the mutant genes required nutritional supplements in their growth medium in order to grow, while strains carrying wild-type genes required no supplements. Thus strain A could only grow on a medium supplemented with the amino acid methionine (*met*) and the vitamin biotin (*bio*) but did not need the amino acids threonine (*thr*) or leucine (*leu*), or the vitamin thiamine (*thi*). In contrast, strain B could only grow on a medium supplemented with *thr*, *leu*, and *thi* but did not require *met* or *bio*. Strains that must have additives in the medium in order to grow are called **auxotrophs**. A strain that is wild-type for all nutritional-requirement genes and thus requires no supplements in its growth medium is called a **prototroph**.

In their experiment, shown in Figure 7.5, Lederberg and Tatum mixed strains A and B together and plated them onto plates of minimal medium (a medium with no growth additives). In control experiments the two strains were plated separately. As expected, neither strain could grow without the appropriate nutritional supplement. The mixed cells, however, gave rise to some prototrophic colonies. Genotypically, these prototrophic strains were *met⁺ bio⁺ thr⁺ leu⁺ thi⁺*, and they arose at a frequency of about 1 in 10 million cells (1×10^{-7}). Since no colonies appeared on the control plates, mutation was ruled out as the origin of the prototrophic colonies. The results suggested that some type of genetic exchange had occurred between the two strains.

To determine whether cell-to-cell contact was needed for the genetic exchange to occur, Bernard

~ FIGURE 7.5

Lederberg and Tatum experiment showing that sexual recombination occurs between cells of *E. coli*. After the cells from group A and group B have been mixed and the mixture plated, a few colonies grow on the minimal medium, indicating that they can now make the essential constituents. These colonies are recombinants produced by an exchange of genetic material between the strains.

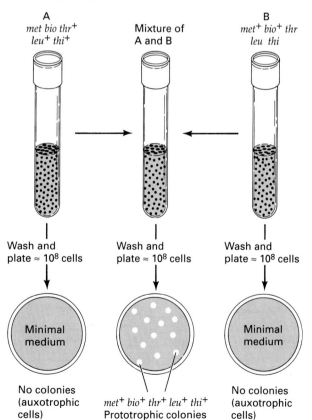

~ FIGURE 7.6

Davis's U tube experiment showing that physical contact between the two bacterial strains of the earlier experiment was needed in order for genetic exchange to occur.

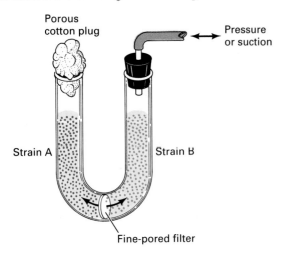

Davis performed an experiment using a U tube apparatus (Figure 7.6). Strains A and B were placed in a liquid medium on either side of the U separated by a fine filter whose pores were too small to allow bacteria to move through it. After several hours of incubation, during which the medium was moved between compartments by alternating suction and pressure, Davis plated the cells on a minimal medium to check for the appearance of prototrophic colonies. No prototrophic colonies appeared. The interpretation was that cell-cell contact was required in order for the previously observed genetic exchange to occur. The genetic exchange could not have resulted from something secreted by the cells. In other words, *E. coli* has a type of mating system, called **conjugation**, in which genetic material can be transferred between bacteria that are transiently connected.

The Sex Factor F

In 1953 William Hayes obtained evidence that the genetic exchange in *E. coli* is unidirectional, not reciprocal. Hayes discovered that one cell acts as a donor and the other cell acts as a recipient. Hayes proposed that the transfer of genetic material between the strains is mediated by a fertility factor called F that the donor cell possesses (F^+) and the recipient cell lacks (F^-). The F factor found in *E. coli* is an example of a **plasmid**, a self-replicating genetic element found in the cytoplasm of bacteria. A wide range of different plasmids occur among bacterial species.

In crosses between F^+ and F^- cells of *E. coli*, recombinants for chromosomal (nonplasmid) gene

markers are rare, but the F factor is transferred from the F^+ to the F^- cell during the conjugation process. The F factor itself is a circular, double-stranded DNA molecule that is distinct from the host cell chromosome. About a fortieth of the size of the host chromosome, the F factor contains a region of DNA, called the **origin** (or O), that is required for its replication within *E. coli* and contains fertility genes, which are required if a cell is to be a donor cell.

When F^+ and F^- cells are mixed, they conjugate. Figure 7.7 shows an electron micrograph of conjugation between an F^+ donor and an F^- recipient. No conjugation occurs between two cells of the same mating type (i.e., two F^+ bacteria or two F^- bacteria). Genetic material is transferred from donor to recipient during conjugation. One strand of the F factor is nicked at the origin, and DNA replication proceeds from that point. As replication proceeds, a copy of the F factor is transferred to the F^- cell as shown in Figure 7.8a. The complementary strand is synthesized in the F^- cell. There is a polarity of transfer; that is, the origin of replication is always transferred first, followed by the rest of the F factor in one particular orientation (clockwise or counterclockwise).

When the complete F factor has been transferred, the F^- cell becomes an F^+ cell as a result of the F factor genes. These genes include ones that specify hairlike cell surface components called **F-pili**, or sex-pili, which allow the physical union of F^+ and F^- cells to take place. Note, though, that while pili are necessary for

~ FIGURE 7.7

Electron micrograph of bacterial conjugation between an F^+ donor and F^- recipient *E. coli* bacterium. The spherical structure on the upper F pilus bridging the two bacteria is a spherical donor-specific RNA phage MS-2. The adsorption of the MS-2 phage blocks conjugation by preventing pilus retraction, thus keeping the cells apart. Magnification 30,000×.

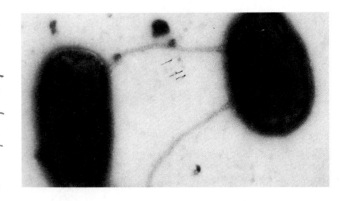

~ FIGURE 7.8

Transfer of genetic material during conjugation in *E. coli;* (a) Transfer of the *F* factor from donor to recipient cell during $F^+ \times F^-$ matings. (b) Production of *Hfr* strain by integration of *F* factor and transfer of bacterial genes from donor to recipient cell during $Hfr \times F^-$ matings.

a) Transfer of the *F* factor

b) Transfer of bacterial genes

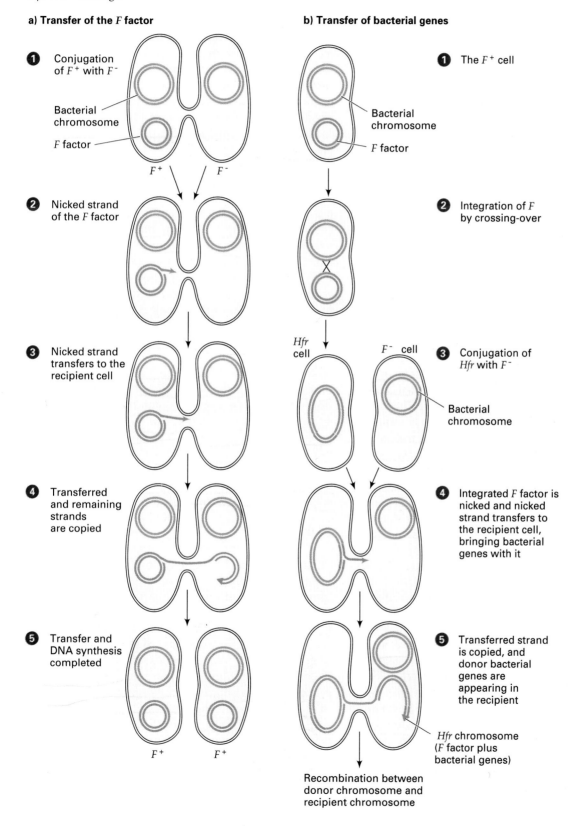

1 Conjugation of F^+ with F^-

Bacterial chromosome

F factor

F^+ F^-

2 Nicked strand of the *F* factor

3 Nicked strand transfers to the recipient cell

4 Transferred and remaining strands are copied

5 Transfer and DNA synthesis completed

F^+ F^+

1 The F^+ cell

Bacterial chromosome

F factor

2 Integration of *F* by crossing-over

Hfr cell F^- cell

3 Conjugation of *Hfr* with F^-

Bacterial chromosome

4 Integrated *F* factor is nicked and nicked strand transfers to the recipient cell, bringing bacterial genes with it

5 Transferred strand is copied, and donor bacterial genes are appearing in the recipient

Hfr chromosome (*F* factor plus bacterial genes)

Recombination between donor chromosome and recipient chromosome

promoting tight pairing of F^+ and F^- cells, there is no evidence that DNA passes through the pilus. In a population of cells, only some of which are F^+ to start with, the end result of the conjugation of F^+ and F^- cells and the subsequent transfer of the F factor is an epidemic conversion of the population to the F^+ state.

𝒦EYNOTE

Some *E. coli* bacteria possess a plasmid, called the *F factor*, that is required for mating. *E. coli* cells with the F factor are F^+ and those without it are F^-. The F^+ cells (donors) can mate with F^- cells (recipients) in a process called conjugation, which leads to the one-way transfer of a copy of the F factor from donor to recipient during replication of the F factor. As a result, both donor and recipient are F^+. None of the bacterial chromosome is transferred during $F^+ \times F^-$ conjugation. ⸻

High-Frequency Recombination Strains

Although geneticists established that F^- cells can be converted to the F^+ state by the transfer of the F factor, they did not know how conjugation resulted in the formation of recombinants for chromosomal genes. This puzzle was solved when William Hayes and Luca Cavalli-Sforza, working separately, isolated unusual donor strains from F^+ stock. In a cross of this donor strain with an F^- strain having different chromosomal gene markers, the frequency of recombinants for the chromosomal genes was about a thousand times greater than normally seen in $F^+ \times F^-$ crosses. This new, highly recombinant type of F^+ donor strain is called an *Hfr*, or **high-frequency recombination**, strain. The two particular strains characterized by Cavalli-Sforza differed in the order with which bacterial genes were transferred; they were later called *Hfr* H and *Hfr* C, after their discoverers. The *Hfr* strains are the strains generally used in mapping *E. coli* genes through conjugation.

The *Hfr* strains originate through a rare single crossover event in which the F factor actually integrates into the bacterial chromosome (Figure 7.8). When the F factor is integrated, it no longer replicates independently; instead, it is replicated as part of the host chromosome. Since the F factor genes are still functional, *Hfr* cells are able to mate with F^- cells. When mating happens, similar events occur as in the $F^+ \times F^-$ matings. The integrated F factor is nicked in one of the two DNA strands and replication begins. During replication

part of the factor first moves into the recipient cell. With time, the donor bacterial chromosome begins to be transferred into the recipient. If there are genetic marker differences between the genes on the donor chromosome that has entered the recipient and the recipient chromosome, recombinants can be isolated. This recombination process occurs by double crossovers between the double-stranded donor DNA that has entered the recipient cell and the double-stranded recipient chromosome, similar to that described for transformation. The recombinant, recipient chromosome is passed on to progeny cells by replication, while the linear DNA fragment is degraded or lost.

In $Hfr \times F^-$ matings the F^- cell almost never acquires the F^+ phenotype. That is, for the recipient cell to become F^+, it must receive a complete copy of the F factor. However, only part of the F factor is transferred at the beginning of conjugation; the rest of the F factor is at the end of the donor chromosome. All of the donor chromosome would have to be transferred in order for a complete functional F factor to be found in the recipient. This is an extremely rare event because all the while the bacteria are conjugating, they are "jiggling" around by Brownian motion, so there is a very high probability that the mating pair will break apart long before the second F factor part is transferred.

The low-frequency recombination of chromosomal gene markers in $F^+ \times F^-$ crosses can be understood when we consider that only about 1 in 10,000 F^+ cells in a population become *Hfr* cells by F factor integration. The reverse process, excision of the F factor, also occurs spontaneously and at low frequency, producing an F^+ cell from an *Hfr* cell. In excision, the F factor loops out of the *Hfr* chromosome, and by a single crossing-over event (just like the integration event), a circular host chromosome and a circular extrachromosomal F factor are generated.

F′ Factors

Occasionally, the F factor excision from the host chromosome is not precise. As a result, the excised F factor may contain a relatively small part of the host chromosome, a section that was adjacent to the F factor when it was integrated. Consider an *E. coli* strain in which the F factor has integrated next to the *lac*$^+$ region, a set of genes required for the breakdown of lactose (Figure 7.9a). If the looping out is not precise in the excision process, then the adjacent *lac*$^+$ host chromosomal genes may be included in the loop (Figure 7.9b). Then by single crossover the looped-out DNA will be separated from the host chromosome (Figure 7.9c). In addition to the F factor, the newly generated plasmid contains the *lac*$^+$ genes of the host chromosome. F factors containing bacterial genes are called *F′* (F prime) factors, and

~ FIGURE 7.9

Production of an F' factor: (a) bacterial chromosome into which the F factor has integrated; (b) the F factor looping out incorrectly, so it includes a piece of bacterial chromosome, the lac^+ genes; (c) excision, in which a single crossover between the looped-out DNA segment and the rest of the bacterial chromosome (i.e., the reverse of integration) results in an F' factor, called F' (lac).

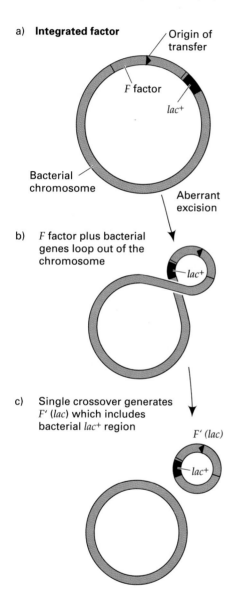

a) **Integrated factor**

Origin of transfer

F factor

lac^+

Bacterial chromosome

Aberrant excision

b) F factor plus bacterial genes loop out of the chromosome

lac^+

c) Single crossover generates F' (lac) which includes bacterial lac^+ region

F' (lac)

lac^+

they are named for the genes they have picked up. For example, an F' with the lac region is called F' (lac).

Cells with F' factors can conjugate with F^- cells since all the F factor functions are present. As in regular conjugation, a copy of the F' factor is transferred to the F^- cell, which then becomes F^+ in phenotype. The recipient also receives a copy of the bacterial gene on the F factor. Typically, the recipient has its own copy of

that gene, so the resulting cell line will be partially diploid, having two copies of one or a few genes and only one copy of all the others. This particular type of conjugation is called **F duction,** or *sexduction.*

Using Conjugation and Interrupted Mating to Map Bacterial Genes

In conjugation, like transformation, one strain acts as the donor and another strain acts as the recipient. In this case, the determining factor for whether a strain is a donor is the presence of an integrated F factor in the chromosome, making the strain an *Hfr* strain. Thus, in planning conjugation experiments to map genes, experimenters will construct an appropriate donor strain by introducing an F factor into it by $F^+ \times F^-$ mating, and then selecting an *Hfr* derivative.

In the late 1950s François Jacob and Elie Wollman studied the transfer of chromosomal genes from *Hfr* strains to F^- cells. Their experimental design was to make an $Hfr \times F^-$ mating and, at various times after conjugation commenced, to break apart the conjugating pairs by mixing the cells in a kitchen blender. The separated cells were then analyzed to determine the relative times at which donor genes entered the recipients and produced genetic recombinants by double crossing-over between the donor fragment and the recipient chromosome (Figure 7.10a; p. 204). That is, if gene a entered the recipient after 9 minutes, and gene b after 15 minutes, genes a and b would be considered to be 6 minutes apart on the genetic map. In this way, on the basis of the appearance of two genes as recombinants in the recipient, the distance between genes was measured as time units (minutes). The genetic map of large chromosomal segments of *E. coli* was constructed in just this way, with map units in minutes, timed from an arbitrarily located origin.

The use of interrupted mating to map bacterial genes may be illustrated by considering one of Jacob and Wollman's experiments, which involved the following cross:

Donor:
   Hfr H str^s thr^+ leu^+ azi^r ton^r lac^+ gal^+

Recipient:
   F^- str^r thr leu azi^s ton^s lac gal

The *Hfr* H strain is prototrophic and is sensitive to the antibiotic streptomycin. The F^- cell carries a streptomycin resistance gene and also a number of mutant genes. These genes cause the F^- to be auxotrophic for threonine (*thr*) and leucine (*leu*), to be sensitive to sodium azide (*azi^s*) and to infection by bacteriophage T1 (*ton^s*), and to be unable to ferment lactose (*lac*) or galactose (*gal*) as a carbon source.

~ FIGURE 7.10

Interrupted-mating experiment involving the cross *Hfr H str^s thr^+ leu^+ azi^r ton^r lac^+ gal^+ × F^- str^r thr leu azi^s ton^s lac gal*. The progressive transfer of donor genes with time is illustrated. Recombinants are generated by an exchange of a donor fragment with the homologous recipient fragment resulting from a double crossover event. (a) At various times after mating commences, the conjugating pairs are broken apart and the exconjugant cells are plated on selective media to determine which genes have been transferred from the *Hfr* to the *F^-*.

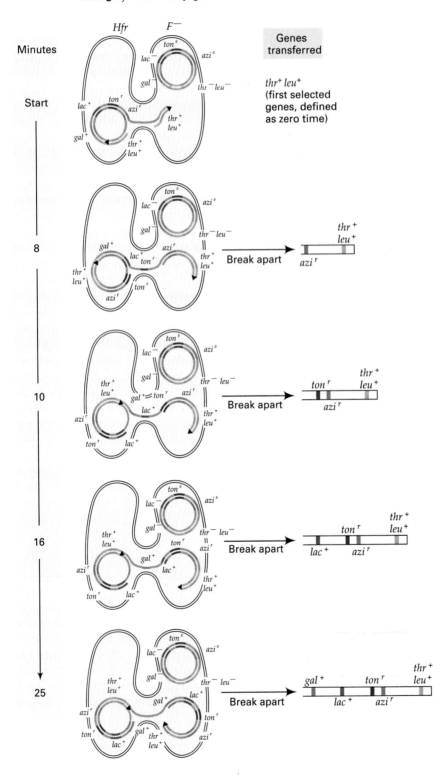

a) Progressive transfer of donor genes to recipient during *Hfr × F^-* conjugation

~ FIGURE 7.10 continued

(b) The graph shows the appearance of donor genetic markers in the F⁻ cells as a function of time; that is, after the selected *thr⁺* and *leu⁺* genes have entered.

b) **Appearance of donor genetic markers in recipient as a function of time**

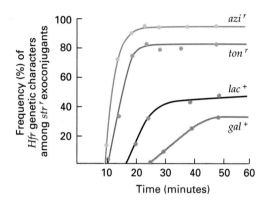

In a conjugation experiment the two cell types are mixed together in a nutrient medium at 37°C, the normal growth temperature for *E. coli*. After a few minutes have passed allowing the *Hfr* H and F⁻ cells to pair, the culture is diluted to prevent the formation of new mating pairs. This procedure ensures some synchrony in the chromosome transfer between cells. Samples are removed from the mating mixture at various times and are then blended to break the pairs apart. These **exconjugants** are plated on a selective medium designed to allow recombinant types to grow and divide while killing both the *Hfr* H and F⁻ parental types.

For this particular cross the medium contains streptomycin to kill the *Hfr* H and lacks threonine and leucine so that the parental F⁻ cannot grow. In this cross the threonine (*thr⁺*) and leucine (*leu⁺*) genes are the first donor chromosomal genes to be transferred to the F⁻, so recombinants formed by the exchange of those genes with the *thr leu* genes of the F⁻ recipient are produced and will grow on the selective medium. Appropriate media can be used to test for the appearance of other (unselected) marker genes among the selected transconjugants.

Figure 7.10b shows an example of the results. The transconjugants here are *thr⁺*, *leu⁺*, and *str^r*. The first unselected marker gene to be transferred is *azi^r*, and recombinants for this gene are seen at about 9 minutes. The second gene transferred is *ton^r* at 10 minutes, followed by *lac⁺* at about 17 minutes and *gal⁺* at about 25 minutes. Figure 7.10b also shows the rates of appearance and the maximum frequencies obtained for each recombinant type. The following conclusions can be

made: (1) As the time between the beginning of conjugation and analysis increases, both the rates of appearance and the maximum frequencies of recombinants decrease; (2) The rate of transfer from mating couple to mating couple is not constant; and (3) The maximal frequency of recombinants becomes smaller the later the gene enters the recipient because with time there is an increasing chance that mating pairs will break apart.

In this experiment, then, each gene from an *Hfr* appears in recombinants at a different but reproducible time after mating commences, and the time intervals between gene appearances are used as a measure of genetic distance. From such data we can conclude that an *Hfr* chromosome is transferred into the F⁻ cell in a linear way. As with F factor transfer, the transfer starts at the origin within the integrated F factor. Genes located far from the origin tend not to be transferred because of the high probability that the conjugating pairs will be broken apart before their location is reached. Thus from the experiment described (*Hfr* H × F⁻) and from the time intervals, the map in Figure 7.11 may be constructed. Again, the map units are in "minutes"; the entire *E. coli* chromosome is about 100 minutes.

Circularity of the *E. coli* Map

Different *Hfr* strains transfer donor genes into the recipient strain with different starting points and polarities of transfer. That is, since the F factor can integrate into the bacterial chromosome at a number of places, and since the F factor itself determines the direction of gene transfer, chromosomal transfer may be in different directions and from different places in the map. Figure 7.12a (p. 206) diagrams the order of chromosomal gene transfer for four different *Hfr* strains H, 1, 2, and 3. Moreover, the genetic distance in time units between

~ FIGURE 7.11

Chromosome map of the genes in the interrupted-mating experiment of Figure 7.10. The marker positions represent the time of entry of the genes into the recipient during the experiment. Thus the map distance is given in minutes. The *azi* marker, for example, appeared about 8 minutes after conjugation began, whereas the *gal* marker appeared after about 25 minutes.

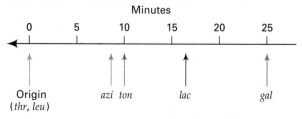

~ FIGURE 7.12

Interrupted-mating experiments with a variety of *Hfr* strains, showing that the *E. coli* linkage map is circular: (a) orders of gene transfer for the *Hfr* strains *H, 1, 2,* and *3*; (b) alignment of gene transfer for the *Hfr* strains; (c) circular *E. coli* chromosome map derived from the *Hfr* gene transfer data. The map is a composite showing various locations of integrated *F* factors. A given *Hfr* strain has only *one* integrated *F* factor.

a) Orders of gene transfer

Hfr strains:

H	origin	*– thr – pro – lac – pur – gal – his – gly – thi*
1	origin	*thr – thi – gly – his – gal – pur – lac – pro*
2	origin	*– pro – thr – thi – gly – his – gal – pur – lac*
3	origin	*– pur – lac – pro – thr – thi – gly – his – gal*

b) Alignments of gene transfer for the *Hfr* strains

H	*thr – pro – lac – pur – gal – his – gly – thi*
1	*pro – lac – pur – gal – his – gly – thi – thr*
2	*lac – pur – gal – his – gly – thi – thr – pro*
3	*gal – his – gly – thi – thr – pro – lac – pur*

c) Circular *E. coli* chromosome map derived from *Hfr* gene transfer data

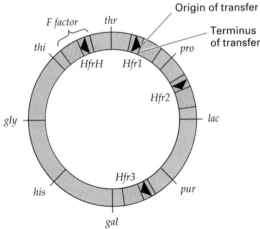

a particular pair of genes is constant, within experimental error, no matter which *Hfr* strain is used as donor; for example, the genetic distance between *thr* and *pro* is the same in *H, 1, 2,* and *3*.

From the data in Figure 7.12a, the best way to construct a genetic map of the chromosome is to try to align the genes transferred by each *Hfr* as shown in Figure 7.12b. In view of the overlap of the genes, the simplest map that can be drawn from these data is a circular one, as shown in Figure 7.12c. This configuration was a truly significant finding, since all previous genetic maps of eukaryotic chromosomes were linear.

KEYNOTE

The *F* factor can integrate into the bacterial chromosome by a single crossover event. Strains in which this integration has happened can conjugate with *F⁻* strains, and transfer of the bacterial chromosome occurs. The strains with the integrated *F* factor are *Hfr* (high-frequency recombination) strains. In *Hfr* × *F⁻* matings the chromosome is transferred in a one-way fashion from the *Hfr* to the *F⁻*, beginning at a specific site called the origin (*O*). The farther a gene is from *O*, the later it is transferred to the *F⁻*, and this result forms the basis for mapping genes by their times of entry into the *F⁻*. Conjugation allows mapping of large chromosome segments.

TRANSDUCTION IN BACTERIA

Transduction is the process by which genetic material is transferred between bacterial strains by bacterial viruses called *bacteriophages* (phages, for short), which are viruses that infect bacteria.

Bacteriophages: An Introduction

Most bacterial strains have phages that are specific for them. For example, *E. coli* is susceptible to infection by DNA-containing phages such as T2, T4, T5, and λ (lambda), among others. As a result of extensive study, most of the genes located on the chromosomes of the commonly used phages have been identified and mapped.

The structure of all phages is simple. A phage contains its genetic material (either DNA or RNA) in a single chromosome surrounded by a coat of protein molecules. Variation in the number and organization of the proteins gives the phages their characteristic appearances. Figure 7.13 presents electron micrographs and diagrams of two phages of great genetic significance, T4 (Figure 7.13a) and λ (Figure 7.13b). Phage T4 is one of a series of similar phages (T2, T4, and T6), collectively called T-even phages. It has a number of distinct structural components: a head (which contains the genetic material DNA), a core, a sheath, a base plate, and tail fibers (the latter two structures enable the phage to attach itself to a bacterium). The head and all other structural components consist of proteins.

~ FIGURE 7.13

Electron micrographs and diagrams of two bacteriophages (1 nm = 10^{-9} m): (a) T4 phage, which is representative of T-even phages; (b) λ phage.

a) T4 phage

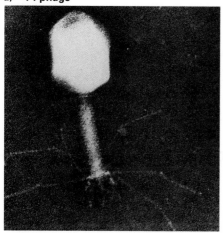

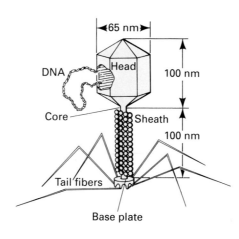

b) λ phage

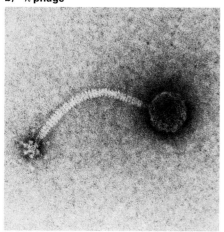

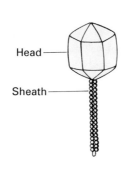

The life cycles of the T4 and the λ phages are not identical. Figure 7.14 (p. 208) presents the stages of the T4 cycle. First (1), the phage particle attaches to the surface of the bacterial cell, and the phage chromosome is injected into the bacterium in a series of events that include a springlike contraction of the phage sheath. Then (2-5), by the action of phage genes, the phage in essence takes over the bacterium, breaking down the bacterial chromosome and directing its growth and reproductive mechanisms to produce progeny phages. Finally (6), the progeny phages are eventually released from the bacteria as the cell is broken open (lysed). The suspension of released progeny phages is called a **phage lysate.** The T4 type of phage life cycle is called the **lytic** cycle, and phages that always follow that cycle when they infect bacteria are called **virulent phages.**

The λ life cycle, shown in Figure 7.15 (p. 209), may be more complex than that of a T4 phage. Phage λ has a structure similar to that of T4. When its DNA is injected into *E. coli,* there are two paths that the phage can follow. One is a lytic cycle, exactly like that of the T phages. The other is the **lysogenic pathway.** In the lysogenic pathway the λ chromosome does not replicate; instead, it inserts (integrates) itself physically into a specific region of the host cell's chromosome, much like F factor integration. In this integrated state the phage chromosome is called a **prophage.** Every time the host cell chromosome replicates, the integrated λ chromo-

~ FIGURE 7.14

Lytic life cycle of a virulent phage, such as T2 or T4.

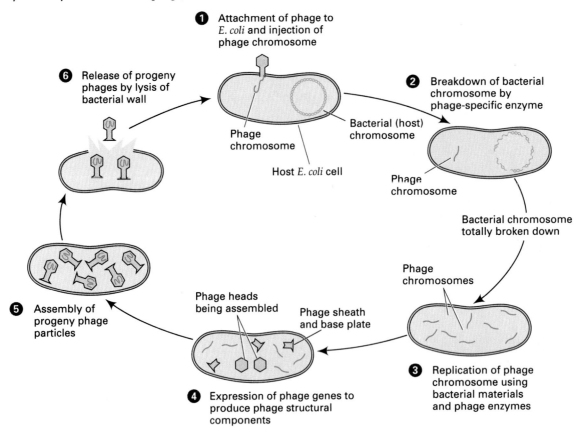

1 Attachment of phage to *E. coli* and injection of phage chromosome

6 Release of progeny phages by lysis of bacterial wall

Phage chromosome

Bacterial (host) chromosome

Host *E. coli* cell

2 Breakdown of bacterial chromosome by phage-specific enzyme

Phage chromosome

Bacterial chromosome totally broken down

Phage chromosomes

5 Assembly of progeny phage particles

Phage heads being assembled

Phage sheath and base plate

4 Expression of phage genes to produce phage structural components

3 Replication of phage chromosome using bacterial materials and phage enzymes

some is also replicated. The bacterium that contains a phage in the prophage state is said to be **lysogenic** for that phage; the phenomenon of the insertion of a phage chromosome into a bacterial chromosome is called **lysogeny.** Phages that have a choice between lytic and lysogenic pathways are called *temperate phages*. Occasionally, the mechanism that maintains the prophage state (which is a function of phage genes) breaks down. Then the lytic cycle is initiated, and progeny λ phages are released. The lytic cycle can also be induced by environmental factors, such as ultraviolet light irradiation.

Transduction Mapping of Bacterial Chromosomes

Transduction, the process in which a phage (called a **phage vector**) carries pieces of bacterial DNA between bacterial strains, is a useful mechanism for mapping bacterial genes. Of necessity, the useful phages are temperate phages, because they lysogenize the bacteria they infect, rather than kill them. Examples are phages P1 and λ for *E. coli* and phage P22 for *Salmonella typhimurium.* Two types of transduction occur: In **generalized transduction** any gene may be transferred between bacteria; in **specialized transduction** only specific genes are transferred.

GENERALIZED TRANSDUCTION. The discovery of generalized transduction is credited to Joshua and Esther Lederberg and Norton Zinder. In 1952 these researchers were testing to see whether a conjugation mechanism occurred in the bacterial species *Salmonella typhimurium,* which causes typhus in the mouse. Their experiment was essentially similar to the one that showed conjugation existed in *E. coli.* They mixed together two multiple auxotrophic strains and looked for the appearance of prototrophs. One strain was *phe+ trp+ tyr+ met his,* which required methionine and histidine, and the other strain was *phe trp tyr met+ his+,* which required phenylalanine, tryp-

~ FIGURE 7.15

Life cycle of a temperate phage, such as λ. When a temperate phage infects a cell, the phage may go through either the lytic or lysogenic cycle.

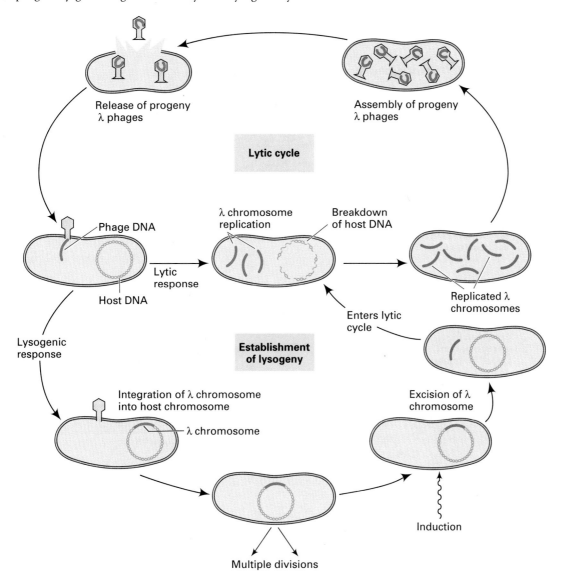

tophan, and tyrosine. When they crossed these two auxotrophic strains, they found prototrophic recombinants phe^+ trp^+ tyr^+ met^+ his^+ at a low frequency.

When Zinder and the Lederbergs used the U tube apparatus (which helped establish conjugation in *E. coli*), they still found prototrophs. This result indicated that recombinants were being produced by a mechanism that did not require cell-to-cell contact. The interpretation was that the agent responsible for the formation of recombinants was a *filterable agent,* since it

could pass through a filter with pores small enough to block bacterial movement. In this particular case the filterable agent was identified as the temperate phage P22.

The mechanism for generalized transduction by P1 is shown in Figure 7.16 (p. 210). Normally, the P1 phage enters the lysogenic state when it infects *E. coli,* existing in the prophage state in these cells (i.e., the phages genes required for taking the phage through the lytic cycle are suppressed). As we have said,

~ FIGURE 7.16

Generalized transduction between strains of *E. coli*: (1) wild-type donor cell of *E. coli* infected with the temperate bacteriophage P1; (2) the host cell DNA is broken up during the lytic cycle; (3) during assembly of progeny phages, some pieces of the bacterial chromosome are incorporated into some of the progeny phages to produce transducing phages; (4) following cell lysis, a low frequency of transducing phages is found in the phage lysate; (5) the transducing phage infecting an auxotrophic recipient bacterium; (6) a double-crossover event results in the exchange of the donor a^+ gene with the recipient mutant a gene; (7) the result is a stable a^+ transductant, with all descendants of that cell having the same genotype.

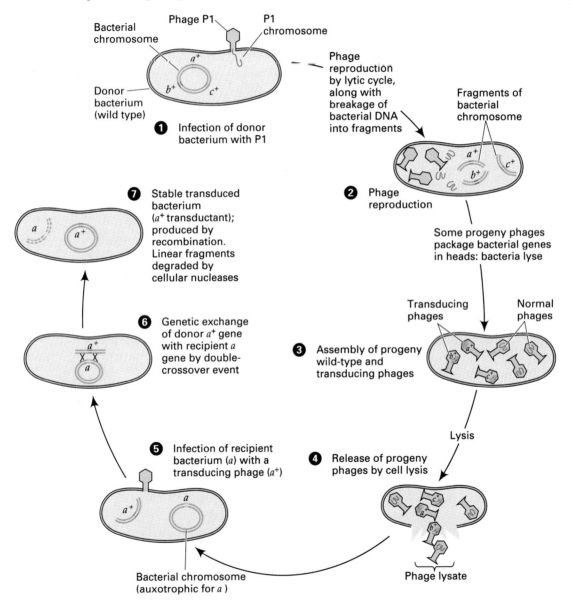

the mechanism that maintains this state breaks down occasionally, and the phage genome goes through the lytic cycle and produces progeny phages. During the lytic cycle, the bacterial DNA is degraded, and on rare occasions, instead of the phage DNA being packaged into the phage head, a piece of bacterial DNA is so packaged (Figure 7.16). Since this event is rare, only a very small proportion of the progeny phages carry bacterial genetic material. These phages are called **transducing phages** since they are the vehicles by which genetic material is shuttled between bacteria. The population of phages in the phage lysate, consisting mostly of normal phages but with some transducing phages present ($1/10^5$ particles), can now be used to lysogenize a new population of bacteria. Recipient bacteria that show genetic recombination of the transferred donor genes are called *transductants*. The recombinants are produced by a double crossover, resulting in an exchange of a donor DNA segment with a homologous segment of the recipient chromosome.

Typically, an experiment like this one is designed so that the donor cell type and the recipient cell type have different genetic markers; then the transduction events can be followed. The strategy behind a transduction experiment is similar to the strategy behind a transformation experiment—and, in fact, behind all genetic-mapping experiments. In the example we are discussing, the donor cell is *thr*$^+$ and the recipient cell is *thr*. Prototrophic transductants are then detectable, since the cell no longer requires threonine in order to grow. In this way, researchers are able to pick out the extremely low frequency of transductants among all the cells present by testing for those cells that are able to do something the nontransduced cells cannot, namely, grow on a minimal medium.

The process we've just described is called *generalized transduction* because the piece of bacterial DNA that the phage erroneously picks up and shuttles to a recipient bacterium is a random piece of the fragmented bacterial chromosome. Thus any genes can be transduced: All that are needed are a temperate phage and bacterial strains carrying different genetic markers. To map genes by using generalized transduction, researchers usually carry out two-factor or three-factor transduction experiments. In a typical transduction experiment, transductants for one of two or more donor markers are selected and these transductants are then analyzed for the presence or absence of other donor markers. Transduction from an $a^+ b^+$ donor to an $a b$ recipient produces various transductants for a^+ and b^+, namely $a^+ b$, $a b^+$, and $a^+ b^+$. If we select for one or other donor markers, we can determine linkage information for the two genes. If we select for a^+ transductants, linkage between genes a and b is given by:

$$\frac{\text{number of single transductants}}{\text{number of total transductants}} \times 100\% =$$

$$\frac{(a^+ b)}{(a^+ b) + (a^+ b^+)} \times 100\%$$

If we select for b^+ transductants, linkage between a and b is given by

$$\frac{(a b^+)}{(a b^+) + (a^+ b^+)} \times 100\%$$

Since the production of a transducing phage is a rare event, and since the amount of bacterial DNA the phage can carry is limited by the capacity of the phage head, the chance of either a^+ or b^+ being included individually in the phage head is proportional to the distance between them. If the genes are close enough to be included in the transducing phage, an $a^+ b^+$ transductant, called a *cotransductant*, can be produced. Thus, **cotransduction** of two or more genes is a good indication that the genes are closely linked.

The map distances between cotransduced genes may be determined by using transduction, and it is by this procedure that fine structure (i.e., detailed) linkage maps of bacterial chromosomes have been constructed. For example, consider the mapping of some *E. coli* genes by using transduction with the temperate phage P1. The donor *E. coli* strain is *leu*$^+$ *thr*$^+$ *azi*^{r} (able to grow on a minimal medium and resistant to the metabolic poison sodium azide). The recipient cell is *leu thr azi*^{s} (needs leucine and threonine as supplements in the medium and sensitive to azide). The P1 phages are grown on the bacterial donor cells, and the phage lysate is used to transduce the recipient bacterial cells. Transductants are selected for any one of the donor markers and then analyzed for the presence of the other unselected markers (in this case, two). Typical data from such an experiment are shown in Table 7.1.

Consider the *leu*$^+$ selected transductants. Of these transductants 50 percent are also transduced for *azi*^{r} and 2 percent for *thr*$^+$. For the *thr*$^+$ transductants 3 percent are *leu*$^+$ and 0 percent are *azi*^{r}. The simplest interpretation is that the *leu* gene is closer to the *thr* gene than is the *azi* gene. The order of genes then is

The transductants are produced by crossing-over between the piece of donor bacterial chromosome brought in by the infecting phage and the homologous region on the recipient bacterial chromosome. The

~ TABLE 7.1

Transduction Data for Deducing Gene Order

SELECTED MARKER	UNSELECTED MARKERS
leu$^+$	50% = *azi*^{r}
	2% = *thr*$^+$
thr$^+$	3% = *leu*$^+$
	0% = *azi*^{r}

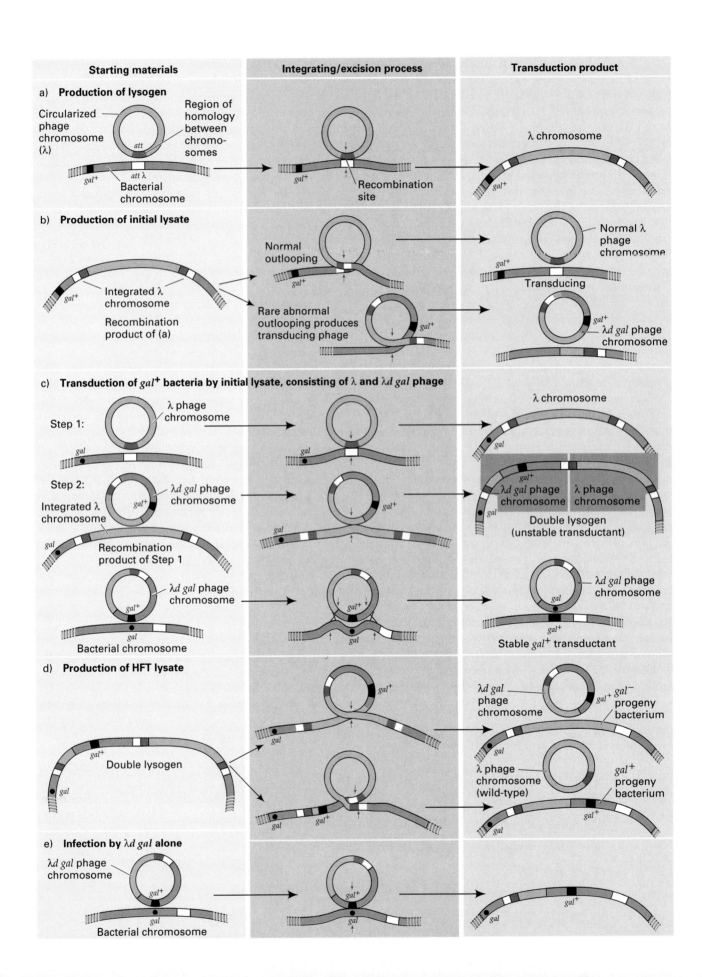

Starting materials

Integrating/excision process

Transduction product

a) **Production of lysogen**

Circularized phage chromosome (λ)

Region of homology between chromosomes

att

att λ

Bacterial chromosome

gal+

Recombination site

λ chromosome

gal+

b) **Production of initial lysate**

Integrated λ chromosome

Recombination product of (a)

gal+

Normal outlooping

Rare abnormal outlooping produces transducing phage

gal+

gal+

Normal λ phage chromosome

gal+

Transducing

λd gal phage chromosome

gal+

c) **Transduction of gal+ bacteria by initial lysate, consisting of λ and λd gal phage**

Step 1:

λ phage chromosome

gal

Step 2:

Integrated λ chromosome

Recombination product of Step 1

λd gal phage chromosome

gal+

gal

λd gal phage chromosome

gal+

Bacterial chromosome

gal

gal

gal

gal+

gal+

gal

λ chromosome

gal

λd gal phage chromosome

gal+

λ phage chromosome

Double lysogen (unstable transductant)

gal+

gal

λd gal phage chromosome

gal+

Stable gal+ transductant

d) **Production of HFT lysate**

Double lysogen

gal

gal+

gal+

gal

gal

gal+

λd gal phage chromosome

gal+   gal− progeny bacterium

gal

λ phage chromosome (wild-type)

gal+ progeny bacterium

gal

gal+

e) **Infection by λd gal alone**

λd gal phage chromosome

gal+

gal

Bacterial chromosome

gal+

gal

gal

gal+

~ FIGURE 7.17 facing page

Specialized transduction by bacteriophage λ: (a) Production of a lysogenic bacterial strain by crossing-over in the region of homology between the circular bacterial chromosome (*att* λ) and the circularized phage chromosome (*att*); (b) Production of initial lysate: induction of the lysogenic bacterium causes out-looping, resulting in many normal λ and a low frequency of a transducing phage, λd *gal*, which includes the bacterial *gal* gene due to incorrect out-looping; (c) Transduction of *gal* bacteria by the initial lysate, producing either unstable transductants by integration of both λ and λd *gal* (resulting in a double lysogen) or stable transductants (single lysogens) by crossing-over around the *gal* region; (d) The unstable double lysogen, which contains both λ and λd *gal*, produces about equal numbers of λd *gal* and wild-type λ phages; the result is a high-frequency transducing (HFT) lysate; (e) Low concentrations of phage from the HFT lysate used to infect bacteria so that a cell can be infected by λd *gal* alone.

infected donor DNA "finds" the region of the recipient chromosome to which it is homologous, and the exchange of parts is accomplished by double (or other even-numbered) crossovers (see Figure 7.16). Map distance can be obtained from these two transduction experiments by determining relative cotransduction frequencies. By extension, similar analyses can be done with data from three-factor or more transduction experiments.

SPECIALIZED TRANSDUCTION. Some temperate bacteriophages can transduce only certain sections of the bacterial chromosome, in contrast to generalized transducing phages, which can carry any part of the bacterial chromosome. An example of such a **specialized transducing phage** is λ, which infects *E. coli*. Figure 7.17 shows the series of events by which phage λ can transduce bacterial genes.

The life cycle of λ was described earlier (see Figure 7.15). In the cell, the λ genome integrates into the bacterial chromosome at a specific site, producing what is called a *lysogen* (Figure 7.17a). That site is called *att* λ, and it is homologous with a site called *att* in the λ DNA. By a single-crossover event the λ chromosome integrates. In the integrated state the phage, now called a *prophage*, is maintained by the action of specific gene products that prevent expression of the λ genes essential for the lytic cycle. The point of integration for this particular phage in the *E. coli* chromosome is between the *gal* (galactose fermentation) and the *bio* (biotin biosynthesis) genes.

The particular *E. coli* strain that λ lysogenizes is *E. coli K12*, and when it contains the λ prophage, it is designated *E. coli K12*(λ). Let us focus just on the *gal* gene and assume that the particular *K12* strain that λ lysogenized is *gal*⁺, that is, it can ferment galactose as a carbon source. This phenotype is readily detectable by plating the cells on a solid medium containing galactose as a carbon source and containing a dye that changes color in response to the products of galactose fermentation. On this medium the *gal*⁺ colonies are pink, while *gal* colonies are white. If we induce the prophage—that is, reverse the inhibition of phage functions—the lytic cycle is initiated. Induction may be accomplished by any procedure that destroys the protein repressor molecules responsible for inhibition (ultraviolet light irradiation is one commonly used procedure).

After the lytic cycle is initiated, the phage chromosome loops out, generating a separate circular λ chromosome by a single-crossover event at the *att* λ/*att* sites (Figure 7.17b). In most cases the excision of the phage chromosome is precise, so that the complete λ chromosome and only the λ chromosome is produced. In rare cases the excision events are not precise: crossing-over between the phage chromosome and the bacterial chromosome at sites other than the homologous recognition sites results in an abnormal circular DNA product. In the case diagrammed, a piece of λ chromosome has been left in the bacterial chromosome, while a piece of bacterial chromosome, including the *gal*⁺ gene, has been added to the rest of the λ chromosome. Since a bacterial gene (or genes) is included in a progeny phage, we have a transducing phage, called, in this case, *λd gal*. The *d* stands for "defective" since not all phage genes are present, and the *gal* indicates that the bacterial host cell *gal* gene has been picked up. This event is similar to *F'* production by defective excision of the *F* factor. The *λd gal* can replicate and lyse the host cell in which it is produced; however, since all *λ* genes are still present, some are on the phage chromosome while the others are in the bacterial chromosome.

Since the abnormal looping-out phenomenon is a rare event, the phage lysate produced from the initial infection diagramed contains mostly normal phages and a few transducing phages ($1/10^5$). The phage lysate can be used to infect bacteria that are *gal*. The phage lysate contains mostly wild-type phages and relatively few (about one in a million) *gal*⁺ transducing phages. Because of the small proportion of transducing phages, the lysate is called a *low-frequency transducing (LFT) lysate*.

Infection of the *gal* bacterial cells with the LFT lysate produces two types of transductants (Figure 7.17c). One type is unstable: The wild-type *λ* integrates at its normal *att λ* site, and then the *λd gal*⁺ phage integrates by a crossing-over event within the common *λ* sequences. In this case both types of phages are integrated in the bacterial chromosome, and the bacterium is heterozygous *gal*⁺/*gal* and hence can ferment galactose. This transductant is unstable because by induction phage growth can be initiated. (This event will be described a little later.) The second type of transductant is stable: These transductants are produced when only a *λd gal*⁺ phage is present (Figure 7.17c). The *gal*⁺ gene carried by the phage may be exchanged for the bacterial *gal* gene by a double-crossover event. Such a transductant is stable because the bacterial chromosome contains only one type of *gal* gene and no phage genes are integrated.

A *high-frequency transducing (HFT) lysate* may be obtained with this system (Figure 7.17d). The starting point is the doubly lysogenic, unstable transductant of Figure 7.17c, in which both a wild-type *λ* and a *λd gal*⁺ genome are integrated side by side in the bacterial chromosome. If this strain is induced, *λd gal*⁺ phages are produced at high frequency. When the HFT lysate is used to infect new nonlysogenic *gal* bacteria so that only one phage infects each bacterium (Figure 7.17e), the defective-phage genome is integrated by a single crossover in the *gal* gene region. This integration produces a *gal*⁺/*gal* transductant that is stable, since it lacks sufficient *λ* genes to lyse the cells. Thus by manipulating the number of phages that infect a bacterium, and by using appropriate selection events, we can detect specialized transduction and cause marker genes to be transduced at high frequency. In this way we can make a detailed analysis of the transduced genes we are studying.

𝒦EYNOTE

Transduction is the process by which bacteriophages mediate the transfer of genetic information from one bacterium (the donor) to another (the recipient). The capacity of the phage particle is limited, so the amount of DNA transferred is usually less than 1 percent of that in the bacterial chromosome. In generalized transduction any bacterial gene can be accidentally incorporated into the transducing phage during the phage life cycle and subsequently transferred to a recipient bacterium. Specialized transduction is mediated by temperate phages (such as λ) whose prophages associate with only one site of the bacterial chromosome. In this case the transducing phage is generated by abnormal excision of the prophage from the host chromosome so that the prophage includes both bacterial and phage genes. Transduction allows the fine-structure mapping of small chromosome segments. ───────

MAPPING GENES IN BACTERIOPHAGES

The same principles used to map eukaryotic genes are used to map phage genes. Crosses are made between phage strains that differ in genetic markers, and the proportion of recombinants among the total progeny is determined. Thus the basic procedure of mapping genes in two-, three-, or four-gene crosses involves the mixed infection of bacteria with phages of different genotypes.

In phages this basic experimental design must be adapted to the phage life cycle. First, we must be able to count phage types. We do so by plating a mixture of phages and bacteria on a solid medium. The concentration of bacteria is chosen so that with growth an entire "lawn" of bacteria forms. Phages are present in much lower concentrations. Each phage infects a bacterium and goes through the lytic cycle. The released progeny phages infect neighboring bacteria, and the lytic cycle is repeated. The result is a cleared patch in the lawn of bacteria. Each clearing is called a **plaque**, and one plaque derives from one of the original bacteriophages that was plated. Figure 7.18 shows plaques of the *E. coli* phages T4 and λ: note the T4 plaque is clear while the λ plaque has a turbid center.

Second, we must be able to distinguish phage phenotypes easily. Since individual phages are visible only under the electron microscope, mutants that affect phage morphology cannot be used. However, several mutants affect the phage life cycle, giving rise to differ-

~ **FIGURE 7.18**

Plaques of the *E. coli* bacteriophages (a) T4 and (b) λ.

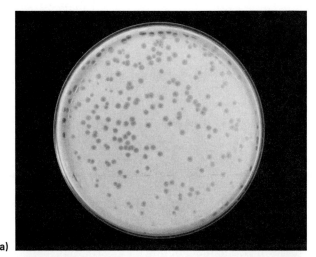

a)

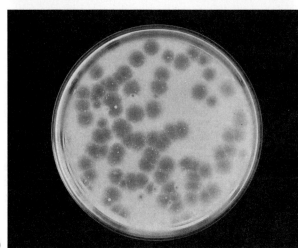

b)

ences in the appearances of plaques on a bacterial lawn. For example, one of the earliest studies of phage recombination was done in the late 1940s by Alfred Hershey and R. Rotman. They used strains of T2 that differed in two phenotypes: *plaque morphology* and *host range* (i.e., which bacterial strain the phage can lyse). One phage strain, genotype *h⁺ r*, was wild-type for the host range gene (*h⁺*, able to lyse only the *B* strain of *E. coli*) and mutant for the plaque morphology gene (*r*, producing large plaques with distinct borders). The other phage strain, *h r⁺*, was mutant for the host range gene (able to lyse both the *B* and the *B/2* strains of *E. coli*) and wild-type for the plaque morphology gene (*r⁺*, producing small plaques with fuzzy borders). In addition, when plated on a lawn containing both the *B* and *B/2* strains, any phage carrying the mutant host range allele *h* (infects both *B* and *B/2*) results in clear plaques, while phages carrying the wild-type *h⁺* allele result in cloudy plaques. The latter characteristic arises because

the *h⁺* allele can only infect the *B* bacteria, leaving a background cloudiness of uninfected *B/2* bacteria.

To map these two genes, we must make a genetic cross. The cross is achieved by adding both types of phages (*h⁺ r* and *h r⁺*) to a culture of bacteria, making sure that there is a high enough concentration of each phage type to ensure that a high proportion of bacteria are infected simultaneously with both phages (Figure 7.19a). Once the two genomes are within the bacterial

~ **FIGURE 7.19**

Schematic of the principles of performing a genetic cross with bacteriophages. (a) Bacteria are coinfected with the two parental bacteriophages, *h⁺ r* and *h r⁺*; (b) Replication of both parental chromosomes; (c) There is pairing of some chromosomes of each parental type, and crossing-over takes place between the two gene loci to produce *h⁺ r⁺* and *h r* recombinants; (d) Progeny phages are assembled and are released into the medium when the bacteria lyse; both parental and recombinant phages are found among the progeny.

a) Coinfect bacteria with the two parental phages, *h⁺ r* and *h r⁺*

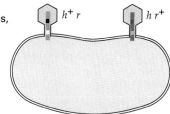

b) Replication of phage chromosomes in cell

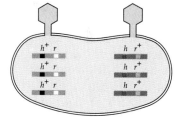

c) Recombination between some parental chromosomes

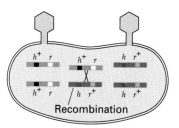

d) Phage assembly, bacterial lysis, and release of progeny phages

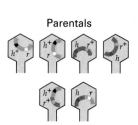

cell, each will replicate (Figure 7.19b). If an h^+ r and an h r^+ chromosome come together, a crossover can occur between the two gene loci to produce h^+ r^+ and h r recombinant chromosomes (Figure 7.19c), which are assembled into progeny phages. When the bacterium lyses, the recombinant progeny are released into the medium, along with non-recombinant (parental) phages (Figure 7.19d).

After the life cycle is completed, the progeny phages are plated onto a bacterial lawn containing a mixture of *E. coli* strains *B* and *B/2*. Four plaque phenotypes are found from this experiment (Figure 7.20), two parental types and two recombinant types. The parental type h r^+ gives a clear and small plaque; the other parental h^+ r gives a cloudy and large plaque. The reciprocal recombinant types give recombined pheno-

types: The h^+ r^+ plaques are cloudy and small, and the h r plaques are clear and large.

Since the four progeny types can be readily analyzed on the solid medium, we can calculate the recombination frequency for recombination events occurring between h and r. In this case the frequency is given by

$$\frac{(h^+ \ r^+) + (h \ r) \text{ plaques}}{\text{total plaques}} \times 100$$

With the same assumption used in mapping eukaryotic genes, namely, that crossovers occur randomly along the chromosome, the recombination frequency can be considered to be the map distance (in map units) between the phage genes. In addition, experiments have shown that the recombination frequency is the same when the parental strains are h^+ r^+ and h r, supporting this conclusion. Soon after these first experiments were performed, a number of other T2 genes were mapped in much the same way. It should be noted, though, that some infected *E. coli* cells will contain only one parental type *or* the other, so the actual frequency of recombination will be underestimated.

FINE-STRUCTURE ANALYSIS OF A BACTERIOPHAGE GENE

All of the genetic-mapping experiments described in this chapter for both bacteria and bacteriophages, and in Chapters 5 and 6, involved the use of mutant alleles of different genes: The recombinational mapping of the distance between genes, called *intergenic mapping* (inter = between), can be used to construct chromosome maps for organisms. The same general principles of recombinational mapping can be applied to mapping the distance between mutational sites within the same gene, a process called *intragenic mapping* (intra = within).

Intragenic mapping is possible because each gene consists of many nucleotide pairs of DNA linearly arranged along the chromosome. This fact was not always known. In the 1950s and early 1960s, Seymour Benzer performed a series of experiments to define mutational and recombinational sites within a gene. In essence, his genetic experiments gave new insights into what alleles were and laid to rest the concept of a gene being a bead on a string. In addition, they revealed much about the relationship between mapping and gene structure.

In his experiments Benzer used phage T4, principally because bacteriophages produce large numbers of progeny, facilitating the potential determination of very low recombination frequencies. Benzer's initial experi-

~ FIGURE 7.20

Plaques produced by progeny of a cross of T2 strains h r^+ × h^+ r. Four plaque phenotypes, representing both parental types and the two recombinants, may be discerned. The parental h r^+ type produces clear, small plaques, and the parental h^+ r type produces cloudy, large plaques. The recombinant h r type produces a clear, large plaque, and the recombinant h^+ r^+ type produces a cloudy, small plaque.

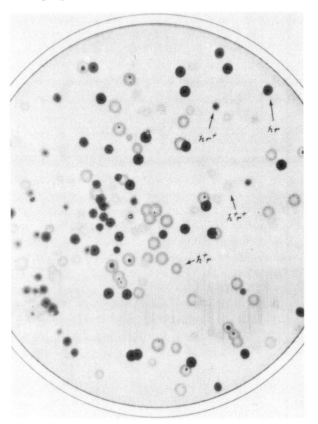

ments involved the detailed genetic mapping of sites within a gene, through **fine-structure mapping.**

When cells of *E. coli* strain *B* or strain *K12*(λ), growing on solid medium in a petri dish, are infected with wild-type T4, small turbid plaques with fuzzy edges are produced as a result of the phage life cycle (Figure 7.21). [Strain *E. coli K12*(λ) contains the λ phage chromosome incorporated into the bacterial chromosome.] On the other hand, when cells of *E. coli* strain *B* are infected with *r* (rapid-lysis) mutants of phage T4, large, clear plaques with distinct edges are produced (Figure 7.21). From intergenic-mapping experiments, the *r* mutations were found to map at several locations in the phage's genome, defining several *r* genes (see Figure 7.31; p. 226).

In his experiments Benzer used particular *r* mutants from the *rII* region of the T4 map. Of special importance to the experiments that we are about to describe was Benzer's 1953 finding, that *rII* mutants, in addition to their *plaque morphology* phenotypic distinction from wild-type (*r*⁺) T4, also have distinct *host range properties.* Specifically, while wild-type T4 is able to grow in and lyse cells of either *E. coli B* or *K12*(λ), *rII* mutants can grow in and lyse cells of *B*, producing large, clear plaques on solid medium (Figure 7.21) but are unable to grow in cells of *K12*(λ). Strain *B* is defined as the *permissive host* for *rII* mutants, while strain *K12*(λ) is the *nonpermissive host.* Thus *rII* mutants are conditional lethal mutants, because they can grow under one set of conditions but not others. The reasons for the inability of *rII* phages to multiply in the *K12*(λ) strain are not known, although it is known that the presence of the incorporated lambda phage genome is necessary for blocking the phage's propagation.

~ FIGURE 7.21

The *r*⁺ and mutant *rII* plaques on a lawn of *E. coli B.* The *r*⁺ plaque is turbid with a fuzzy edge; the *rII* plaque is larger, clear, and has a distinct boundary.

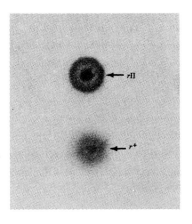

Recombination Analysis of *rII* Mutants

Benzer realized that the growth defect of *rII* mutants on *E. coli K12*(λ) could serve as a powerful selective tool for detecting the presence of a very small proportion of *r*⁺ phages within a large population of *rII* mutants. Between 1953 and 1963 Benzer collected thousands of *rII* mutants, some of which had arisen spontaneously and some of which had been induced by treating *rII* phages with mutagens.

Initially, Benzer set out to construct a fine-structure genetic map of the *rII* region. To do so, he crossed 60 independently isolated *rII* mutants in all possible combinations in *E. coli B* (the permissive host), then collected the progeny phages once the cells had lysed (Figure 7.22; p. 218). (For a phage such as T4, about $10^{10}-10^{11}$ phages per milliliter of phage lysate would be typical.) For each cross he plated a sample of the phage progeny on *E. coli B,* the permissive host, in order to count the total number of progeny phage per milliliter. He plated another sample of the phage progeny on *E. coli K12*(λ), the nonpermissive host, to find the total number of the very rare *r*⁺ recombinants (per milliliter) resulting from genetic recombination between the mutations carried by the two *rII* mutants used in the cross. In this way Benzer was able to calculate the percentage of very rare *r*⁺ recombinants between closely linked genetic sites.

For each cross of two *rII* mutants, such as *rII*1 and *rII*2 (Figure 7.23; p. 219), four genetic classes of progeny were usually found: the two parental *rII* types (*rII*1 and *rII*2) and two recombinant types resulting from a single crossover. One of the recombinant types was the wild type (*r*⁺), and the other was a double mutant carrying both *rII* mutations (*rII*1, 2), which had an *r* phenotype and was phenotypically indistinguishable from one or the other of the parental *rII* mutants. Recall from Chapters 5 and 6 and earlier in this chapter that the map distance between two mutations is given by the percentage of recombinant progeny among all progeny from crosses of the two mutants. For the crosses of *rII* mutants, a single crossover event between the two mutations will give the *r*⁺ and double *rII* mutant recombinants (see Figure 7.23). Therefore the frequencies of the two recombinant classes of phages are expected to be the same. Consequently, the total number of recombinants from each *rII* × *rII* cross is approximated by twice the number of *r*⁺ plaques counted on plates of strain *K12*(λ). The general formula for the map distance between two *rII* mutations is

$$\frac{2 \times \text{number of } r^+ \text{ recombinants}}{\text{total number of progeny}} \times 100\%$$

$$= \text{map distance (in map units)}$$

~ FIGURE 7.22

Benzer's general procedure for determining the number of r^+ recombinants from a cross involving two *rII* mutants of T4.

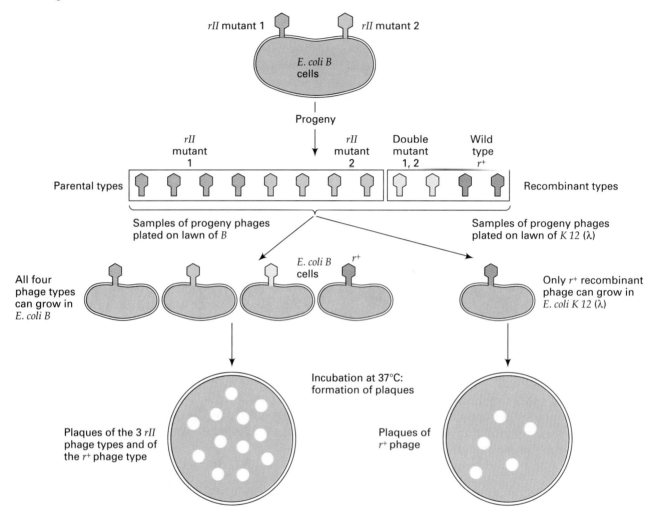

From the recombination data obtained from all possible pairwise crosses of the initial set of 60 *rII* mutants, Benzer was able to construct a linear genetic map (Figure 7.24). Among the mutants some pairs produced no r^+ recombinants when they were crossed. This result was interpreted to mean that those pairs carried mutations at exactly the same site; that is, the same nucleotide pair in the DNA had been changed; there was no possibility of recombination between the mutations. Mutations that change the same nucleotide pair within a gene are termed *homoallelic*. However, most pairs of *rII* mutants did produce r^+ recombinants when crossed, indicating that the mutations they carried had altered different nucleotide pairs in the DNA. Mutations that change different nucleotide pairs within a gene are termed *heteroallelic*. The map showed that the lowest frequency with which r^+ recombinants were formed in any pairwise crosses of *rII* mutants carrying heteroallelic mutations was 0.01 percent.

The minimum map distance figure of 0.01 percent can be used to make a rough calculation of the molecular distance—the distance in base pairs—involved in the recombination event. From extensive mapping experiments with T4, its circular genetic map is known to be about 1500 map units. If two *rII* mutants produce 0.01 percent r^+ recombinants, the mutations are separated by 0.02 map units, or by about $0.02/1500 = 1.3 \times 10^{-5}$ of the total T4 genome. Since the total T4 genome contains about 2×10^5 base pairs, the minimum recombination distance is $(1.3 \times 10^{-5}) \times (2 \times 10^5)$, or about 3 base pairs. That means that genetic recombination can occur in distances of 3 base pairs or less. Since genes were known to consist, typically, of a few hundred base pairs, Benzer's mapping data clearly indicated that recombination could occur *within* a gene. We now know that the nucleotide pair is the *unit of recombination*. Equally important, Benzer's work and its extensions established that the nucleotide pair is

~ **FIGURE 7.23**

Production of parental and recombinant progeny from a cross of two *rII* mutants with
mutation in different sites within the *rII* region. Progeny phages of parental genotype are
produced if no crossing-over occurs, whereas progeny phages with recombinant genotypes
are produced if a single crossover occurs between the sites of the two mutations.

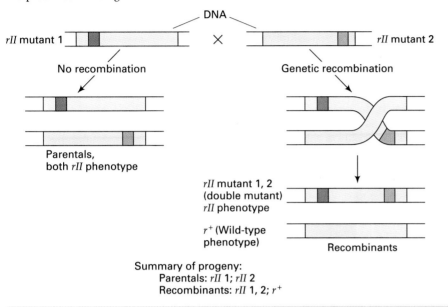

~ **FIGURE 7.24**

Preliminary fine-structure genetic map of the *rII* region of phage T4 derived by Benzer
from crosses of an initial set of 60 *rII* mutants. Lower levels in the figure show finer
detail of the map. In the lowest level, the numbered vertical lines indicate individual *rII*
mutants, and the decimals indicate the percentage of r^+ recombinants found in crosses
between the two *rII* mutants connected by an arrow.

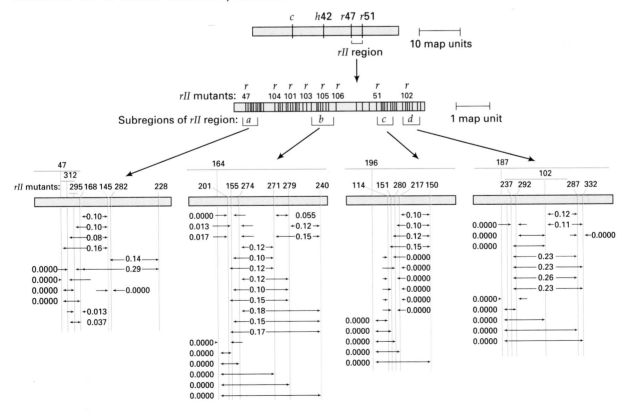

also the *unit of mutation.* These definitions clearly replace the classical ones that the whole gene is the unit of recombination and the unit of mutation.

Deletion Mapping

Following his initial series of crossing experiments, Benzer continued to map the over 3000 *rII* mutants he had so that he could complete his fine-structure map. To map these 3000 mutants would have required approximately 5 million crosses—an overwhelming task even in phages, with which up to 50 crosses can be done per day. Therefore Benzer developed some genetic tricks to simplify his mapping studies. These tricks involved the use of *deletion mapping* to localize unknown mutations.

~ FIGURE 7.25

Segmental subdivision of the *rII* region of phage T4 by means of deletion. Level I shows the whole T4 genetic map. In Level II seven deletions define seven segments of the *rII* region. In Level III three deletions define four subsegments of the *A5* segments. In Level IV three deletions define four subsegments of the *A5c* subsegment. Level V shows the order and spacing of the sites of the *rII* mutations in the *A5c2a2* subsegment, as established by pairwise crosses of seven point mutants. Level VI is a model of the DNA double helix, indicating the approximate scale of the level V map.

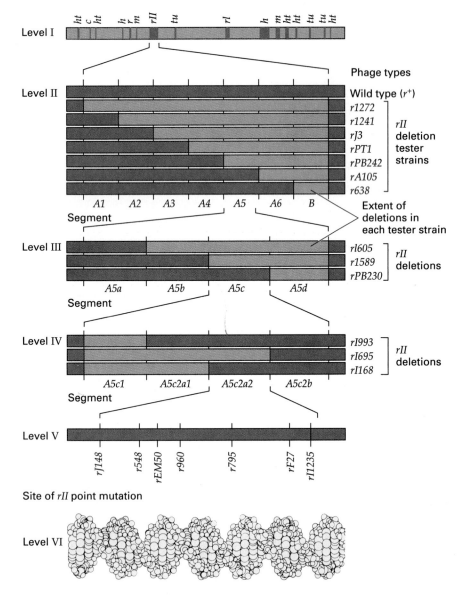

Most of the *rII* mutants isolated by Benzer were **point mutants;** their phenotype resulted from an alteration of a single nucleotide pair. A point mutant can revert to the wild-type state spontaneously or after treatment with an appropriate mutagen. However, some of Benzer's *rII* mutants did not revert, nor did they produce r^+ recombinants in crosses with a number of *rII* point mutants that were known to be located at different places on the *rII* map. These mutants were *deletion mutants,* which involve the loss of a section of DNA. Benzer found a wide range in the extent and location of deleted genetic material was found among the *rII* deletion mutants he studied. Some deletion mutants are shown in Figure 7.25.

In actual practice an unknown *rII* point mutant was first crossed with each of seven standard deletion mutants that defined seven main segments of the *rII* region (segments *A1-A6* and *B* in Figure 7.25). For example, if an *rII* point mutant gave r^+ recombinants when crossed with deletions defining *A6* and *B* (*rA105* and *r638,* respectively) but not with the other five deletions, the point mutant must be in segment *A5.* That is to say, r^+ recombinants can only be produced in crosses with deletions if the deleted segment does not include the region of DNA containing the point mutation. If a point mutant does not produce r^+ recombinants with the deletions *r1272, r1241, rJ3, rPT1,* and *rPB242,*

then the point mutation must be in the segment of DNA that these mutants lack. The fact that r^+ recombinants are produced with *rA105* and *r638* indicates that the point mutation must be in the *A5* region, which is not missing in either deletion mutant.

Once the main segment in which the mutation occurred was known, the point mutant was crossed with each of the relevant secondary set of reference deletions, *r1605, r1589,* and *rPB230* (see Figure 7.25). With segment *A5,* for example, three deletions divide *A5* into the four subsegments *A5a* through *A5d.* The presence or absence of r^+ recombinants in the progeny of the crosses of the *A5 rII* mutant with the secondary set of deletions enabled Benzer to localize the mutation more precisely to a smaller region of the DNA. For example, if the mutation was in segment *A5c,* then r^+ recombinants will be produced with deletion *rPB230* but not with either of the other two deletions. Other deletion mutants defined even smaller regions of each of the four subsegments *A5a* through *A5d;* for example, *A5c* was divided into *A5c1, A5c2a1, A5C2a2,* and *A5c2b* by deletions *r1993, r1695,* and *r1168.*

In three sequential sets of crosses of point mutants with deletion mutants, it was possible to localize any given *rII* point mutant to one of 47 regions defined by the deletions, as shown in Figure 7.26. Then all those point mutants within a given region could be crossed in

~ **FIGURE 7.26**

Map of deletions used to divide the *rII* region into 47 small segments. Here, those segments are shown as small boxes at the bottom of the figure.

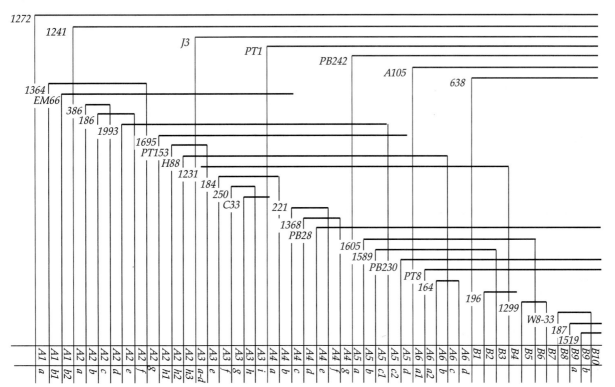

~ **FIGURE 7.27**

Fine-structure map of the *rII* region derived from Benzer's extensive experiments. The number of independently isolated mutations that mapped to a given site is indicated by the number of blocks at the site. Hot spots are represented by a large number of blocks.

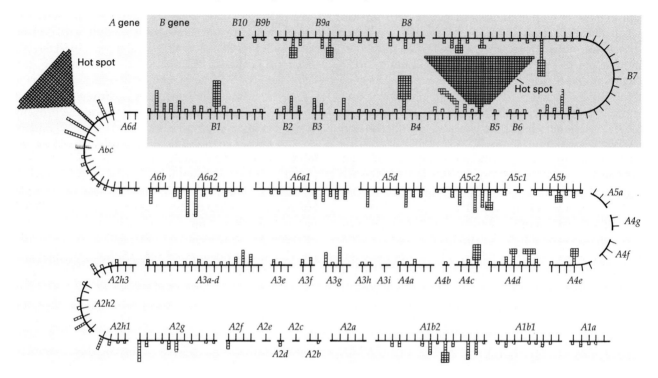

all possible pairwise crosses to construct a detailed genetic map. In this way Benzer used the greater than 3000 *rII* mutants to prove that the *rII* region is subdivisible into more than 300 mutable sites that were separable by recombination (Figure 7.27). The distribution of mutants is not random; certain sites (called *hot spots*) are represented by a large number of independently isolated point mutants.

${K}$EYNOTE

Benzer's detailed recombination analysis of the *rII* region of bacteriophage T4 indicated that the unit of mutation and of recombination is the base pair in DNA. This definition replaced the classical view that genes were indivisible by mutation and recombination. _____

Defining Units of Function by Complementation Tests

From the classical point of view the gene is a *unit of function;* that is, each gene specifies one function. Benzer designed genetic experiments to determine

whether this classical view was indeed true of the *rII* region. To find out whether two different *rII* mutants belonged to the same gene (unit of function), Benzer adapted the **cis-trans** or **complementation test** developed by Edward Lewis to study the nature of the functional unit of the gene in *Drosophila*. For clarification of the following discussion, it will help to know that the complementation tests indicated that the *rII* region actually consists of two genes (units of function), *rIIA* and *rIIB*. A mutation at any point in either gene will produce the *rII* phenotypes [the production of large, clear plaques on *E. coli* B and the inability to grow in *E. coli* K12(λ)]. In other words, there are two distinct genes in the *rII* region, *rIIA* and *rIIB*, each of which specifies a functional product needed for growth in *E. coli* K12(λ).

The complementation test is used to establish how many units of function (genes) define a given set of mutations that express the same mutant phenotypes. In Benzer's work with the *rII* mutants, the nonpermissive strain K12(λ) was infected with a pair of *rII* mutants to see whether the two mutants, each alone unable to grow in strain K12(λ), are able to "work together" to produce progeny phages. If the phages do produce progeny, the two mutants are said to complement each other, meaning that the two mutations must be in different genes (units of function) that specify different functional products. If no progeny phages are produced, the mutants have not complemented, indicating that the mutations are in the same functional unit. (Note that genetic recombination is not necessary for complementation to occur.)

These two situations are diagrammed in Figure 7.28. In the first case (Figure 7.28a), complementation occurs because the *rIIA* mutant still makes a functional *B* product and the *rIIB* mutant makes a functional *A* product. As a result, between the two mutants both products necessary for phage propagation in *E. coli* K12(λ) are generated, so progeny phages are assembled and released. In the second case (Figure 7.28b), no complementation occurs because, while both produce a functional *rIIB* product, the two mutants both lack the *A* function. As a result, phage reproduction in *E. coli* K12(λ) does not occur.

On the basis of the results of complementation tests, Benzer found that *rII* mutants fall into two units of function, *rIIA* and *rIIB* (also called complementation groups, which in this case directly correspond to genes). That is, all *rIIA* mutants complement all *rIIB*

mutants. In contrast, *rIIA* mutants fail to complement other *rIIA* mutants, and *rIIB* mutants fail to complement other *rIIB* mutants. The dividing line between the *rIIA* and *rIIB* units of function is indicated in the fine-structure map of Figure 7.27. Point mutants and deletion mutants in the *rII* region obey the same rules in the complementation tests. The only exceptions are deletions that span parts of both the *A* and the *B* functional units. Such deletion mutants do not complement either *A* or *B* mutants.

For the complementation test examples shown in Figure 7.28, each of the two phages that co-infect the nonpermissive *E. coli* strain K12(λ) carries an *rII* mutation, a configuration of mutations called the *trans* configuration. In this configuration, the two mutations are carried by different phages. As a control, it is usual to co-infect *E. coli* K12(λ) with an *r*[+] (wild-type) phage and an *rII* mutant phage carrying *both* mutations to see whether the expected wild-type function results. When both mutations under investigation are carried on the same chromosome, the configuration is called the *cis* configuration of mutations. (Because of the *cis* and *trans* configurations of mutations used, the comple-

~ **FIGURE 7.28**

Complementation tests for determining the units of function in the *rII* region of phage T4; the nonpermissive host *E. coli* K12(λ) is infected with two different *rII* mutants: (a) complementation occurs; (b) complementation does not occur.

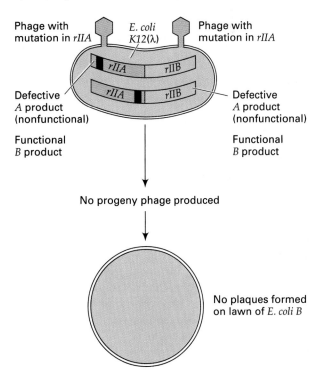

mentation test is also called the *cis-trans* test.) In the *cis* test, the r^+ is expected to be dominant over the two mutations carried by the *rII* mutant phage, so progeny phages will be produced. Thus if the mutants do not complement in *trans,* they are in the same functional unit.

Benzer referred to the genetic unit of function revealed by the cis-trans test as the *cistron.* When it was coined in 1955, the term *cistron* became widely used and replaced the ambiguous term *gene* in many writings. At the present time, *gene* is commonly used and *cistron* is being used less. Nonetheless, *gene* and *cistron* are equivalent in referring to the genetic unit of function. It is appropriate to refer to the *A* and *B* functional units as the *rIIA* and *rIIB* cistrons or genes. Presumably, their two products act in common processes necessary for T4 propagation in strain $K12(\lambda)$. Genetically, the *rIIA* cistron is about 6 map units and 800 base pairs, and the *rIIB* cistron is about 4 map units and 500 base pairs.

The complementation test is commonly used to define the functional units (complementation groups or genes) for mutants with the same phenotype. The principles for performing a complementation test are always the same; only the practical details of performing the test are organism-specific. For example, in yeast one could select two haploid cells that are of different mating types (*a* and *α*) and that carry different mutations conferring the same mutant phenotype. Mating these two types would produce a diploid, which would then be analyzed for complementation of the two mutations. In animal cells two cells, each exhibiting the same mutant phenotype, can be fused together and analyzed for wild-type or mutant phenotype; a wild-type phenotype would indicate that complementation had occurred. Again, in neither of these cases is recombination necessary for complementation to occur.

Let us consider an example of complementation in a diploid organism. (The cross is diagrammed in Figure 7.29, and the example comes from Question 4.21.) Two true-breeding mutant strains of *Drosophila melanogaster* have black body color instead of the wild-type grey-yellow. When the two strains are crossed, all of the F_1 flies have wild-type body color. How can these data be interpreted? The simplest explanation is that complementation has occurred between mutations in two genes, each of which is involved in the body color phenotype. That is, a recessive autosomal gene, ebony (*e*), when homozygous, produces a black body color. On a different autosome is another recessive gene, black (*bl*), which also produces a black body color when homozygous. Since the two parents are homozygotes, they are genotypically *e/e bl⁺/bl⁺* and *e⁺/e⁺ bl/bl,* and each is phenotypically black. The F_1 genotype is *e⁺/e bl⁺/bl,* which is equivalent to the *trans* configura-

~ FIGURE 7.29

Complementation between two black body mutations of *Drosophila melanogaster.*

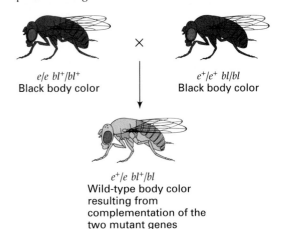

e/e bl⁺/bl⁺
Black body color

×

e⁺/e⁺ bl/bl
Black body color

e⁺/e bl⁺/bl
Wild-type body color
resulting from
complementation of the
two mutant genes

tion of the *rII* cistron experiments. The F_1 has wild-type body color because complementation has occurred. Even without knowing in advance that two independently assorting genes were involved, we would conclude that two cistrons are involved.

What if the F_1 from the cross between two independently isolated, true-breeding black mutant strains were all phenotypically black? The interpretation would be that the two mutations did not show complementation, and therefore the mutations are in the same complementation group; that is, they are different mutant alleles of the same gene.

KEYNOTE

The complementation, or *cis-trans,* test is used to determine how many units of functions (genes) define a given set of mutations expressing the same mutant phenotypes. If two mutants, each carrying a mutation in a different gene, are combined, the mutations will complement and a wild-type function will result. If two mutants, each carrying a mutation in the same gene, are combined, the mutations will not complement and the mutant phenotype will be exhibited. _____

SUMMARY

In this chapter, we have seen how genetic mapping can be accomplished in prokaryotes such as bacteria and bacteriophages. Essentially the same experimental strategy is used for all gene mapping; that is, genetic material is exchanged between strains differing in genetic markers and recombinants are detected and counted. Depending on the particular bacterium, the mechanism of gene transfer may be transformation, conjugation, or transduction.

Transformation is the transfer of genetic material between organisms by means of extracellular pieces of DNA. Conjugation is a process in which there is a uni-directional transfer of genetic information through direct cellular contact between a donor and a recipient bacterial cell. Transduction is a process whereby bacteriophages mediate the transfer of bacterial DNA from one bacterium (the donor) to another (the recipient). Bacteriophage chromosomes may be mapped by infecting bacteria simultaneously with two phage strains and analyzing the resulting phage progeny for parental and recombinant phenotypes. Formally, this method of mapping is the same as that used for mapping genes in haploid eukaryotes. As a summary, two linkage maps

of prokaryotes of genetic significance are shown: *E. coli* (Figure 7.30) and bacteriophage T4 (Figure 7.31; p. 226).

Insights into the relationships between mapping and gene structure were obtained from a fine structure analysis of the bacteriophage T4 *rII* region. The mutational sites within a gene were mapped through intragenic mapping. The results indicated that the unit of mutation and the unit of recombination is the base pair in DNA, replacing the classical definition that genes were indivisible by mutation and recombination.

The method of defining the number of units of function (genes) by complementation tests was discussed. Given a set of mutations expressing the same mutant phenotypes, two mutants are combined and the phenotype is determined. If the phenotype is wild type, the two mutations have complemented and must be in different units of functions. If the phenotype is mutant, the two mutations have not complemented and must be in the same unit of function.

~ FIGURE 7.30

Genetic map of *E. coli* based on conjugation experiments. Units are in minutes timed from an arbitrary origin at 12 o'clock.

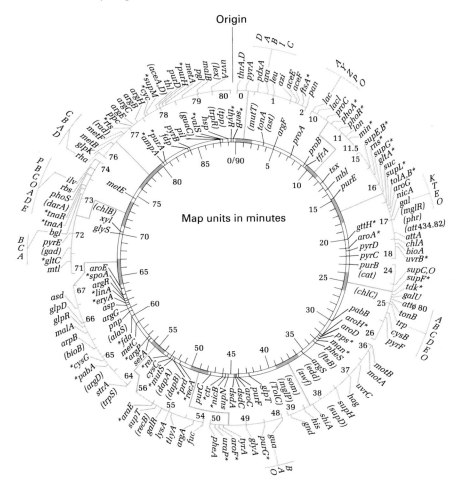

~ FIGURE 7.31

Genetic map of the bacteriophage T4. Note the clustering of genes of related function around the perimeter of the map. Units are in kilobase pairs (kb). (1 kb = 1000 base pairs.)

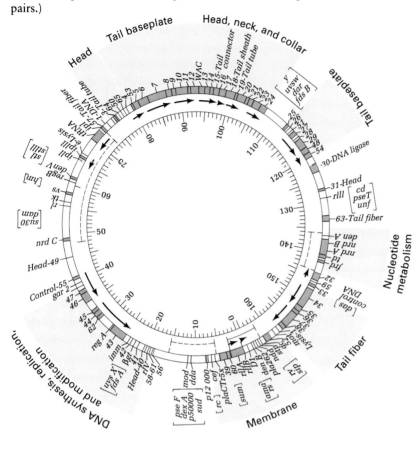

ANALYTICAL APPROACHES FOR SOLVING GENETICS PROBLEMS

Q.1 In a transformation experiment, donor DNA from an $a^+ b^+$ strain was used to transform a recipient strain of genotype $a\ b$. The transformed classes were isolated and their frequencies determined to be:

$a^+ b^+$	307
$a^+ b$	215
$a\ b^+$	278

The total number of transformants was 800. What is the frequency with which the b locus is cotransformed with the a locus?

A.1 The frequency with which b^+ is cotransformed with the a^+ gene is calculated using values for the total number of a^+ transformants, and the number of transformants for both a^+ and b^+. The formula is:

$$\frac{\text{Number of } a^+ b^+ \text{ cotransformants}}{\text{Total number of } a^+ \text{ transformants}} \times 100\%$$

The $a^+ b^+$ cotransformants number 307. The a^+ transformants are represented by two classes: $a^+ b^+$ (307) and $a^+ b$ (215), for a total of 522. The $a\ b^+$ class is irrelevant to the question because they are not transformants for a^+. Thus, the cotransformation frequency for a^+ and b^+ is $307/522 \times 100 = 58.8$ percent.

Q.2 In *E. coli* the following *Hfr* strains donate the markers shown in the order given:

Hfr STRAIN	ORDER OF GENE TRANSFER
1	G E B D N A
2	P Y L G E B
3	X T J F P Y
4	B E G L Y P

All the *Hfr* strains were derived from the same F^+ strain. What is the order of genes in the original F^+ chromosome?

A.2 This question is an exercise in piecing together various segments of the circumference of a circle. The best approach is to draw a circle and label it with the genes transferred from one *Hfr* and then to see which of the other *Hfr*'s transfers an overlapping set. For example, *Hfr 1* transfers E, then B, then D, and so on; and *Hfr 4* transfers B, then E, and so forth. Now we can juxtapose the two sets of genes transferred by the two *Hfr*'s and deduce that the polarities of transfer are opposite:

Hfr 1	G E B D N A
Hfr 4	P Y L G E B

Extending this reasoning to the other *Hfr*'s, we can draw an unambiguous map (see the figure below), with the arrowheads indicating the order of transfer.

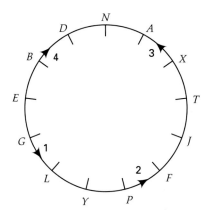

The same logic would be used if the question gave the relative time units of entry of each of the genes. In that case we would expect that the time "distance" between any two genes would be approximately the same regardless of the order of transfer or how far the genes were from the origin.

Q.3 In a transduction experiment, the donor was $c^+ d^+ e^+$ and the recipient was $c\ d\ e$. Selection was for c^+. The four classes of transductants from this experiment were:

CLASS	GENETIC COMPOSITION	NUMBER OF INDIVIDUALS
1	$c^+\ d^+\ e^+$	57
2	$c^+\ d^+\ e$	76
3	$c^+\ d\ e$	365
4	$c^+\ d\ e^+$	2
	Total	500

a. Determine the cotransduction frequency for c^+ and d^+.

b. Determine the cotransduction frequency for c^+ and e^+.

c. Which of the cotransduction frequencies calculated in (a) and (b) represents the greater actual distance between genes? Why?

A.3 a. The analysis is similar to the cotransformation frequency analysis described in Q.1. The formula for the cotransduction frequency for c^+ and d^+ is:

$$\frac{\text{Number of } c^+\ d^+ \text{ cotransductants}}{\text{Total number of } c^+ \text{ transductants}} \times 100\%$$

From the data presented, Classes 1 and 2 are the $c^+ d^+$ cotransductants, and the total number of c^+ transductants is the sum of Classes 1 through 4. Thus the number of c^+ and d^+ transductants is $57 + 76 = 133$, and the cotransductant frequency is $133/500 \times 100 = 26.6$ percent.

b. The analysis is identical in approach to (a). The formula for the cotransduction frequency for c^+ and e^+ is:

$$\frac{\text{Number of } c^+\ e^+ \text{ cotransductants}}{\text{Total number of } c^+ \text{ transductants}} \times 100\%$$

From the data presented, Classes 1 and 4 are the $c^+ e^+$ cotransductants, and the total number of c^+ transductants is the sum of Classes 1 through 4. Thus the number of c^+ and e^+ transductants is $57 + 2 = 59$, and the cotransductant frequency is $59/500 \times 100 = 11.8$ percent.

c. The greater actual distance is for the c^+ and e^+ genes. The principle involved is that the closer two genes are on the chromosome, the greater the chance that they will be cotransduced. Thus, as the distance between genes increases, concomitantly the cotransduction frequency decreases. Since the $c^+ e^+$ cotransduction frequency is 11.8 percent and the $c^+ d^+$ cotransduction frequency is 26.6 percent, genes c^+ and e^+ are farther apart than genes c^+ and d^+.

Q.4 Five different *rII* deletion strains of phage T4 were tested for recombination by pairwise crossing in *E. coli* B. The following results were obtained, where $+ = r^+$

recombinants produced, and 0 = no r^+ recombinants produced:

	A	B	C	D	E
E	0	+	0	+	0
D	0	0	0	0	
C	0	0	0		
B	+	0			
A	0				

Draw a deletion map compatible with these data.

A.4 The principle here is that if two deletion mutations overlap, then no r^+ recombinants can be produced. Conversely, if two deletion mutations do not overlap, then r^+ recombinants can be produced. To approach a question of this kind, we must draw overlapping and nonoverlapping lines from the given data.

Starting with A and B, these two deletions do not overlap since r^+ recombinants are produced. Therefore these two mutations can be represented as follows:

```
      A                    B
  _____            _____
```

The next deletion, C, does not give r^+ recombinants with any of the other four deletions. We must conclude, therefore, that C is an extensive deletion that overlaps all of the other four, with endpoints that cannot be determined from the data given. One possibility is as follows:

```
                  C
  _____

      A                    B
  _____            _____
```

Deletion D does not give r^+ recombinants with A, B, or C, but it does with E. In turn, E gives r^+ recombinants with B and D but not with A or C. Thus D must overlap both A and B but not E, and E must overlap A and C but not B. A compatible map for this situation follows. Other maps can be drawn in terms of the endpoints of the deletions.

```
                  C
  _____

      A                    B
  _____            _____

  E              D
  _____      _____
```

Q.5 Seven different *rII* point mutants (*1* to *7*) of phage T4 were tested for recombination crosses in *E. coli* B with the five deletion strains described in Question 4. The following results were obtained, where $+$ = r^+ recombinants produced and 0 = no r^+ recombinants produced:

	A	B	C	D	E
1	0	+	0	+	+
2	+	0	0	+	+
3	0	+	0	+	0
4	+	+	0	+	0
5	+	0	0	0	+
6	0	+	0	0	+
7	+	+	0	0	+

In which regions of the map can you place the seven point mutations?

A.5 If an r^+ recombinant is produced, the *rII* point mutation must be in the region covered by the deletion mutation with which it was crossed. Thus the matrix of results localizes the point mutations to the regions defined by the deletion mutants. Potentially, the results define the relative extent of deletion overlap. For example, point mutation 7 gives r^+ recombinants with A, B, and E but not with D. Logically, then, 7 is located in the region defined by the part of deletion D that is not involved in the overlap with A and B. Similarly, point mutation 4 gives r^+ recombinants with A, D, and B but not with E. Thus 4 must be in a region defined by a segment of deletion E that does not overlap deletion A. Furthermore, since 4 does not give r^+ recombinants with C either, deletion C must overlap the site defined by point mutation 4. This result, then, refines the deletion map with regard to the E, C, and A endpoints. The map we can draw from the matrix of results is as follows:

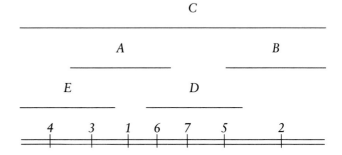

QUESTIONS AND PROBLEMS

*7.1 If an *E. coli* auxotroph A could only grow on a medium containing thymine, and an auxotroph B could only grow on a medium containing leucine, how would you test whether DNA from A could transform B?

7.2 Distinguish among F^-, F^+, F', and *Hfr* strains of *E. coli*.

*7.3 In $F^+ \times F^-$ crosses the F^- recipient is converted to a donor with very high frequency. However, it is rare for a recipient to become a donor in *Hfr* $\times$ F^- crosses. Explain why.

*7.4 With the technique of interrupted mating, four *Hfr* strains were tested for the sequence in which they trans-

mitted a number of different genes to an *F⁻* strain. Each *Hfr* strain was found to transmit its genes in a unique sequence, as shown in the accompanying table (only the first six genes transmitted were scored for each strain).

ORDER OF TRANSMISSION	*Hfr* STRAIN			
	1	2	3	4
First	O	R	E	O
	F	H	M	G
	B	M	H	X
	A	E	R	C
	E	A	C	R
Last	M	B	X	H

What is the gene sequence in the original strain from which these *Hfr* strains derive? Indicate on your diagram the origin and polarity of each of the four *Hfr*'s.

7.5 When *Hfr* donors conjugate with *F⁻* recipients that are lysogenic for phage λ, the recipients (the exconjugant zygotes) usually survive. However, when *Hfr* donors that are lysogenic for λ conjugate with *F⁻* cells that are non-lysogens, the exconjugant zygotes produced from matings that have lasted at least 100 minutes usually lyse, releasing mature λ phage particles. This event is called zygotic induction of λ.

a. Explain zygotic induction.

b. Explain how the locus of the integrated λ prophage can be determined.

7.6 At time zero an *Hfr* strain (strain 1) was mixed with an *F⁻* strain, and at various times after mixing, samples were removed and agitated to separate conjugating cells. The cross may be written as:

$$Hfr\ 1: a^+\ b^+\ c^+\ d^+\ e^+\ f^+\ g^+\ h^+\ str^s$$
$$F^-: a\ b\ c\ d\ e\ f\ g\ h\ str^r$$

(No order is implied in listing the markers.)

The samples were then plated onto selective media to measure the frequency of $h^+\ str^r$ recombinants that had received certain genes from the *Hfr* cell. The graph of the number of recombinants against time is shown in the figure below.

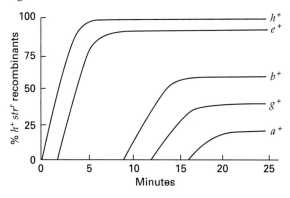

a. Indicate whether each of the following statements is true or false.

   i. All *F⁻* cells that received a^+ from the *Hfr* in the chromosome transfer process are likely also to have received b^+.

   ii. The order of gene transfer from *Hfr* to *F⁻* was a^+ (first), then g^+, then b^+, then e^+, then h^+.

   iii. Most $e^+\ str^r$ recombinants are likely to be *Hfr* cells.

   iv. None of the $b^+\ str^r$ recombinants plated at 15 minutes are also a^+.

b. Draw a linear map of the *Hfr* chromosome indicating
   i. the point of insertion, or origin;
   ii. the order of the genes a^+, b^+, e^+, g^+, and h^+;
   iii. the shortest distance between consecutive genes on the chromosomes.

7.7 Distinguish between the lysogenic and lytic cycles.

7.8 Distinguish between generalized and specialized transduction.

*7.9 Indicate whether each of the following occurs/or is a characteristic of

> Generalized Transduction (GT)
> Specialized Transduction (ST)
> Occurs in Both   (B)
> Occurs in Neither   (N)

a. Phage carries DNA of bacterial or viral DNA origin, never both.

b. Phage carries viral DNA covalently linked to bacterial DNA.

c. Phage integrates into specific site on the host chromosome.

d. Phage integrates at random site on the host chromosome.

e. "Headful" of bacterial DNA encoated.

f. Host lysogenized.

g. Prophage state exists.

h. Temperate phage involved.

i. Virulent phage involved.

7.10 Consider the following transduction data:

DONOR	RECIPIENT	SELECTED MARKER	UNSELECTED MARKER	%
aceF⁺ dhl	*aceF dhl⁺*	*aceF⁺*	*dhl*	88
aceF⁺ leu	*aceF leu⁺*	*aceF⁺*	*leu*	34

Is *dhl* or *leu* closer to *aceF*?

*7.11 Consider the following P1 transduction data:

DONOR	RECIPIENT	SELECTED MARKER	UNSELECTED MARKER	%
$cysB^+$ $trpE$	$cysB$ $trpE^+$	$cysB^+$	$trpE$	37
$cysB^+$ $trpB$	$cysB$ $trpB^+$	$cysB^+$	$trpB$	53

Is $trpE$ or $trpB$ closer to $cysB$?

7.12 Consider the following data with P1 transduction:

DONOR	RECIPIENT	SELECTED MARKER	UNSELECTED MARKER	%
$aroA$ $pyrD^+$	$aroA^+$ $pyrD$	$pyrD^+$	$aroA$	5
$aroA^+$ $cmlB$	$aroA$ $cmlB^+$	$aroA^+$	$cmlB$	26
$cmlB$ $pyrD^+$	$cmlB^+$ $pyrD$	$pyrD^+$	$cmlB$	54

Choose the correct order:

a. $aroA - cmlB - pyrD$
b. $aroA - pyrD - cmlB$
c. $cmlB - aroA - pyrD$

7.13 Order the mutants trp, $pyrF$, and qts on the basis of the following three-factor transduction cross:

$$\begin{array}{rl} \text{Donor} & trp^+ \ pyr^+qts \\ \text{Recipient} & trp \ pyr \ qts^+ \\ \text{Selected Marker} & trp^+ \end{array}$$

UNSELECTED MARKERS	NUMBER
pyr^+ qts^+	22
pyr^+ qts	10
pyr qts^+	68
pyr qts	0

*7.14 Order $cheA$, $cheB$, eda, and $supD$ from the following data:

MARKERS	% COTRANSDUCTION
$cheA$-eda	15
$cheA$-$supD$	5
$cheB$-eda	28
$cheB$-$supD$	2.7
eda-$supD$	0

*7.15 A stock of T4 phage is diluted by a factor of 10^{-8} and 0.1 mL of it is mixed with 0.1 mL of 10^8 $E.$ $coli$ B/mL and 2.5 mL melted agar, and poured on the surface of an agar petri dish. The next day 20 plaques are visible. What is the concentration of T4 phages in the original T4 stock?

*7.16 Wild-type phage T4 grows on both $E.$ $coli$ B and $E.$ $coli$ $K12(\lambda)$, producing turbid plaques. The rII mutants of T4 grow on $E.$ $coli$ B, producing clear plaques, but do not grow on $E.$ $coli$ $K12(\lambda)$. This host range property permits the detection of a very low number of r^+ phages among a large number of rII phages. With this sensitive system it is possible to determine the genetic distance between two mutations within the same gene, in this case the rII locus.

Suppose $E.$ $coli$ B is mixedly infected with $rIIx$ and $rIIy$, two separate mutants in the rII locus. Suitable dilutions of progeny phages are plated on $E.$ $coli$ B and $E.$ $coli$ $K12(\lambda)$. A 0.1-mL sample of a thousandfold dilution plated on $E.$ $coli$ B showed 672 plaques. A 0.2-mL sample of undiluted phage plated on $E.$ $coli$ $K12(\lambda)$ showed 470 turbid plaques. What is the genetic distance between the two rII mutations?

7.17 Construct a map from the following two factor phage cross data (show map distance):

CROSS	% RECOMBINATION
$r_1 \times r_2$	0.10
$r_1 \times r_3$	0.05
$r_1 \times r_4$	0.19
$r_2 \times r_3$	0.15
$r_2 \times r_4$	0.10
$r_3 \times r_4$	0.23

*7.18 The following two-factor crosses were made to analyze the genetic linkage between four genes in phage λ: c, mi, s, and co.

PARENTS	PROGENY
$c +$ $\times$ $+ mi$	1213 c +, 1205 + mi, 84 + +, 75 c mi
$c +$ $\times$ $+ s$	566 c +, 808 + s, 19 + +, 20 c s
$co +$ $\times$ $+ mi$	5162 co +, 6510 + mi, 311 + +, 341 co mi
$mi +$ $\times$ $+ s$	502 mi +, 647 + s, 65 + +, 56 mi s

Construct a genetic map of the four genes.

7.19 Three gene loci in T4 that affect plaque morphology in easily distinguishable ways are r(rapid lysis), m(minute), and tu(turbid). A culture of $E.$ $coli$ is mixedly infected with two types of phage: r m tu and r^+ m^+ tu^+. Progeny phage are collected and the following genotype classes are found:

CLASS	NUMBER
r^+ m^+ tu^+	3,729
r^+ m^+ tu	965
r^+ m tu^+	520
r m^+ tu^+	172
r^+ m tu	162
r m^+ tu	474
r m tu^+	853
r m tu	3,467
	10,342

Construct a map of the three genes. What is the coefficient of coincidence, and what does the value suggest?

*7.20 The rII mutants of bacteriophage T4 grow in $E.$ $coli$ B but not in $E.$ $coli$ $K12(\lambda)$. The $E.$ $coli$ strain B is doubly infected with two *rII* mutants. A 6×10^7 dilution of the lysate is plated on *E. coli* B. A 2×10^5 dilution is plated on *E. coli* K12(λ). Twelve plaques appeared on strain K12(λ) and 16 on strain B. Calculate the amount of recombination between these two mutants.

7.21 Wild-type (r^+) strains of T4 produce turbid plaques, whereas *rII* mutant strains produce larger, clearer plaques. Five *rII* mutations ($a-e$) give the following percentages of wild-type recombinants in two-point crosses:

CROSS	% OF WILD-TYPE RECOMBINANTS
$a \times b$	0.2 percent
$a \times c$	0.9 percent
$a \times d$	0.4 percent
$b \times c$	0.7 percent
$e \times a$	0.3 percent
$e \times d$	0.7 percent
$e \times c$	1.2 percent
$e \times b$	0.5 percent
$b \times d$	0.2 percent
$d \times c$	0.5 percent

What is the order of the mutational sites and what are the map distances between the sites?

7.22 Choose the correct answer in each case.

a. If one wants to know if two different *rII* point mutants lie at exactly the same site (nucleotide pair), one should:
 i. Coinfect *E. coli* K12(λ) with both mutants. If phage are produced, they lie at the same site.
 ii. Coinfect *E. coli* K12(λ) with both mutants. If phage are not produced, they lie at the same site.
 iii. Coinfect *E. coli* B with both mutants and plate the progeny phage on both *E. coli* B and *E. coli* K12(λ). If plaques appear on B but not K12(λ), they lie at the same site.
 iv. Coinfect *E. coli* K12(λ) with both mutants and plate the progeny phage on both *E. coli* B and *E. coli* K12(λ). If plaques appear on K12(λ) but not B, they lie at the same site.

b. If one wants to know if two different *rII* point mutants lie in the same cistron, one should:
 i. Coinfect *E. coli* K12(λ) with both mutants. If phage are produced, they lie at the same cistron.
 ii. Coinfect *E. coli* K12(λ) with both mutants. If phage are not produced, they lie in the same cistron.
 iii. Coinfect *E. coli* B with both mutants and plate the progeny phage on both *E. coli* B and *E. coli* K12(λ). If plaques appear on B but not K12(λ), they lie in the same cistron.
 iv. Coinfect *E. coli* K12(λ) with both mutants and plate the progeny phage on both *E. coli* B and *E. coli* K12(λ). If plaques appear on K12(λ) but not B, they lie in the same cistron.

*7.23 Given the following map with point mutants, and given the data in the table below, draw a topological representation of deletion mutants *r21*, *r22*, *r23*, *r24*, and *r25*. (Be sure to indicate clearly the endpoints of the deletions.)

($+ = r^+$ recombinants are obtained. $0 = r^+$ recombinants are not obtained.)

Map:
$$\underset{\displaystyle \begin{array}{ccccccc} r12 & r16 & r11 & r15 & r13 & r14 & r17 \end{array}}{\rule{6cm}{0.4pt}}$$

DELETION MUTANTS	POINT MUTANTS						
	r11	*r12*	*r13*	*r14*	*r15*	*r16*	*r17*
r21	0	+	0	+	0	+	I
r22	+	+	0	0	+	+	0
r23	0	0	0	+	0	0	+
r24	+	+	0	0	+	+	+
r25	+	+	0	0	0	+	+

7.24 A set of seven different *rII* deletion mutants of bacteriophage T4, *1* through *7*, were mapped, with the following result:

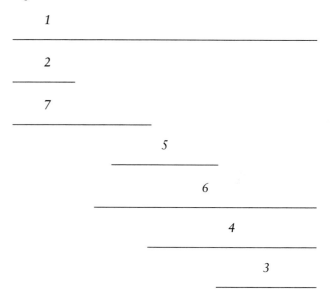

Five *rII* point mutants were crossed with each of the deletions, with the following results, where $+ = r^+$ recombinants were obtained, $0 = $ no r^+ recombinants were obtained:

POINT MUTANTS	DELETION MUTANTS						
	1	2	3	4	5	6	7
a	0	+	+	+	0	0	0
b	0	0	+	+	+	+	0
c	0	+	+	0	0	0	+
d	0	+	0	0	+	0	+
e	0	+	+	+	+	0	0

Map the locations of the point mutants.

*7.25 Given the following deletion map with deletions r31, r32, r33, r34, r35, and r36, place the point mutants r41, r42, and so on, on the map. Be sure you show where they lie with respect to endpoints of the deletions.

	DELETION MUTANTS (+ = r^+ RECOMBINANTS PRODUCED, 0 = NO r^+ RECOMBINANTS PRODUCED)					
POINT MUTANTS	r31	r32	r33	r34	r35	r36
r41	0	0	0	0	+	0
r42	0	0	0	+	0	+
r43	0	0	+	+	+	0
r44	0	0	0	0	+	+
r45	0	+	0	+	+	+
r46	0	0	+	0	+	0

Show the dividing line between the A cistron and the B cistron on your map above from the following data [+ = growth on strain $K12(\lambda)$, 0 = no growth on strain $K12(\lambda)$]:

	COMPLEMENTATION WITH	
MUTANT	rIIA	rIIB
r31	0	0
r32	0	0
r33	0	+
r34	0	0
r35	0	+
r36	0	0
r41	0	+
r42	0	+
r43	+	0
r44	0	+
r45	0	+
r46	0	+

7.26 Mutants in the ade2 gene of yeast require adenine and are pink because of the intracellular accumulation of a red pigment. Diploid strains were made by mating haploid mutant strains. The diploids exhibited the following phenotypes:

CROSS	DIPLOID PHENOTYPES
1×2	pink, adenine-requiring
1×3	white, prototrophic
1×4	white, prototrophic
3×4	pink, adenine-requiring

How many genes are defined by the four different mutants? Explain.

*7.27 In Drosophila, mutants A, B, C, D, E, F, and G all have the same phenotype: the absence of red pigment in the eyes. In pairwise combinations in complementation tests the following results were produced, where + = complementation and − = no complementation:

	A	B	C	D	E	F	G
G	+	−	+	+	+	+	−
F	−	+	+	−	+	−	
E	+	+	−	+	−		
D	−	+	+	−			
C	+	+	−				
B	+	−					
A	−						

a. How many genes are present?
b. Which mutants have defects in the same gene?

7.28 a. A homozygous white-eyed Martian fly (w_1/w_1) is crossed to a homozygous white-eyed fly from a different stock (w_2/w_2). It is well known that wild-type Martian flies have red eyes. This cross produces all white eyed progeny. State whether the following is true or false. Explain your answer.
   i) w_1 and w_2 are allelic genes.
   ii) w_1 and w_2 are non-allelic.
   iii) w_1 and w_2 affect the same function.
   iv) The cross was a complementation test.
   v) The cross was a cis-trans test.
   vi) w_1 and w_2 are allelic by terms of the functional test.

b. The F_1 white-eyed flies were allowed to interbreed, and when you classified the F_2 you found 20,000 white-eyed flies and ten red-eyed progeny. Concerned about contamination, you repeat the experiment and get exactly the same results. How can you best account for the presence of the red-eyed progeny? As part of your explanation give the genotypes of the F_1 and F_2 generation flies.

8 The Beginnings of Molecular Genetics: Gene Function

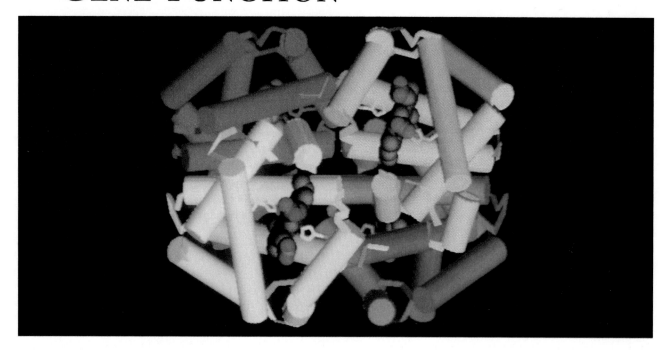

Gene Control of Enzyme Structure
    Garrod's Hypothesis of Inborn Errors of Metabolism
    Genetic Control of *Drosophila* Eye Pigments
    One Gene–One Enzyme Hypothesis

Genetically Based Enzyme Deficiencies in Humans
    Phenylketonuria
    Albinism
    Lesch-Nyhan Syndrome
    Tay-Sachs Disease
    Genetic Counseling

Gene Control of Protein Structure
    Protein Structure
    Sickle-Cell Anemia
    Other Hemoglobin Mutants
    Biochemical Genetics of the Human ABO Blood Groups

PRINCIPAL POINTS

~ There is a specific relationship between genes and enzymes, historically embodied in Beadle and Tatum's "one gene–one enzyme" hypothesis, which states that each gene controls the synthesis or activity of a single enzyme. A more modern maxim is "one gene–one polypeptide."

~ Genetic counseling is the analysis of the probability that individuals have a genetic defect, or of the risk that prospective parents may produce a child with a genetic defect, and, for the latter, the presentation to the individuals involved of any available options for avoiding or minimizing those possible risks. Techniques available to genetic counselors include amniocentesis and chorionic villus sampling.

~ A protein consists of one or more molecular sub-units called polypeptides, which are themselves composed of smaller building blocks called amino acids. The amino acids are linked together in the polypeptide by peptide bonds.

~ The primary amino acid sequence of a protein determines its secondary, tertiary, and quaternary structure, and hence its functional state.

~ From the study of alterations in proteins other than enzymes, convincing evidence was obtained that genes control the structures of all proteins.

*I*n the preceding seven chapters we have explored many aspects of genes. We have learned that genes are defined by mutations, that genes segregate in genetic crosses according to how chromosomes segregate, and that genes can be located—mapped—relative to one another on chromosomes in both prokaryotes and eukaryotes by analyzing the results of appropriate genetic crosses. We have also learned that alleles of genes can have differing relationships, depending on the allele: most are recessive to the wild-type allele, some are dominant to the wild-type allele, and others are co-dominant or incompletely dominant. Throughout all of these discussions we have considered the gene as an abstract entity, a factor that is located on a chromosome and that can give rise to a phenotype. We know, of course, that a gene is a stretch of DNA in a chromosome and that section of DNA contains information that specifies the product for which the gene encodes. In fact, the entire set of genes in the genome specifies a vast array of products that are responsible for all of a cell's and organism's function: that is, *life*. In the popular press, the genetic information contained within an organism's genome is called the "blueprint for life." We now need to describe how a gene functions, that is, how the information in the DNA is used to specify a product and how that product is involved in the production of a phenotype.

We saw in Chapter 7 that complementation tests can be used to assign mutations with the same phenotype to units of function—the genes. In this chapter we will examine gene function: what do genes code for, and how do we know what they code for? We will encounter the classical experiments that demonstrated that genes code for enzymes and for other proteins. In

particular, we examine the involvement of particular sets of genes in directing and controlling a particular biochemical pathway; that is, the series of enzyme-catalyzed steps required to break down or synthesize a particular chemical compound. Instead of viewing the gene in isolation, as in the past few chapters, we will see that the gene must often work in cooperation with other genes in order for cells to function properly. As a whole, the experiments we will discuss were the beginnings of molecular genetics, in that they were designed to understand a gene at the molecular level. In the following chapters our modern understanding of gene structure and function will be developed.

GENE CONTROL OF ENZYME STRUCTURE

Our studies so far have shown how genetic analysis can provide insights about the relationships between mutations and phenotypic change. In this section we analyze some of the classical genetic data that showed that genes code for enzymes and for nonenzymatic proteins.

Garrod's Hypothesis of Inborn Errors of Metabolism

In 1902 Archibald Garrod, an English physician, provided the first evidence of a specific relationship between genes and enzymes. (As we now know, genes specify the amino acid sequence of all proteins, including enzymes.) Garrod studied *alkaptonuria*, a human disease characterized by urine that turns black upon exposure to the air and by the tendency to develop arthritis later in life.

In studying the occurrence of alkaptonuria in families of individuals with the disease, Garrod and his colleague William Bateson discovered two interesting facts that indicated to them that alkaptonuria is a genetically controlled trait: (1) Several members of the same families frequently had alkaptonuria, and (2) the disease was much more common among children of marriages involving first cousins than among children of marriages between unrelated partners. This finding was significant because first cousins have ⅛ of their genes in common and, therefore, the chance is greater for recessive genes to become homozygous in children of first-cousin marriages.

Garrod then found that people with alkaptonuria excrete in their urine all the homogentisic acid (HA) they ingest, whereas normal people excrete none. Moreover, he showed that it is HA in the urine that turns black in the air. This result indicated to Garrod that normal people are able to metabolize HA to its breakdown products but that people with alkaptonuria cannot. That told Garrod that alkaptonurics lack the enzyme that metabolizes HA. Figure 8.1 shows part of the biochemical pathway in which HA is involved and the step blocked in people with alkaptonuria. From his results and the genetic evidence, Garrod concluded that alkaptonuria is a genetic disease that results from the absence of a particular enzyme in the steps in the metabolism of HA to its breakdown products. In Garrod's terms, this disease is an example of an *inborn error of metabolism*. We now know that the mutation responsible for alkaptonuria is recessive, so only people homozygous for the mutant gene express the defect. The gene has been localized to an autosome, but which autosome is unknown.

Garrod also studied three other human genetic diseases that affected biochemical processes, and in each case he was able to conclude correctly that a metabolic pathway was blocked. An important aspect of Garrod's analysis of these human diseases was his understanding that the position of a block in a metabolic pathway can be determined by the accumulation of the chemical compound (HA in the case of alkaptonuria) that precedes the blocked step.

Genetic Control of *Drosophila* Eye Pigments

After Garrod's work the next significant piece of evidence linking genes and enzymes was obtained in 1935. In that year George Beadle and Boris Ephrussi published the results of their experiments on gene-enzyme relationships in a biochemical pathway responsible for the synthesis of eye pigments in the fruit fly *Drosophila melanogaster*. Wild-type *Drosophila* have bright red

~ FIGURE 8.1

Part of the phenylalanine-tyrosine metabolic pathways. (Left) The biochemical steps that operate in normal individuals; (Right) The metabolic block found in individuals with alkaptonuria.

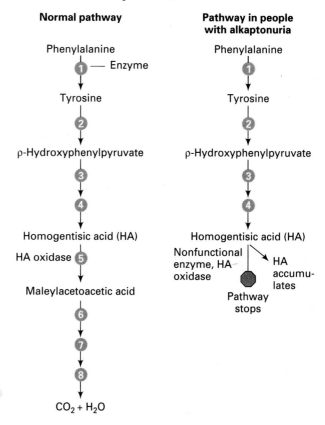

eyes, the result of the blending of two distinct pigments, red and brown.

At the time of these experiments, three genes were known to be involved in the production of brown eye pigment. Mutations in any one of these genes resulted in the absence of brown pigment and, as a consequence, flies with bright orange or scarlet eyes. Mapping experiments had shown that the three genes occupy three distinct loci. In other words, the genes are not clustered, even though they are involved in the same biochemical pathways. Each mutant allele is recessive to its wild-type allele, so the mutant phenotype is only exhibited in flies homozygous for the mutation. The three mutant, brown-pigment genes are designated *st* (scarlet), *cn* (cinnabar), and *v* (vermilion). The wild-type forms of these genes are designated st^+, cn^+, and v^+, respectively.

First, Beadle and Ephrussi isolated two groups of cells, called *imaginal disks*, from *Drosophila* larvae, cells that develop into the two eyes in an adult. When an eye disk is transplanted from a larva into the abdomen of a second larva, the disk will develop into a rec-

~ TABLE 8.1

Results of Eye Disk Transplantation Experiments Involving
Eye Color Mutants of *Drosophila*

SOURCE OF EYE DISK	HOST FLY	COLOR OF TRANSPLANTED EYE AFTER METAMORPHOSIS
+	*v*	+
v	+	+
+	*cn*	+
cn	+	+
cn	*v*	Cinnabar
v	*cn*	+
+	*st*	+
st	+	Scarlet

ognizable eye structure. Once the second larva matures
into an adult, the eye structure can be isolated from its
abdomen. Beadle and Ephrussi transplanted imaginal
eye disks taken from each type of mutant larvae (*st*, *cn*,
and *v*) separately into a wild-type larval host. Once the
wild-type host had become an adult, the researchers
analyzed the eye color of the eye structure produced by
the transplanted disk. They also investigated the fate of
a wild-type disk transplanted into a mutant larval host.
The results of their studies are summarized in
Table 8.1.

~ FIGURE 8.2

Imaginal eye disk transplant experiment in *Drosophila*,
showing autonomous development.

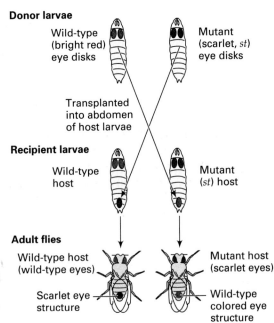

Donor larvae

Wild-type (bright red) eye disks

Mutant (scarlet, *st*) eye disks

Transplanted into abdomen of host larvae

Recipient larvae

Wild-type host

Mutant (*st*) host

Adult flies

Wild-type host (wild-type eyes)

Mutant host (scarlet eyes)

Scarlet eye structure

Wild-type colored eye structure

In the case of the *st* mutant, the *st* disk in the wild-
type host developed into a scarlet eye (Figure 8.2). In
the reciprocal transplant, the wild-type implant developed into a wild-type eye structure with a wild-type,
bright red eye color. These results meant that the wild-
type host was not able to provide substances to the
mutant *st* disk to enable production of the brown pigment. The transplanted *st* disk developed according to
the genetic information in its cells and was not affected
by the environment. When a cell develops in this way, it
is said to undergo **autonomous development**.

The results of similar transplants of mutant *v* and
cn disks (Figure 8.3) illustrate **nonautonomous development**: The development of a cell's phenotype is affected by the environment in which it develops. When
eye disks are transplanted from either *v* or *cn* larvae
into a wild-type host, the disks develop into eye structures with wild-type eye color. In contrast, wild-type
eye disks transplanted into *cn* or *v* hosts exhibit autonomous development, producing wild-type colored eye
structures. The explanation is that the wild-type host is
able to provide substances that the *v* and *cn* disks can
use to bypass the genetic block they have and can,
therefore, make brown pigment. The needed substances
must diffuse into the implanted disk from the host circulation.

Next, Beadle and Ephrussi did reciprocal transplants of eye disks between *v* and *cn* larvae (see
Figure 8.4). These experiments showed that *v* disks
transplanted into *cn* hosts develop nonautonomously
into wild-type, colored eye structures; this result shows
that *cn* larvae can provide the *v* disks with some substance needed for wild-type pigment production (note
that this is an example of complementation: see Chapter 7). In the reciprocal experiment, however, *cn* disks
transplanted into *v* hosts develop autonomously into *cn*
eyes; this result shows that *v* larvae cannot provide *cn*
disks with the material necessary to produce wild-type
eye pigment.

Because of this work, Beadle and Ephrussi concluded that production of the brown pigment necessary
for wild-type bright red eyes involves a biochemical
pathway with at least two precursor pigment substances. Wild-type flies possess both precursors, *cn* flies
have only one, and *v* flies have neither. Figure 8.5
(p. 238) shows biochemical steps and their relationships to their genes involved for wild type, *v*, and *cn*
flies.

In a wild-type fly (Figure 8.5a), the v^+ gene encodes
an enzyme that makes v^+ substance, and the cn^+ gene
encodes an enzyme that converts the v^+ substance to
the cn^+ substance. The cn^+ substance is then converted
to the brown pigment. The brown pigment then combines with a bright orange pigment (synthesized in a
parallel biochemical pathway that involves a number of
other genes) to produce the bright red eye color of the
wild type. In a vermilion-eyed (*v*) mutant fly (Figure

Imaginal eye disk transplant experiments in *Drosophila*, showing nonautonomous development: (a) reciprocal transplant between the wild type and the vermilion mutant *v*; (b) reciprocal transplant between the wild type and the cinnabar mutant *cn*.

a) Vermilion mutant experiment

Donor larvae

Wild-type (bright red) eye disks

Mutant (vermilion, *v*) eye disks

Transplanted into abdomen of host larvae

Recipient larvae

Wild-type host

Mutant (*v*) host

Adult flies

Wild-type host (wild-type eyes)

Mutant host (vermilion eyes)

Wild-type colored eye structure

Wild-type colored eye structure

a) Cinnabar mutant experiment

Donor larvae

Wild-type (bright red) eye disks

Mutant (cinnabar, *cn*) eye disks

Transplanted into abdomen of host larvae

Recipient larvae

Wild-type host

Mutant (*cn*) host

Adult flies

Wild-type host (wild-type eyes)

Mutant host (cinnabar eyes)

Wild-type colored eye structure

Wild-type colored eye structure

Reciprocal imaginal eye disk transplant experiments involving *v* and *cn* mutants of *Drosophila*, showing nonautonomous development of *v* disks in a *cn* host, and autonomous development of *cn* disks in a *v* host.

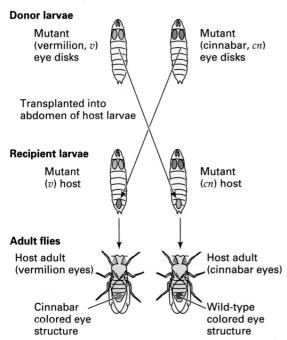

Donor larvae

Mutant (vermilion, *v*) eye disks

Mutant (cinnabar, *cn*) eye disks

Transplanted into abdomen of host larvae

Recipient larvae

Mutant (*v*) host

Mutant (*cn*) host

Adult flies

Host adult (vermilion eyes)

Host adult (cinnabar eyes)

Cinnabar colored eye structure

Wild-type colored eye structure

8.5b), the *v* mutant gene results in an inactive or absent *v* enzyme, so no v^+ substance is produced. The vermilion eye color results from a combination of the independently produced bright orange pigment and the precursor for the brown pigment pathway. In a cinnabar-eyed (*cn*) mutant fly (Figure 8.5c), the *cn* mutant gene results in an inactive or absent *cn* enzyme, so v^+ substance cannot be converted to cn^+ substance. The cinnabar eye color results from a combination of the independently produced bright orange pigment and the v^+ substance of the brown pigment pathway.

   Given these facts, we can now explain how a wild-type colored eye structure develops from a mutant *v* disk transplanted into a *cn* host. The implanted disk comes from a fly with a mutation in the *v* gene so that the v^+ substance cannot be made, but it has the wild-type cn^+ gene. The *cn* host has a mutant *cn* gene, but because it has a normal v^+ gene, it is able to make the v^+ substance. This substance diffuses into the implanted $v\ cn^+$ disk, where it is converted to the cn^+ substance. Since the genes (including the st^+ gene) for the rest of the pathway are also normal in the implanted *v* disk, the cn^+ substance is converted to the brown pigment, and the eye derived from the implanted disk is wild type. Stated in another way, the *cn* host is able to make up for the deficiency of the *v* disk by sup-

Proposed biochemical sequence and relationships in the gene reaction steps for (a) wild-type, (b) *v*, and (c) *cn* flies, as deduced from the results of the Beadle and Ephrussi transplantation experiments diagrammed in Figures 8.2–8.4.

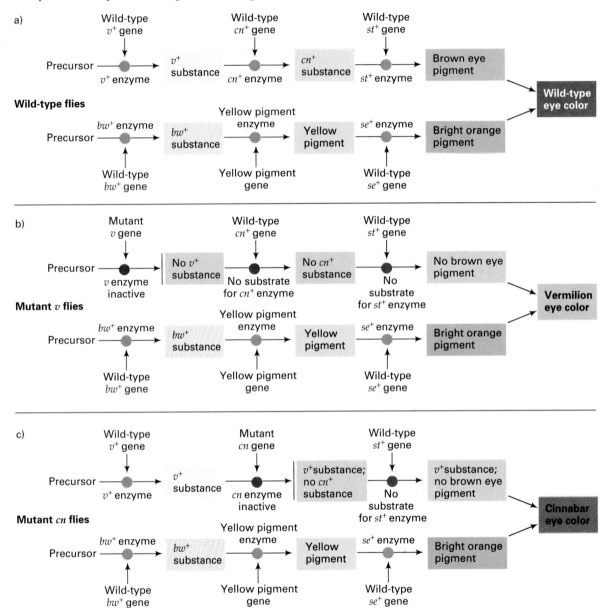

plying it with a diffusible substance so that it can develop into a wild-type colored eye.

In the reciprocal transplant, a *cn* disk implanted into a *v* host cannot convert its own *v*⁺ substance to the *cn*⁺ substance because it possesses the mutant *cn* gene. Because the *v* host is deficient in the production of *v*⁺ substance, the host also cannot produce the *cn*⁺ substance that the implanted *cn* disk would need to give rise to a wild-type, colored eye structure. Thus the re-

sulting eye structure is cinnabar (bright orange), showing autonomous development.

From subsequent biochemical analysis researchers were able to identify the chemical nature of the *v*⁺ and *cn*⁺ substances, to work out the biochemical pathway, and to localize the steps catalyzed by the *v*⁺ and *cn*⁺ gene products (see Figure 8.6). The *v* mutation results in a nonfunctional enzyme; consequently, tryptophan cannot be converted to formylkynurenine. Similarly,

~ FIGURE 8.6

Actual biochemical pathway for the production of the *Drosophila* brown eye pigment from the amino acid tryptophan. The steps catalyzed by the enzymes encoded by the v^+, cn^+, and st^+ genes are indicated.

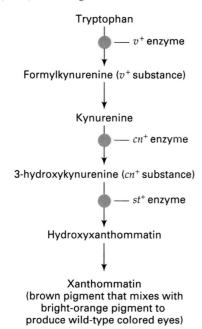

Tryptophan

— v^+ enzyme

Formylkynurenine (v^+ substance)

Kynurenine

— cn^+ enzyme

3-hydroxykynurenine (cn^+ substance)

— st^+ enzyme

Hydroxyxanthommatin

Xanthommatin
(brown pigment that mixes with bright-orange pigment to produce wild-type colored eyes)

the *cn* mutation results in a nonfunctional enzyme that catalyzes a subsequent step in the pathway, so kynurenine cannot be converted to 3-hydroxykynurenine. Lastly, the *st* (scarlet) mutation affects the enzyme that catalyzes the step after the cn^+ step; in *st* mutants, 3-hydroxykynurenine cannot be converted to hydroxyxanthommatin.

In all three mutants no brown pigment is produced because the pathway cannot be completed. At the same time, red pigment is being produced by a separate biochemical pathway controlled by a distinctive set of genes. The eye colors produced in each mutant are slightly different because at each step a slight coloration is added to the red pigment. In other words, the closer the pathway functions toward the brown pigment itself, the more colored the compound is, the more color combines with the red pigment, and the more the eye color resembles the wild-type color.

One Gene–One Enzyme Hypothesis

The work on *Drosophila* eye pigments led to more refined studies on the relationship between genes and enzymes. Those studies are generally regarded as heralding the beginnings of biochemical genetics, a branch of genetics that combines genetics and biochemistry to elucidate the nature of metabolic pathways. The studies were carried out by George Beadle and Edward Tatum with the fungus *Neurospora crassa*. The results showed that there was a direct relationship between genes and enzymes and resulted in the *one gene–one enzyme hypothesis*, a very important landmark in the history of genetics.

ISOLATION OF NUTRITIONAL MUTANTS OF *NEUROSPORA.* In Chapter 5 we discussed the life cycle of *N. crassa* in detail. Briefly, *Neurospora* is a haploid organism that propagates asexually or sexually. One important attribute of *Neurospora* is that it has simple growth requirements. Wild-type *Neurospora*, by definition, is prototrophic, meaning that it can grow and propagate on a *minimal medium* that contains a basic set of ingredients (inorganic salts, a carbon source such as glucose or sucrose, and the vitamin biotin). Beadle and Tatum reasoned that *Neurospora* synthesized the materials it needed for growth (amino acids, nucleotides, vitamins, nucleic acids, proteins, etc.) from the chemicals present in the minimal medium. They also realized that it should be possible to isolate *nutritional mutants* (i.e., auxotrophs) of *Neurospora*: mutant strains that required nutritional supplements in order to grow. These auxotrophic mutants could be isolated because they would not grow on the minimal medium.

Figure 8.7 (p. 240) shows how Beadle and Tatum isolated and characterized nutritional mutants. They treated asexual spores (conidia) with X rays to induce genetic mutants; then they crossed the cultures derived from the surviving spores with a wild-type (prototrophic) strain of the opposite mating type. This sexual cross was done because they wanted to study the genetics of biochemical events, and they had to identify those nutritional mutants that were heritable. By crossing the mutagenized spores with the wild type, they ensured that any nutritional mutant they isolated had segregated in a cross and therefore had a genetic basis rather than a nongenetic reason for requiring the nutrient.

Each progeny spore from the crosses was allowed to germinate in a nutritionally complete medium that contained all necessary amino acids, purines, pyrimidines, and vitamins, in addition to sucrose, salts, and biotin found in minimal medium. Thus any strain that could not make one or more necessary compounds from the basic ingredients found in minimal medium could still grow by using the compounds supplied in the growth medium. Each culture grown from a progeny spore was then tested for growth on minimal medium. Those strains that did not grow were assumed to be auxotrophic mutants; these mutants were, in turn, individually tested for their abilities to grow on various supplemented minimal media. In this screening two media were used: minimal medium plus amino acids, and minimal medium plus vitamins. Theoretically, an

~ **FIGURE 8.7**

Method devised by Beadle and Tatum to isolate auxotrophic mutations in *Neurospora*. Here, the mutant strain isolated is a tryptophan auxotroph.

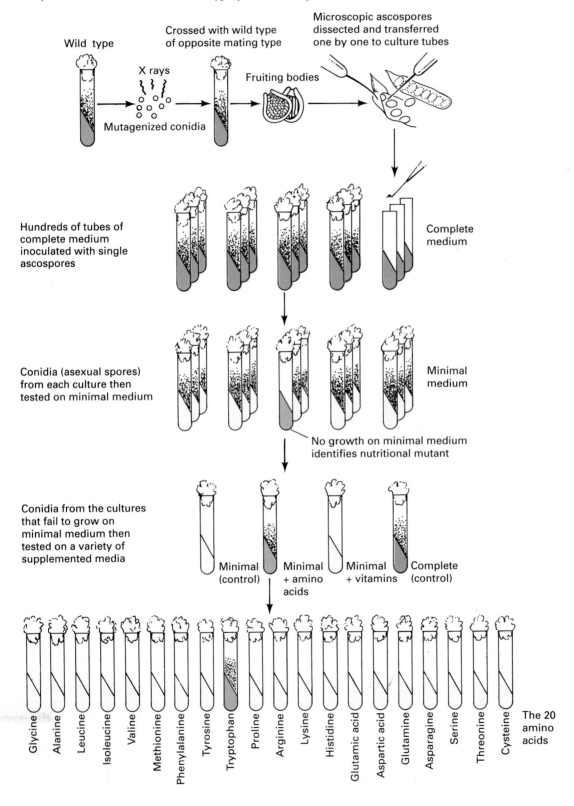

amino acid auxotroph—a mutant strain that requires a particular amino acid in order to grow—would grow on minimal medium plus amino acids but not on either minimal medium plus vitamins or minimal medium. Vitamin auxotrophs would only grow on minimal medium plus vitamins, and so on.

Having categorized the strains into non-auxotrophs, amino acid auxotrophs, and vitamin auxotrophs, Beadle and Tatum next conducted a second round of screening to specify which chemical the strains needed for growth. Let us imagine an amino acid auxotroph. To find out which of the 20 amino acids is required for a particular amino acid auxotroph to grow, the strain would be tested in 20 tubes, each containing minimal medium plus one of the 20 amino acids. For example, a tryptophan auxotroph would be identified if it grows only in the tube containing minimal medium plus tryptophan.

Figure 8.8 shows the method used to confirm that an auxotroph identified by the above procedures has a genetic rather than a nongenetic basis. For a tryptophan auxotroph, the auxotrophic strain is crossed with the wild-type strain. The spores from a single meiosis (a meiotic tetrad) can be isolated and analyzed as described in Chapter 6. If the auxotrophic property is caused by a gene mutation, then half the spores should be wild-type and half should be auxotrophic. This hypothesis is tested by first germinating the spores individually on minimal medium plus tryptophan (so all will grow) and then checking for the ability of each to grow on minimal medium. On the latter medium only the wild-type strains will grow; the auxotrophic strains will not. If, however, the tryptophan auxotroph is not the result of a gene mutation, then all the cultures should grow on minimal medium—or at least there will not be a 4:4 segregation of wild type:auxotroph when tetrads are tested for growth on minimal medium.

~ FIGURE 8.8

Procedure used to confirm the genetic basis of a nutritional defect in *Neurospora*.

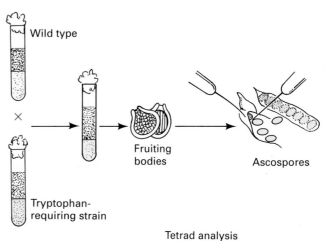

GENETIC DISSECTION OF A BIOCHEMICAL PATHWAY.

Once Beadle and Tatum had isolated and identified nutritional mutants, they set out to investigate the biochemical pathways affected by the mutations. They assumed that *Neurospora* cells, like all cells, function by the interaction of the products of a very large number of genes. Furthermore, they imagined that wild-type *Neurospora* converted the constituents of minimal medium into amino acids and so forth by a series of reactions organized into biochemical pathways. In this way the synthesis of cellular components occurs by a series of small steps, each catalyzed by its own enzyme. As the hypothetical pathway in Figure 8.9 (p. 242) shows, the product of each step is used as the substrate for the next enzyme. Beadle and Tatum called the proposed relationship between an organism's genes and the enzymes necessary for carrying out the steps in a biochemical pathway the one gene–one enzyme hypothesis. That is, a gene controls the production and/or activity of one enzyme. Consequently, mutations that result in the loss of enzyme activity lead to the accumulation of precursors in the pathway (and to possible side reactions), as well as to the absence of the end product of the pathway.

Using the one gene–one enzyme hypothesis, we can make some predictions about the consequences of mutations affecting enzymes in the hypothetical pathway shown. If, for example, a mutation in gene C results in

~ FIGURE 8.9

Hypothetical biochemical pathway for the conversion of a precursor substrate to an end product C in three enzyme-catalyzed steps. Each enzyme is coded for by one gene.

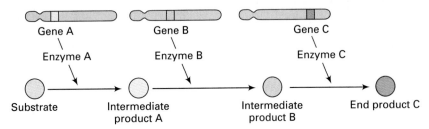

a nonfunctional enzyme C, then intermediate product B in the pathway cannot be converted to the end product C. The consequences to the organism are that substance C must be present in the growth medium in order for the organism to grow. Product B also accumulates in the cells since it cannot be converted to C. Mutations in gene *A* or gene *B* will also result in auxotrophy for compound C.

However, the three mutants *A*, *B*, and *C* are distinguishable on other grounds, namely, their ability to grow or not to grow on various intermediates in the pathway. Gene *C* mutants can grow only if supplemented with C; gene *B* mutants can grow if supplemented with either B or C; and gene *A* mutants can grow on minimal medium plus A, B, or C. A reciprocal pattern of pathway precursor accumulation occurs in these strains. Gene *C* mutants accumulate product B; gene *B* mutants accumulate product A; and gene *A* mutants accumulate the initial substrate for the path-

way. In actual experiments where the pathway is unknown, the logic is carried out in the opposite direction. As a result, from the pattern of growth supplementation and precursor accumulation, the sequence of steps in a pathway can be deduced.

Let us take an actual example in the biochemical pathway for the synthesis of the amino acid arginine in *Neurospora crassa*. Figure 8.10 shows this pathway, along with the genes that code for the enzymes that catalyze each step. The starting point is a set of arginine auxotrophs. Genetic crosses and complementation tests determine that four distinct genes are involved; a mutation in any one of them gives rise to auxotrophy for arginine. These four genes in a wild-type cell are designated $argE^+$, $argF^+$, $argG^+$, and $argH^+$.

Next, the sequence of biochemical steps in the pathway can be deduced by the growth pattern of the mutant strains on media supplemented with presumed arginine precursors. Table 8.2 shows the results ob-

~ FIGURE 8.10

Arginine biosynthetic pathway, showing the four genes in *Neurospora crassa* that code for the enzymes that catalyze each reaction. (The genes are not on the same chromosome).

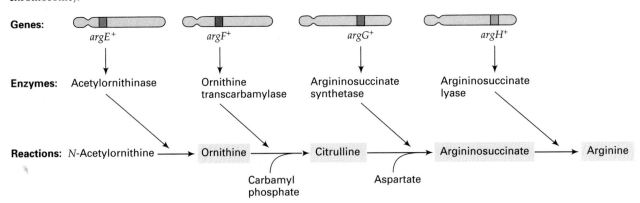

~ TABLE 8.2

Growth Responses of Arginine Auxotrophs

MUTANT STRAIN	GROWTH RESPONSE ON MINIMAL MEDIUM AND . . .				
	NOTHING	ORNITHINE	CITRULLINE	ARGININO-SUCCINATE	ARGININE
Wild-type	+	+	+	+	+
argE	−	+	+	+	+
argF	−	−	+	+	+
argG	−	−	−	+	+
argH	−	−	−	−	+

tained for this set of four mutants. By definition, all four mutant strains can grow on arginine, and none can grow on unsupplemented minimal medium.

As shown in the table, the *argH* mutant strain can grow when supplemented with arginine but not when supplemented with any of the intermediates in the pathway. This result indicates that the *argH* gene codes for the enzyme that controls the last step in the pathway, which leads to the formation of arginine. The *argG* mutant strain grows on media supplemented with arginine or argininosuccinate, the *argF* mutant strain grows on media supplemented with arginine, argininosuccinate, or citrulline, and the *argE* strain grows on arginine, argininosuccinate, citrulline, or ornithine. The later a mutant strain is blocked in a pathway, the fewer intermediate chemicals need to be added to the growth medium for the strain to grow. The earlier a mutant strain is blocked, the greater the number of intermediates that are required for the strain to grow. With these results in mind, we conclude that argininosuccinate must be the immediate precursor to arginine, since argininosuccinate permits all but the *argH* mutant to grow.

By similar reasoning, the pathway can be hypothesized to follow the pattern shown in Figure 8.10. Gene *argF*+ codes for the enzyme that converts ornithine to citrulline. An *argF* mutant strain can, therefore, grow on minimal medium plus citrulline, argininosuccinate, or arginine. As was discussed before, there is an accumulation of the intermediate chemical produced prior to the step blocked by the genetic mutation, and that information can be used to confirm any conclusions about the sequence of steps in a pathway. The *argF* mutant strains, for example, accumulate ornithine, and hence ornithine would be placed prior to citrulline, argininosuccinate, and arginine in the pathway.

With this sort of approach we can dissect a biochemical pathway genetically—that is, we can determine the sequence of steps in the pathway and relate each step to a specific gene or genes. We should not conclude, though, that there is necessarily one gene for each step in a pathway, although that is usually true. In some cases an enzyme may consist of more than one polypeptide chain, each of which is coded for by a specific gene. In that event, more than one gene would specify that enzyme. Therefore Beadle and Tatum's original name for their hypothesis is modified to the *one gene–one polypeptide hypothesis*.

KEYNOTE

A number of classical studies indicated the specific relationship between genes and enzymes, eventually embodied in Beadle and Tatum's one gene–one enzyme hypothesis, which states that each gene controls the synthesis or activity of a single enzyme. Since enzymes may consist of more than one polypeptide, a more modern title for this hypothesis is the one gene–one polypeptide hypothesis.

GENETICALLY BASED ENZYME DEFICIENCIES IN HUMANS

Many human genetic diseases are caused by a single gene mutation that alters the function of an enzyme. In general, an enzyme deficiency caused by a mutation may have simple, or pleiotropic, consequences. Table 8.3 presents several of these diseases (which Garrod would have called inborn errors of metabolism). Studies of these diseases have offered further evidence that some genes code for enzymes.

Phenylketonuria

Like alkaptonuria, *phenylketonuria (PKU)* is a genetic disease most commonly caused by a mutation in the gene for phenylalanine hydroxylase, which results in a block in a metabolic pathway for the metabolism of the amino acids phenylalanine and tyrosine. As a result, phenylalanine cannot be converted to tyrosine (Figure 8.11a and b).

An individual with PKU (a *phenylketonuriac*) cannot make tyrosine, an amino acid required for protein synthesis and for the production of the hormones thyroxine and adrenaline and the skin pigment melanin. This aspect of the phenotype is not very serious

~ Table 8.3

Selected Human Genetic Disorders with Demonstrated Enzyme Deficiencies

Condition	Enzyme Deficiency	Condition	Enzyme Deficiency
Acid phosphatase deficiency	Acid phosphatase	Leigh's necrotizing encephalomelopathy[a]	Pyruvate carboxylase
Albinism	Tyrosinase	Lesch-Nyhan syndrome[a]	Hypoxanthine phosphoribosyl transferase
Alkaptonuria	Homogentisic acid oxidase	Lysine intolerance	Lysine: NAD-oxidoreductase
Ataxia, intermittent[a]	Pyruvate decarboxylase	Male pseudohermaphroditism	Testicular 17,20-desmolase
Disaccharide intolerance	Invertase	Maple sugar urine disease[a]	Keto acid decarboxylase
Fructose intolerance	Fructose-1-phosphate aldolase	Orotic aciduria[a]	Orotidylic decarboxylase
Fructosuria	Liver fructokinase	Phenylketonuria	Phenylalanine hydroxylase
G6PD deficiency (favism)[a]	Glucose-6-phosphate dehydrogenase	Porphyria, acute[a]	Uroporphyrinogen I synthetase
Glycogen storage disease[a]	Glucose-6-phosphatase	Porphyria, congenital erythropoietic[a]	Uroporphyrinogen III cosynthetase
Gout, primary	Hypoxanthine phosphoribosyl transferase	Pulmonary emphysema	α-1-Antitrypsin
Hemolytic anemia	Glutathione peroxidase	Pyridoxine-dependent infantile convulsions	Glutamic acid decarboxylase
Hemolytic anemia	Hexokinase	Pyridoxine-responsive anemia	λ-Aminolevulinic synthetase
Hemolytic anemia	Pyruvate kinase	Kidney tubular acidosis with deafness	Carbonic anhydrase B
Hypoglycemia and acidosis	Fructose-1,6-diphosphatase	Ricketts, vitamin D-dependent	25-Hydroxycholecalciferol 1-hydroxylase
Immunodeficiency[a]	Adenosine deaminase	Tay-Sachs disease[a]	Hexosaminidase A
Immunodeficiency	Purine nucleoside phosphorylase	Thyroid hormone synthesis, defect in	Iodide peroxidase
Immunodeficiency	Uridine monophosphate kinase	Thyroid hormone synthesis, defect in	Deiodinase
Intestinal lactase deficiency (adult)	Lactase	Tyrosinemia	Para-Hydroxyphenylpyruvate oxidase
Ketoacidosis[a]	Succinyl CoA:3-ketoacid CoA-transferase	Xeroderma pigmentosum[a]	DNA-specific endonuclease

[a] Prenatal diagnosis possible or potentially possible. From A. Milunsky, 1976. *N Eng. J. Med.* 295:377. Note that some similar conditions result from various enzyme deficiencies, and single enzyme deficiencies can produce multiple defects.

Source: Modified from J. B. Standbury, J. W. Wyngaarden, and D. S. Fredrickson, eds., 1978. *The Metabolic Basis of Inherited Disease*, 4th ed., Table 1–4. New York: McGraw-Hill.

~ FIGURE 8.11

Phenylalanine-tyrosine metabolic pathways: (a) biochemical steps that operate in normal individuals; (b) metabolic block in the pathway exhibited by individuals with phenylketonuria—phenylalanine cannot be metabolized to tyrosine and unusual metabolites of phenylalanine accumulate; (c) metabolic block in the pathway exhibited by individuals with albinism—no or very little melanin pigment is produced.

a) Normal pathway

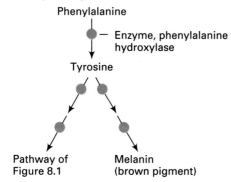

b) Pathway in people with phenylketonuria (PKU)

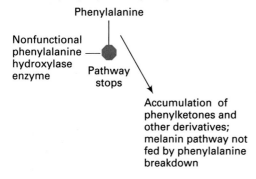

c) Pathway in people with albinism

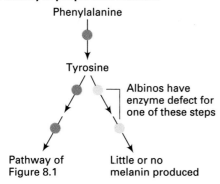

because tyrosine can be obtained from food. Yet food does not normally have a lot of tyrosine. As a result, people with PKU make relatively little melanin, since their bodies can use only tyrosine (rather than phenyl-

alanine and tyrosine) for melanin synthesis. Hence people with PKU tend to have very fair skin and blue eyes (even if they have brown-eye genes). In addition, PKU individuals have relatively low adrenaline levels.

The absence of phenylalanine hydroxylase also results in the accumulation of phenylalanine. However, unlike the accumulated precursor HA for alkaptonuria, which is excreted in the urine, the accumulated phenylalanine in phenylketonuriacs is converted to a number of phenylalanine derivatives. The accumulation of one of these derivatives, phenylpyruvic acid, drastically affects the cells of the central nervous system and produces the PKU's serious symptoms: severe mental retardation, a slow growth rate, and early death.

As with alkaptonuria, PKU is caused by a relatively rare recessive mutation; in Caucasians, only about one in 10,000 newborns are homozygous for the mutant gene and hence have PKU. Heterozygous children born to PKU mothers stand a high risk of having secondary (nongenetic) PKU because the higher levels of phenylalanine in the mother's blood can cause brain damage in fetuses.

In the United States, by law, all newborns must be screened for PKU. In the screening, the infant's blood is assayed for phenylalanine hydroxylase or their wet diapers are treated with ferric oxide. In the latter test, diapers of phenylketonuriacs turn green in the presence of the excreted phenylpyruvic acid and other ketones. A better test is the *inhibition assay test,* which bases its efficacy on the fact that the growth of a bacterial strain is inhibited by phenylalanine. Once PKU is identified, the genetic disease is treated by modifying the infant's diet to limit phenylalanine intake. In this way enough phenylalanine is available for protein synthesis, but it and its derivatives do not accumulate. The dietary correction must be implemented rigidly over a prolonged period to be effective. If the modified diet is dropped after about 5 years of age, for example, there is subsequently a small but significant decline in IQ. In sum, PKU is one of the very few genetic diseases that can be treated in a relatively easy way.

Albinism

Albinism is caused by a recessive mutation, and individuals must be homozygous for the mutation in order to exhibit the condition. About one in 17,000 individuals is albino. The mutation is in a gene for an enzyme used in the pathway from tyrosine to the brown pigment melanin (see Figure 8.11c). Since melanin is not produced, albinos have white skin, white hair, and red eyes (owing to the lack of pigment in the iris). Melanin absorbs light in the UV range and is important in protecting the skin against harmful UV irradiation from the sun. Albinos are therefore very light-sensitive. No

Individual with Lesch-Nyhan syndrome.

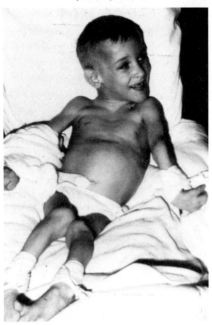

apparent problems result from the accumulation of precursors in the pathway prior to the block.

There are at least two kinds of albinism, since at least two biochemical steps can be blocked to prevent melanin formation. Therefore, if two albinos of different types mate, they can produce normal children as a result of complementation of the two nonallelic mutations.

Lesch-Nyhan Syndrome

Lesch-Nyhan syndrome (Figure 8.12) is a fatal human trait caused by a recessive, single gene mutation on the X chromosome. Only males (who are XY) exhibit Lesch-Nyhan syndrome. This syndrome is the result of a deficiency in the enzyme hypoxanthine guanine phosphoribosyl transferase (HGPRT), an enzyme essential to the utilization of purines (Figure 8.13). When the biosynthetic pathway is highly impaired, as in this case, excess purines accumulate, which are converted to uric acid.

At birth, Lesch-Nyhan individuals are healthy and develop normally for several months. The uric acid

~ FIGURE 8.13

Some steps in the purine biosynthetic pathways, showing the reactions catalyzed by hypoxanthine guanine phosphoribosyl transferase (HGPRT). This enzyme is defective in individuals with Lesch-Nyhan syndrome.

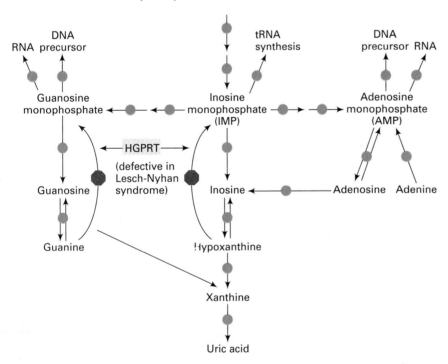

excreted in the urine leads to the deposition of orange uric acid crystals in the diapers, an indication that an infant has this genetic disease. Between three and eight months, delays in motor development lead to weak muscles. Later, the muscle tone changes radically, producing uncontrollable movements and involuntary spasms, seriously affecting feeding activities.

After two or three years, Lesch-Nyhan children begin to show extremely bizarre activity, such as compulsive biting of fingers, lips, and the inside of the mouth. This self-mutilation is difficult to control and is not without severe pain to those afflicted. Typically, behavior toward others becomes aggressive. In intelligence tests the patients score in the severely retarded region, although the difficulties they show in communicating with others may be a major contributing factor to the low scores. Most Lesch-Nyhan individuals die before they reach their twenties, usually from pneumonia, kidney failure, or uremia (uric acid in the blood).

Understandably, the elevated uric acid levels that result from an HGPRT deficiency might give rise to uremia, kidney failure, and mental deficiency. No clear explanation has been advanced, however, of how such a deficiency can lead to self-mutilating behavior.

Tay-Sachs Disease

A number of human diseases are caused by mutations in genes that code for lysosomal enzymes, enzymes that function in intracellular digestion. These diseases, collectively called *lysosomal-storage diseases,* are generally caused by recessive mutations.

The best-known genetic disease of this type is *Tay-Sachs disease* (also called infantile amaurotic idiocy), which is caused by homozygosity for a rare recessive mutation that maps to chromosome 15 (an autosome). While Tay-Sachs disease is rare in the population as a whole, it has a relatively high incidence in Ashkenazi Jews of Central European origin.

The mutant gene codes for the enzyme N-acetyl-hexosaminidase A (hex A), which catalyzes the reaction shown in Figure 8.14. Enzyme hex A cleaves a terminal N-acetylgalactosamine group from a brain chemical called a ganglioside. Tay-Sachs infants are deficient in that enzyme activity, so the uncleaved ganglioside accumulates in the brain cells. This accumulation causes cerebral degeneration and death, usually by age three.

Individuals with Tay-Sachs disease exhibit a number of clinical symptoms, illustrating the pleiotropic

~ FIGURE 8.14

Diagram of the biochemical step for the conversion of the brain ganglioside G$_{M2}$ to the ganglioside G$_{M3}$, catalyzed by the enzyme N-acetylhexosaminidase A (hex A): (a) Normal pathway; (b) Pathway in individuals with Tay-Sachs disease, where the activity of hex A is deficient; the result is an abnormal buildup of ganglioside G$_{M2}$ in the brain cells.

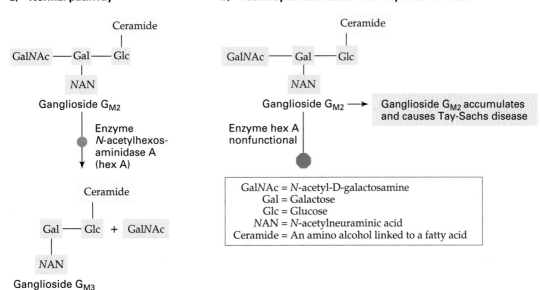

effects of the gene mutation. Typically, the symptom first recognized is an unusually enhanced reaction to sharp sounds. Early diagnosis is also made possible by the presence on the retina of a cherry-colored spot surrounded by a white halo. About a year after birth, there is rapid degeneration as the uncleaved ganglioside accumulates and the brain begins to lose control over normal function and activities. This degeneration involves generalized paralysis, blindness, a progressive loss of hearing, and serious feeding problems. By two years of age the infants are essentially immobile, and death ensues at about three to four years of age, often from respiratory infections.

No cure is known for Tay-Sachs disease, but as we will see in the next section, there are ways to determine whether a fetus has the disease while it is still developing in the uterus.

Genetic Counseling

Genetic counseling is the analysis of the probability that individuals have a genetic defect, or of the risk that prospective parents may produce a child with a genetic defect. In the latter case, genetic counseling involves presenting the available options for avoiding or minimizing those possible risks. Thus, genetic counseling provides people with an understanding of the genetic problems that are or may be inherent in their families or prospective families. Counseling utilizes a wide range of information on human heredity. In many instances, the risk of having a child with a genetic defect may be stated in terms of rather precise probabilities; in others, where the role of heredity is not sufficiently clear, the risk may be estimated only in empirical terms. It is the responsibility of the genetic counselors to supply their clients with clear, unemotional, and non-prescriptive statements based on the family history and on their knowledge of all relevant scientific information and of the probable risks of giving birth to a child with a genetic defect. In sum, genetic counseling is a lot more than the simple presentation of risk facts and figures to patients; it is the prevention and cure of disease, the relief of pain, and the maintenance of health, all of which are goals of the medical profession in general.

Genetic counseling generally starts with pedigree analysis of both families to determine the likelihood that a genetic disease is present in either group. (Pedigree analysis was described in Chapters 2 and 3.) Once evidence is found of a genetic disease in a family or families, prospective parents need to be informed of the probability that they will produce a child with that disease. Early detection of a genetic disease that has a known biochemical basis occurs at one or both of two

levels. One is the detection of heterozygotes (*carriers*) of recessive mutations, and the other is the determination of whether or not the developing fetus shows the biochemical defect. In genetic diseases with known biochemical defects, the two types of procedures are limited to those diseases in which the biochemical defect is expressed in the parents or the developing fetus.

CARRIER DETECTION. *Carrier detection* is the detection of individuals who are heterozygous for a recessive-gene mutation. The heterozygous carrier of a mutant gene is normal in phenotype. Nonetheless, the individual carries a recessive mutation that can be passed on to the progeny. In the case of a recessive mutation that has serious deleterious consequences to a homozygous individual for that mutation, there is great value in determining whether people who are contemplating having a child are both carriers, because in that situation a fourth of the children would get the trait. Carrier detection has been used most widely to detect heterozygotes by biochemical means; that is, in those cases in which a gene product (protein or enzyme) can be assayed and in which a direct correlation exists between the amount of gene product and the number of copies present of the wild-type gene that codes for the product.

Carrier detection by biochemical analysis is limited to those cases where the gene in question is active (that is, producing a measurable protein or enzyme) at the time of analysis. Recently, molecular methods have been developed in which DNA probes are used directly to determine whether the two genes present are both normal, or one is normal and the other is mutant. In this case, the genes need not be active (i.e. no gene products need to be present) at the time of analysis. We will describe the use of DNA probes in diagnosing genetic disease in humans in Chapter 15.

As an example of how carrier detection might proceed in humans, let us suppose that the a^+ gene produces a certain amount of a functional enzyme that can be measured easily and that the mutant gene a is a null allele—it produces no functional enzyme. Homozygous a^+/a^+ individuals might produce twice the amount of enzyme as the heterozygous a^+/a people, and the a/a types would produce none of the enzyme. Therefore, one possible way to identify heterozygotes (the carriers) would be to measure the amount of enzyme they have.

Such measurements, however, are difficult to make. As with many other human characteristics, when one assays for an enzyme among a number of individuals with the same genotype, a range of enzyme activity typically is found. The range of enzyme activity in a^+/a^+ individuals, for example, might be measured by a bell-shaped curve with a high average enzyme activity, while the range of enzyme activity of a^+/a individuals

~ FIGURE 8.15

Theoretical distribution of enzyme activity in populations of a^+/a^+ (homozygous) and a^+/a (heterozygous) individuals.

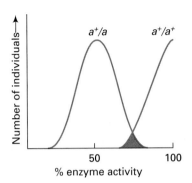

~ FIGURE 8.16

Steps in amniocentesis, a procedure used for prenatal diagnosis of genetic defects.

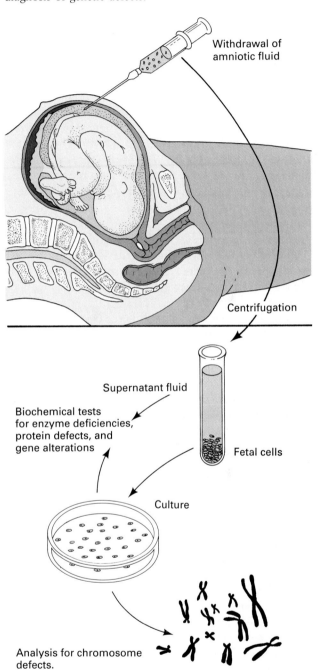

might be demonstrated by a second bell-shaped curve with an average enzyme activity about half that of a^+/a^+ individuals (Figure 8.15). In most cases heterozygotes can be clearly identified, but at the region of overlap it is impossible to make the crucial distinction between a^+/a^+ and a^+/a.

Carriers of Tay-Sachs disease can be tested in this way because carriers have 50 percent of the normal activity for the enzyme hex A. Similarly, Lesch-Nyhan syndrome carriers may be detected by a lower HGPRT enzyme level, and PKU carriers tend to have elevated levels of phenylalanine in their blood.

Identifying carriers is useful since, if both parents are carriers for a recessive genetic disease, the risk is high that they will have a child with the genetic disease. From the principles of genetics the chance is one in four that their progeny will be homozygous for the mutant allele and will have the disease.

AMNIOCENTESIS. A second important aspect of genetic counseling is finding out whether a fetus is normal. This test can be done in many instances by a procedure called **amniocentesis** (Figure 8.16). As a fetus develops in the amniotic sac, it is surrounded by amniotic fluid, which serves to cushion it against shock. In amniocentesis a sample of the amniotic fluid is taken by carefully inserting a syringe needle through the mother's uterine wall and into the amniotic sac. Ultrasound imaging monitors the position of the fetus so that the needle does not damage it. The amniotic fluid contains cells that have sloughed off the fetus's skin; these cells can be cultured in the laboratory. Once cultured, they are examined for chromosomal abnormalities and for the presence or absence of proteins or enzymes. Amniocentesis is possible at any stage of pregnancy, but the quantity of amniotic fluid available and the increased risk to the fetus makes it impractical to perform the

procedure before the 12th week of pregnancy. Also, the time restrictions on abortions make the 18th week usually the latest at which the procedure is done. Typically it is done at about the 16th week. Because amniocentesis is complicated and costly, it is primarily used in high-risk cases.

Amniocentesis analysis can reveal the chromosome composition of the cells and can thus show any obvious chromosomal abnormalities, such as the extra chromosome 21 found in humans with Down syndrome. In general, chromosome abnormalities have serious consequences on the development and function of an individual, and chromosomal analysis offers early detection of such defects. Additionally, chromosomal analysis gives advance information about the sex of the fetus.

Cells can also be examined for protein or enzyme defects. If, for example, both parents are carriers for a known enzyme deficiency disease, and if the enzyme is normally detectable in the fetus, then an enzyme assay on an extract of the cells will clearly reveal whether the fetus will have the disease when it is born. This type of analysis is applicable to Tay-Sachs disease, Lesch-Nyhan syndrome, and any other enzyme deficiency disease in which the enzyme involved is expressed in the developing fetus. Holding increasing promise for the future is the use of DNA probes to diagnose genetic diseases. This technique is limited not by whether a gene product is expressed in the developing fetus, but by whether or not there is a DNA probe that can distinguish the wild-type from the mutant gene.

CHORIONIC VILLUS SAMPLING. Chorionic villus sampling is a relatively new procedure for testing for genetic defects in the developing fetus *in utero*. The procedure can be done between the 8th and 12th week of pregnancy, that is, earlier than amniocentesis, thereby avoiding some of the problems inherent with later sampling. The chorion is a membrane layer surrounding the fetus, and it consists entirely of embryonic tissue. Hence the cells obtained should reflect the genetic makeup of the fetus. A chorionic villus tissue sample may be taken from the developing placenta through the abdomen (as in amniocentesis) or via the vagina using biopsy forceps or a flexible catheter, aided by ultrasound (Figure 8.17). The latter is the preferred method. Once the tissue sample is obtained, the analysis follows that for amniocentesis. An advantage of the technique is that the earlier time of testing permits the parents to learn if the fetus has a genetic defect earlier in the pregnancy than is the case with amniocentesis. Another advantage is that it is not necessary to culture cells to obtain enough to do the biochemical assays. Fetal loss and inaccurate diagnoses due to the presence of maternal cells are more common in chorionic villus sampling than in amniocentesis, however.

Although we can identify carriers of many genetic diseases and can determine if fetuses express genetically-based enzyme deficiencies or are homozygous for mutant genes, in most cases there is no cure for the diseases. Carrier detection, amniocentesis, and chorionic villus sampling serve mainly to inform parents of the risks and probabilities of having a child with the defect. It is then up to the parents to act on the information in the way they see fit. If both are carriers for a

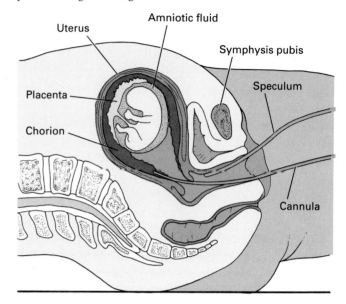

~ FIGURE 8.17

Chorionic villus sampling, a procedure used for early prenatal diagnosis of genetic defects.

serious (perhaps fatal) genetic disease, then they could choose not to have children. If they do conceive a child, they might then have an amniocentesis to see if the fetus has the disease. If it does, then they can prepare themselves to rear the child, or they can opt for an abortion.

KEYNOTE

Genetic counseling is the analysis of the risk that prospective parents may produce a child with a genetic defect, and the presentation to appropriate family members of the available options for avoiding or minimizing those possible risks. Early detection of a genetic disease is done by carrier detection, by amniocentesis, and by chorionic villus sampling. The latter two may also be used to determine whether a fetus exhibits any chromosomal abnormalities.

GENE CONTROL OF PROTEIN STRUCTURE

The preceding section of this chapter provided evidence that genes code for enzymes; this evidence was given because of the historical significance of those studies in advancing our understanding of gene function. While all enzymes are proteins, however, not all proteins are enzymes. To understand completely how genes function, we will look at the experimental evidence that genes are responsible for the structure of nonenzymatic proteins such as hemoglobin. Nonenzymatic proteins are often easier to study than enzymes. Enzymes are usually present in small amounts, while nonenzymatic proteins occur in relatively large quantities in the cell, which makes them easier to isolate and purify. Before discussing the gene control of protein structure, however, we first describe protein structure.

Protein Structure

CHEMICAL STRUCTURE OF PROTEINS. A **protein** is one of a group of high-molecular-weight, nitrogen-containing organic compounds of complex shape and composition. Each cell type has a characteristic set of proteins that give that cell type its functional properties. A protein consists of one or more molecular subunits called **polypeptides,** which are themselves composed of smaller building blocks, the **amino acids,** linked together to form long chains. Each molecule of a given protein consists of the same number and kind of polypeptide chains and each of these in turn is composed of the same number, kind, and sequence of amino acids. It is the sequence of amino acids in a polypeptide that gives the polypeptide its three-dimensional shape and its properties in the cell.

The basic chemical structural units of proteins are amino acids. With the exception of proline, the amino acids have a common structure, which is shown in Figure 8.18. The structure consists of a central carbon atom (α-carbon) to which is bonded an amino group (NH_2), a carboxyl group (COOH), and a hydrogen atom. At the pH commonly found within cells, the NH_2 and COOH groups of the free amino acids are in a charged state; that is, $-NH_3^+$ and $-COO^-$, respectively.

Bound to the α-carbon, each amino acid has an additional chemical group, called the *radical,* or

R group. It is the R group that varies from one amino acid to another and gives each amino acid its distinctive properties. Since different proteins have different sequences and proportions of amino acids, the organization of the R groups gives a protein its structural and functional properties.

There are 20 naturally occurring amino acids; their names, three-letter abbreviations, and chemical structures are shown in Figure 8.19 (p. 252). The 20 amino acids are divided into subgroups, based on whether the R group is acidic (e.g., aspartic acid), basic (e.g., lysine), neutral-polar (e.g., leucine), or neutral-nonpolar (e.g., serine).

The amino acids of a polypeptide are held together by a **peptide bond,** a covalent bond that joins the carboxyl group of one amino acid to the amino group of another amino acid. The formation of the peptide bond is the result of the reaction depicted in Figure 8.20 (p. 253). A polypeptide, then, is a linear, unbranched molecule that consists of many amino acids (usually 100 or more) joined by peptide bonds. Every polypeptide has a free α-amino group at one end (called the N terminus, or the N terminal end) and a free α-carboxyl group at the other end (called the C terminus, or the C terminal end). Polypeptides have polarity: by convention and because the polypeptide is constructed that way, the N terminal end is defined as the beginning of a polypeptide chain.

MOLECULAR STRUCTURE OF PROTEINS. The molecular structure of a protein is relatively complex; it has four levels of structural organization, as shown in Figure 8.21 (p. 254).

1. The primary structure of the polypeptide chain that constitutes a protein is the amino acid sequence (Figure 8.21a).
2. The secondary structure of a protein refers to the folding and twisting of a single polypeptide chain into a variety of shapes. A polypeptide's secondary structure is the result of weak bonds (e.g., electrostatic or hydrogen) that form between NH and CO groups of amino acids that are relatively near each other on the chain. One type of secondary structure found in regions of many polypeptides is the α-helix, a structure discovered by Linus Pauling and Robert Corey in 1951. Figure 8.21b shows the α-helix and diagrams the hydrogen bonding between the NH group of one amino acid (i.e., an NH group that is part of a peptide bond) and the CO group (also part of a peptide bond) of an amino acid that is four amino acids away in the chain. The repeated formation of this bonding results in the helical coiling of the chain. The α-helix content of proteins varies.

~ FIGURE 8.18

General structural formula for an amino acid.

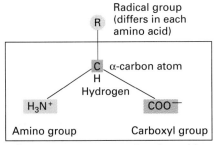

Structures common to all amino acids

~ FIGURE 8.19

Structures of the 20 naturally-occurring amino acids.

Acidic

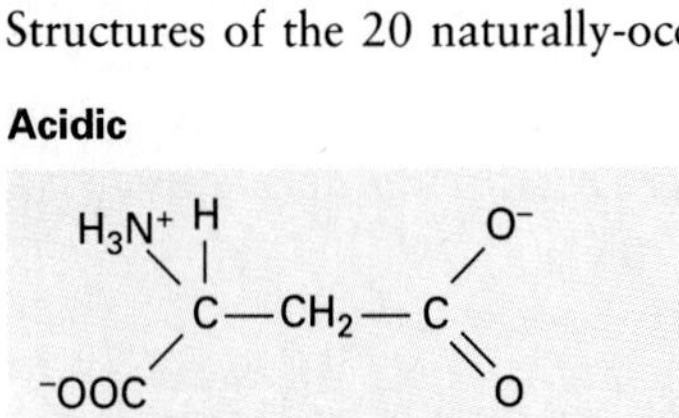

Aspartic acid (Asp)

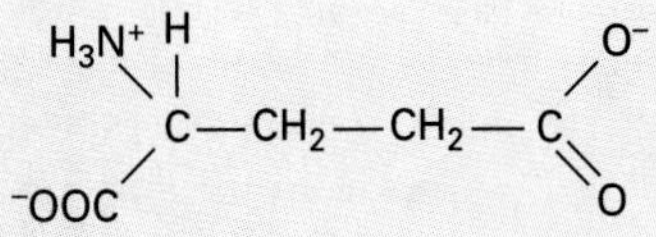

Glutamic acid (Glu)

Neutral, nonpolar

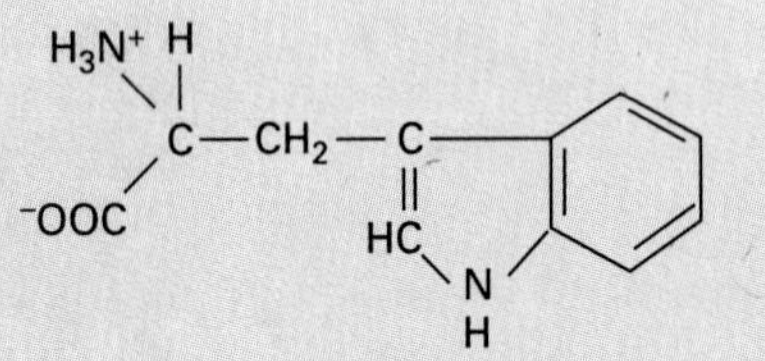

Tryptophan (Trp)

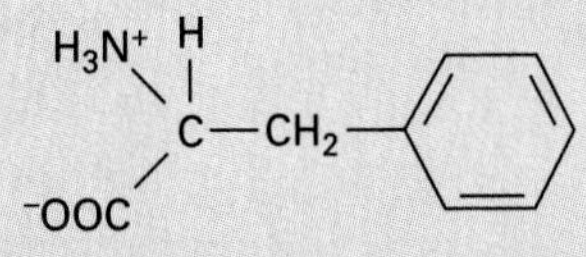

Phenylalanine (Phe)

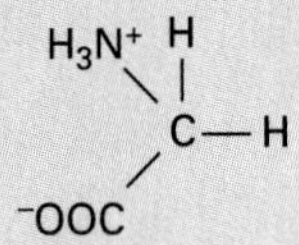

Glycine (Gly)

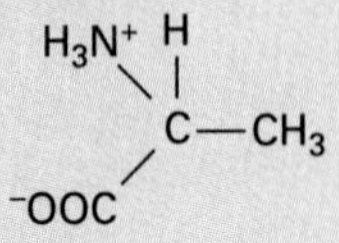

Alanine (Ala)

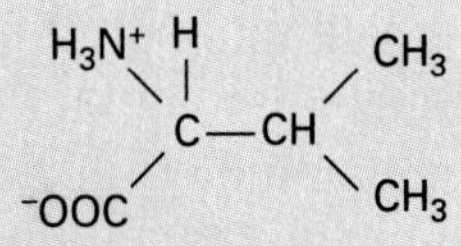

Valine (Val)

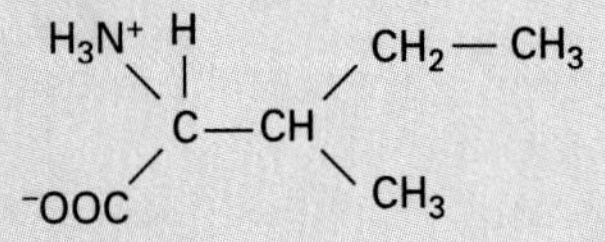

Isoleucine (Ile)

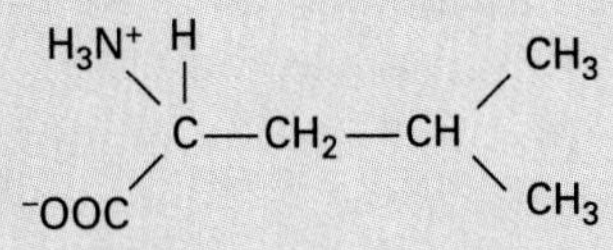

Leucine (Leu)

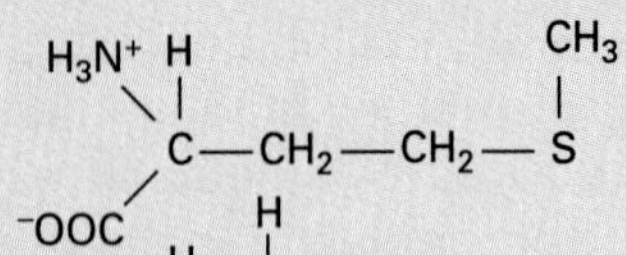

Methionine (Met)

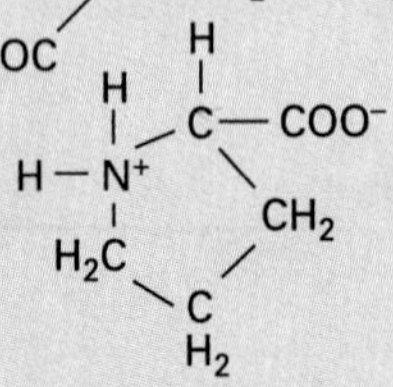

Proline (Pro)

Basic

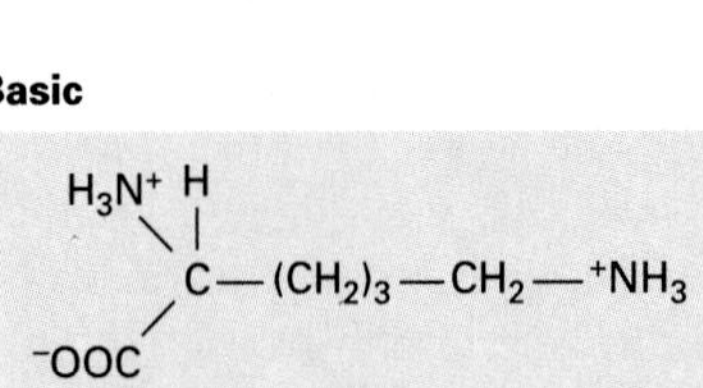

Lysine (Lys)

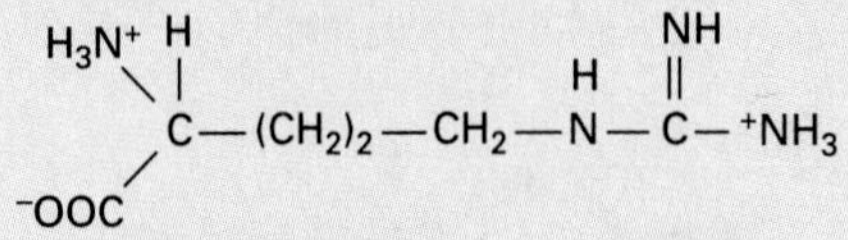

Arginine (Arg)

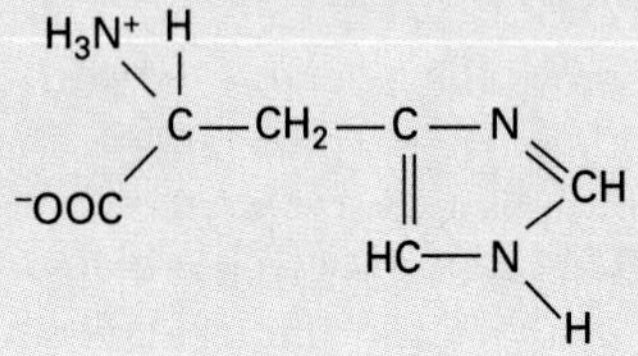

Histidine (His)

Neutral, polar

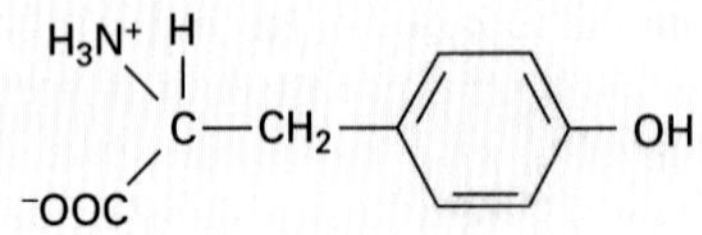

Tyrosine (Tyr)

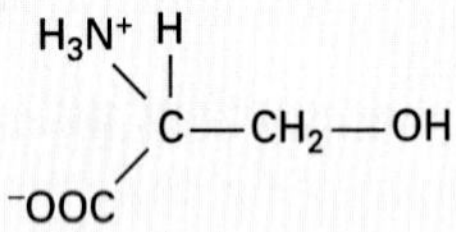

Serine (Ser)

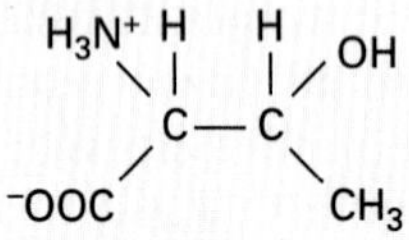

Threonine (Thr)

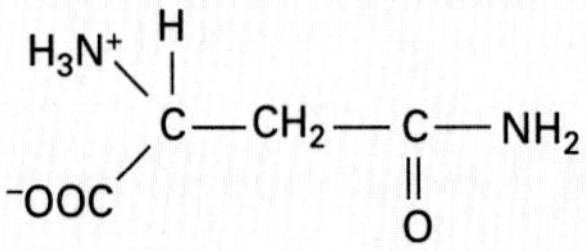

Asparagine (Asn)

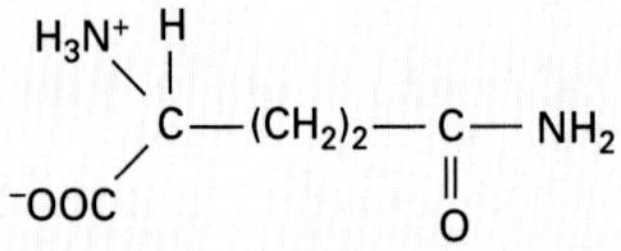

Glutamine (Gln)

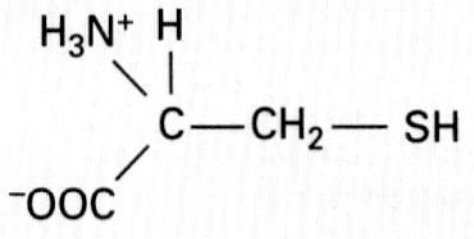

Cysteine (Cys)

~ FIGURE 8.20

Mechanism for peptide bond formation between the carboxyl group of one amino acid and the amino group of another amino acid.

Another type of secondary structure is the β-pleated sheet (Figure 8.21b). Also discovered by Pauling and Corey in 1951, the β-pleated sheet involves a polypeptide chain or chains folded in a "zig-zag" way with parallel regions or chains linked by hydrogen bonds. Most proteins contain a mixture of α-helical and β-pleated sheet regions.

3. A protein's tertiary structure (Figure 8.21c) is the three-dimensional structure into which the helices and other parts of a polypeptide chain are folded. The three-dimensional shape of a polypeptide is often called its *conformation*. Tertiary folding is a direct property of the amino acid sequence of the chain and, hence, is related to the relative frequency and distribution of the R groups along the chain. Shown in Figure 8.21c is the tertiary structure of myoglobin, an oxygen-binding protein found in muscle.

4. Quaternary structure of a protein is shown in Figure 8.21d. Note that primary, secondary, and tertiary structures all refer to single polypeptide chains. As we said earlier, however, proteins may consist of more than one polypeptide chain; such proteins are called multimeric ("many subunits") proteins. Quaternary structure is found only in proteins having more than one polypeptide chain. The term refers to how polypeptides are packaged into the whole protein molecule. Shown in Figure 8.21d is the quaternary structure of a well-known example of a multimeric protein, the oxygen-carrying protein hemoglobin, which consists of four polypeptide chains (two α polypeptides and two β polypeptides), each of which is associated with a heme group (involved in the binding of oxygen). The α and β polypeptides have different amino acid sequences. The α polypeptide contains 141 amino acids, and the β polypeptide contains 146 amino acids. In the quaternary structure of hemoglobin, each α chain is in contact with each β chain; however, little interaction occurs between the two α chains or between the two β chains.

KEYNOTE

A protein consists of one or more molecular subunits called polypeptides, which are themselves composed of smaller building blocks, the amino acids, linked together by peptide bonds to form long chains. The primary amino acid sequence of a protein determines its secondary, tertiary, and quaternary structure and hence its functional state.

Sickle-Cell Anemia

Sickle-cell anemia is a genetically-based human disease that results from an amino acid change in the hemoglobin protein that transports oxygen in the blood.

SYMPTOMS AND BASIS OF THE DISEASE. Sickle-cell anemia was first described in 1910 by J. Herrick. He found that red blood cells from individuals with the disease lose their characteristic shape in the absence of oxygen and assume the shape of a sickle (Figure 8.22; p. 255). The sickled red blood cells are more fragile than normal red blood cells and tend to break sooner. In addition, sickled cells are not as flexible as normal cells and therefore tend to jam up in the capillaries rather than squeeze through them. As a consequence, blood circulation is impaired, and tissues served by the capillaries become deprived of oxygen. Although oxygen deprivation occurs particularly at the extremities, the heart, lungs, brain, kidneys, gastrointestinal tract, muscles, and joints can also suffer from oxygen deprivation and its subsequent damage. The afflicted individual may therefore suffer from a variety of health problems including heart failure, pneumonia, paralysis, kidney failure, abdominal pain, and rheumatism.

Sickle-cell anemia is caused by homozygosity for a mutation in the gene for the β polypeptide of hemoglobin. The sickle-cell mutation β^S is codominant with the wild-type allele β^A. Thus people who are heterozygous

~ FIGURE 8.21

Four levels of protein structure: (a) Primary, the sequence of amino acids in a polypeptide chain. (b) Secondary, the folding and twisting of a single polypeptide chain into a variety of shapes. Shown are two types of secondary structures: the α-helix and β-pleated sheet. Both structures are stabilized by hydrogen bonds. (c) Tertiary, the specific three-dimensional folding of the polypeptide chain. Shown here is the main polypeptide chain of myoglobin, a 153-amino acid, heme-containing polypeptide that carries oxygen in muscles. (d) Quaternary, the specific aggregate of polypeptide chains. Shown here is hemoglobin, which carries oxygen in the blood; it consists of two α chains, two β chains, and four heme groups. The tertiary structures of the α- and β-chains closely resemble that of myoglobin.

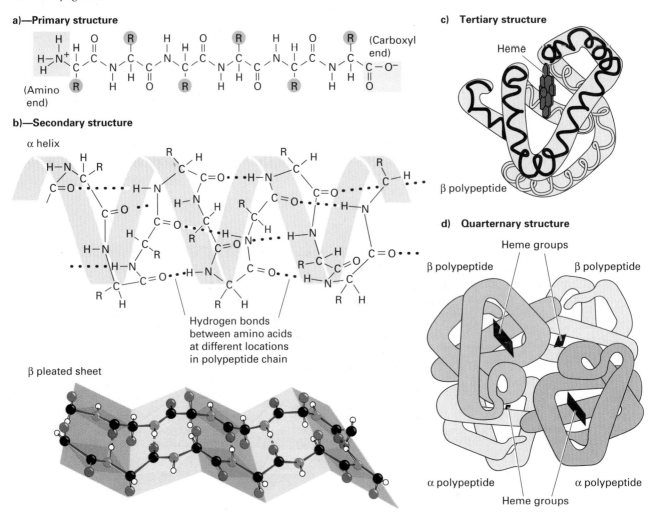

a)—Primary structure

b)—Secondary structure

α helix

Hydrogen bonds between amino acids at different locations in polypeptide chain

β pleated sheet

c) Tertiary structure

Heme

β polypeptide

d) Quaternary structure

Heme groups

β polypeptide   β polypeptide

α polypeptide   α polypeptide

Heme groups

$\beta^A\beta^S$ make two types of hemoglobins. One, called Hb-A (hemoglobin A), is completely normal, with two normal α chains and two normal β chains specified by two wild-type α genes and one wild-type β gene (β^A). The other, called Hb-S, is the defective hemoglobin, with two normal α chains specified by wild-type α genes and two abnormal β chains specified by the mutant β gene β^S. The $\beta^A\beta^S$ heterozygotes are said to

have *sickle-cell trait*, although typically they show few symptoms of the disease.

The evidence that an abnormal hemoglobin molecule was present in individuals with sickle-cell anemia was obtained by Linus Pauling and his co-workers in the 1950s, using the procedure of electrophoresis to separate the electrically charged protein molecules in an electric field. In this technique, proteins of the same

~ FIGURE 8.22

Scanning electron micrograph of (left) normal and (right) sickled red blood cells.

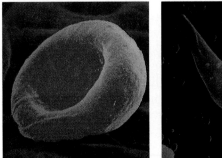

~ FIGURE 8.23

Electrophoresis of hemoglobin found in (left) normal $\beta^A\beta^A$ individuals, (center) in $\beta^A\beta^S$ individuals with sickle-cell trait, and (right) in $\beta^S\beta^S$ individuals who exhibit sickle-cell anemia. The two hemoglobins migrate to different positions in an electric field and hence must differ in electric charge.

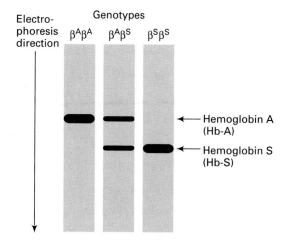

molecular weight but different charge will migrate at different rates. Pauling isolated hemoglobin from the red blood cells of three groups: normal individuals, individuals with sickle-cell trait, and individuals with sickle-cell anemia. He then subjected each sample to electrophoresis. Figure 8.23 shows the results he obtained. Under the conditions he used, the hemoglobin from normal people (Hb-A) migrated more slowly than the hemoglobin from people who had sickle-cell anemia (Hb-S). The hemoglobin from sickle-cell trait individuals behaved in electrophoresis like a 1:1 mixture of Hb-A and Hb-S. Pauling concluded that sickle-cell anemia results from a mutation that alters the chemical structure of the hemoglobin molecule. This experiment was one of the first rigorous proofs that protein structure is controlled by a unique gene.

The precise molecular change in the hemoglobin molecule of sickle-cell anemia individuals was determined by J. Ingram in 1957. He determined the sequence of the α and β polypeptide chains of the two types of hemoglobin, Hb-A and Hb-S, and found that the α chain was identical in the two. In the β chain, however, Ingram found a single amino acid substitution at the sixth position from the N terminal end. Figure 8.24 shows this substitution, which is the replacement of the acidic amino acid glutamic acid with the neutral amino acid valine. The simplest explanation for this change is that the β^S mutation involves a base-pair charge in the DNA so that valine is encoded instead of glutamic acid. This finding was undeniable proof that genes specify the amino acid sequence of

~ FIGURE 8.24

The first seven N terminal amino acids in normal and sickled hemoglobin β polypeptides, showing the single amino acid change from glutamic acid to valine at the amino acid in the sixth position.

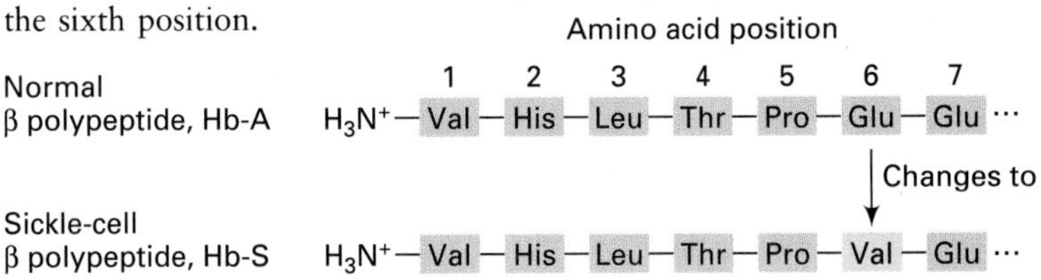

Amino acid position

Normal
β polypeptide, Hb-A H_3N^+ — Val — His — Leu — Thr — Pro — Glu — Glu ···

Changes to

Sickle-cell
β polypeptide, Hb-S H_3N^+ — Val — His — Leu — Thr — Pro — Val — Glu ···

proteins. Furthermore, since the hemoglobin protein consists of two types of polypeptides, this experiment gave additional support for the one-gene–one-polypeptide hypothesis.

The substitution of valine for glutamic acid results in the pleiotropic consequences characteristic of sickle-cell anemia in the following way. The amino acid at position 6 of the β chain is on the outside of the molecule. In Hb-A (the normal hemoglobin) the glutamic acid at that position is a negatively charged, hydrophilic ("water-loving") molecule. In the folding process that occurs (usually in an aqueous solution), hydrophilic regions of a polypeptide (in this case the β polypeptide) find themselves on the outside of the molecule, while hydrophobic ("water-hating") regions generally are folded into the interior of the protein, away from water. Valine is a hydrophobic amino acid, so it tends to be found in the interior of the molecule, rather than on the surface. As a consequence, the Hb-S molecules in individuals with sickle-cell anemia tend to align side by side to form long tubules. This tubular organization of the hemoglobin molecules is responsible for the sickling of the red blood cells.

OCCURRENCE OF SICKLE-CELL ANEMIA IN HUMAN POPULATIONS. The mutant gene responsible for sickle-cell anemia does not have a random distribution among human populations. Rather, it occurs with relatively high frequency in individuals whose ancestors lived for many generations in parts of the world where malaria was present, such as Africa, India, and southern Europe.

Populations in which the mutant gene is common have significant numbers of people with sickle-cell anemia. In the United States these individuals tend to die at a relatively young age, usually about 40 years; in less developed countries, afflicted individuals die much earlier. In addition, females who have sickle-cell anemia and who reach childbearing age tend to spontaneously abort, because insufficient oxygen is available to the developing fetus.

The connection between sickle-cell anemia (an inherited blood disorder) and malaria (a disease that results from a parasite injected into the blood stream by a mosquito) is not as obscure as it might seem. People who are homozygous normals are much more likely to die from malaria than are heterozygous carriers for sickle-cell anemia because the malarial parasite cannot thrive in abnormal red blood cells. As a result, in areas where malaria is endemic heterozygous individuals will survive in higher numbers than those with normal red blood cells. Consequently, the mutant gene continues to survive in the population. For this reason the sickle-cell anemia gene in the American black population,

whose origins lie in Africa, has a much higher frequency than in Caucasian or Oriental populations in the United States.

Other Hemoglobin Mutants

Many other mutant hemoglobins have been detected in general screening programs in which hemoglobin is isolated from red blood cells and subjected to electrophoresis. Altered hemoglobins were detected by a difference in the electrophoretic mobility compared with the Hb-A molecule. Over two hundred hemoglobin mutants have been detected through electrophoresis; Figure 8.25 lists some of these mutants along with the amino acid substitutions that have been identified. Some mutations affect the α chain and others the β chain, and there is a wide variety in the types of amino acid substitutions that occur. From the codon changes that are assumed to be responsible for the substitutions, it is apparent that a single base change is involved in each case.

~ **FIGURE 8.25**

Examples of amino acid substitutions found in (a) α and (b) β polypeptides of various human hemoglobin variants.

a) **α-chain**

	1	2	16	30	57	58	68	141
Normal	Val	Leu	Lys	Glu	Gly	His	Asn	Arg

Hb variants:

	1	2	16	30	57	58	68	141
HbI	Val	Leu	Asp	Glu	Gly	His	Asn	Arg
Hb-G Honolulu	Val	Leu	Lys	Gln	Gly	His	Asn	Arg
Hb Norfolk	Val	Leu	Lys	Glu	Asp	His	Asn	Arg
Hb-M Boston	Val	Leu	Lys	Glu	Gly	Tyr	Asn	Arg
Hb-G Philadelphia	Val	Leu	Lys	Glu	Gly	His	Lys	Arg

b) **β-chain**

	1	2	3	6	7	26	63	67	125	150
Normal	Val	His	Leu	Glu	Glu	Glu	His	Val	Glu	His

Hb variants:

	1	2	3	6	7	26	63	67	125	150
Hb-S	Val	His	Leu	Val	Glu	Glu	His	Val	Glu	His
Hb-C	Val	His	Leu	Lys	Glu	Glu	His	Val	Glu	His
Hb-G San Jose	Val	His	Leu	Glu	Gly	Glu	His	Val	Glu	His
Hb-E	Val	His	Leu	Glu	Glu	Lys	His	Val	Glu	His
Hb-M Saskatoon	Val	His	Leu	Glu	Glu	Glu	Tyr	Val	Glu	His
Hb Zurich	Val	His	Leu	Glu	Glu	Glu	Arg	Val	Glu	His
Hb-M Milwaukee-1	Val	His	Leu	Glu	Glu	Glu	His	Glu	Glu	His
Hb-D β Punjab	Val	His	Leu	Glu	Glu	Glu	His	Val	Gln	His

Not all the identified hemoglobin mutants have as drastic an effect as the sickle-cell anemia mutant because the consequences of the substitution depend both on the amino acids involved and on the location of the mutant amino acid in the chain. For example, in the Hb-C hemoglobin molecule there is a change in the β chain from a glutamic acid to a lysine. Like the Hb-S change, this substitution involves the sixth amino acid from the N terminal end of the polypeptide. But unlike the Hb-S change, this change is not a serious defect because both amino acids are hydrophilic. Therefore people homozygous for the β^C mutation that results in the abnormal β chain of the Hb-C hemoglobin molecule experience only a mild form of anemia.

Biochemical Genetics of the Human ABO Blood Groups

The mechanism of inheritance of the human ABO blood groups was mentioned in Chapter 4. To summarize, the blood group depends on the presence of chemical substances called antigens on the red blood cell surface. When antigens are injected into a host organism, they may be recognized as foreign and removed from the circulation by the host's antibodies. The molecular basis of the ABO blood groups further supports the direct relationship between genes and amino acid sequences in proteins.

An individual generally has antibodies circulating in blood serum that are directed toward those antigens that the individual is not carrying on his or her red blood cells. People of blood group A (genotypes I^A/I^A or I^A/I^O) carry the A antigen on their blood cells, and in their serum they have antibodies against the B antigen (the β antibody). Similarly, people of blood group B (genotypes I^B/I^B or I^B/I^O) carry the B antigen and have antibodies against the A antigen (the α antibody). An individual with AB blood type (genotype I^A/I^B), however, has both the A and B antigens but neither α nor β circulating antibodies. Finally, an individual of blood group O (genotype I^O/I^O) has neither the A nor the B antigens on their blood cells and so has both the α and β antibodies in circulation. As we learned in Chapter 4, I^A, I^B, and I^O are all alleles of one gene (the ABO locus) and hence constitute a multiple allelic series.

The ABO locus is involved in the production of enzymes that are used in the biosynthesis of polysaccharides. The enzymes, called *glycosyltransferases*, act specifically to add sugar groups to a preexisting polysaccharide. The polysaccharides important here are those that attach to lipids to produce compounds called glycolipids. The glycolipids then associate with red blood cell membranes to form the blood group antigens.

Figure 8.26 (p. 258) shows the terminal sugar structures of the polysaccharide components of the glycolipids involved in the ABO blood type system. Most individuals produce a glycolipid called the H antigen, which has the terminal sugar sequence shown in the figure. The I^A allele produces a glycosyltransferase enzyme called α-N-acetylgalactosamyl transferase, which recognizes the H antigen and adds the sugar α-N-acetylgalactosamine to the end of the polysaccharide to produce the A antigen. The I^B allele, on the other hand, produces an α-D-galactosyltransferase, which also recognizes the H antigen but adds galactose to its polysaccharide to produce the B antigen. (Note that this small difference in the structure of the A and B antigens is sufficient to induce an antibody response.) In both cases some H antigen remains unconverted.

For the I^A/I^B heterozygote, both enzymes are produced, so some H antigen is converted to the A antigen and some to the B antigen. The blood cell has both antigens on the surface, so the individual is of blood type AB. It is the presence of both surface antigens that provides the molecular basis of the codominance of the I^A and I^B alleles.

People who are homozygous for the I^O allele produce no enzymes to convert the polysaccharide component of the H antigen glycolipid. As a consequence, their red blood cells carry neither the A nor the B antigen, although they do carry the H antigen. The H antigen does not elicit an antibody response in people of other blood groups because its polysaccharide component is the basic component of the A and the B antigens as well, so it is not exactly a foreign substance. Lastly, people who are heterozygous for the I^O allele will have the blood type of the other allele. For example, in I^B/I^O individuals the I^B allele will result in the conversion of some of the H antigen to the B antigen, determining the blood type of the individual.

Lastly, the H antigen is produced by the action of the dominant *H* allele at a locus that is distinct from the ABO locus. Individuals homozygous for the recessive mutant allele, *h*, do not make the H antigen, and therefore, regardless of the presence of I^A or I^B alleles at the ABO locus, no A or B antigens can be produced. Thus *h/h* individuals are blood type O, in the sense that they lack A and B antigens; they are said to have the Bombay blood type. Note, however, that the blood of people who have the Bombay blood type is not identical to that of people with blood group O, since Bombay blood individuals have anti-O antibodies. As a consequence of this phenomenon, it is possible for two blood type O individuals to produce a child who has blood type A or B. That is, if one parent is *h/h* $I^B/-$ (i.e., *h/h* I^B/I^B or *h/h* I^B/I^O), and the other is *H/H* I^O/I^O, the child could be *H/h* I^B/I^O, and, therefore, be blood type B.

~ FIGURE 8.26

Terminal sugar sequences in the polysaccharide chains of glycolipids involved in the human ABO blood types (a) H antigen; (b) A antigen; (c) B antigen.

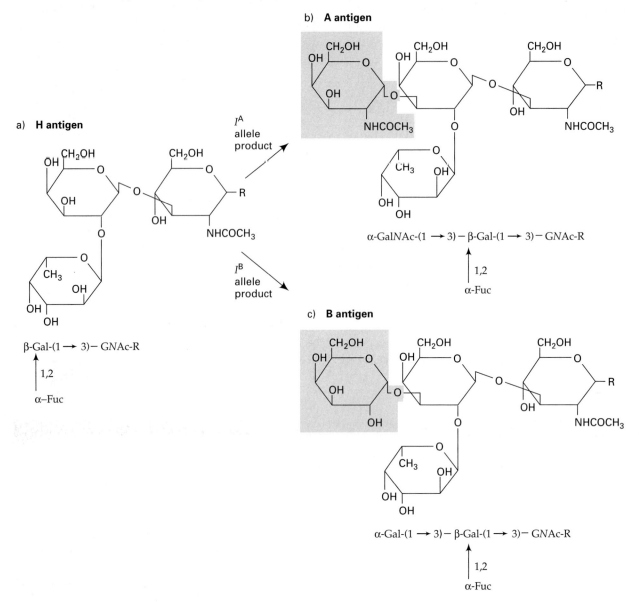

KEYNOTE

From the study of alterations in proteins other than enzymes, such as the mutations responsible for sickle-cell anemia, convincing evidence was obtained that genes control the structures of all proteins.

SUMMARY

In this chapter we began to think of the gene in molecular, rather than abstract terms. Specifically, we discussed gene function—what genes code for, and what is the evidence for that coding. We considered a number of classical experiments that showed that genes coded for enzymes and for nonenzymatic proteins. A number of these experiments were done prior to the unequivocal proof that DNA is genetic material and the elucidation of DNA structure.

As early as 1902 there was evidence for a specific relationship between genes and enzymes. This evidence was obtained by Archibald Garrod in his investigations of "inborn errors of metabolism," human genetic diseases that resulted in enzyme deficiencies. The specific relationship between genes and enzymes is historically embodied in the "one gene–one enzyme" hypothesis proposed by Beadle and Tatum to explain their data on auxotrophic mutants of *Neurospora*. The "one gene–one enzyme" hypothesis states that each gene controls the synthesis or activity of a single enzyme. Since some enzymes and nonenzymatic proteins have more than one polypeptide (e.g. hemoglobin), a more modern maxim is "one gene–one polypeptide." Beadle and Tatum's experiments are landmarks in the development of molecular genetics and their methods for dissecting a biochemical pathway genetically—that is, to determine the sequence of steps in the pathway and relating each step to a specific gene or genes—are classical.

In this chapter we also discussed a number of human examples of genetically-based enzyme deficiencies that give rise to genetic diseases. Many of these are the result of homozygosity for recessive mutant genes. The severity of the disease depends on the effects of the loss of function of the particular enzyme involved. Thus, albinism is a relatively mild genetic disease, while Tay-Sachs disease is lethal. Given the knowledge we currently have about a large number of genetically based enzyme and protein deficiencies in humans, it is possible to make some predictions about the existence of certain genetic diseases in individuals or in a family. Genetic counseling is the analysis of the probability that individuals have a genetic defect, or of the risk that prospective parents may produce a child with a genetic defect, and the presentation to the individuals involved any available options for avoiding or minimizing those possible risks. Techniques available to genetic counselors include amniocentesis and chorionic villus sampling.

To make our understanding of gene function more complete, we examined experimental evidence that genes control the structure of nonenzymatic proteins as well as enzymes. That is, while all enzymes are proteins, not all proteins are enzymes. At the molecular level, a protein consists of one or more macromolecular subunits called polypeptides, which are themselves composed of smaller building blocks called amino acids. The amino acids are linked together in the polypeptide by peptide bonds. The primary amino acid sequence of a protein determines its secondary, tertiary, and quaternary structure, and hence its functional state.

ANALYTICAL APPROACHES FOR SOLVING GENETICS PROBLEMS

Q.1 k^+, l^+, and m^+ are independently assorting genes that control the production of a red pigment. These three genes act in a biochemical pathway as follows:

$$\text{colorless 1} \xrightarrow{k^+} \text{colorless 2} \xrightarrow{l^+} \text{orange} \xrightarrow{m^+} \text{red}$$

The mutant alleles that give abnormal functioning of these genes are k, l, and m; each is recessive to its wild-type counterpart. A red individual homozygous for all three wild-type alleles is crossed with a colorless individual that is homozygous for all three recessive mutant alleles. The F_1 is red. The F_1 is then selfed to produce the F_2 generation.

a. What proportion of the F_2 are colorless?
b. What proportion of the F_2 are orange?
c. What proportion of the F_2 are red?

A.1 a. There are two ways to answer this question. One is to determine all of the genotypes that can specify the colorless phenotypes, and the other is to use subtractive logic, in which the proportions of orange and red are first calculated and these proportions are subtracted from one to give the proportion of colorless progeny in the F_2. We will first consider the second method.

To produce an orange phenotype three things are needed: 1) at least one wild-type k^+ allele must be present so that the colorless 1 to colorless 2 step can occur; 2) at least one wild-type l^+ allele must be present so that the colorless 2 to orange step can occur; and 3) the individual must be m/m so that the orange to red step cannot proceed. Thus, an orange phenotype results from the genotype $k^+/-\ \ l^+/-\ \ m/m$. From an $F_1 \times F_1$ self, the probability of getting an individual with that genotype is $3/4 \times 3/4 \times 1/4 = 9/64$.

To produce a red phenotype three things are needed: 1) at least one wild-type k^+ allele must be present so that the colorless 1 to colorless 2 step can occur; 2) at least one wild-type l^+ allele must be present so that the colorless 2 to orange step can occur; and 3) at least one m^+ allele must be present so that the orange to red step can proceed. Thus, a red phenotype results from the genotype $k^+/- \; l^+/- \; m^+/-$. From an $F_1 \times F_1$ self, the probability of getting an individual with that genotype is $3/4 \times 3/4 \times 3/4 = 27/64$.

By default, all other F_2 individuals are colorless. The proportion of F_2 individuals that are colorless is, therefore, $1 - 9/64 - 27/64 = 1 - 36/64 = 28/64$.

Using the first approach to calculate the proportion of F_2 colorless, we must determine all genotypes that will give a colorless phenotype and then add up the probabilities of each occurring. The genotypes and their probabilities are as follows:

GENOTYPES	PROBABILITIES	
$k/k \; l/l \; m/m$	1/64	(Cannot convert colorless 1)
$k/k \; l/l \; m^+/-$	3/64	(Cannot convert colorless 1)
$k/k \; l^+/- \; m/m$	3/64	(Cannot convert colorless 1)
$k/k \; l^+/- \; m^+/-$	9/64	(Cannot convert colorless 1)
$k^+/- \; l/l \; m/m$	3/64	(Cannot convert colorless 2)
$k^+/- \; l/l \; m^+/-$	9/64	(Cannot convert colorless 2)
TOTAL	28/64	

b. The proportion of the F_2 was calculated in (a), above, i.e. $9/64$.

c. The proportion of the F_2 was calculated in (a), above, i.e. $27/64$.

Q.2 A number of auxotrophic mutant strains were isolated from wild-type, haploid yeast. These strains responded to the addition of certain nutritional supplements to minimal culture medium either by growth (+) or no growth (0). The following table gives the responses for single gene mutants:

MUTANT STRAINS	SUPPLEMENTS ADDED TO MINIMAL CULTURE MEDIUM				
	B	A	R	T	S
1	+	0	+	0	0
2	+	+	+	+	0
3	+	0	+	+	0
4	0	0	+	0	0

Diagram a biochemical pathway which is consistent with the data, indicating where in the pathway each mutant strain is blocked.

A.2 The type of data to be analyzed are very similar to those discussed in the text for Beadle and Tatum's analysis of *Neurospora* auxotrophic mutants, from which they proposed the one gene–one enzyme hypothesis. Recall that an important principle is that the later in the pathway a mutant is blocked, the fewer nutritional supplements need be added to allow growth. In the data given, we must assume that the nutritional supplements are not listed necessarily in the order in which they appear in the pathway.

Analysis of the data indicates that all four strains will grow if given R, and that none will grow if given S. From this we can conclude that R is likely to be the end product of the pathway (all mutants should grow if given the end product) and that S is likely to be the first compound in the pathway (none of the mutants should grow given the first compound in the pathway). Thus, the pathway as deduced so far is:

$$S \longrightarrow [B, A, T] \longrightarrow R$$

where the order of B, A, and T are as yet undetermined.

Now let us consider each of the mutant strains, and see how their growth phenotypes can help define the biochemical pathway:

Strain 1 will grow only if given B or R. Therefore, the defective enzyme in strain 1 must act somewhere prior to the formation of B and R, and after the substances A, T and S. Since we have deduced that R is the end product of the pathway, we can propose that B is the immediate precursor to R, and that strain 1 cannot make B. The pathway so far is:

$$S \longrightarrow [A, T] \xrightarrow{1} B \longrightarrow R$$

Strain 2 will grow on all compounds except S, the first compound in the pathway. Thus, the defective enzyme in strain 2 must act to convert S to the next compound in the pathway, which is either A or T. We do not know yet whether A or T follows S in the pathway, but the growth data at least allow us to conclude where strain 2 is blocked in the pathway, i.e.:

$$S \xrightarrow{2} [A, T] \xrightarrow{1} B \longrightarrow R$$

Strain 3 will grow on B, R and T, but not on A or S. We know that R is the end product and S is the first compound in the pathway. This mutant strain allows us to determine the order of A and T in the pathway. That is, since strain 3 grows on T but not A, T must be later in the pathway than A, and the defective enzyme in 3 must be blocked in the yeast's ability to convert A to T. The pathway now is:

$$S \xrightarrow{2} A \xrightarrow{3} T \xrightarrow{1} B \longrightarrow R$$

Strain 4 will grow only if given the deduced end product R. Therefore, the defective enzyme produced by the mutant gene in strain 4 must act before the formation of R, and after the formation of A, T and B from the first compound S. The mutation in 4 must be blocked in the last step of the biochemical pathway in the conversion of B to R. The final deduced pathway, and the positions of the mutant blocks are:

$$S \xrightarrow{2} A \xrightarrow{3} T \xrightarrow{1} B \xrightarrow{4} R$$

Q.3 The ABO blood group locus determines the production of blood group antigens. People known as secretors have the same AB blood group antigens in their saliva as are found on the surfaces of the red blood cells. Nonsecretors do not have these antigens in their saliva. The secretion phenotype is controlled by a single pair of alleles: *Se* (secretor) and *se* (nonsecretor). The *Se* allele is completely dominant to the *se* allele. The secretor locus is unlinked to the ABO locus.

A child has the blood group O, M and is a secretor. Five matings are indicated in the following list. Explain for each case whether or not the child could have resulted from the mating.

a. O, M secretor × O, MN, secretor
b. A, MN, nonsecretor × B, MN, secretor
c. A, M, secretor × A, M, secretor
d. AB, MN, secretor × O, M, nonsecretor
e. A, M, nonsecretor × A, M, nonsecretor

A.3 The child must be $I^O/I^O\ L^M/L^M\ Se/-$, where the secretor locus can be either homozygous or heterozygous for the dominant *Se* allele.

a. The genotypes of the parents are $I^O/I^O\ L^M/L^M\ Se/-$ and $I^O/I^O\ L^M/L^N\ Se/-$, so it is possible to produce the child described. There are no major assumptions that we have to make, since it does not matter whether the parents are homozygous or heterozygous for the secretor locus.

b. The genotypes of the parents are $I^A/-\ L^M/L^N\ se/se$ and $I^B/-\ L^M/L^N\ Se/-$, so it is possible to produce the child. The assumptions here are that both parents carry the I^O allele and that the secretor parent is heterozygous at that locus.

c. The parents are $I^A/-\ L^M/L^M\ Se/-$ and $I^A/-\ L^M/L^M\ Se/-$. On the assumption that both parents are heterozygous I^A/I^O, it is possible to produce the child. Since both parents are secretors, the genotype is no problem with respect to producing a secretor child.

d. The parents are $I^A/I^B\ L^M/L^N\ Se/-$ and $I^O/I^O\ L^M/L^M\ se/se$. The child could not be produced from this mating since one parent is AB in blood group, meaning that the child must be either A or B blood group. There is no problem with either of the other two loci.

e. The parents are $I^A/-\ L^M/L^M\ se/se$ and $I^A/-\ L^M/L^M\ se/se$. The child could not be produced from this mating since both parents are nonsecretors and hence are homozygous *se/se*. Therefore, all children must be *se/se* and be nonsecretors. There is no problem with either of the other two loci.

QUESTIONS AND PROBLEMS

8.1 Phenylketonuria (PKU) is a heritable metabolic disease of humans; its symptoms include mental deficiency. The gross phenotypic effect is due to:
a. Accumulation of phenylketones in the blood
b. Accumulation of maple sugar in the blood
c. Deficiency of phenylketones in the blood
d. Deficiency of phenylketones in the diet

*8.2 If a person were homozygous for both PKU (phenylketonuria) and AKU (alkaptonuria), what would you expect his or her phenotype to be? Refer to the pathway below.

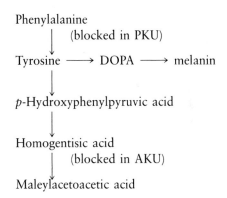

8.3 Refer to the pathway shown in Question 8.2. What effect, if any, would you expect PKU (phenylketonuria) and AKU (alkaptonuria) to have on pigment formation?

*8.4 a^+, b^+, c^+, and d^+ are independently assorting Mendelian genes controlling the production of a black pigment. The alternate alleles that give abnormal functioning of these genes are a, b, c, and d. A black $a^+/a^+\ b^+/b^+\ c^+/c^+\ d^+/d^+$ is crossed with a colorless $a/a\ b/b\ c/c\ d/d$ to give a black F_1. The F_1 is then selfed. Assume that a^+, b^+, c^+, and d^+ act in a pathway as follows:

$$\text{colorless} \xrightarrow{a^+} \text{colorless} \xrightarrow{b^+} \text{colorless} \xrightarrow{c^+} \text{brown} \xrightarrow{d^+} \text{black}$$

a. What proportion of the F_2 are colorless?
b. What proportion of the F_2 are brown?

8.5 Using the genetic information given in Problem 8.4, now assume that a^+, b^+, and c^+ act in a pathway as follows:

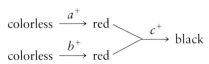

Black can be produced only if both red pigments are present, i.e. c^+ converts the two red pigments together into a black pigment.

a. What proportion of the F_2 are colorless?

b. What proportion of the F_2 are red?

c. What proportion of the F_2 are black?

*8.6a. Three genes on different chromosomes are responsible for three enzymes that catalyze the same reaction in corn:

$$\text{colorless compound} \xrightarrow{a^+, b^+, c^+} \text{red compound}$$

The normal functioning of any one of these genes is sufficient to convert the colorless compound to the red compound. The abnormal functioning of these genes is designated by a, b, and c respectively. A red $a^+/a^+ \, b^+/b^+ \, c^+/c^+$ is crossed by a colorless $a/a \, b/b \, c/c$ to give a red F_1, $a^+/a \, b^+/b \, c^+/c$. The F_1 is selfed. What proportion of the F_2 are colorless?

b. It turns out that another step is involved in the pathway. It is controlled by gene d^+, which assorts independently of a^+, b^+, and c^+.

$$\begin{array}{ccccc} \text{colorless} & \xrightarrow{d^+} & \text{colorless} & \xrightarrow{a^+, b^+, c^+} & \text{red} \\ \text{compound 1} & & \text{compound 2} & & \text{compound} \end{array}$$

The inability to convert colorless 1 to colorless 2 is designated d. A red $a^+/a^+ \, b^+/b^+ \, c^+/c^+ \, d^+/d^+$ is crossed by a colorless $a/a \, b/b \, c/c \, d/d$. The F_1 are all red. The red F_1s are now selfed. What proportion of the F_2 are colorless?

8.7 In hypothetical diploid organisms called mongs, the recessive bw causes a brown eye, and the (unlinked) recessive st causes a scarlet eye. Organisms homozygous for both recessives have white eyes. The genotypes and corresponding phenotypes, then, are as follows:

$bw^+/-$	$st^+/-$	red eye
bw/bw	$st^+/-$	brown
$bw^+/-$	st/st	scarlet
bw/bw	st/st	white

Outline a hypothetical biochemical pathway that would give this type of gene interaction. Demonstrate why each genotype shows its specific phenotype.

*8.8 The Black Riders of Mordor in *Lord of the Rings* ride steeds with eyes of fire. As a geneticist, you are very interested in the inheritance of the fire red eye color. You discover that the eyes contain two types of pigments, brown and red, that are usually bound to core granules in the eye. In wild-type steeds precursors are converted by these granules to the above pigments, but in steeds homozygous for the recessive X-linked gene w (white eye), the granules remain unconverted and a white eye results. The metabolic pathways for the synthesis of the two pigments are shown in Figure 8.A. Each step of the pathway is controlled by a gene: A mutation v gives vermilion eyes; cn gives cinnabar eyes; st gives scarlet eyes; bw gives brown eyes; and se gives black eyes. All the mutations are recessive to their wild-type alleles and all are unlinked. For the following genotypes, show the phenotypes and proportions of steeds that would be obtained in the F_1 of the given matings.

a. $w/w \, bw^+/bw^+ \, st/st \times w^+/Y \, bw/bw \, st^+/st^+$

b. $w^+/w^+ \, se/se \, bw/bw \times w/Y \, se^+/se^+ \, bw^+/bw^+$

c. $w^+/w^+ \, v^+/v^+ \, bw/bw \times w/Y \, v/v \, bw/bw$

d. $w^+/w^+ \, bw^+/bw \, st^+/st \times w/Y \, bw/bw \, st/st$

*8.9 Upon infection of *E. coli* with bacteriophage T4, a series of biochemical pathways result in the formation of

FIGURE 8.A

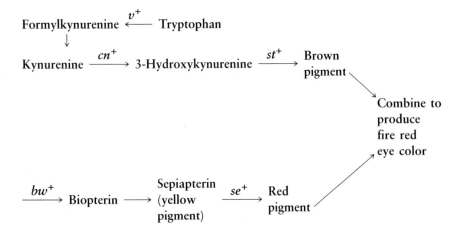

mature progeny phages. The phages are released following lysis of the bacterial host cells. Let us suppose that the following pathway exists:

$$A \xrightarrow[\text{enzyme}]{} B \xrightarrow[\text{enzyme}]{} \text{mature phage}$$

Let us also suppose that we have two temperature-sensitive mutants that involve the two enzymes catalyzing these sequential steps. One of the mutations is cold-sensitive (*cs*) in that no mature phages are produced at 17°C. The other is heat-sensitive (*hs*) in that no mature phages are produced at 42°C. Normal progeny phages are produced when phages carrying either of the mutations infect bacteria at 30°C. However, let us assume that we do not know the sequence of the two mutations. Two models are therefore apparent:

$$(1) \; A \xrightarrow{hs} B \xrightarrow{cs} \text{phage}$$
$$(2) \; A \xrightarrow{cs} B \xrightarrow{hs} \text{phage}$$

Outline how you would experimentally determine which model is the correct model without artificially breaking phage-infected bacteria.

8.10 Four mutant strains of *E. coli* (*a, b, c,* and *d*) all require substance X in order to grow. Four plates were prepared, as shown below. In each case the medium was minimal, with just a trace amount of substance X, to allow a small amount of growth of mutant cells. On plate *a*, cells of mutant strain *a* were spread over the agar, and grew to form a thin lawn. On plate *b* the lawn is composed of mutant *b* cells, and so on. On each plate, cells of the four mutant types were inoculated over the lawn, as indicated by the circles. Dark circles indicates luxuriant growth occurred. That is, this experiment tests whether the bacterial strain spread on the plate can "feed" the four strains inoculated on the plate, allowing them to grow. What do these results show about the relationship of the four mutants to the metabolic pathway leading to substance X?

*8.11 The following table indicates what enzyme is deficient in six different complementing mutants of *E. coli*, none of which can grown on minimal medium. All of them will grow if tryptophan (Trp) is added to the medium.

MUTANT	ENZYME MISSING
trpE	anthranilate synthetase
trpA	tryptophan synthetase
trpF	IGP synthetase
trpB	tryptophan synthetase
trpD	PPA transferase
trpC	PRA isomerase

Each of the plates below shows the results of streaking three of the mutants on minimal medium with just a trace of added tryptophan. Heavy shading indicates regions of heavy growth, indicating where a strain can be fed by the strain streaked next to it on the plate. In what order do the enzymes listed above act in the tryptophan synthetic pathway?

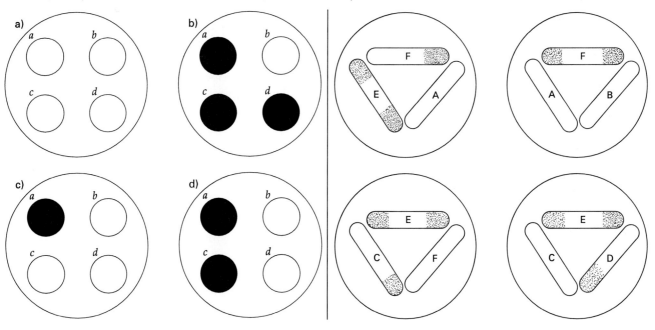

8.12 Referring to the list of mutants and enzymes given in Problem 8.11, explain how it can be that two different complementing mutants (*trpA* and *trpB*) can affect the activity of the same enzyme. How will two such mutants be related in terms of their position within the metabolic pathway?

*8.13 Two mutant strains of *Neurospora* lack the ability to make compound Z. When crossed, the strains usually yield asci of two types: (1) those with spores that are all mutant and (2) those with four wild-type and four mutant spores. The two types occur in a 1:1 ratio.

a. Let *c* represent one mutant, and let *d* represent the other. What are the genotypes of the two mutant strains?

b. Are *c* and *d* linked?

c. Wild-type strains can make compound Z from the constituents of the minimal medium. Mutant *c* can make Z if supplied with X but not if supplied with Y, while mutant *d* can make Z from either X or Y. Construct the simplest linear pathway of the synthesis of Z from the precursors X and Y, and show where the pathway is blocked by mutations *c* and *d*.

8.14 The following growth responses (where + = growth and 0 = no growth) of mutants *1–4* were seen on the related biosynthetic intermediates A, B, C, D, and E. Assume all intermediates are able to enter the cell, that each mutant carries only one mutation, and that all mutants affect steps after B in the pathway.

	GROWTH ON				
MUTANT	A	B	C	D	E
1	+	0	0	0	0
2	0	0	0	+	0
3	0	0	+	0	0
4	0	0	0	+	+

Which of the schemes in the figure fits best with the data with regard to the biosynthetic pathway?

a)

$$B \longrightarrow A \underset{\searrow E}{\overset{\nearrow C}{\longrightarrow} D}$$

b)

$$B \longrightarrow A \longrightarrow D \underset{\searrow E}{\overset{\nearrow C}{}}$$

c)

$$B \longrightarrow A \underset{\searrow C \longrightarrow D}{\overset{\nearrow E}{}}$$

d)

$$B \longrightarrow A \longrightarrow E \underset{\searrow C}{\overset{\longrightarrow D}{}}$$

8.15 Four strains of *Neurospora*, all of which require arginine but have unknown genetic constitution, have the following characteristics. The nutrition and accumulation characteristics are as follows:

	GROWTH ON				
STRAIN	MINIMAL MEDIUM	ORNITHINE	CITRULLINE	ARGININE	ACCUMULATES
1	–	–	+	+	Ornithine
2	–	–	–	+	Citrulline
3	–	–	–	+	Citrulline
4	–	–	–	+	Ornithine

The pairwise complementation tests of the four strains gave the following results (+ = growth on minimal medium and 0 = no growth on minimal medium):

	4	*3*	*2*	*1*
1	0	+	+	0
2	0	0	0	
3	0	0		
4	0			

Crosses among mutants yielded prototrophs in the following percentages:

1 × 2: 25 percent
1 × 3: 25 percent
1 × 4: none detected among 1 million ascospores
2 × 3: 0.002 percent
2 × 4: 0.001 percent
3 × 4: none detected among 1 million ascospores

Analyze the data and answer the following questions.

a. How many distinct mutational sites are represented among these four strains?

b. In this collection of strains, how many types of polypeptide chains (normally found in the wild type) are affected by mutations?

c. Give the genotypes of the four strains, using a consistent and informative set of symbols.

d. Give the map distances between all pairs of linked mutations.

e. Give the percentage of prototrophs that would be expected among ascospores of the following types: (1) strain 1 × wild type; (2) strain 2 × wild type; (3) strain 3 × wild type; (4) strain 4 × wild type.

*8.16 Proteins are (choose the correct answer):

a. Branched chains of nucleotides

b. Linear, folded chains of nucleotides

c. Linear, folded chains of amino acids

d. Invariable enzymes

8.17 A breeder of Irish setters has a particularly valuable show dog that he knows is descended from the famous bitch Rheona Didona, who carried a recessive gene for atrophy of the retina. Before he puts the dog to stud, he must ensure that it is not a carrier for this allele. How should he proceed?

*__8.18__ Suppose you were on a jury to decide the following case. What would you conclude?

The Jones family claims that Baby Jane, given to them at the hospital, does not belong to them but to the Smith family, and that the Smith's baby Joan really belongs to the Jones's family. It is alleged that the two babies were accidentally exchanged soon after birth. The Smiths deny that such an exchange has been made. Blood group determinations show the following results:

> Jones mother, AB
> Jones father, O
> Smith mother, A
> Smith father, O
> Baby Jane, A
> Baby Joan, O

8.19 Referring back to Figure 8.25, what would you expect the phenotype to be in individuals heterozygous for the following two hemoglobin mutations?
a. Hb Norfolk and Hb-S
b. Hb-C and Hb-S

8.20 In evaluating my teacher, my sincere opinion is that:
a. He (she) is a swell person whom I would be glad to have as a brother-in-law/sister-in-law.
b. He (she) is an excellent example of how tough it is when you don't have either genetics or environment going for you.
c. He (she) may have okay DNA to start with, but somehow all the important genes got turned off.
d. He (she) ought to be preserved in tissue culture for the benefit of other generations.

9 THE STRUCTURE OF GENETIC MATERIAL

THE NATURE OF GENETIC MATERIAL: DNA AND RNA
The Discovery of DNA as Genetic Material
The Discovery of RNA as Genetic Material

THE CHEMICAL COMPOSITION OF DNA AND RNA
The Physical Structure of DNA: The Double Helix
Other DNA Structures
DNA in the Cell
Bends in DNA

PRINCIPAL POINTS

~ DNA and RNA are macromolecules composed of smaller building blocks called nucleotides. Each nucleotide consists of a 5-carbon sugar (deoxyribose in DNA, ribose in RNA) to which is attached one of four nitrogenous bases and a phosphate group.

~ In DNA the four nitrogenous bases are adenine, guanine, cytosine, and thymine, while in RNA the four bases are adenine, guanine, cytosine, and uracil.

~ The DNA molecule consists of two polynucleotide chains joined by hydrogen bonds between pairs of bases (A and T, G and C) in a double helix. The diameter of the helix is 2 nm, and there are 10 base pairs in each complete turn (3.4 nm).

~ Three major types of double-helical DNA have been identified by X-ray diffraction analysis, the right-handed A− and B−DNAs, and the left-handed Z−DNA. The common form found in cells is B−DNA. A−DNA probably does not exist in cells. Z−DNA may exist in cells in stretches of DNA that are particularly rich in guanine and cytosine. The functional significance of Z−DNA is unknown.

*I*n the previous eight chapters we have learned much about how traits are inherited from one generation to another, how variations in allele organization along the chromosome arise (through recombination), and how the location of chromosomal genes governing certain traits has been determined (by establishing linkage and by gene mapping). In all our discussions, however, we have taken as givens the existence and the function of the genetic material—genes and the chromosomes in which they are located. In the next several chapters we will explore the molecular structure and function of genetic material—either DNA (deoxyribonucleic acid) or RNA (ribonucleic acid)—and we will examine in detail the mechanisms by which genetic information is transmitted from generation to generation. We will see exactly what the genetic message is, and we will learn how DNA directs the manufacture of proteins, including enzymes.

In this chapter we begin our study of the molecular aspects of genetics by discussing the qualities that the genetic material possesses, the evidence that DNA and RNA are genetic material, and the structure of DNA and RNA molecules.

THE NATURE OF GENETIC MATERIAL: DNA AND RNA

Long before DNA and RNA were proved to carry genetic information, geneticists recognized that specific molecules must fulfill that function. They postulated that the material responsible for inheritable traits would have to have three principal characteristics:

1. It must contain, in a stable form, *the information* for an organism's cell structure, function, development, and reproduction.
2. It must *replicate accurately* so that progeny cells have the same genetic information as the parental cell.
3. It must be capable of *variation*. Without variation (such as through mutation and recombination), organisms would be incapable of change and adaptation, and evolution could not occur.

From experiments in the mid to late 1800s and the early 1900s, scientists suspected that genetic material was composed of protein. Chromosomes were first seen under the microscope in the latter part of the nineteenth century. Chemical analysis over the first 40 years of this century revealed that the nucleus contained a unique molecular constituent, deoxyribonucleic acid (DNA). It was not understood, however, that DNA was the chemical constituent of genes. Rather, DNA was believed to be a molecular framework for a theoretical class of special proteins that carried genetic information. In the 1940s, George Beadle and Edward Tatum obtained evidence from experiments with the orange bread mold, *Neurospora crassa,* that genes function by controlling the synthesis of specific enzymes (this was called the "one gene–one enzyme" hypothesis; see Chapter 8 for a detailed discussion), thereby proving a hypothesis

~ FIGURE 9.1

EM of the bacterium *Diplococcus pneumoniae*. (The cell is in the process of dividing.)

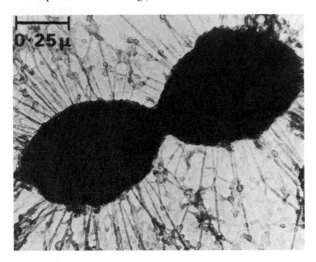

~ FIGURE 9.2

Colonies of smooth (*S*) (left) and rough (*R*) (right) strains of *Diplococcus pneumoniae*.

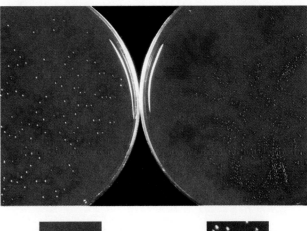

proposed by Archibald Garrod in 1902 (see Chapter 8). Evidence gathered from experiments in the middle of this century showed unequivocally that the genetic material of living organisms consisted of one of two types of nucleic acids: in most organisms it is DNA; in a few it is RNA.

The Discovery of DNA as Genetic Material

One of the first studies, in 1928, that ultimately led to the identification of DNA as the genetic material involved the two strains of bacterium *Diplococcus pneumoniae* (also called pneumococcus) (Figure 9.1). The smooth (*S*) strain is infectious (virulent) and results in the death of the infected animal. Although this was not known in 1928, each *S* type bacterial cell is surrounded by a polysaccharide coat that gives the strain its infectious properties and results in the smooth, shiny appearance of *S* colonies. The rough (*R*) strain is noninfectious (avirulent) because *R* cells lack the polysaccharide coat; colonies of this strain are nonshiny. Figure 9.2 shows the colony phenotypes of the two strains.

The *S* strains can be classified further into *IIS* and *IIIS*, which have clearly defined differences in the chemical composition of the polysaccharide coat. The differences between *IIS* and *IIIS* are genetically determined, as are the *S* and *R* states. In about one *S* cell out of 10^7 there will be a mutational change from *S* to *R*.

Rarely will the *R* colony that results give rise to an *S* cell by a second mutation. The *S* colony that develops has the same capsule type as the *S* strain from which the *R* mutant was originally derived. That is, if the original was *IIS*, then the reversion of the *R* strain that derives from it will also be *IIS*, not *IIIS*.

In 1928 Frederick Griffith injected mice with *R* bacteria, produced by mutation of type *IIS* bacteria (Figure 9.3). Genetically these bacteria lacked the ability to make the polysaccharide coat and, *if* a coat could be made at all, had the genes for making a type II polysaccharide coat. The *R* bacteria did not affect the mice, and after a while the bacteria disappeared from the animals' bloodstreams. Griffith also injected mice with living type *IIIS* bacteria. Those animals died, and living type *IIIS* bacteria could be isolated from their blood. If the type *IIIS* bacteria were heat-killed before injecting them, however, the mice survived. These two experiments showed that the bacteria had to be alive and had to possess the polysaccharide coat in order for them to be infectious.

Lastly, Griffith infected mice with a mixture of living *R* bacteria (derived from *IIS*) and heat-killed type *IIIS* bacteria. In this case the mice died, and living *S* bacteria were present in the blood. These bacteria were all of type *IIIS* and therefore could not have arisen by mutation of the *R* bacteria, since mutation would have produced *IIS* colonies. Rather, Griffith concluded that some *R* bacteria had somehow been transformed into smooth, infectious type *IIIS* cells by interaction with

~ **FIGURE 9.3**

Griffith's transformation experiment. Mice injected with type *IIIS* pneumococcus died, whereas mice injected with either type *IIR* or heat-killed type *IIIS* bacteria survived. When injected with a mixture of living type *IIR* and heat-killed type *IIIS* bacteria, however, the mice died.

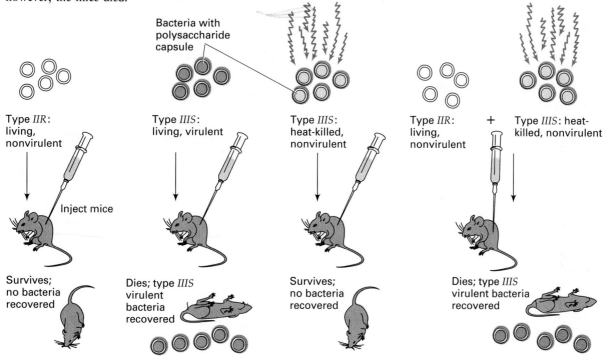

THE NATURE OF THE TRANSFORMING PRINCIPLE.

the dead type *IIIS* cells. The transformed *IIIS* cells retained their infectious properties and capsule type in successive generations, indicating that the transformation was stable. Griffith believed that the unknown agent responsible for the change in genetic material was a protein. He referred to the agent as the *transforming principle*. (See Chapter 7 for a discussion of bacterial transformation.)

From a modern molecular perspective, the data indicate that the genetic material of the virulent bacteria was not destroyed by the heat treatment and that it was released from the heat-killed cells and entered the living avirulent cells where recombination occurred to generate a virulent transformant (Figure 9.4; p. 270). Subsequent research has verified this genetic hypothesis. That is, the *S* gene codes for a key enzyme needed for the synthesis of the carbohydrate-containing capsule. The *R* allele of the gene produces no active enzyme, so the capsule is not made.

THE NATURE OF THE TRANSFORMING PRINCIPLE. In 1944, Oswald T. Avery, Colin M. MacLeod, and Maclyn McCarty showed that the transforming principle was not protein but was instead DNA. They used a system in which the transformation of bacteria from *R* to *S* type was done in the test tube. In their experiments they lysed type *IIIS* cells and separated the lysate into the various cellular macromolecular components such as lipids, polysaccharides, proteins, and the nucleic acids RNA and DNA (Figure 9.5; p. 271). They tested each component to see whether it contained the transforming principle by checking whether or not it could transform living *R* bacteria derived from *IIS*. The nucleic acids (at this point not separated into DNA and RNA) were the only components of the type *IIIS* cells that could transform the *R* cells into *IIIS*.

Avery and his colleagues then used specific **nucleases**—enzymes that degrade nucleic acids—to determine whether DNA or RNA was the transforming prin-

~ FIGURE 9.4

Transformation of a genetic characteristic of a bacterial cell (*Diplococcus pneumoniae*) by addition of heat-killed cells of a genetically different strain. In the example, an *R* cell receives a chromosome fragment containing the *S* gene. Since most *R* cells receive other chromosomal fragments, the efficiency of transformation for a given gene is usually less than 1 percent.

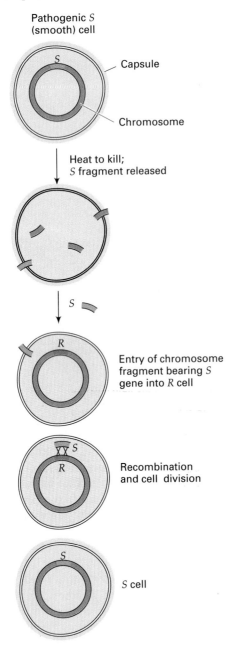

Pathogenic *S* (smooth) cell

Capsule

Chromosome

Heat to kill; *S* fragment released

S

Entry of chromosome fragment bearing *S* gene into *R* cell

Recombination and cell division

S cell

tion resulted. These results strongly suggested that DNA was the genetic material. Although Avery's work was important, it was criticized because the nucleic acids isolated from the bacteria were not completely pure and, in fact, contained some protein contamination.

CONSTANCY OF CHROMOSOMAL DNA AMOUNT. Support for the hypothesis that DNA was the genetic material came from the results of experiments that showed that DNA was located almost exclusively in the nucleus of eukaryotic cells but never in cell locations where chromosomes, the carriers of genetic information, were absent. In addition, Alfred Mirsky and Hans Ris in 1949 demonstrated by cytochemical analysis that the amount of DNA per diploid set of chromosomes was constant for a given organism. Moreover, the diploid DNA amount was shown to be equal to twice the amount of DNA found in the haploid sperm. Different organisms were shown to have different amounts of DNA. We now know that the DNA content of certain cells of an organism can vary depending on the tissue of origin. In general, though, with the exception of spontaneous chromosome loss or breakage, the DNA content is usually a multiple of the DNA content of a zygote cell. For example, the DNA content of human cells may be from two to four times that of a diploid cell, and the DNA content of the root nodule cells of leguminous plants such as the pea is characteristically double that of the cells of the rest of the plant.

Three other aspects of DNA strongly supported its role as the genetic material. (1) DNA is metabolically stable, that is, it is not rapidly synthesized and degraded like many other cellular molecules. (2) The amount of DNA per cell is approximately related to the complexity of the organism. Thus cells of higher organisms contain considerably more DNA than do bacterial cells. (3) The amount of DNA in gametes is approximately one-half that found in 2N cells of the same species.

THE HERSHEY AND CHASE BACTERIOPHAGE EXPERIMENTS. In 1953, Alfred D. Hershey and Martha Chase reported experimental results conclusively showing that DNA is genetic material. They were studying the replication of bacteriophage T2 to see which phage components were needed to complete the life cycle. (See Figure 7.14 for the life cycle of this bacteriophage.) At the time it was known that T2 infected *Escherichia coli* by first attaching itself with its tail fibers, then injecting into the bacterium some genetic material that somehow controlled the production of progeny T2 particles. Ultimately, the bacterium lysed, releasing 100–200 phages that could infect other bacteria. However, the nature of T2's genetic material was not known.

ciple (Figure 9.6; p. 272). When they treated the nucleic acids with **ribonuclease (RNase)**, which degrades RNA but not DNA, the transforming activity was still present. However, when they used **deoxyribonuclease (DNase)**, which degrades only DNA, no transforma-

~ FIGURE 9.5

Chemical method used in the original isolation of a transforming agent.

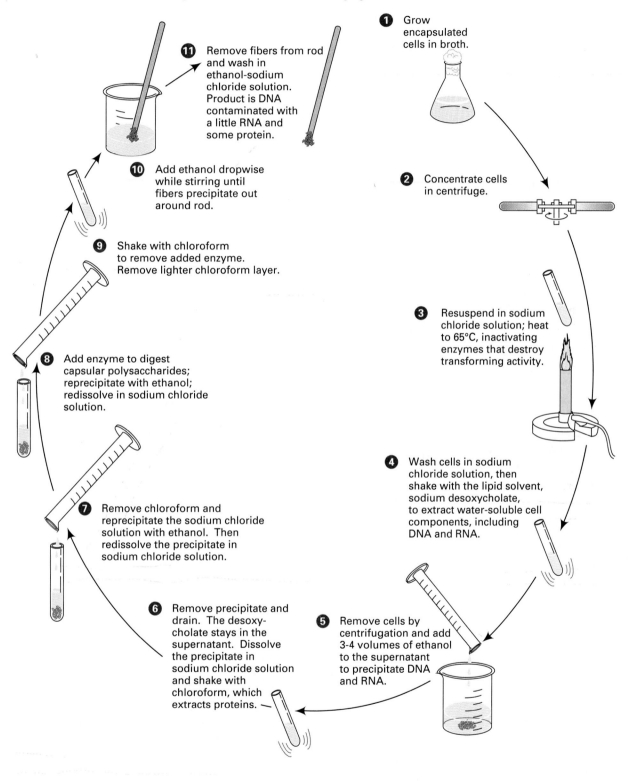

1 Grow encapsulated cells in broth.

2 Concentrate cells in centrifuge.

3 Resuspend in sodium chloride solution; heat to 65°C, inactivating enzymes that destroy transforming activity.

4 Wash cells in sodium chloride solution, then shake with the lipid solvent, sodium desoxycholate, to extract water-soluble cell components, including DNA and RNA.

5 Remove cells by centrifugation and add 3-4 volumes of ethanol to the supernatant to precipitate DNA and RNA.

6 Remove precipitate and drain. The desoxycholate stays in the supernatant. Dissolve the precipitate in sodium chloride solution and shake with chloroform, which extracts proteins.

7 Remove chloroform and reprecipitate the sodium chloride solution with ethanol. Then redissolve the precipitate in sodium chloride solution.

8 Add enzyme to digest capsular polysaccharides; reprecipitate with ethanol; redissolve in sodium chloride solution.

9 Shake with chloroform to remove added enzyme. Remove lighter chloroform layer.

10 Add ethanol dropwise while stirring until fibers precipitate out around rod.

11 Remove fibers from rod and wash in ethanol-sodium chloride solution. Product is DNA contaminated with a little RNA and some protein.

~ **FIGURE 9.6**

Experiment that showed DNA, and not RNA, was the transforming principle. When the nucleic acid mixture of DNA and RNA was treated with ribonuclease (RNase), transformants still resulted. However, when the DNA and RNA mixture was treated with deoxyribonuclease (DNase), no transformants resulted.

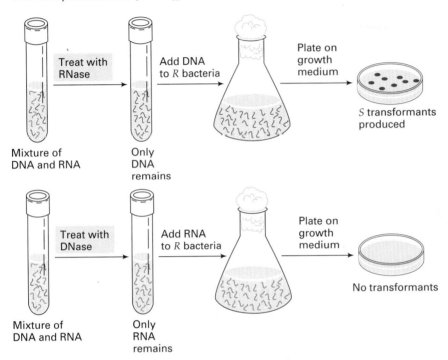

Hershey and Chase grew cells of *E. coli* in media containing either a radioactive isotope of phosphorus (^{32}P) or a radioactive isotope of sulfur (^{35}S) (Figure 9.7a). They used these isotopes because DNA contains phosphorus but no sulfur, and protein contains sulfur but no phosphorus. They infected the bacteria with T2 and collected the progeny phages that were produced. At this point, Hershey and Chase had two batches of T2: One had the proteins radioactively labeled with ^{35}S, and the other had the DNA labeled with ^{32}P.

Since they knew that each T2 phage consisted of approximately half DNA and half protein, they knew that one of these two classes of molecules must be the genetic material. To determine which class it was, they next infected unlabeled *E. coli* with the two types of radioactively labeled T2 and obtained the interesting results shown in Figure 9.7b. When the infecting phage was ^{32}P-labeled, most of the radioactivity could be found within the bacteria soon after infection. Very little could be found in protein parts of the phage (the phage ghosts) released from the cell surface by mixing the cells in a kitchen blender. After lysis, some of the ^{32}P was found in the progeny phages. In contrast, after infecting *E. coli* with ^{35}S-labeled T2, virtually none of the radioactivity appeared within the cell, and none was found in the progeny phage particles. Most of the

radioactivity could be found in the phage ghosts released after treating the cultures in the kitchen blender.

Since genes serve as the blueprint for making the progeny virus particles, it was also presumed that the blueprint must get into the bacterial cell in order for new phage particles to be built. Therefore, since *it was DNA and not protein that entered the cell*, as evidenced by the presence of ^{32}P and the absence of ^{35}S, Hershey and Chase reasoned that *DNA must be the material responsible for the function and reproduction of phage T2*. The protein, they hypothesized, provided a structural framework to contain the DNA and the specialized structures required to inject the DNA into the bacterial cell.

The Discovery of RNA as Genetic Material

Most of the organisms discussed in this book (such as humans, *Drosophila*, yeast, *E. coli*, and phage T2) have DNA as their genetic material. However, some bacterial viruses and some animal and plant viruses have RNA as their genetic material. No prokaryotic or eukaryotic organism has RNA as its genetic material. The following classical experiment showed that RNA is the genetic material of the *tobacco mosaic virus* (TMV).

~ FIGURE 9.7

Hershey-Chase experiment: (a) the production of T2 phages either with (1) ^{32}P-labeled DNA or with (2) ^{35}S-labeled protein; (b) the experimental evidence showing that DNA is the genetic material in T2: (1) The ^{32}P is found within the bacteria and appears in progeny phages, (2) while the ^{35}S is *not* found within the bacteria and is released with the phage ghosts.

a) Preparation of radioactively labeled T2 bacteriophage

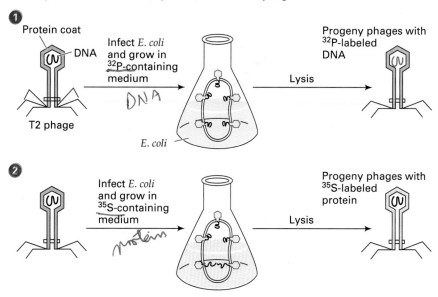

b) Experiment that showed DNA to be the genetic material of T2

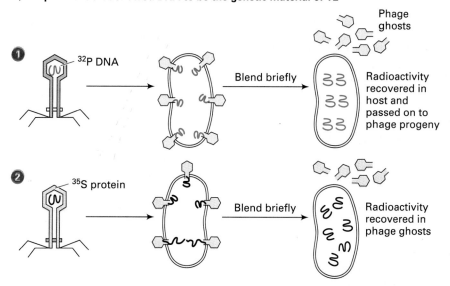

Like phage T2, TMV contains two chemical components, in this case RNA and protein in a spiral (helical) configuration. The protein surrounds the RNA core, protects it from attack by nucleases (Figure 9.8; p. 274), and along with the RNA core, functions in the infection of the plant cells. Many varieties of TMV are known; they differ in the plants they infect and in the extent to which they affect these plants.

In 1956, A. Gierer and G. Schramm showed that when tobacco plants were inoculated with the purified RNA of TMV (i.e., RNA without the protein coat), they developed typical virus-induced lesions. This result indicated strongly that RNA was the genetic material of TMV. This conclusion was supported by the observation that no lesions were produced when the RNA had been degraded by treatment with ribonuclease and then injected into the plant.

In an experiment in 1957, Heinz Fraenkel-Conrat and B. Singer confirmed Gierer and Schramm's conclusions. They isolated the RNA and protein components

~ FIGURE 9.8

Typical tobacco mosaic virus (TMV) particle. The helical RNA core is surrounded by a helical arrangement of protein subunits.

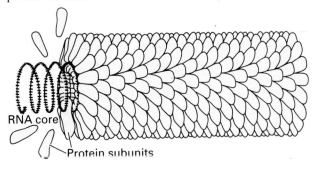

RNA core

Protein subunits

of two distinct TMV strains and reconstituted the RNA of one type with the protein of the other type, and vice versa (Figure 9.9). They then infected the tobacco leaves with the two hybrid viruses. The progeny viruses isolated from the resulting lesions were of the type spec-

ified by the RNA and not by the protein. These results conclusively showed that RNA is the genetic material of TMV.

KEYNOTE

Genetic material must contain all the information for the cell structure and function of an organism and must replicate accurately so that progeny cells have the same genetic information as the parental cell. In addition, genetic material must be capable of variation, one of the bases for evolutionary change. A series of experiments proved that the genetic material of organisms consists of one of two types of nucleic acids, DNA or RNA. Of the two, DNA is most common; only certain viruses have RNA as their genetic material. _____

~ FIGURE 9.9

Demonstration that RNA is the genetic material in tobacco mosaic virus (TMV). Hybrid particles were made from the protein subunits of one TMV strain and the RNA of a different TMV strain. Tobacco leaves were infected with the reconstituted hybrid viruses, and the progeny viruses isolated from the resulting leaf lesions were analyzed. The progeny viruses always had protein subunits specified by the RNA component; that is, the character of the protein coat cannot be transmitted from the hybrid particles in their progeny.

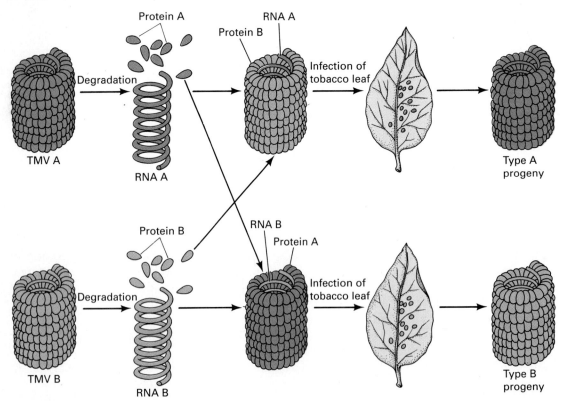

THE CHEMICAL COMPOSITION OF DNA AND RNA

Both DNA and RNA are **macromolecules,** which means that they have a molecular weight of at least a few thousand daltons (1 dalton is equivalent to a twelfth of the mass of the carbon 12 atom, or 1.67×10^{-24} g). In this respect they differ markedly from many other molecules important to cell function, such as sugars and amino acids, which weigh from a hundred to a few hundred daltons.

Both DNA and RNA are polymeric molecules made up of monomeric units called **nucleotides.** Each nucleotide consists of three distinct parts: (1) a **pentose** (5-carbon) **sugar,** (2) a **nitrogenous** (nitrogen-containing) **base,** and (3) a **phosphate group.** Because they can be isolated from nuclei, and because they are acidic, these macromolecules are called nucleic acids.

For RNA the pentose sugar is **ribose,** and for DNA the sugar is **deoxyribose** (Figure 9.10). The two sugars differ by the chemical groups attached to the 2′ carbon: a hydroxyl group (OH) in ribose, a hydrogen group (H) in deoxyribose. (The carbon atoms in the pentose sugars are numbered 1′ to 5′ to distinguish them from the carbon and nitrogen atoms in the rings of the bases.)

The nitrogenous bases fall into two classes, the **purines** and the **pyrimidines.** In DNA the purines are **adenine** (A) and **guanine** (G), and the pyrimidines are **thymine** (T) and **cytosine** (C). The RNA molecule also contains adenine, guanine, and cytosine, but thymine is replaced by **uracil** (U). The chemical structures of these five bases are given in Figure 9.11. Note that thymine contains a methyl group (CH_3) not found in uracil.

In DNA and RNA, bases are always attached to the 1′ carbon of the pentose sugar by a covalent bond. The purine bases are bonded at the 9 nitrogen, while the pyrimidines bond at the 1 nitrogen. The phosphate group (PO_4^{2-}) is attached to the 5′ carbon of the sugar in both DNA and RNA. Examples of a DNA nucleotide (a **deoxyribonucleotide**) and an RNA nucleotide (a **ribonucleotide**) are shown in Figure 9.12a (p. 276). For future discussions, remember that a *nucleotide,* the basic building block of the DNA and RNA molecules,

~ FIGURE 9.10

Structures of ribose and deoxyribose, the pentose sugars of RNA and DNA, respectively. The difference between the two sugars is highlighted.

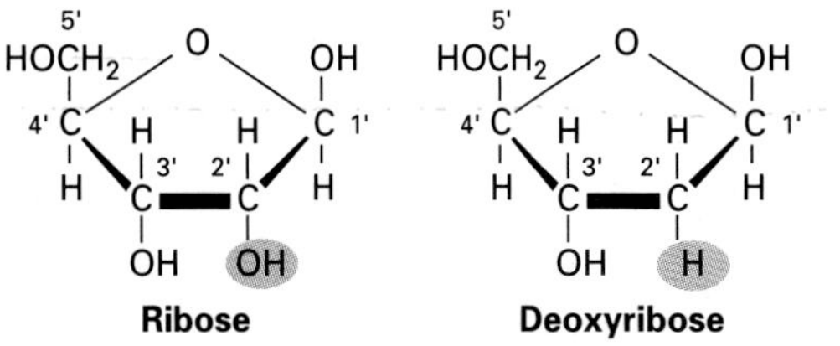

~ FIGURE 9.11

Structures of the nitrogenous bases in DNA and RNA. The parent compounds are purine (top), and pyrimidine (bottom). Differences between the bases are highlighted.

DNA
purines
adenine (A)
guanine (G)
pyrimidine
thymine (T)
cytosine (C)

RNA
purines
A
G
pyrimidine
C
Uracil

thymine is replaced by uracil

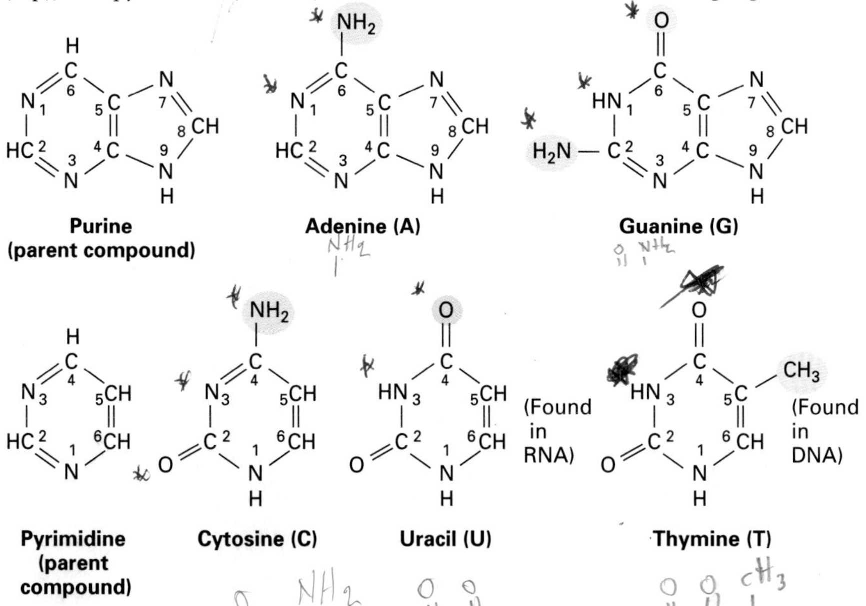

~ FIGURE 9.12

Chemical structures of DNA and RNA: (a) Basic structures of DNA and RNA nucleosides (sugar plus base) and nucleotides (sugar plus base plus phosphate group), the basic building blocks of DNA and RNA molecules. Here, the phosphate groups are orange, the sugars are red, and the bases are brown. (b) A segment of a polynucleotide chain, in this case a single strand of DNA. The deoxyribose sugars are linked by phosphodiester bonds between the 3' carbon of one sugar and the 5' carbon of the next sugar.

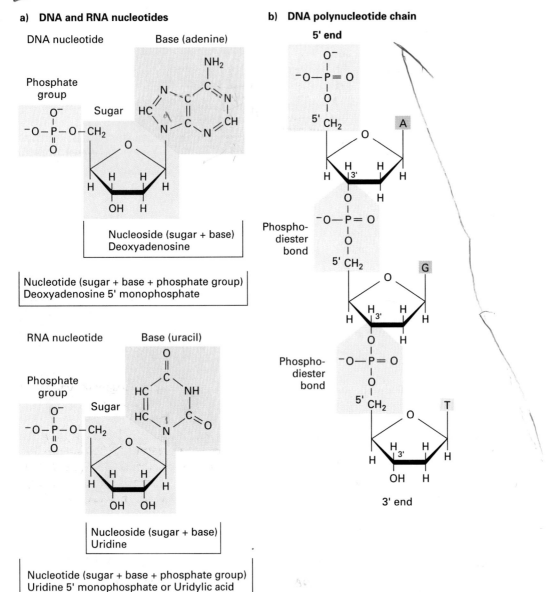

a) DNA and RNA nucleotides

DNA nucleotide   Base (adenine)

Phosphate group   Sugar

Nucleoside (sugar + base)
Deoxyadenosine

Nucleotide (sugar + base + phosphate group)
Deoxyadenosine 5' monophosphate

RNA nucleotide   Base (uracil)

Phosphate group   Sugar

Nucleoside (sugar + base)
Uridine

Nucleotide (sugar + base + phosphate group)
Uridine 5' monophosphate or Uridylic acid

b) DNA polynucleotide chain

5' end

Phospho-diester bond

Phospho-diester bond

3' end

consists of the sugar, a base, and a phosphate group. The sugar plus base only is called a nucleoside, so a nucleotide is also called a **nucleoside phosphate.** A complete listing of the names for the bases, nucleosides, and nucleotides is presented in Table 9.1. The four deoxyribonucleotide subunits of DNA are deoxyadenosine 5'-monophosphate (dAMP), deoxyguanosine 5'-monophosphate (dGMP), deoxycytidine 5'-monophosphate (dCMP), and deoxythymidine 5'-monophosphate (dTMP). The four ribonucleotide subunits of RNA are adenosine 5'-monophosphate (AMP), guanosine 5'-monophosphate (GMP), cytidine 5'-monophosphate (CMP), and uridine 5'-monophosphate (UMP).

~ TABLE 9.1

Names of the Base, Nucleoside, and Nucleotide Components Found in DNA and RNA

		BASE: PURINES (PU)		BASE: PYRIMIDINES (PY)		
		ADENINE (A)	GUANINE (G)	CYTOSINE (C)	THYMINE (T) (DEOXYRIBOSE ONLY)	URACIL (U) (RIBOSE ONLY)
DNA	Nucleoside: deoxyribose + base	Deoxy- adenosine (dA)	Deoxy- guanosine (dG)	Deoxycytidine (dC)	Thymidine (dT)	
	Nucleotide: deoxyribose + base + phos- phate group	Deoxyadenylic acid or deoxy- adenosine monophos- phate (dAMP)	Deoxyguanylic acid or deoxy- guanosine monophos- phate (dGMP)	Deoxycytidylic acid or deoxy- cytidine mono- phosphate (dCMP)	Thymidylic acid or thymi- dine mono- phosphate (TMP)	
RNA	Nucleoside: ribose + base	Adenosine (A)	Guanosine (G)	Cytidine (C)		Uridine (U)
	Nucleotide: ribose + base + phosphate group	Adenylic acid or adenosine monophos- phate (AMP)	Guanylic acid or guanosine monophos- phate (GMP)	Cytidylic acid or cytidine monophos- phate (CMP)		Uridylic acid or uridine monophos- phate (UMP)

Because DNA contains the pentose sugar deoxyribose and the pyrimidine thymine, while RNA contains the pentose sugar ribose and the pyrimidine uracil, these two nucleic acids have different chemical and biological properties. The presence of the 2' OH, for example, enables the RNA to be degraded with alkali, while DNA is resistant to that treatment. Also, the cellular enzymes that catalyze nucleic acid synthesis (polymerases) and nucleic acid degradation (nucleases) are usually DNA-specific or RNA-specific. The differences between the two molecules permit them to be separated and purified relatively simply for study in the laboratory.

To form polynucleotides of either DNA or RNA, nucleotides are linked together by a covalent bond between the phosphate group (which is attached to the 5' carbon of the sugar ring) of one nucleotide and the 3' carbon of the pentose sugar of another nucleotide. These 5'-3' phosphate linkages are called **phosphodiester bonds**. A short polynucleotide chain is diagrammed in Figure 9.12b. The phosphodiester bonds are relatively strong, and as a consequence, the repeated sugar-phosphate-sugar-phosphate backbone of DNA and RNA is a stable structure.

To understand how a polynucleotide chain is synthesized (which we will study in a later chapter), we must be aware of one more feature of the chain: It has polarity; that is, it has a 5' carbon (usually phosphorylated) at one end and a 3' carbon (usually hydroxylated) at the other end, as shown in 9.12b.

KEYNOTE

DNA and RNA occur in nature as macromolecules composed of smaller building blocks called nucleotides. Each nucleotide consists of a 5-carbon sugar (deoxyribose in DNA, ribose in RNA) to which is attached a phosphate group and one of four nitrogenous bases, adenine, guanine, cytosine, and thymine (in DNA) or adenine, guanine, cytosine and uracil (in RNA).

~ FIGURE 9.13

James Watson (left) and Francis Crick (right) with the 1953 model of DNA structure.

The Physical Structure of DNA: The Double Helix

In 1953 James D. Watson and Francis H. C. Crick (Figure 9.13) published a paper in which they proposed a model for the physical and chemical structure of the DNA molecule. According to their model, most DNA consists of two polynucleotide chains wound around each other in a right-handed (clockwise) helix. In generating their model, Watson and Crick used three main pieces of evidence:

1. The DNA molecule was known to be composed of bases, sugars, and phosphate groups linked together as a **polynucleotide** (deoxyribonucleotide) chain.
2. By chemical treatment Erwin Chargaff had hydrolyzed the DNA of a number of organisms and had quantified the purines and pyrimidines released.

His studies showed that in all the DNAs the amount of the purines was equal to the amount of the pyrimidines. More importantly, the amount of adenine (A) was equal to that of thymine (T), and the amount of guanine (G) was equal to that of cytosine (C). These equivalences have become known as Chargaff's rules. In comparisons of DNAs from different organisms, the A/T and G/C ratios are always the same, although the (A + T)/(G + C) ratio (typically presented as %GC) varies (see Table 9.2).

3. Rosalind Franklin, working with Maurice H. F. Wilkins, studied isolated fibers of DNA by using the X-ray diffraction technique, a procedure in which a beam of parallel X rays is directed on a regular, repeating array of atoms. The beam is diffracted ("broken up") by the atoms in a pattern that is characteristic of the atomic weight and the atom's spatial arrangement. The diffracted X rays are recorded on a photographic plate. By analyzing the photograph, Franklin could obtain information about the molecule's atomic structure. The analysis of X-ray diffraction patterns is extremely complicated. As a result, given diffraction patterns can usually be interpreted in more than one way, and models built of the analyzed molecules may not be accurate. Moreover, since the experiments usually use molecules in a crystalline or fiber formation, the structures deduced may not precisely reflect the form of the molecules in the cell.

The diffraction patterns obtained by directing X rays along the length of drawn-out fibers of DNA indicated that the molecule is organized in a highly ordered, helical structure. An example of DNA's X-ray diffraction pattern and the method by which it was obtained are illustrated in Figure 9.14. Franklin interpreted these kinds of data to mean that DNA was a helical structure which had two distinctive regularities of 0.34 nm and 3.4 nm along the axis of the molecule. (1 nanometer [nm] = $1/10^9$ meter = 10 Ångstrom units [Å].)

~ TABLE 9.2

Base Compositions of DNAs from Various Organisms

DNA ORIGIN	PERCENTAGE OF BASE IN DNA				RATIOS		
	A	T	G	C	A/T	G/C	(A + T)/(G + C)
Human (sperm)	31.0	31.5	19.1	18.4	0.98	1.03	1.67
Corn (*Zea mays*)	25.6	25.3	24.5	24.6	1.01	1.00	1.04
Drosophila	27.3	27.6	22.5	22.5	0.99	1.00	1.22
Euglena nucleus	22.6	24.4	27.7	25.8	0.93	1.07	0.88
Escherichia coli	26.1	23.9	24.9	25.1	1.09	0.99	1.00

~ FIGURE 9.14

X-ray diffraction analysis of DNA: The X-ray diffraction pattern of DNA that Watson and Crick used in developing their double-helix model. The dark areas that form an X shape in the center of the photograph indicate the helical nature of DNA. The dark crescents at the top and bottom of the photograph indicate the 0.34-nm distance between the base pairs.

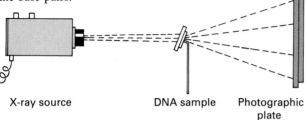

X-ray source   DNA sample   Photographic plate

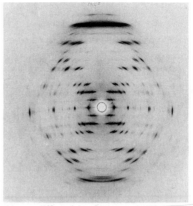

X-ray diffraction pattern

Watson and Crick considered all the evidence just described and began to build three-dimensional models for the structure of DNA. The model they devised, which fit all the known data on the composition of the DNA molecule, is the now-famous double-helix model for DNA. Figure 9.15a shows a three-dimensional model of the DNA molecule, and Figure 9.15b is a diagram of the DNA molecule, showing the arrangement of the sugar-phosphate backbone and base pairs in a stylized way.

~ FIGURE 9.15

Molecular structure of DNA: (a) Three-dimensional molecular model of DNA as prepared by Watson and Crick; (b) Stylized representation of the DNA double helix; (c) Schematic showing the major and minor grooves in double-helical DNA.

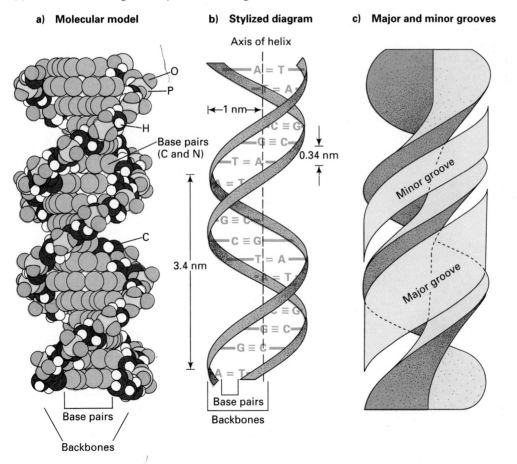

a) Molecular model   b) Stylized diagram   c) Major and minor grooves

The double-helical model of DNA proposed by Watson and Crick has the following main features:

1. The DNA molecule consists of two polynucleotide chains wound around each other in a right-handed double helix; that is, viewed on end, the two strands wind around each other in a clockwise (right-handed) fashion.
2. The diameter of the helix is 2 nm.
3. The two chains are *antiparallel* (= show *opposite polarity*); that is, the two strands are oriented in opposite directions with one strand oriented in the 5′ to 3′ way, while the other strand is oriented 3′ to 5′.
4. The sugar-phosphate backbones are on the outsides of the double helix, while the bases are oriented toward the central axis (see Figure 9.15). The bases of both chains are flat structures oriented perpendicularly to the long axis of the DNA; that

is, the bases are stacked like pennies on top of one another (except for the "twist" of the helix).

5. The bases of the opposite strands are bonded together by relatively weak hydrogen bonds. The only specific pairings observed are A with T (two hydrogen bonds) and G with C (three hydrogen bonds) (Figure 9.16). The weak hydrogen bonds make it relatively easy to separate the two strands of the DNA, for example, by heating. The A–T and G–C base pairs are the only ones that can fit the physical dimensions of the helical model, and they are totally in accord with Chargaff's rules. The specific A–T and G–C pairs are called *complementary base pairs,* so the nucleotide sequence in one strand dictates the nucleotide sequence of the other. For instance, if one chain has the sequence 5′-TATTCCGA-3′ on one chain, the opposite, antiparallel chain must bear the sequence 3′-ATAAGGCT-5′.
6. The base pairs are 0.34 nm apart in the DNA helix. A complete (360°) turn of the helix takes 3.4 nm; therefore, there are 10 base pairs per turn. Each base pair, then, is twisted 36° clockwise with respect to the previous base pair.
7. The two bonds that attach a base pair to its sugar rings are not directly opposite each other. Because of this, the two sugar-phosphate backbones of the double helix are not equally spaced along the helical axis, resulting in grooves between the backbones. These grooves are not of equal size, so they are called the *major groove* and the *minor groove* (Figure 9.15c).

~ FIGURE 9.16

Structures of the complementary base pairs found in DNA. In both cases a purine pairs with a pyrimidine: (a) The adenine-thymine bases, which pair through two hydrogen bonds; (b) The guanine-cytosine bases, which pair through three hydrogen bonds.

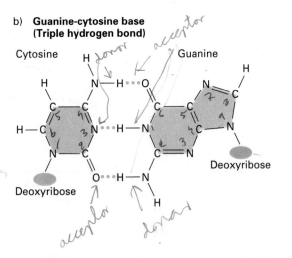

a) **Adenine-thymine base (Double hydrogen bond)**

b) **Guanine-cytosine base (Triple hydrogen bond)**

KEYNOTE

The DNA molecule usually consists of two polynucleotide chains joined by hydrogen bonds between pairs of bases (A and T, G and C) in a double helix. The diameter of the helix is 2 nm, and there are 10 base pairs in each complete turn (3.4 nm). The double-helix model was proposed by Watson and Crick from chemical and physical analyses of DNA.

Other DNA Structures

A more thorough analysis of the early X-ray diffraction patterns of DNA fibers revealed that DNA can exist in different forms, depending on the conditions. When humidity is relatively high, DNA is in what is called the B form (= the double-helical form deduced by Watson

~ FIGURE 9.17

Schematic of A—DNA (left) and B—DNA (right) based on analyses of DNA fibers. The sugar-phosphate backbones of the double helix are represented as ribbons and the runglike base pairs connecting them as planks. In A—DNA, the base pairs are tilted and are pulled away from the axis of the double helix. In B—DNA, the base pairs sit astride the helix axis and are perpendicular to it. (© Irving Geiss.)

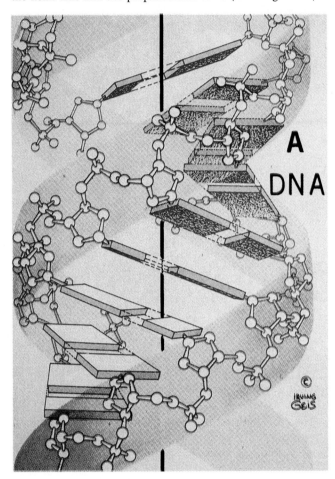

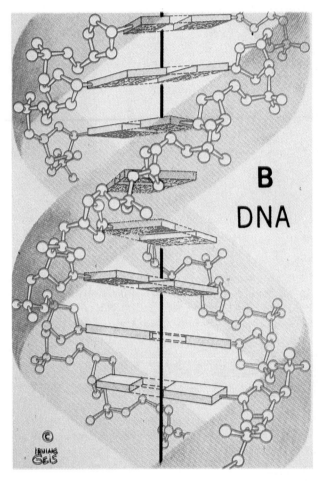

and Crick). When the humidity is relatively low, DNA is found in the A form. Figure 9.17 shows schematic "ball-and-stick" molecular drawings of A—DNA and B—DNA based on analyses of DNA fibers. Both of these DNA forms involve right-handed helices. Note that in A—DNA the base pairs are tilted and pulled away from the axis of the double helix. In contrast, in B—DNA the helix axis passes through the base pairs which are, themselves, oriented perpendicular to that axis.

An inherent problem in analyzing DNA fibers by X-ray diffraction is that the DNA strands are generally not regularly ordered around and along the fiber axis. Thus the DNA models generated from the resulting data show averaged helical structures. Moreover, the data cannot reveal any localized structural variations that might result from particular base sequences. This limitation has been overcome in recent years because methods have been developed to synthesize short DNA molecules (called **oligomers** [oligo = few]) of defined sequences. The pure DNA oligomers can be crystallized and analyzed by single-crystal X-ray diffraction. These more modern approaches have made possible a more detailed analysis of A—DNA and B—DNA. In addition, other forms of DNA have been identified in the lab, the most intriguing of which is Z—DNA because of its totally unexpected structure and because of the evidence for its (rare) existence in cells. Z—DNA has a left-handed helix and a zig-zag sugar-phosphate backbone. (The latter property gave this DNA form its "Z"

~ FIGURE 9.18

Space-filling models of (a) A−DNA, (b) B−DNA, and (c) Z−DNA.

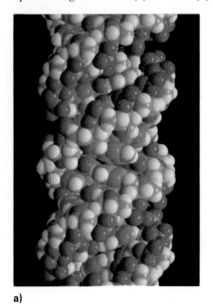

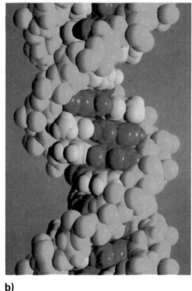

a)   b)   c)

designation.) Figure 9.18 shows space-filling models, and Figure 9.19 shows perspective drawings of A−, B−, and Z−DNA, based on single-crystal studies. Some of the key structural features of each DNA type as deduced from the single-crystal studies are given in Table 9.3.

FEATURES OF A−DNA and B−DNA.

A−DNA and B−DNA are right-handed double helices with 10.9 and 10.0 base pairs per 360-degree turn of the helix, respectively. These values correspond closely with the data derived from DNA fiber studies. The bases are inclined away from the perpendicular from the helix axis (which would be 0 degrees) by 13 degrees in A−DNA and 2 degrees in B−DNA.

The single crystal studies confirmed the fiber studies' conclusion that the A−DNA double helix is short and broad with a deep major groove and a broad and shallow minor groove. The B−DNA double helix is thinner and taller for the same number of base pairs than A−DNA, with a wide major groove and a narrow minor groove; both grooves are of similar depths.

~ TABLE 9.3

Properties of A−DNA, B−DNA, and Z−DNA

PROPERTY	A−DNA	B−DNA	Z−DNA
Helix direction	Right-handed	Right-handed	Left-handed
Base pairs per helix turn	10.9	10.0	12.0
Overall morphology	Short and broad	Longer and thinner	Elongated and thin
Major groove	Extremely narrow and very deep	Wide and of intermediate depth	Flattened out on helix surface
Minor groove	Very broad and shallow	Narrow and of intermediate depth	Extremely narrow and very deep
Helix axis location	Major groove	Through base pairs	Minor groove
Base inclination from helix axis (degrees)	13.0	2.0	8.8
Sugar-base bond conformation*	*anti*	*anti*	*anti* at C; *syn* at G

*The words *anti* and *syn* refer to the two conformations of the C−N bond connecting each base to its sugar ring.

~ FIGURE 9.19

Comparison of (a) A−, (b) B−, and (c) Z−DNA. Short segments of each DNA type are shown. The bases are red and the backbones are in blue. Note the zig-zag course of the sugar-phosphate backbone chain in the Z−DNA. (© Irving Geiss.)

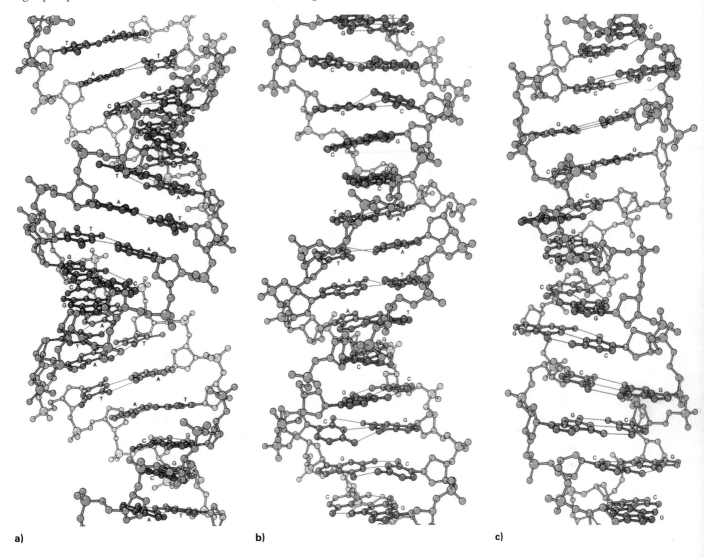

a)   b)   c)

FEATURES OF Z−DNA.

Z−DNA was discovered when single crystals of an alternating cytosine-guanine DNA oligomer were subjected to X-ray diffraction analysis. DNA oligomers containing A−T base pairs do not form Z−DNA. Z−DNA is a left-handed helix generated because of a different bond conformation (a glycosidic bond) between one of its bases and the deoxyribose sugar than is found in A− and B−DNA. Figure 9.20 (p. 284) shows the two conformations of the C−N glycosidic bond connecting each base to its sugar. The *anti* conformation occurs for all bases in A− and B−DNA and at cytosines in Z−DNA. The *syn* conformation is found at the guanines in Z−DNA. The alternation of a purine (G) and a pyrimidine (C) along a Z−DNA double strand produces the *anti-syn* alternation that gives the backbone of Z−DNA a zigzag appearance (see Figures 9.18 and 9.19).

In Z−DNA there are 12.0 base pairs per complete helical turn. The bases are inclined away from the perpendicular with the helix axis by 8.8 degrees. The base pairs have smaller mean propeller twist values than A− or B−DNA, that is, 4.4 degrees. The Z−DNA helix is thin and elongated with a deep, minor groove. The major groove is pushed to the surface of the helix so that it is not really evident as a distinct groove. The helix axis passes through the minor groove in Z−DNA, through the major groove in A−DNA, and through the base pairs in B−DNA (see Figure 9.20).

~ FIGURE 9.20

Two conformations of the C−N bond connecting each base to its sugar ring are shown. (right) The *anti* conformation appears in all A− and B−DNA and at the cytosines in Z−DNA. (left) The *syn* conformation appears at the guanines in Z−DNA. The three points labeled A, B, and Z indicate for each type of helix the location of the helix axis with respect to a base pair. The axis is in the major groove in A−DNA, passes through the base pair in B−DNA, and is in the minor groove in Z−DNA.

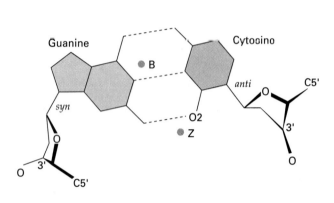

DNA in the Cell

In solution, DNA usually is found in the B form and hence B−DNA is the form typically found in cells. Since A−DNA is found only under relatively low humidity conditions, it is unlikely that any lengthy sections of A−DNA exist within cells. Whether Z−DNA truly exists within cells is currently a matter of significant debate among scientists. There is evidence that certain DNA segments rich in guanine and cytosine may exist in the cell as Z−DNA. However, it is argued by critics that the Z−DNA is an artifact of the analytical procedures rather than being a natural feature of the chromosome. Since the structure of the Z−DNA is strikingly different from B−DNA, it may be speculated that such Z−DNA regions could be an as yet undefined regulatory signal. Along these lines, there is evidence for the existence of proteins that bind specifically to possible Z−DNA regions in the chromosome of SV40 (an animal virus). The binding region is positioned at an important viral control region. However, a direct role for the Z−DNA binding proteins in gene regulation has not been demonstrated in SV40. Moreover, Z−DNA binding proteins also bind to B−DNA, although with less affinity.

Interestingly, in the nematode *Caenorhabditis elegans* several proteins have been found that bind Z−DNA 1,000 times better than B−DNA (a greater affinity than the SV40 Z−DNA binding proteins). The proteins were identified as yolk proteins and, evidently the reason these proteins bind Z−DNA is that Z−DNA mimics the shape and charge distribution of the phospholipids to which the proteins normally bind. These observations illustrate how cautiously conclusions must be made from biochemical experiments; had the researchers in question not been very familiar with yolk proteins of *C. elegans*, they may have concluded that they had definitive proof for Z−DNA binding proteins. Clearly, more definitive data are needed before we have an understanding of what role, if any, Z−DNA has in cellular function.

𝒦EYNOTE

X-ray diffraction analysis of single crystals of DNA oligomers of known sequence has produced detailed molecular models of three DNA types: A−DNA, B−DNA, and Z−DNA. A−DNA and B−DNA are both right-handed double helices while Z−DNA is a left-handed double helix. The number of base pairs per helical turn are 10.9, 10.0, and 12.0 for A−, B−, and Z−DNA, respectively. A−DNA is a short and broad molecule, B−DNA is a longer and thinner molecule, and Z−DNA is elongated and thin. The three DNA forms differ in a number of molecular attributes, including dimensions of the major and minor grooves and location of the helix axis. B−DNA is the form found predominantly in cells. Z−DNA has also been found in the chromosomes of some organisms but its role in cellular function, if any, is not clear. Since A−DNA forms only under conditions of low humidity, it appears unlikely that this DNA type exists in a stable form in cells.

Bends in DNA

Even though the DNA molecule is drawn in textbooks as a very straight, helical molecule, implying that it is a rigid rodlike molecule, this is actually not so. In solution, B−DNA is quite a flexible molecule, undergoing shape changes as it responds to the local environmental ionic and temperature conditions. It is possible that particular bends in DNA might favor the binding of specific protein molecules, and this could be part of the gene regulation process. While specific information on this point is only just coming to light, it is known that the ability of B—DNA to bend slightly is enhanced by the presence of specific base-pair sequences. The sequence CAAAAAT, for example, particularly if it is repeated three or four times separated by 10 or so base pairs, has been shown to cause DNA to bend.

SUMMARY

In this chapter we have learned that DNA or RNA can be the genetic material. All prokaryotic and eukaryotic organisms, and most viruses have DNA as their genetic material, while some viruses have RNA as their genetic material. The form of the genetic material varies among organisms and their viruses. In prokaryotic and eukaryotic organisms, the DNA is always double-stranded, while in viruses the genetic material may be double- or single-stranded DNA or RNA, depending on the virus.

Chemical analysis revealed that DNA and RNA are macromolecules composed of smaller building blocks called nucleotides. Each nucleotide consists of a 5-carbon sugar (deoxyribose in DNA, ribose in RNA) to which is attached one of four nitrogenous bases and

a phosphate group. In DNA the four nitrogenous bases are adenine, guanine, cytosine, and thymine, while in RNA the four bases are adenine, guanine, cytosine, and uracil. Thymine and uracil differ only by the presence of a methyl group in thymine that is absent in uracil. From chemical and physical analysis, it was determined that the DNA molecule consists of two polynucleotide chains joined by hydrogen bonds between pairs of bases (A and T, G and C) in a double helix. (The double helix model was first proposed by Watson and Crick.) The diameter of the helix is 2 nm, and there are 10 base pairs in each complete turn of the helix (3.4 nm); each base pair is thus twisted 36° relative to the preceding base pair in the molecule.

A number of different types of double-helical DNA have been identified by X-ray diffraction analysis. The three major types are the right-handed A— and B—DNAs, and the left-handed Z—DNA. The common form found in cells is B—DNA (the form analyzed by Watson and Crick). A—DNA does not exist in cells. Z—DNA may exist in cells in stretches of DNA that are particularly rich in guanine and cytosine. The functional significance of Z—DNA is unknown.

ANALYTICAL APPROACHES FOR SOLVING GENETICS PROBLEMS

Q.1 The linear chromosome of phage T2 is 52 μm long. The chromosome consists of double-stranded DNA, with 0.34 nm between each base pair. The average weight of a base pair is 660 daltons. What is the molecular weight of the T2 molecule?

A.1 This question involves the careful conversion of different units of measurement. The first step is to put the lengths in the same units: 52 μm is 52 millionths of a meter, or $52,000 \times 10^{-9}$ m, or 52,000 nm. One base occupies 0.34 nm in the double helix, so the number of base pairs in this chromosome is 52,000 divided by 0.34, or 152,941 base pairs. Each base pair, on the average, weighs 660 daltons, and therefore the molecular weight of the chromosome is $152,941 \times 660 = 1.01 \times 10^8$ daltons, or 101 million daltons.

The average length of the double helix in a human chromosome is 3.8 cm, which is 3.8 hundredths of a meter, or 38 million nm—substantially longer than the T2 chromosome! There are over 111.7 million base pairs in the average human chromosome.

Q.2 The accompanying table lists the relative percentages of bases of nucleic acids isolated from different species.

For each one, what type of nucleic acid is involved? Is it double- or single-stranded? Explain your answer.

SPECIES	ADENINE	GUANINE	THYMINE	CYTOSINE	URACIL
(i)	21	29	21	29	0
(ii)	29	21	29	21	0
(iii)	21	21	29	29	0
(iv)	21	29	0	29	21
(v)	21	29	0	21	29

A.2 This question focuses on the base-pairing rules and the difference between DNA and RNA. In analyzing the data, we should determine first whether the nucleic acid is RNA or DNA, and then whether it is double- or single-stranded. If the nucleic acid has thymine, it is DNA; if it has uracil, it is RNA. Thus species (i), (ii), and (iii) must have DNA as their genetic material, and species (iv) and (v) must have RNA as their genetic material.

Next, the data must be analyzed for strandedness. Double-stranded DNA must have equal percentages of A and T and of G and C. Similarly, double-stranded RNA must have equal percentages of A and U and of G and C.

Hence species (i) and (ii) have double-stranded DNA, while species (iii) must have single-stranded DNA since the base-pairing rules are violated, with A = G and T = C but A ≠ T and G ≠ C. As for the RNA-containing species, (iv) contains double-stranded RNA since A = U and G = C, and (v) must contain single-stranded RNA.

Q.3 Yeast DNA, which is double-stranded, has a 31% A, 18% G, 18% C and 31% T. In this instance would you expect the (A + G)/(C + T) and (A + T)/(G + C) ratios to be the same, as they are here? Explain.

A.3 In double-stranded DNA, purines = pyrimidenes because of complementary base pairing; as a result, (A + G)/(C + T) = 1 in all cases. However, while A = T and G = C, the relative proportions of AT and GC base pairs varies with the species. This relationship, often expressed as %GC, is the base ratio of the DNA. Only by chance would the %AT = %GC to give (A + T)/(G + C) = 1 as it is in yeast. Thus, it is not *expected* that (A + G)/(C + T) = (A + T)/(G + C), but it will occur in a few species.

Q.4 Here are four characteristics of one strand (the "original" strand) of a particular long double-stranded DNA molecule:
1. 35% of the adenine-containing nucleotides (As) have guanine-containing nucleotides (Gs) on their 3′ sides;
2. 30% of the As have Ts as their 3′ neighbors;
3. 25% of the As have Cs; and
4. 10% of the As have As as their 3′ neighbors. Use this information to answer the following questions, explaining your reasoning in each case. (Some of the questions may be unanswerable without more information.)
   a. In the complementary DNA strand, what will be the frequencies of the various bases on the 3′ side of A?
   b. In the complementary strand, what will be the frequencies of the various bases on the 3′ side of T?
   c. In the complementary strand, what will be the frequency of each kind of base on the 5′ side of T?
   d. Why is the % of A not equal to the % of T (and the % of C not equal to the % of G) among the 3′ neighbors of A in the original DNA strand described?

A.4 a. This cannot be answered without more information. Although we know that the As neighbored by Ts in the original strand will correspond to As neighbored by Ts in the complementary strand, there will be additional As in the complementary strand about whose neighbors we know nothing.
   b. This also cannot be answered. All the As in the original strand correspond to Ts in the complementary strand, but we know only about the 5′ neighbors of these Ts, not the 3′ neighbors.

c. 10% will be T, 30% will be A, 25% will be G and 35% will be C.
d. The A = T and G = C rule only applies when considering both strands of a double stranded DNA. Here we are considering only the original single strand of DNA.

QUESTIONS AND PROBLEMS

9.1 In the 1920s while working with *Diplococcus pneumoniae*, the agent that causes pneumonia, Griffith discovered an interesting phenomenon. In the experiments mice were injected with different types of bacteria. For each of the following bacteria type(s) injected, indicate whether the mice lived or died:
a. type *IIR*;   c. heat-killed *IIIS*;
b. type *IIIS*;   d. type *IIR* + heat-killed *IIIS*.

9.2 Several years after Griffith described the transforming principle, Avery, MacLeod, and McCarty investigated the same phenomenon.
a. Describe their experiments.
b. What did their experiments demonstrate beyond Griffith's?
c. How were enzymes used as a control in their experiments?

*9.3 By differentially labeling the coat protein and the DNA of phage T2, Hershey and Chase demonstrated that (choose the correct answer)
a. only the protein enters the infected cell.
b. the entire virus enters the infected cell.
c. a metaphase chromosome is composed of two chromatids, each containing a single DNA molecule.
d. the phage genetic material is most probably DNA.
e. the phage coat protein directs synthesis of new progeny phage.

*9.4 Hershey and Chase showed that when phages were labeled with ^{32}P and ^{35}S, the ^{35}S remained outside the cell and could be removed without affecting the course of infection, whereas the ^{32}P entered the cell and could be recovered in progeny phages. What distribution of isotope would you expect to see if parental phages were labelled with isotopes of
a. C?
b. N?
c. H?
Explain your answer.

9.5 What is the evidence that the genetic material of TMV (tobacco mosaic virus) is RNA?

*9.6 In DNA and RNA, which carbon atoms of the sugar molecule are connected by a phosphodiester bond?

9.7 Which base is unique to DNA, and which base is unique to RNA?

9.8 How do nucleosides and nucleotides differ?

9.9 What chemical group is found at the 5′ end of a DNA chain? At the 3′ end of a DNA chain?

9.10 What evidence do we have that in the helical form of the DNA molecule that the base pairs are composed of one purine and one pyrimidine?

*__9.11__ What evidence is there to substantiate the statement: "There are only two base pair combinations in DNA, A−T and C−G"?

9.12 How many different kinds of nucleotides are there in DNA molecules?

*__9.13__ What is the base sequence of the DNA strand that would be complementary to the following single-stranded DNA molecules:
a. 5′ AGTTACCTGATCGTA 3′
b. 5′ TTCTCAAGAATTCCA 3′

*__9.14__ Is an adenine-thymine or guanine-cytosine base pair harder to break apart? Explain your answer.

9.15 The double-helix model of DNA, as suggested by Watson and Crick, was based on a variety of lines of evidence gathered on DNA by other researchers. The facts fell into the following two general categories; give three examples of each:
a. chemical composition;
b. physical structure.

*__9.16__ For double-stranded DNA, which of the following base ratios always equals 1?
a. (A + T)/(G + C)   **d.** (G + T)/(A + C)
b. (A + G)/(C + T)   **e.** A/G
c. C/G

9.17 If the ratio of (A + T) to (G + C) in a particular DNA is 1.00, does this result indicate that the DNA is most likely constituted of two complementary strands of DNA or a single strand of DNA, or is more information necessary?

9.18 Explain whether the (A + T)/(G + C) ratio in double-stranded DNA is expected to be the same as the (A + C)/(G + T) ratio.

*__9.19__ The percent cytosine in a double-stranded DNA is 17. What is the percent of adenine in that DNA?

9.20 Upon analysis, a DNA molecule was found to contain 32% thymine. What percent of this same molecule would be made up of cytosine?

9.21 A sample of double-stranded DNA has 62% G−C content. What is the percentage of A in the DNA?

9.22 A double-stranded DNA polynucleotide contains 80 thymidylic acid and 110 guanylic acid residues. What is the total nucleotide number in this DNA fragment?

9.23 Analysis of DNA from a bacterial virus indicates that it contains 33% A, 26% T, 18% G, and 23% C. Interpret the data.

9.24 What is a DNA oligomer?

9.25 The genetic material of bacteriophage ΦX174 is single-stranded DNA. What base equalities or inequalities might we expect for single-stranded DNA?

9.26 If a virus particle contains double-stranded DNA with 200,000 base pairs, how many complete 360° turns would occur in this molecule?

*__9.27__ A double-stranded DNA molecule is 100,000 base pairs (100 kilobases) long.
a. How many nucleotides does it contain?
b. How many complete turns are there in the molecule?
c. How long is the DNA molecule?

*__9.28__ If nucleotides were arranged at random in a single-stranded RNA 10⁶ nucleotides long, and if the base composition of this RNA was 20% A, 25% C, 25% U and 30% G, how many times would you expect the specific sequence (5′)-GUUA-(3′) to occur?

10 THE ORGANIZATION OF DNA IN CHROMOSOMES

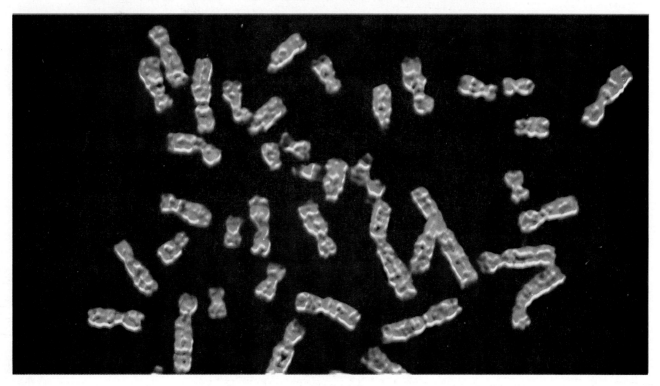

PRINCIPAL POINTS

~ In prokaryotes and viruses in which the genetic material is DNA, the chromosome consists of DNA with a number of proteins associated with it.

~ In bacteria the chromosome is a circular, double-stranded DNA molecule that is compacted by supercoiling of the DNA helix. In the *E. coli* chromosome about 100 independent looped domains of supercoiled DNA have been identified.

~ The complete set of metaphase chromosomes in a eukaryotic cell is called its karyotype. The karyotype is species-specific.

~ The amount of DNA in a prokaryotic chromosome or the haploid amount of DNA in a eukaryotic cell is called the C value. There is not a direct relationship between the C value and the structural or organizational complexity of an organism.

~ The nuclear chromosomes of eukaryotes are complexes of DNA and histone and nonhistone chromosomal proteins. Each chromosome consists of one linear, unbroken, double-stranded DNA molecule running throughout its length. The histones are constant from cell to cell within an organism, while the nonhistones vary significantly between cell types.

~ The large amount of DNA present in the eukaryotic chromosome is compacted by its association with histones in nucleosomes, and by higher levels of folding of the nucleosomes into chromatin fibers. Each chromosome contains a large number of looped domains of 30-nm chromatin fibers attached to a protein scaffold.

~ The centromere region of each eukaryotic chromosome is responsible for the accurate segregation of the replicated chromosome to the daughter cells during mitosis and meiosis. The DNA sequences of centromeres are species-specific.

~ The ends of chromosomes, the telomeres, consist of simple, relatively short, tandemly repeated sequences that are species-specific.

~ Prokaryotic genomes consist mostly of unique-sequence DNA, with only a few repeated sequences and genes. Eukaryotes have both unique and repetitive sequences in the genome. The spectrum of complexity of repetitive DNA sequences among eukaryotes is extensive.

*I*n Chapter 9 we discussed the structures of DNA. In this chapter we will examine how the DNA is organized into chromosomes. The chromosomes of eukaryotes are much more complex than the chromosomes of prokaryotes. Eukaryotic chromosomes consist of a highly ordered complex of DNA and proteins, with special regions—centromeres and telomeres—that are of special importance for chromosome function. Ultimately, geneticists will need an even greater understanding of chromosome structure than we have now so that a complete picture can be obtained of how gene expression is regulated.

THE STRUCTURAL CHARACTERISTICS OF PROKARYOTIC AND VIRAL CHROMOSOMES

The Watson and Crick double-helix model of DNA structure does not by itself describe the structural characteristics of chromosomes. In the following section we will examine the organization of DNA molecules in chromosomes of bacteria and viruses. An organism's genes are discrete regions of longer molecules of DNA (perhaps a thousand to over a million base pairs). It is important to understand how DNA is organized in chromosomes in order to understand how genes function and are regulated.

Bacterial Chromosomes

In Chapter 7 we learned that the DNA of the bacterium *Escherichia coli* is located in a central region called the nucleoid (see Figure 7.1). If an *E. coli* cell is lysed gently, the DNA is released in a highly folded state. The double-stranded DNA is present as a single chromosome, approximately 1100 μm long [4×10^3 kb (1 kb = 1 kilobase pairs = 1000 base pairs)], which contains all the genes necessary for the bacterium to grow and survive in a variety of natural and laboratory environments. What are the properties of this chromosome?

THE *E. COLI* SUPERCOILED CHROMOSOME. The amount of DNA in the *E. coli* chromosome is approximately 1000 times the length of the *E. coli* cell. How is all that DNA packaged into the nucleoid region of the cell? To answer that question we must learn about the properties of linear and circular forms of DNA.

An early prediction was that, when DNA isolated from a bacterium or virus is centrifuged, all of the DNA would sediment at the same rate, forming a band in the centrifuge tube at the end of the experiment. However, in 1963, Jerome Vinograd obtained an unexpected result. When he centrifuged circular DNA from polyoma virus, two bands, not one, were observed. His investigations of this finding led to an understanding that circular DNA can exist in a *relaxed* or a *supercoiled* form. These forms are explained in the following.

Let us consider a 200-base-pair linear piece of DNA in the B-form (Figure 10.1a). Such a molecule has two free ends and, since there are 10.0 base pairs per helical turn of B-form DNA, there are 20 helical turns

in this molecule. If we now simply join the two ends, we have produced a circular DNA molecule; that is, a closed structure without free ends (Figure 10.1b). This circular DNA molecule that we have produced is said to be *relaxed*. Alternatively, if we first untwist one end of the linear DNA molecule by two turns (Figure 10.1c), and then join the two ends, the circular DNA molecule produced will have 18 helical turns and a small unwound region (Figure 10.1d). Such a structure is not energetically favored, and will switch to a structure with 20 helical turns and two superhelical turns (Figure 10.1e). This structure is said to be *supercoiled*; that is, the DNA has supercoils introduced into it by having the double helix twisted in space about its own axis. It is important to remember that supercoiling can only occur in circular DNA or in linear DNA in instances in which the two ends are anchored in some

~ FIGURE 10.1

Illustration of DNA supercoiling. (a) A linear B-form DNA with 20 helical turns (200 base pairs); (b) Relaxed circular DNA produced by joining the two ends of the linear molecule of (a); (c) The linear DNA molecule of (a) unwound from one end by two helical turns; (d) A possible circular DNA molecule produced by joining the two ends of the linear molecule of (c) showing 18 helical turns and a short unwound region; (e) The more energetically favored form of (d), a supercoiled DNA with 20 helical turns and two superhelical turns.

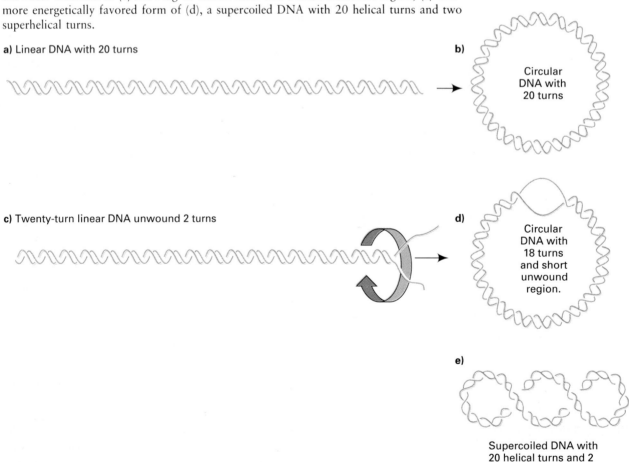

a) Linear DNA with 20 turns

b) Circular DNA with 20 turns

c) Twenty-turn linear DNA unwound 2 turns

d) Circular DNA with 18 turns and short unwound region.

e) Supercoiled DNA with 20 helical turns and 2 superhelical turns

way. Further, the very act of supercoiling produces tension in the DNA molecule. Therefore, if a break (= a nick) is introduced into the sugar-phosphate backbone of a supercoiled DNA molecule, untwisting the molecule produces a relaxed DNA circle. Figure 10.2 pictures relaxed and supercoiled circular DNA of a plasmid.

There are two types of supercoiling, *negative supercoiling* and *positive supercoiling*. In negative supercoiling, the DNA is twisted about its axis in the *opposite* direction from the clockwise turns of the right-handed double helix. DNA with negative supercoils is said to be *underwound*. Positive supercoiling is the opposite of negative supercoiling; that is, the DNA is supercoiled in the *same* direction as the clockwise turns of the helix. Positive supercoiling therefore winds the double helix even more tightly and the DNA is said to be *overwound*.

An important biological function of DNA supercoiling is that it makes the molecule more compact than a relaxed molecule of the same length (compare Figure 10.1e with Figure 10.1b). Therefore, when relaxed and supercoiled DNA molecules of the same length are centrifuged, the supercoiled form will sedi-

~ FIGURE 10.2

Electron micrographs of plasmid DNA: (a) relaxed (nonsupercoiled) DNA; (b) supercoiled DNA. Both molecules are shown at the same magnification.

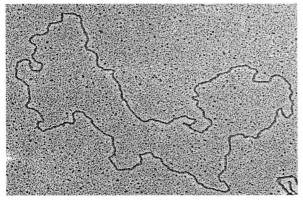

a)

b)

ment more rapidly than the relaxed form. This was what Vinograd observed in his polyoma DNA experiment. Presumably, the native polyoma DNA is a supercoiled DNA, and the relaxed form in the experiment is interpreted to have resulted from some of the supercoiled DNA becoming nicked on one strand.

Given this information, we can explain the packaging of the *E. coli* chromosome into the nucleoid region of the cell as follows: the *E. coli* chromosome is circular and is extensively supercoiled (= negatively supercoiled) resulting in a highly compact molecule.

The supercoiling of the DNA is controlled by a class of enzymes called **topoisomerases.** These enzymes convert one topological form of DNA into another. Two topoisomerases have been identified and characterized in both prokaryotes and eukaryotes. In prokaryotes the two enzymes, called topoisomerase I and II, have been best characterized in *E. coli.* Topoisomerase II (also called *DNA gyrase*) untwists DNA in the left-handed direction to produce negative supercoils. Topoisomerase I does the opposite; that is, it converts negatively supercoiled DNA to the relaxed state. In the presence of both enzymes, then, DNA can be interchanged between the negatively supercoiled and relaxed states. Both types of topoisomerases must be present and functional in cells for DNA to move between supercoiled and relaxed states.

Both topoisomerases function by making a nick or nicks in the DNA chain. The broken ends produced are held together by the topoisomerase enzyme itself so that the ends are unable to rotate freely about the other continuous strand. If that were to happen, only relaxed DNA molecules would result from topoisomerase action. Topoisomerase II functions to introduce two negative turns in the DNA by passing DNA chains through transient double-strand breaks (Figure 10.3; p. 292). Topoisomerase I introduces one positive turn by passing DNA chains through transient single-strand breaks (Figure 10.4; p. 293). Most topoisomerases relax the supertwisting of supercoiled DNA, but only topoisomerase II catalyzes the formation of negatively supercoiled DNA.

PROTEINS COMPLEXED TO BACTERIAL CHROMOSOMES. As we will see, eukaryotic DNA is complexed with a number of discrete structural proteins called histones, which serve to compact the DNA into the chromosome structures characteristic of eukaryotic nuclei.

Are there structural proteins associated with bacterial chromosomes? When *E. coli* cells are lysed (broken open), the DNA of the nucleoid region is released into the medium (Figure 10.5; p. 293). That DNA is organized into loops. Moreover, the DNA in the loops is folded into a more compact shape than would be the case for extended DNA molecules. The explanation for this structure is that the DNA, which is an acidic (nega-

~ FIGURE 10.3

Model for the action of topoisomerase II (DNA gyrase). Starting with DNA that has one positive supercoil, this enzyme introduces two negative turns, resulting in DNA with one negative supercoil.

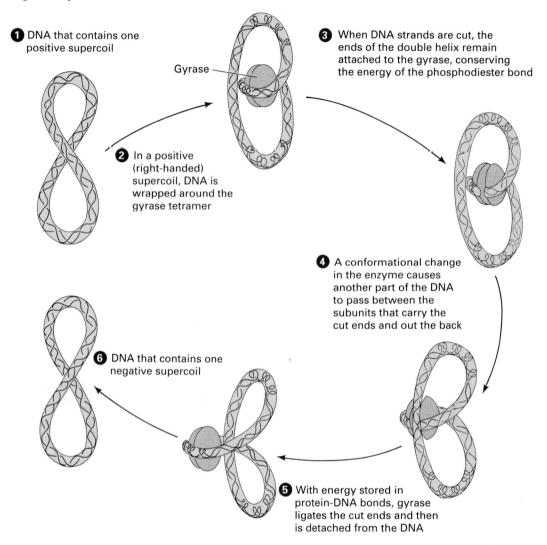

① DNA that contains one positive supercoil

Gyrase

② In a positive (right-handed) supercoil, DNA is wrapped around the gyrase tetramer

③ When DNA strands are cut, the ends of the double helix remain attached to the gyrase, conserving the energy of the phosphodiester bond

④ A conformational change in the enzyme causes another part of the DNA to pass between the subunits that carry the cut ends and out the back

⑥ DNA that contains one negative supercoil

⑤ With energy stored in protein-DNA bonds, gyrase ligates the cut ends and then is detached from the DNA

tively-charged) molecule, is associated with basic (positively-charged) structural proteins. Several of these DNA-binding, structural proteins have been isolated from *E. coli* and characterized. Two of these proteins— HU and H—resemble two of the histone structural proteins found associated with eukaryotic DNA. Based on analyses of the *E. coli* proteins and of the properties of nucleoid DNA, a model for the structure of the bacterial chromosome has been proposed (Figure 10.6). In the model, the bacterial chromosome has about 100 independent *domains*. Each domain consists of a loop of about 40,000 base pairs (40-kilobase pairs [kb])(= 13 μm) of supercoiled DNA. The ends of each domain are held (by unknown means) such that the

supercoiled DNA state in that domain is not affected by events that influence supercoiling of DNA in the other domains.

PLASMIDS IN BACTERIA. In addition to the main chromosome, bacterial cells often contain DNA in plasmids, double-stranded DNA circles that are much smaller than the circular chromosome. In natural populations of bacteria the amount of plasmid DNA may approach 1 or 2 percent of the total amount of cellular DNA. These plasmids are supercoiled and they replicate autonomously, that is, they replicate independently of the main chromosome. For this replication to occur, the plasmids must contain genes that control

~ FIGURE 10.4

Model for the action of topoisomerase I. As a result of the action of this enzyme, the DNA has one less negative supercoil.

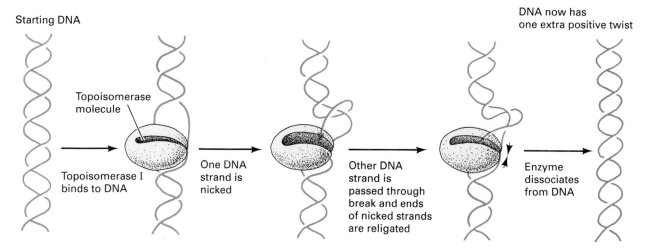

Starting DNA

Topoisomerase molecule

Topoisomerase I binds to DNA

One DNA strand is nicked

Other DNA strand is passed through break and ends of nicked strands are religated

Enzyme dissociates from DNA

DNA now has one extra positive twist

their replication and maintain a balance in the ratio of the number of plasmids to the main chromosome.

T-Even Phage Chromosomes

The T2, T4, and T6 bacteriophages infect *E. coli,* and we have discussed some aspects of their genetics in Chapter 7. (To review their structures, refer to Figure 7.13.) The characteristics of these phages, along with features of other viruses we will discuss, are shown in Table 10.1 (p. 294). Like all viruses, T2, T4, and T6 (collectively called the T-even phages) contain a single chromosome surrounded by a protein coat. The T-even phages are virulent, and infection by them almost always results in the lysis of the infected bacterial cell (see Figure 7.14).

From the results of genetic-mapping experiments, geneticists suggested that the genetic maps of phages T2 and T4 are circular and that the phage chromosomes might have ends at different points around the circular

~ FIGURE 10.5

Nucleoid region released from a lysed *E. coli* cell.

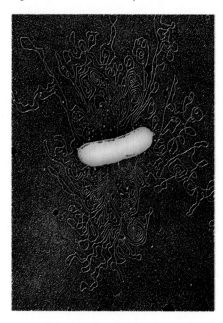

~ FIGURE 10.6

Model for the structure of a bacterial chromosome. The chromosome is organized into looped domains, the bases of which are anchored in an unknown way.

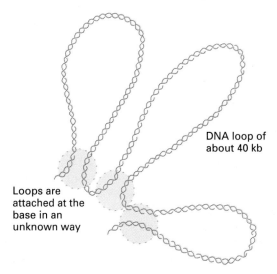

DNA loop of about 40 kb

Loops are attached at the base in an unknown way

~ TABLE 10.1

Characteristics of the Chromosomes of Selected Viruses

VIRUS	HOST	STRUCTURE AND TYPE OF GENETIC MATERIAL	CHROMOSOME DESCRIPTION	LENGTH OF CHROMOSOME (μm)	%GC	
T-even phages	*E. coli*	Double-stranded DNA	Linear; circularly permuted and terminally redundant	60	35	
T7	*E. coli*	Double-stranded DNA	Linear; unique sequence	12	48	
λ	*E. coli*	Double-stranded DNA	Linear; single-stranded "sticky ends"	16	49	
P22	*Salmonella typhimurium*	Double-stranded DNA	Linear; unique sequence	14	48	
ΦX174	*E. coli*	Single-stranded DNA	Circular	1.8	A25 T33	G24 C18
Qβ	*E. coli*	Single-stranded RNA	Linear	1.4	A22 U29	G24 C25
Reovirus	Mammals	Double-stranded RNA	Several pieces	8.3	A38 U28	G17 C17
SV40	Human	Double-stranded DNA	Supercoiled ring	1.7	41	
Murine leukemia	Mouse	Single-stranded RNA	Several pieces		A25 U23	G25 C27
Tobacco mosaic virus (TMV)	Tobacco	Single-stranded RNA	Linear		A30 U26	G25 C19

maps, with the ends overlapping. The phage DNA itself was proposed to be a population of linear chromosomes with ends at different places around the circle, an arrangement referred to as *circularly permuted* chromosomes (Figure 10.7a). The experimental demonstration of the existence of circularly permuted chromosomes is shown in Figure 10.7b.

Here is the logic behind the experiment. If a double-stranded DNA molecule is heated, it denatures into single-stranded molecules because the hydrogen bonds between the bases are broken. If the heated solution is cooled, complementary regions of DNA strands come together again (*renature* or *reanneal*) to form double-stranded molecules held together by complementary-base pairing. If the experiment involves a circularly permuted population of DNA molecules (such as those extracted from the phage progeny in a bacterium that has been infected with a single phage), the terminal region of one strand will be complementary to the middle part of another strand. Therefore renaturation of these two strands will produce a molecule that is double-stranded in the middle and single-stranded on each end. The single-stranded ends have complementary sequences of bases that will renature to form a circle of double-stranded DNA with a circumference that is the same as the length of the linear T2 chromosomes. When this experiment is performed with T2 and T4 chromosomes, circles are seen under the electron microscope, indicating that the chromosomes of these phages are circularly permuted. This fact correlates very well with the fact that the genetic maps of T2 and T4 are circular, as we saw in Chapter 7.

There is also evidence that each linear T2 or T4 DNA molecule has the same sequence of nucleotides at

~ FIGURE 10.7

Circularly permuted DNA molecules: (a) Examples of double-stranded DNA molecules with sequences that are circular permutations of each other. (b) Experimental procedure to test whether a population of chromosomes is circularly permuted. The double-stranded DNAs are denatured by heat into single strands. When they cool, random reassociation of double-stranded DNA will occur, and because of circular permutation, molecules will be produced with complementary single-stranded ends, forming circles that are visible by electron microscopy.

a) **Circular permutations of complementary DNA strands**

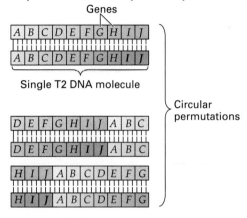

b) **Experimental procedure**

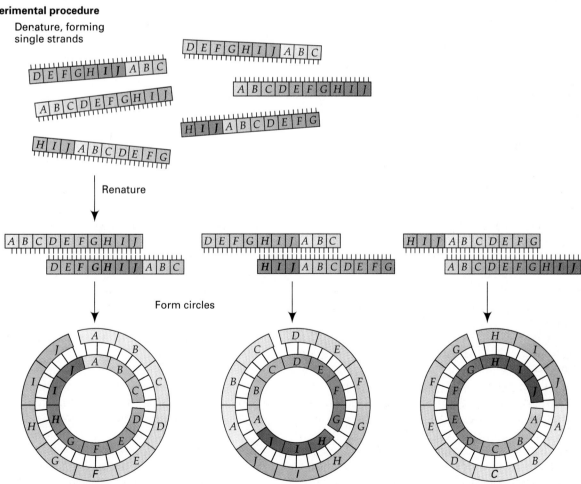

~ FIGURE 10.8

Terminally redundant DNA molecules: (a) Examples of terminally redundant double-stranded DNA molecules. (b) Experimental procedure to test whether a linear, double-stranded DNA molecule is terminally redundant. The molecules are digested to a limited extent with exonuclease III, which removes deoxyribonucleotides from the 3′ ends. If the molecules are terminally redundant, complementary single-stranded ends are exposed. When paired, these form circles that are visible through an electron microscope.

a) **Terminal redundancy and circular permutation**

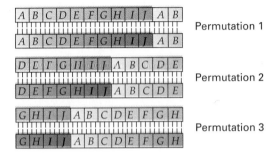

b) **Experimental procedure**

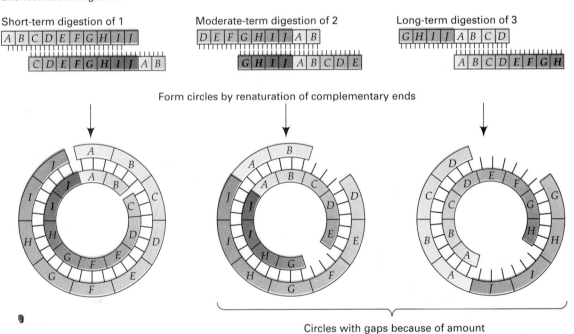

the two ends of the molecule. This *terminal redundancy,* as it is called, is illustrated in Figure 10.8a. (Because of circular permutation, the two ends of each DNA molecule are different in sequence from molecule to molecule.) Terminal redundancy in a linear molecule can be shown experimentally as diagramed in Figure 10.8b. The DNA is treated with the enzyme exonuclease III (an exonuclease is an enzyme that digests DNA at the ends), which removes nucleotides one-by-one from the 3′ ends of each strand of the double-

helical DNA. If the molecule is terminally redundant, the enzyme treatment leaves complementary single-stranded ends, which can pair to form circles.

How are the native, circularly permuted and terminally redundant T2 and T4 chromosomes generated from a single parental phage? The answer is found in the mechanism used to package DNA into the phage heads (Figure 10.9). After the phage chromosome is injected into the host bacterium, it replicates several times. This replication produces a number of chromo-

~ FIGURE 10.9

Circularly permuted and terminally redundant T4 progeny chromosomes produced experimentally by making concatameric molecules and cleaving them into pieces large enough to fill the phage heads.

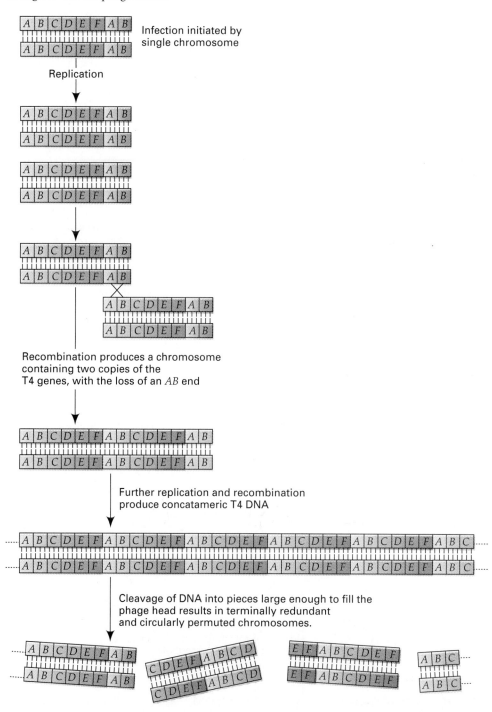

somes, all of which have the same terminally redundant sequence as the parental DNA. Molecular recombination occurs next between the DNA molecules at the terminally redundant ends, splicing the chromosomes together into very long molecules, called *concatamers*. These structures repeat the base sequence of the origi-

nal unit phage chromosome in a tandem fashion. The concatamers undergo several rounds of replication, and late in infection the DNA is packaged into the phage heads that have been produced concurrently. The packaging is done by the *headful;* the length of DNA that is put into each head is determined by the volume of the head. Since the head can hold a little more than a genome's worth of DNA, we see how each phage chromosome contains a terminally redundant region. In addition, the successive clipping of the concatamer molecules into headful lengths leads to the circular permutation of the population of progeny phage chromosomes. In other words, each phage contains the same amount of DNA, with the same sequences represented and with terminal redundancy. In a population of phage chromosomes, individual chromosomes will have different terminal sequences as a result of different permutations of the same sequence.

Bacteriophage ΦX174 Chromosome

The virulent phage ΦX174 is a DNA phage that infects *E. coli*. It has attracted the attention of geneticists because it is smaller than other phages. An electron micrograph of ΦX174 and diagrams of its structure are shown in Figure 10.10. The ΦX174 phage is an icosahedron consisting of protein subunits surrounding the genetic material. At each vertex of the protein coat, there is a spike, which is involved in the infection process. Unlike the T-even phages and the λ phage, ΦX174 does not have a contractile sheath, baseplate, or tail fibers. The study of this phage has provided valuable information about the molecular biology of prokaryotic DNA replication.

In 1959 Robert Sinsheimer found that the DNA of ΦX174 has a base composition that does not fit the complementary-base-pairing rules. That is, instead of having A = T and G = C as is the case for double-stranded DNA, the ratio of bases is 25A:33T:24G:18C. The interpretation was that ΦX174 chromosome is not double-stranded but single-stranded. In addition, if the chromosome is treated with an exonuclease (an enzyme which digests DNA from free ends), the DNA is not affected. The simplest explanation is that the ΦX174 chromosome is circular, a fact that has been corroborated by electron microscopy. Further, the sequence of nucleotides in the chromosome has been determined in its entirety, and it is known to contain 5386 nucleotides. Eleven genes can be identified in this sequence. How the 11 genes are arranged in the chromosome is an illustration of efficient use of DNA, since in some cases parts or even all of one gene is found within the sequence of another gene.

Bacteriophage λ Chromosome

The structure and life cycle of the temperate bacteriophage λ was described in Chapter 7. This phage has been studied extensively for a number of years, so many of its gene functions are well understood.

The phage λ chromosome is double-stranded DNA, is linear, and has no structural proteins associated with it. The two ends of the λ DNA molecule are single-stranded, and complementary. The single-stranded terminus is 12 nucleotides long; its sequence is shown in Figure 10.11a.

Regardless of whether λ goes through the lytic or the lysogenic cycle, the first step after the λ DNA is injected into the host cell is the conversion of the linear molecule into a circular molecule (Figure 10.11a). The complementary ends ("sticky ends") pair and the single-stranded gaps are bonded in a reaction catalyzed by the enzyme DNA ligase (Figure 10.11a). The paired ends are called the *cos* sequence. In the lysogenic cycle

~ FIGURE 10.10

Bacteriophage ΦX174; (a) Electron micrographs of ΦX174 phage particles; (b) Models of the ΦX174 phage particle: top—view along axis of 2-fold symmetry; bottom—view along axis of 5-fold symmetry; (c) single-stranded circular DNA chromosome of ΦX174.

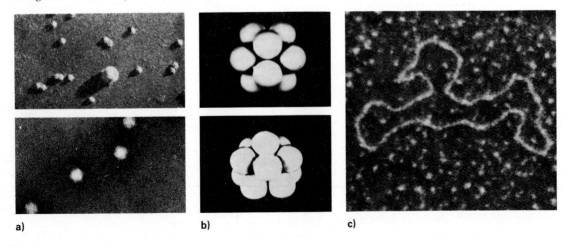

a)   b)   c)

~ FIGURE 10.11

Chromosome structure varies at stages of lytic infection of *E. coli:* (a) parts of the λ chromosome showing the nucleotide sequence of the two single-stranded, complementary ("sticky") ends, and the chromosome circularizing after infection by pairing of the ends, with the single-stranded gaps filled in to produce a covalently closed circle; (b) Generation of the "sticky" ends of the λ DNA during the lytic cycle. During replication of the λ chromosome, a giant concatameric DNA molecule is produced; it contains tandem repeats of the λ genome. The diagram shows the "join" between two adjacent λ chromosomes and the extent of the *cos* sequence. The *cos* sequence is recognized by the *ter* gene product, an endonuclease that makes two cuts at the sites shown by the arrows. These cuts produce one λ chromosome (shaded and unshaded in the diagram) from the concatamer.

a) **Linear λ chromosome (~48,000 base pairs) forms circular λ chromosome**

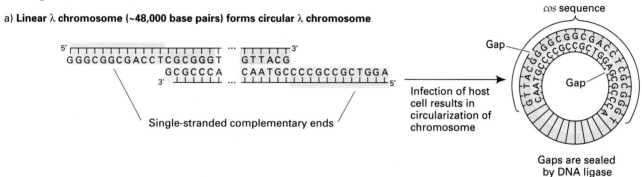

b) **Production of progeny, linear λ chromosomes from concatamers (multiple copies linked end-to-end at complementary ends)**

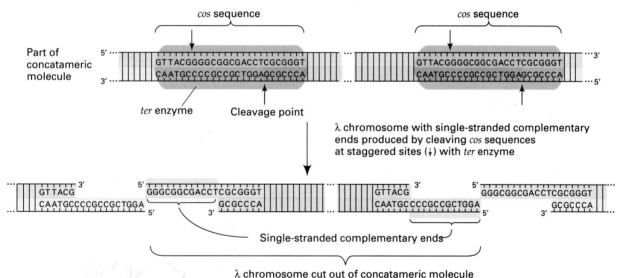

the circular DNA finds a particular site in the *E. coli* chromosome, and by a crossing-over event the DNA is integrated into the main chromosome (see Chapter 7).

In the lytic cycle the DNA replicates and produces a long concatameric molecule similar to the one for T2 and T4. It is from this concatameric structure that progeny phage λ chromosomes are generated as follows: The phage λ chromosome has a gene called *ter* (for "terminus-generating activity": Figure 10.11b), the product of which is a DNA endonuclease (an enzyme

that digests a nucleic acid chain from somewhere along its length rather than at the termini). The endonuclease recognizes the *cos* sequence. Once *ter* is aligned on the DNA at the *cos* site, the endonuclease makes a staggered cut such that linear λ chromosomes with the correct complementary, 12-base-long, single-stranded ends are produced. The chromosomes are then packaged in the assembled phage heads, and progeny λ phages are assembled, then released from the cell when it lyses.

KEYNOTE

In prokaryotes in which the genetic material is DNA, the chromosome consists of DNA with a number of proteins associated with it. In bacteria the chromosome is a circular, double-stranded DNA molecule that is compacted by supercoiling of the DNA helix. In the *E. coli* chromosome, about 100 independent looped domains of supercoiled DNA have been identified. In bacteriophages the genetic material may be double-stranded or single-stranded DNA or RNA. Phage chromosomes can be linear or circular.

THE STRUCTURAL CHARACTERISTICS OF EUKARYOTIC CHROMOSOMES

Now we will examine the general structure of eukaryotic chromosomes, which differ substantially from prokaryotic chromosomes. In addition, we will study how the chromosome is organized at a fundamental molecular level. This will help us understand how the chromosomes replicate, recombine, condense, and become extended again as they go through the cell cycle.

The Eukaryotic Chromosome Complement

A fundamental difference between prokaryotes and eukaryotes is that prokaryotes have but one chromosome, while most eukaryotes have the diploid number of chromosomes in almost all somatic cells. Chromosomes in the interphase nucleus are invisible under the light microscope. During mitosis, as discussed in Chapter 1, they become highly condensed and, when stained, are readily visible under the light microscope. In metaphase, each of the two sister chromatids of each chromosome is in its most highly condensed state. The sister chromatids are attached at one point along their length, the centromere. Almost all cytogenetic analysis of chromosomes has been done with condensed metaphase chromosomes. A photograph of a stained set of metaphase chromosomes from a human male is shown in Figure 10.12. There are 46 chromosomes (each with an X shape representing a pair of sister chromatids joined at the centromere), which vary in size and in the position of the centromere.

The Karyotype

A complete set of all the metaphase chromosomes in a cell is called its **karyotype** (literally, "nucleus type"). For most organisms, all cells have the same karyotype. However, the karyotype is species-specific, so a wide range of number, size, and shape of metaphase chromo-

~ FIGURE 10.12

Mitotic metaphase chromosomes of a human male.

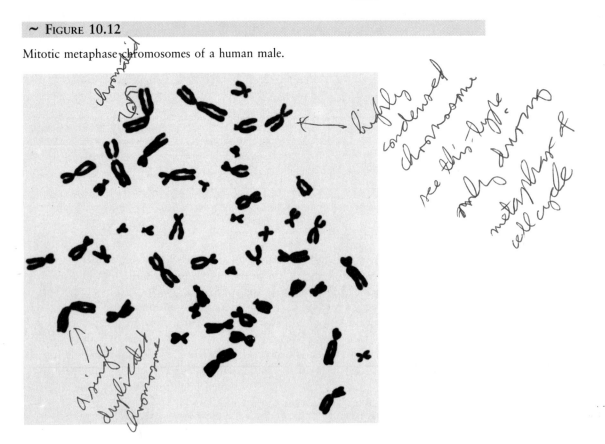

somes is seen among eukaryotic organisms. Even closely-related organisms may have quite different karyotypes. Table 10.2 gives the chromosome numbers of selected eukaryotes. The diploid number of chromosomes in humans is 46, in chimpanzees it is 48, in cats it is 38, and in garden peas it is 14. In the fruit fly, *Drosophila melanogaster*, it is 8. In haploid organisms the number of chromosomes is, for example, 16 in the

~ TABLE 10.2

Chromosome Numbers in Eukaryotic Organisms

SCIENTIFIC NAME	COMMON NAME	NUMBER OF CHROMOSOMES*
DIPLOID ORGANISMS		
ANIMALS		
Homo sapiens	Human	46
Pan troglodytes	Chimpanzee	48
Equus caballus	Horse	64
Bos taurus	Cattle	60
Canis familiaris	Dog	78
Felis domesticus	Cat	38
Oryctolagus cuniculus	Rabbit	44
Rattus norvegicus	Rat	42
Mus musculus	House mouse	40
Meleagris gallopavo	Turkey	82
Xenopus laevis	Toad	36
Caenorhabditis elegans	Nematode	11♂/12♀
Musca domestica	Housefly	12
Drosophila melanogaster	Fruit fly	8
PLANTS		
Pinus ponderosa	Ponderosa pine	24
Pisum sativum	Garden pea	14
Solanum tuberosum	Potato	48
Nicotiana tabacum	Tobacco	48
Triticum aestivum	Bread wheat	42
Zea mays	Corn	20
Vicia faba	Broad bean	12
HAPLOID ORGANISMS		
Chlamydomonas reinhardi	A unicellular alga	16
Neurospora crassa	Orange bread mold	7
Aspergillus nidulans	A green bread mold	8
Saccharomyces cerevisiae	Brewer's yeast	17

*For the diploid organisms, the diploid number of chromosomes is given.

Source: Data adapted from P. L. Altman and D. S. Dittmer (eds.), 1972. *Biology data book*, 2nd ed., vol. 1. Bethesda, MD: Federation of American Societies for Experimental Biology and from other sources.

green alga *Chlamydomonas reinhardi,* 17 in baker's yeast *Saccharomyces cerevisiae,* and 7 in the pink bread mold *Neurospora crassa.*

Figure 10.13 shows the karyotype for the cell of a normal human male. It is customary, particularly with human chromosomes, to arrange chromosomes in order according to size. This karyotype shows 46 chromosomes: 2 pairs of each of the 22 autosomes and 1 of each of the 2 sex chromosomes. (In a karyotype of a human female there are also 46 chromosomes: 2 pairs of each of the 22 autosomes and 2 X chromosomes.) The number of chromosomes is 46 because humans are diploid (2N) organisms, possessing one haploid (N) set of chromosomes (23 chromosomes) from the egg, and another haploid set from the sperm. (In general, the gametes of sexually reproducing organisms are haploid in their chromosome content.)

A knowledge of the size, overall morphology, and banding patterns (discussed later in this chapter) of chromosomes permits geneticists to identify certain chromosome aberrations that correlate with congenital abnormalities or dysfunctions. For example, on rare occasions chromosomes undergo changes in morphology, such as by gaining or losing a portion of a chromosome or by exchanging pieces with a nonhomologous chromosome. In addition, changes in the number of chromosomes in the karyotype may occur because of an error in cell division, for example, as a result of chromosome nondisjunction. In humans, some congenital syndromes (such as Down syndrome or cri-du-chat syndrome) result in characteristic features that correlate with an extra chromosome 21 (trisomy 21), in most cases of Down syndrome, or with the loss of the distal portion of chromosome 5, in the case of cri-du-chat syndrome. Extra or missing chromosomes and large additions and deletions to chromosomes can be observed under the light microscope and will be discussed more fully in Chapter 19.

~ FIGURE 10.13

Human male metaphase chromosomes arranged as a karyotype. Note that groups of chromosomes with similar morphologies are arranged under letter designations (A through G). This arrangement is based on the size of the condensed chromosomes. In the male all chromosomes except the X and Y sex chromosomes are present in pairs.

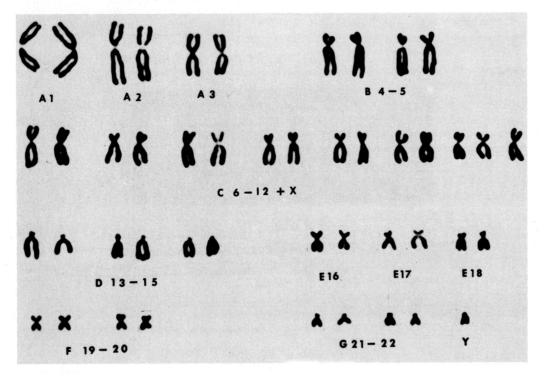

In a human karyotype, the chromosomes are numbered for easy identification. Conventionally, the largest pair of homologous chromosomes is designated 1, the next largest 2, and so on. In humans chromosomes 1 (largest) through 22 (smallest) are called autosomes as distinguished from the pair of sex chromosomes. Formally, the sex chromosomes constitute pair 23 even though, in humans at least, they do not fit properly in the size scale. As shown in Figure 10.13, the X chromosome is a large metacentric chromosome, and the Y chromosome is a much smaller chromosome.

Chromosomal Banding Patterns

The human metaphase chromosomes shown in Figure 10.13 were seen after staining the chromosomes with Feulgen stain. This material stains the chromosomes uniformly, making it difficult to distinguish chromosomes that are similar in size and general morphology. For this reason, in the figure the human chromosomes are grouped into subsets; within each subset one cannot determine which chromosome is which on the basis of size alone.

Fortunately, a number of techniques have been developed that stain certain regions or *bands* of the chromosomes more intensely than other regions. Banding patterns are specific for each chromosome, enabling each chromosome in the karyotype to be distinguished clearly. This makes it easier to distinguish between chromosomes of similar sizes and shapes such as those in the C group in Figure 10.13.

One of these staining techniques is called **Q banding.** In it, metaphase chromosomes are stained with quinacrine mustard, producing what are known as Q bands on the chromosomes. Q bands can be discerned by their fluorescence under ultraviolet light. A metaphase karyotype of human chromosomes showing the Q bands is presented in Figure 10.14. The Q bands are unique for each chromosome in the human set. Although Q banding is a useful diagnostic tool, it is limited by the fact that the fluorescent bands are not permanent.

A more useful technique involves first treating the chromosomes with mild heat or proteolytic enzymes (enzymes that digest proteins) and then staining with Giemsa stain (a permanent DNA dye) to produce a *G banding* pattern (Figure 10.15; p. 305). G and Q bands have the same locations and reflect regions of DNA that are rich in adenine and thymine. In humans, approximately 2000 G bands can be identified. The G bands are stable and are even visible in scanning electron micrographs of chromosomes as constrictions at

the banding regions. The drawings and map of human chromosomes in Figure 6.20 are derived from analysis of G-banded chromosomes.

Cellular DNA Content and the Structural or Organizational Complexity of the Organism

The total amount of DNA in the haploid genome is characteristic of each living species, prokaryote or eukaryote, and is known as its **C value.** Table 10.3 (p. 304) gives the C values for a selection of prokaryotes and eukaryotes. The length of the DNA in each case is also presented as calculated from the number of base pairs. Figure 10.16 (p. 306) presents a summary of the range of C values found in different phyla. Within each phylum, there is an increase in minimum genome size as the complexity of the organisms increases.

~ **FIGURE 10.14**

Q banding in a karyotype of metaphase chromosomes from a normal human male. The Q-banding patterns were produced by staining the chromosomes with the fluorescent dye quinacrine.

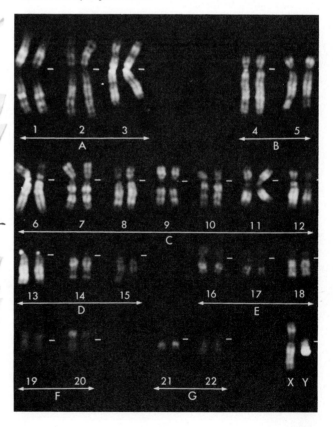

~ TABLE 10.3

Haploid DNA Content (the C Value) of Selected Prokaryotes and Eukaryotes

ORGANISM	C VALUE		
	DALTONS	BASE PAIRS (BP)	LENGTH*
PROKARYOTES			
Bacteria:			
Escherichia coli	2.8×10^9	4.1×10^6	1.4 mm
Salmonella typhimurium	8×10^9	1.1×10^7	3.8 mm
Viruses and phages:			
ΦX174 (double-stranded form)	3.2×10^6	5386	1.8 μm
λ	3.3×10^7	4.65×10^4	16 μm
T2	1.2×10^8	1.75×10^5	60 μm
SV40	3.5×10^6	5226	1.7 μm
EUKARYOTES			
Vertebrates:			
Human	1.9×10^{12}	2.75×10^9	94 cm
Mouse	1.45×10^{12}	2.2×10^9	75 cm
Frog	1.4×10^{13}	2.25×10^{10}	7.65 m
Invertebrates:			
Sea urchin	5×10^{11}	8×10^8	27.2 cm
Drosophila melanogaster	1.2×10^{11}	1.75×10^8	5.95 cm
Caenorhabditis elegans	5×10^{10}	8×10^7	2.7 cm
Plants:			
Lilium longiflorum (lily)	2×10^{14}	3×10^{11}	100 m
Zea mays (maize)	4.4×10^{12}	6.6×10^9	2.24 m
Fungi:			
Neurospora crassa	1.85×10^{10}	2.7×10^7	9.18 mm
Saccharomyces cerevisiae	1.2×10^{10}	1.75×10^7	5.95 mm

*Calculated from the formula 10 bp = 3.4 nm.

The C value data show that the amount of DNA found in the different phyla varies considerably (compare, for example, SV40 with the lily in Table 10.3). The relatively simple, unicellular eukaryotes such as yeast (*Saccharomyces cerevisiae*) has a genome size of about 1.75×10^7 base pairs (bp), that is, about 4–5 times the genome size of *E. coli*. (Any organism having a small genome, as yeast does, makes that organism especially useful for certain kinds of studies of DNA structure and function.) One of the simplest multicellular organisms, the nematode *Caenorhabditis elegans*, has a genome of 8×10^7 bp, which is about four times that of yeast. From such examples, it seems that there is a good relationship between DNA content and the complexity of the organism. However, the relationship becomes less significant as we look at other phyla. In fact, there is also no rule that will hold true for the variation of C values found within a phylum. For example, mammals, birds, and reptiles each show little variation, while amphibians, insects, and plants each vary over a wide range, often tenfold or so.

~ FIGURE 10.15

Morphologies of human chromosomes after G banding. Shown are autosomes 1 through 22 and an X chromosome. The scanning electron micrographs exhibit constriction at the site where the G bands are located, and the light micrographs (insets) show the G banding.

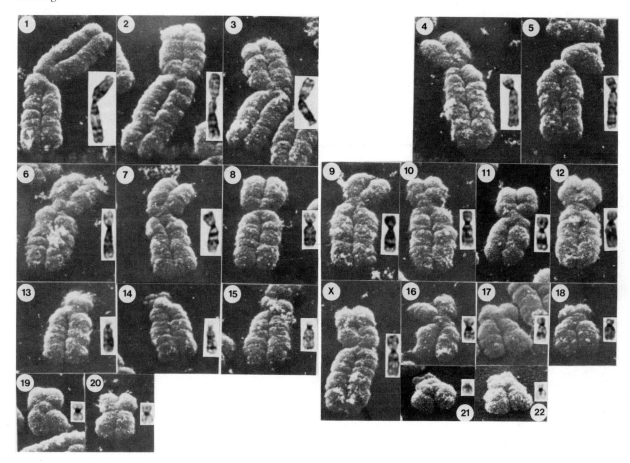

The lack of a direct relationship between the C value and the structural or organizational complexity of the organism is called the _C-value paradox._ We have no complete explanation of this paradox. On one level, the wide range of DNA amounts within groups of organisms suggests that some of the DNA in certain organisms may not encode vital functions. At least one reason for variation in DNA amounts in related organisms may be the presence on chromosomes of different numbers of repeated segments of DNA that appear not to be expressed.

𝒦EYNOTE

In eukaryotes the complete set of metaphase chromosomes in a cell is called its karyotype. The karyotype is species-specific. The amount of DNA in a prokaryotic chromosome or the haploid (N) amount of DNA in a eukaryotic cell is called the C value. The amount of genetic material varies greatly among prokaryotes and eukaryotes. There is not a direct

~ FIGURE 10.16

The amount of DNA per haploid chromosome set in a variety of animals. Variation within certain phyla is quite wide (as in insects and amphibians).

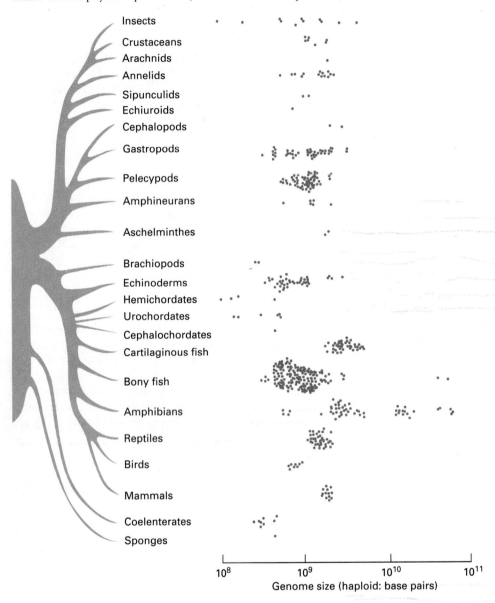

relationship between the *C* value and the structural or organizational complexity of the organism.

The Molecular Structure of the Eukaryotic Chromosome

In comparison with a prokaryotic cell, a eukaryotic cell contains a large amount of DNA in its nucleus. A human cell, for example, has more than a thousand times as much DNA as does *E. coli*. We saw earlier that the *E. coli* chromosome (about 1 mm long in its fully extended state) is supercoiled and supertwisted so that its 4×10^6 base pairs of DNA can be packaged into the nucleoid region. By contrast, the human cell has 5.5×10^9 base pairs of DNA in its diploid nucleus. Without the compacting of this DNA, accomplished by the specific association between DNA and proteins in the chromosomes, the DNA of the chromosomes of a single human cell would be over 6 feet long (about 200 cm) if the molecules were placed end to end!

It is difficult (although not impossible) to isolate intact eukaryotic chromosomes for study. During the processes of releasing the chromosomes from a cell and purifying them, some breakage usually occurs. Studies of carefully isolated chromosomes have shown that *each chromosome consists of one linear, unbroken, double-stranded DNA molecule running throughout its length.*

When eukaryotic chromosomes (that is chromosome fragments) from mitotic metaphase cells are isolated and analyzed, there is about twice as much protein as DNA. The same ratio is found for DNA isolated from the nuclei of interphase cells. The mixture of DNA and protein obtained from these cell types is called **chromatin.** The fundamental structure of chromatin is essentially identical in all eukaryotes.

There are two types of chromatin: **euchromatin** and **heterochromatin.** Euchromatin is chromatin that is condensed during division but becomes uncoiled during interphase. Heterochromatin, on the other hand, refers to chromatin that remains condensed throughout the cell cycle, as evidenced by its staining properties. Functionally, euchromatin is genetically active, whereas heterochromatin is genetically inactive, either because it contains no genes or because the genes it does contain are repressed. Characteristically, heterochromatin replicates later in the S phase of the cell cycle than euchromatin, a result of the higher degree of chromosome condensation in heterochromatin.

Heterochromatin is found in all eukaryotic species near centromeres, at *telomeres* (the ends of the chromosomes), and elsewhere in a species-specific manner. (Telomeres will be discussed later in this chapter.) Two classes of heterochromatin can be distinguished. **Constitutive heterochromatin** is always genetically inactive, and is found at homologous sites on chromosome pairs. Centromeric and telomeric heterochromatin are examples of constitutive heterochromatin. **Facultative heterochromatin** is chromatin that may become heterochromatic. It may contain genes that are inactivated when the chromatin becomes condensed. Barr bodies (inactivated X chromosomes in female mammals—see Chapter 3) are examples of facultative heterochromatin.

There are two major types of proteins associated with DNA in chromatin: **histones** and **nonhistones.** Histones and nonhistones play an important role in determining the physical structure of the chromosome. As we will see, the DNA is wrapped around a core of histone molecules, and the nonhistones are somehow associated with that complex. Various studies have shown that some of the nonhistones have a structural role in the chromosomes: If the histones are removed from the chromosome, the DNA unravels and is displaced from the complex, but a skeleton of nonhistone proteins in the shape of the chromosome remains.

THE HISTONES. The histones are the most abundant proteins associated with chromosomes. Their role is to bind to the negatively charged DNA in the chromosome. The histones are relatively small basic proteins; that is, at the normal pH of a cell, the histones have a net positive charge, thus facilitating their binding to the DNA. This positive charge is found mainly on the amino groups ($-NH_3^+$) of the basic amino acids lysine and arginine (see Figure 8.19 for the chemical structures of these amino acids). In fact, histone molecules consist of approximately 25 percent arginine and lysine. This is a higher percentage than is found in most other proteins.

Five main types of histones are associated with eukaryotic DNA: H1, H2A, H2B, H3, and H4. Weight for weight there is about an equal amount of histone and of DNA in chromatin. The amount and proportions of these histones are constant from cell to cell in all eukaryotic organisms. The properties of the histones from calf thymus DNA are given in Table 10.4. The amino acid sequences of the four histones H2A, H2B, H3, and H4 have been determined for a wide variety of organisms. Comparison of the sequences has indicated, in some cases, a high degree of conservation in the sequences among distantly related species. For example, only two amino acid differences exist in the H4 proteins of cows and peas, and there is only one amino acid difference between sea urchin and calf thymus histone H3. (However, there are 11 amino acid differences in the histone H3 of yeast and *Tetrahymena,* a protozoan.) In fact these four histones are among the most highly conserved of all known proteins. The remarkable degree of amino acid sequence conservation is a strong indicator that histones perform the same basic role in organizing the DNA in the chromosomes of all eukaryotes.

There are several different but closely related H1 histones in each cell. This class of histones has a central

~ **TABLE 10.4**

Characteristics of Histones from Calf Thymus DNA

HISTONE TYPE	BASIC AMINO ACIDS		NUMBER OF AMINO ACIDS	MOLECULAR WEIGHT (DALTONS)
	LYS	ARG		
H1	29%	1%	215	23,000
H2A	11%	9%	129	13,960
H2B	16%	6%	125	13,775
H3	10%	13%	135	15,340
H4	11%	14%	102	11,280

Source: S. C. R. Elgin and H. Weintraub, 1976, *Annu. Rev. Biochem.* 44:725–776.

core region of amino acids that has been conserved through evolution, but the rest of the molecule has been much less conserved. Moreover, in some tissues H1 is not present, being replaced by yet another histone protein. For example, in avian red blood cells (which, unlike mammalian red blood cells, have nuclei), a histone designated H5 is present instead of H1.

NONHISTONE CHROMOSOMAL PROTEINS.

Nonhistones are the second type of protein found in chromatin. By definition, the nonhistone chromosomal proteins are all proteins associated with DNA apart from the histones. There are a very large number of different types of nonhistone chromosomal proteins. As a group, they constitute a relatively smaller proportion of the mass of chromatin than histones. Thus, any individual nonhistone chromosomal protein is present in chromatin in much smaller amounts than any histone. Some nonhistones play a structural role, while others are involved in the regulation of gene expression. For example, RNA polymerase, the enzyme that catalyzes the synthesis of transcription of RNA from a DNA template, may be considered to be an important nonhistone protein. As a class, the nonhistones are very different from the histones. They are usually acidic proteins—that is, proteins with a net negative charge—so they are likely to bind to positively charged histones in the chromatin. Each eukaryotic cell has many different nonhistones in the nucleus, and weight for weight there may be as much nonhistone protein present as DNA and histone proteins combined.

In contrast to the histones, the nonhistone proteins differ markedly in number and type from cell type to cell type within an organism, at different times in the same cell type, and from organism to organism.

KEYNOTE

The nuclear chromosomes of eukaryotes are complexes of DNA, histone proteins, and nonhistone chromosomal proteins. Each chromosome consists of one linear, unbroken, double-stranded DNA molecule running throughout the length of the chromosome. These are five main types of histones—H1, H2A, H2B, H3, and H4—which are constant from cell to cell within an organism. Nonhistones, of which there are a large number, vary significantly between cell types both within and between organisms as well as with time in the same cell type.

NUCLEOSOMES.

The DNA helix of each chromosome is coiled within the nucleus in a nonrandom way. It is known that there are several levels of packing that enable chromosomes several millimeters or even centimeters long to fit into a nucleus that is a few micrometers in diameter. The simplest level of packing involves the winding of DNA around a core of histones in a structure called a **nucleosome,** and the most highly packed state involves higher-order coiling and looping of the DNA-histone complexes, as exemplified by the metaphase chromosomes.

Under the electron microscope the DNA-protein complex is seen as 10-nm fibers (10-nm **nucleofilaments**), that is fiber with a diameter of about 10 nm. Since DNA has a diameter of 2 nm, we must assume that in nucleofilaments the DNA is complexed with histone and nonhistone proteins. In its most unraveled state, the chromatin fiber of a nucleofilament has the appearance of "beads on a string," where the beads are the nucleosomes (Figure 10.17). The thinner "thread" connecting the "beads" is naked DNA.

Nucleosomes were first discovered in the 1970s by A. Olins and D. Olins, and by C. L. F. Woodcock. Their data suggested that the nucleosomes were flattened spheres about 10 nm in diameter, in which the DNA was somehow associated with a core of histones. Current knowledge indicates that the nucleosome consists of DNA associated with the five histones, H1,

~ FIGURE 10.17

Electron micrograph of unraveled chromatin showing the nucleosomes in a "beads on a string" morphology.

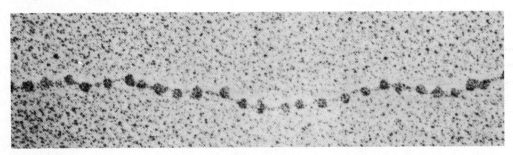

~ **FIGURE 10.18**

Organization of DNA and histones in a chromatin fiber. Gentle digestion of the chromatin fiber with nuclease releases the nucleosomes, consisting of about 200 base pairs of DNA associated with the five histones. More extensive digestion with nuclease removes the linker DNA tails, giving rise to an 11-nm diameter nucleosome core particle that consists of 146 base pairs of DNA associated with the histone octamer (two molecules each of histones H2A, H2B, H3 and H4) and a free molecule of H1 histone. Dissociation of the nucleosome core particle with high salt releases the DNA from the histones and disassembles the histone octamer into individual molecules.

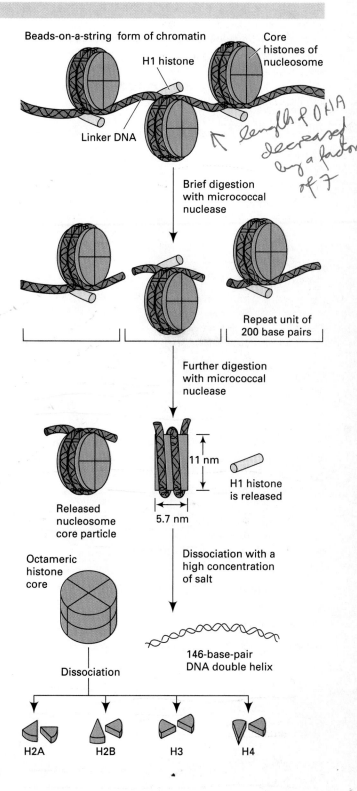

Beads-on-a-string form of chromatin

H1 histone

Core histones of nucleosome

Linker DNA

length of DNA decreased by a factor of 7

Brief digestion with microccocal nuclease

Repeat unit of 200 base pairs

Further digestion with microccocal nuclease

Released nucleosome core particle

11 nm

5.7 nm

H1 histone is released

Octameric histone core

Dissociation with a high concentration of salt

146-base-pair DNA double helix

Dissociation

H2A   *H2B*   *H3*   *H4*

H2A, H2B, H3, and H4. The amount of DNA associated with each nucleosome is about 200 base pairs, with a range of about 150 to 250 base pairs, depending upon the organism and tissue being studied. This regularity was demonstrated by digesting chromatin lightly with micrococcal nuclease, an enzyme that digests DNA to which it has access (Figure 10.18). The *linker DNA* that connects adjacent nucleosomes is most accessible to the nuclease, so the enzyme cuts that DNA, releasing DNA fragments that are about 200-base-pairs long and are still bound to the five histones. In essence, the "beads-on-a-string" are released as individual beads.

With more extensive digestion, the enzyme chews away the tails of linker DNA. The result is a particle with a tightly-protected, 146-bp DNA fragment associated with two each of the core histones H2A, H2B, H3, and H4 (a *histone octamer*). This particle is called a **nucleosome core particle.** (The single molecule of H1 that was originally associated with each nucleosome core particle is released when the linker DNA is degraded by the enzyme.) The nucleosome core particle can be disassociated further in the presence of a high concentration of salt to produce the 146-bp DNA fragment and the eight individual histones.

Figure 10.19 (p. 310) presents a model of a nucleosome. There are about equal masses of DNA and protein. While the core histones are relatively tightly bound to one another and to the DNA, the H1 histone is more loosely associated. As a result it can be released from the nucleosome relatively easily.

Nucleosome core particles have been shown by relatively low resolution X-ray diffraction to be flat-ended, cylindrical particles of dimensions 11 × 11 × 5.7 nm. The DNA is wrapped around the histone core with approximately 80 base pairs per turn, so the 146 base pairs of DNA in the nucleosome core particle involve one-and-three-quarters turns of the DNA. The way the DNA is wound around the histone core causes negative supercoiling (twisting against the clockwise direction of helix formation) of the DNA strand, and altogether the length of the DNA is decreased by a factor of about seven by this coiling.

~ Figure 10.19

Model for the relationship between DNA and histones in the nucleosome.

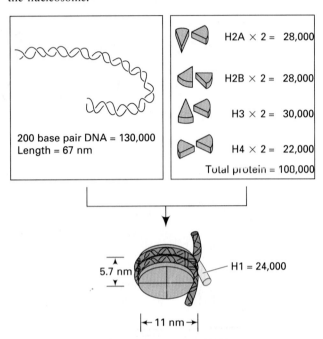

200 base pair DNA = 130,000
Length = 67 nm

H2A × 2 = 28,000

H2B × 2 = 28,000

H3 × 2 = 30,000

H4 × 2 = 22,000

Total protein = 100,000

5.7 nm

H1 = 24,000

11 nm

~ Figure 10.20

(a) Photograph of a wooden model showing the molecular organization of the histone octamer. The model is painted to show the organization of subunits; (b) Interpretative diagram of the photograph in (a).

a)

b)

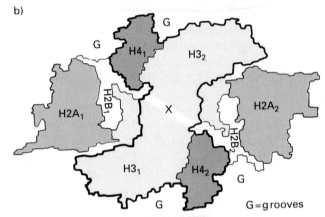

G
G
H4₁
H3₂
H2B₁
H2A₁
X
H2A₂
H2B₂
H3₁
H4₂
G
G
G = grooves

The structure of the histone octamer has been determined using high-resolution X-ray crystallographic techniques. Figure 10.20a shows a photograph of a wooden model of the octamer, and Figure 10.20b is a schematic of the model, showing the different histone domains. The model shows the histone octamer to be a prolate ellipsoid 11 nm long and 6.5 and 7.0 nm in diameter—its general shape is that of a lemon. There is a central $(H3-H4)_2$ tetramer (i.e., two molecules each of H3 and H4) flanked by two H2A–H2B dimers. The DNA helix, when it is placed around the histone octa-

mer in a path suggested by the surface features of the model (Figure 10.21), appears like a spring that holds the H2A–H2B dimers at either end of the $(H3-H4)_2$ tetramer. A more schematic drawing of the nucleosome core particle is presented in Figure 10.22.

~ Figure 10.21

Photographs of a wooden model of the histone octamer showing the path of DNA around the structure: (a) View from the front; (b) View from the back.

a)

b)

~ **FIGURE 10.22**

Schematic of the nucleosome core particle, showing the path of the DNA around the histone octamer.

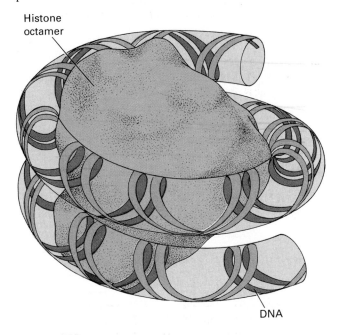

Histone octamer

DNA

In sum, the nucleosome is the fundamental structure of chromatin. The 10-nm fiber, then, may be interpreted as a continuous string of nucleosomes, like tuna cans stacked end to end, or perhaps tilted slightly away from the long axis (Figure 10.23).

HIGHER-ORDER STRUCTURES IN CHROMATIN. In the living cell, chromatin typically will not exist in a "beads-on-a-string" nucleofilament; rather, it is kept in a more highly compacted state. The details of these higher levels of folding, which are found predominantly in the chromosomes, are still incompletely understood. Moreover, the details of the changes in chromatin folding that take place as a cell goes from interphase to mitosis (or meiosis) and back to interphase are not well

~ **FIGURE 10.23**

Model for the organization of nucleosomes in the 10-nm chromatin fiber. The nucleosomes are arranged in a linear array, either stacked end-to-end or tilted slightly away from the long axis of the fiber.

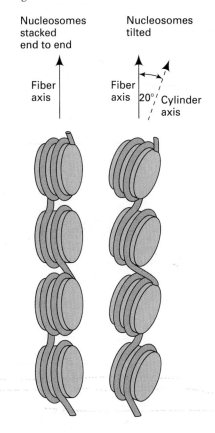

Nucleosomes stacked end to end

Nucleosomes tilted

Fiber axis

Fiber axis   20°   Cylinder axis

understood. There is general agreement, derived from microscopic observations, that the next level of packing above the nucleosome is the *30-nm chromatin fiber* (Figure 10.24). One major problem that has hindered investigations is that as chromatin compaction proceeds, the individual nucleosomes can no longer be distinguished, and hence their arrangement in the 30-nm

~ **FIGURE 10.24**

Electron micrograph of a 30-nm chromatin fiber.

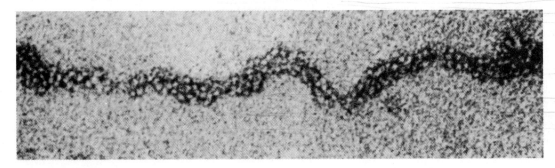

~ FIGURE 10.25

A model for the packaging of nucleosomes into the 30-nm chromatin fiber.

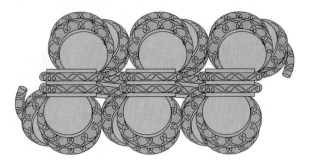

~ FIGURE 10.26

Model for the stages in folding of a nucleosomal chain (the relaxed zigzag ribbon) via the compact zigzag ribbon into the helical ribbon proposed to be the basic structural form of the 30-nm fiber.

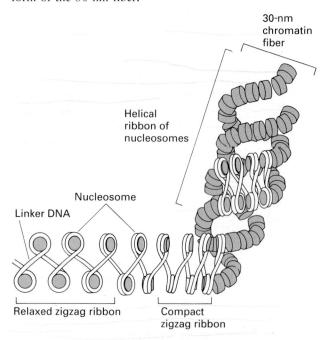

fiber becomes much more difficult to define. Nonetheless, a number of models for the 30-nm fiber have been proposed based on biophysical and biochemical data derived from studies of both compact and relaxed (unraveled) chromatin, one of which is shown in Figure 10.25. Histone H1 must play an important role in the formation of the 30-nm fiber, since H1-depleted chromatin can form 10-nm fibers, but not 30-nm fibers. However, the location(s) of histone H1 in the 30-nm fiber is (are) currently unknown.

A quite different model for the 30-nm fiber involves a zigzag ribbon of nucleosomes which is twisted to generate the 30-nm fiber. The stages in folding of a nucleosomal chain, as indicated in this model, are shown in Figure 10.26. The least compact form is the relaxed zigzag ribbon, which undergoes a transition to a compact zigzag ribbon. This transition involves no major changes in basic structure, especially the loss or addition of nucleosome-nucleosome contacts. The 30-nm fiber is then constructed by folding the compact zigzag ribbon in such a way as to minimize changes in nucleosome-nucleosome contacts. This relatively simple coiling of the compact zigzag ribbon creates a double-helical ribbon with a range of possible diameters and with the nucleosomes in face-to-face contact.

The 30-nm fiber does not provide sufficient packing of the chromosomes to explain the degree to which chromosomes are condensed in the cell nucleus. That is, an average human chromosome would extend approximately 1 mm as a 30-nm fiber, and, as such, would be 200 times longer than the diameter of the nucleus (about 5 μm). Thus, yet further folding of the chromatin fiber must occur. However, the next levels of packing beyond the 30-nm chromatin fiber are not clearly understood. An interphase chromosome may have a diameter of 300 nm while a metaphase chromosome may have a diameter of 700 nm.

A major step in modeling the structure of metaphase chromosomes came when data were obtained showing that metaphase chromosomes from which the histones have been removed still retain a residual folded structure with DNA *looped domains* of 30-nm fiber DNA extending from a condensed protein lattice. The central structure is called the *chromosome scaffold*. The scaffolding actually exhibits the shape of the metaphase chromosome (Figure 10.27), a shape that remains even when the DNA is digested away by nucleases.

Figure 10.28 presents a schematic of a series of looped domains of DNA in a chromosome. The looped domains extend at an angle from the main chromosome axis and are anchored to the protein scaffold. (Recall that bacterial chromosomal DNA is also organized into looped domains.) The amount of DNA in the loops ranges from 10,000 to 90,000 bp, representing an average length of about 0.5 μm of 30-nm fiber. An average human chromosome could contain approximately 2000 looped domains. Figure 10.29 presents a scanning electron micrograph of a region of a mitotic chromosome showing the tips of the looped domains. There is evidence that the loops correspond to the units of replication of the chromosomes.

In highly purified metaphase chromosomes, only two nonhistone scaffold proteins are found. One of these proteins is eukaryotic topoisomerase II. This enzyme, like its bacterial counterpart (see Figure 10.3)

~ FIGURE 10.27

Electron micrographs of a metaphase chromosome depleted of histones: (a) The chromosome maintains its general shape by a nonhistone protein scaffolding from which loops of DNA protrude; (b) Higher magnification of a section of (a), showing the attachment of the DNA loops to the scaffold.

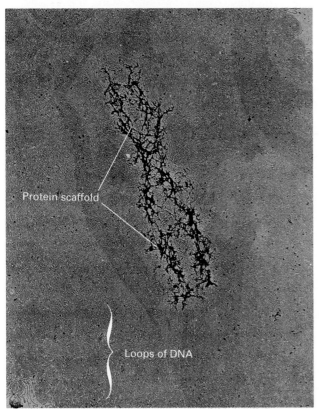

a)

the histone proteins are strip away

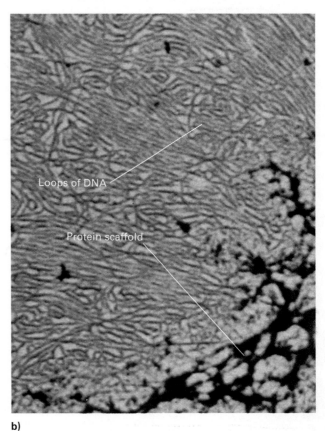

b)

~ FIGURE 10.28

Schematic model for the organization of 30-nm chromatin fiber into looped domains that are anchored to a nonhistone protein scaffold.

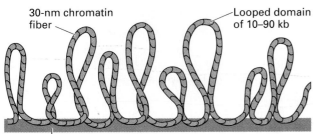

~ FIGURE 10.29

Scanning electron micrograph of a region of a mitotic chromosome showing the tips (knoblike projections) of looped domains of 30-nm chromatin fibers.

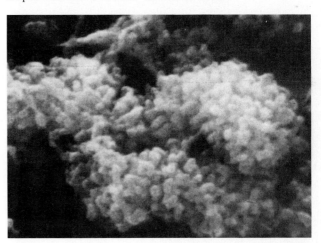

~ FIGURE 10.30

Schematic drawing of the many different orders of chromatin packing that are thought to give rise to the highly condensed metaphase chromosome.

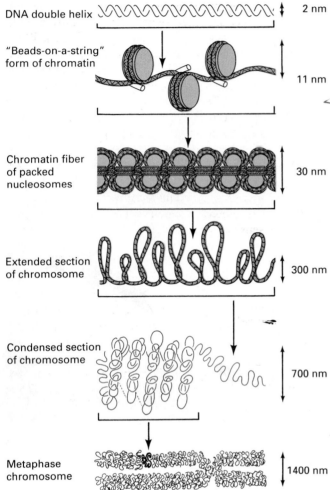

DNA double helix — 2 nm

"Beads-on-a-string" form of chromatin — 11 nm

Chromatin fiber of packed nucleosomes — 30 nm

Extended section of chromosome — 300 nm

Condensed section of chromosome — 700 nm

Metaphase chromosome — 1400 nm

causes negative supercoiling of the DNA, meaning that it twists DNA in the opposite direction from the way in which the double helix is wound. Thus, as with bacterial chromosomes, the existence of looped domains permits localized supercoiling of the DNA without influencing the DNA in other looped domains.

Finally, Figure 10.30 shows the different orders of DNA packing that could give rise to the highly condensed metaphase chromosome. Interphase chromosomes would be packed less tightly than metaphase chromosomes. Very little is known about the mechanisms involved in the transitions between interphase and metaphase chromosome morphologies, or about the more subtle transitions in localized chromosome regions when genes are activated for transcription and are turned off.

KEYNOTE

The large amount of DNA present in the eukaryotic chromosome is compacted by its association with histones in nucleosomes and by higher levels of folding of the nucleosomes into chromatin fibers. The functional state of the chromosome is related to the extent of coiling: The more condensed a part of a chromosome is, the less likely it is that the genes in that region will be active.

Centromeres and Telomeres

So far in this chapter we have described eukaryotic chromosomes as linear structures each containing a single linear DNA molecule wrapped around histones and associated with nonhistone proteins. In terms of size, number, and morphology the chromosome complement of an organism is species-specific. Nonetheless, as described in Chapter 1, all chromosomes behave similarly at the time of cell division. In mitosis, for example, the centromeres of all the chromosomes become aligned at the metaphase plate, the chromatids separate at the centromeres, and one chromatid (now daughter chromosome) of each pair is distributed to each daughter cell. The behavior of chromosomes in mitosis and meiosis is a function of the centromeres, the sites at which chromosomes attach to the mitotic and meiotic spindles, and *telomeres*, the ends of the chromosomes. Centromeres and telomeres are discussed in the following two sections.

CENTROMERES. The centromere region of each chromosome is responsible for the accurate segregation of the replicated chromosomes to the daughter cells during mitosis and meiosis. The error frequency for this process is low but significant, and varies from organism to organism. In yeast, for example, segregation errors (i.e., nondisjunction—see Chapter 3) occur at a frequency of 10^{-5} or less.

In humans nondisjunction occurs much more frequently. However, most zygotes derived from nondisjunctional gametes die early *in utero* (trisomy 21 and some X chromosome abnormalities are exceptions). We can calculate an estimated frequency of nondisjunction in humans as follows: The nondisjunction frequency for chromosome 21 is $1/600 \times 2 = 1/300$. That is, 1/600 is the frequency of Down syndrome births, and the factor 2 is to account for the reciprocal nondisjunction class, that is, the gamete produced with no chromosome 21. When the latter fuses with a normal gamete, the zygote is monosomic for chromosome 21, and that condition is lethal. If we now assume that nondisjunction is an equally probable event for all chromosome pairs (ignoring the sex chromosomes), the expected frequency of nondisjunction in humans is $1/300 \times 22$ (for 22 pairs of autosomes) = 22/300 = approx. 7%.

If the centromere is absent, the chromosome loses its ability to attach to the spindle and thus can only migrate through the cell randomly during the cell division process. Also, acentromeric chromosomes often are degraded so that they are "lost" to the segregation process. That is, the chromosome will replicate but the two chromosome copies will not always segregate properly to the daughter cells.

In many higher eukaryotes the centromere is seen as a constricted region at one point along the chromosome. The region contains the kinetochore (see Chapter 1) to which the spindle fibers attach during cell division, and gives the chromosome its characteristic appearance at metaphase. In most eukaryotes several spindle fibers attach to each centromere, whereas in the yeast *Saccharomyces cerevisiae*, the organism in which centromeres have been best characterized, only one spindle fiber attaches to each centromere.

The DNA sequences (called *CEN* sequences, after the *cen*tromere) of at least 10 *S. cerevisiae* centromeres have been determined. While each centromere has the same function, each differs in specific DNA sequence. Nonetheless the known *CEN* regions are quite similar to one another in nucleotide sequence and organization. Figure 10.31a presents the key sequence information for four yeast centromeres, and Figure 10.31b gives the consensus sequence (i.e., the sequence giving

the base most commonly found at each position) for the known yeast centromeres. The common core centromere region consists of three sequence domains. Element II, a 78–86 bp region of high A + T content (>90 percent A + T bases) is the largest domain. Flanking to one side is element I, which is a conserved PuTCACPuTG sequence (where Pu is a purine, i.e., either A or G), and to the other side is element III, a 26 bp conserved sequence domain that is A–T rich. Interestingly, while the centromere has the same function in all eukaryotes, centromere sequences are quite different in different organisms.

The chromatin organization of *S. cerevisiae* centromeres has been analyzed by mapping chromatin sites that are hypersensitive (i.e., more sensitive than average) to digestion by either micrococcal nuclease or DNase I. The data obtained showed that the core centromere regions I, II, and III are contained within a 220–250 bp protected region that is flanked by highly nuclease sensitive sites (Figure 10.32; p. 316). The protected region is surrounded by an array of nuclease cleavage sites spaced at the usual yeast nucleosomal interval of 160 bp, indicating a normal nucleosomal organization in those areas. As diagrammed in Figure 10.32, *CEN* element III is near the center of the 220–250 bp protected region, suggesting that this element might be directly involved with spindle fiber binding. Note in Figure 10.32 that the centromere core diameter closely matches the spindle fiber microtubule diameter.

Finally, it is clear that the complex structure of the centromere and the attachment of the spindle microtubules must be mediated by proteins, some of which

~ FIGURE 10.31

(a) Regions of sequence homology between *CEN3, CEN4, CEN6,* and *CEN11* in the centromere DNA of yeast, *Saccharomyces cerevisiae*. The sequence elements I–III are set up in a spatial arrangement that is nearly identical in the centromeres of four different chromosomes; (b) Consensus sequence of *S. cerevisiae* centromeres.

a) Specific centromere sequences

Regions:	I	II	III
CEN 3	GTCACATG CAGTGTAC	← 84 bp (93% AT) →	TGTATTTGATTTCCGAAAGTTAAAAA ACATAAACTAAAGGCTTTCAATTTTT
CEN 4	GTCACATG CAGTGTAC	← 78 bp (93% AT) →	TGTTTATGATTACCGAAACATAAAAC ACAAATACTAATGGCTTTGTATTTTG
CEN 6	ATCACGTG TAGTGCAC	← 84 bp (94% AT) →	AGTTTTTGTTTTCCGAAGATGTAAAA TCAAAAACAAAAGGCTTCTACATTTT
CEN 11	GTCACATG CAGTGTAC	← 84 bp (94% AT) →	TGTTCATGATTTCCGAACGTATAAAA ACAAGTACTAAAGGCTTGCATATTTT
	← 8 bp →		← 26 bp →

b) Consensus centromere sequences (in at least 7 of 11 centromeres analyzed)

PuTCACPuTG PyAGTGPyAC	← 78-86 bp(>90% AT) →	TGTTTATGNTTTCCGAAANNNAAAAA ACAAATACNAAAGGCTTTNNNTTTTT
← 8 bp →		← 26 bp →

~ FIGURE 10.32

~ FIGURE 10.32

Model for a centromere and adjacent regions in a yeast chromosome.

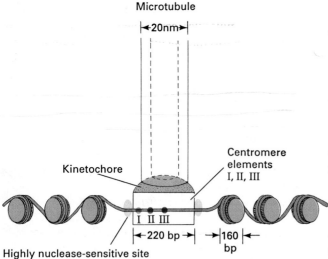

presumably interact directly with centromeric DNA. Some progress is being made in identifying and characterizing centromere-binding proteins. One of the major problems to be solved is the mechanism of attachment of the centromere to the spindle.

TELOMERES. A telomere is the region of DNA at the end of a linear chromosome. It is required for replication and stability of that chromosome. Telomeric regions of chromosomes are characteristically, but not necessarily, heterochromatic. In most organisms that have been examined, the telomeres are positioned just under the nuclear envelope, and are often found associated with each other as well as with the nuclear envelope.

The organization of telomeres is of considerable interest because "broken" chromosomes are highly reactive. That is, they stick readily to other chromosomes, giving broken chromosomes the potential to cause chromosomal aberrations such as translocations. Thus, telomeres must function to stabilize chromosomes against such changes. There is molecular evidence that all telomeres in a given species share a common sequence. Telomeric sequences may be divided into two types:

1. Regions near the ends of chromosomes often contain repeated but still complex DNA sequences extending for many thousands of base pairs from the molecular end of chromosomal DNA. These sequences are called **telomere-associated sequences.**
2. Sequences at, or very close to, the extreme ends of the chromosomal DNA molecules of a number of lower eukaryotes have been determined and have

~ TABLE 10.5

Tandemly Repeated Telomeric DNA Sequences of Eukaryotes

ORGANISM	$5' \rightarrow 3'$ SEQUENCE*
HOLOTRICHOUS CILIATES	
Tetrahymena	...TTGGGG
Glaucoma	
Paramecium	
HYPOTRICHOUS CILIATES	
Stylonchia	...TTTTGGG
Oxytricha	
FLAGELLATES	
Trypanosoma	...TTAGGG
Leptomonas	
Leishmania	
Crithidia	
SLIME MOLDS	
Physarum	...T_nAGGG
Dictyostelium	AG_{1-8}
FUNGUS	
Saccharomyces cerevisiae	$(TG)_{1-3}TG_{2-3}$
HIGHER EUKARYOTES	
Humans	TTAGGG
Arabidopsis	TTTAGGG

* The 5' to 3' strand sequence is directed from the interior of the linear chromosome to its end.

been shown to consist of simple, tandemly repeated DNA sequences. These so-called **simple telomeric sequences** are the essential functional components of telomeric regions in that they are sufficient to supply a chromosomal end with stability. Table 10.5 gives the DNA sequences for the simple telomeric sequences found in a number of lower eukaryotes. In the ciliate *Tetrahymena*, for example, reading from the interior of the chromosome to its end, the repeated sequence consists of elements

~ FIGURE 10.33

Telomere sequence at the ends of *Tetrahymena* chromosomes.

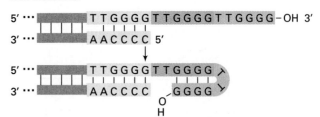

that are 5'-TTGGGG-3' elements, and in the flagellate *Trypanosoma,* the repeated sequence is TTAGGG. Figure 10.33 shows the telomere sequence at the end of a *Tetrahymena* chromosome. Note the unusual G—G base pairing. Much remains to be learned about these sequences in general and the properties of chromosomal ends.

KEYNOTE

The centromere region of each eukaryotic chromosome is responsible for the accurate segregation of the replicated chromosome to the daughter cells during both mitosis and meiosis. The DNA sequences of yeast centromeres (*CEN* sequences) exhibit individual sequence differences while having a high degree of organizational homology. The common core centromere region consists of three sequence domains, one of which has been postulated to serve as a spindle fiber attachment point.

Facts about the ends of chromosomes, the telomeres, are also becoming known. Often the telomeres are associated with the nuclear membrane. Telomeres in a species often share a common sequence. Characteristically, sequences at or very close to the extreme ends of the chromosomal DNA consist of simple, relatively short, tandemly repeated sequences. Repeated, often complex, DNA sequences, called telomere-associated sequences, are found farther in from the chromosome ends. ⎯⎯⎯

SEQUENCE COMPLEXITY OF EUKARYOTIC DNA

Now that we know about the basic structure of DNA and its organization in chromosomes, we can examine some of the distribution of certain sequences in the genomes of prokaryotes and eukaryotes. These data are obtained through denaturation-renaturation analysis of DNA.

Denaturation-Renaturation Analysis of DNA

The two strands of linear, double-helical DNA are held together by the relatively weak hydrogen bonds between the bases. The A—T base pair has two such bonds, while the G—C base pair has three. If a solution of DNA is heated slowly, eventually the hydrogen bonds will break and a population of single-stranded DNA molecules will result; this procedure is called *denaturation* or *melting* of the DNA.

If thermally denatured DNA is cooled rapidly, the single-stranded molecules stay single-stranded. However, if the molecules are cooled slowly, the strands start to come together again, and complementary base pairing occurs to produce some double-stranded molecules. Such *renaturation,* or *reannealing,* of denatured DNA depends on two sequential events. The first is the collision of DNA strands carrying complementary sequences and the proper alignment of these sequences. The second is the formation of hydrogen bonds between the base pairs once the complementary sequences have matched. A complete match is not needed: all that is required is a high degree of complementarity between the two sequences. The study of the kinetics of DNA renaturation, which was pioneered by Roy Britten and David Kohne, gives us useful information about the distribution of sequences—the sequence organization—of DNA in the genome (genomic DNA) and points out yet one more distinct difference between prokaryotes and eukaryotes, as we shall see.

THEORY OF DNA RENATURATION. The rate of renaturation of denatured DNA into double-helical DNA is a function of the concentration of two reactants, the two populations of dissociated single-stranded DNAs. We can write a simple equation for it:

$$S1 + S2 \xrightarrow{k} D$$

Here $S1$ and $S2$ are the two complementary single strands, D is the renatured double-helical DNA, and k is the rate constant for the reaction.

Experimentally, we start out with a known amount of DNA in single-stranded form, and with time (on the order of seconds to hours) an increasing proportion of it becomes double-stranded by DNA–DNA renaturation. It does so as a function of the rate constant k. The rate equation for this experiment is called the C_0t equation and is usually written as

$$\frac{C}{C_0} = \frac{1}{1 - kC_0t}$$

where

k is the rate constant, as before.
t is the time of the reaction, in seconds (s).
C_0 is the initial concentration of single-stranded DNA at time zero, in moles of nucleotides per liter (mol/L).
C is the concentration of single-stranded DNA that remains in the reaction, in moles per liter, after time t has elapsed.

~ FIGURE 10.34

Ideal time course for the renaturation of DNA as seen in a C_0t (initial DNA concentration × time) plot. In the initial state the DNA is single-stranded, and in the final state it is all double-stranded. Note that 80 percent of the renaturation occurs over a 2 log C_0t interval.

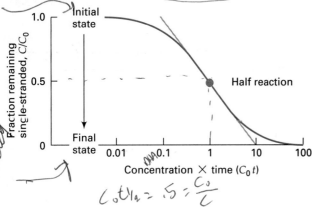

$$C_0t_{1/2} = .5 = \frac{C_0}{C}$$

The equation shows that the fraction of single-stranded DNA remaining in a renaturation reaction (C/C_0) is a function of C_0t, the product of the initial DNA concentration (C_0) and the elapsed time of the reaction (t). Thus the kinetics of DNA renaturation is usually graphed as a **C_0t plot.** An ideal C_0t plot is shown in Figure 10.34. In this plot the fraction of DNA in single-stranded form is graphed as a logarithmic function of the initial concentration of DNA (C_0) multiplied by the time (t), in seconds. The assumption is that all single-stranded DNA fragments in the reaction mix had an equal probability of forming a renatured double-stranded DNA. In other words, we assume that all DNA fragments are *unique sequences*, that is none is represented more than once. On the Y (vertical) axis the initial state for C/C_0—that is, all single-stranded DNA—is indicated by the value of 1, and the final state—that is, all double-stranded DNA—is indicated by the value 0. A convenient value that can be obtained from the curve is $C_0t_{1/2}$, which is the value of C_0t at which the reaction has been half completed (i.e., $C/C_0 = 1/2$).

C_0T CURVES AND THE SEQUENCE ORGANIZATION OF DNA.
In the production of a C_0t curve, DNA is sheared to a length of about 400 base pairs, the fragments are denatured, and renaturation is allowed to proceed. Samples of the reaction are taken at various times, and the relative proportions of single-stranded and double-stranded DNA are determined. Figure 10.35a shows the C_0t curves for several prokaryotic and viral DNAs. The shape of each curve closely parallels the theoretical curve, indicating that all sequences in the DNA of these prokaryotic organisms and viruses are in equal concentration. In fact, the DNAs in question consist mostly of unique sequences.

~ FIGURE 10.35

C_0t plots for a variety of DNAs: (a) C_0t plots showing the renaturation of DNAs from organisms with small genomes: the bacterium *E. coli*, the bacterial viruses T2 and λ, and the animal virus SV40. The dashed line is a theoretical plot for renaturation of unique-sequence DNA from an organism with a genome size equal to that of *E. coli*. (b) Kinetics of renaturation of DNA from calf thymus and *E. coli* as seen in a C_0t plot. The *E. coli* DNA consists almost entirely of unique sequences. However, the shape of the C_0t curve for the calf DNA is very different from that of *E. coli* and indicates that there are some sequences (toward the right of the curve) that renature much more slowly and some (toward the left of the curve) that renature much more quickly than the bacterial DNA sequences.

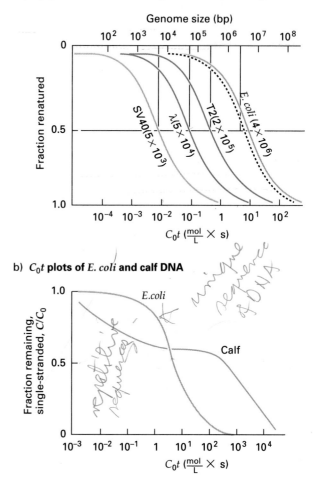

a) C_0t **plots of DNAs from organisms with small genomes**

b) C_0t **plots of *E. coli* and calf DNA**

Note that the curves are not all in the same place on the graph. They occur at different places because the rate of DNA renaturation is related to the amount of DNA in the genome (the genome size) of the organism. Imagine the analogy of a population of nuts and bolts that we are trying to match by randomly trying pairs together. If the population is 100 nuts and 100 bolts of 100 different-sized nut-and-bolt pairs, once a nut is picked, it will take a long time to find the matching bolt, since only 1 out of 100 bolts will fit. However, if only 10 different sizes of nuts and bolts are in the population, when one nut is picked, the chance of finding a matching bolt is 1 in 10, and the process would occur more quickly.

This analogy is equivalent to a renaturation experiment in which there is a fixed weight of DNA, but one genome is a tenth of the size of the other, so the larger genome has one set of sequences and the smaller genome has ten sets of its sequences. As the genome size becomes larger, the $C_0t_{1/2}$ value becomes larger. Further, the $C_0t_{1/2}$ value is related directly to the genome size, so determining the $C_0t_{1/2}$ experimentally allows the genome size to be calculated.

UNIQUE-SEQUENCE AND REPETITIVE-SEQUENCE DNA. If a particular base sequence is present in multiple copies, those sequences will reassociate more rapidly than will unique sequences. Continuing the nuts-and-bolts analogy, if most of the nuts and bolts in the set are unique, then it will take a lot of time to find a matching nut-and-bolt set: however, if all of the nut-and-bolt pairs are the same (similar to highly repetitive sequences), then matched pairs can be formed very quickly. If some nut-and-bolt types are repeated and some are unique, then pairs of the former will be produced more rapidly than pairs of the latter.

Figure 10.35b presents an illustration of the effects of repetitive sequences on DNA renaturation kinetics. The figure shows a C_0t curve for DNA from a calf alongside that for *E. coli*. which, as we have already seen, has only unique-sequence DNA. The curve for calf DNA is very different. Some of the DNA fragments renature at very low C_0t values, indicating that the calf DNA contains repetitive sequences, present in many copies. As a result, the probability of their collision in the solution is much higher than it is for the unique-sequence fragments.

At the other end of the scale some of the calf DNA renatures at very high C_0t values, much higher than those for *E. coli*. These higher C_0t values probably represent unique-sequence DNA, which (because of the much larger genome size in the calf) constitute a smaller proportion and are thus present in low concentration. As a result the calf DNA renatures more slowly than the *E. coli* DNA. In between these two extremes there are apparently a variety of sequences that are repeated to various degrees in the genome, resulting in the broadness of the calf C_0t curve.

By following the kinetics of DNA–DNA renaturation and making C_0t curves, geneticists can determine the abundance of particular types of sequences in DNA. For convenience, we speak of three different types of sequences: **unique** (or single-copy), **moderately repetitive,** and **highly repetitive.** Note that because we are studying randomly sheared pieces of DNA, we are not looking directly at genes but rather at DNA sequences that may or may not contain intact genes.

Unique sequences (sometimes called single-copy sequences), as already indicated, are present as single copies in the genome. (Thus there are two copies per diploid cell.) In current usage the term usually applies to sequences that have one to a few copies per genome. Moderately repetitive sequences are reiterated from a few to as many as 10^3 to 10^5 times in the genome, while highly repetitive sequences are repeated between 10^5 and 10^7 times in the genome.

We could discuss the relative proportions of these unique and repetitive sequences. However, since there is a continuum in the number of repeated sequences within and between organisms, such information would only reflect the somewhat arbitrary boundaries set between the types. Suffice it to say that there is a fair degree of variation between the relative proportion of unique and repetitive sequences among eukaryotic organisms. To give but one concrete example, the distribution of unique and repeated sequences in mouse DNA is shown in Figure 10.36. About 70 percent of the DNA is single-copy material. The repeated sequences fall into two main groups, the highly repeated and the moderately repeated types. About 10 percent of the DNA consists of the highly repeated sequences, where the number of copies is approximately a million. The remaining DNA, that is, about 20 percent of the genome, consists of the heterogeneous class of moderately repeated sequences; these sequences are repeated between 1000 and 100,000 times in the genome.

~ **FIGURE 10.36**

Frequency distribution of unique and repeated nucleotide sequences in mouse DNA. The number of copies of each type of sequence is estimated from the C_0t plot for the mouse.

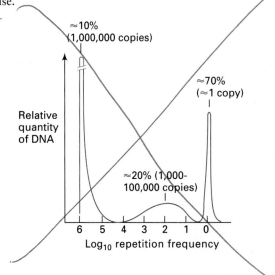

Unique-sequence DNA. Unique sequences take a relatively long time to reassociate after denaturation. Most of the genes that we know about—those that code for proteins in the cell—are in the unique-sequence kinetic class of DNA. Conversely, not all unique-sequence material contains protein-coding sequences. We say this because the amount of DNA present in the unique class would code for more proteins than we believe are produced by the cell.

Moderately repetitive DNA sequences. The moderately repetitive DNA sequence class is heterogeneous in terms of the frequencies of the repeated sequences represented. Some known genes are found in this class, notably the genes for ribosomal RNAs (rRNAs), transfer RNAs (tRNAs), and the histone genes. For example, there are about 200 copies of the major class of rRNA genes in *Neurospora crassa*, 450 copies in the clawed toad (*Xenopus laevis*), 160–200 copies in humans, 260 copies in the sea urchin, 3600 copies in Douglas fir, and 3900 copies in the garden pea.

In most cases these repeated genes are found in one or more clusters in the genome, with each cluster consisting of an uninterrupted tandem array of genes. Sequences that are repeated one after another in a row are called **clustered repeated sequences.** Clusters also occur for some tRNA genes and, in some organisms, for the multiple copies of the histone genes. Only part of each repeated unit of a tandemly repeated cluster is transcribed. In general, in long tandem arrays of complex repeated sequences, portions are transcriptionally active and portions are not.

Also found among the moderately repeated sequences are *interspersed repeated sequences,* which are repeated sequences that are scattered, not clustered, throughout the genome. These sequences are characteristic of most eukaryotic DNAs, although no single unifying description of their arrangement can be applied to all the known examples. The name *interspersed repeated sequences* is applied to those moderately repeated (10^3–10^5, usually) sequence families that are dispersed among unique sequences in the genome.

Two common patterns of interspersion have been found. In the first, the so-called *short-period pattern*—exemplified by that found in *Xenopus* DNA—100–300 base-pair repeated sequences are interspersed with longer unique sequences of about 1000–2000 base pairs in length. Between 50 and 80 percent of DNA typically shows this pattern. Note, though, that other arrangements can also be found in the same genomes—for example, interspersed repeated sequences not much greater than 100–300 base pairs long, long tandem repeated sequences not interspersed with unique-sequence DNA, and long unique-sequence stretches of DNA not interspersed with short repeated sequences.

In the second, or *long-period pattern*—exemplified by that found in *Drosophila* DNA—about 5000 base-pair repeated sequences are dispersed among unique sequences that may be up to 35,000 base pairs or more in length. Recent evidence indicates that these long repeated sequences, even though they consist of different sequence families when the genome is considered as a whole, share a common structure in which the 5000 base-pair repeated sequences are flanked on either side by shorter, direct repeated sequences of about 255–400 base pairs.

A more detailed analysis at the molecular level has revealed additional complexity for the interspersed repeated sequences. On the basis of size and relative abundance, two different classes of interspersed and moderately repeated sequences have been described. One class consists of dispersed families with unit lengths of under 500 base pairs and with as many as hundreds of thousands of copies. These families are called **SINEs,** for **short interspersed repeated sequences.** One example of a SINEs family is the *Alu* sequence family in mammals. These involve 150 to 300 bp moderately repetitive sequences. The DNA sequences of about 50 of the human *Alu* family have been determined. No two sequences are exactly alike, but there are enough similarities to show clearly that they are related. There is evidence that at least some of the *Alu* sequences are located in regions of the chromosomes that are transcribed. However, there does not appear to be any specific function for the *Alu* sequences in controlling transcription.

Other dispersed families of sequences in mammals are several thousand base pairs in length, occurring >20,000 times in the genome. These families are called **LINEs** for **long interspersed repeated sequences.** Only a single LINEs family has been identified in each mammalian species investigated. Families in different species have been shown to be homologous to one another over considerable lengths of DNA. At least some LINEs in a family appear to code for proteins, although the proteins themselves have not been characterized.

All eukaryotic organisms have SINEs and LINEs, although the relative proportions vary considerably. *Drosophila* and birds, for example, have mostly LINEs, while humans and frogs have mostly SINEs.

Highly repetitive DNA sequences. Among the highly repetitive class of DNA sequences are simple clustered repeated sequences, some with repeated units no more than 6 base pairs long repeated as many as 10^6 to 10^7 times in the genome and others hundreds of base pairs long and repeated millions of times. They renature at low C_0t values, thus facilitating their purification. Since these sequences often have a %GC content (base ratio) different from that of the majority of the DNA in the genome, they will usually form a distinct band (or bands) when centrifuged in a solution that separates DNA on the basis of %GC content; such bands are called *satellite bands,* and the DNA contained therein is known as **satellite DNA.**

The amount of satellite DNA found in an organism varies considerably. For example, a small percentage of the human genome consists of satellite DNA, whereas more than 50 percent of the kangaroo rat's genome is satellite DNA. Commonly, the satellite DNA sequences are associated with heterochromatin and thus are localized around centromeres and at the telomeres (ends) of the chromosomes.

It was once thought that satellite DNAs had simple, redundant, repeated units of short base-pair sequences. It is now known that the repeat length can be from a few to several thousands of base pairs, indicating tremendous complexity of these sequences. There is no experimental evidence for the function of satellite DNA sequences. They are not transcribed in the cell; that is, they do not code for a functional product.

$\mathcal{K}$EYNOTE

Prokaryotic genomes consist mostly of unique-sequence DNA, with only a few sequences and genes repeated; eukaryotes have both unique and repetitive sequences in the genome. Particularly in higher eukaryotes, the spectrum of complexity of the repetitive DNA sequences is extensive. The highly repetitive sequences tend to be localized to heterochromatic regions around centromeres and chromosome ends, whereas the unique and moderately repetitive sequences tend to be interspersed. ‎————

SUMMARY

In this chapter we have learned some aspects of the molecular organization of prokaryotic, viral, and eukaryotic chromosomes. Such structural information is important to know in order to understand gene regulation mechanisms more completely.

The chromosomes of prokaryotic organisms consist of circular, double-stranded DNA molecules that are complexed with a number of proteins. The chromosome is compacted within the cell by supercoiling of the DNA helix and the formation of looped domains of supercoiled DNA. Examples of chromosome organization in bacteriophages were discussed, including circular permutation and terminal redundancy of double-stranded DNA in T-even phages, sticky ends in λ double-stranded DNA and their involvement in the phage life cycle, and circular, single-stranded DNA in ΦX174. Each chromosome form is replicated in a particular way to generate progeny chromosomes.

A distinguishing feature of eukaryotic chromosomes is that DNA is distributed among a number of chromosomes. The set of metaphase chromosomes in a cell is called its karyotype. We saw in an earlier chapter how karyotype analysis is useful for the diagnosis of certain human chromosome aberrations. This analysis is aided by the ability to stain human chromosomes so that each chromosome is distinguishable by its banding pattern.

The DNA in the genome specifies an organism's structure, function, and reproduction. The amount of DNA in a genome (made up of a prokaryotic chromosome or of the haploid set of chromosomes in eukaryotes) is called the C value. There is no direct relationship between the C value and the structural or organizational complexity of an organism.

How DNA is organized in eukaryotic chromosomes was detailed in this chapter. The nuclear chromosomes of eukaryotes are complexes of DNA with histone and nonhistone chromosomal proteins. Each chromosome consists of one linear, double-stranded DNA molecule that extends along its length. As a class, the histones are constant from cell to cell within an organism and have been evolutionarily conserved. Nonhistone chromosomal proteins, on the other hand, vary significantly between cell types, and among organisms. The DNA in a eukaryotic chromosome is highly compacted by its association with histones to form nucleosomes and by the several higher levels of folding of nucleosomes into chromatin fibers. The most highly compacted chromosome structure is seen in metaphase chromosomes, while the least compacted structure is seen in interphase chromosomes. The factors controlling the transitions between different levels of chromosome folding as cells go through the cell division cycle are not known. Like prokaryotic chromosomes, eukaryotic chromosomes are organized into a large number of looped domains. These loops consist of 30-nm chromatin fibers and are attached to a protein scaffold.

Important elements for the chromosome segregation in mitosis and meiosis are centromeres and telomeres. Significant progress has been made in defining the sequences of centromeres and telomeres; for example, the DNA sequences of centromeres are relatively complex and are species-specific, and the DNA sequences of telomeres are simple, relatively short, tandemly repeated sequences that are also species-specific. Thus, common chromosome activities in different eukaryotes are controlled by quite different DNA sequences.

By denaturing sheared pieces of genomic DNA and following the kinetics of renaturation of the single-stranded fragments into double-stranded fragments, information can be obtained about genome size and about the frequency with which particular sequences occur in the genome. From such studies, we know that prokaryotic and viral genomes consist mostly of unique-sequence DNA, with only a few repeated sequences. The genomes of eukaryotes contain both unique-sequence and repetitive-sequence DNA, the latter usually being classified into moderately repetitive and highly-repetitive types. Genes are found in unique-sequence DNA, but not all unique-sequence DNA contains genes. Moderately repetitive DNA includes repeated genes such as RNA and tRNA genes, and the histone genes. Highly repetitive DNA generally does not include genes. There is a wide spectrum of complexity of repetitive DNA sequences among eukaryotes.

ANALYTICAL APPROACHES FOR SOLVING GENETICS PROBLEMS

Q.1 When double-stranded DNA is heated to 100°C, the two strands separate because the hydrogen bonds between the strands break. When cooled again, the two strands typically do not come back together again because they do not collide with one another in the solution in such a way that complementary base sequences can form. In other words, heated DNA tends to remain as single strands. Consider the DNA double helix,

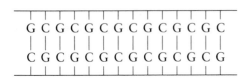

If this DNA is heated to 100°C and then cooled, what might be the structure of the single strands? Assume in formulating your answer that the concentration of DNA is so low that the two strands never find one another.

A.1 This question serves two purposes. First, it reinforces certain information about double-stranded DNA, and, second, it poses a problem that can be solved by simple logic.

In the chapter we examined the structure of DNA and some of the ways to investigate the sequence complexity of the genome, particularly in eukaryotes. One of the experimental approaches discussed was DNA–DNA renaturation in which double-stranded DNA was denatured into single strands and the kinetics of renaturation was measured. Those molecules that renatured rapidly were assumed to be represented in the genome a large number of times, and so on. In the question posed here the double-stranded DNA has been melted to single strands by heat treatment. From the text we know that the renaturation of strands is a function of the product of the concentration at time zero and the time (in minutes)—the C_0t. Conditions are set up here so that the concentration is so low that the two strands do not renature. We are left with an analysis of the base sequences themselves to see whether there is anything special about them in order to avoid an answer of "nothing significant happens." The DNA is a 14-base-pair segment of alternating GC and CG base pairs. By examining just one of the strands, we can see that there is an axis of symmetry at the midpoint such that it is possible for the single strand to form a double-stranded DNA molecule by intrastrand ("within strand") base pairing. The result is a double-stranded hairpin structure, as shown in the following diagram (from the top strand; the other strand will also form a hairpin structure):

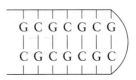

Q.2 An organism has a haploid genome of 10^{10} nucleotide pairs, of which 70% is unique-sequence DNA with a copy number of one, 20% is moderately-repetitive DNA with an average copy number of 1000, and 10% is highly-repetitive DNA with an average copy number of 10^6. Assuming that an average DNA sequence is 10^3 nucleotide pairs, how many different sequences are in each of the three DNA classes?

A.2 a. *Unique-sequence DNA:* the total DNA in this class is 70% of 10^{10} nucleotide pairs = $0.7 \times 10^{10} = 7 \times 10^9$. Only one of each sequence exists so the number of different sequences = $7 \times 10^9 \div 10^3 = \underline{7 \times 10^6}$.

b. *Moderately-repetitive DNA:* DNA in this class is 20% of 10^{10} bp = $0.2 \times 10^{10} = 2 \times 10^9$. Given an average DNA sequence length of 10^3 bp, there are $2 \times 10^9 \div 10^3 = 2 \times 10^6$ sequences in this class. Since the average copy number is 1000, the number of *different* sequences $= 2 \times 10^6 \div 10^3 = 2 \times 10^3$ (2000).

c. *Highly-repetitive DNA:* DNA in this class is 10% of 10^{10} bp $= 0.1 \times 10^{10} = 1 \times 10^9$. Given an average DNA sequence length of 10^3 bp, there are $1 \times 10^9 \div 10^3 = 1 \times 10^6$ sequences in this class. Since the average copy number is 10^6, the number of different sequences $= 1 \times 10^6 \div 10^6 = \underline{1}$.

QUESTIONS AND PROBLEMS

*10.1 The nucleotide sequences of two DNA molecules from a population of T2 DNA molecules are as shown:

1   T A G C T C C →   3   G C T C C T A →

and

2   A T C G A G G ←   4   C G A G G A T ←

These molecules were heat-denatured and then the separated strands were allowed to renature. Diagram the structures of the renatured molecules most likely to appear when (a) strand 2 renatures with strand 3 and (b) strand 3 renatures with strand 4. Mark the strands and indicate sequences and polarity.

10.2 Capital letters represent regions in the chromosome of phage T4. A particular *E. coli* cell was infected by a single T4 chromosome with the sequence ABCDEFAB, but before this chromosome could replicate it suffered a deletion of the E region. This did not interfere with phage replication. What would you expect to be the chromosome sequence(s) of the progeny phage produced upon lysis of this cell? Explain your reasoning.

*10.3 a. If you were to denature and renature a population of normal T4 chromosomes, what kinds of structures would form?
   b. How would the results differ if the chromosomes were from T7?
   c. From λ?

10.4 You are given a single-strand exonuclease (an enzyme that can digest single-stranded DNA from a free end but that cannot digest double-stranded DNA). What would be the results of incubating each of the following phages with single-strand exonuclease and then using the phage to infect *E. coli*?
a. ΦX174   c. T4
b. λ   d. T7

10.5 What are topoisomerases?

*10.6 In typical human fibroblasts in culture, the G1 period of the cell cycle lasts about 10 h, S lasts about 9 h, G2 takes 4 h, and M takes 1 h. Imagine you were to do an experiment in which you added radioactive (^{3}H) thymidine to the medium and left it there for 5 min (pulse), and then washed it out and put in ordinary medium (chase).
a. What % of cells would you expect to become labeled by incorporating the ^{3}H-thymidine into their DNA?
b. How long would you have to wait after removing the ^{3}H medium before you would see labeled metaphase chromosomes?
c. Would one or both chromatids be labeled?
d. How long would you have to wait if you wanted to see metaphase chromosomes containing ^{3}H in the regions of the chromosomes that replicated at the beginning of the S period?

10.7 Assume you did the experiment in Question 10.6, but left the radioactive medium on the cells for 16 h instead of 5 min. How would your answers to the above questions change?

10.8 Karyotype analysis performed on cells cultured from an amniotic fluid sample reveals that the cells contain 47 chromosomes. The Feulgen stained chromosomes are classified into groups and the arrangement shows 6 chromosomes in A, 4 in B, 16 in C, 6 in D, 6 in E, 4 in F, and 5 in the G group. Based on the above information
a. What could be the genotype of the fetus? If more than one possibility exists, give all.
b. How would you proceed to distinguish between the possibilities?

10.9 What is the relationship between cellular DNA content and the structural or organizational complexity of the organism?

*10.10 Match the DNA type with the chromatin type. (More than one DNA type may match a given chromatin type.)

DNA TYPE	CHROMATIN TYPE
Barr body (inactivated DNA)	Euchromatin
Centromere	Facultative heterochromatin
Telomere	Constitutive heterochromatin
Most expressed genes	

10.11 Eukaryotic chromosomes contain (choose the best answer)
a. protein.
b. DNA and protein.
c. DNA, RNA, histone, and nonhistone protein.
d. DNA, RNA, and histone.
e. DNA and histone.

10.12 List four major features that distinguish eukaryotic chromosomes from prokaryotic chromosomes.

10.13 Discuss the structure and role of nucleosomes.

* **10.14** What are the main molecular features of yeast centromeres?

10.15 What are telomeres?

* **10.16** Would you expect to find most protein coding genes in unique-sequence DNA, in moderately repetitive DNA, or in highly repetitive DNA?

10.17 Would you expect to find tRNA genes in unique-sequence DNA, in moderately repetitive DNA, or in highly repetitive DNA?

10.18 The data plotted in the accompanying graph are obtained in a renaturation study of a DNA sample.

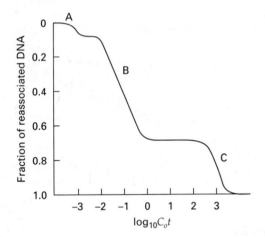

State whether each of the following is true or false. Explain your answer.
a. Satellite DNA is found in C fraction.
b. RNA and histone genes are found in B.
c. Telomere and centromere sequences are found in A.
d. DNA coding for mRNA is found in A.
e. The source of the DNA used in this experiment could be eukaryotic or prokaryotic.

* **10.19** A particular virus has a genome consisting of 10^5 bp of double stranded DNA. When this DNA is denatured and renatured, it reaches 50% renaturation at a C_0t of 3×10^{-1}, and a monophasic renaturation curve is seen. In contrast, when the DNA of a particular amphibian is denatured and renatured, a biphasic renaturation curve is seen. About 50% of this DNA reaches 50% renaturation at a C_0t of 3, while the other half of the DNA reaches 50% renaturation at a C_0t of 3×10^4.
a. What is the size of the amphibian genome?
b. What is the significance of the biphasic renaturation curve shown by the amphibian DNA?
c. How many copies of the C_0t 3 species are present in the amphibian genome?
d. How big are the repeats?

11 DNA REPLICATION AND RECOMBINATION

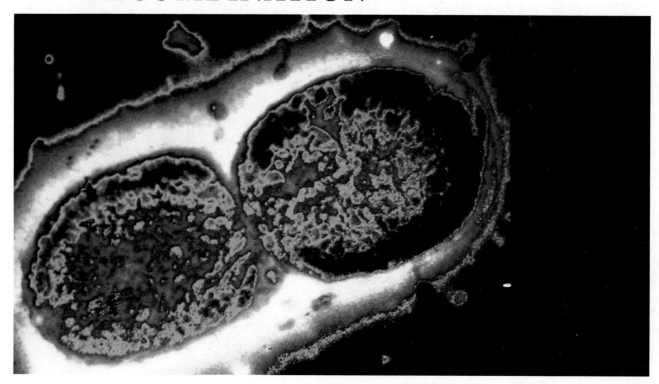

DNA REPLICATION IN PROKARYOTES
Early Models for DNA Replication
The Meselson-Stahl Experiment
DNA Synthesis Enzymes
Molecular Details of DNA Replication
Bacterial DNA Replication and the Cell Cycle

DNA REPLICATION IN EUKARYOTES
Molecular Details of DNA Synthesis in Eukaryotes
Assembly of New DNA into Nucleosomes
Genetics of the Eukaryotic Cell Cycle

DNA RECOMBINATION
Crossing-Over: Breakage and Rejoining of DNA
The Holliday Model for Recombination
Mismatch Repair and Gene Conversion

PRINCIPAL POINTS

~ DNA replication in prokaryotes occurs by a semi-conservative mechanism in which the two strands of a DNA double helix separate and a new complementary strand of DNA is synthesized on each of the two parental template strands. This mechanism ensures the faithful copying of the genetic information at each cell division.

~ The enzymes that catalyze the synthesis of DNA are called DNA polymerases. All DNA polymerases carry out the same DNA synthesis reaction; new strands are made in the 5'-to-3' direction using deoxyribonucleoside 5'-triphosphate (dNTP) precursors.

~ DNA polymerases are incapable of initiating the synthesis of a new DNA strand. All new DNA strands need a short primer of RNA, the synthesis of which is catalyzed by the enzyme DNA primase.

~ DNA replication in *E. coli* requires two of the DNA polymerases and several other enzymes and proteins. Synthesis of DNA on one template strand is continuous while it is discontinuous on the other template strand, a process called semidiscontinuous replication.

~ In eukaryotes, DNA replication occurs in the S phase of the cell cycle and is essentially similar to the replication process in prokaryotes. When the DNA replicates, the chromosome itself must duplicate. That is, histones are assembled into nucleosomes as the replication forks migrate.

~ In both prokaryotes and eukaryotes, genetic recombination involves the breakage and rejoining of homologous DNA double helices.

~ Numerous models have been proposed to describe the molecular events involved with genetic recombination. Some of the enzymes involved are also involved in DNA replication.

~ Mismatch repair of heteroduplex DNA that is an intermediate in genetic recombination can result in a non-Mendelian segregation of alleles, typically a 3:1 or 1:3 ratio rather than the expected 2:2 ratio. This is called gene conversion.

*I*n Chapter 9 we learned that one of the essential properties of genetic material is that it must replicate accurately so that progeny cells have the same genetic information as the parental cell. In this chapter we will learn about the mechanics of DNA replication and chromosome duplication in prokaryotes and eukaryotes and about some of the enzymes and other proteins required for replication. Also, we will learn some of the basic molecular details of DNA recombination.

DNA REPLICATION IN PROKARYOTES

Early Models for DNA Replication

In Chapter 9 we presented Watson and Crick's double-helix model for DNA. In the next paper after the one they published, they reasoned that replication of the DNA would be straightforward if their model was correct. That is, by unwinding the DNA molecule and separating the two strands, each strand could be a template for the synthesis of a new, complementary strand of DNA. As the DNA double helix is progressively unwound from one end, the base sequence of the new

strand is determined by the base sequence of the template strand, following complementary base-pairing rules. When replication is completed, there are two progeny DNA double helices, each consisting of one parental DNA strand and one new DNA strand. This model for DNA replication is known as the **semiconservative model** since each progeny molecule retains one of the parental strands (Figure 11.1a).

At the time, two other models for DNA replication were proposed, the **conservative model** (Figure 11.1b) and the **dispersive model** (Figure 11.1c). In the conservative model, the two parental strands of DNA remain together and as a whole serve as a template for the synthesis of new progeny DNA double helices. Thus, one of the two progeny DNA molecules is actually the parental double-stranded DNA molecule, and the other consists of totally new material. In the dispersive model, the parental double helix is cleaved into double-stranded DNA segments which act as templates for the synthesis of new double-stranded DNA segments. Somehow, the segments reassemble into complete DNA double helices, with parental and progeny DNA segments interspersed. Thus, while the two progeny DNAs are identical with respect to base-pair sequence, the parental DNA has actually become dispersed throughout both progeny molecules.

~ FIGURE 11.1

Three models for the replication of DNA: (a) The semiconservative model (the correct model); (b) The conservative model; (c) The dispersive model. The parental strands are shown in taupe and the newly synthesized strands are shown in red.

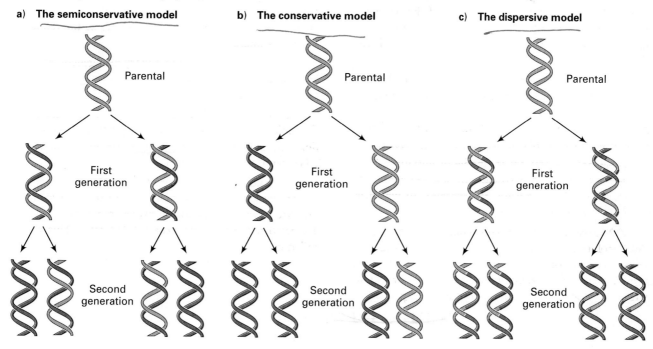

a) The semiconservative model

Parental

First generation

Second generation

b) The conservative model

Parental

First generation

Second generation

c) The dispersive model

Parental

First generation

Second generation

The Meselson-Stahl Experiment

Five years after Watson and Crick developed their model, Matthew Meselson and Frank Stahl obtained experimental evidence that the semiconservative replication model was the correct one. To examine the process of DNA replication, Meselson and Stahl used the bacterium *Escherichia coli* because it can be easily and quickly grown in a minimal medium. In Meselson and Stahl's experiment (Figure 11.2; p. 328) *E. coli* was grown for several generations in a minimal medium in which the only nitrogen source was $^{15}NH_4Cl$ (ammonium chloride). In this compound the normal isotope of nitrogen, ^{14}N, is replaced with ^{15}N, the heavy isotope. (Note: Density is weight/volume so ^{15}N, with one extra neutron in its nucleus, is 1/14 more dense than ^{14}N.) As a result, all the bacteria's cellular nitrogen-containing compounds, including its DNA, contained ^{15}N instead of ^{14}N. ^{15}N DNA can be separated from ^{14}N DNA by using equilibrium density gradient centrifugation (described in Box 11.1; p. 329). Briefly, in this technique a solution of cesium chloride (CsCl) is centrifuged at high speed, causing the cesium chloride to form a density gradient. If DNA is present in the solution, it will band to a position where its buoyant density is the same as that of the surrounding cesium chloride.

As the next step in Meselson and Stahl's experiment, the ^{15}N-labeled bacteria were transferred to a medium containing nitrogen in the normal ^{14}N form. The bacteria were allowed to replicate in the new conditions for several generations. Throughout the period of incubation in the ^{14}N medium, samples of *E. coli* were lysed to release the cellular contents, and the DNA was analyzed in cesium chloride density gradients (Figure 11.2). After one generation in ^{14}N medium, all the DNA had a density that was exactly intermediate between that of totally ^{15}N DNA and totally ^{14}N DNA. After two generations, half the DNA was of the intermediate density and half was of the density of DNA containing entirely ^{14}N. These observations, presented in Figure 11.2, and those for subsequent generations were exactly what the semiconservative model predicted. If the conservative model of DNA replication had been correct, after one generation two bands of DNA would be seen (see Figure 11.1b). One band would be in the heavy-density position of the gradient, containing parental DNA molecules, and both strands would consist of ^{15}N-labeled DNA only. The other band would be in the light-density position, containing progeny DNA molecules with both strands totally ^{14}N-labeled. In subsequent generations, the heavy parental DNA band would be seen at each generation, and in the amount found at the start of the experiment. All new DNA molecules would have both strands totally labeled with ^{14}N. Hence, the amount of DNA in the light-density position would increase with each generation.

~ FIGURE 11.2

The Meselson-Stahl experiment: The demonstration of semiconservative replication in *E. coli*. Cells were grown in ^{15}N-containing medium for several generations, and then transferred to ^{14}N-containing medium. At various times over several generations, samples were taken; the DNA was extracted and analyzed by CsCl equilibrium density gradient centrifugation. Shown in the figure are a schematic interpretation of the DNA composition at various generations, photographs of the DNA bands, and a densitometric scan of the bands.

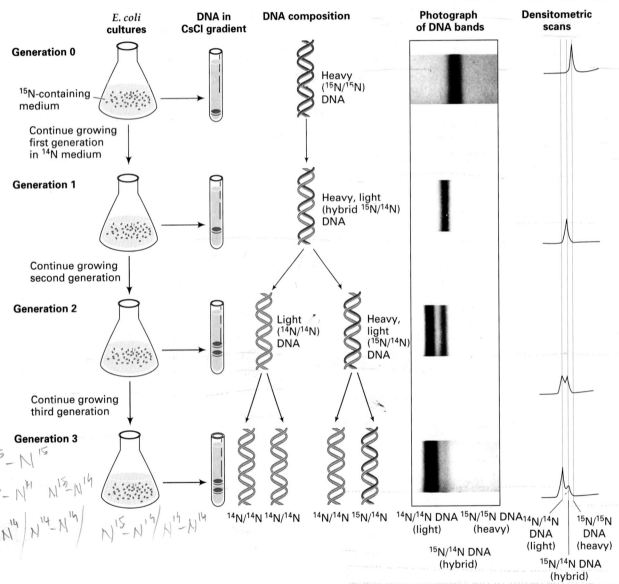

In the conservative model of DNA replication, the most significant prediction was that *at no time would any DNA of intermediate density have been found*. The fact that intermediate-density DNA *was* found ruled out the conservative model.

In the dispersive model for DNA replication, the parental DNA is scattered in double-stranded segments throughout the progeny DNA molecules (see Figure 11.1c). According to this model, all DNA present after one generation in ^{14}N-containing medium would be of intermediate density, and this was seen in the Meselson-Stahl experiment. Thus, the dispersive model could not be ruled out after just one generation of replication. After a second generation in ^{14}N-containing medium, the dispersive model predicted that DNA segments from the first generation would be dispersed through the DNA double helices produced. Thus, the $^{15}N-^{15}N$ DNA segments dispersed among new $^{14}N-^{14}N$ DNA after one generation would then be distributed among twice as many DNA molecules after

Box 11.1
Equilibrium Density
Gradient Centrifugation

In experiments involving equilibrium density gradient centrifugation, a concentrated solution of cesium chloride (CsCl) is centrifuged at high speed. The opposing forces of sedimentation and diffusion produce a stable, linear concentration gradient of the CsCl. The actual densities of CsCl at the extremes of the gradient are related to the CsCl concentration that is centrifuged.

For example, to examine DNA of density 1.70 g/cm^3 (a typical density for DNA), one makes a gradient that spans that density, for example from 1.60 to 1.80 g/cm^3. If the DNA is mixed with the CsCl when the density gradient is generated in the centrifuge, the DNA will come to equilibrium at the point in the gradient where its buoyant density equals the density of the surrounding CsCl (see Box Figure 11.1). The DNA is said to have banded in the gradient. If DNAs are present that have different densities, as is the case with ^{15}N-DNA and ^{14}N-DNA, then they will band (i.e., come to equilibrium) in different positions.

In the experiments the gradients contain the stain ethidium bromide, which complexes with DNA. The ethidium bromide fluoresces under ultraviolet light; therefore the bands of DNA can be visualized.

The method can also be used experimentally to purify DNAs on the basis of their different densities. It has been used, for example, to fractionate nuclear and mitochondrial DNA in extracts of eukaryotic cells and to separate bacterial plasmid DNA from the host bacterium's DNA.

~ BOX FIGURE 11.1

Schematic diagram for separating DNAs of different buoyant densities by equilibrium centrifugation in a cesium chloride density gradient. Illustrated is the separation of satellite DNA (containing highly repetitive sequences) and main band DNA (the remainder of the DNA) from a eukaryote.

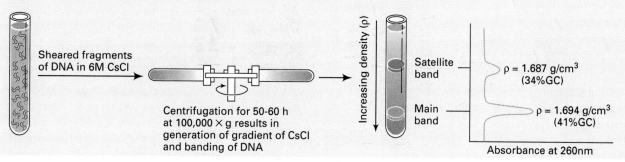

two generations. As a result, the DNA molecules would be found at one band located halfway between the intermediate-density and light-density band positions. With subsequent generations, there would continue to be one band and it would become lighter in density with each generation. Such a slow shift in DNA density was not seen in the results of the Meselson-Stahl experiment and therefore the dispersive model was ruled out.

each of the two parental template strands. Semiconservative replication results in two double-stranded DNA molecules, each of which has one strand from the parent molecule and one newly synthesized strand. This mechanism ensures the faithful copying of the genetic information at each cell division.

KEYNOTE

DNA replication in *E. coli* and other prokaryotes occurs by a semiconservative mechanism in which the strands of a DNA double helix separate and a new complementary strand of DNA is synthesized on

DNA Synthesis Enzymes

In 1955 Arthur Kornberg and his colleagues set out to find the enzymes that were necessary for the DNA replication so that he could describe the reactions involved in detail. His approach was to identify all the necessary ingredients required for the synthesis of *E. coli* DNA in vitro. The first successful synthesis of DNA was accomplished in a reaction mixture containing DNA fragments, a mixture of four deoxyribonucleoside 5'-triphosphates (dATP, dGTP, dTTP, and dCTP) that we will abbreviate as dNTP, and a lysate prepared from *E. coli* cells. So that he could measure the very small amount of DNA expected to be synthesized in the reaction, Kornberg used radioactively labeled dNTPs in the reaction mixture.

Realizing that a crucial component (or components) for DNA synthesis must be present in the *E. coli* lysate, Kornberg next analyzed the lysate in order to find that component. In doing so, he isolated the enzyme that was responsible for DNA synthesis. This enzyme was originally called the *Kornberg enzyme*, but it is now most commonly called **DNA polymerase I**. (By definition, the enzymes that catalyze the synthesis of DNA are called **DNA polymerases**.) Once DNA polymerase I was purified, more detailed information could be obtained about DNA synthesis *in vitro*. The first things studied were the various ingredients necessary for the synthesis of DNA *in vitro*. Researchers found that four components were needed for the in vitro synthesis of DNA. If any one of the following four components were omitted, DNA synthesis would not occur:

1. all four dNTPs (If any one dNTP was missing, no synthesis occurred.)
2. magnesium ions (Mg^{2+})
3. a fragment of DNA to act as a template
4. DNA polymerase I

Subsequent experiments showed that the new DNA that was made *in vitro* was a faithful base-pair for base-pair copy of the original DNA. That is, the original DNA acted as a template for the new DNA synthesis.

ROLE OF DNA POLYMERASES. All DNA polymerases catalyze the polymerization of nucleotides into a DNA chain. The reaction that they catalyze is

$$(dNMP)_n + dNTP \xrightleftharpoons{\text{DNA polymerase}} (dNMP)_{n+1} + PP_i$$

Here the DNA is represented as a string of a number (*n*) of deoxyribonucleoside 5'-monophosphates (dNMP). The next nucleotide to be added is the precursor, dNTP. The DNA polymerase catalyzes a reaction in which one nucleotide is added to the existing DNA chain with the release of inorganic pyrophosphate (PP_i). The enzyme then repeats its action until the new DNA chain is complete.

The action of DNA polymerase in synthesizing a DNA chain is shown at the molecular level in Figure 11.3a. The same reaction in shorthand notation is shown in Figure 11.3b. The reaction has three main features:

1. At the growing end of the DNA chain, DNA polymerase catalyzes the formation of a **phosphodiester bond** between the 3'-OH group of the deoxyribose on the last nucleotide and the 5'-phosphate of the deoxyribonucleoside 5'-triphosphate (dNTP) precursor. The formation of the phosphodiester bond results in the release of two of three phosphates from the dNTP. The important concept here is that *the lengthening DNA chain acts as a primer in the reaction.*

2. The addition of nucleotides to the chain is not random; each deoxyribonucleotide is selected by the DNA polymerase, which is always bound to the DNA and which moves along the template strand as the polynucleotide chain is lengthened. The polymerase finds the precursor (dNTP) that can form a complementary base-pair with the nucleotide on the template strand of DNA. Since the DNA polymerase is bound to the template DNA, it ensures that the correct precursor has been chosen. This does not occur with 100 percent accuracy, but the error frequency is extremely low.

3. The direction of synthesis of the new DNA chain is only from 5' to 3' because of the properties of DNA polymerase.

All known DNA polymerases carry out the same reaction and will make new DNA copies from any DNA added to the reaction mixture, provided all the necessary ingredients are present. Thus, if given human DNA, *E. coli* DNA polymerase can replicate it faithfully, and vice versa.

One of the best-understood systems of DNA replication is that of *E. coli*. For several years after the discovery of the Kornberg enzyme (DNA polymerase I), scientists believed that this enzyme was the only DNA replication enzyme in *E. coli*. However, evidence was obtained indicating that the enzyme was not solely responsible for DNA synthesis *in vivo*.

One such indication came from genetic studies in which *E. coli* strains with mutations in the gene for DNA polymerase I were isolated. One way to study the action of a particular enzyme *in vivo* is to induce a mutation in the gene that codes for that particular enzyme. In this way the phenotypic consequences of the mutation can be compared with the wild-type phenotype. A mutation in the gene coding for an enzyme that is as essential to cell function as DNA polymerase, for instance, would be lethal. The first DNA polymerase I mutant, *polA1,* was isolated in 1969 by Peter DeLucia and John Cairns. The defect in the mutant was a base-pair change in the gene that resulted in a shorter-than-normal protein that was non-functional. This type of mutation is called a nonsense mutation (see Chapter 18). Unexpectedly, *E. coli* cells carrying the *polA1*

~ FIGURE 11.3

DNA chain elongation catalyzed by DNA polymerase. (a) Mechanism at molecular level;
(b) The same mechanism, using a shorthand method to represent DNA.

a) Mechanism of DNA elongation

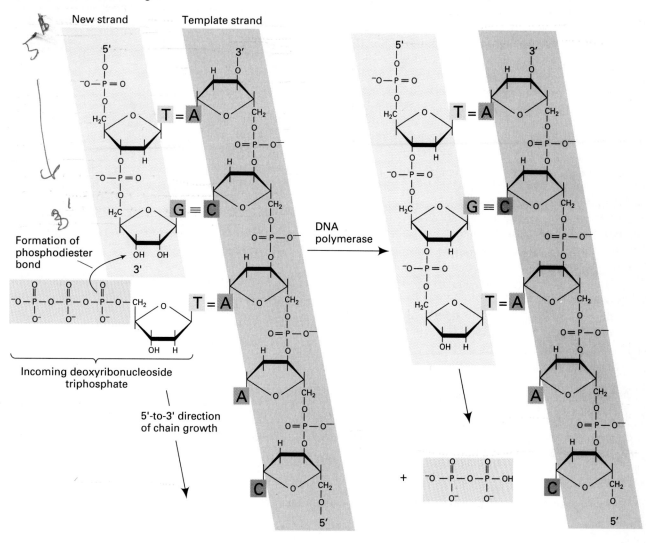

b) Shorthand notation

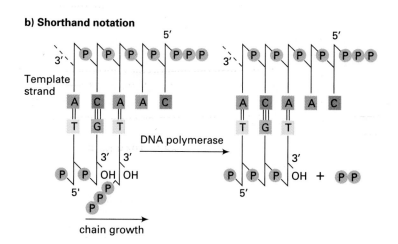

mutation grew and divided normally, showing that DNA polymerase I is not essential to cell function.

Subsequently, DNA polymerase I mutants that were temperature-sensitive were isolated. To study the consequences of mutations in genes coding for essential proteins and enzymes, geneticists find it easiest to work with **conditional mutants,** that is, mutant organisms that are normal under one set of conditions but that become seriously impaired or die under other conditions (Figure 11.4). The most common types of conditional mutants are those that are temperature-sensitive— mutant organisms that function normally until the temperature is raised past some threshold level, at which time some temperature-sensitive defect is manifested.

It was expected that the temperature-sensitive DNA polymerase I mutants would die at elevated temperatures. At *E. coli*'s normal growth temperature of 37°C, the temperature-sensitive *polA* mutant strains produce DNA polymerase I with normal catalytic activity. At 42°C, however, the mutant strains produce DNA polymerase I molecules that have only about 1 percent of the normal catalytic activity of the wild-type enzyme.

Like *polA1*, the temperature-sensitive *polA* mutant cells multiplied at the same rate as the parental (wild-type) strain, even though the temperature was elevated, and DNA polymerase I activity was virtually non-existent. These results indicated that there must be other DNA-polymerizing enzymes in the cell, and a search for these enzymes in extracts of *E. coli* began. In 1970, Martin Gefter, Rolf Knippers, and C. C. Richardson, all working independently, discovered DNA polymerase II, and Tom Kornberg and M. L. Gefter, working together, discovered DNA polymerase III in 1971. In *E. coli*, DNA polymerase II is encoded by the *polB* gene. DNA polymerase III consists of many subunits. Three of these subunits constitute a "core enzyme" that is capable of synthesizing DNA: (1) α (130,000 Daltons [Da]; *polC* [*dnaE*] gene); (2) β (40,000 Da; *dnaN* gene); and; θ [theta] (10,000 Da; no gene yet identified). The activity of these subunits is enhanced by the presence of four other subunits (see Table 11.1 footnote). In every cell there are about 400, 50–100, and 10–20 molecules of DNA polymerase I, II, and III, respectively. The number of copies of DNA polymerase II per cell is not known precisely, but there are far fewer than DNA polymerase I. Indeed, since enzymes II and III are present in the cell in much lower amounts than DNA polymerase I, their activities had been masked previously. A comparison of the important properties of the three *E. coli* DNA polymerases is given in Table 11.1.

~ FIGURE 11.4

The effect of a conditional mutation. (a) A normal gene, when transcribed and translated, produces a protein that functions normally at normal and high temperatures. (b) A gene with conditional mutation produces a slightly altered protein that functions normally at normal temperatures but at high temperatures changes its shape so that function is lost or reduced significantly.

a) **Function of a normal gene**

b) **Function of a gene with a conditional (temperature-sensitive) mutation**

~ TABLE 11.1

Comparison of the Structural and Functional Characteristics of the *E. coli* DNA
Polymerases I, II, and III

DNA POLYMERASE	PROPERTIES						
	POLYMERIZA-TION: $5' \rightarrow 3'$	EXONUCLEASE: $3' \rightarrow 5'$	EXONUCLEASE: $5' \rightarrow 3'$	MOLECULAR WEIGHT (DALTONS)	MOLECULES PER CELL (APPROXI-MATELY)	STRUCTURAL GENES	CONDITIONAL LETHAL MUTANTS
I	yes	yes	yes	109,000	400	*polA*	yes
II	yes	yes	no	120,000	Not known	*polB*	no
III	yes	yes	yes	>250,000[a] many different subunits	10–20	*polC (dnaE), dnaN, dnaQ, dnaZX,* and 2 others	yes

[a]Polymerase III consists of several subunits: a core of α (130,000; *dnaE*), β (40,000; *dnaN*), and θ (theta: 10,000), and ϵ (epsilon: 25,000; *dnaQ*), δ (delta: 32,000), γ (gamma: 52,000; *dnaZX*) and τ (tau: 71,000; *dnaZX*).

Both DNA polymerase I and III are known to be involved in DNA replication, but the role of DNA polymerase II is unknown at this time.

As Table 11.1 shows, all three *E. coli* DNA polymerases have $3' \rightarrow 5'$ exonuclease activity; that is, they can catalyze the removal of nucleotides from the 3' end of a DNA chain. So, in addition to selecting the correct nucleotide precursor to add to the primer DNA strand, the DNA polymerases also check the accuracy of the base pair linking the primer strand terminus and the template strand. If the wrong base pair has been generated in error, hydrogen bonding between complementary bases cannot occur and DNA synthesis cannot proceed. In such cases, the $3' \rightarrow 5'$ exonuclease activity catalyzes the excision of the erroneous nucleotide on the primer strand. The polymerase activity then catalyzes the formation of the correct base pair. Thus in DNA replication, $3' \rightarrow 5'$ exonuclease activity is a **proofreading** mechanism that helps keep the frequency of DNA replication errors very low.

In addition, DNA polymerases I and III have $5' \rightarrow 3'$ exonuclease activities and can remove nucleotides from the 5' end of a DNA strand. This activity, also important in DNA replication, will be examined later in this chapter.

KEYNOTE

The enzymes that catalyze the synthesis of DNA are called DNA polymerases. Three DNA polymerases, I, II, and III, have been identified in *E. coli*: I and III are known to be involved in DNA replication. The three enzymes differ in a number of properties, including size, number of molecules per cell, and proofreading ability.

Molecular Details of DNA Replication

INITIATION OF DNA REPLICATION. The initiation of DNA replication in prokaryotes requires the local denaturation of the DNA at a specific DNA sequence called an **origin of replication.**

Origin of replication. Replication of prokaryotic and viral DNA usually starts at a specific site on the chromosome, the **origin.** At that site the double helix denatures into single strands, exposing the bases for the synthesis of new strands. In a circular chromosome, such as is found in *E. coli*, the local denaturing of the DNA produces what is called a **replication bubble.** The segments of untwisted single strands upon which the new strands are made (following complementary base pairing rules) are called the **template strands.** In *E. coli* there is a single origin, or replication, called *oriC* from which replication proceeds bidirectionally.

Initiation of DNA synthesis. Figure 11.5 (p. 334) shows a model for the formation of a replication bubble at a replication origin in *E. coli*, and the initiation of the new DNA strand. This model is based on *in vitro* studies of DNA replication. An initiator protein binds to the parental DNA molecule at the origin of replication sequence (Figure 11.5,1). The two DNA strands then separate in that region; this involves a

~ FIGURE 11.5

Schematic model for the formation of a replication bubble at a replication origin in *E. coli* and the initiation of the new DNA strand. [From Alberts 2nd ed., Figure 5.50, p. 235]

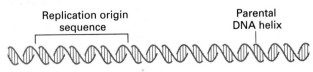

Replication origin sequence | Parental DNA helix

❶ Initiator protein binds to replication origin

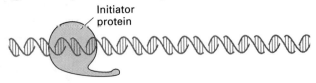

Initiator protein

❷ DNA helicase binds to initiator protein

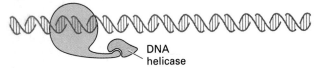

DNA helicase

❸ Helicase loads onto DNA

❹ Helicase denatures helix and binds with DNA primase to form primosome

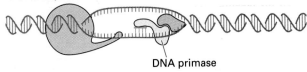

DNA primase

❺ Primase synthesizes RNA primer which is extended as DNA chain by DNA polymerase

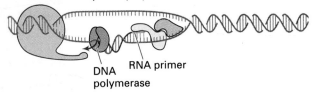

DNA polymerase | RNA primer

localized denaturation and requires the physical untwisting of the DNA. This untwisting is catalyzed by the enzyme **DNA helicase** (product of the *rep* gene), which first becomes bound to the initiator protein (Figure 11.5,2), then is loaded onto the DNA itself (Figure 11.5,3). The untwisting reaction requires energy derived from the hydrolysis of ATP. Next, another enzyme, **DNA primase**, binds to the helicase and the denatured DNA (Figure 11.5,4). Primase is derived from **RNA polymerase** (the enzyme responsible

for RNA synthesis in the cell) by the addition of at least two other proteins.

What is the function of primase? No known DNA polymerases can initiate the synthesis of a DNA strand; they can only catalyze the addition of deoxyribonucleotides (dNTPs) to a preexisting strand. The initiation of DNA synthesis involves the synthesis of a short **RNA primer**, catalyzed by primase (Figure 11.5,5). The complex of the primase, helicase, and perhaps other proteins with the DNA is called the **primosome**. Only very short RNA primers made up of from three to five nucleotides, are made. In a given organism all RNA primers that are synthesized have very similar sequences, suggesting that primases synthesize RNA at specific sequences along the template DNA strand. The primers function as a preexisting polynucleotide chain to which new deoxyribonucleotides can be added by reactions catalyzed by DNA polymerase (Figure 11.5,5). The RNA primers themselves do not remain as part of the new DNA chain; they are removed and replaced with DNA by the action of DNA polymerase, as we shall discuss later.

At this point it is important to make a clear distinction between **template** and **primer** with respect to DNA replication. A template strand is one upon which the new strand is made, following complementary base pairing rules; thus, the base sequence of the template strand directs the synthesis of a new strand with a complementary base sequence. A primer is a short segment of nucleotides bound to the template strand. The primer nucleotides serve as a substrate for the action of DNA polymerase, which extends the primer as a new DNA strand, the sequence of which is complementary to the template strand.

KEYNOTE

No DNA polymerase can initiate the synthesis of a new DNA chain. Instead, the initiation of DNA synthesis first involves the denaturation of double-stranded DNA at an origin of replication, catalyzed by DNA helicase. Next, DNA primase binds to the helicase and the denatured DNA and synthesizes a short RNA primer. The RNA primer is extended by DNA polymerase as new DNA is made. The RNA primer is later removed. _____

SEMIDISCONTINUOUS DNA REPLICATION. When a double-stranded DNA molecule unwinds to expose the two single-stranded template strands for DNA replication, a Y-shaped structure is formed, called a **repli-**

cation fork. A replication fork moves in one direction. This unidirectional movement poses a problem, since DNA polymerases can only catalyze DNA synthesis in the 5'-to-3' direction, yet the two DNA strands are of opposite polarity. As the DNA helix unwinds to provide templates for new synthesis, the new DNA cannot be polymerized continuously on the 3'-to-5' strand.

The work of Reiji and Tuneko Okazaki and their colleagues suggested a mechanism to explain this result. They added a radioactive DNA precursor (^{3}H-thymidine) to cultures of *E. coli* for 0.5 percent of a generation time. Next, they added a large amount of nonradioactive thymidine to prevent the incorporation of any more radioactivity into the DNA. They followed what happened to the radioactively labeled thymidine. At various times they extracted the DNA and determined the size of the newly labeled molecules. At intervals soon after the labeling period, most of the radioactive ^{3}H-thymidine was present in relatively low-molecular-weight DNA about 100–1000 nucleotides long. As time increased, a greater and greater proportion of the labeled molecules was found in high-molecular-weight DNA. These results indicated that DNA replication normally involves the synthesis of short DNA segments, called **Okazaki fragments,** which are subsequently linked together by the action of DNA polymer-

ase to remove the RNA primers. Removal is followed by the action of an enzyme called **DNA ligase,** which catalyzes formation of the final phosphodiester bond between Okazaki fragments to form a long polynucleotide chain. In other words, Okazaki's group had shown that DNA replication is **discontinuous.**

We can relate the discontinuous nature of DNA synthesis to the replication of the circular DNA chromosome in *E. coli.* One location on the chromosome serves as the point of origin for DNA replication. At this point the DNA denatures to expose the two template strands. Since denaturation occurs in the middle of a circular DNA molecule rather than at the end, the result is the two Y-shaped structures linked head-to-head at the points of the Ys. These structures act as two replication forks, with DNA synthesis taking place in both directions (bidirectionally) away from the origin point.

The early stages of **bidirectional replication** are shown in Figure 11.6. This figure also shows how discontinuous replication occurs. As the replication fork migrates, synthesis of one new strand (the **leading strand**), is continuous since the 3'-to-5' template strand (the leading strand template) is being copied. Synthesis of the other new strand (the **lagging strand**) must be discontinuous because the helix must unwind to expose

~ **FIGURE 11.6**

Bidirectional DNA replication. Synthesis of DNA is initiated at the origin and proceeds in the 5'-to-3' direction while the two replication forks are migrating in opposite directions. Replication is continuous on the leading strand and discontinuous on the lagging strand. These events occur in the replication of all DNA in prokaryotes and eukaryotes.

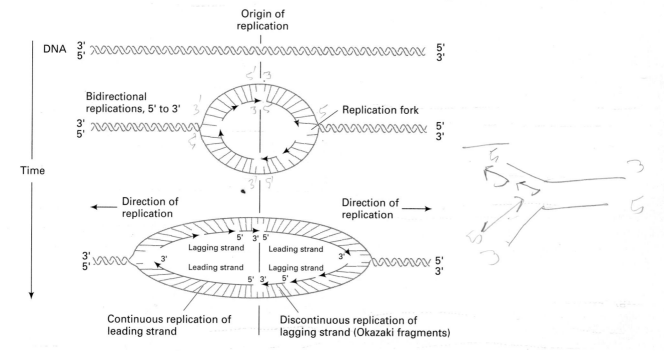

~ FIGURE 11.7

Model for the events occurring around the replication fork of the *E. coli* chromosome. (a) Untwisting; (b) Initiation; (c) Further untwisting and elongation of the new DNA strands; (d) Further untwisting and continued DNA synthesis; (e) Removal of the primer by DNA polymerase I; (f) Joining of adjacent DNA fragments by the action of DNA ligase. (From James E. Watson, et al., *Molecular Biology of the Gene*, 4th ed. Copyright 1965, 1970, 1976, 1987, by the Benjamin/Cummings Publishing Company, Inc. Reprinted by permission.)

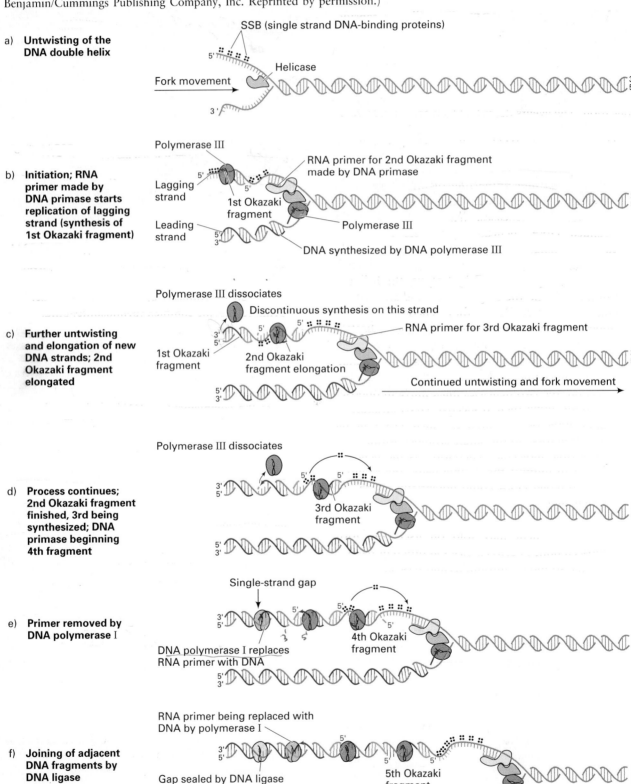

a) **Untwisting of the DNA double helix**

b) **Initiation; RNA primer made by DNA primase starts replication of lagging strand (synthesis of 1st Okazaki fragment)**

c) **Further untwisting and elongation of new DNA strands; 2nd Okazaki fragment elongated**

d) **Process continues; 2nd Okazaki fragment finished, 3rd being synthesized; DNA primase beginning 4th fragment**

e) **Primer removed by DNA polymerase I**

f) **Joining of adjacent DNA fragments by DNA ligase**

a new segment of 3'-to-5' template (the lagging strand template) on which the new strand can be made. Since one new DNA strand is synthesized continuously and the other discontinuously, DNA replication as a whole is considered to occur in a **semidiscontinuous** fashion.

DNA REPLICATION MODEL. We will now consider a model for DNA replication which incorporates all of the facts we have just discussed (Figure 11.7). The model involves a single replication fork; for circular chromosomes with bidirectional replication, two replication forks are formed on either side of the origin of replication. The events described here for a single replication fork apply to each of those two replication forks.

The key steps in DNA replication are as follows:

1. *Denaturation and untwisting of the double helix* (Figures 11.5 and 11.7a). These events are catalyzed by DNA helicases. When bound to single-stranded DNA, helicase hydrolyzes ATP. This causes a change in shape of the enzyme molecule, enabling the enzyme to move along a DNA single strand. By repeated ATP hydrolysis, the helicases can move along the single strand and untwist any double-stranded DNA they encounter. There appear to be two helicases bound at the replication fork, one bound to the lagging-strand template strand and the other to the leading-strand template strand. Since the replication fork moves in one direction, the two helicases must move in opposite directions along their respective DNA strands.

2. *Stabilization of the single-stranded DNA in the replication fork.* In the single-stranded form, DNA is a flexible molecule. Therefore, as the helicases proceed to untwist the double helix, the resulting single-stranded DNA could potentially reform the original hydrogen bonds to reestablish a double-helical molecule. Alternatively, each single-stranded segment may fold on itself and form hydrogen bonds between bases at different parts of the same strand (= intramolecular base pairing). In either case, the double-stranded regions would impede the path of DNA polymerase as it synthesizes a new strand on the exposed DNA template. These two scenarios are avoided by the action of **single-strand DNA-binding (SSB) proteins,** also called **helix-destabilizing proteins.** SSB proteins bind to the single-stranded DNA without covering the bases so that they are still readable by DNA polymerase (Figures 11.7 and 11.8). In binding to the single-stranded DNA, the SSB proteins help the DNA unwinding process by stabilizing the single-stranded DNA. In *E. coli* the SSB protein is a 177-amino acid polypeptide that exists as a tetramer

~ **FIGURE 11.8**

Role of single-strand DNA-binding proteins in stabilizing the single strands of DNA at the replication fork. [From Alberts 2nd ed., Figure 5.45, p. 232]

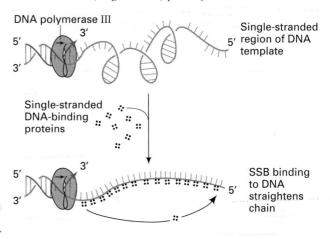

and that binds to a 32-nucleotide segment of DNA. Over 200 of the proteins bind to each replication fork.

3. *Initiation of synthesis of new DNA strands.* The primase-enzyme complex (primosome) binds to the single-stranded DNA and synthesizes the RNA primer (Figures 11.5 and 11.7b). A schematic of the reaction catalyzed by primase is shown in (Figure 11.9). The RNA primers are lengthened by the action of DNA polymerase III, which synthesizes the complementary DNA chains to the template strand (Figure 11.7b and 11.7c). To maintain

~ **FIGURE 11.9**

Drawing of the reaction catalyzed by primase. The enzyme's direction of movement is indicated by the arrow; the RNA primer is synthesized in the 5'-to-3' direction. After a short RNA primer has been synthesized, the enzyme stops, leaving the 3' end of the primer available for elongation by DNA polymerase III.

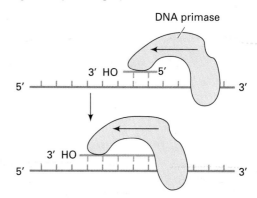

the 5'-to-3' polarity of DNA synthesis, the direction of DNA synthesis is different on the two template strands, depending upon the direction in which the replication fork moves. That is, the leading strand is synthesized in the same direction as the direction of fork movement while the lagging strand is synthesized in the opposite direction. The new piece of DNA synthesized on the lagging strand template is an Okazaki fragment.

4. *Elongation of the new DNA strands.* The DNA is untwisted further by the helicases (Figure 11.7c). On the leading strand template (bottom strand in Figure 11.7), the new leading strand is synthesized continuously. Since DNA synthesis can only proceed in the 5'-to-3' direction, however, the DNA polymerase synthesis reaction on the lagging-strand template (top strand in Figure 11.7) has gone as far as it can. For DNA replication to continue on that strand, a new initiation of DNA synthesis must occur on the single-stranded template that has been produced by the unwinding of the double helix. As before, an RNA primer is made, and this occurs close to the replication fork, catalyzed by the primase which is still bound to the helicase. The primer is lengthened by the action of DNA polymerase III, which displaces SSB proteins as it synthesizes the new Okazaki fragment. Figure 11.10 is a drawing of the replication fork, showing the locations of the major proteins. The figure captures the moment of synthesis of the RNA primer for a new Okazaki fragment.

In Figure 11.7d the process repeats itself: The DNA untwists, continuous DNA synthesis occurs on the leading-strand template, and discontinuous DNA synthesis occurs on the lagging-strand template.

5. *Joining of Okazaki fragments on the lagging strand to make a continuous strand.* Eventually, the unconnected Okazaki fragments on the lagging-strand template are synthesized into a continuous DNA strand. That process requires the activities of two enzymes, DNA polymerase I and **DNA ligase.** Consider two adjacent Okazaki fragments. The 3' end of the newer DNA fragment is adjacent to, but not joined to, the primer at the 5' end of the previously made fragment. The DNA polymerase III dissociates from the DNA, and DNA polymerase I takes over. This enzyme continues the 5'-to-3' synthesis of the newer DNA fragment made by DNA polymerase III, simultaneously removing the primer section of the older fragment (Figure 11.7e). The removal of the RNA primer takes place nucleotide by nucleotide through the action of the 5'→3' exonuclease activity of DNA polymerase I.

When DNA polymerase I has completed replacement of RNA primer nucleotides with DNA nucleotides, a single-stranded gap exists between adjacent nucleotides on the DNA strand between the two fragments. The two fragments are joined into one continuous DNA strand by the enzyme DNA ligase. The result is a longer DNA strand (Figure 11.7f). The catalytic reaction of DNA ligase is diagrammed in Figure 11.11. The whole process is repeated until all the DNA is replicated.

The complicated process of DNA replication in *E. coli* requires many different proteins, some of which function in other cellular processes such as the repair of

---

~ **FIGURE 11.10**

Schematic representation of the replication fork, showing the locations of the major proteins primase, helicase, SSB proteins, and DNA polymerase III. The complex of helicase, primase, and DNA is the primosome.

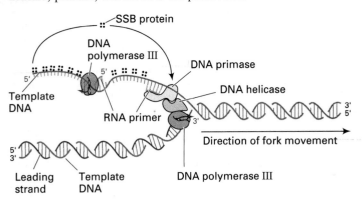

~ FIGURE 11.11

Action of DNA ligase in sealing the gap between adjacent DNA fragments (e.g., Okazaki fragments) to form a longer, covalently continuous chain. The DNA ligase catalyzes the formation of a phosphodiester bond between the 3'-OH and the 5'-phosphate groups on either side of a gap, sealing the gap.

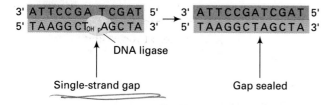

damaged DNA and genetic recombination. The roles of some of the genes that have been identified as necessary for DNA replication are given in Table 11.2.

The picture that has emerged from extensive studies of the proteins involved with replication is that the key proteins in the process are closely associated to form a **replication machine** or **replisome** (Figure 11.12; p. 340). This close association is believed to increase the efficiency of replication significantly. Figure 11.12 is essentially Figure 11.10 but with the lagging-strand DNA folded so that the DNA polymerase III of that strand is complexed with the DNA polymerase III of the leading strand. The folding also brings the 3' end of each completed Okazaki fragment near the site where the next Okazaki fragment will start. The primase-helicase complex (the primosome) moves with the fork, synthesizing new RNA primers as it proceeds. Similarly, with the lagging strand polymerase complexed with the other replication proteins at the fork, that enzyme can be continually reused at the same replication fork, synthesizing a string of Okazaki fragments as it moves with the rest of the replication machine. In sum, the complex of replication proteins that forms at the replication fork moves as a unit along the DNA, and enables new DNA to be synthesized efficiently on both the leading-strand template and lagging-strand template.

PROOFREADING: CORRECTING ERRORS IN DNA

REPLICATION. Occasionally an error is made in DNA replication and an incorrect nucleotide is incorporated into the DNA chain being synthesized. The mismatched base has a very high probability of being excised by the $3' \rightarrow 5'$ exonuclease activity of the DNA polymerase before the next base in the chain is added. This process, or proofreading ability, results in an extremely low error frequency in inserting the wrong base during DNA replication. It can be conjectured that proofread-

~ TABLE 11.2

Functions of Some of the Genes Involved in DNA Replication in *E. coli*

GENE PRODUCT AND/OR FUNCTION	GENE
DNA polymerase I	*polA*
DNA polymerase II	*polB*
DNA polymerase III	*polC* (*dnaE, N, Z*)
Initiation of chromosomal replication	*dnaA*
DNA replication	*dnaB*
DNA replication	*dnaC*
Primase—make primer for extension by DNA polymerase	*dnaG*
DNA replication	*dnaI*
DNA replication	*dnaJ*
DNA replication	*dnaK*
DNA replication	*dnaL*
Initiation of chromosomal replication	*dnaP*
Origin of chromosomal replication	*ori*
Helicase—unwinding activity to generate single-stranded arms of replication fork	*rep*
Single-stranded binding (SSB) protein—stabilize single-stranded arms of replication fork	*ssb*
RNA polymerase, α subunit	*rpoA*
RNA polymerase, β subunit	*rpoB*
RNA polymerase, β' subunit	*rpoC*
RNA polymerase, σ subunit	*rpoD*
DNA ligase—seal single-stranded gaps; join Okazaki fragments	*lig*
Terminus of chromosomal replication	*ter*

~ FIGURE 11.12

Model for the "replication machine," or replisome, the complex of key replication proteins, with the DNA at the replication fork.

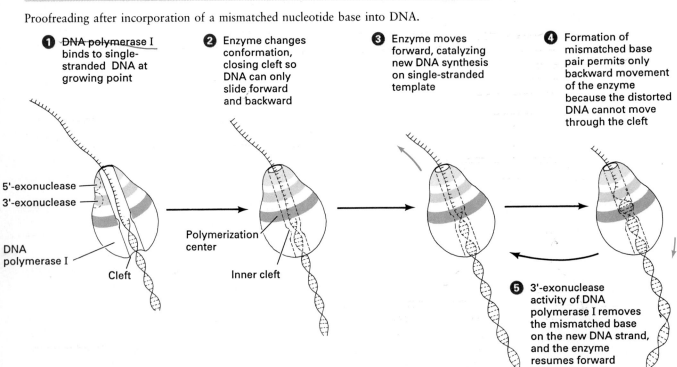

[handwritten margin notes: replication machine = replisome. DNA poly III, DNA ligase, DNA helicase]

ing might occur in the following manner: If an incorrect base is incorporated so that hydrogen bonding between the new base and the base in the template strand cannot form properly, the DNA may assume a distorted shape and be unable to slip easily through a 2-nm cleft in the DNA polymerase to which DNA binds. However, if the DNA were able to move backward and come into con-

tact with the $3' \rightarrow 5'$ exonuclease (proofreading) region of the DNA polymerase, the incorrect base could be removed and polymerization continued (Figure 11.13). As such, the proposed mechanism resembles a correcting typewriter, where a backspace is used to erase the incorrect character and the forward direction is resumed to insert the next correct character.

~ FIGURE 11.13

Proofreading after incorporation of a mismatched nucleotide base into DNA.

1. ~~DNA polymerase I~~ binds to single-stranded DNA at growing point

2. Enzyme changes conformation, closing cleft so DNA can only slide forward and backward

3. Enzyme moves forward, catalyzing new DNA synthesis on single-stranded template

4. Formation of mismatched base pair permits only backward movement of the enzyme because the distorted DNA cannot move through the cleft

5. 3'-exonuclease activity of DNA polymerase I removes the mismatched base on the new DNA strand, and the enzyme resumes forward movement

REPLICATION OF CIRCULAR DNA AND THE TWISTING PROBLEM. In *E. coli* the parental DNA strands remain in a circular form throughout the replication cycle. This is true of many, but not all circular DNA molecules. During replication, these circular DNA molecules exhibit a theta-like (θ) shape because of a replicating bubble's initiation at the replication origin (Figure 11.14). In the case of the *E. coli* chromosome and a number of other circular chromosomes, replication proceeds bidirectionally from the origin. For other circular chromosomes—for example, certain plasmids—replication is unidirectional.

Since the two DNA strands in a circular chromosome must untwist for replication to occur, there is a problem that must be solved as replication proceeds. That is, if both strands remain circular, then as a section of DNA double helix is untwisted to make the replication bubble, positive supercoils will form elsewhere in the molecule. By analogy, if you take a circular piece of double-stranded rope and try to separate the two strands at one point, the rope will become tightly supercoiled at the opposite side of the circle. In the DNA, every 10 base pairs replicated at the fork represent one complete turn of the helix. For the replication fork to move, all of the chromosome ahead of fork would have to rotate. Given a rate of movement of the replication fork of 500 nucleotides per second, at 10 base pairs per turn, the helix ahead of the fork would have to rotate at 50 revolutions per second, or 3000 rpm! (Figure 11.15; p. 342)

The twisting problem is solved by the action of topoisomerases (see Chapter 10), enzymes that introduce positive or negative supercoils into DNA. Topoisomerases play an important role in the replication process by preventing excessively twisted DNA from forming and thereby allowing both parental strands to remain intact during the replication cycle as the replication fork migrates. That is, the unreplicated part of the theta structure ahead of the replication fork repeatedly has negatively supercoils introduced into it by the action of topoisomerase II, relieving the positive supercoiling that occurs as DNA is untwisted during replication. Figure 11.16 (p. 342) shows twisted replicating molecules of the circular DNA of the animal virus SV40: topoisomerases function in this system to prevent excessive twisting.

ROLLING CIRCLE REPLICATION OF DNA. The rolling circle model of DNA replication applies to the replication of several viral DNAs such as Φ X174 and to the replication of the *E. coli* F factor during conjugation and transfer of donor DNA to a recipient (see Chapter 7). The rolling circle model is shown in Figure 11.17 (p. 343). The first step is the generation of a specific cut (nick) in one of the two strands at the origin of replication. The 5′ end of the cut strand is then displaced from the circular molecule. This creates a replication fork structure and leaves a single-stranded stretch of DNA that serves as a template for the addition of deoxyribonucleotides to the free 3′ end. This DNA synthesis is catalyzed by DNA polymerase III.

As the 5′ cut end continues to be displaced from the circular molecule, new DNA is synthesized; that is, this is the leading strand of the previous replication fork diagrams. As replication proceeds, the 5′ end of the cut

~ **FIGURE 11.14**

Bidirectional replication of circular DNA molecules.

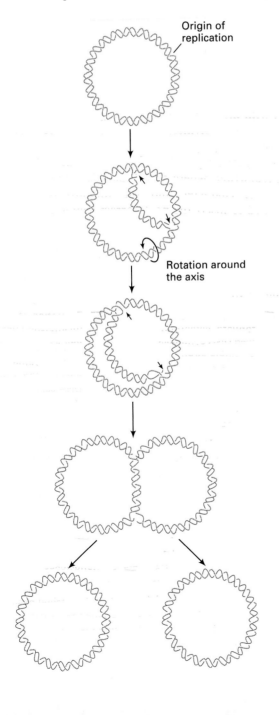

Origin of replication

Rotation around the axis

~ FIGURE 11.15

Illustration of the twisting problem in DNA replication. If the replication fork moves at 500 nucleotides per second, the DNA helix ahead of the fork rotates at 50 revolutions per second, or 3000 rpm.

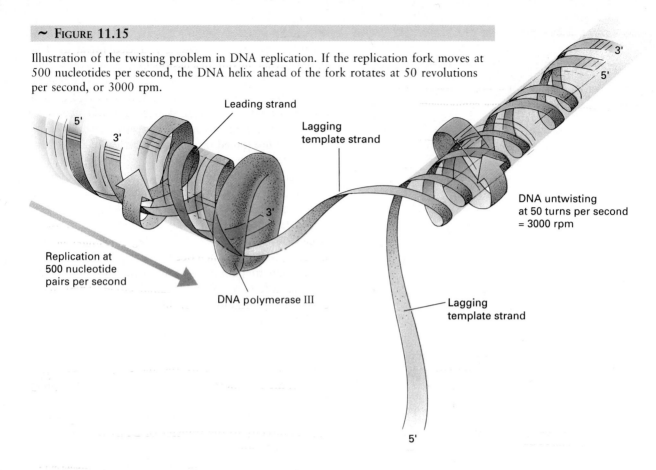

~ FIGURE 11.16

Twisted replicating molecule of circular SV40 DNA. (a) Electron micrograph of an early phase of replication; (b) Diagram showing the unreplicated, supercoiled parent strands and the portions already replicated.

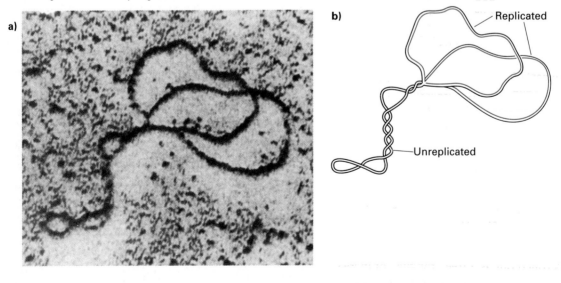

~ FIGURE 11.17

The replication process of double-stranded circular DNA molecules through the rolling circle mechanism. The active force that unwinds the 5′ tail is the movement of the replisome propelled by its helicase components.

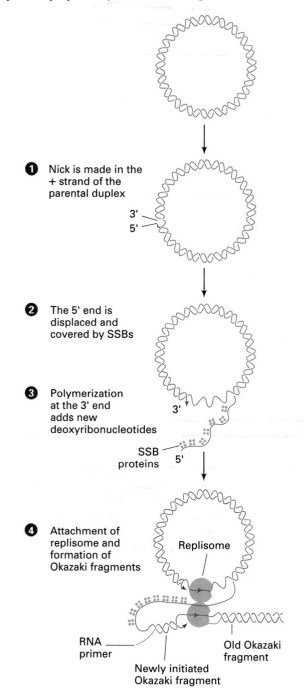

❶ Nick is made in the + strand of the parental duplex

3′
5′

❷ The 5′ end is displaced and covered by SSBs

❸ Polymerization at the 3′ end adds new deoxyribonucleotides

3′

SSB proteins   5′

❹ Attachment of replisome and formation of Okazaki fragments

Replisome

RNA primer

Old Okazaki fragment

Newly initiated Okazaki fragment

DNA strand is rolled out as a free "tongue" of increasing length. (This is analogous to pulling out the end of a roll of kitchen towels.) This single-stranded DNA tongue becomes covered with SSB proteins.

As the circle continues to roll, the single-stranded tongue converts to a double-stranded form, as illustrated earlier in our discussion of lagging-strand synthesis (see Figure 11.7). That is, primase synthesizes short RNA primers that are extended as DNA (Okazaki fragments) by DNA polymerase III. The RNA primers are ultimately removed and adjacent Okazaki fragments are joined through the action of DNA ligase. As the single-stranded DNA tongue rolls out, DNA synthesis continues on the circular DNA template.

Since the parental DNA circle can continue to roll, it is possible to generate a linear double-stranded DNA molecule that is longer than the circumference of the circle. For example, in the later stages of phage lambda DNA replication, which occurs by rolling circle replication, linear tongues are produced that are many times the circumference of the original circle. These molecules, then, contain many head-to-tail copies of the lambda chromosome. How is this multimeric molecule processed to produce unit-length lambda chromosomes? The long molecules are cut into individual lambda chromosomes by the *ter* enzyme (see Figure 10.11); the chromosomes are then packaged into phage heads.

The rolling circle model for DNA replication applies directly to the transfer of DNA from donor to recipient cell during conjugation of *E. coli* cells (Figure 11.18). In this case leading strand DNA synthesis serves to maintain a complete double-stranded copy of the *F* factor or *Hfr* donor chromosome in the donor cell. The displaced 5′ tongue passes through the conjugation tube as a single-stranded molecule, entering the recipient cell in that form. It is not known, however, whether the single-stranded molecule is first converted to double-stranded DNA by lagging strand synthesis and then undergoes recombination with the recipient's chromosome, or whether it recombines with the recipient's chromosome as a single-stranded molecule.

~ FIGURE 11.18

Transfer of single-stranded DNA from a donor bacterium into a recipient, via the rolling circle mechanism.

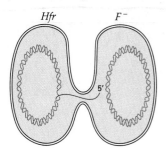

Hfr   *F⁻*

5′

KEYNOTE

Replication of DNA in *E. coli* requires at least two of the DNA polymerases and several other enzymes and proteins. The DNA helix is untwisted to provide templates for the synthesis of new DNA. The untwisting is catalyzed by DNA helicase. Since new DNA is made in the 5′-to-3′ direction, chain growth is continuous on one strand and discontinuous (i.e., in segments that are later joined) on the other strand. This semidiscontinuous model is applicable to many other prokaryotic replication systems, each of which differs in the number and properties of the enzymes and proteins required. In the replication of circular DNA molecules, topoisomers act to prevent the DNA from tangling ahead of the replication fork as the DNA untwists. ————————

Bacterial DNA Replication and the Cell Cycle

In a rapidly growing bacterium such as *E. coli*, DNA replication occurs throughout the cell cycle. At the same time, cell mass is increasing, and the cell is becoming longer. At the beginning of the cell cycle the DNA molecule is already partially replicated (Figure 11.19a). About halfway through the cell cycle, the chromosome has completely replicated, and the two progeny DNA molecules are attached to the cell membrane, their attachment points quite close to one another (Figure 11.19b). (Note that, while the DNA appears to be attached to the cell membrane during the cell division events, none of the stages of DNA replication requires the presence of any membrane component.) As the cell grows, new membrane material is laid down in the narrow zone between the two attachment points. As that zone grows, the attachment points and the DNA segregate to different ends of the cell (Figure 11.19c). During this stage, another round of DNA replication begins. Lastly, a wall is formed across the cell at its midpoint, and two progeny cells are generated by fission. Each progeny cell has a complete replicating chromosome still attached to the cell membrane (Figure 11.19d).

In a nutrient-rich medium, *E. coli* can divide every 20 minutes. In such cells, it takes 40 minutes for a replication fork to go from the origin to the point where the two replication forks meet; that is, with the two replication forks of bidirectional replication, the chromosome is replicated in 20 minutes.

DNA REPLICATION IN EUKARYOTES

The biochemistry of DNA replication is similar in prokaryotes and eukaryotes. However, the added complication in eukaryotes is that DNA is not found in only one chromosome but is distributed among many chromosomes, each of which is a complex aggregate of DNA and proteins. In each cell division cycle, each of these chromosomes must be faithfully duplicated and a copy of each distributed to each of the two progeny cells.

The cell cycle is qualitatively the same from eukaryote to eukaryote, although there are significant differences both in the relative amount of time spent in each phase of the cycle and in the total time spent in one cycle. Among higher eukaryotes, for example, some cells divide once every 3 hours and some cells divide once every 200 hours; human cells in culture divide approximately once every 24 hours.

Recall from Chapter 1 that the cell cycle in most somatic cells of higher eukaryotes is divided into four stages: gap 1 (G_1), synthesis (S), gap 2 (G_2), and mitosis (M) (see Figure 1.18). The DNA replicates and the chromosomes duplicate during the S phase, and the progeny chromosomes segregate into daughter cells during the M phase. Although DNA is neither replicating nor being segregated during G_1 and G_2, the cell is metabolically active and is growing. Events involved in the subsequent synthesis and mitotic activity also occur during G_1 and G_2.

Collectively, the G_1, S, and G_2 phases of the eukaryotic cell cycle constitute interphase, the time between divisions, while division (mitosis or meiosis) itself occurs within the M phase.

The significant events that occur within the G_1, S, G_2, and M phases are as follows:

1. In the G_1 phase, the morphology of the chromosomes changes from the condensed state characteristic of mitosis to the extended state of interphase cells; the change is due to a transition to a lower order of coiling of the chromatin fibers. In addition, during G_1 the cell becomes prepared biochemically for the S phase. The G_1 phase is a period during which there is very active synthesis of both RNA and protein. In particular, proteins needed for the replication and subsequent segregation of DNA are made in G_1.

   During G_1 normal cells must decide whether or not to proliferate. The point at which the decision is made is called the *restriction point*, and once the biochemical events associated with the restriction point occur, the cell is irreversibly programmed to

~ FIGURE 11.19

Model for chromosome segregation to daughter cells of *E. coli*, as proposed by Jacob, Brenner, and Cuzin. (a) A newly produced bacterial cell: the DNA is already partially replicated and is attached to the membrane; (b) Partway through the division cycle: replication of the chromosome has been largely completed, and the two DNAs are attached to the membrane; (c) Near the end of the cycle, when new membrane growth has moved the chromosome attachment points again; (d) The end of the cycle, with the bacterium divided in two (each daughter cell receives one chromosome).

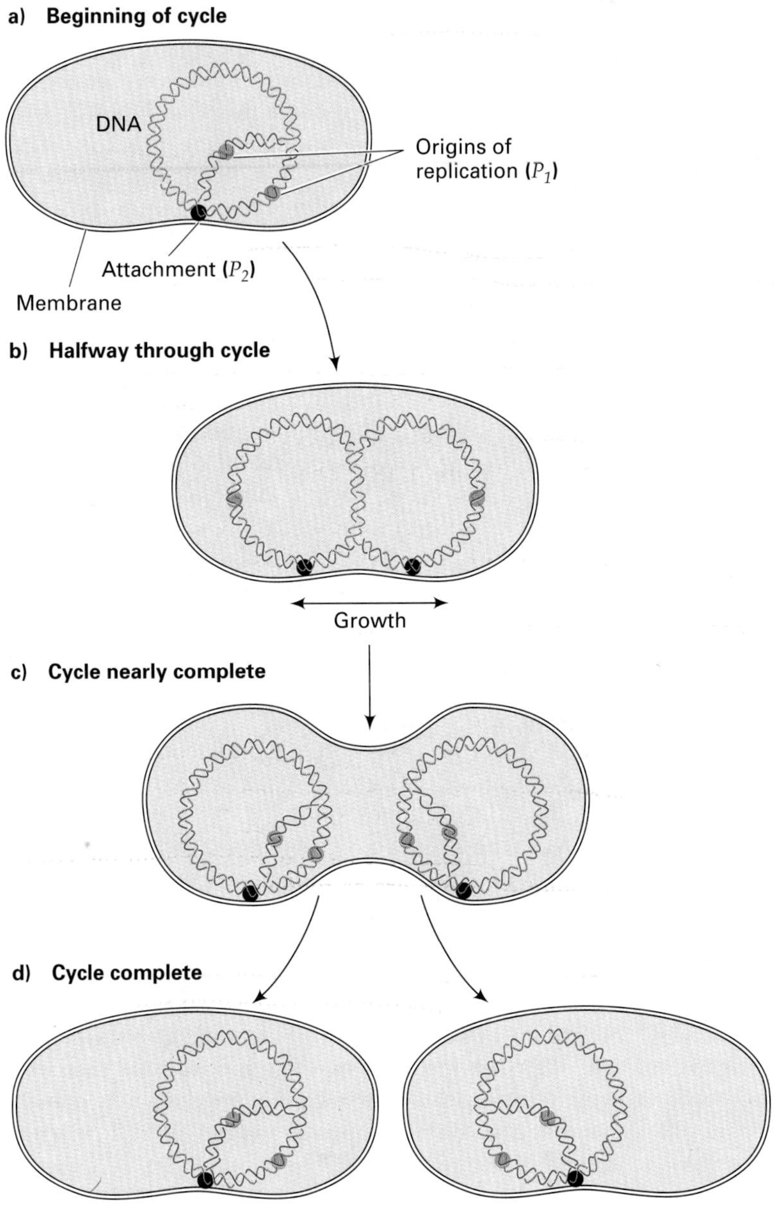

a) **Beginning of cycle**

b) **Halfway through cycle**

c) **Cycle nearly complete**

d) **Cycle complete**

initiate DNA replication and consequently to divide. The precise molecular signals that operate at the restriction point are not known, although it is clear that the cell must reach a certain mass (i.e.,

accumulate sufficient cellular protein) before it can commit itself to the rest of the cell cycle.

If a cell is unable to grow and proliferate—for example, as a result of an exhaustion of nutrients, of contact inhibition (a phenomenon in which cells in contact somehow inhibit each other from dividing further), or of the addition of inhibitory drugs—it becomes quiescent. The quiescent state can be reversed by a change in conditions, such as adding nutrients, diluting the cells, or removing the drugs, respectively.

The G_1 phase is absent in some eukaryotes, such as the slime mold *Physarum polycephalum* and the fission yeast *Schizosaccharomyces pombe*. Events required for the initiation of DNA synthesis in these cells probably occur in the G_2 phase, preceding mitosis.

2. During the S phase, DNA replication takes place. Replication involves the coordination of the mechanism for untwisting the DNA of each chromosome and producing two progeny DNA double helices and the mechanism for duplicating the nucleosome organization of the eukaryotic chromosome.

3. The G_2 phase follows the S phase and occupies perhaps 10 to 20 percent of the cell cycle. During G_2 the chromosomes begin condensing in preparation for mitosis. Both RNA and protein are actively synthesized during this phase. One of the important proteins made during G_2 is *tubulin*. Tubulin is a component of the mitotic spindle apparatus, which is used to segregate the chromosomes into daughter cells during mitosis. The end of the G_2 phase is delineated by the beginning of mitosis. It is not clear what events trigger mitosis.

4. In the last stage of the cell cycle, the M phase, the chromosomes are segregated into the progeny cells. Depending on the cell type, chromosome segregation will occur by mitosis or meiosis; both processes were described in Chapter 1.

Molecular Details of DNA Synthesis in Eukaryotes

As described in Chapter 10, the eukaryotic chromosome is a complex of DNA and proteins (histones and nonhistones) organized into nucleosomes. Thus the replication of the eukaryotic chromosome must involve the replication of the DNA and of the histone core of the nucleosome as well as a doubling of the nonhistones.

As we saw earlier, many of the enzymes and proteins involved in prokaryote DNA replication have been identified. Less is known about the enzymes and proteins involved in eukaryotic DNA replication. It is clear, however, that the sequential steps described for DNA synthesis in prokaryotes also occur for DNA synthesis in eukaryotes, namely, denaturation of the DNA double helix and the semiconservative, semidiscontinuous replication of the DNA.

SEMI-CONSERVATIVE DNA REPLICATION IN EUKARYOTES. Semi-conservative replication of DNA in eukaryotic chromosomes can be visualized using a staining procedure that results in what are called harlequin chromosomes (Figure 11.20). Chromosomes are allowed to go through two rounds of replication. The experimental system is Chinese Hamster ovary (CHO) cells growing in tissue culture, where the constituents of the medium can be manipulated by the investigator. In this case the chemical 5-bromodeoxyuridine (BUdR) is added at the beginning of the experiment. BUdR is a *base analogue,* that is, it has a structure very similar to that of one of the normal bases found in DNA, in this case thymine. During replication, wherever a T is called for in the new strand, BUdR is incorporated instead. If semi-conservative replication occurs, two progeny double helices (sister chromatids), each with one T-containing strand and one BUdR-containing strand will be produced from each parental double helix. By a second round of semiconservative replication in the presence of BUdR, the T-BUdR hybrid DNA gives rise to one sister chromatid with *both* strands of DNA labeled with BUdR and the other sister chromatid with one strand of DNA labeled with BUdR and the other with T. When these mitotic chromosomes are then stained with a fluorescent dye and Giemsa stain, chromatids with two BUdR-labeled strands stain less intensely than the chromatids with one T-labeled and one BUdR-labeled strand. The dark-light staining pattern brings to mind Harlequins, hence the term harlequin chromosome (Figure 11.20).

INITIATION OF DNA REPLICATION.

Replicons. Each eukaryotic chromosome consists of one linear DNA double helix. If there was only one origin of replication per chromosome, the replication of each chromosome would take many, many hours. For example, there are 2.75×10^9 base pairs of DNA in the haploid human genome (23 chromosomes), and the average chromosome is roughly 10^8 base pairs long. With a replication rate of 2 kilobases (2000 bases) per minute in human cells, it would take approximately 830 hours to replicate one chromosome. If each cell cycle were at least that long for a developing human embryo, the gestation period would be many years instead of 9 months.

Actual measurements show that the chromosomes in eukaryotes replicate much faster than would be the case with only one origin of replication per chromosome. The diploid complement of chromosomes in *Drosophila* embryos, for example, replicates in 3 minutes. This is 6 times faster than the replication of the *E. coli* chromosome, even though there is about 100 times more DNA in *Drosophila* than there is in *E. coli*.

The rapidity of *Drosophila* replication is possible because DNA replication is initiated at many origins of

~ FIGURE 11.20

Visualization of semiconservative DNA replication in eukaryotes. Shown are harlequin chromosomes in Chinese Hamster ovary cells that have been allowed to go through two rounds of DNA replication in the presence of the base analogue 5-bromodeoxyuridine, followed by staining.

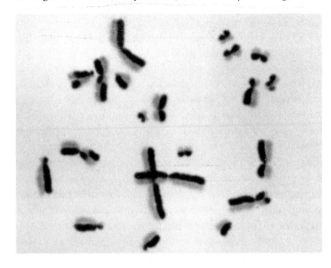

replication throughout the genome, a state of affairs that is general among eukaryotes.

At each origin of replication, the DNA denatures (as in *E. coli*). Replication proceeds bidirectionally, and the DNA double helix opens to expose single strands that act as templates for new DNA synthesis. Eventually, each replication fork will run into an adjacent replication fork, initiated at an adjacent origin or replication. In eukaryotes the stretch of DNA from the origin of replication to the two termini of replication (where adjacent replication forks fuse) on each side of the origin is called a **replicon** or a **replication unit**. The many and relatively small replication origins greatly decrease the overall time for duplication of the chromosomal DNA.

Table 11.3 presents the number of replicons and their average size, and the rate of replication fork movement for a number of organisms. Note that the replicon size is much smaller, and the rate of fork movement is much slower in eukaryotic organisms than in bacteria. In general, each replicating eukaryotic chromosome has many replicons. Also, the number of origins and the average size of the replicon can vary with the nature of the cell, for example, its growth rate and tissue of origin.

Figure 11.21a (p. 348) presents an electron micrograph showing a large number of replicons on a piece of *Drosophila* DNA; Figure 11.21b is an interpretive drawing of that micrograph. The piece of DNA shown is about 500 kb long (about a tenth the size of the *E. coli* chromosome), and many replicons are present.

Replication of DNA does not occur simultaneously in all the replicons in an organism's genome. Instead, a temporal ordering of initiation events occurs during the replication phase. For many cell types a particular temporal ordering of initiation is characteristic of the cell type; that is, the same pattern occurs, cell generation after cell generation. The molecular events that control which replicons replicate early and which replicate late are unknown. The diagram in Figure 11.22 (p. 348) represents the temporal ordering of initiation phenomena from autoradiographic studies. The figure shows one segment of one chromosome in which there are three replicons that always begin replicating at distinct times. When the replication forks fuse at the margins of adjacent replicons, the replicated chromosome has replicated into two sister chromatids. In general, replication of a segment of chromosomal DNA occurs following the synchronous activation of a cluster of origins.

Replication origins. Are there specific sequences that function as origins of replication in eukaryotes? In the yeast, *Saccharomyces cerevisiae*, specific sequences have been identified which, when they are included as part of an extrachromosomal, circular DNA molecule, confer upon that molecule the ability to replicate autonomously within the yeast cell. These sequences are called **autonomously replicating sequences,** or **ARS** elements. A variety of sequences from other eukaryotic organisms are also able to function as ARS elements in yeast.

~ TABLE 11.3

Comparison of Bacterial and Eukaryote Replicons

ORGANISM	NO. OF REPLICONS	AVERAGE LENGTH	FORK MOVEMENT
Bacterium (*E. coli*)	1	4,200 kb	50,000 bp/min
Yeast (*S. cerevisiae*)	500	40 kb	3,600 bp/min
Fruit fly (*D. melanogaster*)	3,500	40 kb	2,600 bp/min
Toad (*X. laevis*)	15,000	200 kb	500 bp/min
Mouse (*M. musculus*)	25,000	150 kb	2,200 bp/min
Plant (*V. faba*)	35,000	300 kb	

~ FIGURE 11.21

Replicating DNA of *Drosophila melanogaster:* (a) Electron micrograph showing replication units (replicons); (b) Diagram, an interpretation of the electron micrograph shown in (a).

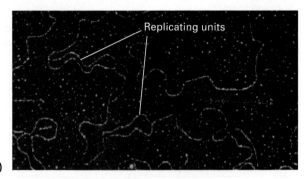

a)

Replicating units

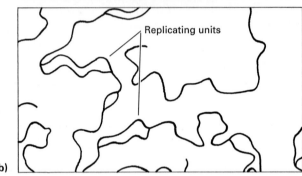

Replicating units

b)

It is believed that ARS elements do correspond to origins of replication that function within the cell and, indeed, replication initiation has been shown unambiguously at some elements in the cell. On the other side of the coin, it appears that not all ARS elements that are known to be located along a well-mapped chromosomal segment are used for replication initiation. Comparison of the sequences of a number of yeast ARS elements has indicated the presence of the following 11-bp consensus sequence (A **consensus sequence** is a sequence indicating the base pair [or base] most commonly found at each position among the different DNAs [or RNAs] analyzed):

5′   A T T T A T Pu T T T A   3′
3′   T A A A T A Py A A A T   5′

The A−T-richness of the sequence fits well with the function of an origin of replication, namely the denaturation of the DNA in that region, because A−T base pairs, with two hydrogen bonds, are easier to disrupt than G−C base pairs, with three hydrogen bonds.

Denaturation of DNA. As in prokaryotes, topoisomerases are used to facilitate DNA untwisting during the replication process. In *E. coli*, enzymes called helicases catalyze the untwisting of the DNA double helix to expose single strands of DNA that act as templates in DNA replication. Untwisting enzymes appear to be involved in eukaryotic DNA replication as well. For example, helicases have been identified in a variety of eukaryotic systems including human cells, mouse, rat, calf, and higher plants. The activity of these

~ FIGURE 11.22

Temporal ordering of DNA replication initiation events in replication units of eukaryotic chromosomes.

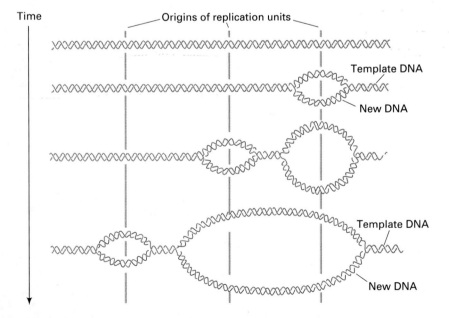

Time

Origins of replication units

Template DNA

New DNA

Template DNA

New DNA

enzymes increases about twenty-fold when a cell begins DNA synthesis, which indicates that they might have a role in DNA replication.

Once the helix unwinds and single-stranded regions are exposed, we expect that proteins similar to the SSB (single-strand binding) proteins in prokaryotes might attach and prevent the DNA from renaturing. Since a large number of proteins have the ability to stick to DNA (such as histones and DNA polymerases), it is difficult to identify exactly which proteins are SSBs.

Single-strand binding proteins *have* been isolated from a number of eukaryotic cells such as yeast, fungi, amphibians, rodents, calf, and human cells. In mammalian cells the SSB proteins have been shown to bind to the single-stranded DNA and leave the bases exposed while simultaneously keeping the DNA in an extended configuration and increasing the affinity of the DNA polymerase for the template. Thus these SSB proteins stimulate polymerase activity (at least in mammals) as measured by the amount of DNA synthesized.

REPLICATION ENZYMES AND PROTEINS. Four different DNA polymerases have been identified in mammalian cells, designated α, β, δ, and γ. Their properties are summarized in Table 11.4. The α, β, and δ polymerases are located in the nucleus, while the γ polymerase is found in the mitochondrion. DNA polymerases α and δ are the two enzymes responsible for nuclear DNA replication. They function essentially like the *E. coli* DNA polymerase III. DNA polymerase β serves a DNA repair function, and DNA polymerase γ catalyzes mitochondrial DNA replication. Both nuclear replication polymerases α and δ can use RNA primers for the initiation of DNA synthesis; δ can also use a DNA primer. A DNA primase activity has been shown to be associated with the α polymerase. The mitochondrial polymerase γ uses a DNA primer for DNA initiation; that primer is synthesized by a primase enzyme, not by a DNA polymerase enzyme. All four enzymes are presumed to have 5'-to-3' exonuclease activity, although no direct evidence is available on that point. Only α and δ have been shown to have proofreading (3'-to-5' exonuclease) activity.

Two DNA polymerases have also been identified and studied in yeast. Yeast polymerase I is equivalent to mammalian polymerase α and yeast polymerase III is equivalent to mammalian polymerase δ. In general, lower-eukaryotic DNA polymerases more closely resemble prokaryotic polymerases than higher-eukaryotic polymerases.

KEYNOTE

In eukaryotes DNA replication occurs in the S phase of the cell cycle and is similar to the replication

~ TABLE 11.4

Properties of DNA Polymerases in Mammalian Cells

PROPERTY	DNA POLYMERASE			
	α	β	δ	γ
Location in cell:				
Nucleus	+	+	+	−
Mitochondrion	−	−	−	+
Polymerization activity:				
5'-to-3'	+	+	+	+
Synthesis from:				
RNA primer	+	−	+	−
DNA primer	−	−	+	+
DNA primase associated with enzyme	+	−	−	−
Exonuclease activity:				
3'-to-5' proofreading	+	−	+	−

process in prokaryotic cells. Synthesis of DNA is initiated by RNA primers, occurs in the 5'-to-3' direction, is catalyzed by DNA polymerases, requires a large number of other enzymes and proteins, and is a semiconservative and semidiscontinuous process. Replication of DNA is initiated at a large number of sites throughout the chromosomes. _____

Assembly of New DNA into Nucleosomes

Eukaryotic DNA is complexed with histones in nucleosome structures, which are the basic units of chromosomes (see Chapter 10). Therefore, when the DNA is replicated, the histone complement must be doubled so that all nucleosomes are duplicated. This involves two processes: the synthesis of new histone proteins and the assembly of new nucleosomes.

HISTONE SYNTHESIS. For many years geneticists thought that the synthesis of histones was coordinated with DNA synthesis in the cell cycle and that histones were made only during the S phase. More recent evidence indicates that histone synthesis is not coordinated tightly with DNA synthesis.

The first line of evidence comes from experiments that examined at which point in the cell cycle the histone messenger RNAs (mRNAs) are made (mRNAs are RNA molecules that are "read" in the cytoplasm of the cell in order to produce proteins). The experimental system was the HeLa cell (a human cell). The results of the experiment showed that histone mRNA is made throughout the cell cycle. Another piece of evidence came from experiments with mouse and Chinese hamster ovary (CHO) cells. Cells in G_1 were separated from

~ FIGURE 11.23

Model for nucleosome assembly at the replication fork of a eukaryotic replicon. RNA-primed Okazaki fragments are on the lagging-strand template. A helicase is at the origin of the fork itself, and DNA polymerases are on both template strands.

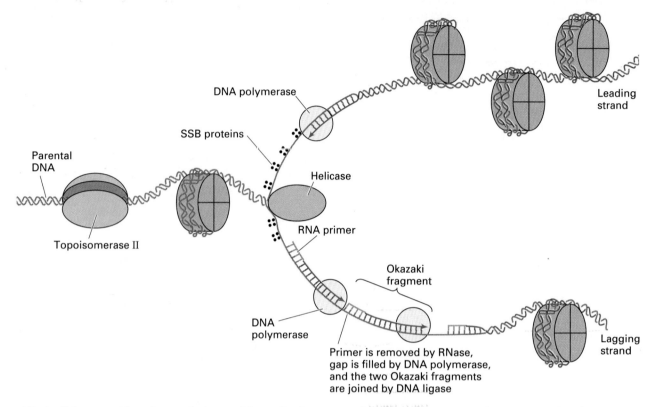

cells in S by centrifugation techniques. The cells then were incubated in the presence of radioactive amino acids, which were, in turn, incorporated into the proteins being manufactured. The proteins were then examined for the presence of radioactivity: Cells from both the G_1 and S phases had synthesized histones. Thus it appears that histones are made at least in G_1 and S and perhaps in G_2 also. The key event that only occurs in the S phase is the association of the new histones with each other and with DNA to form new nucleosomes.

NUCLEOSOMES AND NEWLY SYNTHESIZED DNA.

Newly replicated DNA from *Drosophila* has been examined by electron microscopy to determine the distribution of nucleosomes. The results of these studies show that the diameter of the nucleosomes and their spacing along the DNA fibers produces the identical, beaded chromatin appearance in both unreplicated DNA and newly replicated DNA. In other words, new DNA is assembled into nucleosomes virtually immedi-
... ly.

... ow do histones in nucleosomes segregate from ... ation to cell generation? That is, are new ... made from an all-new set of histones, or ... histones become mixed in both the old

and new nucleosomes? This question has been investigated by using density labeling to distinguish old nucleosomes from new in a way analogous to the Meselson-Stahl experiment. Cells were first grown in a medium containing a high-density label, so that the nucleosomes became "heavy." The cells were then shifted to a new growth medium containing a low-density label, as well as a radioactive amino acid label to enable newly-synthesized histones to be identified. Then, newly-synthesized chromatin was isolated, and the composition of the histones in the nucleosomes was analyzed.

The results of these kinds of experiments have shown that almost all H3 and H4 histones enter chromatin during S phase, while H2A and H2B histones can enter chromatin during S phase or during interphase. The fact that the H2A and H2B histones can enter chromatin during interphase means that those histones can enter old chromatin, perhaps exchanging with histones in nucleosomes of that chromatin. In other words, it is not absolutely clear how nucleosomes segregate from cell generation to cell generation. Figure 11.23 presents a model for new nucleosome assembly at the replication fork of a eukaryotic replicon. Once chromosome duplication has been completed, the cell completes the cell cycle by going through the G_2 and M phases.

KEYNOTE

Eukaryotic nuclear DNA is complexed with histones and organized into nucleosomes, the basic units of chromosomes. When the DNA replicates, new histones are synthesized and are assembled into nucleosomes. It is not known precisely how nucleosomes segregate from cell generation to cell generation.

Genetics of the Eukaryotic Cell Cycle

Useful information can be obtained from studies of mutants that affect a biological process. With respect to the cell division cycle, relatively few mutants are known. A number of eukaryotic cell cycle mutants have been obtained through the use of selection techniques. In general, these mutants were isolated by selection for temperature-sensitive growth. Such cells do not grow either at high temperatures (heat-sensitive mutants) or low (cold-sensitive mutants). To understand the utility of temperature-sensitive cells, remember that cell cycle functions are essential ones. So if one of the functions is lost by mutation, the cell dies—or at least it is not able to reproduce, and therefore there are not high numbers of cells exhibiting the defect for study. However, if a cell is temperature-sensitive for a cell cycle function, then it survives and grows at the normal (permissive) temperature. Only at the nonpermissive temperature is it defective. By switching the temperature experimentally, then, we can get the cells to manifest their defect.

Many temperature-sensitive cell cycle mutants have been isolated in eukaryotes, both in lower eukaryotes and in tissue cultures of cells of higher eukaryotes. Examples of some of the temperature-sensitive mutants that show conditional defects in cell cycle events are listed in Table 11.5. As the data in the table indicate,

~ TABLE 11.5

Examples of Cell Division Cycle Mutants in Various Eukaryotic Organisms

CELL TYPE	MUTANT SYMBOL	CELL CYCLE FUNCTION AFFECTED
Saccharomyces cerevisiae	cdc1	Growth
	cdc2, cdc8, cdc3, cdc10	DNA synthesis; establishment of S phase
	cdc11, cdc12	Mitosis; cytokinesis
	cdc4, cdc6, cdc7	Initiation of DNA synthesis; entry into S phase
	cdc5, cdc14, cdc15	Late nuclear division
	cdc9, cdc13	Medial nuclear division
	cdc25	Start; growth
	cdc28	Start; continuing growth
Ustilago maydis	ts-220	Elongation phase of DNA synthesis
	ts-207	DNA synthesis (S phase)
	pol 1-1	DNA polymerase A
Chinese hamster ovary (CHO) cells	K-12	G_1; entry into S phase blocked
	13B11	DNA synthesis (S phase)
	MS1-1	Mitosis; cytokinesis
Hamster BHK-21 cells	ts AF8	G_1; entry into S blocked
	ts HJ4	Late G_1; possibly initiation of S
	NW1	Mitosis; cytokinesis
Hamster HM-1 cells	ts-655	Mitosis; prophase
	ts-546	Mitosis; metaphase
Mouse leukemia cells	ts 2	Mitosis; cytokinesis
Mouse CAK cells	ts B54	G_1; entry into S blocked
Mouse Ba1B/C cells	ts 2	DNA synthesis (S phase)
Mouse L cells	ts A1S9	Elongation of replication units in DNA synthesis
	ts-C1	DNA synthesis (S phase)
Murine L5178Y cells	ts 2	Mitosis
African green monkey Bsc-1 cells	ts3, ts5, ts9	DNA synthesis (S)

Source: After R. Sheinin et al., 1978. *Annu. Rev. Biochem.* 47:277–316, and G. Simchen, 1978. *Annu. Rev. Genet.* 12:161–191.

the temperature-sensitive defect can be expressed in a number of reactions. Some mutants affect the elongation stage of DNA synthesis, some affect the function of the DNA polymerases, and some affect essentially every discrete stage of the cell division cycle.

The yeast mutants are perhaps the best-studied, and they are particularly useful since virtually all stages of the cell division cycle have mutational defects. Studies have been done of the interrelationships of each of the temperature-sensitive events. For example, genetic studies have shown that the initiation and elongation stages of DNA synthesis must be completed before actual nuclear division occurs, and only then can cytokinesis and cell separation occur, in that order. While we could have predicted some of these dependencies, it is always useful to have experimental confirmation, and this system shows the power of genetic analysis in providing such basic information.

𝒦EYNOTE

Studies of mutants that affect the cell division cycle have helped define how genes control that process. Most cell division cycle mutants are temperature sensitive; that is, they function normally at normal growth temperatures, but do not function, or function poorly at higher or lower temperatures.

DNA RECOMBINATION

In Chapters 5, 6, and 7, we discussed genetic recombination and the mapping of genes. We learned that crossing-over involves the physical exchange of homologous chromosome parts. Now that we have an understanding of DNA structure (Chapter 9), the organization of DNA into chromosomes (Chapter 10), and DNA replication (this chapter), we return to a consideration of genetic recombination, this time discussing some molecular aspects.

Crossing-Over: Breakage and Rejoining of DNA

Crossing-over in eukaryotes occurs in meiosis at the point when the chromosomes have duplicated; there are four chromatids for each pair of synapsed chromomes. Crossing-over involves the breaking and rejoin- DNA. No base pairs are added or deleted to the   result of the breakage and rejoining events.   rly step in recombination is the genera-

tion of a break in one double helix. One possible sequence of events is that cellular enzymes introduce breaks along a pair of synapsed chromosomes in meiosis, thereby stimulating the recombination process. Some evidence for this is that the artificial introduction of breaks in DNA, such as by UV- or X-irradiation, results in a significant stimulation of crossing-over. However, molecular details of the initiation of recombination are not known at this time; as a result, it is conceivable that an entirely different sequence of events occurs than that just described.

𝒦EYNOTE

In both prokaryotes and eukaryotes, genetic recombination involves the breakage and rejoining of homologous DNA double helices. ─────────

The Holliday Model for Recombination

In the mid-1960s Robin Holliday proposed a model for reciprocal recombination. Since then, the *Holliday model* has been refined by other geneticists, particularly Matthew Meselson and Charles Radding.

The Holliday model is diagramed in Figure 11.24 for genetically distinguishable homologous chromosomes, one with alleles a^+ and b^+ at opposite ends, and the other with alleles a and b. The two DNA double helices in the figure participate in the recombination event. The first stage of the recombination process is *recognition and alignment* (Figure 11.24, part 1), in which two homologous DNA double helices become precisely aligned so that no genetic material is added or deleted to either DNA in the subsequent exchange events. (In eukaryotes, this synapsis is accompanied by the formation of a synaptinemal complex, which mediates the meiotic exchange of genetic information. See Chapter 1.) In the second stage, one strand of each double helix breaks; each broken strand then invades the opposite double helix and base pairs with the complementary nucleotides of the invaded helix (Figure 11.24, part 2). Each of these steps is catalyzed by specific cellular enzymes. The process leaves gaps that are closed by the activities of DNA polymerase and DNA ligase, producing what is called a *Holliday intermediate*, with an internal branch point (Figure 11.24, part 3). The two DNA double helices in the Holliday intermediate can rotate, causing the branch point to move to the right or to the left. Figure 11.24, part 4 shows a branch migration event that has occurred to the right. The four-armed structure for the DNA strands is produced simply by pulling the four chromo-

~ FIGURE 11.24

Holliday model for reciprocal genetic recombination. Shown are two homologous DNA double-helices that participate in the recombination process.

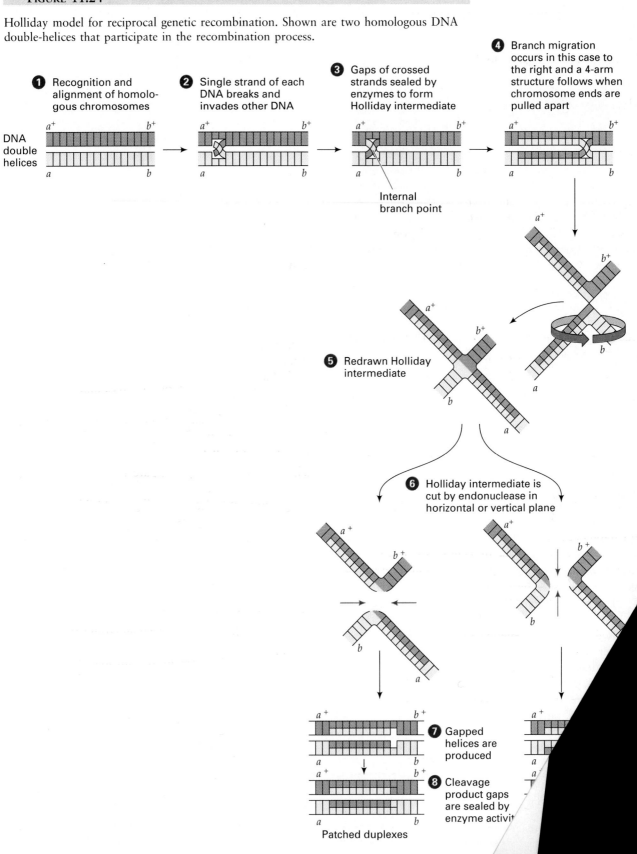

1 Recognition and alignment of homologous chromosomes

2 Single strand of each DNA breaks and invades other DNA

3 Gaps of crossed strands sealed by enzymes to form Holliday intermediate

4 Branch migration occurs in this case to the right and a 4-arm structure follows when chromosome ends are pulled apart

DNA double helices

Internal branch point

5 Redrawn Holliday intermediate

6 Holliday intermediate is cut by endonuclease in horizontal or vertical plane

7 Gapped helices are produced

8 Cleavage product gaps are sealed by enzyme activit

Patched duplexes

~ FIGURE 11.25

Electron micrograph of a Holliday intermediate with some single-stranded DNA in the branch point region.

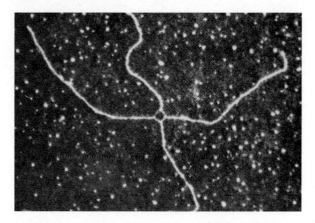

some ends apart. Holliday intermediate structures have been demonstrated to be very stable, with no unpaired bases, even in the region of the branch. Figure 11.25 shows an electron micrograph of a Holliday intermediate with some single-stranded DNA in the branch point region.

The result of branch migration is the generation of complementary regions of hybrid DNA in both double helices (diagramed as stretches of DNA helices with two different colors in Figure 11.24, parts 5 through 8). Careful measurements have indicated that branch migrations can occur very rapidly, perhaps moving as much as 50 base pairs per second in eukaryotes. Thus, regions of hybrid DNA could be formed in the few that might elapse between the formation of a ermediate and the cleavage and ligation odel that occurs next.

 and ligation phase of recombination if the Holliday intermediate is DNA strand passes over or under Thus, if the four-armed Holliday Figure 11.24, part 4 is taken as lower two arms are rotated e ms, the structure shown in of ced.
clea
left) diate is cut by enzymes
11.24, d DNA region of the
 The cuts can be in
 , with both kinds
 . Endonuclease
 11.24, part 6,
 n in Figure
 -stranded

gap. The gaps are sealed by DNA ligase to produce the double helices shown in Figure 11.24, part 8, left. Since each of the resulting helices contains a segment of single-stranded DNA from the other helix, flanked by nonrecombinant DNA, these double helices are called *patched duplexes.*

If the endonuclease cleavage in Figure 11.24, part 5 is in the vertical plane (Figure 11.24, part 6, right), the gapped double helices of Figure 11.24, part 7, right are produced. In this case, there are segments of hybrid DNA in each duplex, but they are formed by what looks like a splicing together of two helices, each with a very long sticky end of one-to-several thousand base pairs. The result of this activity is the double helices in Figure 11.24, part 8, right, which are called *spliced duplexes.*

In the example diagramed in 11.24, the parental duplexes contained different genetic markers at the ends of the molecules. One parent was $a^+ b^+$ and the other was $a\ b$, as would be the case for a doubly heterozygous parent. However, the reciprocal recombination events shown in Figure 11.24 result in different products. In the patched duplexes (Figure 11.24, part 8, left) the markers are $a^+ b^+$ and $a\ b$, which is the parental configuration. On the other hand, the spliced duplexes (Figure 11.24, part 8, right) have the markers in the recombinant configuration, that is, $a^+ b$ and $a\ b^+$. Since the enzymes that cut in the branch region during the final cleavage and ligation phase (Figure 11.24, part 6, left and right) most likely act randomly with respect to the pair of strands they cut, the Holliday model predicts that a physical exchange between two gene loci on homologous chromosomes should result in the genetic exchange of the outside chromosome markers about half of the time.

It is important to note that the Holliday model is only one model to explain the events of recombination. While its basic features are generally accepted, a number of other models have been proposed that attempt to explain recombination either at a more detailed level or recombination in special systems. Discussion of these models is beyond the scope of this text.

KEYNOTE

The Holliday model for genetic recombination involves a precise alignment of homologous DNA sequences of two parental double helices, endonucleolytic cleavage, invasion of the other helix by each broken end, ligation to produce the Holliday intermediate, branch migration, and finally, cleavage and ligation to resolve the Holliday intermediate into the recombinant double helices. Depending on the orientation of the cleavage events, the resulting double helices will be patched duplexes or spliced duplexes. If the recombination events occur between two heterozygous loci, the patched duplexes produced are parental, whereas the spliced duplexes are recombinant for the two loci.

Mismatch Repair and Gene Conversion

In Chapter 6 we discussed tetrad analysis, in which all products of each meiotic event can be isolated and analyzed. Tetrad analysis is limited to relatively few organisms; among them are fungi such as *Saccharomyces cerevisiae* (yeast) and *Neurospora crassa*. Recall that from a cross of $a\ b \times a^+\ b^+$ in which a and b are linked on the same chromosome, but still reasonably distant, three types of tetrads can result: parental ditype (PD), $a\ b$, $a\ b$, $a^+\ b^+$, $a^+\ b^+$; tetratype (T), $a\ b^+$, $a\ b$, $a^+\ b^+$, $a^+\ b$; and the rare nonparental ditype (NPD), $a\ b^+$, $a\ b^+$, $a^+\ b$, $a^+\ b$. For each tetrad type, there is a 2:2 segregation of the alleles.

Occasionally, the Mendelian law of 2:2 segregation of alleles is not obeyed in that 3:1 and 1:3 ratios are seen. Tetrads showing these unusual ratios are proposed to derive from a process called *gene conversion*. Any pair of alleles is subject to gene conversion. Consider, for example, the cross $a^+\ m\ b^+ \times a\ m^+\ b$, where gene m is located between genes a and b. The generation by mismatch repair of a tetrad showing gene conversion is diagrammed in Figure 11.26. The starting point is parental homologous chromosomes synapsed in meiotic prophase (Figure 11.26, part 1). If a recombination event occurs between the two inner chromatids (Figure 11.26, part 2), a patched duplex can be produced with two mismatches (heteroduplexes involving m and m^+ sequences). Both mismatches can be excised and repaired as shown in Figure 11.26, part 3. Following the subsequent two meiotic divisions, a tetrad is produced, showing a 3:1 gene conversion for the + allele of m, while the outside genetic markers a^+/a and b^+/b retain their parental configuration (Figure 11.26, part 4).

Lastly, if two markers are very close together, they can undergo *co-conversion* so that both will exhibit a 1:3 or 3:1 segregation in a tetrad, while outside markers will segregate in a normal 2:2 ratio.

~ **Figure 11.26**

Gene conversion by mismatch repair at two sites. (1) Parent homologues; (2) Recombination between inner two chromatids produces a patched duplex with two mismatches; (3) Both mismatches are repaired by excision and DNA synthesis; (4) Two meiotic divisions produce a tetrad with a 3:1 conversion for the m^+ allele.

❶ Parent homologues in meiotic prophase

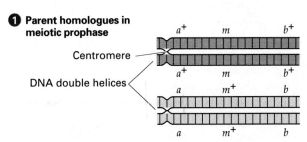

❷ Recombination event generates two mismatches

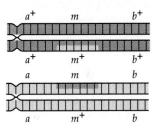

❸ Excision and repair by DNA synthesis

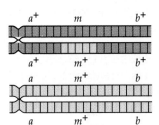

❹ Tetrad produced showing 3:1 gene conversion for m^+

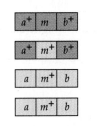

KEYNOTE

In organisms in which all four products of meiosis can be analyzed, mismatch repair of heteroduplex DNA can result in a non-Mendelian segregation of alleles, typically a 3:1 or 1:3 ratio rather than a 2:2 segregation. This phenomenon is known as gene conversion. Co-conversion of alleles can occur if the two alleles are very close together. ────────

SUMMARY

In this chapter we discussed DNA replication and DNA recombination. Many aspects of DNA replication are similar in prokaryotes and eukaryotes, for example, a semiconservative and semi-discontinuous mechanism, synthesis of new DNA in the 5'-to-3' direction, and the use of RNA primers to initiate DNA chains. The enzymes that catalyze the synthesis of DNA are the DNA polymerases. In *E. coli* there are three DNA polymerases, two of which are known to be involved in DNA replication along with several other enzymes and proteins. The DNA polymerases have 3'-to-5' exonuclease activity, which permits proofreading to take place during DNA synthesis if an incorrect nucleotide is inserted opposite the template strand. In eukaryotes, two DNA polymerases are involved in nuclear DNA replication. Neither has associated proofreading activity, that function presumably being the property of a separate protein.

In prokaryotes, DNA replication begins at specific chromosomal sites. Such sites are known also for yeast. A prokaryotic chromosome has one initiation site for DNA replication, while a eukaryotic chromosome has several initiation sites dividing the chromosome into replication units, or replicons. It is not clear whether the initiation sites used are the same sequences from cell generation to cell generation. The existence of replicons means that DNA replication of the entire set of chromosomes in a eukaryotic organism can proceed relatively quickly, in some cases faster than the single *E. coli* chromosome, despite the presence of orders of magnitude more DNA.

Replication of DNA in eukaryotes occurs in the S phase of the cell division cycle. Eukaryotic chromosomes are complexes of DNA with histones and nonhistone chromosome proteins, so not only must DNA be replicated but the chromosome structure must also be duplicated. In particular, the nucleosome organization of chromosomes must be duplicated as the replication forks migrate. It is known that nucleosomes are "mature" soon after a replication fork passes, although it is not known precisely how nucleosomes segregate from cell generation to cell generation.

Some of the same enzymes used for DNA replication are also used in the genetic recombination process. A number of models have been proposed to describe the molecular events involved in the breakage and rejoining of DNA during crossing-over. One model, developed by Holliday, is described in this chapter. All molecular models for genetic recombination involve new DNA synthesis in small regions participating in crossing-over. In a later chapter (Chapter 18) we will see that DNA synthesis may also be involved in the repair of certain genetic damage. Thus, three major cellular processes involve DNA synthesis: DNA replication, genetic recombination, and DNA repair.

ANALYTICAL APPROACHES FOR SOLVING GENETICS PROBLEMS

Q.1 a. Meselson and Stahl used ^{15}N-labeled DNA to prove that DNA replicates semiconservatively. The method of analysis was cesium chloride equilibrium density gradient centrifugation, in which bacterial DNA labeled in both strands with ^{15}N (the heavy isotope of nitrogen) bands to a different position in the gradient than DNA labeled in both strands with ^{14}N (the normal isotope of nitrogen). Starting with a mixture of ^{15}N-containing and ^{14}N-containing DNA, then, two bands result after CsCl density gradient centrifugation. If both DNAs are first heated to 100°C and cooled slowly before centrifuging, the result is different. Two bands are seen in exactly the same positions as before, and a new, third band, is seen at a position halfway between the other two bands. From its position relative to the other two bands, the new band is interpreted to be intermediate in density between the other two bands. Explain the existence of the three bands in the gradient.

b. DNA from *E. coli* containing ^{15}N in both strands is mixed with DNA from another bacterial species, *Bacillus subtilis* containing ^{14}N in both strands. Two bands are seen after CsCl density gradient centrifugation. If the two DNAs are mixed, heated to 100°C, slowly cooled, and then centrifuged, only two bands result. These bands are in the same positions as in the unheated DNA experiment. Explain these results.

A.1 a. When DNA is heated to 100°C, the DNA is denatured to single strands (c.f. Chapter 10). If denatured DNA is allowed to cool slowly, reassociations of complementary strands occurs to produce double-stranded DNA again. Thus, when mixed, denatured $^{15}N-^{15}N$ DNA and $^{14}N-^{14}N$ DNA from the same species is cooled slowly, the single strands pair randomly during renaturation so that $^{15}N-^{15}N$, $^{14}N-^{14}N$, and $^{15}N-^{14}N$ double-stranded DNA is produced. The latter DNA will have a density intermediate between those of the two other DNA types, accounting for the third band. Theoretically, if all DNA strands pair randomly, there should be a distribution of 1:2:1 of $^{15}N-^{15}N$, $^{15}N-^{14}N$, and $^{14}N-^{14}N$ DNAs, and this should be reflected in the relative intersities of the bands.

b. DNAs from different bacterial species have different sequences. In other words, DNA from one species typically is not complementary to DNA from another species. Therefore only two bands are seen because only the two *E. coli* DNA strands can renature to from $^{15}N-^{15}N$ DNA and only the two *B. sublilis* DNA strands can renature to form $^{14}N-^{14}N$ DNA. No $^{15}N-^{14}N$ hybrid DNA can form, so in this case there is no third band of intermediate density.

Q.2 What would be the effect upon chromosome replication in *E. coli* strains carrying deletions of the following genes:
a. *polC*
b. *polA*
c. *dnaG*
d. *lig*
e. *ssb*
f. *ori*
g. *rep*

A.2 When genes are deleted, the function encoded by those genes is lost. All of the genes listed in the question are involved in DNA replication in *E. coli,* and their functions are briefly described in Table 11.2, and discussed further in the text.

a. *polC* encodes DNA polymerase III, the principal DNA polymerase in *E. coli* that is responsible for elongation of DNA chains. A deletion of the *polC* gene would result in the loss of the ability to synthesize DNA strands from RNA primers; hence, synthesis of new DNA strands could not occur, and there would be no chromosome replication.

b. *polA* encodes DNA polymerase I, which is used in DNA synthesis to extend DNA chains made by DNA polymerase III while simultaneously excising the RNA primer by 5'-to-3' exonuclease activity. As discussed in the text, temperature-sensitive mutants of *polA* made inactive DNA polymerase I at the nonpermissive temperature, but chromosome replication occurred nor-

mally. Thus, chromosome replication would occur normally in an *E. coli* strain carrying a deletion of *polA*.

c. *dnaG* encodes DNA primase, the enzyme that synthesized the RNA primer on the DNA template. Without the synthesis of the short RNA primer, DNA polymerase III cannot initiate DNA synthesis and, therefore, chromosome replication will not take place.

d. *lig* encodes DNA ligase, the enzyme that catalyzes the ligation of Okazaki fragments together. In a strain carrying a deletion of *lig*, DNA synthesis would occur, but stable progeny chromosomes would not result because the Okazaki fragments could not be ligated together, so the lagging strand synthesized discontinuously on the lagging strand template would be in fragments.

e. *ssb* encodes the single-strand binding proteins which bind to and stabilize the single-stranded DNA regions produced as the DNA is unwound at the replication fork. In the absence of single-strand binding proteins, impeded or absent DNA replication would result because the replication bubble could not be kept open.

f. *ori* is the origin of replication region in *E. coli*, that is, the location at which chromosome replication initiates. Without the origin, the initiator protein cannot bind, no replication bubble can form, and therefore chromosome replication cannot take place.

g. *rep* encodes helicase, the enzyme that untwists the DNA at the replication fork. Without helicase, the old chromosome cannot be untwisted and is therefore not accessible for replication; no chromosome replication will occur.

QUESTIONS AND PROBLEMS

11.1 Compare and contrast the conservative and semiconservative models for DNA replication.

11.2 Describe the Meselson and Stahl experiment, and explain how it showed that DNA replication is semiconservative.

*11.3 In the Meselson and Stahl experiment ^{15}N-labeled cells were shifted to ^{14}N medium, at what we can designate as generation 0.
a. For the semiconservative model of replication, what proportion of $^{15}N-^{15}N$, $^{15}N-^{14}N$, and $^{14}N-^{14}N$ would you expect to find at generations 1, 2, 3, 4, 6, and 8?
b. Answer the above question in terms of the conservative model of DNA replication.

11.4 Suppose *E. coli* cells are grown on an ^{15}N medium for many generations. Then they are quickly shifted to an ^{14}N medium, and DNA is extracted from the samples taken after one, two, and three generations. The extracted DNA is subjected to equilibrium density gradient centrifugation in CsCl. In the figure below, using the reference positions of pure ^{15}N and pure ^{14}N DNA as guides, indicate where the bands of DNA would equilibrate if replication were semiconservative or conservative.

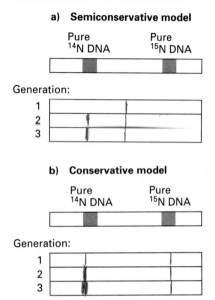

11.5 A spaceship lands on earth and with it a sample of extraterrestrial bacteria. You are assigned the task of determining the mechanism of DNA replication in this organism.

You grow the bacteria in unlabeled medium for several generations, then grow it in presence of ^{15}N exactly for one generation. You extract the DNA and subject it to CsCl centrifugation. The banding pattern you find is shown in the figure below.

It appears to you that this is evidence that DNA replicates in the semiconservative manner, but this result does not *prove* that this is so. Why? What other experiment could you perform (using the same sample and technique of CsCl centrifugation) that would further distinguish between semiconservative and dispersive modes of replication?

11.6 Assume you have a DNA molecule with the base sequence T-A-T-C-A going from the 5′ to the 3′ end of one of the polynucleotide chains. The building blocks of the DNA are drawn as in the following figure. Use this shorthand system to diagram the completed double-stranded DNA molecule, as proposed by Watson and Crick.

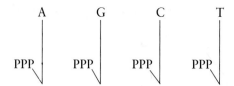

****11.7** List the components necessary to make DNA *in vitro* by using the enzyme system isolated by Kornberg.

****11.8** Give two lines of evidence that the Kornberg enzyme is not the enzyme involved in the replication of DNA for the duplication of chromosomes in growth of *E. coli*.

11.9 Base analogs are compounds that resemble the natural bases found in DNA and RNA, but are not normally found in those macromolecules. Base analogs can replace their normal counterparts in DNA during *in vitro* DNA synthesis. Four base analogs were studied for their effects on *in vitro* DNA synthesis using the *E. coli* DNA polymerase. The results were as follows, with the amounts of DNA synthesized expressed as percentages of that synthesized from normal bases only.

	NORMAL BASES SUBSTITUTED BY THE ANALOG			
ANALOG	A	T	C	G
A	0	0	0	25
B	0	54	0	0
C	0	0	100	0
D	0	97	0	0

Which bases are analogs of adenine? of thymine? of cytosine? of guanine?

11.10 Describe the semidiscontinuous model for DNA replication. What is the evidence showing that DNA synthesis is discontinuous on at least one template strand?

****11.11** Distinguish between a primer strand and a template strand.

****11.12** The length of the *E. coli* chromosome is about 1100 μm.

a. How many base pairs does the *E. coli* chromosome have?

b. How many complete turns of the helix does this chromosome have?

c. If this chromosome replicated unidirectionally and if it completed one round of replication in 60 minutes, how many revolutions per minute would the chromosome be turning during the replication process?

d. The *E. coli* chromosome, like many others, replicates bidirectionally. Draw a simple diagram of a replicating *E. coli* chromosome that is halfway through the round of replication. Be sure to distinguish new and old DNA strands.

11.13 In *E. coli* the replication fork moves forward at 500 nucleotide pairs per second. How fast is the DNA ahead of the replication fork rotating?

11.14 A diploid organism has 4.5×10^8 base pairs in its DNA. This DNA is replicated in 3 minutes. Assuming all replication forks move at a rate of 10^4 base pairs per minute, how many replicons (replication units) are present in this organism's genome?

11.15 The following events, steps or reactions occur during *E. coli* DNA replication. For each entry in Column A, select the appropriate entry in Column B. Each entry in A may have more than one answer, and each entry in B can be used more than once.

COLUMN A	COLUMN B
_____ **a.** Unwinds the double helix	A Polymerase I
	B Polymerase III
_____ **b.** Prevents reassociation of complementary bases	C Helicase
	D Primase
	E Ligase
_____ **c.** Is a RNA polymerase	F SSB protein
	G Gyrase
_____ **d.** Is a DNA polymerase	H None of these
_____ **e.** Is the "repair" enzyme	
_____ **f.** Is the major elongation enzyme	
_____ **g.** A $5' \rightarrow 3'$ polymerase	
_____ **h.** A $3' \rightarrow 5'$ polymerase	
_____ **i.** Has $5' \rightarrow 3'$ exonuclease function	
_____ **j.** Has $3' \rightarrow 5'$ exonuclease function	
_____ **k.** Bonds free 3'-OH end of a polynucleotide to a free 5' monophosphate end of polynucleotide	
_____ **l.** Bonds 3'-OH end of a polynucleotide to a free 5' nucleotide triphosphate	
_____ **m.** Separates daughter molecules and causes supercoiling	

11.16 Compare and contrast the three *E. coli* DNA polymerases with respect to their enzymatic activities.

* **11.17** In *E. coli,* distinguish between the activities of primase; single-stranded, binding protein; helicase; DNA ligase; DNA polymerase I; and DNA polymerase III in DNA replication.

11.18 Describe the molecular action of the enzyme DNA ligase. What properties would you expect an *E. coli* cell to have if it had a temperature-sensitive mutation in the gene for DNA ligase?

11.19 Chromosome replication in *E. coli* commences from a constant point, called the origin of replication. It is known that DNA replication is bidirectional. Devise a biochemical experiment to prove that the *E. coli* chromosome replicates bidirectionally. (*Hint:* Assume that the amount of gene product is directly proportional to the number of genes.)

11.20 What property of DNA replication was indicated by the presence of Okazaki fragments?

11.21 A space probe returns from Jupiter and brings with it a new microorganism for study. It has double-stranded DNA as its genetic material. However, studies of replication of the alien DNA reveal that, while the process is semiconservative, DNA synthesis is continuous on both the leading and the lagging strand templates. What conclusion(s) can you make from that result?

* **11.22** Compare and contrast eukaryotic and prokaryotic DNA polymerases.

11.23 Draw a eukaryotic chromosome as it would appear at each of the following cell cycle stages. Show both DNA strands, and use different line styles for old and newly synthesized DNA.
a. G_1
b. anaphase of mitosis
c. G_2
d. anaphase of meiosis I
e. anaphase of meiosis II

* **11.24** Autoradiography is a technique which allows radioactive areas of chromosomes to be observed under the microscope. The slide is covered with a photographic emulsion, which is exposed by radioactive decay. In regions of exposure the emulsion forms silver grains upon being developed. The tiny silver grains can be seen on top of the (much larger) chromosomes. Devise a method for finding out which regions in the human karyotype replicate during the last 30 min of the S period. (Assume a cell cycle in which the cell spends 10 h in G_1, 9 h in S, 4 h in G_2 and 1 h in M.)

11.25 When the eukaryotic chromosome duplicates, the nucleosome structures must duplicate. Discuss the synthesis of histones in the cell cycle, and discuss the model for the assembly of new nucleosomes at the replication forks.

* **11.26** How is mismatch repair related to recombination and repair of DNA?

* **11.27** Crosses were made between strains, each of which carried one of three different alleles of the same gene, *a*, in

yeast. For each cross, some unusual tetrads resulted at low frequencies. Explain the origin of each of these tetrads:

$$Cross: \quad a1\ a2^+ \quad a1\ a3^+ \quad a2\ a3^+$$
$$\times \qquad \times \qquad \times$$
$$a1^+\ a2 \quad a1^+\ a3 \quad a2^+\ a3$$

$$Tetrads:\ a1^+\ a2 \quad a1^+\ a3 \quad a2^+\ a3$$
$$a1^+\ a2^+ \quad a1^+\ a3 \quad a2^+\ a3^+$$
$$a1\ \ a2^+ \quad a1^+\ a3^+ \quad a2\ \ a3^+$$
$$a1\ \ a2^+ \quad a1\ \ a3^+ \quad a2\ \ a3^+$$

11.28 From a cross of $y1\ y2^+ \times y1^+\ y2$, where $y1$ and $y2$ are both alleles of the same gene in yeast, the following tetrad type occurs at very low frequencies:

$$y1^+\ y2$$
$$y1\ \ y2$$
$$y1\ \ y2$$
$$y1\ \ y2^+$$

Explain the origin of this tetrad at the molecular level.

11.29 In *Neurospora* the *a*, *b*, and *c* loci are situated in the same arm of a particular chromosome. *a* is near the centromere, *b* is near the middle, and *c* is near the telomere of the arm. Among the asci resulting from a cross of $ABC \times abc$, the following ascus was found (the 8 spores are indicated in the order in which they were arranged in the ascus): *ABC, ABC, ABc, ABc, aBC, aBC, abc, abc.* How might this ascus have arisen?

11.30 In the population of asci produced in question 11.29 an ascus was found containing, in this order, the spores *ABC, ABC, ABc, Abc, aBC, aBC, abc, abc.* How could this ascus have arisen?

12 TRANSCRIPTION

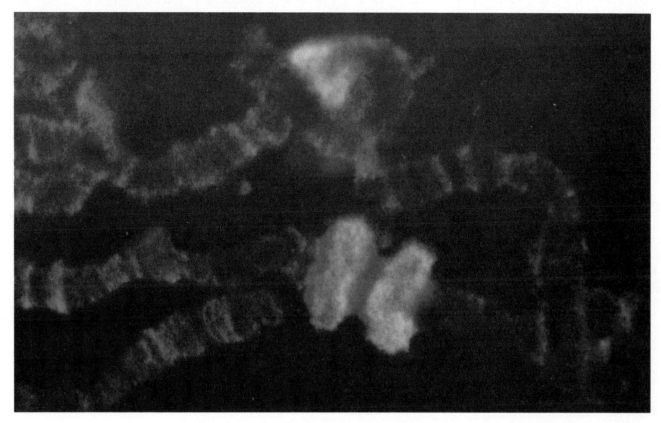

PRINCIPAL POINTS

~ Transcription is the process of reading DNA base-pair sequences into RNA base sequences (mRNA, tRNA, and rRNA in prokaryotes and eukaryotes, and snRNA additionally in eukaryotes).

~ Transcription is similar in prokaryotes and eukaryotes. The DNA denatures, and an RNA polymerase catalyzes the synthesis of an RNA molecule in the 5'-to-3' direction. Only one strand of the double-stranded DNA is transcribed into the single-stranded RNA molecule. Promoter and terminator sequences determine where transcription starts and stops, respectively.

~ The E. coli RNA polymerase transcribes all classes of RNA in the cell; namely, mRNA, tRNA and rRNA. In eukaryotic nuclei there are three RNA polymerases which have different functions: RNA polymerase I transcribes three of the four rRNAs, RNA polymerase II transcribes mRNAs and some snRNAs, and RNA polymerase III transcribes the fourth rRNA, the tRNAs, and the remaining snRNAs.

~ Unlike the E. coli RNA polymerase, none of the three eukaryotic RNA polymerases directly binds to DNA during initiation of RNA synthesis. Rather, the initiation of transcription by these enzymes is mediated by transcription factors that are specific for each polymerase and that recognize the appropriate promoter sequence.

~ The promoter for protein-coding genes is located upstream of the transcription initiation point and consists of different combinations of promoter elements. Specific transcription factors and regulatory factors bind to these elements and facilitate transcription initiation by RNA polymerase II. Further away from the promoter are enhancers, which bind regulatory factors that function to facilitate maximum gene transcription.

~ The promoters for genes transcribed by RNA polymerase III are located within the RNA coding sequences of the genes themselves. The promoters consist of different combinations of functional domains, depending on the class of RNA being transcribed. The domains are recognized by transcription factors, which facilitate transcription initiation by the polymerase.

~ The 18S, 5.8S and 28S rRNAs are transcribed from a single transcription unit to produce a single precursor RNA molecule. Most eukaryotes contain tandem arrays of the transcription units, each of which is separated by a nontranscribed spacer (NTS) sequence. The promoter for the transcription unit is located within the NTS and functions to bind specific transcription factors that facilitate the initiation of transcription by RNA polymerase I.

As was stated in Chapter 9, genetic material has three requirements:

1. It must contain—in a stable form—all the information for an organism's structure, function, development, and reproduction.
2. It must replicate accurately, so that progeny cells have the same genetic information as the parental cell.
3. It must be capable of variation.

In the previous four chapters we examined the structure of genetic material and the evidence that proteins are specified by genes. We also learned how DNA is replicated, so that it can be passed from one cell generation to the next and from a parental to a progeny organism. Now we will begin a discussion of the processes by which genetic material directs the synthesis of proteins. Since all cell and organism functions involve the action of specific proteins, all cell functions are ultimately directed by genetic material.

Each protein consists of one or more chains of amino acids, and the sequence of those amino acids is coded for by a specific base-pair sequence in DNA. When a protein is needed in the cell, the genetic code for that protein's amino acid sequence must be read from the DNA and processed into the finished protein. Two major steps occur in the process of protein synthesis: transcription and translation.

Transcription is the transfer of information from a double-stranded, template DNA molecule to a single-stranded RNA molecule. **Translation** (protein synthesis) is the conversion, in the cell, of the messenger RNA (mRNA) base sequence information into the amino acid sequence of a polypeptide. While replication typically occurs during only part of the cell cycle (at least in eukaryotes), transcription and translation generally occur throughout the cell cycle. In this chapter we will examine the transcription process itself, and in Chapter 13 we will discuss the structures and properties of the different RNA classes and the processing of RNA precursor molecules.

OVERVIEW OF TRANSCRIPTION

In 1956, three years after Watson and Crick proposed their double helix model for DNA, Crick gave the name the **Central Dogma to** the two-step process (transcription followed by translation) for the synthesis of proteins encoded by DNA. It was generally understood at that time that there was a *one-way flow of information* from DNA to protein. That is, all RNA molecules are produced from DNA templates by transcription, and all proteins are produced from RNA templates by translation. Further, it was recognized that all new DNA was produced from old DNA by replication.

Before we begin a detailed discussion of transcription, a brief overview of the process is in order. The genome of an organism consists of specific sequences of base pairs distributed among a species-specific number of chromosomes. All base pairs in the genome are replicated during the DNA synthesis phase of the cell cycle, but only *some* of the base pairs are transcribed into RNA. The specific sequences of base pairs that are transcribed are called *genes*, and thus the transcription process is also referred to as *gene expression*. Around the beginning and end of each gene are base-pair sequences called **gene regulatory elements,** which are involved in the regulation of gene expression. These will be discussed in a later chapter.

Four different major classes of RNA molecules or transcripts are produced by transcription: **messenger RNA (mRNA), transfer RNA (tRNA), ribosomal RNA (rRNA),** and **small nuclear RNA (snRNA)** (Figure 12.1). mRNAs, tRNAs and rRNAs are found in both prokaryotes and eukaryotes, while snRNAs are only found in eukaryotes. The structures of the RNAs and their functions in protein synthesis are described in Chapter 13. For now, note that only the mRNA molecule is translated to produce a protein molecule. A gene that codes for an mRNA molecule, and hence for a protein, is called a **structural gene,** or a protein-coding gene. In this and later chapters we will be discussing primarily structural genes. There are other genes, whose RNA transcripts (the products of transcription) are the final products of gene expression; they do not function as messenger molecules between DNA and protein; that is they are never translated into proteins. The genes that specify tRNA, rRNA, and snRNA molecules are examples of such genes.

Genes are almost always located in the chromosomes, yet they must be able to control functions at various locations throughout the cell. Through transcription, the genes produce RNA transcripts that can diffuse away from the chromosomes and be involved in protein synthesis elsewhere in the cell. While DNA is a very stable structure that endures cell and organism reproduction virtually unchanged, the mRNA tran-

~ **FIGURE 12.1**

Four types of RNA transcripts: messenger RNA (mRNA), transfer RNA (tRNA), ribosomal RNA (rRNA), and small nuclear RNA (snRNA). The latter is found in eukaryotes only, while the other three types are found in both prokaryotes and eukaryotes.

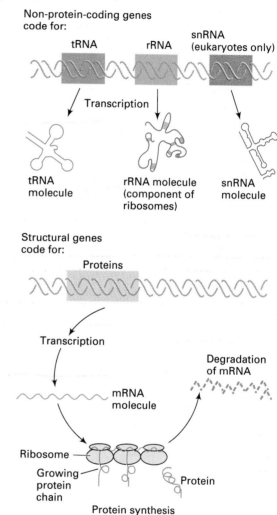

scripts of genes typically have a limited existence, being degraded by cellular enzymes called **ribonucleases.** The transient existence of the RNA transcripts is a key element in the processes that regulate gene expression. This will be discussed when the regulation of gene expression is described in a later chapter.

BASIC PROPERTIES OF TRANSCRIPTION

In both prokaryotes and eukaryotes, transcription occurs by a process that is catalyzed by an enzyme called *RNA polymerase* (Figure 12.2; p. 364). The DNA dou-

~ FIGURE 12.2

Transcription process. The DNA double helix is denatured by the action of RNA polymerase, which then catalyzes the synthesis of a single-stranded RNA chain, beginning at the "start of transcription" point. The RNA chain is made in the $5' \rightarrow 3'$ direction, using only one strand of the DNA as a template to determine the base sequence.

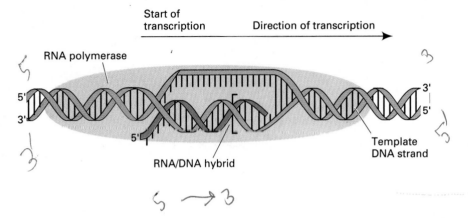

ble helix must unwind before transcription can begin. Only one of the two DNA strands is transcribed into an RNA. This strand is called the *template strand*. The *non-template strand*—the strand complementary to the template strand—has *the same polarity as the RNA strand*.

If each of the two DNA strands of a gene was transcribed to produce an RNA molecule, each gene would produce two RNA products that were complementary in sequence. Translation of the two RNAs would produce two very different proteins. However, genetic evidence indicates that each structural gene encodes but a single protein, as would be expected if only one RNA molecule were transcribed from a given gene.

Some evidence that only one of the two strands of DNA is copied into RNA in a given region came from the work of Julius Marmur and his colleagues in their studies of the RNA species produced when bacteriophage SP8 infects the bacterium *Bacillus subtilis*. The two DNA strands of this double-stranded DNA phage have different densities from each other and from the double-stranded form, and can easily be separated by cesium chloride (CsCl) equilibrium density gradient centrifugation (Figure 12.3).

Marmur and his colleagues isolated SP8 DNA from the phages and separated the two strands into the "heavy" and "light" single-stranded species by CsCl density gradient centrifugation (Figure 12.4). Separately, they allowed SP8 to infect *Bacillus subtilis* growing in a medium containing ^{32}P so that all mRNA species transcribed from the phage genome during the life cycle became radioactively labeled. They reasoned that the mRNAs produced would complement the DNA strand from which they were transcribed and could hybridize to that DNA. They found that hybrids were formed only between the radioactive RNA and the heavier of the two strands, from which they concluded that only one of the two DNA strands was used for transcription (Figure 12.4).

In the SP8 case it was fortunate that only one of the two DNA strands of the chromosome is used for the transcription of *all* genes. In other systems, while it is

~ FIGURE 12.3

Demonstration that the two strands of DNA in phage SP8 have different buoyant densities. The figure shows the locations of DNA bands separated by CsCl equilibrium density gradient centrifugation. Reference DNA, double-stranded SP8 DNA, and the "heavy" and "light" single strands of SP8 are distinguished on the gradient.

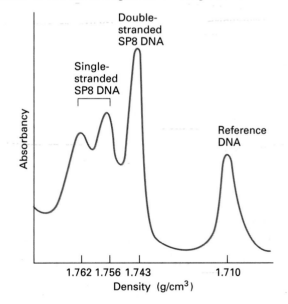

~ FIGURE 12.4

Use of DNA-RNA hybridization to show that an mRNA is complementary to only one of the two strands of its DNA template. Phage SP8 DNA was denatured to single strands by heat, and the two strands were separated based on buoyant differences by CsCl equilibrium density gradient centrifugation (see Figure 12.3). The two strands were separately isolated from the gradient and hybridized with ^{32}P-labeled, phage-encoded mRNAs. The mRNAs hybridized with the heavy DNA strand but not the light DNA strand.

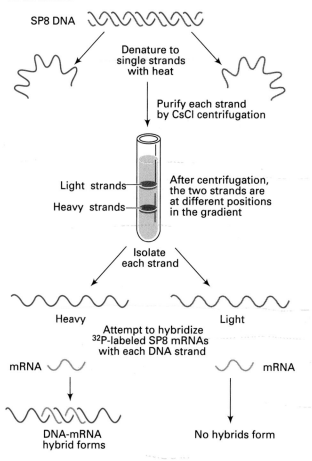

Conclusion: SP8 mRNAs are transcribed from only one DNA strand.

true that only one of the two strands is transcribed for a given gene, *which* DNA strand functions as the template strand for RNA transcription depends upon the gene.

In transcription, the RNA precursors are the ribonucleoside triphosphates ATP, GTP, CTP, and UTP, collectively called NTPs. The enzyme that catalyzes the polymerization reaction is **RNA polymerase**; this enzyme is specific for RNA synthesis and will only use NTPs that contain the ribose sugar rather than the deoxyribose sugar found in DNA precursors.

The RNA polymerization reaction is shown in Figure 12.5 (p. 366); this reaction is very similar to the DNA polymerization reaction. The next nucleotide to be added to the chain is selected by the RNA polymerase for its ability to pair with the exposed base on the DNA template strand. Recall that RNA chains contain nucleotides with the base uracil instead of thymine. Therefore where an A nucleotide occurs on the DNA template chain, a U nucleotide is placed in the RNA chain.

As in DNA replication, RNA is synthesized in the $5' \rightarrow 3'$ direction, and hence the template DNA strand is copied in the $3' \rightarrow 5'$ direction. If the DNA template strand reads

$$3'\text{-A T A C T G G A C-}5'$$

then the RNA chain produced will read

$$5'\text{-U A U G A C C U G-}3'$$

The DNA non-template strand is

$$5'\text{-T A T G A C C T G-}3'$$

TRANSCRIPTION OF BACTERIAL AND BACTERIOPHAGE GENES

RNA Polymerase in Bacteria

Each bacterial species encodes RNA polymerase. The action of RNA polymerase is similar to that of DNA polymerases. It requires a DNA template, magnesium ions, and the four ribonucleoside triphosphates, ATP, UTP, CTP, and GTP. If any of these components is missing, transcription cannot proceed. RNA polymerase differs from DNA polymerases in two important respects: (1) RNA polymerase does not have any proofreading function; (2) RNA polymerase is able to initiate new RNA chains.

The most extensively studied bacterial RNA polymerase is that found in *Escherichia coli*. *E. coli* RNA polymerase (Figure 12.6; p. 366) is a complex enzyme that synthesizes all three classes of RNA: transfer, messenger, and ribosomal.

The enzyme consists of five polypeptide subunits: two alpha (α) subunits, one beta (β) subunit, one beta-prime (β') subunit, and one omega (ω) subunit. The active enzyme is often referred to as the **core enzyme** and can be written $\alpha_2\beta\beta'\omega$; its molecular weight is about 450,000. For the core to begin transcription at the proper place on the chromosome, another polypeptide of weight 85,000 (the sigma factor, or σ) must become associated with it to form what is called the *holoenzyme*, or *complete enzyme*.

~ FIGURE 12.5

Chemical reaction involved in the RNA polymerase-catalyzed synthesis of RNA on a DNA template strand.

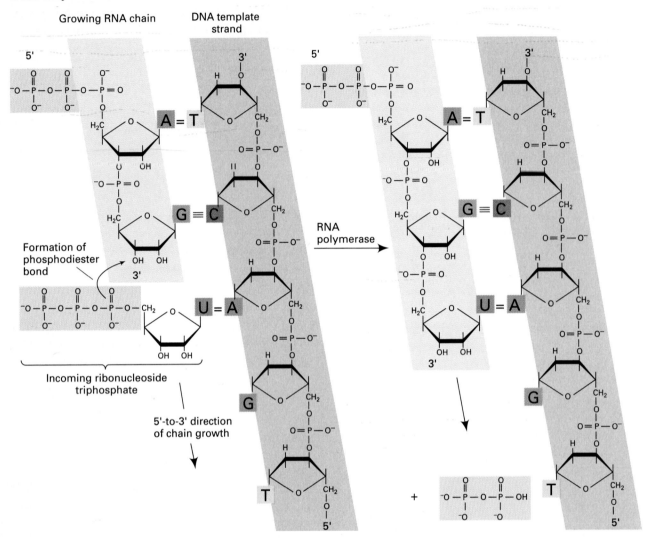

~ FIGURE 12.6

Schematic picture of the *E. coli* RNA polymerase, showing the five subunits of the core enzyme and the dissociable sigma factor.

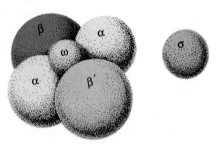

Through studies of mutants, geneticists have identified the chromosomal genes that encode the RNA polymerase subunits. They are called the *rpo* genes: *rpoA* codes for the α subunits; *rpoB* and *rpoC* code for the β and β' subunits respectively; and *rpoD* codes for the sigma factor. The gene for the ω subunit has yet to be found.

To initiate transcription of a gene, RNA polymerase binds to a transcription-controlling sequence adjacent to the start of the gene; this sequence is called the **promoter.** In the cell the initiation of transcription is preceded by the binding of a core enzyme–sigma factor complex to the promoter. The sigma factor is essential for promoter recognition; if sigma is absent, the core enzyme initiates transcription randomly. Once transcription is initiated by the core enzyme–sigma factor complex, (the holoenzyme) the sigma factor dissociates from the core enzyme.

Transcription ceases when a controlling element called a **transcription terminator sequence,** or more simply, a **terminator,** is encountered at the end of a gene. After termination, the core enzyme is released and is reused on the same or on a different gene.

Conceptually, then, a gene may be divided into three sequences with respect to its transcription (Figure 12.7). The promoter (transcription start) sequence specifies where transcription will begin. The RNA coding sequence is the sequence of base pairs transcribed by RNA polymerase into the single-stranded mRNA transcript. The terminator is a sequence of base pairs that indicates where transcription is to stop. For purposes of discussion, the promoter is considered to be "upstream" from the gene, and the terminator is "downstream."

An RNA polymerase enzyme performs several functions, each of which requires different activities of the enzyme:

1. It recognizes a promoter on the double-stranded DNA.
2. It causes the DNA to denature and unwind into single strands at the promoter (similar to the initiation of DNA replication).
3. As a result of reading the promoter sequence, the RNA polymerase orients itself properly and transcribes the entire template strand of the gene.
4. Lastly, it stops transcribing when it reaches and recognizes the terminator.

Figure 12.8 (p. 368) shows the general processes of initiation, elongation, and termination of transcription. These processes will now be discussed in more detail.

~ FIGURE 12.7

Organization of a gene in terms of transcription into three regions: promoter, RNA coding sequence, and terminator.

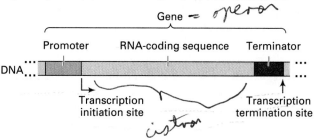

KEYNOTE

Transcription, the process of reading DNA base-pair sequences into RNA base sequences (mRNA, tRNA, and rRNA), is similar in prokaryotes and eukaryotes. The DNA denatures, and an RNA polymerase catalyzes the synthesis of an RNA molecule in the 5′-to-3′ direction. Only one strand of the double-stranded DNA is transcribed into an RNA molecule. Specific base-pair sequences in the DNA determine where transcription begins (promoter) and ends (terminator).

Initiation and Elongation

PROMOTER RECOGNITION AND TRANSCRIPTION INITIATION. Promoter recognition is a function of the sigma factor in association with the core enzyme (Figure 12.8,2). After binding correctly to the DNA, the core enzyme slides along the DNA toward the gene. As it moves, it catalyzes the denaturation of the two DNA strands, thereby exposing the single-stranded DNA strand for use as a template for RNA synthesis. Immediately following denaturation, RNA synthesis commences. In *E. coli* the interval from the time the promoter is recognized to the time RNA synthesis is begun is about 0.2 seconds.

In *E. coli,* two sequences in the promoter are critical for gene expression. These sequences are at two distinct locations upstream from the first base pair to be transcribed into an RNA chain. Conventionally, this initial base pair is designated as +1; base pairs upstream from it (such as promoters) are given negative numbers, while those downstream from it are given positive numbers. It is also conventional to give the sequences of promoters and other controlling elements for the non-template strand—that is, the DNA strand that has the same polarity as the RNA strand. For the *E. coli* RNA polymerase, the two regions of a promoter generally are found at −35 and −10, that is, centered at 35 and 10 base pairs upstream from the base pair at which transcription starts. The −10 region is also called the **Pribnow box** (after the researcher who first discovered it). From examination of the promoters of a large number of genes, the **consensus sequence** (i.e., the most prevalent base found in each position for a large number of promoters that have been analyzed) for the −35 region is

5′-T T G A C A-3′

~ FIGURE 12.8

Transcription of a gene in *E. coli*. The process involves the activities of the RNA polymerase core enzyme, the sigma factor, and in this case, the ρ factor. For the gene shown, RNA chain termination requires the ρ factor. This is called ρ-dependent termination. For many other genes, RNA polymerase alone can read the termination sequence—this is called ρ-independent termination.

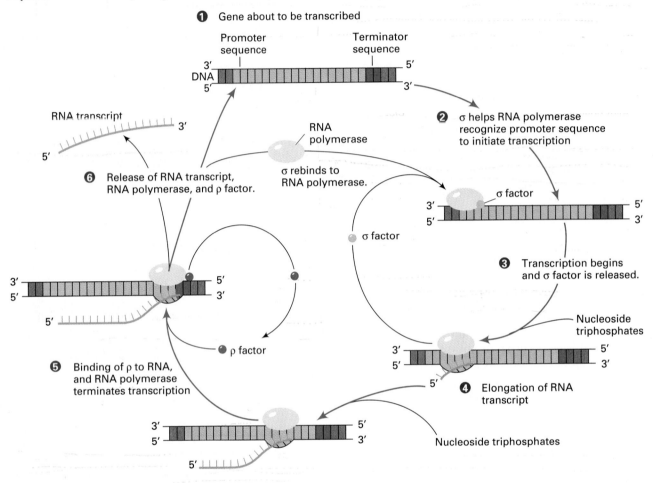

and the consensus sequence for the Pribnow box at −10 is

$$5'\text{-T A T A A T-}3'$$

The *E. coli* RNA polymerase binds to a promoter in two distinct steps. First, it binds loosely to the −35 region of the promoter while the DNA is still in double-helical form (Figure 12.9a). The second step involves a shift to a tighter binding between RNA polymerase and DNA. This shift accompanies a local untwisting of about 17 base pairs of the DNA centered around the −10 region (Figure 12.9b). This untwisting represents almost two complete helical turns of DNA. Note that the −10 region is all AT base pairs; having only two hydrogen bonds such base pairs are easier to break

apart than GC base pairs, which have three. Once the RNA polymerase is bound at the −10 region, it is correctly oriented to begin transcription at the right nucleotide (Figure 12.9b).

Since promoters differ in sequence, the efficiency of RNA polymerase binding varies considerably. As a result, the rate at which transcription is initiated is not constant, a fact that helps to explain why different genes have different rates of expression at the RNA level.

Elongation of the RNA chain

RNA polymerase-catalyzed elongation of the RNA chain occurs as shown in Figure 12.5. All events take place in a region of the DNA that has denatured to

~ FIGURE 12.9

Action of *E. coli* RNA polymerase in the initiation and elongation stages of transcription. (a) In initiation, the enzyme first binds loosely to the promoter at the −35 region; (b) As initiation continues, RNA polymerase binds more tightly to the promoter at the −10 region, accompanied by a local untwisting of about 17 bp centered around the −10 region. At this point, the RNA polymerase is correctly oriented to begin transcription at +1; (c) After a few polymerizations have occurred, the sigma factor dissociates from the core enzyme; (d) As the RNA polymerase elongates the new RNA chain, the enzyme untwists the DNA ahead of it; as the double helix reforms behind the enzyme, the RNA is displaced away from the DNA.

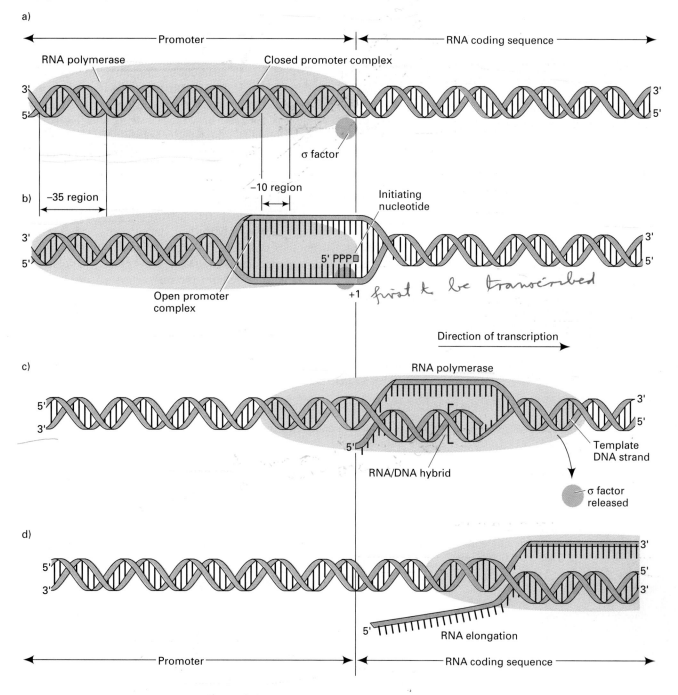

form a transcription bubble that is similar to a DNA replication bubble. Once a few polymerizations have been completed, the sigma factor dissociates from the core enzyme (Figure 12.8,3 and Figure 12.9c) and can be used again in other transcription initiation reactions.

As the RNA polymerase moves along, it untwists the DNA double helix ahead of it, while the helix reforms behind it (Figure 12.8,4 and 12.9d). About 17 base pairs of the DNA are kept unpaired as transcription continues, about the same as are initially untwisted. Within the untwisted region, about 12 bases of RNA are bonded to the DNA in an RNA-DNA hybrid; the rest of the RNA is displaced away from the DNA (Figure 12.9d).

As with the untwisting of DNA during replication, the supercoiling caused by untwisting is counteracted by negative supercoiling introduced into the DNA double helix by DNA topoisomerase II (see Ch. 10).

Termination of Transcription

Termination of the transcription of a prokaryotic gene is signaled by controlling elements called *terminators*. Terminators vary considerably in efficiencies of termination and in their dependence on proteins for their action. One important protein involved in the termination of transcription of some genes is *rho* (ρ) (Figure 12.8,5). The terminators of such genes are called *rho-dependent terminators* (Figure 12.8,5). At many other terminators the core RNA polymerase itself can carry out the termination events. Those terminators are called *rho-independent terminators.* Possibly, as-yet unidentified proteins are needed for *rho*-independent termination.

Rho-independent terminators consist of sequences with twofold symmetry that are about 15 to 20 base pairs before the end of the gene, followed by a string of about six AT base pairs (Figure 12.10). The AT base pairs are transcribed into a string of Us at the end of the RNA. Potentially, the transcript of the region with twofold symmetry can form a hairpin loop. Since mutations that disrupt the twofold symmetry (and therefore affect the transcript's ability to form a hairpin) decrease or prevent termination (see Figure 12.10), it is argued that the hairpin not only forms *in vivo* but also plays an important role in termination. Similarly, the existence of mutations that disrupt the string of AT's in the terminator sequence of the template cause its inactivation. While it is not clear how the AT string and the hairpin loop facilitate *rho*-independent termination, it is possible that the rapid formation of the hairpin loop destabilizes the RNA-DNA hybrid in the terminator region, which in turn leads to the release of the RNA and to transcription termination.

Rho-dependent terminators lack the AT string found in *rho*-independent terminators, and many cannot form hairpin structures. While the mechanism of *rho*-dependent termination is not clearly understood, it is known to require the interaction of *rho* and the nascent RNA chain upstream of the terminator, at regions that are relatively unstructured and untranslated. Following this, the RNA transcript, RNA polymerase, and *rho* factor are released.

~ FIGURE 12.10

Sequence of a ρ-independent terminator and structure of the terminated RNA. The taupe-shaded mutations partially or completely prevent termination.

In sum, three key events occur at a terminator: (1) RNA synthesis stops, (2) the RNA chain is released from the DNA, and (3) RNA polymerase is released from the DNA (Figure 12.8,6).

KEYNOTE

The end product of transcription is an RNA molecule. The three general classes of RNA transcripts that are made in bacteria are messenger RNA, transfer RNA, and ribosomal RNA. Each of these transcripts is made in the 5'→3' direction, is read from the DNA using similar promoter and terminator sequences, and is synthesized by using the catalytic activity of RNA polymerase core enzyme alone or in association with the sigma factor (for initiation). A second protein factor, the *rho* protein, is required for the release of the RNA transcript from the DNA for some genes; termination of transcription of other genes occurs independently of *rho*.

RNA Polymerases and Phage Transcription

In general, small bacteriophages use the RNA polymerase of the host cell for their transcription. Larger phages, however, often have a gene that codes for an RNA polymerase to transcribe phage genes. For example, phage T7 has a gene, *gene1*, that codes for an RNA polymerase. The host *E. coli* RNA polymerase transcribes the part of the phage chromosome that includes *gene1*. The *gene1* transcript is translated in the bacterial cell to produce the single-polypeptide, T7 RNA polymerase, which then transcribes the rest of the phage chromosome. At the same time some of the other phage gene transcripts are translated to produce proteins that shut off the activity or production of the host RNA polymerase. The net effect of all this enzyme activity is that the transcription of the phage chromosome is favored over transcription of the host chromosome. This happens because the phage promoters have a special affinity for the phage-coded RNA polymerase, while the *E. coli* promoter sequences have a low affinity for the phage-coded enzyme and because the *E. coli* RNA polymerase has been inactivated.

The T4 phage exploits the transcription process in a bacterial cell in a different way. Unlike phage T7, phage T4 does not code for its own RNA polymerase. Instead, T4 has a gene that codes for proteins that interact with the *E. coli* RNA polymerase and alter its subunits. The altered host polymerase then transcribes phage genes rather than bacterial genes.

TRANSCRIPTION IN EUKARYOTES

In general, transcription in eukaryotes proceeds in a way similar to that process in prokaryotes. The two main differences are that more than one RNA polymerase enzyme occurs in eukaryotes and that the transcription sequences (promoters and terminators) are different.

RNA Polymerases in Eukaryotes

While one RNA polymerase is found in bacterial cells, typically three RNA polymerases are found in the nucleus of a eukaryotic cell. Their properties are listed in Table 12.1. Each enzyme is involved in the synthesis of a different type of RNA. **RNA polymerase I**, located exclusively in the nucleolus, catalyzes the synthesis of the two large ribosomal RNA (rRNA) molecules (18S and 28S rRNAs) and one of the two small rRNAs, the 5.8S rRNA. All of these molecules are found in the cytoplasmic ribosomes, the organelles responsible for protein synthesis. Ribosome assembly, which we will

~ TABLE 12.1

Properties of Eukaryotic RNA Polymerases

RNA POLYMERASE	LOCATION	PRODUCTS	α-AMANITIN SENSITIVITY
I	Nucleolus	28S, 18S, 5.8S rRNAs	Insensitive
II	Nucleus	hnRNA mRNA some snRNAs	Highly sensitive
III	Nucleus	tRNA 5S rRNA some snRNAs	Intermediate sensitivity

examine in Chapter 13, occurs in the nucleolus. **RNA polymerase II**, found only in the nucleoplasm of the nucleus, is involved in the synthesis of messenger RNAs (mRNAs) and some snRNAs, some of which are involved in RNA precursors. **RNA polymerase III** (also found only in the nucleoplasm) synthesizes the following: (1) the transfer RNAs (tRNAs), which bring amino acids to the ribosome; (2) the 5 S rRNAs, one molecule of which is found in each ribosome and (3) the remaining small nuclear RNAs (snRNA) not made by RNA polymerase II, some of which are involved in RNA processing events.

Another characteristic helps distinguish between the three RNA polymerases: their different sensitivities to inhibition by α-amanitin, a product of the poisonous mushroom *Amanita phalloides*. RNA polymerase I is insensitive to inhibition by α-amanitin, RNA polymerase II exhibits the greatest sensitivity to inhibition, and RNA polymerase III shows intermediate sensitivity to α-amanitin inhibition. These properties have made it possible to isolate and purify the three different RNA polymerase classes.

All eukaryotic RNA polymerases consist of several subunits (two large and at least four small). The subunits of any given RNA polymerase type (I, II, or III) have apparently been conserved through evolution, since they are very similar in size and activity (where that is known) in eukaryotes ranging from yeast to humans.

Compared with what we know about the structure and function of the RNA polymerases in *E. coli* and a number of other prokaryotes, relatively little is known about the structure and function of eukaryotic RNA polymerases. One reason for this lack of knowledge is that few RNA polymerase mutants are known in eukaryotes. Another reason is that the amount of RNA polymerase in a eukaryotic cell is relatively low, thus making purification of the polymerase very difficult. In calf thymus, for example, RNA polymerases constitute only 0.05 percent of the total cellular protein, whereas in *E. coli* RNA polymerase accounts for 1 percent of the total cellular protein.

In addition to the RNA polymerases located in the nucleus, RNA polymerases are also found in the organelles of eukaryotes. A specific RNA polymerase found in mitochondria transcribes mitochondrial genes. Similarly, an RNA polymerase, different from the mitochondrial enzyme, is found in the chloroplast, where it transcribes chloroplast genes.

KEYNOTE

In *E. coli*, a single RNA polymerase synthesizes mRNA, tRNA, and rRNA. Eukaryotes have three distinct nuclear RNA polymerases, each of which transcribes different gene types: RNA polymerase I transcribes the genes for the large ribosomal RNAs; RNA polymerase II transcribes mRNA genes and some snRNA genes; and RNA polymerase III transcribes genes for the small 5S rRNAs, the tRNAs, and the remaining snRNAs.

Transcription of the Different Classes of Eukaryotic Genes

Like prokaryotic genes, eukaryotic genes have promoters and terminators that play important roles in directing the transcriptional machinery where to start and stop. The three different eukaryotic RNA polymerases have different requirements for transcription: they recognize different promoter sequences, they require different arrays of proteins, called *transcription factors*, that are involved in the transcription process, and they have different requirements for the termination of transcription. In addition, unlike the *E. coli* RNA polymerase, none of the three eukaryotic RNA polymerases directly binds to DNA during the initiation of transcription. Rather, the initiation of transcription by these enzymes is mediated by transcription factors that are specific for each polymerase. Once transcription has initiated, the RNA polymerases are then bound to DNA.

TRANSCRIPTION OF PROTEIN-CODING GENES BY RNA POLYMERASE II. RNA polymerase II transcribes those genes that code for proteins, producing mRNA molecules. RNA polymerase II also transcribes some snRNA genes. A protein-coding gene may have a large assortment of regulatory elements for transcription by RNA polymerase II located both upstream and downstream (that is, within the transcribed sequence of the gene) of the RNA initiation site for the gene. Each gene in an animal cell, for example, has a particular combination of positive regulatory elements (for activating transcription) and negative regulatory elements (for repressing transcription) adjacent to it that are uniquely arranged as to number, type, and arrangement on the DNA. These regulatory elements are DNA sequences that are binding sites for specific transcription factors and **regulatory factors**, that is, proteins involved in the activation or repression of transcription of the gene. Generally, the regulatory elements are located within several hundred base pairs from the site of initiation of transcription, usually upstream from that point. However, some regulatory elements are 1000 to 30,000 base pairs away. By definition the regulatory region immediately adjacent to the transcription start site is called the *promoter*, while the more distant regulatory elements are called *enhancers*.

Promoters. The promoters of protein-coding genes have been analyzed in two principal ways. One way was to isolate single-base-pair mutations at every base pair position for a 100 or so base pairs upstream from the start point of transcription and to examine those mutants in terms of their effects on transcription. It was presumed that mutations that caused significant effects on transcription would define important promoter elements. The second way was to compare the DNA sequences upstream of a number of protein-coding genes to see if there were any regions with similar sequences. The results of these experiments indicated that the promoter is arranged as a series of **promoter elements, or modules.** Starting closest to the transcription initiation site, the promoter elements are the **TATA box** or **TATA element** (also called the **Goldberg-Hogness box** after its discoverers), the **CAAT element,** and the **GC element** (Figure 12.11). The elements are named for the general DNA base sequences they contain. Not all promoters have all three of the elements: some lack a TATA element, others lack a CAAT element or a GC element.

The TATA element has the consensus sequence TATAAA (reading 5'-to-3' on the non-template strand). In higher eukaryotes, the TATA element is almost always located at position −30 (25 to 35 base pairs upstream from the first base pair transcribed into RNA). In yeast, the distance between the TATA element and the mRNA initiation site ranges from 40 to 120 bp, depending on the promoter of some genes. Mutations in the TATA element result in relatively little decrease in transcription, although the RNA initiation point often is changed. In promoters that lack a TATA element, typically there is no unique initiation point for RNA synthesis. Thus, it seems that the TATA element functions to specify a particular start point for RNA transcription.

The CAAT element has the consensus sequence 5'-GGCCAATCT-3'. The CAAT element is found at approximately −80 in many genes, but it can also function at a number of other locations and in either orientation with respect to the RNA start point. Mutations in the CAAT element cause a very marked reduction in transcription, indicating that the CAAT element plays a very important role in the initiation of transcription.

The GC element has the consensus sequence 5'-GGGCGG-3'. Often there is more than one copy of a GC element in a promoter, and the elements may function in either orientation. In Figure 12.11, two GC elements are shown, both in opposite orientation on either side of the CAAT element. The GC elements appear to help bind the RNA polymerase near the transcription start point.

Enhancers. While the promoter elements are crucial for determining whether transcription can occur, **enhancers,** or **enhancer elements,** are required for maximal transcription of the gene to occur. An enhancer element helps control transcription from the promoter linked to it, even though the enhancer may be some distance away. Enhancer elements function in either orientation. Also, the enhancer elements function at a large distance from the gene, often over 1000 base pairs from the promoter. In animal cells, enhancer elements can activate genes either when they are upstream or downstream from the RNA initiation site and even when they are within the coding sequence. In most cases, though, the enhancer elements are found upstream of the gene. Similar elements that have essentially the same properties as enhancer elements, except that they repress rather than activate gene transcription, are called **silencer elements.**

There is no consensus sequence for a eukaryotic enhancer element, and the mechanism of enhancer action is not well understood. It is clear that regulatory proteins bind to the enhancer elements, and which regulatory proteins bind depends on the DNA sequence of the enhancer element. A model for how the enhancers affect transcription from a distance is that specific regulatory proteins bind to the enhancer element and the DNA then forms a loop so that the enhancer-bound regulatory proteins interact with the regulatory proteins and transcription factors bound to the promoter elements. Through the interactions of all of the pro-

~ **FIGURE 12.11**

Promoter elements (modules) for a eukaryotic protein-coding gene transcribed by RNA polymerase II. Each promoter element has a different function in transcription. The DNA sequences between the elements are not important for the transcription process. Transcription factors bind to the elements when the gene is transcribed.

teins, transcription is either activated (enhancers) or repressed (silencers).

In yeast, there are elements functionally similar to enhancers that are called **upstream activator sequences (UAS)**. Like enhancers, UASs can function in either orientation and at variable distances upstream of the promoter. However, unlike enhancers, UASs cannot function when located downstream of the promoter.

Transcription factors. None of the three eukaryotic RNA polymerases are able to recognize their respective promoter sequences *in vitro*. Instead, specific proteins, called **transcription factors (TFs)**, are needed to facilitate the initiation of transcription by RNA polymerases. The transcription factors are named for the RNA polymerase with which they work: TFI for RNA polymerase I, TFII for RNA polymerase II, and TFIII for RNA polymerase III. A number of different transcription factors are involved with transcription by each polymerase, and thus the TFs are lettered A, B, C, etc.

For protein-coding genes, some transcription factors bind to specific DNA sequences of the promoter, while others appear to bind to the RNA polymerase II when it initiates transcription. Figure 12.12 is a schematic of events that may occur during the initiation and elongation stages of transcription, catalyzed by RNA polymerase II. During initiation (Figure 12.12a) the key transcription factor is TFIID, which binds to the TATA element; hence, it is also called the **TATA factor.** TFIID is also bound to TFIIA, which is not bound to DNA, and to an upstream DNA-binding protein that is bound to the enhancer sequence. The binding between TFIID and the upstream DNA-binding protein brings the enhancer sequence close to the promoter sequence and results in the DNA forming a loop between the two sequences. Since the binding of two proteins is the driving force in bringing the two sequences together, we can begin to see how the enhancer sequence can function at a great distance from the promoter and in either orientation. When TFIID is bound to the TATA element, RNA polymerase II can bind and, along with transcription factors TFIIB, TFIIF and TFIIE, forms a *transcription initiation complex.* The last three TFs mentioned are presumably required to stimulate transcription or facilitate transcription initiation correctly. Other transcription factors bind to the CAAT and GC promoter elements although their exact roles in transcription initiation remain to be determined.

Once transcription has been initiated, and the elongation phase is occurring (Figure 12.12b), the distribution of transcription factors changes. TFIIB and TFIIE

~ **FIGURE 12.12**

Schematic of the events that may occur during the initiation and elongation stages of transcription catalyzed by RNA polymerase II.

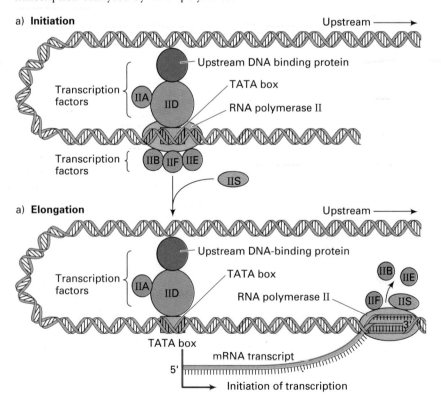

a) **Initiation**

a) **Elongation**

are released from the polymerase, leaving TFIIF bound. Another transcription factor, TFIIS becomes bound to the polymerase during the elongation phase. At the promoter, TFIID remains bound to the TATA element and to TFIIA and the upstream DNA binding protein. This primes the gene's promoter for binding by other RNA polymerase II molecules, facilitating repeated transcription of the gene until regulatory signals specify the repression of transcription.

In view of its relationship to RNA processing events, we will discuss the termination of transcription of protein-coding genes in Chapter 13, when we discuss mRNA synthesis.

$\mathcal{K}$EYNOTE

Protein coding genes in eukaryotes are transcribed by RNA polymerase II. The promoter for these genes is located within a few hundred base pairs upstream of the transcription initiation point and consists of different combinations of promoter elements or modules, depending on the gene. These promoter elements are crucial for determining whether transcription can occur. They are the sites for interaction of transcription factors and regulatory factors. Another important element associated with genes is the enhancer, which, through interaction with regulatory factors, functions to facilitate maximum transcription of the gene with which it is associated. Often, enhancers exert their action at a large distance from the gene, either upstream or downstream of the transcription initiation site.

TRANSCRIPTION BY RNA POLYMERASE III. RNA polymerase III is responsible for the transcription of genes for a variety of small RNAs: transfer RNAs, 5S ribosomal RNAs, and some small nuclear RNAs. We will focus on transcription of 5S rRNA genes (5S rDNA) and tRNA genes (tDNA) in this discussion.

The 5S rRNA genes encode the 120-nucleotide 5S rRNA; one 5S rRNA is found in each large ribosomal subunit. The 5S rRNA genes were the first eukaryotic genes for which the promoter structure and transcription factor requirements were determined. Based on the knowledge of upstream promoters for bacterial genes, the expectation was that the 5S rRNA genes would also have an upstream promoter. The promoter location was investigated by making a series of overlapping deletions that removed either various segments of the upstream sequences and of the gene itself, or of the

~ **FIGURE 12.13**

Use of overlapping deletions for determining the location of the promoter for 5S rRNA genes.

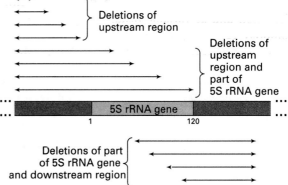

downstream sequences (Figure 12.13). The effect of the deletions on the ability of the gene to be transcribed was then determined. The unexpected result was that the promoter for RNA polymerase III in these genes is *within the gene itself*. More specifically, the experiments showed that removal of any part of a 40-to-50 nucleotide pair sequence from position 50 to 97 of the gene reduced transcription markedly. This internal promoter has been called the **internal control region (ICR)**. The tRNA genes, which encode the 75-90 nucleotide tRNAs, also have internal promoters (ICRs).

What are the similarities and differences between the 5S rDNA and tDNA ICRs? Careful studies of the effects of individual base pair changes of the 5S rDNA ICR have revealed that there are two functional domains, *boxA* and a second domain consisting of an intermediate element and *boxC* (Figure 12.14; p. 376). Similarly the tDNA ICR also has two domains, *boxA* and *boxB* (Figure 12.14). The *boxA* is highly conserved in sequence for both types of ICRs.

How do the ICRs function in initiating transcription? There are three transcription factors that have been shown to be associated with initiation of transcription by RNA polymerase III. They are designated TFIIIA, TFIIIB, and TFIIIC. (TFIIIA was the first eukaryotic TF to be isolated and characterized.) All three TFIIIs are required for transcription of 5S rDNA, but only TFIIIB and C are needed for rDNA transcription. A model for formation of a transcription initiation complex on a 5S rDNA ICR is found in Figure 12.15; p. 376. TFIIIA first binds to *boxC* of the ICR; this facilitates binding of TFIIIC to DNA in the *boxA* region. TFIIIB is the next component to be added, and it apparently binds only to the other TFs and not to DNA. TFIIIB appears to be the crucial protein that correctly positions the RNA polymerase III on the gene. That is, TFIIIB functions as a *transcription initiation factor*.

~ FIGURE 12.14

Organization of the internal control regions, or ICRs (promoters), for 5S rDNA and tDNA. The base-pair coordinates for the ICR elements are approximate, since they vary a little between different genes.

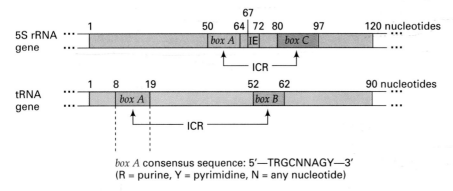

box A consensus sequence: 5'—TRGCNNAGY—3'
(R = purine, Y = pyrimidine, N = any nucleotide)

~ FIGURE 12.15

Model for the formation of a transcription initiation complex on a 5S rDNA ICR.

1 TFIIIA binds to *Box C*

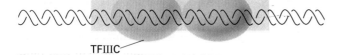

2 This facilitates binding of TFIIIC to *box A*

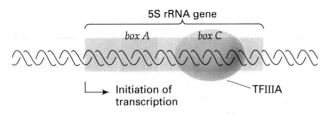

3 TFIIIB binds to other TFs, but not to DNA

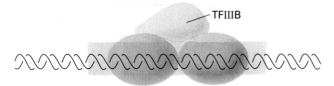

4 TFIIIB positions RNA polymerase III on gene

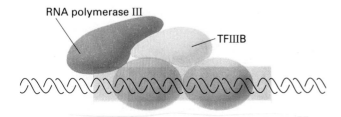

The polymerase then initiates transcription 50 base pairs upstream from the beginning of *boxA*. (Recall that eukaryotic RNA polymerases do not bind to DNA during the initiation of transcription, only to initiation complexes.) Once formed, the initiation complex is stable and facilitates repeated cycles of transcription of the gene by RNA polymerase III.

Termination of transcription for both types of genes involves simple sequences at the 3'-ends of the genes. For 5S rDNA, on the nontemplate strand there is a cluster of four or more T nucleotides surrounded by GC-rich sequences. For tDNA, a cluster of T nucleotides is used to signal transcription termination. Evidence for this has come, for example, from studies of the effects of mutations on termination. If a string of 4-to-6 T nucleotides is added within the gene sequence, transcription will terminate prematurely at that point.

While ICRs are essential sequences for transcription factors to bind in order to facilitate the initiation of transcription by RNA polymerase, other sequences can be involved. Such sequences can be entirely outside of the coding sequence of the gene, usually upstream of it. Mutations or deletions of these sequences usually diminish or, in some cases, almost abolish transcription. A few cases are known of sequences downstream of the gene which affect transcription. It is presumed that proteins, perhaps regulatory proteins, bind to these upstream and downstream sequences and, in some fashion, modulate the level of gene transcription.

𝒦EYNOTE

RNA polymerase III transcribes 5S rRNA genes, tRNA genes, and some snRNA genes. The promoter for RNA polymerase III is located within the gene itself. The internal promoter is called the internal control region and consists of different combinations of functional domains depending on the class of gene involved. The domains are binding sites for transcription factors required for transcription by RNA polymerase III. It is believed that the level of transcription of the genes is regulated by the binding of regulatory proteins to sequences outside of the RNA coding sequences themselves. _____

TRANSCRIPTION OF RIBOSOMAL RNA GENES (rDNA) BY RNA POLYMERASE I. As we will discuss in more detail in Chapter 13, the sequences for 18S, 5.8S, and 28S rRNAs are located adjacent to one another in the DNA and are transcribed onto a single precursor RNA molecule. The sequences in the DNA are said to constitute a *transcription unit* (Figure 12.16). In most eukaryotes, there exists a tandem array of transcription units, the number depending on the organism. Each transcription unit is transcribed by RNA polymerase I to produce a **precursor-rRNA (pre-rRNA) molecule**. The pre-rRNA contains the three rRNA sequences and sequences between, and flanking those three sequences. The latter are called **spacer sequences**, and they are removed from the pre-rRNA in subsequent processing steps. Between transcription units in the rDNA is a region called the **non-transcribed spacer (NTS) sequence**, which is not transcribed. The important sequences that control transcription of the rDNA are within the NTS.

While RNA polymerases II and III each transcribe a wide variety of genes, RNA polymerase I is unique in that it only transcribes the pre-rRNA transcription units. The promoter sequences for RNA polymerase I lie upstream of the transcription initiation site (Figure 12.17). In the human rDNA promoter, there are two domains: 1) a core promoter element that overlaps the start of the rRNA transcript extending from +7 to −45; and 2) an upstream control element (UCE from −107 to −186). For comparison, the *Xenopus laevis*

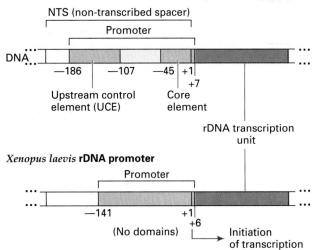

~ **FIGURE 12.17**

The rDNA promoters for humans and for *Xenopus laevis.*

(South African clawed toad) rDNA promoter has no well-defined domains. This promoter extends from +6 to −141. Surprisingly, the promoter sequences are largely species-specific. That is, RNA polymerase I can only transcribe rRNA genes from closely-related species in *in vitro* transcription systems. This species-specificity is determined by the transcription factors that are required for rDNA transcriptions.

RNA polymerase I itself does not bind to the promoter; instead, specific transcription factors bind and form a complex to which RNA polymerase I binds. In humans, two transcription factors have been identified that are necessary for transcription to occur (Figure 12.18; p. 378). Human upstream binding factor (hUBF) is a sequence-specific DNA-binding protein that binds to both elements of the human rDNA promoter to activate transcription. A second protein, SL1, is needed for promoter recognition and initiation of transcription by RNA polymerase I. SL1 does not bind to DNA, but interacts with the promoter through its interaction with

~ **FIGURE 12.16**

Tandemly arranged eukaryotic rDNA transcription units.

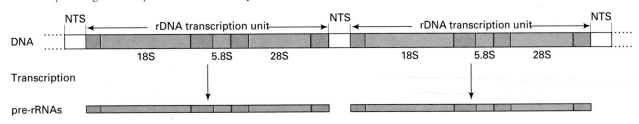

~ FIGURE 12.18

Involvement of transcription factors in human rDNA transcription.

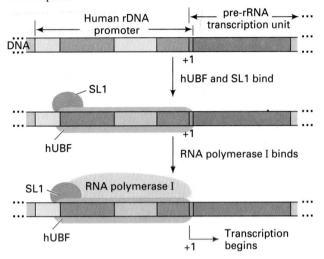

the hUBF-DNA complex. Once hUBF and SL1 are bound to the promoter, RNA polymerase I binds and transcription can begin.

Termination of transcription of the pre-rRNA varies with the organism. In *Xenopus*, for example, there are two important transcription termination sites: T2 and T3. T2 is about 235 base pairs downstream of the 3′ end of the 28S rRNA sequence, and T3 is about 200 base pairs upstream from the transcription initiation site of the next rDNA transcription unit in the array (Figure 12.19). T2 and T3 contain an identical seven-nucleotide sequence, 5′-GACTTGC-3′. T2 is the primary site used for terminating the pre-rRNA transcript, while T3 appears to function as a fail-safe termination site. Since the rDNA transcription units are arranged in a tandem array, the fail-safe termination system allows RNA polymerase molecule potentially to start at one transcription unit and to transcribe that and the following ones in the array.

~ FIGURE 12.19

Terminators for rDNA transcription.

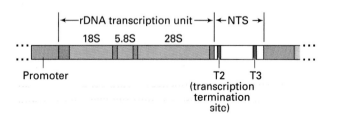

KEYNOTE

RNA polymerase I transcribes the 18S, 5.8S, and 28S rRNA sequences into a single precursor molecule. A tandem array of 18S + 5.8S + 28S rRNA transcription units is found in most eukaryotic organisms. Each transcription unit is separated from the next by a nontranscribed spacer (NTS) sequence. The promoter and terminator for each transcription unit are located within the NTS. As for the other eukaryotic RNA polymerases, specific transcription factors are required for the initiation of transcription by RNA polymerase I.

SUMMARY

We learned in a previous chapter that the DNA contains all of the information for the structure, function, and reproduction of an organism. That information is organized into discrete units called genes. When a gene is expressed, the DNA base pair sequence is transcribed into the base sequence of an RNA molecule. Four major classes of RNA transcripts are produced by transcription of four classes of genes: messenger RNA (mRNA), transfer RNA (tRNA), ribosomal RNA (rRNA) and small nuclear RNA (snRNA). snRNA is found only in eukaryotes, while the other three classes are found in both prokaryotes and eukaryotes. Only mRNA is subsequently translated to produce a protein molecule.

When a gene is transcribed, only one of the two DNA strands is copied. The direction of RNA synthesis is 5′ to 3′ and the reaction is catalyzed by RNA polymerase. In addition to the coding sequences, genes contain other sequences important for the regulation of transcription, including promoter sequences and terminator sequences. Promoter sequences specify where transcription of the gene is to begin, and terminator sequences specify where transcription is to stop. In bacteria there is only one type of RNA polymerase, hence all classes of RNA are transcribed by the same enzyme. Consequently, the promoters for all three classes of genes are very similar. The promoter is recognized by a complex between the RNA polymerase core enzyme and a protein factor called sigma. Once transcription is initiated correctly, the sigma factor dissociates from the enzyme and is reused in other transcription initiation events. Termination occurs in one of two ways. In *rho*-independent termination, a two-fold symmetrical sequence of the DNA enables the resulting RNA transcript to form secondary structures that signal where the RNA polymerase is to stop and release the RNA. In *rho*-dependent termination, a protein factor called *rho* interacts with the new RNA chain upstream of the terminator sequence and somehow facilitates termination of transcription at the correct site.

In eukaryotes there are three different RNA polymerases located in the nucleus. RNA polymerase I, located exclusively in the nucleus, transcribes the 18S, 5.8S and 28S rRNA sequences. These rRNAs are part of ribosomes. RNA polymerase II, located in the nucleoplasm, transcribes protein-coding genes (into mRNAs) and some snRNA genes. RNA polymerase III, located in the nucleoplasm, transcribes tRNA genes, 5S rRNA genes, and the remaining snRNA genes. None of the three eukaryotic RNA polymerases directly binds to promoters. Rather, the promoter sequences for the genes they transcribe are first recognized by specific transcription factors. The transcription factors bind to the DNA and facilitate binding of the polymerase to the transcription factor–DNA complex for correct initiation of transcription.

The promoters for the three types of polymerase differ. For genes transcribed by RNA polymerase II, the promoter consists of a number of sequence modules located within a relatively short distance upstream of the transcription initiation point. These modules serve as binding sites for transcription factors and for regulatory proteins that function to regulate transcription. Other regulatory elements called enhancers are also associated with these genes: they may be upstream or downstream of the transcription initiation site and are required for maximal transcription of the gene to occur. Enhancers function through interaction with specific regulatory proteins. When the regulatory proteins are bound to the enhancer, the DNA forms a loop so that those proteins can interact with the regulatory proteins and transcription factors bound to the promoter.

For genes transcribed by RNA polymerase III, the promoter is located within the gene itself. Transcription factors recognize the promoter, and then the polymerase binds and begins transcription at the correct site. Other sequences upstream of the transcription initiation site function to modulate the rates of transcription initiation by the polymerase.

The 18S, 5.8S and 28S rRNA sequences transcribed by RNA polymerase I are organized into a single transcription unit so that transcription produces a precursor RNA molecule containing all three rRNAs plus extra RNA material. The rDNA transcription units typically are arranged in a tandem array. The promoter for these rRNAs is located upstream of the 18S rRNA coding sequence (the first sequence to be transcribed) in the nontranscribed spacer (NTS) region, which is located between the rDNA transcription units. Again, transcription factors first recognize the promoter; then the polymerase binds to initiate transcription. Promoters for RNA polymerase I are quite species-specific, despite the marked evolutionary conservation of rRNA sequences. In some systems, enhancers have been shown to be associated with rRNA transcription units.

In sum, while the transcription process is very similar in prokaryotes and eukaryotes, the molecular components of the transcription process itself differ considerably. Even within eukaryotes, three different promoters have evolved, along with three distinct RNA polymerases. Much remains to be learned about the associated transcription factors and regulatory proteins before we have a complete understanding of transcription processes and their regulation.

ANALYTICAL APPROACHES FOR SOLVING GENETICS PROBLEMS

Q.1 If two RNA molecules have complementary base sequences, they can hybridize to form a double-stranded structure just as DNA can do. Imagine that in a particular region of the genome of a certain bacterium one DNA strand is transcribed to give rise to the mRNA for protein A, while the other DNA strand was transcribed to give rise to the mRNA for protein B.

a. Would there be any problem in expressing these genes?

b. What would you see in protein B if a mutation occurred which affected the structure of protein A?

A.1 a. mRNA A and mRNA B would have complementary sequences, so they might hybridize with each other and not be available for translation.

b. Every mutation in gene A would also be a mutation in gene B, so protein B might also be abnormal.

Q2. Compare and contrast the following two events in terms of what their consequences would be. Event #1: an incorrect nucleotide is inserted into the new DNA strand during replication, and not corrected by the proofreading or repair systems before the next replication. Event #2: an incorrect nucleotide is inserted into an mRNA during transcription.

A.2 Event #1 would result in a mutation, assuming it occurred within a gene. The mistake would be inherited by future generations, and would affect the structure of all mRNA molecules transcribed from the region and therefore all molecules of the corresponding protein could be affected.

Event #2 would produce a single aberrant mRNA. This could produce a few aberrant protein molecules. Additional normal protein molecules would exist because other, normal, mRNAs would have been transcribed. The abnormal mRNA would soon be degraded. The mRNA mutation would not be hereditary.

QUESTIONS AND PROBLEMS

*12.1 Describe the differences between DNA and RNA.

12.2 Compare and contrast DNA polymerases and RNA polymerases.

12.3 On page 363 you will find the sentence "All base pairs in the genome are replicated during the DNA synthesis phase of the cell cycle, but only *some* of the base pairs are transcribed into RNA." How is it determined *which* base pairs of the genome are transcribed into RNA?

12.4 Discuss the structure and function of the *E. coli* RNA polymerase. In your answer, be sure to distinguish between RNA core polymerase and RNA core polymerase-sigma factor complex.

*12.5 Discuss the similarities and differences between the *E. coli* RNA polymerase and eukaryotic RNA polymerases.

12.6 The RNA polymerases bind to promoter sequences in the DNA.
a. Compare and contrast promoter sequences from prokaryotes and eukaryotes.
b. Apart from the sequence information available for promoters for a number of genes, what evidence is there that promoters do not all have the same base sequence?

12.7 Discuss the molecular events involved in the termination of RNA transcription in prokaryotes.

*12.8 Which classes of RNA do each of the three eukaryotic RNA polymerases synthesize? What are the functions of the different RNA types in the cell?

12.9 What is the Pribnow box? The Goldberg-Hogness box (TATA element)?

*12.10 What is an enhancer element?

12.11 A piece of mouse DNA was sequenced as follows (a space is inserted after every 10th base for ease in counting; "......" means a lot of unspecified bases):

AGAGGGCGGT   CCGTATCGGC   CAATCTGCTC
ACAGGGCGGA   TTCACACGTT   GTTATATAAA
TGACTGGGCG   TACCCCAGGG   TTCGAGTATT
CTATCGTATG   GTGCACCTGA   CT(.......)
GCTCACAAGT   ACCACTAAGC........

What can you see in this sequence to indicate it might be all or part of a transcription unit?

13 RNA Molecules and RNA Processing

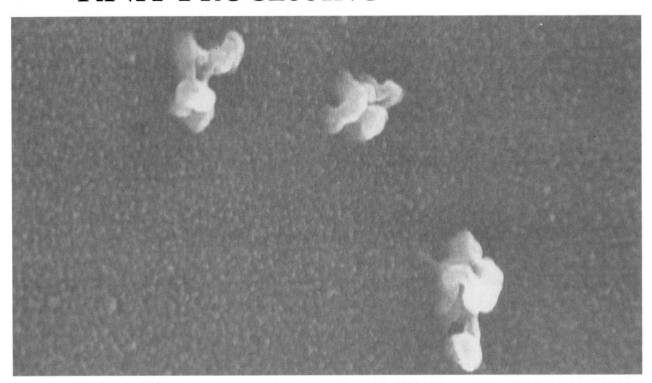

PRINCIPAL POINTS

~ The transcripts of protein-coding genes are linear mRNAs, the lengths of which vary over a wide range. Prokaryotic mRNAs are modified little once they are transcribed, while eukaryotic mRNAs are modified by the addition of a 5' cap and a 3' poly(A) tail.

~ Most eukaryotic mRNAs contain sequences called introns, which do not code for amino acids. The introns are removed as the primary transcript is processed to produce the mature, functional mRNA molecule. Intron removal occurs in complexes of the pre-mRNA with small nuclear ribonucleoprotein particles (snRNPs). The complexes are called spliceosomes. The snRNPs consist of small nuclear RNAs (snRNAs) and several proteins.

~ Transfer RNA (tRNA) molecules bring amino acids to the ribosomes where the amino acids are polymerized into a protein chain. All tRNAs are about the same length, contain a number of modified bases, and have similar three-dimensional shapes.

~ tRNAs are made as pre-tRNA molecules containing 5'-leader and 3'-trailer sequences, which are removed by enzyme activity. Some eukaryotic pre-tRNAs contain introns, which are removed in processing steps that are different from pre-mRNA processing steps.

~ Ribosomes are the cellular organelles on which protein synthesis takes place. In both prokaryotes and eukaryotes, ribosomes consist of two unequally-sized subunits. Each subunit consists of a complex between one or more ribosomal RNA (rRNA) molecules and many ribosomal proteins. Eukaryotic ribosomes are larger and more complex than prokaryotic ribosomes.

~ The 18S, 5.8S, and 28S eukaryotic rRNAs are transcribed from the rDNA onto a single pre-rRNA molecule. The pre-rRNA also contains spacer sequences located at the ends of the molecule and between the rRNA sequences. The spacers are removed as the pre-rRNA is processed in the nucleolus to produce the mature rRNAs and the functional ribosomal subunits.

~ In some precursor rRNAs there are introns, the RNA sequences of which fold into a secondary structure that excises itself, a process called self-splicing. These introns are called group I introns, and the self-splicing reaction requires guanosine but does not involve any proteins.

*I*n the previous chapter we discussed the process of transcribing genetic information into an RNA molecule. In prokaryotes and eukaryotes, messenger RNA, transfer RNA, and ribosomal RNA result from transcription. In eukaryotes only, a fourth RNA, small nuclear RNA is also produced. In this chapter we will discuss the structures and properties of these RNA molecules.

As we shall see in this chapter, the initial RNA transcripts (the **primary transcripts**) of most genes must be modified before the biologically active (mature) RNA molecules are produced. These primary transcripts are called **precursor RNA molecules (pre-RNAs)**, and their processing may involve the addition and/or removal of bases, the chemical modification of some bases, or the removal of sequences from the precursor. Both prokaryotic and eukaryotic tRNA and rRNA precursors undergo processing. Prokaryotic mRNA is not processed, while many eukaryotic mRNAs are generated by the processing of precursor molecules.

OVERVIEW OF TRANSLATION

Although we will be examining the molecular events that comprise the reading of the genetic information coded in mRNA and the transfer of that information into the amino acid sequence of a protein (a process called **translation**) in Chapter 14, we will better understand the functions of each of the major RNAs if we begin this chapter with a general overview of the translation process.

1. **Messenger RNA,** the RNA molecule that contains the coded information for the amino acid sequence of a protein, migrates to a **ribosome,** a cellular organelle that consists of proteins and three (prokaryotes) or four (eukaryotes) rRNA molecules. The ribosome translates the mRNA in the 5'-to-3' direction, starting at a specific start sequence near the 5' end.
2. **Transfer RNA** molecules bring amino acids to the ribosome, where they are matched to the tran-

scribed message on the mRNA. This mRNA message is read in sequential, non-overlapping groups of three bases (called codons).

3. At the ribosome, two tRNAs can bind to the mRNA at one time. A ribosomal enzyme catalyzes the formation of a bond between the two amino acids carried by two adjacent tRNAs, thus beginning the synthesis of a protein chain.

4. The ribosome continues to move along toward the 3′ end of the mRNA, and as each new tRNA matches its attached amino acid to the next mRNA sequence, the protein chain lengthens. The process continues until the ribosome recognizes a sequence of bases in the mRNA that indicates the protein is complete. The completed protein is then released.

THE STRUCTURE AND FUNCTION OF MESSENGER RNA

Figure 13.1 shows the general structure of the mature, biologically active mRNA as it exists in both prokaryotic and eukaryotic cells. mRNAs are transcribed from *structural genes* (protein-coding genes).

The mRNA molecule has three main parts. At the 5′ end is a **leader sequence**, which varies in length from mRNA to mRNA. Within this leader sequence is the coded information that the ribosome reads to orient it correctly for beginning protein synthesis; none of the bases of the leader sequence are translated into amino acids. Following the 5′ leader sequence is the actual **coding sequence** of the mRNA; this sequence determines the amino acid sequence of a protein during translation. The coding sequence varies in length, depending on the length of the protein for which it codes. Following the amino acid–coding sequence in the mRNA, and constituting the rest of the mRNA at the 3′ end of the molecule, is a **trailer sequence**. Like the leader sequence, the trailer sequence is not translated and varies in length from mRNA to mRNA.

~ FIGURE 13.1

General structure of mature prokaryotic and eukaryotic mRNA molecules.

mRNA 5′ [ ] 3′

5′ leader sequence — Protein-coding sequence — 3′ trailer sequence

The production of functioning mRNA is fundamentally different in prokaryotes and eukaryotes. In prokaryotes (Figure 13.2a; p. 384) the RNA transcript functions directly as the mRNA molecule for translation, but in eukaryotes (Figure 13.2b) the primary RNA transcript must be modified in the nucleus by a series of events known as *RNA processing* in order to produce the mature mRNA. In addition, in prokaryotes an mRNA begins to be translated on ribosomes before it has been completely transcribed. This process is called the "coupling" of transcription and translation. An electron micrograph of coupled transcription and translation in *E. coli* is shown in Figure 13.3 (p. 384). In eukaryotes, however, the mRNA must migrate from the nucleus to the cytoplasm (where the ribosomes are located) before it can be translated. A eukaryotic mRNA is always completely transcribed and processed before it is translated.

Production of Prokaryotic mRNA

In prokaryotes the initial transcript of a protein-coding gene is processed very little, if at all, to produce the mature mRNA molecule. That is, an exact point-by-point relationship exists between the order of base pairs in the gene and the order of the corresponding bases in the mature mRNA. This is referred to as *colinearity* between a gene and its primary transcript.

Production of Mature Messenger RNA in Eukaryotes

Unlike prokaryotic mRNAs, which typically are not modified, eukaryotic mRNAs are usually modified at both the 5′ and 3′ ends. These *posttranscriptional modifications* are catalyzed by specific enzymes. In addition, most protein-coding genes have insertions of non-amino-acid-coding sequences called **introns** between amino acid-coding sequences, or **exons.** Both exons and introns are copied into the primary mRNA transcript— the **precursor mRNA (pre-mRNA)**—from which they are removed in the processing of pre-mRNA to the mature mRNA molecule. The synthesis and processing of pre-mRNA is discussed in this section.

5′ CAPPING. The 5′ end of the eukaryotic mRNA is modified by the addition of a *cap* in a process called **5′ capping.** 5′ capping involves the addition of a guanine nucleotide (most commonly 7-methyl guanosine {m^7G}) to the terminal 5′ nucleotide by an unusual 5′-to-5′ linkage (as opposed to the usual 5′-to-3′ linkage) and the addition of two methyl groups (CH_3) to the

~ FIGURE 13.2

Processes for synthesis of functional mRNA in prokaryotes and eukaryotes. (a) In prokaryotes the mRNA synthesized by RNA polymerase does not have to be processed before it can be translated by ribosomes. Also, since there is no nuclear membrane, translation of the mRNA can begin while transcription continues, resulting in a coupling of the transcriptional and translational processes. (b) In eukaryotes, the primary RNA transcript is a precursor-mRNA (pre-mRNA) molecule, which is processed in the nucleus [addition of 5' cap and 3' poly(A) tail, and removal of introns] to produce the mature, functioning mRNA molecule. Only when that mRNA is transported to the cytoplasm can translation occur.

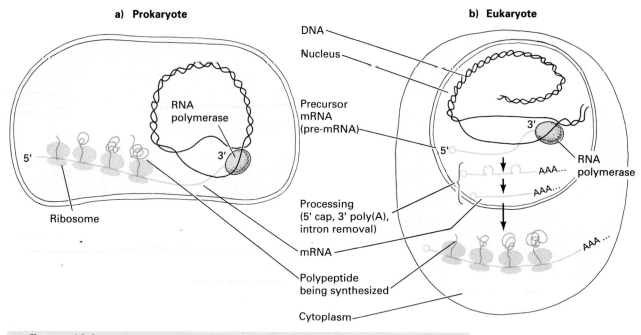

~ FIGURE 13.3

Electron micrograph of coupled mRNA transcription and translation in *E. coli*. From the faint DNA strand, a number of mRNA molecules are emerging. Since the length of the mRNA molecules increases from left to right, we assume that this strand is a single gene with a promoter to the left of the photograph. The globular structures on the mRNAs are ribosomes, which are translating the message into a protein before the synthesis of the mRNA is finished.

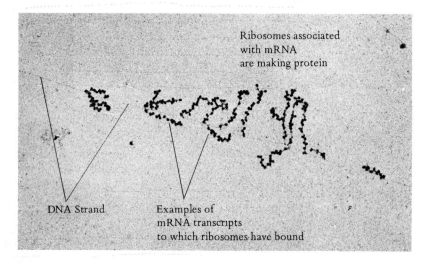

~ FIGURE 13.4

Cap structure at the 5′ end of a eukaryotic mRNA. The cap results from the addition of a guanine nucleotide and two methyl groups.

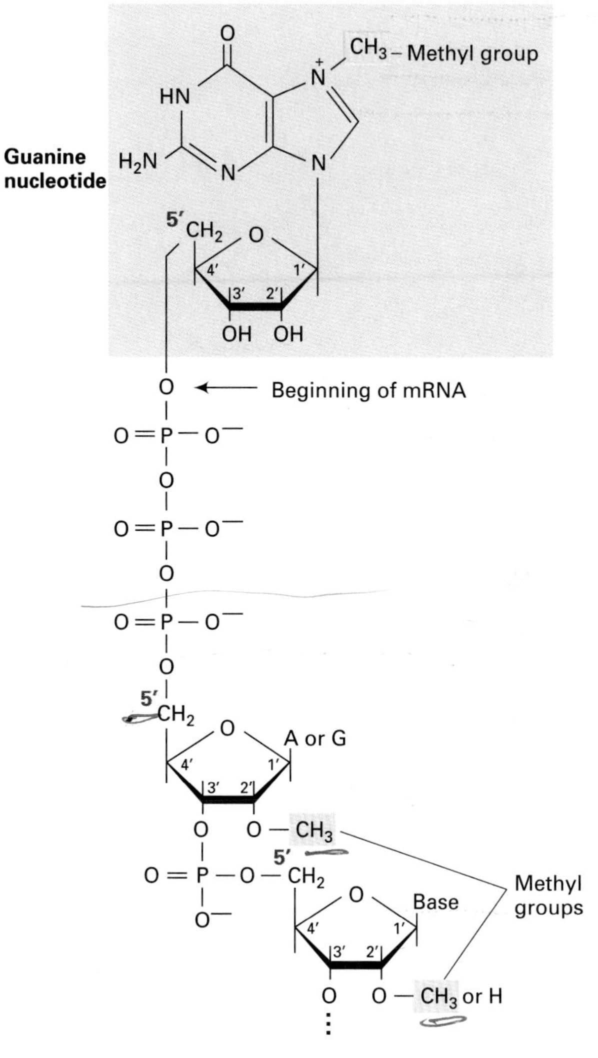

first two nucleotides of the RNA chain (Figure 13.4). 5′ capping takes place as follows. RNA polymerase II starts transcription at the base to which the cap is added. (That DNA site is called the *cap site*. There is no DNA template for the 5′ cap.) When each growing primary transcript is about 20 to 30 nucleotides long, a methylated cap structure is added at the 5′ end of the transcript. The cap is essential for the ribosome to bind to the 5′ end of the mRNA, an initial step of translation.

ADDITION OF THE 3′ POLY(A) TAIL. The posttranscriptional modification of the 3′ ends of eukaryotic mRNAs is the addition of a sequence of about fifty to two hundred and fifty adenine nucleotides. This se-

quence is called a **poly(A) tail.** There is no DNA template for the poly(A) tails, and they are not added to tRNA or rRNA molecules. Poly(A) tails are found on most of the mRNAs of all eukaryotic species. At least in mammalian cells, histone mRNAs are exceptional by virtue of the absence of poly(A) tails.

The addition of the 3′ poly(A) tail provides a mechanism for signaling how the ends of eukaryotic mRNAs are determined. In contrast, in prokaryotes, specific transcription termination sequences exist to specify the end of an mRNA molecule. In eukaryotes, however, there are no specific transcription termination sequences near the end of an mRNA molecule. Instead, mRNA transcription continues, in some cases, for hundreds or thousands of nucleotides, past the so-called **poly(A) addition site,** where cleavage by an RNA-specific endonuclease generates a 3′OH end to which the poly(A) tail is added.

The sequences responsible for controlling poly(A) addition have been identified. About 10 to 30 nucleotides upstream from the poly(A) site there is the strongly conserved sequence AAUAAA (Figure 13.5; p. 386). The AAUAAA sequence and the sequences downstream from the poly(A) site function together to signal the location of the poly(A) site, which leads to cleavage of the primary transcript and addition of the poly(A).

Production of the 3′ poly(A) tail is catalyzed by the enzyme **poly(A) polymerase** in the following reaction. (The energy for the reaction is supplied when the bonds of ATP molecules are broken.)

$$mRNA + nATP \xrightarrow{\text{poly(A) polymerase}} mRNA\text{-}(A)n + nPP_i$$

where n equals 50-250.

What is the function of the poly(A) tail? A number of studies have shown that the poly(A) tail is important for determining the stability of the mRNA. For example, mRNAs can be injected into frog oocytes where they will be translated. Globin mRNA, with its normal poly(A) tail, remains active in frog oocytes for a much longer time than globin mRNA from which the poly(A) tail has been removed because the non-poly(A) mRNA is more rapidly degraded.

INTRONS. The primary transcripts of eukaryotic protein-coding genes often contain long insertions of non-amino-acid-coding sequences. These noncoding sequences are copies from regions of a gene that are called **introns,** or **intervening sequences.** Introns must be excised from each primary transcript (pre-mRNA) in order to convert the transcript into a mature mRNA

~ FIGURE 13.5

Sequences near poly(A) in four eukaryotic mRNAs. Each contains the sequence AAUAAA about twenty nucleotides upstream of the poly(A) addition site.

Human
α-globin
UGGUC UUUG AAUAAAG UC UGA GUGGGC GGC —poly(A)

Human
β-globin
UGCC U AAUAAA AAAC AUUUAUUUUC AUUGC —poly(A)

Mouse
immunoglobulin
light chain
UGGUC UUUG AAUAAAG UC UGAGUGAGUGGC —poly(A)

Chicken
ovalbumin
CC UUUAAUC AU AAUAAA AAC AUGUUUAAGC —poly(A)

molecule that can be used for the synthesis of a complete protein. Those parts of the gene that correspond to the mature mRNA molecule (and therefore code for amino acids) are called **exons**, or **coding sequences.**

Discovery of introns. The first report of genes with introns was given by Alan Jeffreys and Richard Flavell in 1977. They discovered an intron estimated to be 600-base-pairs (bp) long in the gene for the 146-amino acid β-globin chain in rabbits. (The β-globin chain is part of a hemoglobin molecule.) At about the same time Philip Leder's group discovered a similar intron estimated to be 550 bp long, that interrupted each of two nonallelic β-globin genes in the mouse.

Leder's group studied the β-globin genes in cultured mouse cells. Because these cells produce β-globin in large quantities, it is relatively easy to isolate the mRNA for this protein. They analyzed the β-globin mRNA by sucrose gradient centrifugation (Box 13.1), an experimental procedure that measures the size of the macromolecules or organelles in terms of their sedimentation coefficient (given in S values, which relate to the rate of sedimentation in a centrifuge). The data indicated that the β-globin mRNA has a size of about 10S, which is about the size expected if virtually all of the mRNA codes for the 147-amino-acid β-globin protein. Sequence analysis indicated the mature mRNA is 0.7-kb long.

With the β-globin mRNA in hand, geneticists investigated the organization of the gene encoding that mRNA in **R-looping** experiments. The R-looping technique was developed by M. Thomas, Ray White, and Ronald Davis and is diagramed in Figure 13.6. The purified mRNA is hybridized with the double-stranded DNA, which encodes the RNA under conditions in which RNA-DNA hybrids are more stable than DNA-DNA hybrids. The mRNA binds to its complementary sequence in the template DNA strand, displacing a loop of single-stranded (non-template) DNA called an *R*

~ FIGURE 13.6

The R-looping procedure. Double-stranded DNA is heated in the presence of single-stranded RNA under conditions such that the RNA can form a hybrid with a DNA strand and displace the non-template strand of DNA. The resulting R loop—the displaced single-stranded non-template DNA strand—can be seen under an electron microscope.

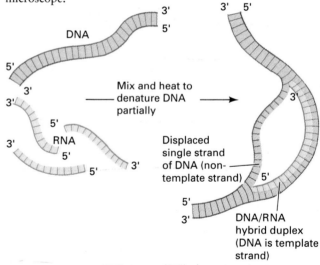

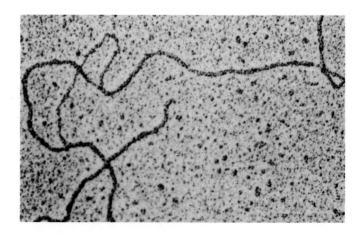

Box 13.1
Sucrose Gradient Centrifugation

Sucrose gradient centrifugation is usually used to separate the components present in a mixture. With this procedure we can measure the rates at which the components sediment in the gradient under centrifugal forces. In sucrose gradient centrifugation the sedimentation rates are converted to **Svedberg units**, using a formula for which the derivation is beyond the scope of this book. Svedberg units, or simply **S values**, are then used as a rough indication of relative sizes of the components being analyzed.

The rate of sedimentation of a component in a sucrose gradient is related both to the molecular weight of the component and to its three-dimensional configuration. Two components of the same molecular weight, for example, will have different S values if one is highly compact and thus sediments rapidly, while the other has a more extended shape and sediments relatively slowly. Sucrose gradient centrifugation is typically used to separate, or estimate, the sizes of RNA molecules, ribosomal subunits, ribosomes, proteins, or various cellular organelles.

The sucrose gradient centrifugation method involves a supporting column of sucrose in which the concentration of sucrose increases toward the bottom of the tube. For the separation of different-sized RNA molecules, for example, a continuous gradient of sucrose concentration, ranging from 10 percent at the top to 30 percent at the bottom, is prepared in a centrifuge tube (Box Figure 13.1a). Next, a small amount of the sample, in this case a solution containing the RNAs, is very carefully layered on top of the gradient (Box Figure 13.1b). The sucrose gradient is then centrifuged at high speeds for several hours, during which time the RNA molecules move through the gradient at different rates depending on their molecular weight and configuration (Box Figure 13.1c). At the completion of centrifugation, the RNAs of similar S values are located in discrete zones or bands in the gradient.

To collect the different RNA types, the bottom of the tube is punctured with a needle, and the drops that run out are collected by using a fraction collector (Box Figure 13.1d). Fractions containing the RNAs are identified by measuring the degree to which each fraction absorbs ultraviolet light. Fractions with RNA will absorb ultraviolet light, while those without RNA will not. The relative positions the RNAs occupy in the gradient indicate the S values of the RNAs: Those with higher S values (larger and/or more compact) are found closer to the bottom of the gradient than those with lower S values.

Note that sucrose gradient centrifugation separates molecules on the basis of their relative rates of sedimentation. Hence it differs from cesium chloride equilibrium density gradient centrifugation, in which molecules are separated on the basis of buoyant density and not size.

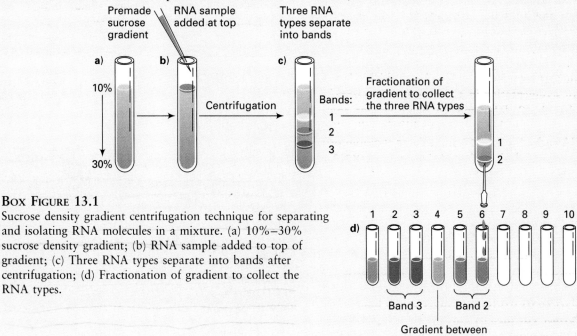

BOX FIGURE 13.1
Sucrose density gradient centrifugation technique for separating and isolating RNA molecules in a mixture. (a) 10%–30% sucrose density gradient; (b) RNA sample added to top of gradient; (c) Three RNA types separate into bands after centrifugation; (d) Fractionation of gradient to collect the RNA types.

loop. The DNA outside the region of complementarity between the RNA and the DNA remains double-stranded. The displaced R loops are visualized by electron microscopy. If the mRNA is colinear with its gene (as is the case for prokaryotic mRNAs), a single R loop is seen, as diagramed in Figure 13.6.

Figure 13.7a shows the result of hybridizing the 0.7-kb β-globin mRNA with DNA containing the β-globin gene. Two R loops were seen to flank a loop interpreted to be double-stranded DNA. This result indicated that there are two sequences in the β-globin gene that are complementary to the two ends of the mRNA. Between these two sequences, though, is a sequence that remains as double-stranded DNA

because it is *not* complementary to any part of the mRNA. The conclusion was that the mouse β-globin gene contained an intron.

Researchers knew at the time of discovering introns that the nucleus contains a large population of RNA molecules of various sizes. These RNA molecules are called **heterogeneous nuclear RNA,** or **hnRNA,** and it was thought that hnRNAs included precursors to mature mRNAs. Using procedures outside the scope of our discussion, a 15S (1.5-kb) RNA molecule was isolated from nuclear hnRNA that was the β-globin pre-mRNA, that is, the primary transcript of the β-globin gene. Like the mature mRNA, the pre-mRNA has a 5' cap and a 3' poly(A) tail. When the 1.5-kb pre-mRNA

~ FIGURE 13.7

Demonstration that the mouse β-globin gene contains an intron: (a) R loops formed between 10S mature β-globin mRNA and the β-globin gene. (b) R loop formed between 15S β-globin pre-mRNA and the β-globin gene. Interpretative diagrams are alongside the micrographs.

a) **10S β-globin mRNA**

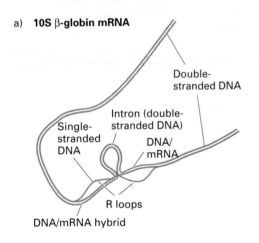

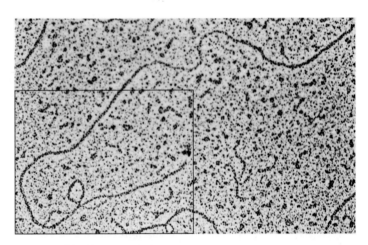

b) **15S β-globin pre-mRNA**

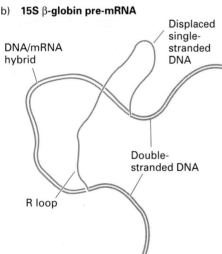

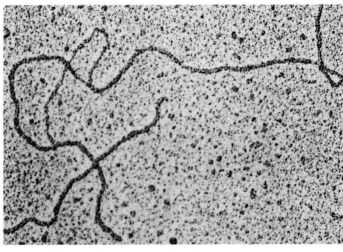

was hybridized to DNA containing the β-globin gene in an R-looping experiment, one continuous R loop was seen (Figure 13.7b). This result indicated that the pre-mRNA was colinear with the gene that encoded it. The interpretation of this series of experiments was that the β-globin gene contains an intron of about 800 nucleotide pairs. Transcription of the gene results in a 1.5-kb pre-mRNA containing both exon and intron sequences. This RNA is found only in the nucleus. By processing events the intron sequence is excised, and the flanking exon sequences are spliced together to produce a 0.7-kb mature mRNA, which is transported to the cytoplasm where it is translated.

As an aside, we raise the question: What is a gene? Up until the work just described, geneticists assumed that a gene was a contiguous stretch of DNA base pairs that was transcribed into a mature mRNA. That mRNA, in turn, was translated into an amino acid sequence. Prokaryotic genes fit this definition closely. However, we have now seen that in the case of the mouse β-globin gene the primary 15S mRNA transcript is not the molecule that is translated. Before it can be translated, a large segment of its genetic material must be excised, and the adjacent segments must be spliced together to produce the mature 10S mRNA. If we define a gene as only the amino acid–coding regions of the DNA, then the presence of introns clearly means that eukaryotic genes are in pieces. On the other hand, if we define a gene as that region of DNA corresponding to the primary RNA transcript, then the whole stretch of coding sequences (exons) plus introns constitute a gene. Both definitions may be encountered in your studies, and it is important to keep the distinctions between the two in mind.

Occurrence of genes with introns. Most eukaryotic protein-coding genes contain introns. Moreover, higher eukaryotes tend to have more genes with introns and longer introns than do lower eukaryotes. We have already seen that early research with the β-globin gene revealed one intron. In subsequent experiments, a second, smaller intron was found.

To give just two additional examples, the gene for the protein vitellogenin (a precursor to two ovary storage proteins) in *Xenopus* has 33 introns, and the chicken ovalbumin gene has 7 introns. Figure 13.8a shows the exon and intron organization of the chicken ovalbumin gene, and Figure 13.8b shows the looping out of the introns when the mature mRNA is hybridized in the gene. Of the 7700 base pairs of the chicken ovalbumin gene, over three-quarters consists of introns; the introns range from 250 to 1600 base pairs in length.

Some eukaryotic genes do not have introns, and prokaryotic genes do not usually contain introns. Some bacteriophage genes have been shown to contain

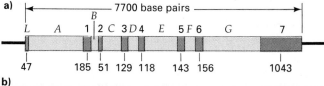

~ **FIGURE 13.8**

The organization of the chicken ovalbumin gene.
(a) Arrangement of exons (L, 1 → 7) and introns (A → G) in the gene. Numbers represent the nucleotide pairs in the exons. (b) Pattern resulting from hybridization of the ovalbumin gene with mature ovalbumin mRNA: the seven introns loop out from the DNA-RNA hybrid.

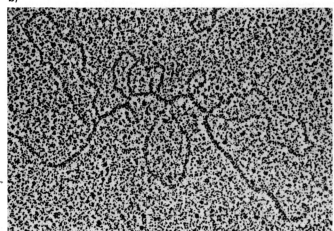

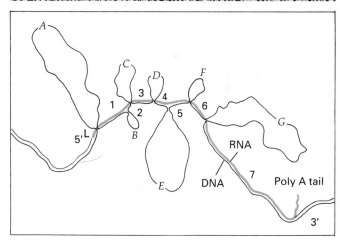

introns. For example, there is a single intron in the bacteriophage T4 thymidylate synthetase gene.

KEYNOTE

The transcripts of protein-coding genes are messenger RNAs. These molecules are linear and vary widely in length in correspondence to the variation in the size of the polypeptides they specify. Prokaryotic mRNAs are seldom modified once they are transcribed, whereas eukaryotic mRNAs are modified by the addition of a cap at the 5′ end and a poly(A) tail at the 3′ end. Many eukaryotic mRNAs contain non-amino-acid-coding sequences called introns, which must be excised from the mRNA transcript to make a mature, functional mRNA molecule. The amino-acid-coding segments separated by introns are called exons. ──────────────

Processing of pre-mRNA to produce mature mRNA. A model for mRNA production from genes with introns is diagramed in Figure 13.9. The sequence of steps is the same for genes without introns, except for the step involving intron removal. In brief, the steps are the transcription of the gene by RNA polymerase II, the addition of the methylated 5′-cap and then the poly(A) tail to produce the pre-mRNA molecule, and finally the processing of the pre-mRNA in the nucleus to remove the introns and splice the exons together to produce the mature mRNA. 5′ capping and 3′ poly(A) addition were described earlier. Here we focus on intron removal during pre-mRNA processing.

Introns in pre-mRNAs are looped out; the loop is then removed by nuclease cleavage. The now-adjacent coding sequences (the exons) are ligated together to generate a contiguous molecule. These events are called **mRNA splicing.** Specific nucleotide signals indicate exon-intron junctions. That is, in RNA introns typically begin with a 5′ GU and end with a 3′ AG. More than just those nucleotides are needed to specify a splicing junction: the 5′ splice junction probably involves at least seven nucleotides and the 3′ splice junction involves at least ten nucleotides of intron sequence.

~ FIGURE 13.9

General sequence of steps in the formation of eukaryotic mRNA. Not all steps are necessary for all mRNAs.

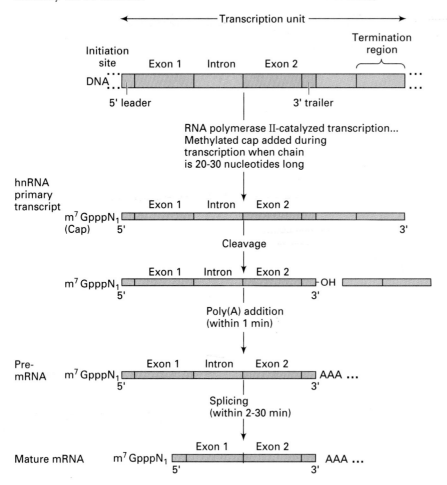

Figure 13.10 diagrams the sequence of events involved in splicing two exons (1 and 2) together with the elimination of an intron. The organization of the pre-mRNA is shown in Figure 13.10a. The first step in splicing is a cleavage at the 5′ splice junction that results in the separation of exon 1 from an RNA molecule that contains the intron and exon 2. The free 5′ end of the intron loops and becomes joined to an A nucleotide that is part of a sequence called a **branch-point sequence**, which is located upstream of the 3′ splice junction (Figure 13.10b). Because of its resemblance to the rope cowboys use to catch cattle, the looped back structure is called an *RNA lariat structure.* For those unfamiliar with the word "lariat," think of the structure as a "p." In mammalian cells, the branch-point consensus sequence is YNCURAY (where Y is a pyrimidine, R is a purine, and N is any base). The key A nucleotide in the sequence is located 18 to 38 nucleotides upstream of the 3′ splice junction. In yeast, there is a more rigid sequence requirement for the branch-

~ FIGURE 13.10

Details of intron removal from a pre-mRNA molecule. At the 5′ end of an intron is the sequence GU and at the 3′ end is the sequence AG. Eighteen to 38 nucleotides upstream from the 3′ end of the intron is an A nucleotide located within the branch-point sequence, which, in mammals is YNCYR⬚A⬚Y, where Y = a pyrimidine, N = any base, R = purine, and A = adenine. (a) Intron removal begins by a cleavage event at the first exon-intron junction. The G at the released 5′ of the intron folds back and forms an unusual 2′ → 5′ bond with the A of the branch-point sequence; (b) This reaction produces a lariat-shaped intermediate. (c) Cleavage at the 3′ intron-exon junction and ligation of the two exons completes the removal of the intron. Inset: Branch-point junction structure involving an unusual 2′ → 5′ phosphodiester bond.

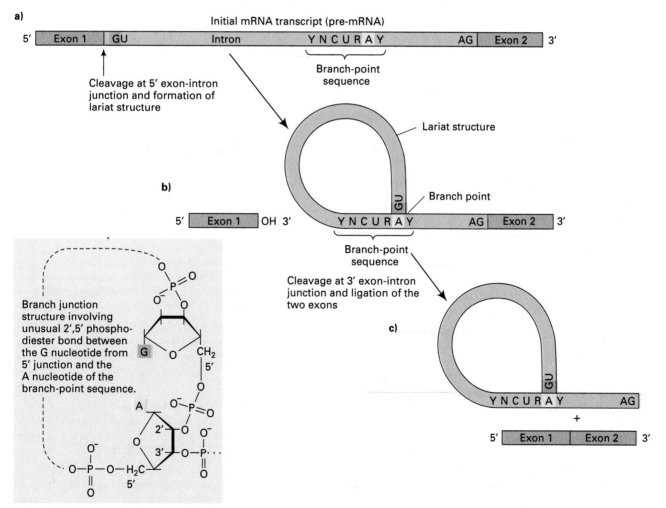

point sequence, although its position is more variable: its sequence is UACUAAC (the underlined A is where the 5′ end of the intron bonds).

The branch point in the RNA that produces the lariat structure involves an unusual 2′-5′ phosphodies-ter bond formed between the 2′ OH of the adenine nucleotide in the branch-point sequence and the 5′ phosphate of the guanine nucleotide at the end of the intron (Figure 13.10 inset). The A itself remains in normal 3′-5′ linkage with its adjacent nucleotides of the intron.

Next, mRNA and a precisely excised intron, the latter still in lariat shape, appear simultaneously as a result of a cleavage event at the 3′ splice junction and ligation of the two coding sequences together (Figure 13.10c). The lariat RNA is subsequently converted to a linear molecule by the action of a de-branching enzyme and is then degraded.

𝒦EYNOTE

Introns are removed from pre-mRNAs in a series of well-defined steps. Introns typically begin with a 5′ GU and end with a 3′ AG. Intron removal begins with the cleavage of the pre-mRNA at the 5′ splice junction. The free 5′ end of the intron loops back and bonds to an A nucleotide in the branch-point consensus sequence, which is located upstream of the 3′ splice junction. Cleavage at the 3′ splice junction releases the intron, which is shaped like a lariat: it is subsequently degraded. The exons that flanked the intron are spliced together once the intron is excised.

The processing of pre-mRNA molecules occurs exclusively in the nucleus. The processing events occur in complexes of the pre-mRNA with small nuclear RNAs (snRNAs), which themselves are associated with proteins. The snRNA-protein complexes are called **small nuclear ribonucleoprotein particles (snRNPs)**. The splicing complexes are called **spliceosomes**. Spliceosomes have been studied most extensively in mammalian cells and yeast. An electron micrograph of a spliceosome from mammalian cells is shown in Figure 13.11. Key components of spliceosomes are the snRNPs. There are six principal snRNAs (named U1-U6) in the nucleus, and they are associated with six to ten proteins to form the snRNPs. Some of the proteins are specific to particular snRNPs, while others are

~ **Figure 13.11**

Electron micrograph of a mammalian spliceosome. [Source: Jack Griffith]

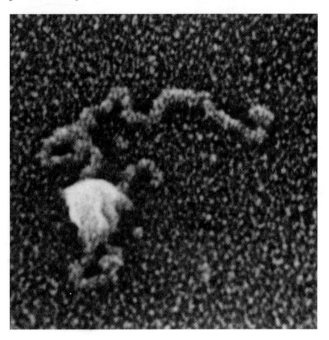

common to all snRNPs. U4 and U6 snRNAs coexist within the same snRNP (U4/U6 snRNP), while the others are found within their own special snRNPs. Each snRNP type is abundant in the nucleus, with at least 10^5 copies per cell. The sequence and proposed secondary structure of human U1 snRNA, which binds to the 5′ splice junction, is shown in Figure 13.12.

A model for the formation of spliceosomes is shown in Figure 13.13. The steps are as follows:

1. U1 snRNP binds to the 5′ splice site. This binding is primarily the result of base pairing of the U1 snRNPs to the 5′ splice site sequence.
2. U2 snRNP binds to the branch-point region.
3. A pre-assembled U4/U6/U5 particle joins the complex. This occurs through association of the particle with the bound U1 and U2 snRNPs.
4. U4 snRNP dissociates from the complex, and this results in the formation of the active spliceosome.

While the steps of spliceosome *assembly* are fairly well understood in mammalian cells and yeast, it is less clear how the spliceosome *functions* in intron removal. One possibility is that the snRNPs help fold the pre-mRNA into a structure such that the pre-mRNA can catalyze its own splicing. Alternatively, the snRNPs may play catalytic roles in splicing.

~ FIGURE 13.12

Sequence and probable secondary structures for human U1 snRNA.

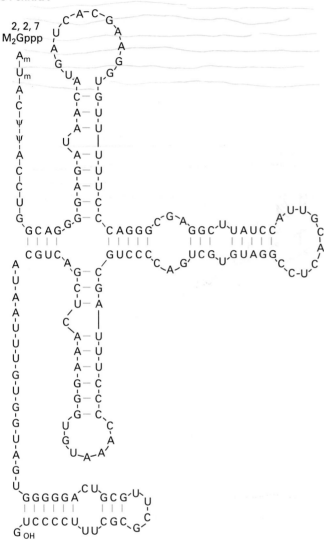

~ FIGURE 13.13

Model for spliceosome assembly in mammalian cells. (From M. R. Green, 1989. *Curr. Opinion Cell Biol.* 1:519–525)

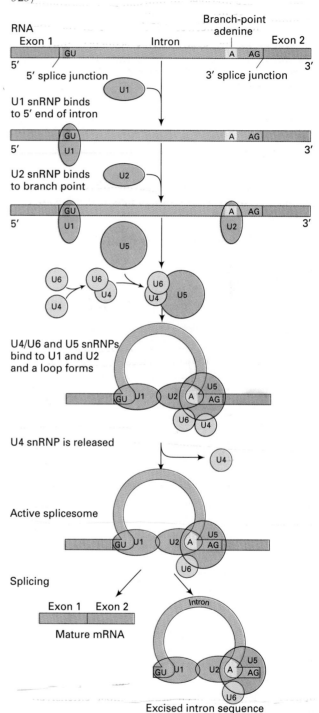

𝒦EYNOTE

The removal of introns from eukaryotic pre-mRNA occurs in the nucleus in complexes called spliceosomes. The spliceosome consists of several small nuclear ribonucleoprotein particles (snRNPs) bound specifically to each intron. The snRNPs themselves consist of usually one but in one case two snRNAs complexed with several proteins. The spliceosome may function to fold the pre-mRNA into a structure in which the pre-mRNA itself acts catalytically to remove the intron. Alternatively, the snRNPs in the spliceosome may catalyze intron removal and exon splicing.

THE STRUCTURE AND FUNCTION OF TRANSFER RNA

In a cell, transfer RNAs (tRNAs) bring amino acids to the ribosome-mRNA complex, where they are polymerized into protein chains in the translation (protein synthesis) process. The sequence of amino acids in the protein is directed by the sequence of codons in the mRNA molecule.

Molecular Structure of tRNA

Transfer RNA molecules constitute between 10 and 15 percent of the total cellular RNA in both prokaryotes and eukaryotes. They have a size of about 4S, and they consist of a single chain of 75 to 90 nucleotides, the sequence of which varies from one tRNA molecule to another. The differences in nucleotide sequences explain the ability of a particular tRNA molecule to bind a particular amino acid.

Primary transcripts of tRNA genes in both prokaryotes and eukaryotes, the **precursor tRNAs** (pre-tRNAs), are extensively modified as they mature. Two types of modification occur: the addition of a 5′ C-C-A 3′ sequence to the 3′ end of some of the tRNAs, and the extensive chemical modification of a number of nucleotides at many places within the chain. Neither of these modifications occur in pre-mRNAs or pre-rRNAs. The type and extent of the modifications vary from tRNA to tRNA, but typically include the addition of methyl groups to specific bases, the reduction of certain uridines, say, to dihydrouridine, or the rearrangement of some uridines to produce pseudouridine (Ψ). Figure 13.14 gives examples of the modified bases found in tRNAs. These modifications are brought about by specific enzyme action, and the modifications themselves cause the particular two- and three-dimensional configurations of tRNAs, which in turn determine the functions of those molecules (e.g., their ability to pick up a specific amino acid and bind to ribosomes).

CLOVERLEAF MODEL FOR THE STRUCTURE OF tRNA. Many tRNAs have been sequenced from a number of organisms, and all the sequences can be arranged into what is called a *cloverleaf model of tRNA*. This model is a secondary structure in that only two dimensions are used to display the sequence (although, as we shall see, tRNA actually has a well-defined three-dimensional structure). Figure 13.15a shows the general features of the cloverleaf model, and Figure 13.15b shows the complete nucleotide sequence of yeast alanine tRNA. The cloverleaf itself results from complementary-base pairing between different sections

~ FIGURE 13.14

Some modified bases found in tRNA molecules.

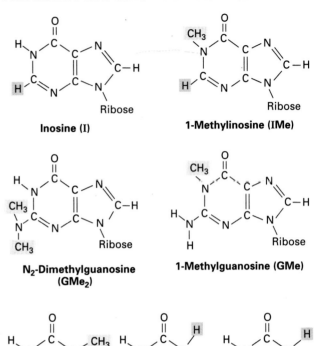

Inosine (I)

1-Methylinosine (IMe)

N₂-Dimethylguanosine (GMe₂)

1-Methylguanosine (GMe)

Ribothymidine (T)

Dihydrouridine (D)

Pseudouridine (ψ)

of the molecule. Four base-paired "stems" appear in the molecule, with the number of base pairs in each stem varying from tRNA to tRNA.

All tRNAs have three unpaired loops, I, II, and IV. Loop I usually contains 9 bases. Loop II has 7 bases and contains within it the three-nucleotide sequence called the **anticodon**, which pairs with a codon (three-nucleotide sequence) in mRNA by complementary-base pairing during translation. This codon-anticodon pairing is crucial for adding the correct amino acid (as specified by the mRNA) to the growing polypeptide chain. Some, but not all, tRNAs have a small loop of variable length, called loop III. Loop IV has 7 bases, and within this loop the sequence 5′ T-ψ-C 3′ appears to be universal among tRNAs. Another universal feature of tRNAs is the 5′ C-C-A 3′ sequence at the 3′ end of the molecule.

THREE-DIMENSIONAL STRUCTURE OF tRNA. Like all molecules, the tRNA has a three-dimensional structure. Because tRNA molecules are so small, they

~ FIGURE 13.15

(a) Schematic of a tRNA molecule shown in the two-dimensional cloverleaf configuration;
(b) The complete nucleotide sequence of yeast alanine tRNA, showing the unusual bases
and anticodon position.

a) **Schematic of tRNA molecule**

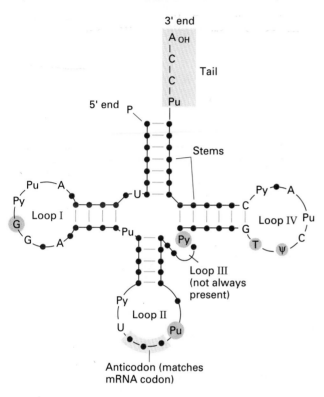

b) **Yeast alanine tRNA**

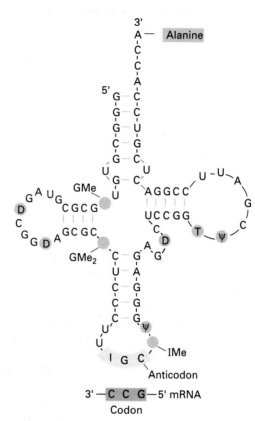

Transfer RNA Genes

can be crystallized, and X-ray crystallography can be used to develop a three-dimensional model. Figure 13.16 (p. 396) shows the tertiary-structure model for yeast tRNA.Phe (this terminology indicates the amino acid specified by the anticodon of the tRNA, in this case phenylalanine). This tRNA was the first to be analyzed using this technique: Figure 13.16a presents a schematic drawing, Figure 13.16b is a photograph of a backbone model, and Figure 13.16c is a photograph of a space-filling molecular model. All other tRNAs that have been examined through X-ray crystallography show similar three-dimensional structures.

From the data they obtained, the crystallographers concluded that all hydrogen-bonded stem structures proposed in the cloverleaf model do exist in the tRNA molecule. In addition, the results showed that there is other hydrogen bonding that folds the cloverleaf into a more compact shape, like an upside-down L. In this L-shaped structure the 3′ end of the tRNA (the end to which the amino acid attaches) is at the opposite end of the L from the anticodon loop.

A three-base sequence (codon) in an mRNA specifies each amino acid to be added to a polypeptide chain. While only 20 different amino acids can be used to make a protein, actually 61 different codons can be used in an mRNA to specify amino acids. Three additional codons do not specify amino acids but instead are used as termination signals for protein synthesis. (These concepts will be further elaborated in the next chapter.) Since each amino acid–specifying codon must be matched by an appropriate anticodon on a tRNA molecule, theoretically any cell could have 61 different tRNA types. And eukaryotes commonly have a number of different tRNA molecules with the same anticodon. So that the instructions for the manufacture of such large numbers of tRNAs can be coded, the genome has many tRNA genes.

THE tRNA GENES OF *E. coli*. Examples of virtually all possible sorts of arrangements of tRNA genes

~ FIGURE 13.16

(a) Schematic of yeast phenylalanine tRNA as determined by X-ray diffraction of tRNA crystals. Note the characteristic L-shaped structure. (b) Photograph of a backbone model and (c) photograph of a space-filling molecular model of yeast phenylalanine tRNA. The CCA end of the molecule is at the upper right, and the anticodon loop is at the bottom.

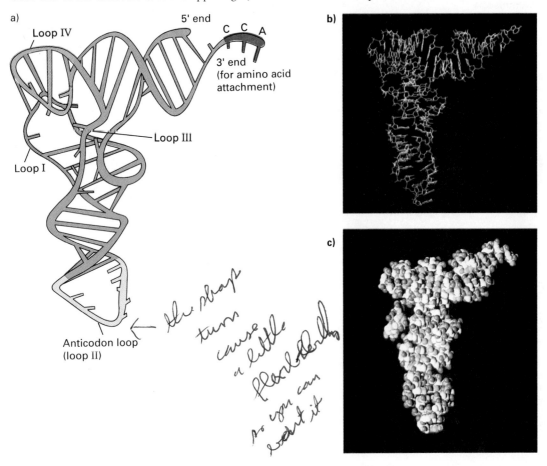

occur in *E. coli*. Some tRNA genes occur only once on the *E. coli* chromosome, while others are present two or more times, a situation called **gene redundancy**. In some cases redundant genes are not linked on the chromosome, but in other cases they are linked, side by side, in what is called a *redundant gene cluster*. Unique tRNA genes can also be linked side by side on the chromosome in a *nonredundant tandem gene cluster*.

EUKARYOTIC tRNA GENES. In general, many more tRNA genes occur in eukaryotes than in prokaryotes. In yeast, for example, about 400 tRNA genes occur in the genome. Higher eukaryotes have a tendency toward even greater redundancy of tRNA genes. *Xenopus laevis*, for example, has over 200 copies of each tRNA gene per genome. As a class, then, tRNA genes are found in the moderately-repetitive kinetic class of DNA (see Chapter 10).

As in *E. coli*, all possible arrangements of tRNA genes occur in eukaryotes: some genes are found singly, whereas others are found in clusters, which can be redundant or nonredundant tandem clusters. There is also evidence of tandem clusters in which some genes are a single copy while others are redundant.

At least some tRNA genes from many eukaryotic organisms (c.f. protein-coding genes) contain introns. For example, about 10 percent of the 400 tRNA genes in yeast have introns. Depending on the tRNA, the intron is 14 to 60 base pairs long and is almost always located between the first and second nucleotides 3′ to the anticodon. The anticodon itself often, but not always, pairs with intron sequences in the pre-tRNA. For reasons that are not understood, introns are found only in tRNAs for the amino acids tyrosine, phenylalanine, tryptophan, lysine, proline, serine, leucine, and isoleucine.

The first tRNA gene in which an intron was found was the yeast gene for tRNA.Tyr, the gene for the tRNA that carries the amino acid tyrosine to the ribosome. Figure 13.17 shows the nucleotide sequence for the initial transcript of the tRNA gene, the pre-tRNA (Figure 13.17a), and the mature tRNA (Figure 13.17b), as well as the organization of the two sequences into cloverleaf models. The 14-base intron (i.e., the tRNA copy of the intron in the gene) is located in the pre-tRNA just to the 3' side of the anticodon. The presence of the transcript of the intron sequence in the pre-tRNA results in a significant change in the anticodon loop, as Figure 13.17a shows. Removal of the intron involves cleavage with a specific endonuclease. The RNA pieces generated by intron removal are then spliced together by an enzyme called **RNA ligase.**

Biosynthesis of Transfer RNAs

The transcription of tRNA genes was described in Chapter 12. In prokaryotes the tRNA genes are transcribed by the same RNA polymerase that transcribes all the genes. In eukaryotes the tRNA genes are transcribed by RNA polymerase III.

As in the case with mRNAs, precursor tRNAs are longer than mature tRNAs and have a 5' leader sequence and a 3' trailer sequence. Unlike mRNA, however, these sequences are removed during the processing of pre-tRNA into the mature tRNA. In eukaryotes the pre-tRNA processing occurs in the nucleus, so only mature tRNAs are found in the cytoplasm. Figure 13.18 (p. 398) shows the nucleotide sequence of a prokaryotic pre-tRNA and the leader and trailer

~ FIGURE 13.17

Cloverleaf models for yeast precursor tRNA.Tyr and mature tRNA.Tyr. (a) The 14-base intron in the precursor, located adjacent to the anticodon-coding region. (b) The mature tRNA.Tyr molecule produced after the intron has been removed.

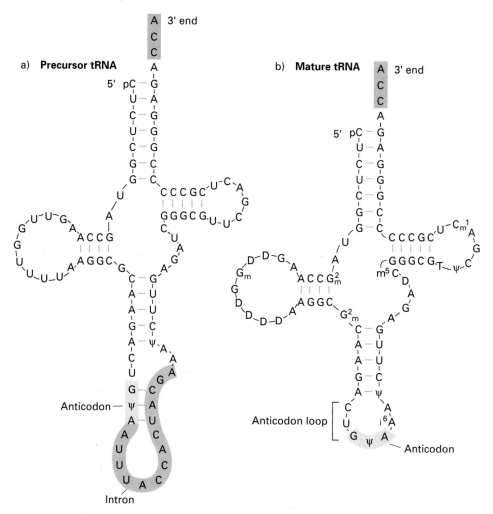

~ FIGURE 13.18

~ FIGURE 13.18

Nucleotide sequence of a prokaryotic precursor tRNA (*E. coli* pre-tRNA.Tyr), showing the nucleotides removed during the processing steps.

sequences that are removed during the processing steps. A similar arrangement exists in eukaryotic systems.

As we mentioned earlier, tRNA genes may exist in clusters. In prokaryotes, but not in eukaryotes, there is evidence that a cluster of tRNA genes may be transcribed to produce a single RNA transcript containing a number of tRNA sequences. The general organization of the multi-tRNA pre-tRNA molecules is

$$5'\text{-leader}-(\text{tRNA}-\text{spacer})n-\text{tRNA}-\text{trailer-}3'$$

where *n* is a number of tRNA-spacers characteristic of a cluster. Figure 13.19 gives an example of a pre-tRNA *E. coli* molecule that contains two tRNA sequences. In such cases the leader, the trailer, and the spacer sequences must be removed during the processing of the pre-tRNA to produce two discrete, mature tRNAs.

The non-tRNA sequences are removed from the pre-tRNA molecules by the action of specific enzymes. In *E. coli* at least two enzymes are needed. The enzyme RNase P catalyzes removal of the 5' leader sequences, and RNase Q catalyzes the removal of the 3' trailer sequence. RNase P has been shown to be a ribonucleoprotein enzyme, that is a complex between an RNA and a protein. Another enzyme (or enzymes) is involved in removing the spacer region between clustered tRNAs.

~ FIGURE 13.19

A tRNA precursor from *E. coli* that contains two tRNA sequences, one for tRNA.Ser and the other for tRNA.Thr. The nucleotides removed during processing are shown in blue.

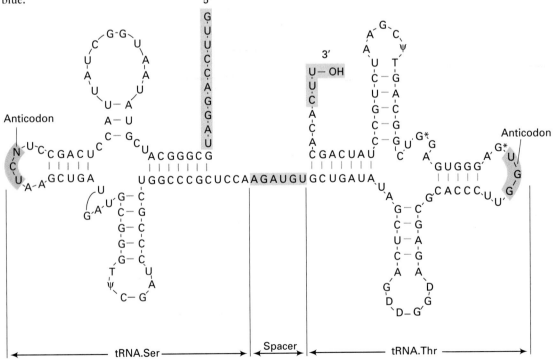

Little is known about the enzymes needed for the removal of the terminal noncoding sequences of eukaryotic pre-tRNAs, except that they must be located in the nucleus, since pre-tRNAs are not found in the cytoplasm.

𝒦EYNOTE

Molecules of transfer RNA bring amino acids to the ribosomes, where the amino acids are polymerized into a protein chain. All the tRNA molecules are between 75 and 90 nucleotides long, contain a number of modified bases, and have similar three-dimensional shapes. A CCA sequence is found at the 3′ end of all tRNAs. tRNAs are made as pre-tRNA molecules containing 5′-leader and 3′-trailer sequences, both of which are removed by enzyme activity. Some eukaryotic pre-tRNAs contain introns, which are removed in processing steps that are different from pre-mRNA processing steps.

RIBOSOMAL RNA

Ribosomes are the organelles within the cell on which protein synthesis takes place. Ribosomes bind to mRNA and facilitate the binding of the anticodon on the tRNA to the codon on the mRNA so that the poly-peptide chain can be synthesized. Ribosomes are complex structures, and we have yet to understand fully how they function. In both prokaryotes and eukaryotes the ribosomes consist of two unequally sized subunits, (the large and small ribosomal subunits), each of which consists of a complex between RNA molecules and proteins. Each subunit contains at least one **ribosomal RNA (rRNA)** molecule and a large number of **ribosomal proteins**.

Structure of the Prokaryotic Ribosome in E. coli

The *E. coli* ribosome is used as a model of a prokaryotic ribosome. Figure 13.20 shows the general structure and composition of this organelle.

The *E. coli* ribosome has a size of 70S, and the sizes of the two subunits are 50S (large subunit) and 30S (small subunit). Figure 13.21 (p. 400) presents electron micrographs of ribosomes and ribosomal subunits of *E. coli*, along with interpretive models that have been constructed from the micrographs. Clearly the two subunits have distinct and recognizable three-dimensional shapes.

Approximately two-thirds of the *E. coli* ribosome consists of ribosomal RNA (rRNA), the rest of the mass consisting of ribosomal proteins. The large 50S subunit has 34 different proteins and two rRNA molecules, one with a size of 23S (2904 nucleotides) and the other with a size of 5S (120 nucleotides). The small 30S ribosomal subunit has 20 different proteins and one rRNA molecule, which has a size of 16S (1542 nucleotides).

~ FIGURE 13.20

Structure of the *E. coli* 70S ribosome. Each of the two subunits (50S and 30S) contains rRNA and ribosomal protein molecules.

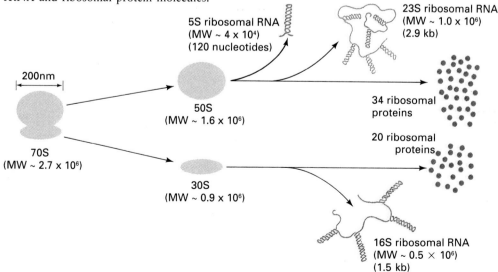

200nm

70S
(MW ~ 2.7 × 10⁶)

50S
(MW ~ 1.6 × 10⁶)

30S
(MW ~ 0.9 × 10⁶)

5S ribosomal RNA
(MW ~ 4 × 10⁴)
(120 nucleotides)

23S ribosomal RNA
(MW ~ 1.0 × 10⁶)
(2.9 kb)

34 ribosomal proteins

20 ribosomal proteins

16S ribosomal RNA
(MW ~ 0.5 × 10⁶)
(1.5 kb)

~ FIGURE 13.21

Electron micrographs of *E. coli* small (30S) ribosomal subunits, large (50S) ribosomal subunits, and complete (70S) ribosomes, shown with photographs of interpretive models.

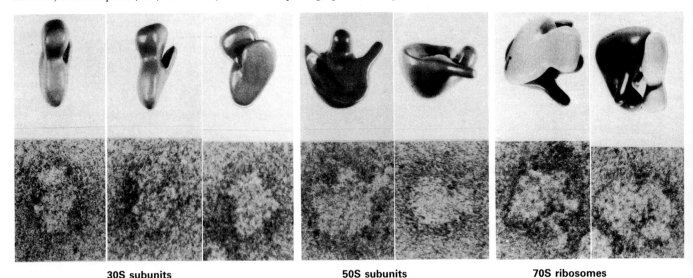

30S subunits   **50S subunits**   **70S ribosomes**

The rRNA Genes of *E. coli*

In prokaryotes and eukaryotes the regions of DNA that contain the genes for rRNA are called **ribosomal DNA (rDNA)**. In *E. coli*, production of equal amounts of the three rRNAs is ensured by the transcription of the three adjacent genes for 16S, 23S, and 5S in rDNA into a single **precursor rRNA (pre-rRNA)** molecule. One transcription unit consists of one gene each for the three rRNAs, and *E. coli* has seven such transcription units (*rrn* regions) positioned apart from each other on the chromosome.

Figure 13.22 gives the general organization of a transcription unit. In each transcription unit, the three rRNA genes are arranged in the order 16S-23S-5S; so the arrangement of the rRNAs on the pre-rRNA transcript is 5′ 16S-23S-5S 3′. In all seven transcription units, one or two tRNA genes are also found in the internal spacer between the 16S and 23S rRNA coding sequences and another one or two tRNA genes are found in the 3′ spacer between the end of the 5S rRNA gene and the 3′ end of the transcription unit in three of the seven transcription units. During RNA synthesis from the transcription units, the tRNA genes are transcribed as part of the pre-rRNA molecule. The tRNAs are then removed from the precursor by the action of the specific pre-tRNA processing enzymes we have already discussed.

~ FIGURE 13.22

General organization of an *E. coli* rRNA transcription unit (an *rrn* region).

E. coli rRNA transcription unit

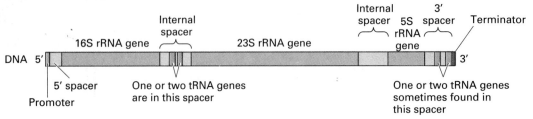

The Ribosomal Protein Genes of *E. coli*

The *E. coli* ribosome has one gene for each of the 54 ribosomal proteins. The genes are located at many places on the *E. coli* genetic map. Some of the ribosomal genes are found clustered on the map; for example, 27 ribosomal protein genes are in the *strA* (streptomycin) region (at 72 min on the map), and 4 ribosomal protein genes are in the *rif* (rifampicin) region (at 88 min). Each of these two regions also contains other genes, such as genes for subunits of RNA polymerase.

Biosynthesis of the *E. coli* Ribosome

In 1969, Masayasu Nomura and his colleagues investigated the question of the assembly of the *E. coli* ribosome. They dissociated the 30S subunit into its component rRNA and ribosomal proteins and then attempted to reconstitute the subunit *in vitro*. They found that at 37°C (the normal growth temperature for *E. coli*), and in the presence of an ideal ionic environment, the parts could be reconstituted into a completely functional 30S subunit. This phenomenon was called **self-assembly.** The result was significant because it indicated that all the information needed to form a functional subunit was somehow embedded into the structure of individual components (rRNA and proteins). A few years later, K. Nierhaus and F. Dohme were able to reconstitute the 50S *E. coli* subunit from its component parts.

So, how is the *E. coli* ribosome assembled *in vivo*? Figure 13.23 shows the scheme for the production of mature *E. coli* rRNAs. Each transcription unit is transcribed into a 30S precursor (p30S) RNA, which contains a 5′ leader sequence, the 16S, 23S, and 5S rRNA sequences (each separated by spacer sequences), and a 3′ trailer sequence.

The p30S molecule is cleaved by RNase III to produce three precursor molecules, one each of the p16S, p23S, and p5S molecules. In normal cells, RNase III cleaves the transcript of the rRNA transcription unit to produce the three rRNA precursors while transcription is still occurring. As a result of this rapid processing, the p30S is not seen in normal cells. However, in mutants that have a temperature-sensitive RNase III activity, the p30S molecule accumulates at the high temperature. Through studies of these mutants, the pathway of p30S synthesis and processing has been discovered. The mature 16S, 23S, and 5S rRNAs are produced by the action of other specific RNases.

As the rRNA genes are being transcribed, the transcript rapidly becomes associated with ribosomal proteins. The cleavage of the transcript takes place, then, within a complex formed between the rRNA transcript (as it is being transcribed) and ribosomal proteins. It is

~ FIGURE 13.23

Scheme for the processing of a precursor rRNA (p30S) to the mature 16S, 23S, and 5S rRNAs of *E. coli*. (The tRNAs have been omitted from the diagram.)

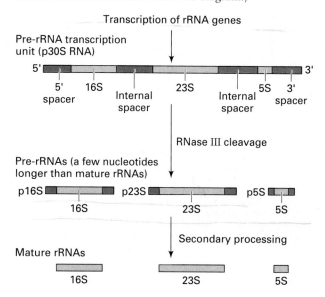

most probably directed by the conformational state of the rRNA-ribosomal protein particle at the time of the cleavage.

Structure of the Eukaryotic Ribosome

In general, eukaryotic ribosomes are larger and more complex than their prokaryotic counterparts. The size of the ribosome and the molecular weights of the rRNA molecules differ from organism to organism. Lower eukaryotes have the smallest ribosomes (although they are larger than ribosomes found in *E. coli*), while mammalian ribosomes are the largest. Nonetheless, all eukaryotic ribosomes have many common structural and chemical features. We will use the mammalian ribosome as a model for discussion, and we will occasionally compare it with the yeast ribosome.

Mammalian ribosomes have a size of 80S (versus 70S for the bacterial ribosomes) and have two subunits, a large one with a size of 60S and a small one with a size of 40S (Figure 13.24; p. 402). The 40S subunit is quite similar in size and mass throughout the eukaryotes, but the 60S subunit varies quite a lot.

The 80S mammalian ribosome consists of about equal weights of rRNA and ribosomal proteins. There are four rRNA types, with sizes of 18S (~1900 nucleotides), 28S (~4700 nucleotides) (the large rRNAs), 5.8S (156 nucleotides), and 5S (120 nucleotides) (the small rRNAs) (Figure 13.24). The ribosomal proteins in eukaryotes have not been characterized as clearly as those for prokaryotes, but in general, the ribosomes of

~ FIGURE 13.24

Composition of whole ribosomes and of ribosomal subunits in mammalian cells.

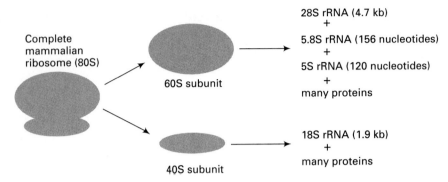

Complete mammalian ribosome (80S)

60S subunit

28S rRNA (4.7 kb)
+
5.8S rRNA (156 nucleotides)
+
5S rRNA (120 nucleotides)
+
many proteins

40S subunit

18S rRNA (1.9 kb)
+
many proteins

higher eukaryotes have 70–80 different ribosomal proteins with about 30 in the small subunit and about 50 in the large subunit. The ribosomes of yeast have 60–70.

In all eukaryotes the 40S ribosomal subunit contains only one rRNA type, 18S rRNA. The large 60S ribosomal subunit contains the other three types of rRNA, the 28S, 5.8S, and 5S rRNAs. The 5.8S rRNA is hydrogen-bonded to the 28S rRNA in the functional ribosome.

Ribosomal RNA Genes in Eukaryotes

LOCATION AND ORGANIZATION OF rRNA GENES IN THE GENOME. Within the nucleus of all eukaryotes, the rRNA genes are located in regions of DNA called ribosomal DNA (rDNA), and around these rDNA regions organelles, called *nucleoli*, form within the nucleus. The nucleus of every eukaryote has one or more nucleoli within which rRNA is transcribed. The resulting rRNAs associate with ribosomal proteins to form the two ribosomal subunits. Because the ribosomal proteins are made in the cytoplasm, ribosome production requires the ribosomal proteins to migrate back into the nucleus and into the nucleolus. Very little information is available on such directional movement of proteins. Once the ribosomal subunits are made, they must be transported to the cytoplasm before they can function in protein synthesis. The ribosomal subunits do not carry out protein synthesis in the nucleus.

Most eukaryotes that have been examined have a large number of copies of the genes for each of the four rRNA species. The genes for 18S, 5.8S, and 28S rRNAs are found in one or more clusters in the genome, and around each cluster a nucleolus is formed. The extent of gene repetition varies from eukaryote to eukaryote and is in the range of 100 to 1000, with a few exceptions. Thus the rRNA genes are representative of the moderately-repetitive DNA in the genome (see Chapter 10). For example, yeast has 140 rRNA gene sets, *Drosophila* has 260, and HeLa (human) cells have 1250.

In most organisms the 5S rRNA genes are located in the genome at a site or sites distinct from the sites for the other rRNA genes. In humans, for example, the 5S genes are clustered, whereas in *Xenopus* they are in clusters, which are scattered throughout the genome. The number of 5S genes relative to the other genes also sets no pattern. The organism may have more of them, the same number, or less of them than the other rRNA genes. In some organisms, such as yeast (*Saccharomyces cerevisiae*) and the cellular slime mold (*Dictyostelium discoideum*), the 5S rRNA genes are interspersed with the other rRNA gene sets.

At the chromosome level the 18S, 5.8S, and 28S rRNA genes are usually found adjacent to one another in the order 18S-5.8S-28S, with each set of three genes repeated many times to form repeated tandem arrays of rRNA genes called **rDNA repeat units.** As in *E. coli,* each set of three rRNA genes is transcribed onto a single pre-rRNA molecule (Figure 13.25). (This is an exception to the general rule that eukaryotic genes are transcribed separately.) The rDNA repeat units of four eukaryotes are diagrammed in Figure 13.26. In general, variation in transcribed spacer lengths is largely responsible for variation in length of the transcribed region (the rDNA transcription unit) of the rDNA repeat unit.

At the gene level, two types of transcribed spacer sequences have been identified (Figure 13.25). The external transcribed spacers (ETS) are DNA sequences that are transcribed into the pre-rRNA and are located immediately adjacent to the 5′ end of the 18S sequence and to the 3′ end of the 28S sequence in the pre-rRNA molecule. The internal transcribed spacers (ITS) are transcribed into the pre-rRNA and are located on either side of the 5.8S sequence. Finally, another class of rDNA spacer sequences, nontranscribed spacer (NTS) sequences, is found at the extreme 5′ and 3′ ends of

~ FIGURE 13.25

Generalized diagram of a eukaryotic, ribosomal DNA-repeating unit. The coding sequences for 18S, 5.8S, and 28S rRNAs are indicated in light brown.

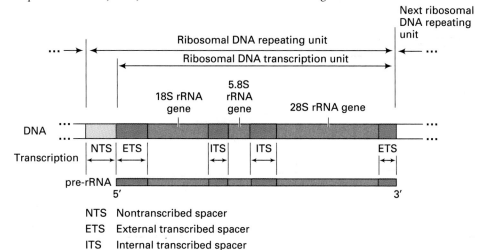

NTS   Nontranscribed spacer
ETS   External transcribed spacer
ITS   Internal transcribed spacer

each tandem rRNA repeating unit in the rDNA. As the name indicates, these regions are not transcribed into the pre-rRNA molecule. NTS sequences vary considerably in length among eukaryotes, from ~3–9 kb in frogs to ~30 kb in humans.

Introns have been found in the rRNA genes of only a few organisms (e.g., *Drosophila*, the slime mold *Physarum*, and the protozoan *Tetrahymena*). Thus, they are certainly not as widespread as the introns in mRNA genes. The introns are excised from the pre-rRNA transcript as it is processed to produce the mature rRNAs. The splicing reactions involved in this case are distinct from those involved in removing introns from pre-mRNA and pre-tRNA. The removal of introns from *Tetrahymena* pre-rRNAs occurs in a very special way and will be discussed later.

𝒦EYNOTE

Ribosomes, the organelles within the cell in which protein synthesis takes place, consist of two unequally sized subunits in both prokaryotes and eukaryotes. Each subunit contains ribosomal RNA and ribosomal proteins. Prokaryotic ribosomes contain three distinct rRNA molecules, whereas eukaryotic cytoplasmic ribosomes (larger and more complex) contain four.

Biosynthesis of Eukaryotic Ribosomes

Eukaryotic ribosomes are assembled in a process similar to that found in prokaryotes. Ribosomal proteins

~ FIGURE 13.26

The ribosomal transcription units of four eukaryotes. The 18S, 5.8S, and 28S regions code for the rRNAs found in all ribosomes. The transcribed spacer regions vary in length and result in large differences in the lengths of the transcription units. Variation in the length of the nontranscribed spacer regions is also considerable.

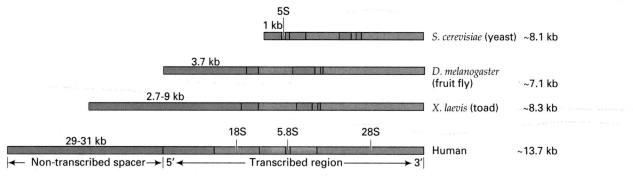

~ FIGURE 13.27

Electron micrograph of transcription of pre-rRNA molecules from rRNA genes in an oocyte from the spotted newt, *Triturus viridescens*. Transcription of the rRNA gene is from left to right.

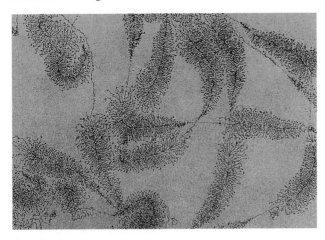

~ FIGURE 13.28

Processing of 45S pre-rRNA in HeLa cells to produce the mature 18S, 5.8S, and 28S rRNAs.

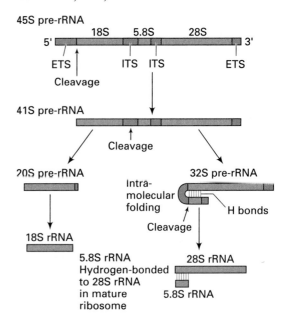

are made in the cytoplasm from mRNAs that are transcribed by RNA polymerase II. The 5S rRNA is transcribed by RNA polymerase III in the nucleus directly from individual 5S rRNA genes. Ribosomal proteins and the 5S rRNA molecule both migrate to the nucleolus in the cell's nucleus by unknown means.

In the nucleolus the 18S, 5.8S, and 28S rRNA genes are transcribed by RNA polymerase I into a large pre-rRNA molecule [such as the 45S pre-rRNA (~13,700 nucleotides) found in human HeLa cells]. Figure 13.27 shows an electron micrograph of pre-rRNA from an rDNA repeat unit of the newt *Triturus viridescens*.

The pre-rRNA is cleaved at specific sites by specific RNases to remove ITS and ETS sequences. Unlike processing for prokaryotes, processing the pre-rRNA does not occur until the molecule is completely synthesized. The discrete, intermediate rRNAs are finally processed into mature 18S, 5.8S, and 28S rRNAs. As an example, Figure 13.28 shows the pre-rRNA-processing pathway in HeLa cells. The pathway is similar in all eukaryotes that have been examined. The first cleavage removes the 5′ ETS sequence and produces a precursor molecule that still contains all three rRNA sequences. In the next step an internal cleavage occurs to produce the precursor to 18S rRNA and the precursor to 28S and 5.8S rRNAs. For the former, the generation of 18S rRNA involves the removal of the ITS sequence to the 3′ side of the sequence. For the latter the molecule folds so that the 5.8S sequence becomes hydrogen-bonded to the 28S sequence. Then the ITS sequences are removed from the 5′ end and from between the 5.8S and 28S sequences, and the ETS is removed from the 3′ end.

All the pre-rRNA-processing events take place in complexes formed between the pre-rRNA, 5S rRNA, and the ribosomal proteins. As the pre-rRNA is cleaved, conformational changes take place in the complexes so that when the processing is finished, the 60S and 40S ribosomal subunits are produced, each with the appropriate mature rRNAs and ribosomal proteins. Once assembled, the two ribosomal subunits migrate out of the nucleolus and into the cytoplasm, where they associate with mRNAs and tRNAs and begin the process of protein synthesis.

KEYNOTE

Three of the four rRNAs in eukaryotic ribosomes, the 18S, 5.8S and 28S rRNAs, are transcribed from the rDNA onto a single pre-rRNA molecule. In addition to the rRNA coding sequences, the pre-rRNA contains spacer sequences located at the ends of the molecule (external transcribed spacers) and internally between the rRNA sequences (internal transcribed spacers). The spacers are removed as the pre-rRNA is processed in the nucleolus to produce the mature rRNAs and the functional ribosomal subunits.

SELF-SPLICING OF INTRONS IN *TETRAHYMENA* PRE-rRNA.

In some species of *Tetrahymena*, for example *T. thermophila*, the genes for the 28S rRNA are all interrupted by a 413-bp intron. The intron sequence is removed during processing of the pre-rRNA in the nucleolus. The excision of the intron unexpectedly was shown to occur by a *protein-independent reaction* in which the RNA intron becomes folded into a secondary structure that promotes its own excision. This process is called **self-splicing** and was discovered by Tom Cech and his research group. The self-splicing of the *Tetrahymena* pre-rRNA intron was the first example of what is now called **group I-intron self-splicing.** Other self-splicing group I introns have been found in rRNA genes and some mRNA genes in the mitochondria of yeast, *Neurospora*, and other fungi, in rRNA genes, and some mRNA and tRNA genes in bacteriophages.

Figure 13.29 diagrams the self-splicing reaction for the group I intron in *Tetrahymena* pre-rRNA. The steps are as follows:

1. The pre-rRNA is cleaved at the 5′ splice junction as a guanosine is added to the 5′ end of the intron. (It is the addition of a G that characterizes group I introns. Another class of self-splicing introns, the group II introns, involve a reactive A in the intron removal process.)
2. The intron is cleaved at the 3′ splice junction.
3. The two exons are spliced together.
4. The excised intron is cleaved again and a splicing event produces a circular RNA molecule and a short piece of RNA.

Self-splicing only occurs in a limited number of systems. In the processing of *Tetrahymena* pre-rRNA, two separate events occur: (1) The precursor is cleaved to remove spacer sequences from the mature rRNAs, and, (2) the intron in the 28S rRNA sequence is removed and the two 28S rRNA parts spliced together. The two processes are not identical. *Removal of the spacer sequences releases rRNAs that remain separate. Intron removal, by contrast, results in the splicing together of the RNA sequences that flanked the intron.*

The self-splicing activity of the intron RNA sequence can not be considered an enzyme activity. That is, while it catalyzes the reaction, it is not regenerated in its original form at the end of the reaction, as is

~ FIGURE 13.29

Self-splicing reaction for the group I intron in *Tetrahymena* pre-rRNA.

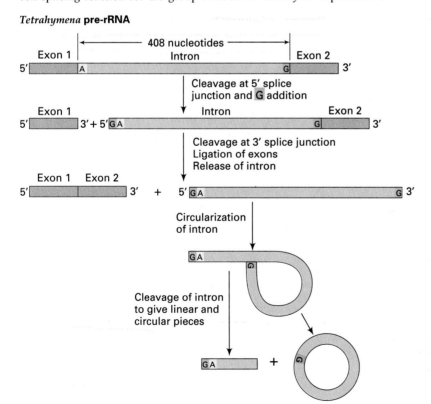

the case with protein enzymes. Recently, however, the *Tetrahymena* intron RNA has been modified in the lab so that it can function as an enzyme in the true sense. This **RNA enzyme** is called a **ribozyme**. Ribozymes can be used to cleave RNA molecules at specific sequences; a complete discussion of ribozymes is beyond the scope of this text.

*K*EYNOTE

In some precursor rRNAs there are introns, the RNA sequences of which fold into a secondary structure that excises itself, a process called self-splicing. The self-splicing reaction requires guanine but does not involve any proteins.

SUMMARY

In this chapter we discussed the structure, synthesis, and function of mRNA, tRNA, and rRNA. Each mRNA encodes the amino acid sequence of a polypeptide chain. The nucleotide sequence of the mRNA is translated into the amino acid sequence of the polypeptide chain on ribosomes. Ribosomes consist of two unequal-sized subunits, each of which contains both rRNA and protein molecules. The amino acids that are assembled into proteins are brought to the ribosome attached to tRNA molecules.

mRNAs have three main parts: a 5' leader sequence, the amino acid coding sequence, and the 3' trailer sequence. Since all three sequences vary from mRNA to mRNA, as a group mRNAs show extensive variability in length. In prokaryotes the primary gene transcript functions directly as the mRNA molecule, while in eukaryotes the primary RNA transcript must be modified in the nucleus by RNA processing events to produce the mature mRNA. These processing events are the addition of a 5' methylated cap, the addition of a 3' poly(A) tail, and the removal of any introns (internal sequences that do not code for amino acids) that are present. The latter is known as RNA splicing and involves specific interactions with snRNPs in molecular structures called spliceosomes. Only when all processing events have been completed is the mRNA functional; at that point, it leaves the nucleus and can be translated in the cytoplasm.

All tRNAs carry out the same function, that is, bringing amino acids to the ribosomes. Thus, all tRNAs are very similar in length (75–90 nucleotides) and in secondary and tertiary structure. tRNAs are usually synthesized as precursor molecules that have extra nucleotides at both the 5' and 3' ends. The extra nucleotides are removed by specific processing events. In addition a number of the bases are modified during the maturation process. A CCA sequence is found at the 3' end of all tRNAs. Some tRNA genes in eukaryotes contain introns; these are removed from the transcripts of those genes by splicing events that are different from those used for intron removal from precursor mRNAs.

rRNAs are important structural components of ribosomes, the cellular organelles on which protein synthesis takes place. In both prokaryotes and eukaryotes, ribosomes consist of two unequal-sized subunits: 50S and 30S in prokaryotes, and 60S and 40S in eukaryotes. The subunits consist of both rRNA and ribosomal proteins. In each case the smaller subunit contains one rRNA molecule, 16S in prokaryotes, and 18S in eukaryotes. The prokaryotic larger subunit contains 23S and 5S rRNAs, while the eukaryotic larger subunit contains 28S, 5.8S, and 5S rRNAs.

In prokaryotes the genes for the 16S, 23S, and 5S rRNAs are transcribed into a single pre-rRNA molecule. There are seven copies of the so-called rRNA transcription unit in *E. coli*. tRNA genes are found in the spacer regions of each transcription unit. The pre-rRNA molecules are processed in a number of enzymatically-catalyzed steps to remove the non-coding sequences found at the ends of the molecules and between the rRNA sequences. At the same time the tRNAs are released by specific enzyme reactions. All of the processing events occur while the pre-rRNA is associating with the ribosomal proteins so that, when processing is complete, the functional 50S and 30S subunits have been assembled.

In eukaryotes there are many copies of the genes for each of the four rRNAs. The 18S, 5.8S, and 28S rRNA genes comprise a transcription unit, and many such transcription units are organized in a tandem array. The 5S rRNA genes are also present in many copies but they are located elsewhere in the genome. The 18S, 5.8S, and 28S sequences are transcribed into pre-rRNA molecules which, in addition to the rRNA sequences, contain spacer sequences at the ends and between the rRNA sequences. As in prokaryotes, the spacer material is removed by specific processing events that take place while the rRNA sequences are associated with ribosomal proteins. The 5S rRNA becomes associated with the assembling 60S subunit so that, at the end of the processing steps, functional 60S and 40S ribosomal subunits have been produced. All of these events occur in the nucleolus. The mature ribosomal subunits then exit the nucleus and participate in protein synthesis in the cytoplasm.

In the pre-rRNA of *Tetrahymena* the 28S rRNA sequence is interrupted by an intron. The intron is removed during processing of the pre-rRNA in the nucleolus. The excision of this particular intron (and a few other examples in other systems) occurs by a protein-independent reaction in which the RNA sequence of the intron folds into a secondary structure that promotes its own excision. This process is called self-splicing.

In sum, the functional RNA molecules of the cell are typically transcribed as precursor molecules that include extra nucleotides that are removed by specific processing events. In prokaryotes the processing events are confined to the tRNAs and rRNAs. In eukaryotes, RNA processing is much more complex, particularly with mRNAs. Mature eukaryotic mRNAs typically have extra nucleotides compared with their precursors, namely, a 5′ cap and 3′poly(A) tail.

ANALYTICAL APPROACHES FOR SOLVING GENETICS PROBLEMS

Q.1 You are given four different RNA samples. Sample I has a short lifetime; sample II has a homogeneous molecular weight; sample III is produced by processing of a larger precursor RNA; and sample IV has an additional sequence added onto the original transcript. For each sample, state whether the RNA could be rRNA, mRNA, or tRNA. If it is not one of those, state what it might be. Note that for each sample more than one RNA could apply. Give reasons for your choices.

A.1 Sample I: A short lifetime is characteristic of mRNAs in prokaryotes, and many mRNAs in eukaryotes. Heterogeneous nuclear RNA of eukaryotes also has a short lifetime. Sample II: rRNA and tRNA species have homogeneous molecular weights since they carry out specific functions within the cell for which their length and three-dimensional configuration are important. Messenger RNA and heterogeneous nuclear RNA are heterogeneous in length. Sample III: rRNA, tRNA, and mRNA are all produced by the processing of a larger precursor RNA molecule. Sample IV: Both tRNA and eukaryotic mRNA have additional sequences added after they are transcribed. For tRNA this sequence is the CCA at the 3′ end, and for mRNA this sequence is the poly(A) tail at the 3′ end.

QUESTIONS AND PROBLEMS

13.1 Compare and contrast the structures of prokaryotic and eukaryotic mRNAs.

*13.2 Compare the structures of the three classes of RNA found in the cell.

13.3 Many eukaryotic mRNAs, but not prokaryotic mRNAs, contain introns. What is the evidence for the presence of introns in genes? Describe how these sequences are removed during the production of mature mRNA.

*13.4 Discuss the posttranscriptional modifications that take place on the primary transcripts of tRNA, rRNA, and protein-coding genes.

13.5 Distinguish between leader sequence, trailer sequence, coding sequence, intron, spacer sequence, nontranscribed spacer sequence, external transcribed spacer sequence, and internal transcribed sequence. Give examples of actual molecules in your answer.

13.6 Describe the organization of the ribosomal DNA repeating unit of a higher-eukaryotic cell.

13.7 In what way(s) is the principle of colinearity between gene and transcript violated in eukaryotes? In what sense is this principle maintained?

*13.8 Which of the following kinds of mutations would be likely to be recessive lethals in humans? Explain your reasoning.
  a. deletion of the U1 genes
  b. deletion within intron 2 of β-globin
  c. deletion of 4 bases at the end of intron 2 and 3 bases at the beginning of exon 3 in β-globin.

13.9 The diagram in Figure 13.A shows the transcribed region of a typical eukaryotic protein-coding gene:

FIGURE 13.A

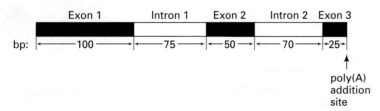

What is the size (in bases) of the fully processed, mature mRNA? Assume in your calculations a poly(A) tail of 200 As.

*13.10 Which of the following could occur in a single mutational event in a human? Explain.
  a. Deletion of 10 copies of the 5S ribosomal RNA genes only
  b. Deletion of 10 copies of the 18S rRNA genes only
  c. Simultaneous deletion of 10 copies of the 18S, 5.8S, and 28S rRNA genes only
  d. Simultaneous deletion of 10 copies each of the 18S, 5.8S, 28S, and 5S rRNA genes.

13.11 During DNA replication in a mammalian cell a mistake occurs: 10 wrong nucleotides are inserted into a 28S rRNA gene, and this mistake is not corrected. What will likely be the effect on the cell?

14 THE GENETIC CODE AND THE TRANSLATION OF THE GENETIC MESSAGE

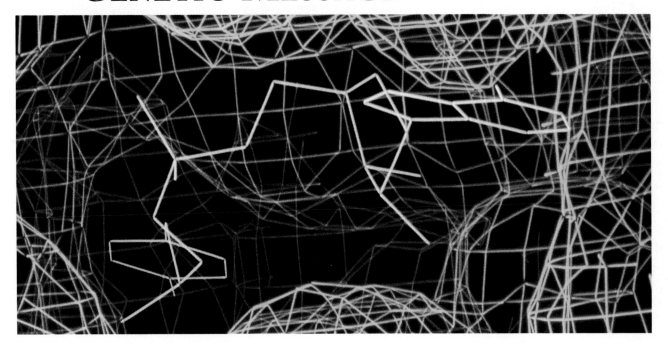

THE NATURE OF THE GENETIC CODE
Deciphering the Genetic Code
Nature and Characteristics of the Genetic Code

TRANSLATION OF THE GENETIC MESSAGE
Aminoacyl-tRNA Molecules
Initiation of Translation
Elongation of the Polypeptide Chain
Termination of Translation

PROTEIN TRANSPORT AND COMPARTMENTALIZATION
Proteins Distributed by the Endoplasmic Reticulum
Proteins Transported into Mitochondria and Chloroplasts
Proteins Transported into the Nucleus

PRINCIPAL POINTS

~ The genetic code is a triplet code in which each three-nucleotide codon in an mRNA specifies one amino acid. Some amino acids are represented by more than one codon. The code is almost universal, and it is read without gaps and in successive, nonoverlapping codons.

~ Translation of the mRNA into a protein chain occurs on ribosomes. Amino acids are brought to the ribosome on tRNA molecules. The correct amino acid sequence is achieved by the specific binding between the codon of the mRNA and the complementary anticodon of the tRNA, and by the specific binding of each amino acid to its specific tRNA.

~ In prokaryotes and eukaryotes, AUG (methionine) is the initiator codon for the start of translation. Elongation of the protein chain involves peptide bond formation between the amino acid attached to the tRNA in the A site of the ribosome and the growing polypeptide attached to the tRNA in the P site. Once the peptide bond has formed, the ribosome translocates one codon along the mRNA in preparation for the next tRNA with its bound amino acid to bind to the available codon.

~ Translation continues until a chain-terminating codon (UAG, UAA, or UGA) is reached in the mRNA. These codons are read by one or more release factor proteins and then the polypeptide is released from the ribosome and the other components of the protein synthesis machinery dissociate.

~ In eukaryotes, proteins are found free in the cytoplasm, as well as in the various cell compartments such as the nucleus, mitochondria, chloroplasts, and secretory vesicles. Proteins destined to enter the cell compartments have specialized amino acid sequences that indicate to the cellular machinery the path they should follow in the cell. For example, proteins to be secreted have N-terminal signal sequences that facilitate their entry into the endoplasmic reticulum for later sorting in the Golgi complex and beyond. Proteins destined for the nucleus, mitochondria, or chloroplasts each have specific sequences that program their localization to those compartments.

As we learned earlier, one of the three requirements for genetic material is that it must contain, in a stable form, all the information for an organism's structure, function, development, and reproduction. All these properties of an organism involve the action of specific proteins, and the information for the proteins is coded in the structural genes of the cell's genetic material. The expression of a protein-coding gene occurs in two major steps: transcription and translation. In **transcription** or **RNA synthesis** (discussed in Chapter 12), the base-pair sequence of the DNA segment constituting the gene is read, and that information is transferred to the base sequence of a single-stranded mRNA molecule. This base sequence carries the specific information for the amino acid sequence of a polypeptide. The conversion in the cell of the mRNA base sequence information into an amino acid sequence of a polypeptide is called **translation.** The DNA base-pair information that specifies the amino acid sequence of a polypeptide is called the **genetic code.**

In this chapter we will study the translation of the genetic message and how the information for the amino acid sequence of proteins is encoded in the nucleotide sequence of messenger RNA. We will see that three classes of RNA—messenger RNA, transfer RNA, and ribosomal RNA—are involved in the translation process.

THE NATURE OF THE GENETIC CODE

In Chapter 13 we learned that nucleotides in the mRNA molecule are read in groups of three (called **codons**) to specify the amino acid sequence in proteins. Since four different nucleotides (A, C, G, U) occur in the message and 20 different amino acids occur in proteins, we deduce that the genetic code for each amino acid must be at least three letters. If it were a one-letter code, only four amino acids could be encoded. If it were a two-letter code, then only $4 \times 4 = 16$ amino acids could be encoded. A three-letter code, however, generates $4 \times 4 \times 4 = 64$ possible codes, more than enough to code for the 20 amino acids. The assumption of a three-letter code also suggests that some amino acids may be specified by more than one codon, which is, in

~ FIGURE 14.1

Experimental details of a cell-free, protein-synthesizing system.

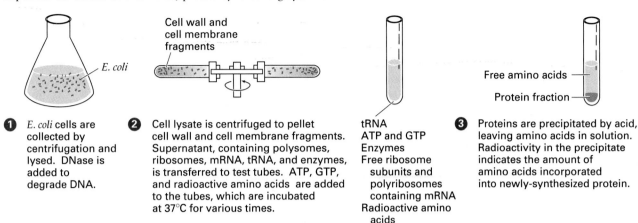

1 E. coli cells are collected by centrifugation and lysed. DNase is added to degrade DNA.

2 Cell lysate is centrifuged to pellet cell wall and cell membrane fragments. Supernatant, containing polysomes, ribosomes, mRNA, tRNA, and enzymes, is transferred to test tubes. ATP, GTP, and radioactive amino acids are added to the tubes, which are incubated at 37°C for various times.

tRNA
ATP and GTP
Enzymes
Free ribosome subunits and polyribosomes containing mRNA
Radioactive amino acids

3 Proteins are precipitated by acid, leaving amino acids in solution. Radioactivity in the precipitate indicates the amount of amino acids incorporated into newly-synthesized protein.

fact, the case. Experimental evidence for the three-letter genetic code was obtained by F. Crick, L. Barnett, S. Brenner, and R. Watts-Tobin in the early 1960s.

Deciphering the Genetic Code

The exact relationship of the 64 codons to the 20 amino acids was determined by experiments done mostly in the laboratories of Marshall Nirenberg and Ghobind Khorana. Essential to these experiments was the use of **cell-free, protein-synthesizing systems,** which contained ribosomes, tRNAs with amino acids attached, and all the necessary protein factors for polypeptide synthesis. These cell-free, synthesizing systems were assembled from components isolated and purified from *Escherichia coli* (Figure 14.1). Protein synthesis in these systems is inefficient, in that very little protein is made. Therefore, radioactively labeled amino acids have to be used to measure the incorporation of amino acids into new proteins.

Figure 14.2 shows the typical time course for the incorporation of radioactive amino acids into proteins in an *in vitro* system. Protein synthesis takes place rapidly for about 15 minutes, then stops gradually because the mRNA that was in the E. coli extract (the endogenous mRNA) degrades. If fresh mRNA is added, protein synthesis resumes and proceeds until the new mRNA (the exogenous mRNA) is degraded. Thus, once the E. coli extract has been made and been incubated to allow the system's original endogenous mRNA to degrade, it becomes an excellent system for testing the coding capacity of synthetic mRNAs. The system's usefulness lies in the fact that a single type of mRNA molecule can be added to the system, enabling the investigators to analyze the result of this addition on the composition of the polypeptide produced. As we shall

see, the mRNAs used were synthetically made in the laboratory; their relative simplicity generated valuable information about the relationships between codons and amino acids that would not have been possible with the very complex populations of natural mRNAs.

~ FIGURE 14.2

Typical time course for the incorporation of radioactive amino acids into proteins in an *in vitro* system. Initially radioactivity is incorporated into proteins until the endogenous mRNA breaks down. Incorporation of radioactive amino acids into proteins resumes when exogenous mRNA is added and continues until that mRNA breaks down. (From Watson, et al., *Molecular biology of the gene.* vols. I and II. 4th ed. Copyright 1987 Benjamin Cummings Publishing Company, Inc. Reprinted by permission.)

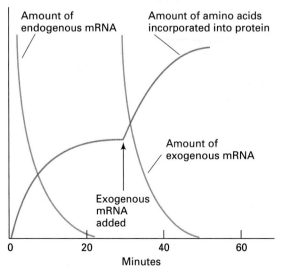

In one approach to establish which codons specify which amino acids, synthetic mRNAs containing one, two, or three different types of bases were made and added to cell-free, protein-synthesizing systems. The polypeptides made in these systems were then analyzed. When the synthetic mRNA contained only one type of base, the results were unambiguous. Synthetic poly(U) mRNA, for example, directed the synthesis of a polypeptide consisting of phenylalanines (a polyphenylalanine chain). Since the genetic code is a triplet code, this result indicated that UUU is a codon for phenylalanine. Similarly, a synthetic poly(A) mRNA directed the synthesis of a polylysine, and poly(C) directed the synthesis of polyproline, indicating that AAA is a codon for lysine and CCC is a codon for proline. The results from poly(G) were inconclusive since the poly(G) folds up upon itself, so it cannot be translated *in vitro*.

Synthetic mRNAs made by the random incorporation of two different bases (called *random copolymers*) were also analyzed in the cell-free, protein-synthesizing systems. When mixed copolymers are made, the bases are incorporated into the synthetic molecule in a random way. Thus poly(AC) molecules can contain eight different codons (CCC, CCA, CAC, ACC, CAA, ACA, AAC, and AAA), and the proportions of each depend on the A-to-C ratio used to make the polymer. In the cell-free, protein-synthesizing system, poly(AC) synthetic mRNAs caused the incorporation of asparagine, glutamine, histidine, and threonine into polypeptides, in addition to the lysine expected from AAA codons and the proline expected from CCC codons. The proportions of asparagine, glutamine, histidine, and threonine incorporated into the polypeptides produced depended on the A/C ratio used to make the mRNA, and these observations were used to deduce information about the codons that specify the amino acids. For example, since an AC random copolymer containing much more A than C resulted in the incorporation of many more asparagines than histidines, researchers concluded that asparagine is coded by two As and one C and histidine by two Cs and one A. With experiments of this kind, the base composition (*but not the base sequence*) of the codons for a number of amino acids was determined.

Another experimental approach also used copolymers, but these copolymers had been synthesized so that they had a known sequence, not a random one. For example, a repeating copolymer of U and C gives a synthetic mRNA of UCUCUCUCUC. When this copolymer is tested in a cell-free, protein-synthesizing system, the resulting polypeptide had a repeating amino acid pattern of leucine-serine-leucine-serine. . . . From this result, researchers conclude that UCU and CUC specify leucine and serine, although they could not determine from this information alone which coded for which.

Yet another approach used a *ribosome-binding assay,* developed in 1964 by Nirenberg and Philip Leder. This assay depended on the fact that, in the absence of protein synthesis, specific tRNA molecules will bind to complexes formed between ribosomes and mRNAs. For example, when synthetic mRNA poly(U) is mixed with ribosomes, it forms a poly(U)-ribosome complex, and only tRNA.Phe (i.e., the tRNA that will bring phenylalanine to an mRNA and that has the appropriate anticodon, AAA, for the UUU codon) will bind to the UUU codon. Importantly, the specific binding of the appropriate tRNA to the mRNA-ribosome complex does not require the presence of long mRNA molecules, since the binding of a trinucleotide (i.e., a codon-length piece of RNA) will suffice. The discovery of the trinucleotide-binding property made it possible to determine relatively easily the specific relationships between many codons and the amino acids for which they code. Note that in this particular approach, *the specific nucleotide sequence of the codon is determined*, not merely the nucleotide composition. With this approach, many of the ambiguities that had arisen from other approaches were resolved. For example, UCU was found to promote the binding of a tRNA.Ser, that is, a tRNA that brings the amino acid serine to the message. Thus UCU must code for serine. Similarly, CUC causes tRNA.Leu binding, so CUC codes for leucine. All in all, about 50 codons were clearly identified by using this approach.

In sum, no single approach produced an unambiguous set of codon assignments, but information obtained through all the approaches enabled all codon assignments to be deduced with high degrees of certainty.

Nature and Characteristics of the Genetic Code

From the types of experiments just described, all 64 codons were assigned with the result shown in Figure 14.3. Note that each codon is written as it appears in mRNA and reads in a 5′-to-3′ direction. The characteristics of the genetic code are as follows:

1. *The code is a triplet code.* Each mRNA codon that specifies an amino acid in a polypeptide chain consists of three nucleotides.
2. *The code is comma-free.* The mRNA is read continuously, three nucleotides (one codon) at a time, without skipping any nucleotides of the message.
3. *The code is nonoverlapping.* The mRNA is read in successive groups of three nucleotides. A message of AAGAAGAAG . . . in the cell would be read as lysine-lysine-lysine . . . ,which is what the AAG specifies. Theoretically, three readings are possible from this message, depending on where the reading

~ FIGURE 14.3

The genetic code. Of the 64 codons, 61 specify 20 amino acids. One of those 61 codons, AUG, is used in the initiation of protein synthesis. The other 3 codons are chain-terminating codons and do not specify any amino acid.

Second letter

First letter	U	C	A	G	Third letter
U	UUU Phe / UUC / UUA Leu / UUG	UCU / UCC Ser / UCA / UCG	UAU Tyr / UAC / UAA Stop / UAG Stop	UGU Cys / UGC / UGA Stop / UGG Trp	U C A G
C	CUU / CUC Leu / CUA / CUG	CCU / CCC Pro / CCA / CCG	CAU His / CAC / CAA Gln / CAG	CGU / CGC Arg / CGA / CGG	U C A G
A	AUU / AUC Ile / AUA / AUG Met	ACU / ACC Thr / ACA / ACG	AAU Asn / AAC / AAA Lys / AAG	AGU Ser / AGC / AGA Arg / AGG	U C A G
G	GUU / GUC Val / GUA / GUG	GCU / GCC Ala / GCA / GCG	GAU Asp / GAC / GAA Glu / GAG	GGU / GGC Gly / GGA / GGG	U C A G

■ = Chain termination codon (stop)
▨ = Initiation codon

is begun (i.e., the *reading frame* in which message is read); namely, the repeating AAG, the repeating AGA, and the repeating GAA. Later in this chapter we will examine the mechanisms in the cell that ensure that the translation of the genetic code in an mRNA begins at the correct point.

4. *The code is almost universal.* All organisms share the same genetic language. Thus, for example, lysine is coded for by AAA or AAG in the mRNA of all organisms, arginine by CGU, CGC, CGA, CGG, AGA, and AGG, and so on. Hence we can isolate an mRNA from one organism, translate it by using the machinery isolated from another organism, and produce the protein as if it had been translated in the original organism. The code, however, is not completely universal. For example, the mitochondria of some organisms, such as mammals, have minor changes in the code, as does the nuclear genome of the protozoan, *Tetrahymena* (discussed in Chapter 21).

5. *The code is degenerate.* With two exceptions [AUG (methionine) and UGG (tryptophan)], more than one codon occurs for each amino acid. This multiple coding is called the **degeneracy** of the code. A

close examination of Figure 14.3 reveals particular patterns in this degeneracy. Thus, when the first two nucleotides in a codon are identical and the third letter is U or C, the codon often codes for the same amino acid. For example, UUU and UUC specify phenylalanine; similarly, CAU and CAC specify histidine. Also, when the first two nucleotides in a codon are identical and the third letter is A or G, the same amino acid is often specified. For example, UUA and UUG specify leucine, and AAA and AAG specify lysine. In some cases, with identity in the first two positions the base in the third position may be U, C, A, or G and the same amino acid will be specified. An example is CUU, CUC, CUA, and CUG, all for leucine.

Even though there is degeneracy of the code, that does not mean that all codons are used equally. Studies have shown that codon usage is not random. Rather, some codons are used repeatedly, while others are almost never used. Figure 14.4 (p. 414) illustrates the results of a study of a number of genes in animals. This study showed, for example, a significant bias toward the use of certain codons for particular amino acids. For example, UUC is the more frequently used codon for phenylalanine, followed by UUU.

6. *The code has start and stop signals.* Specific start and stop signals for protein synthesis are contained in the code. In both eukaryotes and prokaryotes, AUG (methionine) is most commonly used as a start codon for protein synthesis, although in rare cases GUG may also be used. Thus in examining the sequence of a particular mRNA for the location of the amino acid–coding sequence, we look near the 5′ end for the occurrence of the AUG start codon and begin reading the amino acid sequence from there.

As Figure 14.3 shows, only 61 of the 64 codons specify amino acids; these codons are called the **sense codons.** The other three codons—UAG (amber), UAA (ochre), and UGA (opal)—do not specify an amino acid, and no tRNAs in normal cells carry the appropriate anticodons. These three codons are the **stop codons;** also called **nonsense codons,** or **chain-terminating codons.** They are used singly or in tandem groups (UAG UAA, for example) to specify the end of the translation process of a polypeptide chain. Again, when we read a particular mRNA sequence, we look for the presence of a stop codon *in the same reading frame* as the AUG start codon to determine where the amino acid–coding sequence for the polypeptide ends. We will examine starting and stopping protein synthesis when we study the details of the translation process.

~ FIGURE 14.4

Codon usage in a number of genes of animals. The numbers next to the codons are the instances of the codons in 1001 codons tabulated.

Second letter

	U	C	A	G	
U	13 UUU Phe 28 UUC Phe 2 UUA Leu 9 UUG Leu	16 UCU 18 UCC 9 UCA Ser 2 UCG	10 UAU Tyr 23 UAC Tyr UAA Stop UAG Stop	10 UGU Cys 13 UGC Cys UGA Stop 12 UGG Trp	U C A G
C	9 CUU 27 CUC Leu 7 CUA 47 CUG	14 CCU 17 CCC Pro 10 CCA 5 CCG	10 CAU His 21 CAC His 10 CAA Gln 28 CAG Gln	8 CGU 11 CGC Arg 4 CGA 5 CGG	U C A G
A	11 AUU 24 AUC Ile 4 AUA 16 AUG Met	15 ACU 28 ACC Thr 11 ACA 6 ACG	8 AAU Asn 28 AAC Asn 19 AAA Lys 49 AAG Lys	12 AGU Ser 21 AGC Ser 8 AGA Arg 10 AGG Arg	U C A G
G	9 GUU 21 GUC Val 5 GUA 33 GUG	28 GCU 38 GCC Ala 14 GCA 6 GCG	16 GAU Asp 24 GAC Asp 21 GAA Glu 34 GAG Glu	22 GGU 32 GGC Gly 16 GGA 11 GGG	U C A G

First letter (left) · Third letter (right)

7. *Wobble occurs in the anticodon.* Since 61 sense codons specify amino acids in mRNA, a total of 61 tRNA molecules could have the appropriate anticodons. Theoretically, though, the complete set of 61 sense codons can be read by fewer than 61 distinct tRNAs because of the wobble in the anticodon. The **wobble hypothesis,** proposed by Francis Crick, is shown in Table 14.1. Sequence analysis had shown that the base at the 5′ end of the anticodon (complementary to the base at the 3′ end of the codon; i.e., the third letter) is not as constrained as the other two bases. This feature allows for less exact base pairing so that the base at the 5′ end of the anticodon can potentially pair with one of three different bases at the 3′ end of the codon; it can wobble. As Table 14.1 shows, no single tRNA molecule can recognize four different codons. But if the tRNA molecule contains the modified nucleoside inosine (see Figure 13.14) at the 5′ end of the anticodon, then three different codons can be read by that one tRNA. Figure 14.5a gives an example of how a single leucine tRNA can read two different leucine codons by base-pairing wobble, and Figure 14.5b shows how a single glycine tRNA can read three different glycine codons by wobble pairing.

KEYNOTE

The genetic code is a triplet code in which each codon (a contiguous set of three bases) in an mRNA specifies one amino acid. Since 64 code words are possible and since 20 amino acids exist, some amino acids are specified by more than one codon. The genetic code in mRNA is read without gaps and in successive, nonoverlapping codons. The code is universal: The same codons specify the same amino

~ TABLE 14.1

Wobble in the Genetic Code

NUCLEOTIDE AT 5′ END OF ANTICODON		NUCLEOTIDE AT 3′ END OF CODON
G	can pair with	U or C
C	can pair with	G
A	can pair with	U
U	can pair with	A or G
I (inosine)	can pair with	A, U, or C

~ FIGURE 14.5

Example of base-pairing wobble. (a) Two different leucine codons (CUC, CUU) can be read by the same leucine tRNA molecule, contrary to regular base-pairing rules. (b) Three different glycine codons (GGU, GGC, GGA) can be read by the same glycine tRNA molecule by base-pairing wobble involving inosine in the anticodon.

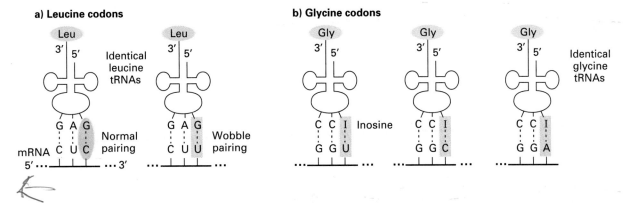

acids in most systems. Protein synthesis is typically initiated by codon AUG (methionine), and it is terminated by three codons singly or in combination: UAG, UAA, UGA (these codons do not code for any amino acid). _____

TRANSLATION OF THE GENETIC MESSAGE

Since we now know the types of RNA made in the cell, the structure of ribosomes (from Chapter 13), and the genetic code, we are ready to examine the details of the translation of the genetic message (also called protein synthesis). In brief, protein synthesis takes place on ribosomes, where the genetic message encoded in mRNA is translated. The mRNA molecule is translated in the 5'-to-3' direction (the same direction in which it is made), and the polypeptide is made in the N-terminal to C-terminal direction. Amino acids are brought to the ribosome bound to tRNA molecules. The correct amino acid sequence is achieved as a result of (1) the specific binding between the codon of the mRNA and the complementary anticodon in the tRNA, and (2) the specific binding of each amino acid to its own specific tRNA.

That the mRNA codon recognizes the tRNA anticodon and not the amino acid carried by the tRNA was proved by G. von Ehrenstein, B. Weisblum, and S. Benzer. They attached cysteine to tRNA.Cys *in vitro*; then they chemically converted the attached cysteine to alanine. The resulting alanyl-tRNA.Cys, was used in the *in vitro* synthesis of hemoglobin. *In vivo*, the α and β chains of hemoglobin each contain one cysteine.

When the hemoglobin made *in vitro* was examined, however, the amino acid alanine was found in both chains at the positions normally occupied by cysteine. This result could only mean that the alanyl-tRNA.Cys had read the cysteine codon and had inserted the amino acid it carried, in this case alanine. Therefore, the researchers concluded that the *specificity of codon recognition lies in the tRNA molecule and not in the amino acid it carries.*

The three basic stages of protein synthesis—*initiation, elongation,* and *termination*—are similar in prokaryotes and eukaryotes. In the following sections we will discuss each of these stages in turn, concentrating, as before, on the processes in *E. coli*, with which most work has been done. In the discussions we will note where significant differences in translation occur in prokaryotes and eukaryotes.

Aminoacyl-tRNA Molecules

The important function of tRNA molecules in protein synthesis is that they bring specific amino acids to the mRNA-ribosomal complex so that the correct polypeptide chain can be assembled. In this section we will see how an amino acid becomes attached to its appropriate tRNA molecule to produce an aminoacyl-tRNA molecule.

ATTACHMENT OF AMINO ACID TO TRNA. The correct amino acid is attached to the tRNA by an enzyme called an **aminoacyl-tRNA synthetase**. Since 20 different amino acids exist, 20 types of aminoacyl-tRNA synthetases also exist. Further, since degeneracy is common in the genetic code (i.e., a single amino acid may be specified by more than one codon), all the tRNAs that are specific for a particular amino acid must have a common recognition site or identity site for the aminoacyl-tRNA synthetase for that amino acid. The aminoacyl-tRNA synthetases themselves are highly specific in their function, so few errors are made in the bonding of an amino acid to its tRNA.

In general, the sequence differences between the tRNAs specify which enzyme will recognize the tRNA. However, no single rule exists for recognition by the aminoacyl tRNA synthetases. For *E. coli* tRNA.Val and tRNA.Met, the recognition site is the anticodon itself, but this obvious location is not the recognition site for all tRNAs. In other *E. coli* tRNAs, the recognition sites have been localized to various positions, for example to a single base pair in the acceptor stem (the stem to which the amino acid attaches) or to a combination of nucleotides in different parts of a tRNA.

Figure 14.6 shows how an amino acid is attached to a tRNA molecule to produce an **aminoacyl-tRNA,** in this case seryl-tRNA, and presents the generalized structure of an aminoacyl-tRNA molecule. The amino acid attaches at the 3′ end of the tRNA by a linkage between the carboxyl group of the amino acid and the 3′-OH or 2′-OH group of the ribose of the adenine nucleotide found at the end of every tRNA (the amino acid is attached to the 3′-OH in the figure). The act of adding the amino acid to the tRNA is called **charging,** and the product is commonly referred to as a **charged tRNA.**

$\mathcal{K}$EYNOTE

Protein synthesis occurs on ribosomes, where the genetic message encoded in mRNA is translated. Amino acids are brought to the ribosome on charged tRNA molecules. The correct amino acid sequence is achieved as a result of (1) the specific binding between the codon of the mRNA and the complementary anticodon in the tRNA and (2) the specific binding of each amino acid to its own specific tRNA. ─────────

Initiation of Translation

INITIATOR CODON AND INITIATOR tRNA. In both prokaryotes and eukaryotes, translation begins at the AUG initiator codon in the mRNA, which specifies methionine. As a result, newly made proteins in both types of organisms begin with methionine; in some cases this methionine is subsequently removed.

Prokaryotes. In prokaryotes, the initiator methionine is actually a specially modified form of methionine, called **formylmethionine** (abbreviated fMet), in which a formyl group has been added to the methionine's amino group. The fMet is brought to the ribosome attached to a special tRNA, called tRNA.fMet, which has the anticodon 5′-CAU-3′. This tRNA is special, since it is involved specifically with the initiation process of protein synthesis. The aminoacylation of tRNA.fMet occurs as follows: First, methionyl-tRNA synthetase catalyzes the addition of methionine to the tRNA. Then an enzyme called *transformylase* catalyzes the addition of the formyl group to the methionine. The resulting molecule is designated fMet-tRNA.fMet (this nomenclature indicates that the tRNA is specific for the attachment of fMet and that, in fact, fMet is attached to it).

The tRNA.fMet molecule is only used in the initiation of translation. When an AUG codon in an mRNA molecule is encountered at a position other than at the start of the amino acid–coding sequence, another species of tRNA is used to insert methionine at that point in the polypeptide chain. This tRNA is called tRNA.Met, and it is aminoacylated by the same aminoacyl-tRNA synthetases as is tRNA.fMet to produce Met.tRNA.Met. However, tRNA.Met and tRNA.fMet are coded for by different genes and have different primary nucleotide sequences.

Eukaryotes. In eukaryotes, protein synthesis initiation occurs in much the same way. That is, AUG is the usual initiation codon, and hence the N-terminal amino acid in polypeptides is methionine. The N-terminal methionine is not formylated since the enzymes needed to carry out this function do not exist in the eukaryotic cytoplasm. Nonetheless, a special methionine tRNA is used for initiation of translation in eukaryotes, and a distinct species of tRNA.Met is used to read AUG codons elsewhere in an mRNA molecule.

INITIATION SEQUENCE. In addition to the AUG intiation codon, the initiation of protein synthesis in prokaryotes requires other information coded in the base sequence of an mRNA molecule upstream (to the 5′ side) of the initiation codon. These sequences serve to align the ribosome on the message in the proper reading frame so that polypeptide synthesis can proceed correctly. To identify these aligning sequences, researchers sequenced a number of mRNAs in the region upstream from the AUG initiation codon. To do so, they allowed ribosomal subunits to bind to the mRNA under conditions in which protein synthesis could not commence. Under these conditions the 30S ribosomal subunits freeze on the mRNA message at the **ribosome-binding site,** the site at which the ribosome

~ FIGURE 14.6

Molecular details of the attachment of an amino acid to a tRNA molecule. (a) In the first step the amino acid (serine here) reacts with ATP to produce an aminoacyl-AMP complex. This reaction is catalyzed by an aminoacyl-tRNA synthetase (seryl-tRNA synthetase). (b) In the second step (also catalyzed by aminoacyl-tRNA synthetase) the aminoacyl-AMP complex then reacts with the appropriate tRNA molecule (tRNA.Ser) to produce an aminoacyl-tRNA (Ser-tRNA.Ser). (c) In the general molecular structure of an aminoacyl-tRNA molecule (charged tRNA), the carboxyl group of the amino acid is attached to the 3'-OH or 2'-OH group of the 3' terminal adenine nucleotide of the tRNA.

a) First step

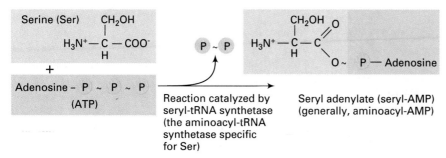

b) Second step

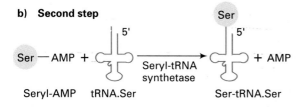

c) General structure

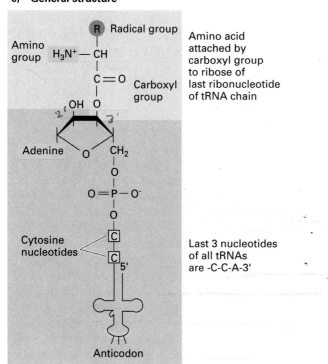

~ FIGURE 14.7

Sequences of some prokaryotic ribosome-binding sites. The initiation codon, AUG, is shown in yellow. (Note that GUG can be used as an initiator codon in some systems.) The purple regions indicate the regions of contiguous complementarity between mRNA and the 3' end of 16S rRNA.

		Binding site sequences					Initiation codon			
Phage R17 A protein	UCC	UAG	GAG	GUU	UGA	CCU	AUG	CGA	GCU	UUU
Phage Qβ replicase	UAA	CUA	AGG	AUG	AAA	UGC	AUG	UCU	AAG	ACA
Phage λ Cro	AUG	UAC	UAA	GGA	GGU	UGU	AUG	GAA	CAA	CGC
Phage φX174 A	AAU	CUU	GGA	GGC	UUU	UUU	AUG	GUU	CGU	UCU
E. coli trpB	AUA	UUA	AGG	AAA	GGA	ACA	AUG	ACA	ACA	UUA
E. coli lacZ	UUC	ACA	CAG	GAA	ACA	GCU	AUG	ACC	AUG	AUU
E. coli RNA polymerase β	AGC	GAG	CUG	AGG	AAC	CCU	AUG	GUU	UAC	UCC

becomes oriented in the correct reading frame for the initiation of protein synthesis. When the subunits are bound to the mRNAs, the message is protected from RNase attack, and the protected regions can be isolated and studied.

Figure 14.7 shows the sequences of some prokaryotic ribosome-binding sites. The AUG codon can be clearly identified. Most of the binding sites have a purine-rich sequence about 8 to 12 nucleotides upstream from the initiation codon. Evidence from the work of J. Shine and L. Dalgarno indicates that this purine-rich sequence, and other nucleotides in this region, are complementary to a pyrimidine-rich region (which always contains the sequence CCUCC) at the 3' end of 16S rRNA (Figure 14.8). The mRNA region that

binds in this way has become known as the **Shine-Dalgarno sequence**. Apparently the formation of complementary base pairs between the mRNA and 16S rRNA allows the ribosome to locate the true initiator regions of the message.

Eukaryotic rRNAs do not have the CCUCC sequence found in prokaryotic 16S rRNA. Instead, the eukaryotic ribosome uses another way to initiate protein synthesis on mRNA. First, a **eukaryotic initiator factor** eIF4A, a multimer of several proteins including the *cap binding protein* (CBP), recognizes and binds to the cap at the 5' end of the mRNA (see Chapter 13). The 40S ribosomal subunit then binds, along with other eIFs, and migrates along the mRNA, scanning for the first AUG codon, to which it binds. The 60S ribosomal subunit then binds, displacing the eIFs, and protein synthesis is initiated at the AUG codon.

~ FIGURE 14.8

Sequences involved in the binding of ribosomes to the mRNA in the initiation of protein synthesis in prokaryotes. (a) Nucleotide sequence at the 3' end of E. coli 16S rRNA; (b) Example of how the 3' end of 16S rRNA can base-pair with the nucleotide sequence 5' upstream from the AUG initiation codon.

a) **Sequence at 3' end of 16S rRNA**

3'...AUUCCUCCAUAG...5'

b) **Example of mRNA leader and 16S rRNA pairing**

Shine-Dalgarno sequence   Initiation codon

5'...AUGUACUAAGGAGGUUGUAUGGAACAACGC...3'
                |||||||||
          3'...AUUCCUCCAUAG...5'
              16S rRNA 3' end

FORMATION OF THE INITIATION COMPLEX.

Prokaryotes. As Figure 14.9 shows, translation in *E. coli.* commences with the formation of a 30S initiation complex. At the beginning of this process, three protein **initiation factors**, IF1, IF2, and IF3, are bound to the 30S ribosomal subunit along with a molecule of GTP (guanosine triphosphate). The fMet-tRNA.fMet and the mRNA then attach to the 30S-IF-GTP complex to form the *30S initiation complex.* IF3 is released as a result of this process. Next, the 50S subunit binds, leading to GTP hydrolysis and the release of IF1 and IF2. The final complex is called the *70S initiation complex.* All three IF molecules are recycled for use in other initiation reactions. The 70S ribosome has two binding sites for aminoacyl-tRNA, the peptidyl (P), and aminoacyl (A) sites; the fMet-tRNA.fMet is bound to the mRNA in the P site.

~ FIGURE 14.9

Initiation of protein synthesis in prokaryotes. A 30S ribosomal subunit, complexed with initiation factors and GTP, binds to mRNA and fMet-tRNA.fMet to form a 30S initiation complex. Next, the 50S ribosomal subunit binds, forming a 70S initiation complex. During the latter event, the initiation factors are released and GTP is hydrolyzed.

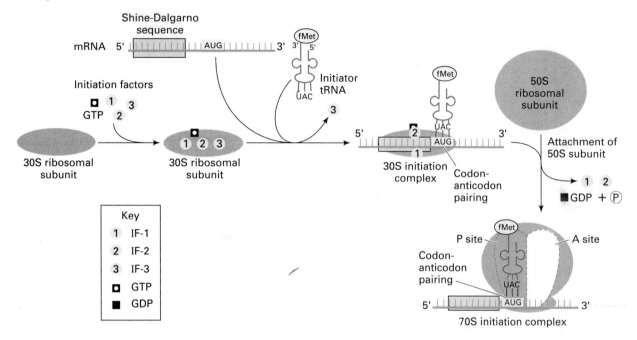

Eukaryotes.

Eukaryotes. The formation of the eukaryotic 80S initiation complex is generally similar to that of prokaryotes, except that the methionine on the initiator tRNA is not formylated, and more initiation factors (*eukaryotic initiation factors, or eIFs*) are involved. At least nine eIFs have been identified in mammalian cells. The initiator AUG codon is recognized first by the 40S ribosomal subunit scanning from the 5' cap to the first AUG. The 60S subunit then binds to form the *80S initiation complex.*

Elongation of the Polypeptide Chain

After the initiation events have been completed, the elongation phase of translation begins. This phase has three steps:

1. the binding of aminoacyl-tRNA to the ribosome;
2. the formation of a peptide bond; and
3. the movement (translocation) of the ribosome along the mRNA, one codon at a time. Figure 14.10 (p. 420) diagrams these events.

BINDING OF AMINOACYL-tRNA. At the outset of the elongation phase, in the 70S initiation complex, the fMet-tRNA.fMet is hydrogen-bonded to the AUG initiation codon in the peptidyl (P) site of the ribosome.

fMet-tRNA.fMet is the only tRNA to bind before the complete ribosome is formed. The orientation of this tRNA-codon complex exposes the next codon in the mRNA in the aminoacyl (A) site (Figure 14.10,1). In Figure 14.10 this codon (UCC) specifies serine (Ser).

The next step in the translation process is the binding of the appropriate aminoacyl-tRNA (in this case Ser-tRNA.Ser) to the newly exposed codon (Figure 14.10,2). The aminoacyl-tRNA is brought to the ribosome complexed with the protein elongation factor EF-Tu and a molecule of GTP. The association of the aminoacyl-tRNA with the codon in the aminoacyl site is accompanied by the hydrolysis of GTP to GDP and Ⓟ, and the release of EF-Tu to which the GDP is bound.

The EF-Tu is recycled in the following steps. First, a second elongation factor, EF-Ts, binds to EF-Tu and displaces the GDP. Next, GTP binds to the EF-Tu–EF-Ts complex to generate an EF-Tu–GTP complex simultaneously with the release of EF-Ts. The aminoacyl-tRNA binds to the EF-Tu–GTP, and that complex can then, if the correct codon-anticodon pairing forms, bind to the A site in the ribosome.

PEPTIDE BOND FORMATION. In Figure 14.10,2 fMet-tRNA.fMet is bound in the P site, and the second specified aminoacyl-tRNA (in this case Ser-tRNA.Ser)

~ FIGURE 14.10

Elongation stage of polypeptide synthesis in prokaryotes.

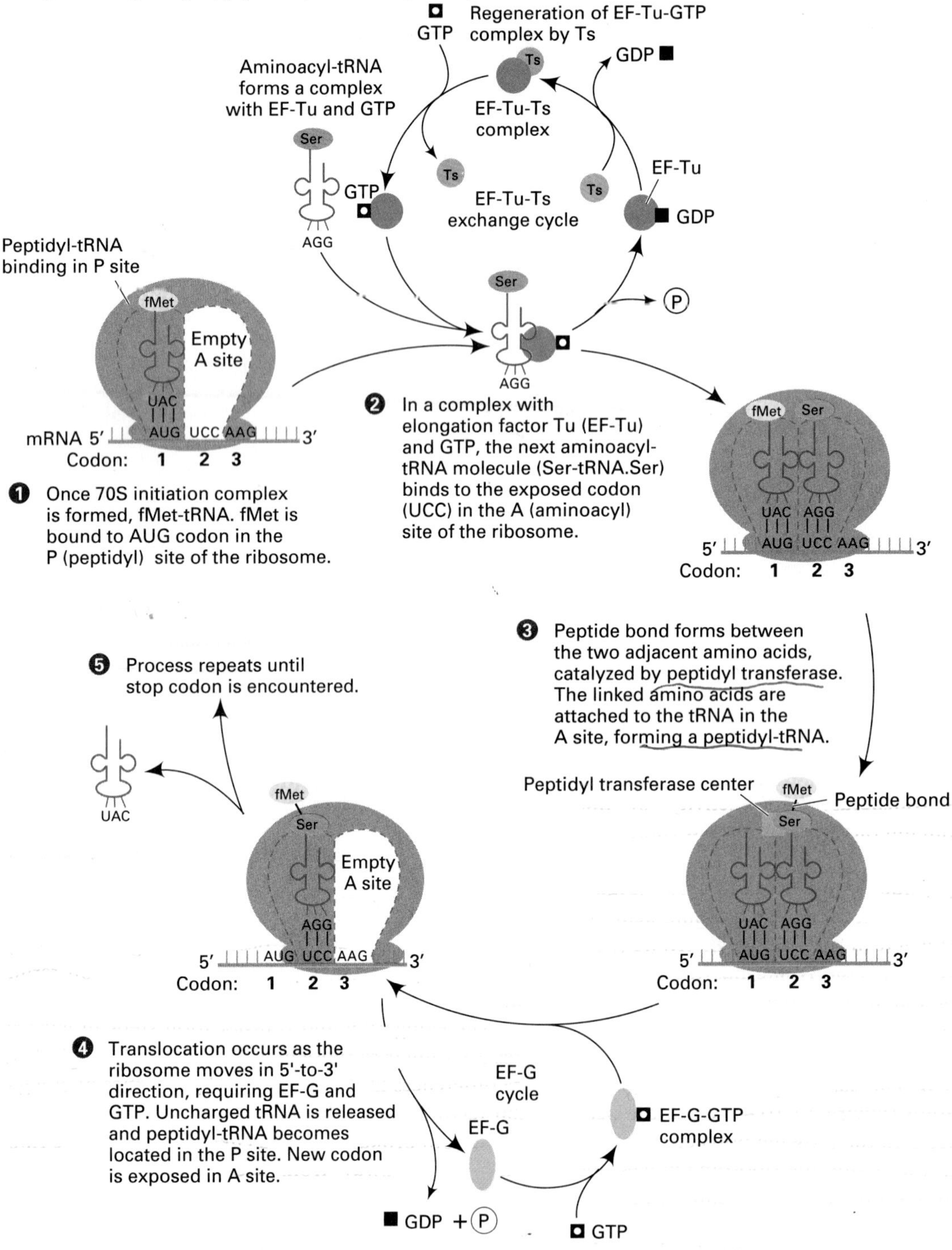

GTP
Regeneration of EF-Tu-GTP complex by Ts
GDP
Aminoacyl-tRNA forms a complex with EF-Tu and GTP
Ser
EF-Tu-Ts complex
Ts
GTP
EF-Tu-Ts exchange cycle
EF-Tu
GDP
AGG
Peptidyl-tRNA binding in P site
fMet
Empty A site
UAC
mRNA 5′ AUG UCC AAG 3′
Codon: 1 2 3
P
1 Once 70S initiation complex is formed, fMet-tRNA. fMet is bound to AUG codon in the P (peptidyl) site of the ribosome.
2 In a complex with elongation factor Tu (EF-Tu) and GTP, the next aminoacyl-tRNA molecule (Ser-tRNA.Ser) binds to the exposed codon (UCC) in the A (aminoacyl) site of the ribosome.
fMet Ser
UAC AGG
5′ AUG UCC AAG 3′
Codon: 1 2 3
3 Peptide bond forms between the two adjacent amino acids, catalyzed by peptidyl transferase. The linked amino acids are attached to the tRNA in the A site, forming a peptidyl-tRNA.
5 Process repeats until stop codon is encountered.
UAC
Peptidyl transferase center
Peptide bond
fMet
Ser
Empty A site
AGG
5′ AUG UCC AAG 3′
Codon: 1 2 3
4 Translocation occurs as the ribosome moves in 5′-to-3′ direction, requiring EF-G and GTP. Uncharged tRNA is released and peptidyl-tRNA becomes located in the P site. New codon is exposed in A site.
EF-G cycle
EF-G
EF-G-GTP complex
GDP + P
GTP

is bound in the A site. The ribosome maintains the two aminoacyl-tRNAs in the correct configuration so that a peptide bond can form between the two amino acids (Figure 14.10,3).

The two steps involved in the formation of a peptide bond are shown in Figure 14.11. The first step is the breakage of the bond between the carboxyl group of the amino acid and the tRNA in the P site. In this case the breakage is between fMet and its tRNA. The second step is the formation of the peptide bond between the now-freed fMet and the Ser attached to the tRNA in the A site. This reaction is catalyzed by the

~ FIGURE 14.11

The formation of a peptide bond between the first two amino acids (fMet and Ser) of a polypeptide chain is catalyzed on the ribosome by peptidyl transferase. (a) Adjacent aminoacyl-tRNAs bound to the mRNA at the ribosome; (b) Following peptide bond formation, an uncharged tRNA is in the P site and a dipeptidyl-tRNA is in the A site.

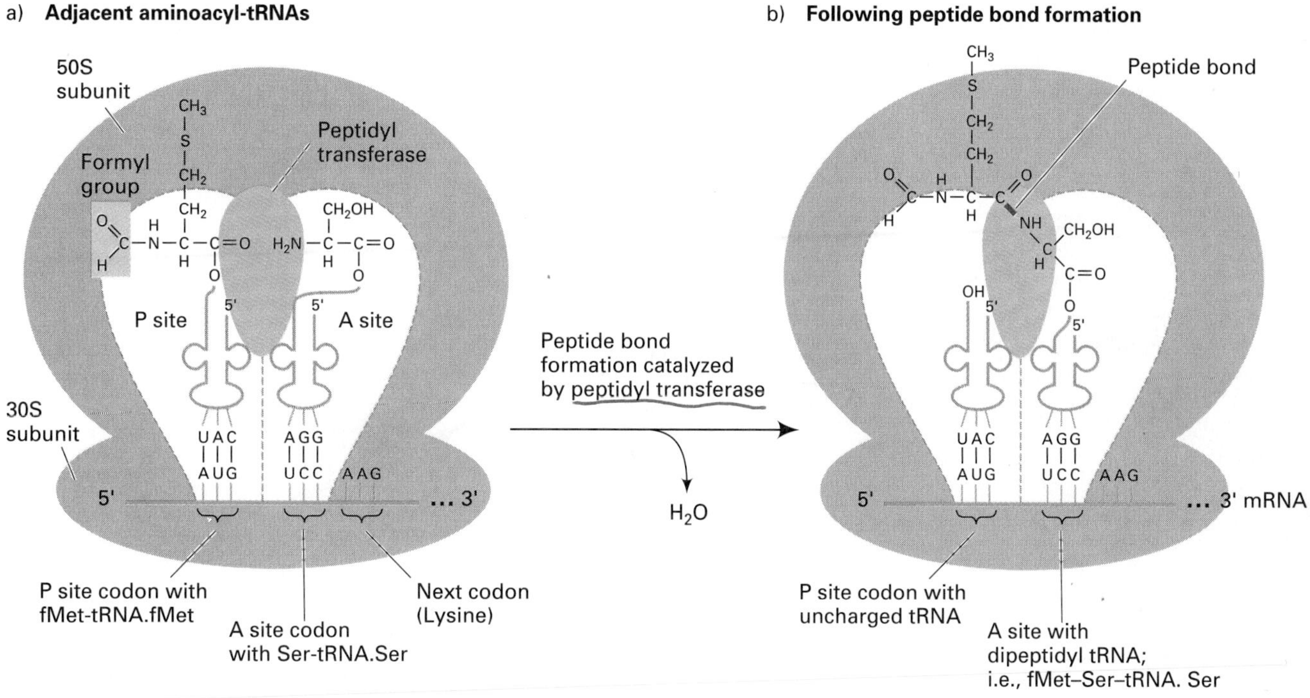

a) **Adjacent aminoacyl-tRNAs**

b) **Following peptide bond formation**

enzyme **peptidyl transferase.** The activity of this enzyme results from interaction among a few ribosomal proteins of the 50S ribosomal subunit.

Once the peptide bond has formed (Figure 14.10,3), a tRNA without an attached amino acid (an uncharged tRNA) is left in the P site. The tRNA (now called peptidyl-tRNA) in the A site has the first two amino acids of the polypeptide chain attached to it, in this case fMet-Ser.

TRANSLOCATION. The last step in the elongation cycle is **translocation** (Figure 14.10,4). Once the peptide bond is formed and the growing polypeptide chain is on the tRNA in the A site, the ribosome moves one codon along the mRNA toward the 3' end. Translocation requires the activity of another protein elongation factor, EF-G. An EF-G–GTP complex binds to the ribosome, and translocation then takes place along with ejection of the uncharged tRNA from the P site. The EF-G is then released in a reaction requiring GTP hydrolysis; EF-G can then be reused. During the translocation step the peptidyl-tRNA remains attached to its codon. And since the ribosome has moved, the peptidyl-tRNA is now located in the P site (hence the name,

the *peptidyl* site). The exact mechanism for the physical translocation of the ribosome is not known.

After translocation is completed, the A site is vacant. An aminoacyl-tRNA with the correct anticodon binds to the newly exposed codon in the A site, using the process already described (Figure 14.10,5). The whole process is repeated until translation terminates at a stop codon.

The elongation and translocation steps in eukaryotes are similar to those in prokaryotes, although there are differences in the number and properties of elongation factors and in the exact sequences of events.

In both prokaryotes and eukaryotes, once the ribosome moves away from the initiation site on the mRNA, the initiation site is open for another initiation event to occur. Thus many ribosomes may simultaneously be translating each mRNA. The complex between an mRNA molecule and all the ribosomes that are translating it simultaneously is called a **polyribosome** or **polysome** (Figure 14.12; p. 422). An average mRNA may have eight to ten ribosomes synthesizing protein from it. Simultaneous translation enables a large amount of protein to be produced from each mRNA molecule.

~ FIGURE 14.12

Electron micrograph and diagram of a polysome, a number of ribosomes each translating the same mRNA sequentially.

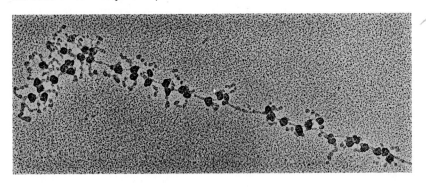

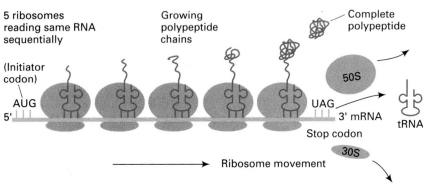

KEYNOTE

The AUG (methionine) initiator codon signals the start of translation in prokaryotes and eukaryotes. Elongation proceeds when a peptide bond forms between the amino acid attached to the tRNA in the A site of the ribosome and the growing polypeptide attached to the tRNA in the P site. Translocation occurs when the now-uncharged tRNA in the P site is released from the ribosome and the ribosome moves one codon down the mRNA. _____

Termination of Translation

Elongation continues until the polypeptide coded for in the mRNA is completed. The completion is signaled by one of three stop codons, UAG, UAA, and UGA, which are the same in prokaryotes and eukaryotes. The stop codons do not code for any amino acid, and so no tRNAs in the cell have anticodons for them. The ribosome recognizes a chain termination codon only with the help of proteins called **termination factors,** or **release factors (RF),** which read the chain termination codons and then initiate a series of specific termination events (Figure 14.13).

E. coli has three RFs—RF1, RF2, and RF3—and each is a single polypeptide. Factor RF1 recognizes UAA and UAG, while RF2 recognizes UAA and UGA. Thus these two release factors have overlapping specificity in codon recognition. Factor RF3, which does not recognize any of the stop codons, plays a stimulatory role in the termination events. In eukaryotes only one release factor, called eukaryotic release factor (eRF), is needed. eRF recognizes all three stop codons. No stimulatory factor analogous to RF3 has been detected in eukaryotes.

Since no amino acid exists in the A site for the polypeptide (attached to the tRNA in the P site) to be transferred to, the polypeptide is released from the tRNA in the P site of the ribosome in a reaction catalyzed by peptidyl transferase. The polypeptide and the tRNA to which it was attached are then released from the ribosome. The ribosome dissociates into the two subunits, which are recycled and used in further protein synthesis events.

~ FIGURE 14.13

Termination of translation. The ribosome recognizes a chain termination codon (UAG) with the aid of release factors. The release factor reads the stop codon, and this initiates a series of specific termination events leading to the release of the completed polypeptide.

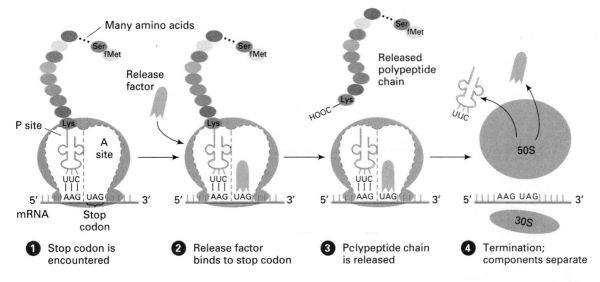

| ❶ Stop codon is encountered | ❷ Release factor binds to stop codon | ❸ Pclypeptide chain is released | ❹ Termination; components separate |

𝒦EYNOTE

Protein synthesis continues until a chain-terminating codon is located in the A site of the ribosome. These codons are read by one or more release factor proteins. Then the polypeptide and its tRNA are released from the ribosome, and the ribosome disengages from the reading frame of the mRNA.

PROTEIN TRANSPORT AND COMPARTMENTALIZATION

In eukaryotes, some proteins may be secreted, depending on cell type, while other proteins need to be located in different cell compartments in order to function. That is, eukaryotic cells are compartmentalized, and the examination of the contents of these compartments (such as the nucleus, mitochondria, chloroplasts, and lysosomes) shows that each contains a specific set of proteins. Therefore proteins must end up in the compartments in which they are to function. The sorting of proteins to their appropriate compartments is under genetic control. Similarly, in prokaryotes, certain proteins become localized in the membrane and others are secreted. A complete discussion of protein localization

mechanisms in eukaryotes and prokaryotes is beyond the scope of this text. Therefore, we will focus on eukaryotes, and briefly describe protein transport into the ER, the mitochondria, the chloroplasts, and the nucleus.

Proteins Distributed by the Endoplasmic Reticulum

Electron microscopic studies showed that secretory cells, such as those found in the liver and pancreas, have an extensive membrane structure called the endoplasmic reticulum (ER). Some of the ER appears rough, due to the presence of ribosomes on the membrane. This observation has resulted in the terms *membrane-bound* and *free* ribosomes. Proteins synthesized on the rough ER are extruded into the space between the two membranes of the ER (the cisternal space) and are then transferred to the Golgi complex. Proteins to be secreted are packaged into secretory vesicles. From these vesicles the proteins are secreted to the outside of the cell by the fusion of the vesicles with the cell membrane. Figure 14.14 (p. 424) is a simple diagram of all these events. For our purposes, it is necessary to understand that initially a common mechanism translocates proteins destined to be secreted from the cell, proteins that will become embedded in the plasma membrane, and proteins that are packaged into lysosomes into the ER cisternal space. Thereafter, sorting to their final destinations occurs primarily at the Golgi complex.

~ FIGURE 14.14

Movement of secretory proteins. These proteins move from their site of synthesis on the rough endoplasmic reticulum, through the Golgi apparatus, into vesicles, and then into the extracellular space. The arrow indicates the general path of the proteins to be secreted.

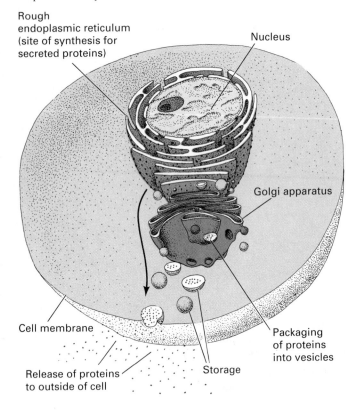

Rough endoplasmic reticulum (site of synthesis for secreted proteins)

Nucleus

Golgi apparatus

Cell membrane

Packaging of proteins into vesicles

Release of proteins to outside of cell

Storage

In 1975, Günther Blobel and his colleagues found that all the major proteins secreted by the pancreas initially contain extra amino acids at the amino terminal end. As a result of their studies, Blobel and B. Dobberstein proposed the **signal hypothesis,** which states that the secretion of proteins out of a cell occurs through the binding of a hydrophobic, amino terminal extension (the **signal sequence**) to the membrane and the subsequent removal and degradation of the extension in the cisternal space of the ER.

Our current understanding of the signal hypothesis is as follows (Figure 14.15): The growing proteins (i.e., those being synthesized on ribosomes) that are to be secreted from the cell or inserted into membranes all have an N-terminal extension of about 15 to 30 amino acids, called the *signal sequence.* Only proteins that have signal sequences can be transferred across or inserted into the membrane of the rough ER while translation is taking place.

When a protein destined for the ER exposes its signal sequence, a cytoplasmic receptor particle called **the signal recognition particle** (SRP: a complex of the small RNA molecule 7SL RNA, with six proteins) recognizes the signal sequence, binds to it, and blocks further translation of the mRNA. The temporary halt in protein synthesis brought about by SRP occurs when the polypeptide chain is long enough so that the signal sequence has completely emerged from the ribosome and can be recognized by SRP. Translation stops until the nascent polypeptide-SRP-ribosome-mRNA complex reaches and binds to the ER. The SRP recognizes an integral membrane protein of the ER called the **docking protein** (also called the *signal recognition protein* receptor). The association of the SRP and docking protein facilitates the binding of polypeptide's signal sequence and associated ribosomes to the ER. Translation resumes, and the SRP is released. The growing polypeptide (with its signal sequence) now is translocated through the membrane into the cisternal space of the ER. The transport of the protein molecule into the ER simultaneously with its synthesis is an example of **cotranslational transport.**

Once the signal sequence is fully into the cisternal space of the ER, it is removed from the polypeptide by the action of an enzyme called **signal peptidase.** When the complete polypeptide is entirely within the ER cisternal space, it is typically modified further by the addition of specific carbohydrate groups to produce *glycoproteins.* The glycoproteins must now be sorted for their final destinations.

The Golgi complex is the main distribution center where most of the sorting decisions are made. Proteins targeted for the plasma membrane are carried from the Golgi complex to the plasma membrane in transport vesicles. Proteins destined to be secreted are packaged into secretory storage vesicles, which form by budding from the Golgi complex. The secretory vesicles migrate to the cell surface where they fuse with the plasma membrane and release their packaged proteins to the outside of the cell. Lysosomal proteins are collected into a specific class of Golgi export vesicles, which eventually become functional lysosomes.

Currently, significant progress is being made in defining in yeast the precise steps involved in protein translocation across the ER membrane and the subsequent sorting events through the study of defined genetic mutants called *sec* (secretion) mutants, which affect different stages of those processes. This kind of combined genetic-cell biology approach will contribute significant new knowledge to this area in the future.

Proteins Transported into Mitochondria and Chloroplasts

A large number of proteins find their way into mitochondria and chloroplasts. These proteins include the ribosomal proteins of the organellar ribosomes, some

~ Figure 14.15

Model for the translocation of proteins into the endoplasmic reticulum in eukaryotes.

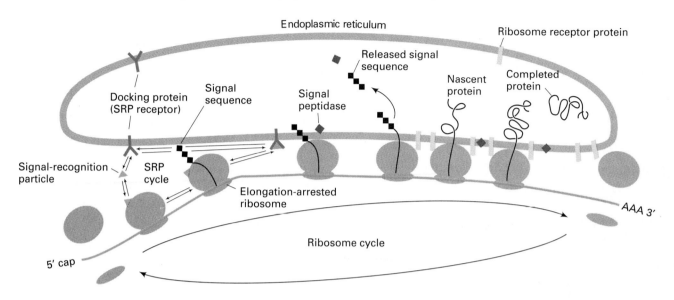

enzymes of the electron transport chain in mitochondria, and some photosynthesis enzymes in chloroplasts.

In contrast to proteins that are translated simultaneously with entry into the ER, proteins destined for the mitochondria and chloroplasts are synthesized completely before the import process begins. That is, ribosomes that make these proteins do not become associated with either organelle's outer membranes. Transport in which synthesis of the protein is completed before import into the organelle takes place is called **posttranslational transport.**

Proteins imported into mitochondria and chloroplasts are made as precursor proteins, which contain extra sequences at their N-terminal ends called **transit sequences** that are both necessary and sufficient for transport. The precursors associate with specific receptors on the outer membrane of the appropriate organelle. They are then transported into the organelle. Once inside the organelle, the transit sequence is removed from the protein by an enzyme, **transit peptidase.**

Proteins Transported into the Nucleus

Many proteins find their way into the nucleus, including the ribosomal proteins that are needed for ribosome assembly, histone and nonhistone chromosomal proteins, gene regulatory proteins, and the many enzymes needed for DNA replication, DNA repair, recombination, transcription, and RNA processing. Except for small proteins (<60 kDa), the nucleus is extremely selective about which proteins it allows in. Like the other proteins we have discussed, nuclear proteins have

simple sequences—signal sequences—that specify their translocation into the nucleus.

Like mitochondrial and chloroplast proteins, proteins destined for the nucleus are transported by a posttranslational transport mechanism. The recognition of the signal sequence may be by a component of the nuclear membrane. The evidence indicates that nuclear proteins enter the nucleus via the nuclear pore complex.

Unlike the other compartmentalized proteins we have discussed, the signal sequences of many nuclear proteins are not removed once they enter the nucleus. The reason for this is that each time the cell divides, the nuclear envelope is degraded, then reforms prior to cytokinesis. Since, during cell division, the nuclear proteins are free in the cytoplasm, they must retain their ability to reenter the nucleus selectively once the nuclear envelope reforms.

𝒦eynote

Eukaryotic proteins that enter the endoplasmic reticulum, the mitochondrion, the chloroplast, or the nucleus are distinguished from proteins that remain free in the cytoplasm by the presence of specific targeting sequences. Proteins that enter the ER, for instance, have signal sequences at their N-terminal ends. Characteristically, the signal sequences contain a significant number of hydrophobic amino acids. Proteins destined for the ER are translocated into the cisternal space of the ER, where the signal sequence is removed by signal peptidase. They are then sorted to their final destinations by the Golgi complex.

Proteins destined for the mitochondrion, the chloroplast, and the nucleus also have specific sequences to direct them into their respective organelles. Once mitochondrial or chloroplast proteins are inside the respective organelle, the special sequence called a *transit sequence,* is removed. For nuclear proteins, however, the signal sequence remains, to allow those proteins to become localized in the nucleus again after cell division. ———————

SUMMARY

In this chapter we discussed the features of the genetic code and the process of translation. From a number of experiments using synthetic mRNAs, the genetic code was found to have the following features: it is a triplet code, it is comma-free, it is nonoverlapping, it is universal, and, with two exceptions, it is degenerate. Of the 64 codons, 61 specify amino acids, while the other three codons are chain-terminating codons. One codon, AUG (for methionine) is typically used to specify the first amino acid in a protein chain.

Protein synthesis occurs on ribosomes, where the genetic message encoded in mRNA is translated. Amino acids are brought to the ribosome on charged tRNA molecules. Each tRNA has an anticodon, which binds specifically to a codon in the mRNA. As a result, the correct amino acid sequence is achieved by (1) the specific binding between the codon of the mRNA and the complementary anticodon in the rRNA and (2) the specific binding of each amino acid to its own specific tRNA.

In both prokaryotes and eukaryotes an AUG codon is the initiator codon for the start of protein synthesis. In prokaryotes, the initiation of protein synthesis requires an upstream sequence of the AUG codon, to which the small ribosomal subunit binds. This sequence is the Shine-Dalgarno sequence, and it binds specifically to the 3′ end of the 16S rRNA of the small ribosomal subunit, thereby associating the small subunit with the mRNA. No functionally equivalent sequence occurs in eukaryotic mRNAs; instead, the ribosomes load onto the mRNA at the 5′ end of the mRNA and scan towards the 3′ end, initiating translation at the first AUG codon they encounter.

In both prokaryotes and eukaryotes, the initiation of protein synthesis requires protein factors called initiation factors. Initiation factors are bound to the ribosome-mRNA complex during the initiation phase and dissociate once the polypeptide chain has been initiated. During the elongation phase, during which the polypeptide chain is elongated one amino acid at a time concomitantly with the movement of the ribosome towards the 3′ end of the mRNA one codon at a time, protein factors called elongation factors play important catalytic roles. The signal for polypeptide chain growth to stop is the presence of a chain-terminating codon (UAG, UAA, or UGA) in the mRNA. No naturally-occurring tRNA has an anticodon that can read a chain-terminating codon. Instead, specific protein factors called release factors read the stop codon and initiate the events characteristic of protein synthesis termination; namely, the release of the completed polypeptide from the ribosome, the release of the tRNA from the ribosome, and the dissociation of the two ribosomal subunits from the mRNA.

Eukaryotic proteins that enter the endoplasmic reticulum, the mitochondrion, the chloroplast, or the nucleus are different from proteins that remain free in the cytoplasm in that they possess specific targeting sequences, that is, sequences that are required for localizing them into the appropriate cellular compartment. Proteins destined for the ER, for example, have specific hydrophobic signal sequences at their amino ends. When a protein being synthesized exposes its signal sequence, a signal recognition particle (SRP) binds to it and blocks further translation of the mRNA. Translation is blocked until the nascent polypeptide-SRP-ribosome-mRNA complex reaches and binds to the ER by interaction between the SRP and an integral membrane protein of the ER called the docking protein. Resumption of translation then results in the growing polypeptide being translocated into the cisternal space of the ER where the signal sequence is removed. The protein is then sorted for its final location by the Golgi complex.

Proteins destined for the mitochondrion, chloroplast, or nucleus also have specific sequences to direct them into their respective organelles. These sequences are called transit sequences. Once mitochondrial or chloroplast proteins are inside the respective organelle, this transit sequence is removed. For nuclear proteins, however, the signal sequence remains attached so that those proteins can become transported again to the nucleus after cell division.

In sum, protein synthesis is a complex process involving the interaction between three major classes of RNA (mRNA, tRNA, and rRNA) and a large number of accessory protein factors that act catalytically in the process. With very few exceptions the genetic code which specifies the amino acids for each mRNA codon is the same in every organism, prokaryotic and eukaryotic. By repeated translation of an mRNA molecule by a string of ribosomes (producing a polysome), a large number of protein molecules can be produced. Thus, from a single gene, large quantities of a protein can be produced by two amplification steps: 1) the production of multiple mRNAs from the gene; and 2) the production of many protein molecules by repeated translation of each mRNA.

ANALYTICAL APPROACHES FOR SOLVING GENETICS PROBLEMS

Q.1 a. How many of the 64 codon permutations can be made from the three nucleotides A, U, and G?

b. How many of the 64 codon permutations can be made from the four nucleotides A, U, G, and C, with one or more C's in each codon?

A.1 a. This question involves probability. There are four bases, so the probability of a cytosine at the first position in a codon is 1/4. Conversely, the probability of a base other than cytosine in the first position is $(1 - 1/4) = 3/4$. These same probabilities apply to the other two positions in the codon. Therefore the probability of a codon without a cytosine is $(3/4)^3 = 27/64$.

b. This question involves the relative frequency of codons that have one or more cytosines. We have already calculated the probability of a codon *not* having a cytosine, so all the remaining codons have one or more cytosines. The answer to this question, therefore, is $(1 - 27/64) = 37/64$.

Q.2 Random copolymers were used in some of the experiments directed toward deciphering the genetic code. For each of the following ribonucleotide mixtures, give the expected codons and their frequencies, and give the expected proportions of the amino acids that would be found in a polypeptide directed by the copolymer in a cell-free, protein-synthesizing system.
a. 2U:1C
b. 1U:1C:2G

A.2 a. The probability of a U at any position in a codon is 2/3, and the probability of a C at any position in a codon is 1/3. Thus, the codons, and their relative frequencies, and the amino acids for which they code are:

$$UUU = (2/3)(2/3)(2/3) = \frac{8}{27} = 0.296 = 29.6\% \quad Phe$$

$$UUC = (2/3)(2/3)(1/3) = \frac{4}{27} = 0.148 = 14.8\% \quad Phe$$

$$UCC = (2/3)(1/3)(1/3) = \frac{2}{27} = 0.0743 = 7.43\% \quad Ser$$

$$UCU = (2/3)(1/3)(2/3) = \frac{4}{27} = 0.148 = 14.8\% \quad Ser$$

$$CUU = (1/3)(2/3)(2/3) = \frac{4}{27} = 0.148 = 14.8\% \quad Leu$$

$$CUC = (1/3)(2/3)(1/3) = \frac{2}{27} = 0.0743 = 7.43\% \quad Leu$$

$$CCU = (1/3)(1/3)(2/3) = \frac{2}{27} = 0.0743 = 7.43\% \quad Pro$$

$$CCC = (1/3)(1/3)(1/3) = \frac{1}{27} = 0.037 = 3.7\% \quad Pro$$

In sum, 44.4% Phe, 22.23% Ser, 22.23% Leu, 11.13% Pro. (Does not quite add up to 100% because of rounding off errors.)

b. The probability of a U at any position in a codon is 1/4, the probability of a C at any position in a codon is 1/4, and the probability of a G at any position in a codon is 1/2. Thus, the codons and their relative frequencies, and the amino acids for which they code are:

$$UUU = (1/4)(1/4)(1/4) = \frac{1}{64} = 1.56\% \quad Phe$$

$$UUC = (1/4)(1/4)(1/4) = \frac{1}{64} = 1.56\% \quad Phe$$

$$UCU = (1/4)(1/4)(1/4) = \frac{1}{64} = 1.56\% \quad Ser$$

$$UCC = (1/4)(1/4)(1/4) = \frac{1}{64} = 1.56\% \quad Ser$$

$$CUU = (1/4)(1/4)(1/4) = \frac{1}{64} = 1.56\% \quad Leu$$

$$CUC = (1/4)(1/4)(1/4) = \frac{1}{64} = 1.56\% \quad Leu$$

$$CCU = (1/4)(1/4)(1/4) = \frac{1}{64} = 1.56\% \quad Pro$$

$$CCC = (1/4)(1/4)(1/4) = \frac{1}{64} = 1.56\% \quad Pro$$

$$UUG = (1/4)(1/4)(1/2) = \frac{2}{64} = 3.13\% \text{ Leu}$$

$$UGU = (1/4)(1/2)(1/4) = \frac{2}{64} = 3.13\% \text{ Cys}$$

$$UGG = (1/4)(1/2)(1/2) = \frac{4}{64} = 6.25\% \text{ Trp}$$

$$GUU = (1/2)(1/4)(1/4) = \frac{2}{64} = 3.13\% \text{ Val}$$

$$GUG = (1/2)(1/4)(1/2) = \frac{4}{64} = 6.25\% \text{ Val}$$

$$GGU = (1/2)(1/2)(1/4) = \frac{4}{64} = 6.25\% \text{ Gly}$$

$$GGG = (1/2)(1/2)(1/2) = \frac{8}{64} = 12.5\% \text{ Gly}$$

$$CCG = (1/4)(1/4)(1/2) = \frac{2}{64} = 3.13\% \text{ Pro}$$

$$CGC = (1/4)(1/2)(1/4) = \frac{2}{64} = 3.13\% \text{ Arg}$$

$$CGG = (1/4)(1/2)(1/2) = \frac{4}{64} = 6.25\% \text{ Arg}$$

$$GCC = (1/2)(1/4)(1/4) = \frac{2}{64} = 3.13\% \text{ Ala}$$

$$GCG = (1/2)(1/4)(1/2) = \frac{4}{64} = 6.25\% \text{ Ala}$$

$$GGC = (1/2)(1/2)(1/4) = \frac{4}{64} = 6.25\% \text{ Gly}$$

$$UCG = (1/4)(1/4)(1/2) = \frac{2}{64} = 3.13\% \text{ Ser}$$

$$UGC = (1/4)(1/2)(1/4) = \frac{2}{64} = 3.13\% \text{ Cys}$$

$$CUG = (1/4)(1/4)(1/2) = \frac{2}{64} = 3.13\% \text{ Leu}$$

$$CGU = (1/4)(1/2)(1/4) = \frac{2}{64} = 3.13\% \text{ Arg}$$

$$GUC = (1/2)(1/4)(1/4) = \frac{2}{64} = 3.13\% \text{ Val}$$

$$GCU = (1/2)(1/4)(1/4) = \frac{2}{64} = 3.13\% \text{ Ala}$$

In sum, 3.12% Phe, 6.25% Ser, 9.38% Leu, 6.25% Pro, 6.26% Cys, 6.25% Trp, 12.5% Val, 25% Gly, 12.5% Arg, 12.5% Ala.

QUESTIONS AND PROBLEMS

*14.1 The form of genetic information used directly in protein synthesis is (choose the correct answer):
a. DNA   c. rRNA
b. mRNA   d. Ribosomes

14.2 The process in which ribosomes engage is (choose the correct answer):

a. Replication   d. Disjunction
b. Transcription   e. Cell division
c. Translation

14.3 What are the characteristics of the genetic code?

14.4 Base-pairing wobble occurs in the interaction between the anticodon of the tRNAs and the codons. On the theoretical level, determine the minimum number of tRNAs needed to read the 61 sense codons.

14.5 Antibiotics have been very useful in elucidating the steps of protein synthesis. If you have an artificial messenger of the sequence of AUGUUUUUUUUUUUUU . . . , it will produce the following polypeptide in a cell-free, protein-synthesizing system: fMet-Phe-Phe-Phe. . . . In your search for new antibiotics you find one called putyermycin, which blocks protein synthesis. When you try it with your artificial mRNA in a cell-free system, the product is fMet-Phe. What step in protein synthesis does putyermycin affect? Why?

14.6 Describe the reactions involved in the aminoacylation (charging) of a tRNA molecule.

14.7 Compare and contrast the following in prokaryotes and eukaryotes: (a) protein synthesis initiation; (b) protein synthesis elongation; (c) protein synthesis termination.

14.8 Discuss the two species of methionine tRNA, and describe how they differ in structure and function. In your answer, include a discussion of how each of these tRNAs binds to the ribosome.

*14.9 Random copolymers were used in some of the experiments that revealed the characteristics of the genetic code. For each of the following ribonucleotide mixtures, give the expected codons and their frequencies, and give the expected proportions of the amino acids that would be found in a polypeptide directed by the copolymer in a cell-free, protein-synthesizing system:
a. 4 A:6 C   c. 1 A:3 U:1 C
b. 4 G:1 C   d. 1 A:1 U:1 G:1 C

*14.10 Other features of the reading of mRNA into proteins being the same as they are now (i.e., codons must exist for 20 different amino acids), what would the minimum WORD (CODON) SIZE be if the number of different bases in the mRNA were, instead of four:
a. Two
b. Three
c. Five

14.11 Suppose that at stage A in the evolution of the genetic code only the first two nucleotides in the coding triplets led to unique differences and that any nucleotide could occupy the third position. Then, suppose there was a stage B in which differences in meaning arose depending upon whether a purine (A or G) or pyrimidine (C or T) was present at the third position. Without reference to the number of amino acids or multiplicity of tRNA molecules, how many triplets of different meaning can be constructed out of the code at stage A? At stage B?

*14.12 A gene makes a polypeptide 30 amino acids long containing an alternating sequence of phenylalanine and tyrosine. What are the sequences of nucleotides corresponding to this sequence in the following:
a. The DNA strand which is read to produce the mRNA, assuming Phe = UUU and Tyr = UAU in mRNA
b. The DNA strand which is not read
c. tRNA

14.13 A segment of a polypeptide chain is Arg-Gly-Ser-Phe-Val-Asp-Arg. It is encoded by the following segment of DNA:

Which strand is the template strand? Label each strand with its correct polarity (5' and 3').

*14.14 Two populations of RNAs are made by the random combination of nucleotides. In population A the RNAs contain only A and G nucleotides (3A:1G), while in population B the RNAs contain only A and U nucleotides (3A:1U). In what ways *other than amino acid content* will the proteins produced by translating the population A RNAs differ from those produced by translating the population B RNAs?

14.15 In *E. coli* a particular tRNA normally has the anticodon 5'-GGG-3', but because of a mutation in the tRNA gene, the mutant tRNA has the anticodon 5'-GGA-3'.
a. What amino acid would this tRNA carry?
b. What codon would the normal tRNA recognize?
c. What codon would the mutant tRNA recognize?
d. What would be the effect of the mutation on the proteins in the cell?

*14.16 A particular protein found in *E. coli* normally has the N-terminal sequence Met-Val-Ser-Ser-Pro-Met-Gly-

Ala-Ala-Met-Ser. . . . In a particular cell a mutation alters the anticodon of a particular tRNA from 5'-GAU-3' to 5'-CAU-3'. What would be the N-terminal amino acid sequence of this protein in the mutant cell? Explain your reasoning.

14.17 The gene encoding an *E. coli* tRNA containing the anticodon 5'-GUA-3' mutates so that the anticodon now is 5'-UUA-3'. What will be the effect of this mutation? Explain your reasoning.

*14.18 The normal sequence of the coding region of a particular mRNA is shown in Figure 14.A, along with several mutant versions of the same mRNA. Indicate what protein would be formed in each case. (.... = many [a multiple of 3] unspecified bases.)

14.19 The normal sequence of a particular protein is given below, along with several mutant versions of it. For each mutant, explain what mutation occurred in the coding sequence of the gene.

> Normal: Met-Gly-Glu-Thr-Lys-Val-Val-....-Pro
> Mutant 1: Met-Gly
> Mutant 2: Met-Gly-Glu-Asp
> Mutant 3: Met-Gly-Arg-Leu-Lys
> Mutant 4: Met-Arg-Glu-Thr-Lys-Val-Val-....-Pro

14.20 In the recessive condition in humans known as Sickle Cell Disease (SCD), the β-globin moiety of hemoglobin is found to be abnormal. The only difference between it and the normal β-globin is that the 6th amino acid from the N-terminal is valine, whereas the normal β-globin has glutamic acid at this position. Explain how this occurred.

14.21 Antibiotics have been useful in determining whether cellular events depend on transcription or translation. For example, actinomycin D is used to block transcription, and cycloheximide (in eukaryotes) is used to block translation. In some cases, though, surprising results are obtained after antibiotics are administered. The addition of actinomycin D, for example, may result in an increase, not a decrease, in the activity of a particular enzyme. Discuss how this result might come about.

FIGURE 14.A

normal: AUGUUCUCUAAUUAC(....)AUGGGGUGGGUGUAG
mutant a: AUGUUCUCUAAUUAG(....)AUGGGGUGGGUGUAG
mutant b: AGGUUCUCUAAUUAC(....)AUGGGGUGGGUGUAG
mutant c: AUGUUCUCGAAUUAC(....)AUGGGGUGGGUGUAG
mutant d: AUGUUCUCUAAAUAC(....)AUGGGGUGGGUGUAG
mutant e: AUGUUCUCUAAUUC(....)AUGGGGUGGGUGUAG
mutant f: AUGUUCUCUAAUUAC(....)AUGGGGUGGGUGUGG

15 RECOMBINANT DNA TECHNOLOGY AND THE MANIPULATION OF DNA

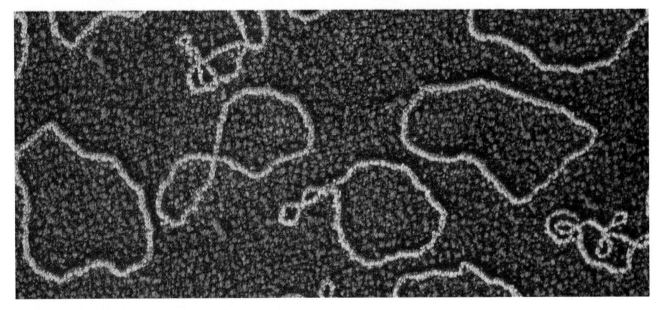

POLYMERASE CHAIN REACTION (PCR)
APPLICATIONS OF RECOMBINANT DNA TECHNOLOGY
Analysis of Biological Processes
Diagnosis of Human Genetic Diseases
DNA Fingerprinting
Human Gene Therapy
Human Genome Project
Commercial Products
Genetic Engineering of Plants

PRINCIPAL POINTS

~ Restriction enzymes recognize specific nucleotide pair sequences in DNA and cleave the DNA either within the recognition sequence, or at a site nearby. Because they cut DNA at specific sites, restriction enzymes are generally used in recombinant DNA cloning experiments.

~ Different kinds of cloning vectors (DNA molecules capable of replication in a host organism) have been developed to construct and clone recombinant DNA molecules. All cloning vectors must replicate within their host organism, they must have one or more restriction sites at which foreign DNA can be inserted, and they must have one or more dominant selectable markers to detect those cells which contain the vectors. The main classes of vectors are plasmids, bacteriophages, and cosmids. Shuttle vectors have the above properties but, in addition, are able to replicate in more than one type of host.

~ Genomic libraries are collections of clones that contain one copy of every DNA sequence in an organism's genome. Genomic libraries are useful, for example, for isolating specific genes and for studying the organization of the genome.

~ Once eukaryotic mRNAs have been isolated from the cell, it is possible to make DNA copies of those mRNA molecules. The resulting DNA copy is called a cDNA and the cDNAs can be cloned using appropriate cloning vectors.

~ Specific sequences in genomic libraries and cDNA libraries can be identified using a number of approaches. These approaches include the use of specific DNA or cDNA probes, heterologous probes, complementation of mutations, and oligonucleotide probes.

~ Genes and cloned DNA sequences can be analyzed to determine the arrangement and specific locations of restriction sites, a process called restriction mapping.

Gene transcripts can be analyzed qualitatively and quantitatively using recombinant DNA procedures.

~ Rapid methods have been developed for determining the sequence of a cloned piece of DNA. One method, the Maxam-Gilbert procedure, uses specific chemicals to modify and cleave the DNA chain at specific nucleotides. The other method, the dideoxy or Sanger procedure, uses enzymatic synthesis of a new DNA chain on a cloned template DNA strand. With this procedure, synthesis of new strands is stopped by the incorporation of a dideoxy analog of the normal deoxyribonucleotide. Using four different dideoxy analogs, the new strands stop at all possible nucleotide positions, thereby allowing the complete DNA sequence to be determined.

~ The polymerase chain reaction (PCR) uses synthetic oligonucleotides to amplify a specific segment of DNA many thousand fold in an automated procedure. A major benefit of PCR is that small amounts of DNA are needed. Thus, DNA from a single cell can be amplified using PCR. PCR is finding increasing applications both in research and in the commercial arena, including generating specific DNA segments for cloning or for sequencing, and for amplifying DNA for analysis for the presence of specific genetic defects.

~ There are many applications for recombinant DNA technology and related procedures. For example, with appropriate probes, detection of specific genetic diseases is possible, and many products in the clinical, veterinary, and agricultural areas have been developed. A project is underway to determine the complete sequence of the human genome. The knowledge obtained will expand greatly our understanding of human genetics. With continued advances in genetic engineering of plants, it is expected that many types of improved crops (increased yields, disease resistance) will be forthcoming.

Recently, experimental procedures have been developed that allow researchers to construct **recombinant DNA molecules** in test tubes. This process entails combining genetic material from two different sources into a single DNA molecule. The technology has opened the way for new and exciting research possibilities and affirms the plausibility of **genetic engineering,** that is, the alteration of the genetic constitution of cells or individuals by directed and selective modification, insertion, or deletion of an individual gene or genes. In some cases, novel gene combinations are made by joining DNA fragments from different organisms. In this chapter we discuss recombinant DNA techniques and present some examples of how **recombinant DNA technology** is furthering our knowledge of the structure and function of the prokaryotic and eukaryotic genomes, how cloned DNA sequences are being used for genetic diagnosis, and how new routes for producing commercial products have been opened because of the ability to manipulate DNA sequences.

Recombinant DNA typically is made by taking a piece of DNA from an organism and splicing it into a **cloning vector.** A cloning vector is a DNA molecule capable of replication in a host organism, such as a bacterium, and into which a gene or a piece of DNA can be specifically inserted at known positions. The resulting molecule is introduced into a host such as *E. coli*, yeast, animal cell, or plant cell. Reproduction of the host cell results in the replication (**molecular cloning**) of the recombinant DNA molecule, thus producing many copies for analysis and/or manipulation.

RESTRICTION ENDONUCLEASES

Recombinant DNA molecules could not be constructed without the use of **restriction endonucleases** (or **restriction enzymes**), which are able to cleave double-stranded DNA molecules at specific nucleotide pair sequences called *restriction sites,* or restriction enzyme recognition sequences. Restriction enzymes are used to produce a pool of DNA molecules to be cloned, and to assemble recombinant DNA molecules from the DNA fragments and cloning vectors, as well as to analyze the organization of any piece of DNA. No similar enzymes have been identified in eukaryotes. Restriction enzymes are named for the organism from which they are isolated. Conventionally, a three-letter system is used, italicized, followed by a roman numeral. Additional letters are sometimes added to signify a particular strain of the bacterial species from which the enzymes were obtained. For example, *Bgl*II is from *Bacillus globigi; Eco*RI is from *E. coli* strain RY13, and *Hin*dIII is from *Haemophilus influenzae* strain R_d. The names are pronounced in particular ways but those pronunciations

follow no set pattern. For example, *Bam*HI is "bam-H-one," *Bgl*II is "bagel-two," *Eco*RI is "echo-R-one," *Hin*dIII is "hin-D-three," *Hha*I is "ha-ha-one," and *Hpa*II is "hepa-two." The pronunciation of other enzymes will be given when they are introduced.

Restriction enzymes fall into two classes. Type I restriction enzymes recognize a specific nucleotide pair sequence in DNA and then cleave the DNA at a nonspecific site away from that sequence. Since the cut site is not nucleotide pair specific, Type I restriction enzymes are not very useful for the construction or analysis of recombinant DNA molecules.

Type II restriction enzymes also recognize a specific nucleotide pair sequence in DNA and, in this case, they cleave the DNA within that sequence. Since all DNA fragments generated by cleavage with a particular Type II restriction enzyme have been cut at the same sequence, these enzymes are very valuable for constructing recombinant DNA molecules.

The recognition sequences for Type II restriction enzymes have an axis of symmetry through the midpoint of the recognition sequence: the base sequence from 5′ to 3′ on one DNA strand is the same as the base sequence from 5′ to 3′ on the complementary DNA strand (Figure 15.1). Thus, the sequences are said to have *twofold rotational symmetry.* Because of this property we can determine the complete restriction enzyme sequence if we know only the first three nucleotides, starting from the 5′ end. So if we are given the sequence 5′ C T G 3′, we can deduce that there must be a symmetrical sequence on the opposite DNA strand because of two-fold rotation symmetry, giving

$$5' \text{ C T G ? ? ? } 3'$$
$$3' \text{ ? ? ? G T C } 5'$$

Filling in the rest by the base pairing rules, we get

$$5' \text{ C T G C A G } 3'$$
$$3' \text{ G A C G T C } 5'$$

Type II restriction enzymes have been isolated from a large number of different bacterial strains. Since each of the enzymes cuts DNA at an enzyme-specific nucleotide pair sequence, the number of cuts the enzyme

---

~ FIGURE 15.1

Restriction enzyme recognition sequence in DNA, showing twofold rotational symmetry of the sequence. Shown here is the recognition sequence for the restriction enzyme *Eco*RI.

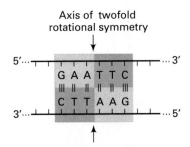

makes in a particular DNA molecule or molecules depends on the number of times the particular recognition/cleavage sequence is present in the DNA. In some cases, restriction enzymes (called *isoschizomers*) isolated from different bacteria recognize and cleave the same nucleotide pair sequences.

The cleavage sites for a number of restriction enzymes are shown in Table 15.1 along with the number

~ TABLE 15.1

Characteristics of Some Restriction Endonucleases

	ENZYME NAME	PRONUN- CIATION	ORGANISM IN WHICH ENZYME IS FOUND	RECOGNITION SEQUENCE AND POSITION OF CUT*	NUMBER OF CLEAVAGE SITES IN DNA FROM	
					λ	pBR322
ENZYMES WITH 6-bp RECOGNITION SEQUENCES	BamHI	"bam-H-one"	Bacillus amyloliquefaciens H	5' G G A T C C 3' 3' C C T A G G 5'	5	1
	BglII	"bagel-two"	Bacillus globigi	A G A T C T T C T A G A	5	0
	EcoRI	"echo-R-one"	E. coli RY13	G A A T T C C T T A A G	5	1
	HaeII	"hay-two"	Haemophilus aegyptius	R G C G C Y Y C G C G R	>30	11
	HindIII	"hin-D-three"	Haemophilus influenzae R_d	A A G C T T T T C G A A	6	1
	PstI	"P-S-T-one"	Providencia stuartii	C T G C A G G A C G T C	18	1
	SalI	"sal-one"	Streptomyces albus G	G T C G A C C A G C T G	2	1
	SmaI	"sma-one"	Serratia marcescens	C C C G G G G G G C C C	3	0
ENZYMES WITH 4-bp RECOGNITION SEQUENCES	HaeIII	"hay-three"	Haemophilus egyptius	G G C C C C G G	>50	22
	HhaI	"ha-ha-one"	Haemophilus hemolyticus	G C G C C G C G	>50	31
	HpaII	"hepa-two"	Haemophilus parainfluenzae	C C G G G G C C	>50	26
	Sau3A	"sow-three-A"	Staphylococcus aureus 3A	G A T C C T A G	116	22
ENZYMES WITH 8-bp RECOGNITION SEQUENCES	NotI	"not-one"	Nocardia otitidis-caviarum	G C G G C C G C C G C C G G C G	0	0

* In this column the two strands of DNA are shown with the sites of cleavage indicated by arrows. Since there is an axis of two-fold rotational symmetry in each recognition sequence, the DNA molecules resulting from the cleavage are symmetrical.

Key: R = purine; Y = pyrimidine.

of cuts each enzyme makes in phage λ DNA and in the cloning vector pBR322. This information can be used as a guideline for choosing an enzyme for a particular application. The most commonly-used restriction enzymes recognize four nucleotide pairs (e.g., *Hae*III {"hay-three"}, *Hha*I) or six nucleotide pairs (e.g., *Bam*HI, *Eco*RI). A number of other enzymes recognize and cleave a five-nucleotide-pair sequence (e.g., *Rsa*I {"rosa-one"}). And, recently two enzymes have been found that recognize eight-nucleotide-pair sequences (e.g., *Not*I {"not-one"}).

For DNA with random distribution of nucleotide pairs, there is a clear relationship between the number of nucleotide pairs in the recognition sequence and the frequency of cutting the DNA. That is, an enzyme that recognizes a four-nucleotide-pair sequence will cut more frequently than one that recognizes a five-nucleotide-pair sequence, and so on. Let us consider this more mathematically.

Consider DNA with a random distribution of nucleotide pairs. For that DNA, there is an equal chance of finding one of the four possible nucleotide pairs $\frac{G}{C}$, $\frac{C}{G}$, $\frac{A}{T}$, or $\frac{T}{A}$ at any one position. Consider the restriction enzyme *Hae*III which recognizes the sequence $\frac{5'\ GGCC\ 3'}{3'\ CCGG\ 5'}$.

The probability of this sequence occurring in DNA is computed as follows:

1st nucleotide pair: $\frac{G}{C}$, probability = 1/4

2nd nucleotide pair: $\frac{G}{C}$, probability = 1/4

3rd nucleotide pair: $\frac{C}{G}$, probability = 1/4

4th nucleotide pair: $\frac{C}{G}$, probability = 1/4

The probability of finding any one of the nucleotide pairs is independent of the probability of finding one of the other nucleotide pairs. Therefore the probability of finding the *Hae*III restriction site in DNA with a random distribution of nucleotide pairs is $1/4 \times 1/4 \times 1/4 \times 1/4 = 1/256$. In short, the recognition sequence for *Hae*III occurs on the average once every 256 base pairs in such a piece of DNA.

In general, the probability of occurrence of a recognition sequence for a restriction enzyme (called a *restriction site*) is given by the formula $(1/4)^n$ where n is the number of nucleotide pairs in the recognition sequence. These values are given in Table 15.2.

It is important to note, though, that DNA from living organisms does not have a random distribution of base pairs. Rather, some DNA is GC-rich and other

~ TABLE 15.2

Occurrence of Restriction Sites for Restriction Enzymes in DNA with Randomly-distributed Nucleotide Pairs

NUCLEOTIDE PAIRS IN RESTRICTION SITE	PROBABILITY OF OCCURRENCE
4	$(1/4)^4$ = 1 in 256 bp
5	$(1/4)^5$ = 1 in 1024 bp
6	$(1/4)^6$ = 1 in 4096 bp
8	$(1/4)^8$ = 1 in 65,476 bp
n	$(1/4)^n$

DNA is AT-rich. The latter, for example, would have few sites for *Hae*III, so this enzyme would cut AT-rich DNA much less frequently than the table indicates. Thus, you should use the predicted frequencies of cutting in Table 15.2 only as a rough guide in planning experiments. Note that enzymes with eight-nucleotide-pair restriction sites cut, on the average, only once every 65,476 bp. Because they cut relatively infrequently, they are useful for cutting large genomes—e.g., human—into relatively large pieces.

~ FIGURE 15.2

Examples of how restriction enzymes cleave DNA. (a) *Sma*I results in blunt ends; (b) *Bam*HI results in overhanging ("sticky") 5' ends; (c) *Pst*I results in overhanging ("sticky") 3' ends.

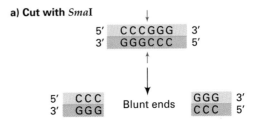

a) Cut with *Sma*I

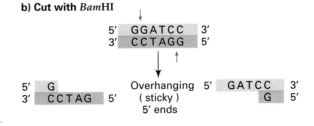

b) Cut with *Bam*HI

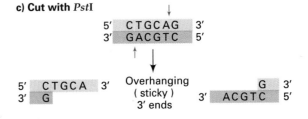

c) Cut with *Pst*I

As Table 15.1 indicates, some enzymes, such as *Hae*III or *Sma*I ("sma-one"), cut both strands of DNA between the same two nucleotide pairs to produce *blunt ends* (Figure 15.2a), while others, such as *Eco*RI, *Bam*HI, and *Hind*III, make staggered cuts in the symmetrical nucleotide pair sequence to produce *sticky* or *staggered ends*. The staggered ends may either have an overhanging 5' end, as in the case of cleavage with *Bam*HI or *Eco*RI (Figure 15.2b), or an overhanging 3' end, as in the case of cleavage with *Hae*II ("hay-two") or *Pst*I ("P-S-T-one") (Figure 15.2c). In each case, the restriction enzyme makes its cut by cleaving the DNA backbone between the 3' carbon of the deoxyribose sugar and the phosphate on the subsequent deoxyribose in the chain (Figure 15.3). Thus, all restriction enzyme-cut DNA fragments have a phosphate on their 5' ends and a hydroxyl group on their 3' ends.

Restriction enzymes that produce staggered ends are of particular value in cloning DNA fragments because every DNA fragment generated by cutting a piece of DNA with the same restriction enzyme has the same base sequence at the two staggered ends. Therefore, if the ends of two pieces of DNA produced by the action of the same restriction enzyme (such as *Eco*RI) come together in solution, perfect base pairing occurs; that is,

the two single-stranded DNA are said to *anneal* (Figure 15.4; p. 436). In the presence of DNA ligase, the sugar-phosphate backbones can then be sealed to produce a whole DNA molecule with a reconstituted restriction enzyme site. Even DNA fragments with blunt ends can be ligated together with DNA ligase at high concentrations of the enzyme. As we will see in the following section on cloning vectors, this ligation of catalyzed ends by DNA ligase is the principle behind the formation of recombinant DNA molecules.

ℋEYNOTE

Restriction enzymes, or restriction endonucleases, are enzymes that recognize specific nucleotide pair sequences in DNA and cleave the DNA either non-specifically at a site away from a twofold symmetrical recognition site (Type I restriction enzymes) or at a specific site within the recognition site (Type II restriction enzymes). Cleavage of the DNA with a Type II restriction enzyme can be staggered (producing DNA fragments with single-stranded, "sticky,"

~ FIGURE 15.3

Cleavage of DNA with a restriction enzyme results in DNA fragments with a phosphate on their 5' ends and a hydroxyl on their 3' ends.

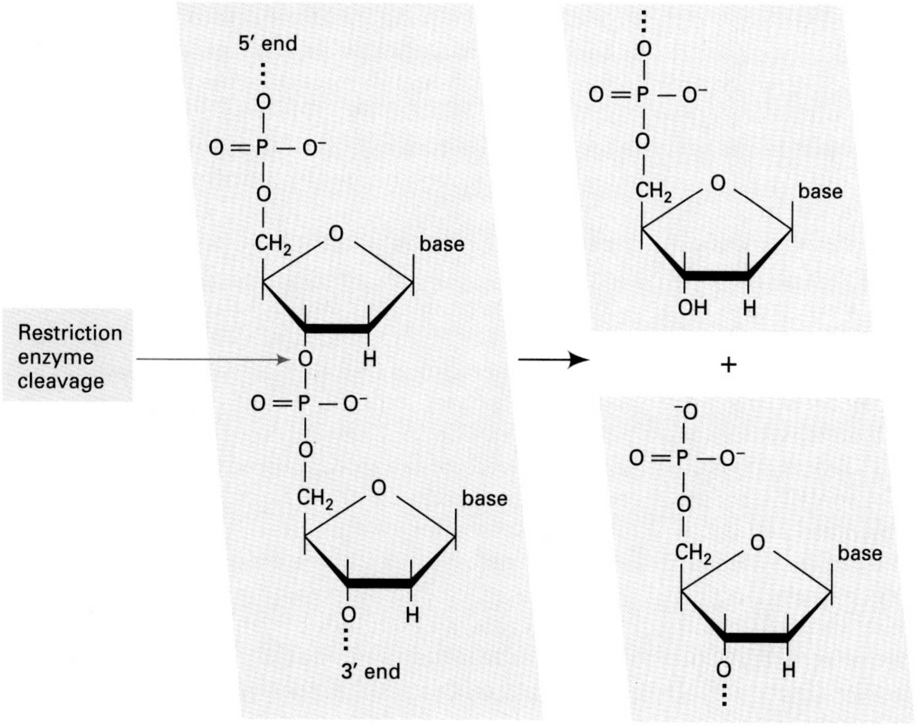

~ FIGURE 15.4

Cleavage of DNA by the restriction enzyme *Eco*RI. *Eco*RI makes staggered, symmetrical cuts in DNA, leaving "sticky" ends. A DNA fragment with a sticky end produced by *Eco*RI digestion can bind by complementary base pairing (anneal) to any other DNA fragment with a sticky end produced by *Eco*RI cleavage. The gaps may then be sealed by DNA ligase.

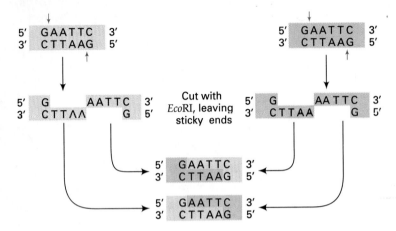

ends) or blunt. Because they cut DNA at specific sites, Type II restriction enzymes are generally used in recombinant DNA cloning experiments.

CLONING VECTORS AND CLONING

Geneticists are interested in understanding the organization of genes, as well as how their expression is regulated. Much can be learned if many copies of the gene are available for study. For example, the sequence can be determined and analyzed for features of interest. But until recently studying individual genes was difficult because of the complexity of the cells of higher eukaryotes. Each higher eukaryotic cell has two copies of each gene. If we isolate the cell's DNA, it will be sheared randomly because the DNA is so long. The gene of interest will be found on DNA fragments of many different sizes. Now, through recombinant DNA technology, it is possible to produce many identical copies of the gene or of any DNA sequence for study. Generating many copies of a DNA sequence is called *cloning*.

To clone a piece of DNA, two things are needed: a disabled host that has no significant chance of surviving outside the laboratory, and a **cloning vector**. A cloning vector is a DNA molecule capable of replication in a host organism and into which a gene or a piece of DNA can be specifically inserted at known positions.

While this chapter concentrates mostly on bacterial hosts, recombinant DNA molecules are also routinely introduced into yeast (an important host these days) and other fungi, into other lower eukaryotes, into animal cells and into plant cells, among other systems. In many cases the host organism is a derivative of the natural form found in the environment.

Three types of vectors are commonly used for cloning DNA sequences in bacterial cells: plasmids, bacteriophages (e.g., λ), and genetic elements called *cosmids*. Each differs in the way it must be manipulated to clone pieces of DNA and in the maximum amount of DNA that can be cloned using it. Each type of vector has been specially constructed in the laboratory to possess the features necessary to make it an efficient cloning vector.

Plasmid Cloning Vectors

Plasmids (see Chapter 7 and the chapter on "Transposable Genetic Elements," Chapter 20) are extrachromosomal genetic elements that replicate autonomously within bacterial cells. Their DNA is circular and double-stranded and carries the sequences (origins or *ori*) required for the replication of the plasmid, as well as for the plasmid's other functions. The particular plasmids used for cloning experiments are all derivatives of naturally occurring plasmids that have been "engineered" to have features that facilitate gene cloning. Because they are most commonly used, here we focus on features of *E. coli* plasmid cloning vectors.

A plasmid cloning vector must have three features (Figure 15.5):

~ FIGURE 15.5

Important features of a plasmid cloning vector.

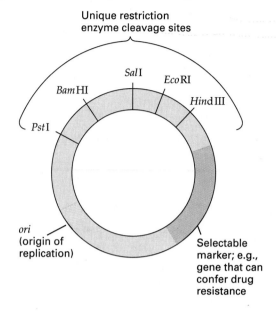

Unique restriction
enzyme cleavage sites

*Sal*I

*Eco*RI

*Bam*HI

*Hind*III

*Pst*I

ori
(origin of
replication)

Selectable
marker; e.g.,
gene that can
confer drug
resistance

1. An *ori* sequence, which directs the replication of the plasmid within the *E. coli* host.
2. A *dominant selectable marker,* which enables *E. coli* cells that carry the plasmid to be easily distinguished from cells that lack the plasmid. The usual selectable marker is a gene that confers an antibiotic resistance phenotype upon the *E. coli* host, for example, the amp^R gene for ampicillin resistance, the tet^R gene for tetracycline resistance, or the cam^R gene for chloramphenicol resistance. When plasmids carrying antibiotic resistance genes such as these are introduced by transformation (see Chapter 7) into a plasmid-free and therefore antibiotic-sensitive *E. coli* host cell, the uptake of an antibiotic-resistance plasmid can be detected easily.
3. *Unique restriction enzyme cleavage sites* for the insertion of the DNA sequences that are to be cloned.

Figure 15.6a (p. 438) diagrams one of the first plasmids that was used extensively for molecular cloning, *pBR322*. By convention, plasmid cloning vectors have names beginning with "p." The following letters may be a person's initials, a university's initials, or a name, and the number's origin varies from plasmid to plasmid. The plasmid pBR322 contains 4361 nucleotide pairs [4.361 kb (kb = kilobase pair = 1000 base pairs)]. By comparison, the *E. coli* chromosome is 4720 kb long. pBR322 contains two antibiotic-resistance genes, amp^R and tet^R, and has scattered, unique restriction enzyme cleavage sites for a number of

enzymes, such as *Eco*RI, *Hind*III, *Bam*HI, *Sal*I, *Pst*I, *Pvu*I ("P-V-U-one"), *Pvu*II ("P-V-U-two"), *Ava*I ("ah-va-one"), and *Cla*I ("clah-one"). These sites are ideal for inserting pieces of DNA for cloning, since cutting the plasmid with any of the enzymes simply converts the circular molecule into a linear molecule. Then, when a piece of DNA generated by cutting with the same restriction enzyme is added to the linear plasmid, a larger circular DNA molecule can be produced by ligating the two DNAs together.

The sites for *Hind*III, *Bam*HI, and *Sal*I lie within the tet^R gene, and the sites for *Pst*I and *Pvu*I are within the amp^R gene. Therefore, DNA fragments inserted into the *Hind*III, *Bam*HI, or *Sal*I site inactivate the tetracycline-resistance gene, and fragments inserted into the *Pst*I or *Pvu*I site inactivate the ampicillin-resistance gene.

A number of other restriction enzymes cut the pBR322 plasmid. Since these other enzymes cleave pBR322 at two or more sites, however, they are not used in cloning experiments. If the plasmid is cleaved into two or more pieces by their action, it is not possible simply to insert a piece of DNA into the plasmid and reseal the pieces into a circle containing the pBR322 parts in the correct order.

More sophisticated and useful plasmid cloning vectors have been developed in recent years. An array of vectors with various features allows a researcher to pick an appropriate vector for the experiments in which he or she is engaged. Figure 15.6b diagrams one of these newer vectors, pUC19 ("puck-19"). This 2,686 bp vector has the following features:

1. Its small size enables more DNA to be cloned than with pBR322.
2. It has a high copy number, approaching 500 copies per cell.
3. Like pBR322, pUC19 has the amp^R selectable marker.
4. Unlike pBR322, where only a few unique restriction sites are scattered around the vector, pUC19 contains a number of unique restriction sites clustered in one region. Such a region of clustered unique restriction sites is called a **polylinker** or **multiple cloning site.**
5. The polylinker is inserted near the 5' end of the *E. coli lacZ*+ gene, which encodes the enzyme β-galactosidase. The insertion was engineered so that functional β-galactosidase is produced when pUC19 is introduced into a *lacZ*− mutant cell that makes nonfunctional β-galactosidase. Then, when a piece of DNA is cloned into the polylinker, the β-galactosidase reading frame is disrupted so functional β-galactosidase can *not* be produced. A simple color test is used to distinguish colonies

~ **FIGURE 15.6**

(a) Restriction map of the plasmid cloning vector pBR322. Unique restriction sites are highlighted in green. (b) pUC19, a newer-generation plasmid cloning vector. This plasmid has a polylinker, an amp^R selectable marker, and a color selection system for distinguishing vectors with and without DNA inserts.

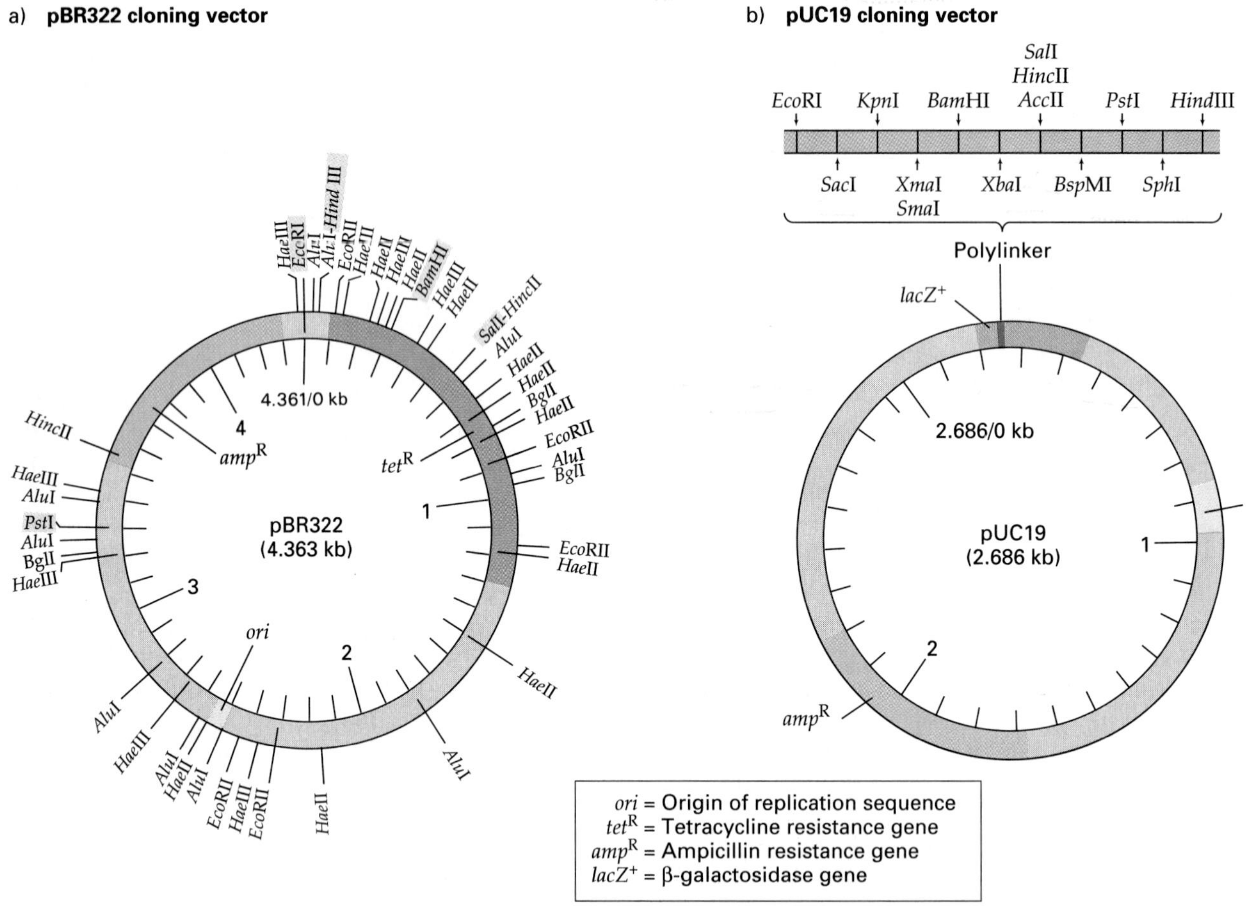

a) **pBR322 cloning vector**

b) **pUC19 cloning vector**

ori = Origin of replication sequence
tet^R = Tetracycline resistance gene
amp^R = Ampicillin resistance gene
$lacZ^+$ = β-galactosidase gene

containing pUC19 with no inserts (blue) from colonies containing pUC19 with inserts (white). This test provides a rapid, visual way to identify cells containing potentially interesting clones.

Similar vectors are available with variants on these features, for example with different arrays of unique restriction sites in the polylinker, and with phage promoters flanking the polylinker-$lacZ^+$ region. The phage promoters, such as those for T7, T3, or SP6 DNA-dependent RNA polymerases, are used to make RNA copies of the cloned DNA sequences *in vitro* in the presence of the appropriate polymerase. If the RNA is made radioactive by the incorporation of ^{32}P-NTPs in the reaction mixture, it can be used as a probe in further analytical techniques. DNA fragments of up to 5 to 12 kb are efficiently cloned in plasmid vectors. Plas-

mids carrying larger DNA fragments are often unstable and tend to lose most of the inserted DNA.

Figure 15.7 illustrates how a piece of DNA can be inserted into a plasmid cloning vector, such as pUC19. In the first step, pUC19 is cut with a restriction enzyme that has a site in the polylinker. Next, the piece of DNA to be cloned is generated by cutting high-molecular-weight DNA with the same restriction enzyme. The DNA fragments are mixed with the cut vector and randomly allowed to join. In some cases, the DNA fragment of interest becomes inserted between the two cut ends of the plasmid. The resulting recombinant DNA plasmid is introduced into an *E. coli* host by transformation. Resulting ampicillin-resistant colonies indicate cells that were transformed by plasmids, and white colonies (versus blue) on color selection plates indicate the recombinant DNA clones.

~ FIGURE 15.7

Insertion of a piece of DNA into the plasmid cloning vector pUC19 to produce a recombinant DNA molecule. pUC19 contains several unique restriction enzyme sites localized in a polylinker. Insertion of a DNA fragment into the polylinker disrupts the β-galactosidase gene, making it possible to distinguish colonies containing pUC19 without an insert (blue) from colonies containing pUC19 with an insert (white).

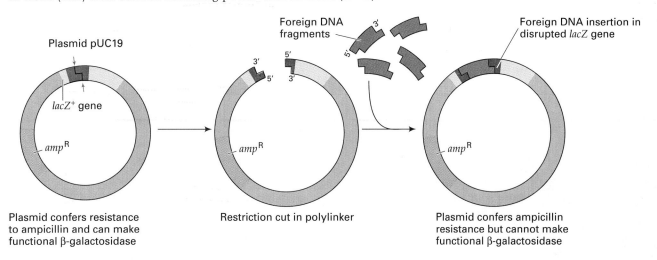

Plasmid pUC19

lacZ⁺ gene

*amp*ᴿ

Plasmid confers resistance to ampicillin and can make functional β-galactosidase

Foreign DNA fragments

5'
3'
5'
3'

*amp*ᴿ

Restriction cut in polylinker

Foreign DNA insertion in disrupted *lacZ* gene

*amp*ᴿ

Plasmid confers ampicillin resistance but cannot make functional β-galactosidase

Bacteriophage Lambda Cloning Vectors

As discussed in Chapter 7, wild-type λ phages have a choice between the lysogenic and lytic cycles when they infect the *E. coli* host cell. In the lytic cycle the λ chromosome replicates, and genes are expressed for the production of progeny phages and the subsequent lysis of the bacterial cell. The derivatives of λ that are used as cloning vectors have all been engineered so that only the lytic cycle is possible. The λ cloning vectors possess restriction enzyme sites so that DNA fragments can be cloned using them. One such vector, the lambda replacement vector (Figure 15.8; p. 440), has a chromosome in which there is a "left" arm and a "right" arm that collectively contain all the essential genes for the lytic cycle. Between the two arms is a disposable segment of DNA ("disposable" since it does not contain any genes needed for phage propagation). The junctions between the disposable central segment and the two arms each have a cleavage site for one restriction enzyme, *Eco*RI; no other *Eco*RI sites are present in the λ replacement vector's DNA.

Cloning a DNA fragment using a λ replacement vector is achieved as follows: First the λ vector is cut with the enzyme *Eco*RI to separate the two arms from the replaceable segment. Next, the DNA fragments to be cloned are generated by cutting high-molecular-weight DNA from a given organism with *Eco*RI. The foreign DNA fragments are then mixed with the λ DNA fragments and the pieces ligated together by DNA ligase.

The only splicing events that produce viable, functional λ chromosomes are those in which the foreign DNA is inserted between the left arm and right arm and in which the total length of the recombined DNA fragments is approximately 45–50 kb. Since only DNA fragments of about 45–50 kb can be packaged into lambda DNA particles, DNA molecules that are either too long or too short are unable to be packaged. Given the number of essential genes that must be present on the left and right arms for phage propagation, only about 15 kb of DNA can be inserted into a λ replacement vector. λ cloning vectors are commonly used for initial cloning of genes of interest.

Cosmid Cloning Vectors

Plasmid and λ cloning vectors can only be used to clone small DNA fragments (5 to 10 kb in plasmids and about 15 kb in λ). Since some genes from higher eukaryotes may be as long as 30 kb, they could not be cloned in one piece using these cloning vectors.

DNA fragments of approximately 35 to 40 kb can be cloned in vectors called **cosmids** (Figure 15.9; p. 440). Unlike plasmid and phage vectors, which are derived from naturally occurring genetic elements, cosmids do not occur naturally—they have been constructed by combining features of both plasmids (e.g., pBR322) and phage λ. That is, a cosmid has an *ori* sequence to permit its replication in *E. coli*, a dominant selectable marker such as *amp*ᴿ, and unique restriction sites for the insertion and cloning of DNA fragments. In

~ FIGURE 15.8

Scheme for using phage λ DNA as a cloning vector. Foreign DNA fragments of approximately 15 kb in length can be cloned in λ vectors.

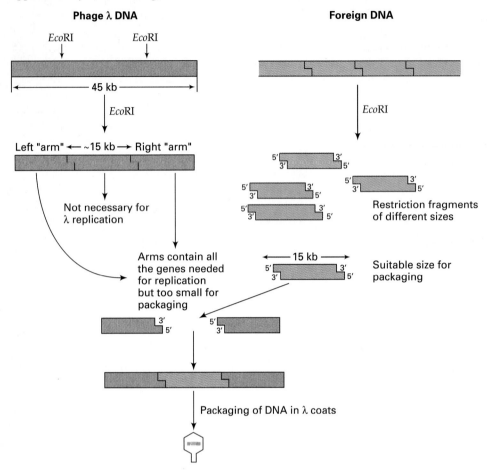

addition, cosmids have a *cos* site that is derived from phage λ. Recall from Chapter 10 that the *cos* site is the site at which multiple copies of the λ genome, attached

in one long piece called a concatamer, are cleaved into 34- to 45-kb pieces to be packaged into phage heads. Recall also that the cut at the *cos* site is staggered and results in a linear DNA molecule with sticky (single-stranded), complementary ends. In fact, all that is needed for packaging DNA into a phage head are *cos* sites that are a specified distance apart. Thus the *cos* sites on a cosmid permit packaging of DNA into a λ phage particle that facilitates the introduction of large DNA molecules into a bacterial cell. This property is absent from plasmid cloning vectors.

λ *cos* sites have been cloned into plasmid cloning vectors to produce cosmids as small as 5 kb. When a DNA fragment of about 40–45 kb is cloned into such a cosmid, the recombinant DNA molecule is then the right size to be packaged into a phage head. The phage is then used to introduce the recombinant cosmid into the *E. coli* host cell, where it replicates as a plasmid does. Recombinant cosmids that are either too small (<35 kb) or too large (> ~ 50 kb) cannot be packaged.

~ FIGURE 15.9

Cosmid cloning vehicle.

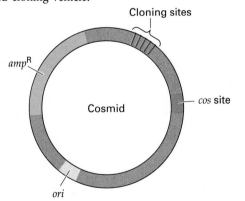

Shuttle Vectors

So far all the cloning vectors we've discussed (plasmids, λ phage, and cosmids) must be used to clone DNA within *E. coli* cells. Other vectors have been developed to introduce recombinant DNA molecules into a variety of prokaryotic and eukaryotic organisms. There are vectors that can be used to transform mammalian cells in culture, as well as vectors to transform other animal cells, and vectors to transform plant cells. However, the most developed vectors are those used to transform yeast cells. A **shuttle vector** is a cloning vector that allows it to replicate in two or more host organisms. Shuttle vectors are used for experiments in which recombinant DNA is to be introduced into organisms other than *E. coli*. Figure 15.10 diagrams the yeast-*E. coli* shuttle vector YEp24, which can be introduced into yeast or *E. coli* cells.

Like *E. coli* cloning vectors, YEp24 has an *ori* sequence that allows it to replicate in *E. coli* and has

dominant selectable marker that confer tetracycline resistance and ampicillin resistance upon *E. coli* cells that contain this vector. YEp24 also contains the selectable marker *URA3* (a wild-type yeast gene for an enzyme required for uracil biosynthesis—in yeast, wild-type genes are symbolized by uppercase letters and the corresponding mutant genes by lowercase letters). This marker enables yeast *ura3* mutant host cells containing YEp24 to be identified. That is, if YEp24 transforms a yeast cell carrying a *ura3* mutation, the yeast cell's phenotype would be changed from uracil-requiring to uracil-independent by the presence of the *URA3* gene in the YEp24 vector. YEp24 also carries a yeast-specific sequence, the two-micron circle (2μ), that allows it to replicate autonomously in a yeast cell. Thus, YEp24 is able to replicate in both yeast and *E. coli*. Not all shuttle vectors have the ability to replicate in the nonbacterial host. Those that do not will typically integrate into a host cell's chromosome and be replicated as that chromosome replicates.

~ FIGURE 15.10

The yeast-*E. coli* shuttle vector YEp24. Unique restriction enzyme cleavage sites are shown with names highlighted in green.

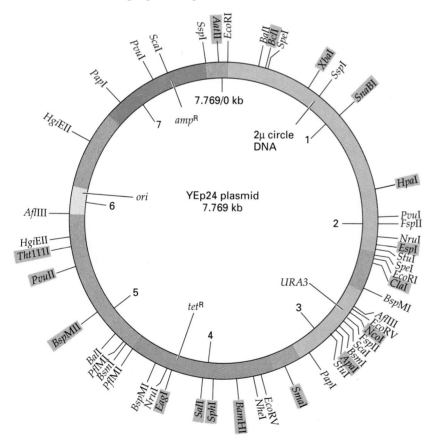

KEYNOTE

Different kinds of vectors have been developed to construct and clone recombinant DNA molecules. To be useful, all vectors must be able to replicate within their host organism, they must have one or more restriction enzyme cleavage sites at which foreign DNA fragments can be inserted, and they must have one or more dominant selectable markers. Three major types of cloning vectors are used to clone recombinant DNA in *E. coli*: plasmids, which are introduced by transformation into the *E. coli* host, where they replicate; and bacteriophage λ and cosmids, both of which are packaged into λ phage particles, which, in turn, inject the DNA into the *E. coli* host. In λ clones, the injected λ DNA replicates and progeny phages are produced, while in cosmid clones, the cosmids replicate like plasmids.

Shuttle vectors are cloning vectors that can also be used to clone recombinant DNA in more than one organism.

CONSTRUCTION OF GENOMIC LIBRARIES, CHROMOSOME LIBRARIES, AND cDNA LIBRARIES

Researchers are interested in cloning individual genes so that those genes can be studied in detail. It is relatively easy to isolate high-molecular-weight DNA from cells of an organism and to cleave that DNA into fragments through the use of restriction enzymes. The restriction fragments collectively represent the entire genome of the organism, and they may be cloned in an appropriate cloning vector. The collection of clones that contains at least one copy of every DNA sequence in the genome is called a **genomic library**. A genomic library can be searched to identify and single out a recombinant DNA molecule that contains a particular gene or DNA sequence of interest.

Alternatively, mRNA molecules can be isolated from cells and DNA copies, called **complementary DNA (cDNA)**, can be made. These cDNA molecules can then be cloned. If a specific mRNA molecule can be isolated, the corresponding cDNA can be made and cloned. The analysis of that cloned cDNA molecule can then provide information about the gene that encoded the mRNA. More typically, the entire mRNA population of a cell is isolated and a corresponding set of cDNA molecules is generated and inserted into a cloning vector to produce what is called a **cDNA library**. The following sections describe the construction of genomic and cDNA libraries.

Genomic Libraries

A genomic library is a collection of clones that, it is hoped, contains one copy of every DNA sequence in the genome. Researchers typically want to study a specific gene rather than a whole chromosome. One approach to obtaining a clone of a gene is to isolate it from a genomic library through the use of a specific probe. In this section we focus on the construction of genomic libraries of eukaryotic DNA.

PRODUCTION OF GENOMIC LIBRARIES. There are two ways to produce genomic libraries:

1. Genomic DNA is completely digested by a restriction enzyme, and the resulting DNA fragments are then cloned in either a plasmid, λ, or a cosmid cloning vector. This technique does have a drawback. If the specific gene the researcher wants to study contains restriction sites for the enzyme, the gene will be split into two or more fragments when the DNA is digested by the restriction enzyme. In this case, the gene would then be cloned in two or more pieces. Another drawback is that the average size of the fragment produced by digestion of eukaryotic DNA with restriction enzymes is relatively small (about 4 kb for restriction enzymes that have six-base-pair recognition sequences: see Table 15.2). Thus, an entire library would need to contain a very large number of recombinant DNA molecules, and screening for the specific gene would be very laborious.

2. The problems of genes split into fragments and the large number of recombinant DNA molecules can be minimized by cloning longer DNA fragments (e.g., in λ or cosmids). Longer DNA fragments can be generated by mechanically shearing high-molecular-weight (usually 100 to 150 kb) DNA. Starting with genomic DNA isolated from a large population of cells, the goal is random fragmentation of the DNA. For example, passage of the DNA through a syringe needle will produce a population of overlapping DNA fragments. However, the DNA fragments produced in this way cannot be inserted directly into a cloning vector since their ends were not generated by cutting with restriction enzymes and, hence, they will not pair with the ends produced by restriction enzyme digestion of a vector. Therefore, to clone these DNA fragments

they first must be subjected to several additional enzymatic manipulations to add appropriate ends to the molecules—a complex procedure.

A more common approach for producing DNA fragments of appropriate size for constructing a genomic library is to perform a *partial digestion* of the DNA with restriction enzymes that recognize frequently occurring four-base-pair recognition sequences (Figure 15.11a). Partial digestion means that only a portion of the available restriction sites is actually cut with the enzyme. This is achieved by limiting the amount of the enzyme used and/or the time of incubation with the DNA. The ideal result of partial digestion is a population of overlapping fragments representing the entire genome. Sucrose gradient centrifugation is then used to collect fragments of the desired size for cloning. Those fragments can be cloned directly since the ends of the fragments were produced by restriction enzyme digestion. For example, if the

~ FIGURE 15.11

Use of a restriction enzyme to produce DNA fragments of appropriate size for constructing a genomic library.

a) **Partial digestion of DNA by a restriction enzyme, e.g., *Sau*3A, generates a series of overlapping fragments, each with identical 5' GATC sticky ends**

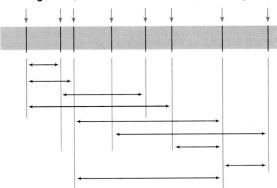

b) **Resulting fragments may be inserted into *Bam*HI site of plasmid cloning vector**

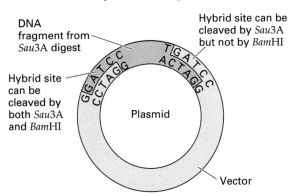

DNA is digested with the enzyme *Sau*3A, which has the recognition sequence $\frac{5'\text{-GATC-}3'}{3'\text{-CTAG-}5'}$, the ends are complementary to the ends produced by digestion of a cloning vector with *Bam*HI, which has the recognition sequence 5'-GGATCC-3' (Figure 15.11b). That is, in

$$5'\text{-GATC-}3'$$
$$3'\text{-CTAG-}5'$$

*Sau*3A cuts to the left of the G to give a 5' overhang with the sequence 5' GATC . . . . 3'. In the sequence:

$$5'\text{-GGATCC-}3'$$
$$3'\text{-CCTAGG-}5'$$

*Bam*HI cuts between the two G nucleotides also to give a 5' overhang with the sequence 5' GATCC . . . . 3'. These two 'sticky' ends can pair to produce a hybrid recognition site. Since the hybrid site contains a 5' GATC 3' sequence, it can be cleaved by *Sau*3A. However, whether it can be cleaved by *Bam*HI depends on the base pair to the 3' side of the C in the *Sau*3A-digested fragment. If it is a C/G nucleotide pair, then the hybrid site will be as follows

$$5'\text{-GGATCC-}3'$$
$$3'\text{-CCTAGG-}5'$$

the recognition site for *Bam*HI. This is the case with the left-hand hybrid site in Figure 15.11b. If any other nucleotide pair is next along the *Sau*3A fragment, the hybrid site will *not* be a *Bam*HI cleavage site (e.g., the right-hand hybrid site in Figure 15.11b).

The recombinant DNA molecules produced by ligating the *Sau*3A-cut fragments and the *Bam*HI-cut vectors together are used to transform *E. coli* cells. In the case of plasmid and cosmid libraries, the transformants are plated on selective medium to clone the sequences. Each colony that is produced represents a different cloned DNA sequence since each bacterium that gave rise to a colony contained a different recombinant DNA molecule.

For λ genomic libraries, the recombinant DNA is packaged into λ particles. Those particles are used to infect a culture of *E. coli*, and each DNA fragment is cloned by the repeated rounds of infectin and lysis that each original λ phage goes through in the culture. Eventually, the culture becomes transparent as all the bacteria have been lysed and a population of progeny λ phages, with a concentration of 10^{10} to 10^{11} phages/ml, is produced, with many, many representatives of each of the recombinant DNA molecules that were originally constructed in the test tube.

The aim of the methods just described is to produce a library of recombinant molecules that is as complete as possible. However, not all sequences of the eukaryotic genome are equally represented in such a library. For example, if the restriction sites in a particular region are very far apart or extremely close together, the chances of obtaining a fragment of clonable size are small. Additionally, some regions of eukaryotic chromosomes may contain sequences that affect the ability of vectors containing them to replicate in *E. coli*: these sequences would then be lost from the library.

Lastly, the probability of having any DNA sequence represented in the genomic library can be calculated from the formula

$$N = \frac{\ln(1 - P)}{\ln(1 - f)}$$

where N is the necessary number of recombinant DNA molecules, P is the probability desired, and f is the fractional proportion of the genome in a single recombinant DNA molecule (i.e., f is the average size, in kilobase pairs, of the fragments used to make the library divided by the size of the genome, in kilobase pairs). For example, for a 99 percent chance that a particular yeast DNA fragment is represented in a library of 40-kb fragments in a cosmid genomic library, where the yeast genome size is 14,000 kb,

$$N = \frac{\ln(1 - 0.99)}{\ln[1 - (40/14,000)]}$$
$$= 1610$$

That is, 1610 recombinant DNA molecules of 40-kb size would be needed to be 99 percent sure that a given DNA sequence was represented in the library.

$\mathcal{K}$EYNOTE

A genomic library is a collection of clones that contains at least one copy of every DNA sequence in an organism's genome. Like regular book libraries, genomic libraries are great resources of information: in this case, the information is about the genome. They are used for isolating specific genes and for studying the organization of the genome, among many other things.

Chromosome Libraries

The genome size of many organisms, including humans, is so large that many thousands of clones are needed to represent the entire genome. This makes searching the library for a gene of interest very time-consuming. One approach that has been used to reduce the screening time of large genomes is to make libraries of the individual chromosomes in the genome. In humans, this gives 23 different libraries. Then, if a gene has been localized to a chromosome by genetic means, researchers can restrict their attention to the library of that chromosome when they search for its DNA sequence.

Individual chromosomes of an organism can be separated if their morphologies and sizes are distinct enough, as is the case for human chromosomes. The most efficient procedure currently used to isolate chromosomes individually is *flow cytometry*. In this procedure, chromosomes from cells in the mitotic phase of the cell cycle are stained with a fluorescent dye. Chromosomes released from cells are passed through a laser beam connected to a light detector. This system sorts and fractionates the chromosomes based on their differences in dye binding and resulting light scattering. Approximately 100 chromosomes can be sorted and isolated per second. As a result of the application of this procedure, libraries of DNA prepared from all human chromosomes are now available to researchers.

cDNA Libraries

In **cDNA cloning**, mRNA molecules are isolated from eukaryotic cells and cDNA copies are made and inserted into a cloning vector. When the entire poly(A)-mRNA population of a cell is isolated, and cDNA copies are made and cloned, the result is a **cDNA library**. Since a cDNA library reflects the gene activity of the cell type at the time the mRNAs are isolated, the construction and analysis of cDNA libraries is useful for comparing gene activities in different cell types of the same organism, since there would be similarities and differences in the clones represented in the cDNA libraries of each cell type.

It is important to recognize that the clones in the cDNA library represent mature mRNAs found in the cell. Since mature mRNAs are processed molecules, the sequences obtained are *not* equivalent to gene clones. In particular, intron sequences are present in gene clones but *not* in cDNA clones. For any mRNA, cDNA clones can be useful for subsequently isolating the corresponding gene for the mRNA.

SYNTHESIS OF cDNA MOLECULES.

In eukaryotes, RNA extracted from cells contains molecules of ribosomal RNA, transfer RNA, messenger RNA, and snRNA. The mRNA in this mixture are unique in that most, but not all, contain a poly(A) tail. These poly(A)$^+$ mRNAs can be purified from the mixture by passing the RNA molecules over a column to which short chains of deoxythymidylic acid, called *oligo(dT) chains*, have been attached. As the RNA molecules pass through an oligo(dT) column, the poly(A) tails on the mRNA molecules form complementary base pairs with the oligo(dT) chains. As a result the mRNAs are captured on the column while the other RNAs pass through. The captured mRNAs are subsequently released and collected, for example, by increasing the ionic strength of the buffer passing over the column so that the hydrogen bonds will be disrupted. This method results in signifi-cant enrichment of poly(A)$^+$ mRNAs in the mixed RNA population to about 50% versus about 3% in the cell.

Once an enriched population of mRNA molecules has been isolated, double-stranded complementary DNA (cDNA) copies are made in vitro, using the enzyme reverse transcriptase (Figure 15.12). First, a short oligo(dT) chain (a short, single-stranded DNA chain containing only T nucleotides) is hybridized to the poly(A) tail at the 3' end of each mRNA strand. The oligo(dT) acts as a primer for **reverse transcriptase** (RNA-dependent DNA polymerase), which makes a complementary DNA copy of the mRNA strand. (The enzyme's name comes from the fact that it catalyzes a reaction that is the reverse of transcription: i.e., RNA → DNA.) Next, RNase H, DNA polymerase I, and DNA ligase are used to synthesize the second DNA

~ FIGURE 15.12

The synthesis of double-stranded complementary DNA (cDNA) from a polyadenylated mRNA using reverse transcriptase, RNase H, DNA polymerase I, and DNA ligase.

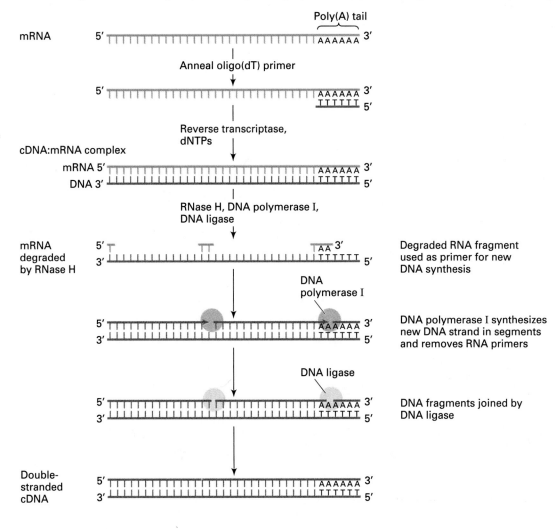

strand. RNase H degrades the RNA strand in the hybrid DNA-mRNA, DNA polymerase I makes new DNA fragments using the partially degraded RNA fragments as primers, and finally DNA ligase ligates the new DNA fragments together to make a complete chain. The result is a double-stranded cDNA molecule, the sequence of which is derived from the original poly(A)-mRNA molecule.

PRODUCTION OF cDNA LIBRARIES. Once cDNA molecules have been synthesized, they must be cloned. This is usually done by adding specific short DNA sequences, called *linkers,* which contain cleavage sites for a particular restriction enzyme so that the cDNAs can be inserted into a cloning vector.

Figure 15.13 illustrates the cloning of cDNA using a **restriction site linker,** or **linker,** which is a relatively short, double-stranded piece of DNA (oligodeoxyribonucleotide) about 8 to 12 nucleotide pairs long. The linker is synthesized by chemical means and contains the cleavage site for a particular restriction

enzyme within its sequence. The linker shown in Figure 15.13 is the *Bam*HI linker (abbreviated in the figure to just the six-base-pair recognition sequence for the enzyme).

Both the cDNA molecules and the linkers have blunt ends, and they can be joined together at high concentrations of T4 DNA ligase. Sticky ends are produced in the cDNA molecule by cleaving the cDNA (with linkers at each end) with *Bam*HI. The resulting DNA is linked into a cloning vector that has also been cleaved with *Bam*HI. The resulting recombinant DNA molecule is transformed into an *E. coli* host cell for cloning.

𝒦EYNOTE

Given a population of mRNAs purified from a cell, it is possible to make DNA copies of those mRNA molecules. First, the enzyme reverse transcriptase makes a single-stranded DNA copy of the mRNA; then RNase H, DNA polymerase I, and DNA ligase are used to make a double-stranded DNA copy called complementary DNA (cDNA). This cDNA can be spliced into cloning vectors using restriction site linkers. ───────

~ FIGURE 15.13

The cloning of cDNA by using *Bam*HI linkers.

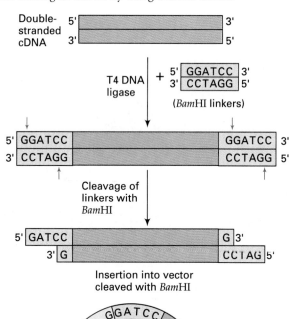

IDENTIFYING SPECIFIC CLONED SEQUENCES IN cDNA LIBRARIES AND GENOMIC LIBRARIES

Identifying Specific Cloned Sequences in a cDNA Library

A cDNA clone can be identified in a cDNA library in a number of ways. One way is to select the cDNA clone that codes for a specific protein (Figure 15.14). This approach requires that a protein can be purified in sufficient quantities so that antibodies can be made against the protein. It also requires that the cDNAs be cloned in a special kind of vector called an *expression vector* (Figure 15.14,1). In such a vector, the cDNA is placed next to a promoter sequence and to translation start signals. In the host an mRNA will be transcribed corresponding to the cDNA and the translation of the mRNA will produce the encoded protein. The expression vector also contains a transcription termination signal downstream of the cDNA sequence.

~ FIGURE 15.14

Identifying specific cDNA plasmids in a cDNA library.

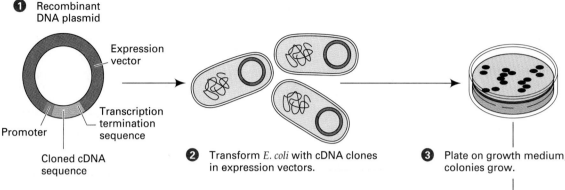

1 Recombinant DNA plasmid

Expression vector

Promoter

Transcription termination sequence

Cloned cDNA sequence

2 Transform *E. coli* with cDNA clones in expression vectors.

3 Plate on growth medium; colonies grow.

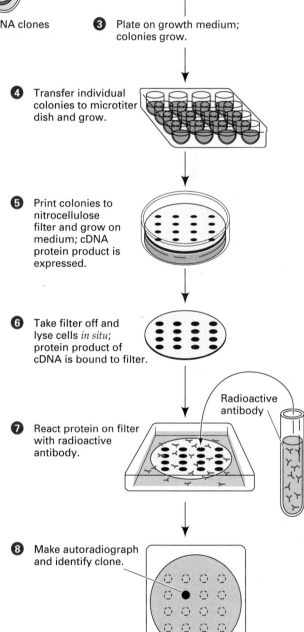

4 Transfer individual colonies to microtiter dish and grow.

5 Print colonies to nitrocellulose filter and grow on medium; cDNA protein product is expressed.

6 Take filter off and lyse cells *in situ*; protein product of cDNA is bound to filter.

7 React protein on filter with radioactive antibody.

Radioactive antibody

8 Make autoradiograph and identify clone.

Theoretically, each individual *E. coli* cell in a population transformed with a cDNA library in an expression vector contains a different cDNA clone (Figure 15.14,2). When the cells are plated, each bacterium will give rise to a colony (Figure 15.14,3). These clones are preserved, for example, by picking each colony off the plate and placing it into the medium in a well of a microtiter dish (Figure 15.14,4). Replicas of the set of clones are placed (printed) onto a nitrocellulose filter that has been placed on a petri plate of selective medium appropriate for the recombinant molecules (Figure 15.14,5). Incubation of the plate produces colonies growing on the filter in the same pattern produced by the clones in the microtiter dish. The filter is peeled from the dish and treated to lyse the cells *in situ* (Figure 15.14,6). The proteins within the cell, including those expressed from the cDNA, become stuck to the filter. The filter is then incubated with an antibody to the protein of interest (Figure 15.14,7).

If the antibody is radioactively labeled, any clones that expressed the protein of interest can be identified by placing the dried filter against X-ray film, leaving it in the dark for a period of time (from one hour to overnight) to produce what is called an *autoradiogram* (Figure 15.14,8). The process is called *autoradiography*. When the film is developed, dark spots are seen wherever the radioactive probe bound to the filter in the antibody reaction (Figure 15.14,9). (The dark spots are the result of the decay of the radioactive atoms changing the silver grains of the film). A cDNA clone can be used as a probe, for example, to analyze the genome of the same or other organisms for homologous sequences, to isolate the nuclear gene for the mRNA from a genomic library, or to quantify mRNA production.

Identifying Specific Cloned Sequences in a Genomic Library

Given the existence of a probe, such as a cloned cDNA probe, it is possible to identify in a genomic library the cloned gene that codes for the mRNA molecule from which the cDNA was made, and then to isolate it for characterization.

Screening plasmid and cosmid libraries. The process of screening for specific DNA sequences in a plasmid or cosmid library is shown in Figure 15.15. The process is similar to that just described for screening a cDNA library. As in that procedure, clones in a microtiter dish (Figure 15.15,1) are printed onto a nitrocellulose filter, on the surface of which the colonies grow (Figure 15.15,2). The filter is placed sequentially on filter papers that have been soaked in different solutions to lyse the bacteria, denature the DNA to single strands, and then bind the single-stranded DNA to the filter (Figure 15.15,3).

Next, the filter is placed in a plastic bag (e.g., Seal-A-Meal bags, which are used for boiling vegetables), and incubated with the cDNA probe (Figure 15.15,4), which has been made radioactive through the incorporation of ^{32}P (Box 15.1; pp. 450–51).

To prepare the radioactive DNA for use as a probe, the DNA is denatured by boiling and then quickly cooled on ice to produce single-stranded DNA molecules. These radioactive molecules are added to the filters to which the denatured (single-stranded) DNA from each colony has been bound. Wherever the two sequences are complementary, DNA-DNA hybrids will form between the probe and the colony DNA. If the cDNA probe is derived from the mRNA for β-globin, for example, that probe will hybridize with DNA bound to the filter that encodes the β-globin mRNA, that is, the β-globin gene. After sufficient time for hybridization has elapsed (approximately 24 hours), the filters are washed to remove unbound probe, dried and subjected to autoradiography (Figure 15.15,5). From the position(s) of the spot(s) on the film (resulting from the decay of the ^{32}P), the locations of the original bacterial culture(s) in the microtiter dish(es) can be determined and the clone(s) of interest isolated for further characterization.

Screening λ genomic libraries. To screen for a clone of interest in a genomic library made in λ, samples of the phage suspension are plated on a bacterial lawn so that plaques are produced (Figure 15.16). Each plaque consists of a pool of progeny phages produced by successive rounds of infection and lysis of bacteria initiated originally by a single phage infection. Each time a bacterium is lysed by a phage infection, not only are progeny phages released, but also released are many

~ Figure 15.15

Screening plasmid and cosmid libraries for specific DNA sequences.

1. Microtiter dish contains plasmid or cosmid library.

2. Library replica plated on nitrocellulose filter on growth medium; colonies produced.

Filter
Colony
Growth medium

DNA

3. Filter removed from culture dish, bacteria lysed, DNA denatured and bound to filter.

Probe DNA hybridized to DNA on filter

4. Radioactive probe solution added to filter in Seal-a-Meal bag.

Autoradiogram

Clone detected by probe

5. Autoradiogram made from washed, dried filter. Dark spots indicate clones detected by probe.

copies of replicated λ DNA that were not packaged into particles. A piece of nitrocellulose filter is laid carefully on top of the plate and left there for a few minutes, permitting the λ DNAs to bind to the filter; this is called making a *plaque lift*. The filter is removed and

~ FIGURE 15.16

Screening a bacteriophage λ library for a specific gene clone.

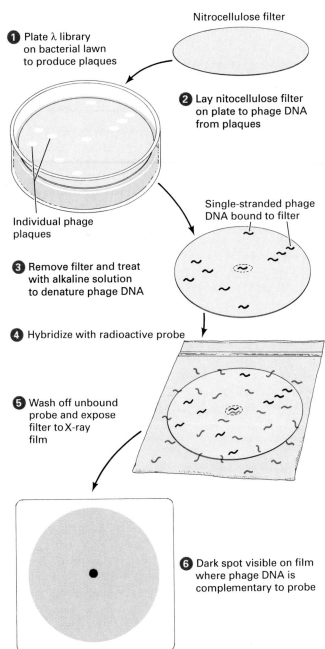

1 Plate λ library on bacterial lawn to produce plaques

Nitrocellulose filter

2 Lay nitocellulose filter on plate to phage DNA from plaques

Individual phage plaques

Single-stranded phage DNA bound to filter

3 Remove filter and treat with alkaline solution to denature phage DNA

4 Hybridize with radioactive probe

5 Wash off unbound probe and expose filter to X-ray film

6 Dark spot visible on film where phage DNA is complementary to probe

placed on blotting paper soaked in an alkaline solution, which denatures the DNA into single strands. After neutralization, the filters are incubated with a radioactive probe, as described for the screening of the plasmid and cosmid libraries.

A number of plaque lifts can be made from each plate of λ plaques, allowing the investigator to screen the same library simultaneously with a number of different probes. Evidence of hybridization of the probe with DNA from a plaque is seen as a spot on the auto-

radiogram. The position of the spot on the film is used to find the plaque on the plate from which the plaque lift was made. Phages with the cloned DNA fragment of interest can be isolated from the plaque and propagated in a bacterial culture to produce large quantities of phage DNA for analysis.

Identifying Specific DNA Sequences in Libraries Using Heterologous Probes

Originally cDNA probes were used to identify and isolate genes. To date, a very large number of genes have been cloned from both prokaryotes and eukaryotes. Thus, it is more common these days to identify genes of interest in a genomic library by using clones of equivalent genes from other organisms as probes. Such probes are called *heterologous probes* and their effectiveness depends upon a good degree of homology between the probes and the genes. For that reason, the greatest success with this approach has come either with highly conserved genes or with probes from a species closely related to the organism from which a particular gene is to be isolated.

Identifying Genes in Libraries by Complementation of Mutations

For those organisms in which genetic systems of analysis have been well developed and for which there are well-defined mutations, it is possible to clone genes by complementation of those mutations. For example, the yeast *Saccharomyces cerevisiae* has been extremely well exploited genetically, a large number of mutations have been generated and characterized, and integrative and replicative transformation systems using *E. coli*-yeast shuttle vectors (see earlier) have been developed.

To clone a yeast gene by complementation, first a genomic library is made of DNA fragments from the wild-type yeast strain in a replicative shuttle vector such as YEp24 (see Figure 15.10). The library is used to transform a host yeast strain carrying a mutation to enable transformants to be selected—*ura3* in the case of YEp24—and a mutation in the gene for which the wild-type gene is to be cloned. Let us consider the cloning of the *ARG1* gene, the wild-type gene for an enzyme needed for arginine biosynthesis (Figure 15.17; p. 451) by complementation of an *arg1* mutation. A yeast strain carrying the *arg1* mutation would have an inactive enzyme for arginine biosynthesis and, hence, a growth requirement for arginine. A genomic library is made using DNA isolated from a wild-type (*ARG1*) yeast strain (Figure 15.17,1 and 2). When a population of *ura3 arg1* yeast cells is transformed with the YEp24 genomic library (Figure 15.17,3), some cells will receive plasmids containing the normal (*ARG1*) gene for the

Box 15.1
Radioactive Labeling of DNA

The radioactive labeling of the cDNA probe may be done by the process of **nick translation** (Box Figure 15.1). In nick translation, the DNA to be labeled is incubated in a buffer containing the enzymes DNase I and DNA polymerase I, and the four deoxyribonucleoside triphosphate DNA precursors (dATP, dGTP, dCTP, and dTTP). In the reaction mixture, one or more of the deoxyribonucleoside triphosphate precursors has ^{32}P in the phosphate group, which is attached to the 5′ carbon of the deoxyribose sugar. (In Box Figure 15.1, the dCTP is ^{32}P-labeled.) This phosphate group is called the α-phosphate because it

is the first in the chain of three; the α-phosphate is used in forming the phosphodiester bonds of the sugar-phosphate backbone. In the reaction mixture, the DNase I enzyme makes "nicks," that is, single-stranded gaps in the backbones of the DNA molecules. DNA polymerase I recognizes the nicks and proceeds to extend the DNA chain from the free 3′ end, simultaneously removing DNA ahead of it by its 5′-to-3′ exonuclease activity. Wherever a G nucleotide is on the intact template strand, the new DNA strand will have a ^{32}P-labeled C nucleotide incorporated. In this way, the DNA becomes radioactively labeled with ^{32}P.

DNA can also be labeled by the newer *random primer* method (Box Figure 15.2). The DNA is denatured to single strands by boiling and quick cooling on ice. DNA primers, six nucleotides long (hexanucleotides), synthetically made by the random incorporation of nucleotides, are annealed to the

~ Box Figure 15.1
Radioactive labeling of DNA fragments by nick translation. Here the reaction is shown occurring on one strand; in actuality both strands become labeled by nick translation.

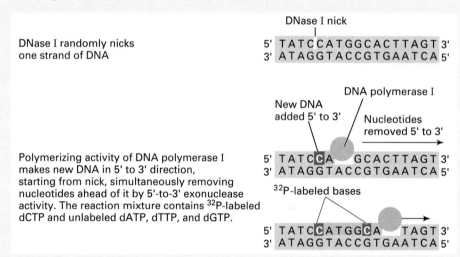

arginine biosynthesis enzyme. The plasmid's *ARG1* gene will be transcribed, and the resultant mRNA will be translated to produce a normal, functional enzyme for arginine biosynthesis. The cell will be able to grow on minimal medium—that is, in the absence of arginine—despite the presence of a defective *arg1* gene on the cell's chromosomes (Figure 15.17,4). The *ARG1* gene is said to overcome the functional defect of the *arg1* mutation by *complementation* of that mutation (Figure 15.17,5 and 6). In sum, the cells transformed by the *ARG1*-containing plasmid vector are isolated in

this experiment by selection on a growth medium lacking arginine (and lacking uracil to select for transformation). The plasmid is then isolated from the cells and the cloned gene is characterized.

Identifying Genes or cDNAs in Libraries Using Oligonucleotide Probes

A number of genes have been isolated from libraries by using synthetically-made oligonucleotide probes. (Oligonucleotides—literally, few nucleotides—are rela-

DNA. The *hexanucleotide random primers* will pair with complementary sequences in the DNA, and this will occur at many locations because all possible hexanucleotide sequences are present. The primers are elongated by the Klenow fragment of DNA polymerase I that uses radioactively-labeled precursors (dNTPs). (The Klenow fragment lacks 3'-to-5' exonuclease activity which would otherwise remove the short primer.)

~ **BOX FIGURE 15.2**
Random primer method of radioactively labeling DNA.

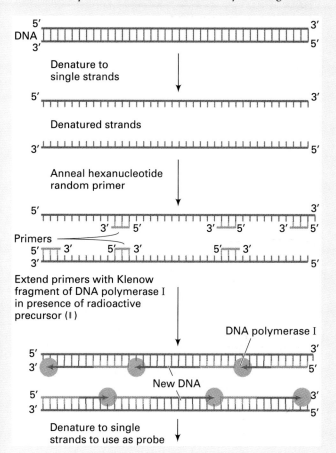

~ **FIGURE 15.17**

Example of cloning a gene by complementation of mutations: the cloning of the yeast *ARG1* gene.

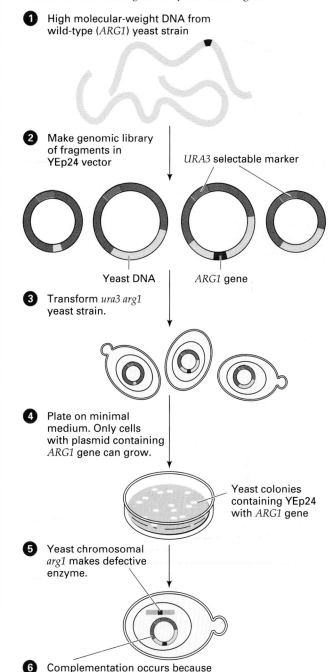

1 High molecular-weight DNA from wild-type (*ARG1*) yeast strain

2 Make genomic library of fragments in YEp24 vector

URA3 selectable marker

Yeast DNA   *ARG1* gene

3 Transform *ura3 arg1* yeast strain.

4 Plate on minimal medium. Only cells with plasmid containing *ARG1* gene can grow.

Yeast colonies containing YEp24 with *ARG1* gene

5 Yeast chromosomal *arg1* makes defective enzyme.

6 Complementation occurs because *ARG1* in vector produces functional enzyme.

tively short, single-stranded pieces of DNA.) This method requires that at least some of the amino acid sequence is known for the protein encoded by the gene. Then, since the genetic code is universal, oligonucleotides about 20 nucleotides long can be designed which, if translated, would give the known amino acid sequence. Because of the degeneracy of the genetic code—up to six different codons can specify a given amino acid—a number of different oligonucleotides are made, all of which could encode the targeted amino acid sequence. These mixed oligonucleotides are used as probes to search the libraries with the hope that at least one of the oligonucleotides will detect the gene or cDNA of interest. While not successful all of the time, the overall success rate makes this a viable method to use where nucleotide sequence information is unavailable for a gene or mRNA.

KEYNOTE

Specific sequences in cDNA libraries and genomic libraries can be identified using a number of approaches, including the use of specific antibodies, cDNA probes, heterologous probes, complementation of mutations, and mixed oligonucleotide probes.

TECHNIQUES FOR THE ANALYSIS OF GENES AND GENE TRANSCRIPTS

Restriction Enzyme Analysis of Genes

As part of the analysis of genes, it is often useful to determine the arrangement and specific locations of restriction sites. This information is useful, for example, for comparing homologous genes in different species, for analyzing intron organization, or for planning experiments to clone parts of a gene, such as its promoter or controlling sequences, into a vector. The arrangement of restriction sites in a gene can be analyzed without actually cloning the gene by using a cDNA probe or a closely related heterologous gene probe. The process of analysis proceeds as follows:

1. Samples of high-molecular-weight DNA are cut with different restriction enzymes (Figure 15.18,1 and 2), each of which will produce DNA fragments of different lengths (depending on the locations of the restriction sites on the DNA molecules).

2. The DNA restriction fragments are separated according to their molecular size by *agarose gel electrophoresis* (Figure 15.18,3). The gel is a rectangular, horizontal slab of agarose (a firm, gelatinous material), which has a matrix of pores through which DNA passes in an electric field. Each gel is cast by boiling a buffered agarose solution and allowing it to cool in a mold. A toothed comb is used to divide the gel into a number of lanes so that a number of different samples can be analyzed simultaneously. Since DNA is negatively charged due to its phosphates, the DNA migrates toward the positive pole. Because small DNA fragments can "squirm" more readily through the small pores in the gel, the small fragments move through the gel more rapidly than large DNA fragments. Thus, the smallest fragments migrate the farthest distance while the largest fragments migrate the least.

After electrophoresis, the DNA is stained with ethidium bromide so that it can be seen under ultraviolet light illumination. When total cellular DNA is digested with a restriction enzyme, the result is usually a continuous smear of fluorescence down the length of the gel lane due to the fact that the enzyme produces fragments of all sizes.

3. The DNA fragments are transferred to a nitrocellulose filter so that they are in exactly the same relative position on the filter as they were on the gel (Figure 15.18,4). The transfer to the nitrocellulose filter is done by the **Southern blot technique** (named after its inventor, Edward Southern), diagramed in Figure 15.19 (p. 454). In brief, the gel is soaked in alkali to denature the double-stranded DNA into single strands. The gel is neutralized and placed on a piece of blotting paper that spans a glass plate. The ends of the paper are in a container of buffer and act as wicks. A piece of nitrocellulose filter is laid down so that it covers the gel. Additional sheets of blotting paper (or paper towels) and a weight are stacked on top of the nitrocellulose filter. The buffer solution in the bottom tray is wicked up by the blotting paper, passed through the gel, through the nitrocellulose filter, and finally into the stack of blotting paper. During this process, the DNA fragments are picked up by the buffer and transferred from the gel to the nitrocellulose, to which they adhere because of nitrocellulose's chemical properties. The fragments on the filter are arranged in exactly the same way as they were in the gel (Figure 15.18,5).

4. Once the Southern blot technique is completed, the single-stranded DNA fragments are fixed to the nitrocellulose filter by baking it in a vacuum at 80°C for 2 hours. The filter is then placed in a hybridization buffer, and any radioactive probe added to the nitrocellulose filter at this point will bind to any complementary DNA fragment (see Figure 15.18,6).

5. After the hybridization of the probe and the DNA fragments, the filter is washed to remove radioactive probes that have not hybridized. The filter is dried, and an autoradiograph is prepared to determine the position(s) of the hybrids. If DNA fragment size markers (e.g., λ DNA cut with *Hin*dIII or with *Hin*dIII and *Eco*RI) are separated in a different lane in the agarose gel electrophoresis process, the sizes of the genomic restriction fragments that hybridized with the probe can be calculated. From the fragment sizes obtained, a **restriction map** can be generated to show the relative positions of the restriction sites. Suppose, for example, that using only *Bam*HI produces a DNA fragment of 3 kb that hybridizes with the radioactive probe. If a combination of *Bam*HI and *Pst*I is then used and produces two DNA fragments, one of 1 kb and the

~ FIGURE 15.18

Procedure for the analysis of cellular DNA for the presence of sequences complementary to a radioactive probe, such as a cDNA molecule made from an isolated mRNA molecule. The hybrids, shown as three bands in this theoretical example, are visualized by autoradiography.

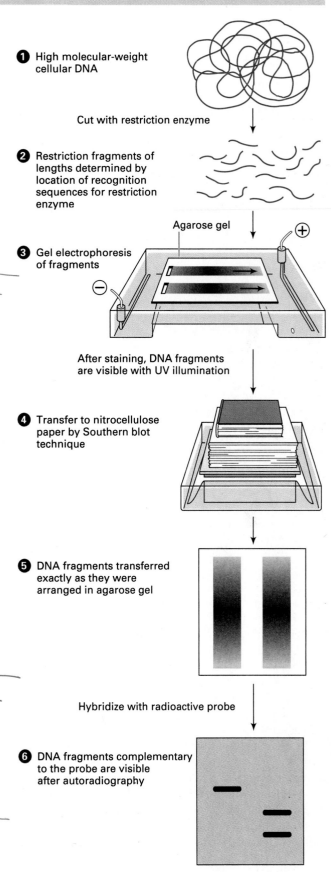

1 High molecular-weight cellular DNA

Cut with restriction enzyme

2 Restriction fragments of lengths determined by location of recognition sequences for restriction enzyme

Agarose gel

3 Gel electrophoresis of fragments

After staining, DNA fragments are visible with UV illumination

4 Transfer to nitrocellulose paper by Southern blot technique

5 DNA fragments transferred exactly as they were arranged in agarose gel

Hybridize with radioactive probe

6 DNA fragments complementary to the probe are visible after autoradiography

other of 2 kb, we would deduce that the 3-kb *Bam*HI fragment contains a *Pst*I restriction site 1 kb from one end and 2 kb from the other end. Further analysis with other enzymes, individually and combined, enables the researcher to construct a map of all the enzyme sites relative to all other sites.

Restriction Enzyme Analysis of Cloned DNA Sequences

Cloned DNA sequences are also analyzed to determine the arrangement and specific locations of restriction sites for essentially the same reasons as those for determining gene restriction sites. Because cloned DNA sequences represent a homogeneous population of relatively low-molecular-weight DNA molecules, restriction enzyme cleavage of cloned DNA sequences produces a relatively small number of discretely sized DNA fragments rather than a continuous array of different sized DNA fragments as is the case with cleavage of genomic DNA. These DNA fragments are readily visualized following agarose gel electrophoresis and ethidium bromide staining, permitting restriction maps to be constructed without the need for hybridization with a radioactive probe and autoradiography.

EXAMPLE OF RESTRICTION MAPPING. Let us consider the construction of a relatively simple restriction map (Figure 15.20; p. 455). We have cloned a 5.0-kb piece of DNA and wish to construct a restriction map of it (Figure 15.20,1). Three samples of the linear DNA are digested with restriction enzymes as follows: one sample is digested with *Eco*RI, a second is digested with *Bam*HI, and the third is digested with both *Eco*RI and *Bam*HI. The results of each reaction are analyzed on an agarose gel alongside a control of the same DNA *not* cut with any enzyme (Figure 15.20,2). Included are size markers so that the sizes of the DNA fragments can be computed. After electrophoresis, the gel is stained with ethidium bromide and photographed under ultraviolet light. From the photograph the distance each

~ FIGURE 15.19

Southern blot technique for the transfer of DNA fragments from an electrophoretic gel to nitrocellulose paper.

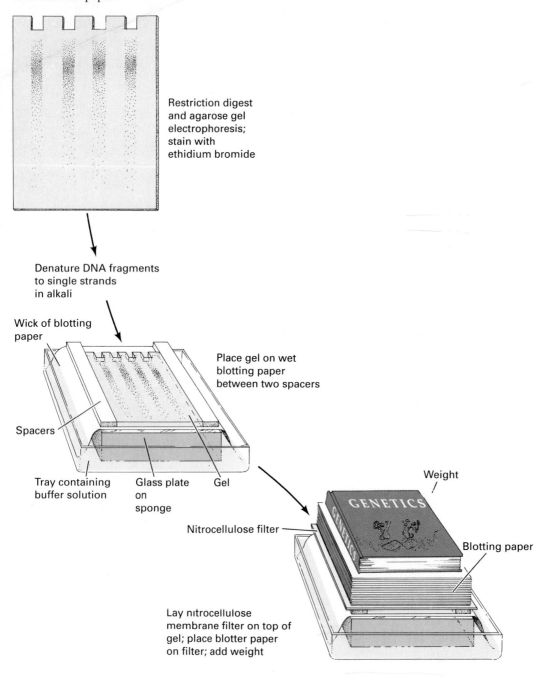

Restriction digest and agarose gel electrophoresis; stain with ethidium bromide

Denature DNA fragments to single strands in alkali

Wick of blotting paper

Place gel on wet blotting paper between two spacers

Spacers

Tray containing buffer solution

Glass plate on sponge

Gel

Weight

GENETICS

Nitrocellulose filter

Blotting paper

Lay nitrocellulose membrane filter on top of gel; place blotter paper on filter; add weight

DNA band migrated can be measured. In the case of the marker, the molecular sizes of each DNA band is known, so a calibration curve can be drawn of DNA size (in kb) vs. migration distance (in mm) (Figure 15.20,3). The migration distances for the DNA bands from our uncut and cut DNA are then used with the calibration curve to determine the molecular sizes of the DNA fragments in the bands (Figure 15.20,4). For our theoretical example the results are (Figure 15.20,5):

SAMPLE	FRAGMENT SIZES
Uncut DNA	5.0 kb
DNA cut with *Eco*RI	4.5 kb, 0.5 kb
DNA cut with *Bam*HI	3.0 kb, 2.0 kb
DNA cut with *Eco*RI and *Bam*HI	2.5 kb, 2.0 kb, 0.5 kb

~ **FIGURE 15.20**

Construction of a restriction map for *Eco*RI and *Bam*HI in a DNA fragment.

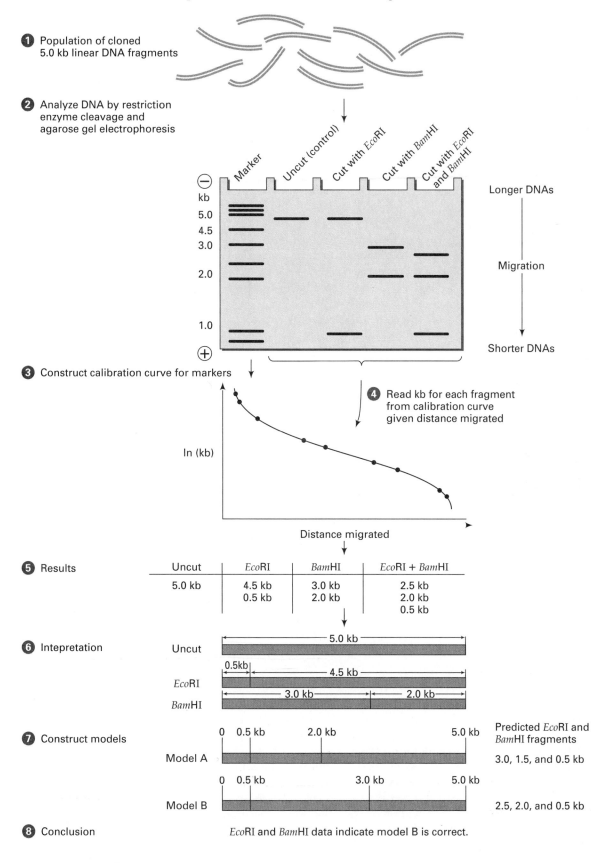

1 Population of cloned 5.0 kb linear DNA fragments

2 Analyze DNA by restriction enzyme cleavage and agarose gel electrophoresis

Marker
Uncut (control)
Cut with *Eco*RI
Cut with *Bam*HI
Cut with *Eco*RI and *Bam*HI

⊖
kb
5.0
4.5
3.0
2.0

1.0

⊕

Longer DNAs

Migration

Shorter DNAs

3 Construct calibration curve for markers

4 Read kb for each fragment from calibration curve given distance migrated

ln (kb)

Distance migrated

5 Results

Uncut	*Eco*RI	*Bam*HI	*Eco*RI + *Bam*HI
5.0 kb	4.5 kb	3.0 kb	2.5 kb
	0.5 kb	2.0 kb	2.0 kb
			0.5 kb

6 Interpretation

Uncut
5.0 kb

*Eco*RI
0.5kb | 4.5 kb

*Bam*HI
3.0 kb | 2.0 kb

7 Construct models

Model A
0   0.5 kb   2.0 kb   5.0 kb

Model B
0   0.5 kb   3.0 kb   5.0 kb

Predicted *Eco*RI and *Bam*HI fragments

3.0, 1.5, and 0.5 kb

2.5, 2.0, and 0.5 kb

8 Conclusion

*Eco*RI and *Bam*HI data indicate model B is correct.

The results are analyzed as follows (Figure 15.20,6–8):

1. The uncut DNA is confirmed to be 5.0 kb long.
2. When the DNA is cut with EcoRI, two DNA fragments result. This indicates that there is only one restriction site for EcoRI in the DNA. The molecular sizes indicate that this site is 0.5 kb from one end of the molecule.
3. Using similar logic, there is one restriction site for BamHI located 2 kb from one end of the molecule.
4. At this point, we know there is one restriction site for each enzyme, but we don't know the relationship between the two. From the information we have, we can make two models (see Figure 15.20,7). In model A, the EcoRI site is 0.5 kb from one end and the BamHI site is 2.0 kb from that same end. In model B, the EcoRI site is 0.5 kb from one end and the BamHI site is 3.0 kb from that end (i.e., 2.0 kb from the other end). Model A predicts that cutting with EcoRI and BamHI will produce three fragments of 0.5, 1.5, and 3.0 kb (going from left-to-right along the DNA). In model B, cutting with both enzymes will produce three fragments of 0.5, 2.5, and 2.0 kb. The actual data show three fragments with sizes 2.5, 2.0 and 0.5 kb indicating that model B is correct.

In real situations, restriction mapping typically involves data that are much more complicated, for example involving more restriction enzymes and a number of sites for each enzyme. Analysis may be done with a complete plasmid (which is circular) or with the cloned sequence or part of it cut out of the plasmid and purified by *preparative agarose gel electrophoresis*. In this technique, large quantities of DNA are cut with appropriate restriction enzymes, and the fragments are separated by agarose gel electrophoresis. After staining with ethidium bromide, the bands can be visualized under an ultraviolet light. At that time the bands can be physically cut out of the gel and the DNA extracted for analysis.

MAPPING RESTRICTION SITES BY PARTIAL DIGESTION. If there are many restriction sites for a given enzyme on a DNA fragment, another restriction mapping procedure can be used. This method involves partial digestion of a DNA fragment with a restriction enzyme. The DNA fragment is labeled at its end so that all fragments *starting with that end* can be visualized by autoradiography. The method is illustrated for a theoretical DNA fragment in Figure 15.21.

First, a cloned, linear piece of DNA is labeled at the 5′ ends with ^{32}P (Figure 15.21,1). The DNA fragments are cut with a restriction enzyme (enzyme A) that has a restriction site located asymmetrically along the DNA

so that the two resulting fragments can easily be distinguished by size (Figure 15.21,2). The two fragments are separated by gel electrophoresis and the fragments are purified separately from the gel (Figure 15.21,3). Both fragments can be mapped separately for restriction sites by partial digestion with enzyme B, a restriction enzyme known to have multiple sites in the fragments (Figure 15.21,4). (Partial digestion is done by reducing the amount of enzyme in the reaction and the time of the reaction is short so that only a few of the sites for the enzyme are cut.) As a result, all possible combinations of cuts at the restriction sites will occur. By using end-labeling, though, our analysis is confined to only those fragments that start with the labeled end when autoradiography is used to identify fragments after electrophoresis (Figure 15.21,5). From the migration distances of the fragments and a marker calibration curve, the sizes of the fragments are computed. This gives directly the locations of the restriction sites from the labeled end. For example, if there were six labeled fragments detected with sizes (reading from the top of the gel) 2.3 kb, 2.0 kb, 1.2 kb, 0.9 kb, 0.6 kb, and 0.1 kb, we would conclude that the following restriction map is the case:

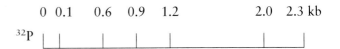

$$0 \quad 0.1 \qquad 0.6 \quad 0.9 \quad 1.2 \qquad\qquad 2.0 \quad 2.3 \text{ kb}$$

Analysis of Gene Transcripts

Geneticists raise a number of questions about gene transcripts, including: Is a gene active in a particular cell type? How much of the transcript is present in different tissues or at different phases of the cell cycle, or at different developmental stages, etc.? Here we will consider one method of transcript analysis.

NORTHERN BLOT ANALYSIS. A related blotting technique to the Southern blot technique (called **northern blot analysis**) has been developed to analyze RNA rather than DNA. (In this case, the name is derived, not from a person, but to indicate that it is a technique essentially opposite to the Southern technique.) In northern blotting, RNA extracted from a cell is separated by size using gel electrophoresis, and the RNA molecules are transferred and bound to a filter in a procedure that is essentially identical to Southern blotting. After hybridization with a radioactive probe, an autoradiograph is taken of the bands corresponding to the RNA species that were complementary to the probe. If appropriate RNA size markers are present in one of the gel lanes, the sizes of the RNA species identified with the probe can be determined.

~ FIGURE 15.21

Mapping the locations of multiple restriction sites for an enzyme in a DNA fragment by partial digestion.

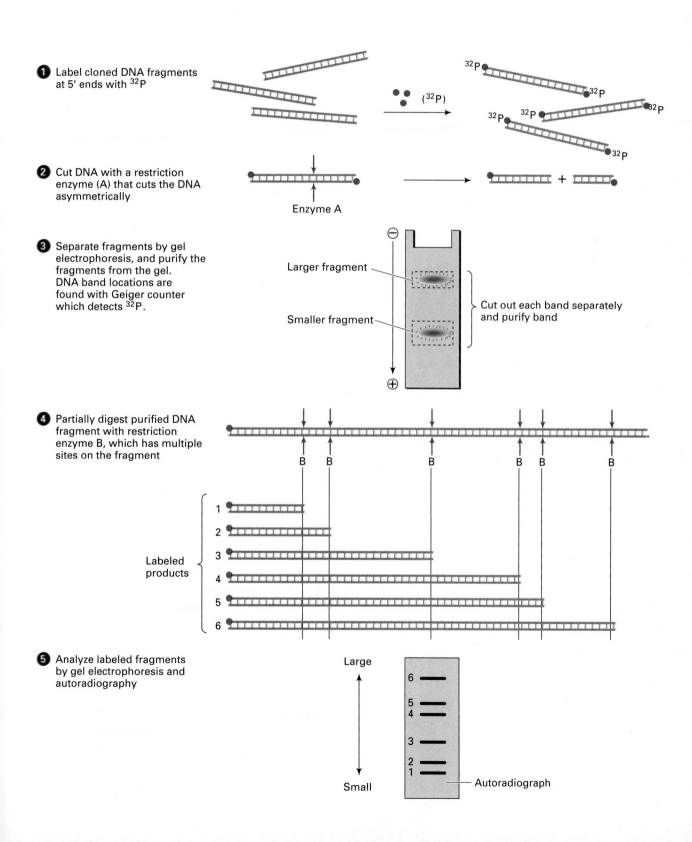

1 Label cloned DNA fragments at 5' ends with ^{32}P

2 Cut DNA with a restriction enzyme (A) that cuts the DNA asymmetrically

Enzyme A

3 Separate fragments by gel electrophoresis, and purify the fragments from the gel. DNA band locations are found with Geiger counter which detects ^{32}P.

Larger fragment

Smaller fragment

Cut out each band separately and purify band

4 Partially digest purified DNA fragment with restriction enzyme B, which has multiple sites on the fragment

Labeled products

5 Analyze labeled fragments by gel electrophoresis and autoradiography

Large

Small

Autoradiograph

Northern blotting is useful in several kinds of experiments. For example, northern blotting can reveal the size or sizes of the specific mRNA encoded by a gene. In some cases, a number of different mRNA species encoded by the same gene have been identified in this way, suggesting that either different promoter sites can be used, different terminator sites can be used, or alternative mRNA processing can occur. Northern blotting can also be used to investigate whether or not an mRNA species is present in a cell type or tissue, and how much of it is present. This type of experiment is useful for determining levels of gene activity, for instance, during development, in different cell types of an organism, or in cells before and after they are subjected to specific natural or experimentally-induced physiological stimuli.

ⱩEYNOTE

Genes and cloned DNA sequences are often analyzed to determine the arrangement and specific locations of restriction sites. The analytical process involves cleavage of the DNA with restriction enzymes, followed by separation of the resulting DNA fragments by agarose gel electrophoresis, and staining of the DNA fragments with ethidium bromide so that they may be visualized with ultraviolet light. The DNA fragments produced by cleavage of cloned DNA sequences can be seen as discrete bands, enabling restriction maps to be constructed based on the calculated molecular lengths of the DNA in the bands. DNA fragments produced by cleavage of genomic DNA show a wide range of sizes, resulting in a continuous smear of DNA fragments in the gel. In this case, specific gene fragments can only be visualized by transferring the DNA fragments to a nitrocellulose filter in a procedure called the Southern blot technique, hybridizing a specific radioactive probe with the DNA fragments, and detecting the hybrids by autoradiography. At that point, a restriction map can be made.

Gene transcripts can be analyzed by separating mRNA species by gel electrophoresis, transferring the mRNAs to a filter by the northern blot technique, and hybridizing with a specific radioactive probe.

DNA SEQUENCE ANALYSIS

Cloned DNA fragments may be analyzed to determine the nucleotide pair sequence of the DNA. This information is useful, for example, for identifying gene sequences and controlling sequences within the fragment, and for comparing the sequences of homologous genes from different organisms.

Two techniques for the *rapid sequencing of DNA molecules* were developed in the 1970s. One method, developed by Allan Maxam and Walter Gilbert, uses chemicals to cleave DNA at different sites and is called **Maxam-Gilbert sequencing,** after its developers. The other method, called **dideoxy sequencing** or **Sanger sequencing,** was developed by Fred Sanger and involves enzymatic extension of a short primer. The dideoxy method is more commonly used today. Methods for sequencing DNA are continually being improved and rapid automated procedures are destined to be developed in the future. Such procedures will be essential as research teams proceed to determined the complete sequences of various genomes such as yeast, the nematode *Caenorhabditis elegans*, mouse, and humans. Sequencing the complete genome of humans has been named the **human genome project.**

Maxam-Gilbert DNA Sequencing

The Maxam-Gilbert technique uses specific chemical reactions to break the DNA at specific nucleotides. The starting point is a homogeneous population of DNA molecules of about two hundred to a thousand base pairs in length. In our hypothetical example, however, we will determine the sequence of a 10-base-pair DNA fragment.

In brief, the steps for DNA sequencing by the Maxam-Gilbert technique are as follows:

STEP ONE: LABELING THE DNA. Each strand of the double-stranded DNA molecule is radiolabeled with ^{32}P at the 5' or 3' end of a chain. The DNA is then either denatured into single strands, each of which is labeled at only one end, or cut with an enzyme somewhere along its length to generate two double-stranded fragments, each of which has only one strand with a terminal ^{32}P molecule.

STEP TWO: CHEMICAL MODIFICATION AND CLEAVAGE OF THE DNA. Consider a hypothetical homogeneous population of single-stranded DNA molecules ten base pairs long and ^{32}P-labeled at the 5' end:

$$^{32}P\ 5'\ \frac{?}{1}\ \frac{?}{2}\ \frac{?}{3}\ \frac{?}{4}\ \frac{?}{5}\ \frac{?}{6}\ \frac{?}{7}\ \frac{?}{8}\ \frac{?}{9}\ \frac{?}{10}\ 3'$$

The sample is divided into four fractions, and each fraction is treated chemically in a different way to modify and remove one of the four bases, thus breaking the DNA backbone at that point.

The first fraction is then treated to indicate the position of every G in the strand by cleaving the DNA at each G. Because this reaction works to a lesser extent on As in the chain, it is designated G > A; that is, this reaction works to identify G more than it does A. The second fraction is treated to cleave the DNA at As. Since this reaction cleaves Gs to a lesser extent, it is designated A > G. The third fraction is treated to cleave the DNA at Ts and Cs, and the fourth fraction is treated to cleave only at Cs in the chain.

STEP THREE: ELECTROPHORETIC ANALYSIS OF DNA FRAGMENTS:

The cleaved chains from each reaction are then separated according to their length by *polyacrylamide gel electrophoresis*. In this technique, a jellylike slab of buffered *polyacrylamide* is used to separate DNA fragments in an electric field. (This is very similar to the separation of DNA fragments in agarose gels, except polyacrylamide gels are used to separate much smaller-sized DNA fragments.) The single-stranded DNA molecules migrate at a rate related to their length (the smallest fragment migrates fastest, the largest migrates slowest). After electrophoresis the DNA fragments are visualized by autoradiography. Figure 15.22 diagrams the autoradiograph expected of the four groups. Figure 15.23 is a photograph of an actual autoradiograph from a Maxam-Gilbert DNA sequencing experiment.

The first lane of the autoradiograph in Figure 15.22 shows the result of the G > A reaction where the fragments generated by cleavage at Gs appear as thicker bands than those generated by cleavage at As. The results may thus be interpreted to mean that the second and seventh bases were Gs and that the first, sixth, and eighth were As.

The second lane shows the A > G results, which are exactly the opposite of those in the first lane, confirming the location of A's and G's.

~ **FIGURE 15.22**

Autoradiograph expected for the four Maxam-Gilbert sequencing reactions with the DNA fragments 5'-A-G-C-C-T-A-G-A-C-T-3'.

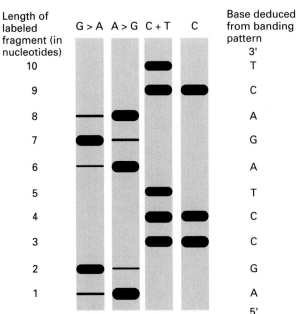

~ **FIGURE 15.23**

Photograph of an autoradiograph from a Maxam-Gilbert DNA sequencing experiment.

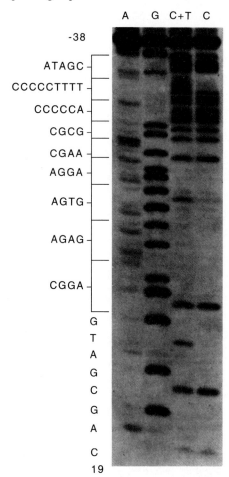

The last two lanes locate the C and T bases in the chain. Lane 3 shows the C + T positions, and lane 4 shows the C positions. Comparing of the two lanes enables the C and T positions to be deduced: A band appearing in both lanes indicates a C; a band appearing only in lane 3 indicates a T.

The DNA sequence is read from the band positions, starting at position 1 (the smallest fragment) and deducing the base from the banding patterns at each step of the ladder. In the example, position 1 has two bands, the one in the A > G lane being denser than the one in the G > A lane, thereby indicating an A. Position 2 has a dense band in G > A and a faint band in A > G, indicating a G. Positions 3 and 4 have bands in both the C + T and C lanes, indicating Cs. Position 5 has a band in the C + T lane but no band in the C lane, indicating a T. Following similar logic, positions 6 through 10 are A, G, A, C, and T, respectively.

Dideoxy (Sanger) DNA Sequencing

Dideoxy DNA sequencing, like Maxam-Gilbert sequencing, requires a clone of the DNA to be produced so that a homogeneous population of molecules is available for analysis. Consider a segment of DNA cloned into a restriction site of a plasmid cloning vector (Figure 15.24,1). The plasmid is made linear by cutting with restriction enzyme II, and the two strands are denatured by heat to single strands (Figure 15.24,2). (Techniques are also used to anneal the oligonucleotide to circular plasmid DNA without making it linear.) Next, a short oligonucleotide, one complementary to the plasmid sequence adjacent to one of the I cloning sites, is annealed to the DNA—it will anneal to only one of the two DNA strands (Figure 15.24,3). The oligonucleotide is designed so that its 3' end will be next to the cloned DNA piece. The oligonucleotide acts as a

~ FIGURE 15.24

Preparation of a template DNA for dideoxy sequencing.

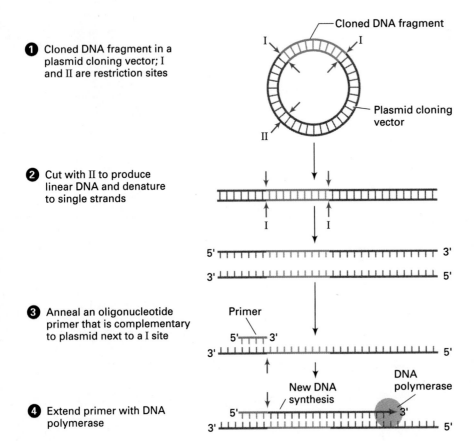

1   Cloned DNA fragment in a plasmid cloning vector; I and II are restriction sites

2   Cut with II to produce linear DNA and denature to single strands

3   Anneal an oligonucleotide primer that is complementary to plasmid next to a I site

4   Extend primer with DNA polymerase

primer for DNA synthesis, and the 5'-to-3' orientation chosen ensures that the DNA made is a *complementary copy of the cloned DNA* cloned from both ends (Figure 15.24,4).

Figure 15.25 illustrates the use of dideoxy sequencing to determine the same sample we discussed for Maxam-Gilbert sequencing in Figure 15.22. For each DNA fragment to be sequenced, four different reactions are set up with single-stranded DNA to which the primer has been annealed. Each reaction contains the four normal precursors of DNA, that is dATP, dTTP, dCTP, and dGTP, and DNA polymerase. The precursors are radioactively labeled with ^{32}P so that newly-synthesized DNA can be easily detected. The reactions differ in the presence of a different modified nucleotide call a **dideoxy nucleotide** (Figure 15.26). The only difference between a dideoxy nucleotide and the deoxynucleotides normally used in DNA synthesis is dideoxy nucleotide has 3'-H on the deoxyribose sugar rather than a 3'-OH. If a dideoxy nucleoside triphosphate (ddNTP) is used in a DNA synthesis reaction, the dideoxy nucleotide can be incorporated into the growing DNA chain. However, once that happens, no further DNA synthesis can then occur because the *absence of a 3'-OH prevents the formation of a phosphodiester bond with an incoming DNA precursor.*

The four sequencing reactions, then, have *one of* ddATP, ddGTP, ddCTP, and ddTTP, in addition to the four normal precursors dATP, dGTP, dCTP, and dTTP. So that some DNA synthesis occurs in the dideoxy sequencing reactions, only a small proportion of the precursors are dideoxy precursors. Generally, the dideoxy precursor is present in about 1/100th the amount of the normal precursor. The primer is extended by DNA polymerase, and when a particular nucleotide is specified by the template strand, there is a small chance that the dideoxy nucleotide will be incorporated instead of the normal nucleotide in the appropriate reaction mixture. For example if an A is specified

~ FIGURE 15.25

Dideoxy DNA sequencing of a theoretical DNA fragment.

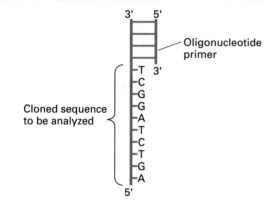

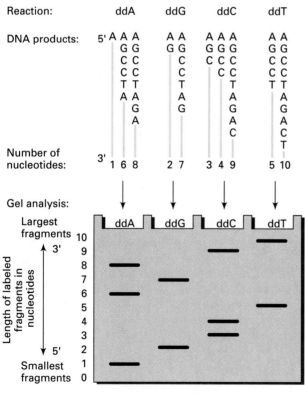

Sequence deduced from banding pattern of autoradiograph made from gel:

5' A-G-C-C-T-A-G-A-C-T 3'

~ FIGURE 15.26

A dideoxy nucleotide DNA precursor.

Dideoxynucleoside triphosphate

(Normal DNA precursor has OH at 3' position)

by the template strand a ddA could be incorporated rather than dA in the reaction mixture. Once a dideoxy nucleotide is incorporated, elongation of the chain stops. In a population of molecules in the same DNA synthesis reaction, then, new DNA chains will stop at all possible positions where the nucleotide is required because of the incorporation of the dideoxy nucleotide. In the ddA reaction, the many different chains that are produced all end with ddA. All chains end with ddG in the ddG reaction, all chains end with ddC in the ddC reaction, and all chains end with ddT in the ddT reaction. The DNA chains in each reaction mixture are separated by polyacrylamide gel electrophoresis and the locations of the DNA bands are revealed by autoradiography.

An example of a dideoxy sequencing gel result is shown in Figure 15.27. There the various DNA chains that are synthesized in each of the dideoxy reactions are given to illustrate the principle involved in dideoxy sequencing. As with Maxam-Gilbert sequencing, the 5′-to-3′ DNA sequence is read from the bottom to the top of the gel since the small fragment migrates the fastest.

KEYNOTE

Rapid methods have been developed for determining the sequence of a cloned piece of DNA. One method, the Maxam-Gilbert procedure, uses specific chemicals to modify and cleave the DNA chain at specific nucleotides. The other method, the dideoxy or Sanger procedure, uses enzymatic synthesis of a new DNA chain on a cloned template DNA strand. With this procedure, synthesis of new strands is stopped by the incorporation of a dideoxy analog of the normal deoxyribonucleotide. Using four different dideoxy analogs, the new strands stop at all possible nucleotide positions, thereby allowing the complete DNA sequence to be determined. _____

Analysis of DNA Sequences

Sequences determined by either sequencing method are usually entered into computer data bases. The data bases are analyzed, for example, to compare a variety of sequences for homologous regions, controlling site similarities, and so on. Appropriate computer programs can search DNA sequences for possible protein-coding regions by looking for an initiator codon in frame with a chain terminating codon (called an **open reading frame**, or ORF). (Note that the discovery of a possible ORF does not mean that that DNA sequence encodes a protein in the cell. Many experiments would have to be done to see if that is the case. And, the failure to detect an ORF does not necessarily mean that the DNA sequence has no function; it may represent a regulatory region.) Other programs can be used to translate a cloned DNA sequence theoretically into an amino acid sequence and to predict secondary and tertiary structures of the polypeptide.

POLYMERASE CHAIN REACTION (PCR)

The **polymerase chain reaction** (PCR) is a method for selectively and repeatedly replicating defined DNA sequences from a DNA mixture. The starting point for

~ FIGURE 15.27

Autoradiogram of a dideoxy sequencing gel.

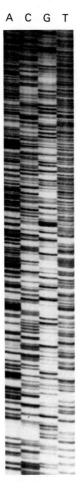

A C G T

PCR is the DNA mixture containing the DNA sequence to be amplified, and a pair of oligonucleotide primers that flank that DNA sequence. The primers are made synthetically, so it is necessary that some sequence information be available about the DNA sequence of interest so that it can be amplified. The PCR procedure is as follows:

1. Denature the DNA to single strands by incubating at 94°C (Figure 15.28,1; p. 464).
2. Anneal the specific pair of primers that flank the targeted DNA sequence (Figure 15.28,2).
3. Extend the primers by DNA polymerase (Figure 15.28,3). For this, a special DNA polymerase called *Taq* ("tack") *polymerase* is used. This polymerase, isolated from the bacterium *Thermus aquaticus*, has a very high heat tolerance and hence is not destroyed by the repeated heat-denaturation that occurs of the PCR process, as *E. coli* DNA polymerase would be. *Thermus aquaticus* lives in hot springs, hence it was reasonable to expect to find thermotolerant enzymes in that organism.
4. Repeat the steps for heat denaturation, annealing of primers, and primer extension for as many cycles as desired (Figure 15.28,4).

Using PCR, the amount of new DNA generated increases geometrically (Figure 15.29; p. 465). Starting with one molecule of DNA, one cycle of PCR produces two molecules, two cycles produces four molecules, three cycles produces eight molecules, and so on. Ten cycles produces 1026 copies of the original DNA; in 20 cycles there will be 1,048,576 copies! The procedure is rapid, each cycle taking only a few minutes using a *thermal cycler*, a machine that automatically cycles through the temperature changes in a programmed way.

One disadvantage of the PCR procedure is that *Taq* polymerase does not have proofreading properties, so that errors are introduced into the DNA copies at low frequencies. Of course, should an error be introduced in an early cycle of PCR, then all subsequent copies made from that DNA will also have that error. PCR is also susceptible to contamination. That is, if the reaction mixture becomes contaminated with DNA to which the primers can bind, then that DNA can also be amplified as well as the targeted DNA.

There are many applications for PCR, including amplifying DNA for cloning, amplifying DNA from genomic DNA preparations for sequencing without cloning, and amplifying DNA for disease diagnosis. In each case, some DNA sequence information must be available so that appropriate pairs of primers can be synthesized. In disease diagnosis, for example, PCR can be used to detect bacterial pathogens or viral pathogens such as HIV (human immunodeficiency virus, the cau-

sation agent of AIDS), human cytomegalovirus (a common virus infecting immunocompromised patients), and hepatitis B virus. PCR can also be used in genetic disease diagnosis, which is discussed in the next section. This makes it easier to detect single-copy sequences in total DNA isolated from cells.

One other interesting application of PCR is amplifying "ancient" DNA for analysis. That is, samples of tissues preserved many hundreds or thousands of years ago are available, for example from mummies, woolly mastodons, and insects in amber. PCR makes it possible to amplify selected DNA sequences, then analyze the sequences of those DNA molecules for comparison with contemporary DNA samples. Such comparisons enable us to make evolutionary comparisons between ancient forebears and present-day descendants, such as elephants.

KEYNOTE

The polymerase chain reaction (PCR) uses specific oligonucleotides to amplify a specific segment of DNA many thousand fold in an automated procedure. PCR is finding increasing applications both in research and in the commercial arena, including generating specific DNA segments for cloning or for sequencing, and for amplifying DNA for analysis for the presence of specific genetic defects.

APPLICATIONS OF RECOMBINANT DNA TECHNOLOGY

Recombinant DNA technology has many applications, including the diagnosis of human genetic diseases such as sickle-cell anemia; the synthesis of commercially important products such as human insulin, human growth hormone, and interferon; *in vitro* modifications of genes; and genetic engineering of plants. Some of the important applications will be briefly outlined in this section.

Analysis of Biological Processes

One of the fundamental and widespread applications of recombinant DNA methodologies is in basic research to explore biological functions. Questions in most areas of biology are being addressed with these techniques. In genetics, researchers are investigating such things as the

~ FIGURE 15.28

The polymerase chain reaction (PCR) for selective amplification of DNA sequences.

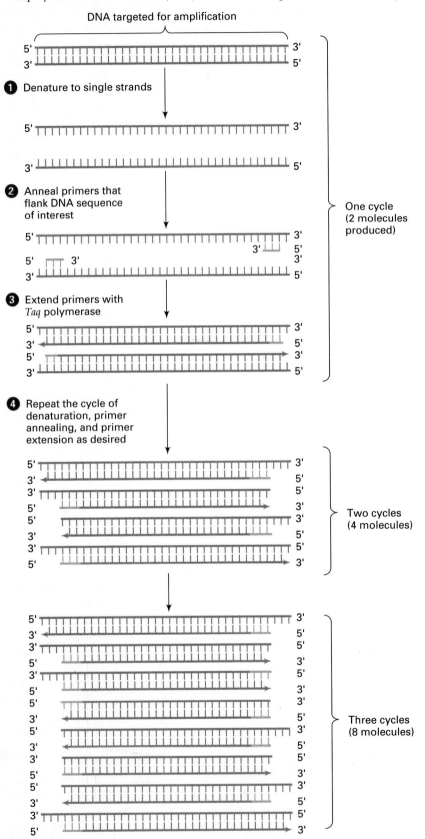

~ FIGURE 15.29

Geometric amplification of DNA by PCR.

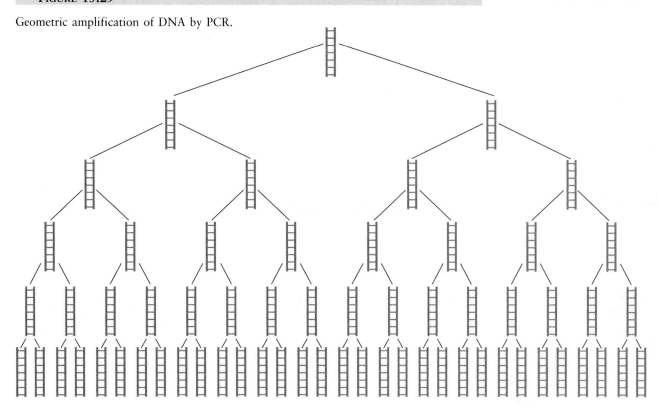

functional organization of genes and the regulation of gene expression. In developmental biology, key regulatory genes and target genes responsible for developmental events are being discovered and analyzed. In evolutionary biology, DNA sequence analysis is adding new information about the evolutionary relationships between organisms. It is rare today to find a research scientist who is not aware of advances being made in his or her field through the use of recombinant DNA techniques.

Diagnosis of Human Genetic Diseases

In Chapter 8, we discussed biochemical approaches to diagnosing genetic diseases. Recombinant DNA techniques can be used to investigate an increasing number of mutated genes that cause human diseases such as sickle-cell anemia, Huntington's chorea, and diabetes. In a number of instances this research has led to simple diagnostic procedures that can be used for carrier detection and for the analysis of the genetic condition of the fetus. Most of the basic techniques used in these procedures were described in Chapter 8.

Recombinant DNA approaches to detecting genetically based human disease (or genetically based disease in other organisms, for that matter) require cellular DNA as the starting point. Cellular DNA can be isolated from white blood cells in parental blood (in the case of carrier detection) or from fetal cells isolated by amniocentesis or chorionic villus sampling and cultured (in the case of fetal diagnosis) (see Figures 8.16 and 8.17). The DNA is digested with a restriction enzyme, producing restriction fragments of lengths determined by the locations of the restriction sites along the DNA molecules. The restriction fragments are then analyzed as described earlier: the fragments are separated according to size by agarose gel electrophoresis, then Southern blotted to a nitrocellulose filter for hybridization with a specific ^{32}P-labeled DNA probe.

These analytical procedures are most useful when the genetic mutation that causes a disease is associated with a change in the number or distribution of restriction sites, either within the gene or in a flanking region. That is, in the human genome, and the genomes of other eukaryotes, different restriction maps may be found among individuals for the same region of a chromosome (i.e., detected by the same probe). The different restriction maps result from different patterns of distribution of restriction sites and are called **restriction fragment length polymorphisms,** (or **RFLPs**), because they are detected by the presence of restriction fragments of different lengths on gels.) RFLPs arise, for example, by the addition or deletion of DNA between restriction sites, or by base-pair changes that create or abolish a restriction site sequence. A restriction map is independent of gene function, so an RFLP is detected whether the DNA sequence change responsible affects a detectable phenotype or not. An RFLP can be used as a genetic marker in the same way as the "conventional" genetic markers we have discussed previously. In this case we assay the DNA—that is the genotype—directly in the form of a restriction map. No dominance exists, so both parental types are seen in heterozygotes.

RFLPs are useful both for mapping chromosome regions as well in human disease diagnosis. For the latter there are many cases of an RFLP being associated with a gene known to cause a disease, as the following example illustrates. The genetic disease *sickle-cell anemia* (discussed in more detail in Chapter 8) results from a single base-pair change in the gene for the hemoglobin's β-globin polypeptide, resulting in an abnormal form of hemoglobin, Hb-S, instead of the normal Hb-A form. This change from AT to TA results in the substitution of a valine for a glutamic acid in the sixth amino acid of the polypeptide, which, in turn, produces abnormal associations of hemoglobin molecules, sickling of the red blood cells, tissue damage, and sometimes death.

Using a cDNA probe for human β-globin, an RFLP has been shown for the restriction enzyme *Hpa*I. Figure 15.30a shows the arrangement of *Hpa*I sites in normal and sickle-cell gene segments of the chromosome, and Figure 15.30b is a schematic of the DNA bonding patterns for normal individuals, sickle-cell anemia individuals, and heterozygotes on an electrophoretic gel. Specifically, following digestion with *Hpa*I, Southern blotting, and probing with cDNA probe (as described earlier), normal individuals who produce Hb-A will display fragments on the autoradiograph of either 7.6 or 7.0 kb. In contrast, people with the sickle-cell anemia mutation, who produce Hb-S, will display an *Hpa*I fragment of 13 kb. Heterozygotes (individuals with sickle cell trait) can be detected because they will have both the 13 kb and 7.6 or 7.0 kb (normal) DNA. Note that in this case the RFLP is only tightly linked to the gene and does not result from the base-pair change in the gene's coding region.

A different RFLP is directly related to the base pair change that results in sickle cell anemia. That base pair change produces a new *Dde*I ("d-d-e-one") restriction site that is not present in DNA of normal individuals. In normal individuals, there are two relevant *Dde*I sites: one is upstream of the start of the β-globin gene, and the other is within the coding sequence itself (Figure 15.31). When DNA from normal individuals is cut with *Dde*I, and the fragments separated by gel electrophoresis, blotted and probed, a 376-base pair fragment is

~ FIGURE 15.30

Restriction fragment length polymorphism (RFLP) associated with the DNA flanking the β-globin gene. (a) Diagrams of DNA segments showing the RFLPs. (b) Schematic drawing of the results of gel electrophoretic analysis of DNA cut with *Hpa*I.

a) **Normal genes; product found in Hb-A molecule**

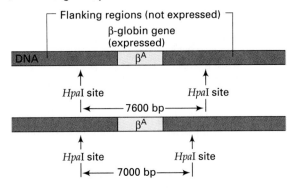

Sickle-cell mutant gene; product found in Hb-S molecule

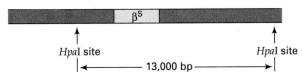

b) **Schematic of gel electrophoretic analysis of DNA cut with *Hpa*I**

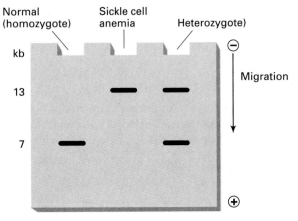

seen. DNA from individuals with sickle cell anemia analyzed in the same way gives two fragments of 175 bp and 201 bp by virtue of the additional *Dde*I site. Heterozygotes can be detected by the presence of three bands of 376 bp, 201 bp, and 175 bp.

An RFLP that is related to the genetic disease PKU (see Chapter 8) has been identified. Recall that PKU results from a deficiency in the activity of the enzyme phenylalanine hydroxylase. Following digestion of genomic DNA with *Hpa*I, Southern blotting, and probing with a cDNA probe derived from phenylalanine

~ FIGURE 15.31

Detection of sickle-cell gene by the *Dde*I restriction fragment length polymorphism.

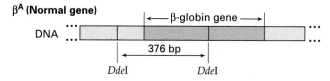

β^A (Normal gene)

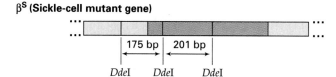

β^S (Sickle-cell mutant gene)

hydroxylase mRNA, different-sized restriction fragments are produced from DNA isolated from PKU individuals and from DNA isolated from homozygous normal individuals. Like the first sickle-cell anemia example, this RFLP results from a difference outside the coding region of the gene, in this case to the 3′ side of the gene. The RFLP can be used for diagnosing PKU in fetuses following amniocentesis or chorionic villus sampling.

Other examples of human genetic diseases for which recombinant DNA technology can or will soon provide early diagnosis include four types of thalassemia (hemoglobin diseases resulting in anemia), α-antitrypsin deficiency (which is a deficiency of a serum protein), hemophilia A, hemophilia B, Huntington's Chorea, cystic fibrosis, and Duchenne muscular dystrophy (a progressive disease resulting in muscle atrophy and muscle dysfunction).

The beauty of the recombinant DNA approach is that it directly assays for a DNA genotype (i.e., RFLP), so detection does not depend on the expression of the gene. Thus this approach is not limited to detecting genetic diseases in which the gene involved is active in the parent or the fetus at the time of analysis. For example, phenylalanine hydroxylase, the enzyme defective in individuals with PKU, is found in the liver but is not found either in blood serum or in fibroblast cells, the cells usually cultured following amniocentesis. The DNA of fibroblast cells can be analyzed, however, for RFLPs, and both individuals with the disease *and carriers* can be detected.

DNA Fingerprinting

Everyone is familiar with the use of fingerprints in forensic science. The principle is that no two individuals have the same fingerprints so that fingerprints left at the scene of a crime are important evidence in a crimi-

nal investigation. Similarly, no two human individuals (except identical twins) have exactly the same genome, base pair for base pair, and this has led to the development of DNA techniques for use in forensic science and in paternity and maternity testing. Such a use might seem unfounded, at first, since it is clear that the similarities in the DNA of different individuals greatly exceed the differences, as we would expect of a genome which specifies the human species. In fact, estimates indicate that about 1 base pair in 1000 base pairs is the site of polymorphisms among humans. For **DNA fingerprinting**, or the use of DNA analysis to identify an individual, scientists have identified highly polymorphic markers scattered throughout the genome. Each marker consists of short, identical segments of DNA tandemly arranged head to tail. Differences between individuals result from a great variation in the number of tandem repeats, called *variable number of tandem repeats,* or VNTRs. Probes have been constructed to analyze the VNTRs so that, in most cases, an unambiguous identification of an individual, based on a DNA sample, is possible.

DNA fingerprinting proceeds as follows (Figure 15.32; p. 468): DNA is obtained from the crime scene and the suspect (Figure 15.32,1 and 2). The crime scene source of the DNA may be obtained from blood, semen, other bodily tissues or fluids, or hair roots. In some cases, PCR may be used to amplify the sample to produce enough DNA for analysis. The DNA is cut with the appropriate restriction enzyme(s) (Figure 15.32,1), the resulting fragments are separated by electrophoresis (Figure 15.32,2), Southern blotted to a filter (Figure 15.32,3), and probed with a radioactive VNTR probe (Figure 15.32). The DNA banding pattern on an autoradiogram made from the filter is then analyzed to compare the two samples (Figure 15.32,5). For example, if the DNA banding pattern for a semen sample taken from a rape victim matches exactly the DNA banding pattern for samples taken from a suspect, this is excellent evidence that the suspect is the rapist. The odds of two people having exactly the same DNA banding patterns are extremely low (with the exception of the patterns of identical twins). DNA fingerprinting is being used more and more frequently in U.S. criminal cases, although there remain many legal challenges to the method. In the United Kingdom, DNA fingerprinting has become an accepted method to resolve immigration cases.

Human Gene Therapy

Is it possible to treat genetic diseases? The diseases result from the phenotypes caused by the mutant gene in somatic cell. Theoretically, if a wild-type copy of the gene could be introduced into somatic cells where the gene's activity is required, then the genetic disease

~ FIGURE 15.32

Procedure for DNA fingerprinting.

1 DNA is obtained from crime scene (e.g., from semen samples in the case of a rape victim) and from the suspect (e.g., from blood sample). In separate analyses, the DNA is cut into fragments with restriction enzyme.

Semen sample from crime scene

Blood sample from suspect

2 Gel electrophoresis of DNAs from each sample and of standards

Semen

Standards Suspect

Smears

3 Southern blot prepared from the gel

4 Filter from the blot is incubated with a radioactive DNA probe. DNA probe binds to specific DNA sequences on the filter.

5 Excess probe is washed away, leaving radioactive probe on filter.

Probe solution

Filter with bound DNA

6 Autoradiograms prepared. The banding pattern for each sample is a DNA fingerprint. If the fingerprints are identical, as here, the prosecution has a strong piece of evidence against the suspect.

Standards Suspect

Semen

could be treated. However, since germline cells would not be similarly treated, the mutant gene could still be passed on to the progeny.

Experiments that could lead to human gene therapy are in a relatively early stage, and many scientific, ethical, and legal questions must be addressed before we will see its implementation. Nonetheless, the results with animal model systems are encouraging, so it seems likely that human gene therapy will be used in the future. The general procedures involved are as follows. First, a sample of a patient's mutant, affected cells is taken. Next, normal, wild-type copies of the mutant gene are introduced into the cell by transformation. Lastly, the "engineered" cells are reintroduced into the patient where, it is hoped, their division and establishment in the patient can lead to the loss or reduction of the symptoms of the genetic disease. Likely human genetic diseases that could be early candidates for gene therapy are, for example, sickle cell anemia, thalassemias, phenylketonuria, Lesch-Nyhan Syndrome, and Tay-Sachs syndrome.

Human Genome Project

Currently underway is an extensive effort to map all of the estimated 50,000 to 100,000 human genes, and to obtain the sequence of the complete 3 billion (3×10^9) nucleotide pairs of the genome. This **human genome project** is being coordinated by James Watson. Associated with the project are parallel efforts to obtain gene maps and complete sequences of the genomes of a number of other model organisms, including yeast, *Drosophila,* the nematode *Caenorhabditis elegans,* and mouse. These efforts will obtain basic information about the organization and functional aspects of the targeted organism's genomes. In particular, the project, which is the largest biological research project ever planned, will expand immensely our knowledge of human genetics and of human genetic diseases. Directly and indirectly we can expect that knowledge to spawn efforts to devise ways to treat and/or cure genetic diseases. The project will take up to 20 years to complete and will cost at least $3 billion.

Commercial Products

Many new biotechnology companies have emerged since the mid-1970s, when basic recombinant DNA techniques were developed. The first was Genentech. With older, established pharmaceutical and chemical companies, these companies are focusing on using recombinant DNA technology to make a wide array of commercial products that are useful in businesses in the human health industry, and agriculture sector, as well as in the general marketplace. Some examples are as follows:

1. Tissue plasminogen activator (TPA)—used to prevent or reverse blood clots, therefore preventing strokes, heart attacks, or pulmonary embolisms.
2. Human growth hormone—used to treat pituitary dwarfism.
3. Tissue growth factor-beta (TGF-β)—promotes new blood vessel and epidermal growth; hence, is potentially useful for wound healing and burns.
4. Human blood clotting factor VIII—to treat hemophiliacs.
5. Human insulin ("humulin")—to treat insulin-dependent diabetes.
6. Bovine growth hormone—to increase cattle and dairy yields.
7. Recombinant vaccines—for treatment of human and animal viral diseases.
8. Genetically engineered bacteria and other microorganisms for improved production of, for example, industrial enzymes (e.g., amylases to break down starch to glucose), citric acid (flavoring), and ethanol.
9. Genetically engineered bacteria that can accelerate the degradation of oil pollutants, or of certain chemicals in toxic wastes (e.g., dioxin).
10. Genetically engineered bacteria that, when sprayed on fields of crops such as strawberries or potatoes, lower by a degree or two the temperature at which the leaves will freeze, thus providing some protection against frost damage compared with untreated plants.

Genetic Engineering of Plants

For many centuries the traditional genetic engineering of plants involved selective breeding experiments in which plants with desirable traits were used as parents for the next generation in order to reproduce offspring with those traits. As a result, humans have produced a hardy variety of plants (e.g., corn, wheat, oats) and have been successful in breeding varieties with increased yields, all by using standard plant breeding techniques. (Similar experiments have also been done with animals, e.g., dogs and horses, to produce desired breeds.) In this section, we will discuss briefly and selectively the application of recombinant DNA technology to plant breeding.

TRANSFORMATION OF PLANT CELLS. Historically, introducing genes into plant cells has been a problem that has placed plant genetic engineering's rate of progress behind bacterial, fungal, and animal genetic engineering. This problem has now been solved. Currently, one solution that has been found is to exploit features of a soil bacterium, *Agrobacterium tumefaciens*, which

infects many kinds of plants. Techniques are now being developed to use the bacterium to introduce genes into its host plants.

Agrobacterium tumefaciens causes crown gall disease, characterized by tumors (the gall) at wounding sites. Most dicotyledonous plants (called *dicots*) are susceptible to crown gall disease, but monocotyledonous plants are not. (A plant is said to be a dicot if the embryo of the seed has two cotyledons; examples are potatoes, petunias, and apples. In monocots, the embryo of the seed has a single cotyledon; examples are corn, wheat, and grass.) *Agrobacterium tumefaciens* transforms plant cells at the wound site, causing the cells to grow and divide autonomously and, therefore, to produce the tumor.

The transformation of plant cells is mediated by a plasmid in the *Agrobacterium* called the *Ti plasmid* (the Ti stands for tumor-inducing) (Figure 15.33; p. 470). Ti plasmids are circular DNA plasmids analogous to pBR322, but, in comparison, Ti plasmids are enormous in size (about 200 kb vs. 4.36 kb for pBR322).

The interaction between the infecting bacterium and the plant cell of the host stimulates the bacterium to excise a 30-kb region of the Ti plasmid called *T-DNA* (so-called because it is transferred DNA). T-DNA is flanked by two repeated 25-bp sequences. Both of these sequences are required for the transfer of tDNA to plant cells. Excised tDNA, once transferred from the bacterium to the plant cell, then integrates into the nuclear genome. Integration of the tDNA occurs at random chromosomal sites. As a result of the integration of tDNA, the plant cell acquires the genes found on the T-DNA, including the genes for plant cell transformation. However, the genes needed for the excision, transfer, and integration of the T-DNA into the host plant cell are not part of the T-DNA. Instead, they are found elsewhere on the Ti plasmid, in a region called the *vir* (for virulence) region.

Using recombinant DNA approaches, researchers have found that excision, transfer, and integration of the T-DNA require only the 25-bp terminal repeat sequences, because the alteration or removal of the rest of the T-DNA does not affect those processes. As a result, the Ti plasmid and the T-DNA it contains is a useful vector for introducing new DNA sequences into the nuclear genome of somatic cells from susceptible plant species. A variety of transformation vectors have been derived from the Ti plasmid and T-DNA for the efficient introduction of genes into plants.

APPLICATIONS OF PLANT GENETIC ENGINEERING. In the near future we can expect a range of plants to be produced that have been engineered with recombinant DNA techniques. Of particular value will be crop plants that have increased yield (e.g., through more effi-

~ **FIGURE 15.33**

Formation of tumors (crown galls) in plants by infection with certain species of *Agrobacterium*. Tumors are induced by the Ti plasmid, which is carried by the bacterium and which integrates some of its DNA (the T, or transforming DNA) into the plant cell's chromosome.

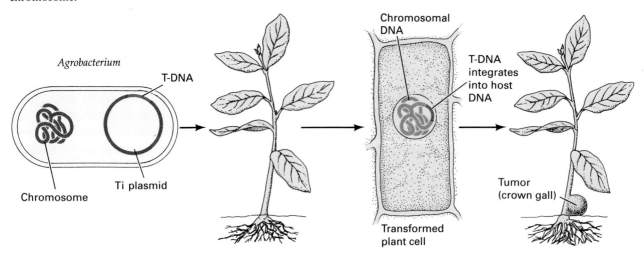

cient nitrogen fixation), insect pest resistance, and herbicide resistance (to enable fields to be sprayed to kill weeds but not the crop). With more sophisticated approaches, some of which are already available, it will be possible to introduce genes into plants and/or control their expression in different tissues. An example would be controlling the time at which fruit ripens. If all fruits in a field ripen at the same time, harvesting will become easier and spoilage during shipping will be reduced.

KEYNOTE

Recombinant DNA technology is finding ever-increasing applications in the world. In the basic research laboratory, biological processes across all of biology are being studied. With appropriate probes, a number of genetic diseases can now be diagnosed using recombinant DNA methods, and DNA finger-printing is being used in forensic science. In addition, many products in the clinical, veterinary, and agricultural areas can be synthesized in commercial quantities using recombinant DNA procedures. Transformation of plants with vectors based on the Ti plasmid has opened the way for genetic engineering of plants using recombinant DNA technology. It is expected that many types of improved crops will result from applications of this new technology.

SUMMARY

In this chapter we have discussed some of the procedures involved in recombinant DNA technology and the manipulation of DNA. Collectively, these procedures are also referred to as genetic engineering, at least in the popular press. We have seen how it is possible to cut DNA at specific sites using restriction enzymes, how DNA can be cloned into specially-constructed vectors, and how the cloned DNA can be analyzed in various ways. Through the construction of genomic libraries and cDNA libraries, and the application of screening procedures to those libraries, a large number of genes have been cloned and identified from a wide variety of organisms.

Restriction mapping analysis has provided detailed molecular maps of genes and chromosomes analogous to the genetic maps constructed on the basis of recombination analysis. Rapid DNA sequencing methods that have been developed for use with cloned DNA fragments have given us an enormous amount of information about the DNA organization of genes, both their coding sequences and regulatory sequences. The amount of DNA sequence information available is growing at an extremely rapid rate and computer databases of such sequences are available for researchers to analyze. For example, when a new gene is sequenced, it is of interest to determine if it has any DNA sequences in common with other sequenced genes that are in the database. In this way, potential families of proteins or parts of proteins have been identified. Such groups possibly have related functions in the organisms from which they derive.

Recently, a new technique called polymerase chain reaction (PCR) has been developed. Given some sequence information about a DNA fragment, synthetic oligonucleotides can be made which can be used to amplify large amounts of the DNA fragment from the genome in a repeated cycle of DNA denaturation, annealing of oligonucleotides to act as primers, and extension of the primers with a special DNA polymerase. Numerous applications have been rapidly found for PCR, including cloning rare pieces of DNA, preparing DNA for sequencing without cloning, and genetic disease diagnosis.

Recombinant DNA technology is being widely applied in both basic research and commercial areas. There is probably not an area of basic biology in which recombinant DNA techniques have not been applied to the questions being asked. The human genome project has the mandate to generate a complete map of all of the genes in the human genome, and to obtain the complete sequence of the human genome. The knowledge obtained from this ambitious, long-range project will contribute markedly to our understanding of human genetics. In the commercial area, recombinant DNA techniques are being used, for example, to develop new pharmaceuticals (including drugs, other therapeutics, and vaccines), to develop new tools for diagnosis of infectious and genetic diseases, for human gene therapy, in forensic analysis (e.g., analyzing DNA from a crime scene to match with a suspect), and in agricultural areas (e.g., improving disease resistance and yields of livestock and crops). While products generated by genetic engineering have been slower coming to the market than originally expected, there is an undiminished enthusiasm for the genesis and commercialization of a wide array of useful products in the present decade.

ANALYTICAL APPROACHES FOR SOLVING GENETICS PROBLEMS

While this is a rather descriptive area, it is often necessary to interpret data derived from restriction enzyme analysis of DNA fragments in order to generate a restriction map, that is, a map of the locations of restriction enzymes. The logic used for this type of analysis is very similar to that used in generating a genetic map of loci from two-point mapping crosses.

Q.1 A piece of DNA 900 bp long is cloned and then cut out of the vector for analysis. Digestion of this linear piece of DNA with three different restriction enzymes singly and in all possible combinations of pairs gave the following restriction fragment size data:

ENZYME(S)	RESTRICTION FRAGMENT SIZES
*Eco*RI	200 bp, 700 bp
*Hin*dIII	300 bp, 600 bp
*Bam*HI	50 bp, 350 bp, 500 bp
*Eco*RI + *Hin*dIII	100 bp, 200 bp, 600 bp
*Eco*RI + *Bam*HI	50 bp, 150 bp, 200 bp, 500 bp
*Hin*dIII + *Bam*HI	50 bp, 100 bp, 250 bp, 500 bp

Construct a restriction map from these data.

A.1 The approach to this kind of problem is to consider a pair of enzymes and to analyze the data from the single and double digestions. First let us consider the *Eco*RI and *Hin*dIII data. Cutting with *Eco*RI produces two fragments, one of 200 bp and the other of 700 bp, while cut-ting with *Hin*dIII also produces two fragments, one of 300 bp and the other of 600 bp. Thus, we know that both restriction sites are asymmetrically located along the linear DNA fragment with the *Eco*RI site 200 bp from an end and the *Hin*dIII site 300 bp from an end. When we consider the *Eco*RI + *Hin*dIII data we can determine the positions of these two restriction sites relative to one another. If, for example, the *Eco*RI site is 200 bp from the fragment end, and the *Hin*dIII site is 300 bp from that same end, then we would predict that cutting with both enzymes would produce three fragments of sizes 200 bp (end to *Eco*RI site), 100 bp, (*Eco*RI site to *Hin*dIII site), and 600 bp (*Hin*dIII site to other end). On the other end, if the *Eco*RI site is 200 bp from one fragment end and the *Hin*dIII site is 300 bp from the other fragment end, cutting with both enzymes would produce three fragments of sizes 200 bp (end to *Eco*RI site), 400 bp (*Eco*RI site to *Hin*dIII site), and 300 bp (*Hin*dIII site to end). The actual data support the first model.

Now we pick another pair of enzymes, for example, *Hin*dIII and *Bam*HI. Cutting with *Hin*dIII produces fragments of 300 bp and 600 bp as we have seen, and cutting with *Bam*HI produces three fragments of sizes 50 bp, 350 bp, and 500 bp, indicating that there are two *Bam*HI sites in the DNA fragment. Again the double digestion products are useful in locating the sites. Double digestion with *Hin*dIII and *Bam*HI produces four fragments of 50 bp, 100 bp, 250 bp, and 500 bp. The simplest interpretation of the data is that the 300-bp *Hind*III fragment is cut into the 50-bp and 250-bp fragments by *Bam*HI, and that the 600-bp *Hind*III fragment is cut into the 100-bp and 500-bp fragments by *Bam*HI. Thus, the restriction map shown in the accompanying figure can be drawn:

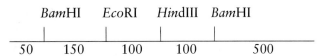

The *Bam*HI + *Eco*RI data are compatible with this model.

Q.2 The recessive allele, *bw*, when homozygous, results in brown eyes in *Drosophila*, in contrast to the wild-type bright red eye color. A restriction fragment length polymorphism (RFLP) for a particular DNA region results in either two restriction fragments (type I) or one restriction fragment (type II) when *Drosophila* DNA is cut with restriction enzyme C, the fragments separated by DNA electrophoresis, blotted to nitrocellulose, and probed with a particular radioactive DNA probe.

A true-breeding brown-eyed fly with type I DNA was crossed with a true-breeding, wild-type fly with type II DNA. The F$_1$ flies had wild-type eye color and exhibited both type I and type II DNA patterns. The F$_1$ flies were crossed with true-breeding brown-eyed flies with type I DNA, and the progeny were scored for eye color and RFLP type. The results were as follows:

CLASS	PHENOTYPES	NUMBER
1	red eyes, type I and II DNA	184
2	red eyes, type I DNA	21
3	brown eyes, type I DNA	168
4	brown eyes, type I and II DNA	27
		400 total progeny

Analyze these data.

A.2 The eye color mutation is a familiar genetic marker. The restriction fragment length polymorphisms (RFLPs) are also genetic markers and can be analyzed just like any gene marker. If we symbolize the type I DNA as I and the type II DNA as II, the F$_1$ cross is:

$$\frac{bw^+ \; \text{II}}{bw \; \; \text{I}} \times \frac{bw \; \; \text{I}}{bw^+ \; \text{II}}$$

(The cross is drawn as if the markers were linked; of course, we have yet to show this.) This is a testcross with the exception that DNA markers do not exhibit dominance or recessiveness. If the eye color and RFLP markers are unlinked, the result would be equal numbers of the four progeny classes. However, the data show a great excess of these two classes: (a) red eyes, type I and II DNA; and (b) brown eyes, type I DNA. Their origin was the pairing of F$_1$ bw^+ II and bw I gametes with bw I gametes to give bw^+ II/bw I and bw I/bw I progeny geno-

types, respectively. Similarly, the progeny (a) red eyes, type I DNA and (b) brown eyes, type I and II DNA derive from pairing bw^+ I and bw II gametes with bw I gametes to give bw^+ I/bw I and bw II/bw I progeny, respectively. These two latter classes occur with about equal frequency, that is a frequency much lower than those for the other two classes. The simplest explanation is that the brown eye color gene and the RFLP marker are linked, so the F$_1$ cross was a mapping cross, much like those we analyzed in Chapter 5. Classes 1 and 3 are the parentals, and classes 2 and 4 are the recombinants. The map distribution between *bw* and the RFLP is, therefore:

$$\frac{21 + 27}{\text{Total}} \times 100\%$$

$$= \frac{48}{400} \times 100\%$$

$$= 12 \text{ map units}$$

QUESTIONS AND PROBLEMS

*15.1 A new restriction endonuclease is isolated from a bacterium. This enzyme cuts DNA into fragments that average 4096 base pairs long. Like all other known restriction enzymes, the new one recognizes a sequence in DNA that has twofold rotational symmetry. From the information given, how many base pairs of DNA constitute the recognition sequence for the new enzyme?

15.2 An endonuclease called *Avr*II ("a-v-r-two") cuts DNA whenever it finds the sequence $\begin{array}{l} 5'\text{-CCTAGG-}3' \\ 3'\text{-GGATCC-}5' \end{array}$ About how many cuts would *Avr*II make in the human genome, which is about 3×10^9 base pairs long and about 40% GC?

15.3 About 40% of the base pairs in human DNA are GC. On the average, how far apart (in terms of base pairs) will the following sequences be?
a. Two *Bam*HI sites?
b. Two *Eco*RI sites?
c. Two *Not*I sites?
d. Two *Hae*III sites?

15.4 What are the features of plasmid cloning vectors that make them useful for constructing and cloning recombinant DNA molecules?

*15.5 Genomic libraries are important resources for isolating genes of interest and for studying the functional organization of chromosomes. List the steps you would use to make a genomic library of yeast in a lambda vector.

15.6 The human genome contains about 3×10^9 bp of DNA. How many 40 kb pieces would you have to clone into a library if you wanted to be 90% certain of including a particular sequence?

*15.7 Suppose you wanted to produce human insulin (a peptide hormone) by cloning. Assume that this could be done by inserting the human insulin gene into a bacterial host, where, given the appropriate conditions, the human gene would be transcribed then translated into human insulin. Which do you think it would be best to use as your source of the gene, human genomic insulin DNA or a cDNA copy of this gene? Explain your choice.

15.8 You are given a genomic library of yeast prepared in a bacterial plasmid vector. You are also given a cloned cDNA for human actin, a protein which is conserved in protein sequences among eukaryotes. Outline how you would use these resources to attempt to identify the yeast actin gene.

15.9 Restriction endonucleases are used to construct restriction maps of linear or circular pieces of DNA. The DNA is usually produced in large amounts by recombinant DNA techniques. The generation of restriction maps is similar to the process of putting the pieces of a jigsaw puzzle together. Suppose we have a circular piece of double-stranded DNA that is 5000 base pairs long. If this DNA is digested completely with restriction enzyme I, four DNA fragments are generated: fragment *a* is 2000 base pairs long; fragment *b* is 1400 base pairs long; *c* is 900 base pairs long; and *d* is 700 base pairs long. If, instead, the DNA is incubated with the enzyme for a short time, the result is incomplete digestion of the DNA; not every restriction enzyme site in every DNA molecule will be cut by the enzyme, and all possible combinations of adjacent fragments can be produced. From an incomplete digestion experiment of this type, fragments of DNA were produced from the circular piece of DNA, which contained the following combinations of the above fragments: *a-d-b*, *d-a-c*, *c-b-d*, *a-c*, *d-a*, *d-b*, and *b-c*. Lastly, after digesting the original circular DNA to completion with restriction enzyme I, the DNA fragments were treated with restriction enzyme II under conditions conducive to complete digestion. The resulting fragments were: 1400, 1200, 900, 800, 400, and 300. Analyze all the data to locate the restriction enzyme sites as accurately as possible.

*15.10 A piece of DNA 5000 bp long is digested with restriction enzymes A and B, singly and together. The DNA fragments produced were separated by DNA electrophoresis and their sizes were calculated, with the following results:

DIGESTION WITH

A	B	A + B
2100 bp	2500 bp	1900 bp
1400 bp	1300 bp	1000 bp
1000 bp	1200 bp	800 bp
500 bp		600 bp
		500 bp
		200 bp

Each A fragment was extracted from the gel and digested with enzyme B, and each B fragment was extracted from the gel and digested with enzyme A. The sizes of the resulting DNA fragments were determined by gel electrophoresis, with the following results.

A FRAGMENT	FRAGMENTS PRODUCED BY DIGESTION WITH B	B FRAGMENT	FRAGMENT PRODUCED BY DIGESTION WITH A
2100 bp ⟶	1900, 200 bp	2500 bp ⟶	1900, 600 bp
1400 bp ⟶	800, 600 bp	1300 bp ⟶	800, 500 bp
1000 bp ⟶	1000 bp	1200 bp ⟶	1000, 200 bp
500 bp ⟶	500 bp		

Construct a restriction map of the 5000 bp DNA fragment.

*15.11 Draw the banding pattern you would expect to see on a DNA-sequencing gel if you applied the Maxam and Gilbert DNA-sequencing method to the following single-stranded DNA fragment (which is labeled at the 5' end with ^{32}P):

$5'^{32}$-P-A-A-G-T-C-T-A-C-G-T-A-T-A-G-G-C-C-3'.

15.12 DNA was prepared from small samples of white blood cells from a large number of people. Ten different patterns were seen when these DNAs were all digested with EcoRI, then subjected to electrophoresis and Southern blotting. Finally, the blot was probed with a radioactively labeled cloned human sequence. The figure below shows the ten DNA patterns taken from ten people.

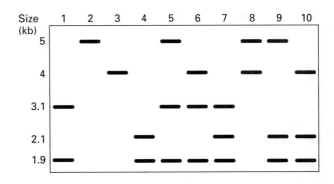

a. Explain the hybridization patterns seen in the ten people in terms of variation in EcoRI sites.
b. If the individuals whose DNA samples are in lanes 1 and 6 on the blot were to produce offspring together, what bands would you expect to see in DNA samples from these offspring?

*15.13 Filled symbols in the pedigree (Figure 15.A) indicate people with a rare autosomal dominant genetic disease. DNA samples were prepared from each of the individuals in the pedigree. The samples were restricted, electrophoresed, blotted, and probed with a cloned human

FIGURE 15.A

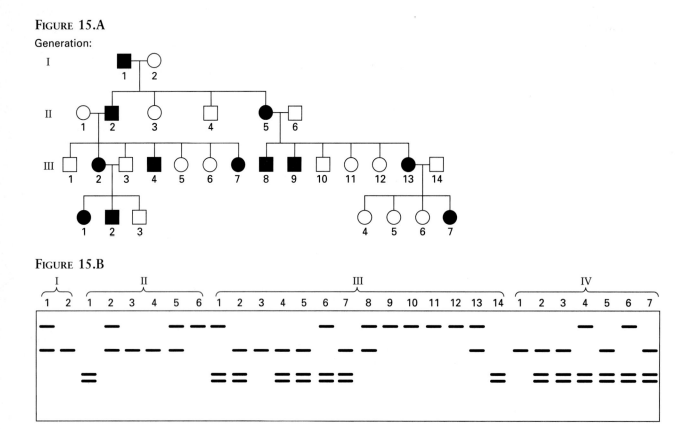

FIGURE 15.B

sequence called DS12-88, with the results shown in Figure 15.B. Do the data in these two figures support the hypothesis that the locus for the disease that is segregating in this family is linked to the region homologous to DS12-88? Make your answer quantitative, and explain your reasoning.

15.14 Imagine that you have been able to clone the structural gene for an enzyme in a catecholamine biosynethetic pathway from the adrenal gland of rats. How could you use this cloned DNA as a probe to determine whether this same gene functions in the brain?

15.15 Imagine that you find an RFLP in the rat genomic region homologous to your cloned catecholamine synthetic gene from 15.14, and that in a population of rats displaying this polymorphism there is also a behavioral variation. You find that some of the rats are normally calm and placid, but others are hyperactive, nervous, and easily startled. Your hypothesis is that the behavioral difference seen is caused by variations in your gene. How could you use your cloned sequence to test this hypothesis?

*15.16 The PCR technique was used to amplify two genomic regions (homologous to probes A and B) in DNA from *Neurospora*. Two strains of opposite mating type (strains J and K) were found to differ from each other in

FIGURE 15.C

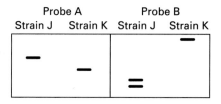

both amplified regions, as shown in Figure 15.C, when the amplified DNA was cut with *Eco*RI. Strain J was crossed to strain K, and 100 asci resulting from the cross were dissected, and the PCR reaction was done on DNA from the individual spores in each ascus. Six different patterns of distribution of DNA types within asci were seen as shown in Figure 15.D. Only four patterns are shown for each ascus. Remember that spores 1 and 2 in the ascus are ordinarily identical, as are spores 3 and 4, 5 and 6, and 7 and 8. In Figure 15.D the band pattern displayed by spores 1 and 2 is indicated in the lane designated "1". The pattern shown by spores 3 and 4 is in lane 2, the pattern shown by spores 5 and 6 is in lane 3, and the pattern shown by spores 7 and 8 is in lane 4. Draw a map showing the relationships among the region detected by probe A, the region homologous to probe B, and any relevant centromeres.

FIGURE 15.D

Spore:	48 asci	2 asci	2 asci	12 asci	32 asci	4 asci

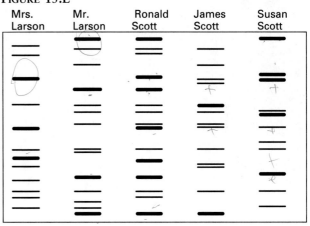

15.17 One application of DNA fingerprinting technology has been to identify stolen children and return them to their parents. Bobby Larson was taken from a supermarket parking lot in New Jersey in 1978, when he was 4 years old. In 1990, a sixteen-year-old boy called Ronald Scott was found in California, living with a couple named Susan and James Scott, who claimed to be his parents. Authorities suspected that Susan and James might be the kidnappers, and that Ronald Scott might be Bobby Larson. DNA samples were obtained from Mr. and Mrs. Larson, and from Ronald, Susan and James Scott. Then DNA fingerprinting was done, using a probe for a particular VNTR family, with the results shown in Figure 15.E. From the information in the figure, what can you say about the parentage of Ronald Scott? Explain.

FIGURE 15.E

16 REGULATION OF GENE EXPRESSION IN BACTERIA AND BACTERIOPHAGES

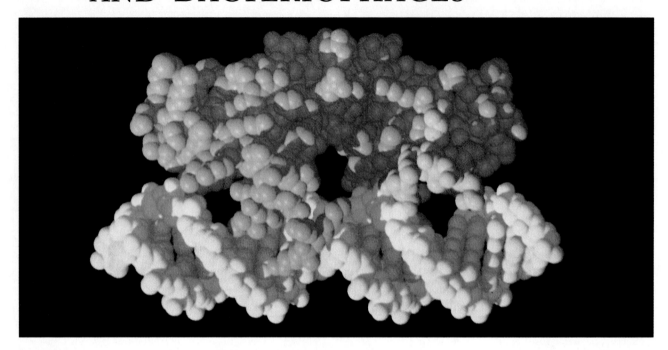

PRINCIPAL POINTS

~ From studies of the regulation of expression of lactose-utilizing genes in *E. coli*, a model was generated that is the basis for the regulation of gene expression in a large number of bacterial and bacteriophage systems. That is, the genes for the enzymes are contiguous in the chromosome and are adjacent to a controlling site (an operator) and a single promoter. This complex constitutes a transcriptional regulatory unit called an operon. A regulatory gene is associated with the operon, but may or may not be nearby.

~ In the lactose operon, the addition of lactose to the cell results in the coordinate synthesis of three enzymes, the genes for which are located contiguously on the chromosome. Enzyme induction occurs as follows: a lactose metabolite binds with a repressor protein, inactivating it and preventing it from binding to the operator. As a result, RNA polymerase can bind to the promoter and transcribe the three genes onto a single polygenic mRNA.

~ Expression of a number of bacterial amino acid synthesis operons is accomplished by a repressor-operator system and through attenuation at a second controlling site called an attenuator. The repressor-operator system functions essentially like that for the *lac* operon, except that the addition of amino acid to the cell activates the repressor, thereby turning the operon off. The attenuator is located downstream from the operator; it

is a partially effective transcription termination site that allows only a fraction of RNA polymerases to transcribe the rest of the operon. Attenuation involves a coupling between transcription and translation, and the formation of particular RNA secondary structures that signal whether or not transcription can continue.

~ Under harsh conditions, a bacterium experiences stress. As a result, the cell responds with a stringent response, that is, a rapid shutdown of essential cellular activities. The stringent response is very complex, and involves the synthesis of a specific molecule under stress conditions that somehow signals the inhibition of major cellular biochemical pathways.

~ Bacteriophages such as T4 and lambda are especially adapted for undergoing reproduction within a bacterial host. Many of the genes related to the production of progeny phages, or to the establishment or reversal of lysogeny of temperate phages, are organized into operons. These operons, like bacterial operons, are controlled through the interaction of regulatory proteins with operators that are adjacent to clusters of structural genes. T4 has been an excellent model for studying the genetic control of development, and lambda has been an excellent model for studying the genetic switch that controls the choice between lytic and lysogenic pathways in a lysogenic phage.

Organisms do not live and reproduce in a constant environment. Through evolutionary processes, they have developed ways to compensate for environmental changes and hence to function in a variety of environments. One way that an organism can adjust to a new environment is to alter its gene activity so that gene products appropriate to the new conditions are synthesized, with the result that the organism is optimally adjusted to grow and reproduce in that environment. This is particularly evident in free-living bacteria.

A change in the array of gene products synthesized in a prokaryotic cell involves regulatory mechanisms that control gene expression—that is, the synthesis of the product of a gene by transcription and, in the case of protein-coding genes, translation. In general, genes whose activity is controlled in response to the needs of a cell or organism are called **regulated genes**. An organism also possesses a large number of genes whose products are essential to the normal functioning of a growing and dividing cell, no matter what the life-supporting

environmental conditions are. These genes are always active in growing cells and are known as **constitutive genes;** examples include genes that code for the synthesis of proteins, and enzymes needed for protein synthesis and glucose metabolism. It is important to realize, though, that all genes are regulated on some level. If the environment suddenly becomes no longer conducive for normal cell function, for example, the expression of all genes, including constitutive genes, will be reduced by specific and/or general regulatory mechanisms. Thus, the distinction between regulated and constitutive genes is a somewhat arbitrary one.

In this chapter we examine the mechanisms by which gene expression is regulated in prokaryotic organisms, particularly in bacteria and bacteriophages. In the wild a bacterium lives in a potentially unstable environment that does not ensure a constant food supply or access to the same types of nutrients at all times. To survive, then, a bacterium must be able to adapt physiologically and adjust its cellular biochemical activities to changing conditions. To adapt to alterations in their environments, bacteria have evolved several regulatory mechanisms for turning off genes not needed in new environmental conditions and for turning on different genes needed for survival in those new conditions. In this chapter we learn in detail about some of the basic gene regulation mechanisms in bacteria and how this regulation relates to the organization of genes in the bacterial genome.

Since bacteriophages are parasites, relying on the activities of a host bacterium to reproduce, they must direct their own reproductive cycle in a carefully programmed way; this cycle is directed through the regulation of gene expression. In this chapter we examine specifically the gene regulation events required for the production of progeny phages and, in the case of phage lambda, for controlling the choice between the lytic and lysogenic pathways.

REGULATED AND CONSTITUTIVE GENES

When gene expression is turned on in a bacterium by the addition of a substance (such as lactose) to the medium, the genes are said to be *inducible*. The regulatory substance that brings about this gene induction is called an **inducer,** and it is a member of a class of small molecules, called **effectors** or effector molecules, that are involved in the control of expression of many regulated genes.

To make the distinction clear, Figure 16.1 diagrams the difference between an inducible gene and a constitutive gene. For an inducible gene the transcription of the gene occurs only in response to a particular molecular event occurring at a specific sequence of nucleotide pairs near the gene, a location called a **controlling site.** The molecular event typically involves an inducer and a regulatory protein. When the molecular event occurs, RNA polymerase binds to the promoter (usually adjacent to the controlling site) upstream from the gene(s) and initiates transcription of the gene. The gene is "turned on," mRNA is made, and the synthesis of the enzyme coded for by the gene(s) is induced under the regulatory effect of the controlling site. The controlling site itself does not code for any product. The phenomenon of producing a gene product only in response to an inducer is called **induction.**

Constitutive genes, on the other hand, are always expressed in a growing and dividing cell. Their promoter is always accessible to RNA polymerase binding, so transcription of the gene(s) is continuous and there is a constant supply of the essential gene products. Thus constitutive genes are said to be under *promoter control.* Constitutive genes may also have controlling sites associated with them; such sites are involved in regulatory systems for controlling the rates of transcription of these genes. For example, under certain conditions, it is necessary to reduce or block the expression of the constitutive genes. Such conditions include heat shock (a temperature pulse that imposes a significant physiological stress on the cell) and the onset of starvation conditions (for example the exhaustion of an amino acid in the medium in which is growing an auxotroph for that amino acid).

GENE REGULATION OF LACTOSE UTILIZATION IN *E. COLI*

Lactose as a Carbon Source for *E. coli*

Escherichia coli is able to grow in a simple medium containing salts (including a nitrogen source) and a carbon source such as glucose. These chemicals provide

---

~ **FIGURE 16.1**

Inducible and constitutive gene systems.

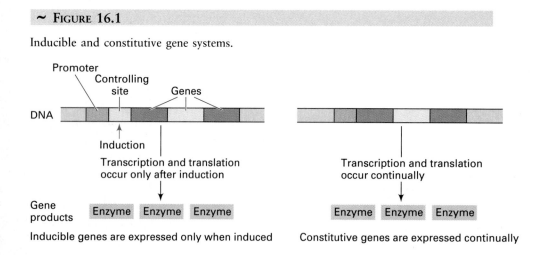

Inducible genes are expressed only when induced   Constitutive genes are expressed continually

molecules that can be manipulated by the cell's enzymatic machinery to produce everything the cell needs to grow and reproduce, such as nucleic acids, proteins, and lipids. The energy for these biochemical reactions comes from the glucose metabolism, a process that is of central importance to the functioning of a bacterial cell and of cells of all organisms. The enzymes required for the glucose metabolism are coded for by constitutive genes.

If lactose, or one of several other sugars, is provided to *E. coli* as a carbon source instead of glucose, a number of enzymes are rapidly synthesized; these enzymes are needed for the metabolism of this particular sugar. (The same series of events, each involving a "sugar specific" set of enzymes, would be triggered by other sugars as well.) The enzymes are synthesized because the genes that code for them become actively transcribed in the presence of the sugar; these same genes are inactive if the sugar is absent. Moreover, the mRNAs for the enzymes have a relatively short half-life, so the transcripts must be made continually in order for the enzymes to be produced. In other words, the genes are regulated genes whose products are needed only at certain times.

When lactose is the sole carbon source in the growth medium, three proteins are synthesized, all of which are needed for lactose utilization in *E. coli*:

1. *β-galactosidase* has two functions that are shown in Figure 16.2. It catalyzes the isomerization ("conversion to a different form") of lactose to *allolactose,* a compound important in the regulation of expression of the lactose utilization genes. It also catalyzes the breakdown of lactose into its two component monosaccharides, *glucose* and *galactose.* In the growing cell the galactose is converted to glucose through the action of enzymes encoded by a gene system specific for galactose catabolism. The glucose is utilized by constitutively produced enzymes.
2. *Lactose permease* (also called *M protein*) is found in the *E. coli* membrane and is needed for the active transport of lactose from the growth medium into the cell.
3. *Transacetylase* is a protein whose function is poorly understood.

In a wild-type *E. coli* that is growing in a medium containing glucose (or another carbon source) but no

~ FIGURE 16.2

Reactions catalyzed by the enzyme β-galactosidase. Lactose brought into the cell by the permease is either converted to glucose and galactose (top) or to allolactose (bottom), the true inducer for the lactose operon of *E. coli.*

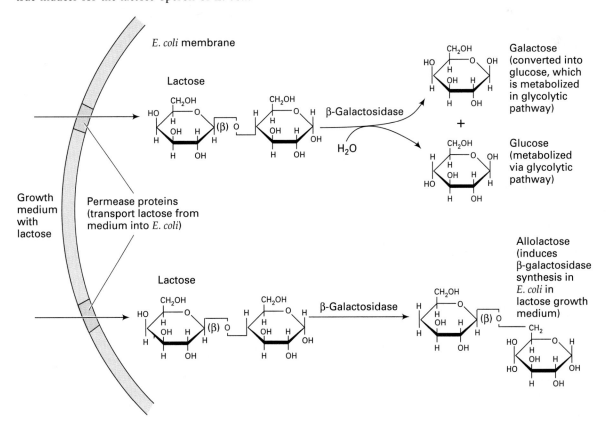

lactose, only a few molecules of each of the three proteins are produced, indicating a low level of expression of the three genes that code for the proteins. If lactose but no glucose is present in the growth medium, however, the number of molecules of each of the three proteins increases coordinately about a thousandfold, indicating that the three previously (essentially) inactive genes are now being actively transcribed and translated. This process is called **coordinate induction.** The inducer molecule directly responsible for the increased production of the three proteins is actually allolactose, which is produced from lactose as a result of one of the activities of the β-galactosidase enzyme.

Experimental Evidence for the Regulation of the *lac* Genes

Our basic understanding of the organization of the genes, the controlling sites involved in lactose utilization, and the control of expression of the *lac* genes came largely from the genetic experiments of Francois Jacob and Jacques Monod, for which they received the Nobel Prize.

MUTANTS OF THE PROTEIN-CODING GENES. Mutants of the protein-coding genes were obtained following treatment with mutagens (chemicals that induce mutations); they were identified on the basis of enzyme assays, since in every case the activity of a particular enzyme was reduced greatly as a consequence of a mutation in the structural gene coding for that enzyme.

From these mutation studies, the β-galactosidase gene was named *lacZ*, the permease gene *lacY*, and the transacetylase gene *lacA*. The *lacZ⁻*, *lacY⁻*, and *lacA⁻* mutations obtained were used to map the locations of the three genes, and from these experiments it was established that the order of the structural genes in the genome was *lacZ-lacY-lacA* and that the three genes were tightly linked in a cluster.

In Chapter 14 we studied the types of mutations that affect the reading of the genetic message. For *lac* genes, missense mutations, which result in the substitution of one amino acid for another in a polypeptide, only affect the expression of the gene in which the mutation maps. A *lacZ⁻* missense mutation results in a nonfunctional or partially functional β-galactosidase, but permease and transacetylase are totally normal. Similarly, a *lacY⁻* missense mutation results in a nonfunctional or partially functional permease, but β-galactosidase and transacetylase are completely functional, and a *lacA⁻* missense mutation results in a nonfunctional transacetylase, but β-galactosidase and permease are completely functional.

It was expected that nonsense (chain-terminating) mutations would act similarly for the *lac* genes. That was not the case. Nonsense mutations mapping in the *lacZ* gene not only resulted in loss of function of β-galactosidase, but also in the loss of permease and transacetylase activities. *lacY* nonsense mutations resulted in non-functional permease and no transacetylase activity, but β-galactosidase activity was unaffected. Lastly, *lacA* nonsense mutants exhibited no transacetylase activity, but normal β-galactosidase and permease activities. That is, nonsense mutations in the cluster of three genes involved in lactose utilization have different effects, depending on where they are located within the cluster. In other words, the nonsense mutations exhibit *polar effects*, and the phenomenon is called **polarity.** (Nonsense mutations that show polar effects are often called *polar mutations*.)

The interpretation of the polar effects of nonsense mutations in the *lac* structural genes is that all three genes are transcribed onto a single mRNA molecule—called a **polygenic mRNA** or **polycistronic mRNA**—rather than onto three separate mRNAs. Unequivocal evidence now indicates that RNA polymerase initiates transcription at a single promoter and that a polygenic mRNA is synthesized with the gene transcripts in the order 5'-*lacZ⁺*-*lacY⁺*-*lacA⁺*-3'. Translation of this mRNA begins near the 5' end and proceeds toward the 3' end (Figure 16.3a). So a ribosome will

1. Synthesize β-galactosidase;
2. Slide along the mRNA until it recognizes the initiation sequence for permease;
3. Synthesize permease;
4. Slide along the mRNA until it recognizes the initiation sequence for transacetylase;
5. Synthesize transacetylase; and
6. Dissociate from the mRNA.

A nonsense mutation in the *lacZ* gene exerts its effect as shown in Figure 16.3b. The ribosome begins to translate the *lacZ* sequence and stops at the premature nonsense codon; a partially completed and non-functional β-galactosidase is released. The ribosome continues to slide along the polygenic mRNA but typically dissociates before the start codon for permease. Thus no ribosomes (or at least, very few ribosomes) translate the downstream permease and transacetylase sequences, so no functional enzymes can be produced. One can see by extension how a permease nonsense mutation will affect downstream transacetylase production but not affect upstream β-galactosidase translation. An essential part of this explanation is that ribosomes *only* load onto the polygenic mRNA at the 5' end and not at the internally located start codons. Because polygenic mRNAs are characteristic of bacterial operons, nonsense mutations in structural genes in an operon have the potential to prevent expression of normal genes that are further along the polygenic mRNA.

~ FIGURE 16.3

Translation of the polygenic mRNA encoded by *lac* utilization genes in (a) wild-type *E. coli* and (b) a mutant strain with a nonsense mutation in the β-galactosidase (*lacZ*) gene.

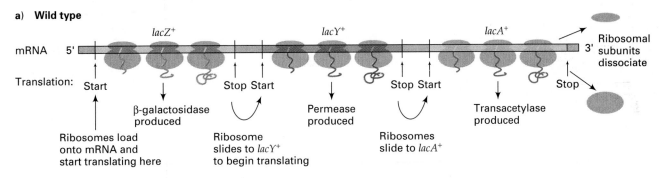

a) Wild type

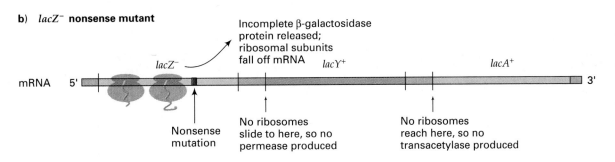

b) *lacZ⁻* nonsense mutant

REGULATORY MUTANTS. Of special interest to Jacob and Monod were mutants that affected the regulation of the three lactose utilization genes. Recall that in the wild-type *E. coli* the three gene products are not made in the absence of lactose. On the other hand, in the presence of lactose all three genes are induced coordinately. Jacob and Monod isolated a number of mutants in which all gene products of the lactose operon were synthesized *constitutively*; that is, all three proteins were synthesized in both the presence or absence of lactose. Jacob and Monod hypothesized that mutants were regulatory mutants that did not affect the functions of the enzymes themselves. Instead, the mu-

tants affected the cellular mechanisms responsible for regulating the normal expression of the genes that code for the enzymes. As a result of their mapping experiments, Jacob and Monod recognized the existence of two classes of constitutive mutants: One class mapped to a relatively small DNA region adjacent to the *lacZ* gene. They called this class the **operator** (*lacO*). The other class mapped to a gene-sized DNA region a short distance away, which they called the *lacI* gene or *lac* **repressor** gene. Figure 16.4 diagrams the organization of the *lac* structural gene cluster and the associated regulatory elements. We will discuss this complex, called the *lac* operon, later in this chapter.

~ FIGURE 16.4

Organization of the *lac* genes of *E. coli* and the associated regulatory elements, the operator, promoter, and regulatory gene. The region defined as the *lac* operon is shown.

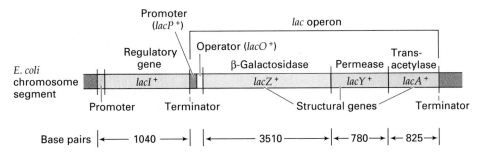

Operator mutants. The mutations of the operator were called operator-constitutive, or $lacO^c$, mutations. All $lacO^c$ mutants exhibit the phenotype of synthesis of the lactose utilization enzymes in the presence *or* absence of lactose. Through the use of partial-diploid strains (i.e., F′ strains; see Figure 7.10), Jacob and Monod were able to define better the role of the operator in regulating the expression of the *lac* genes. One such partial diploid was $lacO^+$ $lacZ^-$ $lacY^+$/ $lacO^c$ $lacZ^+$ $lacY^-$ (both gene sets have a normal promoter, and the *lacA* gene is omitted because it is not germane to our discussions).

One of the chromosomes in the partial diploid has a normal operator ($lacO^+$), a mutant β-galactosidase gene ($lacZ^-$), and a normal permease gene ($lacY^+$). The other chromosome has a constitutive operator mutant ($lacO^c$), a normal β-galactosidase gene ($lacZ^+$), and a mutant permease gene ($lacY^-$). This partial diploid was tested for the production of β-galactosidase (from the $lacZ^+$ gene) and of permease (from the $lacY^+$ gene), both in the presence and the absence of the inducer lactose.

In the absence of inducer, the β-galactosidase is synthesized, but permease is not. Only when lactose is added to the culture does permease synthesis occur. In other words, in this partial diploid the $lacZ^+$ gene is *constitutively expressed* (i.e., the gene is active in the presence or absence of lactose) whereas the $lacY^+$ gene is under normal inducible control (i.e., *the gene is inactive in the absence of lactose and active in the presence of lactose*). The interpretation of these results was that *$lacO^c$ mutations affect only those genes that are adjacent to it on the same chromosome strand.* Furthermore, the $lacO^+$ region only controls those genes adjacent to it and has no effect on the genes on the other chromosome strand. This phenomenon of a gene or DNA sequence controlling only genes that are on the same, contiguous piece of DNA is called **cis-dominance.** Thus the $lacO^c$ mutation is cis-dominant since the defect affects the adjacent genes only and cannot be overcome (complemented) by a normal $lacO^+$ region elsewhere in the genome.

Jacob and Monod also concluded from these results that the operator region does not produce a diffusible product that functions in the cell. That is, if it did, then in the $lacO^+$/$lacO^c$ diploid state, one or the other of the regions would have controlled all the lactose utilization genes wherever they were.

The *lacI* gene regulatory mutants. The second class of *lac* constitutive mutants defined the *lacI* gene. The study of these $lacI^-$ mutants in partial-diploid strains was instrumental in obtaining an understanding of the normal regulation of the *lac* genes.

In the partial diploid $lacI^+$ $lacO^+$ $lacZ^-$ $lacY^+$/ $lacI^-$ $lacO^+$ $lacZ^+$ $lacY^-$ (in which both gene sets have normal operators and normal promoters), one chromosome has a normal *lacI* gene ($lacI^+$), a mutant β-galactosidase gene ($lacZ^-$), and a wild-type permease gene ($lacY^+$): the other chromosome has a constitutive mutant *lacI* gene ($lacI^-$), a normal β-galactosidase gene ($lacZ^+$), and a mutant permease gene ($lacY^-$). As before this partial diploid was tested for structural gene expression in the absence and the presence of the inducer lactose. The result was that no β-galactosidase or permease was produced in the absence or lactose, but both were synthesized in the presence of lactose. In other words, the expression of *both* genes was inducible. This means that the $lacI^+$ gene in the cell can overcome the defect of the $lacI^-$ mutation. Since the two *lacI* genes are located on different chromosomes (that is, they are in a *trans* configuration) the $lacI^+$ is said to be **transdominant** to $lacI^-$.

Since the $lacI^+$ gene controlled the genes on the other chromosome strand, the *lacI* gene must produce a diffusible product. Jacob and Monod proposed that the $lacI^+$ gene produces a functional **repressor molecule** (hence, the *lacI* gene is also called the *lac* **repressor gene**) and that no functional repressor molecules are produced in $lacI^-$ mutants. Thus in a haploid bacterial strain that has a $lacI^-$ mutation, the *lac* operon is constitutive. In a partial diploid with both a $lacI^+$ and a $lacI^-$, however, the functional repressor molecules produced by the $lacI^+$ gene control the expression of both *lac* operons present in the cell, making both operons inducible.

Promoter mutants. The promoter for the structural genes (located at the *lacZ* gene end of the cluster of *lac* genes; see Figure 16.4) is also susceptible to mutation. Most of the known promoter mutants ($lacP^-$) affect all three structural genes. Even in the presence of lactose, the lactose utilization enzymes are not made—or are made only at very low rates. Since the promoter is the recognition sequence for RNA polymerase and does not code for any product, the effect of a $lacP^-$ mutation is confined to the genes that it controls on the same chromosomal strand. The $lacP^-$ mutations are another example of cis-dominant mutations.

Jacob and Monod's Operon Model for the Regulation of the *lac* Genes

Based on their studies of genetic mutants affecting the regulation of the synthesis of the lactose utilization enzymes, Jacob and Monod proposed their now-classical **operon model.** By definition, an **operon** is a *cluster of genes, the expressions of which are regulated together by operator-regulator protein interactions, plus the* *operator region itself and the promoter.* (The *lac* operon is diagrammed in Figure 16.4.) The order of the controlling elements and genes is promoter-operator-*lacZ-lacY-lacA*. For this operon, the regulatory gene, *lacI*, is located essentially adjacent to the structural genes, just upstream of the promoter. The *lacI* gene has its own promoter and terminator. *lacI* is under promoter control, so this is another example of a constitutive gene.

The following description of the Jacob-Monod model for the regulation of the *lac* operon has been embellished with more up-to-date molecular information. Figure 16.5 diagrams the state of the *lac* operon in wild-type *E. coli* growing in the absence of lactose. In this case RNA polymerase molecules bind to the repressor gene (*lacI*$^+$) promoter and transcribe the *lacI* gene. Translation produces a polypeptide consisting of 360 amino acids. Four of these polypeptides associate together to form the repressor protein. Because synthesis of the repressor is under promoter control, the gene product is synthesized constitutively.

The repressor has affinity for the base-pair sequence of the operator to which it binds. When the repressor is bound to the operator, RNA polymerase can bind to the operon's promoter but cannot initiate transcription; hence, transcription of the three protein-coding genes in the operon cannot occur. The *lac* operon is said to be under *negative control* since the binding of the repressor at the operator site blocks transcription of the structural genes. (The low level of transcription of the genes that results in the presence of a few molecules of each protein, even in the absence of

lactose, occurs because repressors do not just bind and stay; they bind and unbind. In the split second when one repressor unbinds and before another binds, an RNA polymerase can initiate transcription of the operon, even in the absence of lactose.)

When wild-type *E. coli* grows in the presence of lactose as the sole carbon source (Figure 16.6; p. 484), some of the lactose transported into the cell is converted by existing molecules of β-galactosidase into allolactose, which, in turn, induces production of the *lac* operon enzymes. (Thus, allolactose, not lactose, is the actual inducer of the *lac* operon protein-coding genes.) Besides having a recognition site for the *lac* operator, the *lac* repressor protein also has a recognition site for allolactose. When allolactose binds to the repressor, it changes the shape of the repressor. As a result, the repressor loses its affinity for the *lac* operator, and it dissociates from the site. Free repressor proteins are also altered so that they cannot bind to the operator.

In the absence of repressor, RNA polymerase is now able to bind to the operon's promoter and initiate the synthesis of a single polygenic mRNA molecule that contains the transcripts for the *lacZ*$^+$, *lacY*$^+$, and *lacA*$^+$ genes. The polygenic mRNA for the *lac* operon is translated by a string of ribosomes to produce the three proteins specified by the operon. This efficient mechanism ensures the coordinate (simultaneous) production of proteins of related function.

EFFECT OF *lacO*^{c} MUTATIONS. The *lacO*^{c} mutations lead to constitutive production of the *lac* operon genes and are cis-dominant to *lacO*$^+$. The explanation

~ FIGURE 16.5

Functional state of the *lac* operon in wild-type *E. coli* growing in the absence of lactose.

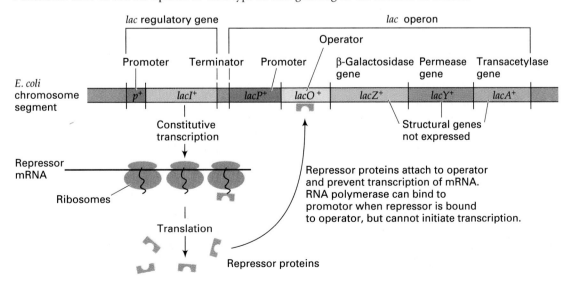

~ FIGURE 16.6

Functional state of the *lac* operon in wild-type *E. coli* growing in the presence of lactose as the sole carbon source.

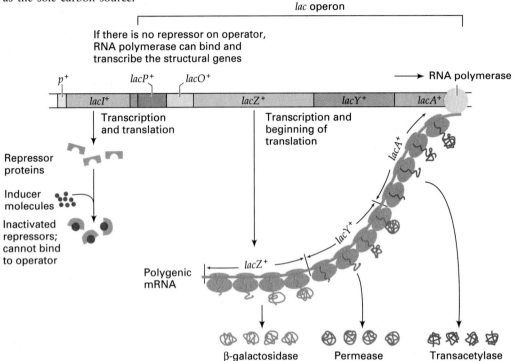

of the phenotype of *lacO^c* mutations is that base-pair alterations of the operator DNA sequence make it unrecognizable by the repressor protein. Since the repressor cannot bind, the structural genes become constitutively expressed. The cis dominance of *lacO^c* is due to the fact that repressor cannot bind to a *lacO^c* mutant, but can bind to another wild-type operator. The cis dominance of a *lacO^c* mutant is illustrated for the partial diploid described earlier, *lacI^+ lacO^+ lacZ^- lacY^+/lacI^+ lacO^c lacZ^+ lacY^-*, growing in the absence (Figure 16.7a) and presence (Figure 16.7b) of inducer.

EFFECTS OF *lacI* GENE MUTATIONS. The effect of *lacI^-* mutations on *lac* operon expression is shown in Figure 16.8a (p. 486). The *lacI^-* mutations map within the repressor structural gene and result in amino acid changes in the repressor. The repressor's shape is changed, and it cannot now recognize and bind to the operator. As a consequence, transcription cannot be prevented, even in the absence of lactose, and there is constitutive expression of the *lac* operon.

The dominance of the *lacI^+* (wild-type) gene over *lacI^-* mutants is illustrated for the partial diploid described earlier, *lacI^+ lacO^+ lacZ^- lacY^+/lacI^- lacO^+ lacZ^+ lacY^-*, in Figure 16.8b and c. In the absence of the inducer (Figure 16.8b; p. 846), the defective *lacI^-* repressor is unable to bind to either normal operator

(*lacO^+*) in the cell. But sufficient normal repressors, produced from the *lacI^+* gene, are diffusing through the cell; they will bind to the two operators and hence block transcription of both operons. In the presence of inducer (Figure 16.8c; p. 847) the wild-type repressors become inactivated, so both operons are transcribed. One produces a defective β-galactosidase and a normal permease, while the other produces a normal β-galactosidase and a defective permease; between them, active β-galactosidase and permease are produced. Thus, in *lacI^+/lacI^-* partial diploids, both operons present in the cell are under inducible control.

Other classes of *lacI* gene mutants have been identified since the time Jacob and Monod studied the *lacI^-* class of mutants. One of these classes, the *lacI^s* (*super-repressor*) mutants, shows no production of *lac* enzymes in the presence or absence of lactose. In partial diploids with a *lacI^+/lacI^s* genotype, the *lacI^s* allele is trans-dominant, affecting both chromosomal strands. Figure 16.9 (p. 488) diagrams the effect of *lacI^s* mutations in the partial diploid. Our interpretation here is that the mutant repressor gene produces a polypeptide product since, if there were no product, the *lacI^s* mutants would not be dominant to the *lacI^+* gene. The superrepressor protein is not altered in its ability to bind to the operator region but, instead, cannot recognize the inducer allolactose. Therefore the mutant su-

~ FIGURE 16.7

Cis-dominant effect of *lacO*ᶜ mutation in a partial-diploid strain of *E. coli*. (a) In the absence of the inducer, the *lacO*⁺ operon is turned off, while the *lacO*ᶜ operon produces functional β-galactosidase from the *lacZ*⁺ gene and nonfunctional permease molecules from the *lacY*⁻ gene; (b) In the presence of the inducer, the functional β-galactosidase and defective permease are produced from the *lacO*ᶜ operon, while the *lacO*⁺ operon produces nonfunctional β-galactosidase from the *lacZ*⁻ gene and functional permease from the *lacY*⁺ gene. Between the two operons in the cell, functional β-galactosidase and permease are produced.

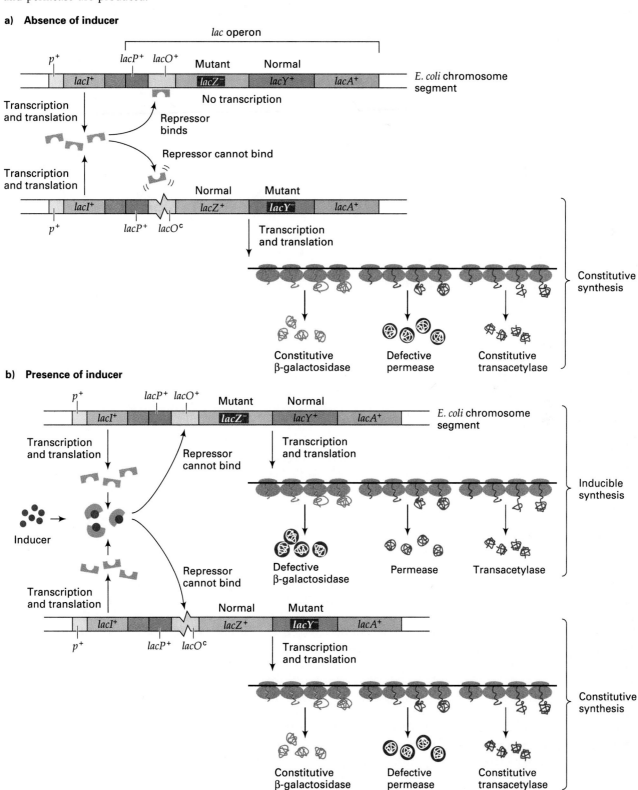

~ FIGURE 16.8

Effect of a *lacI⁻* mutation: (a) On *lac* operon expression in a haploid cell, where mutant, inactive repressor molecules that cannot bind to the operator *lacO⁺* are produced, so the structural genes are transcribed constitutively. In a partial diploid strain *lacI⁺ lacO⁺ lacZ⁻ lacY⁺/lacI⁻ lacO⁺ lacZ⁺ lacY⁻* in (b) the absence or (c) presence of inducer.

a) **Haploid strain (in presence or absence of inducer)**

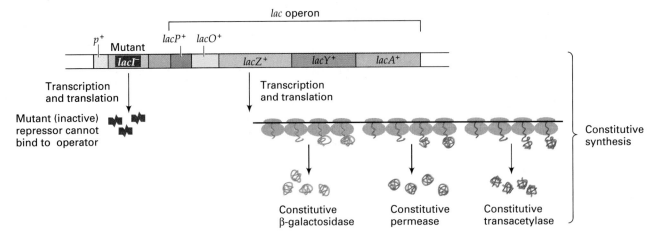

b) **Partial diploid in the absence of inducer**

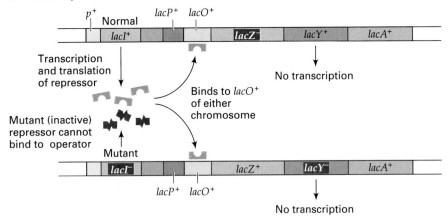

perrepressors get stuck on the operators, and transcription of the operons can never occur. The presence of normal repressors in the cell has no effect on this situation, since once a *lacI*ˢ repressor is on the operator, nothing can remove it. As a consequence, cells with a *lacI*ˢ mutation cannot use lactose as a carbon source.

A third type of repressor gene mutation is the *lacI⁻ᵈ* (*dominance*) class. These mutations are clustered toward the 5′ end of the *lacI* gene. In haploid cells the *lacI⁻ᵈ* mutants have a constitutive phenotype like the *lacI⁻* mutants; the *lac* enzymes are made in the presence or absence of lactose. Unlike the *lacI⁻* mutations, the *lacI⁻ᵈ* mutations are trans-dominant to *lacI⁺* in *lacI⁻ᵈ/lacI⁺* partial diploids, so *lac* enzymes are produced constitutively even in the presence of the *lacI⁺* gene.

The dominance of *lacI⁻ᵈ* mutants relates to the structure of the *lac* repressor. The repressor protein has four identical polypeptides. There are relatively few, perhaps a dozen, repressor molecules in the cell. In the

lacI⁻ᵈ mutants, the repressor subunits do not combine normally, so no complete repressor is formed, and no operator-specific binding is possible. The *lacI⁻ᵈ/lacI⁺* diploids have a mixture of normal and mutant polypeptides, which combine randomly to form repressor tetramers. The presence of one or more defective polypeptides in the repressor is enough to block normal binding to the operator. So, there is a good chance that no normal repressor proteins will be produced, since there are so few molecules per cell. As a result of the absence (or virtual absence) of complete, functional repressors, a constitutive enzyme phenotype results.

Lastly, some mutations in the repressor gene promoter affect the expression of the repressor gene itself. We mentioned earlier that the extent of transcription of a gene is a function of the affinity of that gene's promoter for RNA polymerase molecules. Clearly, since relatively few repressor molecules are synthesized in the wild-type *E. coli* cells, the repressor gene promoter

~ FIGURE 16.8 continued

c) Partial diploid in the presence of inducer

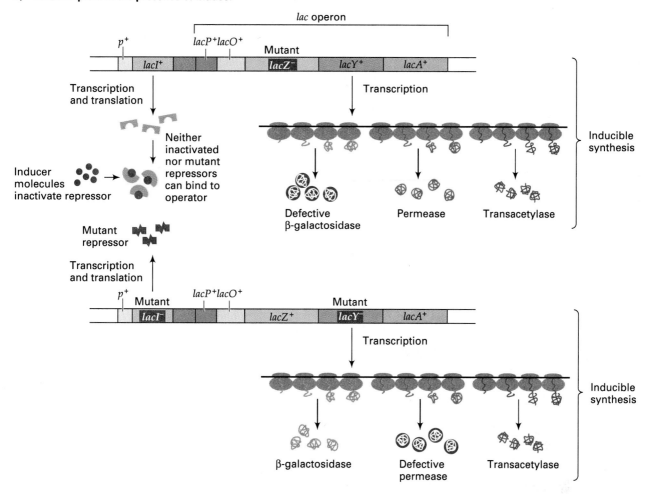

must be of low affinity. As in any other region of the DNA, the repressor gene promoter is subject to base-pair changes by mutation. Base-pair mutations have been found that decrease and that increase transcription rates. (Remember that the repressor gene is a constitutively expressed gene under promotor control.) The most useful mutants in this regard are $lacI^Q$ and $lacI^{SQ}$ mutants (where Q stands for "quantity" and SQ for "superquantity"). Both mutations result in an increase in the rate of transcription of the repressor gene, with the $lacI^{SQ}$ mutants giving the greater increase. These mutants were useful historically because they produce large numbers of repressor molecules, which facilitated their isolation and purification and, consequently, the determination of the amino acid sequence of the constituent polypeptide subunit. Since $lacI^Q$ and $lacI^{SQ}$ mutants produce more repressor molecules than the wild type, these mutants reduce the efficiency of induction of the *lac* operon. They can be induced, however, at very high lactose concentrations.

The mutants of the *lacI* gene point out the functions of the product of that gene, the repressor. Specifi-

cally, the repressor is involved in three recognition interactions, any one of which can be affected by mutation: (1) binding of the repressor to the operator region; (2) binding of the inducer to the repressor; and (3) binding of individual repressor polypeptides to each other to form the active repressor tetramer.

Positive Control of the *lac* Operon

In the previous sections we have learned that in the *lac* operon the repressor protein functions to prevent the transcription of the operon's structural genes by binding to the operator, providing the inducer is not present. The repressor is exerting a negative effect on the expression of the *lac* operon, and this operon is said to be under *negative control*. Several years after Jacob and Monod proposed their operon model, researchers also found a positive control system that regulates the *lac* operon, a system that functions to turn on the expression of the operon. This system is used to ensure that the *lac* operon will be expressed if lactose is the sole carbon source *but not if glucose is present as well*.

~ FIGURE 16.9

Dominant effect of *lacI*^{s} mutation over wild-type *lacI*$^+$ in a *lacI*$^+$ *lacO*$^+$ *lacZ*$^+$ *lacY*$^+$ *lacA*$^+$/*lacI*^{s} *lacO*$^+$ *lacZ*$^+$ *lacY*$^+$ *lacA*$^+$ partial-diploid cell growing in the presence of lactose.

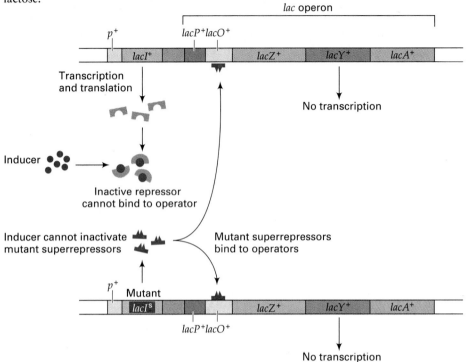

~ FIGURE 16.10

Role of cyclic AMP (cAMP) in the functioning of glucose-sensitive operons such as the lactose operon of *E. coli*.

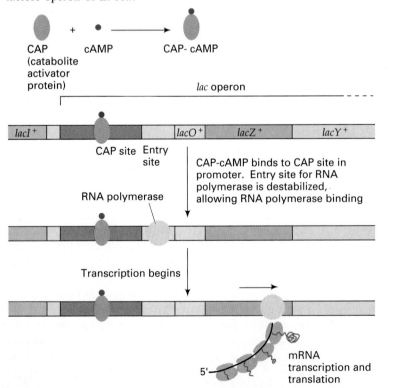

We have carefully noted all along in our discussion of the *lac* operon that expression of the protein-coding genes occurs when lactose is added as the sole carbon source. Recall also that glucose is preferred as a carbon source to all other sugars. Lactose is converted to glucose and galactose (which is, in turn, converted to glucose).

If both glucose and lactose are present in the medium, the glucose is used preferentially and the *lac* operon is not expressed. The *lac* operon is repressed under these conditions because the concentration of a *positive regulator* that binds to the *lac* operon to facilitate transcription is reduced in the presence of glucose. Figure 16.10 shows the mechanism involved. First, a protein called *CAP* (catabolite activator protein) binds with *cAMP* (cyclic AMP, or cyclic adenosine 3′,5′-monophosphate; see Figure 16.11), to form a CAP–cAMP complex. This complex is the positive-regulator molecule. The CAP protein itself is a dimer that consists of two identical polypeptides. Next, the CAP–cAMP complex binds to a specific site in the DNA, called the *CAP binding site*, and this facilitates RNA polymerase binding to the promoter sequence in the presence of an inducer. The polymerase then moves toward the structural genes and begins transcription near the start of the β-galactosidase gene.

In the presence of glucose, **catabolite repression** (**glucose effect**) occurs. A breakdown product—a catabolite—of glucose causes a rapid decrease in the cellular levels of cAMP. (See Figure 16.11 for the steps involved in the synthesis and degradation of cAMP.) Insufficient CAP–cAMP complex is then available to facilitate polymerase binding to the *lac* operon promoter, and transcription is blocked, even though repressors are removed from the operator by the presence of allolactase.

Catabolite repression occurs in a number of other bacterial operons related to catabolism of sugars other than glucose. In all instances, if glucose is present, catabolite repression occurs and, through a similar mechanism of CAP–cAMP binding to a CAP binding site in DNA blocks the expression of those operons. It has been shown that the CAP binding site is not located at a constant distance from the transcription start site among the operons, indicating that direct protein-protein interaction between CAP and RNA polymerase is unlikely.

Molecular Details of *lac* Operon Regulation

From DNA- and RNA-sequencing experiments much information is now available concerning the nucleotide sequences of the significant *lac* operon regulatory sequences. One general approach to obtain this information has been to purify the protein known to bind to a regulatory site and to let it bind to isolated *lac* operon

~ **FIGURE 16.11**

Structure of cyclic AMP (cAMP, or cyclic adenosine 3′,5′-monophosphate). cAMP is synthesized from ATP in a reaction catalyzed by adenylcyclase, and it is broken down in a reaction catalyzed by phosphodiesterase.

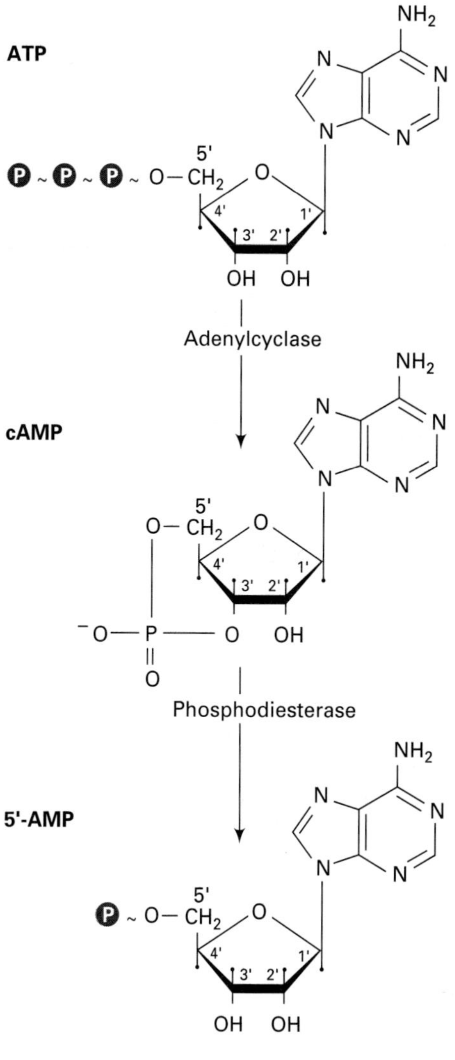

DNA *in vitro*. For example, if the repressor is bound to the *lac* operator, it will protect that region of the operon from deoxyribonuclease digestion. If DNase is allowed to digest the rest of the DNA, the operator sequence can be isolated, cloned by using recombinant DNA technology, and sequenced.

PROMOTER REGION OF THE *lac* REPRESSOR GENE (*lacI*). Figure 16.12 (p. 490) shows the nucleotide pair sequence for the *lacI* gene promoter region, the sequence for the 5′ end of the repressor mRNA, and the first few amino acids of the repressor protein itself. The nucleotide sequence of the repressor mRNA can be aligned with this promoter sequence, with its start approximately in the middle. As usual with all gene transcripts, translation does not start right at the end of the

~ FIGURE 16.12

Base-pair sequences of the *lac* operon *lacI*⁺ gene promoter and of the 5′ end of the repressor mRNA. Also shown is the amino acid sequence of the first part of the repressor protein itself.

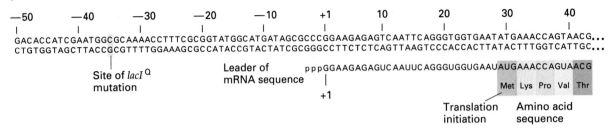

mRNA molecule: The translation start sequence (ribosome binding site) is a Shine-Dalgarno sequence (AGGG) at 12 to 9 bases upstream from the AUG start codon (Chapter 14). Here the first AUG codon (the actual start codon) is nucleotides 29–31 from the 5′ end of the messenger. Figure 16.12 also shows the single base-pair change found for a particular *lacI*^Q mutant; this change, from CG to TA, brings about a tenfold increase in repressor production.

Lac OPERON CONTROLLING SITES. Figure 16.13 shows the nucleotide pair sequence of the *lac* operon controlling sites. The orientation of this sequence was put together from several different pieces of information. First, the amino acid sequences of the repressor protein and of β-galactosidase are completely known,

and this information enables us to identify the coding regions of the *lacI* gene and of the *lacZ*⁺ gene. Then the other regions were identified on the basis of "protection" experiments of the kind described a little earlier. Here CAP−cAMP complex, RNA polymerase, and repressor protein were used separately to bind to the DNA, and DNase-resistant regions were then sequenced. (Although studies indicate the results that follow, we should perhaps be cautious before we definitely assume that the discrete boundaries between the various parts of the controlling region are real.)

The beginning of the promoter region is defined as position −84 in the figure (i.e., 84 base pairs upstream from the mRNA initiation site), immediately next to the stop codon for the *lacI* gene. Nucleotide pairs −54 to −58, and −65 to −69 are the consensus sequence

~ FIGURE 16.13

Base-pair sequence of the controlling sites, promoter and operator, for the lactose operon of *E. coli*. Also shown are locations of some known *lacO*^c mutations (indicated by arrows).

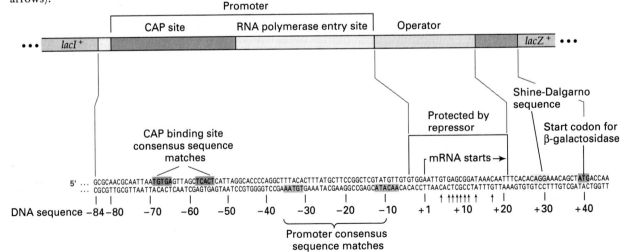

matches for the CAP—cAMP binding site. The fact that this location is upstream from the polymerase binding site (the region spanned by nucleotide pairs -47 to -8, including -10 and -35 consensus sequence matches) fits in well with the model that CAP—cAMP binding destabilizes the DNA in order to facilitate RNA polymerase binding (see Figure 16.10). Together, the region from -8 to -84 that includes the CAP protein and the RNA polymerase interaction sites (including a Pribnow box) essentially define the *lac* operon promoter region.

Immediately next to the promoter region is the operator. The region protected by the repressor protein is the area containing nucleotide pairs -3 to $+21$. When the repressor is bound to the operator, it is impossible for polymerase to bind to the DNA, and thus it cannot transcribe the genes.

The β-galactosidase mRNA has a *leader region* before the start codon is encountered. The actual start of the mRNA here is nucleotide pair $+1$ in Figure 16.13, which is very close to the beginning of the repressor binding site. Transcription of the *lac* operon includes a large proportion of the operator region in addition to the protein-coding genes themselves. The start codon for β-galactosidase, which defines the beginning of the *lacZ* gene, is at nucleotide pairs $+39$ to $+41$. Thus the first 38 bases of the *lac* mRNA are not translated.

Figure 16.13 also shows the base-pair substitutions that have been identified for some of the *lacO^c* mutations that have been studied. In each case a single base-pair change is responsible for the altered control of the *lac* operon.

In conclusion, the lactose operon has proved to be a model system for understanding gene regulation in prokaryotic organisms. Jacob and Monod's original work on this system had a great impact on further studies. As the first molecular model for the regulation of gene expression in any organism, it sparked numerous studies in both prokaryotes and eukaryotes to see whether operons were generally the case. We now know that operons are prevalent in bacteria and bacteriophages but that they do not exist in eukaryotes. Each operon is regulated through the interaction of a regulatory protein with an operator, and this regulation takes place in several different ways. In the *lac* operon, for example, the repressor exerts a negative effect on structural gene expression in the absence of lactose; the genes are turned on when the inducer removes this negative effect by binding to the repressor. Other inducible operons use a similar mechanism; yet others (for example, the arabinose operon) are turned on by an inducer binding to a regulatory protein, altering it so that it binds to a controlling site, thereby facilitating operon transcriptions. The latter operons are said to be positively-controlled operons.

KEYNOTE

Studies of the synthesis of the lactose-utilizing enzymes of *E. coli* generated a model that is the basis for the regulation of gene expression in a large number of bacterial and bacteriophage systems. In the lactose system the addition of lactose to the cells bring about a rapid synthesis of three enzymes required for lactose utilization. The genes for these enzymes are contiguous on the *E. coli* chromosome and are adjacent to a controlling site (an operator) and a single promoter. The genes, the operator, and the promoter constitute an operon, which is transcribed as a single unit. In the absence of lactose the operon is turned off.

TRYPTOPHAN OPERON OF *E. COLI*

Just as glucose may not always be available in a bacterial growth medium for use as a carbon source, all necessary amino acids may not be present in a growth medium to enable bacteria to assemble proteins and to produce compounds requiring amino acids as precursors or as donators of specific chemical groups. If an amino acid is missing, a bacterium has certain operons and other gene systems that enable it to manufacture that amino acid so that it may grow and reproduce. Each step in the biosynthetic pathway through which amino acids are assembled is catalyzed by a specific enzyme coded by a specific gene.

When all 20 amino acids are present, the genes encoding the enzymes for the amino acid biosynthetic pathways are turned off. If an amino acid is not present in the medium, however, then the genes must be turned on in order for the biosynthetic enzymes to be made. Unlike the case in the *lac* operon, where gene activity is induced when a chemical (lactose) is added to the medium, in this case there is a repression of gene activity when a chemical (an amino acid) is added. We refer to amino acid biosynthesis operons controlled in this way as *repressible operons*. One repressible operon in *E. coli* that has been extensively studied is the operon for the biosynthesis of the amino acid tryptophan (Trp). In *E. coli*, tryptophan is used sparingly in proteins, perhaps occurring once in every hundred amino acids or so. Although the regulation of the *trp* operon shows some basic similarities to the regulation of the classical *lac* operon, some intriguing differences also appear to be common among similar, repressible, amino acid biosynthesis operons in bacteria.

~ **FIGURE 16.14**

Organization of controlling sites and the structural genes of the *E. coli* tryptophan operon. Also shown are the steps catalyzed by the products of the structural genes *trpA*, *trpB*, *trpC*, *trpD*, and *trpE*.

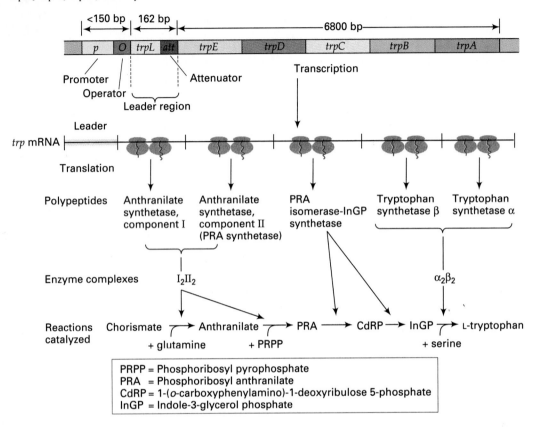

PRPP = Phosphoribosyl pyrophosphate
PRA = Phosphoribosyl anthranilate
CdRP = 1-(*o*-carboxyphenylamino)-1-deoxyribulose 5-phosphate
InGP = Indole-3-glycerol phosphate

Gene Organization of the Tryptophan Biosynthesis Genes

Figure 16.14 shows the organization of the controlling sites and of the genes that code for the tryptophan biosynthetic enzymes and how they relate to the biosynthetic steps. Much of the work that we will discuss is that of Charles Yanofsky and his collaborators.

Five structural genes (*A–E*) occur in the tryptophan operon. The promoter and operator regions are closely integrated in the DNA and are upstream from the *trpE* gene. Between the promoter-operator region and *trpE* is a 162-base-pair region called *trpL*, or the leader region. Within *trpL*, relatively close to *trpE*, is an *attenuator site* (*att*) that plays an important role in the regulation of the tryptophan operon, as we will see later.

The entire tryptophan operon is approximately 7000 base pairs long. Transcription of the operon results in the production of a polygenic mRNA containing the transcripts for the five structural genes. Each of these transcripts is translated to an equal extent.

Regulation of the *trp* Operon

An expanded schematic of the regulatory elements of the tryptophan operon is shown in Figure 16.15. The initiation of transcription of the *trp* operon is regulated at the operator. A regulatory gene *trpR* for the operon is unlinked to the protein-coding genes (and therefore does not appear in Figure 16.14). The product of *trpR* is an *aporepressor protein*, which, alone, has no affinity

~ **FIGURE 16.15**

Regulatory elements of the *E. coli* tryptophan operon.

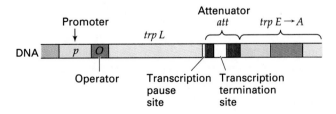

for the operator. However, the aporepressor has two binding sites for tryptophan and is converted to an active repressor by binding with tryptophan. The active complex has affinity for the operator. When the repressor-tryptophan complex is bound to the operator, the promoter is not accessible to RNA polymerase, and initiation of transcription cannot occur. (Tryptophan is an example of an effector molecule, just as allolactose is the effector molecule for the *lac* operon.) As a result of repression, transcription of the *trp* operon can be reduced about seventy-fold. In sum, this regulatory system responds to the amount of tryptophan in the cell.

A secondary regulatory mechanism is involved in the expression of the *trp* operon, usually under conditions of relatively severe tryptophan starvation. In brief, not all RNA transcripts initiated at the promoter site are continued through the protein-coding genes, and the proportion that are is related to the amount of tryptophan that is present. As we will see, this system responds to the level of charged tRNA.Trp in the cell. Within the *trpL* region is the attenuator, a transcription termination site (Figures 16.14 and 16.15). Those RNA transcripts that do not extend into the structural genes are terminated at the attenuator in a process called **attenuation.** Attenuation can reduce transcription of the *trp* operon by 8-fold to 10-fold. Thus repression and attenuation together can regulate the transcription of the *trp* operon over a range of about 560-fold to 700-fold. The details of these regulatory mechanisms are more fully elaborated next.

EXPRESSION OF THE *trp* OPERON IN THE PRESENCE OF TRYPTOPHAN. When tryptophan is abundant in the growth medium, the repressor protein is activated by tryptophan and binds to the operator, thereby preventing the initiation of transcription of the *trp* operon protein-coding genes by RNA polymerase. As a result, the tryptophan biosynthesis enzymes are not produced.

EXPRESSION OF THE *trp* OPERON IN THE ABSENCE OF OR IN THE PRESENCE OF LIMITING AMOUNTS OF TRYPTOPHAN. In the absence of tryptophan (or if tryptophan is present in very low levels) in the growth medium, a tryptophan-repressor complex is not bound to the operator, and transcription of the operon is initiated at the promoter. Following the binding of RNA polymerase, transcription can proceed toward the structural genes. The first region to be transcribed is the leader region (*trpL*), a region of 162 base pairs that does not code for any of the tryptophan biosynthesis enzymes. After the *trpL* region is transcribed, the protein-coding genes are transcribed, in order, onto a polygenic mRNA.

When very low levels of tryptophan are present, all the RNAs produced contain transcripts of the protein-coding genes. However, when some, but not a great deal, of tryptophan is present, *E. coli* produces two types of mRNA transcripts of the tryptophan operon. One transcript is of the entire operon—the leader region plus the protein-coding genes. The second transcript is an mRNA transcript of only the first 140 base pairs of the operon, which includes most of the leader region. The relative proportions of the two transcripts depend on the amount of tryptophan present—the more tryptophan there is present, the greater the number of short transcripts.

The existence of a number of partial transcripts suggests that a transcription termination site exists within the *trpL* region at the 140-base-pair position. The regulation that occurs at this site is *attenuation,* and the site itself is the attenuator (*att*).

The DNA sequence of the whole tryptophan operon is known. Inspection of the sequence for the leader region revealed that the transcript of this region has a ribosome binding site with an AUG start codon in frame with a UGA chain termination codon such that translation of the transcript would produce a 14-amino acid-long polypeptide. This polypeptide has never been isolated from cells, but it is clear that it is made and that it is involved in the attenuation process. Of particular interest is the existence of two adjacent codons for tryptophan near the stop codon of the leader transcript. If cells are starved for tryptophan, then the amount of Trp-tRNA.Trp molecules drops dramatically, because very few tryptophan molecules are available for the aminoacylation of the tRNA.Trp molecules. A ribosome translating the leader transcript would stall at the Trp codons, because the next specified amino acid in the peptide could not be added to the growing chain. As a result, the leader peptide could only be completed if sufficient tryptophan were present.

The sequence data therefore suggested that the position of the ribosome on the leader transcript has an important role in the regulation of transcription termination at the attenuator. In Chapter 13 we learned that transcription and translation are closely coupled in prokaryotes, so this hypothesis was by no means out of line. In fact, evidence from other systems suggests that the tandem arrangement of the Trp codons in this case is not a chance event. Figure 16.16 (p. 494) shows the predicted leader peptides of a number of other amino acid biosynthetic operons of *E. coli* and of *Salmonella typhimurium*, which also have an attenuation mechanism for regulation of gene expression. In every case a significant number of the codons specify the amino acid for which the operon specifies the biosynthetic enzymes. For example, the histidine operon of *E. coli*

~ **FIGURE 16.16**

Predicted amino acid sequences of the leader peptides of a number of attenuator-controlled bacterial operons. Shown are the peptides for the *pheA, his, leu, thr,* and *ilv* operons of *E. coli* or *Salmonella typhimurium*. The amino acids that regulate the respective operons are capitalized.

pheA: Met – Lys – His – Ile – Pro – Phe – Phe – Phe – Ala – Phe – Phe – Phe – Thr – Phe – Pro

his: Met – Thr – Arg – Val – Gln – Phe – Lys – His – His – His – His – His – His – Pro – Asp

leu: Met – Ser – His – Ile – Val – Arg – Phe – Thr – Gly – Leu – Leu – Leu – Leu – Asn – Ala – Phe
          Ile – Val – Arg – Gly – Arg – Pro – Val – Gly – Ile – Gln – His

thr: Met – Lys – Arg – Ile – Ser – Thr – Thr – Ile – Thr – Thr – Thr – Ile – Thr – Ile – Thr – Thr
          Gly – Asn – Gly – Ala – Gly

ilv: Met – Thr – Ala – Leu – Leu – Arg – Val – Ile – Ser – Leu – Val – Val – Ile – Ser – Val – Val
          Val – Ile – Ile – Ile – Pro – Pro – Cys – Gly – Ala – Ala – Leu – Gly – Arg – Gly – Lys – Ala

has a string of 7 histidines in the leader peptide, and 7 of the 15 amino acids in the leader for the phenylalanine A operon of *E. coli* are phenylalanine.

MOLECULAR MODEL FOR ATTENUATION. How does amino acid starvation and the possible stalling of the ribosome at the Trp codons control transcription by RNA polymerase? Yanofsky proposed a model involving different secondary structures of the leader region's transcript.

Analysis of the nucleotide sequence of the leader peptide mRNA indicated that there were four regions that could form secondary structures by complementary base pairing (Figure 16.17). As we will see, pairing

of regions 1 and 2 results in a *pause signal*, pairing of 3 and 4 is a *termination of transcription signal*, and pairing of 2 and 3 is an *antitermination signal* for transcription.

Figure 16.18 shows the possible alternative secondary structure of nucleotides 50–141 of the *trp* leader transcript. Figure 16.18a shows the two hairpin loops that are formed by hydrogen bonding between complementary base pairs. The four regions that are involved in the pairing are designated as 1–4 in Figure 16.18 (see inset in the figure). Thus, in this form regions 1 and 2 are paired, as are regions 3 and 4. The two Trp codons are located in region 1, starting at position 54. Figure 16.18b shows the alternative secondary structure that could form. In this case region 2 is hydrogen-bonded to region 3.

The model proposed by Yanofsky is shown in Figure 16.19 (p. 496). In the following discussion a ribosome is considered to be translating the RNA transcript a little way behind the point at which RNA polymerase is at work extending the transcript—transcription and translation are very tightly coupled in this regulatory system. This coupling is accomplished by a pause of the RNA polymerase just after the RNA of the 1:2 structure has been synthesized. The pausing is a transient halt of leader RNA synthesis. The pause lasts long enough for the ribosome to load and to begin translating the leader peptide, resulting in the tight coupling of transcription and translation.

The position of the ribosome on the leader transcript determines which of the other two RNA secondary structures forms. If the cells are starved for tryptophan (Figure 16.19a), the ribosome stalls on the Trp codons. The stall allows region 2 to pair with region 3 once the latter is synthesized. As a result, when region 4 is synthesized, it cannot pair with region 3 because region 3 is already paired with region 2. With this secondary structure of the leader transcript, RNA poly-

~ **FIGURE 16.17**

Four regions of the tryptophan operon leader mRNA that can form alternate secondary structures by complementary base pairing.

Leader peptide coding region:

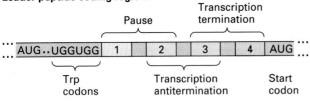

Alternative RNA structures:

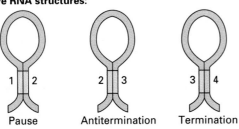

~ FIGURE 16.18

Secondary structure alternatives for the *trp* operator leader transcript; the inset (center) shows the numbering of the four stems that can participate in hydrogen bonding: (a) Regions (1 + 2) and (3 + 4) are hydrogen bonded together in a structure assumed to terminate transcription at the attenuator; (b) Regions 2 and 3 are hydrogen-bonded together, while regions 1 and 4 are unbonded; the 2:3 binding prevents transcription termination at the attenuator.

a) **Regions 1 + 2 and 3 + 4 paired**   b) **Regions 2 and 3 paired**

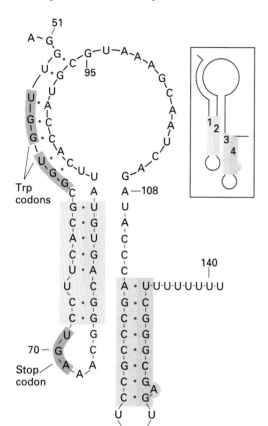
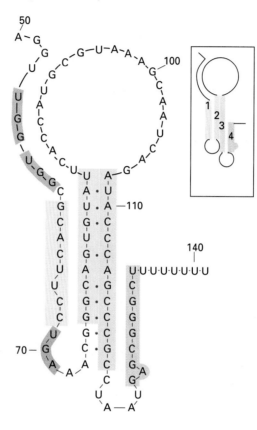

merase is able to continue past the attenuator, and the structural genes are transcribed. The resulting polygenic mRNA is translated into the biosynthetic enzymes necessary to begin tryptophan synthesis.

If, instead, sufficient tryptophan is present (Figure 16.19b) so that the ribosome can read the Trp codons, the ribosome stalls at the stop codon for the leader peptide at positions 69–71 in the leader transcript. As a result, region 2 is unable to pair with region 3, and the latter is then able to pair with region 4 when that region is synthesized. This secondary structure somehow stimulates RNA polymerase to terminate transcription. This structure is referred to as the attenuator. That is, the RNA polymerase is sensing the secondary structure of the RNA it is making. It is also important to realize that the RNA polymerase and ribosome are moving

together and that the secondary structure loops are forming as they are synthesized. The key signal for attenuation is the level of Trp-tRNA.Trp in the cell, since that determines whether or not ribosomes will stall on the leader transcript.

Genetic evidence for the attenuation model has been obtained through the study of mutants. One type of mutant shows less efficient transcription termination at the attenuator, increasing structural gene expression. The mutations involved are single base-pair changes leading to the base changes in the leader transcript shown in Figure 16.20 (p. 497). In each case the change is in the regions of 3:4 pairing; each causes a disruption of the pairing so that the structure is less stable. In the less stable state, the structure is less able to prevent transcription from proceeding into the structural genes.

~ FIGURE 16.19

Models for attenuation in the tryptophan operon of *E. coli*; the taupe structures are ribosomes that are translating the leader transcript: (a) tryptophan-starved cells; (b) non-tryptophan-starved cells.

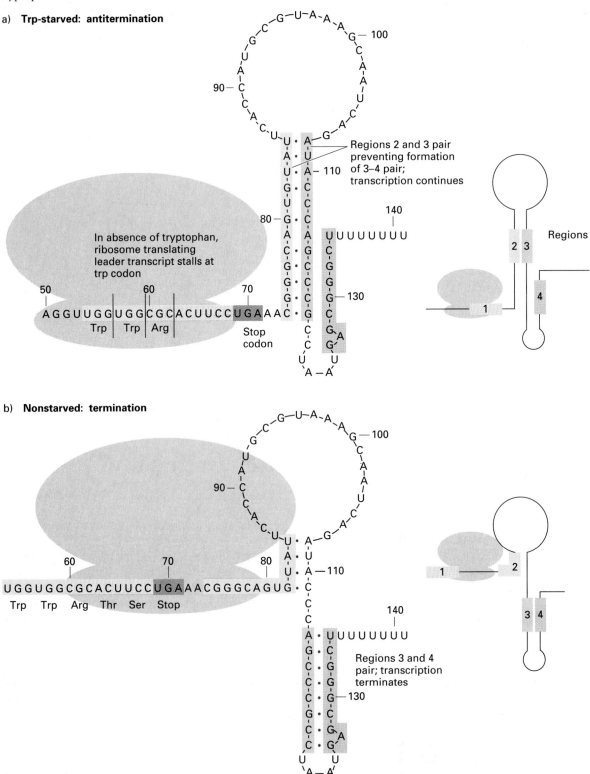

a) **Trp-starved: antitermination**

Regions 2 and 3 pair preventing formation of 3–4 pair; transcription continues

In absence of tryptophan, ribosome translating leader transcript stalls at trp codon

A G G U U G G U G G C G C A C U U C C U G A A A C
Trp | Trp | Arg

Stop codon

Regions

b) **Nonstarved: termination**

U G G U G G C G C A C U U C C U G A A A C G G G C A G U G
Trp Trp Arg Thr Ser Stop

Regions 3 and 4 pair; transcription terminates

~ FIGURE 16.20

In the *trpL* region, mutation sites that show less efficient transcription at the attenuator site. The mutations map to DNA regions that correspond to regions 3 and 4 in the RNA.

Further direct evidence for the attenuation model came from DNA manipulations in which the DNA sequences for the two Trp codons were changed to encode another amino acid. In those mutant strains, attenuation was *not* seen in response to changing levels of tryptophan, but it *was* seen in response to changing levels of the amino acid now specified by the codons.

FINE-SCALE REGULATION OF THE *trp* OPERON. Everything we have said so far relates to regulation at the gene expression level in the *trp* operon. One other mechanism operates in addition to this regulation to give an overall regulatory system for gene expression. This mechanism is the degradation of the polygenic mRNA transcript of the protein-coding genes. As is the case with the lactose operon and with most other prokaryotic genes, the transcripts have a relatively short half-life, so mRNA must continually be made in order for the enzymes to be produced. The short half-life allows the cell to change its physiological state when the environment changes. When tryptophan is added to the medium, transcription termination at the attenuator and repression at the operator block further transcription of the structural genes, and the structural gene mRNAs already made are rapidly degraded ("turned over"). In this way the cell does not make unnecessary materials.

Another part of the overall regulatory system is a fine-tuning system. Gene regulation (turning genes on and off) involves control processes that function in response to long-term changes in the environment.

Cells must also respond to short-term changes that occur. If, for example, tryptophan being made by the tryptophan biosynthetic pathway accumulates at a rate faster than it is being used, a mechanism called *feedback inhibition* (also called end-product inhibition) is triggered.

In feedback inhibition the end product of a biosynthetic pathway can often be recognized by the first enzyme in the biosynthetic pathway. When too much tryptophan is produced, for example, the tryptophan binds to the first enzyme in the tryptophan biosynthetic pathway, thereby altering the enzyme's three-dimensional conformation. (The technical term for this configurational change is *allosteric shift*, and the phenomenon is called *allostery*.) When the shape of the enzyme is changed, the function of the enzyme is impaired, because the substrate for the enzyme can no longer bind to the enzyme. As a result, the tryptophan pathway is temporarily turned off at the first step in the biosynthetic pathway. The enzyme remains nonfunctional until the tryptophan level in the cell drops and the tryptophan dissociates from the enzyme. This step reverses the feedback inhibition of the enzyme's function, and the pathway is restarted. This feedback mechanism is a common attribute of many biosynthetic pathways in cells of all kinds (both prokaryotic and eukaryotic), serving to halt the flow of intermediates in a pathway.

Regulation of Other Amino Acid Biosynthesis Operons

Other amino acid biosynthetic operons also have attenuator mechanisms. The molecular information that has been gathered for those operons indicates that the attenuation mechanism is very similar in each case. Interestingly, some of the operons use only attenuation and not repression. For example, no histidine repressor protein, or any mutation that gives a phenotype that we would expect of an altered repressor protein, has been found for the histidine operon of *Salmonella typhimurium*.

Attenuation has also been shown to regulate a number of genes not involved with amino acid biosynthesis, for example, the ribosomal protein gene S10, rRNA operons (*rrn*), and the *ampC* gene of *E. coli*, as well as some animal virus and animal cell genes. Among all these examples, a number of attenuation mechanisms have been found to be involved.

KEYNOTE

Expression of the tryptophan (*trp*) operon of *E. coli* is accomplished primarily through a repressor-operator system, which responds to free tryptophan levels, and through attenuation at a second controlling site called an attenuator, which responds to Trp-tRNA.Trp levels. The attenuator is located downstream from the operator in the leader region. The attenuator is a partially effective termination site that allows only a fraction of RNA polymerases to transcribe the rest of the operon. In the presence of tryptophan enough Trp-tRNA.Trp is present so that the ribosome can move past the attenuator and allow the leader transcript to form a secondary structure that causes transcription to be blocked. In the absence of tryptophan the ribosomes stall at the attenuator, and the leader transcript forms a secondary structure that permits continued transcription activity. _____

SUMMARY OF OPERON FUNCTION

There are numerous examples of operons in bacteria and their bacteriophages. The following generalizations can be made about the regulation of operons:

1. A regulator protein (e.g., the *lac* repressor) plays a key role in the regulation process since it is able to bind to a controlling site in the operon, the operator.
2. The transcription of a set of clustered structural genes is controlled by an adjacent operator through interaction with a regulator protein.
3. The trigger for changing the state of an operon from off to on and vice versa is an effector molecule (e.g., allolactose), which controls the conditions under which the regulator protein will or will not bind to the operator.

It should be pointed out, however, that not in all cases are genes for a related function clustered in the prokaryotic genome.

GLOBAL CONTROL OF TRANSCRIPTION

As we have seen, specific mechanisms allow the control of gene expression for individual catabolic and biosynthetic pathways. In addition, the bacteria are able to respond at a more general level to major environmental changes. Clearly, it is advantageous in the evolutionary sense for a bacterium to be able to withstand difficult

conditions. Such conditions can be mimicked in the laboratory. For example, an amino acid auxotroph starved for that amino acid suffers severe hardship. A trauma may be induced by shifting cells from a nutrient-rich medium to a nutrient-poor medium in what is called a nutritional shift-down experiment. The cell responds to this stress with a **stringent response** (or **stringent control**), that is, a rapid shutdown of essential cellular activities.

The study of the stringent response has shown that regulatory mechanisms in bacteria coordinate protein synthesis with a number of other basic cellular activities. Starvation conditions, for example, result in a rapid inhibition of glycolysis and transport mechanisms and of the syntheses of proteins, rRNAs, tRNAs, mRNAs, lipids, carbohydrates, and nucleotides. Were these events not to occur, the cell would grow in an uncontrolled fashion, with all macromolecular components (except proteins) accumulating in large amounts. The stringent response is very complex, and we shall only deal with some of the information known.

Mutants have been isolated that do not show the stringent response under starvation conditions. These mutants are called *relaxed* (*relA*) mutants, and their study has helped us understand the stringent response. When starved, wild-type cells rapidly accumulate an unusual nucleotide, ppGpp (guanosine 5′-diphosphate-3′-diphosphate) (Figure 16.21), which *relA* cells do not accumulate. The ppGpp is synthesized by a ribosomal protein enzymatic activity that is activated on ribosomes that are starved of aminoacylated tRNA molecules. The ppGpp is an example of an effector mole-

~ FIGURE 16.21

Structure of the unusual nucleotide ppGpp (guanosine tetraphosphate, or guanosine 5′-diphosphate-3′-diphosphate).

cule, a regulatory molecule that responds to the state of the cell. Apparently, the ppGpp brings about a change in the preference that the RNA polymerase has for promoters; as a result, different genes are transcribed. This reaction occurs because of its direct interaction with RNA polymerase, resulting in a lowered affinity of RNA polymerase for DNA. The RNA polymerase sigma factor plays some role in this change in promoter affinity, but its exact involvement is not understood.

$\mathcal{K}$EYNOTE

Under harsh conditions, a bacterium experiences stress. As a result, the cell responds with a stringent response, that is, a rapid shutdown of essential cellular activities. The stringent response is very complex, and involves the synthesis of a specific molecule under stress conditions that somehow signals the inhibition of major cellular biochemical pathways.

GENE REGULATION IN BACTERIOPHAGES

Because bacteriophages exist by parasitizing bacteria, they themselves do not need to code for the required enzymes for all the biosynthetic pathways needed to make progeny bacteriophages. Instead, the essential components for phage reproduction are provided by the host cell, and those components are directed by the products of phage genes. Most genes of a phage, then, code for products that control the production of progeny phage particles. Since there is a temporal sequence of events for phage particle assembly, there must be specific regulatory pathways to turn genes on and off at particular times. In this section we will discuss selected examples of the regulation of gene expression in two phages that we have discussed before: T4 and lambda (λ).

Regulation of Gene Expression in Phage T4

In Chapter 1 we saw that the T4 phage is a complex array of different proteins: head proteins, core proteins, baseplate proteins, and tail fiber proteins. In the mature phage the genetic material (double-stranded DNA) is packaged within the head.

FUNCTIONAL ORGANIZATION OF THE T4 GENOME. Figure 16.22 (p. 500) shows the gene organization of the T4 chromosome. (Recall that although the chromosome in the phage particle is linear, the

genetic map is circular, owing to terminal redundancy and circular permutation: Chapter 10.) As indicated on the map, genes with related functions are often found in clusters, many of which are operons. The genes required for DNA replication, for example, are clustered in two principal areas: one between 5 and 6 o'clock on the map, and the other between 8 and 11 o'clock. These two clusters are separated by a gene cluster (from about 6 to 7 o'clock) that controls tail fiber production. The gene clusters found throughout the rest of the genome direct the synthesis of the various components of the phage progeny and direct the synthesis of enzymes required for the lysis of the host cell when mature progeny phages have accumulated within. It has been well established that the arrangement of related genes in clusters is important in regulating gene expression.

GENETIC CONTROL OF T4 PHAGE ASSEMBLY. Applying the genetic approach described in Chapter 8 for dissecting the steps in a biochemical pathway, Robert Edgar and William Wood delineated the morphogenesis (genetic control of morphology) of phage T4. They isolated conditional lethal mutants of T4 with mutations in genes essential to the assembly of mature phages. Two types of mutants were identified: nonsense mutants and temperature-sensitive mutants.

Edgar and Wood infected *E. coli* cells with these phage mutants under nonpermissive conditions so that the genetic defects would be manifested. For each mutant the phage assembly proceeded until the step at which an essential protein (coded for by the mutated gene) was absent or inactive. They found that depending on the protein that was missing, different precursor phage particles accumulated in the bacterial cell. These particles were isolated by artificially lysing the cells and centrifuging the contents, then examining those contents under the electron microscope. By correlating the precursor particle with the genetic mutant, they constructed a morphogenetic pathway that involved the functions of about 50 genes, as shown in Figure 16.23 (p. 501). The figure indicates that the pathway has three different parts. One of the branches is concerned with the assembly of the head and the packaging of the DNA within it. The second is involved with tail assembly, and the third with tail fiber production. In the final step of the pathway the head attaches to the tail, after which the tail fibers are added.

Regulation of Gene Expression in Phage Lambda

As we learned in Chapter 7, phage lambda is a temperate phage. When it infects *E. coli,* it can go through a lytic cycle (like that of T4), in which DNA replication and phage assembly occur, or it can establish a lyso-

~ FIGURE 16.22

T4 genetic map, showing the clustering of genes with related functions. In most cases, a number of genes are transcribed from a single promoter onto a single polygenic mRNA, indicated by arrows on the inside of the circle. (Shaded blocks indicate the relative sizes of the genes.)

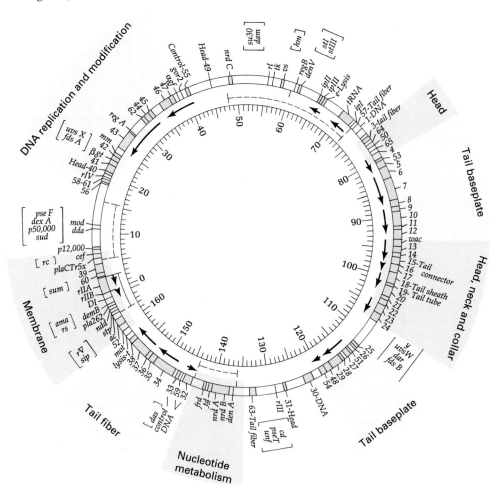

genic state, in which the phage chromosome integrates into the bacterial chromosome and is then called the prophage. When the phage is in the lytic cycle, phage genes for reproduction are active. Those same genes are repressed when the phage is in the lysogenic (prophage) state. In this section we will examine the proposed regulatory mechanisms that determine whether a λ phage enters the lytic or lysogenic cycle.

FUNCTIONAL ORGANIZATION OF THE λ GE-NOME. Figure 16.24 (p. 502) shows the genetic map of the λ phage. Like the T4 chromosome, the mature λ chromosome is linear, in this case with complementary "sticky" ends. Once free in the host cell, the λ chromosome circularizes. Thus it is conventional to show the genetic map in a circular form. The point of joining of the two ends is at 12 o'clock in the figure.

In λ, functionally related genes are clustered in the genome. The genes for DNA replication that are active during the lytic cycle are clustered in an operon at about 1 o'clock (the early right operon), whereas the genes for lysogeny (e.g., those controlling integration and excision of the λ chromosome) are clustered in an operon at 10 to 12 o'clock (the early left operon). The genes for the various structural components of the phage particles (active only in the lytic cycle)—namely, the heads and the tails—are clustered together in the late operon (from 2 to 7 o'clock).

EARLY TRANSCRIPTION EVENTS. When λ infects *E. coli*, it chooses between the lytic and lysogenic pathways soon after the λ chromosome is injected into the host cell. This decision involves a sophisticated *genetic switch*. If the lytic pathway is followed, progeny phages

~ FIGURE 16.23

Morphogenetic pathway of T4 phage production, which has three branches and involves over 50 genes (indicated by numbers in the diagram).

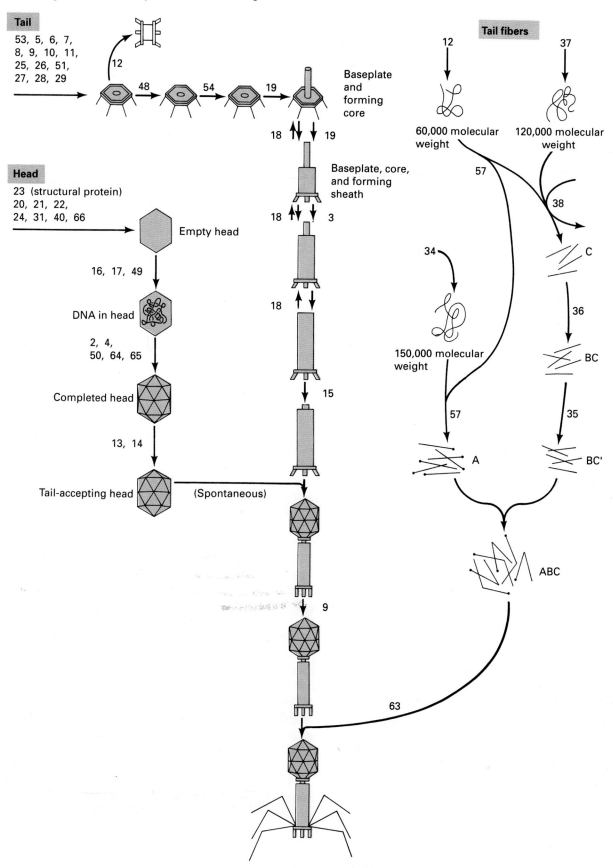

~ FIGURE 16.24

A map of phage λ, showing the major genes.

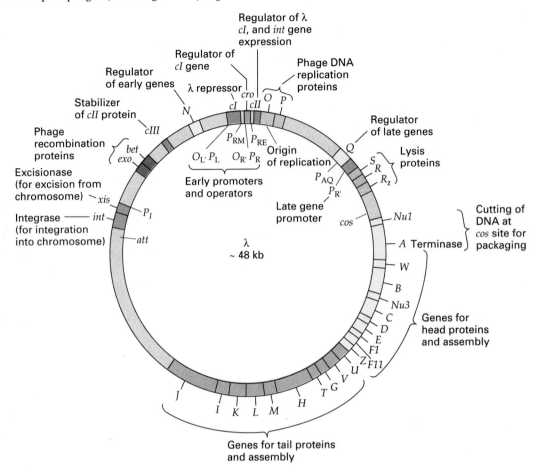

are assembled, the bacterial cells are lysed, and the phages are released. If the lysogenic pathway is followed, the λ chromosome becomes integrated into the *E. coli* chromosome at a specific site and no progeny phages are produced. In this integrated, prophage state, the lytic pathway genes are repressed and the λ genome replicates as part of the *E. coli* chromosome replication.

The two pathways are followed as the result of the expression of different genes. When the λ chromosome first infects a cell, however, some stages of phage growth are the same, regardless of whether the lytic or lysogenic pathway is chosen. First, the linear chromosome becomes circular through the pairing of the 12-nucleotide-long complementary ends and the subsequent action of DNA ligase to form a covalent circle. Since no repressor molecules have been made to prevent expression of the genes for the lytic pathway, transcription begins at promoters P_L and P_R (Figure 16.25, part 1). (P_L is the promoter for leftward transcription of the left early operon, and P_R is the promoter for rightward transcription of the right early operon.) Promoter

P_L is on a different DNA strand from the P_R promoter, and hence P_L is oriented in an opposite direction from P_R. As a result, transcription occurs in opposite directions, counterclockwise for P_L and clockwise for P_R.

From P_R, the first gene to be transcribed is *cro* (control of repressor and other), the product of which is the Cro protein. This protein plays an important role in setting the genetic switch to the lytic pathway, as will be discussed later. From P_L, the first gene to be transcribed is N. The resulting N protein is a transcription *antiterminator* that allows RNA synthesis to proceed past certain transcription terminators. This process is called *antitermination* or *readthrough transcription*. The action of N protein requires the presence of sites in the DNA called *nut* (N utilization) sites. N protein binds to these sites, then alters RNA polymerases that are transcribing the region so that almost all transcription terminators that follow are ignored by the enzyme.

N protein acts to allow RNA polymerase to proceed leftward of gene N, and rightward of gene *cro*, thereby including all of the early genes (Figure 16.25,

~ FIGURE 16.25

Expression of λ genes after infecting *E. coli* and the transcriptional events that occur when either the lysogenic or lytic pathways are followed.

1 Phage growth begins when RNA polymerase binds early promoters P_L and P_R, making mRNA for *N* and *cro* genes. (Throughout figure, ☐ are inactive promoters and ☐ are active promoters.)

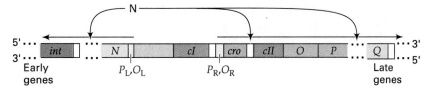

2 N protein, acting at 3 sites, extends RNA synthesis to other genes; O and P proteins permit DNA synthesis to begin; cII protein is made and acts as shown below.

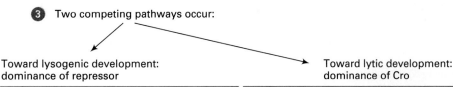

3 Two competing pathways occur:

Toward lysogenic development: dominance of repressor

Toward lytic development: dominance of Cro

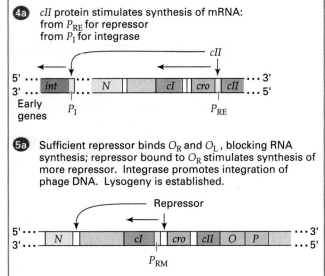

4a *cII* protein stimulates synthesis of mRNA: from P_{RE} for repressor from P_I for integrase

5a Sufficient repressor binds O_R and O_L, blocking RNA synthesis; repressor bound to O_R stimulates synthesis of more repressor. Integrase promotes integration of phage DNA. Lysogeny is established.

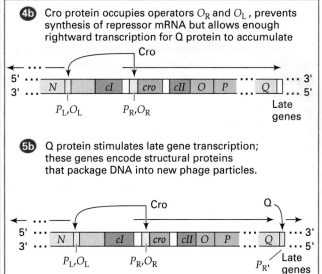

4b Cro protein occupies operators O_R and O_L, prevents synthesis of repressor mRNA but allows enough rightward transcription for Q protein to accumulate

5b Q protein stimulates late gene transcription; these genes encode structural proteins that package DNA into new phage particles.

part 2). In the right early operon, genes transcribed as a result of the action of N protein include *cII*, the product of which stimulates the synthesis of λ repressor (the product of the *cI* gene, to be discussed shortly), *O* and *P*, which are genes for DNA replication, and *Q*, which is needed to turn on late genes for lysis and phage proteins. The Q protein is a positive regulatory protein that functions as an antiterminator protein, similar to the function of N protein. That is, Q protein binds to a site

called *qut* (*Q* utilization), alters RNA polymerase, transcribing the right operon. This permits readthrough transcription into the late genes, which are involved in the lytic pathway. However, only when the lytic pathway is chosen and transcription continues from P_R for a sufficient time does enough Q protein accumulate to function effectively.

The antiterminator activity of N protein in the left early operon leads to the expression of genes *cIII* through *int* (see Figure 16.24). The cIII protein stabilizes the cII protein which, as already mentioned, stimulates the synthesis of λ repressor, and *int* encodes an integrase enzyme that is needed for integration of the λ genome at the bacterial λ attachment site.

THE LYSOGENIC PATHWAY. After the early transcription events, either the lysogenic or lytic pathway is followed (Figure 16.25, part 3). The choice for the lysogenic pathway is made in the following way.

The integration of the prophage into the *E. coli* chromosome depends on protein products coded by genes in the left early operon. These proteins are synthesized after transcription termination is overcome by the action of the N protein. The establishment of the lysogenic state requires the protein products of the *cII* (right early operon) and *cIII* (left early operon) genes. The cII protein (stabilized by cIII protein) activates transcription of the *cI* gene (located between the P_L and P_R promoters) (Figure 16.25, part 4a). As a result of the activation event, *cI* transcription is initiated from a promoter called P_{RE} (promoter for *repressor establishment*), located to the right of the *cI-cro* region. This transcription takes place in a counterclockwise direction. The product of the *cI* gene, the λ repressor protein, acts to block the expression of particular λ genes.

The λ repressor is a diffusible molecule that maintains the lysogenic state of the integrated λ prophage by preventing the expression of the genes necessary for chromosome replication, progeny phage assembly, and cell lysis.

From the work of Mark Ptashne, the λ repressor is known to be a protein of 236 amino acids. This protein folds into two nearly equal-sized domains, connected by a string of 40 amino acids (Figure 16.26a). The

~ FIGURE 16.26

(a) The λ repressor showing the amino (N) and carboxyl (C) domains and the short connecting segment; (b) Repressor monomers form dimers, which can dissociate to monomers.

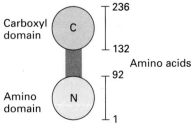

a) λ **repressor**

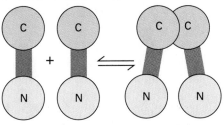

b) **Dimerization of λ repressor**

domains represent the amino and carboxyl parts of the polypeptide. The functional λ repressor is a dimer of the polypeptide, formed largely by contacts between the carboxyl domains (Figure 16.26b). Almost all of the repressor in the cell is in the dimeric form. As soon as enough λ repressor is produced within the cell, it binds to two operator regions O_L and O_R (Figure 16.24), whose sequences overlap the P_L and P_R promoters,

~ FIGURE 16.27

(a) The DNA and part of the protein sequences of right operator (O_R) and parts of the adjacent genes (*cI* and *cro*) and (b) DNA sequence of the left operator O_L and adjacent gene (*N*) regions of phage λ. (The orientation of O_L has been reversed from that in the chromosome to allow comparison with O_R.) The brackets indicate the six binding sites for the λ repressor. The starting points for transcription of genes *N*, *cro*, and *cI* are indicated.

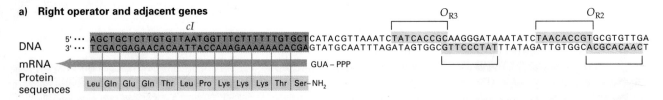

a) **Right operator and adjacent genes**

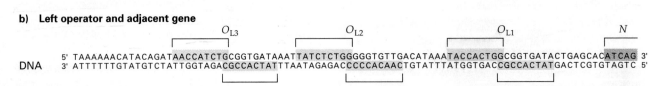

b) **Left operator and adjacent gene**

respectively. The binding of the λ repressor thus prevents the further initiation of transcription by RNA polymerase at the early operons controlled by P_L and P_R. As a result, transcription of the N and cro genes is blocked, and since the two proteins specified by these two genes are unstable, the levels of these two proteins in the cell drop dramatically. The binding of the λ repressor also blocks the N protein–controlled readthrough transcription of the O, P, and Q genes in the right early operon. Furthermore, a repressor bound to O_R stimulates the synthesis of more repressor mRNA from a different promoter, P_{RM}, thereby maintaining its concentrations in the cell (Figure 16.25, part 5a). Thus in the presence of enough λ repressors, the lysogenic state is established by the binding of the repressor to the O_L and O_R operator regions and integration of λ DNA catalyzed by integrase. With recombinant DNA techniques the O_L and O_R operators have been cloned and sequenced (see Figure 16.27). What is most interesting about the sequences is that there are three 17-nucleotide-long repressor binding sites for both the O_L and the O_R regions: O_{L1}, O_{L2}, and O_{L3}, and O_{R1}, O_{R2}, and O_{R3}, respectively.

Binding studies using purified repressor show that the relative binding strength (binding affinity) of the repressor for each of the operator binding sites in O_L is $O_{L1} > O_{L2} > O_{L3}$; and in O_R it is $O_{R1} > O_{R2} > O_{R3}$. Let us consider the events that occur at the right operator (O_R) sites, which are major sites where the genetic switch controlling the choice between the lysogenic and lytic pathways operates; O_L sites are not part of the switch.

Figure 16.28 shows the arrangement of the O_R repressor binding sites relative to the cI (λ repressor) and cro (control of $repressor$ and $other$) genes. Note

~ FIGURE 16.28

Part of the chromosome, showing the two promoters P_{RM} and P_R and the right operator (O_R), which overlaps them. Transcription from P_{RM} occurs leftward to transcribe the repressor gene (cI). Transcription from P_R proceeds rightward to transcribe the cro gene.

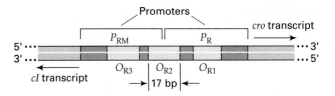

that the O_{R1} site overlaps the P_R promoter and that the O_{R3} site overlaps the P_{RM} (RM = repressor maintenance) promoter. Because of the different binding affinities of the three binding sites for repressor, if we started with a repressor-free operator, a repressor dimer would bind first to O_{R1} (Figure 16.29a). Once a repressor becomes bound to O_{R1}, the affinity of O_{R2} for the second repressor increases because the second repressor dimer not only binds to O_{R2}, but it also contacts the repressor that is bound at O_{R1} (Figure 16.29b). The presence of the initial repressor is enhanced by the binding of a second repressor, a mech-

~ FIGURE 16.27 continued

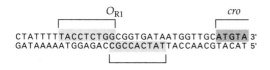

~ FIGURE 16.29

λ repressor binds to three sites in the right operator (O_R). (a) The affinity of repressor for O_{R1} is about 10 times higher than for O_{R2} or O_{R3}; (b) Once repressor is bound to O_{R1}, a second repressor rapidly binds to O_{R2}; (c) Repressor binds to O_{R3} only at high repressor concentrations.

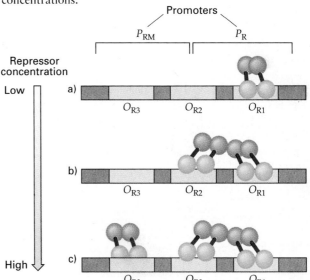

anism called *cooperativity of repressor binding*. At higher repressor concentrations, repressor will also bind to O_{R3} (Figure 16.29c). The affinity of the repressor for this operator site is lower than for O_{R2} because in this case the repressor must bind independently without also contacting a previously bound repressor.

For determining the lysogenic state, then, repressor is typically present at concentrations such that repressor dimers are bound to both O_{R1} and O_{R2}, thereby preventing RNA polymerase from binding to P_R (see Figure 16.24 and Figure 16.25, part 5a) and further transcribing the *cro* gene. However, since no repressor is bound to O_{R3}, RNA polymerase can bind to P_{RM} and the repressor gene *cI* can be transcribed (Figure 16.30). As repressor concentration increases as a result of this transcriptional activity, repressor dimers bind also to the weak O_{R3} site. This blocks repressor synthesis by preventing RNA polymerase from binding and transcribing the *cI* gene. The repressor concentration drops because no new repressor transcripts are being made and existing repressor molecules are being degraded. Once the repressor concentration drops sufficiently, the O_{R3} site is freed again, and RNA polymerase can then bind to P_{RM}, and *cI* transcription can resume. In other words, the λ repressor plays a negative regulatory role by blocking transcription of the *cro* gene, and a positive regulatory role by maintaining optimum repressor levels in the cell by modulating RNA polymerase binding at P_{RM}.

Repressor binding to O_R is not permanent; instead, it is a dynamic event in which repressor dimers continuously fall off the operator binding sites to be replaced by other repressor molecules. The concentration of repressor molecules in the cell is sufficiently high to ensure that O_{R1} and O_{R2} are likely to have repressor bound to them most of the time, enabling the lysogenic state to be maintained. As we will see, this state changes dramatically when the genetic switch is thrown and the lytic pathway is induced.

THE LYTIC PATHWAY. The lysogenic pathway is favored when enough λ repressor is made so that early promoters are turned off, thereby repressing all the genes needed for the lytic pathway. An important gene in this regard is *Q*, since the Q protein is a positive regulatory protein required for the production of lysis proteins and phage coat proteins.

Let us consider the induction of the lytic pathway caused by ultraviolet light irradiation. Inducers such as ultraviolet light typically damage the DNA, and this somehow causes a change in the function of the bacterial protein RecA (product of the *recA* gene). Normally RecA functions enzymatically in DNA recombination events, but when DNA is damaged, it changes its function to that of a protease, which cleaves λ repressor monomers within the linker segment, thereby separating the amino and carboxyl domains (Figure 16.31). This event causes the chain reaction shown in Figure 16.32: the cleaved monomers are unable to form dimers and, as a result, when repressors fall off the operator through the process of normal degradation, they are not replaced.

The absence of repressor at O_{R1} that results leads to RNA polymerase binding at P_R and the *cro* gene is then further transcribed. The resulting Cro protein then controls the subsequent lytic growth. Cro protein action results in a decrease in RNA synthesis from P_L and P_R, thereby reducing synthesis of cII protein, the

repressors are not cut when they are at the operons

~ **FIGURE 16.30**

In the lysogenic state, repressors are bound to O_{R1} and O_{R2}, but not to O_{R3}, which allows RNA polymerase to bind to P_{RM} and transcribe the *cI* (repressor) gene. Polymerase cannot bind to P_R since that promoter is covered by repressors; hence the *cro* gene is not transcribed.

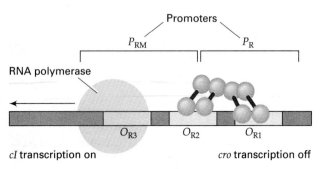

cI transcription on   *cro* transcription off

~ **FIGURE 16.31**

Cleavage of repressor by RecA protein. Repressor is cleaved in the linker between the C and N domains of the repressor.

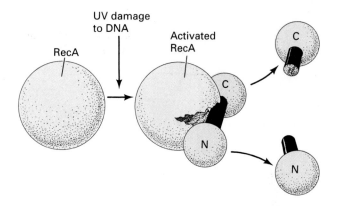

~ FIGURE 16.32

Activated RecA protein cleaves repressor which, then, cannot dimerize. As a result, when repressors dissociate from operators they are not replaced, *cro* is transcribed, and induction of lytic growth follows.

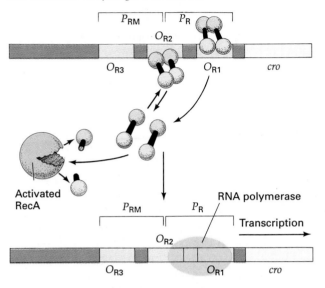

regulator of λ repressor synthesis, and blocking synthesis of λ repressor mRNA from P_{RM} (Figure 16.25, part 4b). At the same time, transcription of the right early operon genes from P_R is decreased, but enough Q proteins are accumulated to stimulate transcription (by antitermination) of the late genes for lytic growth (Figure 16.25, part 5b).

How does Cro protein work? Cro is a 66-amino acid protein that is folded into one globular domain; virtually all of the monomeric Cro polypeptides exist in the cell in dimers (Figure 16.33a), and each dimer can bind to the 17-nucleotide pair operator binding sites (Figure 16.33b). Cro dimers bind independently to the three O_R sites, although the order of affinity for the three sites is *exactly opposite* that of the λ repressor

dimer; that is, the strongest affinity of Cro protein is for O_{R3} (Figure 16.34a), and after that the affinity is approximately equal for O_{R2} and O_{R1} (Figure 16.34b). Unlike the λ repressor, which has both a positive and negative regulatory action, Cro acts only as a negative regulator. The first Cro synthesized binds to O_{R3}, blocking the ability of RNA polymerase to bind to P_{RM}, thereby stopping further repressor synthesis. At this point, the switch has been reset and the lytic pathway is followed. Through continued P_R function, *cro* transcription continues, along with transcription of the right early operon genes that are needed for the early stages of the lytic pathway. As with the repressor protein, the Cro protein controls its own transcription. At high Cro protein concentrations, all three O_R binding sites become occupied by Cro protein (Figure 16.34c), blocking transcription initiation at the P_R promoter. This results in decreased transcription of the *cro* gene, and of the other early lytic pathway genes.

In summary, lambda uses complex regulatory systems to direct either the lytic or the lysogenic pathway. The choice is made between the two pathways, using an elaborate genetic switch, most of the details of which are now in hand. This switch involves two regulatory proteins with opposite affinities for three binding sites within each of the two major operators that control the λ life cycle.

~ FIGURE 16.34

Cro protein binds to the same three sites in the right operator (O_R) as λ repressor, but with different affinities for the three sites. (a) The affinity of Cro protein for O_{R3} is about 10 times higher than for O_{R2} or O_{R1}; (b) Once a Cro protein is bound to O_{R3}, the second dimer binds with equal chance to O_{R2} or O_{R1}; (c) At high Cro dimer concentration, all three sites are occupied.

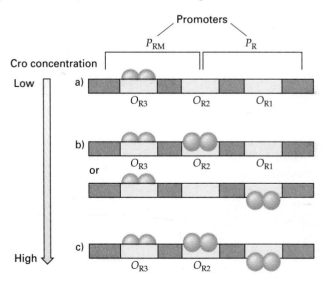

~ FIGURE 16.33

(a) Cro protein. This globular polypeptide is found in the cell as dimers; (b) Cro dimer bound to a 17-bp O_R site.

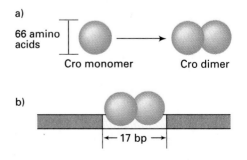

𝒦EYNOTE

Bacteriophages are especially adapted for undergoing their reproduction within a bacterial host. Many of the genes related to the production of progeny phages, or to the establishment or reversal of lysogeny in temperate phages, are organized into operons. These operons (like bacterial operons) are controlled through the interaction of regulatory proteins with operators that are adjacent to clusters of structural genes.

SUMMARY

Bacteria typically live in environments that can change rapidly. To function efficiently in the changing environment, bacteria have evolved special systems for regulating gene expression in response to the dictates of conditions. Generally, when a particular function is needed, the genes for that function (e.g., the lactose utilization genes when lactose is present as a carbon source, or the tryptophan biosynthesis genes when tryptophan is absent from the medium) are coordinately transcribed so that the relevant enzymes are coordinately produced. Conversely, when the conditions change so that function is not needed, gene transcription is turned off.

From studies of the regulation of expression of the three lactose-utilizing genes in *E. coli* bacteria, Jacob and Monod developed a model that is the basis for the regulation of gene expression in a large number of bacterial and bacteriophage systems. That is, the genes for the enzymes are contiguous in the chromosome and are adjacent to a controlling site (an operator) and a single promoter. This complex constitutes a transcriptional regulatory unit called an operon. A regulatory gene, which may or may not be nearby, is associated with the operon. The addition of lactose to the cell results in the coordinate induction of the *lac* genes. Induction occurs as follows: lactose binds with a repressor protein that is encoded by the regulatory gene, inactivating it and preventing it from binding to the operator. As a result, RNA polymerase can bind to the promoter and transcribe the three genes onto a single polygenic mRNA. As long as lactose is present, mRNA continues to be produced and the enzymes are made. When lactose is no longer present, the repressor protein is no longer inactivated, and it binds to the operator, thereby preventing RNA polymerase from transcribing the *lac*

genes. If glucose is present as well as lactose, the lactose operon is not induced. This is the case because less energy is required to metabolize glucose than lactose. This phenomenon is called catabolite repression.

The genes for a number of bacterial amino acid biosynthesis pathways are arranged in operons. Expression of these operons is accomplished by a repressor-operator system and, in some cases, also through attenuation at a second controlling site called an attenuator. The tryptophan operon is an example of an operon with both types of transcription regulation systems. The repressor-operator system functions essentially like that for the *lac* operon, except that the addition of amino acid to the cell activates the repressor, thereby turning the operon off. This makes sense since the role of the gene products is to make the amino acid. The attenuator is located downstream from the operator in a leader region that is translated. The attenuator is a partially effective transcription termination site that allows only a fraction of RNA polymerases to transcribe the rest of the operon. Attenuation involves a coupling between transcription and translation, and the formation of particular RNA secondary structures that signal whether or not transcription can continue.

Key to the attenuation phenomenon is the presence of multiple copies of codons for the amino acid the synthesis of which the operon controls. When enough of the amino acid is present in the cell, enough charged tRNAs are produced so that the ribosome can read the key codons in the leader region, and this causes the RNA that is being made by the RNA polymerase ahead of the ribosome to assume a secondary structure which signals transcription to stop. However, when the cell is starved for that amino acid, there are insufficient charged tRNAs to be used at the key codons so that the ribosome stalls at that point. As a result the RNA ahead of that point assumes a secondary structure that permits continued transcription into the structural genes. The combination of repressor-operator regulation and attenuation control permits a fine degree of control of transcription of the operon.

Normally, bacteria are continually adapting to the changing environment by turning on and off appropriate operons. Under harsh conditions, such as starvation of an amino acid auxotroph for that amino acid, bacteria experience stress. As a result, the cells respond with a stringent response, that is, a rapid shutdown of essential cellular activities, including DNA synthesis, RNA synthesis, protein synthesis, and other primary metabolic pathways. The stringent response is very complex, and involves the synthesis of a specific molecule under stress conditions that somehow signals the rapid and reversible inhibition of major cellular biochemical pathways.

Operons are also extensively used by bacteriophages to coordinate the synthesis of proteins with related functions. As we know, bacteriophages such as T4 and lambda are especially adapted for undergoing their reproduction within a bacterial host. Many of the genes related to the production of progeny phages, or to the establishment or reversal of lysogeny of temperate phages, are organized into operons. These operons, like bacterial operons, are controlled through the interaction of regulatory proteins with operators that are adjacent to clusters of structural genes. T4 has been an excellent model for studying the genetic control of development, since the genes with related developmental functions (e.g., the head genes, tail fiber genes, etc.) are often clustered. Bacteriophage lambda has been an excellent model for studying the elaborate genetic switch that controls the choice between lytic and lysogenic pathways in a lysogenic phage. The switch involves two regulatory proteins, the lambda repressor and the *cro* protein, which have opposite affinities for three binding sites within each of the two major operators that control the lambda life cycle.

In sum, operons are commonly encountered in bacteria and their viruses. They provide a simple way to coordinate the transcription of genes with related function at the transcriptional level. Generally, there is an interaction between an effector molecule (such as lactose or tryptophan) and a repressor molecule. That interaction may inactivate (lactose) or activate (tryptophan) the repressor causing it to be released from or bind to the operator, respectively. If no repressor is bound to the operator, RNA polymerase can bind to the adjacent promoter and the genes can be transcribed. Operons have not been found in eukaryotic systems.

ANALYTICAL APPROACHES FOR SOLVING GENETICS PROBLEMS

Q.1 In the laboratory you are given ten strains of *E. coli* with the following lactose operon genotypes:

(1) *lacI⁺ lacP⁺ lacO⁺ lacZ⁺*;
(2) *lacI⁻ lacP⁺ lacO⁺ lacZ⁺*;
(3) *lacI⁺ lacP⁺ lacOᶜ lacZ⁺*;
(4) *lacI⁻ lacP⁺ lacOᶜ lacZ⁺*;
(5) *lacI⁺ lacP⁺ lacOᶜ lacZ⁻*;
(6) F' *lacI⁺ lacP⁺ lacOᶜ lacZ⁻/lacI⁺ lacP⁺ lacO⁺ lacZ⁺*;
(7) F' *lacI⁺ lacO⁺ lacZ⁻/lacI⁺ lacP⁺ lacOᶜ lacZ⁺ lacP⁺*;
(8) F' *lacI⁺ lacP⁺ lacO⁺ lacZ⁺/lacI⁻ lacP⁺ lacO⁺ lacZ⁻*;
(9) F' *lacI⁺ lacP⁺ lacOᶜ lacZ⁻/lacI⁻ lacP⁺ lacO⁺ lacZ⁺*;
(10) F' *lacI⁻ lacP⁺ lacO⁺ lacZ⁻/lacI⁻ lacP⁺ lacOᶜ lacZ⁺*.

(Note: In the partial-diploid strains (6–10) one copy of the *lac* operon is in the host chromosome and the other copy is in the *F* episome.)

For each strain, predict whether β-galactosidase will be produced (a) if lactose is absent from the growth medium, and (b) if lactose is present in the growth medium.

A.1 The answers are as follows, where + = β-galactosidase produced and − = it is not produced:

GENOTYPE	NON-INDUCED: LACTOSE ABSENT	INDUCED: LACTOSE PRESENT
(1)	−	+
(2)	+	+
(3)	+	+
(4)	+	+
(5)	−	−
(6)	−	+
(7)	+	+
(8)	−	+
(9)	−	+
(10)	+	+

To answer this question completely, we need a good understanding of how the lactose operon is regulated in the wild type and of the consequences of particular mutations on the regulation of the operon.

Strain (1) is the standard wild-type operon. No enzyme is produced in the absence of lactose, since the repressor produced by the *lacI* gene binds to the operator (*lacO*) and blocks the initiation of transcription. When lactose is added, it binds to the repressor, changing its conformation so that it no longer can bind to the *lacO* region, thereby facilitating transcription of the structural genes by RNA polymerase.

Strain (2) is a haploid strain with a mutation in the *lacI* gene (*lacI⁻*). The consequence is that the repressor protein cannot bind to the (normal) operator region, so there is no inhibition of transcription, even in the absence of lactose. This strain, then, is constitutive, meaning that the functional enzyme is produced by the *lacZ⁺* gene in the presence or absence of lactose.

Strain (3) is the other possible constitutive mutant. In this case the repressor gene is a wild type and the β-galactosidase gene *lacZ⁺* is a wild type, but there is a mutation in the operator region (*lacO^c*). Therefore repressor protein cannot bind to the operator, and the transcription occurs in the presence or absence of lactose.

Strain (4) carries both of the regulatory mutations of the previous two strains. Functional repressor is not produced, but even if it were, the operator is changed so that it cannot bind. The consequence is the same: constitutive enzyme production.

Strain (5) produces functional repressor, but the operator (*lacO^c*) is mutated. Therefore transcription cannot be blocked, and lactose polygenic mRNA is produced in the presence or absence of lactose. However, because there is also a mutation in the β-galactosidase gene (*lacZ⁻*), no functional enzyme is generated.

In the partial-diploid strain (6) one lactose operon is completely wild type, and the other carries a constitutive operator mutation and a mutant β-galactosidase gene. In the absence of lactose no functional enzyme is produced. For the wild-type operon the functional repressor binds to the operator and blocks transcription. For the operon with the two mutations the operator region is mutated and cannot bind repressor, so the mRNA for the mutated operon is produced, since transcription cannot be inhibited, and the *lacZ* gene is mutated so that functional enzyme cannot be produced. In the presence of lactose, functional enzyme is produced, since repression of the wild-type operon is relieved so that the *lacZ⁺* gene can be transcribed. This type of strain provided one of the pieces of evidence that the operator region does not produce a diffusible substance.

In diploid (7) functional enzyme will be produced in the presence or absence of lactose, because one of the operons has a *lacO^c* mutation that does not respond to a repressor and that is linked to a wild-type *lacZ⁺* gene. That operon is transcribed constitutively. The other operon is inducible, but because there is a *lacZ⁻* mutation, no functional enzyme is produced.

Diploid (8) has a wild-type operon and an operon with a *lacI⁻* regulatory mutation and a *lacZ⁻* mutation. The *lacI⁺* gene product is diffusible so that it can bind to the *lacO⁺* region of both operons, thereby putting both operons under inducer control. This strain gives a classic demonstration that the *lacI⁺* gene is trans-dominant to a

lacI⁻ mutation. In this case the particular location of the one *lacZ⁻* mutation is irrelevant: The same result would have been obtained had the *lacZ⁺* and *lacZ⁻* been switched between the two operons. In this diploid, β-galactosidase is not produced unless lactose is present.

In strain (9) β-galactosidase is produced only when the lactose is present, because the *lacO^c* region controls only those genes that are adjacent to it on the same chromosome (cis dominance), and in this case one of the adjacent genes is *lacZ⁻*, which codes for a nonfunctional enzyme. The diploid is heterozygous *lacI⁺/lacI⁻*, but *lacI⁺* is trans-dominant, as discussed for strain (8). Thus the only normal *lacZ⁺* gene is under inducer control.

Diploid (10) has defective repressor protein as well as an *lacO^c* mutation adjacent to a *lacZ⁺* gene. On the latter ground alone this diploid is constitutive. The other operon is also constitutively transcribed, but because there is a *lacZ⁻* mutation, no functional enzyme is generated from it.

QUESTIONS AND PROBLEMS

16.1 How does lactose bring about the induction of synthesis of β-galactosidase, permease, and transacetylase? Why does this event not occur when glucose is also in the medium?

16.2 Operons produce polygenic mRNA when they are active. What is a polygenic mRNA? What advantages, if any, do they confer on a cell in terms of its function?

16.3 If an *E. coli* mutant strain synthesizes β-galactosidase whether or not the inducer is present, what genetic defect(s) might be responsible for this phenotype?

16.4 Distinguish the effects you would expect from (a) a missense mutation and (b) a nonsense mutation in the *lacZ* (β-galactosidase) gene of the *lac* operon.

16.5 The elucidation of the regulatory mechanisms associated with the enzymes of lactose utilization in *E. coli* was a landmark in our understanding of regulatory processes in microorganisms. In formulating the operon hypothesis as applied to the lactose system, Jacob and Monod found that results from particular partial-diploid strains were invaluable. Specifically, in terms of the operon hypothesis, what information did the partial diploids provide that the haploids could not?

16.6 For the *E. coli lac* operon, write the partial-diploid genotype for a strain that will produce β-galactosidase constitutively and permease by induction.

16.7 Mutants were instrumental in the elaboration of the model for the regulation of the lactose operon.

a. Discuss why $lacO^c$ mutants are cis-dominant but not trans-dominant.

b. Explain why $lacI^s$ mutants are trans-dominant to the wild-type $lacI^+$ allele but $lacI^-$ mutants are recessive.

c. Discuss the consequences of mutations in the repressor gene promoter as compared with mutations in the structural gene promoter.

*16.8 This question involves the lactose operon of *E. coli.* Complete Table 16.A, using + to indicate if the enzyme in question will be synthesized and − to indicate if the enzyme will *not* be synthesized.

16.9 A new sugar, sugarose, induces the synthesis of two enzymes from the *sug* operon of *E. coli.* Some properties

TABLE 16.A

GENOTYPE	INDUCER ABSENT		INDUCER PRESENT	
	β-GALACTOSIDASE	PERMEASE	β-GALACTOSIDASE	PERMEASE
a. $lacI^+ \ lacP^+ \ lacO^+ \ lacZ^+ \ lacY^+$				
b. $lacI^+ \ lacP^+ \ lacO^+ \ lacZ^- \ lacY^+$				
c. $lacI^+ \ lacP^+ \ lacO^+ \ lacZ^+ \ lacY^-$				
d. $lacI^- \ lacP^+ \ lacO^+ \ lacZ^+ \ lacY^+$				
e. $lacI^s \ lacP^+ \ lacO^+ \ lacZ^+ \ lacY^+$				
f. $lacI^+ \ lacP^+ \ lacO^c \ lacZ^+ \ lacY^+$				
g. $lacI^s \ lacP^+ \ lacO^c \ lacZ^+ \ lacY^+$				
h. $lacI^+ \ lacP^+ \ lacO^c \ lacZ^+ \ lacY^-$				
i. $lacI^{-d} \ lacP^+ \ lacO^+ \ lacZ^+ \ lacY^+$				
j. $lacI^- \ lacP^+ \ lacO^+ \ lacZ^+ \ lacY^+/$ $lacI^+ \ lacP^+ \ lacO^+ \ lacZ^- \ lacY^-$				
k. $lacI^- \ lacP^+ \ lacO^+ \ lacZ^+ \ lacY^-/$ $lacI^+ \ lacP^+ \ lacO^+ \ lacZ^- \ lacY^+$				
l. $lacI^s \ lacP^+ \ lacO^+ \ lacZ^+ \ lacY^-/$ $lacI^+ \ lacP^+ \ lacO^+ \ lacZ^- \ lacY^+$				
m. $lacI^+ \ lacP^+ \ lacO^c \ lacZ^- \ lacY^+/$ $lacI^+ \ lacP^+ \ lacO^+ \ lacZ^+ \ lacY^-$				
n. $lacI^- \ lacP^+ \ lacO^c \ lacZ^+ \ lacY^-/$ $lacI^+ \ lacP^+ \ lacO^+ \ lacZ^- \ lacY^+$				
o. $lacI^s \ lacP^+ \ lacO^+ \ lacZ^+ \ lacY^+/$ $lacI^+ \ lacP^+ \ lacO^c \ lacZ^+ \ lacY^+$				
p. $lacI^{-d} \ lacP^+ \ lacO^+ \ lacZ^+ \ lacY^-/$ $lacI^+ \ lacP^+ \ lacO^+ \ lacZ^- \ lacY^+$				
q. $lacI^+ \ lacP^- \ lacO^c \ lacZ^+ \ lacY^-/$ $lacI^+ \ lacP^+ \ lacO^+ \ lacZ^- \ lacY^+$				
r. $lacI^+ \ lacP^- \ lacO^+ \ lacZ^+ \ lacY^-/$ $lacI^+ \ lacP^+ \ lacO^c \ lacZ^- \ lacY^+$				
s. $lacI^- \ lacP^- \ lacO^+ \ lacZ^+ \ lacY^+/$ $lacI^+ \ lacP^+ \ lacO^+ \ lacZ^- \ lacY^-$				
t. $lacI^- \ lacP^+ \ lacO^+ \ lacZ^+ \ lacY^-/$ $lacI^+ \ lacP^- \ lacO^+ \ lacZ^- \ lacY^+$				

of deletion mutations affecting the appearance of these enzymes are as follows (here, + = enzyme induced normally, i.e., synthesized only in the presence of the inducer; C = enzyme synthesized constitutively; 0 = enzyme cannot be detected):

MUTATION OF	ENZYME 1	2
Gene *A*	+	0
Gene *B*	0	+
Gene *C*	0	0
Gene *D*	C	C

a. The genes are adjacent in the order *ABCD*. Which gene is most likely to be the structural gene for enzyme 1?
b. Complementation studies using partial-diploid (*F'*) strains were made. The episome (*F'*) and chromosome each carried one set of *sug* genes. The results were as follows (symbols are the same as in previous table):

GENOTYPE OF *F'*	CHROMOSOME	ENZYME 1	2
A⁺ B⁻ C⁺ D⁺	*A⁻ B⁺ C⁺ D⁺*	+	+
A⁺ B⁻ C⁻ D⁺	*A⁻ B⁺ C⁺ D⁺*	+	0
A⁻ B⁺ C⁻ D⁺	*A⁺ B⁻ C⁺ D⁺*	0	+
A⁻ B⁺ C⁺ D⁺	*A⁺ B⁻ C⁺ D⁻*	+	+

From all the evidence given, determine whether the following statements are true or false:

1. It is possible that gene *D* is a structural gene for one of the two enzymes.
2. It is possible that gene *D* produces a repressor.
3. It is possible that gene *D* produces a cytoplasmic product required to induce genes *A* and *B*.
4. It is possible that gene *D* is an operator locus for the *sug* operon.
5. The evidence is also consistent with the possibility that gene *C* could be a gene that produces a cytoplasmic product *required* to induce genes *A* and *B*.
6. The evidence is also consistent with the possibility that gene *C* could be the controlling end of the *sug* operon (end from which mRNA synthesis presumably commences).

16.10 Four different polar mutations, *1*, *2*, *3*, and *4*, in the *lacZ* gene of the lactose operon were isolated following mutagenesis of *E. coli*. Each caused total loss of β-galactosidase activity. Two revertant mutants, due to suppressor mutation in genes unlinked to the *lac* operon, were isolated from each of the four strains: Suppressor mutations of polar mutation *1* or *1A* and *1B*; those of polar mutation *2* or *2A* and *2B*; and so on. Each of the

eight suppressor mutations was then tested, by appropriate crosses, for its ability to suppress each of the four polar mutations; the test involved examining the ability of a strain carrying the polar mutation and the suppressor mutation to grow with lactose as the sole carbon source. The results follow (+ = growth on lactose and − = no growth):

POLAR MUTATION	SUPPRESSOR MUTATION							
	1A	*1B*	*2A*	*2B*	*3A*	*3B*	*4A*	*4B*
1	+	+	+	+	+	+	+	+
2	+	−	+	+	+	+	−	−
3	+	−	+	−	+	+	−	−
4	+	+	+	+	+	+	+	+

A mutation to a UAG codon is called an amber nonsense mutation, and a mutation to a UAA codon is called an ochre nonsense mutation. Suppressor mutations allowing reading of UAG and UAA are called amber and ochre suppressors, respectively.

a. Which of the polar mutations are probably amber? Which are probably ochre?
b. Which of the suppressor mutations are probably amber suppressors? Which are probably ochre suppressors?
c. How would you explain the anomalous failure of suppressor *2B* to permit growth with polar mutation *3*? How could you test your explanation most easily?
d. Explain precisely why ochre suppressors suppress amber mutants but amber suppressors do not suppress ochre mutants.

*16.11 What consequences would a mutation in the catabolite activator protein (CAP) gene of *E. coli* have for the expression of a wild-type *lac* operon?

16.12 The lactose operon is an inducible operon, whereas the tryptophan operon is a repressible operon. Discuss the differences between these two types of operons.

16.13 In the bacterium *Salmonella typhimurium* seven of the genes coding for histidine biosynthetic enzymes are located adjacent to one another in the chromosome. If excess histidine is present in the medium, the synthesis in all seven enzymes is coordinately repressed, whereas in the absence of histidine all seven genes are coordinately expressed. Most mutations in this region of the chromosome result in the loss of activity of only one of the enzymes. However, mutations mapping to one end of the gene cluster result in the loss of all seven enzymes, even though none of the structural genes have been lost. What is the counterpart of these mutations in the *lac* operon system?

*16.14 What is stringent control, and how does this regulatory system work?

16.15 Bacteriophage λ, upon infecting an *E. coli* cell, has a choice between the lytic and lysogenic pathways. Discuss the molecular events that determine which pathway is taken.

16.16 How do the lambda repressor protein and the cro protein regulate their own synthesis?

*****16.17** If a mutation was induced in the *cI* gene of phage lambda such that the resulting *cI* gene product was nonfunctional, what phenotype would you expect the phage to exhibit?

16.18 Bacteriophage λ can form a stable association with the bacterial chromosome because the virus manufactures a repressor. This repressor prevents the virus from replicating its DNA, making lysozyme and all the other tools used to destroy the bacterium. When you induce the virus with UV light, you destroy the repressor, and the virus goes through its normal lytic cycle. This repressor is the product of a gene called the *cI* gene and is a part of the wild-type viral genome. A bacterium that is lysogenic for λ^+ is full of repressor substance, which confirms immunity against any λ virus added to these bacteria. These added viruses can inject their DNA, but the repressor from the resident virus prevents replication, presumably by binding to an operator on the incoming virus. Thus this system has many analogous elements to the lactose operon. We could diagram a virus as shown in the figure. Several mutations of the *cI* gene are known. The c_i mutation results in an inactive repressor.

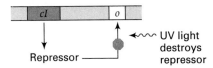

a. If you mix λ containing a c_i mutation, can it lysogenize (form a stable association with the bacterial chromosome)? Why?
b. If you infect a bacterium simultaneously with a wild-type c^+ and a c_i mutant of λ, can you obtain stable lysogeny? Why?
c. Another class of mutants called c^{IN} makes a repressor that is insensitive to UV destruction. Will you be able to induce a bacterium lysogenic for c^{IN} with UV light? Why?

17 Regulation of Gene Expression and Development in Eukaryotes

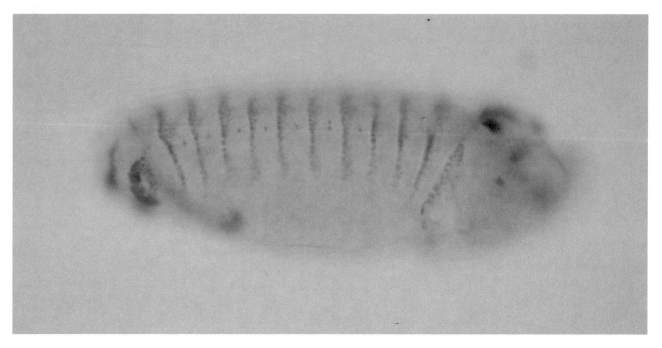

PRINCIPAL POINTS

~ In eukaryotes, gene expression is regulated at a number of distinct levels. That is, regulatory systems appear to exist for the control of transcription, of precursor-RNA processing, transport of the mature RNA out of the nucleus, translation of the mRNAs, degradation of the mature RNAs, and degradation of the protein products.

~ Eukaryotic protein-coding genes contain both promoter elements and enhancer elements. Some promoter elements are required for transcription to begin. Other promoter elements have a regulatory function; these are specialized with respect to the gene they control, binding specific regulatory proteins that modulate expression of the gene. Specific regulatory proteins bind also to the enhancer elements and activate transcription through their interaction with proteins bound to the promoter elements. Enhancer elements and promoter elements appear to bind many of the same proteins. This implies that both types of regulatory elements affect transcription by a similar mechanism, probably involving interactions of the regulatory proteins.

~ Transcriptionally active chromatin is more sensitive to digestion with DNase I than transcriptionally inactive chromatin. This sensitivity is a result of a loosened DNA-protein structure in the areas of transcriptional activity. For active genes, certain sites—called hypersensitive sites—are highly sensitive to DNase I digestion. These hypersensitive sites probably correspond to the binding sites for RNA polymerase and regulatory proteins.

~ The DNA of most eukaryotes has been shown to be methylated at a certain proportion of the bases. Methyl groups are added mostly to cytosines. In those eukaryotes, transcriptionally active genes generally exhibit lower levels of DNA methylation than transcriptionally inactive genes.

~ Despite extensive searching, no operons have been found in eukaryotes. However, genes for related functions in some biochemical pathways are sometimes linked. More often, genes that are coordinately regulated are dispersed throughout the genome.

~ In higher eukaryotes, well-studied systems of short-term gene regulation are those for the control of enzyme synthesis by steroid hormones. Some hormones (the steroid hormones) exert their action by binding directly to the cell's genome in a hormone-receptor complex. Other hormones (polypeptide hormones) act at the cell surface, activating a system to produce cyclic

AMP, a substance that acts as a second messenger to control gene activity. The specificity of hormone action results from the presence of hormone receptors in only certain cell types and by interactions of steroid-receptor complexes with cell-type-specific regulatory proteins.

~ Gene expression in eukaryotes can be regulated at the level of RNA processing. This type of regulation operates to determine the production of mature RNA molecules from precursor-RNA molecules. Two regulatory events that exemplify this level of control are choice of poly(A) site and choice of splice site. In both cases different types of mRNAs are produced, depending upon the choices made. Examples are known of both types of regulation in developmental systems.

~ mRNA transport from the nucleus to the cytoplasm is another important control point. A spliceosome retention model has been proposed for this regulation. In this model, spliceosome assembly on a precursor-mRNA competes with nuclear export of the molecule. Thus, during intron removal, the spliceosome serves to keep the RNA in the nucleus but, when all introns have been removed and the spliceosome has dissociated, the free mRNA can interact with the nuclear pore and exit.

~ Gene expression is regulated also by mRNA translation control and by mRNA degradation control. The latter is believed to be a major control point in the regulation of gene expression. Structural features of individual mRNAs have been shown to be responsible for the range of mRNA degradation rates observed, although the precise roles of cellular factors and enzymes have yet to be determined.

~ Long-term regulation is involved in controlling events that activate and repress genes during development and differentiation. Those two processes result from differential gene activity of a genome that contains a constant amount of DNA, rather than from a programmed loss of genetic information.

~ *Drosophila* has become a relatively tractable model system in which to study the genetic control of development. *Drosophila* body structures result from specific gradients in the egg, and the subsequent determination of embryo segments that directly correspond to adult body segments. Both processes are under genetic control as defined by mutations that disrupt the development events. Studies of the mutants indicate that *Drosophila* development is directed by a temporal regulatory cascade.

*I*n prokaryotes, gene expression is commonly regulated in a unit of protein-coding genes and adjacent controlling sites (promoter and operator) collectively called an operon. The molecular mechanisms in operon function provide a relatively simple means of controlling the coordinate synthesis of proteins with related functions. The discovery of operons in prokaryotes stimulated searches for similar regulatory systems in eukaryotes, but no operons have been found. Therefore, gene regulation in eukaryotes must occur by means other than the use of operons. In this chapter we look at some of the well-studied regulatory systems in eukaryotes.

Eukaryotic gene regulation falls into two categories. Short-term regulation involves regulatory events in which gene sets are quickly turned on or off in response to changes in environmental or physiological conditions in the cell's or organism's environment. Long-term gene regulation involves regulatory events other than those required for rapid adjustment to local environmental or physiological changes; that is, those events that are required for an organism to develop and differentiate. Both of these categories of gene regulation are addressed in this chapter.

LEVELS OF CONTROL OF GENE EXPRESSION IN EUKARYOTES

Prokaryotic organisms are unicellular and free-living. As we discussed in Chapter 16, the control of gene expression in prokaryotes occurs primarily at the transcriptional level, with some control at the translational level. Transcriptional control is accomplished through the interaction of regulatory proteins with upstream regulatory DNA sequences. A rapid transition between protein synthesis and no protein synthesis is achieved by both switching off gene transcription and rapid degradation of the mRNA molecules involved.

Eukaryotes fall into two major groups: unicellular organisms and multicellular organisms. In both eukaryote groups the control of gene expression is more complicated than is the case with prokaryotes. Figure 17.1 diagrams a number of levels at which gene expression can be regulated in eukaryotes. These levels of control are *transcriptional control, RNA processing control, transport control, mRNA translation control, RNA degradation control,* and *protein degradation control.* We will consider each of these in turn with the exception of the last, since that is less relevant to the study of gene expression than the others. Keep in mind that these control processes help coordinate the generation of new proteins in different cells at different times.

~ FIGURE 17.1

Levels at which gene expression can be controlled in eukaryotes.

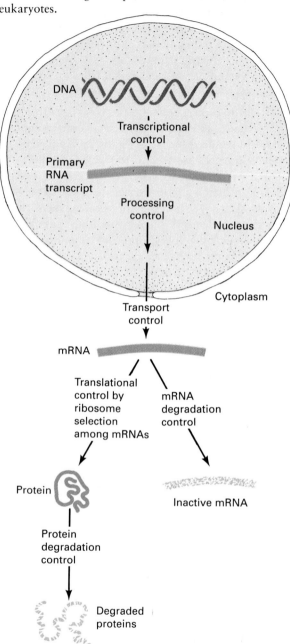

Transcriptional Control

GENERAL ASPECTS OF TRANSCRIPTIONAL CONTROL. **Transcriptional control** regulates whether or not a gene is to be transcribed and the rate at which transcripts are produced.

As we discussed in Chapter 12, eukaryotic genes that are protein-coding contain both promoter elements and enhancers. The promoter elements are found just upstream of the transcription initiation site, while the enhancers are usually some distance away, usually much farther upstream. We can think of the different promoter elements as modules that function in the regulation of expression of the gene with which they are associated. Certain promoter elements, such as the TATA element, are required for transcription to begin. Other promoter elements function to determine whether or not transcription of the gene occurs. These latter promoter elements are DNA sequences to which specific regulatory proteins bind (Figure 17.2). If transcription is activated, the promoter element is a *positive regulatory element* and the protein is a *positive regulatory protein*. If transcription is turned off, the promoter element is a *negative regulatory element* and the protein is a *negative regulatory protein*.

A regulatory promoter element is specialized with respect to the gene (or genes) it controls because it binds a signalling molecule that is involved in the regulation of that gene's expression. Depending on the particular gene, there can be one, a few, or many regulatory promoter elements, since under various conditions there may be one, a few, or many regulatory proteins that control the gene's expression. The remarkable specificity of regulatory proteins in binding to their specific regulatory element in the DNA and to no others ensures careful control of which genes are activated and which are repressed.

While promoter elements are crucial for determining whether transcription can occur, enhancer elements determine that maximal transcription of the gene occurs. Recall that an enhancer element, by definition, activates transcription from the promoter linked to it, even though the enhancer may be some distance away (see Chapter 12). Regulatory proteins bind to the enhancer elements, and which regulatory proteins bind depends on the DNA sequence of the enhancer element. A model for how the enhancers affect transcription from a distance sets out this pattern. Specific regulatory proteins bind to the enhancer element; the DNA then forms a loop so that the enhancer-bound regulatory proteins interact with the regulatory proteins and transcription factors bound to the promoter elements. Through the interactions of the proteins, transcription is either activated or repressed.

Both promoters and enhancers are important in regulating transcription of a gene. Each regulatory promoter and enhancer element binds a corresponding regulatory protein. Some regulatory proteins are found in most or all cell types, while others are found in only a limited number of cell types. Some of the gene regulatory proteins activate transcription when they bind to the enhancer or promoter element, while others repress transcription. Therefore, the net effect of a regulatory element on transcription depends on the combination of different proteins bound. If positive regulatory proteins are bound at both the enhancer and promoter elements, the result is activation of transcription. However, if a negative regulatory protein binds to the enhancer and a positive regulatory protein binds to the promoter element, the result will depend on the interaction between the two regulatory proteins. If the negative regulatory protein has a strong effect, the gene will be repressed. In this case, the enhancer is acting as a silencer element.

Enhancer elements and promoter elements appear to bind many of the same proteins. This implies that both types of regulatory elements affect transcription

---

~ FIGURE 17.2

Schematics of (a) positive regulation of gene transcription and (b) negative regulation of gene transcription.

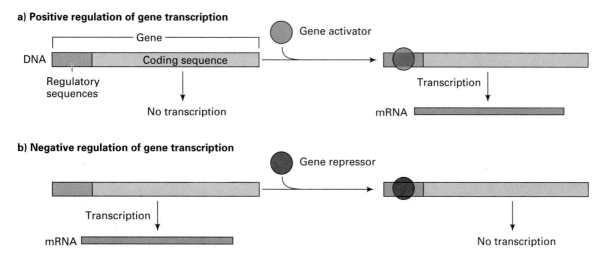

a) Positive regulation of gene transcription

by a similar mechanism, probably involving interactions of the regulatory proteins as described before. Interestingly, there appears to be a relatively small number of regulatory proteins that control transcription. Therefore, by combining relatively few regulatory proteins in particular ways, the transcription of different arrays of genes is regulated, and a large number of cell types is specified. This is called **combinatorial gene regulation.**

KEYNOTE

Eukaryotic protein-coding genes contain both promoter elements and enhancer elements. Some promoter elements are required for transcription to begin. Other promoter elements have a regulatory function; these are specialized with respect to the gene they control, binding specific regulatory proteins that modulate expression of the gene. Specific regulatory proteins bind also to the enhancer elements and activate transcription through their interaction with proteins bound to the promoter elements. Enhancer elements and promoter elements appear to bind many of the same proteins. This implies that both types of regulatory elements affect transcription by a similar mechanism, probably involving interactions of the regulatory proteins.

CHROMOSOME CHANGES AND TRANSCRIPTIONAL CONTROL. As we learned in Chapter 10, the eukaryotic chromosome consists of DNA complexed with histones and nonhistone chromosomal proteins. Ultimately, the regulation of gene transcription must involve particular DNA nucleotide sequences that are recognized by the appropriate regulatory molecules.

In the evolutionary sense, the histones are extremely conserved; their amino acid sequence, structure, and function are all very similar among all eukaryotic organisms. Furthermore, from cell to cell in an organism, the five different types of histones are arranged in an essentially constant way along the DNA. Clearly, this uniform distribution of histones along the DNA gives a system nowhere near the specificity needed for controlling the expression of thousands of genes. Nonetheless, histones do become modified, primarily through acetylation and phosphorylation, thereby altering their ability to bind to the DNA. If the gene that they cover is to be expressed, the histones must be modified to loosen their grip on the DNA so that the DNA strands can interact with RNA polymer-

ase or regulatory proteins. In essence, then, histones act primarily as negative controlling elements in eukaryotic gene regulation—that is, they are gene repressors.

If histones are not the molecules responsible for turning specific genes on, then this process must be controlled by the nonhistones, which not only are more numerous (there are hundreds of different types) but also vary significantly in number and distribution from one cell to another. Since it is to be expected that cells with different functions will have different arrays of genes being expressed, the variations in nonhistones from cell to cell strongly suggest that nonhistones play an important role in controlling which genes are expressed and therefore in controlling cell differentiation and organism development.

Evidence for a specific regulatory role of nonhistones has come from chromatin reconstitution experiments. Chromatin from a cell may be dissociated into its component DNA, histones, and nonhistones, and it may be reconstituted, a process that makes it possible to "mix and match" chromatin components from cells of different types. If the DNA and histones of transcriptionally inactive cells are mixed with nonhistones of transcriptionally active cells, transcriptional activity will occur.

Evidence indicates that transcriptionally active chromatin is in a much looser configuration than transcriptionally inactive chromatin. Thus the DNA's highly compact secondary and tertiary structures maintained in the nucleosomes unwind to allow the DNA to be accessible to regulatory proteins and to RNA polymerase. Over the past few years a number of studies have revealed significant physical differences between active and inactive genes, involving either the packing of protein along the DNA or the folding and supercoiling of the DNA-protein structure.

DNase sensitivity and gene expression. When a gene becomes transcriptionally active, the region of the chromosome containing the gene becomes much more sensitive to digestion by DNase I (an endonuclease) than transcriptionally inactive regions. The interpretation for this general result is that the DNA-protein structure of the chromosome becomes loosened in regions where genes are active so that DNase I can access the DNA and degrade it.

One well-studied system involves globin and ovalbumin synthesis in chickens. In the nuclei of **erythroblasts** (red blood cell precursors) from chick embryos, the genes for globin are susceptible to digestion by DNase I, whereas the sequences encoding ovalbumin, which is not produced in erythroblasts, are resistant to DNase digestion. Conversely, in hen oviduct nuclei, in which ovalbumin mRNA is actively transcribed but globin mRNA is not produced, the ovalbumin gene sequences are much more sensitive to DNase digestion than are the globin gene sequences.

The experimental approach used to demonstrate changes in DNase I sensitivity of DNA sequences involves the use of specific cloned genes as probes and the Southern blotting technique (see Chapter 15). Figure 17.3 shows how this approach was used for the work with the chicken globin gene and ovalbumin gene just described. Chromatin was extracted from erythroblast nuclei and, as a control, from cells growing in culture that were not making globin. Samples were treated with different concentrations of DNase I; DNA was then extracted from each chromatin sample and digested with the restriction enzyme *Bam*HI (see Chapter 15), which cuts the DNA at sites on either side of the globin gene sequence. The two *Bam*HI sites are 4.5 kb apart on the chromosomal DNA. The resulting DNA fragments were separated by agarose gel electrophoresis and transferred to nitrocellulose filters using the Southern blotting technique. The filters were hybridized with a radioactive globin DNA probe to determine the sizes of the globin gene DNA fragments that were present following DNase digestion.

The results were as follows: For chromatin extracted from erythroblasts, a gradual disappearance of the 4.5-kb *Bam*HI-*Bam*HI, globin gene-containing DNA fragments was seen with increasing DNase concentration (Figure 17.3a), indicating that the chromatin segment containing the globin gene was sensitive to DNase digestion. That is, the chromatin organization was loosened sufficiently in the globin gene region such that the globin DNA sequence was accessible to DNase I. By contrast, in chromatin isolated from control cells not making globin, the 4.5-kb *Bam*HI DNA fragment was not digested at any DNase concentration. This result indicated that, in cells not making globin, the chromatin was still highly condensed; as a result, the globin gene sequence was inaccessible to DNase so no degradation occurred. Moreover, analysis of the ovalbumin gene (which is also flanked on either side by *Bam*HI sites so that cutting with *Bam*HI produces a 20-kb DNA fragment) in chromatin from erythroblasts using a cloned ovalbumin gene probe showed that the ovalbumin gene was not sensitive to digestion by DNase I, as evidenced by the presence of the 20-kb *Bam*HI-*Bam*HI DNA fragment at all DNase I concentrations (Figure 17.3b). The latter result is explained by the fact that ovalbumin is not made in erythroblasts, so that region of the genome has not become loosened.

Many experiments of this sort lead to the conclusion that almost all transcriptionally active genes have an increased DNase I sensitivity. The extent of the

~ **FIGURE 17.3**

Relationship of gene expression and sensitivity to DNase. (a) In erythroblasts, (cells making globin) globin DNA is sensitive to DNase I. (b) In erythroblasts, which do not make ovalbumin, ovalbumin DNA is resistant to DNase.

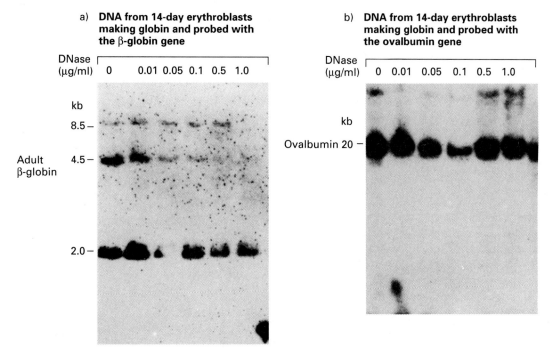

DNase sensitive region varies with the gene in question, ranging from a few kilobase pairs around the gene to as much as 20 kb of flanking DNA.

DNase hypersensitive sites. Careful analysis of the regions of DNA around transcriptionally active genes has shown that certain sites, called **hypersensitive sites** or **hypersensitive regions** are even more highly sensitive to digestion by DNase I. These sites or regions are typically the first to be cut with DNase I. Most, but not all, DNase-hypersensitive sites are in the regions upstream from the start of transcription, probably corresponding to the DNA sequences where RNA polymerase and other gene regulatory proteins bind.

𝒦EYNOTE

Transcriptionally active chromatin is characterized by a sensitivity to digestion with DNase I as a result of the loosened DNA-protein structure in those areas. By contrast, transcriptionally inactive chromatin is resistant to DNase I digestion. For transcriptionally active genes, certain sites, probably corresponding to the binding sites for RNA polymerase and regulatory proteins, are especially sensitive to DNase I digestion. These sites are called hypersensitive sites.

DNA METHYLATION AND TRANSCRIPTIONAL CONTROL. Once DNA has been synthesized within the eukaryotic cell, a small proportion of bases become chemically modified through the action of enzymes. Potentially, modified DNA bases could act as signals in the processes in which DNA is involved, such as DNA replication, DNA repair, and DNA expression, although in most cases we have no information about this activity. A particular modified base has received significant attention since it has been correlated with gene activity in a number of cases. This base, *5-methylcytosine* (5mC), is generated by the modification of DNA shortly after replication by the action of the enzyme DNA methylase (Figure 17.4).

The percentage of 5mC in eukaryotic DNA varies over a wide range. In mammalian DNA, for example, about 3 percent of cytosines are present as 5mC. The DNAs of less complex eukaryotes, such as *Drosophila, Tetrahymena,* and yeast, appear to contain very little, if any, 5mC. There is no clear relationship, however, between the amount of 5mC and the complexity of an organism, evolutionarily speaking.

~ FIGURE 17.4

Production of 5-methylcytosine from cytosine in DNA by the action of the enzyme DNA methylase.

Cytosine (in DNA)   **5-Methylcytosine (5mC)**

The distribution of 5mC is nonrandom, with most of the 5mC (greater than 90 percent in mammalian DNA) being found in the sequence CG. Since this sequence has twofold rotational symmetry in DNA, the cytosines are symmetrically located on opposite strands of the DNA molecule. These CG sequences also form part of restriction enzyme recognition sites, which allow the use of restriction enzymes to investigate the extent of methylation of a segment of DNA.

Certain restriction enzymes are used as tools for studies of 5mC methylation, since many enzymes with cytosine in their recognition sequence fail to cleave the DNA when this base is methylated. The enzyme *Hpa*II ("hepa-two"), for example, will cleave DNA at the sequence 5'-CCGG-3' (the opposite strand is complementary to this sequence, of course) but not if the internal cytosine is methylated, that is, if it is 5'-C^{m}CGG-3'. The enzyme *Msp*I ("M-S-P-one") also will cleave the CCGG sequence recognized and cleaved by *Hpa*II, but unlike *Hpa*II, it will cleave the methylated sequence C^{m}CGG.

The use of *Hpa*II and *Msp*I in analyzing the extent of methylation of a segment of DNA is illustrated by the following theoretical example. Consider a fragment of genomic DNA that contains three CCGG sequences (Figure 17.5). If the sequences are not methylated (Figure 17.5a), then both *Hpa*II and *Msp*I will cleave the DNA at these sequences to produce four DNA fragments of discrete sizes. These DNA fragments can be identified following (1) electrophoresis of the digested DNA on agarose gels, (2) Southern blotting to nitrocellulose, and (3) probing the fragments with a radioactively labeled DNA probe that will hybridize with any and all of the fragments. In this case DNA fragments of identical sizes will be produced from both digests. If one of the CCGG sequences is methylated—for example, the first one from the left (Figure 17.5b)—then all three

~ FIGURE 17.5

Effect of 5-methylcytosine on cleavage of DNA with *Hpa*II and *Msp*I: (a) unmethylated (CCGG) sequence; (b) methylated (C^mCGG) sequence.

a) **Unmethylated**

DNA fragment, no methylated CCGG sequences; all sites can be cleaved with *Hpa*II and *Msp*I

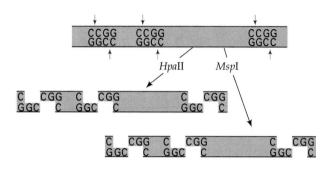

b) **Methylated**

DNA fragment with one methylated C^mCGG site that can be cleaved by *Msp*I but not by *Hpa*II

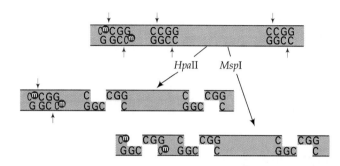

Agarose gel electrophoresis, Southern blotting to nitrocellulose, probing with labeled DNA, autoradiographing

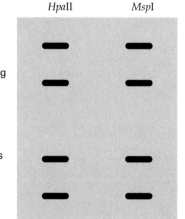

*Hpa*II and *Msp*I results identical

Patterns different for the two enzymes since *Hpa*II was unable to cut the C^mCGG sequence

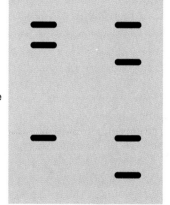

sequences will be cleaved by *Msp*I, whereas only the two unmethylated sequences will be cleaved by *Hpa*II. As a result, the array of DNA fragments produced by digestion of the same DNA piece with the two enzymes will be different, as indicated in the stylized blot result. In actual practice, a piece of DNA is digested separately by the two enzymes *Hpa*II and *Msp*I, and if the results show that the *Msp*I has digested the DNA into more pieces than *Hpa*II, the interpretation is that one of the CCGG sequences is methylated; that is, C^mCGG.

With the *Hpa*II/*Msp*I approach a large number of systems have been examined in an attempt to determine whether there is a relationship between gene methylation and transcriptional activity of the gene. For at least thirty genes examined a negative correlation exists between DNA methylation and transcription; there is a low level of methylated DNA (hypomethylated DNA) in transcriptionally active genes in comparison with transcriptionally inactive genes. We must exercise cau-

tion in taking this broad generalization too far, because not all methylated CCGG sequences in a gene region become demethylated when the gene is expressed and because, as was stated earlier, not all organisms have been shown to have significant amounts of methylated C in their DNA. Moreover, the fact that a correlation exists between the level of DNA methylation and the transcriptional state of a gene provides no information about cause and effect. In other words, at this time we do not know whether methylation changes are prerequisites for the onset of transcriptional activity (in those organisms in which there is significant DNA methylation) or whether the changes are secondary consequences of the transcriptional activity initiated by other means. Some experiments, for example, suggest that the level of methylation of a gene does not by itself determine whether the gene will be transcribed.

In summary, in at least some systems (notably mammals and many other animals) 5-methylcytosine appears to function as part of a gene-silencing system. In a number of cases removal of methyl groups is correlated with increases of transcriptional activity. The 5-methylcytosine changes cannot explain all changes in transcriptional activity, however, since not all eukaryotic organisms have significant amounts of 5mC in their DNA.

𝒦EYNOTE

The DNA of most eukaryotes has been shown to be methylated, with most methylations occurring on cytosine residues. Transcriptionally active genes exhibit lower levels of DNA methylation than transcriptionally inactive genes. ────────────

EXAMPLES OF TRANSCRIPTIONAL CONTROL OF GENE EXPRESSION IN EUKARYOTES. In this section we shall look at some examples of the short-term regulation of gene expression in eukaryotes. This regulation involves rapid responses to changes in environmental or physiological conditions. The responses occur primarily at the transcriptional level. As we examine these systems, keep in mind how short-term regulation of gene expression can occur in bacteria through regulatory units called operons so that you can compare the molecular events involved.

Galactose-utilizing genes of yeast. A well-studied regulatory system is the galactose-utilizing system of yeast. Three genes, GAL1, GAL7, and GAL10, encode enzymes that function in a pathway to metabolize the sugar galactose (Figure 17.6a). The enzymes are galactokinase (GAL1 gene), galactose transferase (GAL7 gene), and galactose epimerase (GAL10 gene). Once D-glucose 6-phosphate is made, it enters the glycolytic pathway, where it is acted upon by enzymes that are always present in the cell as a result of constitutive gene expression. In the absence of galactose, the GAL genes are not transcribed. When galactose is added, there is a rapid, coordinate induction of transcription of the GAL genes, and therefore a rapid production of the three galactose-utilizing enzymes.

Genetic studies have shown that the GAL1, GAL7, and GAL10 genes are located near each other, but do not constitute an operon (Figure 17.6b). Genetic studies have also identified a regulatory gene, GAL4, which encodes the GAL4 protein. GAL4 binds to a specific sequence in the DNA called upstream activator sequence–galactose (UAS_G). A UAS_G is located between the divergently-transcribed GAL1 and GAL10 genes. The UAS_G consists of four similar, 17 bp sequences, each of which shows two-fold symmetry,

~ FIGURE 17.6

The galactose metabolizing pathway of yeast. (a) The steps catalyzed by the enzymes encoded by the GAL genes; (b) The organization of the GAL protein-coding genes. The arrows indicate the direction of transcription. (UDP = uridine diphosphate)

a) Pathway

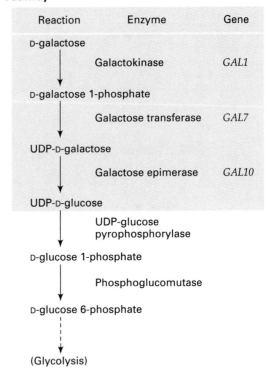

b) Gene organization

and each of which binds GAL4. GAL4 binds to each 17-bp sequence as a dimer (see Chapter 16, on lambda repressor and its target sequences). If only one GAL4 dimer is bound, transcription is activated at a low level; maximal expression requires all sites to be occupied by GAL4 dimers.

GAL4 is similar in functional organization to lambda repressor; that is, there is a DNA-binding domain and a domain required for activating transcription. This was shown in a "domain-swapping" experiment. Using recombinant DNA techniques, different DNA fragments from the GAL4 genes were fused to DNA fragments from bacterial repressors, enabling protein hybrids to be made. One protein hybrid had the DNA binding domain of a bacterial repressor and the GAL4 activator domain. This hybrid could not function in yeast because there are no sequences recognized by the bacterial repressor DNA binding domain. However, if a DNA sequence to which the bacterial

~ FIGURE 17.7

Model for the activation of the *GAL* genes of yeast.

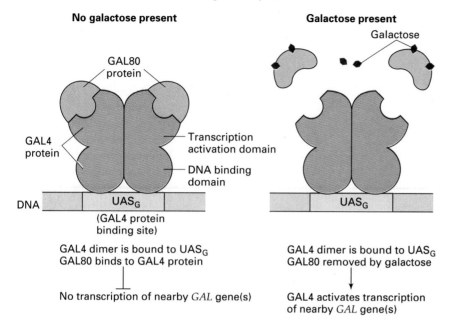

No galactose present

GAL80 protein

GAL4 protein

Transcription activation domain

DNA binding domain

DNA   UAS$_G$
(GAL4 protein binding site)

GAL4 dimer is bound to UAS$_G$
GAL80 binds to GAL4 protein

No transcription of nearby *GAL* gene(s)

Galactose present

Galactose

UAS$_G$

GAL4 dimer is bound to UAS$_G$
GAL80 removed by galactose

GAL4 activates transcription of nearby *GAL* gene(s)

repressor DNA binding domain can bind is placed next to a gene and introduced into yeast, the hybrid protein can bind to it, and if galactose is added, transcription of the gene adjacent to the bacterial repressor sequence is induced.

Another protein, GAL80, is involved in *GAL* gene regulation. Encoded by the *GAL80* gene, the GAL80 protein responds to the presence of galactose and interacts with the GAL4 protein. Figure 17.7 presents a model for the regulation of transcription of the *GAL* protein-coding genes. In the absence of galactose, a GAL4 protein dimer is bound to a UAS$_G$. GAL80 protein is bound to the transcription activation domain of GAL4, preventing transcription. In the presence of galactose, interaction with galactose prevents GAL80 from binding to GAL4 protein's transcription activation domain. As a result, transcription of the associated *GAL* genes is activated by an as yet unknown mechanism. Thus, in this system, GAL4 protein acts as a positive regulator (activator), while GAL80 protein acts as a negative regulator.

KEYNOTE

No operons have been found in eukaryotes, although the genes for related functions in some biochemical pathways may be linked. More often, genes that are coordinately regulated are dispersed throughout the eukaryotic genome. ———————————————

Steroid hormone regulation of gene expression in animals. Short-term regulation enables an organism or cell to adapt rapidly to changes in its physiological environment so that it can function as optimally as possible. While lower eukaryotes exhibit some cell specialization, higher eukaryotes are differentiated into a number of cell types, each of which carries out a specialized function or functions. The cells of higher eukaryotes are not exposed to rapid changes in environment as are cells of bacteria and of microbial eukaryotes. The environment to which most cells of higher eukaryotes are exposed, the intercellular fluid, is relatively constant in the nutrients, ions, and other important molecules it supplies. The intercellular fluid, in a sense, is analogous to a bacterium's growth medium. The constancy of the cell's environment is, in turn, maintained through the action of chemicals called *hormones,* which are secreted by various cells in response to signals and which circulate in the blood until they stimulate their target cells. Elaborate feedback loops control the amount of hormone secreted, modulating the response so that appropriate levels of chemicals in the blood and tissues are maintained.

A hormone, then, is an effector molecule that is produced in low concentrations by one cell and that causes a physiological response in another cell. As shown in Figure 17.8 (p. 524), some hormones—e.g., steroid hormones—exert their action by binding directly to the cell's genome and causing changes in gene expression. Other hormones—e.g., polypeptide

~ FIGURE 17.8

Mechanisms of action of polypeptide hormones and steroid hormones.

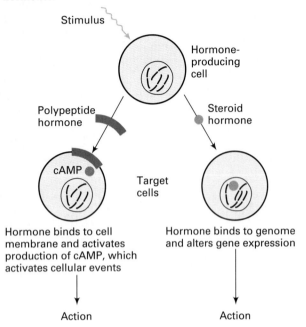

hormones—act at the cell surface to activate a membrane-bound enzyme, adenyl cyclase, which produces cyclic AMP (cAMP) from ATP (see Chapter 16). The cAMP acts as a second messenger molecule to activate the cellular events normally associated with the release of the hormone involved. The exact molecular mechanisms of these reactions are not known.

A variety of types of molecules can act as hormones, such as polypeptides, fatty acid derivatives, and steroids. Since the action of steroid hormones on gene expression has been well studied, we shall concentrate on steroid hormones here. Steroid hormones have been shown to be important to the development and physiological regulation of organisms ranging from fungi to humans. Figure 17.9 gives the structures of four of the most common mammalian steroid hormones. All have a common four-ring structure; the differences in the side groups are responsible for their different physiological effects.

A key to hormone action is that hormones act on specific target cells that have receptors capable of recognizing and binding that particular hormone. For most of the polypeptide hormones (e.g., insulin, ACTH, vasopressin), the receptors are on the cell sur-

~ FIGURE 17.9

Structures of some mammalian steroid hormones. (a) Hydrocortisone, which helps regulate carbohydrate and protein metabolism; (b) Aldosterone, which regulates salt and water balance; (c) Testosterone, which is used for the production and maintenance of male sexual characteristics; (d) Progesterone, which, with estrogen, prepares and maintains the uterine lining for embryo implantation.

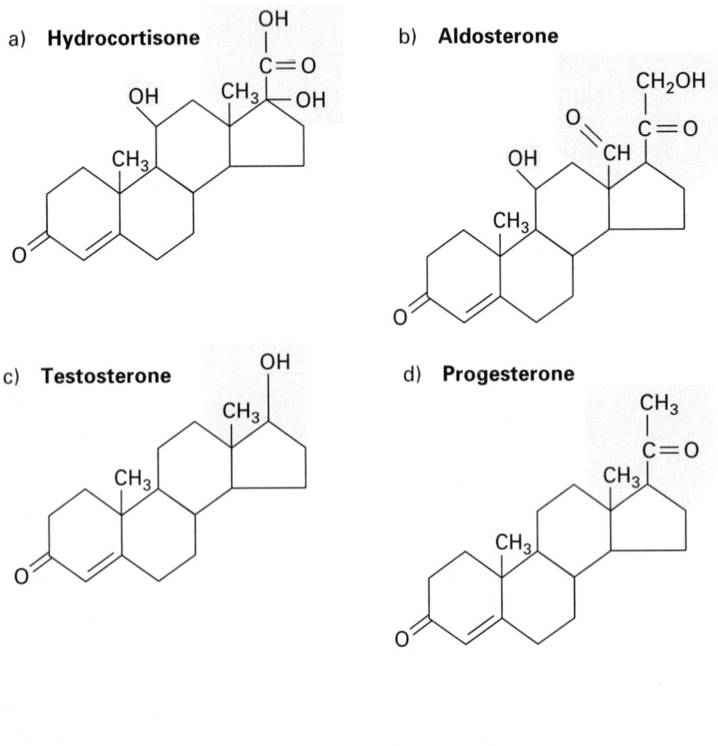

a) **Hydrocortisone**

b) **Aldosterone**

c) **Testosterone**

d) **Progesterone**

face, whereas the receptors for steroid hormones are inside the cell.

Table 17.1 presents examples of the induction of specific proteins by selected steroid hormones. As can be seen, there are tissue-specific effects. The specificity of the response to steroid hormones is controlled by the hormone receptors. Thus, only target tissues that contain receptors for a particular steroid hormone can respond to that hormone, and this directly controls the sites of the hormone's action. With the exception of receptors for the steroid hormone glucocorticoid, which are widely distributed among tissue types, steroid receptors are found in a limited number of target tissues. While steroid hormones have well-substantiated effects on transcription, steroids also can affect the stability of mRNAs, and possibly the processing of mRNA precursors.

Mammalian cells contain between 10,000 and 100,000 steroid receptor molecules, which are proteins with a very high affinity for their respective steroid hormones. The receptor molecules are located in the cytoplasm. For example, in the absence of glucocorticoid hormone, the glucocorticoid hormone receptor is found in the cytoplasm in a complex with a protein called Hsp90. When a steroid hormone enters a cell it binds to its specific receptor molecule to form a steroid-receptor complex (Figure 17.10; p. 526). In the case of the glucocorticoid hormone, the hormone displaces Hsp90 from the receptor-Hsp90 complex, forming a glucocorticoid hormone-receptor complex. The steroid-receptor complex diffuses to the nucleus where it binds to specific DNA sequences, activating the transcription of specific mRNA molecules. The new mRNAs can appear within minutes after a steroid hormone encounters its target cell, enabling new proteins to be produced rapidly.

All the genes activated by a specific steroid hormone have in common a DNA sequence to which the steroid-receptor complex binds. The binding regions are called **steroid response elements** or **REs.** An additional letter is added to the acronym to indicate the specific steroid involved. Thus, GRE is the glucocorti-

~ TABLE 17.1

Examples of Proteins Induced by Steroid Hormones

HORMONE	TISSUE	INDUCED PROTEIN
Estrogen	Oviduct (hen)	Conalbumin[a] Lysozyme[a] Ovalbumin[a] Ovomucoid[a]
	Liver (rooster)	Vitellogenin Apo-very low density lipoprotein (apoVLDL)
	Liver (frog)	Vitellogenin
	Pituitary gland (rat)	Prolactin
Glucocorticoids	Liver (rat)	Tyrosine aminotransferase Tryptophan oxygenase
	Kidney (rat)	Phosphoenolypyruvate carboxykinase
	Pituitary gland (rat)	Growth hormone
Progesterone	Oviduct (hen)	Avidin[a] Conalbumin[a] Lysozyme[a] Ovomucoid[a]
	Uterus (rat)	Uteroglobin
Testosterone	Liver (rat)	α-2u Globulin
	Prostate gland (rat)	Aldolase

[a]Egg white proteins.

~ FIGURE 17.10

Model for the action of steroid hormones in vertebrate cells.

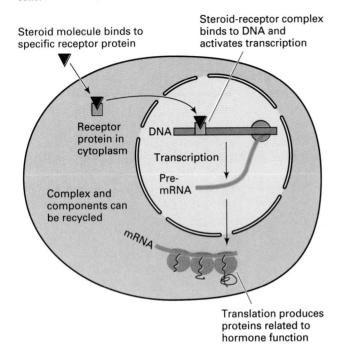

Steroid molecule binds to specific receptor protein

Steroid-receptor complex binds to DNA and activates transcription

Receptor protein in cytoplasm

Complex and components can be recycled

DNA

Transcription

Pre-mRNA

mRNA

Translation produces proteins related to hormone function

coid response element and ERE is the estrogen response element. The REs are located, often in multiple copies, in the enhancer region of genes.

How the hormone-receptor complexes, once bound to the correct REs, regulate transcriptional levels is not completely known. Potentially, functional interactions arise among the hormone-receptor complexes (recall that multiple REs are present for many genes and therefore multiple hormone-receptor complexes can bind to each gene) and with transcription factors. These interactions may facilitate the initiation of transcription by RNA polymerase II. It is presumed that each steroid hormone regulates its specific transcriptional activation by the same general mechanism. The unique action of each type of steroid results from the different receptor proteins and REs involved.

Lastly, it is of particular interest that in different types of cells the same steroid hormone may activate different sets of genes, even though the various cells have the same steroid hormone receptor. This is explained by the fact that many regulatory proteins bind to both promoter elements and enhancers to facilitate gene activation (see the discussions earlier in this chapter and Chapter 12). Thus a steroid-receptor complex can activate a gene only if the correct array of other regulatory proteins is present. Since these latter regulatory proteins are cell-type specific, different gene activation patterns can result.

In sum, steroid hormones act as positive effector molecules, and receptors act as regulatory molecules. When the two combine, the resulting complex activates gene transcription, resulting in a large (and specific) increase in cellular mRNA levels. The specific responses characteristic of each steroid hormone result from the fact that receptors are found in only certain cell types, and each of those cell types contains different arrays of other cell-type-specific regulatory proteins that interact with the steroid-receptor complex to activate specific sets of genes.

*K*EYNOTE

In higher eukaryotes one of the well-studied systems of short-term gene regulation is the control of enzyme synthesis by hormones. Some hormones (the steroid hormones) exert their action by binding directly to the cell's genome in a hormone-receptor complex, while others act at the cell surface, activating a system to produce cAMP, which acts as a second messenger to control gene activity. The specificity of hormone action is caused by the presence of hormone receptors in only certain cell types and by interactions of steroid-receptor complexes with cell-type-specific regulatory proteins. ———

Hormone control of gene expression in plants. Many chemicals have been identified that play important roles in controlling growth and development in plants; these chemicals are called *plant hormones*. Plant hormones fall into five main classes, namely ethylene, abscisic acid, auxins, cytokinins, and gibberellins (Figure 17.11). These extracellular hormones are responsible for many activities. Ethylene, for example, induces fruit ripening, flower maturation, and plant senescence, among other things. The molecular actions of plant hormones are generally ill-defined.

Among the best-characterized plant hormones are *gibberellins*. These hormones have actions similar to those of steroid hormones in mammals. That is, gibberellins stimulate transcription and thereby result in an increase in the production of specific proteins. Those proteins are responsible for profound changes in plant cell form and cell differentiation.

Gibberellins have been shown to have many effects when applied to plants, such as making certain mutant dwarf plants grow tall or making normal plants grow taller. Gibberellins are also important because they stimulate germination of some seeds; for example, those of barley (Figure 17.12). In this process the

~ FIGURE 17.11

Chemical structures of five plant hormones.

Ethylene

Abscisic acid

Indoleacetic acid (an auxin)

Zeatin (a cytokinin)

Gibberellic acid (a gibberellin)

~ FIGURE 17.12

Effect of gibberellins on the germination of barley seeds.

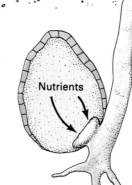

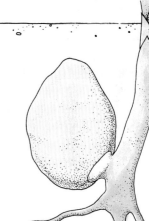

Sunlight

Approximate soil level

Aleurone layer

Endosperm

Gibberellins

Embryo

Seed coat

α-amylase

Nutrients

Day 1:
Gibberellins are produced by the embryo and diffuse to the aleurone layer, the outermost layer of the endosperm of barley.

Day 2:
The aleurone layer now secretes the enzyme α-amylase, which breaks down starch in the endosperm, releasing nutrients.

Day 3:
Using the nutrients released by the breakdown of the endosperm, the embryo has grown considerably.

Day 6:
The endosperm has been exhausted, but the plant can now rely on photosynthesis to obtain nutrients.

embryo of the seed produces gibberellins that diffuse to the aleurone layer, the outermost layer of the endosperm. As a result of the effect of the gibberellins on gene expression, particularly on the gene for the enzyme α-amylase, the aleurone layer synthesizes and secretes α-amylase, which breaks down the endosperm, releasing nutrients that enable the embryo to grow. By the time the endosperm has been exhausted, the young, developing plant can rely on photosynthesis to obtain nutrients for continued growth.

Even though the action of some gibberellins has been well described, no gibberellin receptor has been detected to date, and the molecular aspects of their action are poorly understood.

RNA Processing Control

A second level of control is **RNA processing control,** which regulates the production of mature RNA mole-

cules from precursor-RNA molecules. As we discussed in Chapter 13, all three major classes of RNA molecules, which include mRNA, tRNA, and rRNA, can be synthesized as precursor molecules. Perhaps the most significant range of control is exhibited by the mRNA class and involves control of processing of mRNA precursor molecules. We will discuss two types of RNA processing control: choice of alternative poly(A) sites (Figure 17.13a) and choice of alternative splice sites (Figure 17.13b). Small nuclear ribonucleoprotein particles (snRNAs) may well play a role in the differential processing process.

CHOICE OF A POLY(A) SITE. There are a number of examples of regulation by choice of poly(A) site. One example concerns the production of immunoglobulin M (IgM), one of the immunoglobulin molecules involved in the immune response system. (The functioning of the immune system is discussed later in this

~ **FIGURE 17.13**

Models for control of the processing of mRNA precursors in eukaryotes by choice of (a) poly(A) sites or (b) splice sites. (p = transcription initiation site [promoter]; t = transcription termination site.)

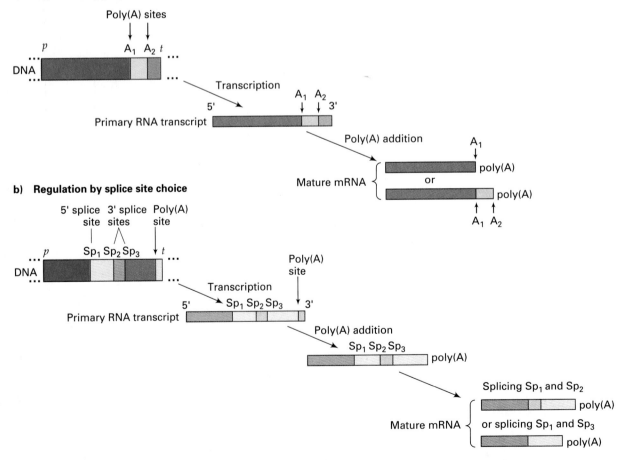

chapter.) Five copies of the monomer aggregate to form the functional pentameric IgM molecule. These molecules, each of which consists of two copies of a small protein (the light chain) and two copies of a large protein (the heavy chain), are manufactured throughout the life of a cell (Figure 17.14). IgM's configuration and its properties vary according to the cell's stage of development. If the cell is in an early developmental stage, the IgM produced associates with the cell membrane. If the cell is in a later developmental stage, the IgM is secreted from the cell. These variations are the result of variant forms of the heavy chain. Encoded by the μ gene, the transcribed heavy chain produces a pre-mRNA that can be processed at different sites. The site that is utilized reflects the cell's developmental stage. As a result, the membrane-bound IgM form has a longer heavy chain than the secreted form.

CHOICE OF SPLICE SITE. A well-studied example of gene regulation by differential splicing of pre-mRNA involves sex determination in *Drosophila* (see Chapter 3). Recall that sex is determined by the X-chromosome:

autosome (X:A) ratio. A number of mutations disrupt normal sex determination. Their study has led to a *regulation cascade model* for sex determination in *Drosophila* (Figure 17.15; p. 530). First, the X:A ratio is read during development, in an as-yet-unknown way. For wild-type *Drosophila*, the ratio for females (XX) is 2X:2 autosome sets, and the ratio for males (XY) is 1:2. This information is transmitted to the sex determination genes, which make the choice between the alternative female and male developmental pathways, starting with the master regulatory gene *Sxl* (*sex-lethal*).

Sxl responds to the dosage of *sis* elements, which act early in embryogenesis to make either active (in females) or no (in males) Sxl proteins. The next gene in the regulatory cascade is *tra* (*transformer*). In the presence of active Sxl protein (in females), the *tra* gene is spliced in a productive way, resulting in an active feminizing product. In males, where Sxl protein is absent, *alternative splicing* occurs (i.e., the default splicing takes place), producing mRNAs with no open reading frames.

~ **FIGURE 17.14**

Structure of an IgM monomer. Two light (L) chains and two heavy (H) chains are linked by disulfide bridges. Each type of chain has a variable domain (V_L and V_H) characterized by variations in amino acid sequences among IgM molecules. The light chains have one constant domain (C_L), while the heavy chains have four constant domains (C_H). In all IgM molecules, constant domains have the same amino acid sequence. The variable domains collectively constitute the antigen-binding sites. Functional IgM is a pentamer of the structure shown here.

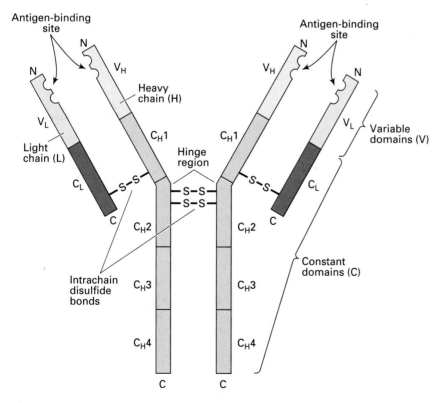

~ FIGURE 17.15

Regulatory cascade for sex determination in *Drosophila*. In XX (female) flies, *Sxl* is active, leading to functional *tra* protein and female splicing of *dsx* transcripts, to give *dsx* female products. In XY (male) flies *Sxl* is inactive, leading to no *tra* protein and default splicing of *dsx* transcripts, to give *dsx* male products.

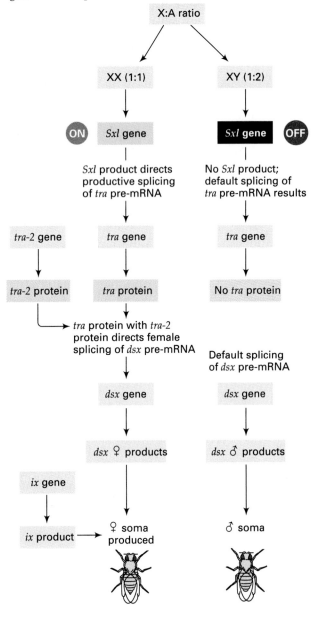

Next, in XX flies, the *tra* gene product, together with the *tra*-2-encoded protein, activates female-specific splicing of the *dsx* (*doublesex*) mRNAs. The resulting *dsx* ♀ products, with *ix* (*intersex*) product, specify the determination of female somatic cells. In XY flies, no products are generated from *tra*, because of the default mRNA splicing. As a result, no functional

products are generated to activate female splicing of the *dsx* transcript. Instead, alternative, default splicing occurs, giving rise to dsx ♂ products and leading to the development of male somatic cells.

𝒦EYNOTE

Gene expression in eukaryotes can be regulated at the level of RNA processing. This type of regulation operates to determine the production of mature RNA molecules from precursor-RNA molecules. Two regulatory events that exemplify this level of control are choice of poly(A) site and choice of splice site. In both cases different types of mRNAs are produced, depending upon the choices made. There are known examples of both types of regulation in developmental systems. _____

Transport Control

The next level of control is **transport control,** which is the regulation of the number of transcripts that exit the nucleus to the cytoplasm. A profound difference between prokaryotes and eukaryotes is the presence of a nucleus with a surrounding membrane in the latter. This nuclear membrane potentially is a control point in gene expression.

We know that primary transcripts are processed extensively in the nucleus. Experiments have also shown that perhaps one half of the primary transcripts of protein-coding genes (the hnRNA population) never leaves the nucleus, but is degraded there. How is transport of mature mRNA from the nucleus regulated? It seems that the mRNA exits through the nuclear pores, but very little is known about the export process itself or of the signals required for nuclear retention and export. There is some evidence that snRNPs are important for retaining mRNAs in the nucleus. For example, in mutants of yeast that prevent spliceosome assembly, mRNA export from the nucleus is facilitated. This has lead to a *spliceosome retention* model. In this model, spliceosome assembly competes with nuclear export. Thus, when pre-mRNAs are in spliceosomes undergoing processing, the RNA is retained in the nucleus, unable to interact with the nuclear pore. However, when processing is complete, the mRNA dissociates from the spliceosome, which is still associated with the intron. The free mRNA is able to interact with the nuclear pore, but the intron cannot. It is still not clear whether mRNAs require a specific export signal, or whether they are exported by default.

$\mathcal{K}$EYNOTE

The transport of mRNAs from the nucleus to the cytoplasm is another important control point. A spliceosome retention model has been proposed for this regulation. In this model, spliceosome assembly on a precursor-mRNA competes with nuclear export of the molecule. Thus, during intron removal, the spliceosome serves to keep the RNA in the nucleus but, when all introns have been removed and the spliceosome has dissociated, the free mRNA can interact with the nuclear pore and exit.

mRNA Translation Control

Messenger RNA molecules are subject to **translational control** by ribosome selection among mRNAs. Thus, differential translation can greatly affect gene expression. For example, mRNAs are stored in many vertebrate and invertebrate unfertilized eggs. In the unfertilized state, the rate of protein synthesis is very slow; however, protein synthesis increases significantly after fertilization. Since this increase occurs without new mRNA synthesis, it can be concluded that translational control is responsible. Lastly, the resulting proteins are subject to **protein degradation control,** in which the rate of protein degradation is regulated.

mRNA Degradation Control

Once in the cytoplasm, all RNA species are subjected to **degradation control,** in which the rate of RNA breakdown (also called RNA turnover) is regulated. Usually, both rRNA (in ribosomes) and tRNA are very stable species. By contrast, mRNA molecules exhibit a diverse range of stability, with some mRNA types known to be stable for many months while others degrade within minutes. Examples of changes in the stability of mRNA molecules have been described. For example, the addition of a regulatory molecule to a cell type can lead to an increase in synthesis of a particular protein or proteins. This is accomplished by both an increase in the rate of transcription of the gene(s) involved and/or an increase in the stability of the mRNAs produced. Table 17.2 presents examples of systems in which changes in mRNA stability for a number of cell types occur in the presence and absence of specific effector molecules.

mRNA degradation, in fact, is believed to be a major control point in the regulation of gene expression. The diversity of mRNA degradation rates has been shown to be due to the recognition of structural features in individual mRNAs. Specifically, the selectivity of mRNA degradation is largely the result of interactions between nucleases and internal mRNA structures. For example, a group of AU-rich sequences in the 3'-*un*translated *r*egion (UTR) of many short-lived mRNAs is responsible for their metabolic instability. It is not known how the AU-rich sequences act to cause mRNA instability, but possibly they cause destabiliza-

~ TABLE 17.2

Examples of Tissues or Cells in Which Regulation of mRNA Stability Occurs in Response to Specific Effector Molecules[a]

			HALF-LIFE OF MRNA	
mRNA	TISSUE OR CELL	REGULATORY SIGNAL (=EFFECTOR MOLECULE)	WITH EFFECTOR	WITHOUT EFFECTOR
Vitellogenin	Liver (frog)	Estrogen	500 h	16 h
Vitellogenin	Liver (hen)	Estrogen	~24 h	<3 h
Apo-very low density lipoprotein (apoVLDL)	Liver (hen)	Estrogen	~20–24 h	<3 h
Ovalbumin, conalbumin	Oviduct (hen)	Estrogen, progesterone	>24 h	2–5 h
Casein	Mammary gland (rat)	Prolactin	92 h	5 h
Prostatic steroid-binding protein	Prostate (rat)	Androgen	Increases 30×	

[a]Note that the effector molecule in each case results in an increase in transcription as well as stabilization of the mRNA.

tion of the poly(A). Clearly, a lot remains to be learned about this important control point, since there is little information about how cellular factors affect the mRNA's susceptibility to degradation and how nucleases are involved.

KEYNOTE

Gene expression is regulated also by mRNA translation control and by mRNA degradation control. The latter is believed to be a major control point in the regulation of gene expression. Structural features of individual mRNAs have been shown to be responsible for the range of mRNA degradation rates, although the precise roles of cellular factors and enzymes have yet to be determined. In sum, in prokaryotes, control of gene expression occurs mainly at the transcriptional level, in association with rapid turnover of mRNA molecules. In eukaryotes, gene expression is regulated at a number of distinct levels. Regulatory systems appear to exist for the control of transcription, a precursor-RNA processing, transport out of the nucleus, degradation of mature RNA species, translation of the mRNA, and degradation of the protein product. ———————————————

GENE REGULATION IN DEVELOPMENT AND DIFFERENTIATION

Most higher eukaryotes have many different cells, tissues, and organs with specialized functions (e.g., in animals, liver cells and nerve cells perform different functions; in plants, root cells and leaf cells perform different functions), and yet all cells of the same organism have the same genotype. During an organism's growth throughout its life cycle, cells which were genetically identical become physiologically and phenotypically differentiated in terms of their structure and function.

Two terms are used in the discussion of long-term gene regulation. **Development** refers to the process of regulated growth that results from the genome's interaction with cytoplasm and the environment and that involves a programmed sequence of phenotypic events that are typically irreversible. The total of the phenotypic changes constitutes the life cycle of an organism. **Differentiation**, the most spectacular aspect of development, involves the formation of different types of cells,

tissues, and organs from a zygote through the processes of specific regulation of gene expression; differentiated cells have characteristic structural and functional properties.

At a very general level we can safely assume that the processes in differentiation and development are the result of a highly programmed pattern of gene activation and gene repression. But while we know a great deal about the molecular events that turn the bacterial *lac* operon on and off, we are a long way from understanding the molecular bases for the differentiation and development events in higher eukaryotes. We can describe these events well at the morphological level, and we can describe them to some extent at the biochemical level. We are only in the relatively early stages of knowing the details of how the complex activation-repression patterns are programmed, because of the greater complexity of the processes in comparison with the processes of bacterial operons.

In the following, we will discuss only selected aspects of gene regulation in development and differentiation, since much of this area belongs more appropriately in a developmental biology course.

Gene Expression in Higher Eukaryotes

Compared with the prokaryotic genome, the genome of higher eukaryotes is much larger, and as discussed in Chapter 10, it has a large amount of highly repetitive and moderately repetitive DNA. Highly repetitive DNA constitutes perhaps 20 to 40 percent of the genomic DNA of higher eukaryotes and does not appear to encode structural information. The remaining 60 to 80 percent of the DNA is distributed between the moderately repetitive and unique-sequence DNA.

What is the function of the moderately repetitive and unique-sequence DNA? In his studies of the sea urchin, Eric Davidson isolated the species' total mRNA, and by determining the total percentage of unique-sequence DNA to which this RNA hybridized, he was able to quantify the amount of DNA that encodes structural information (proteins). Instead of confining his observations to a specific stage in sea urchin development, where he expected only a subset of the total structural information to be expressed, Davidson extended his hybridizations to mRNAs obtained from populations of sea urchins in all stages of development and from mature sea urchin tissues. The results indicated the percentage of unique-sequence DNA to which RNA hybridized at various stages of development. Maximal transcription was found to occur in the oocyte, with about 6 percent of the unique-sequence DNA hybridized. Transcriptional activity remains high during the blastula, gastrula, and larval stages of development (>2 percent unique-sequence DNA hybridized) and then drops to low levels (<0.8 percent unique-sequence DNA hybridized) in mature tissue such as the intestine and tubefoot. These results mean that at any one time a maximum of about 6 percent of the unique-sequence DNA is transcriptionally active and producing proteins.

Because higher eukaryotes are more complex than prokaryotes, it was expected that a larger proportion of their DNA would encode structural information. However, it was shown that the majority of the sea urchin's DNA must have other functions. We now know that the structural genes (which are transcribed) are located within long stretches of DNA that are not transcribed. The function of the large amount of non-transcribed DNA in higher eukaryotes (less is found in lower eukaryotes) has long been controversial. It could be "junk" DNA left over from evolutionary changes, for example. Or it might have important regulatory roles as yet undetermined, or a structural role.

Constancy of DNA in the Genome During Development

In early studies of the genetic processes associated with differentiation and development, an important area of research focused on whether differentiation and development involve a *loss* of genetic information (i.e., of DNA), or whether these processes involve a programmed sequence of gene activation and repression events in a genome that is unchanged from the sequence found in the nucleus of the zygote, from which the mature organism develops. The most direct way to decide is to determine whether the genome of a differentiated cell contains the same genetic information found in the zygote. Two important studies were especially instrumental in answering this question: One involved the study of carrots; the other focused on frogs.

REGENERATION OF CARROT PLANTS FROM MATURE SINGLE CELLS. The first study was done in the 1950s by Frederick Steward, who worked with carrot plants. He took a carrot, dissociated the tissue so that the cells were separated, and then attempted to culture new carrot plants from those cells by using plant tissue culture techniques (Figure 17.16). He was successful, and mature plants with edible carrots were produced (cloned) from phloem cells. That the mature cells had the potential to act as zygotes and develop into complete plants indicated that mature cells had all the DNA found in zygotes. Steward's findings supported the notion that the DNA content of a cell remains constant during development and differentiation and provided evidence against the notion that development and differentiation do not involve losses of genetic information.

NUCLEAR TRANSPLANTATION EXPERIMENTS WITH *Xenopus*. An objection to the general applicability of the results of the carrot experiments is that plants are much more able to propagate themselves vegetatively than are animals. For example, horticulturists have long been able to regenerate plants from cuttings. Researchers also had to determine whether the DNA content of animal cells remained constant during development. This result was indicated by a study involving the South African clawed toad, *Xenopus laevis*. This work, done in 1964 by John Gurdon, was based on earlier work done in the early 1950s by Robert Briggs and Thomas J. King with another amphibian, the leopard frog, *Rana pipiens*.

Gurdon's experiments (Figure 17.17; p. 534) tested whether or not a nucleus taken from a tadpole (one differentiated stage of the *Xenopus* life cycle) could direct the development of an egg into a new tadpole—or even into an adult toad. The nucleus that he used was tagged with a genetic marker so that he could be

~ FIGURE 17.16

Cloning of a mature carrot plant from a cell of a mature carrot.

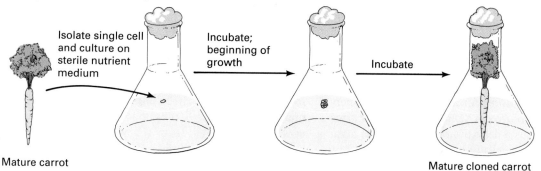

Mature carrot   Isolate single cell and culture on sterile nutrient medium   Incubate; beginning of growth   Incubate   Mature cloned carrot

~ FIGURE 17.17

Representation of Gurdon's experiments, which showed the totipotency of the nucleus of a differentiated cell of *Xenopus laevis*.

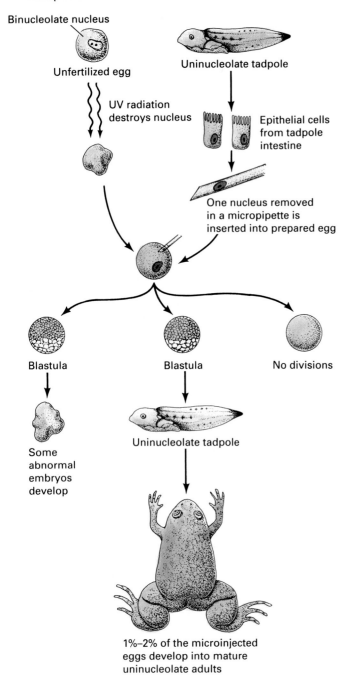

Binucleolate nucleus

Unfertilized egg

Uninucleolate tadpole

UV radiation destroys nucleus

Epithelial cells from tadpole intestine

One nucleus removed in a micropipette is inserted into prepared egg

Blastula

Blastula

No divisions

Some abnormal embryos develop

Uninucleolate tadpole

1%–2% of the microinjected eggs develop into mature uninucleolate adults

cells have no nucleoli. Since the nucleolus organizer region normally contains the ribosomal RNA genes, homozygous *O-nu/O-nu* organisms cannot make their own ribosomes, so this genotype is ultimately lethal.

Heterozygous *+/O-nu* individuals develop in a normal fashion, and their nuclei have one nucleolus per nucleus instead of the two found in *+/+* cells. Since it is easy to determine whether a nucleus has one or two nucleoli by the use of a microscope, Gurdon used *+/O-nu* (uninucleolate) tadpoles as the source for donor nuclei and *+/+* (binucleolate) eggs as recipients.

The experiment required some very skillful manipulations of cells and nuclei under the microscope. First, intestinal epithelial cells (i.e., cells of the gut lining) were isolated from uninucleolate tadpoles, and their nuclei were removed by using a micropipette larger than the diameter of the nucleus but smaller than the diameter of a whole cell. Gentle sucking of a cell caused the cell to break, and the nucleus could be sucked into the pipette. Second, an unfertilized egg from a binucleolate strain was isolated and then irradiated with ultraviolet light to destroy the genetic information it contained. Next, the intestinal *+/O-nu* nucleus was introduced into the recipient egg, which was then incubated to see whether development would occur.

In some cases no divisions occurred; in others abnormal embryos developed; and in 1 to 2 percent of the cases a fertile, adult *+/O-nu* toad developed. Most frequently, however, swimming tadpoles were generated that later died. Because the nuclei of the tadpoles and adult toads produced were always uninucleolate, Gurdon concluded that the donor tadpole nucleus must have contained all the genetic information needed to specify an adult toad. The nucleus from the *Xenopus* intestine is said to be totipotent, where **totipotency** is the capacity of a nucleus to direct a cell through all the stages of development and therefore produce a normal adult. In this particular study, only a few cells (1 to 2 percent) were totipotent.

The results of these and many other studies of various plants and animals indicate that differentiation and development in most cases do not involve a loss of genetic information from the genome. Thus, because there is constancy of DNA in the genome over an organism's life cycle, differentiation and development must result from regulatory processes affecting gene expression. (As always, there are some rare exceptions. Certain organisms, for example, have somatic cells that do not contain all of the genetic information that is found in germ line cells, while other organisms display mature differentiated cells that have genetic material arranged differently from the way it is found in its embryonic precursor cells.)

sure that it was responsible for any of the developmental changes that occurred. *Xenopus* has a mutation called *O-nu*, which is a deletion for the nucleolus organizer region, and the nuclei of homozygous *O-nu/O-nu*

~ FIGURE 17.18

~ FIGURE 17.18

Molecular organization of the human α-globin and β-globin genes.

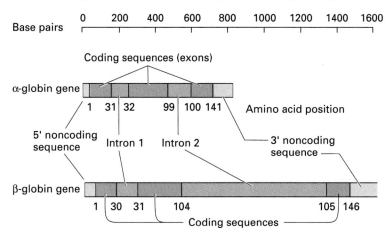

KEYNOTE

Long-term regulation is involved in controlling the events that activate and repress genes during development and differentiation. Development and differentiation result from differential gene activity of a genome that contains a constant amount of DNA, from the zygote stage to the mature organism stage. Thus these events do not result from a loss of genetic information.

Differential Gene Activity in Tissues and During Development

Specialized cell types have different cell morphologies; for example, nerve cells are clearly different from intestinal epithelial cells. Different organs and tissues can also be distinguished. Since researchers have shown that the amount of DNA remains constant during development, the simplest hypothesis is that the observable phenotypic differences reflect differential gene activity, and the following examples discuss some of the evidence supporting this hypothesis.

HEMOGLOBIN TYPES AND HUMAN DEVELOPMENT. In this book so far, human adult hemoglobin Hb-A has been examined in many contexts. Hb-A is a tetrameric protein made up of two α and two β polypeptides, where each type of polypeptide is coded by a separate gene, α and β. The two genes appear to have arisen during evolution by duplication of a single ancestral gene, followed by alteration of the base sequences

in each gene. The organization of the genes (Figure 17.18) shows that each contains two introns (intron 1 and intron 2), which are transcribed with the coding sequences but are removed to produce a mature mRNA transcript. While the introns are of different sizes in the two genes, they are placed in very similar positions in the two genes.

Hemoglobin Hb-A is only one type of hemoglobin found in humans. Genetic studies have shown that several distinct genes code for α- and β-like globin polypeptides that are assembled in specific combinations to form different types of hemoglobin. The various types are synthesized and function at different times during human development. Figure 17.19 shows

~ FIGURE 17.19

Comparison of synthesis of different globin chains at given stages of embryonic, fetal, and postnatal development.

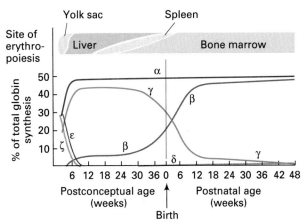

the globin chains synthesized at different stages of human development. In the human embryo the hemoglobin initially made in the yolk sac is a tetramer of two ζ (zeta) polypeptides and two ϵ (epsilon) polypeptides. From comparisons of the amino acid sequences, ζ is an α-like polypeptide, and ϵ is a β-like polypeptide. After about three months of development, synthesis of embryonic hemoglobin ceases (i.e., the ζ and ϵ genes are no longer transcriptionally active), and the site of hemoglobin synthesis shifts to the liver and the spleen. Here, true α polypeptides are made, and the two other polypeptide chains in the tetramer are either β-like γA chains or β-like γG chains, producing two forms of fetal hemoglobin. Each type of γ (gamma) chain is coded for by distinct genes. The hemoglobin formed here is called *fetal hemoglobin* (Hb-F).

Fetal hemoglobin is made until just before birth, at which time synthesis of the two types of γ chains stops, and the site of hemoglobin switches to the bone marrow, where α and true β polypeptides are made, along with some β-like δ (delta) polypeptides. In the newborn through adult human, most of the hemoglobin is our familiar $\alpha_2\beta_2$ tetramer (Hb-A), with about one in forty molecules having the constitution $\alpha_2\delta_2$. Thus, there is clear evidence for switching of globin gene types during human development, and this switching must involve a sophisticated gene regulatory system that turns appropriate globin genes on and off over a long time period.

The α-like genes (α, ζ) are all located on chromosome 16, while the β-like genes (ϵ, γA, γG, δ, and β) are all on chromosome 11 (Figure 17.20). Note that there is an ζ gene and two α genes—all of these genes are transcribed at the appropriate times. The α-like genes are arranged in the same orientation on the chromosome with regard to their transcription; that is, the promoter for each is to the left of the gene in the figure,

and transcription is from left to right. Between the ζ and α genes is a sequence that closely resembles the base-pair sequence of the α-globin gene. Careful scrutiny of this sequence reveals that it does not have all the features required for producing a functional polypeptide product. Sequences such as this one, which are highly related to known functional genes but which themselves are defective (they cannot produce a functional product), are called *pseudogenes*. Pseudogenes are thought to be nonfunctional relics of gene duplication and gene modification that occurred some time ago in evolutionary time.

The β-like genes are also each arranged in the same orientation with respect to transcription. This cluster has a pseudogene located between the γ genes and the δ gene. Most intriguing is the fact that for both clusters the order of the genes exactly parallels the order in which the genes become transcriptionally active during human development. Recall that embryonic hemoglobin consists of ζ and ϵ polypeptides, and these genes are the first functional genes at the left of the clusters. Next, the α and γ genes are transcribed to produce fetal hemoglobin (Hb-F), and if one moves from left to right, these genes are the next functional genes that can be transcribed from the clusters. Lastly the δ and β polypeptides are produced, and these genes are last in line in the β-like globin gene cluster. Although such an arrangement surely must occur by more than coincidence, there is no insight as yet about how the gene order relates to the regulation of expression of these genes during development.

POLYTENE CHROMOSOME PUFFS DURING DIPTERAN (TWO-WINGED FLY) DEVELOPMENT.

Polytene chromosomes are a special type of chromosome that consists of a bundle of chromatids. These

~ FIGURE 17.20

Linkage maps of human globin gene clusters. (Map is not to scale.)

chromatid bundles result from repeated cycles of chromosome duplication without nuclear division or chromosome segregation (a process called *endoreduplication*), and they are readily visible after staining under the light microscope. Polytene chromosomes are characteristic of certain tissues of some insects in the Order Diptera, for example, the salivary glands in the larval stages, and they may be 1000 times the thickness of corresponding chromosomes at meiosis or in the nuclei of ordinary somatic cells. After staining, distinct and characteristic bands, called *chromomeres*, can be seen along each chromosome (Figure 17.21). Each chromo-

~ FIGURE 17.21

Diagram of the complete set of *Drosophila* polytene chromosomes in a single salivary gland cell. There are four chromosome pairs, but each pair is tightly synapsed so that only a single chromosome is seen for each pair. The four chromosome pairs are linked together by regions near their centromeres to produce a large *chromocenter*.

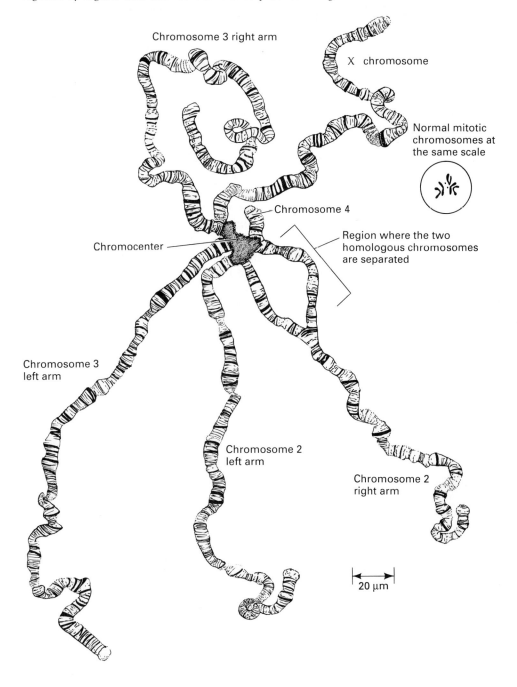

Chromosome 3 right arm

X chromosome

Normal mitotic chromosomes at the same scale

Chromosome 4

Chromocenter

Region where the two homologous chromosomes are separated

Chromosome 3 left arm

Chromosome 2 left arm

Chromosome 2 right arm

20 µm

some band contains one or more genes, and it is likely that at least some of the interband regions also contain genes.

Specific chromomeres have been shown to exhibit *puffs* at characteristic and reproducible times during development (Figure 17.22). In other words, the puffs appear and disappear in specific patterns at certain chromosomal loci as development proceeds, so it is fair to say that they are developmentally controlled. (Note that puffing occurs as a result of very high levels of gene expression. Most genes that are expressed do so at low levels and do not puff.)

While we do not know what molecular event is responsible for the induction of a puff, hormonal control is involved in many cases. Dipteran larvae go through different stages, each separated by a molt. The molting process is controlled by the molting hormone *ecdysone*, which is made in the larvae's prothoracic gland. The prothoracic gland is, in turn, stimulated to produce ecdysone by the action of a brain hormone synthesized by neurosecretory cells. The levels of ecdysone increase and decrease periodically during larval development. Ecdysone induces the transcription of some genes, the protein products of which turn on and/or off other genes that are important for larval molting and pupation. If we inject ecdysone into larvae

that are between molts, it causes the formation of the same sequence and types of puffs that are characteristically seen in larvae undergoing molting under endogenous ecdysone control. Thus, we can conclude that hormone control of gene expression plays a role in insect development.

Electron microscopic studies have shown that puffing involves a loosening of the tightly coiled DNA into long, looped structures more accessible to RNA polymerase. At the molecular level this event must involve an uncoiling of the supercoiled DNA-protein structure of the chromosome.

Evidence that the puffs are the visual manifestation of transcriptionally active genes has come from radioactive tracer experiments. If radioactive uridine (an RNA precursor) is added to developing flies, it is incorporated into the RNA being synthesized. If salivary glands are dissected from larvae and their polytene chromosomes are prepared and autoradiographed, radioactivity is localized at the puff sites, indicating that RNA, presumably transcripts, is associated with the puffs. That is, when a polytene chromosome gene is to be transcribed to produce a protein required for salivary gland activities during development, the chromosome structure loosens in order to permit efficient transcription of that region of the DNA. When transcription is completed, the chromosome reassumes its compact configuration and the puff disappears.

In sum, polytene puffs, the sites of RNA synthesis in dipteran insects, are directly related to the developmental processes in the organism, and at least some of the puffs are regulated by hormone action. We can conclude, therefore, that at least some gene regulation of developmental processes in dipteran flies is done by hormones, and that this regulation occurs at the transcriptional level.

Immunogenetics

In this section we will discuss how antibodies, the proteins that are of central importance in the immune response, are encoded in the DNA. We will see that during development, DNA rearrangements occur in cells that will produce the genes from which antibody polypeptides are transcribed.

Synopsis of the Immune System

1. All vertebrates have an *immune system*. The immune system provides protection against infectious agents such as viruses, bacteria, fungi, and protozoans.
2. The immune system distinguishes between molecules that are *foreign* from molecules that are *self*.

~ **Figure 17.22**

Light micrograph of a polytene chromosome from *Chironomus*, showing two puffs that result from localized uncoiling of the chromosome structure and indicate transcription of those regions.

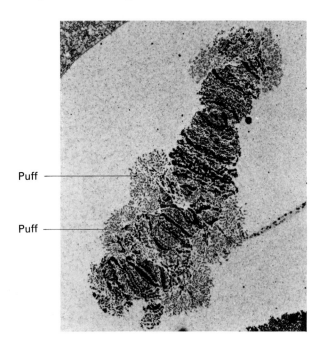

Puff

Puff

3. Any substance that elicits an immune response is called an **antigen** (*anti*body *gen*erator).

4. The cells that are responsible for immune specificity are *lymphocytes,* specifically *T cells* and *B cells.* We will focus our discussion on B cells. B cells develop in the adult bone marrow. They secrete **antibodies**—specialized proteins called **immunoglobulins**—which circulate in the blood and lymph and which are responsible for the humoral ("humor" = fluid) immune responses, i.e., the antibodies bind to foreign antigens they encounter.

5. A key feature of the immune response is that, once an organism has been exposed to a particular antigen, it becomes *immune* to the effects of a new infection. That is, when an individual encounters a foreign antigen, the immune system mounts a response by committing cells to making antibodies against that foreign antigen. The next time the same antigen is encountered, the individual "remembers," and sufficient antibodies can be synthesized rapidly to respond to the new invasion.

6. The establishment of immunity against a particular antigen results from **clonal selection,** a process whereby cells that already have antibodies specific to the antigen on their surfaces are stimulated to proliferate and secrete that antibody. During development each lymphocyte becomes committed to react with a particular antigen, even though the cell has never seen the antigen. For the humoral response system, there is a population of B cells, *each of which can recognize a single antigen.* A B cell recognizes an antigen because the cell has made antibody molecules, which are attached to the outer membrane of the cell and which act as receptor molecules. When an antigen is presented to a B cell that has the appropriate antibody receptor capable of binding to the antigen, that B cell is stimulated selectively to proliferate, producing a clonal population of plasma cells, each of which makes and secretes the identical antibody. It is important to note that *any given cell makes only one specific kind of antibody towards one specific antigen.* The actual immune response, though, may involve the binding of many different antibodies to an array of antigens on such invaders as an infecting virus or bacterium. The end result is inactivation of the infecting agent.

ANTIBODY MOLECULES. All antibody molecules made by a given B cell are identical; that is, they have the same protein chains and bind the same antigen. Considering the organism as a whole, millions of different antibody types can be produced, each with a different amino acid sequence and a different antigen-binding site.

As a group, antibodies are proteins called *immunoglobulins (Ig's).* A stylized antibody (immunoglobulin) molecule is shown in Figure 17.23a (p. 540; also see Figure 17.14), and a model of an antibody molecule based on X-ray crystallography is shown in Figure 17.23b. Both figures show the molecule's two short polypeptide chains, called *light (L) chains,* and two long polypeptide chains, called *heavy (H) chains.* (All antibody molecules also have carbohydrates attached to the regions of H chains not involved in binding with L chains.) The two H chains are held together by disulfide $(-S-S-)$ bonds, and an L chain is bonded to each H chain by disulfide bonds. Other disulfide bonds within each L and H chain cause intrachain folding.

The overall structure resembles a Y, with the two arms containing the two antigen-binding sites. The two L chains in each Ig molecule are identical, as are the two H chains, so the two antigen-binding sites are identical. The hinge region (see Figure 17.23a) allows the two arms to move in space, which facilitates the antigen-binding reaction. Also, one arm can then bind an antigen on, say, one virus, while the other arm binds the same antigen on a different virus. Such crosslinking of antibody molecules in solution helps inactivate infecting agents.

In mammals there are five classes of antibodies: IgA, IgD, IgE, IgG, and IgM. (IgM was diagrammed in Figure 17.14.) They have different H chains, i.e., α (alpha), δ (delta), ϵ (epsilon), γ (gamma), μ (mu), respectively. Two types of L chains are found: κ (kappa) and λ (lambda). Both L chain types are found in all Ig classes, but a given antibody molecule will have either two identical κ chains or two identical λ chains. A complete discussion of the functions of the five Ig classes is beyond the scope of this text. For here, we need to be aware that the major class of immunoglobulin in the blood is IgG, and IgM plays an important role in the early stages of an antibody response to a previously-unseen antigen. We will focus on these two antibody classes from now on.

Each polypeptide chain in an antibody is organized into domains of about 110 amino acids (Figure 17.23a). Each L chain (κ or λ) has two domains, and the H chain of IgG (the γ chain) has four domains, while the IgM's H chain (the μ chain) has five domains. The N-terminal domains of the H and L chains have highly variable amino acid sequences that constitute the antigen-binding sites. These domains, representing in IgG the N-terminal half of the L chain and the N-terminal quarter of the H chain, are termed the *variable,* or V, regions, and are symbolized generically as V_L for the light and V_H for the heavy chain. The amino acid sequence of the rest of the L chain is constant for all antibodies (with the same L chain type, i.e., μ or λ) and is termed C_L. Similarly, the amino acid sequence of the

~ FIGURE 17.23

Antibody molecule. (a) Diagram showing the two heavy and two light chains, and the antigen-binding sites. The heavy and light chains are held together by disulfide bonds; (b) Model of antibody molecule based on X-ray crystallography.

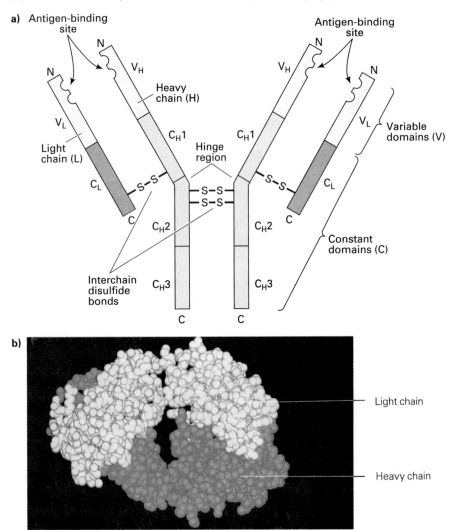

rest of the H chain is constant and is termed C_H. Thus, the production of antibody molecules involves synthesizing polypeptide chains, one part of which varies from molecule to molecule, and the other part of which is constant. How this occurs is discussed in the following section.

ASSEMBLY OF ANTIBODY GENES FROM GENE SEGMENTS DURING B CELL DEVELOPMENT. It is thought that a mammal may produce $10^6 - 10^8$ different antibodies. Since each antibody molecule consists of one kind of L chain and one kind of H chain, then $10^6 - 10^8$ antibodies theoretically would require $10^3 - 10^4$ different L chains and $10^3 - 10^4$ different H chains, if L and H chains paired randomly. The

dilemma is that the human genome is thought to contain perhaps only 10^5 genes. Therefore, how can the observed diversity be generated? The answer is that variability in L and H chains results from particular DNA rearrangements that occur during B cell development. These rearrangements involve the joining of different gene segments to form a gene that is transcribed to produce an Ig chain; the process is called *somatic recombination* and the many permutations of recombinations that are possible are responsible largely for the diversity of antibody structure. The process will now be illustrated for mouse immunoglobulin chains.

Light chain gene recombination. In mouse germ-line DNA, there are a number of gene segments that encode

~ FIGURE 17.24

Production of the light chain gene in mouse by recombination of V, J, and C gene segments during development. The rearrangement shown is only one of many possible recombinations.

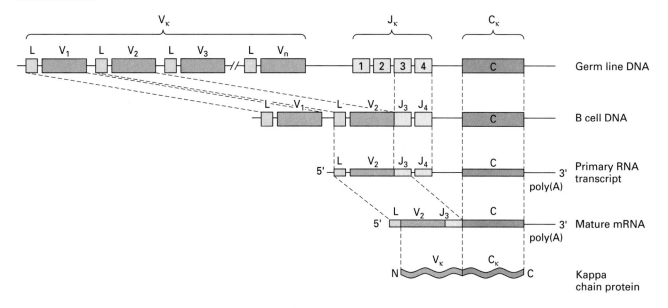

the κ light chain (Figure 17.24). There are three types of segments:

1. L-V_κ segments, which consist of a leader sequence (L) and a sequence, V_κ, which varies from segment to segment. The V_κ segment encodes most of the amino acids of the L chain variable domain. The L sequence also encodes a signal sequence (see Chapter 14) required for secretion of the Ig molecule—the signal sequence is subsequently removed and is not part of the functional antibody molecule;
2. A C_κ segment, which specifies the constant domain of the kappa L chain; and
3. J_κ segments, which are *joining* segments used to join V_κ and C_κ segments in the production of a functional κ light chain gene.

In the pre-B cell, the L-V_κ, J_κ and C_κ segments, in that order, are widely dispersed on the chromosome. Each L-V_κ segment is about 400 nucleotide pairs long; L-V_κ segments are tandemly arranged on a chromosome with about 7 kb of DNA between them. The J_κ segments are about 30 nucleotide pairs long, and they are tandemly arranged about 20 kb apart. An intron of about 2 to 4 kb separates the J_κ segments and the single C_κ segment. As the B cell develops, a particular L-V_κ segment will become associated with one of the J_κ segments and with the C_κ segment (see Figure 17.24). In the example L-$V_{\kappa 2}$ has recombined next to $J_{\kappa 4}$. After

transcription and intron removal, the mature mRNA has the organization L-$V_{\kappa 2}J_{\kappa 4}C_\kappa$; translation and leader removal produces the κ light chain that the B cell is committed to make.

In mouse there are about 350 L-V_κ gene segments, four functional J_κ segments, and one C_κ gene segment. Thus, the number of possible κ chain variable regions that can be produced by this mechanism is $350 \times 4 = 1400$. A similar mechanism exists for mouse λ light chain gene assembly. In this case there are only two L-V_λ gene segments, and four C_λ gene segments, each with its own J_λ gene segment. Thus, fewer λ variable regions can be produced than is the case for κ chains.

Further diversity of κ chains results from imprecise joining of the V_κ and J_κ gene segments. That is, the 3′ ends of V_κ gene segments and the 5′ ends of J_κ gene segments have specific recognition sequences that signal the recombination of a V_κ with a J_κ. These recognition sequences are presumed to be recognized by as-yet-unidentified enzymes that catalyze the recombination event. During this process a few nucleotide pairs from V_κ and a few nucleotides from J_κ are lost from the DNA at the $V_\kappa J_\kappa$ joint, generating significant diversity in sequence at that point. Thus, diversity of κ light chains results from: (1) variability in the sequences of the multiple V_κ gene segment; (2) variability in the sequences of the four J_κ gene segments; and (3) variability in the number of nucleotide pairs deleted at V_κ-J_κ joints.

~ FIGURE 17.25

Production of heavy chain genes in mouse by recombination of V, D, J, and C gene segments during development. Depending upon the C_H segment used, the resulting antibody molecule will be IgM, IgD, IgG, IgE, or IgA. Shown here is the assembly of an IgG heavy chain. This rearrangement is only one of the many thousands possible.

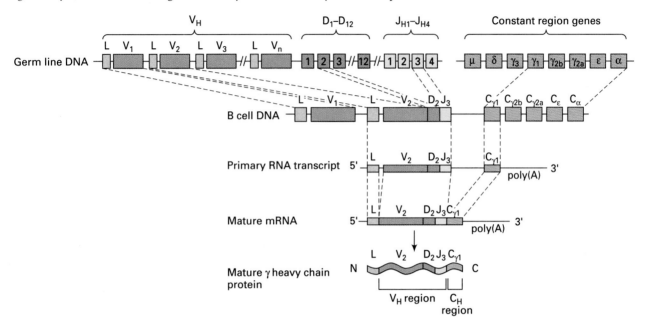

Heavy chain gene recombination. The immunoglobulin heavy chain gene is also encoded by V_H, J_H, and C_H segments. In this case additional diversity is provided by a third gene segment, D (diversity), which is located between the V_H segments and the J_H segments (Figure 17.25). As the figure shows, in the germ line of mouse there is a tandem array of L-V_H segments, then a gap, then 12 D segments, then a gap, and then 4 J_H segments. After another gap, the constant region gene segments are arranged in a cluster which, in mouse, has the order μ, δ, γ (four different ones for four different, but similar, IgG H chain constant domains), ϵ, and α for the H chain constant domains of IgM, IgD, IgG, IgE and IgA, respectively. As with L chain gene rearrangements, further antibody diversity results from imprecise joining of the gene segments that comprise the chain's variable region. In Figure 17.25, an IgG heavy chain is assembled.

𝒦EYNOTE

Antibodies are specialized proteins called immunoglobulins, which bind specifically to antigens. Immunity against a particular antigen results from clonal selection, in which cells already making the required antibody are stimulated to proliferate. Antibody molecules consist of two light (L) chains and two heavy (H) chains. The amino acid sequence of one domain of each type chain is variable; this variation is responsible for the different antigen-binding site on different antibody molecules. The other domain(s) of each chain are constant in amino acid sequence. In germ-line DNA the coding regions for immunoglobulin chains are scattered in tandem arrays of gene segments. Thus, for light chains, there are many variable region (V) gene segments, a few joining (J) gene segments, and one constant region (C) gene segment. During development, somatic recombination occurs to bring particular gene segments together into a functional L chain gene. A large number of different L chain genes result from the many possible ways in which the gene segments can recombine. Similar rearrangements occur for H chain genes, but with the addition of several D (diversity) segments that are between V and J, increasing the possible diversity of H chain genes.

GENETIC REGULATION OF DEVELOPMENT IN *DROSOPHILA*

There are a few systems in which significant progress is being made in relating gene activity to developmental events. One such system is *Drosophila* development, and this section is a brief overview of what is known.

Drosophila Developmental Stages

The production of an adult *Drosophila* from a fertilized egg involves a well-ordered sequence of developmentally programmed events under strict genetic control (Figure 17.26). About 24 hours after fertilization, a *Drosophila* egg hatches into a larva, which undergoes three molts, after which it is called a pupa. The pupa metamorphoses into an adult fly. The whole process from egg to adult fly takes about nine days.

Embryonic Development

Development commences with a single fertilized egg, giving rise to cells that have different developmental fates. What follows is a brief discussion of the information that has been obtained about the relationship between the developmental events in the egg and the determination of adult body parts.

Before a mature egg is fertilized, particular molecular gradients are established within it. The posterior end is indicated by the presence of a region called the *polar cytoplasm* (Figure 17.27a; p. 544). At fertilization, the two parental nuclei are roughly centrally located in the egg. The two nuclei fuse to produce a 2N zygote nucleus (Figure 17.27b). For the first nine divisions, only the nuclei divide in a common cytoplasm—cytokinesis does not occur—to produce what is called a multinucleate *syncytium* (Figure 17.27c). (After seven divisions, some nuclei migrate into the polar cytoplasm, where they become precursors to germ-line cells.) Next, the nuclei migrate and divide to form a layer at the surface of the egg, producing the *syncytial blastoderm* (Figure 17.27d). After four more divisions, membranes form around the nuclei to produce somatic cells, about 4000 of which make up the *cellular blastoderm* (Figure 17.27e).

Subsequent development of body structures depends upon two processes (Figure 17.28; p. 544): (1) Two gradients of molecules form, one along the posterior-anterior axis and the other along the dorsal-ventral axis of the egg. Somehow, a nucleus is aware of its position with reference to the molecular concentration in the two gradients; (2) Regions are determined in the embryo that correspond to adult body segments. In

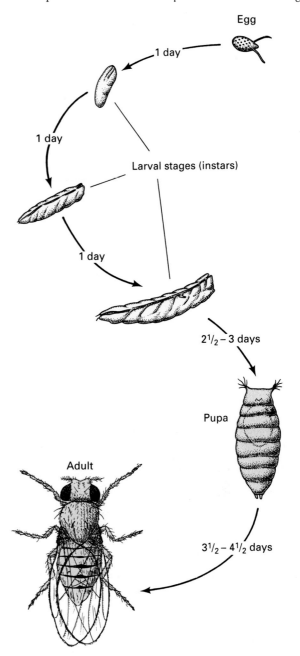

~ **FIGURE 17.26**

Development of an adult *Drosophila* from a fertilized egg.

Egg

1 day

1 day

Larval stages (instars)

1 day

2½ – 3 days

Pupa

Adult

3½ – 4½ days

the cellular blastoderm stage, where the regions are not well defined, they are called *parasegments*. In subsequent stages, where they are visible, they are called *segments*, forming a striped pattern along the anterior-posterior axis of the embryo. The embryonic segments give rise to the segments of the adult fly.

Genes involved in regulating *Drosophila* development are defined by mutations that have a lethal pheno-

~ FIGURE 17.27

Embryonic development in *Drosophila*. (a) The fertilized egg, with the two parental nuclei. Polar cytoplasm indicates the posterior end. (b) The two parental nuclei fuse to produce a diploid zygote nucleus. (c) The nucleus undergoes nine divisions in a common cytoplasm to produce a multinucleate syncytium. (d) Nuclei migrate and divide, producing a layer at the periphery of the egg. This stage is the syncytial blastoderm. (e) Nuclei divide four times, a membrane then forms around each to produce the somatic cells of the cellular blastoderm.

a)   Fertilized egg with two parental nuclei

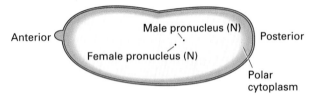

b)   Parental nuclei fuse and produce a diploid zygote nucleus

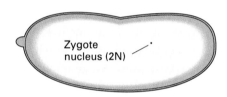

c)   The nucleus divides for nine divisions in a common cytoplasm to give a multinucleate syncytium

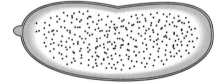

d)   Syncytial blastoderm nuclei migrate and divide, producing a layer at the periphery of the egg

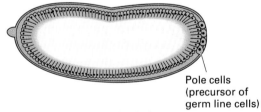

e)   Cellular blastoderm. Nuclei divide four times; membranes form around them and produce somatic cells

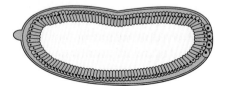

~ FIGURE 17.28

Drosophila development results from gradients in the egg that define parasegments in the cellular blastoderm, and segments in the embryo and adult. The adult segment organization directly reflects the segment pattern of the embryo.

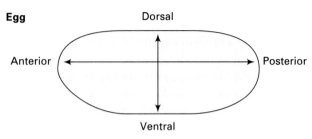

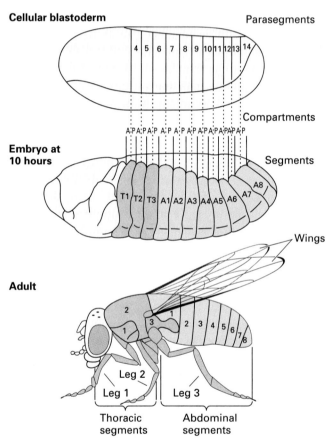

type early in development or that result in the development of abnormal structures (e.g., embryos with abnormal striping). Three major classes of such genes have been defined:

1. *Maternal genes:* These are expressed by the mother during oogenesis. Their products function to specify the gradients in the egg and to determine spatial organization in early development. Examples are *bicoid* (*bcd:* required for normal anterior develop-

~ FIGURE 17.29

Functions of segmentation genes as defined by mutations.

Gene	Normal larva and affected parts	Effect of mutation	Time of expression
Gap (*Krüppel*)		Adjacent segments missing	< 11 divisions
Pair-rule (*even-skipped*)		Deletion in every other segment	11–12 divisions
Segment polarity gene (*gooseberry*)		Segments replaced by mirror images	13 divisions

ment), *tudor* (*tud:* posterior segment development), and *torso* (*tor:* terminal posterior segment development).

2. *Segmentation genes:* These are expressed after fertilization. Mutations in these genes affect the number or polarity of body segments. At least 20 gene loci have been shown to affect segmentation. The loci are subclassified on the basis of the size of the unit affected (Figure 17.29). That is, mutations in *gap genes* (e.g., *Krüppel, hunchback*) result in several adjacent segments being deleted; mutations in *pair rule genes* (e.g., *even-skipped, fushi tarazu*) result in the same part of the pattern deleted in every other segment; and mutations in *segment polarity genes* (e.g., *gooseberry, engrailed*) have segments replaced by mirror images.

In brief, then, genes control *Drosophila* development in a temporal regulatory cascade. First, the maternal genes define the anterior-posterior gradients in the egg; then the gap genes are transcribed, functioning to define four major regions of the egg. Next, pair rule genes are transcribed, and their products are confined to pairs of segments. Last, segment polarity genes are expressed; their products act at the individual segment level.

3. *Homeotic genes:* These genes control the *identity* of a segment, but do not affect the number, polarity, or size of segments. Homeotic mutants, which act after segmentation genes, cause one body part to develop into a different body part. Examples of homeotic mutants will be described later.

KEYNOTE

Development of *Drosophila* body structures results from gradients along the posterior-anterior and dorsal-ventral axes of the egg and from the subsequent determination of regions in the embryo that directly correspond to adult body segments. As defined by mutations, genes control *Drosophila* development in a temporal regulatory cascade. First, maternal genes specify the gradients in the egg, then segmentation genes (gap genes, pair rule genes, and segment genes) determine the segments of the embryo and adult, and homeotic genes next specify the identity of the segments. _____

Imaginal Discs

Two types of cells are specified by cellular blastoderm cells: (1) those that will produce larval tissues; and (2) those that will develop into the adult tissues and organs. For the latter, certain groups of undifferentiated cells form larval structures called **imaginal discs,** each of which differentiate into a specific organ in the adult fly. That is, a number of *Drosophila* adult structures develop from imaginal discs, which are determined early in larval development. The adult structures are appendages and related structures, for example the eyes and external genitalia. Other structures, such as the nervous system, gut, and cuticle do not develop from imaginal discs. A determined state that is so programmed is very stable, although rate changes do occur. For instance, there is a clear distinction between the stage when a phenotype is determined and the stage when the differentiation processes that give rise to the phenotype occur.

Thus larvae contain two groups of cells: Cells of one group are involved exclusively with larval development and function, while cells of the other group are grouped into the clusters of cells that comprise imaginal discs. Imaginal disc cells remain in an embryonic state throughout larval development, even though larval cells differentiate around them. Each disc eventually differentiates into a particular adult structure. Figure 17.30a shows the positions of some imaginal discs in a mature larva and the adult structures that develop from them.

From the genetic point of view imaginal discs are excellent subjects for study because each disc develops during the first larval stage, and when it consists of about 20−50 cells, it is already programmed to specify its given adult structure—its fate is determined. From then on the number of cells in each disc increases by mitotic division, until by the end of the larval stages, there are many thousand cells per disc.

The nature of the disc determination process is unknown, although at a simple level it probably involves a programming of those genes that are accessible to hormone (ecdysone)-stimulated activation during the pupal stage. Evidence suggests that the determination process is a remarkably stable event; some mechanism apparently keeps all the cells in a particular disk programmed in the same way. This evidence comes from transplantation experiments (c.f. Chapter 8), in which discs or parts of discs are transplanted from a larva into the abdomen of an adult fly. In the adult abdomen the disc gets larger as its cells grow and divide, but the disc does not differentiate into its adult structure because the adult abdomen lacks the necessary hormones to induce such differentiation. After the disc has grown, it can be reisolated from the abdomen and part of it transplanted into the abdomen of another adult. Continued growth, isolation, and transplantation to a new host (called *serial transplantation*) at least 150 times does not change the programming of the disc. That is, when, at the end of the series of transplantations, the disc is transplanted into a larval host (in which the necessary hormones for differentiation exist), the disc will still, with very rare exceptions, develop into the same adult structure as the original disc would have.

~ Figure 17.30

Locations of imaginal discs in a mature *Drosophila* larva and the adult structures derived from each disc.

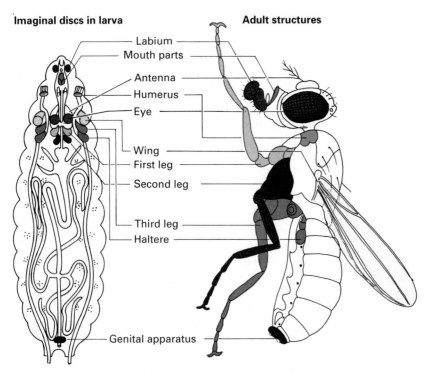

Imaginal discs in larva　　　　**Adult structures**

- Labium
- Mouth parts
- Antenna
- Humerus
- Eye
- Wing
- First leg
- Second leg
- Third leg
- Haltere
- Genital apparatus

TRANSDETERMINATION. Rarely, the determined state of an imaginal disc does change, as the transplantation studies of Ernst Hadorn showed. The disc does not totally dedifferentiate but switches to another determined path in a process called **transdetermination.** For example, an eye disc may become altered so that it specifies a wing structure, or an antenna disc may become altered so that it specifies a leg structure. Since the position of the discs in the larvae relates to the positions of adult structures, the consequence of such transdetermination events is that the adult fly has familiar structures at unusual places, such as a leg where an antenna should be. Significantly, the *entire disc changes its determined state;* no studies have reported, for instance, that half a disc becomes transdetermined while the other half does not. The results of such transdetermination would be easy to detect by visual scrutiny of adult flies since they would be hybrid structures.

Figure 17.31, which shows the pathways of transdetermination in *Drosophila* that have been detected by transplantation experiments such as those of Hadorn, indicates that certain discs transdetermine only into certain other disc types. Leg discs, for example, transdetermine into wing discs, but the opposite transdetermination occurs far less frequently.

⨔EYNOTE

A number of *Drosophila* adult structures develop from imaginal discs, which are determined early in larval development. Rarely, the determined state of an imaginal disc changes so that a different adult structure is produced; this is called transdetermination. ———————————

Homeotic Genes

Once the basic segmentation pattern has been laid down, the homeotic genes give a specific developmental identity to each of the segments. The homeotic genes have been defined by mutations that affect the development of the fly. That is, **homeotic mutations** alter the identity of particular segments, transforming them into copies of other segments. The principal pioneer of genetic studies of homeotic mutants is Edward Lewis, and the more recent molecular analysis has been done in many laboratories, including Thomas Kaufman's, Walter Gehring's, W. McGinnis's, Matthew Scott's, and Welcome Bender's.

Lewis's pioneering studies involved mutations in a cluster of homeotic genes called the *bithorax* complex (BX-C). BX-C determines the posterior identity of the fly, namely thoracic segment T3 and abdominal segments A1-A8. BX-C contains three complementation groups called *Ultrabithorax (Ubx), abdominal-A (abd-A),* and *Abdominal-B (Abd-B),* each of which constitutes one protein-coding transcription unit. Mutations in these homeotic genes are often lethal, so the fly typically does not survive past embryogenesis. Some nonlethal mutations have been characterized, however, which permit development into an adult fly. Figure 17.32 (p. 548) shows the abnormal adult structures that can result from *bithorax* mutations. A diagram showing the segments of a normal adult fly is in Figure 17.32a: note that the wings are located on segment Thorax 2 (T2), while the pair of halteres (rudimentary wings used as "balancers" in flight) are on segment T3. A photograph of a normal adult fly clearly showing the wings and halteres is presented in Figure 17.32b. Figure 17.32c shows one type of developmental abnormality that can result from nonlethal

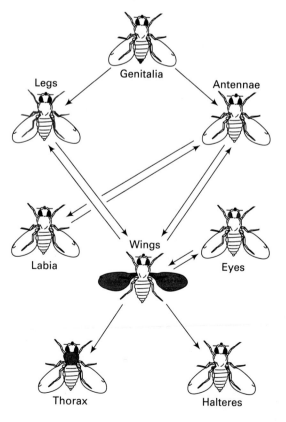

~ FIGURE 17.32

(a) Drawing of a normal fly. T = thoracic segment. A = abdominal segment. The haltere (rudimentary wing) is on T3 (see Figure 17.34). (b) Photograph of a normal fly with a single set of wings. (c) Photograph of a fly homozygous for three mutant alleles (bx^3, abx, and pbx) that results in the transformation of segment T3 into a structure like T2: namely, a segment with a pair of wings. These flies therefore have two sets of wings.

a)

Haltere (rudimentary wing)

b)

Haltere

Wing

c)

homeotic mutations in BX-C: shown is a fly that is homozygous for three separate mutations in the *Ubx* complementation group, *abx*, bx^3, and *pbx*. Collectively these mutations transform segment T3 into an adult structure similar to T2. The transformed segment exhibits a fully developed set of wings. The fly lacks halteres, however, because no normal T3 segment is present.

Another well-studied group of mutations defines another large cluster of homeotic genes called the *Antennapedia* complex (ANT-C). ANT-C determines the anterior identity of the fly, namely the head and thoracic segments T1 and T2. ANT-C contains at least four genes, called *Deformed (Dfd)*, *fushi tarazu (ftz)*, *Sex combs reduced (Scr)*, and *Antennapedia (Antp)*. Most ANT-C mutations are lethal. Among the nonlethal mutations is a group of mutations in *Antp* that result in leg parts instead of an antenna growing out of the cells near the eye during the development of the eye disc (Figure 17.33a and b). Note that the leg has a normal structure, but it is obviously positioned in an abnormal location. A different mutation in *Antp*, called *Aristapedia*, has a different effect: only the distal part of the antenna, the arista, is transformed into a leg (Figure 17.33c).

The homeotic genes ANT-C and BX-C, therefore, encode products that are involved in controlling the normal development of the relevant adult fly structures.

The *Antennapedia* complex (ANT-C) and the bithorax complex (BX-C) have been cloned. Both complexes are very large. In ANT-C, for example, the *Antp* gene is 103-kb long, with many introns; this gene encodes a mature mRNA of only a few kilobase pairs. BX-C covers more than 300 kb of DNA and contains only three protein-coding regions: *Ubx*, *abd-A*, and *Abd-B* (Figure 17.34; p. 550). At least *Ubx* and *abd-A* have introns. Other RNA products are known to be transcribed from BX-C, but they do not code for proteins, and they have no known functions. If we count the base pairs required to encode the proteins, more than 250 kb of DNA remains in the complex. This other DNA is believed to consist of regulatory regions of significant size and complexity, the functions of which are to control the expression of the protein-coding genes.

Because the ANT-C and BX-C protein-coding genes have similar functions but are located in different places in the genome, Lewis predicted that the genes would have related structures. Analysis of the DNA sequences for the genes revealed the presence of similar sequences of about 180 bp that has been named the **homeobox.** The homeobox is part of the protein-coding sequence of each gene, and the corresponding 60-amino-acid part of each protein is called the **homeodomain.**

Homeoboxes have been found in over 20 *Drosophila* genes, most of which regulate development. All homeodomain-containing proteins appear to be located in the nucleus, and there is a similarity in structure between the homeodomain and other known DNA binding proteins and transcription factors. Thus, it seems that homeodomain-containing proteins play a role in transcriptional regulation through interaction

~ FIGURE 17.33

(a) Scanning electron micrograph (left) and drawing (right) of the antennal area of a wild-type fly; (b) Scanning electron micrograph (left) and drawing (right) of the antennal area of the homeotic mutant of *Drosophila*, *Antennapedia*, in which the antenna is transformed into a leg; (c) Scanning electron micrograph (left) and drawing (right) of the homeotic mutant of *Drosophila*, *Aristapedia*, in which the arista is transformed into a leg.

a) **Normal**

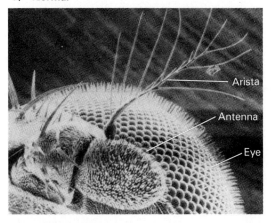

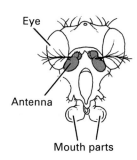

b) **Antennapedia**

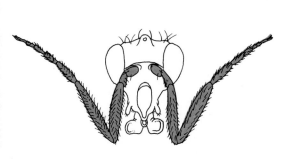

c) **Aristapedia**

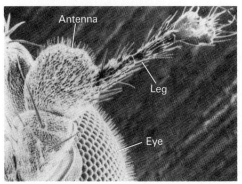

~ FIGURE 17.34

Organization of the *bithorax* complex (BX-C). The DNA spanned by this complex is 300 kb long. The transcription units for *Ubx*, *abdA*, and *AbdB* are shown below the DNA: the exons are shown by colored blocks and the introns by bent, dotted lines. All three genes are transcribed from right to left. Shown above the DNA are regulatory mutants that affect the development of different fly segments.

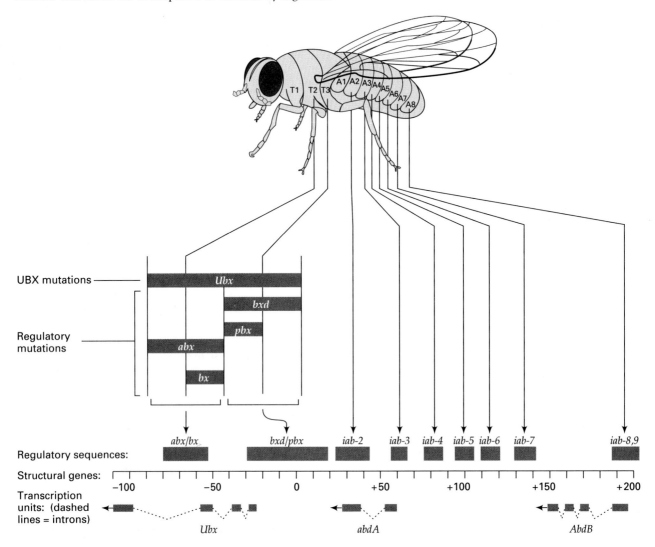

with specific DNA sequences. This role has been definitely established in a few cases.

Homeoboxes have also been found in many higher eukaryotes, including humans, mice, chickens, and *Xenopus,* and in arthropods, annelids, ascidians, echinoderms, brachiopods, tapeworms, molluscs, and other chordates. Homeoboxes have not yet been found in coelenterates, sponges, flatworms, slime molds, filamentous fungi, or bacteria. It is hoped that study of the homeoboxes will provide a more general understanding of transcriptional regulation of genes involved in development.

KEYNOTE

Homeotic genes in *Drosophila* determine the developmental identity of a segment, once the basic segmentation pattern has been laid down. Homeotic genes have in common similar DNA sequences called homeoboxes. The homeobox is part of the protein-coding sequence of each gene and the corresponding approximately 60-amino-acid part of the protein is called the homeodomain. Homeoboxes have been found in developmental genes in other organisms, and homeodomains are thought to play a role in regulating transcription by binding to specific DNA sequences. ——————————

GENETIC REGULATION OF DEVELOPMENT IN THE NEMATODE *CAENORHABDITIS ELEGANS*

The nematode worm *Caenorhabditis elegans (C. elegans)* is another excellent model system for studying the genetic control of development. The reasons are as follows:

1. Adult worms are about 1 mm long, and the life cycle is about 3 days.
2. Sex determination is similar to that for *Drosophila,* except XX individuals are hermaphrodites and XYs are males (see Chapter 3). Genetic crosses and selfings can readily be done.
3. The genome is about 8×10^4 kb, which is relatively small. Efforts are underway to obtain its complete sequence so all potential genes will be identified at the sequence level.
4. The worm's body is transparent, enabling all cells to be seen under the microscope.
5. Development is programmed in an invariant way. That is, each adult hermaphrodite has exactly 945 cells, and these cells are traceable back to the zygote along particular cell pathways or *cell lineages.*

Particularly because of the defined cell lineages and available molecular tools, great strides are being made in dissecting the development of *C. elegans* using genetic, cellular, and molecular approaches.

SUMMARY

In this chapter we have considered a number of examples of gene regulation in eukaryotes. The general picture is one of much greater complexity than prokaryotes. Genes are not organized into operons, since genes of related function are often scattered around the genome; nonetheless, genes are regulated coordinately. The chromosome organization in eukaryotes also makes for greater complexity in regulating gene expression.

Three main topics were discussed in the chapter: the levels of control of gene expression, gene regulation during development and differentiation, and genetic regulation of development in *Drosophila* and *C. elegans.* Given the information available, each of these topics could be the subject of its own chapter.

Levels of Control of Gene Expression

Gene expression in eukaryotes is regulated at a number of distinct levels. Regulatory systems have been found for: (1) the control of transcription; (2) of precursor-RNA processing; (3) of transport of the mature RNA out of the nucleus; (4) of translation of the mRNAs; (5) of degradation of the mature RNAs; and (6) of degradation of the protein products. All but the last were discussed in this chapter.

The control of transcription itself involves a number of elements. At the DNA level, transcription of protein-coding genes involves the interaction of transcription factors with promoter elements, and of regulatory proteins with both promoter elements and enhancer elements. Depending on the element and the protein that binds to it, the effect on transcription may be positive or negative. Activation of transcription through protein binding at the distant enhancer elements is thought to involve a looping of the DNA, brought about by interaction of the regulatory proteins bound at the enhancer and promoter elements. While unique regulatory proteins certainly bind to promoter elements and to enhancer elements, some regulatory proteins are shared by the two, indicating that both types of regulatory elements affect transcription by a similar mechanism.

At the chromosome level, the regulation of transcription must deal with the interaction of histones and nonhistones with the DNA. Generally, we can view the eukaryote chromosome as repressed for transcription by this interaction, so activation of transcription can be seen as a depression phenomenon that results from a loosening of the DNA-protein structure in the neighborhood of a gene being activated. Relevant to this, we know that transcriptionally active chromatin is more sensitive to digestion with DNase I than transcriptionally inactive chromatin. Moreover, active genes contain DNase-hypersensitive sites, which are highly sensitive to DNase I digestion. These sites probably correspond to the binding sites for RNA polymerase and regulatory proteins. Gene activation in many eukaryotes is accompanied in many instances by a decrease in the level of DNA methylation, although the relationship, if any, to chromosome organization changes is not clear.

Gene expression can be regulated at the level of RNA processing. This type of regulation operates to determine the production of mature RNA molecules from precursor-RNA molecules. Two regulatory events that exemplify this level of control are choice of poly(A) site and choice of splice site. In both cases different types of mRNAs are produced, depending upon the choice made. Examples are known of both types of regulation in developmental systems. For example, different classes of immunoglobulin molecules can result from poly(A) site selection, and choice of splice site plays a key role in the *Drosophila* sex determination system.

mRNA transport from the nucleus to the cytoplasm is another important control point. A spliceosome retention model has been proposed for this regulatory system. In this model, spliceosome assembly on a precursor-mRNA competes with nuclear export of the molecule. It is argued that, during intron removal, the spliceosome that is complexed with the precursor-mRNA prevents the RNA molecule from leaving the nucleus. Then, when all introns have been removed and the spliceosome has dissociated from the now-mature mRNA, the free mRNA molecule can interact with the nuclear pore and move to the cytoplasm.

Gene expression is also regulated by mRNA translation control and by mRNA degradation control. The latter is believed to be a major control point in the regulation of gene expression, as evidenced by the wide range of mRNA stabilities found within organisms. It is clear that nucleases are ultimately responsible for the degradation of the RNAs, and the signals for the differential mRNA stabilities seem to be a property of the structural features of individual mRNAs. For example, a group of AU-rich sequences in the 3′-untranslated regions of short-lived mRNAs is responsible for their instability, although how this signal is read by the cell and transmitted to the cellular factors and enzymes that must be directly involved in the mRNA degradation remains under investigation.

Gene Regulation During Development and Differentiation

This topic is a vast one and brings together material in the domains of the geneticist, the cell biologist, the embryologist, the developmental biologist, and the molecular biologist. Classical experiments showed that development and differentiation result from differential gene activity of a genome that contains a constant amount of DNA, rather than from a programmed loss of genetic information. Development and differentiation, then, must involve regulation of gene expression, using the levels of control we have just discussed. Adding to the complexity is the fact that we must consider the regulation of a large number of genes for each developmental process, and communication between differentiating tissues, as well as systems for timing the activation and repression of genes during those important events. For example, the structure and function of a cell are often determined early, even though the manifestations of this determination process are not seen until later in development. Such early determination events may involve some preprogramming of genes that will be turned on later. Neither the nature of these determination events, nor the mechanism of timing in developmental processes are well understood, although progress is being made in understanding the genetic regulation of development in certain organisms such as *Drosophila*.

In sum, a great deal of information has been learned in the past decade or so about gene regulation in eukaryotes. We have merely scratched the surface in this chapter. Thousands of researchers are currently working to elaborate the molecular details of gene regulation in model systems. Much of our advancing knowledge has been made possible by the application of recombinant DNA and related technologies, and we can look forward to significant increases in our understanding of eukaryotic gene regulation by the turn of the century.

ANALYTICAL APPROACHES FOR SOLVING GENETICS PROBLEMS

Q.1 A region of the yeast chromosome specifies three enzyme activities in the histidine biosynthesis pathway; these activities are synthesized coordinately. How would you distinguish among the following three models?

a. Three genes are not organized into an operon. They code for three discrete mRNAs that are translated into three distinct enzymes.

b. Three genes are arranged in an operon. The operon is transcribed to produce a single polygenic mRNA, whose translation produces three distinct enzymes.

c. One gene (a supergene) is transcribed to produce a single mRNA, whose translation produces a single polypeptide with three different enzyme activities.

A.1 A key feature of an operon (model b) is that a contiguously arranged set of genes is transcribed onto a single polygenic mRNA. Thus a nonsense mutation in a structural gene will result in the loss of not only the enzyme activity coded for by that gene but also the enzyme activities coded for by the structural genes that are more distant from the promoter (see Chapter 16). If there is a supergene coding for a single polypeptide with three different enzyme activities (model c), polar effects of nonsense mutations will also be seen. However, if there are three genes that are closely linked but each with its promoter (model a), transcription will produce three distinct mRNAs, so a nonsense mutation will affect only the gene in which it is located and no other gene. Therefore, if nonsense mutations are shown to have polar effects, then either model b or model c is correct, but model a cannot be correct. If nonsense mutations do not have polar effects, no matter in which gene they are located, then model a must be correct.

Characterization of the enzyme activities coded for by the three genes would enable us to distinguish between models b and c. That is, in the operon model b, three distinct polypeptides would be produced. These polypeptides could be isolated and purified individually by using standard techniques.

Note that it would be particularly important to make sure that inhibitors of protein-degrading enzymes, proteases, are present during cell disruption so that if there is a trifunctional polypeptide present, it is not cleared by the proteases to produce three separable enzyme activities.

Thus if the operon model b is correct, we could show that there are three distinct polypeptides, each exhibiting only one of the enzyme activities—that is, three polypeptides and three enzyme activities. On the other hand, if the supergene model c is correct, it should *only* be possible to isolate a large polypeptide with all three enzyme activities (again assuming careful isolation and purification procedures); no polypeptides with only one of the enzyme activities should exist.

QUESTIONS AND PROBLEMS

*17.1 Eukaryotic organisms have a large number of copies (usually more than a hundred) of the genes that code for ribosomal RNA, yet they have only one copy of each gene that codes for each ribosomal protein. Explain why.

*17.2 The human α, β, γ, δ, ϵ and ζ globin genes are transcriptionally active at various stages of development. Fill in the following table, indicating whether the globin gene in question is sensitive (S) or resistant (R) to DNase I digestion at the developmental stages listed.

	TISSUE		
GLOBIN GENE	EMBRYONIC YOLK SAC	SPLEEN	ADULT BONE MARROW
α			
β			
γ			
δ			
ζ			
ϵ			

17.3 A cloned DNA sequence was used to probe a Southern blot. There were two DNA samples on the blot, one from white blood cells and the other from a liver biopsy of the same individual. Both samples had been digested with *Hpa*II. The probe bound to a single 2.2 kb band in the white blood cell DNA, but bound to two bands (1.5 and 0.7 kb) in the liver DNA.

a. Is this difference likely to be due to a somatic mutation in a *Hpa*II site? Explain.

b. How would it affect your answer if you knew that white blood cell and liver DNA from this individual both showed the 2 band pattern when digested with *Msp*I?

17.4 What is a hormone?

17.5 How do hormones participate in the regulation of gene expression in eukaryotes?

*17.6 The following figure shows the effect of the hormone estrogen on ovalbumin synthesis in the oviduct of 4-day-old chicks. Chicks were given daily injections of estrogen ("Primary Stimulation") and then after 10 days the injections were stopped. Two weeks after withdrawal (25 days), the injections were resumed ("Secondary Stimulation").

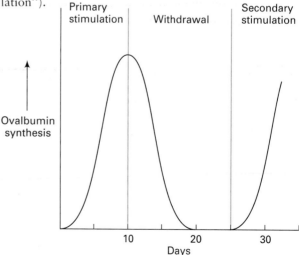

Provide possible explanations of these data.

17.7 Distinguish between the terms *development* and *differentiation*.

17.8 What is totipotency? Give an example of the evidence for the existence of this phenomenon.

17.9 Discuss some of the evidence for differential gene activity during development.

*17.10 The enzyme lactate dehydrogenase (LDH) consists of four polypeptides (a tetramer). Two genes are known to specify two polypeptides, A and B, which combine in all possible ways (A_4, A_3B, A_2B_2, AB_3, and B_4) to produce five LDH isozymes. If, instead, LDH consisted of three polypeptides (i.e., it was a trimer), how many possible isozymes would be produced by various combinations of polypeptides A and B?

17.11 Discuss the expression of human hemoglobin genes during development.

17.12 Discuss the organization of the hemoglobin genes in the human genome. Is there any correlation with the temporal expression of the genes during development?

17.13 In humans, β-thalassemia is a disease caused by failure to produce sufficient β-globin chains. In many cases, the mutation causing the disease is a deletion of all or part of the β-globin structural gene. Individuals homozygous for certain of the β-thalassemia mutations are able to survive because their bone marrow cells produce γ-globin chains. The γ-globin chains combine with α-globin chains to produce fetal hemoglobin. In these people, fetal hemoglobin is produced by the bone marrow cells throughout life, whereas normally it is produced in the fetal liver. Use your knowledge about gene regulation during development to suggest a mechanism by which this expression of γ-globin might occur in β-thalassemia.

17.14 What are the polytene chromosomes? Discuss the molecular nature of the puffs that occur in polytene chromosomes during development.

17.15 Puffs of regions of the polytene chromosomes in salivary glands of *Drosophila* are surrounded by RNA molecules. How would you show that this RNA is single-stranded and not double-stranded?

*17.16 In experiment A, ^{3}H-thymidine (a radioactive precursor of DNA) is injected into larvae of *Chironomus*, and the polytene chromosomes of the salivary glands are later examined by autoradiography. The radioactivity is seen to be distributed evenly throughout the polytene chromosomes. In experiment B, ^{3}H-uridine (a radioactive precursor of RNA) is injected into the larvae, and the polytene chromosomes are examined. The radioactivity is first found only around puffs; later, radioactivity is also found in the cytoplasm. In experiment C, actinomycin D (an inhibitor of transcription) is injected into larvae and then

^{3}H-uridine is injected. No radioactivity is found associated with the polytene chromosomes, and few puffs are seen. Those puffs that are present are much smaller than the puffs found in experiments A and B. Interpret these results.

*17.17 The following figure shows the percentage of ribosomes found in polysomes in unfertilized sea urchin oocytes (0 h) and at various times after fertilization:

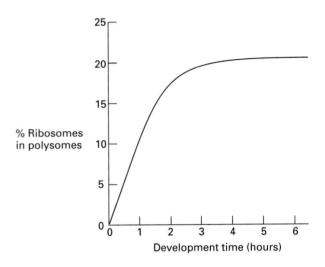

In the unfertilized egg, less than 1% of ribosomes are present in polysomes, while at 2 h post-fertilization, about 20% of ribosomes are present in polysomes. It is known that no new mRNA is made during the time period shown. How may the data be interpreted?

17.18 The mammalian genome contains about 10^5 genes. Mammals can produce about 10^6 to 10^8 different antibodies. Explain how it is possible for both of the above sentences to be true.

17.19 Define *imaginal disc, homeotic mutant,* and *transdetermination.*

*17.20 Imagine that you observed the following mutants (a through e) in *Drosophila*. Based on the characteristics given, assign each of the mutants to one of the following categories: maternal gene, segmentation gene, or homeotic gene.

a. Mutant *a*: In homozygotes phenotype is normal, except wings are oriented backwards.

b. Mutant *b*: Homozygous females are normal but produce larvae that have a head at each end and no distal ends. Homozygous males produce normal offspring (assuming the mate is not a homozygous female).

c. Mutant *c*: Homozygotes have very short abdomens, which are missing segments AB2 through AB4.

d. Mutant *d*: Affected flies have wings growing out of their heads in place of eyes.

e. Mutant *e*: Homozygotes have shortened thoracic regions and lack the second and third pair of legs.

*17.21 If actinomycin D, an antibiotic that inhibits RNA synthesis, is added to newly fertilized frog eggs, there is no significant effect on protein synthesis in the eggs. Similar experiments have shown that actinomycin D has little effect on protein synthesis in embryos up until the gastrula stage. After the gastrula stage, however, protein synthesis is significantly inhibited by actinomycin D, and the embryo does not develop any further. Interpret these results.

*17.22 It is possible to excise small pieces of early embryos of the frog, transplant them to older embryos, and follow the course of development of the transplanted material as the older embryo develops. A piece of tissue is excised from a region of the late blastula or early gastrula that would later develop into an eye and is transplanted to three different regions of an older embryo host (see a in Figure 17.A). If the tissue is transplanted to the head region of the host, it will form eye, brain, and other material characteristic of the head region. If the tissue is transplanted to other regions of the host, it will form organs and tissues characteristic of those regions in normal development (e.g., ear, kidney, etc.). In contrast, if tissue destined to be an eye is excised from a neurula and transplanted into an older embryo host to exactly the same places as used for the blastula/gastrula transplants, in every case the transplanted tissue differentiates into an eye (see b in Figure 17.A). Explain these results.

FIGURE 17.A

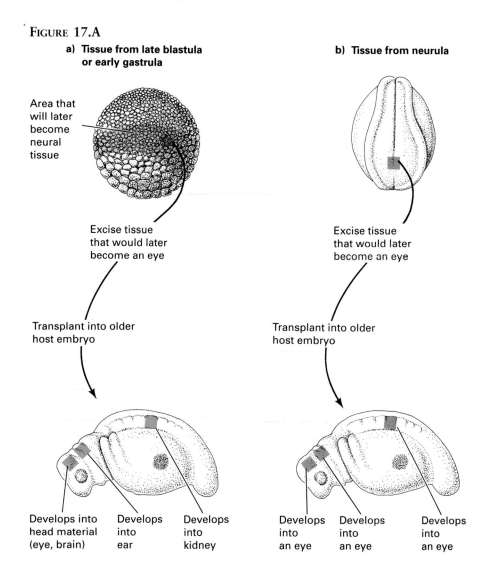

a) Tissue from late blastula or early gastrula

Area that will later become neural tissue

Excise tissue that would later become an eye

Transplant into older host embryo

Develops into head material (eye, brain)

Develops into ear

Develops into kidney

b) Tissue from neurula

Excise tissue that would later become an eye

Transplant into older host embryo

Develops into an eye

Develops into an eye

Develops into an eye

18 GENE MUTATIONS

PRINCIPAL POINTS

~ A mutation is any heritable change in the genetic material that is not caused by genetic recombination. Mutations of genes may occur spontaneously or may be induced experimentally by the application of mutagens.

~ Mutations can occur at the level of the chromosome or at the level of a gene. The former are called chromosomal aberrations; the latter are called gene mutations. Gene mutations may occur, for example, by a substitution of one base pair for another or by the addition or deletion of one or more base pairs.

~ The consequences of a gene mutation to an organism depend upon a number of factors, especially the extent to which the amino acid coding information is changed. For example, missense mutations cause the substitution of one amino acid for another, and nonsense mutations cause premature termination of synthesis of the polypeptide.

~ The effects of a gene mutation can be reversed either by reversion of the gene sequence to its original state, or by a mutation at a site distinct from that of the original mutation. The latter is called a suppressor mutation. Suppressor mutations that occur within the same gene as the original mutation are intragenic suppressors. They act either by altering a different nucleotide in the same codon affected by the original mutation, or by altering a nucleotide in a different codon. Suppressor mutations that occur in a different gene (the suppressor gene) from the original mutation are called intergenic suppressors. Often, suppression by this class of suppressors involves a tRNA with an altered anticodon.

~ Gene mutations may be caused by exposure to radiation. Radiation may cause genetic damage by breaking chromosomes, by producing chemicals that interact with DNA, or by causing unusual bonds between DNA bases. Certain kinds of genetic damage caused by radiation, but not chromosome breakage, can be corrected by repair systems that exist within the cell.

~ Gene mutations may also be caused by exposure to certain chemicals, called chemical mutagens. A variety of chemical mutagens is known, and they act in various ways. Base analogs physically replace the proper bases during DNA replication and then shift chemical form so that their base-pairing properties change. Base modifiers cause chemical changes to existing bases, thereby altering their base-pairing properties. Base-pair substitution mutations are the common result of treatment with base analogs and base modifiers. Intercalating agents are inserted between existing adjacent bases during replication resulting in single base-pair additions or deletions, that is, frameshift mutations.

~ Both prokaryotic and eukaryotic cells have a number of repair systems that deal with different kinds of DNA damage. All of the systems use enzymes to make the corrections. Some of the systems directly correct the mutational lesion while others first excise the lesion creating a single-stranded gap and then synthesize new DNA for the resulting gap. Without such repair systems mutational lesions would accumulate and ultimately be lethal to the cell or organism. Not all mutational lesions are repaired, and hence mutations do occur at relatively low frequencies.

~ Geneticists have made great progress in understanding how cellular processes take place by studying mutants that have defects in those processes. With microorganisms, a number of screening procedures enrich selectively for mutants of interest from a heterogeneous mixture of cells in a mutagenized population of cells.

Throughout the book so far, especially the first seven chapters, we used the terms *normal, wild type, variants,* and *mutants* rather loosely. And in Chapters 8–14 we learned about the *normal* functions of genes at the molecular level; specifically, what genes are, how genes are replicated, and how genes are expressed through transcription and translation. Our picture of gene functions shows these processes occurring with great accuracy and without variation. The existence of mutants, though, indicates that changes *do* occur in genes and that such changes often have such significant consequences to the organism that normal function is no longer possible (see Figure 18.1; p. 558). Furthermore, were it not for the production of variant organisms with the potential to adapt to new environments and new situations, evolution could not occur.

In this chapter and the next we will examine some mechanisms by which genetic variations can arise; they can arise through changes at the base-pair level or at the chromosomal level. (*A mutation is defined as any heritable change in the genetic material.*) Along with genetic recombination, which we studied in Chapters 5–7, these mechanisms are the principal ways by which genetic variants are produced in all organisms.

~ FIGURE 18.1

Concept of a mutation.

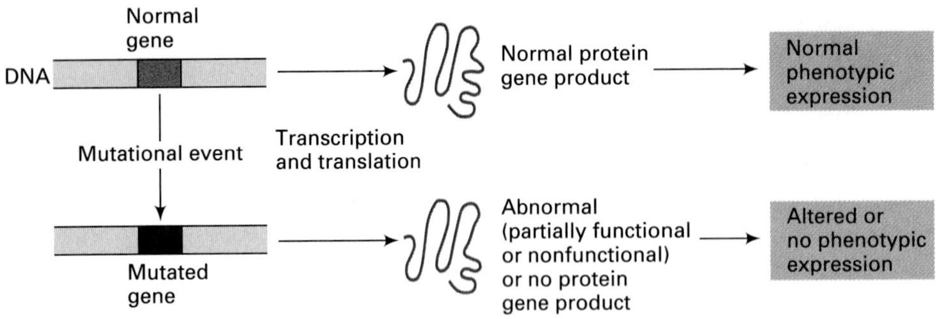

In this chapter we focus on some of the mechanisms that result in genetic changes at the gene level—changes in the base-pair sequences of genes and the phenotypic consequences of these changes. We also examine some of the methods used to select for genetic mutants, methods that are important in the geneticist's approach to understanding a process. As we have seen throughout our studies thus far, geneticists often study a number of organisms having mutations that affect a particular process in order to determine the differences between the normal and mutant organisms. These comparative investigations may take place at a number of levels, but most frequently they involve physiological comparisons and investigations of biochemical and/or molecular events, because such defects in the mutants may be defined in some detail. From the information gained from studying the mutant process, geneticists can extrapolate the function of the normal gene.

GENE MUTATIONS DEFINED

A **mutation** is any detectable and heritable change in the genetic material not caused by genetic recombination. A mutation can be transmitted to daughter cells and even to succeeding generations, thereby giving rise to mutant cells or mutant individuals. If a mutant cell gives rise only to somatic cells (in multicellular organisms), a mutant spot or area is produced, but the mutant characteristic is not passed on to the succeeding generation. This type of mutation is called a **somatic mutation.** However, mutations in the germ line of sexually reproducing organisms may be transmitted by the gametes to the next generation, giving rise to an individual with the mutant state in both its somatic and germ line cells. Such mutations are called **germ-line mutations.** Since the genetic material is usually DNA, a mutation may be the result of any detectable, unnatural change that affects DNA's chemical or physical consti-

tution, its replication, its phenotypic function, or the sequence of one or more DNA base pairs (base pairs may, for example, be added, deleted, substituted, reversed in order or inverted, or transposed to new positions).

Types of Mutations

Mutations can occur at the level of the chromosome or at the level of the gene. A change in the organization of a chromosome or chromosomes is called a **chromosomal aberration,** also called a **chromosomal mutation** (see Chapter 19). When a chromosomal mutation involves a change in the number of sets of chromosomes in the genome, the mutation is called a **genome mutation.** A mutation that occurs at the level of a gene is called a **gene mutation,** and it can involve any one of a number of alterations of the DNA sequence of the gene, including base-pair substitutions and additions or deletions of one or more base pairs. Those gene mutations that affect a single base pair of DNA are called **point mutations.**

Mutations can occur spontaneously, but they can also be induced experimentally by the application of a **mutagen,** any physical or chemical agent that significantly increases the frequency of mutational events above the spontaneous mutation rate. Mutations that result from treatment with mutagens are called **induced mutations;** naturally occurring mutations are **spontaneous mutations.** There are no qualitative differences between spontaneous and induced mutations. The manifestation of a mutant phenotype is typically the result of a change in DNA that results in the altered function of a protein.

With the above background we can now move to the definitions of several terms related to gene mutation. These terms are listed in Figure 18.2. Keep in mind that since the tertiary structure of a polypeptide is a function of its primary amino acid sequence (which is

~ **FIGURE 18.2**

Types of gene mutations.

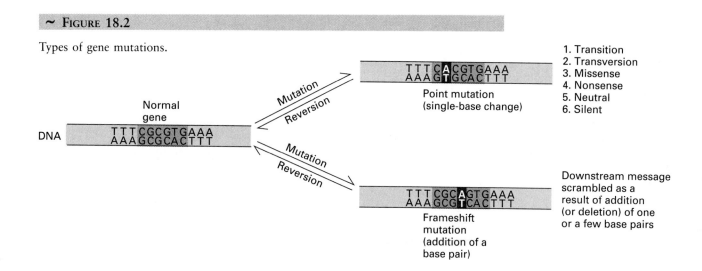

1. Transition
2. Transversion
3. Missense
4. Nonsense
5. Neutral
6. Silent

Point mutation (single-base change)

Downstream message scrambled as a result of addition (or deletion) of one or a few base pairs

Frameshift mutation (addition of a base pair)

coded for by a gene), a polypeptide synthesized by a mutant strain may be structurally different from the wild-type polypeptide. If so, the mutant strain will be partially functional, nonfunctional, or not produced.

A **base-pair substitution mutation** (point mutation) is a change in a gene such that one base pair is replaced by another base pair. For example, an AT may be replaced by a GC pair.

A **transition mutation** is a specific type of base-pair substitution mutation that involves a change in the chemical structure of the DNA from one purine-pyrimidine base pair to the other purine-pyrimidine base pair at a particular site. The four types of transition mutations are AT to GC, GC to AT, TA to CG, and CG to TA.

A **transversion mutation** is another specific type of base-pair substitution mutation that involves a change in the chemical structure of the DNA from a purine-pyrimidine base pair to a pyrimidine-purine base pair at the same site. Examples of transversion mutations are AT to TA, GC to CG, AT to CG, and GC to TA.

A **missense mutation** is a gene mutation in which a base-pair change in the DNA causes a change in an mRNA codon with the result that a different amino acid is inserted into the polypeptide in place of the one specified by the wild-type codon. A base-pair substitution mutation can result in a missense mutation. Suppose, for example, that a mutation occurs that changes a nucleotide pair, perhaps from GC to AT (a transition mutation). If this mutation occurs in the amino acid–coding region of a gene, then when the mutated gene is transcribed, the mRNA will have a different codon from the one specified in the normal or wild-type gene. If the mutated codon causes a wrong amino acid to be incorporated into the protein, the function of the protein may be impaired, and the mutant organism might function differently from a wild-type organism.

Whether a mutant phenotype is readily detectable depends on the particular amino acid substitution. In humans, for example, a particular single nucleotide pair change in the β-globin gene leads to an amino acid substitution in the β-hemoglobin chain. If the individual is homozygous for that mutation, he or she will have sickle-cell anemia. Note that a mutation may not produce a phenotypic change in those cases where the insertion of an incorrect amino acid in the polypeptide chain does not lead to an appreciable change in the function of the protein.

A **nonsense mutation** is a base-pair change in the DNA that results in the change of an mRNA codon from one that specifies an amino acid to a chain-terminating (nonsense) codon (UAG, UAA, or UGA). For example, a mutation in the DNA template strand from 3′-TTC-5′ to 3′-ATC-5′ would change the mRNA codon from 5′-AAG-3′ (lysine) to 5′-UAG-3′, which is a nonsense codon (Figure 18.3; p. 560). Because a nonsense mutation gives rise to chain termination at an incorrect place in a polypeptide, the mutation prematurely ends the polypeptide. Instead of complete polypeptides, polypeptide fragments (usually nonfunctional) are released from the ribosomes.

A **neutral mutation** is a base-pair change in the gene that changes a codon in the mRNA such that the resulting amino acid substitution produces no change in the function of the protein translated from that message. A neutral mutation occurs when the new codon codes for a different amino acid but one that is chemically equivalent to the original and hence does not affect the protein's function. An example would be a change from the codon AGG to AAG, which would substitute the amino acid lysine for the amino acid arginine. Both amino acids are basic amino acids and are sufficiently similar in properties so that the protein's function may well not be significantly diminished.

~ FIGURE 18.3

Nonsense mutation.

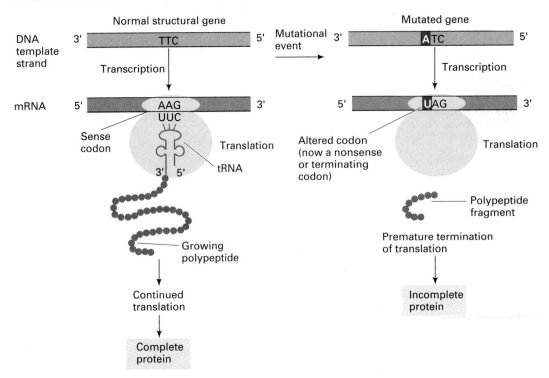

A **silent mutation** is a special case of a neutral mutation. That is, a silent mutation is a base-pair change in the gene that alters a codon in the mRNA such that the *same* amino acid is inserted in the protein. The resulting protein has wild-type function because its amino acid sequence is unchanged. An example of a silent mutation would be a change from mRNA codon AGG to AGA, both of which specify arginine.

A **frameshift mutation** results from the addition or deletion of a base pair in a gene. Such additions or deletions shift the mRNA's reading frame by one base so that incorrect amino acids are added to the polypeptide chain after the mutation site. A nonfunctional polypeptide results.

𝒦EYNOTE

A mutation is any detectable and heritable change in the genetic material that is not caused by genetic recombination. Mutations of genes may occur spontaneously or may be induced experimentally by the application of mutagens. Mutations that affect a single base pair of DNA are called base-pair substitution mutations or point mutations. ────────

Reverse Mutations and Suppressor Mutations

Point mutations generally fall into two classes in terms of their effects on the phenotype in comparison to the wild type. **Forward mutations** are mutations that occur in the direction wild-type to mutant, and **reverse mutations (or reversions)** occur in the direction mutant to wild-type.

A reversion can restore the partially functional or nonfunctional polypeptide produced by a mutant gene to its full function or to partial function. Reversion of a nonsense mutation, for instance, occurs when a base-pair change in DNA corresponds to the incorrect nonsense codon in the mRNA so that the codon again specifies a different amino acid. The change codes for either the original amino acid of the wild type (in which case the mutation is a **true reversion**) or to some other amino acid (in which case the mutation is a *partial reversion*, since complete function is unlikely to be restored). Reversion of missense mutations can occur in the same way.

The effects of a mutation may be diminished or abolished by a mutation at another site. The latter is called a **suppressor mutation**, which, by definition, is a mutation at a second site (also called a secondary or **second-site mutation**) that totally or partially restores a function lost because of a primary mutation at another site. A suppressor mutation does not result in a reversal of the original mutation; instead, it masks or compensates for the effects of the primary mutation.

There are two major classes of suppressor mutations: those which occur within the same gene as the original mutations but at a different site (called **intragenic** [intra = within] **suppressors**); and those occurring in a different gene (called **intergenic** [inter = between] **suppressors**). Both intragenic and intergenic suppressors operate to allow production of functional or partially functional copies of the protein rendered inactive by the original deleterious mutation. Thus, function can be restored *only* when both the original mutation and the suppressor mutation are present together in the same cell.

Suppressors allow the synthesis of some functional protein in different ways for the intragenic and intergenic classes. Intragenic suppressors act in one of two ways: by altering a different nucleotide in the same codon in which the original mutation occurred, or by altering a nucleotide in a different codon. Table 18.1 illustrates the first situation. Here, a three base-pair DNA sequence in the wild type specifies the mRNA codon CGU, which is read as arginine. The original (first) mutation is a CG-to-AT transversion at the first base pair resulting in the mRNA codon AGU, which specifies serine. The suppressor (second) mutation is a TA-to-AT transversion at the third position, giving the mRNA codon AGA, which is an arginine codon. Thus, with both mutations, the protein will be completely functional in cells. This type of intragenic suppressor can also correct frameshift mutations that result when an extra nucleotide is added or a nucleotide is deleted. Suppression here, for example, occurs when a nucleotide is removed from the same codon in which an earlier mutation inserted a nucleotide. In the second type of intragenic suppressor, a second mutation alters a nucleotide in a different codon. Most often, this type of

second mutation suppresses a frameshift mutation; that is, adding a nucleotide downstream of a nucleotide deletion to restore reading frame, or deleting a nucleotide downstream of an addition.

Intergenic suppression is the suppression of a mutational defect by a second mutation in another gene. Genes that cause suppression of mutations in other genes are called **suppressor genes**. Suppressor genes do not act by changing the nucleotide sequence of a mutant gene. Rather, they act by changing the way the mRNA is read. A number of different suppressor genes have been characterized in several systems, particularly *E. coli* and yeast. Each suppressor gene can suppress the effects of only one nonsense, missense, or frameshift mutation; hence, suppressor genes can suppress only a small proportion of the point mutations that theoretically can occur within a gene. On the other hand, a given suppressor gene will suppress *all* mutations for which it is specific, whatever gene the mutation is in.

As mentioned above, intergenic suppressors act on nonsense, missense, and frameshift mutations. The suppressors of nonsense mutations have been particularly well characterized. Such suppressors often result when particular tRNA genes are mutated so that (in contrast to what occurs with wild-type tRNAs) their anticodons recognize a chain-terminating codon and put an amino acid into the chain. Thus instead of polypeptide chain synthesis being stopped prematurely, the altered tRNA inserts an amino acid at that position, and that amino acid may restore full or partial function to the polypeptide. Since there are three types of nonsense codons, there are three classes of nonsense suppressors: one for UAG, one for UAA, and one for UGA. If, for example, a gene for tRNA.Tyr (which has the anticodon 5'-GUA-3') is mutated so that the tRNA has the anticodon 5'-CUA-3', the mutated tRNA (which still has Tyr attached) will read the nonsense codon 5'-UAG-3' (Figure 18.4; p. 562). But instead of terminating the chain, the mRNA will insert its tyrosine at that point in the chain. The extent to which the now-completed protein will be functional will then depend on the effects of the inserted tyrosine in the protein. If it is an important part of the protein, for example its active site, then the incorrect amino acid may not restore function to a significant degree. If it is in a less crucial area, the protein may have a significant degree of function.

If we have changed this particular class of tRNA.Tyr so that its anticodon can now read a nonsense codon, it cannot read the original codon that specifies the amino acid it carries. Thus nonsense suppressor tRNAs are typically produced by mutation of tRNA genes that are redundant in the genome; that is, for which several different genes all specifying the same

~ TABLE 18.1

Example of an Intragenic Suppressor Mutation Altering a Nucleotide in the Same Codon as the Original Mutation

	DNA SEQUENCE	MRNA CODON SPECIFIED	AMINO ACID CODED FOR
Wild-type condition	5'-CGT-3' 3'-GCA-5'	5'-CGU-3'	Arg
Original mutation	5'-AGT-3' 3'-TCA-5'	5'-AGU-3'	Ser
After suppressor mutation	5'-AGA-3' 3'-TCT-5'	5'-AGA-3'	Arg

~ FIGURE 18.4

Mechanism for action of an intergenic nonsense suppressor mutation that results from mutation of a tRNA gene. In this example, a tRNA.Tyr gene has mutated so that the tRNA's anticodon is changed 5'-GUA-3' to 5'-CUA-3', which can read a UAG nonsense codon, inserting tyrosine in the polypeptide chain at that codon.

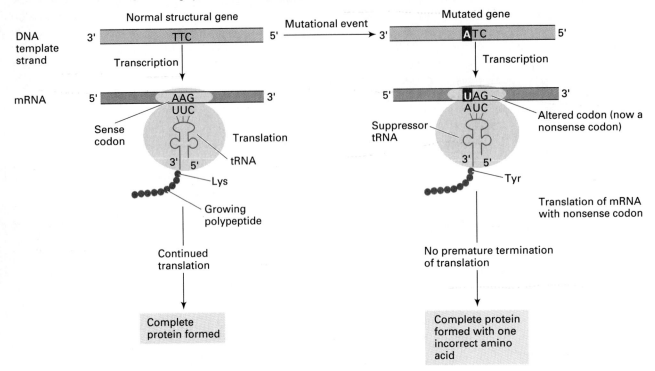

tRNA base sequence exist. Therefore if there is a mutation in one of the redundant genes (usually one that codes for minor amounts of tRNA) so that the tRNA will read, say, UAG, the other genes specifying the same tRNA produce a tRNA molecule that will read the normal Tyr codon.

We can view nonsense suppression as a competition between the normal release factor binding to the nonsense codon and the suppressor tRNA. UAG and UGA suppressor tRNAs do well in this competition, since they succeed in reading over one-half of the nonsense codons. By contrast, for reasons that are not clear, UAA suppressor tRNAs are only one to five percent efficient. Given these figures, one might suppose that suppression of the *normal* chain-terminating codons—which are frequently UAG and UGA—would also occur, producing longer-than-normal proteins. Such proteins are known as *read-through proteins*. However, for reasons that are not clear, UAG and UGA suppressor tRNAs do not produce large numbers of read-through proteins.

𝒦EYNOTE

A suppressor mutation is a mutation at a second site that completely or partially restores a function lost because of a primary mutation at another site.

Intragenic suppressors are suppressor mutations that occur within the same gene as the original mutation but at a different site. They act either by altering a different nucleotide in the same codon as that in which the original mutation occurred, or by altering a nucleotide in a different codon. Intergenic suppressors are suppressor mutations that occur in a different gene (called a suppressor gene) than that in which the original mutation occurred. Intergenic suppressors are known for nonsense, missense, and frameshift mutations. Typically, suppression involves a tRNA with an altered anticodon. Thus, nonsense mutations are suppressed by mutated tRNAs which now can read the chain-terminating codon and insert an amino acid at the mutation site. _____

CAUSES OF MUTATION

Now that we have studied the various types of gene mutations, we are ready to discuss how mutations occur. The two general processes we will consider are spontaneous mutations and induced mutations.

Spontaneous Mutations

Spontaneous mutations are mutations that occur in nature, without the apparent intervention of humans. All types of point mutations described in the previous section occur spontaneously. For a long while geneticists thought that spontaneous mutations were produced by mutagens indigenous to the environment, such as radiation and chemicals. However, evidence indicates that although the rate at which spontaneous mutations appear is extremely low, that rate is nonetheless too high to be accounted for by indigenous mutagens alone.

There are attempts to quantify the occurrence of mutations. Two different terms are often used. **Mutation rate** presents the probability of a particular kind of mutation as a function of time, e.g., number per nucleotide pair per generation, or number per gene per generation. **Mutation frequency,** on the other hand, is the number of occurrences of a particular kind of mutation in a population of cells or individuals, e.g., number per 100,000 organisms or number per one million gametes.

In *Drosophila*, for example, the spontaneous mutation rate for individual genes is about 10^{-4} to 10^{-5} per gene per generation. In humans the rate varies between 10^{-4} and 4×10^{-6} per gene per generation. For eukaryotes in general, the spontaneous mutation rate is 10^{-4} to 10^{-6} per gene per generation and for bacteria and phages, the rate is 10^{-5} to 10^{-7} per gene per generation. The spontaneous mutation frequencies at specific loci for various organisms are presented in Table 23.5. That the spontaneous mutation rate is affected by the genetic constitution of the organism is shown by the fact that male and female *Drosophila* of the same strain have identical mutation rates, while different strains may exhibit different mutation rates.

Spontaneous mutations can result from any one of a number of events, including errors in DNA replication and spontaneous chemical changes in DNA. Spontaneous mutations can also result from the movement of what are called transposable genetic elements, which will be discussed in Chapter 20.

DNA REPLICATION ERRORS. A mismatched base pair such as A–C might form during DNA replication. When the DNA with the A–C mismatch replicates again, one progeny helix will result with a GC base pair in place of the original AT base pair; that is, a transition mutation is produced. The other progeny helix will still have an AT base pair in that position. Because the bases themselves are able to exist in alternate chemical forms, called **tautomers,** erroneous base pairs form. When each base is in its rare form, different hydrogen bondings become possible and lead to mismatched base pairs (Figure 18.5). As an example, a rare form of cytosine can pair with adenine (Figure 18.5a), and a rare form of

---

~ FIGURE 18.5

Examples of mismatched bases in DNA. (a) Mismatched bases resulting from rare forms of pyrimidines; (b) mismatched bases resulting from rare forms of purines.

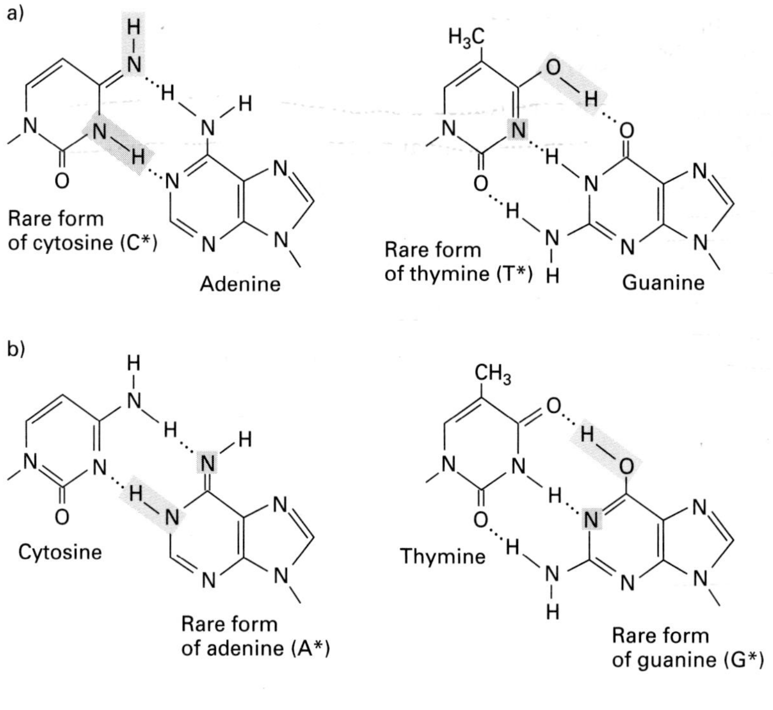

~ FIGURE 18.6

Production of a mutation as a result of a tautomeric shift in a DNA base. A guanine on the top strand (a) shifts to its rare form (*G) during DNA replication (b). In its rare form, thymine on the new strand pairs with it. (c) and (d) The guanine shifts back to its more stable normal form during the next DNA replication. The mismatched thymine specifies an adenine on the new strand and the overall result is a GC-to-AT transition mutation for the DNA strand lineage complementary to that in which the tautomeric shift occurred. The other DNA stand lineages produce no mutations.

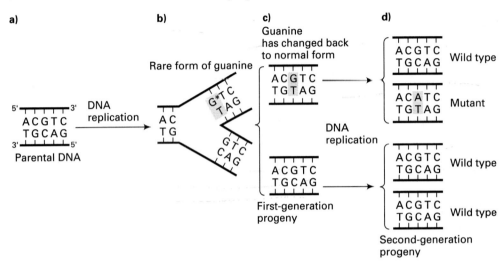

guanine can pair with thymine (Figure 18.5b). The change in the chemical form of a base is called a **tautomeric shift.** Figure 18.6 illustrates how a GC to AT transition mutation can be produced as a result of a tautomeric shift. Note that many more mutations would result from tautomeric shifts than are actually observed if it were not for the proofreading activity of DNA polymerases, which recognizes many of the mismatches, excises them, and replaces them with the correct base pair (see Chapter 11).

Small additions and deletions can occur spontaneously during replication (Figure 18.7). The mechanism for introducing such errors involves erroneous looping out of bases from one or the other DNA strand. If DNA loops out from the template strand, a deletion mutation results; if DNA loops out from the strand being synthesized, an addition mutation results. Such additions or deletions in the DNA, unless they happen in multiples of three, will generate frameshift mutations.

SPONTANEOUS CHEMICAL CHANGES. Two of the most common chemical events that occur to produce spontaneous mutations are *depurination* and *deamination* of particular bases. In depurination a purine, either adenine or guanine, is removed from the DNA when the bond breaks between the purine and the deoxyribose (Figure 18.8). Thousands of purines are lost by depurination in a typical generation time of a mamma-

lian cell in tissue culture. If such lesions are not repaired, there is no base to specify a complementary base during DNA replication. Instead, a randomly chosen base is inserted, which may produce a mismatched base pair. The overall result may be a genetic mutation, as described in the previous section.

Deamination is the removal of an amino group from a base. As an example, a susceptible base for this event is cytosine, the deamination of which produces uracil (Figure 18.9; p. 566). Uracil is not a normal base in DNA. A repair system described later in this chapter removes most of the uracils produced by deamination of cytosine, thereby minimizing the mutational consequences of this event. However, if the uracil is not repaired (see later), it will direct that an adenine be incorporated in the new DNA strand during replication, ultimately resulting in the conversion of a CG base pair to a TA base pair, that is, a transition mutation. Later discussion of the deaminating effects of nitrous acid presents other examples of mutations resulting from deamination. DNA of both prokaryotes and eukaryotes contains relatively small amounts of the modified base 5-methylcytosine (5mC) (Figure 18.10 [p. 566]; see also Figure 17.5) in place of the normal base cytosine. 5mC is also subject to deamination, the product of which is thymine (Figure 18.10). Since thymine is a normal nucleotide in DNA, there are no repair mechanisms that can detect and correct such

~ FIGURE 18.7

Spontaneous generation of addition and deletion mutants by DNA looping-out errors during replication.

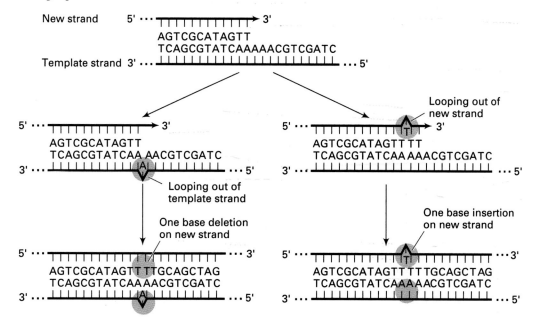

~ FIGURE 18.8

Depurination (loss of a purine) from a single strand of DNA. The sugar-phosphate backbone is unbroken.

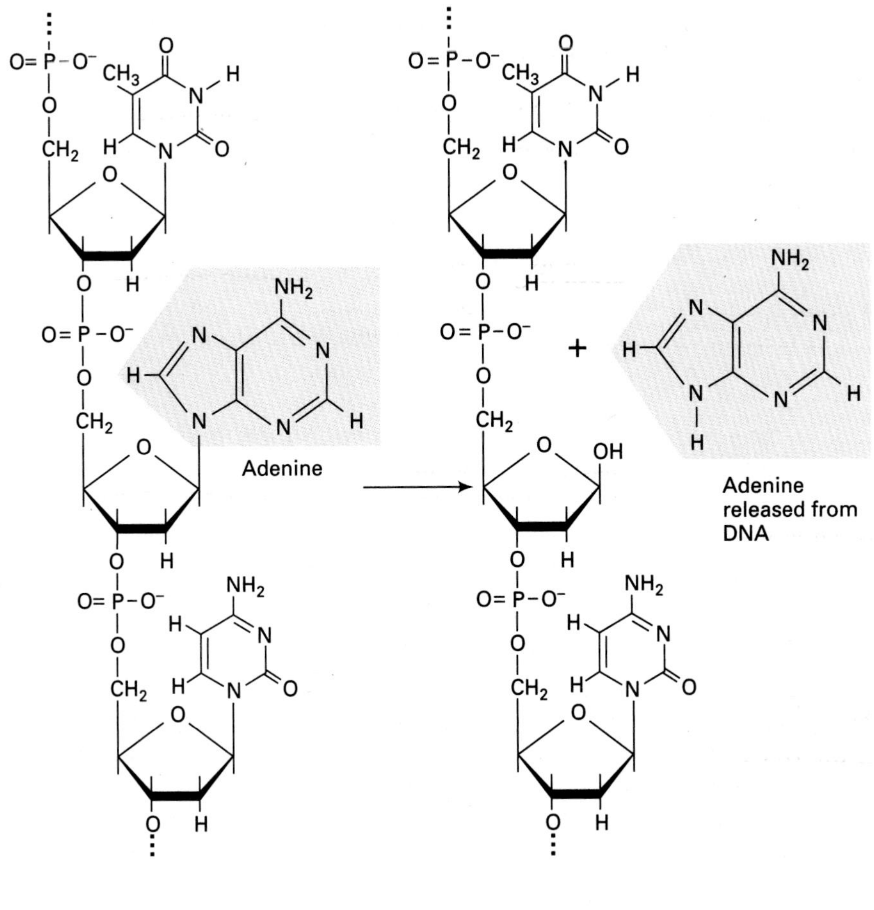

~ FIGURE 18.9

Deamination of cytosine to uracil.

Cytosine   **Uracil**

~ FIGURE 18.10

Deamination of 5-methylcytosine (5mC) to thymine.

5-methylcytosine (5mC)   **Thymine (T)**

mutations. As a consequence, deamination of 5mC results in 5mCG to TA transitions. Because significant proportions of other kinds of mutations are corrected by repair mechanisms but 5mC deamination mutations are not, locations of 5mC in the genome often appear as *mutational hot spots*, that is, nucleotides where a higher than average frequency of mutation occurs (Figure 18.11).

Induced Mutations

Since the rate of spontaneous mutation is so low, genetic researchers generally use mutagens to enhance mutation frequency so that a significant number of organisms have mutations in the gene being studied. Two classes of mutagens are used—radiation and chemical—both of which involve specific mechanisms of action.

RADIATION. The part of the electromagnetic spectrum with wavelengths shorter than visible light (i.e.,

less than 1 μm) consists of non-ionizing radiation and ionizing radiation. The former is exemplified by ultraviolet light, and the latter includes X-rays, gamma rays, and cosmic rays. Because energy increases as wavelength decreases, these three types of radiation have higher energy than visible light, with ionizing radiation having greater energy than non-ionizing radiation, and gamma rays having greater energy than X-rays, and so on. As a result of its energy, ionizing radiation can penetrate tissues, hence the use of X-rays as a diagnostic tool. High-energy irradiation also produces ions by colliding with atoms and releasing electrons, which in turn collide with other atoms, releasing more electrons, and so on. Therefore, along the track of each high-energy ray a string of ions is formed, which can initiate a number of chemical reactions, including mutations. Both gene mutation and chromosome breakage can be induced by ionizing radiation.

Ultraviolet light (UV) rays are non-ionizing; that is, they have insufficient energy to induce ionizations.

~ FIGURE 18.11

Distribution of spontaneous GC to AT transition mutations to stop codons in the *lac* repressor gene of *E. coli*, showing the hot spots for mutation at 5mC nucleotides.

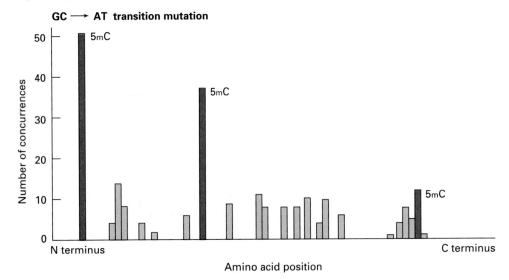

Nonetheless, ultraviolet light is a useful mutagen, and at high enough doses it can kill cells. For geneticists, other scientists and medical personnel, this property is a useful one: UV light is used as a sterilizing agent in some applications. Ultraviolet light causes mutations because the purine and pyrimidine bases in DNA absorb light very strongly if it has a wavelength of 254–260 nm, that is, in the ultraviolet range. At this wavelength UV light induces gene mutations primarily by causing photochemical (light-induced chemical) changes in the DNA. This mutagenic activity is attenuated to some extent because UV radiation has lower-energy wavelengths than X-rays and, hence, has very limited penetrating power.

Our sun is a very powerful source of UV radiation, but much of the UV light is screened out by the atmospheric layers. Nonetheless, significant amounts of UV radiation are present in sunlight, as evidenced by the tanning (or burning) of humans who sunbathe.

One of the effects of UV radiation on DNA is the formation of abnormal chemical bonds between adjacent pyrimidine molecules in the same strand or between pyrimidines on the opposite strands of the double helix. This bonding is induced mostly between adjacent thymines on the same or opposite strands of the DNA, forming what are called *thymine dimers* (Figure 18.12), usually designated T̂T. This unusual pairing disrupts the normal pairing of the Ts with the corresponding As on the opposite strand, causes a bulge in the DNA strand, and weakens the hydrogen bonds between T and A on opposite strands.

Usually mutations are produced by UV radiation because of the persistence of the thymine dimers or their imperfect repair. The thymine dimers, however, can be repaired completely by specific cellular systems (see p. 576). Generating viable mutants requires a special host enzyme system that has been characterized in bacteria, and is presumed to exist in eukaryotic cells. This system, called the *SOS repair system* (also *error*

prone repair), is induced as an emergency response to prevent cell death when significant DNA damage has been incurred. The action of the SOS system is not confined to UV-induced damage; it is also activated after different kinds of DNA damage.

The following describes how SOS repair works. When a replication fork encounters significant DNA damage, such as a thymine dimer induced by UV light, the fork stops and reinitiates DNA synthesis some distance away. This leaves a gap in the new DNA strand to which a specific protein called the RecA protein (encoded by the *recA* gene) binds. In other circumstances the RecA protein functions in recombinational repair; but when it binds to single-stranded DNA (i.e., in the gap), a different function of the RecA protein is activated. The RecA protein destroys a repressor that prevents expression of the SOS genes, a group of about 15 genes responsible for the SOS system. As a result, the SOS genes are activated, and the gap is repaired through a variety of mechanisms. The key to producing mutations is that the SOS repair system facilitates the insertion of a base in the new DNA strand, even though no template base exists. This process, normally prevented in DNA synthesis, has been named *bypass synthesis*. Since a random base will be inserted, both transition and transversion mutagens may result from SOS-induced bypass synthesis.

𝒦EYNOTE

Mutations may be induced by the use of radiation or chemical mutagens. Radiation may cause genetic damage by breaking chromosomes, by producing chemicals that affect the DNA (as in the case of X-rays), or by causing the formation of unusual bonds between DNA bases, such as thymine dimers (as in the case of ultraviolet light). If DNA is dam-

~ FIGURE 18.12

Production of thymine dimers by ultraviolet light irradiation. The two components of the dimer are covalently linked in such a way that the DNA double helix is distorted at that position.

aged significantly (e.g., by UV light), mutations are produced as a result of the activation of SOS genes, the function of which is to permit the insertion of bases in new DNA where no base exists on the template strand. The SOS process is called bypass synthesis or error-prone repair. If the damage is not repaired, it may be lethal to the cell. ────────

CHEMICAL MUTAGENS

Base analogs. Base analogs are chemicals that have molecular structures that are extremely similar to the bases normally found in DNA. The base analog 5-bromouracil (5BU), for example, has a bromine residue instead of the methyl group of thymine. These base analogs require DNA replication for their incorporation, and because they often can form base pairs with more than one of the usual bases, mutations can result. We shall consider two commonly used base-analog mutagens, 5-bromouracil and 2-aminopurine.

The mutagen 5-bromouracil exists in two states. In the normal state it resembles thymine and will pair only with adenine in DNA (Figure 18.13a). In its alternate, rarer state it pairs only with guanine (Figure 18.13b). Mutagen 5BU induces mutations by switching between

---

~ FIGURE 18.13

Mutagenic effects of the base analog 5-bromouracil (5BU). (a) In its normal state 5BU pairs with adenine; (b) In its rare state, 5BU (indicated by highlighted 5BU) pairs with guanine; (c) The two possible mutation mechanisms. 5BU induces transition mutations when it incorporates into DNA in one state, then shifts to its alternate state during the next round of DNA replication.

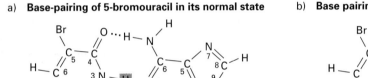

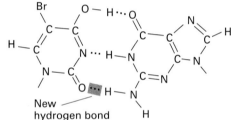

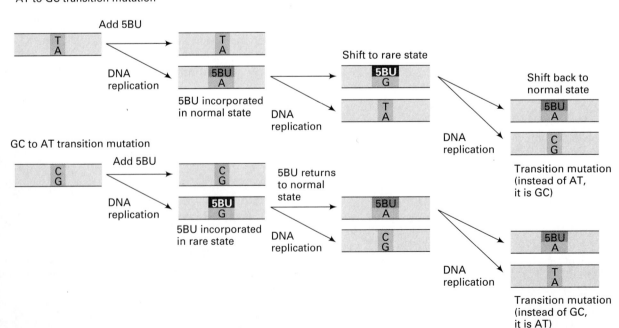

the two forms once the base analog has been incorporated into the DNA (Figure 18.13c). This is essentially similar to the way spontaneous mutations occur by DNA replication errors. The difference here is that base analogs exist much more frequently in the rare state than do normal bases. Hence, the frequency of mutations induced by base analogs is much higher than the frequency of spontaneous mutations. If, for example, 5BU is incorporated in its normal state, it pairs with adenine. If it changes into its rare state during replication, then guanine instead of adenine will be put in as the complementary base pair. In the next round of replication the G–5BU base pair will be resolved into a G–C base pair instead of the A–T base pair. By this process a transition mutation is produced, in this case from AT to GC. The 5BU will induce a mutation from GC to AT if it is first incorporated into DNA in its rare state and then switches to the normal state during replication. Thus 5BU-induced mutations can be reverted by a second treatment of 5BU.

The base analog 2-aminopurine (2AP) acts as a mutagen in essentially the same way as 5BU does, and like 5BU, 2AP exists in normal and rare states (Figure 18.14a and b). In its normal state 2AP resembles adenine but has an amino group at a different position on the purine ring than does adenine; in its normal state it base-pairs with thymine. In its rare state 2AP resembles guanine and base-pairs with cytosine. As with 5BU, 2AP induces transition mutations, which can be reverted by a second application of 2AP. The mutation will be either AT to GC or GC to AT, depending on its state during its initial incorporation into the DNA and upon its state during replication. Note that since the mutations involved are transitions in both cases, 5BU can revert mutations induced by 2AP and vice versa.

Finally, one of the approved drugs given to AIDS patients, AZT (azidothymidine), is a base analog. It works as follows. The AIDS virus, HIV-I (human immunodeficiency virus-I) is a retrovirus. That is, its genome is RNA, but when the virus enters a cell, the viral enzyme reverse transcriptase (see Chapter 15) makes a DNA copy of the RNA (i.e., a cDNA). That DNA can then be incorporated into the cell's genomic DNA, from where it can direct new viral synthesis. AZT can be incorporated into DNA as an analog of thymidine. AZT (as a triphosphate derivative) is a substrate for reverse transcriptase in the viral RNA → DNA step. However, AZT is *not* a good substrate for cellular DNA polymerases. Thus, AZT acts as a selective poison by inhibiting the production of the viral cDNA, therefore blocking new viral synthesis, which would derive from integrated viral DNA copies.

BASE-MODIFYING AGENTS. Unlike base analogs, which actually replace bases in the DNA and require DNA replication for their incorporation, a number of chemicals act as mutagens by directly modifying the chemical structure and properties of the bases. Figure 18.15 (p. 570) shows the action of three types of mutagens that work in this way: a deaminating agent, a hydroxylating agent, and an alkylating agent.

Nitrous acid, HNO_2 (Figure 18.15a), is a deaminating agent that removes amino groups ($-NH_2$) from the bases guanine, cytosine, and adenine. Treatment of guanine with nitrous acid produces xanthine, but since this purine base has the same pairing properties as guanine, only a neutral mutation results (Figure 18.15a, part 1). However, when cytosine is treated with nitrous acid, uracil (which pairs with adenine) is produced (Figure 18.15a, part 2). The deamination of

~ **FIGURE 18.14**

Base-pairing properties of the normal and rare states of the base-analog mutagen 2-aminopurine (2AP). (a) In its normal state 2AP pairs with thymine; (b) In its rare state 2AP pairs with cytosine.

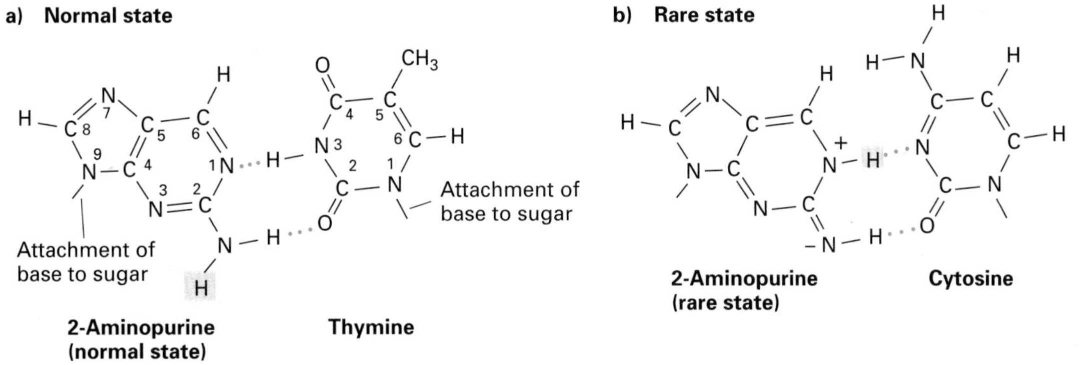

~ FIGURE 18.15

Action of three base-modifying agents. (a) Nitrous acid (HNO$_2$) modifies guanine, cytosine, and adenine. The cytosine and adenine modifications result in mutations, while the guanine modification does not. (b) Hydroxylamine (NH$_2$OH) reacts only with cytosine. (c) Methylmethane sulfonate (MMS), an alkylating agent, alkylates guanine and thymine.

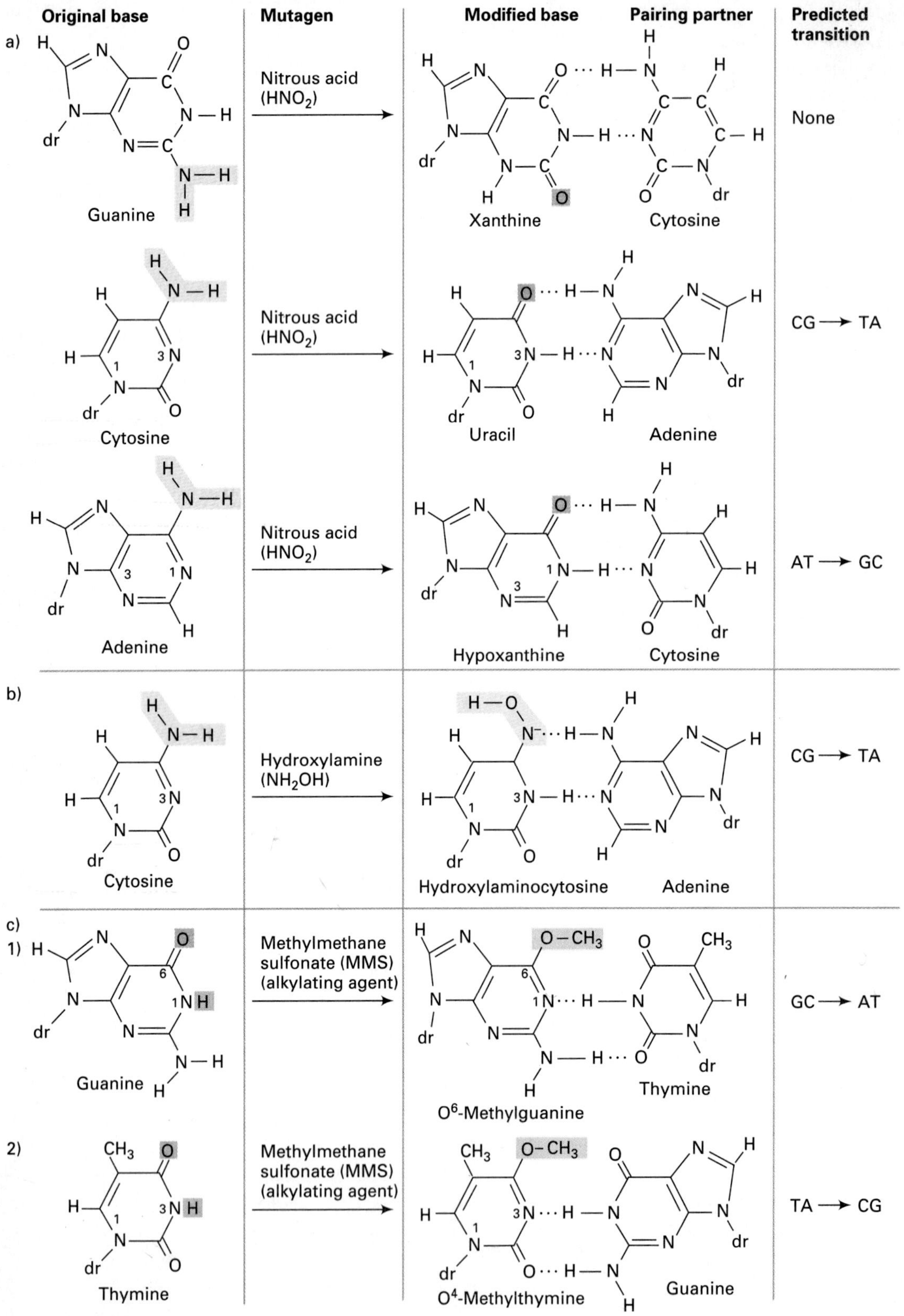

cytosine by nitrous acid, then, produces a CG-to-TA transition mutation during replication. Likewise, nitrous acid modifies adenine to produce hypoxanthine, a base that pairs with cytosine rather than thymine, thus resulting in an AT-to-GC transition mutation (Figure 18.15a, part 3). A nitrous acid-induced mutation can be reverted by a second treatment with nitrous acid.

Another base-modifying mutagen is hydroxylamine, NH_2OH (Figure 18.15b), which reacts specifically with cytosine, modifying it by adding a hydroxyl group (OH) so that it can pair only with adenine instead of with guanine. The mutations induced by hydroxylamine are, therefore, CG-to-TA transitions. Since only a one-way transition can occur with this mutagen, hydroxylamine-induced mutations cannot be reversed by a second treatment with this chemical. They can be reverted, however, by treatment with other mutagens (such as 5BU, 2AP, and nitrous acid) that cause TA-to-CG transition mutations.

The last base-modifying agent we will consider is methylmethane sulfonate, MMS (Figure 18.15c). This agent is one of a diverse group of alkylating agents that introduce alkyl groups (e.g., $-CH_3$, $-CH_2CH_3$) onto the bases at a number of places. An alkylating agent alters one base so that it pairs with the wrong base. As shown in Figure 18.15c, after treatment with MMS, the now methylated guanine pairs with thymine, giving GC-to-AT transitions (Figure 18.15c, part 1), and the methylated thymine pairs with guanine, giving TA-to-CG transitions (Figure 18.15c, part 2).

Intercalating agents. The intercalating mutagens (including proflavin, acridine, ethidium bromide, dioxin and ICR-170) act by inserting themselves (*intercalating*) between adjacent bases in one or both strands of the DNA double helix (Figure 18.16). The chemical structures of proflavin and acridine orange are shown in Figure 18.16a.

If the intercalating agent inserts between adjacent base pairs of the DNA strand that is the template for new DNA synthesis (Figure 18.16b), then an extra base (chosen at random: G in the figure) must be inserted in the new DNA strand opposite the intercalating agent. After one more round of replication, accompanied by loss of the intercalating agent, the overall result is a frameshift mutation due to the insertion of one base pair (CG in Figure 18.16b). If, in place of a base, the intercalating agent inserts into the new DNA strand (Figure 18.16c), then when that DNA double helix replicates after the intercalating agent is lost, the result is a frameshift mutation due to the deletion of one base pair (TA in Figure 18.16c).

~ **FIGURE 18.16**

Intercalating mutations. (a) Structures of representative intercalating agents, proflavin and acridine orange: (b) Frameshift mutation by addition, when agent inserts into template strand; (c) Frameshift mutation by deletion, when agent inserts into newly synthesizing strand.

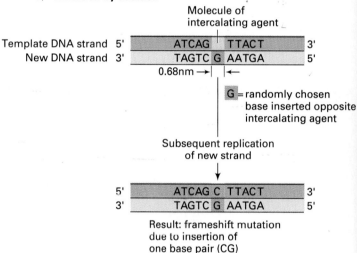

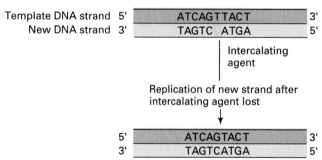

So, following the insertion of an intercalating agent into a strand of DNA, the consequence is that when the DNA replicates, a base is either added or deleted, resulting in a frameshift mutation. As a consequence of the frameshift mutation, all the amino acids past the mutation point are likely to be incorrect, and the resulting protein will most likely be nonfunctional. Frameshift mutations induced by intercalating agents can be reverted by treatment with these agents.

KEYNOTE

Chemical mutagens act in a variety of ways. Base analogs physically replace the proper bases during DNA replication, then shift form so that their base-pairing capabilities change. Base-modifying agents cause chemical changes in existing bases, thereby altering their base-pairing properties. Base analogs and base-modifying agents result in base-pair substitution mutations. Intercalating agents are inserted between existing adjacent bases during replication, causing single base-pair additions or deletions, depending on whether the insertion is in the parental or new strand. The result is a frameshift mutation.

SITE-SPECIFIC *IN VITRO* MUTAGENESIS OF DNA.

Spontaneous and induced mutations do not occur in specific genes; rather they are scattered essentially randomly throughout the genome. However, most geneticists wish to study the effects of mutations in genes in which they are particularly interested. After mutagenizing cells or organisms with radiation or chemical mutagens, the surviving population must be screened to identify those with the mutation(s) of interest. (Some of these screening methods are described later in this chapter.) With the advent of recombinant DNA technology, it is possible to clone genes and produce large amounts of that DNA for analysis and manipulation. That means it is now possible to alter (that is, to mutate) DNA in the test tube, reinsert the DNA back into the cell by infection or transformation, and examine the effects of the mutation. Such techniques enable geneticists to make mutations at specific nucleotide positions within the gene. This procedure is called *site-specific in vitro mutagenesis of DNA*.

Site-specific mutagenesis can be used to create point mutations or small deletions or additions. Figure 18.17 diagrams how a point mutation can be created. The starting point is a clone of a gene or piece of gene. The clone is denatured to single strands, then an oligonucleotide (a short, synthetic segment of DNA) is annealed to one of the strands. The oligonucleotide is synthesized to be complementary to part of the cloned DNA, except where the point mutation is desired. In the example, we wish to change a GC base pair to an AT base pair. Therefore, the oligonucleotide has the sequence TAGACTAT. The underlined A is the mutant site, an incorrect partner for the C on the single-stranded cloned molecule. Once the oligonucleotide has

~ FIGURE 18.17

Site-specific *in vitro* mutagenesis of DNA to introduce a point mutation at a specific site.

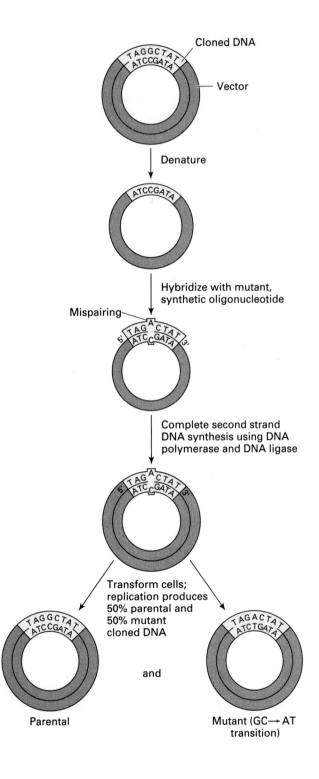

annealed, it serves as a primer for DNA polymerase to make a complete complementary copy of the circular template strand. DNA ligase then catalyzes the circularization of the new strand. At this point we have a circular, double-stranded molecule with a mismatched AC base pair at one point. The DNA is introduced into cells by transformation. When the DNA replicates, two types of molecules are produced: one is the original recombinant molecule with GC at the target site, and the other is the mutant with AT at the target site. This procedure can be used to generate single base-pair deletions or additions, using synthetic oligonucleotides with one fewer or one more base than the complementary cloned sequence.

With this approach, a researcher can, for example, mutate a gene of unknown function, introduce the mutant gene into a cell and investigate the resulting phenotypic changes, thereby gaining insight into the normal gene's function. This type of experimental approach is called reverse genetics. That is, while genetics traditionally has worked from a phenotype back to the gene responsible for the phenotype, reverse genetics works forward from the gene to the function that gene specifies. Site-specific mutagenesis has also been used to generate specific changes in key regulatory sequences to investigate which base pairs are important and which are not. Our understanding of promoter and enhancer elements has come, in part, from site-specific mutagenesis of sequences upstream or downstream from a gene.

THE AMES TEST: A SCREEN FOR POTENTIAL MUTAGENS AND CARCINOGENS

All the mutagens (radiation and chemicals) we have examined so far are those commonly used to induce mutations during genetic research. Presently, scientists are aware that many chemicals in our environment might have mutagenic effects. For example, about one-half of the natural chemicals tested in laboratory animals were carcinogens, that is they caused cancer. (Natural chemicals are those that are normally indigenous in such things as food.) About the same proportion of synthetic chemicals was shown to be carcinogenic in laboratory animals. (Synthetic chemicals include pesticides, drugs, and food additives.) Nonetheless, recent data indicate that mitogenesis (induced cell division) plays a dominant role in carcinogenesis. Many carcinogens, for instance, are *not* mutagens. Instead, carcinogens induce mitogenesis.

Several methods are used to test new or old environmental chemicals for carcinogenic effects. One method is to test a chemical directly for carcinogenicity by setting up carefully controlled experiments with inbred lines of mice or rats. The inbred lines are genetically identical so that a statistical analysis of treated versus sham-treated animals (i.e., animals treated in exactly the same way as the experimental animals but without use of the potential carcinogen) for the incidence of cancers provides information about the probability that a chemical is carcinogenic. Although these studies are expensive and time-consuming, they are done routinely by research laboratories, often under contract to federal agencies or to private companies.

In the early 1970s a more rapid test was developed by Bruce Ames. The Ames test uses the bacterium *Salmonella typhimurium* as the test organism. Two strains are used, both of which are auxotrophic for histidine. [An auxotrophic mutation affects an organism's ability to make a particular molecule which is essential for growth. Histidine (*his*) auxotrophs, then, require the presence of histidine in the growth medium in order to grow; normal (*his*+) individuals do not.] In one *his* strain the *his* mutation can be reverted to *his*+ by a base-pair substitution, and in the other strain the *his* mutation can be reverted by a frameshift mutation. (These strains also carry other mutations that make them particularly suitable for the experimental manipulations used, but a discussion of these features is beyond the scope of this text.)

The procedure for the Ames test is summarized in Figure 18.18 (p. 574). Rat livers are obtained, homogenized, and centrifuged to sediment the cellular debris. The enzymes of the rat liver are collected from the supernatant and are added to a liquid culture of the auxotrophic *Salmonella typhimurium*, along with the chemical being tested. At the same time a control experiment is run, in which no potential carcinogen is present.

Enzymes from the rat liver are used in these experiments because in the living organism, as in humans, the liver enzymes detoxify and, in certain cases, as will become evident, actually toxify various chemicals, including many potential mutagens. Their presence allows the investigators to determine whether a chemical that by itself is not mutagenic can become mutagenic when processed in the liver.

The mixture is then plated onto a medium containing no histidine. Revertants of a mutant *Salmonella typhimurium* strain containing the base-pair substitution mutation or of a strain containing a frameshift mutation are sought. The control mixture is plated in the same way.

~ FIGURE 18.18

Ames test for potential mutagens.

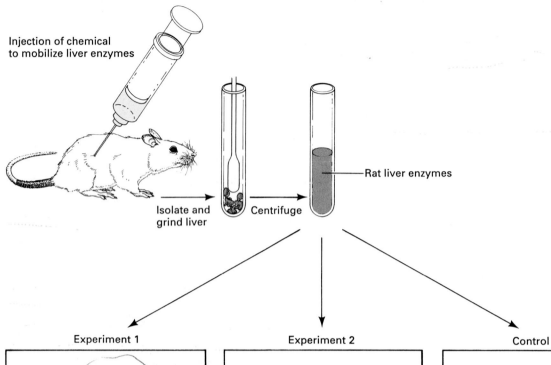

Injection of chemical to mobilize liver enzymes

Isolate and grind liver

Centrifuge

Rat liver enzymes

Experiment 1

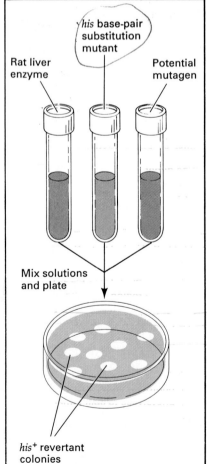

Rat liver enzyme

his base-pair substitution mutant

Potential mutagen

Mix solutions and plate

his+ revertant colonies

Experiment 2

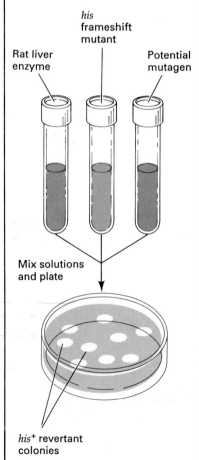

Rat liver enzyme

his frameshift mutant

Potential mutagen

Mix solutions and plate

his+ revertant colonies

Control

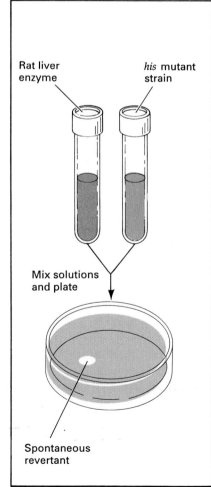

Rat liver enzyme

his mutant strain

Mix solutions and plate

Spontaneous revertant

Conclusion: chemical solution is a mutagen

The *his*⁺ revertants are detected, since they will produce colonies on the histidine-lacking plates. If there are significantly more *his*⁺ revertants on the plates spread with the chemical-treated mixtures than on the control plate, the chemical is a mutagen. That is, the number of colonies on the control plate indicates the spontaneous reversion rate of the test bacteria. If more colonies are seen on the experimental plates, it indicates that the chemical induced mutations. Whether or not the chemical is a carcinogen cannot be determined directly by this test. To date, the Ames test has identified a large number of mutagens among environmental chemicals, such as hair dye additives, vinyl chloride, and particular food colorings, and among natural compounds.

DNA REPAIR MECHANISMS

Both prokaryotic and eukaryotic cells have a number of repair systems to deal with damage to DNA (e.g., the SOS system). All of the systems use enzymes to make the correction. Some of the systems directly correct the mutational lesion while others first excise the lesion creating a single-stranded gap and then synthesize new DNA for the resulting gap.

Direct Correction of Mutational Lesions

REPAIR BY DNA POLYMERASE PROOFREADING.
In bacterial genes, the frequency of base-pair substitutions varies from 10^{-7} to 10^{-11} errors per replication event. On the other hand, DNA polymerase makes errors in inserting nucleotides while it is synthesizing the new DNA strand perhaps at a frequency of 1 in 10. The discrepancy between the initial base substitution frequency by DNA polymerase and the observed mutation rate is accounted for by proofreading activity of the polymerase itself (see Chapter 11). That is, bacterial DNA polymerases, in addition to having 5′-to-3′ polymerizing activity, also have a 3′-to-5′ exonuclease activity. The latter is necessary to ensure accurate DNA replication. When an incorrect nucleotide is inserted, the error is often (but not always) detected by the polymerase, perhaps by the fact that the erroneous base pair results in a bulge in the double helix. Or, since the incorrect base cannot form a hydrogen bond with the complementary base, perhaps the polymerase will not add a nucleotide to the growing 3′-OH end unless that nucleotide is properly hydrogen-bonded. Thus, DNA synthesis stalls and cannot proceed until the wrong nucleotide is excised and the proper one added, forming the necessary hydrogen bond. The incorrect nucleotide is excised by the 3′-to-5′ exonuclease activity as the enzyme reverses along the template strand. The polymerase then resumes its 5′-to-3′ polymerizing activity.

While 3′-to-5′ exonuclease activity is a property of bacterial DNA polymerases, no such enzyme activities are found in eukaryotic DNA polymerases. Proofreading does take place in eukaryotes, although the proofreading activity is not located on the DNA polymerase itself. Presumably it is located on other proteins.

That 3′-to-5′ exonuclease activity of DNA polymerase is important for maintaining a low mutation rate is shown nicely by the existence of *mutator* mutations in *E. coli*. Strains carrying mutator mutations show a much higher than normal mutation frequency for all genes. These mutations have been shown to affect proteins, substances whose normal functions are required for accurate DNA replication. Relevant to this discussion is the *mutD* mutator of *E. coli*, which results in an altered ϵ (epsilon) subunit of DNA polymerase III. This enzyme is the primary replication enzyme of *E. coli*. *Mut D* causes a defect in 3′-to-5′ proofreading activity, so that many incorrectly inserted nucleotides are unrepaired.

PHOTOREACTIVATION OF UV-INDUCED PYRIMIDINE DIMERS.
Another example of direct correction of mutational lesions involves the repair of UV-light-induced thymine (or other pyrimidine) dimers. This system is called **photoreactivation** or **light repair** (Figure 18.19). By this process the dimers are reverted directly to the original form by exposure to visible light in the wavelength range of 320–370 nm (blue light). Photoreactivation is catalyzed by an enzyme called

~ FIGURE 18.19

Repair of a thymine dimer by photoreactivation.

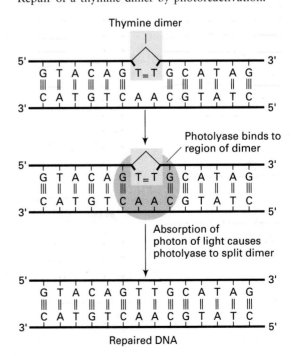

photolyase (encoded by the *phr* gene), which, when activated by a photon of light, splits the dimers apart. Since photolyase has been found in all organisms that have been studied, it may be a ubiquitous enzyme. Presumably, it functions by scavenging along the double helix, seeking the bulges that result when thymine dimers are present. Photolyases are apparently very effective since few thymine dimers (and, hence, mutations) are left after photoreactivation.

REPAIR OF ALKYLATION DAMAGE. Earlier in this chapter, we discussed the use of base-modifying agents as mutagens (see Figure 18.15). Some of these mutagens—the alkylating agents—transfer alkyl groups (usually methyl or ethyl groups) to reactive sites on the bases or phosphates. A particularly reactive site is the oxygen of carbon 6 in the guanine, as shown in Figure 18.15c. The alkylated product can mispair with thymine and result in a GC to AT transition when the DNA is replicated. Alkylation damage like this can be removed by specific DNA repair enzymes. In this particular case, an enzyme encoded by the *ada* gene, called O^6-*methylguanine methyltransferase*, recognizes the O^6-methylguanine in the DNA and removes the methyl group, thereby correcting it back to its original form. Note that in this repair system the modified base is not removed from the DNA.

Repair Involving Excision of Base Pairs

EXCISION REPAIR. A second repair process for UV-light induced pyrimidine dimers is called **excision repair** or **dark repair,** since it does not depend on the presence of light. The experiments that proved the existence of repair processes that did not require light did so by determining processes that could *only* take place in the presence of light. This repair mechanism was discovered in 1964, both by R. P. Boyce and P. Howard-Flanders and by R. Setlow and W. Carrier. They isolated some UV-sensitive mutants of *E. coli.* After UV-irradiation, the mutants exhibited a higher-than-normal rate of induced mutation in the dark. These mutants were *uvrA* mutants (uvr means "UV repair"). The *uvrA* mutants can repair dimers only with the input of light; that is, they lack the dark-repair system. Since wild-type organisms can repair dimers in the dark, the wild-type structures are labeled *uvrA$^+$*.

The excision repair system in *E. coli* corrects not only pyrimidine dimers but also other serious damage-induced distortions of the DNA helix. The system works as diagrammed in Figure 18.20. The helix distortions are recognized by the UvrABC endonuclease, a multi-subunit enzyme encoded by the three genes *uvrA,* *uvrB,* and *uvrC.* This enzyme makes one cut in the damaged DNA strand eight nucleotides to the 5' side of

~ FIGURE 18.20

Excision repair of pyrimidine dimer and other damage-induced distortions of DNA catalyzed by the UvrABC endonuclease.

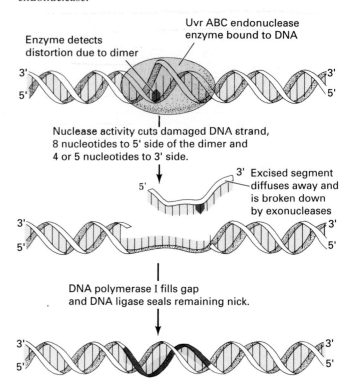

the damage (e.g., the dimer) and four nucleotides to the 3' side of the damage, thereby releasing a 12-nucleotide stretch of single-stranded DNA containing the damaged base(s). The 12-nucleotide gap is filled by DNA polymerase I and sealed by DNA ligase. The excision repair system is found in most organisms that have been studied, and its mechanism of action is thought to be essentially like the one found in *E. coli.* Occasionally, errors are introduced in the repair synthesis of the DNA, and such errors are another source of mutations resulting from UV radiation. The most common cause of the errors is incorrect pairing of new nucleotides with those in the template strand.

REPAIR BY GLYCOSYLASES. Damaged bases can also be excised in another way, involving the action of a glycosylase enzyme that detects an individual unnatural base and catalyzes its removal from the deoxyribose sugar to which it is attached (Figure 18.21,1). This catalytic activity leaves a hole in the DNA from which the base was removed. This is called an *AP site* (for apurinic, where there is no A or G, or apyrimidinic, where there is no C or T). Such holes can also occur

~ FIGURE 18.21

Excision repair of damaged bases involving the action of glycosylase. Glycosylase detects the damaged base and catalyzes its removal. A few adjacent bases are removed by AP endonuclease, and the resulting single-stranded gap is repaired by DNA polymerase I and DNA ligase.

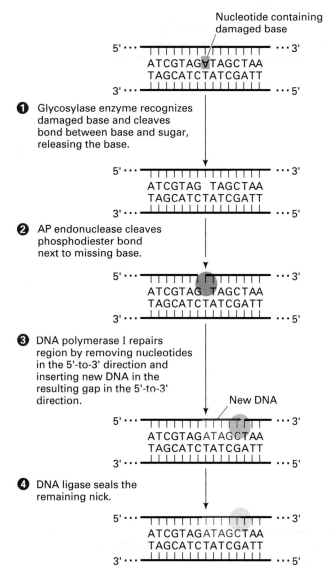

Nucleotide containing damaged base

5′ ··· ATCGTAGATAGCTAA ··· 3′
TAGCATCTATCGATT
3′ ··· ··· 5′

❶ Glycosylase enzyme recognizes damaged base and cleaves bond between base and sugar, releasing the base.

5′ ··· ATCGTAG TAGCTAA ··· 3′
TAGCATCTATCGATT
3′ ··· ··· 5′

❷ AP endonuclease cleaves phosphodiester bond next to missing base.

5′ ··· ATCGTAG TAGCTAA ··· 3′
TAGCATCTATCGATT
3′ ··· ··· 5′

❸ DNA polymerase I repairs region by removing nucleotides in the 5′-to-3′ direction and inserting new DNA in the resulting gap in the 5′-to-3′ direction.

New DNA

5′ ··· ATCGTAGATAGCTAA ··· 3′
TAGCATCTATCGATT
3′ ··· ··· 5′

❹ DNA ligase seals the remaining nick.

5′ ··· ATCGTAGATAGCTAA ··· 3′
TAGCATCTATCGATT
3′ ··· ··· 5′

through natural, spontaneous losses of bases, discussed earlier in the chapter. The hole is recognized by a specific enzyme called *AP endonuclease*, which cuts the backbone besides the missing base (Figure 18.21,2). This leaves a primer end from which DNA polymerase I (in *E. coli*) initiates repair synthesis, using its $5′ \rightarrow 3′$ exonuclease activity to remove a few nucleotides ahead of the missing base, and using its polymerizing activity to fill in the gap (Figure 18.21,3). DNA ligase seals the

remaining nick (Figure 18.21,4). DNA polymerase I activity is displayed in this repair system, in which the enzyme simultaneously polymerases DNA and removes nucleotides ahead of the growing chain. This process is called *nick translation* (see Chapter 15).

REPAIR BY MISMATCH CORRECTION. As should be apparent by now, the replication process does not exhibit perfect fidelity in inserting nucleotides in the new chain, so a number of error correction systems have evolved to reduce the frequency of errors (= mutations) in the genome. While proofreading by DNA polymerase is an efficient way of correcting many errors soon after they are made, a significant number of errors still remain uncorrected after replication has been completed. Such errors usually involve mismatched base pairs which, in the next round of replication, will result in spontaneous mutations.

A rough estimate is that DNA polymerase leaves about one error per 10^8 base pairs per generation in *E. coli*. The measured rate of mutation, however, is one error per 10^{10}–10^{11} base pairs per generation. This difference results from another system of repair called *mismatch correction*. This system involves a *mismatch correction enzyme* encoded by three genes, *mutH*, *mutL*, and *mutS* (Figure 18.22; p. 578). The enzyme searches newly replicated DNA for mismatched base pairs. Once identified, it catalyzes the removal of a single-stranded segment of DNA that includes the mismatched base. DNA polymerase and DNA ligase then fill in the gap and seal it, respectively.

How does the enzyme know which of the mismatched bases in the pair is the erroneous one introduced during replication? It determines this by establishing which of the two DNA strands is the parental strand and which is the newly-synthesized strand. The signal that it uses is methylation of a specific DNA sequence near the mismatch. This sequence is GATC, in which the A is usually methylated by the action of *dam*-methylase, an enzyme encoded by the *dam* gene. The GATC sequence is palindromic: that is, the same sequence is read 5′-to-3′ on both DNA strands (see Chapter 15). Thus, both A nucleotides in the DNA segment are methylated. However, the A nucleotide in the GATC sequence of a *newly replicated DNA strand* is not methylated until a short time after its synthesis. Therefore, for a short while after replication, the parental strand has a methylated GATC sequence, while the new strand has an unmethylated GATC sequence. This difference is detected by mismatch correction enzyme which then removes the mismatch only from the new strand. In addition to correcting base-pairing errors, the enzyme can correct small base-pair deletions or additions (see Figure 18.7), thereby also reducing the frequency of frameshift mutations.

~ FIGURE 18.22

Mechanism of mismatch correction repair. The mismatch correction enzyme recognizes which strand the base mismatch is on by reading the methylation state of a nearby GATC sequence. If the sequence is unmethylated, a segment of that DNA strand containing the mismatch is excised and new DNA is inserted.

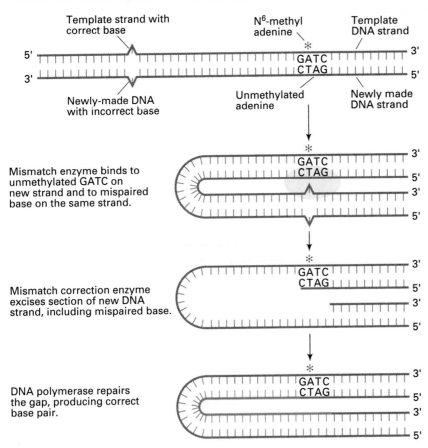

$\mathcal{K}$EYNOTE

Both prokaryotes and eukaryotes have a number of repair systems that deal with different kinds of DNA damage. Without such repair systems mutational lesions would accumulate and be lethal to the cell or organism. Not all mutational lesions are repaired; hence, mutations do appear, but at relatively low frequencies. _____

Human Genetic Diseases Resulting from DNA Replication and Repair Errors

Human cells may carry some naturally occurring genetic diseases in which the defect has been attributed to defects in DNA replication. Some of these mutants are listed in Table 18.2. Perhaps the most well-known is *xeroderma pigmentosum*, which is caused by homozygosity for a recessive mutation. People with this lethal affliction are photosensitive, and portions of their skin that have been exposed to light show intense pigmentation, freckling, and warty growths that may become malignant. The function affected in these people is excision repair of damage caused by ultraviolet light, X-rays, or gamma radiation or by chemical treatment. Thus individuals with xeroderma pigmentosum are unable to repair radiation damage to DNA and eventually die, often as a result of malignancies that arise from the damage. Since the disease is inherited, the defect must result from a mutation in a gene coding for a protein involved in the repair of DNA damage.

~ TABLE 18.2

Examples of Some Naturally Occurring Human Cell Mutants That Are Defective in DNA Replication

DISEASE	SYMPTOMS	FUNCTION AFFECTED
Xeroderma pigmentosum (XP)	Lethal; skin freckling, cancerous growths on skin	Repair of DNA damaged by UV irradiation or chemicals
Ataxia telangiectase (AT)	Muscle coordination defect; propensity for respiratory infection; radiation sensitive; cancer-prone; broken chromosomes	Repair replication of DNA
Fanconi's anemia (FA)	Aplastic anemia[a]; pigmentary changes in skin; malformation of heart, kidney, and extremities; leukemia	Repair replication of DNA, UV dimers and chemical adducts not removed from DNA
Bloom's syndrome (BS)	Dwarfism; sun-sensitive skin disorder; chromosome breaks	Elongation of DNA chains intermediate in replication

[a] Individuals with aplastic anemia make no, or very few, red blood cells.
Source: After R. Sheinin et al., 1978. *Annu. Rev. Biochem.* 47:227–316.

SCREENING PROCEDURES FOR THE ISOLATION OF MUTANTS

Geneticists have made great progress over the years in understanding how normal processes take place by studying mutants that have defects in those processes. Researchers use mutagens to induce mutations at a greater rate than the rate at which spontaneous mutations occur. As we have mentioned, however, mutagens are not directed in their action toward particular genes. Instead, they change base pairs at random, without regard to the positions of the base pairs in the genetic material. Thus a mutagen's effect can only be studied if the mutation affects the function of a gene governing a phenotype researchers can observe. Mutations in regulatory or protein-coding genes usually result in a phenotypic change, unless the mutation is a neutral one.

Once mutations have occurred, they must be detected if they are to be studied. Mutations of haploid organisms are readily detectable because there is only one copy of the genome. Mutations of diploid organisms, however, may be more difficult to detect. Let us consider mutations of *Drosophila*, an experimental organism in which genetic crosses can be made as desired. In such an organism, dominant mutations will be readily detectable. Recessive mutations are less readily detectable. Sex-linked recessive mutations can be detected because they are expressed in one-half of the sons of a mutated, heterozygous female. Autosomal recessive mutations can only be detected if the mutation is homozygous. To simplify detection of different types of mutations, special tester strains of *Drosophila* have been constructed to which strains carrying potentially new mutations are crossed. Analysis of the progeny quickly establishes whether a new mutation has been made. The discussion of these tester strains is beyond the scope of this text.

The detection of mutations in humans is much more difficult than in *Drosophila* because geneticists cannot make controlled crosses. Dominant mutations can be readily detected, of course, but other types of mutations may be revealed only by pedigree analysis or by direct biochemical or molecular probing. Pedigree analysis can detect sex-linked recessive mutations by mother-son transmission patterns. Rare autosomal recessive mutations, however, will rarely be homozygous, so few pedigrees will have individuals exhibiting such a trait.

Thus, the detection of mutations is not necessarily a simple matter. Fortunately, for some organisms of genetic interest, particularly microorganisms, screening procedures have been developed to help geneticists obtain particular mutants of interest from among a heterogeneous mixture in a mutagenized population. What follows are descriptions of a few screening procedures for the isolation of particular types of mutations.

Visible Mutations

Visible mutations affect the morphology or physical appearance of an organism. Examples of visible mutations are eye-color or wing-shape mutants of *Drosophila*, coat-color mutants of animals (e.g., albino organisms), colony-size mutants of yeast, and plaque morphology mutants of bacteriophages. Since visible mutations, by definition, are readily apparent, screening is done by inspection.

Nutritional Mutations

An **auxotrophic mutation** (also called a **nutritional** or **biochemical mutation**) affects an organism's ability to make a particular molecule essential for growth. Auxotrophic mutations are most readily detected in microorganisms such as *E. coli*, yeast, *Neurospora*, or some unicellular algae that grow on simple and defined growth media from which they synthesize the enzymes used to make all the molecules essential to their growth. A number of screening procedures can isolate auxotrophic mutants, and some of them will be described now.

In 1952 Esther and Joshua Lederberg developed a procedure to screen for auxotrophic mutants of any microorganism that grows in discrete colonies on solid medium. This procedure is called **replica plating**; Figure 18.23 diagrams how replica plating can be used to isolate arginine auxotrophs of a microorganism.

In the replica-plating technique, samples from a culture of a colony-forming organism or cell type that has or has not been mutagenized are plated onto a complete medium that contains all possible nutrients. Upon incubation, colonies grow wherever a cell has landed on the medium. These colonies consist of clones of the original cell; that is, they are genetically identical copies of the cell that initiated the cloning. The population of colonies will be a mixture of prototrophic (wild-type) colonies and auxotrophic mutant colonies. The pattern of the colonies is transferred onto sterile velveteen cloth. Replicas of the original colony pattern on the cloth are then made by gently pressing new plates onto the velveteen. If the new plate contains minimal

~ FIGURE 18.23

Replica-plating technique to screen for mutant strains of a colony-forming microorganism.

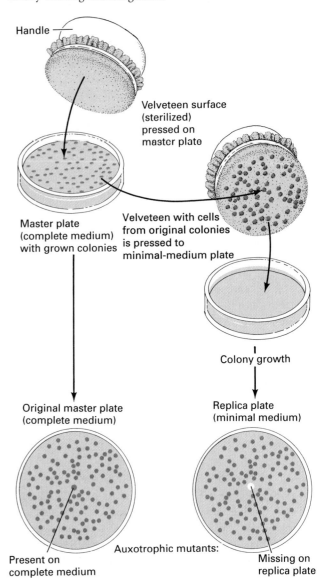

Handle

Velveteen surface (sterilized) pressed on master plate

Master plate (complete medium) with grown colonies

Velveteen with cells from original colonies is pressed to minimal-medium plate

Colony growth

Original master plate (complete medium)

Replica plate (minimal medium)

Auxotrophic mutants:

Present on complete medium

Missing on replica plate

with those on the minimal-medium replica plate, researchers can readily identify the potential auxotrophic colonies. If another plate contains minimal medium *plus* arginine, then the colonies that grow on this medium but not on the minimal medium are the arginine auxotrophs, that is, the mutants that require arginine in order to grow. They can then be selected from the original master plate and cultured for further study.

A useful replica-plating technique to screen for a range of mutant types employs an antibiotic to kill many nonmutant cells but not many or any mutant cells. For example, the antibiotic nystatin (named after New York state) may be used to "enrich" for mutants of yeast; that is, to increase the proportion of mutant to nonmutant cells in the population. Nystatin is an effective killer of growing yeast cells.

Suppose we wish to enrich selectively for adenine auxotrophic *(ade)* mutants of yeast. We incubate the mutagenized population of cells in a culture medium that lacks adenine but contains nystatin. Under these conditions, wild-type cells grow and are killed by the nystatin, while the *ade* (adenine-requiring) cells that either do not grow or grow very slowly are selectively spared. If the exposure to the antibiotic is carefully limited, the population of surviving cells has a much higher proportion of *ade* mutant cells than is found in the original, unselected population of cells. This method can be modified to select for a number of other classes of mutants. An essentially identical method with penicillin as the selective antibiotic may be used with *E. coli*.

Conditional Mutations

The products of many genes—DNA polymerase and RNA polymerase, for example—are important for the growth and division of cells, and most mutations in these genes result in a lethal phenotype. The structure and function of this class of genes may be studied by inducing *conditional mutations* in the genes. A common type of conditional mutation to study is a heat-sensitive mutation that is characterized by normal function at the normal growth temperature and no or severely impaired function at a higher temperature. In yeast, for instance, normal growth temperature is 23°C; heat-sensitive mutations typically are isolated at 36°C.

Essentially the same procedures are used to screen for heat-sensitive mutations of microorganisms as for auxotrophic mutations. Replica plating can select for temperature-sensitive mutants when the replica plate is incubated at a higher temperature than the master plate. Nystatin selection could also be used to enrich for heat-sensitive mutations of yeast that do not grow at 36°C.

KEYNOTE

Geneticists have made great progress in understanding how cellular processes take place by studying mutants that have defects in those processes. With microorganisms, a number of screening procedures enrich selectively for mutants of interest from a heterogeneous mixture of cells in a mutagenized population of cells.

SUMMARY

In this chapter we have seen that there are a number of different kinds of mutations at the DNA level. Mutations occur spontaneously at a low rate. The rate of induction of mutations can be increased through the use of mutagens like irradiation and certain chemicals. Typically, mutagens are used by researchers so that a mutant of interest is more likely to be found in a population of cells. Chemical mutagens work in a number of different ways, such as by acting as base analogs, by modifying bases, or by intercalating into the DNA. The latter results in frameshift mutations, while the others result in base-pair substitution mutations. Bruce Ames has devised a test—the Ames test—to determine if a chemical (e.g., an environmental or commercial chemical) has the potential to cause mutations in humans. A large number of potential human carcinogens have been found in this way.

Cells possess a number of repair mechanisms that function to correct at least some mutations. These repair mechanisms include: (1) repair by DNA polymerase proofreading, in which a base-pair mismatch in DNA being synthesized is immediately repaired by 3'-to-5' excision; (2) photoreactivation of pyrimidine dimers induced by UV light; (3) excision repair, in which pyrimidine dimers and other DNA damage that distorts the DNA helix are excised and replaced with new DNA; (4) repair of damaged bases by glycosylases and AP endonuclease; and (5) repair by mismatch correction. The latter involves a mechanism whereby the methylation state of a DNA sequence signals which DNA strand is newly synthesized so that the mismatched base on that strand is corrected. The collective array of repair enzymes results in a reduction of mutation rates for spontaneous errors of several orders of magnitude. Such repair mechanisms cannot cope, however, with the extensive number of mutations that result from the use of chemical mutagens or UV irradiation.

Lastly, we considered some examples of how a mutagenized population of cells can be screened for particular mutants of interest, e.g., auxotrophs and conditional mutations. Over the decades, spontaneous and induced mutants have proved invaluable for the genetic analysis of biological function.

Analytical Approaches for Solving Genetics Problems

Q.1 Five strains of *E. coli* containing mutations that affect the tryptophan synthetase A polypeptide have been isolated. Figure 18.A shows the changes produced in the protein itself in the indicated mutant strains.

In addition, *A23* can be further mutated to insert Ile, Thr, Ser, or the wild-type Gly into position 210.

a. Using the genetic code (see Figure 14.4), explain how the two mutations *A23* and *A46* can result in two different amino acids being inserted at position 210. Give the nucleotide sequence of the wild-type gene at that position and the two mutants.

b. Can mutants *A23* and *A46* recombine? Why or why not?

c. From what you can infer of the nucleotide sequence in the wild-type gene, indicate, for the codons specifying amino acids 48, 210, 233, and 234, whether or not a nonsense mutant could be generated by a single nucleotide substitution in the gene.

A.1 a. There are no simple ways to answer questions like this one. The best approach is to scrutinize the genetic code dictionary and use a pencil and paper to try to define the codon changes that are compatible with all the data. The number of amino acid changes in position 210 of the polypeptide is helpful in this case. The answer is that the original codon for glycine in the wild type was GGA; also, *A23* is AGA, and *A46* is GAA. In other words, the two different amino acid substitutions in the two mutants result from different, single base-pair changes in the part of the gene that corresponds to this codon.

b. The answer to this question follows from the answer deduced in part a. Mutants *A23* and *A46* can recombine since the mutations in the two mutant strains are in different base pairs. The results of a single recombination event at position 210 are a wild-type GGA codon (Gly) and a double mutant AAA codon (Lys).

c. Amino acid 48 had a Glu-to-Val change. This change must have involved GAA to GUA or GAG to GUG. In either case the Gly codon can mutate with a single base-pair change to a nonsense codon, that is, UAA or UAG, respectively.

Amino acid 210 in the wild type has a GGA codon, as we have already discussed. This gene could mutate to the UGA nonsense codon with a single base-pair change.

Amino acid 233 had a Gly-to-Cys change. This change must have involved either GGU to UGU or GGC to UGC. In either case the Gly codon cannot mutate to a nonsense codon with one base-pair change.

Amino acid 234 had a Ser-to-Leu change. This change was either UCA to UUA or UCG to UUG. If the Ser codon was UCA, it changed to UGA in one step. But if the Ser codon was UCG, it cannot change to a nonsense codon in one step.

Q.2 The chemically induced mutations, *a*, *b*, *c*, and *d* show specific reversion patterns when subjected to treatment by the following mutagens: 2-aminopurine (AP), 5-bromouracil (BU), proflavin (pro), hydroxylamine (HA), and ethylmethane sulfonate (EMS). The reversion patterns are shown in the following table:

	MUTAGENS TESTED IN REVERSION STUDIES				
MUTATION	AP	BU	PRO	HA	EMS
a	−	−	+	−	−
b	−	+	−	+	+
c	+	−	−	−	−
d	−	−	−	−	+

(*Note:* + indicates many reversions to wild type were found; − indicates no reversions, or very few, to wild type were found.)

Figure 18.A

Mutant number			*A3* *A23* *A46* *A78* *A169*	

Amino acid position in chain   N terminus   C terminus

Amino acid in the wild type   48 210 233 234 — Glu Gly Gly Ser

Amino acid change in mutant   Val Arg Glu Cys Leu

Indicate, for each original mutation (i.e., a^+ to a, b^+ to b, etc.), the probable base-pair change (i.e., AT to GC, deletion of GC, etc.) and the mutagen that was most probably used to induce the original change.

A.2 This question tests knowledge of the base-pair changes that can be induced by the various mutagens used. Mutagen AP induces mainly AT-to-GC changes and can cause GC-to-AT changes also. Thus AP-induced mutations can be reverted by AP. Mutagen BU induces mainly GC-to-AT changes and can cause AT-to-GC changes, so BU-induced mutations can be reverted by BU. Proflavin causes single base-pair deletions or additions with no specificity. Proflavin-induced changes can be reverted by a second treatment with proflavin, as we discovered in the discussion of the experiments used to prove the three-letter basis of the genetic code. Mutagen HA causes one-way transitions from GC to AT, and so HA-induced mutations cannot be reverted by HA. Mutagen MMS alkylates G, which results in its loss from the DNA; it is replaced at random by one of the four bases. If the replacing base is G, there is no mutation, and the effect of MMS is not detected. Substitution with any one of the other three bases, though, results in a mutation. So a third of the time there will be a GC-to-AT change; a third of the time there will be a GC-to-TA change; and a third of the time there will be a GC-to-CG change. With these mutagen specificities in mind, we can answer the questions for each mutation in turn.

Mutation a^+ to a: The a mutation was reverted only by proflavin, indicating that it was a deletion or an addition. Therefore the original mutation was induced by proflavin, since that is the only mutagen in the list that can cause an addition or a deletion.

Mutation b^+ to b: The b mutation was reverted by BU, HA, or MMS. A key here is that HA only causes GC-to-AT changes. Therefore b must be GC, and the original b^+ must have been an AT. Thus the mutational change of b^+ to b must have been caused by treatment with AP or BU, since these are the two mutagens in the list that are capable of inducing that change.

Mutation c^+ to c: The c mutation was reverted only by AP. Since it could not be reverted by HA, c must be an AT and c^+ a GC. The mutational change from c^+ to c therefore involved a GC-to-AT transition and could have resulted from treatment with BU, HA, or MMS.

Mutation d^+ to d: The d mutation was reverted only by MMS, indicating that something other than a transition mutation was involved in the d^+-to-d mutational event. Mutagen MMS causes GC-to-CG transversions, so the most likely event in d^+ to d was a GC-to-CG change (or CG to GC). Hence the reversion was a CG-to-GC change (or GC to CG). Thus the original mutagen could only have been MMS.

QUESTIONS AND PROBLEMS

*18.1 Mutations are (choose the correct answer):
a. Caused by genetic recombination.
b. Heritable changes in genetic information.
c. Caused by faulty transcription of the genetic code.
d. Usually but not always beneficial to the development of the individuals in which they occur.

18.2 Answer true or false: Mutations occur more frequently if there is a need for them.

*18.3 The following is not a class of mutation (choose the correct answer):
a. Frameshift.
b. Missense.
c. Transition.
d. Transversion.
e. None of the above (i.e., all are classes of mutation).

*18.4 Ultraviolet light usually causes mutations by a mechanism involving (choose the correct answer):
a. One-strand breakage in DNA.
b. Light-induced change of thymine to alkylated guanine.
c. Induction of thymine dimers and their persistence or imperfect repair.
d. Inversion of DNA segments.
e. Deletion of DNA segments.
f. All of the above.

18.5 For the middle region of a particular polypeptide chain, the normal amino acid sequence and the amino acid sequence of several mutants were determined, as shown below (.... indicates additional, unspecified amino acids). For each mutant, say what DNA level change has occurred, and whether the change is a point mutation (transversion or transition, missense or nonsense) or a frame shift. (Refer to the codon dictionary in Figure 14.3.)
a. Normal: .... Phe Leu Pro Thr Val Thr Thr Arg Trp
b. Mutant 1: .... Phe Leu His His Gly Asp Asp Thr Val
c. Mutant 2: .... Phe Leu Pro Thr Met Thr Thr Arg Trp
d. Mutant 3: .... Phe Leu Pro Thr Val Thr Thr Arg
e. Mutant 4: .... Phe Pro Pro Arg
f. Mutant 5: .... Phe Leu Pro Ser Val Thr Thr Arg Trp

*18.6 In mutant strain X of *E. coli*, a Leu tRNA which recognizes the codon 5'-CUG-3' in normal cells has been altered so that it now recognizes the codon 5'-GUG-3'. A missense mutation, which affects amino acid 10 of a particular protein, is suppressed in mutant X cells.
a. What are the anticodons of the two Leu tRNAs, and what mutational event has occurred in mutant X cells?
b. What amino acid would normally be present at position 10 of the protein (without the missense mutation)?

c. What amino acid would be put in at position 10 if the missense mutation is not suppressed (i.e., in normal cells)?

d. What amino acid is inserted at position 10 if the missense mutation is suppressed (i.e., in mutant X cells)?

18.7 In any kind of chemotherapy, the object is to find a means to kill the invading pathogen or cancer cell without killing the cells of the host. To do this successfully, one must find and exploit biological differences between target organisms and host cells. Explain the nature of the biological difference between host cells and HIV-I virus that permits the use of azidothymidine for chemotherapy.

*****18.8** The mutant *lac z-1* was induced by treating *E. coli* cells with acridine, while *lac z-2* was induced with 5BU. What kinds of mutants are these likely to be? Explain. How could you confirm your predictions by studying the structure of the β-galactosidase in these cells?

*****18.9 a.** The sequence of nucleotides in an mRNA is:

$$5'-AUGACCCAUUGGUCUCGUUAG-3'$$

How many amino acids long would you expect the polypeptide chain made with this messenger to be?

b. Hydroxylamine is a mutagen that results in the replacement of an AT base pair for a GC base pair in the DNA; that is, it induces a transition mutation. When applied to the organism that made the mRNA molecule shown in part a, a strain was isolated in which a mutation occurred at the 11th position of the DNA that coded for the mRNA. How many amino acids long would you expect the polypeptide made by this mutant to be? Why?

18.10 In a series of 94,075 babies born in a particular hospital in Copenhagen, 10 were achondroplastic dwarfs (this is an autosomal dominant condition). Two of these 10 had an achondroplastic parent. The other 8 achondroplastic babies each had two normal parents. What is the apparent mutation rate at the achondroplasia locus?

*****18.11** Three of the codons in the genetic code are chain-terminating codons for which no naturally occurring tRNAs exist. Just like any other codons in the DNA, though, these codons can change as a result of base-pair changes in the DNA. Confining yourself to single base-pair changes at a time, determine which amino acids could be inserted in a polypeptide by mutation of these chain-terminating codons: (a) UAG; (b) UAA; (c) UGA. (The genetic code is listed in Figure 14.3.)

18.12 The amino acid substitutions in the figure occur in the α and β chains of human hemoglobin. Those amino

acids connected by lines are related by single nucleotide changes. Propose the most likely codon or codons for each of the numbered amino acids. (Refer to the genetic code listed in Figure 14.3.)

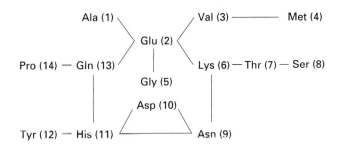

*****18.13** Yanofsky studied the tryptophan synthetase of *E. coli* in an attempt to identify the base sequence specifying this protein. The wild type gave a protein with a glycine in position 38. Yanofsky isolated two *trp* mutants, *A23* and *A46*. Mutant *A23* had Arg instead of Gly at position 38, and mutant *A46* had Glu at position 38. Mutant *A23* was plated on minimal medium, and four spontaneous revertants to prototrophy were obtained. The tryptophan synthetase from each of four revertants was isolated, and the amino acids at position 38 were identified. Revertant 1 had Ile, revertant 2 had Thr, revertant 3 had Ser, and revertant 4 had Gly. In a similar fashion, three revertants from *A46* were recovered, and the tryptophan synthetase from each was isolated and studied. At position 38 revertant 1 had Gly, revertant 2 had Ala, and revertant 3 had Val. A summary of these data is given in the figure. Using the genetic code shown in the figure below, deduce the codons for the wild type, for the mutants *A23* and *A46*, and for the revertants, and place each designation in the space provided in the following figure.

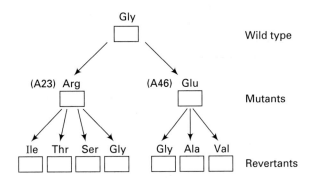

18.14 Consider an enzyme chewase from a theoretical microorganism. In the wild-type cell the chewase has the following sequence of amino acids at positions 39 to 47 (reading from the amino end) in the polypeptide chain:

$$-Met-Phe-Ala-Asn-His-Lys-Ser-Val-Gly-$$
$$\quad 39 \quad 40 \quad 41 \quad 42 \quad 43 \quad 44 \quad 45 \quad 46 \quad 47$$

A mutant of the organism was obtained; it lacks chewase activity. The mutant was induced by a mutagen known to cause single base-pair insertions or deletions. Instead of making the complete chewase chain, the mutant makes a short polypeptide chain only 45 amino acids long. The first 38 amino acids are in the same sequence as the first 38 of the normal chewase, but the last 7 amino acids are as follows:

$$-\text{Met}-\text{Leu}-\text{Leu}-\text{Thr}-\text{Ile}-\text{Arg}-\text{Val}-$$
$$\quad 39 \quad\; 40 \quad\;\; 41 \quad\; 42 \quad\;\; 43 \quad\; 44 \quad\; 45$$

A partial revertant of the mutant was induced by treating it with the same mutagen. The revertant makes a partly active chewase, which differs from the wild-type enzyme only in the following region:

$$-\text{Met}-\text{Leu}-\text{Leu}-\text{Thr}-\text{Ile}-\text{Arg}-\text{Gly}-\text{Val}-\text{Gly}-$$
$$\quad 39 \quad\; 40 \quad\;\; 41 \quad\; 42 \quad\;\; 43 \quad\; 44 \quad\; 45 \quad\; 46 \quad\; 47$$

Using the genetic code given in Figure 14.3, deduce the nucleotide sequences for the mRNA molecules that specify this region of the protein in each of the three strains.

18.15 Two mechanisms in *E. coli* were described for the repair of DNA damage (thymine dimer formation) after exposure to ultraviolet light: photoreactivation and excision (dark) repair. Compare and contrast these mechanisms, indicating how each achieves repair.

*18.16 After a culture of *E. coli* cells was treated with the chemical 5-bromouracil, it was noted that the frequency of mutants was much higher than normal. Mutant colonies were then isolated, grown, and treated with nitrous acid; some of the mutant strains reverted to wild type.
a. In terms of the Watson-Crick model, diagram a series of steps by which 5BU may have produced the mutants.
b. Assuming the revertants were not caused by suppressor mutations, indicate the steps by which nitrous acid may have produced the back mutations.

18.17 A single, very hypothetical strand of DNA is composed of the base sequence indicated in this figure.

$$5'-\text{T}-\text{HX}-\text{U}-\text{A}-\text{G}-\text{BU-enol}-2\text{AP}-\text{C}-\text{BU}-\text{X}-2\text{AP-imino}-3'$$

In the figure, A indicates adenine, T indicates thymine, G indicates guanine, C denotes cytosine, U denotes uracil, BU is 5-bromouracil, 2AP is 2-aminopurine, BU-enol is a tautomer of 5BU, 2AP-imino is a rare tautomer of 2AP, HX is hypoxanthine, and X is xanthine; 5' and 3' are the numbers of the free, OH-containing carbons on the deoxyribose part of the terminal nucleotides.
a. Opposite the bases of the hypothetical strand, and using the shorthand of the figure, indicate the sequence of bases on a complementary strand of DNA.

b. Indicate the direction of replication of the new strand by drawing an arrow next to the new strand of DNA from part a.
c. When postmeiotic germ cells of a higher organism are exposed to a chemical mutagen before fertilization, the resulting offspring expressing an induced mutation are almost always mosaics for wild-type and mutant tissue. Give at least one reason that in the progenies of treated individuals these mosaics are found and not the so-called complete or whole-body mutants.

The following information applies to Problems 18.18 through 18.22. A solution of single-stranded DNA is used as the template in a series of reaction mixtures. It has the base sequence as follows:

A = adenine, G = guanine, C = cytosine, T = thymine, H = hypoxanthine, and HNO$_2$ = nitrous acid. For Problems 18.18 through 18.22, use this shorthand system and draw the products expected from the reaction mixtures. Assume that a primer is available in each case.

*18.18 The DNA template + DNA polymerase + dATP + dGTP + dCTP + dTTP + Mg^{2+}.

18.19 The DNA template + DNA polymerase + dATP +dGMP + dCTP + dTTP + Mg^{2+}.

18.20 The DNA template + DNA polymerase + dATP +dHTP + dGMP + dTTP + Mg^{2+}.

*18.21 The DNA template is pretreated with HNO$_2$ + DNA polymerase + dATP + dGTP + dCTP + dTTP + Mg^{2+}.

18.22 The DNA template + DNA polymerase + dATP + dGMP + dHTP + dCTP + dTTP + Mg^{2+}.

18.23 A strong experimental approach to determining the mode of action of mutagens is to examine the revertibility of the products of one mutagen by other mutagens. The following table represents collected data on revertibility of various mutagens on *rII* mutations in phage T2; + indicates majority of mutants reverted, − indicates virtually no reversion; BU = 5-bromouracil, AP = 2-aminopurine, NA = nitrous acid, and HA = hydroxylamine. Fill in the empty spaces.

Mutation induced by	Proportion of mutations reverted by				Base-pair substitution inferred
	BU	AP	NA	HA	
BU	+	_	_	−	_____
AP	_	−	_	+	_____
NA	+	+	_	+	_____
HA	_	_	+	−	GC → AT

*18.24 Three *ara* mutants of *E. coli* were induced by mutagen X. The ability of other mutagens to cause the reverse change (*ara* to *ara*$^+$) was tested, with the results shown in Table 18.A.

Assume all *ara*$^+$ cells are true revertants. What base changes were probably involved in forming the three original mutations? What kind(s) of mutations are caused by mutagen X?

TABLE 18.A

Frequency of *ara*$^+$ Cells Among Total Cells After Treatment

MUTANT	MUTAGEN				
	NONE	BU	AP	HA	FRAMESHIFT
ara-1	1.5×10^{-8}	5×10^{-5}	1.3×10^{-4}	1.3×10^{-8}	1.6×10^{-8}
ara-2	2×10^{-7}	2×10^{-4}	6×10^{-5}	3×10^{-5}	1.6×10^{-7}
ara-3	6×10^{-7}	10^{-5}	9×10^{-6}	5×10^{-6}	6.5×10^{-7}

19 CHROMOSOME ABERRATIONS

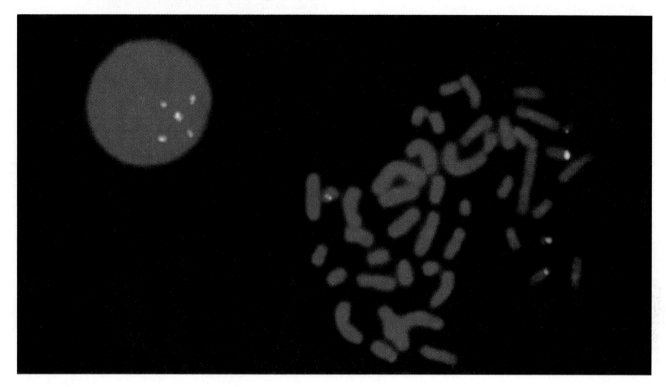

TYPES OF CHROMOSOME CHANGES

VARIATIONS IN CHROMOSOME STRUCTURE
Deletion
Duplication
Inversion
Translocation

VARIATIONS IN CHROMOSOME NUMBER
Changes in One or a Few Chromosomes
Changes in Complete Sets of Chromosomes

CHROMOSOME ABERRATIONS AND HUMAN TUMORS
Chronic Myelogenous Leukemia and the Philadelphia Chromosome
Chromosomal Aberration and Burkitt's Lymphoma

CHROMOSOME REARRANGEMENTS THAT ALTER GENE EXPRESSION
Amplification or Deletion of Genes
Inversions That Alter Gene Expression
Transpositions That Alter Gene Expression

PRINCIPAL POINTS

~ Chromosome aberrations are variations from the wild-type condition in either chromosome number or chromosome structure. Chromosome aberrations can occur spontaneously, or they can be induced by chemical or radiation treatment.

~ Chromosome aberrations may involve parts of chromosomes rather than whole chromosomes. The four major types of such aberrations are deletions (loss of a DNA segment), duplications (duplication of a DNA segment), inversions (change in orientation of a DNA segment within a chromosome without loss of DNA material), and translocations (movement of a DNA segment to another location in the genome without loss of DNA material).

~ Variations in the chromosome number of a cell or an organism give rise to aneuploidy, monoploidy, and polyploidy. Aneuploidy is the state in which the cell has one, two, or a few whole chromosomes more or less than the basic number of the species involved. Monoploidy is the state in which a normally-diploid cell has only one set of chromosomes, and polyploidy is the state in which there are more than the normal number of sets of chromosomes present.

~ Any or all of these chromosome number or chromosome structure aberrations may have serious, even lethal, consequences to the organism. The abnormal phenotypes result typically from problems with chromosome segregation during meiosis, with gene disruptions at chromosome breakage sites, or with gene expression levels when gene dosage is altered. Some human tumors, for example, are associated with chromosome aberrations, either through a change in the number of chromosomes, or through a change in chromosome structure.

~ Some changes in chromosome organization occur naturally as mechanisms of altering gene expression, often as part of a developmental program. Examples of such changes are amplification of genes, deletion of a chromosome or part of a chromosome, inversion of a chromosome segment that results in a switch in transcription from one gene or set of genes to an alternative set, and transposition of a gene from a silent location in the genome to an active location where transcription then occurs.

Chapter 18 discussed gene mutations, that is, heritable changes that occur within the limits of a single gene. This chapter discusses another kind of genetic change, namely chromosome aberrations, which are changes in normal chromosome structure or chromosome number. In practice, discriminating between gene mutations and chromosomal structural changes, especially minute ones, is frequently impossible. However, many chromosome aberrations are significant enough to be visible under the microscope. Thus the study of chromosome behavior in normal cells and of how this behavior is altered in cells with chromosome aberrations involves both genetic and cytological approaches and hence falls into the field of *cytogenetics*. In this chapter we also briefly discuss some examples of changes in chromosome organization that affect gene expression.

many chromosomes as somatic cells. Moreover, the organization and the number of genes on the chromosomes of an organism are the same from cell to cell. These characteristics of chromosome number and gene organization typically hold for all members of the same species. Exceptions to these tenets do occur and are known as **chromosome aberrations.** Chromosome aberrations will be defined here as *variations from the wild-type condition in either chromosome structure or chromosome number.* Chromosome aberrations can arise spontaneously or can be induced experimentally by chemical or radiation mutagens, and in many instances they can be detected cytologically during mitosis and meiosis. The ability to detect different kinds of chromosome aberrations depends on the size, structure, and number of chromosomes concerned and the ease with which they can be handled.

TYPES OF CHROMOSOME CHANGES

Cells of the same organism characteristically have the same number of chromosomes, with the exception of gametes, which (in diploid organisms) have half as

VARIATIONS IN CHROMOSOME STRUCTURE

This type of chromosome aberration involves changes in parts of individual chromosomes rather than whole chromosomes or sets of chromosomes in a genome.

There are four types of structural alterations: *deletions* and *duplications* (both of which involve a change in the amount of genetic material on a chromosome), *inversions* (which involve a change in the arrangement of a chromosomal segment), and *translocations* (which involve a change in the location of a chromosomal segment).

Much information about changes in chromosome structure has been gained from the study of **polytene chromosomes,** a special type of chromosome found in certain insects that consists of a bundle of chromatids (see Figure 17.21 and Figure 19.9a; p. 593). Recall that these chromatid bundles result from repeated cycles of chromosome duplication without nuclear or cell division, and they are easily detectable through microscopic examination. Polytene chromosomes may be a thousand times the size of corresponding chromosomes at meiosis or in the nuclei of ordinary somatic cells. In Diptera, homologous chromosomes are somatically paired, and therefore the number of polytene chromosomes per cell corresponds to half the diploid chromosome of somatic cells. The number of chromatids per polytene chromosome is species-specific.

As a result of the intimate pairing of the multiple copies of chromatids, characteristic banding patterns are easily observed, enabling cytogeneticists to identify unambiguously any segment of a chromosome and therefore to see clearly where chromosome changes have occurred. In *Drosophila melanogaster,* for example, over 5000 bands and interbands can be counted in the four polytene chromosomes. For a long time each band was thought to represent a single protein-coding gene and the region between bands to represent intergenic DNA. It is now known that each band contains on the average 30,000 base pairs (30 kb) of DNA, which encode several average-sized proteins. Careful analysis involving cloning and sequencing has shown that many bands contain a number of genes (up to seven) that are transcribed independently. Sometimes the genes in a band are related in function.

Deletion

A **deletion** is a chromosome aberration resulting in the loss of a segment of the genetic material and the genetic information contained therein from a chromosome (Figure 19.1). The deleted segment may be located anywhere along a chromosome. (Deletion of a chromosome end results in an unstable chromosome unless the end gets capped with a telomere.) A deletion is initiated by breaks in chromosomes induced by agents such as heat, radiation (especially ionizing radiation: see Chapter 18), viruses, and chemicals or by errors in recombination enzymes.

~ **FIGURE 19.1**

Deletion with a chromosome segment (here, *D*) missing.

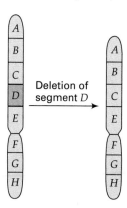

The consequences of the deletion depend on the genes or parts of genes that have been removed. In diploid organisms the effects may be lessened by the presence of another set of the deleted genes in the homologous chromosome. However, if the homolog contains recessive genes with deleterious results, then the consequences can be severe. We are most knowledgeable, therefore, about the deletions that cause a detectable phenotypic change. If the deletion involves the loss of the centromere, for example, the result is an acentric chromosome, which is usually lost during meiosis, thus leading to a deletion of an entire chromosome from the genome producing a monosomic cell.

In organisms in which karyotype analysis is practical, deletions can be detected by that procedure, if the deficiencies are large enough so that one can see an apparently mismatched pair of homologous chromosomes, with one shorter than the other. In heterozygous individuals, deletions result in unpaired loops when the two homologous chromosomes pair at meiosis. Unless the deletion is large, such loops are only readily observable in the polytene chromosomes of somatic cells of certain insects that are heterozygous for a deletion (i.e., one homolog is normal and one homolog has a deletion) (Figure 19.2; p. 590).

A number of human disorders are caused by deletions. In many, but not all, cases the disorders are found in heterozygous individuals; that is, homozygotes for deletions are often lethal if the deletion is large enough. One human disorder caused by a heterozygous deletion is cri-du-chat syndrome, which results from an observable deletion of part of the short arm of chromosome 5, one of the larger human chromosomes (Figure 19.3; p. 590). Children with cri-du-chat syndrome are severely mentally retarded, have a number of physical abnormalities, and cry with a sound like the mew of a cat (hence the name, which is French for "cry of the cat").

~ FIGURE 19.2

Cytological effects at meiosis of heterozygosity for a deletion. Paired *Drosophila* salivary gland polytene chromosomes in a strain with a heterozygous deletion, showing the unpaired region; numbers refer to bands, which are presumed to indicate genes.

Paired polytene chromosomes

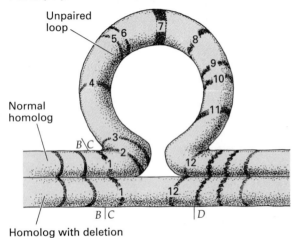

Duplication

A **duplication** is a chromosome aberration that results in the doubling of a segment of a chromosome (Figure 19.4). Duplications may or may not be lethal to an organism. The size of the duplicated segment may vary considerably, and duplicated segments may occur at different locations in the genome or in a tandem configuration (i.e., adjacent to each other). When the order of genes in the duplicated segment is the opposite of the order of the original, it is a *reverse tandem duplication*;

~ FIGURE 19.3

Cri-du-chat syndrome results from the deletion of part of one of the copies of human chromosome 5.

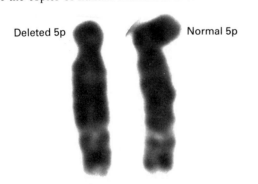

~ FIGURE 19.4

Duplication, with a chromosome segment (here, *BC*) repeated.

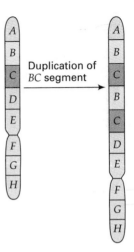

when the duplicated segments are tandemly arranged at the end of a chromosome, it is a *terminal tandem duplication* (Figure 19.5). Cytologically, heterozygous duplications result in unpaired loops similar to those described for chromosome deletions.

Evidence suggests that duplications of particular genetic regions can have unique phenotypic effects, as in the *Bar* mutant on the X chromosome of *Drosophila melanogaster* that we discussed in Chapter 3. In strains carrying the *Bar* mutation (not to be confused with the

~ FIGURE 19.5

Forms of chromosome duplications are tandem, reverse tandem, and terminal tandem duplications.

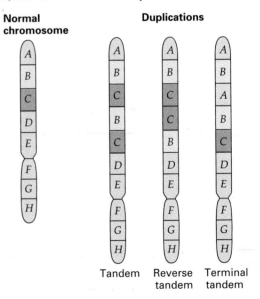

~ FIGURE 19.6

Chromosome constitutions of *Drosophila* strains, showing the relationship between duplications of region 16A of the X chromosome and the production of reduced-eye size phenotypes: (a) wild type; (b) bar mutant; (c) double-bar mutant.

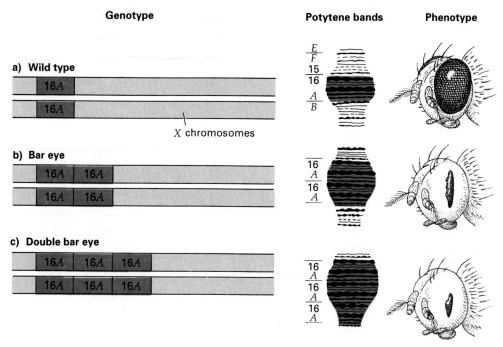

Barr body discussed in Chapter 3 and later in this chapter) the number of facets of the compound eye is fewer than for the normal eye (shown in Figure 19.6a), giving the eye a bar-shaped (slit-like) rather than an oval appearance. *Bar* acts like an incomplete dominant mutation since females heterozygous for *bar* have more facets and hence a somewhat larger bar-shaped eye (sometimes referred to as "kidney" eye) than do females homozygous for *bar*. Males hemizygous for *bar* have very small eyes like those of homozygous *bar* females. Cytological examination of polytene chromosomes has shown that the *bar* trait is the result of a duplication of a small segment of the X chromosome (Figure 19.6b). Some flies have been found that have three copies of that segment instead of the normal one copy, a condition called *double bar* (Figure 19.6c). These flies have very small, slitlike eyes.

A duplication such as *bar* probably arose as a result of a process called *unequal crossing-over* (see Figure 19.7). Unequal crossing-over may result when homologous chromosomes pair inaccurately, perhaps because similar DNA sequences occur in neighboring regions of chromosomes. Crossing-over in the mispaired region will result in gametes with a duplication or a deletion. Figure 19.7 shows how the duplicated 16A segment of

the X chromosome of *Drosophila* that is associated with *Bar* (Figure 19.6b) could have arisen.

Duplications have played an important role in the evolution of multigene families. For example, the α-globin and β-globin genes are each arranged in an array (see Chapter 17). The sequences of the genes in the α-globin family are all very similar, as are the sequences of the β-globin family of genes. It is thought that each family evolved from a different ancestral gene by duplication and divergence.

~ FIGURE 19.7

Duplications and deletions of chromosome segments that arise because of unequal crossing-over.

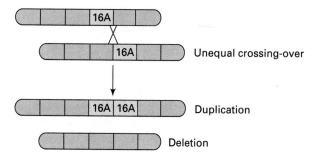

~ FIGURE 19.8

Pericentric and paracentric inversions.

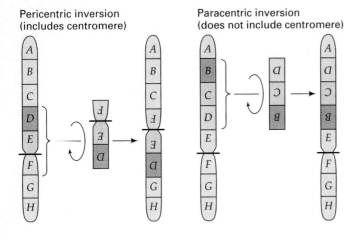

Inversion

An **inversion** is a chromosome aberration that results when a segment of a chromosome is excised and then reintegrated in an orientation 180° from the original orientation (Figure 19.8). When the inverted segment occurs on one chromosome arm and does not include the centromere, the inversion is called a **paracentric inversion.** When the inverted segment includes the parts of both chromosome arms and, hence, includes the centromere, the inversion is called a **pericentric inversion.**

In general, genetic material is not lost when an inversion takes place, although there can be phenotypic consequences when the break points (inversion points) occur within genes or within regions that control gene expression. Additionally, since gene order (i.e., the physical relationships between genes) may affect gene regulation, inversions can disrupt gene regulation. Homozygous inversions can be detected because of the unusual linkage relationships that result for the genes within the inverted segment and the genes that flank the inverted segment. For example, if the normal chromosome is *ABCDEFGH* and the *BCD* segment is inverted, the gene order will now be *ADCBEFGH* (see Figure 19.8).

The meiotic consequences of a chromosome inversion depend on whether the inversion occurs in a homozygote or a heterozygote. If the inversion is homozygous (e.g., *ADCBEFGH/ADCBEFGH*, where the *BCD* segment is the inverted segment), then meiosis will take place normally and there are no problems related to gene duplications or deletions. The situation differs if the individual is heterozygous with a normal chromosome (e.g., *ABCDEFGH/ADCBEFGH*, where one chromosome has a normal gene order and the other has an inverted *BCD* segment). In inversion heterozygotes, as these individuals are called, the homologous chromosomes attempt to pair so that the best possible match of the base-pair sequences is made. Because of the inverted segment on one homolog, pairing of homologous chromosomes necessitates the formation of loops containing the inverted segments, called inversion loops (Figure 19.9a; also see Figure 19.10; p. 594). Inversion heterozygotes, then, may be identified by looking for loops in cells undergoing meiosis or, where appropriate, in polytene chromosomes.

Heterozygous inversions can also be identified in genetic experiments, since recombination is reduced significantly or suppressed. Actually, the frequency of crossing-over is not diminished in inversion heterozygotes in comparison with normal cells, but gametes derived from recombined chromatids are inviable, as illustrated in Figure 19.9b, which shows the effects of a single crossover in the inversion loop of an individual heterozygous for a paracentric inversion. During the first meiotic anaphase, the two centromeres migrate to opposite poles of the cell. As a result of the crossover between genes *B* and *C* in the inversion loop, one recombinant chromatid becomes stretched across the cell as the two centromeres begin to migrate, forming a **dicentric bridge.** As the two centromeres continue to migrate to opposite poles in the cell, the dicentric bridge, under physical tension, breaks randomly. The other recombinant product of the crossover event, a chromosome without a centromere (an acentric fragment), is unable to continue through meiosis and is lost (i.e., it is not found in the resultant gametes).

In the second meiotic division the centromeres divide, and the chromosomes are segregated to the four gametes. Two of the gametes have complete sets of genes and are viable: the gamete with the normal order of genes (*ABCDEFGH*) and the gamete with the inverted segment (*ADCBEFGH*). The other two gametes are inviable because they both are deficient for many of the genes of the chromosome. In general, the only gametes produced that can give rise to viable progeny are those containing the chromosomes that were not involved in the crossover event.

Figure 19.10 shows the consequences of a single crossover in the inversion loop of an individual heterozygous for a pericentric inversion. The results of the crossover event and the ensuing meiotic divisions are two viable gametes, with the nonrecombinant chromosomes *ABCDEFGH* (normal) and *ADCBEFGH* (inversion), and two recombinant gametes that are inviable, each as a result of the deletion of some genes and the duplication of other genes.

~ FIGURE 19.9

Consequences of a paracentric inversion: (a) photomicrograph of an inversion loop in polytene chromosomes of a strain of *Drosophila melanogaster* that is heterozygous for a paracentric inversion; (b) meiotic products resulting from a single crossover within a heterozygous, paracentric inversion loop. Crossing-over occurs at the four-strand stage involving two nonsister homologous chromatids.

a)

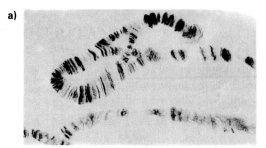

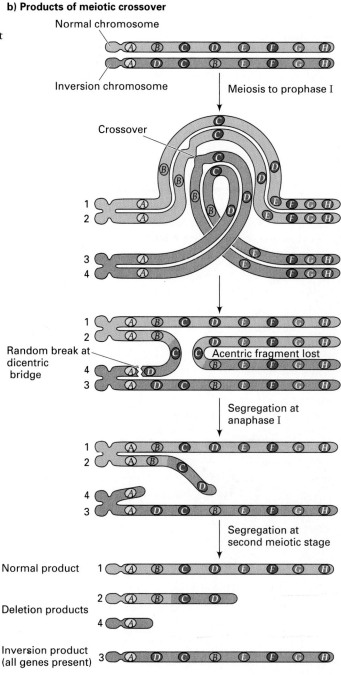

It should be noted that not all crossover events within an inversion loop lead to inviable recombinants. One exception occurs if a double crossover takes place in the inversion loop and the same two chromatids are involved in each crossover (i.e., it is a two-strand double crossover; see Chapter 6). A second exception occurs when the duplicated and deleted segments of the recombinant chromatids do not affect gene expression and hence viability to a significant degree, as when the chromosome segments involved are very small.

Translocation

A **translocation** (also called a **transposition**) is a chromosome aberration in which there is a change in position of chromosome segments and the gene sequences

~ FIGURE 19.10

Meiotic products resulting from a single crossover within a heterozygous, pericentric inversion loop. Crossing-over occurs at the four-strand stage, involving two nonsister homologous chromatids.

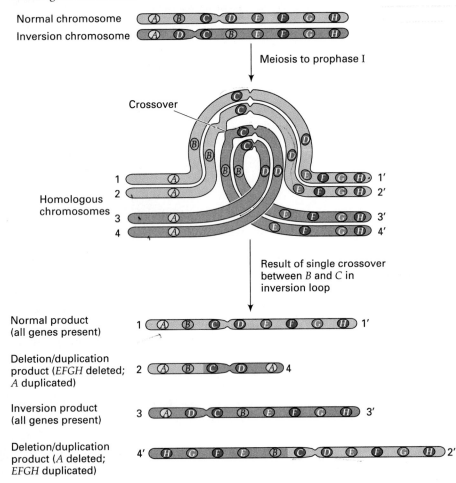

they contain (see Figure 19.11). Simple translocations may be classified as follows:

1. *Intrachromosomal* (within a chromosome) translocations (Figure 19.11a) involve a change in position of a chromosome segment within the same chromosome, either from one chromosome arm to the other or within the same chromosome arm.
2. *Interchromosomal* (between chromosomes) translocations involve the transfer of a chromosome segment from one chromosome into a nonhomologous chromosome (called a nonreciprocal translocation; see Figure 19.11b) or the reciprocal exchange of segments between two nonhomologous chromosomes (called a reciprocal translocation; see Figure 19.11c). A variety of nonreciprocal and reciprocal translocation types occur, depending on

the locations of the chromosome segment or segments involved.

In homozygotes for the translocations, the genetic consequence of translocations is an alteration in the linkage relationships of genes, in comparison with strains with normal chromosomes. For example, in the nonreciprocal translocation shown in Figure 19.11a, the *BC* segment has moved to the other chromosome arm and has become inserted between the *F* and *G* segments. As a result, genes in the *F* and *G* segments are now farther apart than they are in the normal strain, and genes in the *A* and *D* segments are now more closely linked. Similarly, in reciprocal translocations new linkage relationships are produced.

The consequences of translocations to the production of meiotic products depend on the type of translo-

~ FIGURE 19.11

Translocations: (a) nonreciprocal intrachromosomal; (b) nonreciprocal interchromosomal; (c) reciprocal.

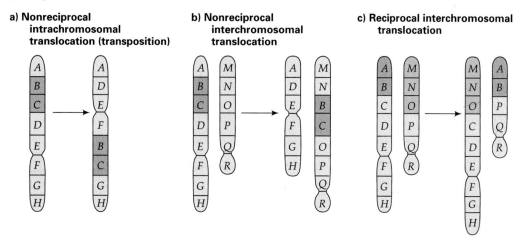

a) Nonreciprocal intrachromosomal translocation (transposition)

b) Nonreciprocal interchromosomal translocation

c) Reciprocal interchromosomal translocation

cation involved. In many cases, some of the gametes produced have duplications and/or deletions and, hence, in many cases are inviable. There are exceptions. For example, familial Down syndrome results from a duplication stemming from a translocation. This is discussed later in the chapter. We concentrate here on reciprocal translocations, since they are the most frequent and the most important to genetics.

In strains homozygous for a reciprocal translocation, meiosis proceeds normally, since all the chromosome pairs can synapse to produce bivalents. In strains heterozygous for a reciprocal translocation, however, as a result of the requirements of pairing between homologous chromosomes, crosslike configurations are seen in meiotic prophase I (Figure 19.12; p. 596). These crosslike figures consist of four associated chromosomes, each chromosome being partially homologous to two other chromosomes in the group. Depending on how the chromosomes segregate, three possible paths produce the ring and figure eight configurations seen at metaphase I. Segregation at anaphase I may occur in three different ways, producing six types of gametes. Of these six gametes, two are functional because alternate disjunction produces germ cells which contain a complete set of genes, no more, no less (one gamete contains a pair of normal chromosomes and the other contains the pair of reciprocally translocated chromosomes). The other four types of gametes are often nonfunctional, since each type contains duplications and deletions of chromosome segments.

Therefore, if we assume that the chromosomes in a reciprocal-translocation heterozygote migrate randomly in pairs to opposite poles in meiosis, we expect that two-thirds of the gametes generated will be nonfunctional because of duplications and deletions. Our assumption is not quite valid, though, in the sense that not all possible segregation patterns occur with the same frequency. The nonfunctional right pair of gametes diagramed in Figure 19.12 actually occur infrequently. In actuality, then, approximately half the gametes produced from a heterozygous reciprocal translocation are nonfunctional—hence the term *semisterility* is applied to this event. (This term is also used for inversion heterozygotes.)

In practice, animal gametes that have duplicated or deleted chromosome segments may function, but the zygotes formed by such gametes typically die. The gametes may function normally and viable offspring may result if the duplicated and deleted chromosome segments are small. In plants, pollen grains with duplicated or deleted chromosome segments typically do not develop completely and hence are nonfunctional.

$\mathcal{K}$EYNOTE

Chromosome aberrations may involve parts of individual chromosomes rather than whole chromosomes or sets of chromosomes. The four major types of structural alterations are deletions and duplications (both of which involve a change in the amount of DNA on a chromosome), inversions (which involve no change in the amount of DNA on a chromosome but rather a change in the arrangement of a chromosomal segment), and translocations (which also involve no change in DNA amount but involve a change in location of one or more DNA segments).

~ FIGURE 19.12

Meiosis in a translocation heterozygote in which no crossover occurs.

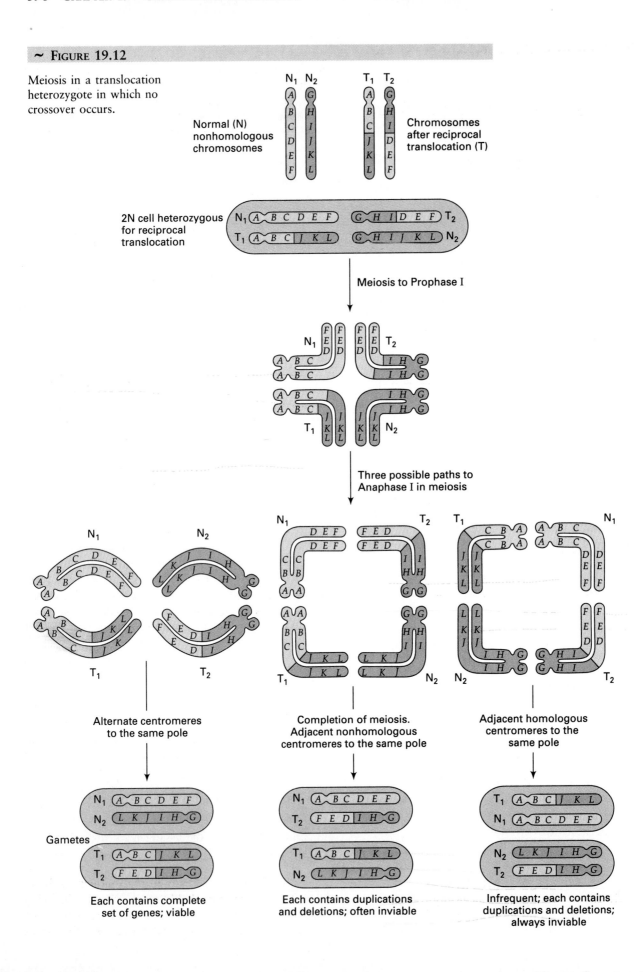

VARIATIONS IN CHROMOSOME NUMBER

Most eukaryotic organisms are diploid, meaning that they have two sets of chromosomes in their somatic cells; the gametes of these organisms are haploid, since they have only one set of chromosomes. The following discussions are related to a diploid organism.

Variations may occur from the wild-type state in the total amount of genetic material in a cell. Such changes may occur spontaneously or may be experimentally induced. Variations in chromosome number, as opposed to variations in parts of chromosomes, give rise to aneuploidy, monoploidy, and polyploidy.

Changes in One or a Few Chromosomes

ANEUPLOIDY. **Aneuploidy** describes the abnormal case in which one to a few whole chromosomes are lost from or added to the normal set of chromosomes (Figure 19.13). In most cases, aneuploidy is lethal in animals, so it is detected mainly in aborted fetuses. Aneuploidy is encountered more frequently, however, in plants because it is less likely to be lethal. The condition can occur, for example, from the loss of individual

chromosomes in mitosis or meiosis. In this case nuclei are formed with less than the normal chromosome numbers. Aneuploidy may also result from nondisjunction, the irregular distribution to the cell poles of sister chromatids in mitosis or of homologous chromosomes in meiosis (see Chapter 3). In nondisjunction, one progeny nucleus with more and one with less than the normal number of chromosomes are produced.

In the case of diploid organisms, aneuploid variations take four main forms (see Figure 19.13):

1. **Nullisomy** (a nullisomic cell) involves a loss of one homologous chromosome pair; that is, the cell is 2N − 2.
2. **Monosomy** (a monosomic cell) involves a loss of a single chromosome; that is, the cell is 2N − 1.
3. **Trisomy** (a trisomic cell) involves a single extra chromosome; that is, the cell has three copies of one chromosome type and two copies of every other chromosome type. A trisomic cell is 2N + 1.
4. **Tetrasomy** (a tetrasomic cell) involves an extra chromosome pair, resulting in the presence of four copies of one chromosome type and two copies of every other chromosome type. A tetrasomic cell is 2N + 2.

Aneuploidy may involve the loss or the addition of more than one specific chromosome or a chromosome pair. Such situations are characterized, for example, by the term *doubly monosomic* (2N − 1 − 1), *doubly tetrasomic* (2N + 2 + 2), and so forth.

All forms of aneuploidy have serious consequences in meiosis. Monosomics, for example, will produce two kinds of haploid gametes, N and N − 1. Alternatively, the odd, unpaired chromosome in the 2N − 1 cell may be lost during anaphase and not be included in either daughter nucleus, thereby producing two N − 1 gametes. In the following sections we examine some examples of aneuploidy as they are found in the human population.

Trisomy-21. Trisomy-21 occurs when there are three copies of the chromosome 21, as shown in the karyotype in Figure 19.14a (p. 598). Trisomy-21 occurs with a frequency of about 3510 per million conceptions and about 1500 per one million live births. Individuals with trisomy-21 have **Down syndrome** and exhibit such abnormalities as low IQ, epicanthal folds over the eyes, short and broad hands, and below-average height. Figure 19.14b is a photograph of a Down syndrome individual. Trisomy of other chromosomes in humans is rarely found with the exception of the sex chromosomes, which we shall study in the following subsection. Trisomy-21 is probably more common in living humans than other trisomics because chromosome 21 is a very small chromosome with fewer genes than most

~ FIGURE 19.13

Normal (theoretical) set of metaphase chromosomes in a diploid (2N) organism *(top)* and examples of aneuploidy *(bottom)*.

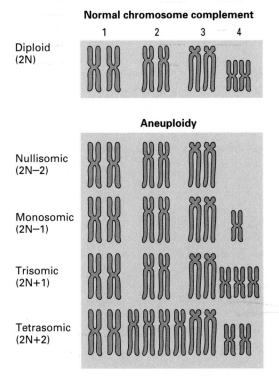

Normal chromosome complement

Diploid (2N)

Aneuploidy

Nullisomic (2N−2)

Monosomic (2N−1)

Trisomic (2N+1)

Tetrasomic (2N+2)

~ FIGURE 19.14

Trisomy-21 (Down syndrome): (a) karyotype; (b) individual.

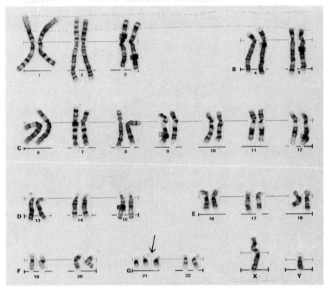

a)

b)

of the other chromosomes. The disruptions caused by the extra dose of chromosome 21 genes are less severe than would be the case with, say, the largest autosome (nonsex chromosome). Trisomics are produced for all autosomes since they are detected among fetuses that are aborted spontaneously, and it is from such fetuses that we have obtained knowledge about this sort of chromosome aberrancy. Such studies have also revealed that all autosomal monosomics in humans are lethal.

The relationship between the age of the mother and the probability of her having a trisomy-21 individual is indicated in Table 19.1. (The correlation with age of the father is much more tenuous.) Prior to the birth of a female child, the eggs in the ovary go through meiosis and arrest at diplotene. When the female becomes fertile, each month at ovulation, the nucleus of a secondary oocyte (see Chapter 1) begins the second meiotic division but progresses only to metaphase, when division is again arrested. If a sperm penetrates the secondary oocyte, the second meiotic division is completed. Apparently, as an egg matures the probability of nondisjunction increases with the storage time of the egg in the ovary. It is important, then, that older mothers-to-be consider testing to determine whether the fetus has a normal complement of chromosomes, e.g., by amniocentesis or chorionic villus sampling (see Chapter 8). Alternatively, the karyotype can be determined by chorionic villus sampling (see Chapter 8).

Down syndrome individuals can also result from a different sort of chromosomal aberrancy called a

Robertsonian translocation, which results in three copies of the long arm of chromosome 21. This form of Down syndrome is called *familial Down syndrome*. A Robertsonian translocation is a type of nonreciprocal translocation in which the long arms of two nonhomologous acrocentric chromosomes (chromosomes with centromeres near their ends) become attached to a single centromere (Figure 19.15). The short arms of the two acrocentric chromosomes become attached to form the reciprocal product, which typically contains nonessential genes and is usually lost within a few cell divisions. In humans, when a Robertsonian translocation joins the long arm of chromosome 21 with the long arm

~ TABLE 19.1

Relationship Between Age of Mother and Risk of a Trisomy-21

AGE OF MOTHER	RISK OF TRISOMY-21 IN CHILD
<29	1/3000
30–34	1/600
35–39	1/280
40–44	1/70
45–49	1/40
All mothers combined	1/665

~ FIGURE 19.15

Production of a Robertsonian translocation by chromosome breakage and rejoining involving two acrocentric chromosomes. The breakage points are indicated by arrows.

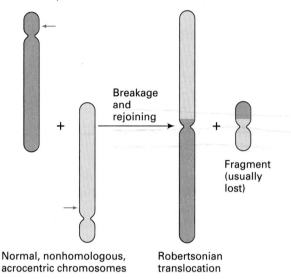

Normal, nonhomologous,   Robertsonian
acrocentric chromosomes   translocation

~ FIGURE 19.16

The three segregation patterns of a heterozygous Robertsonian translocation involving the human chromosomes 14 and 21. Fusion of the resulting gametes with gametes from a normal parent produces zygotes with various combinations of normal and translocated chromosomes.

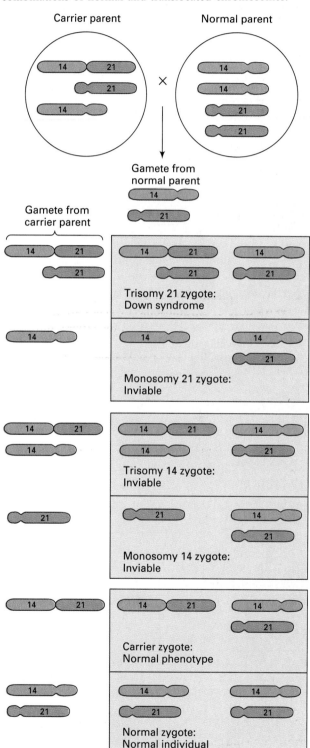

of chromosome 14 (or 15), the heterozygous carrier is phenotypically normal, since there are two copies of all major chromosome arms and hence two copies of all essential genes involved.

However, there is a high risk of Down syndrome among the offspring of these carriers. Figure 19.16 shows the gametes produced by the heterozygous carrier parent and by the normal parent and the chromosome constitutions of the offspring generated. The normal parent produces gametes with one copy each of the relevant chromosomes 14 and 21. The heterozygous carrier parent produces three reciprocal pairs of gametes, each pair being produced as a result of different segregation of the three chromosomes involved. As Figure 19.16 shows, the zygotes produced by pairing of these gametes with gametes of normal chromosomal constitution are as follows: one-sixth include normal chromosomes 14 and 21; one-sixth are heterozygous carriers like the parent and are phenotypically normal; three-sixths are inviable because of monosomy for chromosome 21 or chromosome 14, or trisomy for chromosome 14; and one-sixth are trisomy-21 and therefore give rise to a Down syndrome individual. (Note that these latter individuals actually have 46 chromosomes but because of the Robertsonian translocation, they have three copies of the long arm of chromosome 21, which is sufficient to give the Down syndrome symptoms. Similarly, the trisomy-14 individuals shown in Figure 19.16 have 46 chromosomes, but they have three copies of the long arm of chromosome 14.

Apparently the dosage of the genes involved on this larger chromosome is more critical, and consequently, these individuals are inviable.) In sum, one-half of the zygotes produced are inviable and one-third of the viable zygotes will give rise to a Down syndrome individual—a much higher risk than we discussed earlier with respect to the mother's age.

There are very few other examples of surviving trisomic individuals in humans, and these generally involve the small chromosomes. Many other human traits also have their basis in chromosome aberrancies such as partial deletions of chromosomes or rearrangements of chromosomes. More drastic changes in chromosome complement such as polyploidy, monosomy, and trisomy of the larger chromosomes have been detected in aborted fetuses, reinforcing the notion that correct gene dosage is very important to human development and function.

Aneuploidy in sex chromosomes. Aberrancies in the number of the X and Y chromosomes are much more readily found among living humans than are aberrancies in the number of autosomes. As we discussed in Chapter 3, the reason they are more readily found is that a *dosage compensation* mechanism exists in mammals by which the X chromosomes in excess of one are inactivated by becoming highly condensed masses of chromatin called Barr bodies (see Figure 3.16). Thus, normal XX females have one Barr body and normal XY males have no Barr bodies. However, individuals with a greater-than-normal number of X chromosomes have all but one X chromosome present as Barr bodies. The inactivation of an X chromosome to produce a Barr body is called *lyonization*, described earlier in Chapter 3.

Some well-studied human syndromes resulting from sex chromosome aneuploidy were discussed in Chapter 3, for example Turner syndrome (45,XO females with no Barr bodies; see Figure 3.14), Klinefelter syndrome (e.g., 47,XXY males with one Barr body; see Figure 3.15), 47,XYY males (no Barr bodies), and triplo-X females (47,XXX with two Barr bodies). About 500 live births out of 1 million have XO or XXX (female) sex chromosome abnormalities, and about 1700 live births out of 1 million have XXY, XYY or other male sex chromosome abnormalities.

The effects manifested in individuals with the X chromosome aberrancies are probably the result of the activities of the X chromosome genes during embryonic development up until the time that all but one of the X chromosomes are deactivated. Other hypotheses include the possibility that the Barr body does not represent a *completely* inactivated X chromosome. Furthermore, in some gonadal tissues, X inactivation does not occur.

Table 19.2 presents a summary of various aneuploid abnormalities for autosomes and for sex chromosomes.

~ TABLE 19.2

Aneuploid Abnormalities in the Human Population

CHROMOSOMES	SYNDROME	FREQUENCY AT BIRTH
AUTOSOMES		
Trisomic 21	Down	1/700
Trisomic 13	Patau	1/5000
Trisomic 18	Edwards	1/10,000
SEX CHROMOSOMES, FEMALES		
XO, monosomic	Turner	1/5000
XXX, trisomic XXXX, tetrasomic XXXXX, pentasomic		1/700
SEX CHROMOSOMES, MALES		
XYY, trisomic	Normal	1/1000
XXY, trisomic XXYY, tetrasomic XXXY, tetrasomic XXXXY, pentasomic XXXXXY, hexasomic	Klinefelter	1/500

Changes in Complete Sets of Chromosomes

Monoploidy and **polyploidy** involve variations from the normal state in the number of complete sets of chromosomes. Monoploidy and polyploidy are lethal for most animal species, but are tolerated more readily by plants. Both have played significant roles in plant speciation.

MONOPLOIDY. If an organism is usually diploid, a monoploid individual has only one set of chromosomes (Figure 19.17a). Monoploidy is sometimes called haploidy, although haploidy is typically used to describe the chromosome complement of gametes.

Monoploidy occurs only rarely. Owing to the presence of recessive lethal mutations in the chromosomes of many diploid eukaryotic organisms, which are usually counteracted by dominant alleles in heterozygous individuals, many monoploids probably do not survive and thus are undetected. Certain species have monoploid organisms as a normal part of their life cycle. Some male wasps, ants, and bees, for example, are normally monoploid because they develop from unfertilized eggs.

Currently, monoploids are used extensively in plant-breeding experiments, in which it is possible to isolate monoploid cells from the normal haploid products of meiosis in plant anthers. These monoploids are then induced to grow and are studied for genetic traits of interest.

Cells of a monoploid individual can also be mutagenized. The mutants can then be isolated for study without researchers having to take into account the masking of recessive mutations, as they must do when they are studying diploid cells.

POLYPLOIDY. *Polyploidy* is when a cell or organism has more than its normal number of sets of chromosomes (Figure 19.17b). Again the best examples of polyploids come from plants. In terms of a normal diploid cell, a cell with three sets of chromosomes is called a *triploid*, and one with four sets of chromosomes is called a *tetraploid*. Polyploids, which may arise spontaneously or may be induced experimentally, often occur as a result of a breakdown in the spindle apparatus in one or more meiotic divisions or in mitotic divisions. For example, the drug colchicine inhibits the formation of the mitotic spindle, so administering this drug can

~ **FIGURE 19.17**

Variations in number of complete chromosome sets: (a) Monoploidy (only one set of chromosomes instead of two); (b) Polyploidy (more than the normal number of chromosomes).

Normal chromosome complement

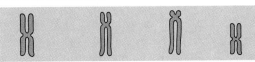

Diploid (2N)

a) **Monoploidy**
(only one set of chromosomes)

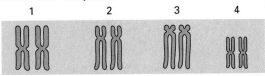

Monoploid (N)

b) **Polyploidy**
(more than the normal number of sets of chromosomes)

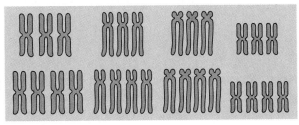

Triploid (3N)

Tetraploid (4N)

produce somatic tissue with twice the normal sets of chromosomes.

In **autopolyploidy** all the sets of chromosomes are of the same species. The condition probably results from a defect in meiosis, which in turn leads to diploid or triploid gametes. For example, if a diploid gamete resulting from an error in meiosis fuses with a normal haploid gamete, the zygote and the organism that develops from it will have three sets of chromosomes: it will be triploid.

In **allopolyploidy** the sets of chromosomes involved differ; that is, they come from different, though usually related, species. This situation can arise in the following way: First, two different species interbreed producing an organism in which each cell has one haploid set of chromosomes from each parent. Then both chromosome sets are doubled.

For example, fusion of haploid gametes of two diploid plants that can intercross may produce an N_1 + N_2 hybrid plant. That is, the hybrid will have the haploid set of chromosomes from plant 1 and the haploid set from plant 2. These plants are usually sterile, owing to the differences between the two chromosome sets. As a result, no viable gametes are produced at meiosis. Rarely, through a division error, the complete set of chromosomes doubles, producing tissues of $2N_1$ + $2N_2$ genotype. That is, the cells in the tissue have a diploid set of chromosomes from plant 1 and a diploid set from plant 2. Each diploid set can function normally in meiosis, so that gametes produced from the ($2N_1$ + $2N_2$) plant are (N_1 + N_2). Fusion of two gametes like this can produce fully fertile, allotetraploid, $2N_1$ + $2N_2$ plants.

There are two general classes of polyploids: those that have an *even* number of chromosome sets and those that have an *odd* number of sets. Polyploids with an even number of chromosome sets have a better chance of being at least partially fertile, since there is the potential for chromosomes to come together in pairs during meiosis. However, polyploids with an odd number of chromosome sets always have an unpaired chromosome for each chromosome type, so the probability of a balanced gamete is extremely low. As a result, such organisms are usually sterile.

An example of an important polyploid with an even number of chromosome sets is the cultivated bread wheat plant *Triticum aestivum*, which is hexaploid (6N). This plant is fertile and produces gametes with 21 chromosomes. The cultivated banana is an example of a polyploid organism with an odd number of chromosome sets; it is triploid. As a result, the gametes have a variable number of chromosomes and few fertile seeds are set, thereby making the banana a highly palatable fruit to eat. Because of the triploid state, cultivated bananas are propagated vegetatively.

In general, the development of "seedless" fruits relies on "odd-number" polyploidy.

While there are a number of polyploid animal species (e.g., North American suckers, a freshwater fish), animals appear to have evolved to the point that the consequences of more than two sets of chromosomes in a diploid organism are generally severe or lethal. (The molecular basis for the severe or lethal effects is not well understood.) Thus the evidence for polyploidy in animals often comes from studies of spontaneously aborted fetuses, so we know that polyploidy is one of the major bases for early, spontaneous abortions of human or other animal fetuses. In contrast, plants are much more likely to tolerate polyploidy, although, as we have seen, they may be sterile.

KEYNOTE

Variations in the chromosome number of a cell or an organism give rise to aneuploidy, monoploidy, and polyploidy. In aneuploidy a cell or organism has one, two, or a few whole chromosomes more or less than the basic number of the species under study. In monoploidy an organism that is usually diploid has only one set of chromosomes. And in polyploidy an organism has more than its normal number of sets of chromosomes. Any or all of these abnormal conditions may have serious consequences to the organism.

CHROMOSOME ABERRATIONS AND HUMAN TUMORS

A number of human tumors have been shown to be associated with chromosome aberrations, either through a change in the number of chromosomes through nondisjunction, or through a change in chromosome structure(s) involving deletions, duplications, inversions, and translocations. It is not clear whether the tumor state is caused by chromosomal aberrations, or whether the chromosomal aberration results from the growth activities of the tumor cell. A number of examples, however, support the first hypothesis. In some cases, for example, a tumor that exhibits a chromosomal abnormality early in its inception will develop other chromosomal aberrations in time, and this often correlates with a progression to an uncontrolled growth state. Examples of tumors associated with a consistent chromosome aberration are chronic myelogenous leukemia (reciprocal translocation involving chromosomes 9 and 22), Burkitt's lymphoma (reciprocal translocation involving chromosomes 8 and 14), constitutional retinoblastoma (deletion of part of chromosome 13), and meningioma (monosomy, chromosome 22).

Chronic Myelogenous Leukemia and the Philadelphia Chromosome

Chronic myelogenous leukemia is an invariably fatal cancer involving uncontrolled replication of myeloid stem cells. Ninety percent of chronic myelogenous leukemia patients have a chromosome aberration in the leukemic cells called the *Philadelphia chromosome* (Ph^1). The Philadelphia chromosome results from a reciprocal translocation event in which a part of the long arm of chromosome 22 (the second smallest human chromosome) is translocated to chromosome 9, and a small part from the tip of chromosome 9 is translocated to chromosome 22 (Figure 19.18). Thus, chronic myelogenous leukemia actually results from two chromosomal aberrations, one involving chromosome 22 and the other involving chromosome 9. This reciprocal translocation event apparently activates genes (called *oncogenes*: see Chapter 20), which initiate the transition from a differentiated cell to a tumor cell with an uncontrolled pattern of growth.

Chromosomal Aberration and Burkitt's Lymphoma

Burkitt's lymphoma, a particularly common disease in Africa, is a viral induced tumor that affects cells of the immune system called B cells. Characteristically, the tumorous B cells secrete antibodies. Ninety percent of tumors in Burkitt's lymphoma patients are associated with a reciprocal translocation involving chromosomes 8 and 14 (Figure 19.19). As in the case for chronic myelogenous leukemia and the Philadelphia chromo-

~ FIGURE 19.18

Origin of the Philadelphia chromosome in chronic myelogenous leukemia by a reciprocal translocation involving chromosome 9 and 22.

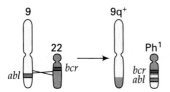

~ FIGURE 19.19

The common translocation involving chromosomes 8 and 14 seen in Burkitt's lymphoma. Diagrams of normal and aberrant chromosomes are shown. The approximate locations of the c-*myc* and immunoglobulin genes are shown.

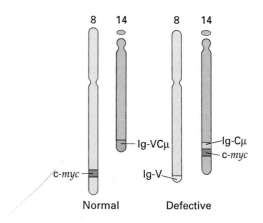

some, there is evidence that a cancer-causing oncogene (in this case c-*myc*) becomes activated as a result of the reciprocal translocation event. The c-*myc* oncogene is located on chromosome 8. In translocations involving chromosome 14, the oncogene plus a distal part of the chromosome moves to chromosome 14 near DNA encoding parts of immunoglobulin chains. This translocation event results in altered transcription of the gene and is associated with the transition of the cell in which the translocation has occurred to the tumor state. In addition, the translocation event results in the activation of transcription of immunoglobulin genes near the chromosome 14 breakpoint. This transcription activation is responsible for the observed secretion of antibodies (immunoglobulin molecules) associated with the disease.

CHROMOSOME REARRANGEMENTS THAT ALTER GENE EXPRESSION

Up to this point in the chapter we have discussed chromosomal aberrations; that is, abnormal alterations in the organization of genes in the chromosomes. In this section we will describe some chromosomal rearrangements that occur in nature as mechanisms of altering gene expression. Many of such chromosomal rearrangements are part of the normal developmental program of an organism, and thus belong more in the area of developmental biology than genetics. Here, we will only give a brief list of the types of rearrangements that alter gene expression.

Amplification or Deletion of Genes

The programmed amplification or deletion of genes is found widely in nature. Amplification may involve a large part or all of the genome, or only a small set of genes. The former is exemplified by polytenization of chromosomes in insect salivary glands and the latter by the amplification of the 18S, 5.8S, and 28S rRNA transcription units in frog eggs. Programmed deletions also vary in extent. They may involve a systematic loss of whole chromosomes (e.g., the loss of an X chromosome in some organisms as a way to achieve the same dosage compensation that other organisms acquire through X-chromosome inactivation), or the loss of one or a few genes.

Gene amplification generally leads to the synthesis of more gene product, not to a change in gene expression of the genes involved. The effects of deletions on gene expression vary, however. If a gene is completely deleted, then the expression of that gene is abolished. However, deletions can also affect regulatory regions associated with genes, which also can result in an increase or decrease of gene expression.

Inversions That Alter Gene Expression

Inversion as a mode of regulation of gene expression is rare, only involving a few genes in bacteria and their viruses. One way that an inversion can function as a regulator is as follows: A single promoter flanks a piece of DNA with genes A and B in opposite orientations. If the A gene is near the promoter, then the A gene is expressed. If the segment of DNA inverts, the B gene is brought next to the promoter, and the B gene is expressed instead of the A gene. Thus, the inversion event acts as a genetic switch. This type of inversion control of gene expression is seen in phage Mu, where two types of tail fibers can be made, which enables the phage to infect more than one type of bacterial host.

Transpositions That Alter Gene Expression

Transpositions are movements of DNA segments from one location to another in the genome. Programmed transpositions that alter gene expression are known for a number of systems, two of which will be mentioned here.

1. *Antigenic variation in Trypanosomes.* Trypanosomes are parasitic protozoa that can multiply for long periods in the bloodstream of mammalian hosts. The parasite evades the host immune defense system by repeatedly altering its surface coat so that it avoids attack by antibodies. The surface coat consists of a single glycoprotein species, and changes result from transposition of a silent copy of a new surface protein gene to the expression site, where it is transcribed. With at least 10^3 silent surface protein genes in the genome, there is an extremely large repertoire of possible surface coats.

2. *Mating type switching in yeast.* Yeast has two mating types, *a* and *α* (see Chapter 6). Mating type is controlled by the DNA sequence present at the mating types locus, *MAT*. Under genetic control, haploid cells may frequently switch mating type during growth. Figure 19.20a diagrams the organization of mating-type genes on chromosome III of an *α* cell. The *α* phenotype is specified by a 2.5-kb segment of the *MATα* locus. An identical copy of the 2.5-kb, *α*-specific sequence is present on chromosome III at a locus called *HMLα*. This DNA is not expressed because associated with it is a regulatory sequence *HMLE* that represses transcription from the two, divergent promoters; that is, *HMLα* is a "silent," unexpressed copy of the *MATα* sequence. The repressing sequence has properties similar to enhancers (see Chapter 12 and Chapter 17); that is, it works in either orientation and at various distances from *HMLα*. However, this sequence represses rather than activates, so this regulatory sequence is called a **silencer**. On the opposite side of the active *MATα* locus is a locus called *HMRa*, which contains all of the DNA necessary for the mating type *a* phenotypes. *HMRa* also has two divergent promoters, and transcription from them is repressed by an associated silencer, *HMRE*.

Mating-type switching in yeast, then, involves replacing the mating-type sequence at the active *MAT* locus with a *copy* of a sequence with opposite mating-type information. Thus, in an *α* cell, a copy of the *HMRa* sequence will replace the active sequence at *MAT* to produce a mating-type *a* cell (Figure 19.20b). The next switch of the cell would replace the active *a* sequence at *MAT* with a copy of the *HMLα* sequence (Figure 19.20c).

$\mathcal{K}$EYNOTE

Some chromosome organizational changes occur in nature as mechanisms for altering gene expression. Often these changes function as genetic switches during development. Examples of such changes are amplification of the entire genome, amplification of

~ FIGURE 19.20

(a) Organization of mating-type genes on chromosome III of a mating-type α yeast cell. The *HMLE* sequence represses transcription of *HMLα*, and the *HMRE* sequence represses transcription *HMRa*; (b) Mating-type switching from α to a; (c) Mating-type switching from a to α.

a)

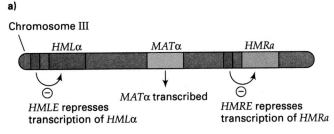

Chromosome III

HMLα   MATα   HMRa

⊖   ↓   ⊖
HMLE represses   MATα transcribed   HMRE represses
transcription of HMLα   transcription of HMRa

b)

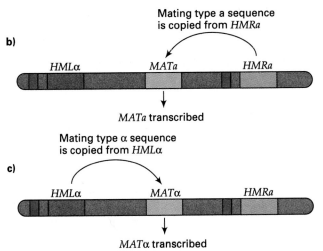

Mating type a sequence
is copied from HMRa

HMLα   MATa   HMRa

↓
MATa transcribed

Mating type α sequence
is copied from HMLα

c)

HMLα   MATα   HMRa

↓
MATα transcribed

one or a few genes, deletion of a chromosome or part of a chromosome, inversion of a chromosome segment that switches transcription from one gene to another, and transposition of a gene from a silent location to an active location, where transcription can occur.

SUMMARY

In this chapter we have considered the kinds of chromosome aberrations that occur. Chromosome aberrations are defined as variations from the wild-type conditions in either chromosome structure or chromosome number. They may occur spontaneously or their frequency can be increased by exposure to radiation or to chemical mutagens. Variations in chromosome structure are exemplified by those aberrations where changes from the normal state occur in parts of individ-ual chromosomes rather than whole chromosomes or sets of chromosomes in the genome. The four major types of structural aberrations are (1) deletion, in which a chromosome segment is lost; (2) duplication, in which a chromosome segment is present in more copies than normal; (3) inversion, in which the orientation of a chromosome segment is opposite that of the wild type; and (4) translocation, in which a chromosome segment has moved to a new location in the genome. The consequences of these kinds of structural aberrations depend on the specific aberration involved. Firstly, each kind of aberration involves one or more breaks in a chromosome. If a break occurs within a gene, then effectively a gene mutation has been produced, the consequence of which will depend on the function of the gene and when it is expressed. In the case of deletions, of course, genes may be lost and multiple mutant phenotypes may result. In some cases deletions and duplications result in lethality or severe defects as a consequence of a deviation from the normal gene dosage.

Secondly, chromosomal aberrations in the heterozygous condition can result in production of some gametes that are inviable because of duplications and/or deficiencies. Commonly this is seen following meiotic crossing-overs that produce inversions and translocations in heterozygotes.

Variations in chromosome number involve departure from the normal diploid (or haploid) state of the organism. For diploids, three classes of such aberrations are observed: (1) aneuploidy, in which one to a few whole chromosomes are lost from or added to the normal chromosome set; (2) monoploidy, in which only one set of chromosomes is present; and (3) polyploidy, in which an integral number greater than two of the haploid number of chromosomes is present. The consequences of these chromosome aberrations depend on the organism. In general, plants are more tolerant than animals of variations in the number of chromosome sets; for example, wheat is hexaploid. While some animal species are naturally polyploid, in the majority of instances monoploidy and polyploidy are lethal, probably because gene expression problems occur when abnormal numbers of gene copies are present. Even in viable individuals, viable gametes may not result because of segregation problems during meiosis.

Lastly, we mentioned a few examples of natural systems in which changes in chromosome organization are associated with altered gene expression. These changes include amplification of chromosomes or chromosome segments, deletions, inversions, and transpositions. A number of these changes function as genetic switches during development.

ANALYTICAL APPROACHES FOR SOLVING GENETICS PROBLEMS

Q.1 Diagram the meiotic-pairing behavior of the four chromatids in an inversion heterozygote *abcdefg/ a'b'f'e'd'c'g'*. Assume that the centromere is to the left of gene *a*. Next, diagram the early anaphase configuration if a crossover occurred between genes *d* and *e*.

A.1 This question requires a knowledge of meiosis and the ability to draw and manipulate an appropriate inversion loop. Part a of the figure below gives the diagram for the meiotic pairing.

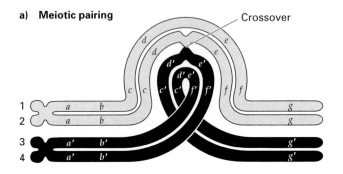

a) Meiotic pairing

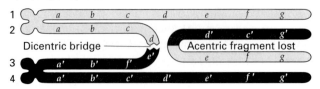

b) Early anaphase

Note that the lower pair of chromatids (*a'*, *b'*, etc.) must loop over in order for all the genes to align; this looping is characteristic of the pairing behavior expected for an inversion heterozygote.

Once the first diagram has been constructed, answering the second part of the question is straightforward. We diagram the crossover, then trace each chromatid from the centromere end until the other end is reached. It is convenient to distinguish maternal and paternal genes, perhaps by *a'* versus *a*, and so on, as we did in part a in the figure. The result of the crossover between *d* and *e* is shown in part b.

In anaphase I of meiosis the two centromeres, each with two chromatids attached, migrate toward the opposite poles of the cell. At anaphase the noncrossover chromatids (top and bottom chromatids in the figure) segregate to the poles normally. As a result of the single crossover between the other two chromatids, however, unusual chromatid configurations are produced, and these

configurations are found by tracing the chromatids from left to right. If we begin by tracing the second chromatid from the top, we get ○ *a b c d e' f' b' a'* ○ (where ○ is a centromere); in other words, we have a single chromatid attached to two centromeres. This chromatid also has duplications and deletions for some of the genes. Thus during anaphase this so-called dicentric chromosome becomes stretched between the two poles of the cells as the centromeres separate, and the chromosome will eventually break at a random location. The other product of the single crossover event is an acentric fragment that can be traced starting from the *right* with the second chromatid from the top. This chromatid is *g f e d' c' g'*, which contains neither a complete set of genes nor a centromere—it is an acentric fragment that will be lost as meiosis continues.

Thus the consequence of a crossover event within the inversion in an inversion heterozygote is the production of gametes with duplicated or deleted genes. Hence these gametes often will be inviable. Viable gametes are produced, however, from the noncrossover chromatids: One of these chromatids (1 in part b of the figure) has the normal gene sequence, and the other (4 in part b of the figure) has the inverted gene sequence.

Q.2 Eyeless is a recessive gene (*ey*) on chromosome 4 of *Drosophila melanogaster*. Flies homozygous for *ey* have tiny eyes or no eyes at all. A male fly trisomic for chromosome 4 with the genotype *+/+/ey* is crossed with a normal, eyeless female of genotype *ey/ey*. What are the expected genotypic and phenotypic ratios that would result from random assortment of the chromosomes to the gametes?

A.2 To answer this question, we must apply our understanding of meiosis to the unusual situation of a trisomic cell. Regarding the *ey/ey* female, only one gamete class can be produced, namely, eggs of genotype *ey*. Gamete production with respect to the trisomy for chromosome 4 occurs by a random segregation pattern in which two chromosomes migrate to one pole and the other chromosome migrates to the other pole in meiosis I. (This pattern is similar to the meiotic segregation pattern shown in secondary nondisjunction of XXY cells; see Chapter 3.) Three types of segregation are possible in the formation of gametes in the trisomy, as shown in part a of Figure 19.A. The union of these sperm at random with eggs of genotype *ey* occurs as shown in part b.

The resulting genotypic and phenotypic ratios are illustrated in part c.

FIGURE 19.A

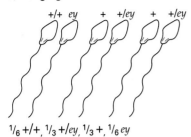

a) Segregation

+/+ ey   +   +/ey   +   +/ey

$\frac{1}{6}$ +/+, $\frac{1}{3}$ +/ey, $\frac{1}{3}$ +, $\frac{1}{6}$ ey

b) Union

		Eggs ey	Phenotype
	+/+	+/+/ey	+
	ey	ey/ey	ey
Sperm	+	+/ey	+
	+/ey	+/ey/ey	+
	+	+/ey	+
	+/ey	+/ey/ey	+

c) Summary of genotypes and phenotypes

Ratios:	Genotypes	Phenotype
	$\frac{1}{6}$ +/+/ey	$\frac{5}{6}$ wild type
	$\frac{1}{3}$ +/ey/ey	$\frac{1}{6}$ eyeless
	$\frac{1}{3}$ +/ey	
	$\frac{1}{6}$ ey/ey	

QUESTIONS AND PROBLEMS

*19.1 A normal chromosome has the following gene sequence:

$$A \quad B \quad C \quad D_{\circ} E \quad F \quad G \quad H$$

Determine the chromosome mutation in each of the following chromosomes:
a. $A \quad B \quad C \quad F \quad E_{\circ} D \quad G \quad H$

b. $A \quad D_{\circ} E \quad F \quad B \quad C \quad G \quad H$

c. $A \quad B \quad C \quad D_{\circ} E \quad F \quad E \quad F \quad G \quad H$

d. $A \quad B \quad C \quad D_{\circ} E \quad F \quad F \quad E \quad G \quad H$

e. $A \quad B \quad D_{\circ} E \quad F \quad G \quad H$

*19.2 Define pericentric and paracentric inversions.

19.3 Very small deletions behave in some instances like recessive mutations. Why are some recessive mutations known not to be deletions?

19.4 What would be the result, in terms of protein structure, if a small inversion were to occur within the coding region of a structural gene?

*19.5 Inversions are said to affect crossing-over. The following homologs with the indicated gene order are given:

$\bullet A \quad B \quad C \quad D \quad E$

$\circ A \quad D \quad C \quad B \quad E$

a. Diagram the way these chromosomes would align in meiosis.
b. Diagram what a single crossover between homologous genes B and C in the inversion would result in.
c. Considering the position of the centromere, what is this sort of inversion called?

19.6 Single crossovers within the inversion loop of inversion heterozygotes give rise to chromatids with duplications and deletions. What happens when, within the inversion loop, there is a two-strand double crossover in such an inversion heterozygote when the centromere is outside the inversion loop?

19.7 An inversion heterozygote possesses one chromosome with genes in the normal order:

$\underset{\circ}{\quad} a \quad b \quad c \quad d \quad e \quad f \quad g \quad h$

It also contains one chromosome with genes in the inverted order:

$\underset{\circ}{\quad} a \quad b \quad f \quad e \quad d \quad c \quad g \quad h$

A four-strand double crossover occurs in the areas $e-f$ and $c-d$. Diagram and label the four strands at synapsis (showing the crossovers) and at the first meiotic anaphase.

*19.8 Mr. and Mrs. Lambert have not yet been able to produce a viable child. They have had two miscarriages and one severely defective child who died soon after birth. Studies of banded chromosomes of father, mother, and child showed all chromosomes were normal except for pair number 6. The number 6 chromosomes of mother, father, and child are shown in Figure 19.B.

FIGURE 19.B

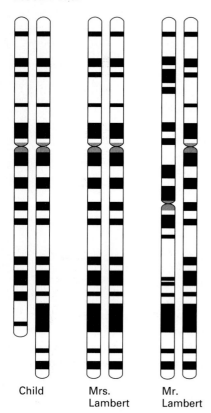

Child   Mrs.   Mr.
         Lambert Lambert

FIGURE 19.C

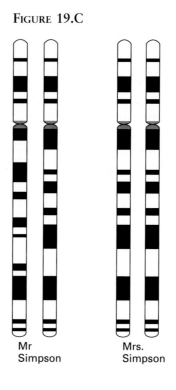

Mr   Mrs.
Simpson   Simpson

a. Does either parent have an abnormal chromosome? If so, what is the abnormality?
b. How did the chromosomes of the child arise? Be specific as to what events in the parents gave rise to these chromosomes.
c. Why is the child not phenotypically normal?
d. What can be predicted about future conceptions by this couple?

19.9 Mr. and Mrs. Simpson have been trying for years to have a child but have been unable to conceive. They consulted a physician, and tests revealed that Mr. Simpson had a very reduced sperm count. His chromosomes were studied, and a testicular biopsy was done as well. His chromosomes proved to be normal, except for pair 12. Figure 19.C shows a normal pair of number 12 chromosomes and Mr. Simpson's number 12 chromosomes.
a. What is the nature of the abnormality in Mr. Simpson's chromosome 12s?
b. What abnormal feature would you expect to see in the testicular biopsy (cells in various stages of meiosis can be seen).
c. Why is Mr. Simpson's sperm count low?
d. What can be done about it?

*19.10 The following gene arrangements in a particular chromosome are found in *Drosophila* populations in different geographical regions. Assuming the arrangement in part a is the original arrangement, in what sequence did the various inversion types most likely arise?
a. *ABCDEFGHI*
b. *HEFBAGCDI*
c. *ABFEDCGHI*
d. *ABFCGHEDI*
e. *ABFEHGCDI*

*19.11 Chromosome I in maize has the gene sequence *ABCDEF*, whereas chromosome II has the sequence *MNOPQR*. A reciprocal translocation resulted in *ABCPQR* and *MNODEF*. Diagram the expected pachytene configuration of the F_1 of a cross of homozygotes of these two arrangements.

19.12 Diagram the pairing behavior at prophase of meiosis I of a translocation heterozygote that has normal chromosomes of gene order *abcdefg* and *tuvwxyz* and has translocated chromosomes *abcdvwxyz* and *tuefg*. Assume that the centromere is at the left end of all chromosomes.

*19.13 Mr. and Mrs. Denton have been trying for several years to have a child. They have experienced a series of miscarriages, and last year they had a child with multiple congenital defects. The child died within days of birth. The birth of this child prompted the Denton's physician to order a chromosome study of parents and child. The results of the study are shown in the figure below. Chromosome banding was done, and all chromosomes were normal in these individuals except some copies of number 6 and number 12. The number 6 and number 12 chromosomes of mother, father, and child are shown in the figure (the number 6 chromosomes are the larger pair).

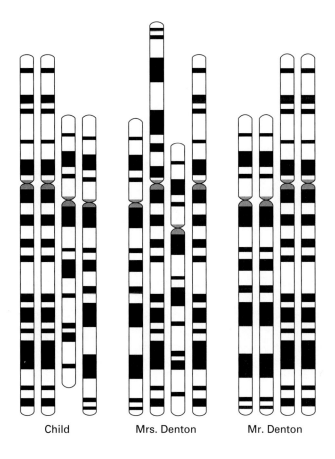

Child   Mrs. Denton   Mr. Denton

a. Does either parent have an abnormal karyotype? If so, which parent has it, and what is the nature of the abnormality?
b. How did the child's karyotype arise (what pairing and segregation events took place in the parents)?
c. Why is the child phenotypically defective?
d. What can this couple expect to occur in subsequent conceptions?
e. What medical help, if any, can be offered to them?

19.14 Define the terms *aneuploidy, monoploidy,* and *polyploidy.*

19.15 If a normal diploid cell is 2N, what is the chromosome content of the following: (a) a nullisomic; (b) a monosomic; (c) a double monosomic; (d) a tetrasomic; (e) a double trisomic; (f) a tetraploid; (g) a hexaploid?

*19.16 In humans, how many chromosomes would be typical of nuclei of cells that are (a) monosomic; (b) trisomic; (c) monoploid; (d) triploid; (e) tetrasomic?

*19.17 An individual with 47 chromosomes, including an additional chromosome 15, is said to be:
a. A triplet
b. Trisomic
c. Triploid
d. Tricycle

*19.18 A color-blind man marries a homozygous normal woman and after four joyful years of marriage they have two children. Unfortunately, both children have Turner syndrome, although one has normal vision and one is color-blind. The type of color blindness involved is a sex-linked recessive trait.
a. For the color-blind child with Turner syndrome, did nondisjunction occur in the mother or the father? Explain your answer.
b. For the Turner child with normal vision, in which parent did nondisjunction occur? Explain your answer.

19.19 In general, polyploids with even multiples of the chromosome set are more fertile than polyploids with odd multiples of the chromosome set. Why?

*19.20 From Mendel's first law genes A and a segregate from each other and appear in equal numbers among the gametes. However obvious that phenomenon may seem now, it was not obvious to Mendel. Mendel did not know that his plants were diploid. In fact, since plants are frequently tetraploid, he could have been unlucky enough to have started with peas that were 4N rather than 2N. Let us assume that Mendel's peas were tetraploid, that every gamete contains two alleles, and that the distribution of alleles to the gamete is random. Suppose we have a cross of $AA\ AA \times aa\ aa$, where A is dominant, regardless of the number of a alleles present in an individual.
a. What will be the genotype of the F_1?
b. If the F_1 is selfed, what will be the phenotypic ratios in the F_2?

19.21 What phenotypic ratio of A to a is expected if $AA\ aa$ plants are testcrossed against $aa\ aa$ individuals? (Assume that the dominant phenotype is expressed whenever at least one A is present, no crossing-over occurs, and each gamete receives two chromosomes.)

19.22 The root tip cells of an autotetraploid plant contain 48 chromosomes. How many chromosomes were contained by the gametes of the diploid from which this plant was derived?

*** 19.23** How many chromosomes would be found in somatic cells of an allotetraploid derived from two plants, one with N = 7 and the other with N = 10?

19.24 Plant species A has a haploid complement of 4 chromosomes. A related species B has N = 5. In a geographical region where A and B are both present, C plants are found that have some characters of both species and somatic cells with 18 chromosomes. What is the chromosome constitution of the C plants likely to be? With what plants would they have to be crossed in order to produce fertile seed?

20 Transposable Genetic Elements, Tumor Viruses, and Oncogenes

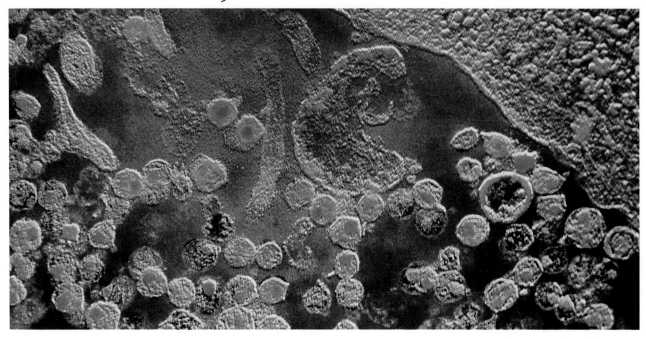

PRINCIPAL POINTS

~ Transposable genetic elements (TGE) are unique DNA segments that can insert themselves at one or more sites in a genome. In bacteria the three major types of transposable genetic elements are insertion sequence (IS) elements, transposons (Tn), and some temperate bacteriophages, such as Mu. IS and Tn elements have repeated sequences at their ends and encode proteins that are responsible for their transposition. When IS and Tn elements integrate into the genome, a short sequence at the target site is duplicated, giving rise to directly repeated sequences flanking the integrated element.

~ Transposable genetic elements in eukaryotes resemble bacterial transposons in general structure and transposition properties. While most transposons move using a DNA-to-DNA mechanism, some eukaryotic transposons such as yeast *Ty* elements transpose via an RNA intermediate (using a transposon-encoded reverse transcriptase). Such transposons resemble retroviruses in genome organization and other properties.

~ The presence of TGEs in a cell is usually detected by the changes they bring about in the expression and activities of the genes at or near the chromosomal sites into which they integrate.

~ Some forms of cancer are caused by tumor viruses. Both DNA and RNA tumor viruses are known. RNA tumor viruses, which contain tumor-inducing genes termed oncogenes, are part of the retroviruses. Both normal cells and non-viral-induced cancer cells contain sequences that are related to the viral oncogenes. The model is that tumor-causing retroviruses have picked up certain normal cellular genes—called proto-oncogenes—while simultaneously losing some of their genetic information. In normal cells, proto-oncogenes function in various ways to regulate cell differentiation. In the retrovirus these genes have been modified or their expression controlled differently so that the protein product, now synthesized under viral control, is both qualitatively and quantitatively changed, as well as being expressed in cells in which the products are not normally found. These gene products, which include growth factors, are directly responsible for the transformation of cells to the cancerous state.

The classic picture of genes is one in which the genes are at fixed loci on a chromosome. We now know, however, that certain genetic elements of chromosomes of both prokaryotes and eukaryotes have the capacity to move from one location to another in the genome. These mobile genetic elements have been given a number of names in the literature, including controlling elements, jumping genes, insertion sequences, and transposons. We shall use the term that has become fairly widely accepted, **transposable genetic elements** (**TGE**), since this generic term reflects the *transposition* events associated with these elements.

In structure and function TGEs are similar in both prokaryotes (in which they are found in bacteria, phages, and plasmids) and in eukaryotes and their viruses. In eukaryotes, TGEs have the property to move to new positions within the same chromosome or to a different chromosome. In both prokaryotes and eukaryotes, TGEs are identified by the changes they caused; for example, they can produce mutations by inserting into genes, they can affect gene expression by inserting into gene regulatory sequences, and they can produce various kinds of chromosome aberrations. They have also probably been important in genome evolution. These effects of TGEs have been established through genetic, cytological, molecular, and recombinant DNA procedures.

In this chapter we also discuss tumor viruses, oncogenes, and cancer. Tumor viruses cause cancers, and both DNA and RNA tumor viruses are known. Here we focus on retroviruses, which are RNA tumor viruses. Retroviruses replicate via a DNA intermediate that becomes integrated into the host cell's chromosome. Continuing the theme of genes that move, tumor-inducing retroviruses cause cancer because of the activities of normal genes that have been picked up from the host genome, inserted into the retroviral genome, and altered in some way. When in the retroviral genome, these genes are called *viral oncogenes*. The cellular counterparts of these genes—*cellular proto-oncogenes*—typically encode proteins involved in growth regulation of cells.

TRANSPOSABLE GENETIC ELEMENTS IN BACTERIA

There are three types of transposable genetic elements in bacteria: insertion sequence (IS) elements, transposons (Tn), and certain temperate bacteriophages.

Insertion Sequences

An **insertion sequence (IS)**, or **IS element,** is the simplest transposable genetic element found in prokaryotes, and is a mobile segment of DNA. An IS element contains genes required for inserting the DNA segment into a chromosome and for mobilizing the element to different locations. IS elements are normal constituents of bacterial chromosomes and plasmids.

DISCOVERY OF IS ELEMENTS. The IS elements were first identified in *E. coli* as a result of their effects on the expression of a set of three genes whose products are needed to metabolize the sugar galactose as a carbon source. A certain set of mutations affecting the expression of these genes did not have properties typical of the classes of gene mutations discussed in Chapter 18. Careful analysis of this unusual set of mutations revealed that the mutant phenotypes resulted from the insertion of an approximately 800-base-pair (bp) DNA segment into a gene. This particular DNA segment is now called insertion sequence 1, or IS1 (Figure 20.1).

PROPERTIES OF IS ELEMENTS. IS1 is one of a family of genetic elements capable of moving around the genome. It integrates into the chromosome at locations with which it has *no* homology. This event is an example of a *transposition event*. If the integration of an IS element places it within a gene, then that gene is usually inactivated. The exact effect depends on the IS element involved, however. For example, the 1327-bp IS2 element decreases gene expression if it integrates into the chromosome in one orientation, but if it integrates in the opposite orientation it increases gene expression. By contrast, the 768-bp IS1 element decreases the expression of the gene into which it integrates no matter what its orientation is.

There are a number of IS elements that have been identified in *E. coli*, including IS1, IS2, IS4, and IS10R,

~ FIGURE 20.1

An insertion sequence (IS) transposable genetic element. The IS element has inverted repeat (IR) sequences at the ends.

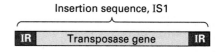

Insertion sequence, IS1

each present in 5–30 copies per genome. Altogether, they constitute approximately 0.3 percent of the cell's genome. Among prokaryotes as a whole, the ISs range in size from 768 bp to over 5000 bp and are normal cell constituents (i.e., they are found in all *E. coli*). Table 20.1 lists the properties of some of the IS elements. All IS elements that have been sequenced end with perfect or nearly perfect inverted terminal repeats of between 9 and 41 bp. This means that the same sequence is found at each end of an IS but in opposite orientations (Figure 20.2; p. 614). For example, if the sequence at one end is: 5'-ACAGTTCAG-3', the sequence at the other end of that strand is 5'-CTGAACTGT-3'. If the stretch of DNA containing an IS is denatured and allowed to reanneal, complementary base pairing can occur within the same strand of DNA (see Figure 20.2). Between the complementary ends is a segment of unpaired DNA, which gives the reannealed molecule a lollipop appearance (called a stem-and-loop structure) under the electron microscope.

IS TRANSPOSITION. Transposition is catalyzed by an enzyme activity called **transposase,** which is encoded by the IS element. The process of IS insertion into a chromosome *target site* is shown in Figure 20.3 (p. 615). First, a staggered cut is made in the host target site, then the IS element is inserted, becoming joined to the single-stranded ends. The gaps are filled in by DNA polymerase and DNA ligase, and this results in an integrated IS

~ TABLE 20.1

Properties of Some IS Elements in Bacteria

INSERTION SEQUENCE	LENGTH (bp)	INVERTED TERMINAL REPEATING SEQUENCE (bp)	DIRECT REPEATS AT TARGET (bp)
IS1	768	18/23	9
IS2	1327	32/41	5
IS4	1400	16/18	11 or 12
IS5	1250	16	4
IS10R	1329	22	9

~ FIGURE 20.2

Consequence of having inverted terminal repeated sequences in DNA. When the DNA is denatured and allowed to renature, the single strands can form stem-and-loop (lollipop) structures. [Note: The base sequences given for the inverted repeating segment are for illustration only and are not the actual sequences found, either in length or sequence.]

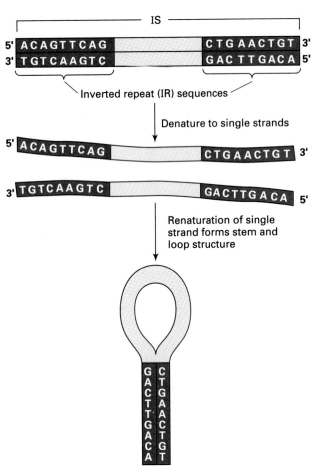

element with two *direct repeats*, or *target repeats*, of the target site sequence flanking the IS element. "Direct" in this case means that the two sequences are repeated in the same orientation (see Figure 20.3). The sizes of direct repeats varies with the IS element, but tend to be small (4–12 bp; Table 20.1).

There are actually two types of transposition events. In *conservative transposition*, the IS element moves from one site to another as a complete physical entity. IS10 and IS50 can move only by conservative transposition. In *replicative transposition*, the IS element is duplicated so that the transposing element is actually a copy of the original element. IS1 and IS903 can transpose by replicative transposition as well as by conservative transposition.

All copies of a given IS element have the same sequence, including that of the inverted terminal repeats. Mutations that affect the inverted terminal repeat sequences of IS elements affect transposition, indicating that the inverted terminal repeat sequences are key sequences recognized by transposase during a transposition event.

The frequency of transposition varies among the IS elements, with an average of 10^{-3} to 10^{-4} per IS element per generation. For a given target site, the integration rate for an IS element is 10^{-5} to 10^{-7} per generation. Integration of a free IS element is not completely random, but there are no general rules. Some IS elements show preference for particular regions (i.e., integration hotspots), while others integrate only at particular sequences.

Transposons

The next most complex type of transposable genetic element in bacteria is the **transposon (Tn)**. Transposons are mobile DNA segments that, like IS elements, contain genes for the insertion of the DNA segment into the chromosome and for the mobilization of the ele-

~ TABLE 20.2

Properties of Some Transposons Found in Bacteria

TRANSPOSON	GENE MARKER(S)	LENGTH (bp)	TERMINAL REPEAT SEQUENCES	MODULE ORIENTATION
Tn3	Resistance to ampicillin	4957	38 bp	Inverted
Tn5	Resistance to kanamycin	5700	IS50L, IS50R	Inverted
Tn9	Resistance to chloramphenicol	2638	IS1	Inverted
Tn10	Resistance to tetracycline	9300	IS10R, IS10L	Inverted
Tn903	Resistance to kanamycin	3100	IS903	Inverted

~ FIGURE 20.3

Schematic of the integration of an IS element into chromosomal DNA. As a result of the integration event, the target site becomes duplicated to produce direct target repeats. Thus, the integrated IS element is characterized by its inverted repeat (IR) sequences flanked by direct target repeat sequences. Integration involves making staggered cuts in the host target site. After insertion of the IS, the gaps that result are filled in with DNA polymerase and DNA ligase. [Note: The base sequences given for the IR are for illustration only and are not the actual sequences found, either in length or sequences.]

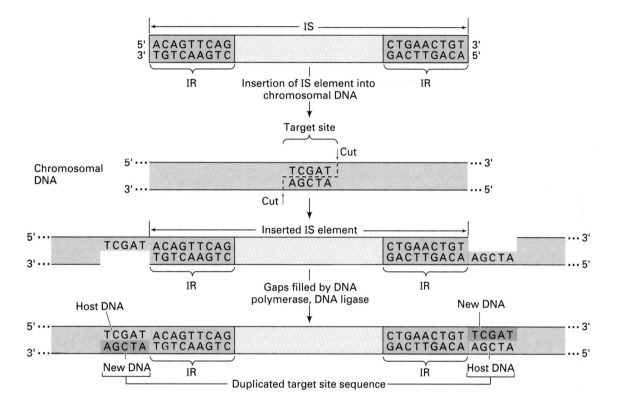

ment to other locations on the chromosome. Unlike IS elements, Tns also contain genes of identifiable function (e.g., for drug resistance). Nonetheless, there is no question that ISs and Tns are related. For example, Tns often end in long (800–1500 bp) direct or inverted repeats that are IS elements themselves, or are derived from them. In fact, many Tns are mobile in the genome because they have IS elements (called *IS modules*) at their ends; the IS modules provide transposition functions. Since IS elements themselves end in short, inverted repeats, a Tn flanked by two IS elements will also be flanked by inverted repeats. In other words, essentially all IS elements and Tns are characterized by terminal inverted repeats. A generalized transposon is shown in Figure 20.4. The properties of a number of transposons are summarized in Table 20.2.

~ FIGURE 20.4

Generalized structure of a transposon. The central region carries genes such as drug resistance genes and is flanked by either direct or inverted repeats. These repeats are IS elements, which themselves have inverted repeats at their ends.

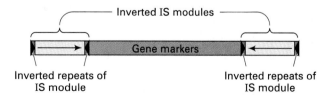

~ FIGURE 20.5

Structure of bacterial transposon Tn10.

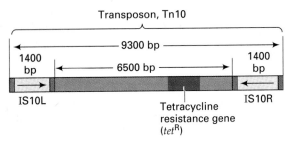

To illustrate the concepts just presented, Figure 20.5 shows the genetic organization of Tn10. This transposon is 9300 bp long and consists of 6500 bp of nonrepeated material flanked at each end with a 1400-bp IS-like module. The IS modules of this transposon are designated IS10L (left) and IS10R (right) and are arranged in an inverted orientation. Since IS10L and IS10R are inverted repeats, then, as we saw for IS elements, when Tn10 is denatured to single strands and allowed to renature, a stem-and-loop structure forms (c.f. Figure 20.2).

Within the 6500-bp segment is the gene for tetracycline resistance. Thus, cells containing Tn10 are resistant to the antibiotic tetracycline. *In vivo*, Tn10 promotes its own transposition, a process by which a copy of the Tn element is placed in a new location on the chromosome while the original insert is retained. These events are catalyzed by enzymes encoded by specific genes in the transposon. Deletion and insertion events also occur as a result of the gene activities of Tn10.

Figure 20.6 shows the structure of another transposon, Tn3. Tn3 has inverted repeat (IR) sequences 38 bp long at each end, and three known genes. One gene encodes β-lactamase, an enzyme that

can degrade ampicillin. Therefore, bacteria carrying the transposon are ampicillin resistant. The two other genes are involved in transposition events: one encodes *transposase*, an enzyme that catalyzes insertion of the Tn into new sites, and the other encodes *resolvase,* an enzyme involved in the particular recombinational events associated with transposition. Resolvase is not found in all transposons.

Transposons move in essentially the same way as IS elements; that is, by conservative transposition and/or by replicative transposition. The method of transposition is directed by the IS modules that a Tn element contains. Tn10, for example, moves only by conservative transposition, as does IS10. Specific models for transposition have been proposed for particular Tns.

Like IS elements, Tn elements produce direct repeats of a host target site sequence when they insert into a chromosome; for example, Tn5, Tn9, and Tn10 produce a 9-bp repeat, while Tn3 produces a 5-bp repeat (Figure 20.7).

Transposition of transposons requires the activity of the enzyme transposase, which recognizes the transposon ends to catalyze movement of the element. Transposition of Tn elements in bacteria is rare, occurring once in 10^7 cell generations for Tn10. This is the case because less than one transposase molecule per cell generation is made by Tn10.

A number of detailed models have been generated for transposition, one of which is shown in Figure 20.8 (p. 618). The first step in this model (proposed in 1979) is a staggered cut, producing a single-stranded nick at each end of the transposable genetic element (either an IS or a Tn). At the same time a staggered cut of opposite polarity is made in the target DNA (Figure 20.8, part 1). Next, one end of the TGE is attached by a single strand to each single-stranded end of the staggered cut on the DNA, thereby generating two replication forks (Figure 20.8, part 2). By semiconservative replication involving

~ FIGURE 20.6

Structure of bacterial transposon Tn3. Tn3 encodes three enzymes: β-lactamase (destroys β-lactam antibiotics like penicillin and ampicillin), transposase, and resolvase.

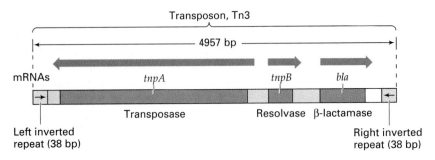

~ **FIGURE 20.7**

DNA sequence of a target site of Tn3: the important 5-nucleotide segment is highlighted. After Tn3 inserts, the same 5-nucleotide sequence is found at each end.

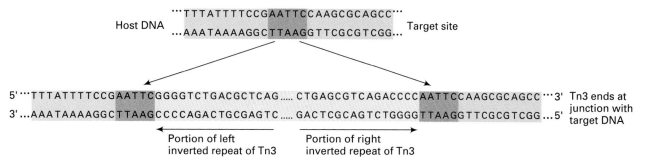

both forks, two transposons are generated, each with one old and one new DNA strand and with short, direct repeating strands representing a duplication of the target site (Figure 20.8, part 3). Thus if *ab* and *cd* were originally part of circular molecules, as is often the case, then *a, b, c,* and *d* are now covalently connected in one large, circular structure called a *cointegrate*. A transposon is present at both junction points between the two "parental" types of DNA. Lastly, crossing-over between the two transposons, catalyzed by resolvase, will resolve the cointegrate into the starting replicon *ab* and the target replicon *cd* (Figure 20.8, part 4). The target DNA replicon now has a copy of the transposable genetic element flanked by a short, direct repeat that was generated by DNA replication to fill in the staggered cut that started the entire process.

IS Elements and Transposons in Plasmids

When we studied bacterial genetics in Chapter 7, we learned that the transfer of genetic material between conjugating *E. coli* is the result of the function of the fertility factor *F*. The *F* factor, a circular double-stranded DNA molecule, is one example of a bacterial **plasmid,** an extrachromosomal genetic element capable of self-replication. Plasmids such as *F* that are also capable of integrating into the bacterial chromosome are called **episomes.**

Figure 20.9a (p. 619) shows the genetic organization of the *E. coli F* factor. It consists of 94,500 bp of DNA that code for a variety of proteins. The important genetic elements are:

1. *tra* (for "transfer") genes, required for the conjugal transfer of the DNA from a donor bacterium to a recipient bacterium;
2. Genes that encode proteins required for the plasmid's replication;

3. Four IS elements: two copies of IS3, one of IS2, and one of an insertion sequence called gamma-delta.

It is because the *E. coli* chromosome has copies of these insertion sequences at various positions that the *F* factor can integrate into the *E. coli* chromosome at different sites and in different orientations. Thus *F* factor integration occurs by conventional recombination between the homologous sequences of the insertion elements. Once integrated into the *E. coli* chromosome, the *tra* genes in the *F* factor direct the conjugal transfer functions.

Another class of plasmids that has medical significance is the *R* plasmid group, which was discovered in Japan in the 1950s. This discovery came about as a result of research into a cure for dysentery, an intestinal disease that is the result of infection by the pathogenic bacterium *Shigella*. The usual treatment of bacterial infections is the administration of antibiotics such as penicillin or ampicillin. However, the *Shigella* strain that occurred in dysentery patients in the Japanese hospitals was found to be resistant not only to penicillin but also to tetracycline, sulfanilamide, chloramphenicol, and streptomycin—all antibiotics that are usually effective in killing bacteria or stopping their growth.

Scientists discovered that the multiple-resistance phenotype was transmissible to other non-resistant strains of *Shigella* as well as to other bacteria commonly found in the human intestine. Subsequently, they found that the genes responsible for the drug resistances were carried on *R* plasmids, which can promote the transfer of genes between bacteria by conjugation, just as the *F* factor can.

Figure 20.9b shows the genetic organization of one type of *R* plasmid. One segment of an *R* plasmid that is homologous to a segment in the *F* factor is the part needed for the conjugal transfer of genes. That segment and the plasmid-specific genes for DNA replication

~ FIGURE 20.8

Transposition of a transposable genetic element (TGE): (1) Single-stranded cuts in the two starting circular DNA molecules, one of which has a TGE and the other does not. (2) The two cut DNAs come together, forming two replication forks. (3) A cointegrate molecule is produced by semiconservative replication. (4) Crossing-over between the two TGEs catalyzed by resolvase resolves the cointegrate into two circular DNA molecules, each with a TGE.

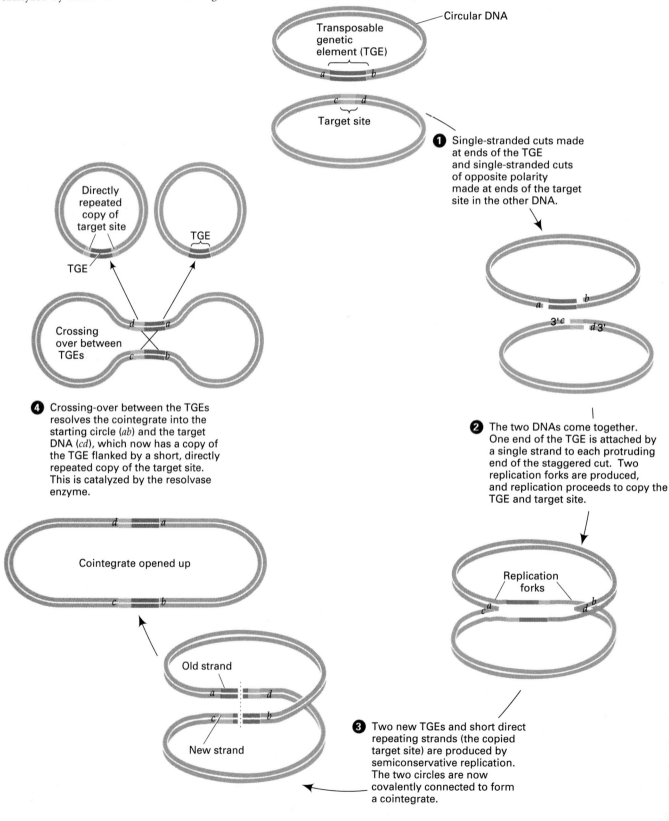

~ FIGURE 20.9

~ FIGURE 20.9

Organizational maps: (a) The *E. coli* F factor. The map shows the locations of genes required for transfer of the F factor during conjugation and for the replication of F; also shown are the insertion elements responsible for F's integration into the bacterial chromosome; (b) A typical R plasmid. The map shows the region of homology with the F factor, the area needed for DNA replication, and the sites of three antibiotic-resistance genes (*amp*, *km*, and *tet*). These genes are flanked by insertion sequences and hence are true transposons.

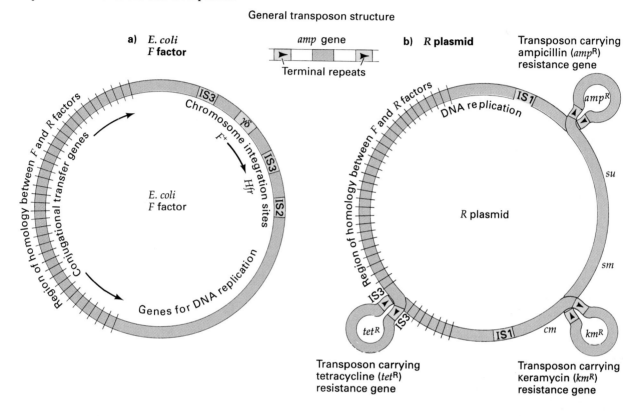

constitute what is called the *RTF* (resistance transfer factor) region. The rest of the R plasmid differs from type to type and includes the antibiotic-resistance genes or other types of genes of medical significance, such as resistance to heavy-metal ions.

The resistance genes in R plasmids are, in fact, transposons; that is, each resistance gene is located between flanking, directly repeated segments such as one of the IS modules. Thus, each transposon with its resistance gene in the R plasmid can be inserted into new locations on other plasmids or on the bacterial chromosome, while at the same time leaving behind a copy of itself in the original position. Different R plasmids contain various combinations of transposons carrying drug-resistance genes. In an individual to whom drugs are being administered, any bacterium containing an R plasmid that is resistant to the drugs will be able to survive and propagate, whereas bacteria without resistance genes will be killed. Thus, there is selection

pressure for new R plasmids to be produced by transposition of transposons that include genes for drug-resistance.

Temperate Bacteriophage Mu

Mu is a temperate bacteriophage that infects *E. coli*. Recall that *temperate bacteriophages* (e.g., λ) can go through the lytic cycle or enter the lysogenic phase (see Chapter 7). Mu is also a transposon, and it can cause mutations when it transposes. In fact, the name Mu stands for mutator.

In the phage particle, the Mu genome is a 37-kb linear piece of DNA consisting mostly of phage DNA, with unequal lengths of host DNA at the two ends (Figure 20.10a; p. 620). When Mu infects *E. coli* and enters the lysogenic state, the Mu genome integrates into the host chromosome by conservative transposition to produce the integrated prophage DNA, flanked

~ FIGURE 20.10

Temperate bacteriophage Mu genome shown (a) in phage particles and (b) integrated into the *E. coli* chromosome as a prophage.

a)   Phage DNA present in virus particles

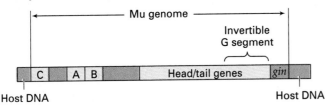

b)   Prophage DNA

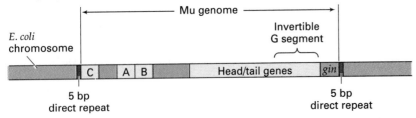

by a 5-bp direct repeat of the host target site sequence (Figure 20.10b). During integration, the flanking host DNA that was present in the particle is lost. In the lysogenic state, a phage-encoded repressor prevents most Mu gene expression (c.f. λ repressor: Chapter 7), and the Mu prophage replicates when the *E. coli* chromosome replicates.

Mu's lytic cycle is different from that of λ. The Mu genome integrates in a different way than for lysogeny and remains integrated throughout the lytic phase. Replication of the Mu genome occurs by replicative transposition. The details of Mu transposition remain to be determined.

Transposition by Mu causes mutations of various kinds. Apart from simple insertions, Mu can also cause deletions, inversions, and translocations. Figure 20.11 shows how Mu (and other multi-copy transposons) can cause deletions (Figure 20.11a) and inversions (Figure 20.11b) by homologous recombination between two identical copies. Deletion occurs if the two Mu prophages or transposons are in the same orientation in the chromosome, while inversion occurs if the two are in opposite orientations.

ℋEYNOTE

Transposable genetic elements are unique DNA segments that can insert themselves at one or more sites in a genome. The presence of TGEs in a cell is usually detected by the changes they bring about in the expression and activities of the genes at or near the chromosomal sites into which they integrate. In prokaryotes the three major types of transposable genetic elements are insertion sequence (IS) elements, transposons (Tn), and temperate bacteriophages such as Mu.

TRANSPOSABLE GENETIC ELEMENTS IN EUKARYOTES

Transposable genetic elements have been identified in many eukaryotes, and they have been studied mostly in yeast, *Drosophila*, corn, and humans. Their structure and function are very similar to those of prokaryotic transposable genetic elements. Eukaryotic transposable genetic elements have been shown to be capable of integrating into chromosomes at a number of sites. Thus, such elements may be able to affect the function of virtually any gene, turning it on or off, depending on the element involved and how it integrates into the gene. The integration events themselves, like those of most prokaryotic TGEs, involve nonhomologous recombination. Many of the eukaryotic TGEs carry genes, and there is good evidence that these genes are transcribed from the integrated elements and that at least some of

~ FIGURE 20.11

Production of deletion or inversion by homologous recombination between two Mu genomes or two transposons. (a) For a deletion to occur, the two elements are in the same orientation; (b) For an inversion to occur, the two elements are in opposite orientations.

a) Deletion by recombination between two Mu genomes or transposons in same orientation

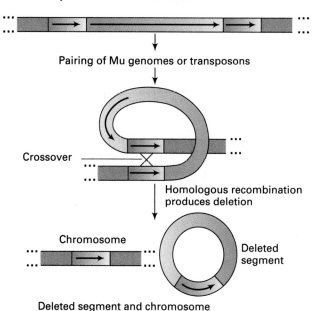

b) Inversion by recombination between two Mu genomes or transposons in opposite orientations

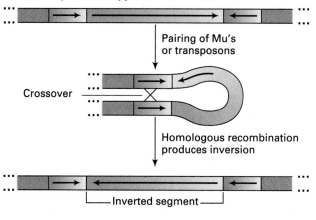

the resulting mRNA is translated. However, in most cases the role for the products of such mRNA is unknown.

Transposons in Plants

MCCLINTOCK'S DISCOVERY OF TRANSPOSONS IN CORN. One of the earliest observations of a transpos-

able genetic element came from Barbara McClintock's work with corn (*Zea mays*) in the 1940s and 1950s. On the basis of her genetic work, she described a number of what she called "controlling elements," which modify or suppress gene activity in corn. One of the characteristics she studied was the extent of pigmentation in the corn kernels. It was known at the time that a number of different genes must function together for the synthesis of red anthocyanin pigment that gives the kernel a purple color. Classical genetic experiments had shown that mutation of any one of these genes causes a kernel to be unpigmented. McClintock studied kernels that, rather then being purple or white, exhibited spots of purple pigment on an otherwise white kernel (Figure 20.12; p. 622). She knew that the phenotype was the result of an unstable mutation. From her careful genetic and cytological studies she came to the conclusion that the spotted phenotype was not the result of any conventional kind of mutation but, rather, was due to a controlling element, which we now know is a transposon.

The explanation for the spotted kernels is as follows: If the corn plant carries a wild-type C gene, the kernel will be purple; mutants in the C gene block purple pigment production, so the kernel is colorless. During kernel development, revertants of the mutation occur, leading to a spot of purple pigment. The genetic nature of the reversion is supported by the fact that descendants of the cell which underwent the reversion also can produce the pigment. The earlier in development the reversion occurs, the larger is the purple spot. McClintock determined that the original c (colorless) mutation resulted from a "mobile controlling element" (in modern terms, a transposon), called Ds for "dissociation," being inserted into the C gene (Figure 20.13a and b; p. 622). We now know this to take place by a transposition event. Another mobile controlling element, called Ac for "activator," is required for transposition of Ds into the gene. Ac can also result in Ds transposing (excising perfectly) out of the c gene, giving a wild-type revertant (Figure 20.13c). The Ac-Ds system will be discussed more later.

The remarkable fact of McClintock's conclusion was that at the time there was no precedent for the existence of transposable genetic elements—indeed, the genome was thought to be very static with regard to gene locations. Only much more recently have transposable genetic elements been widely identified and studied, and only in 1983 was direct evidence obtained for the movable genetic elements proposed by McClintock. For her pioneering studies and model building she was awarded the Nobel Prize for medicine in 1983.

TYPES OF TRANSPOSONS AND THEIR EFFECTS. In recent years, a number of transposons have been identified in corn and other plants. Some of these transpos-

~ FIGURE 20.12

Corn kernels showing spots of pigment produced by cells in which a transposable genetic element had transposed out of a pigment-producing gene, thereby allowing function of that gene to be restored. The cells in the white areas of the kernel lack pigment because a pigment-producing gene continues to be inactivated by the presence of a TGE within that gene.

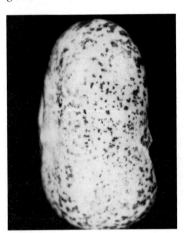  

~ FIGURE 20.13

Kernel color in corn and transposon effects. (a) Purple kernels result from the active C gene. (b) Colorless kernels can result when the *Ac* transposon activates *Ds* transposition and *Ds* inserts into C, producing a mutation. (c) Spotted kernels result by reversion of the *c* mutation during kernel development when *Ac* activates *Ds* transposition out of the C gene.

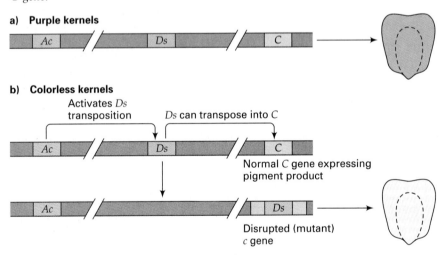

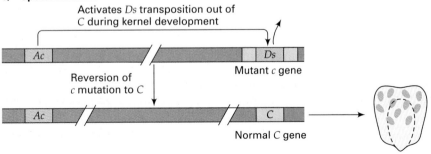

~ TABLE 20.3

Transposable Genetic Elements in Plants

TRANSPOSABLE ELEMENT	PROPERTIES
Ac	In corn, *Ac* (*Activator*) is the autonomous element of the *Ac/Ds* family.
Ds	In corn, *Ds* is a nonautonomous element whose transposition is dependent on the presence of an active *Ac* element. *Ds* (*Dissociation*) refers to the ability of particular *Ds* elements to induce chromosome breakage at their insertion site.
Mu1	*Mu1* (*Mutator 1*) is an autonomous transposable element in corn originally identified in strains with a very high frequency of spontaneous mutations.
Tam	Transposable element in *Antirrhinum majus*.
Tgm	Transposable element in *Glycine max*.

From H.P. Doring and P. Starlinger, 1986. *Annu. Rev. Genet.* 20:175–200.

able elements are presented in Table 20.3. Like the transposons we have already discussed, these transposons have inverted repeated (IR) sequences at their ends and generate short direct repeats of the target site DNA when they integrate.

When a plant transposon inserts into a chromosome, there are a number of possible consequences depending upon the properties of the transposon, including activation or repression of adjacent genes, chromosome aberrations such as duplications, deletions, inversions, translocations, or chromosome breakage. That is, as with bacterial IS elements and transposons, generally, transposition of transposons into genes causes mutations. Disruption of a gene typically results in a *null mutation*, that is, a mutation where the function of a gene is reduced to zero. Transposition into controlling regions of genes can either decrease or increase gene expression. If a transposon moves into a gene's promoter, the efficiency of that promoter can be decreased or obliterated. The transposon may provide promoter activity itself, however, and lead to an increase in gene expression.

Geneticists have identified several families of transposons (controlling elements) in corn, and each family possesses a characteristic array of transposons with respect to numbers, types, and locations. Each family consists of two forms of transposons: *autono-*

mous elements, which can transpose by themselves; and *nonautonomous elements,* which cannot transpose by themselves. When either element is inserted into a gene, a mutant allele of that gene is produced. When an autonomous element is inserted into a gene, the resulting mutant allele is *unstable* because the element can excise and transpose to a new location. The frequency of transposition out of a gene is higher than the spontaneous reversion frequency for a regular point mutation; hence, the allele produced by an autonomous element is referred to as a *mutable* allele.

By contrast, mutant alleles resulting from the insertion of a nonautonomous element in a gene are *stable,* since the element is unable to transpose out of the locus by itself. However, if the autonomous element of its family is also present in the genome, that element can provide transposition factors in *trans,* and the nonautonomous element will now exhibit properties of autonomous elements with respect to transposition around the genome.

THE *Ac–Ds* CONTROLLING ELEMENTS IN CORN. Transposition events in corn are often followed by visible phenotypic changes when the transposon inserts near a gene. Such changes are best seen in heterozygotes when the dominant allele is inactivated by the transposon, resulting in a cell with a recessive phenotype. Mitotic progeny of that cell remain in the same location, giving rise to a tissue spot or sector with a clone of cells that exhibit the recessive phenotype. As we have discussed, tissue sectors are readily seen in kernels as colored patches in a background of a different color.

The *Ac–Ds* family of controlling elements has been studied in detail. McClintock's mapping studies had shown that *Ac* and *Ds* were not conventional genetic loci because they could move to new locations in the genome; that is, they could transpose and cause mutations by inserting into other loci. *Ac* is the autonomous element of the family, and, hence, mutations caused by *Ac* are unstable. *Ds* is the nonautonomous element of the family. *Ds* mutations are stable if only *Ds* is present; they are unstable in the presence of an *Ac* element. In the mid-1950s McClintock demonstrated that at least some *Ds* elements were derived from *Ac* elements.

Ac is 4563 bp long, with 11-bp imperfect terminal inverted repeats (IRs) (Figure 20.14a; p. 624). *Ac* contains a single transcription unit that comprises most of the element's length. A single 3.5-kb mRNA is produced, which encodes an 807-amino acid polypeptide. Upon insertion into a chromosome site, an 8-bp direct duplication of the target site is generated.

Ds does not transpose in the absence of *Ac* and remains as a stable insertion in the chromosome. When an *Ac* element is present, it provides functions in *trans*

~ FIGURE 20.14

The structure of the *Ac* autonomous transposable element of corn, and of several *Ds* nonautonomous elements derived from *Ac*. (a) The *Ac* element is 4563 bp long and has imperfect 11-bp inverted terminal repeats. *Ac* has a single transcription unit, comprising most of the element, and going from left to right. The five exons of the resulting RNA are shown in green within the *Ac* element itself. (b) Several *Ds* elements are shown in the same way as for the *Ac* element. They are derived from *Ac* by internal deletions of various lengths.

a)   Activator element (*Ac*)

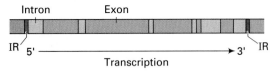

b)   Dissociation elements (*Ds*)

that activate *Ds*, causing it to transpose to a new site or to break the chromosome in which it is located.

Ds elements are heterogeneous in length and sequence; a number of them are diagrammed in Figure 20.14b. All *Ds* elements have the same terminal IRs as *Ac* elements, and many of these *Ds* elements have been generated from *Ac* by deletion of various lengths (e.g., *Ds9, Ds2d1, Ds2d2, Ds6*; Figure 20.14b) or by more complex sequence rearrangements. The deletions and complex sequence rearrangements are responsible for the transposition-defective phenotype of the various *Ds* elements.

Genetic evidence indicates that both the timing and frequency of transposition of the *Ac* and *Ds* elements, as well as the genetic rearrangements associated with the elements, are developmentally regulated. This is a unique property of corn transposons. Transposition of the *Ac* element does not itself involve replication of the element, so that it does not leave a copy of itself in the original chromosomal location when it transposes: that is, *Ac* exhibits conservative transposition. *Ac* transposes only during or after chromosome replication (Figure 20.15). The starting point for discussion is a chromosome with one copy of *Ac* at a site called the *donor site*. Upon replication of the region of the chromosome containing *Ac*, two copies of *Ac* result, one on

each progeny chromatid. There are two possible results of *Ac* transposition, depending on whether transposition occurs to a replicated or unreplicated chromosome site.

First, let us consider transposition to a replicated chromosome site (Figure 20.15a). If one of the two *Ac* elements transposes at this time, an empty donor site is left on one chromatid, while an *Ac* element remains in the homologous donor site on the other chromatid. The transposing *Ac* element inserts into a new, already-replicated recipient site, which is often on the same chromosome. In Figure 20.15a the site is shown on the same chromatid as the parental *Ac* element. Thus, in the case of transposition to an already-replicated site, there is no net increase in the number of *Ac* elements.

Figure 20.15b shows transposition to an unreplicated chromosome site. As with the first case, one of the two *Ac* elements transposes, leaving an empty donor site on one chromatid and an *Ac* element in the homologous donor site on the other chromatid. But now the transposing element inserts into a nearby recipient site that has yet to be replicated. When that region of the chromosome replicates, the result will be a copy of the transposed *Ac* element on both chromatids, in addition to the one original copy of the *Ac* element at the donor site on one chromatid. Thus, in the case of transposition to an unreplicated recipient site, there is a net increase in the number of *Ac* elements.

Transposition of most *Ds* elements occurs in the same way as *Ac* transposition, with functions for transposition supplied by an *Ac* element in the genome.

$\mathcal{K}$EYNOTE

The mechanism of transposition of plant transposons is quite similar to transposition of bacterial transposons. Transposons integrate at a target site by a precise mechanism so that the integrated elements are flanked at the insertion site by a short duplication of target site DNA of characteristic length. For *Ac* and *Ds* element transposition, for example, integration results in 6- to 8-bp direct repeats. Many plant transposons occur in families, the autonomous elements of which are able to direct their own transposition, and the nonautonomous elements of which are able to transpose only when activated by an autonomous element in the same genome. Most nonautonomous elements are derived from autonomous elements by internal deletions or complex sequence rearrangements. _____

~ **FIGURE 20.15**

The *Ac* transposition mechanism. (a) Transposition to an already-replicated recipient site results in no net increase in the number of *Ac* elements in the genome; (b) Transposition to an unreplicated recipient site results in a net increase in the number of *Ac* elements when the region of the chromosome containing the transposed element is subsequently replicated.

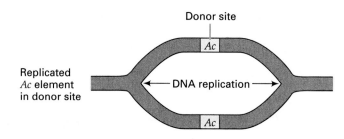

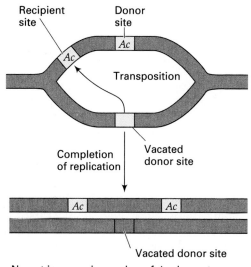

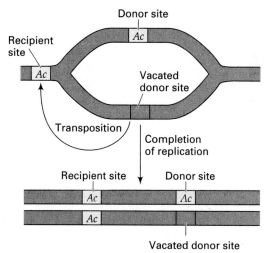

Ty Elements in Yeast

TY STRUCTURE AND PROPERTIES. *Ty* elements are segments of yeast DNA that are capable of transposition within the yeast genome. Figure 20.16 diagrams a *Ty* transposable genetic element of yeast, which is about 6000 bp long and includes two 334-bp-long, directly repeated sequences called long terminal repeats (LTR) or deltas (δ), one at each end of the element. The deltas consist of about 70 percent AT, and each contains a promoter and sequences recognized by transposing enzymes. The *Ty* elements encode a single, 5700-nucleotide mRNA that begins at the promoter in the delta at the 5′ end of the element (see Figure 20.16). The mRNA transcript contains two open reading frames (ORFs); that is, regions with a start codon in frame with a chain-terminating codon, indicating that two proteins could be produced from the mRNA. The

~ **FIGURE 20.16**

The *Ty*-transposable element of yeast. (Courtesy of Dr. Gerry Fink.)

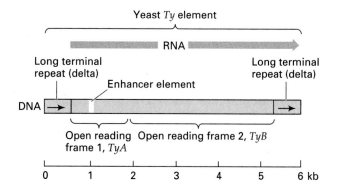

two regions have been designated *TyA* and *TyB*. There are about 35 copies of *Ty* in some yeast strains, although the number of copies of this element varies between strains.

All *Ty* have a number of structural properties in common with bacterial transposons:

1. They are found inserted in many different target sequences that share no obvious homology to each other or to the ends of the transposable genetic elements.
2. Upon insertion the elements generate a duplication of a few base pairs of the target sequence located immediately adjacent to the ends of the element.
3. The elements contain terminal repeated sequences bounding the main body of the element.

Because *Ty* has no genetic marker associated with it, it is much more difficult to follow its movements around the genome than it is to follow the drug-resistance markers of bacterial transposons. In at least twelve cases in which *Ty* element movements have been followed, a 5-bp terminal repeat of the yeast target DNA has been generated upon integration of the element. Since there is a delta at each end of a *Ty* element, a frequent event is recombination between two deltas by homologous recombination (Figure 20.17). This produces a released *Ty* element that has one delta and leaves a delta behind in the yeast chromosome. Studies have shown that there are at least a hundred copies of delta per yeast cell. These delta "droppings" are incapable of transposition, but because they contain transcription initiation signals, they can affect the expression of the genes near their location. Since there are multiple *Ty* elements in a cell, recombination can also occur between *Ty* elements located at different positions in the genome. The result can be translocations, deletions, or inversions.

***Ty* ELEMENTS AND RETROVIRUSES.** *Ty* elements have been shown to transpose via an RNA intermediate in a way very similar to **retroviruses,** single-stranded RNA viruses that replicate via double-stranded DNA intermediates. Because of their similarity with retroviruses, *Ty* elements have been referred to as *retrotransposons*.

In brief, retroviruses have single-stranded RNA genomes. When a retrovirus infects a cell, the RNA genome is used to produce a double-stranded DNA catalyzed by viral reverse transcriptase. The DNA integrates into the host's chromosome where it can be transcribed.

Analysis of yeast *Ty* elements revealed a number of similarities in the organization of those elements and of retroviruses. This analysis led to the hypothesis that the

~ **FIGURE 20.17**

Excision of a *Ty* element by homologous recombination between directly-repeated deltas, resulting in a delta being left behind at the original *Ty* insertion site.

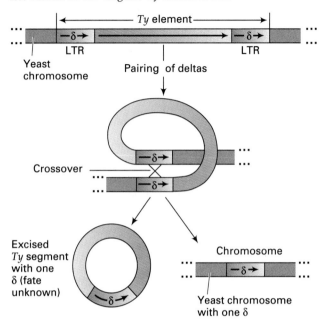

Ty elements move to new chromosomal locations using the same mechanisms as those involved in provirus production and integration of retroviruses. That is, rather than transposing DNA to DNA as is the case with bacterial transposons and most eukaryotic transposons, *Ty* elements are hypothesized to transpose by making an RNA copy of the integrated DNA sequence and then by creating a new *Ty* element by reverse transcription. The new element would then integrate at a new chromosome location.

Evidence substantiating the above hypothesis was obtained through experiments using *Ty* elements that had been constructed using recombinant DNA techniques; these *Ty* elements had special features that enabled their transposition to be monitored easily. One compelling piece of evidence came from experiments in which an intron was placed into the *Ty* element by recombinant DNA techniques (there is none in normal *Ty* elements) and the element was monitored from the beginning through the transposition event. At the new location, the *Ty* element no longer had the intron sequence, a fact that was consistant with the notion that transposition occurred via an RNA intermediate. The intron had been removed by the usual splicing processes that function to remove introns from pre-mRNAs. Subsequently it has been demonstrated that *Ty* elements encode a reverse transcriptase. Moreover, *Ty* viruslike particles (*Ty*-VLPs) containing *Ty* RNA and reverse transcriptase activity have been identified.

~ FIGURE 20.18

Structure of the transposable element *copia*, a retrotransposon found in *Drosophila melanogaster*.

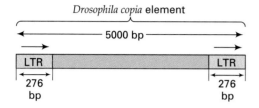

Drosophila copia element

Drosophila Transposons

A number of classes of transposons have been identified in *Drosophila*. In this organism it is estimated that about 15 percent of the genome is mobile.

THE *COPIA* RETROTRANSPOSON. Figure 20.18 diagrams the structure of a *copia* transposable genetic element of *Drosophila*. There are several families of *copia*-like elements. The members of each family are highly conserved and are located at about 5 to 100 widely scattered sites in the genome. The *copia* ele-

ments code for abundant mRNAs found in *Drosophila*. All *copia* elements found in *Drosophila* are capable of transposition, although there are differences among *Drosophila* strains in the number and distribution of these elements.

The structure of the *copia* element is similar to that of the *Ty* elements of yeast. Directly repeated termini (LTRs) of 276 bp are found at either end of a 5000-bp segment of DNA. Inverted repeats of 17 bp are located at the ends of each LTR. Like yeast *Ty* elements, *copia* elements transpose via an RNA intermediate, using a reverse transcriptase catalyzed process. Viruslike particles (VLPs) have also been identified in *copia*. Integration of a *copia* element into the genome, like other transposons, results in a duplication of a short target site sequence, in this case 3 to 6 bp long.

HYBRID DYSGENESIS AND *P* ELEMENTS. Hybrid dysgenesis is the appearance of a series of defects, including mutations, chromosomal aberrations, and sterility, when certain strains of *Drosophila melanogaster* are crossed. For example, hybrid dysgenesis occurs when female laboratory strains are mated to males from natural populations (Figure 20.19a). The laboratory strain is said to be of the *M* cytotype (maternal contributing cell type) and the naturally-occurring

~ FIGURE 20.19

Hybrid dysgenesis, exemplified by the production of sterile flies, results from (a) a cross of M♀ × P♂, but not from (b) a cross of M♂ × P♀.

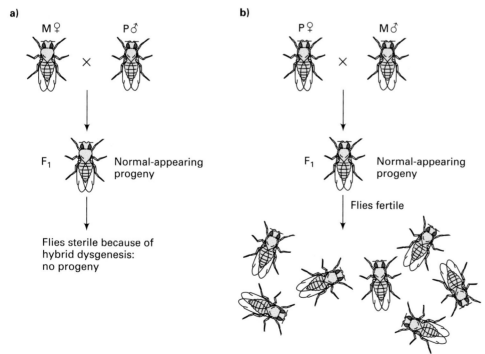

strain is said to be of the *P cytotype* (paternal contributing cell type), thus the cross is M♀ × P♂. In the reciprocal P♀ × M♂ cross, hybrid dysgenesis does not occur and all progeny are fertile (Figure 20.19b).

Hybrid dysgenesis primarily affects germ line cells. That is, F₁ hybrids of a M♀ × P♂ cross have normal somatic tissues, but gonads do not develop, and the flies are sterile. Hybrid dysgenesis results when chromosomes of the P male parent become exposed to the cytoplasm derived from the M female parent. Cytoplasm from P females does not induce hybrid dysgenesis, so progeny of P♀ × P♂ are fully fertile.

The model for the induction of hybrid dysgenesis in the M♀ × P♂ crosses is as follows. The chromosomes of the P male carry 30–50 copies of a transposon called the *P element*. The M strain does not have any *P* elements. *P* elements vary in length from 500–2900 bp; each has 31-bp inverted terminal repeats. The shorter *P* elements (which are nonautonomous elements, like *Ds* in corn) are derived from the longest *P* element by internal elements. The longest *P* elements are autonomous elements like *Ac* in corn; that is, they encode a transposase, which can catalyze its own transposition and the transposition of the shorter *P* elements. In the hybrid flies produced from the M♀ × P♂ crosses, the chromosomes have *P* elements inserted at new sites. This occurs because *P* element transposition has become activated, and such activation occurs only in the germline.

P elements are thought to encode a repressor protein that prevents the transcription of the transposase gene, therefore preventing transposition of *P* elements. Thus, in the P cytotype, the *P* elements are stable in the chromosomes. However, in hybrids from M♀ × P♂, the cytoplasm derives from the M cytotype, which lacks *P* repressors. Transposase is produced, activating *P* element transposition, which leads to the hybrid dysgenesis phenomenon.

Apart from their role in hybrid dysgenesis, *P* elements are used as vectors for transferring genes into the germline of *Drosophila* embryos, allowing genetic manipulation of the organism.

𝒦EYNOTE

Transposable genetic elements in eukaryotes are typically transposons. Transposons can transpose to new sites while leaving a copy behind in the original site, or they may excise themselves from the chromosome. When the excision is imperfect, deletions can occur, and by various recombination events other chromosomal rearrangements such as inversions and duplications may occur. While most transposons move by using a DNA-to-DNA mechanism, some eukaryotic transposons such as yeast *Ty* elements and *Drosophila copia* elements transpose via an RNA intermediate (using a transposon encoded reverse transcriptase), thereby resembling retroviruses.

TUMOR VIRUSES, ONCOGENES, AND CANCER

Occasionally, differentiated cells revert to an undifferentiated state (a process called *dedifferentiation*), and instead of remaining in a nondividing mode, they begin to divide mitotically and give rise to tissue masses called *tumors*. The failure of cells to remain constrained in their growth properties is called **transformation** (not to be confused with transformation of a cell by uptake of DNA). Tumors may be either benign, in which case their cells divide and they increase in size without threat to the life of the organism, or malignant. In a malignant tumor the cells divide and the tumor increases in size, threatening the life of the organism. In some cases, cells break off from the tumor and migrate to adjacent tissues and throughout the organism, resulting in tumor formation at many sites. The spreading of tumor cells throughout the body is called *metastasis*, and metastasizing tumors are called **cancers**. Tumor initiation in an organism is called **oncogenesis** (*onkos*, "mass" or "bulk"; *genesis*, "birth"). There are many causes of cancer, such as spontaneous genetic changes (e.g., spontaneous gene mutations or chromosome aberrations), exposure to mutagens, radiation, or cancer-inducing viruses (tumor viruses). There is also evidence for hereditary predisposition to cancer.

In this section we will focus on tumor viruses and, in particular, on the class of tumor viruses that consists of retroviruses. We will see that tumor-causing retroviruses carry genes that have been acquired from the host's genome. As we said earlier, the yeast *Ty* element and *Drosophila copia* element are transposable genetic elements that have some properties in common with retroviruses.

Tumor Viruses

Transformation of cells to the tumorous state can result from infection with **tumor viruses**, which induce the cells they infect to dedifferentiate and to divide to produce a tumor. Tumor viruses may have DNA or RNA genomes and are widely found in animals. The ability of a tumor virus to transform a cell is the property of a gene or genes in the viral genome called *oncogenes*. By definition, an **oncogene** is a gene that transforms a normal cell to the tumorous (cancerous) state. *DNA tumor viruses,* such as adenoviruses and SV40 (simian virus 40), have oncogenes that have been difficult to study. By contrast, *RNA tumor viruses,* all of which are retroviruses, have been a fruitful source of information about the relationship between viruses, oncogenes, and the host cell. We will focus our attention on retroviruses in the remaining discussions.

Retroviruses

STRUCTURE. The RNA tumor viruses exemplified by *Rous sarcoma* virus, feline leukemia virus, mouse mammary tumor virus, and human immunodeficiency virus (HIV, the causative agent of AIDS) are all retroviruses. A drawing of a stylized retrovirus particle is shown in Figure 20.20. Within a protein core, which often is icosahedral in shape, are two copies of the 7-to-

10-kb single-stranded RNA genome. The core is surrounded by an envelope derived from host membranes and coated with viral-encoded glycoproteins. When the virus infects a cell, the envelope glycoproteins interact with a host-encoded cell surface receptor.

LIFE CYCLE. When a retrovirus infects a cell, the RNA genome does not act as mRNA but serves as a template for the synthesis of a double-stranded DNA copy of the retrovirus (the provirus) catalyzed by reverse transcriptase, an enzyme brought into the cell as part of the virus particle. The double-stranded DNA then integrates into the host chromosome where it can serve as a template for transcription of viral mRNA (i.e., transcription of parts of the inserted DNA) or of progeny genomes (i.e., transcription of the entire inserted DNA).

As an example of a retrovirus, the RNA genome organization of *Rous sarcoma* virus (RSV) is shown in Figure 20.21a (p. 630). Typical retroviruses have three protein-coding genes required for life cycle functions: *gag* encodes a precursor protein that, when cleaved, produces virus particle proteins; *pol* encodes a precur-

~ FIGURE 20.20

Stylized drawing of a retrovirus.

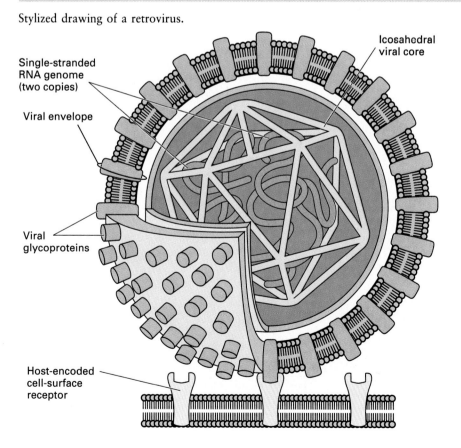

Icosahedral viral core

Single-stranded RNA genome (two copies)

Viral envelope

Viral glycoproteins

Host-encoded cell-surface receptor

~ **FIGURE 20.21**

The Rous sarcoma virus (RSV) RNA genome and the integration of the provirus into the host chromosome. (a) RSV genome RNA; (b) RSV proviral DNA produced by reverse transcriptase; (c) Circularization of the proviral DNA; (d) Staggered nicks are made in viral and cellular DNAs; (e) By an exchange process, the viral ends become joined to the ends of the host cell DNA; (f) The single-stranded gaps are filled in and a complete, double-stranded integrated RSV provirus results.

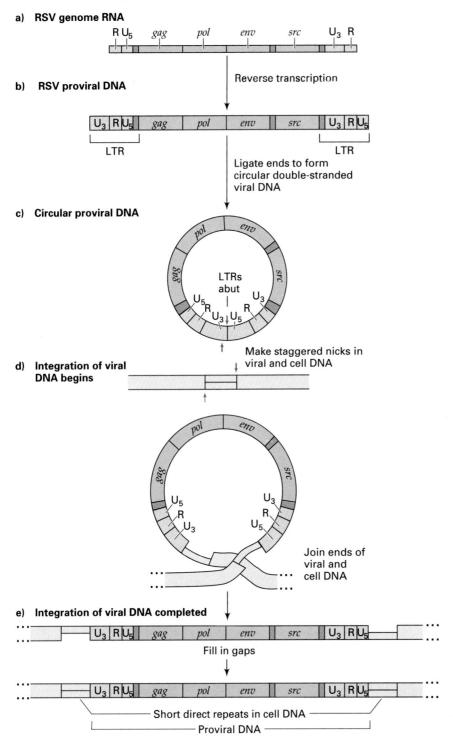

a) RSV genome RNA

b) RSV proviral DNA

c) Circular proviral DNA

d) Integration of viral DNA begins

e) Integration of viral DNA completed

sor protein that is cleaved to produce reverse transcriptase and an enzyme involved in the integration of the proviral DNA into the host cell chromosome; and *env* encodes the precursor to an envelope glycoprotein. The RSV retrovirus also carries an oncogene called *src*, which is not involved in the viral life cycle. (Different retroviruses carry different oncogenes.) The ends of the RNA genome consist of the sequences R and U5 (shown here at left) and U3 and R (shown at right) (Figure 20.21a). When DNA is produced by reverse transcriptase, the end sequences are duplicated to produce long terminal repeats (LTRs: Figure 20.21b). The LTRs contain many of the regulatory signals in the viral sequence (e.g., for initiation of transcription and 3′ cleavage and polyadenylation of mRNAs) (Figure 20.22).

For integration of the proviral DNA into the host's chromosome, the first step is ligation of the ends of the linear molecule to produce a circular, double-stranded molecule (Figure 20.21c). This brings two LTRs next to each other. Staggered nicks are made in both the viral and cellular DNAs (Figure 20.21d). By an exchange process, the viral ends become joined to the ends of the cellular DNA; at this point integration has occurred (Figure 20.21e). Lastly, the single-stranded gaps are filled in. Like the transposition of transposons, the integration of retrovirus proviral DNA results in a duplication of DNA at the target site, producing short, direct repeats in the host cell DNA flanking the provirus.

TYPES OF RETROVIRUSES. There are two types of retroviruses: (1) *Replication-competent retroviruses* can direct their own life cycle but do not harm the cells they infect—that is, they are nontransforming viruses; and (2) *Transformation-competent retroviruses* have the ability to transform an infected cell to the tumorous state. The latter type of retrovirus may or may not be able to self-replicate. In short, all RNA tumor viruses are retroviruses, but not all retroviruses induce tumors.

VIRAL ONCOGENES. In 1910, F. Peyton Rous showed that, when pieces of a particular kind of chicken tumor were transplanted into other chickens from the same stock, the second group of chickens developed the same kind of tumors. Rous showed that cell-free filtrates of the sarcoma inoculated into chickens also resulted in tumor development. These results were interpreted to mean that some kinds of cancer might result from infection with a "filter-passing microbe" or a "virus." We now know that Rous's results (published in 1911) are explained by Rous sarcoma virus (RSV).

It is now understood that tumor induction by RSV results from the activity of a particular gene, called a **viral oncogene,** present in the retroviral genome. The discovery that the product of a single gene was both necessary and sufficient for tumor induction and formation was a significant breakthrough. RSVs transforming viral oncogene is called *src,* after the sarcoma tumor it induces. Viral oncogenes (which will be called v-*onc*'s) are responsible for many different cancers. Only retroviruses that contain a v-*onc* gene are tumor viruses. Such retroviruses are called **transducing retroviruses,** because they have picked up DNA other than *gag, pol,* and *env;* that is, they have picked up an oncogene from the cellular genome. (We will learn how this happens later in this section.) The v-*onc* genes are named for the tumor that the virus causes, with the prefix "v" to indicate that the gene is of viral origin. Thus the v-*onc* gene of RSV is v-*src.* Table 20.4 (p. 632) lists some transducing retroviruses and their viral oncogenes.

Cells infected by RSV rapidly transform into the tumorous state because of the presence of the v-*src* gene. Since RSV contains all the genes necessary for viral replication (i.e., *gag, env,* and *pol*), a RSV-transformed cell produces progeny RSV particles. In this ability RSV is an exception, however, because all other

---

~ FIGURE 20.22

Features of LTRs. Most retroviruses (and many "transposable elements" of *Drosophila* and yeast) have LTRs with these features. [Note: The LTRs have been drawn much larger than they are in order to show their features.]

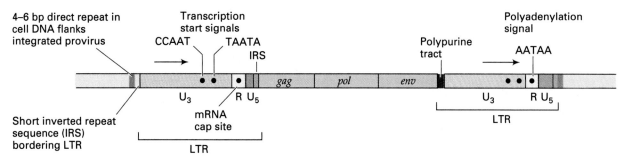

~ TABLE 20.4

Some Acute Transforming Retroviruses and Their Oncogenes

ONCO-GENE	RETROVIRUS ISOLATES	v-onc ORIGIN	v-onc PROTEIN	VIRUS DISEASE
src	Rous sarcoma virus (RSV)	Chicken	$pp60^{src}$	Sarcoma
fps*	Fujinami sarcoma virus (FuSV)	Chicken	$P130^{gag\text{-}fps}$	Sarcoma
	PRCII-avian sarcoma virus (ASV)	Chicken	$P105^{gag\text{-}fps}$	Sarcoma
fes*	Snyder-Theilen feline sarcoma virus (FeSV)	Cat	$P85^{gag\text{-}fes}$	Sarcoma
	Gardner-Arnstein FeSV	Cat	$P110^{gag\text{-}fes}$	Sarcoma
yes	Y73-ASV	Chicken	$P90^{gag\text{-}yes}$	Sarcoma
fgr	Gardner-Rasheed FeSV	Cat	$P70^{gag\text{-}actin\text{-}fgr}$	Sarcoma
ros	UR2-ASV	Chicken	$P68^{gag\text{-}ros}$	Sarcoma
abl	Abelson murine leukemia virus (MLV)	Mouse	$P90\text{-}P160^{gag\text{-}abl}$	Pre-B cell leukemia
	Hardy-Zuckerman (HZ2)-FeSV	Cat	$P98^{gag\text{-}abl}$	Sarcoma
ski	SKV	Chicken	$P110^{gag\text{-}ski\text{-}pol}$	Squamous carcinoma
erbA	Avian erythroblastosis virus (AEV)	Chicken	$P75^{gag\text{-}erb\text{-}A}$	Erythroblastosis and sarcoma
erbB	Avian erythroblastosis virus (AEV)	Chicken	$gp65^{erbB}$	Erythroblastosis and sarcoma
fms	McDonough (SM)-FeSV	Cat	$gp180^{gag\text{-}fms}$	Sarcoma
fos	FBJ (Finkel-Biskis-Jinkins)-MSV	Mouse	$pp55^{fos}$	Osteosarcoma
mos	Moloney MSV	Mouse	$P37^{env\text{-}mos}$	Sarcoma
sis	Simian sarcoma virus (SSV)	Monkey	$P28^{env\text{-}sis}$	Sarcoma
	Parodi-Irgens FeSV	Cat	$P76^{gag\text{-}sis}$	Sarcoma
myc	MC29	Chicken	$P100^{gag\text{-}myc}$	Sarcoma, carcinoma, and myelocytoma
myb	Avian myeloblastosis virus (AMV)	Chicken	$p45^{myb}$	Myeloblastosis
	AMV-E26	Chicken	$P135^{gag\text{-}myb\text{-}ets}$	Myeloblastosis and erythroblastosis
rel	Reticuloendotheliosis virus (REV)	Turkey	$p64^{rel}$	Reticuloendotheliosis
kit	HZ4-FeSV	Cat	$P80^{gag\text{-}kit}$	Sarcoma
raf*	3611-MSV	Mouse	$P75^{gag\text{-}raf}$	Sarcoma
H-ras	Harvey MSV (Ha-MSV)	Rat	$pp21^{ras}$	Sarcoma and erythroleukemia
	RaSV	Rat	$P29^{gag\text{-}ras}$	Sarcoma?
K-ras	Kirsten MSV (Ki-MSV)	Rat	$pp21^{ras}$	Sarcoma and erythroleukemia
ets	AMV-E26	Chicken	$P135^{gag\text{-}myb\text{-}ets}$	See above for myb

* fps, the oncogene of several chicken sarcoma viruses, is derived from a chicken gene, c-fps, that is the chicken equivalent of c-fes, a gene of cats first discovered as the oncogene fes in several feline sarcoma viruses. It was not realized that fes and fps were homologous when they were assigned different names. Note that AMV-E26 has two oncogenes.

SOURCE: R. Weiss, N. Teich, H. Varmus, and J. Coffin (eds.), *RNA Tumor Viruses*, 2d ed. Cold Spring Harbor Laboratory, Cold Spring Harbor, N.Y., 1985.

~ FIGURE 20.23

Structure of four defective transducing viruses (not drawn to scale) (a) Avian myeloblastosis virus (AMV) contains the v-*myb* oncogene, which replaces some of the 3′ end of *pol* and most of *env*; (b) Avian defective leukemia virus (DLV) contains the v-*myc* oncogene, which replaces the 3′ end of *gag*, all of *pol*, and the 5′ end of *env*; (c) Feline sarcoma virus (FeSV) contains the v-*fes* oncogene, which replaces the 3′ end of *gag*, and all of *pol* and *env*; (d) Abelson murine leukemia virus (AbMLV) contains the v-*abl* oncogene, which replaces the 3′ end of *gag*, and all of *pol* and *env*.

a) Avian myeloblastosis virus (AMV) genomic RNA

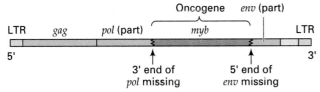

b) Avian defective leukemia virus (DLV) genomic RNA

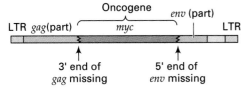

c) Feline sarcoma virus (FeSV) genomic RNA

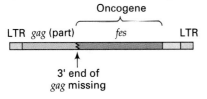

d) Abelson murine leukemia virus (AbMLV) genomic RNA

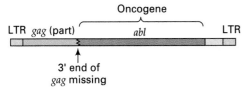

transducing retroviruses are defective in some of their genes (Figure 20.23): they can transform cells but are unable to produce progeny viruses. These retroviruses can produce progeny viral particles, however, if cells containing them are also infected with another virus (a *helper virus*) that can supply the gene products that are missing (Figure 20.24; p. 634).

CELLULAR PROTO-ONCOGENES. In the mid-1970s, J. Michael Bishop and Harold Varmus, along with a number of other researchers, demonstrated that normal animal cells contain genes with DNA sequences very closely related to the viral oncogenes. In the early 1980s, R. A. Weinberg and M. Wigler showed independently that a variety of human tumor cells contain oncogenes. These genes, when introduced into other cells growing in culture, transformed those cells into tumor cells. Human oncogenes were very similar to viral oncogenes that had been characterized earlier, even though there was no evidence for viral induction of the human cancers involved. Moreover, the human oncogenes were shown also to be closely related to genes found in normally growing cells.

For both the animal and human oncogenes, the related genes in normal cells were not identical in sequence but there were long sequences showing high degrees of similarity. That is, all cells contain normal, non-transforming copies of oncogenes called **proto-oncogenes.** When a proto-oncogene is mutated or altered so that it is now a transforming oncogene, it is called a **cellular oncogene,** or a **c-*onc*** (the "c" stands for "cellular," and the *onc* is replaced by the three-letter sequence of the related viral oncogene). Thus, a c-*src* gene for RSV is present in all chickens, both in normal cells and in tumor cells. Transducing retroviruses, then, have picked up cellular proto-oncogenes in a *modified form*. When the transducing retrovirus infects a normal cell, the hitchhiking oncogene (now a v-*onc*) transforms the cell into a tumor cell. (Normal cells can also become transformed into tumor cells in the absence of infection by a tumor virus if [1] a proto-oncogene becomes mutated to a cellular oncogene, or [2] if certain chromosomal aberrations occur [see Chapter 19] such that the proto-oncogene is placed next to the regulatory region of another gene.)

There are many significant differences between a cellular proto-oncogene and its viral oncogene v-*onc* counterpart:

1. Most proto-oncogenes contain introns that are not present in the corresponding v-*onc*. The v-*src* oncogene of RSV, for example, is 1.7 kb long and is transcribed to produce a spliced mRNA containing viral sequences and the *src* sequence. The chicken *src* proto-oncogene gene is over 7 kb long, consisting of 12 exons. The mRNA transcribed from this gene is about 4 kb long with no associated retroviral sequences. The molecular organizations of the *src* cellular proto-oncogene and the RSV that contains v-*src* are shown in Figure 20.25a (p. 635); the mRNA transcripts of RSV are shown in Figure 20.25b.

~ FIGURE 20.24

Transformation of cells by viruses produced by a cell making a defective transforming virus and a helper virus. Top: When a cell is infected by both a defective transforming virus and a nondefective helper virus, a transformed cell results, which produces both types of virus. Middle: When a cell is infected by only a defective transforming virus, transformed virus-nonproducer cells are generated. Transforming viruses can be rescued from these cells by infection with a helper virus. Bottom: When a cell is infected by only a helper virus, a nontransformed, virus-producing cell results.

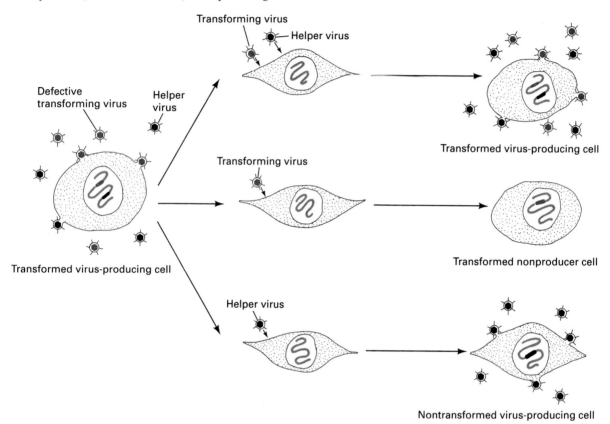

v-*onc*'s do not contain introns. This is not surprising, since viral propagation involves the production of RNA from the integrated proviral DNA, and since any introns present would be removed from that RNA molecule. The RNA is then packaged into progeny virus particles.

2. When a normal cellular proto-oncogene becomes a retroviral oncogene, the retroviral genome is altered significantly. That is, transducing retroviruses typically arise by complex rearrangements of the DNA after integration of a retrovirus near a cellular proto-oncogene (Figure 20.26; p. 636). Usually, the newly acquired genetic information replaces some or all of the *gag, pol,* and *env* genes,

making the new retrovirus defective (and incapable, therefore, of replication).

3. The v-*onc* is transcribed in a different range of cells, and results in larger amounts of mRNA than the corresponding proto-oncogene. This results from the fact that the gene is now under viral control, using retroviral promoter, enhancer, and poly(A) addition signals. Thus, there is a significant quantitative change in the protein encoded by the oncogene. In addition, since the proto-oncogene has usually become mutated when picked up by the retroviruses (e.g., point mutations, additions, deletions, rearrangements), there is typically a qualitative change in that protein, exhibited as functional

~ FIGURE 20.25

(a) Top: Molecular organization of the chicken *src* proto-oncogene. The gene contains twelve exons (shown as purple boxes). Below is the molecular organization of the Rous sarcoma virus RNA genome to indicate the relationship of nucleotide sequences in the cellular *src* proto-oncogene and v-*src*; v-*src* was produced mostly by intron removal; (b) mRNAs produced by transcription of RSV proviral DNA genome.

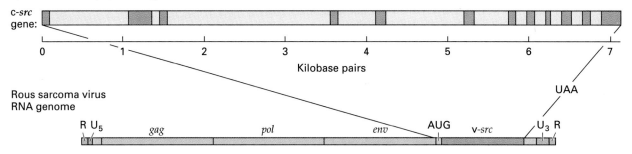

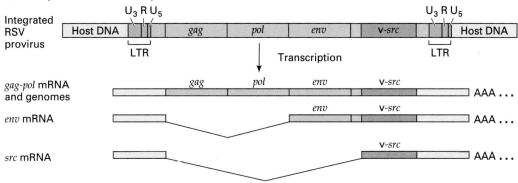

differences from the proto-oncogene product. It is generally accepted that proto-oncogenes (i.e., the *normal* gene counterpart of viral oncogenes) code for products that are essential for normal development and cell function.

PROTEIN PRODUCTS OF PROTO-ONCOGENES. There are at least 30 known proto-oncogenes. Based on DNA sequence similarities, and similarities in amino acid sequences of the protein products, proto-oncogenes fall into several distinct classes, each with a characteristic type of protein product (Table 20.5; p. 637). The major classes of protein products are growth factors (e.g., *sis* product), protein kinases (e.g., *src* prod-

uct), GTP-binding proteins (e.g., H-*ras* product), nuclear proteins (e.g., *myc* product), and hormone receptors (e.g., *erbA* product). In the following we will consider just two examples of protein products, growth factors, and protein kinases.

Growth factors. Based on the effect of oncogenes on cell growth and division, it was hypothesized early that proto-oncogenes might be "switching" genes involved with the control of cell multiplication during differentiation. Evidence supporting this hypothesis came when the product of the viral oncogene v-*sis* was shown to be identical to part of platelet-derived growth factor (PDGF, a factor found in blood platelets in mammals),

~ FIGURE 20.26

Model for the formation of a transducing retrovirus. First a wild-type provirus integrates near a cellular proto-oncogene (c-*onc*). Next, a hypothetical deletion-fusion event fuses a c-*onc* exon into the *gag* region of the retrovirus. The LTR in this fusion directs the synthesis of a transcript that undergoes splicing to produce an mRNA for the *gag-onc* fusion protein. If the cell is also infected with a wild-type (helper) retrovirus, as part of the virus life cycle a wild-type RNA and a *gag-onc* RNA can be packaged into one virus particle. In a virus particle containing the two RNAs, reverse transcriptase can start copying the *gag-onc* RNA, then switch to copying the wild-type RNA to produce a new defective transducing retrovirus. (From James D. Watson, et al., *Molecular biology of the gene*, 4th ed. Copyright 1965, 1970, 1976, 1987 by The Benjamin/Cummings Publishing Company, Inc. Reprinted by permission.)

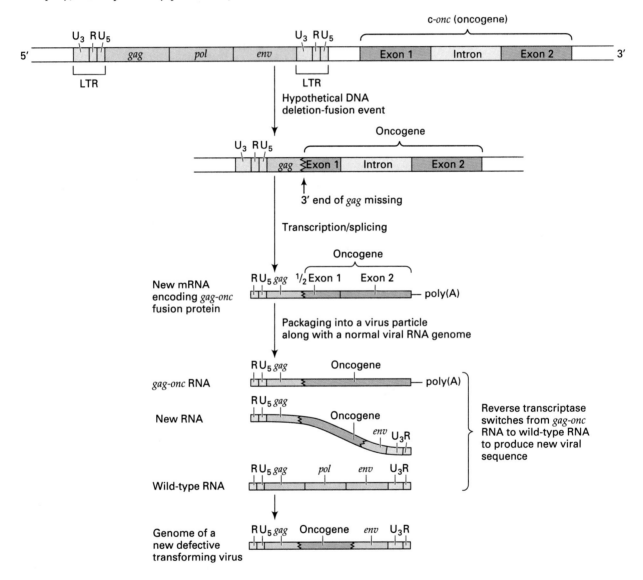

which is released after tissue damage. PDGF affects only one type of cell, fibroblasts, causing them to grow and divide. The fibroblasts are part of the wound-healing system. PDGF itself consists of two polypeptides, one of which has been shown to be encoded by c-*sis*, the cellular oncogene of *sis*. The causal link between PDGF and tumor induction was demonstrated in an experiment in which the cloned PDGF gene was intro-

~ TABLE 20.5

Oncogene Families and Their Protein Products

ONCOGENE	SUBCELLULAR LOCATION OF PROTEIN	PROPERTIES OR NORMAL FUNCTION OF PROTEIN	ONCOGENE FOUND IN ANIMAL RETROVIRUS
Class I: Protein kinases			
src	Plasma membrane	Tyrosine-specific protein kinase	Rous avian sarcoma
yes	Plasma membrane	Tyrosine-specific protein kinase	Yamaguchi avian sarcoma
fgr	?	Tyrosine-specific protein kinase	Gardner-Rasheed feline sarcoma
abl	Plasma membrane	Tyrosine-specific protein kinase	Abelson murine leukemia
fps (fes)	Cytoplasm	Tyrosine-specific protein kinase	Fujinami avian sarcoma (and feline sarcoma)
erbB	Plasma membrane (transmembrane)	EGF receptor/tyrosine-specific protein kinase	Avian erythroblastosis
fms	Plasma membrane (transmembrane)	CSF-1 receptor/tyrosine-specific protein kinase	McDonough feline sarcoma
ros	Plasma membrane (transmembrane)	Tyrosine-specific protein kinase	UR II avian sarcoma
kit	Plasma membrane		Feline sarcoma
mos	Cytoplasm	Serine/Threonine protein kinase	Moloney murine sarcoma
raf (mil)	?	Serine/Threonine protein kinase	3611 Murine sarcoma
Class II: GTP binding proteins			
H-*ras*	Plasma membrane	Guanine nucleotide binding protein with GTPase activity	Harvey murine sarcoma
K-*ras*	Plasma membrane	Guanine nucleotide binding protein with GTPase activity	Kirsten murine sarcoma
Class III: Growth factors			
sis	Secreted	Derived from a gene that encodes platelet derived growth factor (PDGF)	Simian sarcoma
Class IV: Nuclear proteins			
myc	Nucleus		Avian MC29 myelocytomatosis
myb	Nucleus		Avian myeloblastosis
fos	Nucleus		FBJ osteosarcoma
ski	Nucleus		Avian SKV770
Class V: Hormone receptor			
*erb*A	Cytoplasm	Thyroid hormone receptor	Avian erythroblastosis

Adapted from James D. Watson, et al., *Molecular biology of the gene*, 4th ed. Copyright © 1965, 1970, 1976, 1987 by The Benjamin/Cummings Publishing Company, Inc. Reprinted by permission.

duced into a cell that normally does not make PDGF (i.e., a fibroblast); that cell was transformed into a tumor cell.

We can generalize and say that tumor cells can result from the synthesis of growth factors. Such growth factors are not normally produced in those cells, but the introduction of a certain class of transducing retrovirus can cause the growth factor genes to come under viral, not cellular control.

Protein kinases. A large number of proto-oncogenes appear to encode protein kinases. These enzymes catalyze the addition of phosphate groups to proteins, thereby modifying their functions. The *src* gene product, for example, has been shown to be a protein kinase and has been termed pp60^{src}. The viral protein, pp60^{v-src}, and the proto-oncogene, pp60^{c-src}, differ in only a few amino acids, and both proteins are found bound to the inner surface of the plasma membrane.

What is particularly interesting about the *src* protein kinases is that both versions add a phosphate group to the amino acid tyrosine; that is, they are tyrosine protein kinases. Before this discovery, the protein kinases that had been characterized had all been shown to add phosphates only to the amino acids serine or threonine. Since protein phosphorylation was known to be an important event in effecting a multitude of metabolic changes in cells, the *src* discovery was an exciting one. It suggested a possible explanation of how the *src* and other tyrosine protein kinase-coding oncogenes might transform a normal cell into a cancer cell, in which many metabolic differences are evident compared with normal cells. For example, a large class of proteins, including the receptors for growth factors, use protein phosphorylation to transmit their signals through the membrane. In the specific case of the *src* product, the oncogene prompts a cascade of regulatory events. First, the protein kinase catalyzes the phosphorylation of phosphatidyl inositol, which results in an increase in the concentration of diacylglycerol. This compound, in turn, activates the enzyme protein kinase C, which is thought to play a key role in the sequence by which growth factor receptors transmit their message to the nucleus. Thus, the action of protein kinases with regard to oncogene functions appears also to be linked to growth factors and their activities.

𝒦EYNOTE

Some forms of cancer are caused by tumor viruses. Both DNA tumor viruses and RNA tumor viruses are known. The RNA tumor viruses are retroviruses
that contain tumor-inducing genes called oncogenes. Both normal cells and non-viral-induced cancer cells contain sequences that are homologous to the viral oncogenes. Apparently, retroviruses that induce tumors have picked up certain normal cellular genes, called proto-oncogenes, while (in most cases) they simultaneously lose part of their genetic information. The proto-oncogenes function in normal cells in various ways to regulate cell differentiation. In the retrovirus, these genes have become modified so that the cell's protein product, now produced under viral control, is both quantitatively and qualitatively changed, as well as being expressed in cells in which their products are not usually found. These gene products, which include growth factors, are directly responsible for the transformation of cells to the cancerous state.

HIV—The AIDS Virus

Acquired ImmunoDeficiency Syndrome—AIDS—is induced by the retrovirus *HIV* (*human immunodeficiency virus*) (Figure 20.27a). The capsid of HIV is bullet-shaped. HIV contains complete *gag*, *pol*, and *env* genes, so HIV can self-propagate. In addition, HIV contains five or six other genes that are not oncogenes (Figure 20.27b). The exact functions of these genes are not completely understood. HIV infects T lymphocytes, which are part of the immune system, but instead of transforming the cells to the tumorous state, HIV kills them. As a consequence, the immune system is damaged severely or destroyed, leading to immunodeficiency. In the absence of a well-functioning immune system, the patient becomes vulnerable to infections (e.g., bacterial, viral, fungal) and becomes susceptible to numerous cancers. Patients die most frequently from the infections. To date, at least 10 million people in the world are infected with HIV.

SUMMARY

Bacteria and eukaryotic cells contain a variety of transposable genetic elements that have the property of moving from one site to another in the genome. In bacteria there are three types of transposable genetic elements: insertion sequence (IS) elements, transposons (Tn), and certain temperate bacteriophages. The simplest type of transposable element is an IS element. An

~ **FIGURE 20.27**

(a) Schematic drawing of a cross-section through an HIV particle. Note that the capsid of this particular retrovirus is "bullet"-shaped; (b) Organization of the HIV genome.

a)

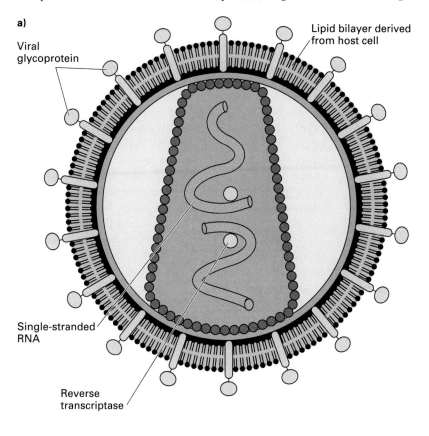

Viral glycoprotein

Lipid bilayer derived from host cell

Single-stranded RNA

Reverse transcriptase

b)

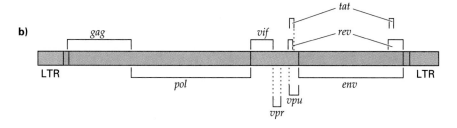

IS element typically consists of inverted terminal repeat sequences flanking a coding region, the products of which provide transposition activity. Tn elements are more complex, consisting essentially of terminally repeated IS elements flanking a DNA segment that often confers an antibiotic resistance phenotype. One or both of the terminal IS elements provides transposition functions for movement of the Tn element. When IS and Tn elements integrate into the genome, a short sequence at the target site typically becomes duplicated, giving rise to directly repeated sequences that flank the integrated element. Plasmids and temperate bacteriophages are much more complex transposable genetic elements. The integration of plasmids such as the *F* factor or the *R* plasmid into the genome is a property of homology between IS elements carried on the plasmids and IS elements present in the bacterial chromosome.

Transposable genetic elements in eukaryotes resemble bacterial transposons in general structure and transposition properties. Corn transposons often occur as families, each family containing an autonomous element (an element capable of transposing by itself) and one or more nonautonomous elements (elements that can only transpose if the autonomous element of the family is also present in the genome). Interestingly, and perhaps uniquely, the timing and frequency of transposition of corn transposons is developmentally regulated. Some eukaryotic transposons, such as yeast *Ty* elements and the *Drosophila copia* elements, transpose via an RNA intermediate (using a transposon-encoded reverse transcriptase). These types of transposons resemble retroviruses in genome organization and other properties, and hence have been called retrotransposons.

The presence of bacterial transposons (IS and Tn elements) and eukaryotic transposons in a cell is usually detected by the changes they bring about in the expression and activities of the genes at or near the chromosomal sites into which they integrate. Gene expression may be increased or decreased if the element inserts into a promoter or other regulatory sequence, mutant alleles of a gene can be produced if an element inserts within the coding sequence of the gene, and various chromosome rearrangements or chromosome breakage events can occur as a result of the mechanics of transposition. Generally, mutant alleles produced by insertion of a transposable element are unstable, since they revert when the transposable element undergoes a new transposition event.

Some forms of cancer are caused by tumor viruses. The RNA tumor viruses are called retroviruses. Retroviruses are single-stranded RNA viruses that may or may not contain tumor-inducing genes termed oncogenes. When a cell is infected by a retrovirus, a double-stranded DNA copy of the RNA genome is produced by reverse transcriptase, an enzyme brought into the cell by the virus. That DNA can integrate into the nuclear genome of the cell, where it is known as a provirus. The provirus directs the synthesis of progeny retroviruses that are released from the cell without causing the death of the cell.

Both normal cells and non-viral-induced cancer cells contain sequences that are related to the viral oncogenes. The model is that tumor-causing retroviruses have picked up certain normal cellular genes—called proto-oncogenes—while simultaneously losing some of their genetic information. In normal cells, proto-oncogenes function in various ways to regulate cell differentiation. In the retrovirus these genes have been modified or their expression controlled differently so that the protein product, now synthesized under viral control, is both qualitatively and quantitatively changed, as well as being expressed in cells in which the products are not normally found. These gene products, which include growth factors, are directly responsible for the transformation of cells to the cancerous state.

ANALYTICAL APPROACHES FOR SOLVING GENETICS PROBLEMS

Q.1 Imagine that you are a corn geneticist and are interested in a gene you call *zma*, which is involved in formation of the tiny hair-like structures on the upper surfaces of leaves. You have a cDNA clone of this gene. In a particular strain of corn that contains many copies of *Ac* and *Ds*, but no other transposable elements, you observe a mutation of the *zma* gene. You want to figure out whether this mutation does or does not involve the insertion of a transposable element into the *zma* gene. How would you proceed? Suggest at least two approaches, and say how your expectations for an inserted transposable element would differ from your expectations for an ordinary gene mutation.

A.1 One approach would be to make a detailed examination of leaf surfaces in mutant plants. Since there are many copies of *Ac* in the strain, if a transposable element has inserted into *zma* it should be able to leave again, so that the mutation of *zma* would be unstable. The leaf surfaces should then show a patchy distribution of regions with and regions without the hair-like structures. A simple point mutation would be expected to be stable.

A second approach would be to digest the DNA from mutant plants and DNA from normal plants with a particular restriction endonuclease, run the digested DNAs on a gel and prepare a Southern blot, and probe the blot using the cDNA. If a transposable element has inserted into the *zma* gene in the mutant plants, then the probe should bind to different molecular weight fragments in mutant as compared to normal DNAs. This would not be the case if a simple point mutation had occurred.

QUESTIONS AND PROBLEMS

20.1 Compare and contrast the types of transposable genetic elements in bacteria.

20.2 What are the properties in common between bacterial and eukaryotic transposons?

20.3 An IS element became inserted into the *lacZ* gene of *E. coli*. Later, a small deletion occurred in this gene, which removed forty base pairs, starting to the left of the IS element. Ten *lacZ* base pairs were removed, including the left copy of the target site, and the thirty left most base pairs of the IS element were removed. What will be the consequence of this deletion?

*20.4 A molecular biologist was studying the DNA of a newly discovered fungus, using denaturation-renaturation experiments to gain information about the amount of single-copy *vs.* repeated sequences. He was surprised to discover that, while most of the DNA of this fungus fit the usual C_ot expectations (renaturation time inversely related to concentration), 2% of the DNA failed to behave this way. This 2% always renatured at the same rate (very quickly) irrespective of concentration (even when the DNA solution was very dilute). Suggest an explanation for this observation.

20.5 A geneticist was studying glucose metabolism in yeast, and had deduced both the normal structure of the enzyme glucose-6-phosphatase (G6Pase) and the DNA sequence of its coding region. She had been using a wild-type strain called A to study another enzyme for many generations, when she noticed a morphologically peculiar mutant had arisen from one of the strain A cultures. She grew the mutant up into a large stock and found that the defect in this mutant involved a markedly reduced G6Pase activity. She isolated the G6Pase protein from these mutant cells and found it was present in normal amounts, but had an abnormal structure. The N-terminal 70% of the protein was normal. The C-terminal 30% was present but altered in sequence by a frame shift reflecting the insertion of 1 base pair, and the N-terminal 70% and the C-terminal 30% were separated by 111 new amino acids unrelated to normal G6Pase. These amino acids represented predominantly the AT rich codons (Phe, Leu, Asn, Lys, Ile, Tyr). There were also two extra amino acids added at the C-terminal end. Explain these results.

*20.6 Consider two theoretical yeast transposons, A and B. Each contains an intron. Each transposes to a new location in the yeast genome and then is examined for the presence of the intron. In the new locations, you find that A has no intron, while B does. What can you conclude about the mechanisms of transposon movement for A and B from these facts?

*20.7 An investigator has found a retrovirus capable of infecting human nerve cells. This is a complete virus, capable of reproducing itself, and it contains no oncogenes. People who are infected suffer a debilitating encephalitis. The investigator has shown that when he infects nerve cells in culture with the complete virus, the nerve cells are killed as the virus reproduces, but if he infects cultured nerve cells with a virus in which he has created deletions in the *env* or *gag* genes, no cell death occurs. The investigator is interested in finding ways to bring about nerve cell growth or regeneration in people who have suffered nerve damage. For example, in a patient with a severed spinal cord, nerve regeneration might relieve paralysis. The investigator has cloned the human nerve growth factor gene, and wants to insert it into the genome of his retrovirus from which he has deleted parts of the *env* and *gag* genes. He would then use the engineered retrovirus to infect cultured nerve cells. Adult nerve cells do not normally produce large amounts of nerve growth factor. If he is successful in inducing growth in them without causing any cell death, he would like to move on to clinical trials on injured patients. When the investigator applied for grant support to do this work, his application was denied on grounds that there were inadequate safeguards in the plan. Why might this work be dangerous? What comparisons can you draw between the virus the investigator wants to create and, for example, Avian myeloblastosis virus?

21 ORGANIZATION AND GENETICS OF EXTRANUCLEAR GENOMES

PRINCIPAL POINTS

~ The genomes of both mitochondria and chloroplasts are circular, double-stranded DNAs. Both classes of organelles contain several nucleoid regions in which the DNA is located, and each nucleoid contains several copies of the DNA molecule.

~ The mitochondrial and chloroplast genomes contain the genes for the rRNA components of the ribosomes that are assembled and function in the organelles, for many (if not all) of the tRNAs used in organellar protein synthesis, and for a few proteins that remain in the organelles and perform functions specific to the organelles. All other proteins are nuclear-encoded, synthesized on cytoplasmic ribosomes, and imported into the organelles. These proteins include the ribosomal proteins for the organellar ribosomes.

~ Many examples of mitochondrial and chloroplast mutants are known. The inheritance of these extranuclear genes follows rules different from those for nuclear genes: no meiotic segregation is involved, uniparental (and often maternal) inheritance is generally exhibited, a trait resulting from an extranuclear mutation persists after nuclear substitution, and extranuclear genes are non-mappable to the known nuclear linkage groups.

~ Maternal effect is defined as the predetermination of gene-controlled traits by the maternal *nuclear* genotype prior to the fertilization of the eggs. Maternal effect is different from extranuclear inheritance. The maternal inheritance pattern of extranuclear genes occurs because the zygote obtains most of its organelles (containing the extranuclear genes) from the female parent. In contrast, in the maternal effect the inherited trait is controlled by the maternal *nuclear* genotype before the fertilization of the egg and does *not* involve extranuclear genes.

~ There are numerous examples of extranuclearly inherited gene mutations, involving genes in mitochondrial DNA or chloroplast DNA, including leaf variegation in four o'clock plants, certain slow-growing mutants in fungi, and certain antibiotic resistance or dependence traits in *Chlamydomonas*.

~ Not all cases of extranuclear inheritance result from genes on mitochondrial DNA or chloroplast DNA. Many other examples in eukaryotes result from infectious heredity, in which symbiotic, cytoplasmically-located bacteria or viruses are transmitted when cytoplasms mix.

To this point in the text, in our discussion of eukaryotic genetics we have analyzed the structure and expression of the genes located on chromosomes in the nucleus and have defined rules for the segregation of nuclear genes. The nucleus is not the only place in the cell in which DNA is located, however. Outside the nucleus, DNA is found in two principal organelles, the mitochondria (found in both animals and plants) and chloroplasts (found only in green plants). The genes in these mitochondrial and chloroplast genomes have become known as extrachromosomal genes, cytoplasmic genes, non-Mendelian genes, organellar genes, or extranuclear genes.

Although *extranuclear* is used in the discussions that follow, the term *non-Mendelian* is also informative because extranuclear genes do not follow the rules of Mendelian inheritance as do nuclear genes in segregation and recombination. Recently, the application of modern molecular biology techniques has led to rapid advances in knowledge about the organization of extranuclear genomes. In this chapter we examine some of these advances and discuss the inheritance patterns that extranuclear genes follow from generation to generation. First, we examine the genetic structure of mitochondria and chloroplasts; we then discuss how their genomes segregate. There are nonchromosomal, nuclear genetic elements that segregate in a non-Mendelian fashion, but they will not be discussed in this text.

ORGANIZATION OF EXTRANUCLEAR GENOMES

Mitochondrial Genome

Mitochondria, organelles found in the cytoplasm of all aerobic animal and plant cells, are the principal sources of energy in the cell (see Figure 1.12). They contain the enzymes of the Krebs cycle, carry out oxidative phosphorylation, and are involved in fatty acid biosynthesis. The presence of DNA in mitochondria has been shown by a variety of techniques, including electron microscopy, autoradiography, and the isolation from mitochondria of a DNA species with a buoyant density different from that of nuclear DNA by virtue of different

~ FIGURE 21.1

Nuclear and mitochondrial DNA. Mitochondrial DNA can be detected under the electron microscope as small circles and can often be separated from nuclear DNA by CsCl density gradient centrifugation, since the two DNAs differ in buoyant density.

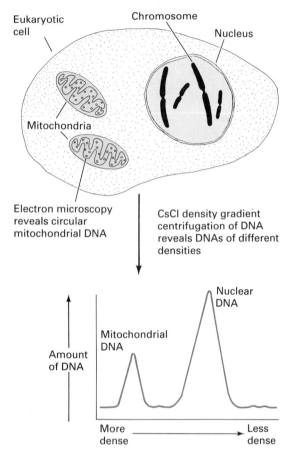

~ FIGURE 21.2

Electron micrograph of mitochondrial DNA.

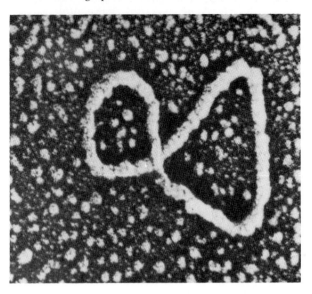

lion base pairs (80 to 800 μm), depending on the species. Corn, for example, has a mitochondrial genome of 600,000 base pairs (200 μm).

Despite the differences in the amount of DNA, mitochondria from all organisms contain about the same amount of unique-sequence DNA; that is, DNA that codes for functional mitochrondrial products. So despite the disparity in mitochondrial size, the only difference between animal mitochondria and that of other organisms is that essentially the entire mitochondrial genomes of animals encode products, whereas the mitochondrial genomes of fungi and plants include a lot of DNA that does not code for products, in addition to coding sequences.

While the informational content of the mitochondrial genome is very small in comparison to the nuclear genome, the relative amount of mitochrondrial DNA is actually quite large. Within mitochondria are *nucleoid regions* (similar to those of bacterial cells), each of which contains several copies of the mitochondrial chromosome. Yeast, for example, has between 4 and 5 mtDNA molecules per nucleoid, and each mitochondrion has 10 to 30 nucleoids. Since each yeast cell has between 1 and 45 mitochondria per cell, there are many mtDNA molecules per cell. As a consequence of the multiplicity of mitochondrial chromosomes, the contribution of mitochondrial activities to overall cellular activities can be relatively significant when compared with nuclear contributions.

REPLICATION. Replication of mtDNA uses mitochondrial DNA polymerases that function independently of nuclear DNA polymerases. As in replication of nuclear DNA, RNA primers are synthesized for initi-

GC percentages (Figure 21.1). Some of the properties of mitochondrial DNA, or mtDNA, are presented in Table 21.1, along with properties of nuclear and chloroplast DNAs.

STRUCTURE. Mitochondrial DNA is a double-stranded, supercoiled, circular chromosome devoid of histones and, therefore, not organized into nucleosomes (Figure 21.2). The size of the chromosome varies from organism to organism but is constant among mitochondria within any particular species. Interestingly, the chromosomes of higher-animal mitochondria are significantly smaller than those of fungi and plants. The mitochondrial chromosome in humans, for example, is 16,569 base pairs, while that from *Drosophila* is about 18,000 base pairs, that from *Neurospora crassa* is about 60,000 base pairs, and that from yeast is about 75,000 base pairs. The mitochondrial genomes of plants are even larger, ranging from 250,000 to 2 mil-

~ TABLE 21.1

Properties of Nuclear, Mitochondrial, and Chloroplast DNAs

	NUCLEAR DNA		MITOCHONDRIAL DNA mtDNA		CHLOROPLAST DNA cpDNA	
	%GC	GENOME SIZE (bp FOR ALL CHROMOSOMES)	%GC	GENOME SIZE (bp PER CHROMOSOME)[a]	%GC	GENOME SIZE (bp PER CHROMOSOME)[b]
ANIMALS						
Human	41	2.75×10^9	41	14×10^3	—	—
Chick	43	1.2×10^9	49	15×10^3	—	—
Sea urchin	39	7.2×10^8	45	13×10^3	—	—
D. melanogaster	42	1.75×10^8	22	18×10^3	—	—
FUNGI						
N. crassa	53	4.3×10^7	42	60×10^3	—	—
S. cerevisiae	44	1.75×10^7	18	75×10^3	—	—
PLANTS						
C. reinhardi	64	1×10^8	71	2×10^5	36	30×10^3
Corn	49	6.6×10^9	48	2.2×10^5	39	13×10^3
Tobacco	38	1.1×10^9	51	2×10^5	47	15×10^3
Sweet pea	35	9.4×10^9	38	9×10^4	46	12×10^3

[a]Each mitochondrion is believed to contain about five copies of the mitochondrial chromosome.

[b]Each chloroplast is believed to contain about twenty copies of the chloroplast chromosome.

SOURCE: Data mainly compiled from "Chloroplast DNA: Physical and genetic studies" by Ruth Sager and Gladys Schlanger and "Mitochondrial DNA" by Margit M. K. Nass, 1976. In *Handbook of genetics*, Robert C. King, ed., v. 5, *Molecular genetics*, p. 371 and p. 477. New York: Plenum Press.

ation. The mitochondrial DNA replication process occurs throughout the cell cycle, without any preference for the S phase of the cell cycle, which is when nuclear DNA replicates. Some differences appear in the details of mtDNA replication among eukaryotes, and the displacement loop (D loop) model (deduced from observing animal mitochondria *in vivo*) is useful as a general scheme. Figure 21.3 shows an electron micrograph of an early stage of D loop replication; Figure 21.4 (p. 646) diagrams the model in more detail.

In most animals the two strands of mtDNA have different densities, so they are called the H (heavy) and L (light) strands. The D loop model of replication shows relative asynchrony in DNA replication for the two complementary H and L strands. In the D loop model, the synthesis of the new H strand is started in one origin (the L strand origin) and forms a D loop structure (which can be seen by electron microscopy). As the new H strand extends, initiation of synthesis of the new L strand at a second origin (the H strand origin) takes place. Both strands are completed by continuous replication. Lastly, the circular DNAs are each converted to a supercoiled form, with approximately one hundred superhelical twists.

Evidence indicates that mitochondria (and chloroplasts) grow and divide rather than assemble from simple components. The classical experiment for rep___ tion of complete mitochondria was done in ___ David Luck, who used density-labeling pro___ study mitochondrial division in *Neuros___* (Figure 21.5; p. 646). In essence, this exp___ ilar to the one done by Meselson and St___ DNA replication is semiconservative ___

~ FIGURE 21.3

Electron micrograph of the D loop structure found in replicating mitochondrial DNA; a small D loop is shown.

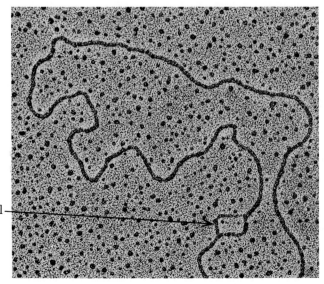

Small D loop

~ FIGURE 21.4

Model for mitochondrial DNA replication that involves the formation of a D loop structure.

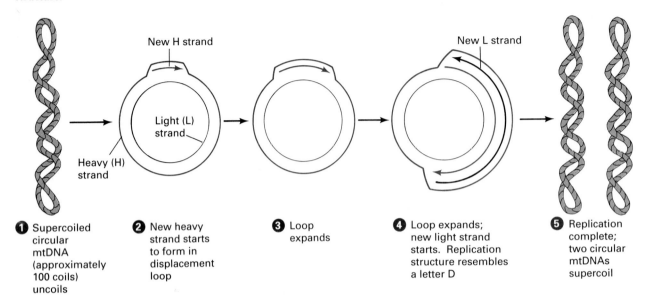

1 Supercoiled circular mtDNA (approximately 100 coils) uncoils

2 New heavy strand starts to form in displacement loop

3 Loop expands

4 Loop expands; new light strand starts. Replication structure resembles a letter D

5 Replication complete; two circular mtDNAs supercoil

Luck grew *N. crassa* on a medium that was very low in protein but high in a lipid precursor, choline. Under these conditions the mitochondria produced were more dense than those found in *Neurospora* cultures grown in normal medium. Next, Luck shifted the *Neurospora* cells to a medium with a high protein content and a low lipid precursor content. After one generation he examined the mitochondria for their density. There are two possible outcomes: If mitochondria grow and divide, then all progeny mitochondria would contain half of the material made in the low-density medium and half of the material made in the high-density medium. These mitochondria would have an intermediate density between light and heavy mitochondria. If, however, mitochondria are made *de novo*, then there would be two populations of mitochondria: those of low density made early in the experiment and those of high density made after the shift to high-density medium. Luck observed mitochondria of intermediate density, indicating that mitochondria grow and divide.

~ FIGURE 21.5

Luck's experiment, which showed that mitochondria arise by division of preexisting mitochondria, not by assembly from simple components.

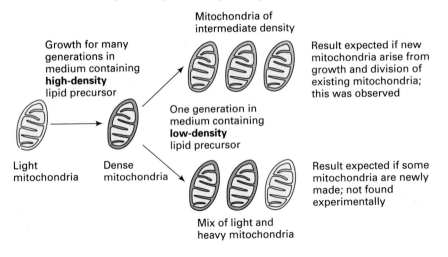

Mitochondria of intermediate density

Growth for many generations in medium containing **high-density** lipid precursor

Result expected if new mitochondria arise from growth and division of existing mitochondria; this was observed

One generation in medium containing **low-density** lipid precursor

Light mitochondria

Dense mitochondria

Mix of light and heavy mitochondria

Result expected if some mitochondria are newly made; not found experimentally

GENE ORGANIZATION OF MTDNA. Mitochondrial DNA contains information for a number of mitochondrial components, such as tRNAs, rRNAs, and proteins. Mitochondria contain ribosomes that are responsible for all protein-synthetic activity that occurs within mitochondria. Messenger RNAs synthesized within the mitochondria remain in the organelle and are translated by mitochondrial ribosomes. The two rRNA molecules found in mitochondrial ribosomes are coded for in the mtDNA, while most, if not all, of the ribosomal proteins of mitochondrial ribosomes are coded for in the nuclear genome. In humans, mtDNA encodes 2 rRNAs, 22 tRNAs, 13 proteins that are part of the inner mitochondrial membrane and 1 required for oxidative metabolism. The mtDNA molecule has been studied in a variety of organisms, and since it is relatively small, a great deal of progress has been made in determining the exact organization of its genes.

Our present-day understanding of the gene organization of mtDNA derives mainly from DNA sequencing experiments. Today the sequence of human (and several other organisms') mtDNA is completely known. Some of its features were unexpected and may be general for mtDNAs. Figure 21.6 presents the map of the genes of human mtDNA. The genes coding for the 16S and 12S rRNAs of the large and small mitochondrial ribosomal subunits are located adjacent to one another on the H strand. The genes for tRNAs are present at various positions around the H strand and the L strand, some in clusters, some individually, and one between the two rRNA genes.

The protein-coding genes are found in both strands. Their positions were identified in two ways. One method involved searching the DNA sequence for possible start and stop signals in the same reading frame (an <u>o</u>pen <u>r</u>eading <u>f</u>rame, or ORF, also called <u>un</u>-

~ FIGURE 21.6

Map of the genes of human mitochondrial DNA. The outer circle shows the genes transcribed from the H (heavy) strand, and the inner circle shows the genes transcribed from the L (light) strand. The origins of replication for H and L strands are indicated. 16S and 12S (blue) are the rRNA genes. tRNA genes are shown in purple and protein-coding genes in yellow. Code: ATPase 6 and 8: components of the mitochondrial ATPase complex. COI, COII, and COIII: cytochrome c oxidase subunits. cyt b: cytochrome b. ND1-6: NADH dehydrogenase components (previously URF1-6, where URF = unidentified reading frame).

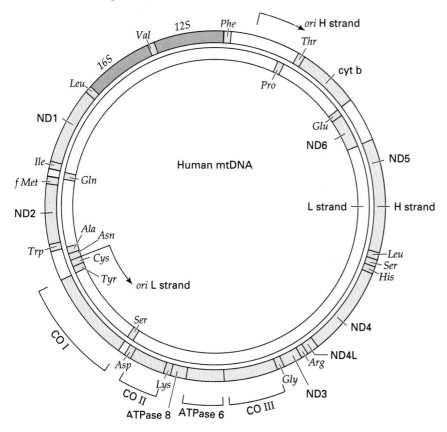

identified reading frame, or URF; see Chapter 17). The signals had to be separated by significant lengths; the sequences had to be able to code for amino acids. This search was accompanied by using computer programs designed for this purpose. The second method (used primarily by G. Attardi and his group) was to align the 5′ and 3′ end proximal sequences of mitochondrial mRNAs with the DNA sequence. (Note that mitochondrial mRNAs have a 3′ poly(A) tail but no 5′ cap.)

At this time the products of all URFs of the human mitochondrial genome have been functionally defined. Careful analysis of the gene sequences revealed some differences in the genetic code used in mitochondria compared with the nuclear genetic code. For example, a UGA codon which signifies chain termination in the nucleus is read as tryptophan in the human mitochondrion. Other differences include ATA in a mitochondrial gene coding for methionine, whereas in the nucleus it codes for isoleucine.

Figure 21.7 shows the human mitochondrial genetic code, the coding properties of the tRNAs, and the total number of codons used in the whole genome. All the codons are used except the chain-terminating codons UAA and UAG, which have the same function in the classical genetic code. No tRNAs have been found with anticodons for AGA and AGG, so perhaps these codons also have a chain-terminating function. Interestingly, the genetic code of yeast mitochondria is not identical with that of human mitochondria, and Table 21.2 shows the points of difference. Differences in the genetic code have also been detected in ciliated protozoans. These differences suggest that the mitochondrial genetic code may be read differently in many eukaryotic organisms. In other words, the genetic code is not as universal as was originally thought.

The tRNA genes in human mtDNA have been identified by using a computer program designed to search the DNA sequence for polynucleotide stretches that could form the classical cloverleaf, two-dimensional structure. However, a number of features thought to be invariable in cytoplasmic tRNAs, in terms of biological function, vary in mitochondrial tRNAs. For example, one serine tRNA has one complete loop missing, yet it is still functional in mitochondria.

For both the cytoplasmic and mitochondrial genetic codes, there are many instances in which the first two bases in the codon are the same and the same

~ FIGURE 21.7

The genetic code of the human mitochrondrial genome. Each box indicates that one tRNA is used to read those codons within. The numbers within each box indicate the total number of those codons found in the whole mitochrondrial DNA molecule. Highlighted are the codons that have different coding properties in the human mitochondrion than in the nuclear genome of all eukaryotic organisms or in the genomes of prokaryotic organisms.

First letter	Second letter: U	C	A	G	Third letter
U	77 UUU Phe	32 UCU Ser	46 UAU Tyr	5 UGU Cys	U
	140 UUC	99 UCC	89 UAC	17 UGC	C
	73 UUA Leu	83 UCA	UAA Stop	93 **UGA** Trp	A
	17 UUG	7 UCG	UAG Stop	10 UGG	G
C	65 CUU Leu	41 CCU Pro	18 CAU His	7 CGU Arg	U
	167 CUC	119 CCC	79 CAC	25 CGC	C
	276 CUA	52 CCA	81 CAA Gln	29 CGA	A
	45 CUG	7 CCG	9 CAG	2 CGG	G
A	125 AUU Ile	51 ACU Thr	33 AAU Asn	14 AGU Ser	U
	196 AUC	155 ACC	130 AAC	39 AGC	C
	166 **AUA** Met	133 ACA	85 AAA Lys	**AGA** Stop	A
	40 AUG	10 ACG	10 AAG	**AGG** Stop	G
G	30 GUU Val	43 GCU Ala	15 GAU Asp	24 GGU Gly	U
	49 GUC	124 GCC	55 GAC	88 GGC	C
	71 GUA	80 GCA	64 GAA Glu	67 CGA	A
	18 GUG	8 GCG	24 GAG	34 GGG	G

~ TABLE 21.2

Differences Between Human and Yeast Mitochondrial Genetic Codes

CODON[a]	NUCLEAR CODE	AMINO ACID MITOCHONDRIAL CODE	
		MAMMAL	YEAST
UGA	Termination	Tryptophan	Tryptophan
AUA	Isoleucine	Methionine	Isoleucine
CUN[b]	Leucine	Leucine	Threonine
AGG, AGA	Arginine	Termination?	Arginine
CGN[b]	Arginine	Arginine	Termination?

[a]All sequences read 5' to 3'.

[b]N = any one of the four bases A, G, U, and C.

~ FIGURE 21.8

Organization of mitochondrial rRNA in HeLa cells and a number of organisms.

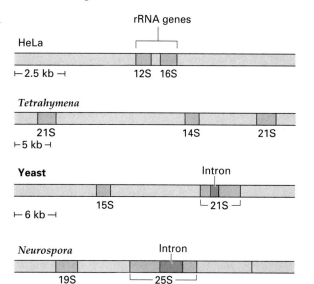

amino acid is specified, whether the third base is U or C (e.g., UUU or UUC = Phe), A or G (e.g., AAA or AAG = Lys), or U, C, A, or G (e.g., GUU, GUC, GUA, or GUG = Val). The term *family box* is used for groups of codons of this kind: UUU or UUC are a family box for phenylalanine (Phe), for example. In the cytoplasm, with base-pairing wobble, a minimum of 32 tRNAs are needed to read all of the 61 sense codons. No cytoplasmic tRNA can read more than three of these related codons, even with wobble. So even when the same amino acid is specified by the first two bases no matter what the third base is (a four-codon family box), the four codons need *two* tRNAs in order for them to be read. Many of the mitochondrial tRNAs, however, apparently have the ability to read all four members in a four-codon family box. Table 21.3 contrasts the two systems. As a result of this extended wobble situation in mitochondria, only 22 tRNA genes are needed in mammalian mitochondria. For all four-codon family boxes,

only one tRNA is used, and this tRNA has U (uracil) in the anticodon at the position corresponding to the third letter of the codon. Similar conclusions have been reached in studies of *Neurospora crassa* and yeast mitochondria.

We know that each mitochondrial genome contains one copy of each of the two rRNA genes. Figure 21.8 diagrams the organization of the two rRNA genes and illustrates the variation in the organization of rRNA genes in selected mitochondrial genomes.

In animals, as exemplified by HeLa (human) cells, the genes for the small (12S) and large (16S) rRNAs are very closely linked. In the spacer separating them, one tRNA gene is found. By contrast, in fungi the two genes

~ TABLE 21.3

Example of Four-Position, Base-Pair Wobble in Mitochondrial tRNAs[a]

CODON	ANTICODON NUCLEAR WOBBLE	MITOCHONDRIAL WOBBLE
GCU		
GCC	GGC	
		UGC
GCA		
GCG	UGC	

[a]All sequences read 5' to 3'.

are separated by a large amount of DNA: 6 kb in *Neurospora* and about 30 kb in yeast. A number of tRNA genes are located in these spacer regions. The rRNA genes of *Neurospora* and yeast both contain introns (Figure 21.8). In the protozoan *Tetrahymena* there is an extra copy of the large rRNA gene, but not of the small rRNA gene. This extra copy has presumably arisen by a duplication event and is not commonly found in mitochondrial genomes.

One of the striking features of human mtDNA is that it is efficiently organized into a compact structure. The reading frames for the protein-coding and rRNA genes are mostly adjacent to tRNA genes, with no, or very few, nucleotides in between. Apparently, very large transcripts are produced during the expression of mitochondrial genes. By processing, the tRNA transcript sequences are spliced out, since they are easily recognized by their cloverleaf structure. With essentially no spacers in between the mRNA transcripts and tRNA transcripts in the large RNA precursor molecules, this process generates mature tRNAs and monogenic mRNAs.

Another aspect of compact organization of genes in the mitochondrial genome is that the AUG initiation codon of a mitochondrial mRNA is right at the 5' end of the molecule. This contrasts with the situation for cytoplasmic mRNAs where the AUG start codon is located internally after a 5' leader sequence (see Chapter 14). Mitochondrial mRNAs also lack the 5' cap structure found in cytoplasmic mRNAs. Since cytoplasmic mRNAs require the 5' cap and an internally located AUG codon for translation, the contrasting features of mitochondrial mRNAs indicate that the mitochondrial protein synthesis machinery must be quite different from that of nuclear DNA—the ribosome and the rest of the initiation mechanism may simply recognize the end of the mRNA without any requirement for a specific ribosome binding site or a 5' cap structure.

mtDNA's mechanism of translation initiation is also different from that for bacterial systems, which requires an internal ribosome-binding site located upstream of the AUG initiation codon.

Another way in which human mtDNA is efficiently organized is that the DNA sequence for most of the mitochondrial mRNAs does not encode a complete chain-terminating codon. Instead, the transcripts end with either U or UA in the same reading frame as the initiation codon, and apparently the polyadenylation (addition of a poly(A) tail at the 3' end) of the mRNA supplies the missing part(s) of the UAA stop codon. While this efficient organization is prevalent for animal mtDNA, the situation is completely different in those organisms with larger mitochondrial genomes. In yeast, which has a genome that is five times larger while coding for about the same number of components, the genes are separated by long AT-rich sequences that have no obvious biological function. Introns are also found in some genes in these organisms, while they are never found in animal mitochondrial genomes.

MITOCHONDRIAL RIBOSOMES. Mitochondrial ribosomes are structurally diverse among organisms, with all of them having a higher protein/RNA ratio than *E. coli* ribosomes. Table 21.4 shows examples of structural characteristics of mitochondrial ribosomes in humans and yeast. In animals, mitochondrial ribosomes are about the same size or perhaps slightly larger than *E. coli* ribosomes in terms of molecular weight and volume. However, owing to the high protein/RNA content, animal ribosomes have a relatively low buoyant density of 55S to 60S. In fungi such as yeast and *Neurospora*, the mt ribosomes have a sedimentation coefficient of 70S to 75S.

We can make some rough generalizations about the rRNA content of mt ribosomes. In most cases the ribosomes lack the 5S and 5.8S rRNA components charac-

~ TABLE 21.4

Relative Size (in S Values) of Mitochondrial (mt) Ribosomes and Cytoplasmic (cyto) Ribosomes in Humans and Yeast

COMPONENT	HUMAN (HeLa)		YEAST	
	mt	cyto	mt	cyto
RIBOSOMES	60S	74S	75S	80S
Large subunit	45S	60S	53S	60S
Small subunit	35S	40S	35S	40S
RIBOSOMAL RNA				
Large ribosomal subunit	16S	28S	21S	26S
Small ribosomal subunit	12S	18S	15S	18S

teristic of cytoplasmic ribosomes. Higher plants do have a 5S rRNA molecule in mt ribosomes but it is distinct from that molecule found in cytoplasmic ribosomes. In other words, most mt ribosomes have two unequal-sized subunits, with a single rRNA molecule in each subunit. In animal mitochondria these rRNA molecules have sedimentation coefficients of 12S and 16S, which are significantly smaller than the large rRNAs of *E. coli* ribosomes.

The number of ribosomal proteins found in mt ribosomes is not well defined. Yeast appears to have 60 to 70 proteins in the mt ribosomes, while animal mitochondrial ribosomes have between 70 and 100 ribosomal proteins.

The proteins of mitochondrial ribosomes appear to be totally distinct from the proteins found in cytoplasmic ribosomes, and with one possible exception in both *Neurospora* and in yeast, mitochondrial proteins are encoded by nuclear genes. Hence, these proteins are synthesized on cytoplasmic ribosomes and subsequently transported into the mitochondria. This process points again to the remarkable abilities of cells to direct the products of protein synthesis to the specific sites where they are needed. We know that cytoplasmic ribosomes in eukaryotes contain about 65 to 70 ribosomal proteins, and lower eukaryotes appear to have 60 to 70 ribosomal proteins in mitochondrial ribosomes. Thus 125 to 140 ribosomal proteins are made on cytoplasmic ribosomes. About half of them subsequently migrate into the nucleus to be used in the assembly of cytoplasmic ribosomes, while the remainder enter the mitochondria, where they are used in the assembly of mitochondrial ribosomes and of transcripts of the two rRNA genes located on the mitochondrial chromosome. But we have much to learn about migration between cellular compartments.

MITOCHONDRIAL PROTEIN SYNTHESIS. Mitochondria have their own protein-synthetic machinery, and the process of protein synthesis is, in some ways, analogous to the process of protein synthesis in bacteria. A species of tRNA.fMet is present in mitochondria, and it appears to function in the initiation of protein synthesis. The initiation factors (IFs) are also very similar to those in bacteria. In a number of cases mitochondrial and bacterial components are interchangeable in an *in vitro*, protein-synthesizing system. For example, *Neurospora crassa* ribosomes are able to recognize, bind, and translocate *E. coli* fMet-tRNA.fMet in response to AUG, and *E. coli* IFs can substitute for mt IFs in catalyzing the initiation of protein synthesis.

Another piece of evidence for the similarity between *E. coli* and mt ribosomes is that the latter are sensitive to most inhibitors of bacterial ribosome function, including streptomycin, spectinomycin, neomycin,

and chloramphenicol. In some cases the degree of sensitivity to these agents, most of which are antibiotics, is quite different from that of *E. coli*. Also, mt ribosomes are generally insensitive to antibiotics or other agents to which cytoplasmic ribosomes are sensitive, such as cycloheximide. By the selective use of antibiotics, the site of synthesis of proteins found in mitochondria can be investigated. For example, those mt proteins that are synthesized in the presence of cycloheximide, an inhibitor of cytoplasmic ribosomes, must be made on mt ribosomes. Conversely, those mt proteins made in the presence of chloramphenicol, an inhibitor of mt ribosomes, must be made on cytoplasmic ribosomes. Using this approach, scientists have discovered that of the seven subunits of the mitochondrial enzyme cytochrome oxidase in yeast mitochondria, four are made in the cytoplasm and three are made in the mitochondria (see Figure 21.9; p. 652).

Chloroplast Genome

Chloroplasts are cellular organelles found only in green plants and are the site of photosynthesis in the cells containing them. Chloroplasts are characterized by a double membrane surrounding an internal, chlorophyll-containing lamellar structure embedded in a protein-rich stroma (see Figure 1.13). Like mitochondria, chloroplasts contain their own genomes, although we do not know as much about the chloroplast (cp) genome as we do about the mitochondrial genome.

STRUCTURE. In many respects the structure of the chloroplast genome is similar to the structure of mitochondrial genomes. Table 21.1 gives some properties of cpDNA. In all cases the DNA is double-stranded, circular, devoid of structural proteins, and supercoiled. An electron micrograph of cpDNA is shown in Figure 21.10a (p. 652). In many cases the GC content of cpDNA differs greatly from that of the nuclear and mitochondrial DNA, which allows cpDNA to be isolated by CsCl equilibrium density gradient centrifugation. Figure 21.10b shows the result of such an experiment using total cellular DNA from the unicellular alga *Chlamydomonas* (Figure 21.11; p. 653). Two discrete bands of DNA are seen (apart from the two marker bands) and indicate the presence of molecular species with different base-pair compositions. Since the cpDNA has a GC content of 36 percent compared with 64 percent for nuclear DNA and 71 percent for mtDNA, the cpDNA produces a *satellite band* that can be isolated and studied further. The relatively large amount of nuclear DNA results in a fairly broad peak that does not permit resolution of the mtDNA peak.

The number of copies of cpDNA per chloroplast varies from species to species. In all cases there are mul-

~ FIGURE 21.9

Synthesis of the multi-subunit protein cytochrome oxidase takes place on both cytoplasmic and mitochondrial ribosomes.

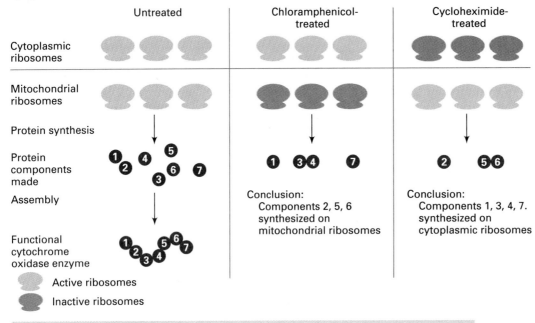

	Untreated	Chloramphenicol-treated	Cycloheximide-treated
Cytoplasmic ribosomes			
Mitochondrial ribosomes			

Protein synthesis

Protein components made

Assembly

Conclusion:
Components 2, 5, 6 synthesized on mitochondrial ribosomes

Conclusion:
Components 1, 3, 4, 7. synthesized on cytoplasmic ribosomes

Functional cytochrome oxidase enzyme

Active ribosomes

Inactive ribosomes

~ FIGURE 21.10

Chloroplast (cp) DNA: (a) electron micrograph of a chloroplast DNA molecule; (b) example of the results of a CsCl equilibrium density gradient centrifugation experiment done with total DNA extracted from the unicellular alga *Chlamydomonas*. The large amount of nuclear DNA results in a broad peak that obscures the mtDNA peak.

a)

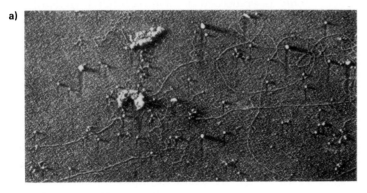

b) Satellite cpDNA band in CsCl gradient

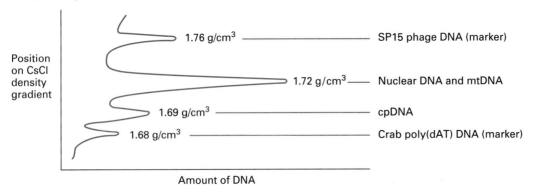

Position on CsCl density gradient

1.76 g/cm^3 ———— SP15 phage DNA (marker)

1.72 g/cm^3 ——— Nuclear DNA and mtDNA

1.69 g/cm^3 ———————— cpDNA

1.68 g/cm^3 ———————— Crab poly(dAT) DNA (marker)

Amount of DNA

~ FIGURE 21.11

Chlamydomonas.

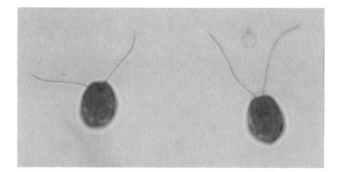

~ FIGURE 21.12

Euglena.

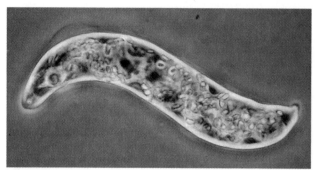

tiple copies per chloroplast, and these copies are found in nucleoid regions that are also present in multiple copies. For example, leaf cells of the garden beet have between 4 and 8 cpDNA molecules per nucleoid, from 4 to 18 nucleoids per chloroplast, and about 40 chloroplasts per cell, giving almost 6000 cpDNA molecules per cell. In *Chlamydomonas* the one chloroplast in a cell generally contains between 500 and 1500 cpDNA molecules.

With a length of about 40 to 45 μm in many plants (62 μm in *Chlamydomonas*), cpDNA is approximately 8 to 9 times longer than animal mtDNA. Although its great length indicates that a potentially large amount of genetic information is coded for in cpDNA, how much of the genome codes for products has yet to be determined.

REPLICATION. The replication of cpDNA is a semiconservative process that probably is similar to the process for mitochondria. It is clear that chloroplast-specific enzymes are used in this replication process. These enzymes are distinct from the enzymes of nuclear replication and of mitochondrial replication. Most cp enzymes are coded for by nuclear genes and transported into the chloroplast. Chloroplasts themselves grow and divide in essentially the same way as do mitochondria.

GENE ORGANIZATION OF cpDNA. The chloroplast genome contains genes for the cp rRNAs and perhaps for some ribosomal proteins, for a number of tRNAs, and for some cp enzymes and membrane proteins that are needed for electron transport in the process of photosynthesis. All of the mRNAs transcribed from chloroplast genes are translated by chloroplast ribosomes.

One of the chloroplast proteins that *has* been well characterized is part of ribulose bisphosphate decarboxylase, the first enzyme used in the pathway for the fixation of carbon dioxide in the photosynthetic pro-

cess. This enzyme is a major protein and is found in the chloroplasts of all plant tissues. Since it constitutes about 50 percent of the protein found in green-plant tissue, it is the most prevalent protein in the world. Ribulose bisphosphate decarboxylase contains eight polypeptides, four identical small ones and four identical large ones. The small polypeptide is coded by a nuclear gene, and the large polypeptide is coded by a chloroplast gene. (This result was determined by using selective antibiotics in essentially the same way as they are used in mitochondrial protein synthesis studies.)

For purposes of comparison, the organizations of the chloroplast genomes in *Chlamydomonas reinhardi*, *Euglena gracilis* (pictured in Figure 21.12), and the garden pea *Pisum sativum* are presented in Figure 21.13 (p. 654). In brief, each cp genome contains essentially the same genes, but their arrangements differ. It is postulated that during evolution the different arrangements of cpDNA arose by inversions of DNA segments.

CHLOROPLAST RIBOSOMES AND PROTEIN SYNTHESIS. Chloroplast protein synthesis uses ribosomes that are completely distinct from mitochondrial ribosomes and cytoplasmic ribosomes. They have a sedimentation coefficient of 70S and consist of two unequal-sized subunits, 50S and 30S—thus they are similar in sedimentation coefficients to prokaryotic ribosomes.

The large subunit of the cp ribosome contains at least two species of rRNA, present in one copy each: 23S and 5S. The small subunit of the cp ribosome contains one copy of a 16S rRNA. Many of the ribosomal proteins are made on cytoplasmic ribosomes and are, therefore, encoded in the nuclear genome. The most detailed study done so far, which used *Euglena*, has shown that 9 ribosomal proteins are made in the chloroplast, and 12 are made in the cytoplasm.

Protein synthesis in chloroplasts is similar to the process in prokaryotes. Formylmethionyl tRNA is used

~ FIGURE 21.13

Organizations of the chloroplast genomes of
(a) *Chlamydomonas reinhardi;* (b) *Euglena gracilis;* and
(c) *Pisum sativum.*

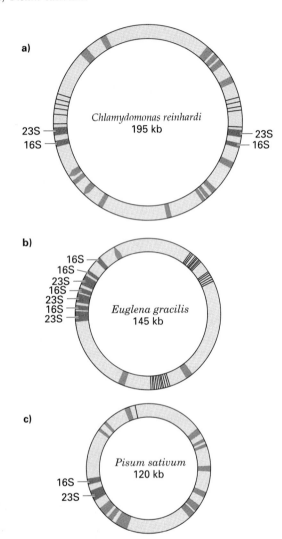

a)

Chlamydomonas reinhardi
195 kb

23S
16S

23S
16S

b)

16S
16S
23S
16S
23S
16S
23S

Euglena gracilis
145 kb

c)

Pisum sativum
120 kb

16S
23S

KEYNOTE

The genomes of both mitochondria and chloroplasts are circular, double-stranded DNAs that contain genes for the rRNA components of the ribosomes of these organelles, for many (if not all) of the tRNAs used in organellar protein synthesis, and for a few proteins that remain in the organelles and perform functions specific to the organelles. At least in mitochondria, the genetic code is different from that found in nuclear protein-coding genes.

RULES OF EXTRANUCLEAR INHERITANCE

Since the pattern of inheritance shown by genes located in organelles differs strikingly from the pattern shown by nuclear genes, the term **non-Mendelian inheritance** is appropriate to use when we are discussing extranuclear genes. In fact, the criteria for establishing that a phenotype is the result of an extranuclear (or extrachromosomal) gene (rather than a nuclear gene) are results that do not conform to results expected from crosses of nuclear genes.

Here are the four main characteristics of *extranuclear inheritance:*

1. In higher eukaryotes, the results of reciprocal crosses involving extranuclear genes are not the same as reciprocal crosses involving nuclear genes. (As was discussed in Chapters 2 and 3, in a reciprocal cross, one selects a pair of contrasting genotypes, then conducts two crosses, reversing the sexes of the parent [the source of the male or female gametes] in each cross. For example, if A and B represent contrasting genotypes, A♀ × B♂ and A♂ × B♀ would represent a pair of reciprocal crosses.)

Extranuclear genes usually exhibit the phenomenon of **uniparental inheritance** from generation to generation; that is, all progeny (both males and females) have the phenotype of only one parent. (The fact that both males and females resemble one of the parents distinguishes uniparental inheritance of an extranuclear gene from that of a sex-linked nuclear gene.) Typically, for higher eukaryotes it is the mother's phenotype that is expressed exclusively, a phenomenon called **maternal inheritance.** Maternal inheritance takes place because the amount of cytoplasm in the female gamete, the egg, usually greatly exceeds that in the

to initiate all proteins, and the formylation reaction is catalyzed by a transformylase localized in the chloroplast. The chloroplast uses IFs and EFs (elongation factors) that are distinct from those of the cytoplasmic protein synthesis system.

Like mitochondrial ribosomes, cp ribosomes are resistant to cycloheximide, an inhibitor of cytoplasmic ribosomes, but are sensitive to virtually all inhibitors known to block prokaryotic protein synthesis. The ribosomes themselves are found both free and membrane-bound in the chloroplast. The free ribosomes appear to make the large subunit of the ribulose bisphosphate decarboxylase enzyme, while ribosomes bound to the membrane are presumed to make the hydrophobic proteins that are important in photosynthesis. Hydrophobic proteins are located in the membranes of the chloroplast.

male gamete, the sperm. Therefore, the zygote receives most of its cytoplasm (containing the extranuclear genes in the organelles; i.e., the mitochondria and, where applicable, the chloroplasts) from the female parent and a negligible amount from the male parent.

In contrast, the results of reciprocal crosses between a wild-type and a mutant strain are ordinarily identical if the genes being transmitted are located on nuclear chromosomes. One exception for nuclear genes occurs when sex-linked genes are involved (see Chapter 3), but even then the results are distinct from those for extranuclear inheritance. In the case of a sex-linked gene (recessive in this example), a cross between a wild-type female and a mutant male will produce F_1s that are all wild type. In the F_1s of the reciprocal cross between a mutant female and a wild-type male, however, the females will be wild type and the males will be mutant.

2. Extranuclear genes cannot be mapped to known nuclear chromosomes. If the nuclear chromosomes of an organism are well mapped, any new nuclear mutations can be located on the genetic linkage map by mapping experiments that follow Mendelian principles of segregation and recombination. If a new mutation does not show linkage to any of the nuclear genes, it is probably an allele of an extranuclear gene.

3. Ratios typical of Mendelian segregation are not found. For nuclear gene mutations the rules of Mendelian segregation predict that all F_1 progeny of a cross between a homozygous mutant (recessive) and a wild type will be wild type in phenotype and that a 3:1 ratio of wild type:mutant phenotypes will be exhibited among the F_2 progeny. Such a segregation ratio is not characteristic of extranuclear genes—the uniparental inheritance pattern characteristic of extranuclear genes is clearly a deviation from Mendelian segregation.

4. Extranuclear inheritance is indifferent to nuclear substitution. When a particular phenotypic characteristic persists after the nucleus is removed and replaced with a nucleus having a different genotypic constitution associated with alternative characteristics, this indicates that the characteristic is likely to be controlled by a genome other than the nuclear genome, such as that of the mitochondrion or the chloroplast.

KEYNOTE

Numerous examples of mutations in mitochondrial and chloroplast genomes are known. The inheritance of these extranuclear genes follows rules different from those for nuclear genes. In particular, no meiotic segregation is involved, generally uniparental (and often maternal) inheritance is exhibited, the trait persists even after nuclear substitution, and extranuclear genes are not mappable to the known nuclear-linkage groups.

MATERNAL EFFECT

The maternal inheritance pattern of extranuclear genes is distinct from the phenomenon of **maternal effect,** which is defined as the predetermination of gene-controlled traits by the maternal nuclear genotype prior to the fertilization of the egg. *Maternal effect does not involve any extranuclear genes and is discussed here to make the distinction from extranuclear inheritance clear.*

Maternal effect is seen, for instance, in the inheritance of the direction of coiling of the shell of the snail *Limnaea peregra* (Figure 21.14). In this snail the direction of coiling of the shell is determined by a single pair of *nuclear* alleles, *D* for coiling to the right (dextral coiling) and *d* for coiling to the left (sinistral coiling). Breeding experiments have shown that the *D* allele is completely dominant to the *d* allele and that the shell-coiling phenotype is always determined by the genotype

~ **FIGURE 21.14**

The snail, *Limnaea peregra.*

~ **FIGURE 21.15**

Inheritance of the direction of shell coiling in the snail *Limnaea peregra* is an example of maternal effect. (a) Cross between true-breeding dextral-coiling female (*D/D*) and sinistral-coiling male (*d/d*); (b) Cross between sinistral-coiling female (*d/d*) and dextral-coiling male (*D/D*).

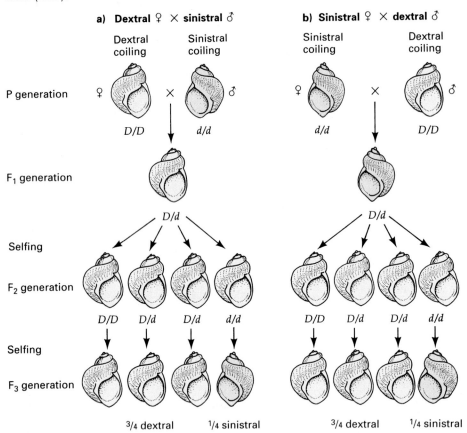

of the mother. Figure 21.15 shows the results of reciprocal crosses between a true-breeding, dextral-coiling, and a sinistral-coiling snail. All the F_1s have the same genotype since a nuclear gene is involved, yet the phenotype is different for the reciprocal crosses. Phenotypically different results from a reciprocal cross are characteristic of extranuclear inheritance. In addition, because in every case the F_1 phenotypes are those of the mothers in the cross, maternal inheritance would *seem* to be indicated.

When we self the dextral F_1s from the cross involving a dextral mother (Figure 21.15a), all the F_2s are phenotypically dextral coiling. From Mendelian principles we predict a 1:2:1 ratio of *D/D*, *D/d*, and *d/d* genotypes in these snails. As a result, three-fourths of these F_2 snails give rise to dextral-coiling F_3 progeny upon selfing, whereas the *d/d* snails, despite being dextral-coiling themselves, give rise to sinistral-coiling F_3 progeny upon selfing; that is, the progeny snails in this case do *not* resemble the mother in snail-coiling

phenotype. This result does *not* fit our criteria for extranuclear inheritance, since if the coil direction phenotype were controlled by an extranuclear gene, the progeny would always exhibit the phenotype of the mother, owing to maternal inheritance. In the reciprocal cross (Figure 21.15b) the sinistral-coiling F_1s are genotypically identical to the dextral-coiling F_1s of Figure 21.15a, and the F_2 and F_3 genotypes and phenotypes are similarly identical. Thus, the coiling phenotype is governed directly by the nuclear genotype of the mother and is an example of maternal effect. The exact molecular basis of the phenotype is not known, although at the cellular level it is known that the coiling results from alternate sloping of the mitotic spindle during cleavage of the developing embryo.

KEYNOTE

Maternal effect is different from extranuclear inheritance. The maternal inheritance pattern of extranuclear genes occurs because the zygote contains most of its organelles (containing the extranuclear genes) from the female parent, whereas in the maternal effect the trait inherited is controlled by the maternal *nuclear* genotype before the fertilization of the egg and does not involve extranuclear genes.

EXAMPLES OF EXTRANUCLEAR INHERITANCE

In this section we shall discuss the properties of a selected number of mutations in extranuclear chromosomes in order to illustrate the principles of extranuclear inheritance.

Leaf Variegation in the Higher Plant *Mirabilis jalapa*

One of the first exceptions to Mendelian inheritance was demonstrated in 1909 through the work of Carl Correns, a plant geneticist and one of the three rediscoverers of Mendelian principles of inheritance. Correns had been studying the inheritance of a number of variegated-leaf forms in a number of flowering plants. Many of these phenotypes were controlled by genes that showed typical Mendelian inheritance. One phenotype that did not follow the expected pattern was the yellowish white patches of a variegated strain of *Mirabilis jalapa* (also called the four o'clock, or the marvel of Peru; Figure 21.16). This strain, called *albomaculata*, also has occasional shoots that are wholly green or wholly yellow-white (see Figure 21.17).

All types of shoots (variegated, green, white) give rise to flowers, so it is possible to make crosses by taking pollen from one type of flower and using it to fertilize a flower on the same or different type of shoot. Table 21.5 (p. 658) summarizes the results of these intercrosses. The flowers on green shoots give only green progeny, regardless of whether the pollen is from green, white, or variegated shoots. Flowers on white shoots give only white progeny, regardless of the source of the pollen. (However, because the whiteness indicates the absence of chlorophyll and hence an inability to carry out photosynthesis, the white progeny die soon after seed germination.) Finally, flowers on variegated shoots all give rise to three types of progeny—completely green, completely white, and variegated—regardless of the nature of the pollen used to fertilize them. No pattern is seen for the relative proportions of these three types. In subsequent generations maternal inheritance is always seen in the same patterns just described.

~ **FIGURE 21.16**

The four o'clock, *Mirabilis jalapa*.

In sum, these breeding experiments with *M. jalapa* show results that indicate maternal inheritance. That is, the *progeny phenotype in each case was the same as that of the maternal parent* (i.e., the color of the prog-

~ **FIGURE 21.17**

Leaf variegation in *Mirabilis jalapa*. Shoots that are all green, all white, and variegated are found on the same plant, and flowers may form on any of these shoots.

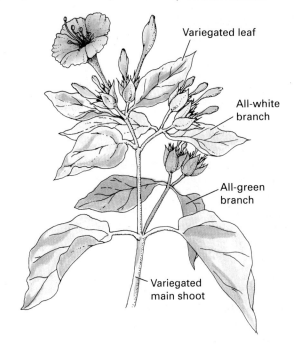

~ TABLE 21.5

Results of Crosses of Variegated Plants of *Mirabilis jalapa*

PHENOTYPE OF BRANCH BEARING ♀ (EGG) PARENT	PHENOTYPE OF BRANCH BEARING ♂ (POLLEN) PARENT	PHENOTYPE OF PROGENY
White	White	White
	Green	White
	Variegated	White
Green	White	Green
	Green	Green
	Variegated	Green
Variegated	White	Variegated, green, or white
	Green	Variegated, green, or white
	Variegated	Variegated, green, or white

eny shoots resembled the color of the parental flower), which is indicative of maternal inheritance. Moreover, the results of reciprocal crosses differed from the expected patterns, and there was a lack of constant proportions of the different phenotypic classes in the segregating progeny. These last two properties are also characteristic of traits showing extranuclear inheritance and are not expected for traits showing Mendelian inheritance.

The basis for the green color of higher plants is the presence of the green pigment chlorophyll in large numbers of chloroplasts. Green shoots in *Mirabilis* have a normal complement of chloroplasts. White shoots have abnormal, colorless chloroplasts called leukoplasts, which lack chlorophyll and hence are incapable of carrying out photosynthesis.

The simplest explanation for the inheritance of leaf color in the *albomaculata* strain of *Mirabilis jalapa* is that the abnormal chloroplasts are defective as a result of a mutant gene in the cpDNA. During plant growth the two types of organelles, chloroplasts and leukoplasts, may segregate so that a particular cell and its progeny cells have *only* chloroplasts (leading to green tissues), *only* leukoplasts (leading to white tissues), or a mixture of chloroplasts and leukoplasts (leading to variegation). This model is shown in Figure 21.18.

In a variegated plant, then, the white shoots derive from cells in which, through segregation, only leukoplasts are present. Green shoots are similarly derived from cells containing only chloroplasts. Variegated shoots are generated from cells that contain chloroplasts and leukoplasts, so by later segregation, patches of white tissue are produced on the shoots and leaves. Let us take this model one step further. The flowers on a green shoot have only chloroplasts, and so through maternal inheritance these chloroplasts form the basis of the phenotype of the next generation. Simi-

lar, logical arguments can be made for the flowers on white or variegated shoots.

Implicit in this simple model are three assumptions. One is that the pollen contributes essentially no cytoplasmic information (i.e., no chloroplasts or leukoplasts) to the egg (a reasonable assumption in view of the far greater size of the egg compared with the pollen). Thus in the zygote the extranuclear genetic determinants in the egg overwhelm those in the pollen. The second assumption is that the chloroplast genome replicates autonomously and that, by growth and division of plastids (the general term for photosynthetic organelles), the wild-type and mutant cpDNA molecules have the potential to segregate randomly to the new plastids so that pure lines can be generated from a mixed line. The third assumption is that segregation of plastids to daughter cells is random, so that some daughters receive chloroplasts, some receive leukoplasts, and some receive mixtures.

*K*EYNOTE

The leaf color phenotypes in a variegated strain of the four o'clock, *Mirabilis jalapa*, show maternal inheritance. The abnormal chloroplasts in white tissue are the result of a mutant gene in the cpDNA, and the observed inheritance patterns follow the segregation of cpDNA. _____

The *poky* Mutant of *Neurospora*

The *poky* mutant of *Neurospora*, which involves a change in mtDNA, illustrates a number of the classical expectations of extranuclear inheritance. The *poky*

~ **FIGURE 21.18**

Model for the inheritance of leaf color in the four o'clock, *Mirabilis jalapa.*

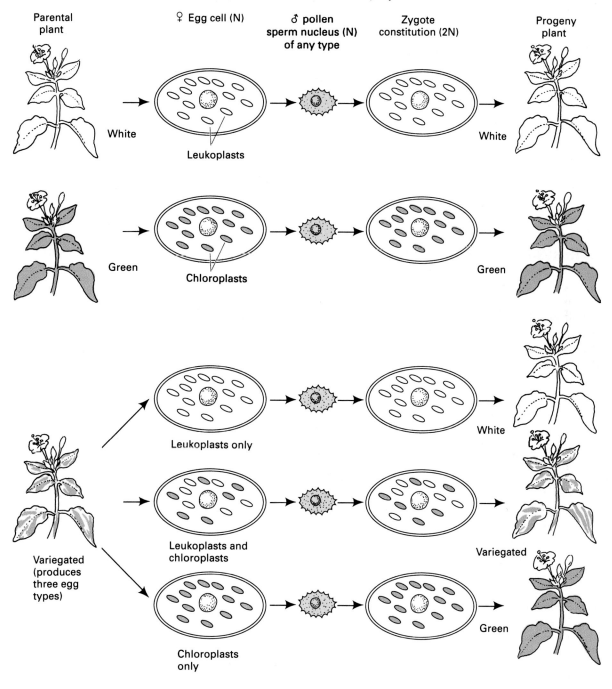

mutant is also called [*mi*-1], where the square brackets indicate an extranuclear gene and *mi* represents *maternal* inheritance. The phenotype of the *poky* mutant is a much slower growth than wild type, either on solid medium or in liquid medium.

Neurospora crassa is an obligate aerobe; that is, it requires oxygen in order to grow and survive, so mito-

chondrial functions are essential for its growth. Biochemical analysis showed that the *poky* mutant is defective in aerobic respiration as a result of changes in the cytochrome complement of the mitochondria. The three principal cytochromes are a + a_3, b, and c. Compared with the wild type, *poky* lacks cytochromes a + a_3 and b; *poky* also has a greater amount of cytochrome c. The change in the cytochrome spectrum affects the ability of the mitochondria to generate sufficient ATP to support rapid growth, so slow growth results.

In *Neurospora* the sexual phase of the life cycle is initiated following a fusion of nuclei from mating type *A* and *a* parents. A sexual cross can be made in one of two ways: either by putting both parents on the crossing medium simultaneously or by inoculating the medium with one strain, and after three or four days at 25°C, adding the other parent. In the latter case the first parent on the medium produces all the *protoperithecia*, the bodies that will give rise to the true fruiting bodies in which are the asci with the sexual spores.

Compared with the conidia, the asexual spores, the protoperithecia have a tremendous amount of cytoplasm and hence can be considered the female parent in much the same way as an egg of a plant or an animal is the female parent. Now by adding conidia of a strain of the opposite mating type to the crossing medium, we have what is called a *controlled cross* in which one strain acts as the female and the other as the male parent. Using a strain to produce the protoperithecia as the female parent and conidia of another strain as a male parent, geneticists can make reciprocal crosses to determine whether any trait shows extranuclear inheritance. This experiment is now illustrated for the *poky* mutant.

Mary and Herschel Mitchell did reciprocal crosses between *poky* and the wild type, with the following results:

poky ♀ × wild type ♂ → all *poky*
wild type ♀ × *poky* ♂ → all wild type

In other words, all the progeny show the same phenotype as the maternal parent, indicating maternal inheritance as a characteristic for the *poky* mutation.

This analysis can be made more refined by using tetrad analysis (Chapter 6) to follow the phenotype more closely. Recall that in *Neurospora* the eight products of a meiosis and subsequent mitosis are retained in linear order within the ascus. The eight ascospores can be removed from the asci, and strains germinated from the spores can be analyzed for the particular phenotype being followed. By doing tetrad analysis for the *poky* ♀ × wild type ♂ cross, we find an 8:0 ratio of *poky*:wild-type progeny (Figure 21.19a) and for the wild type ♀ × *poky* ♂ cross, we get a 0:8 ratio of *poky*:wild-type progeny with regard to the growth phenotype (Figure 21.19b). At the same time nuclear genes show the 4:4 segregation expected of Mendelian inheritance. Again, the simplest explanation for these results is that *poky* is determined by an extranuclear gene (located in this case in the mitochondria) that exhibits a pattern of maternal inheritance.

~ **FIGURE 21.19**

Results of reciprocal crosses of *poky* and normal (wild-type) *Neurospora*. (a) *poky* ♀ × normal ♂; (b) normal ♀ × *poky* ♂.

a)

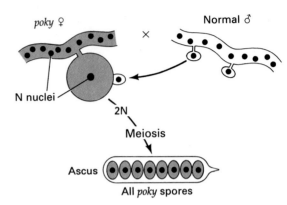

b)

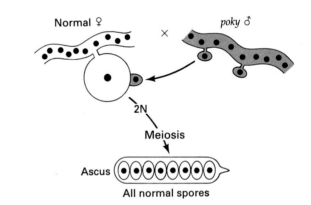

𝒦**EYNOTE**

The slow-growing *poky* mutant of *Neurospora crassa* shows maternal inheritance and deficiencies for some mitochondrial cytochromes. ———

Yeast *petite* Mutants

Yeast grows as single cells rather than as a mycelium like *Neurospora*. Thus, on solid medium, yeast forms discrete colonies consisting of many thousands of individual cells clustered together. Yeast can grow with or without oxygen. In the absence of oxygen yeast obtains energy for cell growth and cellular metabolism through fermentation metabolism, in which the mitochondria are not involved.

CHARACTERISTICS OF *PETITE* MUTANTS. In the late 1940s, Boris Ephrussi and his colleagues studied the growth of yeast cells on a solid medium that allowed growth to occur either by aerobic respiration or by fermentation. Occasionally, they noticed a colony that was much smaller than wild-type colonies. Since Ephrussi was French, the small colonies were called *petites* (French for "small"), and the wild-type colonies were called *grandes* (French for "big"). Ephrussi found that *petite* colonies were small not because the cells were small but because the growth rate of the mutant *petite* strain is significantly slower than that of the wild-type. Thus there are fewer cells in the *petite* colonies.

Biochemical analysis shows that the *petites* are essentially incapable of carrying out aerobic respiration and so must obtain their energy primarily from fermentation, which is a relatively inefficient process.

In an unmutagenized population of cells between 0.1 and 1 percent of the cells spontaneously become *petite*. In the presence of an intercalating agent (a chemical that can wedge itself between adjacent base pairs in DNA: see Chapter 18) such as ethidium bromide, 100 percent of the cells become *petite*! The yeast *petite* system is particularly useful in studies of extranuclear inheritance because there is an abundance of *petite* mutants and because yeast cells that lack mitochondrial functions can still survive and grow. Such mutations to respiratory deficiencies in yeast are automatically conditional mutants. On a medium that can support fermentation, the *petites* grow more slowly than the *grandes*, whereas on a medium that supports only aerobic respiration, *petites* are unable to grow.

NUCLEAR, NEUTRAL, AND SUPPRESSIVE *PETITE* MUTANTS. Let us now turn to the genetics of *petite* mutants. Some *petites* have as their basis a mutation in the nuclear genome—a characteristic that is not surprising since some subunits of some mitochondrial proteins are encoded by nuclear genes. The nuclear gene mutations are called *pet*⁻. When a *pet*⁻ mutant is crossed with the wild type (*pet*⁺), the diploid is a *pet*⁺/*pet*⁻ *grande*. When this *pet*⁺/*pet*⁻ cell goes through meiosis, each resulting tetrad shows a 2:2 segregation of *grande* (*pet*⁺):*petite* (*pet*⁻) phenotype. This result is typical of Mendelian inheritance, so we will consider these *petites* no further, except to say that nuclear *petites* occur much less frequently than extranuclear *petites*.

One class of *petites* that exhibits the traits of extranuclear inheritance is the *neutral petite* class (symbolized [*rho*⁻N]). Figure 21.20 shows the inheritance pattern of [*rho*⁻N] *petites*. When a neutral *petite* is crossed with wild type ([*rho*⁺N]), the resulting diploid strain, [*rho*⁺N]/[*rho*⁻N], has a normal (*grande*) phenotype. When these diploids go through meiosis, all resulting meiotic tetrads show a 0:4 ratio of *petite*:*grande* strains, while at the same time nuclear markers segregate 2:2. This result is a classical example

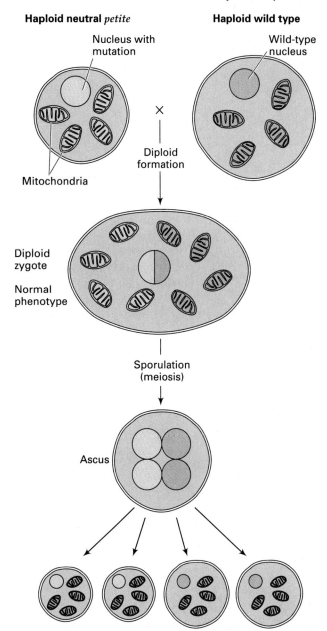

~ **FIGURE 21.20**

Extrachromosomal inheritance of neutral *petite* in yeast.

Haploid neutral *petite*

Haploid wild type

Nucleus with mutation

Wild-type nucleus

×

Diploid formation

Mitochondria

Diploid zygote

Normal phenotype

Sporulation (meiosis)

Ascus

Nuclear gene segregation: 2 neutral *petite* mutants: 2 wild type
Extranuclear segregation: 0 neutral *petite* mutants: 4 wild type

of uniparental inheritance, in which all progeny have the phenotype of only one parent. *We cannot ascribe this phenomenon to maternal inheritance, however, since the two haploid cells that fuse to produce the diploid are the same size and contribute equally to the cytoplasm.*

What is the nature of the neutral *petite* mutation? As we have said, *petites* can be induced readily by treatment with intercalating agents like ethidium bromide. With prolonged treatment the majority of *petites* are of

~ FIGURE 21.21

Extrachromosomal inheritance of suppressive *petite* in yeast.

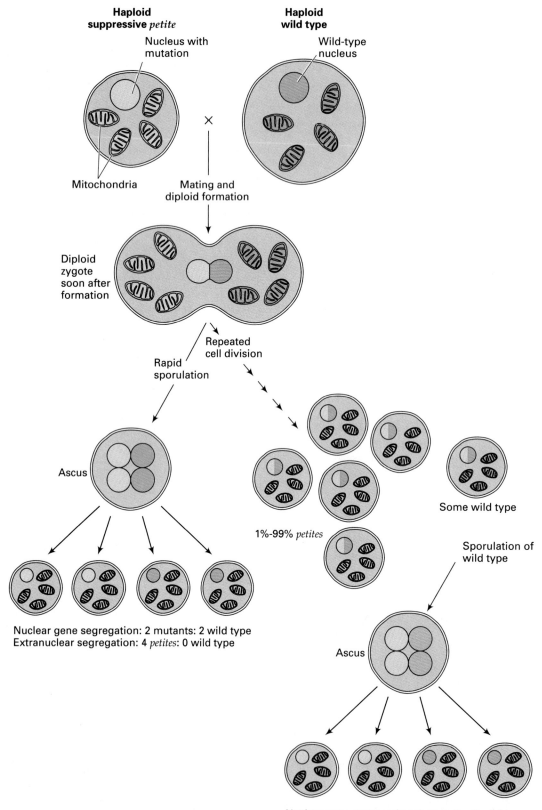

Haploid suppressive *petite*
Nucleus with mutation

Haploid wild type
Wild-type nucleus

Mitochondria

× Mating and diploid formation

Diploid zygote soon after formation

Repeated cell division

Rapid sporulation

Ascus

Some wild type

1%-99% *petites*

Sporulation of wild type

Nuclear gene segregation: 2 mutants: 2 wild type
Extranuclear segregation: 4 *petites*: 0 wild type

Ascus

Nuclear gene segregation: 2 mutants: 2 wild type
Extranuclear segregation: 0 *petites*: 4 wild type

the neutral type. The mitochondrial genome is implicated because cytochromes are altered, because there is evidence for extranuclear inheritance, and because the mitochondria are the only other site in the yeast cell in which genetic material is found.

An examination of the mitochondrial genetic material in the neutral *petites* reveals a remarkable characteristic: Essentially 99–100 percent of the mtDNA is missing. Not surprisingly, then, the neutral *petites* are unable to perform mitochondrial functions. They survive, however, because fermentation processes are localized in the cytoplasm. In genetic crosses with wild-type yeasts normal (wild-type) mitochondria form a population from which new, normal (wild-type) mitochondria are produced in all progeny, and hence the *petite* trait is lost after one generation. The *petite* is a neutral mutation in the sense that it does not affect the normal phenotype.

The suppressive *petites* are different from the neutrals. Most [*rho⁻*] mutants are of the suppressive type. Like the neutrals and the nuclear *petites*, the [*rho⁻S*] *petites* are deficient in mitochondrial protein synthesis, but they differ from the other two classes in that they exert an effect in the diploid zygote when crossed with a normal [*rho⁺*] strain (see Figure 21.21).

When a [*rho⁺*]/[*rho⁻S*] diploid is formed, it has respiratory properties intermediate between those of normal and *petite* strains. If this 2N is sporulated immediately after it is produced, the resulting ascospores show a 4:0 segregation of *petite*:normal. If, instead, the [*rho⁺*]/[*rho⁻S*] diploid is allowed to divide mitotically a number of times, the diploid progeny population will have between 1 and 99 percent *petites*, depending on the particular suppressive *petite* strain. Sporulation of any of the normals present in that population gives 0 *petites*:4 normals.

The suppressive mutations result from changes in the mtDNA. They start out as deletions of part of the mtDNA, and then, by some correction mechanism, sequences that are not deleted become reiterated until the normal amount of mtDNA is restored. During these events rearrangements of the mtDNA sometimes occur. Since the protein-coding genes in the mitochondrial genome are widely scattered, any significant deletion and/or rearrangement of mtDNA is likely to lead to deficiencies in mitochondrial protein synthesis, especially in enzymes involved in aerobic respiration.

KEYNOTE

Yeast *petite* mutants grow slowly and have various deficiencies in mitochondrial functions as a result of alterations in mitochondrial DNA. The *petite* mutants show extranuclear inheritance, with particular patterns of inheritance varying with the type of *petite* involved. ――――――――――

Extranuclear Genetics of *Chlamydomonas*

One of the most thorough analyses of chloroplast genetics was begun by Ruth Sager and her colleagues in their work with the unicellular alga *Chlamydomonas reinhardi*. This motile, haploid organism has two flagella and a single chloroplast that contains many copies of cpDNA. The life cycle of this organism was given in Figure 6.2. There are two mating types, mt^+ and mt^-, usually abbreviated as + and −. In a process known as syngamous mating, a zygote is formed by fusion of two equal-sized cells (which, therefore, contribute an equal amount of cytoplasm), one of each mating type. A thick-walled cyst develops around the zygote. After meiosis, four haploid progeny cells are produced, and since mating type is determined by a nuclear gene, a 2:2 segregation of +:− mating-types results.

From a systematic study of traits affecting this organism, researchers identified a number of traits that showed the expected patterns for extranuclear inheritance. Some of these extranuclear traits were shown to involve resistance to or dependence on antibiotics that affect chloroplast protein synthesis by changing ribosome structure and/or function.

One trait that is inherited in an extranuclear manner confers erythromycin resistance ([*ery⁻ʳ*]) on the organism. Wild-type *Chlamydomonas* cells are erythromycin-sensitive ([*ery⁻ˢ*]). If we perform a cross of + [*ery⁻ʳ*] × − [*ery⁻ˢ*], about 95 percent of the offspring are erythromycin-resistant phenotypes (Figure 21.22a; p. 664). This result is a classic instance of uniparental inheritance, an attribute we have come to expect of extranuclear traits. The reciprocal cross, − [*ery⁻ʳ*] × + [*ery⁻ˢ*] (Figure 21.22b) also shows uniparental inheritance about 95 percent of the time, although here the erythromycin-sensitive phenotype is inherited. Thus even though both parents contribute equal amounts of cytoplasm to the zygote, the progeny always resemble the + parent with regard to chloroplast-controlled phenotypes. Somehow, preferential segregation of one chloroplast type occurs in this organism, or perhaps one parental type is inactivated preferentially.

As we indicated, only 95 percent of the zygotes in the above crosses show uniparental inheritance. The other 5 percent of the zygotes show extranuclear traits from both parents. Such zygotes are said to show **biparental inheritance,** indicating that both types of chloroplast chromosomes are present and active in these zygotes. The genetic condition of these zygotes is defined as a **cytohet,** the term deriving from "*cyto*plasmically *heter*ozygous." In many instances the

~ **FIGURE 21.22**

Uniparental inheritance in *Chlamydomonas*. (a) From a cross of $+ [ery^{-r}] \times - [ery^{-s}]$, 95 percent of the zygotes give tetrads that segregate 2:2 for the nuclear mating type genes, and 4:0 for the extranuclear gene carried by the + parent (here, $[ery^{-r}]$); (b) from the reciprocal cross of a $- [ery^{-r}] \times + [ery^{-s}]$, 95 percent of the zygotes give tetrads segregating $2+:2-$ and $0 [ery^{-r}]:4 [ery^{-s}]$, again showing uniparental inheritance for the extranuclear trait of the + parent.

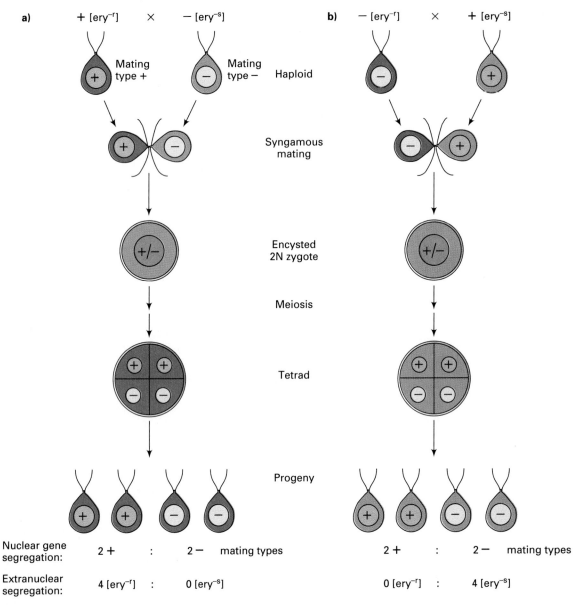

extranuclear traits of biparental zygotes segregate into pure types (i.e., either erythromycin-sensitive or erythromycin-resistant) on successive mitotic divisions. This phenomenon presumably involves the segregation of the different chloroplast chromosomes, and hence the different chloroplasts, into pure types. This situation parallels the situation we discussed for *Mirabilis*.

Geneticists are able to make crosses between strains carrying different extranuclear traits that are chloroplast-controlled. The same principles of inheritance described for the erythromycin resistance trait apply to these traits. That is, in most cases there is uniparental inheritance with progeny resembling the + parent in extranuclear phenotype, and in the remaining cases there is biparental inheritance. Occasionally, a biparental zygote does not segregate the two traits; instead, both traits continue to be expressed in subsequent generations. The explanation is that a recombi-

~ FIGURE 21.23

Maps of *Chlamydomonas* cpDNA: (a) circular genetic map; (b) restriction map.

a) Genetic map

b) Restriction map

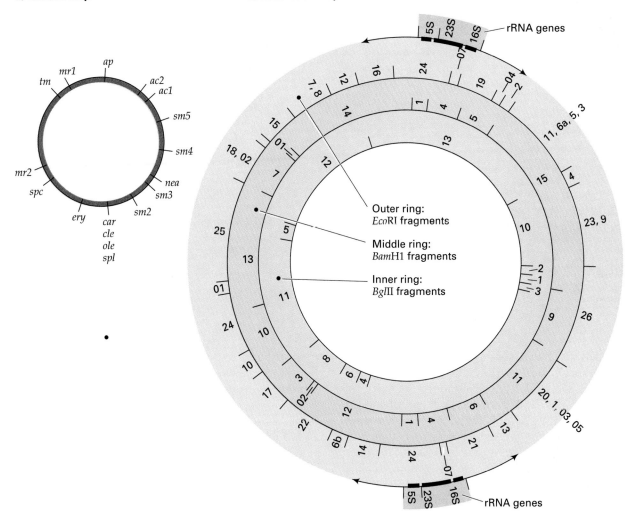

nation event takes place between the two types of cpDNA so that a recombinant chromosome carrying both alleles originally carried by the two parents is produced. Sager and her colleagues have performed extensive analyses of these recombinants and, on the basis of the relative frequencies of recombination, have produced a genetic map (Figure 21.23a) and a restriction map (Figure 21.23b) of the cpDNA of *Chlamydomonas*.

𝒦EYNOTE

A number of genes have been identified and mapped in the *Chlamydomonas* chloroplast genome. These genes are inherited in an extranuclear manner.

Infectious Heredity—Killer Yeast

The examples of extranuclear inheritance we have discussed so far have all resulted from mutations in the mtDNA or cpDNA. There are many other examples of eukaryotic extranuclear inheritance that are due to the presence of cytoplasmic bacteria or viruses coexisting with the eukaryotes in a symbiotic relationship. One example is the killer phenomenon in yeast. Some yeast strains secrete a killer toxin, which will kill sensitive strains of yeast. (Killer strains are, of course, immune to the toxin.) The killer phenomenon results from the presence in the cell's cytoplasm of two types of viruses, L and M (Figure 21.24a; p. 666). Neither appear to produce deleterious effects on the host cell.

The L virus consists of a protein capsid within which is one 4.6-kb double-stranded (ds) RNA genome

~ Figure 21.24

The killer phenomenon in yeast. (a) Killer yeast contains two virus types, L and M, each of which contains a double-stranded RNA genome. L-dsRNA encodes both virus particles and the replication enzyme required for L and M virus replication. M-dsRNA encodes the killer toxin; (b) Sensitive yeast, which can be killed by killer toxin, either have L viruses but no M viruses, or have neither virus type.

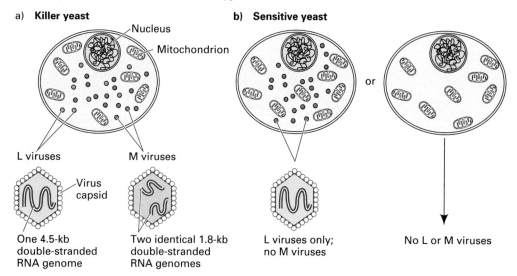

a) **Killer yeast**

Nucleus

Mitochondrion

L viruses

M viruses

Virus capsid

One 4.5-kb double-stranded RNA genome

Two identical 1.8-kb double-stranded RNA genomes

b) **Sensitive yeast**

or

L viruses only; no M viruses

No L or M viruses

called L-dsRNA. L-dsRNA encodes the capsid proteins for both L and M viral particles and the viral polymerase required for viral RNA replication. So, since all viral capsids (the viral particles) are encoded by a L-dsRNA, M viruses are only found in cells if L viruses are also present. The M virus consists of a protein capsid encoded by L-dsRNA, and two identical copies of a 1.8-kb double-stranded RNA genome called M-dsRNA. M-dsRNA encodes the protein killer toxin, which is secreted from the cell. The same protein is responsible for conferring immunity upon the killer cell.

Sensitive yeast cells (cells that can be killed by the M-encoded killer toxin) are of two types (Figure 21.24b). One type has only L viruses, and the other has neither L nor M viruses. In both of these types, no immunity function is produced because killer toxin is not made.

Unlike most viruses, the yeast L and M viruses are not found outside of the cell, so sensitive yeast cells cannot be infected by viruses that invade from outside. Rather, virus transmission from yeast to yeast occurs whenever there is cytoplasmic mixing, most commonly when two yeast cells mate. All progeny of the mating will inherit copies of the viruses in the parental cells, producing an infectious mechanism of cytoplasmic inheritance. For example, if a killer yeast mates with a sensitive yeast that lacks viruses (obviously before killing it), the resulting diploid will be a killer because of the presence of both L and M viruses. Both virus types

are distributed to the progeny, all of which will exhibit the killer phenotype. This case, then, exhibits extranuclear, uniparental inheritance.

Keynote

Not all cases of extranuclear inheritance result from genes on mtDNA or cpDNA. Many other examples in eukaryotes result from infectious heredity in which symbiotic, cytoplasmically-located bacteria or viruses are transmitted when cytoplasms mix. The killer phenomenon in yeast is such an example: it results from cytoplasmically-located viruses. _____

Summary

Both mitochondria and chloroplasts contain DNA, the length of which varies from organism to organism. The genomes of both organelles are naked, circular, double-stranded DNAs. The two genomes contain genes that are not duplicated in the nuclear genome; hence, the organelle genes are contributing different information for the function of the cell. The organellar genes encode the rRNA components of the ribosomes that are assembled and function in the organelles, for many (if not all) of the tRNAs used in organellar protein synthesis, and for a few proteins that remain in the organelles and perform functions specific to them. Many of the proteins function in multi-protein complexes. Some of these proteins are encoded by the organelle; the remainder are nuclear-encoded. Nuclear-encoded proteins found in organelles are synthesized on cytoplasmic ribosomes, then imported into the appropriate organelles. These proteins include the ribosomal proteins for the organellar ribosomes and, in the case of mitochondria, component proteins of cytochromes.

Given that organelles contain genetic material, it is not surprising that there are many examples of mitochondrial and chloroplast mutants. The inheritance of these extranuclear genes follows rules different from those for nuclear genes, and this is how extranuclear genes were originally identified. That is, for extranuclear genes, no meiotic segregation is involved, uniparental (and often maternal) inheritance is generally exhibited, a trait resulting from an extranuclear mutation persists after nuclear substitution, and extra-

nuclear genes cannot be mapped to the known nuclear linkage groups. Not all cases of extranuclear inheritance result from genes on mtDNA or cpDNA. Many other examples result from infectious heredity, in which cytoplasmically located bacteria or viruses are transmitted when cytoplasms mix.

A specific inheritance pattern sometimes observed is the phenomenon in which the mother specifically affects the phenotype of the offspring. This phenomenon, called the maternal effect, is defined as the predetermination of gene-controlled traits by the maternal *nuclear* genotype prior to the fertilization of the eggs. Maternal effect is different from extranuclear inheritance. That is, the maternal inheritance pattern of extranuclear genes occurs because the zygote receives most of its organelles (containing the extranuclear genes) from the female parent, whereas in the maternal effect the trait inherited is controlled by the maternal *nuclear* genotype before the fertilization of the egg and does *not* involve extranuclear genes.

ANALYTICAL APPROACHES FOR SOLVING GENETICS PROBLEMS

Q.1 In *Neurospora*, strains of *poky* have been isolated that have reverted to nearly wild-type growth, although they still retain the abnormal metabolism and cytochrome pattern characteristic of *poky*. These strains are called *fast-poky*. A cross of a *fast-poky* female with a wild-type male gives a 1:1 segregation of *poky*:*fast-poky* ascospores in all asci. Interpret these results, and predict the results you would expect from the reciprocal of the stated cross.

A.1 The *poky* mutation shows maternal inheritance, so when it is used as a female parent, all the progeny ascospores will carry the *poky* determinant. Conventionally, extranuclear genes are designated within square brackets so in this case the cross with respect to the *poky* [*po*] determinant was [*po*] × [*N*], where *N* signifies normal cytoplasm. All progeny are [*po*]. The asci show a 1:1 segregation for the *poky* and *fast-poky* phenotypes. This ratio is characteristic of a nuclear gene segregating in the cross. Therefore the simplest explanation is that the factor that causes *poky* strains to be *fast-poky* is a nuclear gene mutation; we can call it *F* (its actual designated name). The *F* gene segregates in meiosis, as do all nuclear genes. The cross can be rewritten as [*po*]*F* ♀ × [*N*] + ♂, which gives a 1:1 segregation of [*po*] + and [*po*]*F*. The former is *poky* and the latter is *fast-poky*.

With these results behind us, the reciprocal cross may be diagramed as [*N*] + ♀ × [*po*]*F* ♂. All the progeny spores from this cross are [*N*], with half of them being *F* and half of them being +. If *F* has no effect on normal cytoplasm, then these two classes of spores would be phenotypically indistinguishable, which is the case.

Q.2 Four slow-growing mutant strains of *Neurospora crassa*, coded *a*, *b*, *c*, and *d*, were isolated. All have an abnormal system of respiratory mitochondrial enzymes. The inheritance patterns of these mutants were tested in controlled crosses (see p. 660) with the wild type, with the following results:

PROTO-PERITHECIAL (FEMALE) PARENT		CONIDIAL (MALE) PARENT	PROGENY (ASCOSPORES)	
			WILD TYPE	SLOW GROWING
Wild type	×	*a*	847	0
a	×	Wild type	0	659
Wild type	×	*b*	1113	0
b	×	Wild type	0	2071
Wild type	×	*c*	596	590
Wild type	×	*d*	1050	1035

Give a genetic interpretation of these results.

A.2 This question asks us to consider the expected transmission patterns for nuclear genes and for extranuclear genes. The nuclear genes will have a 1:1 segregation in the offspring, since this organism is a haploid organism and hence should exhibit no differences in the segregation patterns, whichever strain is the maternal parent. On the other hand, a distinguishing characteristic of extranuclear genes is a difference in the results of reciprocal crosses. In *Neurospora* this characteristic is usually manifested by all progeny having the phenotype of the maternal parent. With these ideas in mind we can analyze each mutant in turn.

Mutant *a* shows a clear difference in its segregation in reciprocal crosses and is, in fact, a classic case of maternal inheritance. The interpretation here is that the gene is extranuclear; hence, the gene must be in the mitochondrion. The *poky* mutant described in this chapter shows this type of inheritance pattern.

By the same reasoning used to analyze mutant *a*, the mutation in strain *b* must also be extranuclear.

Mutants *c* and *d* segregate 1:1, indicating that the mutations involved are in the nuclear genome. In these cases we need not consider the reciprocal cross since there is no evidence for maternal inheritance. In fact, the actual mutations that are the basis for this question are female and sterile, so the reciprocal cross cannot be done. We can confirm that the mutations are in the nuclear genome by doing mapping experiments, using known nuclear markers. Evidence of linkage to such markers would confirm that the mutations are not extranuclear.

QUESTIONS AND PROBLEMS

21.1 Compare and contrast the structure of the nuclear genome, the mitochondrial genome, and the chloroplast genome.

21.2 How do mitochondria reproduce? What is the evidence for the method you describe?

21.3 What genes are present in the human mitochondrial genome?

21.4 What conclusions can you draw from the fact that most nuclear-encoded mRNAs and all mitochondrial mRNAs have a poly(A) tail at the 3' end?

*21.5** Discuss the differences between the universal genetic code of the nuclear genes and the code found in mammalian and fungal mitochondria. Is there any advantage to the mitochondrial code?

21.6 When the DNA sequences for most of the mRNAs in human mitochondria are examined, no nonsense codons are found at their termini. Instead, either U or UA is found. Explain this result.

21.7 Compare and contrast the cytoplasmic and mitochondrial protein-synthesizing systems.

21.8 Compare and contrast the organization of the ribosomal RNA genes in mitochondria and in chloroplasts.

*21.9** What features of extranuclear inheritance distinguish it from the inheritance of nuclear genes?

21.10 Distinguish between maternal effect and extranuclear inheritance.

21.11 Distinguish between nuclear (segregational), neutral, and suppressive *petite* mutants of yeast.

*21.12** The inheritance of the direction of shell coiling in the snail *Limnaea peregra* has been studied extensively. A snail produced by a cross between two individuals has a shell with a right-hand twist (dextral-coiling). This snail produces only left-hand (sinistral) progeny on selfing. What are the genotypes of the F_1 snail and its parents?

21.13 *Drosophila melanogaster* has a sex-linked, recessive, mutant gene called *maroon-like* (*ma-l*). Homozygous *ma-l* females or hemizygous *ma-l* males have light-colored eyes, owing to the absence of the active enzyme xanthine dehydrogenase, which is involved in the synthesis of eye pigments. When heterozygous *ma-l*⁺/*ma-l* females are crossed with *ma-l* males, all the offspring are phenotypically wild type. However, half the female offspring from this cross, when crossed back to *ma-l* males, give all *ma-l* progeny. The other half of the females, when crossed to *ma-l* males, give all phenotypically wild-type progeny. What is the explanation for these results?

*21.14** When females of a particular mutant strain of *Drosophila melanogaster* are crossed to wild-type males, all the viable progeny flies are females. Hypothetically, this result could be the consequence of either a sex-linked, lethal mutation or a maternally inherited factor that is lethal to males. What crosses would you perform in order to distinguish between these alternatives?

21.15 Reciprocal crosses between two *Drosophila* species, *D. melanogaster* and *D. simulans*, produce the following results:

melanogaster ♀ × simulans ♂ → females only
simulans ♀ × melanogaster ♂ → males, with few or no females

Propose a possible explanation for these results.

21.16 Some *Drosophila* flies are very sensitive to carbon dioxide; they become anesthetized when it is administered to them. The sensitive flies have a cytoplasmic particle called *sigma* that has many properties of a virus. Resistant flies lack *sigma*. The sensitivity to carbon dioxide shows strictly maternal inheritance. What would be the outcome of the following two crosses: (a) sensitive female × resistant male and (b) sensitive male × resistant female?

*21.17 In yeast a haploid nuclear (segregational) *petite* is crossed with a neutral *petite*. Assuming that both strains have no other abnormal phenotypes, what proportion of the progeny ascospores are expected to be *petite* in phenotype if the diploid zygote undergoes meiosis?

21.18 When grown on a medium containing acriflavin, a yeast culture produces a large number of very small (*tiny*) cells that grow very slowly. How would you determine whether the slow-growth phenotype was the result of a cytoplasmic factor or a nuclear gene?

*21.19 In *Neurospora* a chromosomal gene *F* suppresses the slow-growth characteristic of the *poky* phenotype and makes a *poky* culture into a *fast-poky* culture, which still has abnormal cytochromes. Gene *F* in combination with normal cytoplasm has no detectable effect. (Hint: Since both nuclear and extranuclear genes have to be considered, it will be convenient to use symbols to distinguish the two. Thus cytoplasmic genes will be designated in square brackets; e.g., [N] for normal cytoplasm, [po] for *poky*.)

a. A cross in which *fast-poky* is used as the female (protoperithecial) parent and a normal wild-type strain is used as the male parent gives half *poky* and half *fast-poky* progeny ascospores. What is the genetic interpretation of these results?

b. What would be the result of the reciprocal cross of the cross described in part a, that is, normal female × *fast-poky* male?

21.20 Reciprocal crosses between two types of the evening primrose, *Oenothera hookeri* and *Oenothera muricata*, produce the following effects on the plastids:

O. hookeri female × *O. muricata* male → yellow plastids
O. muricata female × *O. hookeri* male → green plastids

Explain the difference between these results, noting that the chromosome constitution is the same in both types.

*21.21 A form of male sterility in corn is maternally inherited. Plants of a male-sterile line crossed with normal pollen give male-sterile plants. Some lines of corn carry a dominant, so-called restorer (*Rf*) gene, which restores pollen fertility in male-sterile lines.

a. If a male-sterile plant is crossed with pollen from a plant homozygous for gene *Rf*, what will be the genotype and phenotype of the F_1?

b. If the F_1 plants of part a are used as females in a testcross with pollen from a normal plant (*rf/rf*), what would be the result? Give genotypes and phenotypes, and designate the type of cytoplasm.

*21.22 A few years ago the political situation in Chile was such that very many young adults were kidnapped, tortured, and killed by government agents. When abducted young women had young children or were pregnant, those children were often taken and given to government supporters to raise as their own. Now that the political situation has changed, grandparents of stolen children are trying to locate and reclaim their grandchildren. Imagine that you are a judge in a trial centering on the custody of a child. Mr. and Mrs. Escobar believe Carlos Mendoza is the son of their abducted, murdered daughter. If this is true, then Mr. and Mrs. Sanchez are the paternal grandparents of the child, as their son (also abducted and murdered) was the husband of the Escobars' daughter. Mr. and Mrs. Mendoza claim Carlos is their natural child. The attorney for the Escobar and Sanchez couples informs you that scientists have discovered a series of RFLPs in human mitochondrial DNA. He tells you his clients are eager to be tested, and ask that you order that Mr. and Mrs. Mendoza and Carlos be tested also.

a. Can mitochondrial RFLP data be helpful in this case? In what way?

b. Do all seven parties need to be tested? If not, who actually needs to be tested in this case? Explain your choices.

c. Assume the critical people have been tested, and you have received the results. How would the results determine your decision?

21.23 The pedigree in the figure below shows a family in which an inherited disease called Leber's optic atrophy is segregating. This condition causes blindness in adulthood. Studies have recently shown that the mutant gene causing Leber's optic atrophy is located in the mitochondrial genome.

a. Assuming II-4 marries a normal person, what proportion of his offspring should inherit Leber's optic atrophy?

b. What proportion of the sons of II-2 should be affected?

c. What proportion of the daughters of II-2 should be affected?

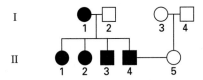

*21.24 Imagine you have discovered a new genus of yeast. In the course of your studies on this organism, you isolate DNA and subject it to CsCl density gradient centrifugation. You observe a major peak at a density of 1.75 g/cm^3 and a minor peak at a density of 1.70 g/cm^3. How could you determine whether the minor peak represents organellar (presumably mitochondrial) DNA, as opposed to a relatively AT-rich repeated sequence in the nuclear genome?

22 QUANTITATIVE GENETICS

PRINCIPAL POINTS

~ Discontinuous traits exhibit only a few distinct phenotypes. Continuous traits display a range of phenotypes.

~ Continuous traits have many phenotypes because they are encoded by many genotypes (are polygenic) and/or because environmental factors cause each genotype to produce a range of phenotypes.

~ The polygene or multiple gene hypothesis of inheritance proposes that quantitative traits are determined by a number of genes, whose effects add together to determine the phenotype.

~ The broad-sense heritability of a trait is the proportion of the phenotypic variance that results from genetic differences among individuals. The narrow-sense heritability is the proportion of the phenotypic variance that results from additive genetic variance.

~ The amount that a trait changes in one generation as a result of selection for the trait is called the response to selection. The magnitude of the response to selection depends upon the selection differential and the narrow-sense heritability.

~ Genetic correlations arise when two traits are influenced by the same genes. When a trait is selected, genetically correlated traits will also exhibit a response to selection.

~ Inbreeding is mating of closely related individuals. Inbreeding leads to an increase in homozygosity.

*I*n the previous chapters, much of our attention focused on the molecular aspects of gene structure, function, and expression. By isolating mutants that affect a particular process and by comparing the mutants with the wild type, geneticists can often describe the molecular basis for the mutant phenotype and can piece together an understanding of how the events for normal cells and organisms proceed. The mutations used in these molecular studies, and in fact most of the traits we have studied up to this point, have been characterized by the presence of a few distinct phenotypes. The seed coats of pea plants, for example, were either grey or white, the seed pods were green or yellow, and the plants were tall or short. In each trait the different phenotypes were distinct, and each phenotype was easily separated from all other phenotypes. A trait such as this, with only a few distinct phenotypes, is called a **discontinuous trait.** The phenotypes of a discontinuous trait can be described in qualitative terms, and all individuals can be placed into a few phenotypic categories, as shown in Figure 22.1. Some additional examples of discontinuous traits are the ABO blood types, coat colors in mice, the prototrophic versus auxotrophic mutants of bacteria, and the presence or absence of extra digits on the hand.

For discontinuous traits, a simple relationship usually exists between the genotype and the phenotype. In most cases, each genotype produces only a single phenotype, and frequently each phenotype results from a single genotype. When dominance or epistasis occurs, the same phenotype may be produced by several differ-

~ **FIGURE 22.1**

Discontinuous distribution of shell color in the snail *Cepaea nemoralus* from a population in England.

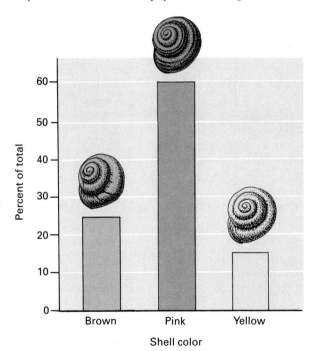

~ FIGURE 22.2

Continuous distribution of height for male students at the University of Minnesota in 1924–25.

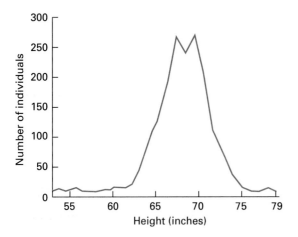

~ FIGURE 22.3

Continuous distribution of birth weight for first-born males born in New York City around 1930.

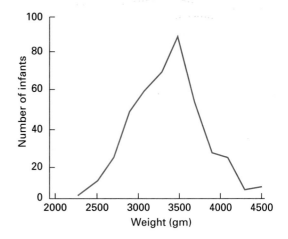

ent genotypes, but the relationship between the genes and the trait remains simple. Because of this simple relationship, genotypes can be inferred by studying the phenotypes of parents and offspring, and in this way Mendel was able to work out the basic principles of heredity.

Not all traits, however, exhibit phenotypes that fall into a few distinct categories. Many traits, such as birth weight and adult height in humans, protein content in corn, and number of eggs laid by *Drosophila,* exhibit a wide range of possible phenotypes. Traits such as these, with a continuous distribution of phenotypes, are called **continuous traits.** The distribution of a continuous trait is illustrated in Figure 22.2. Since the phenotypes of continuous traits must be described in quantitative terms, such traits are also known as **quantitative traits,** and the study of the inheritance of quantitative traits comprises the field of **quantitative genetics.**

Numerous traits have a continuous distribution, and quantitative genetics plays an important role in many areas of basic and applied biology. In agriculture, for example, crop yield, rate of weight gain, milk production, and fat content are all continuous traits which are studied with the techniques of quantitative genetics. In psychology, methods of quantitative genetics are frequently employed for the study of complex behavioral traits, such as IQ, learning ability, and personality. Geneticists also use these methods to study other continuous traits found in humans, such as blood pressure, antibody titer, fingerprint pattern, and birth weight (see Figure 22.3). The traditional techniques of transmis-

sion genetics provide little information about the inheritance of these complex traits, and different, more quantitative methods are required.

THE NATURE OF CONTINUOUS TRAITS

Biologists began developing statistical techniques for the study of continuous traits during the latter part of the 19th century, even before they were aware of Mendel's principles of heredity. Francis Galton and his associate Karl Pearson studied a number of continuous traits in humans, such as height, weight, and mental traits; by demonstrating that the traits of parents and their offspring are statistically associated, they were able to show that these traits are inherited, but they were not successful in determining how genetic transmission occurs. After the rediscovery of Mendel's work, considerable controversy arose over whether continuous traits also follow Mendel's principles, or whether they are inherited in some different fashion. Around 1903, Wilhelm Johannsen conducted a series of experiments on the inheritance of seed weight of beans, and he demonstrated that continuous variation is partly genetic and partly environmental. Several years later, Herman Nilsson-Ehle, working with wheat, proposed that continuous traits are determined by multiple genes, each of which segregates according to Mendelian principles.

Why Some Traits Have Continuous Phenotypes

Continuous traits have many phenotypes. To understand the inheritance of continuous traits, we must first determine why some traits have many phenotypes.

Multiple phenotypes of a trait arise in several ways. Frequently a range of phenotypes occurs because numerous genotypes exist among the individuals of a group—this happens when the trait is influenced by a large number of loci. For example, when a single locus with two alleles determines a trait, three genotypes are present: *AA*, *Aa*, and *aa*. With two loci, each with two alleles, the number of genotypes is $3^2 = 9$ (*AA BB, Aa BB, AA Bb, Aa Bb, AA bb, aa BB, aa Bb, Aa bb,* and *aa bb*). In general, the number of genotypes is 3^n, where *n* equals the number of loci with two alleles; if more than two alleles are present at a locus, the number of genotypes is even greater. As the number of loci influencing a trait increases, the number of genotypes quickly becomes large. Traits that are encoded by many loci are referred to as **polygenic traits**. If each genotype in a polygenic trait encodes a separate phenotype, many phenotypes will be present. And because many phenotypes are present, and the differences between phenotypes are slight, the trait appears to be continuous.

More frequently, several genotypes of a polygenic trait produce the same phenotype. For example, with dominance, only two phenotypes are produced by the three genotypes at a locus (*AA* and *Aa* produce one phenotype and *aa* produces a different phenotype). If dominance occurs among the alleles at each of three loci, the number of genotypes is $3^3 = 27$, but the number of phenotypes is only $2^3 = 8$. When epistatic interactions occur among the alleles at different loci, as discussed in Chapter 4, a number of genotypes may code for the same phenotype. In many polygenic traits, multiple genotypes code for the same phenotype and the relationship between genotype and phenotype is obscured.

A second reason that a trait may have a range of phenotypes is that environmental factors also affect the trait. When environmental factors exert an important influence on the phenotype, each genotype is capable of producing a range of phenotypes. Which phenotype is expressed depends both on the genotype and on the specific environment in which the genotype is found. For most continuous traits, both multiple genotypes and environmental factors influence the phenotype; such a trait is **multifactorial**.

When multiple genes and environmental factors influence a trait, one does not find the simple relationship between genotype and phenotype that exists in discontinuous traits. Therefore, the simple rules of inheritance that we learned in classical genetics provide us with little information about the genes involved in continuous traits. In order to understand the genetic basis of these traits and their inheritance, we must employ special concepts and analytical procedures.

*K*EYNOTE

Discontinuous traits exhibit only a few distinct phenotypes and can be described in qualitative terms. Continuous traits, on the other hand, display a spectrum of phenotypes and must be described in quantitative terms. Many phenotypes are present in a continuous trait because the trait is encoded by many loci, producing many genotypes, and because environmental factors cause each genotype to produce a range of phenotypes.

Questions Studied in Quantitative Genetics

Not only are the methods used in quantitative genetics different from those we have previously studied, but the fundamental nature of the questions asked is also different. In transmission genetics, we frequently determined the probability of inheriting a discontinuous trait. With most continuous traits, however, numerous genes are involved and no simple relationship exists between the genotype and the phenotype; thus, we cannot make precise predictions about the probability of inheriting a continuous trait, as we did for simple discontinuous traits. For the same reason, it is difficult to identify the DNA sequences that code for the trait, so molecular studies of continuous traits are usually impossible. In spite of these limitations, some important questions about the genetic nature of continuous traits can be addressed. What follows is a set of questions frequently studied by quantitative geneticists.

1. To what degree does the observed variation in phenotype result from differences in genotype and to what degree does this variation reflect the influence of different environments? In our study of discontinuous traits, this question assumed little importance, because the differences in phenotype were usually entirely genetic.
2. How many genes determine the phenotype of the trait? When only a few loci are involved and the trait is discontinuous, the number of loci involved can often be determined by examining the phenotypic ratios in genetic crosses. With complex, continuous traits, however, determining the number of loci involved is more difficult.

3. Are the contributions of the determining genes equal? Or do a few genes have major effects on the trait and other genes only modify the phenotype slightly?

4. To what degree do alleles at the different loci interact with one another? Are the effects of alleles additive? Do epistatic interactions occur?

5. When selection occurs for a particular phenotype, how rapidly does the trait change? Do other traits change at the same time?

6. What is the best method for selecting and mating individuals so as to produce desired phenotypes in the progeny?

STATISTICS

With discontinuous traits, we described the phenotypes of a group of individuals by listing the numbers or frequencies of each phenotype; for instance, we might state that three-fourths of the progeny were yellow and one-fourth were green. With continuous traits, such a simple description of the phenotypes is often impossible because a spectrum of phenotypes is present, and usually the phenotype can only be described in terms of a measurement. Furthermore, the relationship between genotype and phenotype is complex, so we do not observe the simple ratios predicted by Mendelian principles. Consequently, quantitative genetics requires an understanding of statistical procedures for describing continuous traits and for answering many of the questions we have listed above.

Samples and Populations

Suppose we want to describe some aspect of a trait in a large group of individuals. For example, we might be interested in the average birth weight of infants born in New York City during the year 1987. Since thousands of babies are born in New York City every year, collecting information on each baby's weight might not be practical. An alternative method would be to collect information on a subset of the group, say birth weights on 100 infants born in New York City during 1987, and then use the average obtained on this subset as an estimate of the average for the entire city. Biologists and other scientists commonly employ this sampling procedure in data collection, and statistics are necessary for analyzing such data. The group of ultimate interest (in our example, all infants born in New York City during 1987) is called the **population,** and the subset used to give us information about the population (our set of 100 babies) is called a **sample.** For a sample to give us reliable information about the population, it must be large enough so that chance differences between the sample and the population are not misleading. If our sample consisted of only a single baby, and that infant was unusually large, then our estimate of the average birth weight of all babies would not be very accurate. The sample must also be a random subset of the population. If all the babies in our sample came from Hope Hospital for Premature Infants, then we would grossly underestimate the true birth weight of the population. Figure 22.3 shows a distribution of birth weight in humans.

KEYNOTE

To describe and study a large group of individuals, scientists frequently examine a subset of the group. This subset is called a sample, and the sample provides information about the larger group, which is termed the population. The sample must be of reasonable size and it must be a random subset of the larger group to provide accurate information about the population. _____

Distributions

When we studied discontinuous traits, we were able to describe the phenotypes found among a group of individuals by stating the proportion of individuals falling into each phenotypic class. As we discussed earlier, continuous traits exhibit a range of phenotypes, and describing the phenotypes found within a group of individuals is more complicated. One means of summarizing the phenotypes of a continuous trait is with a **frequency distribution,** which is a description of the population in terms of the proportion of individuals that have each phenotype.

To make a frequency distribution, classes are constructed to consist of individuals falling within a specified range of the phenotype, and the number of individuals in each class is counted. Table 22.1 presents a

~ TABLE 22.1

Weight of 5494 F_2 Beans (Seeds of *Phaseolus vulgaris*) Observed by Johannsen in 1903

WEIGHT (cg) (MIDPOINT OF RANGE)	5–15 (10)	15–25 (20)	25–35 (30)	35–45 (40)	45–55 (50)	55–65 (60)	65–75 (70)	75–85 (80)	85–95 (90)
Number of beans	5	38	370	1676	2255	928	187	33	2

~ FIGURE 22.4

Frequency histogram for bean weight in *Phaseolus vulgaris* plotted from the data in Table 22.1. A normal curve has been fitted to the data and is superimposed on the frequency histogram.

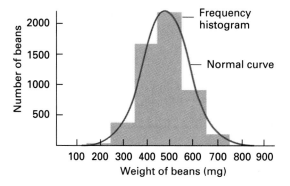

frequency distribution constructed from the data in Johannsen's study of the inheritance of weight in the dwarf bean, *Phaseolus vulgaris*. As shown in the table, Johannsen weighed 5494 beans from the F_2 progeny of a cross and classified them into nine groups or classes, each of which covered a 10-centigram (cg) range of weight. A frequency distribution such as this can be displayed graphically by plotting the phenotypes in a frequency histogram, as shown in Figure 22.4 for Johannsen's beans. In the histogram, the phenotypic classes are indicated along the horizontal axis and the number present in each class is plotted on the vertical axis. If a curve is drawn tracing the outline of the histogram, the curve assumes a shape that is characteristic of the frequency distribution. Several different types of distributions are illustrated in Figure 22.5.

Many continuous phenotypes exhibit a symmetrical, bell-shaped distribution similar to the one shown in Figure 22.5a. This type of distribution is called a **normal distribution.** The normal distribution is produced when a large number of independent factors influence the measurement. Since many continuous traits are multifactorial (influenced by multiple genes and multiple environmental factors), observing a normal distribution for these traits is not surprising. Two other types of distribution are illustrated in Figures 22.5b and 22.5c.

Binomial Theorem

The **binomial distribution** is the theoretical frequency distribution of events that have two possible outcomes. For example, among humans two sexes are present, male and female. In a family of three children, several different combinations of the sexes are possible: three boys, two boys and one girl, one boy and two girls, and three girls. Suppose that we want to know what the probability is that in a family of three children, two will be boys and one will be a girl. The probability is not $\frac{1}{2} \times \frac{1}{2} \times \frac{1}{2} = \frac{1}{8}$, but rather $\frac{3}{8}$. This is because there are three different ways, or three permutations, in which a family of three children can have two boys and one girl: The first two children might be boys and the third child a girl (BBG). Alternatively, the first child might be a boy, followed by a girl and then another boy (BGB), or the first child might be a girl followed by two boys (GBB). In each case, the probability is $\frac{1}{2} \times \frac{1}{2} \times \frac{1}{2} = \frac{1}{8}$ (using the product rule; see Chapter 2), and so the total probability is the sum of these individual probabilities (using the sum rule; see Chapter 2), or $\frac{1}{8} + \frac{1}{8} + \frac{1}{8} = \frac{3}{8}$.

~ FIGURE 22.5

Three different types of distributions. (a) A normal distribution of percent sucrose in sugar beets; (b) A skewed distribution representing coat color in guinea pigs; (c) A bimodal distribution of the size of female rattlesnakes.

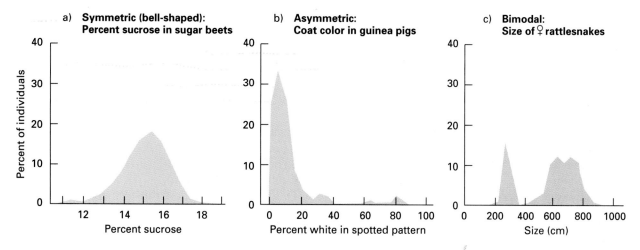

In a family with three children, four different combinations of the sexes are possible: three boys, two boys and a girl, one boy and two girls, and three girls. Three boys can occur in only one way: BBB. Therefore the probability that a family of three children has all boys is $\frac{1}{2} \times \frac{1}{2} \times \frac{1}{2} = \frac{1}{8}$. As we have seen, the probability of two boys and one girl is $\frac{3}{8}$. Following the same reasoning, the probability of two girls and one boy is $\frac{3}{8}$, because three permutations of this combination are possible (GGB, GBG, and BGG). Since three girls can be present in only one way (GGG), the probability of this combination is $\frac{1}{2} \times \frac{1}{2} \times \frac{1}{2} = \frac{1}{8}$. In summary, we have probabilities of $\frac{1}{8}:\frac{3}{8}:\frac{3}{8}:\frac{1}{8}$ for three boys:two boys and one girl:two girls and one boy:three girls.

As you might imagine, such computations get rather laborious when the number of combinations is large. A convenient and efficient way around the tedium is to use the *binomial theorem*. According to this theorem, the frequencies (or probabilities of occurrence) of the various combinations correspond to the terms of the *binomial expansion*.

The general expression for the binomial expansion is $(a + b)^n$, where a is the probability of one event occurring and b is the probability of the alternative event occurring; $a + b$ must equal 1. The term n is the number of trials. The first three binomial expansions are as follows:

$$(a + b)^2 = a^2 + 2ab + b^2$$
$$(a + b)^3 = a^3 + 3a^2b + 3ab^2 + b^3$$
$$(a + b)^4 = a^4 + 4a^3b + 6a^2b^2 + 4ab^3 + b^4$$

Using these general formulas, we could consider a, for example, to be the probability of having a boy and b the probability of having a girl. The exponents of a and b in the expansions correspond, then, to the number of children of each type in the family to which the term applies.

Let us consider the first expansion. The exponent indicates that we are dealing with a two-child family. The term a is the probability of having a boy, namely, $\frac{1}{2}$, and b is the probability of having a girl, also $\frac{1}{2}$. To the right of the equals sign are the terms for the various two-child families that can result, and from these terms the probability of each combination can be determined. The a^2 term, for example, is the probability of having two boys, which is $(\frac{1}{2})^2 = \frac{1}{4}$. The probability of having a boy and a girl is $2ab = 2(\frac{1}{2})(\frac{1}{2}) = \frac{2}{4}$; and so on.

For our purposes we must be able to compute the coefficients for terms in the binomial expansion. As Figure 22.6 shows, there is a symmetry in the coefficients of successive powers of the binomial such that they can be arranged in what is called Pascal's triangle.

~ **FIGURE 22.6**

Coefficients for terms in the binomial expansion $(a + b)^n$ for values of n from 1 to 8.

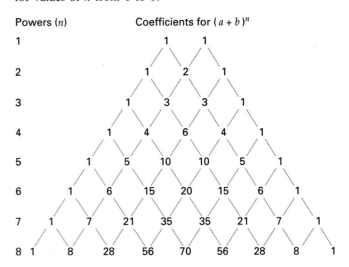

In this triangle each number is the sum of the numbers one place to the left and right in the row immediately above. There is also a symmetry in the exponents of the expansion. Consider, for example, the expansion of $(a + b)^5$:

$$(a + b)^5 = a^5 + 5a^4b + 10a^3b^2 + 10a^2b^3 + 5ab^4 + b^5$$

Note that the exponents of a begin at the power to which the binomial is raised and then decrease incrementally to zero; at the same time the reverse happens to the exponents of b. Also, note that the number of terms in the expansion is always $n + 1$.

Last, we need a reasonably quick method of determining the value of a particular coefficient in an expansion, without searching for the nearest copy of Pascal's triangle. The principle here is as follows: If the probability of occurrence of event A is a, and the probability of occurrence of the alternative event B is b (i.e., $1 - a$), then the probability that in n attempts event A will take place s times and event B will take place t, or $(n - s)$, times is calculated as follows:

$$\frac{n!}{s!t!} a^s b^t$$

The symbol ! in this expression means "factorial" and is the product of all the integers from 1 to the term n involved. For example, 5! is $1 \times 2 \times 3 \times 4 \times 5$. By definition, 0! is 1, not zero.

To illustrate how to use this formula, consider the following problem. In a family of six children, what is the probability that three will be girls and three will be boys? The number of events (n) is 6, because there are six children in the family. The probability of event A is the probability of having a girl, so $a = \frac{1}{2}$, and similarly the probability of event B is the probability of having a boy, so $b = \frac{1}{2}$. The number of times event A takes place is the number of girls in the family; therefore $s = 3$. The number of times event B takes place is the number of boys in the family; therefore $t = 3$. Substituting these values in the formula, we obtain

$$\frac{6!}{3!3!}(\tfrac{1}{2})^3(\tfrac{1}{2})^3$$

$$= \frac{1 \times 2 \times 3 \times 4 \times 5 \times 6}{1 \times 2 \times 3 \times 1 \times 2 \times 3}(\tfrac{1}{2})^3(\tfrac{1}{2})^3$$

$$= \frac{720}{36}(\tfrac{1}{2})^3(\tfrac{1}{2})^3$$

$$= 20(\tfrac{1}{8})(\tfrac{1}{8}) = {}^{20}\!/_{64} = {}^{5}\!/_{16}$$

Later in this chapter, we will use the binomial expansion for determining the phenotypic ratios in crosses involving a polygenic trait (Polygene Hypothesis for Quantitative Inheritance).

The Mean

A distribution of phenotypes can be summarized in the form of two convenient statistics, the mean and the variance. The **mean,** which is also known as the average, gives us information about the center of the distribution of the phenotypes in a sample. The mean of a sample is calculated by simply adding up all the individual measurements (x_1, x_2, x_3, etc.) and dividing by the number of measurements we added (n). We can represent this calculation with the following formula:

$$\text{Mean} = \bar{x} = \frac{x_1 + x_2 + x_3 + \cdots x_n}{n}$$

This equation can be abbreviated by using the symbol Σ, which means the summation, and x_i, which means the ith (or individual) value of x:

$$\text{Mean} = \bar{x} = \frac{\Sigma x_i}{n}$$

When data are taken from a frequency distribution, such as the distribution of bean weights seen in Table 22.1, several individuals have the same phenotype. Here, the mean can be estimated by first multiplying the midpoint of each class, m_i (in this case the middle weight in each class) by the number of individuals in

that class, or f_i. These products are then added up and divided by the total number of measurements:

$$\bar{x} = \frac{\Sigma f_i m_i}{n}$$

Table 22.2 (p. 678) presents a sample calculation of the mean.

The mean is frequently used in quantitative genetics to summarize the phenotypes of a group of individuals. For example, in an early study of continuous variation, Edward M. East examined the inheritance of flower length in several strains of the tobacco plant. He crossed a short-flowered strain of tobacco with a long-flowered strain. Within each strain, however, flower length varied some, so East reported that the mean phenotype of the short strain was 40.4 mm and the mean phenotype of the long strain was 93.1 mm. The F_1 progeny, which consisted of 173 plants, had a mean flower length of 63.5 mm. In this situation, the mean provides a convenient way to quickly summarize the phenotypes of parents and offspring.

The Variance and the Standard Deviation

A second statistic that provides key information about a distribution is the **variance.** The variance is a measure of how much the individual measurements spread out around the mean—how variable the measurements are. Two distributions may have the same mean but very different variances, as shown in Figure 22.7. The variance, symbolized as s^2, is defined as the average squared deviation from the mean.

~ FIGURE 22.7

Graphs showing three distributions with the same mean but different variances.

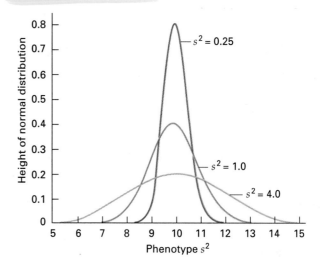

~ TABLE 22.2

Sample Calculations of the Mean, Variance, and Standard Deviation for Body Length of 10 Spotted Salamanders from Penobscot County, Maine

BODY LENGTH x_i (mm)	$(x_i - \bar{x})$	$(x_i - \bar{x})^2$
65	$(65 - 57.1) =\ \ \ 7.9$	$7.9^2 =\ \ \ 62.41$
54	$(54 - 57.1) = -3.1$	$-3.1^2 =\ \ \ 9.61$
56	$(56 - 57.1) = -1.1$	$-1.1^2 =\ \ \ 1.21$
60	$(60 - 57.1) =\ \ \ 2.9$	$2.9^2 =\ \ \ 8.41$
56	$(56 - 57.1) = -1.1$	$-1.1^2 =\ \ \ 1.21$
55	$(55 - 57.1) = -2.1$	$-2.1^2 =\ \ \ 4.41$
53	$(53 - 57.1) = -4.1$	$-4.1^2 =\ \ 16.81$
55	$(55 - 57.1) = -2.1$	$-2.1^2 =\ \ \ 4.41$
58	$(58 - 57.1) =\ \ \ 0.9$	$0.9^2 =\ \ \ 0.81$
59	$(59 - 57.1) =\ \ \ 1.9$	$1.9^2 =\ \ \ 3.61$
$\Sigma x_i = 571$		$\Sigma(x_i - \bar{x})^2 = 112.9$

Mean $= \bar{x} = \dfrac{\Sigma x_i}{n} = \dfrac{571}{10} = 57.1$

Variance $= s_x{}^2 = \dfrac{\Sigma(x_i - \bar{x})^2}{n-1} = \dfrac{112.9}{9} = 12.54$

Standard deviation $= s_x = \sqrt{12.54} = 3.54$

$$\text{Variance} = s^2 = \frac{\Sigma(x_i - \bar{x})^2}{n-1}$$

It can be calculated by first subtracting each individual measurement from the mean. Each value obtained from this subtraction is squared and all the squared values are added up. The sum of this calculation is then divided by the number of original measurements minus one. (For mathematical reasons, which we will not discuss here, the variance is obtained by dividing $n-1$ instead of n.)

Another statistic, closely related to the variance, is the **standard deviation,** which is simply the square root of the variance:

$$\text{Standard deviation} = s = \sqrt{s^2}$$

The standard deviation is often preferred to the variance, because the standard deviation is in the same units as the original measurements, while the variance is in the units squared. Sample calculations for the variance and the standard deviation are presented in Table 22.2.

In a normal distribution about two-thirds of the individual observations have values within one stan-dard deviation above or below ($\pm 1s$) the mean of the distribution (Figure 22.8). About 95 percent of the values fall within two standard deviations ($\pm 2s$) of the mean, and over 99 percent fall within three standard deviations ($\pm 3s$). Therefore, a broad curve implies a considerable variability in the quantity measured and a correspondingly large standard deviation. A narrow curve, in contrast, indicates relatively little variability in the quantity measured and a correspondingly small standard deviation.

The variance and the standard deviation can provide us with valuable information about the phenotypes of a group of individuals. In our discussion of the mean, we saw how East used the mean to describe flower length of parents and offspring in crosses of the tobacco plant. When East crossed a strain of tobacco with short flowers to a strain with long flowers, the F_1 offspring had a mean flower length of 63.5 mm, which was intermediate to the phenotypes of the parents. When he intercrossed the F_1, the mean flower length of the F_2 offspring was 68.8 mm, approximately the same as the mean phenotype of the F_1. However, the F_2 progeny differed from the F_1 in an important attribute that is not apparent if we only examine the means of the phenotype—the F_2 were more variable in phenotype

~ FIGURE 22.8

Normal distribution curve showing the proportions of the data in the distribution that are included within certain multiples of the standard deviation.

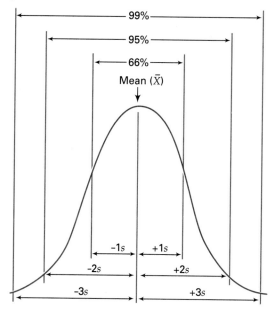

to give the covariance of x and y, where n equals the number of xy pairs:

$$\text{cov}_{xy} = \frac{\Sigma(x_i - \bar{x})(y_i - \bar{y})}{n - 1}$$

An algebraically equivalent equation, which is easier to compute, is

$$\text{cov}_{xy} = \frac{\Sigma x_i y_i - \dfrac{1}{n}(\Sigma x_i \Sigma y_i)}{n - 1}$$

where $\Sigma x_i y_i$ is the sum of each value of x multiplied by each corresponding value of y, Σx_i is the sum of all x values, and Σy_i is the sum of all y values. The correlation coefficient r can then be obtained by dividing the covariance by the product of the standard deviations of x and y.

$$\text{Correlation coefficient} = r = \frac{\text{cov}_{xy}}{s_x s_y}$$

where s_x equals the standard deviation of x and s_y equals the standard deviation of y. Table 22.3 (p. 680) gives a sample calculation of the correlation coefficient between two variables.

The correlation coefficient can range from -1 to $+1$. The sign of the correlation coefficient, whether it is positive or negative, indicates the direction of the correlation. If the correlation coefficient is positive, then an increase in one variable tends to be associated with an increase in the other variable. If seed size and seed number are positively correlated in sunflowers, for example, plants with larger seeds also tend to produce more seeds. Positive correlations are illustrated in Figure 22.9 (p. 681) parts *b, c, d,* and *f.* A negative correlation coefficient indicates that an increase in one variable is associated with a decrease in the other. If seed size and seed number are negatively correlated, plants with large seeds tend to produce fewer seeds on the average than plants with smaller seeds. Figure 22.9, part *e* represents a negative correlation. The absolute value of the correlation coefficient (its magnitude if the sign is ignored) provides information about the strength of the association. When the correlation coefficient is close to -1 or to $+1$, the correlation is strong, meaning that a change in one variable is almost always associated with a corresponding change in the other variable. For example, the x and y variables in Figure 22.9, part *f* are strongly associated, and have a correlation coefficient of 0.9. On the other hand, a correlation coefficient near 0 indicates a weak relationship between the variables as is illustrated in Figure 22.9*b*.

Several important points about correlation coefficients warrant emphasis at this point. First, a correla-

than the F_1. The variance in the flower length of the F_2 was 42.4, whereas the variance in the F_1 was only 8.6. This finding indicated that more genotypes were present among the F_2 progeny than in the F_1. Thus, the mean and the variance are both necessary for fully describing the distribution of phenotypes among a group of individuals.

Correlation

Often two variables are associated or correlated. This means that if one variable changes, the other is also likely to change. For example, head size and body size are correlated in most animals—individuals with large bodies have correspondingly large heads, and those with small bodies tend to have small heads. The **correlation coefficient** is a statistic that measures the strength of the association between two variables. Suppose we have two variables, x and y (x might equal body length and y might equal head width), and we wish to calculate the correlation between them. We begin by obtaining the covariance of x and y; the **covariance** is computed by taking the first x value and subtracting it from the mean of x and then multiplying the result by the deviation of the first y value from the mean of y. This is done for each pair of x and y values and the products are added together. The sum in then divided by $n - 1$

~ TABLE 22.3

Sample Calculation of the Correlation Coefficient for Body Length and Head Width of Tiger Salamanders

BODY LENGTH (mm) x_i	$x_i - \bar{x}$	$(x_i - \bar{x})^2$	HEAD WIDTH (mm) y_i	$y_i - \bar{y}$	$(y_i - \bar{y})^2$	$x_i y_i$
72.00	−7.92	62.67	17.00	−0.75	0.56	1224
62.00	−17.92	321.01	14.00	−3.75	14.06	868
86.00	6.08	37.01	20.00	2.25	5.06	1720
76.00	−3.92	15.34	14.00	−3.75	14.06	1064
64.00	−15.92	253.34	15.00	−2.75	7.56	960
82.00	2.08	4.34	20.00	2.25	5.06	1640
71.00	−8.92	79.51	15.00	−2.75	7.56	1065
96.00	16.08	258.67	21.00	3.25	10.56	2016
87.00	7.08	50.17	19.00	1.25	1.56	1653
103.00	23.08	532.84	23.00	5.25	27.56	2369
86.00	6.08	37.01	18.00	0.25	0.06	1548
74.00	−5.92	35.01	17.00	−0.75	0.56	1258
$\Sigma x_i =$ 959.00		$\Sigma(x_i - \bar{x})^2 =$ 1686.92	$\Sigma y_i =$ 213.00		$\Sigma(y_i - \bar{y})^2 =$ 94.25	$\Sigma x_i y_i =$ 17,385

$\bar{x} = \Sigma x_i/n = 959/12 = 79.92$

$\bar{y} = \Sigma y_i/n = 213/12 = 17.75$

Variance of $x = s_x^2 = \Sigma(x_i - \bar{x})^2/n - 1 = 1686.92/11 = 153.35$

Standard deviation of $x = s_x = \sqrt{s_x^2} = \sqrt{153.35} = 12.38$

Variance of $y = s_y^2 = \Sigma(y_i - \bar{y})^2/n - 1 = 94.25/11 = 8.57$

Standard deviation of $y = s_y = \sqrt{s^2} = \sqrt{8.57} = 2.93$

Covariance $= \text{cov}_{xy} = (\Sigma x_i y_i - 1/n(\Sigma x_i \Sigma y_i))/n - 1$

$\text{cov}_{xy} = \dfrac{(17385 - 1/12(959 \times 213))}{12 - 1}$

$\text{cov}_{xy} = 32.97$

Correlation coefficient $= r = \text{cov}_{xy}/(s_x s_y) = 32.97/(12.38 \times 2.93)$

$r = 0.91$

tion between variables means only that the variables are associated; *correlation does not imply that a cause-effect relationship exists.* The classic example of a noncausal correlation between two variables is the positive correlation that exists between number of Baptist ministers and liquor consumption in cities with population size over 10,000. One should not conclude from this correlation that Baptist ministers are consuming all the alcohol. Alcohol consumption and number of Baptist ministers are associated because both are positively correlated with a third factor, population size; larger cities contain more Baptist ministers and also have higher alcohol consumption. Assuming that two factors are causally related because they are correlated may lead to erroneous conclusions.

Another important point is that *correlation is not the same thing as identity.* Correlation only means that a change in one variable is associated with a corresponding change in the other variable. Two variables can be highly correlated, and yet have very different values. For example, the height of today's college-age males is correlated with the height of their fathers; tall fathers tend to produce tall sons and short fathers tend to produce short sons. This correlation results from the fact that genes influence human height. However, most college-age males today are taller than their fathers, probably because better diet and health care have increased the average height of all individuals in recent years. Thus, fathers and sons exhibit a correlation in height, but they are not the same height.

~ FIGURE 22.9

Scatter diagrams showing the correlation of *x* and *y* variables. Diagrams *b, c, d,* and *f* show positive correlations, whereas diagram *e* shows a negative correlation. The absolute value of the correlation coefficient (*r*) indicates the strength of association. For example, diagram *f* illustrates a relatively strong correlation and diagram *b* illustrates a relatively weak correlation. In diagram *a*, the *x* and *y* variables are not correlated.

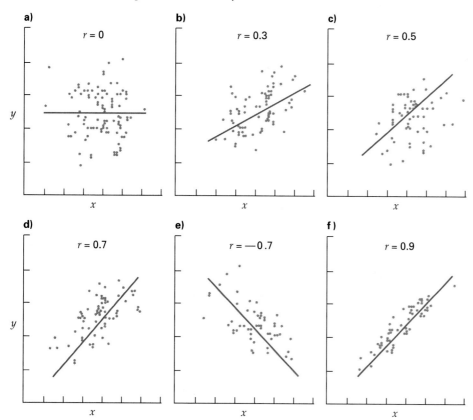

KEYNOTE

The correlation coefficient is a measure of how strongly two variables are associated. A positive correlation coefficient indicates that the two variables change in the same direction; an increase in one variable is usually associated with a corresponding increase in the other variable. When the correlation coefficient is negative, the variables are inversely related; an increase in one variable is most often associated with a decrease in the other. The absolute value of the correlation coefficient provides information about the strength of the association. Correlation does not imply that a cause-effect relationship exists between the two variables. ⎯⎯⎯⎯

Regression

The correlation coefficient tells us about the strength of association between variables and indicates whether the relationship is positive or negative, but it provides little information about the precise relationship between the variables. Often we are interested in knowing how much of a change in one variable is associated with a given change in another variable. Returning to our example of the correlation between heights of father and son, we might ask: If a father is 6 feet tall, what is the most likely height of his son? To answer this question, **regression** analysis is used.

The relationship between two variables can be expressed in the form of a **regression line**, as shown in Figure 22.10 (p. 682) for the relationship between heights of father and son. Each point on the graph represents the actual height of a father (value on the hori-

~ FIGURE 22.10

Regression of son's height on father's height. Each point represents a pair of data for the height of a father and his son. The regression equation is $y = 36.05 + 0.49x$.

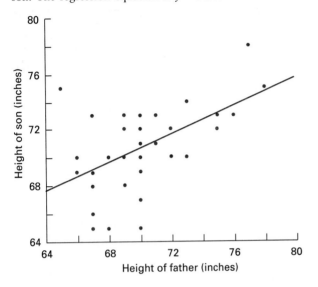

~ FIGURE 22.11

Regression lines with different slopes. The slope indicates how much of a change in the y variable is associated with a change in the x variable.

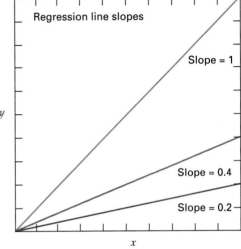

zontal or x axis) and the height of his son (value on the vertical or y axis). The regression line is a mathematically computed line that represents the best fit of a line to the points; what this means is that the squared vertical distance from the points to the regression line is minimized. The regression line can be represented with the equation

$$y = a + bx$$

where x and y represent the values of the two variables (in Figure 22.10 the heights of father and son, respectively), b represents the **slope** (also called the **regression coefficient**), and a is the **y intercept**. The slope can be calculated from the covariance of x and y and the variance of x in the following manner:

$$\text{Slope} = b = \frac{\text{cov}_{xy}}{s^2_x}$$

The slope indicates how much of an increase in the variable on the y axis is associated with a unit increase in the variable on the x axis. For example, a slope of 0.5 for the regression of father and son height would mean that for each 1-inch increase in height of a father, the expected height of the son would increase 0.5 inches. The y intercept is the expected value of y when x is zero (the point at which the regression line crosses the y axis). Examples of regression lines with different slopes are presented in Figure 22.11. Regression analysis is one method that is commonly used for measuring the

extent to which variation in a trait is genetically determined, as is described later in the section on heritability.

Analysis of Variance

One last statistical technique that we will mention briefly is **analysis of variance**. Analysis of variance, often called simply **ANOVA**, is a powerful statistical procedure for determining if differences in means are significant and for dividing the variance into components. We might be interested in knowing, for example, if males with the XYY karyotype (see Chapter 3) differ in height from males with a normal XY karyotype. We would proceed by first calculating the mean height of a sample of XYY males and the mean height of a sample of XY males. Suppose we found the mean height of our sample of XYY males was 74 inches, while the mean height of our sample of XY males was 70. The means are different, but our result could be due to chance differences between the samples, rather than to some factor associated with the extra Y chromosome.

Analysis of variance can provide us with the probability that the difference in means of two samples results from chance. For example, it might indicate that there is less than a 1 percent probability that the difference we observed in the mean heights resulted from chance. We would probably conclude, then, that the difference in mean heights of XYY males and XY males is not due to chance differences in our samples, but results from some significant factor associated with the difference in chromosomes. The analysis of variance can also be used to determine how much of the variation in height is associated with the difference in karyotype. We might find that the difference in the karyotypes is associated with 40 percent of the overall variation in height among the individuals in our samples. Factors other than difference in the number of Y chromosomes (perhaps genes on other chromosomes, diet, health care, etc.) would be responsible for the other 60 percent of the variation.

The calculations involved in analysis of variance are beyond the scope of this book, but the concept of breaking down the variance into components—called partitioning the variance—is important in quantitative genetics. For example, frequently we are interested in knowing how much of the variation in a trait is associated with genetic differences among individuals and how much of the variation is associated with environmental factors. Suppose that we wanted to increase milk production in a herd of dairy cattle. We might use analysis of variance to determine how much of the variation in milk production among the cows results from environmental differences and how much arises because of genetic differences. If much of the variation is genetic, we could increase milk production by selective breeding. On the other hand, if most of the variation is environmental, selective breeding will do little to increase milk production, and our efforts would be better directed toward providing the optimum environment for high production. Analysis of variance is often used in this type of problem.

POLYGENIC INHERITANCE

In the examples of polygenic inheritance described in the following sections, geneticists did not know at first how these traits were inherited, although it was apparent that their pattern of inheritance differed from that of discontinuous traits.

Inheritance of Ear Length in Corn

An organism that has been the subject of genetic and cytological studies for many years is corn, *Zea mays*. Ear length is one of the traits of the corn plant that came under scrutiny in a classic study that demonstrated quantitative inheritance for that trait. In this study, reported in 1913, Rollins Emerson and E. East started their experiments with two pure-breeding strains of corn, each of which displayed little variation in ear length. The two varieties were Black Mexican sweet corn (which had short ears of mean length 6.63 cm) and Tom Thumb popcorn (which had long ears of mean length 16.80 cm).

Emerson and East crossed the two strains and then interbred the F_1 plants. Figure 22.12 (p. 684) presents the results in both photographs and a histogram. Note that the F_1s have a mean ear length of 12.12 cm, which is approximately intermediate between the mean ear lengths of the two parental lines. The parental plants are pure breeding, so we can assume that each is homozygous for whatever genes control the lengths of their ears. Since the two parental plants differ in ear length, though, each must be genetically different. When two pure-breeding strains are crossed, the F_1 plants will be heterozygous for all genes, and all plants should have the same genotype. Therefore the range of ear length phenotypes seen in the F_1 plants must be due to factors other than genetic differences; these other factors are probably environmental, since it is impossible to grow plants in exactly identical conditions.

In the F_2 the mean ear length of 12.89 cm is about the same as the mean for the F_1 population, but the F_2 population has a much larger variation around the mean than the F_1 population has. This variation is easy to see in Figure 22.12b; it can also be shown by calculating the standard deviation s. The standard deviation of the long-eared parent is 0.816, and that of the short-eared parent is 1.887. In the F_1, $s = 1.519$, and in the F_2, $s = 2.252$. These numbers confirm that the F_2 has greater variability, something we could conclude by looking at the data.

Is this variation the result of the effects of environmental factors? Certainly, if the environment was responsible for variation in the parental and the F_1 generations, we have every reason to believe that it would have a similar effect on the F_2. However, we have no reason to suppose that the environment would have a greater influence on the F_2 than on the other two generations, and thus there must be another explanation for the greater variation in ear length in the F_2 generation. A more reasonable hypothesis is that the increased variability of the F_2 results from the presence of greater genetic variation in the F_2.

Setting aside the environmental influence for the moment, the data reveal four observations that apply generally to quantitative-inheritance studies similar to this one:

1. The mean value of the quantitative trait in the F_1 is approximately intermediate between the means of the two true-breeding parental lines.
2. The mean value for the trait in the F_2 is approximately equal to the mean for the F_1 population.
3. The F_2 shows more variability around the mean than the F_1 does.
4. The extreme values for the quantitative trait in the F_2 extend further into the distribution of the two parental values than do the extreme values of the F_1.

~ FIGURE 22.12

Inheritance of ear length in corn: (a) representative corn ears from the parental, F_1, and F_2 generations from an experiment in which two pure-breeding corn strains that differ in ear length were crossed and then the F_1s interbred. (b) Histograms of the distributions of ear length (in centimeters) of ears of corn from the experiment represented in part (a); the vertical axes represent the percentages of the different populations found at each ear length.

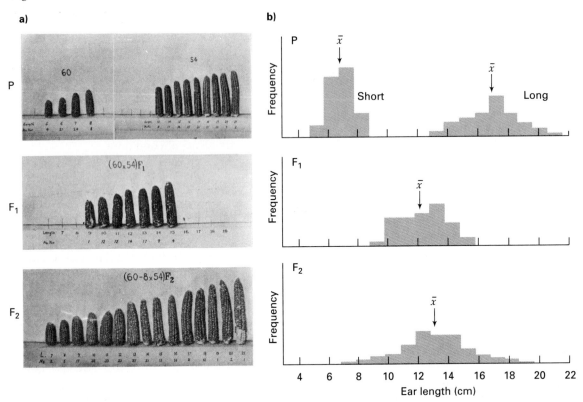

Can the data presented be explained in terms of standard Mendelian genetic principles that govern the inheritance of discontinuous traits? No, the data cannot support such an interpretation. That is, if a single gene were responsible for the two phenotypes of the original parents (AA = homozygous, long; aa = homozygous, short), then the F_1 data could be explained if we assume incomplete dominance. However, crossing the F_1 heterozygote (Aa) should produce a 1:2:1 ratio of AA, Aa, and aa, or long, intermediate, and short phenotypes. Clearly, the data do not fall into such discrete classes.

𝒦EYNOTE

For a quantitative trait, the F_1 progeny of a cross between two phenotypically distinct, pure-breeding parents has a phenotype intermediate between the parental phenotypes. The F_2 shows more variability than the F_1, with a mean phenotype close to that of

the F_1. The extreme phenotypes of the F_2 extend well beyond the range of the F_1 and into the ranges of the two parental values.

Polygene Hypothesis for Quantitative Inheritance

The simplest explanation for the data obtained from Emerson and East's experiments on corn ear length and from other experiments with quantitative traits is that quantitative traits are controlled by not one but by many genes. This explanation, called the **polygene**, or **multiple-gene**, hypothesis for quantitative inheritance, is regarded as one of the landmarks of genetic thought.

The polygene hypothesis can be traced back to 1909 and the classic work of Hermann Nilsson-Ehle, who studied the color of the wheat kernel. Like Mendel and like Emerson and East, Nilsson-Ehle started by crossing true-breeding lines of plants with red kernels and plants with white kernels. The F_1 had grains that were all the same shade of an intermediate color between red and white. At this point he could not rule out incomplete dominance as the basis for the F_1 results. However, when he intercrossed the F_1s, a number of the F_2 progeny showed a ratio of approximately 15 red (all shades) : 1 white kernels, clearly a deviation from a 3 : 1 ratio expected for a monohybrid cross. He recognized four discrete shades of red, in addition to white, among the progeny. He counted the relative number of each class and found a 1 : 4 : 6 : 4 : 1 phenotypic ratio of wheat with dark red, medium red, intermediate red, light red, and white kernels. Note that $\frac{1}{16}$ of the F_2 has a kernel phenotype as extreme as the original red parent, and $\frac{1}{16}$ has a kernel phenotype as extreme as the original white parent.

How can the data be interpreted in genetic terms? Recall from Chapter 4 (Analytical Approaches for Solving Genetics Problems, Q3c, pp. 120–121) that a 15 : 1 ratio of two alternative characteristics resulted from the interaction of the products of two genes, each of which affects the same trait; the two genes are known as duplicate genes. The explanation for the 15 : 1 ratio was that two allelic pairs are involved in determining the phenotypes segregating in the cross. Since several of the F_2 populations from the wheat crosses exhibited a 15 red : 1 white ratio, we can apply that explanation to the kernel trait.

Let us hypothesize that there are two pairs of independently segregating alleles that control the production of red pigment: alleles R (red) and C (crimson) result in red pigment and alleles r and c result in the lack of pigment. Nilsson-Ehle's parental cross and the F_1 genotypes can then be shown as follows:

$$P \quad RR\,CC \quad \times \quad rr\,cc$$
$$\text{(dark red)} \qquad \text{(white)}$$
$$F_1 \qquad Rr\,Cc$$
$$\text{(intermediate red)}$$

When the F_1 is interbred, the distribution of genotypes in the F_2 is that typical of dihybrid inheritance, that is, $\frac{1}{16} RR\,CC + \frac{2}{16} Rr\,CC + \frac{1}{16} rr\,CC + \frac{2}{16} RR\,Cc + \frac{4}{16} Rr\,Cc + \frac{2}{16} rr\,Cc + \frac{1}{16} RR\,cc + \frac{2}{16} Rr\,cc + \frac{1}{16} rr\,cc$. If R and C are dominant to r and c, the 9 : 3 : 3 : 1 phenotypic ratio characteristic of dihybrid inheritance will result. For the wheat kernel color phenotype, then, dominance is not the simple answer, since the observed phenotypic ratio approximates 1 : 4 : 6 : 4 : 1.

Note that these numbers in the phenotypic ratio are the same as the coefficients in the binomial expansion of $(a + b)^4$. A simple explanation is that each dose of a gene controlling pigment production allows the synthesis of a certain amount of pigment. Therefore the intensity of red coloration is a function of the number of R

~ **TABLE 22.4**

~ TABLE 22.4

Genetic Explanation for the Number and Proportions of F_2 Phenotypes for the Quantitative Trait Red Kernel Color in Wheat

GENOTYPE	NUMBER OF GENES FOR RED	PHENOTYPE	FRACTION OF F_2
RR CC	4	Dark red	$\frac{1}{16}$
RR Cc or Rr CC	3	Medium red	$\frac{4}{16}$
RR cc or rr CC or Rr Cc	2	Intermediate red	$\frac{6}{16}$
rr Cc or Rr cc	1	Light red	$\frac{4}{16}$
rr cc	0	White	$\frac{1}{16}$

or C alleles in the genotype; RR CC (term a^4 in the expansion) would be dark red and rr cc (b^4 in the expansion) would be white. Table 22.4 summarizes this situation with regard to the five phenotypic classes observed by Nilsson-Ehle. In other words, the genes represented by capital letters code for products that add to the phenotypic characteristic; for example, each allele, R or C, causes more red pigment to be added to the wheat kernel color phenotype. Alleles that contribute to the phenotype are called **contributing alleles**. The alleles that do not have any effect on the phenotype of the quantitative trait, such as the r and c alleles in the wheat kernel color trait, are called **noncontributing alleles**. Thus the inheritance of red kernel color in wheat is an example of a multiple gene or polygene series of as many as four contributing alleles.

We must be cautious in interpreting the genetic basis of this particular quantitative trait, though. Some F_2 populations show only three phenotypic classes with a 3 : 1 ratio of red to white, while other F_2 populations show a 63 : 1 ratio of red to white, with discrete classes of color between the dark red and the white. These results indicate that the genetic basis for the quantitative trait can vary with the strain of wheat involved. The 3 : 1 case could be explained by a single-gene system with two contributing alleles, while the 63 : 1 case could indicate a polygene series with six contributing alleles. The number of discrete classes in the latter case would be seven, with the proportion of each class following the coefficients in the binomial expansion of $(a + b)^6$, that is, 1 : 6 : 15 : 20 : 15 : 6 : 1.

The multiple-gene hypothesis that fits the wheat kernel color example so well has been applied to other examples of quantitative inheritance, including ear length in corn. In its basic form the multiple-gene hypothesis proposes that a number of the attributes of quantitative inheritance can be explained on the basis of the action and segregation of a number of allelic pairs that have identical (or nearly identical) additive effects on the phenotype and that involve no complete dominance. Simplified, the assumptions of the hypothesis are as follows:

1. In no allelic pair does one allele exhibit dominance over another allele. Instead, a series of contributing and noncontributing alleles is involved.
2. Each contributing allele in a series has an equal effect.
3. The effect of each contributing allele is additive.
4. No genetic interaction (epistasis) occurs among the alleles at different loci in a polygenic series.
5. No genetic linkage is exhibited between the genes in a polygenic series, and therefore they assort independently.
6. There are no environmental effects.

It would be naive to suppose that each of these assumptions is uniformly, or even widely, valid. Since each assumption is open to criticism from theoretical principles, it is surprising that so many examples conform so well to the assumptions. In fact, many polygenic effects do conform to the first four assumptions.

When a large number of polygenes are involved, however, it is hard to imagine that linkage or environmental effects can be ignored, and we have no reason to suppose that evolution has made genes in a small polygenic series unlinked. Also, environmental effects cannot be ignored. In the ear length data, for example, the variability of length in the true-breeding strains is probably the result of environmental effects. With plants, variations in environmental agents such as soil composition, water availability, and local pest concentrations can affect the quantitative-trait phenotypes. Thus we should always remember that an organism's genotype merely specifies the organism's *potential* phenotype and that the genotypic blueprint is interpreted during the organism's development and differentiation in a way that is influenced by internal and external environments.

For the most part the multiple-gene hypothesis is satisfactory as a working hypothesis for interpreting many quantitative traits. The whole picture of quantitative traits is very complicated, though, and there are still many gaps in our understanding of quantitative inheritance. In addition, we have a lot to learn about the molecular aspects of quantitative traits. For example, while the proposal that a number of alleles each function to produce a particular amount of pigment is an attractive hypothesis, what does that hypothesis mean at the molecular level? Is some regulation of product output exerted at the translational level or in a biochemical pathway? In a large polygenic series, how

many biochemical pathways are controlled? Thus quantitative inheritance provides an explanation for the inheritance of continuous traits that is compatible with Mendel's laws, but many aspects of quantitative inheritance are still unknown.

*K*EYNOTE

Quantitative traits are based on genes in a multiple-gene series, or polygenes. The multiple-gene hypothesis assumes that contributing and noncontributing alleles in the series operate so that as the number of contributing alleles increases, there is an additive (or occasionally multiplicative) effect on the phenotype, that no genetic interaction occurs between the gene loci, that the loci are unlinked, and that there are no environmental effects. Since the last three assumptions are not always true, the analysis of multiple-gene series that control quantitative traits is difficult.

Determining the Number of Polygenes for a Quantitative Trait

We must be able to estimate the number of genes in a multiple-gene system so that we can apply our knowledge in a predictive way to genetic crosses involving quantitative traits. This predictive ability is important, for example, to plant and animal breeders who want to breed features into the organisms with which they deal. Such features are often phenotypes controlled by a polygene system or systems, and the breeders need to be able to predict the probability of obtaining progeny with particular combinations of phenotypes from breeding programs. Examples here usually relate to yields, such as number and size of tobacco leaves from a tobacco plant, quantity of milk from a dairy cow, beef from beef cattle, or ears of corn from a corn plant. Related to these factors, of course, is the quality of the yield, which can also be affected by polygene systems.

As you might imagine, it is not an easy task to estimate the number of genes controlling a quantitative trait. If a large number of genes are involved, many different phenotypes are present, and the progeny are not easily placed into distinct phenotypic classes.

Where discrete phenotypic classes can be identified, however, we can estimate the number of genes involved. In Nilsson-Ehle's studies of the inheritance of red kernel color in wheat, for example, all F_2 populations were not identical. While crosses of some red-kerneled varieties gave F_2 progeny that fell into five phenotypic classes, others gave seven phenotypic classes. The latter crosses showed an approximately $1:6:15:20:15:6:1$ ratio of different shades, ranging from dark red to white. Since the relative proportions of the phenotypic classes correspond to the coefficients in the binomial expansion of $(a + b)^6$, it is likely that three pairs of alleles controlled kernel color in this particular cross.

In general, the relationship between the number of independently segregating allelic pairs in a polygene series and the number of quantitative-trait phenotypes in the F_2 is as shown in Table 22.5. When one pair of alleles controls a trait, then $\frac{1}{4}$ of the F_2 should show one phenotypic extreme (e.g., the darkest red) and $\frac{1}{4}$ should show the other phenotypic extreme (e.g., white). The general formula $(\frac{1}{2})^{2n}$ describes the predicted probability for an extreme phenotype when n is the number of pairs of segregating alleles involved. For two genes, and therefore two pairs of alleles, the fraction of the F_2 population with an extreme character is $(\frac{1}{2})^4 = \frac{1}{16}$. For three genes (three pairs of alleles) the fraction is $\frac{1}{64}$, as in the wheat kernel example. As the number of genes increases, the fraction of the F_2 with an extreme phenotype decreases, and it does so very rapidly. For five genes the fraction is $1/1024$, and for ten genes it is $1/1,048,576$.

The number of genotypic classes also increases rapidly as the number of pairs of alleles increases. The beginnings of this increase are shown in Table 22.5. Three pairs of alleles yield 27 possible genotypic classes in the F_2; five pairs of alleles yield 243 genotypes; and ten pairs of alleles yield 59,049 genotypes. In general, the number of genotypic classes in the F_2 is $(3)^n$, where n is the number of pairs of alleles involved.

As the number of pairs of alleles in a multiple-gene series increases, the F_2 population rapidly assumes a continuum of phenotypic variation for which distinctions between classes are impossible. In short, when the number of pairs of alleles is five or more, it is difficult for us to be sure of the number of pairs of alleles involved in controlling a quantitative effect. Other methods are available for determining the number of pairs of alleles; they have as their basis more sophisticated mathematical treatments. Even with these methods, however, we still encounter significant complications caused by the environment, different dominance relationships of the genes in the series, and linkage.

HERITABILITY

Heritability is the proportion of a population's phenotypic variation that is attributable to genetic factors. The concept of heritability is used to examine the relative contributions of genes and environment to variation in a specific trait. As we have seen, continuous traits are frequently influenced by multiple genes and by environmental factors. One of the important questions addressed in quantitative genetics is: How much of the variation in phenotype results from genetic differences among individuals and how much is a product of environmental variation? For example, multifactorial traits such as weight of cattle, number of eggs laid by chickens, and the amount of fleece produced by sheep are important for breeding programs and agricultural management. Many ecologically important traits, such as variation in body size, fecundity, and developmental rate are also multifactorial, and the genetic

~ **TABLE 22.5**

Probability Information for a Quantitative Character Controlled by a Polygenic Series in Which Independently Segregating Allelic Pairs Have Duplicate, Cumulative Effects

NUMBER OF ALLELIC PAIRS	NUMBER OF SEGREGATING ALLELES	FRACTION OF POPULATION SHOWING AN EXTREME EXPRESSION OF THE CHARACTER	NUMBER OF GENOTYPES IN F_2	NUMBER OF PHENOTYPES IN F_2	F_2 PHENOTYPIC RATIOS ARE THE COEFFICIENTS IN THE BINOMIAL EXPANSION OF
1	2	$(\frac{1}{2})^2 = \frac{1}{4}$	$(3)^1 = 3$	3	$(a + b)^2$
2	4	$(\frac{1}{2})^4 = \frac{1}{16}$	$(3)^2 = 9$	5	$(a + b)^4$
3	6	$(\frac{1}{2})^6 = \frac{1}{64}$	$(3)^3 = 27$	7	$(a + b)^6$
4	8	$(\frac{1}{2})^8 = \frac{1}{256}$	$(3)^4 = 81$	9	$(a + b)^8$
n	$2n$	$(\frac{1}{2})^{2n}$	$(3)^n$	$2n + 1$	$(a + b)^{2n}$

contribution to this variation is important for understanding how natural populations evolve. The extent to which genetic and environmental factors contribute to human variation in traits like blood pressure and birth weight is important for health care. Also, the extent to which genes affect human behaviors, such as alcoholism and criminality, may be useful in establishing social policy. Hence, the study of heritability is crucial in a variety of scientific disciplines.

The following section deals with two types of heritability: broad-sense heritability and narrow-sense heritability. To assess heritability, we must first measure the variation in the trait and then we must partition that variance into components attributable to different causes.

Components of the Phenotypic Variance

The **phenotypic variance**, represented by V_P, is a measure of the variability of a trait. It is calculated by computing the variance of the trait for a group of individuals, as was outlined in this chapter's section on statistics. Differences among individuals arise from several factors; therefore, we can partition the phenotypic variance into several components attributable to different sources. First, some of the phenotypic variation arises because of genetic differences among individuals (different genotypes within the group). This contribution to the phenotypic variation is called **genetic variance** and is represented by the symbol V_G. As noted, additional variation often results from environmental differences among the individuals; in other words, different environments experienced by individuals may contribute to the differences in their phenotypes. The **environmental variance** is symbolized by V_E, and by definition, it includes any nongenetic source of variation. Temperature, nutrition, and parental care are examples of obvious environmental factors that may cause differences among individuals. Environmental variance also includes random factors that occur during development, factors that are sometimes referred to as "developmental noise." A third source of phenotypic variance is genetic-environmental interaction, represented by V_{GE}. Genetic-environmental interaction occurs when the relative effects of the genotypes differ among environments. For example, in a cold environment, genotype AA of a plant may be 40 cm in height and genotype Aa may be 35 cm in height. However, when the genotypes are moved to a warm climate, genotype Aa is now 60 cm, as compared with genotype AA, which is only 50 cm in height. This relationship is shown in Figure 22.13. In this example, both genotypes grow taller in the warm environment, but the relative performance of the genotypes switches in the two environments. Therefore, both environmental differences

(temperature) and genetic differences (genotypes) contribute to the phenotypic variance. However, the effects of genotype and environment cannot simply be added together. An additional component that accounts for how genotype and environment interact must be considered, and this is V_{GE}.

The phenotype variance, composed of differences arising from genetic variation, environmental variation, and genetic-environmental variation, can be represented by the following equation:

$$V_P = V_G + V_E + V_{GE}$$

The relative contributions of these three factors to the phenotypic variance depend upon the genetic composition of the population, the specific environment, and the manner in which the genes interact with the environment.

The genetic variance V_G can be further subdivided into components arising from different types of interactions between genes. Some of the genetic variance occurs as a result of the average effects of the different alleles on the phenotype. For example, an allele g may,

~ FIGURE 22.13

Hypothetical example of genetic-environmental interaction. At 15°C, the genotype AA is taller with a height of 40 cm. Genotype Aa is only 35 cm tall. However, at 30°C, Aa is taller (60 cm) while AA is 55 cm tall. Both genotype and environment influence plant height, but the effects of the genotypes differ between environments.

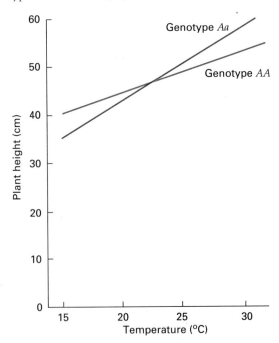

on the average, contribute 2 centimeters in height to a plant, and the allele G may contribute 4 centimeters. In this case, the *gg* homozygote would contribute $2 + 2 = 4$ centimeters in height, the *Gg* heterozygote would contribute $2 + 4 = 6$ centimeters in height, and the *GG* homozygote would contribute $4 + 4 = 8$ centimeters in height. To determine the genetic contribution to height, we would then add the effects of alleles at this locus to the effects of alleles at other loci which might influence the phenotype. Genes such as these are said to have additive effects, and this type of genetic variation is called **additive genetic variance.** You may recall that the genes studied by Nilsson-Ehle, which determine kernel color in wheat, are strictly additive in this way. Some alleles contribute to the pigment of the kernel and others do not; the added effects of all the individual contributing alleles determine the phenotype of the kernel. Thus, the genotypes *AA bb*, *aa BB*, and *Aa Bb* all produce the same phenotype, since each genotype has two contributing alleles. The phenotypic variance arising from the additive effects of genes is the additive genetic variance and is symbolized by V_A.

Some genes may exhibit dominance, and this comprises another source of genetic variance, the **dominance variance** (V_D). Dominance occurs when one allele masks the expression of the other allele at the locus. If dominance is present, the individual effects of the alleles are not strictly additive; we must also consider how alleles at a locus interact. In the presence of dominance, the heterozygote *Gg* would contribute 8 centimeters in height to the phenotype, the same amount as the *GG* homozygote. Finally, epistatic interactions may occur among alleles. Recall that in epistasis, alleles at different loci interact to determine the phenotype. The presence of epistasis adds another source of genetic variation, called epistatic or **interaction variance** (V_I). So we can partition the genetic variance as follows:

$$V_G = V_A + V_D + V_I$$

The total phenotypic variance can then be summarized as

$$V_P = V_A + V_D + V_I + V_E + V_{GE}$$

Partitioning the phenotypic variance into these components is useful for thinking about the contribution of different factors to the variation in phenotype. When determining the narrow-sense heritability of a trait (which is discussed in the following section), for example, we are primarily concerned with the relative contribution of the additive genetic component V_A. However, precisely estimating all, or even some, of these components is frequently a formidable task.

Broad-Sense and Narrow-Sense Heritability

Geneticists frequently partition the phenotypic variance of a trait to determine the extent to which variation among individuals results from genetic differences. Thus, they are interested in how much of the phenotypic variance V_P can be attributed to genetic variance V_G. This quantity, the proportion of the phenotypic variance that consists of genetic variance, is called the **broad-sense heritability** and is expressed as follows:

$$\text{Broad-sense heritability} = H^2 = \frac{V_G}{V_P}$$

The heritability of a trait can range from 0 to 1. A broad-sense heritability of 0 indicates that none of the variation in phenotype among individuals results from genetic differences. A heritability of 0.5 means that 50 percent of the phenotypic variation arises from genetic differences among individuals, and a heritability of 1 would suggest that all the phenotypic variance is genetically based.

Broad-sense heritability includes genetic variation from all types of genes. Frequently, we are more interested in the proportion of the phenotypic variation that results only from additive genetic effects. This is because the additive genes are those that allow us to predict the phenotype of the offspring from the phenotypes of the parents. To understand the reason for this, consider a cross involving a trait that results from the effect of alleles at a single locus. One parent is $A^1 A^1$ and 10 cm tall; the other parent is $A^2 A^2$ and is 20 cm tall. All the offspring from this cross (the F_1) will be $A^1 A^2$. If the alleles are additive and contribute equally to height, the offspring should be 15 cm tall, exactly intermediate between the parents. However, if A^2 is dominant, all the offspring will be 20 cm, resembling only one of the parents. Thus, if dominant genes are important, the offspring will not be intermediate to the parental phenotypes. In a similar fashion, epistatic genes will not always contribute to the resemblance between parents and offspring.

Because the additive genetic variance allows one to make accurate predictions about the resemblance between offspring and parents, quantitative geneticists frequently determine the proportion of the phenotypic variance that results from additive genetic variance, a quantity referred to as the **narrow-sense heritability.** The additive genetic variance is also that variation that responds to selection in a predictable way, and thus the narrow-sense heritability provides information about how a trait will evolve. The narrow-sense heritability is represented mathematically as

$$\text{Narrow-sense heritability} = h^2 = \frac{V_A}{V_P}$$

Understanding Heritability

Despite their utility, heritability estimates have a number of significant limitations. Unfortunately, these limitations are often ignored. As a result, heritability is one of the most misunderstood and widely abused concepts in all of genetics. Before we discuss how heritability is determined, it is important that we list some of the important qualifications and limitations of heritability.

1. Heritability does not indicate the extent to which a trait is genetic. What heritability does measure is the *proportion of the phenotypic variance* among individuals in a population that results from genetic differences. The "proportion of the phenotypic variance among individuals in a population that results from genetic differences" may seem like the same thing as "the extent to which a trait is genetic," but these two statements actually mean quite different things. Genes often influence the development of a trait, and thus the trait may be said to be genetic. However, the differences in phenotype among individuals may not be genetic at all. Let us consider an example. Our ability to learn about the game of football is dependent upon the proper development of our nervous system, which is clearly controlled by genes. Thus we could say that knowledge of the game of football is a genetically influenced trait. However, the differences we see among individuals in their knowledge of football is usually not due to genes but to different environments and personal interests. Since heritability measures the proportion of the phenotypic variation among individuals that results from genetic differences, heritability for knowledge of football would be zero, despite the fact that genes do influence our knowledge of football.

   Consider another situation. Suppose that genes are critical in determining the phenotypes of a trait. If all individuals in a population have identical genes at the loci that control the trait, then the genetic variance is zero ($V_G = 0$). Because heritability equals the proportion of the phenotypic variance that results from genetic variance ($H^2 = V_G/V_P$), and V_G is zero, the heritability must be zero. Although the heritability in this case is zero, it would be incorrect to assume that genes play no role in the development of the trait. Similarly, a high heritability does not negate the importance of environmental factors influencing a trait; a high heritability might simply mean that the environmental factors that influence the trait are relatively uniform among the individuals studied.

2. Heritability does not indicate what proportion of an individual's phenotype is genetic. Since it is based on the variance, which can only be calculated on a group of individuals, heritability is characteristic of a population. An individual does not have heritability—a population does.

3. Heritability is not fixed for a trait. Thus, there is no universal heritability for a trait like human stature. Rather, the heritability value for a trait depends upon the genetic makeup and the specific environment of the population. We could calculate the heritability of stature among Hopi Indians living in Arizona, for example, but heritability calculated for other populations and other environments might be very different.

   To illustrate this point, suppose that we calculated heritability for adult height on individuals living in a small New England town and obtained a value of 0.7. This would indicate that 70 percent of the variation in adult height among these individuals is due to genetic variation. Heritability for other populations might not be the same. Residents of San Francisco, for example, might be more ethnically heterogeneous than the inhabitants of the small New England town; therefore the San Francisco population would have more genetic variation for height. Let us assume that the environmental variance of the two populations was about the same. If the environmental variances of the two populations were similar, but the genetic variance was greater in San Francisco, then heritability calculated for the height of San Francisco residents also would be greater.

   Genes are not the only factor that influences height in humans; diet, an environmental effect, is also a major determinate of height. Since most individuals in our small New England town probably receive an adequate diet, at least in terms of calories, the environmental variance for height would not be large. In a developing nation, however, some individuals might receive adequate nutrition, while the diet of others might be severely deficient. Because greater differences in diet exist, in such a nation the environmental variance for height would be larger, and as a result, the heritability of height would be less. Thus heritability calculated for human height might differ substantially for residents of the small New England town, residents of San Francisco, and residents of a developing country.

   These examples illustrate that heritability can be applied only to a specific group of individuals in a specific environment. If the genetic composition of the group is different, or the environment is different, the original heritability value is no longer valid. Changing groups or environments does not alter the way in which genes affect the trait, but it may change the amount of genetic and environmental variance for the trait, which would then alter the heritability.

4. Even if heritability is high in each of two populations, and the populations differ markedly in a particular trait, one cannot assume that the populations are genetically different. For example, suppose that you obtain some genetically variable mice and divide them into two groups. You feed one group a nutritionally rich diet, and you are careful to provide each mouse with exactly the same amount of food, space, water, and other environmental necessities. The mice grow to a large size because of the rich diet. When you measure heritability for adult body weight, you obtain a high value of 0.93. The high heritability is not surprising since the mice were genetically variable and environmental differences were kept at a minimum. The second group of mice comes from the same genetic stock, but you feed them an impoverished diet, lacking in calories and essential nutrients; again, each mouse gets exactly the same amount of food, space, water, and necessities. Because of the poor diet, the mice of this second group are all smaller than those in the first group. When you calculate heritability for adult weight in the small mice, you again obtain a high value of 0.93, because the mice were genetically variable and the environmental differences were kept to a minimum. Because the heritability of body weight is high in both groups and the mice of the two groups differ in adult weight, some people might suggest that the mice are genetically different in body size. Yet, any claim that the mice of the two groups differ genetically is clearly wrong—both groups came from the same stock. The important point is that heritability cannot be used to draw conclusions about the nature of differences between populations.

5. Traits shared by members of the same family do not necessarily have high heritability. When members of the same family share a trait, the trait is said to be **familial.** Familial traits may arise because family members share genes or because they are exposed to the same environmental factors. Thus, familiality is not the same as heritability.

$\mathcal{K}$EYNOTE

Broad-sense heritability of a trait represents the proportion of the phenotypic variance that results from genetic differences among individuals. The narrow-sense heritability is more restricted—it measures the proportion of the phenotypic variance that results from additive genetic variance. Narrow-sense heritability allows quantitative geneticists to make predictions about the resemblance between parents and offspring, and it represents that part of the phenotypic variance that responds to selection in a predictable manner. _____

How Heritability Is Calculated

A number of different methods are available for calculating heritability. Many of the methods involve comparing related and unrelated individuals, or comparing individuals with different degrees of relatedness. If genes are important in determining the phenotypic variance, then closely related individuals should be more similar in phenotype, since they have more genes in common. Alternatively, if environmental factors are responsible for determining differences in the trait, then related individuals should be no more similar in phenotype than unrelated individuals. An important point to remember is that *the related individuals studied must not share a more common environment.* We assume that if related individuals are more similar in phenotype, it is because they share similar genes. If related individuals also share a more common environment than unrelated individuals, separating the effects of genes and environment is much more difficult and frequently impossible. The absence of common environmental factors among related individuals can often be achieved in domestic plants and animals, and common environments may not exist among family members in the wild; however, this is very difficult to obtain in humans, where family structure and extended parental care create common environments for many related individuals.

HERITABILITY FROM PARENT-OFFSPRING REGRESSION. Some of the methods used to calculate heritability include comparison of parents and offspring, comparison of full and half sibs, comparison of identical and nonidentical twins, and response-to-selection experiments. Here, we will discuss the parent-offspring method; later we will examine the calculation of heritability through response-to-selection.

If additive genes are important in determining the differences among individuals, then we expect that offspring should resemble their parents. To quantify the degree to which genes influence a trait, we must first measure the phenotypes of parents and offspring in a series of families, and then statistically analyze the relationship between their phenotypes. Correlation and regression are appropriate for this type of problem.

We can represent the relationship between offspring phenotype and parental phenotype by plotting the mean phenotype of the parents against the mean phenotype of the offspring, as shown in Figure 22.14. Each point on the graph represents one family. If the points are randomly scattered across the plot, as in Figure 22.14c, then no relationship exists between the traits of parents and offspring. We would conclude that additive genetic differences are not important in determining the phenotypic variability and that the heritability is low. On the other hand, if a definite relationship exists between the phenotypes of parents and offspring, as shown in Figure 22.14a and 22.14b, then additive genetic variance is more important and heritability is high (assuming that no common environmental effects between parents and offspring influence the trait).

In a plot of the parent and offspring phenotypes, the slope of the regression line can provide us with information about the magnitude of the heritability (see section of this chapter on statistics). If the slope is 0, as shown in Figure 22.14c, then the narrow-sense heritability h^2 is zero. When the slope of the parent-offspring regression is 1, as in Figure 22.14a, the mean offspring phenotype is exactly intermediate to the phenotype of the two parents, and genes with additive effects determine all the phenotypic differences. If the slope is less than 1 but greater than zero, as in Figure 22.14b, both additive genes and nonadditive factors (genes with dominance, genes with epistasis, and environmental factors) affect the phenotypic variation. It is possible to show mathematically that when the mean phenotype of the offspring is regressed against the mean phenotype

of the parents, the narrow-sense heritability (h^2) equals the slope of the regression line (b):

$h^2 = b$ (for the regression of mean offspring phenotype and mean parental phenotype)

When the mean phenotype of the offspring is regressed against the phenotype of only one parent, the narrow-sense heritability is twice the slope:

$h^2 = 2b$ (for the regression of mean offspring phenotype and one parental phenotype)

Heritability values for a number of traits in different species are given in Table 22.6. These heritability values are based upon various populations and have been determined using a variety of methods, including the parent-offspring method. Estimates of heritability are rarely precise, and most measured heritability values have large errors. This lack of precision is reflected in the fact that heritabilities calculated for the same trait in the same organism often vary widely. Heritability values calculated for human traits must be viewed with special caution, given the difficulties of separating genetic and environmental influences in humans.

Identification of Genes Influencing a Quantitative Trait

One of the limitations of the traditional approach of quantitative genetics is that it tells us little about the genes involved in quantitative traits. While heritability may indicate how much of the phenotypic variance in a trait is due to genetic differences, it tells us nothing

---

~ **Figure 22.14**

Three hypothetical regressions of mean parental wing length on mean offspring wing length in *Drosophila*. In each case, the slope of the regression line (b) equals the narrow-sense heritability (h^2).

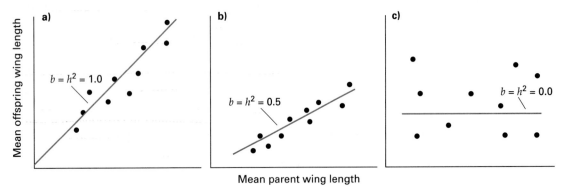

~ TABLE 22.6

Heritability Values for Some Traits in Humans, Domesticated Animals, and Natural Populations

ORGANISM	TRAIT	HERITABILITY	REF*
Humans	Stature	0.65	1
	Serum immunoglobulin (IgG) level	0.45	1
Cattle	Milk yield	0.35	1
	Butterfat content	0.40	1
	Body weight	0.65	1
Pigs	Back-fat thickness	0.70	1
	Litter size	0.05	1
Poultry	Egg weight	0.50	1
	Egg production (to 72 weeks)	0.10	1
	Body weight (at 32 weeks)	0.55	1
Mice	Body weight	0.35	1
Drosophila	Abdominal bristle number	0.50	1
Jewelweed	Germination time	0.29	2
Milkweed bugs	Wing length (females)	0.87	3
	Fecundity (females)	0.50	3
Spring peeper (frog)	Size at metamorphosis	0.69	4
Wood frog	Development rate (mountain population)	0.31	5
	Size at metamorphosis (mountain population)	0.62	5

Note: The estimates given in this table apply to particular populations in particular environments; heritability values for other individuals may differ.

*References: (1) Falconer, D. S. 1981. Introduction to Quantitative Genetics, 2d ed. Longman, New York; (2) Mitchell-Olds, T. 1986. Evolution 40:107–116; (3) Palmer, J. O., and H. Dingle. 1986. Evolution 40:767–777; (4) Travis, J., et al. 1987. Evolution 41:145–156; (5) Berven, K. A. 1987. Evolution 41:1088–1097.

about how many genes are involved, where they are located, or what proteins the genes produce. Identification of individual genes that affect quantitative traits like yield in corn would allow more efficient design of breeding programs to enhance these traits; finding genes in humans that affect quantitative traits such as high blood pressure would allow identification of those at risk for the disease and early application of preventive treatments.

Using genetic crosses and rather complex statistical procedures, it is possible to identify and map genes that affect quantitative traits, just as genes for Mendelian traits are mapped by linkage analysis (see Chapter 5). The approach for quantitative traits makes use of genetic markers. Genetic markers are genes with an observable phenotypic effect and whose chromosome location is known. In specially designed genetic crosses, or in pedigrees with humans, one looks for associations between a set of genetic markers and the quantitative trait of interest. Significant associations indicate that one or more genes affecting the quantitative trait (called *quantitative trait loci* or *QTLs*) are located close to the genetic marker on the same chromosome; in other words, the marker and the QTL are linked. Using this approach, Thoday and his colleagues mapped genes that affected bristle number in *Drosophila*, a trait that had been shown to exhibit quantitative inheritance. They identified five genes that accounted for almost 90 percent of the phenotypic variation in bristle number.

Similar studies have been carried out on a few other organisms, but past attempts to identify QTLs have been limited by the paucity of known genetic markers. To identify QTLs, a number of genetic markers of known chromosome location are required. Ideally, the set of markers should span the lengths of all the chromosomes. Furthermore, the markers must be variable; in other words, each marker locus must possess two or more common alleles. The markers must also code for easily observable phenotypes.

The recent development of methods to identify restriction fragment length polymorphisms (RFLPs; see Chapter 15) provides the technology to identify large numbers of genetic markers that can be used for mapping genes that affect quantitative traits. For example, hundreds of new genetic markers in corn and tomatoes have been uncovered with RFLP methods. These markers now provide the means to map QTLs for a number of agriculturally important traits in these plants. Work on mapping human genes has also provided hundreds of RFLPs with known chromosomal locations; these can be used for mapping QTLs that affect important quantitative traits in humans. Undoubtedly, these new developments will greatly enhance our understanding of how numerous loci interact to determine quantitative traits.

RESPONSE TO SELECTION

Two fields of study in which quantitative genetics has played a particularly important role are plant and animal breeding and evolutionary biology. Both fields are concerned with genetic change within groups of organisms: in the case of plant and animal breeding, genetic change can lead to improvement in yield, hardiness, size, and other agriculturally important qualities; in the case of evolutionary biology, genetic change occurs in natural populations as a result of the forces of nature. **Evolution** can be defined as genetic change that takes place over time within a group of organisms. Therefore, both evolutionary biologists and plant and animal breeders are interested in the process of evolution, and both use the methods of quantitative genetics to predict the rate and magnitude of genetic change that will occur.

Evolution results from a variety of forces, many of which will be explored more thoroughly in the next chapter. Here we concentrate on genetic change that results from selection. **Natural selection** is undoubtedly one of the most powerful forces of evolution in natural populations. First fully described by Charles Darwin, the concept of natural selection can be summarized in four statements:

1. More individuals are produced every generation than can survive and reproduce.
2. Much phenotypic variation exists among these individuals, and some of this variation is heritable.
3. Individuals with certain traits survive and reproduce better than others; these individuals leave more offspring in the next generation.
4. Since individuals with certain traits leave more offspring, and since those traits are heritable, traits that allowed increased survival and reproduction will be more common in the next generation.

The essential element of natural selection is that individuals with certain genotypes leave more offspring than others. In this way, groups of individuals change or evolve over time and become better adapted to their particular environment.

Humans bring about evolution in domestic plants and animals through the similar process of **artificial selection.** In artificial selection, humans, not nature, select the individuals that are to survive and reproduce. If the selected traits have a genetic basis, they too will change over time and evolve, just as traits in natural populations evolve as a result of natural selection. Artificial selection can be a powerful force in bringing about rapid evolutionary change, as evidenced by the extensive variation observed among domesticated plants and animals. For example, all breeds of domestic dogs are derived from one species that was domesticated some 10,000 years ago. The large number of breeds that exist today, encompassing a tremendous variety of sizes, shapes, colors, and even behaviors, has been produced by artificial selection and selective breeding (Figure 22.15 and also Figure 2.1).

Both the process of natural selection, as described by Charles Darwin, and artificial selection practiced by plant and animal breeders depend upon the presence of genetic variation. Only if genetic variation is present within a population of individuals can that population change genetically and evolve. Furthermore, the amount and the type of genetic variation present is extremely important in determining how fast evolution will occur. Both evolutionary biologists and breeders are interested in the question of how much genetic variation for a particular trait exits within a population. As we have seen, quantitative genetics is often employed to answer this question.

Estimating the Response to Selection

When natural or artificial selection is imposed upon a phenotype, the phenotype will change from one generation to the next, provided that genetic variation underlying the trait is present in the population. The amount

~ **FIGURE 22.15**

All breeds of dogs have been produced by artificial selection, a demonstration of the tremendous power of this method for bringing about change.

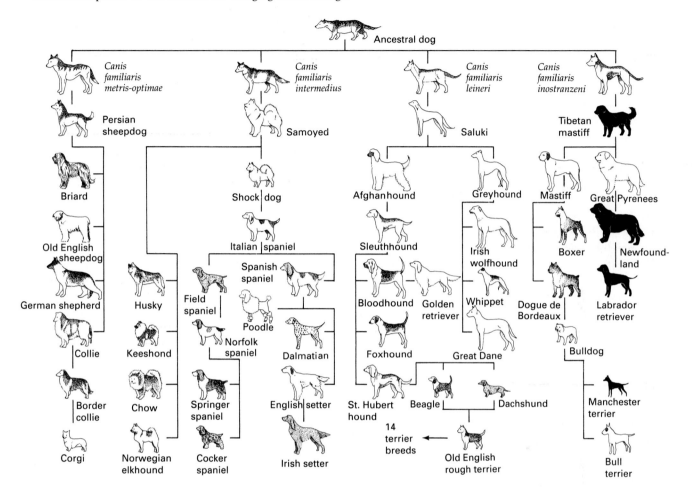

that the phenotype changes in one generation is termed the **selection response.** To illustrate the concept of selection response, suppose a geneticist wishes to produce a strain of *Drosophila melanogaster* with large body size. In order to increase body size in fruit flies, the geneticist would first examine files from a genetically diverse culture and would measure the body size of these *unselected flies.* Suppose that our geneticist found the mean body weight of the unselected flies to be 1.3 mg. After determining the mean body weight of this population, the geneticist would select flies that were endowed with large bodies (assume that the mean body weight of these selected flies was 3.0 mg). He would then place the large, selected flies in a separate culture vial and allow them to interbreed. After the F_1 offspring of these selected parents emerged, the geneticist would measure the body weights of the F_1 flies and compare them with the body weights of the original, unselected population.

What our geneticist has done in this procedure is to apply selection for large body size to the population of fruit flies. If genetic variation underlies the variation in body size of the original population, the offspring of the selected flies will resemble their parents and the mean body size of the F_1 generation will be greater than the mean body size of the original population. If the F_1 flies have a mean body weight of 2.0 mg, which is considerably larger than the mean body weight of 1.3 mg observed in the original, unselected population, a response to selection has occurred.

The amount of change that occurs in one generation, or the selection response, is dependent upon two things: the narrow-sense heritability and the selection differential. The selection differential is defined as the difference between the mean phenotype of the selected parents and the mean phenotype of the unselected population. In our example of body size in fruit flies, the original population had a mean weight of 1.3 mg, and the mean weight of the selected parents was 3.0 mg, so the selection differential is 3.0 mg − 1.3 mg = 1.7 mg. The selection response is related to the selection differential and the heritability by the following formula:

Selection response = narrow-sense heritability × selection differential

When the geneticist applied artificial selection to body size in fruit flies, the difference in the mean body weight of the F_1 flies and the original population was 2.0 mg − 1.3 mg = 0.7 mg, which is the response to selection. We now have values for two of the three parameters in the above equation: the selection response (0.7 mg) and the selection differential (1.7 mg). By rearranging the formula for the selection response, we can solve for the narrow-sense heritability.

Narrow-sense heritability = h^2 = selection response/selection differential

$$h^2 = 0.7 \text{ mg}/1.7 \text{ mg} = 0.41$$

Measuring the response to selection provides another means for determining the narrow-sense heritability, and heritabilities for many traits are determined in this way.

A trait will continue to respond to selection, generation after generation, as long as genetic variation for the trait remains within the population. The results from an actual, long-term selection experiment on phototaxis in *Drosophila pseudoobscura* are presented in Figure 22.16. Phototaxis is a behavioral response to light. In this study, flies were scored for the number of times the fly moved toward light in a total of 15 light-dark choices. Two different response-to-selection experiments were carried out. In one, attraction to light was selected, and in the other, avoidance of light was selected. As can be seen in Figure 22.16, the fruit flies responded to selection for positive and negative phototactic behavior for a number of generations. Eventually, however, the response to selection tapered off, and finally no further directional change in phototactic behavior occurred. One possible reason for this lack of response in later generations is that no more genetic variation for phototactic behavior exists within the population. In other words, all flies at this point are homozygous for all the alleles affecting the behavior. If this were the case, phototactic behavior could not undergo further evolution in this population unless input of additional genetic variation occurred. More often, some variation still exists for the trait, even after the selection response levels off, but the population fails

~ FIGURE 22.16

Selection for phototaxis in *Drosophila pseudoobscura*. The upper graph is the line selected for avoidance of light. The lower graph is the line selected for attraction to light. The phototactic score is the number of times the fly moved toward the light out of a total of 15 light–dark choices. (After Dobzhansky and Spassky, 1969 Proc. Natl. Acad. Sci. *62: 75–80.*)

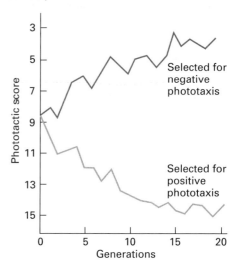

to respond to selection because the genes for the selected trait have detrimental effects on other traits. These detrimental effects occur because of genetic correlations, which are discussed in the next section.

₭EYNOTE

The amount that a trait changes in one generation as a result of selection for the trait is called the selection response or the response to selection. The magnitude of selection response depends on both the intensity of selection, what is termed the selection differential, and the narrow-sense heritability.

Genetic Correlations

The phenotypes of two or more traits may be associated or correlated; this means that the traits do not vary independently. For example, fair skin, blond hair, and blue eyes are often found together in the same individual. The association is not perfect—we sometimes see individuals with dark hair, fair skin, and blue eyes—but the traits are found together with enough regularity for us to say that they are correlated. The **phenotypic** correlation between two traits can be computed by measuring two phenotypes on a number of individuals and then calculating a correlation coefficient for the two traits. (See the section on statistics for a discussion of correlation coefficients.) One reason for a phenotypic correlation among traits is that the traits are influenced by a common set of genes. Indeed, this is the most likely reason for the association among hair color, eye color, and skin color in humans. Rarely do genes affect only a single trait. More commonly, each gene influences a number of traits, and this is particularly true for the polygenes that influence continuous traits. When genes affect multiple phenotypes, we say they are *pleiotropic*, a concept we introduced in Chapter 4. If a gene is pleiotropic, it simultaneously affects two or more traits, and thus the phenotypes of these two traits will be correlated. For example, the genes that affect growth rates in humans influence both weight and height, so these two phenotypes tend to be correlated. When pleiotropy is present and some of the genes influencing two traits are the same, we say that a **genetic correlation** exists for the traits.

Pleiotropy is not the only cause of phenotypic correlations among traits. Environmental factors may also influence several traits simultaneously and may cause nonrandom associations between phenotypes. For example, adding fertilizer to the soil often causes plants both to grow taller and to produce more flowers. If we measured plant height and counted the number of flowers on a group of responsive plants, some of which received fertilizer and some of which did not, we would find that the two traits are correlated; those plants receiving fertilizer would be tall and would have many flowers, while those without fertilizer would be short and have few flowers. However, this phenotypic correlation does not result from any genetic correlation (pleiotropy), but from the common effect of the environmental factor, the fertilizer, on both traits.

Genetic correlations may be positive or negative. A positive correlation means that genes causing an increase in one trait bring about a simultaneous increase in the other trait. In chickens, body weight and egg weight have a positive genetic correlation. If breeders select for heavier chickens, thereby favoring the genes for large body size, the size of the chickens will increase and the mean weight of the eggs produced by these chickens will also increase. This increase in egg weight occurs because the genes that produce heavier chickens have a similar effect on egg weight.

Other traits exhibit negative genetic correlations. In this case, genes that cause an increase in one trait tend to produce a corresponding decrease in the other trait. Egg weight and egg number in chickens have a negative genetic correlation. When breeders select for chickens that produce larger eggs, the average egg size increases, but the number of eggs laid by each chicken decreases.

Genetic correlations are important for predicting the changes in phenotype that result from selection. If two traits are genetically correlated, they evolve together. Any genetic change in one trait occurring as a result of selection will simultaneously alter the other trait, since both are influenced by the same genes. Such correlation often presents problems for plant and animal breeders. As an example, milk yield and butterfat content have a negative genetic correlation in cattle. The same genes that cause an increase in milk production bring about a decrease in butterfat content of the milk. Thus, when breeders select for increased milk yield, the amount of milk produced by the cows may go up, but the butterfat content will decrease at the same time. Negative correlations among traits often place practical constraints on the ability of breeders to select for desirable traits. Knowing about the presence of genetic correlations before undertaking an expensive breeding program is essential in order to avoid the production of associated undesirable traits in the selected stock.

An organism's ability to adapt to a particular environment is also strongly influenced by genetic correlations among traits, and thus genetic correlations are of considerable interest to evolutionary biologists. As an illustration, consider two traits in tadpoles: developmental rate and size at metamorphosis. Most tadpoles are found in small ponds and pools, where fish (potential predators) are absent and food is abundant. A major liability in using this type of aquatic habitat is that ponds often dry up, frequently before the tadpoles have developed sufficiently to metamorphose into frogs and leave the water. One might expect, then, that natural selection would favor a maximum rate of development in tadpoles, so that the tadpoles could quickly metamorphose into frogs. However, many species of tadpoles fail to develop at maximum rates, contrary to what we might expect to evolve under natural selection. One reason for the slow rate of development is that a negative genetic correlation may exist between developmental rate and body size at metamorphosis. Genes that accelerate development also tend to cause metamorphosis at a smaller size, at least in some populations. Thus, selection for fast metamorphosis will also produce smaller frogs, and size is extremely important in determining the survival of young frogs. Small frogs tend to lose water more rapidly in the terrestrial environment, are more likely to be eaten by predators, and have more difficulty finding sufficient food. The negative genetic correlation between developmental rate and body size at metamorphosis places constraints on the frogs' ability to evolve rapid development and on their ability to evolve large body size at metamorphosis.

~ TABLE 22.7

Genetic Correlations among Traits in Humans, Domesticated Animals, and Natural Populations

ORGANISM	TRAITS	GENETIC CORRELATION	REF*
Humans	IgG, IgM	0.07	1
Cattle	Butterfat content, milk yield	−0.38	1
Pigs	Weight gain, back-fat thickness	0.13	1
	Weight gain, efficiency	0.69	1
Chickens	Egg weight, egg production	−0.31	1
	Body weight, egg weight	0.42	1
	Body weight, egg production	−0.17	1
Mice	Body weight, tail length	0.29	1
Jewelweed	Seed weight, germination time	−0.81	2
Milkweed bugs	Wing length, fecundity	−0.57	3
Wood frogs	Developmental rate, size at metamorphosis	−0.86	5
Drosophila	Early life fecundity, resistance to starvation	−0.91	6

Note: The estimates given in this table apply to particular populations in particular environments; genetic correlations for other individuals may differ.

*References: (1) Falconer, D. S. 1981. Introduction to Quantitative Genetics, 2d ed. Longman, New York; (2) Mitchell-Olds, T. 1986. Evolution 40:107–116; (3) Palmer, J. O., and H. Dingle. 1986. Evolution 40:767–777; (4) Travis, J., et al. 1987. Evolution 41:145–156; (5) Berven, K. A. 1987. Evolution 41:1088–1097; (6) Service, P. M., and M. R. Rose. 1985. Evolution 39:943–944.

Knowing about such genetic correlations is important for understanding how animals adapt or fail to adapt to a particular environment. Table 22.7 presents some genetic correlations that have been detected in studies of quantitative genetics.

𝒦EYNOTE

Genetic correlations arise when two traits are influenced by the same genes. When a trait is selected, any genetically correlated traits will also exhibit a selection response. ────────

INBREEDING

Now that we have an understanding of quantitative inheritance, heritability, and selection, we can consider the consequences of **inbreeding, or mating among related individuals.** In addition, we will consider **heterosis,** the phenomenon in which hybrids made between true-breeding (inbred) lines of animals or plants often exhibit greater fitness than the parentals.

For centuries human populations have been aware of the possible serious and unfavorable effects of inbreeding. In fact, laws in many countries prevent marriages between close relatives. The logic behind such thinking is that close relatives (e.g., first cousins) have many genes in common because of common ancestry. As a result, there is a greater chance of bringing deleterious recessive mutations into the homozygous state by a mating of close relatives than by a mating of unrelated individuals.

A substantial amount of data from animal and plant breeders indicates the deleterious effects of inbreeding. However, a number of plants and some animals reproduce predominantly by self-fertilization, and many of these species are quite successful. For example, the land snail, *Rumina decollata*, was introduced into the United States from its native habitat in Europe sometime before 1822. By 1915 it had spread to Florida, Texas, Oklahoma, and Mexico, and it is common today across the southern United States. Genetic studies indicate that *Rumina decollata* is completely homozygous in North America and reproduces only through self-fertilization. This species illustrates that inbreeding is not always harmful.

In other plants and animals, however, inbreeding is deleterious, and mechanisms have evolved to minimize self-fertilization. In hermaphroditic organisms, for example, the two types of gametes often mature at different times, thus forcing cross-fertilization between individuals, or *outbreeding*. In other organisms, genetic systems cause self-incompatibility. Figure 22.17 illustrates the genetic basis of incompatibility for a number of groups of plants, such as evening primroses and tobacco. In this example a polygenic series controls compatibility in sexual reproduction. When pollen of one incompatibility type falls onto a style of the same genetic type, the pollen tube does not grow properly. Thus fertilization depends on the pollen and style having different incompatibility alleles, thereby virtually assuring that only cross-fertilization occurs. These incompatibility systems can be very complicated, with at least forty different alleles being involved in some plants. Incompatibility alleles are generally designated S_1, S_2, S_3 and so on (the S stands for sterility).

In plants where self-fertilization *is* possible, some mechanisms may still operate to minimize that process and facilitate cross-fertilization. For example, orchids have complex flower structures designed so that pollen must be carried from one plant to another by insects.

The consequences of inbreeding can be examined by undertaking an inbreeding program with an organism that does not have an incompatibility system. At the outset the population would show a range of phenotypic variation reflecting the heterogeneity of the alleles the organisms carry. After repeated self-fertilization distinct phenotypic traits would become fixed, or pure, in the lines. In plants, for example, lines would sort out that would be relatively uniform for height and flower size. Some lines would sort out to show deleterious effects, such as lethal white seedlings. Thus a number of lines produced by an inbreeding program will not be able to survive. At the same time the surviving lines would almost certainly show less vigor than the original parental line because of the deleterious effects of some of the homozygous pairs of alleles produced; this is called *inbreeding depression*.

As an example, Figure 22.18 (p. 700) shows the consequences of 30 generations of inbreeding on the grain yield of three different lines of corn, A, B, and C. All three lines derived from the same parental line. Note that in each line yield gradually declines as the number of generations of self-fertilization increases. Eventually, each line becomes fixed at a particular yield level, probably because all the alleles that control grain yield have become homozygous. Since this trait is under polygenic control, the different fixed yields reflect different patterns of homozygosity of the genes involved.

~ **FIGURE 22.17**

Multiple allelic series determining compatibility in sexual reproduction.

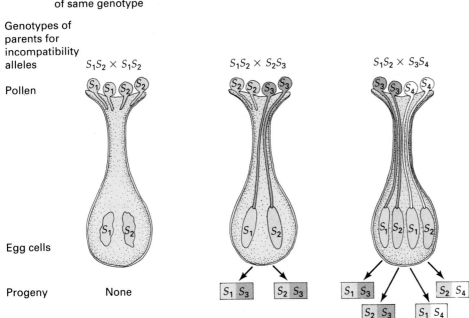

~ FIGURE 22.18

Consequences of continued inbreeding of three different lines of corn, A, B, and C.

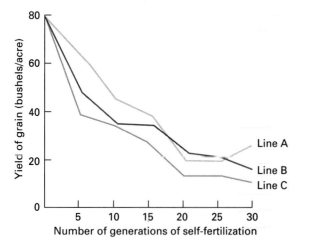

Homozygosity and Inbreeding

The homozygosity brought about by inbreeding is unquestionably responsible for many of the deleterious effects associated with inbreeding, since as inbreeding continues, more and more loci become homozygous. In this situation there is no preference for either dominant or recessive alleles becoming homozygous. However, the deleterious effects of inbreeding are mostly caused by homozygosity for certain recessive alleles. In a genetically heterogeneous individual the effects of deleterious recessive alleles are masked by wild-type alleles. In contrast, deleterious dominant alleles always exhibit their effects, whether they are heterozygous or homozygous. Thus deleterious dominant alleles (which are also rare) are seen directly in the heterogeneous population without resorting to inbreeding. If the deleterious dominant alleles affect vigor significantly, their frequencies are soon reduced in the population by natural selection. Deleterious recessive alleles concern geneticists because the frequencies of such alleles are much higher in populations than the frequencies of dominant deleterious alleles, since the recessive alleles can be maintained in a population in the heterozygous state.

To see how self-fertilization leads to a reduction of heterozygosity, let us consider a cross between a population of two heterozygous individuals, $Aa \times Aa$, the result of which is ¼ AA, ½ Aa, and ¼ aa offspring. In other words, while 100 percent of the population is heterozygous initially, after one generation only 50 percent of the population is heterozygous, and the rest are homozygous. When the progeny are self-fertilized, the AA and aa types breed true, while the Aa individuals give progeny of the same type and in the same pro-

portions as in the original cross. So after a second round of self-fertilization the proportion of heterozygotes is further reduced by a factor of 2; that is, now the proportions of the three genotypes are ⅜ AA, ¼ Aa, and ⅜ aa. Generally speaking, the relative number of heterozygotes in a population is reduced 50 percent for each self-fertilized generation.

Figure 22.19 illustrates how heterozygosity in a theoretical population is reduced by inbreeding over four generations. After the four generations the original number of Aa individuals is reduced by a factor of $(2)^4 = 16$, that is, from 1600 to 100. Note, though, that inbreeding does not favor an increase in the number of either the recessive or the dominant alleles. Instead, the proportion of homozygotes increases and leads specifically to an increase in the relative proportion of the population expressing the recessive trait. Thus, as shown in Figure 22.19, the same number of A and a alleles occurs in each generation, but their distribution among the three genotypic classes has changed. (The effects of inbreeding on genotype frequencies will be discussed in more detail in the next chapter.)

The example just presented is greatly simplified, since it considers only one gene locus. In real situations, unless the population is already inbred, a large number of loci are heterozygous. Nonetheless, the basic effects operate on all gene loci so that both continuous (quantitative) and discontinuous traits are affected.

Many studies of outcrossing organisms (organisms that normally do not engage in inbreeding) provide examples of deleterious expression of recessive traits that result from homozygosity brought about by inbreeding. In Chapter 8, for example, we discussed Tay-Sachs disease, which causes early death in infants homozygous for a particular recessive allele. Tay-Sachs disease is very rare in the general human population, but it occurs with a fairly high frequency in particular Ashkenazi Jewish populations, such as in the New York City area. This high frequency occurs because religious beliefs lead to marriages between individuals in a relatively small population, and since the population has remained discrete for many generations, there are many genes in common among the individuals. As a consequence, the probability of heterozygotes for Tay-Sachs disease is much higher in the Ashkenazi population than in the population at large, and thus marriages between heterozygotes are much more likely in the Ashkenazi population. A fourth of the children of such marriages would be expected to have Tay-Sachs disease.

Similar situations are found in relatively small human populations in which marriages between group members are more common than marriages to individuals from outside the population. The Amish population of Pennsylvania, for example, exhibits a number of

~ **FIGURE 22.19**

Theoretical cascade diagram showing the reduction in heterozygosity over four generations when only self-fertilization is permitted.

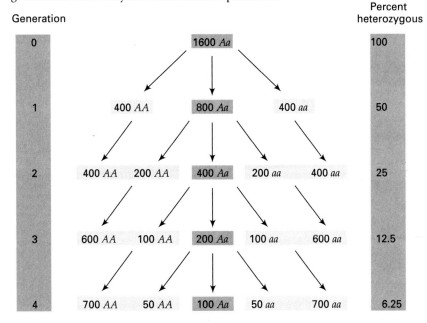

genetic abnormalities attributable to homozygosity for one or more deleterious recessive alleles. It is quite common, for instance, to find Amish people with supernumerary digits (extra fingers or toes).

Inbreeding does not always result in the expression of deleterious recessive alleles. It can bring together particular combinations of dominant alleles with deleterious or beneficial effects or can bring out recessive alleles with nondeleterious effects. The primary result of extensive inbreeding, however, is to fix phenotypes by making the genes controlling them homozygous. The fixation process is a random one (unless the genes involved are linked closely), so a number of pure-breeding lines can be generated from one starting point. Once established, the pure-breeding lines change genetically only in response to mutation. Thus the inbreeding of a heterozygous organism of genotype *Aa Bb* will produce four distinct, pure lines by producing all possible homozygous types, each with a distinct phenotype: *AA BB*, *AA bb*, *aa BB*, and *aa bb*.

The fixation consequence of inbreeding is important in animal and plant breeding (and in evolution, since it reduces the genetic variation in a population, as we will see in Chapter 23). Professional animal and plant breeders make their living by generating stable breeding lines that give rise to homogeneous and predictable offspring. Their breeding programs balance inbreeding and outbreeding to fix desirable characters in the absence of deleterious characters. In crop plants,

for example, it may be important for all plants to have the same height to facilitate machine harvesting or for all plants to bear fruit at the same time to accommodate shipping schedules. Similarly, various breeds of dogs represent the result of extensive breeding programs to accentuate some features (e.g., the flattened face of the English bulldog or the short legs of a dachshund) while minimizing other aspects of the general phenotype. If you are familiar with purebred animals of any type, you may be aware of the physiological problems that arise with fair frequency in purebreds (e.g., hip dysplasia in certain dog breeds). Many of these problems result from breeding programs that must inevitably involve some inbreeding as breeders try to coax along a desirable phenotype that has arisen.

KEYNOTE

Inbreeding involves the mating of closely related individuals of a species. At the beginning of an inbreeding program the population shows much phenotypic variation, since the individuals carry a heterogeneous array of alleles. With continued inbreeding loci become homozygous, resulting in the formation of a variety of pure lines, each with a characteristic set of phenotypes that remain constant with further inbreeding. Inbreeding leads to the fixation of either beneficial, neutral, or deleterious traits, depending on the alleles that become homozygous.

SUMMARY

Quantitative genetics is the field of genetics that studies the inheritance of continuous, or quantitative traits—those traits with a range of phenotypes. Continuous traits usually result from the influence of multiple genes and environmental factors. Statistics, such as the mean, variance, and standard deviation can be used to describe continuous traits; correlation, regression, and analysis of variance are statistical procedures that are used for studying continuous traits.

Polygenic traits are caused by genes at many loci, each of which follows the principles of Mendelian inheritance. The multiple gene hypothesis assumes that the effects of the alleles are additive, that no genetic interaction occurs between gene loci, that the loci are unlinked, and that environmental factors do not influence the trait. Frequently, not all these assumptions hold.

The broad-sense heritability is the proportion of the phenotypic variance in a population that is due to genetic differences. The narrow-sense heritability indicates the proportion of the phenotypic variance that results from additive genetic variance. The narrow-sense heritability can be calculated by regressing the mean phenotype of the offspring against the mean phenotypes of both parents; when this is done the narrow-sense heritability equals the slope of the regression line. If the phenotype of only one parent is used, the narrow-sense heritability equals twice the slope.

The response to selection is the amount a trait changes in one generation as a result of selection; it depends upon the narrow-sense heritability and the selection differential. Genetic correlation occurs when two traits are influenced by the same genes. When a trait responds to selection, any genetically correlated trait will also exhibit a selection response.

Inbreeding is the preferential mating of closely related individuals. Continued inbreeding causes loci to become homozygous.

ANALYTICAL APPROACHES FOR SOLVING GENETICS PROBLEMS

Q.1 Assume that genes *A*, *B*, *C*, and *D* are members of a multiple-gene series that control a quantitative trait. Each of these genes has a duplicate, cumulative effect in that each contributes 3 cm of height to the organism when it is present. Each gene assorts independently. In addition, gene *L* is always present in the homozygous state, and the *LL* genotype contributes a constant 40 cm of height. The alleles *a*, *b*, *c*, and *d* do not contribute anything to the height of the organism. If we ignore height variation caused by environmental factors, an organism with genotype *AA BB CC DD LL* would be 64 cm high, and one with genotype *aa bb cc dd LL* would be 40 cm. A cross is made of *AA bb CC DD LL* × *aa BB cc DD LL* and is carried into the F_2 by selfing of the F_1.
a. How does the size of the F_1 individuals compare with the size of each of the parents?
b. Compare the mean of the F_1 with the mean of the F_2, and comment on your findings.
c. What proportion of the F_2 population would show the same height as the *AA bb CC DD LL* parent?
d. What proportion of the F_2 population would show the same height as the *aa BB cc DD LL* parent?

e. What proportion of the F_2 population would breed true for the height shown by the *aa BB cc DD LL* parent?
f. What proportion of the F_2 population would breed true for the height characteristic of F_1 individuals?

A.1 This question explores our understanding of the basic genetics involved in a multiple-gene series that in this case controls a quantitative trait. The approach we will take is essentially the same as the approach used with a series of independently assorting genes that control distinctly different traits. That is, we make predictions on the basis of genotypes and relate the results to phenotypes, or we make predictions on the basis of phenotypes and relate the results to genotypes.
a. Each allele represented by a capital letter contributes 3 cm of height to the base height of 40 cm, which is controlled by the ever present *LL* homozygosity. Therefore the *AA bb CC DD LL* parent, which has six capital-letter alleles from the *A*-through-*D*, multiple-gene series, is 40 + (6 × 3) = 58 cm high. Similarly, the *aa BB cc DD LL* parent has four capital-letter alleles and therefore is 40 + 12 = 52 cm high. The F_1

from a cross between these two individuals would be heterozygous for the *A*, *B*, and *C* loci and homozygous *D* and *L*, that is, *Aa Bb Cc DD LL*. This progeny has five capital-letter alleles apart from *LL* and hence is 40 + 15 = 55 cm high.

b. The F_2 is derived from a self of the *Aa Bb Cc DD LL* F_1. All the F_2 individuals will be *DD LL*, making them at least 40 + 6 = 46 cm high. Now we must deal with the heterozygosity at the other three loci. What we need to calculate is the relative proportions of individuals with all the various possible numbers of capital-letter alleles. This calculation is equivalent to determining the relative distribution of three independently assorting traits, each showing incomplete dominance. In other words, we must calculate directly the relative frequencies of all possible genotypes for the three loci and collect those with no, one, two, three, four, five, and six capital-letter alleles. Thus the probability of getting an individual with two capital-letter alleles for each locus is ¼, the probability of getting an individual with one-capital letter allele for each locus is ½, and the probability of getting an individual with no capital-letter alleles for each locus is ¼. So the probability of getting an F_2 individual with six capital-letter alleles for the *A*, *B*, and *C* loci is $(¼)^3 = $ ¹⁄₆₄, and the same probability is obtained for an individual with no capital-letter alleles. This analysis gives us a clue about how we should consider all the possible combinations of genotypes that have the other numbers of capital-letter alleles. That is, the simplest approach is to compute the coefficients in the binomial expansion of $(a + b)^6$. The expansion gives a 1:6:15:20:15:6:1 distribution of zero, one, two, three, four, five, and six capital-letter alleles, respectively. Now since each capital-letter allele in the *A*, *B*, and *C* set contributes 3 cm of height over the 46-cm height given by the *DD LL* genotype common to all, then the F_2 individuals would fall into the following distribution:

Number of Capital-Letter Alleles	Height Added to Basic Height of 46 cm for Common *DD LL* Genotype (cm)	Height of Individuals (cm)	Frequency
6	18	64	1
5	15	61	6
4	12	58	15
3	9	55	20
2	6	52	15
1	3	49	6
0	0	46	1

The distribution is clearly symmetrical, giving an average of 55 cm, the same height shown in F_1 individuals.

c. The *AA bb CC DD LL* parent was 58 cm, so we can read the proportion of F_2 individuals that show this same height directly from the table in part b. The answer is ¹⁵⁄₆₄.

d. The *aa BB cc DD LL* parent was 52 cm, and from the table in part b the proportion of F_2 individuals that show this same height is ¹⁵⁄₆₄.

e. We are asked to determine the proportion of the F_2 population that would breed true for the height shown by the *aa BB cc DD LL* parent, which was 52 cm. To breed true, the organism must be homozygous. We have also established that *DD LL* is a constant genotype for the F_2 individuals, giving a basic height of 46 cm. Therefore for a height of 52 cm, two additional, active, capital-letter alleles must be present, apart from those at the *D* and *L* loci. With the requirement for homozygosity, there are only three genotypes that give a 52-cm height; they are *AA bb cc DD LL*, *aa BB cc DD LL*, and *aa bb CC DD LL*. The probability of each combination occurring in the F_2 is ¹⁄₆₄, so the answer to the problem is ¹⁄₆₄ + ¹⁄₆₄ + ¹⁄₆₄ = ³⁄₆₄. (Note that the individual probability for each genotype may be calculated. That is, probability of *AA* = ¼, probability of *bb* = ¼, probability of *cc* = ¼, probability of *DD LL* = 1, giving an overall probability for *AA bb cc DD LL* of ¹⁄₆₄.)

f. We are asked to determine the proportion of the F_2 population that would breed true for the height characteristic of F_1 individuals. Again, the basic height given by *DD LL* is 46 cm. The F_1 height is 55 cm, so three capital-letter alleles must be present in addition to *DD LL* to give that height, since (3 × 3) cm = 9 cm, and 9 cm + 46 cm = 55 cm. However, since an individual must be homozygous to be true breeding, the answer to this question is none, since 3 is an odd number, meaning that at least one locus must be heterozygous in order to get the 55-cm height.

Q.2 Five field mice collected in Texas had weights of 15.5 g, 10.3 g, 11.7 g, 17.9 g, and 14.1 g. Five mice collected in Michigan had weights of 20.2 g, 21.1 g, 20.4 g, 22.0 g, and 19.7 g. Calculate the mean weight and the variance in weight for mice from Texas and for mice from Michigan.

A.2 To answer this question, we use the formulas given in the chapter section on statistics. The formula for the mean is

$$\bar{x} = \frac{\Sigma x_i}{n}$$

The symbol Σ means to add, and the x_i represents all the individual values. We begin by summing up all the weights of the mice from Texas:

$$\Sigma x_i = 15.5 + 10.3 + 11.7 + 17.9 + 14.1 = 69.5$$

Next, we divide this summation by n, which represents the number of values added together. In this case, we added together five weights, so $n = 5$. The mean for the Texas mice is therefore

$$\frac{\Sigma x_i}{n} = \frac{69.5}{5} = 13.9$$

To calculate the variance in weight among the Texas mice, we utilize the formula

$$s^2 = \frac{\Sigma(x_i - \bar{x})^2}{n - 1}$$

We must take each individual weight and subtract it from the mean weight of the group. Each value obtained from this subtraction is then squared, and all squared values are added up, as shown below.

$15.5 - 13.9 =$	1.6	$(1.6)^2 = 2.56$
$10.3 - 13.9 =$	-3.6	$(-3.6)^2 = 12.96$
$11.7 - 13.9 =$	-2.2	$(-2.2)^2 = 4.84$
$17.9 - 13.9 =$	4.0	$(4.0)^2 = 16.0$
$14.1 - 13.9 =$	0.2	$(0.2)^2 = \underline{0.04}$
		36.4

The sum of all the squared values is 36.4. All that remains for us to do is to divide this sum by $n-1$, which is $5 - 1 = 4$.

$$s^2 = \frac{\Sigma(x_i - \bar{x})^2}{n - 1} = \frac{36.4}{4} = 9.1$$

The mean and the variance for the Texas mice are therefore 13.9 and 9.1.

We now repeat these steps for the mice from Michigan.

$$\Sigma x_i = 20.2 + 21.2 + 20.4 + 22.0 + 19.7 = 103.5$$

$$\frac{\Sigma x_i}{n} = \frac{103.5}{5} = 20.7$$

$$s^2 = \frac{\Sigma(x_i - \bar{x})^2}{n - 1}$$

$20.2 - 20.7 =$	-0.5	$(-0.5)^2 = 0.25$
$21.2 - 20.7 =$	0.5	$(0.5)^2 = 0.25$
$20.4 - 20.7 =$	-0.3	$(-0.3)^2 = 0.9$
$22.0 - 20.7 =$	1.3	$(1.3)^2 = 1.69$
$19.7 - 20.7 =$	-1.0	$(-1.0)^2 = \underline{1.0}$
		4.09

$$s^2 = \frac{\Sigma(x_i - \bar{x})^2}{n - 1} = \frac{4.09}{4} = 1.23$$

The mean and the variance for the Michigan mice are 20.79 and 1.23.

We conclude that the Michigan mice are considerably heavier than the Texas mice and the Michigan mice also exhibit less variance in weight.

QUESTIONS AND PROBLEMS

*22.1 The following measurements of head width and wing length were made on a series of steamer-ducks:

SPECIMEN	HEAD WIDTH (cm)	WING LENGTH (cm)
1	2.75	30.3
2	3.20	36.2
3	2.86	31.4
4	3.24	35.7
5	3.16	33.4
6	3.32	34.8
7	2.52	27.2
8	4.16	52.7

a. Calculate the mean and the standard deviation of head width and of wing length for these eight birds.
b. Calculate the correlation coefficient for the relationship between head width and wing length in this series of ducks.
c. What conclusions can you make about the association between head width and wing length in steamer-ducks?

22.2 Using methods described in this chapter, answer the following questions.
a. In a family of 6 children, what is the probability that 3 will be girls and 3 will be boys?
b. In a family of 5 children, what is the probability that 1 will be a boy and 4 will be girls?
c. What is the probability that in a family of 6 children, all will be boys?

*22.3 In flipping a coin, there is a 50 percent chance of obtaining heads and a 50 percent chance of obtaining tails on each flip. If you flip a coin 10 times, what is the probability of obtaining exactly 5 heads and 5 tails?

*22.4 The F_1 generation from a cross of two pure-breeding parents that differ in a size character is usually no more variable than the parents. Explain.

22.5 If two pure-breeding strains, differing in a size trait, are crossed, is it possible for F_2 individuals to have phenotypes that are more extreme than either grandparent (i.e., be larger than the largest or smaller than the smallest in the parental generation)? Explain.

22.6 Two pairs of genes with two alleles each, A/a and B/b, determine plant height additively in a population. The homozygote $AA\ BB$ is 50 cm tall, the homozygote $aa\ bb$ is 30 cm tall.
a. What is the F_1 height in a cross between the two homozygous stocks?

b. What genotypes in the F$_2$ will show a height of 40 cm after an F$_1$ × F$_1$ cross?

c. What will be the F$_2$ frequency of the 40-cm plants?

*22.7 Three independently segregating genes (*A*, *B*, *C*), each with two alleles, determine height in a plant. Each capital-letter allele adds 2 cm to a base height of 2 cm.

a. What are the heights expected in the F$_1$ progeny of a cross between homozygous strains *AA BB CC* (14 cm) × *aa bb cc* (2 cm)?

b. What is the distribution of heights (frequency and phenotype) expected in an F$_1$ × F$_1$ cross?

c. What proportion of F$_2$ plants will have heights equal to the heights of the original two parental strains, *AA BB CC* and *aa bb cc*?

d. What proportion of the F$_2$ will breed true for the height shown by the F$_1$?

22.8 Repeat Problem 22.6, but assume that each capital-letter allele acts to double the existing height; for example, *Aa bb cc* = 4 cm, *AA bb cc* = 8 cm, *AA Bb cc* = 16 cm, and so on.

22.9 Assume three equally and additively contributing pairs of genes control flower length in nasturtiums. A completely homozygous plant with 10 mm flowers is crossed to a completely homozygous plant with 30 mm flowers. F$_1$ plants all have flowers about 20 mm long. F$_2$ plants show a range of lengths from 10 to 30 mm, with about 1/64 of the F$_2$ having 10 mm flowers and 1/64 having 30 mm flowers. What distribution of flower length would you expect to see in the offspring of a cross between an F$_1$ plant and the 30 mm parent?

*22.10 In a particular experiment the mean internode length in spikes of the barley variety *asplund* was found to be 2.12 mm. In the variety *abed binder* the mean internode length was found to be 3.17 mm. The mean of the F$_1$ of a cross between the two varieties was approximately 2.7 mm. The F$_2$ gave a continuous range of variation from one parental extreme to the other. Analysis of the F$_3$ generation showed that in the F$_2$ 8 out of the total 125 individuals were of the *asplund* type, giving a mean of 2.19 mm. Eight other individuals were similar to the parent *abed binder*, giving a mean internode length of 3.24 mm. Is the internode length in spikes of barley a discontinuous or a quantitative trait? Why?

22.11 From the information given in Problem 22.10, determine how many gene pairs involved in the determination of internode length are segregating in the F$_2$.

22.12 Assume that the difference between a type of oats yielding about 4 g per plant and a type yielding 10 g is the result of three equal and cumulative multiple-gene pairs *AA BB CC*. If you cross the type yielding 4 g with the type yielding 10 g, what will be the phenotypes of the F$_1$ and the F$_2$? What will be their distribution?

*22.13 Assume that in squashes the difference in fruit weight between a 3-lb type and a 6-lb type is due to three allelic pairs, *A/A*, *B/b*, and *C/c*. Each capital-letter allele contributes a half pound to the weight of the squash. From a cross of a 3-lb plant (*aa bb cc*) with a 6-lb plant (*AA BB CC*), what will be the phenotypes of the F$_1$ and the F$_2$? What will be their distribution?

22.14 Refer to the assumptions stated in Problem 22.13. Determine the range in fruit weight of the offspring in the following squash crosses: (a) *Aa Bb CC × aa Bb Cc*; (b) *AA bb Cc × Aa BB cc*; (c) *aa BB cc × AA BB cc*.

*22.15 Assume that the difference between a corn plant 10 dm (decimeters) high and one 26 dm high is due to four pairs of equal and cumulative multiple alleles, with the 26-dm plants being *AA BB CC DD* and the 10-dm plants being *aa bb cc dd*.

a. What will be the size and genotype of an F$_1$ from a cross between these two true-breeding types?

b. Determine the limits of height variation in the offspring from the following crosses:
(1) *Aa BB cc dd × Aa bb Cc dd*;
(2) *aa BB cc dd × Aa Bb Cc dd*;
(3) *AA BB Cc DD × aa BB cc Dd*;
(4) *Aa Bb Cc Dd × Aa bb Cc Dd*.

22.16 Refer to the assumptions given in Problem 22.15. But for this problem two 14-dm corn plants, when crossed, give nothing but 14-dm offspring (case A). Two other 14-dm plants give one 18-dm, four 16-dm, six 14-dm, four 12-dm, and one 10-dm offspring (case B). Two other 14-dm plants, when crossed, give one 16-dm, two 14-dm, and one 12-dm offspring (case C). What genotypes for each of these 14-dm parents (cases A, B, and C) would explain these results? Would it be possible to get a plant taller than 18 dm by selection in any of these families?

22.17 A quantitative geneticist determines the following variance components for leaf width in a population of wild flowers growing along a roadside in Kentucky:

Additive genetic variance (V_A)	= 4.2
Dominance genetic variance (V_D)	= 1.6
Interaction genetic variance (V_I)	= 0.3
Environmental variance (V_E)	= 2.7
Genetic-environmental variance (V_{GE})	= 0.0

a. Calculate the broad-sense heritability and the narrow-sense heritability for leaf width in this population of wildflowers.

b. What do the heritabilities obtained in part *a* indicate about the genetic nature of leaf width variation in this plant?

*22.18 Assume all genetic variance affecting seed weight in beans is genetically determined and is additive. From a population where the mean seed weight was 0.88 g, a farmer selected two seeds, each weighing 1.02 g. He planted these and crossed the resulting plants to each other, then collected and weighed their seeds. The mean weight of their seeds was 0.96 g. What is the narrow-sense heritability of seed weight?

*22.19 Members of the inbred rat strain SHR are salt sensitive: they respond to a high salt environment by developing hypertension. Members of a different inbred rat strain, TIS, are not salt sensitive. Imagine you placed a population consisting only of SHR rats in an environment that was variable in regard to distribution of salt, so that some rats would be exposed to more salt than others. What would be the heritability of blood pressure in this population?

*22.20 On his farm in Kansas a farmer is growing a variety of wheat called TK138. He calculates the narrow-sense heritability for yield (the amount of wheat produced per acre) and finds that the heritability of yield for TK138 is 0.95. The next year he visits a farm in Poland and observes that a Russian variety of wheat, UG334, growing there has only about 40 percent as much yield as TK138 grown on his farm in Kansas. Since he found the heritability of yield in his wheat to be very high, he concludes that the American variety of wheat (TK138) is genetically superior to the Russian variety (UG334), and he tells the Polish farmers that they can increase their yield by using TK138. What is wrong with his conclusion?

22.21 Dermatoglyphics are the patterns of the ridged skin found on the fingertips, toes, palms, and soles. (Fingerprints are dermatoglyphics.) Classification of dermatoglyphics is frequently based on the number of triradii; a triradius is a point from which three ridge systems separate at angles of 120°. The number of triradii on all ten fingers was counted for each member of several families, and the results are tabulated below.

FAMILY	MEAN NUMBER OF TRIRADII IN THE PARENTS	MEAN NUMBER OF TRIRADII IN THE OFFSPRING
I	14.5	12.5
II	8.5	10.0
III	13.5	12.5
IV	9.0	7.0
V	10.0	9.0
VI	9.5	9.5
VII	11.5	11.0
VIII	9.5	9.5
IX	15.0	17.5
X	10.0	10.0

a. Calculate the narrow-sense heritability for the number of triradii by the regression of the mean phenotype of the parents against the mean phenotype of the offspring.

b. What does your calculated heritability value indicate about the relative contributions of genetic variation and environmental variation to the differences observed in number of triradii?

*22.22 A scientist wishes to determine the narrow-sense heritability of tail length in mice. He measures tail length among the mice of a population and finds a mean tail length of 9.7 cm. He then selects the ten mice in the population with the longest tails; mean tail length in these selected mice is 14.3 cm. He interbreeds the mice with the long tails and examines tail length in their progeny. The mean tail length in the F_1 progeny of the selected mice is 13 cm.

Calculate the selection differential, the response to selection, and the narrow-sense heritability for tail length in these mice.

22.23 Suppose that the narrow-sense heritability of wool length in a breed of sheep is 0.92, and the narrow-sense heritability of body size is 0.87. The genetic correlation between wool length and body size is −0.84. If a breeder selects for sheep with longer wool, what will be the most likely effects on wool length and on body size?

*22.24 The heights of 10 college-age males and the heights of their fathers are presented below.

HEIGHT OF SON (INCHES)	HEIGHT OF FATHER (INCHES)
70	70
72	76
71	72
64	70
66	70
70	68
74	78
70	74
73	69

a. Calculate the mean and the variance of height for the sons and do the same for the fathers.
b. Calculate the correlation coefficient for the relationship between the height of father and height of son.
c. Determine the narrow-sense heritability of height in this group by regression of the son's height on the height of father.

*22.25 The narrow-sense heritability of egg weight in a particular flock of chickens is 0.60. A farmer selects for increased egg weight in this flock. The difference in the mean egg weight of the unselected chickens and the selected chickens is 10 g. How much should egg weight increase in the offspring of the selected chickens?

22.26 Would a law banning marriages between individuals and their stepparents be founded on genetic principles?

22.27 Is there any circumstance for which inbreeding does not have deleterious consequences?

23 POPULATION GENETICS

PRINCIPAL POINTS

~ The gene pool of a population is the total of all genes within a group of sexually interbreeding individuals. The gene pool can be described in terms of gene and genotypic frequencies.

~ The Hardy-Weinberg law states that in a large, randomly mating population, free from evolutionary forces, the gene frequencies do not change, and the genotypic frequencies stabilize after one generation in the proportions p^2, $2pq$, and q^2, where p and q equal the gene frequencies of the population.

~ The classical, balance, and neutral models are hypotheses to explain how much genetic variation should exist within natural populations and what forces are responsible for the variation observed.

~ Mutation, genetic drift, migration, and natural selection are forces that can alter gene frequencies of a population.

~ Recurrent mutation changes gene frequencies, and at equilibrium the relative rates of forward and reverse mutations determine gene frequencies of a population in the absence of other forces.

~ Genetic drift is random change in gene frequencies due to chance. Genetic drift produces genetic change within populations, genetic differentiation among populations, and loss of genetic variation within populations.

~ Migration, also termed gene flow, involves movement of genes between populations. Migration can alter the gene frequencies of a population, and it tends to reduce genetic divergence among populations.

~ Natural selection is differential reproduction of genotypes. It is measured by fitness, which is the relative reproductive ability of genotypes. Natural selection can produce a number of different effects on the gene pool of a population.

~ Nonrandom mating affects the genotypic frequencies of a population, inbreeding leads to an increase in homozygosity, and outbreeding leads to an increase in heterozygosity.

~ Rates of evolution can be measured by comparing DNA or RNA sequences. Different genes and even different parts of the same gene tend to evolve at different rates.

~ In eukaryotic organisms, genes frequently occur in multiple copies with identical or similar sequences. A group of such genes is termed a multiple gene family.

~ The mitochondrial DNA of some organisms evolves at a faster rate than the nuclear DNA.

~ Concerted evolution is a process that maintains sequence uniformity among multiple copies of the same sequence within a species.

~ Evolutionary relationships among organisms can be revealed by study of DNA and RNA sequences.

The science of genetics can be broadly divided into three major subdisciplines: transmission genetics (also called classical genetics), molecular genetics, and population genetics. Each of these three areas focuses on a different aspect of heredity. **Transmission genetics** is primarily concerned with genetic processes that occur within individuals, and how genes are passed from one individual to another. Thus, the unit of study for transmission genetics is the individual. In **molecular genetics,** we are largely interested in the molecular nature of heredity—how genetic information is encoded within the DNA and how biochemical processes of the cell translate the genetic information into the phenotype. Consequently, in molecular genetics we focus on the cell.

Population genetics, which we introduce in this chapter, is the field of genetics that studies heredity in groups of individuals. Population geneticists investigate the patterns of genetic variation found within groups and how these patterns change, or evolve over time. In this discipline, our perspective shifts away from the individual and the cell, and focuses instead upon a **Mendelian population.** A Mendelian population is a group of *interbreeding* individuals, who share a common set of genes. The genes shared by the individuals of a Mendelian population are called the **gene pool.** To understand the genetics of the evolutionary process, we study the gene pool of a Mendelian population, rather than the genotypes of its individual members. Except for rare mutations, individuals are born and die with the same set of genes; what changes genetically over time (evolves) is the hereditary makeup of a group of individuals, reproductively connected in a Mendelian population.

Questions frequently studied by population geneticists include:

1. How much genetic variation is found in natural populations, and what forces of nature control the amount of variation observed?

2. What evolutionary forces shape the genetic structures of populations?
3. What forces are responsible for producing genetic divergence among populations?
4. How do biological characteristics of a population, such as breeding system, fecundity, and age structure, influence the gene pool of the population?

To answer these questions, population geneticists frequently develop mathematical models and equations to describe what happens to the gene pool of a population under various conditions. An example is the set of equations that describes the influence of random mating on the gene and genotypic frequencies of an infinitely large population, a model called the **Hardy-Weinberg law,** which we discuss later in the chapter. At first, students often have difficulty grasping the significance of models such as these, because the models are frequently simple and require numerous assumptions that are unlikely to be met by organisms in the real world. Beginning with simple models is useful, however, because with such models we can examine the effects of individual evolutionary factors on the genetic structure of a population. Once we understand the results of the simple models, we can incorporate more realistic conditions into the equations. In the case of the Hardy-Weinberg law, the assumptions of infinitely large size and random mating may seem unrealistic, but these assumptions are necessary for simplifying the mathematical analysis. We cannot begin to understand the impact of nonrandom mating and limited population size on gene frequencies until we first know what happens under the simpler conditions of random mating and large population size.

GENOTYPIC AND GENE FREQUENCIES

Genotypic Frequencies

To study the genetic composition of a Mendelian population, population geneticists must first quantitatively describe the gene pool of the group. This is done by calculating genotypic frequencies and allelic frequencies within the population. A frequency is a percent, or a proportion, and always ranges between 0 and 1. If 43 percent of the people in a group have red hair, the frequency of red hair in the group is 0.43. To calculate the **genotypic frequencies** at a specific locus, we count the number of individuals with one particular genotype and divide this number by the total number of individuals in the population. We do this for each of the genotypes at the locus, and all the genotypic frequencies should add up to 1. Consider a locus that determines

the pattern of spots in the scarlet tiger moth, *Panaxia dominula* (Figure 23.1). Three genotypes are present in most populations, and each genotype produces a different phenotype. E. B. Ford collected moths at one locality in England and found the following numbers of genotypes: 452 *BB*, 43 *Bb*, and 2 *bb*, out of a total of 497 moths. The genotypic frequencies are therefore:

$$P = f(BB) = 452/497 = 0.909$$
$$H = f(Bb) = 43/497 = 0.087$$
$$Q = f(bb) = 2/497 = \underline{0.004}$$
$$\text{Total} \quad 1.000$$

Population geneticists often use the capital letters, P, H, and Q to represent the frequencies (f) of the three genotypes at a locus with two alleles. Be careful that you do not confuse these symbols with the small letters p and q which are used to represent the allelic frequencies, as discussed in the next section. Box 23.1 (pp. 710–711) presents a sample calculation of genotypic frequencies for some hemoglobin variants in a human population.

~ FIGURE 23.1

Panaxia dominula, the scarlet tiger moth. The top two moths are normal homozygotes (*BB*), those in rows 2 and 3 are heterozygotes (*Bb*), and the bottom moth is the rare homozygote (*bb*).

Box 23.1
Sample Calculation of Gene and Genotypic Frequencies for Hemoglobin Variants among Nigerians

Hemoglobin Genotypes

AA	AS	SS	AC	SC	CC	Total
2017	783	4	173	14	11	3002

Calculation of Gene Frequencies

$$\text{Gene frequency} = \frac{\text{number of copies of a given allele in the population}}{\text{sum of all alleles in the population}}$$

$$f(S) = \frac{(2 \times \text{number of } SS \text{ individuals}) + (\text{number of } AS \text{ individuals}) + (\text{number of } SC \text{ individuals})}{(2 \times \text{total number of individuals})}$$

$$f(S) = \frac{(2 \times 4) + 783 + 14}{(2 \times 3002)} = \frac{805}{6004} = 0.134$$

$$f(A) = \frac{(2 \times \text{number of } AA \text{ individuals}) + (\text{number of } AS \text{ individuals}) + (\text{number of } AC \text{ individuals})}{(2 \times \text{total number of individuals})}$$

$$f(A) = \frac{(2 \times 2017) + 783 + 173}{(2 \times 3002)} = \frac{4990}{6004} = 0.831$$

$$f(C) = \frac{(2 \times \text{number of } CC \text{ individuals}) + (\text{number of } AC \text{ individuals}) + (\text{number of } SC \text{ individuals})}{(2 \times \text{total number of individuals})}$$

$$f(C) = \frac{(2 \times 11) + 173 + 14}{(2 \times 3002)} = \frac{209}{6004} = 0.035$$

Gene Frequencies

While genotypic frequencies are useful for examining the effects of certain evolutionary forces on a population, in most cases population geneticists use frequencies of alleles to describe the gene pool. The frequencies of alleles at a locus are called the **allelic frequencies** or, more commonly, the **gene frequencies**. The use of gene frequencies offers several advantages over the genotypic frequencies. First, there are always fewer alleles than genotypes, and so the gene pool can be described with fewer parameters when gene frequencies are used. If a locus has three alleles, six genotypic frequencies must be calculated to describe the gene pool, whereas use of the allelic frequencies requires the calculation of only three allelic frequencies. Furthermore, in sexually reproducing organisms genotypes break down to alleles when gametes are formed, and alleles, not genotypes, are passed from one generation to the next. Consequently, only alleles have continuity over time, and the gene pool evolves through changes in the frequencies of alleles.

Gene frequencies may be calculated in either of two ways: from the observed numbers of different genotypes, or from the genotypic frequencies. First, we can calculate the gene frequencies directly from the *numbers* of genotypes. In this method, we count the number of alleles of one type and divide it by the total number of alleles in the population:

$$\text{Gene frequency} = \frac{\text{number of copies of a given allele in the population}}{\text{sum of all alleles in the population}}$$

As an example, imagine a population of 1000 diploid individuals with 353 *AA*, 494 *Aa*, and 153 *aa* individuals. Each *AA* individual has two *A* alleles, while each *Aa* heterozygote possesses only a single *A* allele. Therefore, the number of *A* alleles in the population is (2 × the number of *AA* homozygotes) + (the number of *Aa* heterozygotes), or (2 × 353) + 494 = 1200. Because

Calculation of Genotypic Frequencies

$$\text{Genotypic frequency} = \frac{\text{number of individuals}}{\text{total number of individuals}}$$

$$f(SS) = \frac{4}{3002} = 0.0013$$

$$f(AS) = \frac{783}{3002} = 0.261$$

$$f(AA) = \frac{2017}{3002} = 0.672$$

$$f(SC) = \frac{14}{3002} = 0.0047$$

$$f(AC) = \frac{173}{3002} = 0.058$$

$$f(CC) = \frac{11}{3002} = 0.0037$$

Data are from Livingstone, F. B. 1973. Data on the Abnormal Hemoglobins and Glucose-6-Phosphate Dehydrogenase Deficiency in Human Populations of 1967–1973. Contributions in Human Biology No. 1. Museum of Anthropology, University of Michigan.

every diploid individual has two alleles, the total number of alleles in the population will be twice the number of individuals, or 2 × 1000. Using the equation given above, the gene frequency is 1200/2000 = 0.60.

When two alleles are present at a locus, we can use the following formulas for calculating gene frequencies:

$$p = f(A) = \frac{(2 \times \text{number of } AA \text{ homozygotes}) + (\text{number of } Aa \text{ heterozygotes})}{(2 \times \text{total number of individuals})}$$

$$q = f(a) = \frac{(2 \times \text{number of } aa \text{ homozygotes}) + (\text{number of } Aa \text{ heterozygotes})}{(2 \times \text{total number of individuals})}$$

The frequencies of two alleles, $f(A)$ and $f(a)$, are commonly symbolized as p and q. The gene frequencies for a locus should always add up to 1; therefore, once p is calculated, q can be easily obtained by subtraction: $1 - p = q$. A sample calculation of gene frequencies is presented in Box 23.1.

GENE FREQUENCIES WITH MULTIPLE ALLELES.

Suppose we have three alleles—A^1, A^2, and A^3—at a locus, and we want to determine the gene frequencies. Here, we employ the same rule that we used with two alleles: we add up the number of alleles of each type and divide by the total number of alleles in the population:

$$p = f(A^1) = \frac{(2 \times A^1A^1) + (A^1A^2) + (A^1A^3)}{(2 \times \text{total number of individuals})}$$

$$q = f(A^2) = \frac{(2 \times A^2A^2) + (A^1A^2) + (A^2A^3)}{(2 \times \text{total number of individuals})}$$

$$r = f(A^3) = \frac{(2 \times A^3A^3) + (A^1A^3) \times (A^2A^3)}{(2 \times \text{total number of individuals})}$$

To illustrate the calculation of gene frequencies when more than two alleles are present, we will use data from a study on genetic variation in milkweed beetles. Walter Eanes and his coworkers examined gene frequencies at a locus that codes for the enzyme phosphoglucomutase (PGM). Three alleles were found at this locus; each allele codes for a different molecular variant of the enzyme. In one population, the following numbers of genotypes were collected:

AA	=	4
AB	=	41
BB	=	84
AC	=	25
BC	=	88
CC	=	32
Total	=	274

The frequencies of the alleles are

$$f(A) = p = \frac{(2 \times 4) + 41 + 25}{(2 \times 274)} = 0.135$$

$$f(B) = q = \frac{(2 \times 84) + 41 + 88}{(2 \times 274)} = 0.542$$

$$f(C) = r = \frac{(2 \times 32) + 88 + 25}{(2 \times 274)} = 0.323$$

As seen in these calculations, we add twice the number of homozygotes that possess the allele and one times each of the heterozygotes that have the allele. We then divide by twice the number of individuals in the population, which represents the total number of alleles present. In the top part of the equation, notice that for each allelic frequency, we do not add all the heterozygotes, because some of the heterozygotes do not have the allele; for example, in calculating the gene frequency of *A*, we do not add the number of *BC* heterozygotes in the top part of the equation, because we are adding the number of *A* alleles, and *BC* individuals do not have an *A* allele. We can use the same procedure for calculating gene frequencies when four or more alleles are present.

GENE FREQUENCIES FROM GENOTYPIC FREQUENCIES. A second method for calculating gene frequencies is from the *genotypic frequencies*. This calculation may be quicker if we have already determined the frequencies of the genotypes. For two alleles, the gene frequencies are

$p = f(A) =$ (frequency of the *AA* homozygote) + ($\frac{1}{2}$ × frequency of the *Aa* heterozygote)

$q = f(a) =$ (frequency of the *aa* homozygote) + ($\frac{1}{2}$ × frequency of the *Aa* heterozygote)

The frequency of the homozygote is added to half of the heterozygote frequency because half of the heterozygote's alleles are *A* and half are *a*. If three alleles (*A*^{1}, *A*^{2}, and *A*^{3}) are present in the population, the gene frequencies are

$p = f(A^1) = f(A^1A^1) + \frac{1}{2}f(A^1A^2) + \frac{1}{2}f(A^1A^3)$

$q = f(A^2) = f(A^2A^2) + \frac{1}{2}f(A^1A^2) + \frac{1}{2}f(A^2A^3)$

$r = f(A^3) = f(A^3A^3) + \frac{1}{2}f(A^1A^3) + \frac{1}{2}f(A^2A^3)$

Although calculating gene frequencies from genotypic frequencies may be quicker than calculating them directly from the numbers of genotypes, more rounding error will occur. As a result, calculations from direct counts are usually preferred.

GENE FREQUENCIES AT AN X-LINKED LOCUS. Calculation of gene frequencies at an X-linked locus is slightly more complicated, because males have only a single X-linked allele. However, we can use the same rules we used for autosomal loci. Remember that each homozygous female carries two X-linked alleles; heterozygous females have only one of that particular allele, and all males may have only a single X-linked allele. To determine the number of alleles at an X-linked locus,

we multiply the number of homozygous females by 2, then add the number of heterozygous females and the number of hemizygous males. We next divide by the total number of alleles in the population. When determining the total number of alleles, we add twice the number of females (because each female has two X-linked alleles) to the number of males (who have a single allele at X-linked loci). Using this reasoning, the frequencies of two alleles at an X-linked locus (X^A and X^a) are determined with the following equations:

$$p = f(X^A) = \frac{(2 \times X^AX^A \text{ females}) + (X^AX^a \text{ females}) + (X^AY \text{ males})}{(2 \times \text{number of females}) + (\text{number of males})}$$

$$q = f(X^a) = \frac{(2 \times X^aX^a \text{ females}) + (X^AX^a \text{ females}) + (X^aY \text{ males})}{(2 \times \text{number of females}) + (\text{number of males})}$$

Allelic frequencies at an X-linked locus can be determined from the genotypic frequencies by

$$p = f(X^A) = f(X^AX^A) + \frac{1}{2}f(X^AX^a) + f(X^AY)$$

$$q = f(X^a) = f(X^aX^a) + \frac{1}{2}f(X^AX^a) + f(X^aY)$$

Students should strive to understand the logic behind these calculations, not just memorize the formulas. If you fully understand the basis of the calculations, you will not need to remember the exact equations and will be able to determine gene frequencies for any situation.

THE HARDY-WEINBERG LAW

The **Hardy-Weinberg law** is the most important principle in population genetics, for it explains how reproduction influences gene and genotypic frequencies of a population. The Hardy-Weinberg law is named after the two individuals who independently discovered it in the early 1900s (Box 23.2). We begin our discussion of the Hardy-Weinberg law by simply stating what it tells us about the gene pool of a population. We then explore the implications of this principle, and briefly discuss how the Hardy-Weinberg law is derived. Finally, we present some applications of the Hardy-Weinberg law and test a population to determine if the genotypes are in Hardy-Weinberg proportions.

The Hardy-Weinberg law is divided into three parts—a set of assumptions and two major results. A simple statement of the law is:

Box 23.2
Hardy, Weinberg, and the History of Their Contribution to Population Genetics

Godfrey H. Hardy (1877–1947) a mathematician at Cambridge University, often met R. C. Punnett, the Mendelian geneticist, at the faculty club. One day in 1908 Punnett told Hardy of a problem in genetics that he attributed to a strong critic of Mendelism, G. U. Yule (Yule later denied having raised the problem). Supposedly, Yule said that if the gene for short fingers (brachydactyly) was dominant (which it is) and its allele for normal-length fingers was recessive, then short fingers ought to become more common with each generation. In time virtually everyone in Britain should have short fingers. Punnett believed the argument was incorrect, but he could not prove it.

Hardy was able to write a few equations showing that, given any particular frequency of alleles for short fingers and alleles for normal fingers in a population, the relative number of people with short fingers and people with normal fingers will stay the same generation after generation providing no natural selection is involved that favors one phenotype or the other in producing offspring. Hardy published a short paper describing the relationship between genotypes and phenotypes in populations, and within a few weeks a paper was published by Wilhelm Weinberg (1862–1937), a German physician of Stuttgart, that clearly stated the same relationship. The Hardy-Weinberg law signaled the beginning of modern population genetics.

To be complete, we should note that in 1903 the American geneticist W. E. Castle of Harvard University was the first to recognize the relationship between allele and genotypic frequencies, but it was Hardy and Weinberg who clearly described the relationship in mathematical terms. Thus, the law is sometimes referred to as the Castle-Hardy-Weinberg law.

Part 1: In an infinitely large, randomly mating population, free from evolutionary forces (mutation, migration, natural selection);

Part 2: The frequencies of the alleles do not change over time; and

Part 3: After one generation of random mating, the genotypic frequencies will remain in the proportions p^2 (frequency of AA), $2pq$ (frequency of Aa), and q^2 (frequency of aa), where p is the allelic frequency of A and q is the allelic frequency of a. The sum of the genotypic frequencies should be equal to 1 (that is, $p^2 + 2pq + q^2 = 1$).

In short, the Hardy-Weinberg law explains what happens to a population's gene and genotypic frequencies as the genes are passed from generation to generation in the absence of evolutionary forces.

Assumptions of the Hardy-Weinberg Law

Part 1 of the Hardy-Weinberg law presents certain conditions, or assumptions, which must be present for the law to apply. First, the law indicates that the population must be infinitely large. If a population is limited in size, chance deviations from expected ratios can cause changes in gene frequency. Changes in gene frequency such as these are termed **genetic drift.** It is true that the assumption of infinite size in part 1 is unrealistic—no population has an infinite number of individuals. However, sampling error has a significant effect on gene frequencies only in "fairly small populations." (We will discuss this phenomenon later, when we examine genetic drift in more detail.) At this point, we merely want to stress that populations need not be infinitely large for the Hardy-Weinberg law to hold true.

A second condition of the Hardy-Weinberg law is that mating must be random. **Random mating** refers to matings between genotypes occurring in proportion to the frequencies of the genotypes in the population. More specifically, the probability of a mating between two genotypes is equal to the product of the two genotypic frequencies.

To illustrate random mating, consider the $M-N$ blood types in humans, discussed in Chapter 4. The $M-N$ blood type is due to an antigen on the surface of a red blood cell, similar to the ABO antigens, except that incompatibility at the $M-N$ system does not cause problems during blood transfusion. The $M-N$ blood type is determined by one locus with two codominant alleles, L^M and L^N. In a population of Eskimos, the frequencies of the three $M-N$ genotypes are $L^M/L^M = 0.835$, $L^M/L^N = 0.156$, and $L^N/L^N = 0.009$. If Eskimos interbreed randomly, the probability of a mating between an L^M/L^M male and an L^M/L^M female is equal to the frequency of L^M/L^M times the frequency of $L^M/L^M = 0.835 \times 0.835 = 0.697$. Similarly, the probabilities of other possible matings are equal to the products of the genotypic frequencies when mating is random.

The requirement of random matings for the Hardy-Weinberg law is often misinterpreted. Many students assume, incorrectly, that the population must be interbreeding randomly for all traits for the Hardy-Weinberg law to hold. If this were true, human populations would never obey the Hardy-Weinberg law, because humans do not mate randomly. Humans mate preferentially for height, IQ, skin color, socioeconomic status, and other traits. However, while mating may be nonrandom for some traits, most humans still mate randomly for the $M-N$ blood types; few of us even know what our $M-N$ blood types are. The principles of the Hardy-Weinberg law apply to any locus for which random mating occurs, even if mating is nonrandom for other loci.

A third condition of the Hardy-Weinberg law is that the population must be free from evolutionary forces. Remember, in the Hardy-Weinberg law we are interested in whether heredity alone changes gene frequencies and how reproduction influences genotypic frequencies. Therefore, the influence of other evolutionary factors must be excluded. Later we will discuss these other evolutionary forces and their effect on the gene pool of a population. This condition (that no evolutionary forces act upon the population) applies only to the locus in question—a population may be subject to evolutionary forces acting on some genes, while still meeting the Hardy-Weinberg assumptions at other loci.

Predictions of the Hardy-Weinberg Law

If the conditions in part 1 of the Hardy-Weinberg law are met, the population is in genetic equilibrium and two results are expected. First, the frequencies of the alleles do not change from one generation to the next, and therefore the gene pool is not evolving at this locus. Second, the genotypic frequencies will be in the proportions p^2, $2pq$, and q^2 after one generation of random mating. Also the genotypic frequencies will remain in these proportions as long as all the conditions required by the Hardy-Weinberg law are met. When the genotypes are in these proportions, the population is said to be in **Hardy-Weinberg equilibrium.** An important use of the Hardy-Weinberg law is that it provides a mechanism for determining the genotypic frequencies from the gene frequencies when the population is in equilibrium.

To summarize, the Hardy-Weinberg law makes several predictions about the gene frequencies and the genotypic frequencies of a population when certain conditions are satisfied. The necessary conditions are that the population is large, randomly mating, and free from evolutionary forces. When these conditions are met, the Hardy-Weinberg law indicates that gene fre-

quencies will not change, genotypic frequencies will change only in the first generation, and thereafter the genotypic frequencies will be determined by the gene frequencies, occurring in the proportions p^2, $2pq$, and q^2.

Derivation of the Hardy-Weinberg Law

The Hardy-Weinberg law states that when a population is in equilibrium, the genotypic frequencies will be in the proportions p^2, $2pq$, and q^2. To understand the basis of these frequencies at equilibrium, consider a hypothetical population in which the frequency of allele A is p and the frequency of allele a is q. In producing gametes, each genotype passes on both alleles that it possesses with equal frequency; therefore, the frequencies of A and a in the gametes are also p and q. Table 23.1 shows the combinations of gametes when mating is random. This table illustrates the relationship between the allelic frequencies and the genotypic frequencies, which forms the basis of the Hardy-Weinberg law. We see that when gametes pair randomly, the genotypes will occur in the proportions p^2 (AA), $2pq$ (Aa), and q^2 (aa). These genotypic proportions result from the expansion of the square of the allelic frequencies $(p + q)^2 = p^2 + 2pq + q^2$, and the genotypes reach these proportions after one generation of random mating.

The Hardy-Weinberg law also states that gene and genotypic frequencies remain constant generation after generation if the population remains large, randomly mating, and free from evolutionary forces. This result can be demonstrated by considering a hypothetical, randomly mating population, as illustrated in Table 23.2. In Table 23.2, all possible matings are given. By definition, random mating means that the frequency of mating between two genotypes will be equal to the product of the genotypic frequencies. For example, the frequency of an $AA \times AA$ mating will be equal to p^2 (the frequency of AA) $\times p^2$ (the frequency of AA) = p^4. The frequencies of the offspring produced from each mating are also presented in Table 23.2.

We see that the probability of $AA \times Aa$ (or $2p^3q$) and $Aa \times AA$ (or $2p^3q$) matings is $4p^3q$, and we know

~ **TABLE 23.1**

Possible Combinations of A and a Gametes from Gametic Pools for a Population

		♂	
GAMETES		$p\ A$	$q\ a$
♀	$p\ A$	$p^2\ AA$	$pq\ Aa$
	$q\ a$	$pq\ Aa$	$q^2\ aa$

In sum, $p^2\ AA + 2pq\ Aa + q^2\ aa = 1.00$

~ TABLE 23.2

Algebraic Proof of Genetic Equilibrium in a Randomly Mating Population for Two Alleles in a Population in Which $(p + q)^2 = p^2 + 2pq + q^2 = 1.00$

TYPE OF MATING ♀ ♂	MATING FREQUENCY	OFFSPRING FREQUENCIES		
		AA	Aa	aa
$p^2\ AA \times p^2\ AA$	p^4	p^4	—	—
$\left.\begin{array}{l} p^2\ AA \times 2pq\ Aa \\ 2pq\ Aa \times p^2\ AA \end{array}\right\}$ *	$4p^3q$	$2p^3q$	$2p^3q$	—
$\left.\begin{array}{l} p^2\ AA \times q^2\ aa \\ q^2\ aa \times p^2\ AA \end{array}\right\}$	$2p^2q^2$	—	$2p^2q^2$	—
$2pq\ Aa \times 2pq\ Aa$	$4p^2q^2$	p^2q^2	$2p^2q^2$	p^2q^2
$\left.\begin{array}{l} 2pq\ Aa \times q^2\ aa \\ q^2\ aa \times 2pq\ Aa \end{array}\right\}$	$4pq^3$	—	$2pq^3$	$2pq^3$
$q^2\ aa \times q^2\ aa$	q^4	—	—	q^4
Totals	$(p^2 + 2pq + q^2)^2 = 1$	$p^2(p^2 + 2pq + q^2) = p^2$	$2pq(p^2 + 2pq + q^2) = 2pq$	$q^2(p^2 + 2pq + q^2) = q^2$

Genotype frequencies = $(p + q)^2 = p^2 + 2pq + q^2 = 1$ in each generation afterward
Gene (allele) frequencies = $p(A) + q(a) = 1$ in each generation afterward

* For example, matings between AA and Aa will occur at $p^2 \times 2pq = 2p^3q$ for $AA \times Aa$ and at $p^2 \times 2pq = 2p^3q$ for $Aa \times AA$, for a total of $4p^3q$. Two progeny types, AA and Aa, result in equal proportions from these matings. Therefore offspring frequencies are $2p^3q$ (i.e., ½ × $4p^3q$) for AA and for Aa.

from Mendelian principles that these crosses produce ½ AA and ½ Aa offspring. Therefore, the probability of obtaining AA offspring from these matings is $4p^3q \times$ ½ = $2p^3q$. The frequencies of offspring produced by each type of mating are presented in the body of the table. At the bottom of the table, the total frequency for each genotype is obtained by addition. As we can see, after random mating the genotypic frequencies are still p^2, $2pq$, and q^2 and the gene frequencies remain at p and q. The population can thus be represented in the zygotic and gametic stages as follows:

$$\begin{array}{cc} \text{Zygotes} & \text{Gametes} \\ p^2\ AA + 2pq\ Aa + q^2\ aa & p\ A + q\ a \end{array}$$

Each generation of zygotes produces A and a gametes in proportions p and q. The gametes unite to form AA, Aa, and aa zygotes in the proportions $p^2 + 2pq + q^2$, and the cycle is repeated indefinitely as long as the assumptions of the Hardy-Weinberg law hold. This short proof gives the theoretical basis for the Hardy-Weinberg law.

The Hardy-Weinberg law indicates that at equilibrium, the genotypic frequencies depend upon the frequencies of the alleles. This relationship between allelic frequencies and genotypic frequencies for a locus with two alleles is represented in Figure 23.2. Several aspects

~ FIGURE 23.2

Relation of the frequencies of the genotypes AA, Aa, and aa, and the frequencies of alleles A and a (in values of p and q, respectively) in a randomly mating population, as predicted by the Hardy-Weinberg formula. It is assumed that all three genotypes have equal reproductive success.

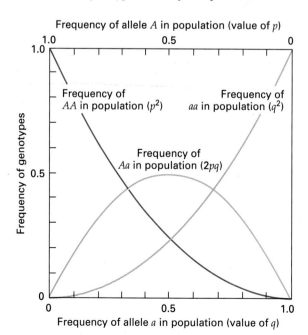

of this relationship should be noted: (1) The maximum frequency of the heterozygote is 0.5, and this maximum value occurs only when the frequencies of A and a are both 0.5; (2) if gene frequencies are between 0.33 and 0.66, the heterozygote is the most numerous genotype; (3) when the frequency of one allele is low, the homozygote for the allele is the rarest of the genotypes. When $f(a) = q = 0.1$, the frequencies of the genotypes are $f(AA) = 0.81$, $f(Aa) = 0.18$, and $f(aa) = 0.01$.

This point is also illustrated by the distribution of genetic diseases in humans, which are frequently rare and recessive. For a rare recessive trait, the frequency of the gene causing the trait will be much higher than the frequency of the trait itself, because most of the rare alleles are in nonaffected heterozygotes. Albinism, for example, is a rare recessive condition in humans. One form of albinism is *tyrosinase-negative albinism*. In this type of albinism, affected individuals have no tyrosinase activity, which is required for normal production of pigment. Among North American whites, the frequency of tyrosinase-negative albinism is roughly 1 in 40,000, or 0.000025. Since albinism is a recessive condition, the genotype of affected individuals is aa. According to the Hardy-Weinberg law, the frequency of the aa genotypes equals q^2. If $q^2 = 0.000025$, then $q = 0.005$ and $p = 1 - q = 0.995$. The heterozygote frequency is therefore $2pq = 2 \times 0.995 \times 0.005 = 0.00995$ (almost 1 percent). Thus, while the frequency of albinos is low (1 in 40,000), individuals heterozygous for albinism are much more common (almost 1 in 100). Heterozygotes for recessive traits can be common, even when the trait is rare.

$\mathcal{K}$EYNOTE

The Hardy-Weinberg law describes what happens to gene and genotypic frequencies of a large population as a result of random mating, assuming that no evolutionary forces act on the population. If these conditions are met, gene frequencies do not change from generation to generation, and the genotypic frequencies stabilize after one generation in the proportions p^2, $2pq$, and q^2, where p and q equal the frequencies of the alleles in the population. ————

Extensions of the Hardy-Weinberg Law

When two alleles are present at a locus, the Hardy-Weinberg law tells us that at equilibrium the frequencies of the genotypes will be p^2, $2pq$, and q^2, which is the square of the allelic frequencies $(p + q)^2$. If three

alleles are present—for example, alleles A, B, and C—with frequencies equal to p, q, and r, the frequencies of the genotypes at equilibrium are also given by the square of the allelic frequencies:

$$(p + q + r)^2 = p^2\ (AA) + 2pq\ (AB) + q^2\ (BB) + 2pr\ (AC) + 2qr\ (BC) + r^2\ (CC)$$

In the blue mussel found along the Atlantic coast of North America, three alleles are common at a locus coding for the enzyme leucine amino peptidase (LAP). Geographic variation in allele frequencies of this locus is shown in Figure 23.3. For a population of mussels inhabiting Long Island Sound, R. K. Koehn and colleagues determined that the frequencies of the three alleles were as follows:

Allele	Frequency
LAP^{98}	$p = 0.52$
LAP^{96}	$q = 0.31$
LAP^{94}	$r = 0.17$

If the population were in Hardy-Weinberg equilibrium, the expected genotypic frequencies would be

Genotype	Expected Frequency		
LAP^{98}/LAP^{98}	p^2	$= (0.52)^2$	$= 0.27$
LAP^{98}/LAP^{96}	$2pq$	$= 2(0.52)(0.31)$	$= 0.32$
LAP^{96}/LAP^{96}	q^2	$= (0.31)^2$	$= 0.10$
LAP^{96}/LAP^{94}	$2qr$	$= 2(0.31)(0.17)$	$= 0.10$
LAP^{94}/LAP^{98}	$2pr$	$= 2(0.52)(0.17)$	$= 0.18$
LAP^{94}/LAP^{94}	r^2	$= (0.17)^2$	$= 0.03$

The square of the gene frequencies can be used in the same way to estimate the expected frequencies of the genotypes when four or more alleles are present at a locus.

If alleles are X-linked, females may be homozygous or heterozygous, but males carry only a single allele for each X-linked locus. For X-linked alleles in females, the Hardy-Weinberg frequencies are the same as those for autosomal loci: p^2 (X^AX^A), $2pq$ (X^AX^B), and q^2 (X^BX^B). In males, however, the frequencies of the genotypes will be p (X^AY) and q (X^BY), the same as the frequencies of the alleles in the population. For this reason, recessive X-linked traits are more frequent among males than among females. To illustrate this concept, consider red-green color blindness, which is an X-linked recessive trait. The frequency of the color-blind allele varies among human ethnic groups; the frequency among American blacks is 0.039. At equilibrium, the expected frequency of color-blind males in this group is $q = 0.039$, but the frequency of color-blind females is only $q^2 = (0.039)^2 = 0.0015$.

~ **FIGURE 23.3**

Geographic variation in allelic frequencies of the locus coding for leucine amino peptidase (*LAP*) in the blue mussel.

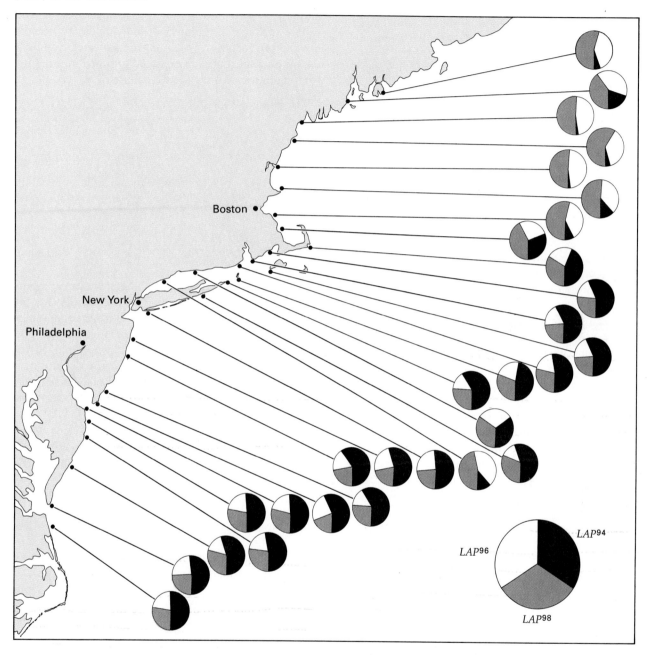

When random mating occurs within a population, the equilibrium genotypic frequencies are reached in one generation. However, if the alleles are X-linked and the sexes differ in allelic frequency, the equilibrium frequencies are approached over several generations. This is because males receive their X chromosome from their mother only, while females receive an X from both the mother and the father. Consequently, the fre-

quency of an X-linked allele in males will be the same as the frequency of that allele in their mothers, whereas the frequency in females will be the average of that in mothers and fathers. With random mating, the allelic frequencies in the two sexes oscillate back and forth each generation, and the difference in gene frequency between the sexes is reduced by half each generation, as shown in Figure 23.4 (p. 718). Once the allelic frequen-

~ **FIGURE 23.4**

Representation of the gradual approach to equilibrium of an X-linked gene with an initial frequency of 1.0 in females and 0 in males.

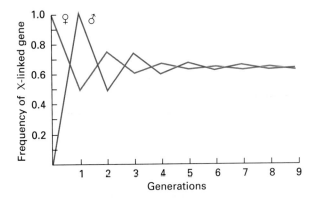

cies of the males and females are equal, the frequencies of the genotypes will be in Hardy-Weinberg proportions after one more generation of random mating.

Testing for Hardy-Weinberg Proportions

To determine whether the genotypes of a population are in Hardy-Weinberg proportions, we first compute p and q from the observed frequencies of the genotypes. Once we have obtained these gene frequencies, we can calculate the expected genotypic frequencies (p^2, $2pq$, and q^2) and compare these frequencies with the actual observed frequencies of the genotypes using a chi-square test (see Chapter 5). The chi-square test gives us the probability that the difference between what we observed and what we expect under the Hardy-Weinberg law, is due to chance.

To illustrate this procedure, consider a locus that codes for transferrin (a blood protein) in the red-backed vole, *Clethrionomys gapperi*. Three genotypes are found at the transferrin locus: *MM*, *MJ*, and *JJ*. In a population of voles trapped in the Northwest Territories of Canada in 1976, 12 *MM* individuals, 53 *MJ* individuals, and 12 *JJ* individuals were found. To determine if the genotypes are in Hardy-Weinberg proportions, we first calculate the gene frequencies for the population using our familiar formula:

$$p = \frac{(2 \times \text{number of homozygotes}) + (\text{number of heterozygotes})}{(2 \times \text{total number of individuals})}$$

Therefore:

$$p = f(M) = \frac{(2 \times 12) + (53)}{(2 \times 77)} = 0.50$$

$$q = f(J) = 1 - p = 0.50$$

Using p and q calculated from the observed genotypes, we can now compute the expected Hardy-Weinberg proportions for the genotypes: $f(MM) = p^2 = (0.50)^2 = 0.25$; $f(MJ) = 2pq = 2(0.50)(0.50) = 0.50$; and $f(JJ) = q^2 = (0.50)^2 = 0.25$. However, for the chi-square test, actual numbers of individuals are needed, not the proportions. To obtain the expected numbers, we simply multiply each expected proportion times the total number of individuals counted (N), as shown below.

	Expected	Observed
$f(MM) = p^2 \times N$ $= 0.25 \times 77 =$	19.3	12
$f(MJ) = 2pq \times N$ $= 0.50 \times 77 =$	38.5	53
$f(JJ) = q^2 \times N$ $= 0.25 \times 77 =$	19.3	12

With observed and expected numbers, we can compute a chi-square value to determine the probability that the differences between observed and expected numbers could be the result of chance. The chi-square is computed using the same formula that we employed for analyzing genetic crosses; that is, d, the deviation, is calculated for each class as (observed − expected); d^2, the deviation squared, is divided by the expected number e for each class; and chi-square (χ^2) is computed as the sum of all d^2/e values. For this example, $\chi^2 = 10.98$. We now need to find this value in the chi-square table (Table 5.2, p. 139) under the appropriate degrees of freedom. This step is not as straightforward as in our previous χ^2 analyses. In those examples the number of degrees of freedom was the number of classes in the sample minus 1. Here, however, while there are three classes, there is only one degree of freedom, because the frequencies of alleles in a population have no theoretically expected values. Thus p must be estimated from the observations themselves. So one degree of freedom is lost for every parameter (p in this case) that must be calculated from the data. Another degree of freedom is lost because, for the fixed number of individuals, once all but one of the classes have been determined, the last class has no degree of freedom and is set automatically. Therefore with three classes (*MM*, *MJ*, and *JJ*), two degrees of freedom are lost, leaving one degree of freedom.

In the chi-square table under the column for one degree of freedom, the chi-square value of 10.98 indicates a P value less than 0.05. Thus the probability that the differences between the observed and expected values is due to chance is very low. That is, the observed numbers of genotypes do not fit the expected numbers under Hardy-Weinberg law.

Using the Hardy-Weinberg Law to Estimate Gene Frequencies

An important application of the Hardy-Weinberg law is the calculation of gene frequencies when one or more alleles is recessive. For example, we have seen that albinism in humans results from an autosomal recessive gene. Normally, this trait is rare, but among the Hopi Indians of Arizona, albinism is remarkably common. Charles M. Woolf and Frank C. Dukepoo visited the Hopi villages in 1969 and observed 26 cases of albinism in a total population of about 6000 Hopis (Figure 23.5). This gave a frequency for the trait of 26/6000, or 0.0043, which is much higher than the frequency of albinism in most populations. Although we have calculated the frequency of the trait, we cannot directly determine the frequency of the albino gene, because we cannot distinguish between heterozygous individuals and those individuals homozygous for the normal allele. Recall that our computation of the gene frequency involves counting the number of alleles:

$$p = \frac{(2 \times \text{number of homozygotes}) + (\text{number of heterozygotes})}{2N}$$

~ FIGURE 23.5

Three Hopi girls, photographed about 1900. The middle child is an albino. Albinism, an autosomal recessive disorder, occurs in high frequency among the Hopi Indians of Arizona.

But, since heterozygotes for a recessive trait such as albinism cannot be identified, this is impossible. Nevertheless, we can determine the gene frequency from the Hardy-Weinberg law if the population is in equilibrium. At equilibrium, the frequency of the homozygous recessive genotype is q^2. For albinism among the Hopis, $q^2 = 0.0043$, and q can be obtained by taking the square root of the frequency of the trait. Therefore, $q = \sqrt{0.0043} = 0.065$, and $p = 1 - q = 0.935$. Following the Hardy-Weinberg law, the frequency of heterozygotes in the population will be $2pq = 2 \times 0.935 \times 0.065 = 0.122$. Thus, one out of eight Hopis, on the average, carries an allele for albinism!

We should not forget that this method of calculating gene frequency rests upon the assumptions of the Hardy-Weinberg law. If the conditions of the Hardy-Weinberg law do not apply, then our estimate of gene frequency will be inaccurate. Also, once we calculate gene frequencies with these assumptions, we cannot then test the population to determine if the genotypic frequencies are in the Hardy-Weinberg expected proportions. To do so would involve circular reasoning, for we assumed Hardy-Weinberg proportions in the first place to calculate the gene frequencies.

GENETIC VARIATION IN NATURAL POPULATIONS

One of the most significant questions addressed in population genetics is how much genetic variation exists within natural populations. The amount of heritable variation within populations is important for several reasons. First, it determines the potential for evolutionary change and adaptation. The amount of variation also provides us with important clues about the relative importance of various evolutionary forces, since some forces increase variation and others decrease it. Furthermore, the manner in which new species arise may depend upon the amount of genetic variation harbored within populations. For all these reasons, population geneticists are interested in measuring genetic variation and in attempting to understand the evolutionary forces that control it.

Models of Genetic Variation

During the 1940s and 1950s, population geneticists developed two opposing hypotheses about the amount of genetic variation within natural populations. These hypotheses, or models of genetic variation, have become known as the **balance model** and the **classical model.** The classical model emerged primarily from the work of laboratory geneticists, who proposed that most natural populations possess little genetic variation.

According to this model, within each population one allele functions best, and this allele is strongly favored by natural selection. As a consequence, almost all individuals in the population are homozygous for this "best" or "wild-type" allele. New alleles arise from time to time through mutation, but almost all are deleterious and are kept in low frequency by selection. Once in a long while, a new mutation arises that is better than the wild-type allele. This new allele increases the survival and reproduction of the individuals that carry it, so the frequency of the allele increases over time because of its selective advantage. Eventually the new allele reaches high frequency and becomes the new wild-type. In this way, a population evolves, but little genetic variation is found within the population at any one time.

In contrast, the balance model predicts that no single allele is best and widespread. The gene pool of a population consists of many alleles at each locus; therefore individuals in the populations are heterozygous at numerous loci. Evolution occurs through gradual shifts in the frequencies and kinds of alleles. To account for the presence of this variation, the proponents of the balance model suggest that natural selection actively maintains genetic variation within a population through balancing selection. Balancing selection is selection that maintains a balance between alleles, preventing any single allele from reaching high frequency. One form of balancing selection is overdominance, in which the heterozygote has higher fitness than either homozygote. The balance view of genetic variation arose among those population geneticists whose backgrounds were in natural history and who studied wild populations.

Measuring Genetic Variation with Protein Electrophoresis

For many years, population geneticists were unable to establish how much variation existed within natural populations. It was difficult to determine which model of genetic variation was correct because no general procedure was available for assessing the amount of genetic variation in nature. To distinguish between the classical and balance models, population geneticists required data on genotypes at many loci of many individuals from multiple species.

Until 1966, no technique existed for obtaining this information. Naturalists recognized that plants and animals in nature frequently differ in phenotype (Figure 23.6), but the genetic basis of most traits is too complex to assign specific genotypes to individuals. A few species do possess simple genetic traits, such as spot patterns in butterflies and shell color in snails, but these isolated cases were too few to provide any general estimate of genetic variation. Indirect methods suggested that many populations possess large amounts of genetic variation, but no direct measure was available.

In 1966, population geneticists began to apply the principle of protein electrophoresis to the study of natural populations. Electrophoresis is a biochemical technique that separates proteins with different molecular structures. (See Box 23.3 [p. 722] for a description of protein electrophoresis.) For population geneticists, electrophoresis provides a technique for quickly genotyping many individuals at many loci. This procedure is now used to examine genetic variation in hundreds of plant and animal species. The amount of genetic variation within a population is commonly measured with two parameters, the **proportion of polymorphic loci** and **heterozygosity.**

A polymorphic locus is any locus that has more than one allele present within a population. The proportion of polymorphic loci (P) is calculated by determining the number of polymorphic loci and dividing by the total number of loci examined. For example, suppose we examined 33 loci in a population of green frogs and found that 18 were polymorphic. The proportion of polymorphic loci would be $18/33 = 0.55$. Heterozygosity (H) is the proportion of an individual's loci that are heterozygous. In our green frogs, suppose we genotyped individuals from one population at a locus coding for esterase and found that the frequency of heterozygotes was 0.09. Heterozygosity for this locus would be 0.09. We would average this heterozygosity with those for other loci and obtain an estimate of heterozygosity for the population.

Table 23.3 (p. 723) presents estimates of the proportion of polymorphic loci and heterozygosity for many species that have been surveyed with electrophoresis. The results of these studies are unambiguous—most species possess large amounts of genetic variation in their proteins, and the classical model is clearly wrong. Actually, the technique of standard electrophoresis misses a large proportion of the genetic variation that is present, because only genetic variants that cause a change in the movement of the protein on a gel will be observed. Therefore, the true amount of genetic variation is even greater than that revealed by this technique.

Studies of electrophoretic variation in natural populations showed that the classical model was incorrect, because populations possess large amounts of genetic variation. However, this did not prove that the balance model was correct. The balance model predicts that much genetic variation is found within natural populations, and this turns out to be the case. But the balance model also proposes that large amounts of genetic variation are maintained by balancing selection,

~ FIGURE 23.6

Extensive phenotypic variation exists in most natural populations, as apparent in the color patterns of the Cuban tree snail, a species treasured for its varicolored beauty.

and this fact was not conclusively demonstrated by electrophoretic studies.

At this point, the nature of the controversy changed. Previously, the central question was how much genetic variation exists. Now, emphasis shifted to the question of what maintains the extensive variation observed. A new hypothesis arose to replace the classical model. This idea, called the **neutral-mutation hypothesis,** acknowledges the presence of extensive genetic variation in proteins but proposes that this vari-

Box 23.3
Analysis of Genetic Variation with Protein Electrophoresis

Gel electrophoresis has been widely used to study genetic variation in natural populations. This process is a biochemical technique that separates large molecules on the basis of size, shape, and charge. To examine genetic variation in proteins with electrophoresis, separate tissue samples, taken from a number of individuals, are ground up, releasing the proteins into an aqueous solution. The individual solutions are then inserted into a gel made of starch, agar, polyacrylamide, or some other porous substance. Electrodes are placed at the ends of the gel, as shown in Box Figure 23.1a, and a direct electrical current is supplied, setting up an electrical field across the gel.

Because proteins are charged molecules, they will migrate within the electrical field. Molecules with different shape, size, and/or charge will migrate at different rates and will separate from one another within the gel over time. If two individuals have different genotypes at a locus coding for a protein, they will produce slightly different molecular forms of the protein, which can be separated and identified with electrophoresis.

After a current has been supplied to the gel for several hours and the proteins allowed to separate, the current is turned off and the proteins visualized. The gel now contains a large number of different enzymes and proteins; to identify the product of a single locus, one must stain for a specific enzyme or protein. This is frequently accomplished by having the enzyme in the gel carry out a specific biochemical reaction. The substrate for the reaction is added to the gel, along with a dye that changes color when the reaction takes place. A colored band will appear on the gel wherever the enzyme is located.

If two individuals have different molecular forms of the enzyme, indicating differences in genotypes, bands for those individuals will appear at different locations on the gel. By examining the pattern of bands produced, it is also possible to determine whether an individual is homozygous or heterozygous for a particular allele. A diagram of the banding pattern is shown in Box Figure 23.1b. Histochemical stains are available for dozens of enzymes, so many individuals may be quickly genotyped for variation at a number of loci.

~ **BOX FIGURE 23.1**
The technique of protein electrophoresis used to measure genetic variation in natural populations. (a) Solutions of proteins from different individuals are inserted into the slots in the gel, and a direct current is supplied to separate the proteins. (b) After electrophoresis, enzyme-specific staining reactions are used to reveal the proteins. The pattern of bands on the gel indicates the genotype of each individual.

a)

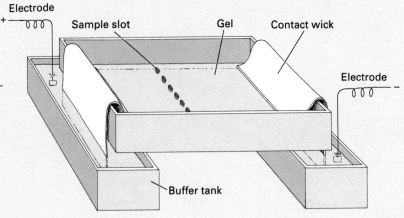

b)

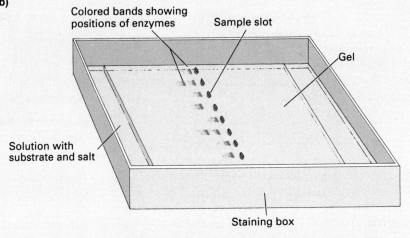

~ **TABLE 23.3**

Genic Variation in Some Major Groups of Animals and in Plants

GROUP	NUMBER OF SPECIES OR FORMS	MEAN NUMBER OF LOCI PER SPECIES	MEAN PROPORTION OF LOCI	
			POLYMORPHIC PER POPULATION	HETEROZYGOUS PER INDIVIDUAL
INSECTS				
Drosophila	28	24	0.529 ± 0.030	0.150 ± 0.010
Others	4	18	0.531	0.151
Haplodiploid wasps	6	15	0.243 ± 0.039	0.062 ± 0.007
MARINE INVERTEBRATES	9	26	0.587 ± 0.084	0.147 ± 0.019
SNAILS				
Land	5	18	0.437	0.150
Marine	5	17	0.175	0.083
FISH	14	21	0.306 ± 0.047	0.078 ± 0.012
AMPHIBIANS	11	22	0.336 ± 0.034	0.082 ± 0.008
REPTILES	9	21	0.231 ± 0.032	0.047 ± 0.008
BIRDS	4	19	0.145	0.042
RODENTS	26	26	0.202 ± 0.015	0.054 ± 0.005
LARGE MAMMALS[a]	4	40	0.233	0.037
PLANTS[b]	8	8	0.464 ± 0.064	0.170 ± 0.031

[a] Human, chimpanzee, pigtailed macaque, and southern elephant seal.

[b] Predominantly outcrossing species; mean gene diversity is 0.233 ± 0.029.

From Selander, R. K. 1976. Genetic variation in natural populations, in F. J. Ayala. *Molecular Evolution.* Sinauer, Sunderland, Mass.

ation is neutral with regard to natural selection. This does not mean that the proteins detected by electrophoresis have no function, but rather that the different genotypes are physiologically equivalent. Therefore, natural selection does not act on the neutral alleles, and random processes such as mutation and genetic drift shape the patterns of genetic variation that we see in natural populations. The neutral-mutation hypothesis proposes that variation at some loci does affect fitness, and natural selection eliminates variation at these loci.

Distinguishing between the balance model and the neutral-mutation hypothesis has been difficult. Population geneticists still argue over the relative merits of the two, and no clear consensus has emerged as to which model of genetic variation is correct. In reality, both models may be partly correct; at some loci genetic variation may be essentially neutral, while at others, it may be acted upon by natural selection.

KEYNOTE

During the 1940s and 1950s, two hypotheses were developed to explain the amount of variation in natural populations. The classical model predicted that little variation would occur, and the balance model proposed that much genetic variation would be maintained in natural populations. Through the use of electrophoresis, population geneticists eventually demonstrated that natural populations harbor much genetic variation, thus disproving the classical model. However, the forces responsible for keeping this variation within populations are still controversial. The neutral-mutation hypothesis proposes that the genetic variation detected by electrophoresis is neutral with regard to natural selection, whereas the balance model proposes that this variation is favored by balancing selection._____

Measuring Genetic Variation with RFLPs and DNA Sequencing

Techniques in molecular genetics now provide the means to directly examine genetic variation in the DNA and to unambiguously determine the amount of genetic variation present in natural populations. One such technique employs restriction enzymes, which were discussed in Chapter 15, for detecting genetic variation. You will recall that restriction enzymes make double-stranded cuts in DNA at specific base sequences. Most restriction enzymes recognize a sequence of four bases or six bases; see Table 15.1. For example, the restriction enzyme *Bam*HI recognizes the sequence $\frac{\text{GGATCC}}{\text{CCTAGG}}$; whenever this sequence appears in the DNA, *Bam*HI will cut the DNA. The resulting fragments can be separated by agarose gel electrophoresis and can be observed by staining the DNA or by using probes for specific genes (see Chapter 15).

Suppose that two individuals differ in one or more nucleotides at a particular DNA sequence and that the differences occur at a site recognized by a restriction enzyme (Figure 23.7). One individual has a DNA molecule with the restriction site, but the other individual does not, because the sequence of DNA nucleotides differ. If the DNA from these two individuals is mixed

~ FIGURE 23.7

DNA from individual 1 and individual 2 differ in one nucleotide, found within the sequence recognized by the restriction enzyme *Bam*HI. Individual 1's DNA contains the *Bam*HI restriction sequence and is cleaved by the enzyme. Individual 2's DNA lacks the *Bam*HI restriction sequence and is not cleaved by the enzyme. When placed on an agarose gel and separated by electrophoresis, the DNAs from 1 and 2 produce different patterns on the gel. This variation is called a restriction fragment length polymorphism.

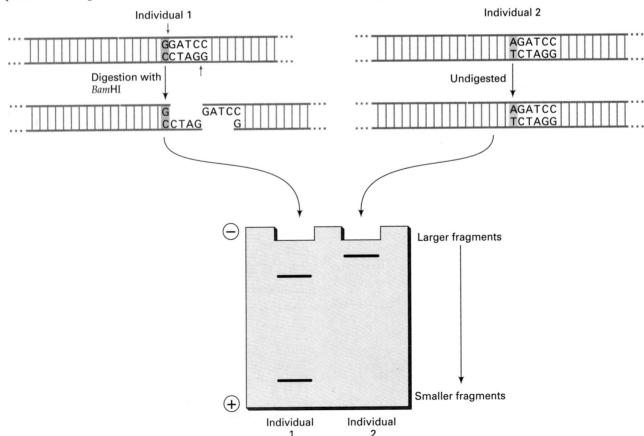

with the restriction enzyme and the resulting fragments are separated on a gel, the two individuals produce different patterns of fragments, as shown in Figure 23.7. The different patterns on the gel are termed Restriction Fragment Length Polymorphisms, or RFLPs. They indicate that the DNA sequences of the two individuals differ. RFLPs are inherited in the same way that alleles coding for other traits are inherited, only the RFLPs do not produce any outward phenotypes; their phenotypes are the fragment patterns produced on a gel when the DNA is cut by the restriction enzyme. RFLPs can be used as genetic markers for mapping genes. They can also provide information about how DNA sequences differ among individuals. Such differences involve only a small part of the DNA, specifically those few nucleo-tides recognized by the restriction enzyme. However, if we assume that restriction sites occur randomly in the DNA, which is not an unreasonable assumption since the sites are not expressed as traits, the presence or absence of restriction sites can be used to estimate the overall differences in sequence.

To illustrate the use of RFLPs for estimating genetic variation, suppose we isolate DNA from 5 wild mice, cut the DNA with the restriction enzyme *Bam*HI, and separate the fragments with agarose gel electrophoresis. We then transfer the DNA to nitrocellulose, using Southern blotting (see Chapter 15), and add a probe that will detect the gene for β hemoglobin. A typical set of restriction patterns that might be obtained is shown in Figure 23.8. Remember that each mouse carries two

~ FIGURE 23.8

Restriction patterns from five mice. The patterns differ due to the presence (+) or absence (−) of a restriction site. Each mouse has two homologous chromosomes, each of which potentially carries the restriction site. Thus a mouse may be +/+ (has the restriction site on both chromosomes), +/− (has the restriction site on one chromosome), or −/− (has the restriction site on neither chromosome). When the restriction site is present, the DNA is broken into two fragments after digestion with the restriction enzyme; the fragments will be detected after electrophoresis. When the restriction site is absent, the DNA fragment is not cut with the restriction enzyme. The pattern of DNA fragments visualized after electrophoresis indicates the chromosome constitution of the mice.

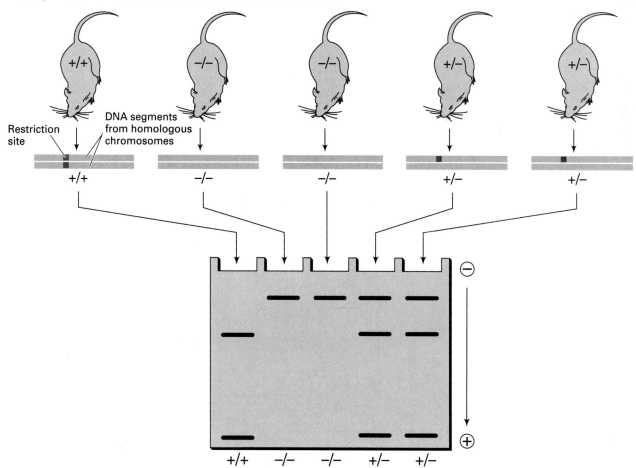

copies of the β hemoglobin gene, one on each homologous chromosome. Thus, a mouse could be $+/+$ (the restriction site is present on both chromosomes), $+/-$ (the restriction site is present on one chromosome and is absent from the other), and $-/-$ (the restriction site is absent from both chromosomes). For the ten chromosomes present among these particular five mice, four have the restriction site and six do not.

To calculate the expected heterozygosity in nucleotide sequence, we use the formula:

$$H_{nuc} = \frac{n(\Sigma c_i) - \Sigma c_i^2}{j(\Sigma c_i)(n - 1)}$$

In this equation, n equals the number of homologous DNA molecules examined. In our example we looked at five mice, each with two homologous chromosomes, so $n = 10$. The quantity j equals the number of nucleotides in the restriction site; in our example this is six, because *Bam*HI recognizes a six-base sequence. For each restriction site i, c_i represents the number of molecules in the sample that were cleaved at that restriction site. In our example, we examined only a single restriction site, so we have a single value of c_i which is four. If additional restriction enzymes were used, we would have a value of c_i for each site recognized by an enzyme. The symbol Σc_i is the total number of cuts at all cleavage sites in all the chromosomes. Since we examined only a single restriction site, and that site was cut in four of the chromosomes, $\Sigma c_i = 4$. Thus we obtain

$$H_{nuc} = \frac{10(4) - (4)^2}{6(4)(10 - 1)} = \frac{40 - 16}{24(9)} = \frac{24}{216} = .11$$

The use of this equation assumes that each RFLP results from a single nucleotide difference.

Nucleotide heterozygosity has been studied for a number of different organisms through the use of restriction enzymes. A few examples are shown in Table 23.4. Nucleotide heterozygosity typically varies from 0.002 to 0.02 in eukaryotic organisms. This means that an individual is heterozygous (contains different nucleotides on the two homologous chromosomes) at about one in every 50 to 500 nucleotides. Another way to interpret nucleotide heterozygosity is that two randomly chosen chromosomes from a population will differ at about one in every 50 to 500 nucleotides.

One disadvantage of using RFLPs for examining genetic variation is that this method reveals variation at only a small subset of the nucleotides that make up a gene. With RFLP analysis, we are taking a small sample of the nucleotides (those recognized by the restriction enzyme) and using this sample to estimate the overall level of variation. DNA sequencing methods, which were described in Chapter 15, provide a method for detecting all nucleotide differences that exist among a set of DNA molecules. For example, Martin Krietman sequenced 11 copies (obtained from different fruit flies) of a 2,659 base pair segment of the alcohol dehydrogenase gene in *Drosophila melanogaster*. Among the 11 copies, he found different nucleotides at 43 positions within the 2,659 base pair segment. Furthermore, only 3 of the 11 copies were identical at all nucleotides examined—thus, there were 8 different alleles (at the nucleotide level) among the 11 copies of this gene! This suggests that populations harbor a tremendous amount of genetic variation in their DNA sequences.

~ TABLE 23.4

Estimates of Nucleotide Heterozygosity for DNA Sequences

DNA SEQUENCES	ORGANISM	H_{nuc}
β-Globin genes	Humans	0.002
Growth hormone gene	Humans	0.002
Alcohol dehydrogenase gene	Fruit fly	0.006
Mitochondrial DNA	Humans	0.004
H4 gene region	Sea urchin	0.019

CHANGES IN GENE FREQUENCIES OF POPULATIONS

We have seen that heredity does not, by itself, generate changes in the gene frequencies of a population. When the population is large, randomly mating, and free from outside forces, no evolution occurs. For many populations, however, the conditions required by the Hardy-Weinberg law do not hold. Populations are frequently small, mating may be nonrandom, and other evolutionary forces may occur. In these circumstances, gene frequencies do change, and the gene pool of the population evolves in response to the interplay of different evolutionary factors. In the following sections we discuss the role of four evolutionary forces—mutation, genetic drift, migration, and natural selection—in changing the gene frequencies of a population.

Mutation

One force that can potentially alter the frequencies of alleles within a population is mutation. As we discussed in Chapter 18, gene mutations consist of heritable changes in the DNA that occur within a locus. Usually a mutation converts one allelic form of the gene to another. The rate at which mutations arise is generally low, but varies between loci and among species (Table 23.5; pp. 728–29). Certain genes affect overall mutation rates, and many environmental factors, such as chemicals, radiation, and infectious agents, may increase the number of mutations.

Evolution is a two-step process: first, genetic variation arises; then, different alleles increase or decrease in frequency in response to evolutionary forces. Mutation is potentially important in both steps.

Ultimately, mutation is the source of all new genetic variation; new combinations of alleles may arise through recombination, but new alleles only occur as a result of mutation. Thus, mutation provides the raw genetic material upon which evolution acts. Most mutations will be detrimental and will be eliminated from the population. A few mutations, however, will convey some advantage to the individuals that possess them and will spread through the population. Whether a mutation is detrimental or advantageous depends upon the specific environment, and if the environment changes, previously harmful mutations may become beneficial. For example, after the widespread use of the insecticide DDT, insects with mutations that conferred resistance to DDT were capable of surviving and reproducing; because of this advantage, the mutations spread and many insect populations quickly evolved resistance to DDT. Mutations are of fundamental importance to the process of evolution, since they provide the genetic variation upon which other evolutionary forces act.

Mutations also have the potential to affect evolution by contributing to the second step in the evolutionary process—by changing the frequency of alleles within a population. Consider a population that consists of 50 individuals, all with the genotype AA. The frequency of $A(p)$ is 1.00 [$(2 \times 50)/100$]. If one A allele mutates to a, the population now consists of 49 AA individuals and 1 Aa individual. The frequency of p is now [$(2 \times 49) + 1$]/100 = 0.99. When another mutation occurs, the frequency of A drops to 0.98. If these mutations continue to occur at a low but steady rate over long periods of time, the frequency of A will eventually decline to zero, and the frequency of a will reach a value of 1.00.

This effect of the mutational process on allelic frequencies is analogous to the following situation: If we have a large jar of black marbles, and every year or so we take out one black marble and replace it with a white marble, the jar eventually will contain only white marbles. Similarly, in populations the slow, persistent pressure of recurrent mutation may lead to significant changes in gene frequency, assuming other evolutionary forces are not present.

The mutation of A to a is referred to as a *forward mutation*. For most genes, mutations also occur in the reverse direction; a may mutate to A. These mutations are called *reverse mutations*, and they typically occur at a somewhat lower rate than forward mutations. To include reverse mutation in our marble jar analogy, suppose we start with a jar of black marbles, but instead of taking out a black marble at each dip and replacing it with a white one, we take out any marble at random and replace it with one of the opposite color. With this sampling system there is no selection against either marble type. At the start we are more likely to pull out black marbles and replace them with white ones, but as white marbles become more common in the jar, we are more and more likely to pull out a white one and replace it with black. At equilibrium, black and white marbles will occur in equal frequency. At this point no further changes in black and white marble frequencies will occur as long as the same method of marble replacement is applied.

In our example with marbles, the forward and reverse mutation rates were equal, since every marble drawn from the jar, whether black or white, was replaced with one of the opposite color. In practice, forward and reverse mutation rates usually differ. The forward mutation rate—the rate at which A mutates to a $(A \rightarrow a)$—is symbolized with u; the reverse mutation rate—the rate at which a mutates to A $(A \leftarrow a)$—is symbolized with v. Consider a hypothetical population in which the frequency of A is p and the frequency of a is q. We assume that the population is large and that no selection occurs on the alleles. Each generation, a proportion u of all A alleles mutates to a. The actual number mutating depends upon both u and the frequency of A alleles. For example, suppose the population consists of 100,000 alleles. If u equals 10^{-4}, one out of every 10,000 A alleles mutates to a. When $p = 1.00$, all 100,000 alleles in the population are A, and free to mutate to a, so $10^{-4} \times 100,000 = 10$ A alleles should mutate to a. However, if $p = 0.10$, only 10,000 alleles are A and free to mutate to a. Therefore, with a mutation rate of 10^{-4} only 1 of the A alleles will undergo mutation. The decrease in the frequency of A resulting from mutation of $A \rightarrow a$ is equal to up; the increase in frequency resulting from $A \leftarrow a$ is equal to vq. As a result of mutation, the amount that A decreases in one generation is equal to the increase in A alleles due to reverse mutations minus the decrease in A alleles due to forward mutations. That is, the change in the frequency of $A(\Delta p)$ is

$$\Delta p = vq - up$$

As we have seen, when p is high, the number of A alleles mutating to a is relatively large, but as more and

~ TABLE 23.5

Spontaneous Mutation Frequencies at Specific Loci for Various Organisms

ORGANISM	TRAIT	MUTATION PER 100,000 GAMETES*
T2 Bacteriophage (virus)	To rapid lysis ($r^+ \rightarrow r$)	7
	To new host range ($h^+ \rightarrow h$)	0.001
E. coli (K12) (bacteria)	To streptomycin resistance	0.00004
	To phage T1 resistance	0.003
	To leucine independence	0.00007
	To arginine independence ($leu^- \rightarrow leu^+$)	0.0004
	To tryptophan independence	0.006
	To arabinose dependence ($ara^+ \rightarrow ara^-$)	0.2
Salmonella typhimurium (bacteria)	To histidine dependence	0.2
	To tryptophan independence	0.005
Diplococcus pneumoniae (bacteria)	To penicillin resistance	0.01
Neurospora crassa	To adenine independence	0.0008–0.029
	To inositol independence	0.001–0.010
	(one inos allele, JH5202)	1.5
Drosophila melanogaster males	y^+ to yellow	12
	bw^+ to brown	3
	e^+ to ebony	2
	ey^+ to eyeless	6
Corn	Wx to waxy	0.00
	Sh to shrunken	0.12
	C to colorless	0.23
	Su to sugary	0.24
	Pr to purple	1.10
	I to i	10.60
	R^r to r^r	49.20

*Mutation rate estimates of viruses, bacteria, Neurospora, and Chinese Hamster somatic cells are based on virus or cell counts.

more A alleles mutate to a, the change in p decreases. At the same time, when q is small, few alleles will be available to undergo reverse mutation, but as q increases, the number of alleles undergoing reverse mutation increases. Eventually, the population achieves equilibrium, in which the number of alleles undergoing forward mutation is exactly equal to the number of alleles undergoing reverse mutation. At this point, no further change in gene frequency occurs, in spite of the fact that forward and reverse mutations continue to take place.

When equilibrium is reached, the change in gene frequency is zero:

$$\Delta p = vq - up = 0$$

Therefore,

$$vq = up$$

and

$$vq = u(1 - q)$$

since $p = (1 - q)$. Solving this equation, we obtain the equilibrium value of q (symbolized by $\hat{q}$):

$$vq = u - uq$$

$$vq + uq = u$$

$$q(v + u) = u$$

$$\hat{q} = \frac{u}{u + v}$$

~ TABLE 23.5 continued

Spontaneous Mutation Frequencies at Specific Loci for Various Organisms[a]

ORGANISM	TRAIT	MUTATION PER 100,000 GAMETES[b]
Mouse	a^+ to *nonagouti*	2.97
	b^+ to *brown*	0.39
	c^+ to *albino*	1.02
	d^+ to *dilute*	1.25
	ln^+ to *leaden*	0.80
	Reverse mutations for above genes	0.27
Chinese hamster somatic cell tissue culture	To azaguanine resistance	0.0015
	To glutamine independence	0.014
Humans	Achondroplasia	0.6–1.3
	Aniridia	0.3–0.5
	Dystrophia myotonica	0.8–1.1
	Epiloia	0.4–1
	Huntington's chorea	0.5
	Intestinal polyposis	1.3
	Neurofibromatosis	5–10
	Osteogenesis imperfecta	0.7–1.3
	Pelger's anomaly	1.7–2.7
	Retinoblastoma	0.5–1.2

[a] Mutations to independence for nutritional substances are from the auxotrophic condition (e.g., *leu*⁻) to the prototrophic condition (e.g., *leu*⁺).

[b] Mutation frequency estimates of viruses, bacteria, *Neurospora*, and Chinese hamster somatic cells are based on particle or cell counts rather than gametes.

From Strickberger, M. W. 1985. *Genetics*, 3d ed. Macmillan, New York.

The equilibrium value for p, $\hat{p}$, is $1 - \hat{q}$. If we solve the above equations for p instead of q, we obtain

$$\hat{p} = \frac{v}{u + v}$$

Consider a population in which the initial gene frequencies are $p = 0.9$ and $q = 0.1$ and the forward and reverse mutation rates are $u = 5 \times 10^{-5}$ and $v = 2 \times 10^{-5}$, respectively. (These values are similar to forward and reverse mutation rates observed for many genes.) In the first generation the change in gene frequency is

$$\Delta p = vq - up$$
$$= (2 \times 10^{-5} \times 0.1) - (5 \times 10^{-5} \times 0.9)$$
$$\Delta p = -0.000043$$

The frequency of A decreases by only four-thousandths of 1 percent. At equilibrium, the frequency of the a allele, $\hat{q}$, equals

$$\hat{q} = \frac{u}{u + v}$$

$$\hat{q} = \frac{5 \times 10^{-5}}{(5 \times 10^{-5}) + (2 \times 10^{-5})} = 0.714$$

If no other forces act on a population, after many generations the alleles will reach equilibrium. Therefore, mutation rates determine the allelic frequencies of the population in the absence of other evolutionary forces. However, because mutation rates are so low, the change in gene frequency due to mutation pressure is exceedingly slow. Furthermore, as gene frequencies approach their equilibrium values, the change in frequency becomes smaller and smaller. Suppose that the mutation rate of $A \rightarrow a$ is $u = 10^{-5}$; suppose also that no reverse mutation occurs. With no reverse mutation, the equilibrium frequency of A is 0.0. If the frequency of A is initially 1.00, 1000 generations are required to change the frequency of A from 1.00 to 0.99. To change the frequency from 0.50 to 0.49, 2000 generations are required, and to change it from 0.1 to 0.09, 10,000 generations are necessary. If some reverse mutation occurs, the rate of change is even slower.

In practice, mutation by itself changes the gene frequencies at such a slow rate that populations are rarely in mutational equilibrium. Other forces have more profound effects on gene frequencies, and mutation alone rarely determines the gene frequencies of a population. For example, achondroplastic dwarfism is an autosomal dominant trait in humans that arises through recurrent mutation. However, the frequency of this disorder in human populations is determined by an interaction of mutation pressure and natural selection.

$\mathcal{K}$EYNOTE

Recurrent mutation can alter the gene frequencies of a population over time, provided that other evolutionary forces are not active. Eventually, a mutational equilibrium is reached, in which the gene frequencies of the population remain constant in spite of continuing mutation; the gene frequencies at this equilibrium are a function of the forward and reverse mutation rates. ───────

Genetic Drift

A major assumption of the Hardy-Weinberg law is that the population is infinitely large. Real populations are not infinite in size, but frequently they are large enough that expected ratios are realized and chance factors have insignificant effects on gene frequencies. Some populations are small, however, and in these groups chance factors may produce random changes in gene frequencies. Random change in gene frequency due to chance is called **genetic drift,** or simply *drift* for short. Sewall Wright (Figure 23.9), a brilliant population geneticist who laid much of the theoretical foundation of the discipline, championed the importance of genetic drift in the 1930s, so sometimes genetic drift is called the *Sewall Wright effect* in his honor.

~ FIGURE 23.9

Sewall Wright, an important figure in the field of population genetics, made a major contribution to our understanding of the role of genetic drift in the evolutionary process.

CHANCE CHANGES IN GENE FREQUENCY. Random changes in gene frequency resulting from chance can be an important evolutionary force in small populations. Imagine a small group of humans inhabiting a South Pacific island. Suppose that this population consists of only 10 individuals, five of whom have green eyes and five of whom have brown eyes. For this example, we assume that eye color is determined by a single locus (actually, eye color is polygenic) and that the allele for green eyes is recessive to brown (*BB* and *Bb* code for brown eyes and *bb* codes for green). The frequency of the allele for green eyes is 0.6 in the island population. A typhoon strikes the island, killing 50 percent of the population; five of the inhabitants perish in the storm. Just by chance, those five individuals who die all have brown eyes. Eye color in no way affects the probability of surviving; the fact that only those with green eyes survive is strictly the result of chance. After the typhoon, the gene frequency for green eyes is 1.0. Evolution has occurred in this population—the frequency of the green-eye allele has changed from 0.6 to 1.0, simply as a result of chance.

Now, imagine the same scenario, but this time with a population of 1000 individuals. As before, 50 percent of the population has green eyes and 50 percent has brown eyes. A typhoon strikes the island and kills half the population. How likely is it that, just by chance, all 500 people who perish will have brown eyes? In a population of 1000 individuals, the probability of this occurring by chance is extremely remote. This example illustrates an important characteristic of genetic drift—chance factors are likely to produce significant changes in gene frequencies only in small populations.

Random factors producing mortality in natural populations, such as the typhoon in the above example, is only one of several ways in which genetic drift arises. Chance deviations from expected ratios of gametes and zygotes also produce genetic drift. We have seen the importance of chance deviations from expected ratios in the genetic crosses we studied in earlier chapters. For example, when we cross a heterozygote with a homozygote ($Aa \times aa$), we expect 50 percent of the progeny to be heterozygous and 50 percent to be homozygous. We do not expect to get exactly 50 percent every time, however, and if the number of progeny is small, the observed ratio may differ considerably from the expected. Recall that the Hardy-Weinberg law is based upon random mating and expected ratios of progeny resulting from each type of mating (Table 23.2). If the actual number of progeny differs from the expected ratio due to chance, genotypes may not be in Hardy-Weinberg proportions. As a result, changes in gene frequencies may occur.

Chance deviations from expected proportions arise from a general phenomenon called **sampling error.** Imagine that a population produces an infinitely large pool of gametes, with alleles in the proportions p and q. If random mating occurs and all the gametes unite to form zygotes, the proportions of the genotypes will be equal to p^2, $2pq$ and q^2, and the frequencies of the alleles in these zygotes will remain p and q. If the number of progeny is limited, however, the gametes that unite to form the progeny constitute a sample from the infinite pool of potential gametes. Just by chance, or by "error," this sample may deviate from the larger pool; the smaller the sample, the larger the potential deviation.

Flipping a coin is analogous to the situation in which sampling error occurs. When we flip a coin, we expect 50 percent heads and 50 percent tails. If we flip the coin 1000 times, we will get very close to that expected fifty-fifty ratio. But, if we flip the coin only four times, we would not be surprised if by chance we obtain 3 heads and 1 tail, or even all tails. When the sample—in this case the number of flips—is small, the sampling error can be large. All genetic drift arises from such sampling error.

MEASURING GENETIC DRIFT. Genetic drift is random, and thus we cannot predict what the gene frequencies will be after drift has occurred. However, since sampling error is related to the size of the population, we can make predictions about the magnitude of genetic drift. Ecologists often measure population size by counting the number of individuals, but not all individuals contribute gametes to the next generation. To determine the magnitude of genetic drift, we must know the **effective population size,** which equals the equivalent number of adults contributing gametes to the next generation. If the sexes are equal in number and all individuals have an equal probability of producing offspring, the effective population size equals the number of breeding adults in the population. However, when males and females are not present in equal numbers, the effective population size is

$$N_e = \frac{4 \times N_f \times N_m}{N_f + N_m}$$

where N_f equals the number of breeding females and N_m equals the number of breeding males.

Students often have difficulty understanding why this equation must be used—why the effective population size is not simply the number of breeding adults. The reason is that males, as a group, contribute half of all genes to the next generation and females, as a group, contribute the other half. Therefore, in a population of 70 females and 2 males, the two males are not genetically equivalent to two females; each male contributes $\frac{1}{2} \times \frac{1}{2} = 0.25$ of the genes to the next generation, whereas each female contributes $\frac{1}{2} \times \frac{1}{70} = 0.007$ of all genes. The small number of males disproportionately influences what alleles are present in the next generation. Using the above equation, the effective population size is $N_e = (4 \times 70 \times 2)/(70 + 2) = 7.8$, or approximately 8 breeding adults. What this means is that in a population of 70 females and 2 males, genetic drift will occur as if the population had only four breeding males and four breeding females. Therefore, genetic drift will have a much greater effect in this population than in one with 72 breeding adults equally divided between males and females.

Other factors, such as differential production of offspring, fluctuating population size, and overlapping generations can further reduce the effective population size. Considering these complications, it is quite difficult to accurately measure effective population size.

The amount of variation among populations resulting from genetic drift is measured by the **variance of gene frequency,** which equals

$$s_p{}^2 = \frac{pq}{2N_e}$$

where N_e equals the effective population size and p and q equal the gene frequencies. A more useful measure is the **standard error of gene frequency**, which is the square root of the variance of gene frequency:

$$s_p = \sqrt{\frac{pq}{2N_e}}$$

The standard error can be used to calculate the 95 percent confidence limits of gene frequency, which indicate the expected range of p in 95 percent of such populations. The 95 percent confidence limits equal approximately $p \pm 2s_p$. Suppose, for example, that $p = 0.8$ in a population with N_e equal to 50. The standard error in gene frequency is $s_p = \sqrt{pq/2N_e} = 0.04$. The 95 percent confidence limits for p are therefore $p + 2s_p = 0.72 \leq p \leq 0.88$. To interpret the 95 percent confidence limits, imagine that 100 populations with N_e of 50 have p initially equal to 0.8. Genetic drift may cause gene frequencies in some populations to change; the 95 percent confidence limits tell us that in the next generation, 95 of the original 100 populations should have p within the range of 0.72 to 0.88. Therefore, if we observe a change in p greater than this, say from 0.8 to 0.68, we know that the probability that this change will occur by genetic drift is less than 0.05. Most likely we would conclude that some force other than genetic drift contributed to the observed change in gene frequency.

CAUSES OF GENETIC DRIFT. All genetic drift arises from sampling error, but there are several ways in which sampling error occurs in natural populations. First, genetic drift arises when population size remains continuously small over many generations. Undoubtedly, this situation is frequent, particularly where populations occupy marginal habitats, or when competition limits population growth. In such populations, genetic drift plays an important role in the evolution of gene frequencies. The effect of genetic drift arising from small population size is seen in an experiment conducted by Buri with *Drosophila melanogaster*. Buri examined the frequency of two alleles, bw^{75} and bw, at a locus that determines eye color in the fruit flies. He set up 107 experimental populations, and the initial frequency of bw^{75} was 0.5 in each. The flies in each population interbred randomly, and in each generation Buri randomly selected 8 males and 8 females to be the parents for the next generation. Thus, effective population size was always 16 individuals. The distribution of allelic frequencies in these 107 populations is presented in Figure 23.10. Notice that the gene frequencies in the early generations were clumped around 0.5, but genetic drift caused the frequencies in the populations to spread out or diverge over time. By generation 19, the frequency of bw^{75} was 0 or 1 in most populations.

~ **FIGURE 23.10**

Results of Buri's study of genetic drift in 107 populations of *Drosophila melanogaster*. Shown are the distributions of allelic frequency among the populations in 19 consecutive generations. Each population consisted of 16 individuals.

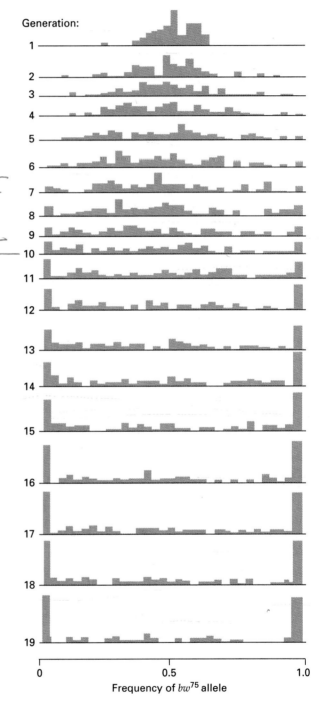

Another way in which genetic drift arises is through **founder effect**. Founder effect occurs when a population is initially established by a small number of individuals. Although the population may subsequently grow in size and later consist of a large number of individuals, the gene pool of the population is derived from the genes present in the original founders. Chance may play a significant role in determining which genes were present among the founders, and this has a profound effect upon the gene pool of subsequent generations.

Many excellent examples of founder effect come from the study of human populations. Consider the inhabitants of Tristan da Cunha, a small, isolated island in the South Atlantic. This island was first permanently settled by William Glass, a Scotsman, and his family in 1817. (Several earlier attempts at settlement failed.) They were joined by a few additional settlers, some shipwrecked sailors, and a few women from the distant island of St. Helena, but for the most part the island remained a genetic isolate. In 1961, a volcano on Tristan da Cunha erupted, and the population of almost 300 inhabitants was evacuated to England. During their two-year stay in England, geneticists studied the islanders and reconstructed the genetic history of the population. These studies revealed that the current gene pool of Tristan da Cunha has been strongly influenced by genetic drift.

Three forms of genetic drift occurred in the evolution of the island's population. First, founder effect took place at the initial settlement. By 1855, the population of Tristan da Cunha consisted of about 100 individuals, but 26 percent of the genes of the population in 1855 were contributed by William Glass and his wife. Even in 1961, these original two settlers contributed 14 percent of all the genes in the 300 individuals of the population. The particular genes that Glass and other original founders carried heavily influenced the subsequent gene pool of the population. Second, population size remained small throughout the history of the settlement, and sampling error continually occurred.

A third form of sampling error, called bottleneck effect, also played an important role in the population of Tristan da Cunha. Bottleneck effect occurs when a population is drastically reduced in size. During such a population reduction, some genes may be lost from the gene pool as a result of chance. Recall our earlier example of the population consisting of 10 individuals inhabiting a South Pacific island. When a typhoon struck the island, the population size was reduced to five, and by chance, all individuals with brown eyes perished in the storm, changing the frequency of green eyes from 0.6 to 1.0. This is an example of bottleneck effect. Bottleneck effect can be viewed as a type of founder effect, since the population is refounded by those few individuals that survive the reduction.

Two severe bottlenecks occurred in the history of Tristan da Cunha. The first took place around 1856 and was precipitated by two events: the death of William Glass and the arrival of a missionary who encour-

aged the inhabitants to leave the island. At this time many islanders emigrated to America and to South Africa, and the population dropped from 103 individuals at the end of 1855 to 33 in 1857. A second bottleneck occurred in 1885. The island of Tristan da Cunha has no natural harbor, and the islanders intercepted passing ships for trade by rowing out in small boats. On November 28, 1885, fifteen of the adult males on the island put out in a small boat to make contact with a passing ship. In full view of the entire island community, the boat capsized and all fifteen men drowned. Following this disaster, only four adult males were left on the island, one of whom was insane and two of whom were old. Many of the widows and their families left the island during the next few years, and the population size dropped from 106 to 59. Both bottlenecks had a major effect on the gene pool of the population. All the genes contributed by several settlers were lost, and the relative contributions of others were altered by these events. Thus, the gene pool of Tristan da Cunha has been influenced by genetic drift in the form of founder effect, small population size, and bottleneck effect.

As we shall see later, when we discuss migration, gene flow among populations increases the effective population size and reduces the effects of genetic drift. Small breeding units that lack gene flow are genetically isolated from other groups and often experience considerable genetic drift, even though surrounded by much larger populations. A good example is a religious sect, known as the Dunkers, found in eastern Pennsylvania. Between 1719 and 1729, fifty Dunker families emigrated from Germany and settled in the United States. Since that time, the Dunkers have remained an isolated group, rarely marrying outside of the sect, and the number of individuals in their communities has always been relatively small.

During the 1950s geneticists studied one of the original Dunker communities in Franklin County, Pennsylvania. At the time of the study, this population had about 300 members, and the population size had remained relatively constant for many generations. The investigators found that some of the gene frequencies in the Dunkers were very different from the frequencies found among the general population of the United States. The Pennsylvania frequencies were also different from the frequencies of the West German population from which the Dunkers descended. Table 23.6 (p. 734) presents some of the gene frequencies at the ABO blood group locus. The ABO allele frequencies among the Dunkers are not the same as those in either the United States population or the West German population. Nor are the Dunker frequencies intermediate between the United States and German frequencies. (Intermediate frequencies might be expected if intermixing of Dunk-

~ TABLE 23.6

Frequencies of Alleles Controlling the ABO Blood Group System in Three Human Populations

POPULATION	ALLELE FREQUENCIES			PHENOTYPE (BLOOD GROUP) FREQUENCIES			
	I^A	I^B	I^O	A	B	AB	O
Dunker	0.38	0.03	0.59	0.593	0.036	0.023	0.348
United States	0.26	0.04	0.70	0.431	0.058	0.021	0.490
West Germany	0.29	0.07	0.64	0.455	0.095	0.041	0.410

ers and Americans had occurred.) The most likely explanation for the unique Dunker gene frequencies observed is that genetic drift has produced random change in the gene pool. Founder effect probably occurred when the original 50 families emigrated from Germany, and genetic drift has most likely continued to influence gene frequencies each generation since 1729, because the population size has remained small.

EFFECTS OF GENETIC DRIFT. Genetic drift produces changes in gene frequencies, and these changes have several effects on the genetic structure of populations. First, genetic drift causes the gene frequencies of a population to change over time. This is illustrated in Figure 23.11. The different lines represent gene fre-

~ FIGURE 23.11

The effect of genetic drift on the frequency (q) of allele A^2 in four populations. Each population begins with q equal to 0.5, and the effective populations size for each is 20. The mean frequency of allele A^2 for the four replicates is indicated by the magenta line. These results were obtained by computer simulation.

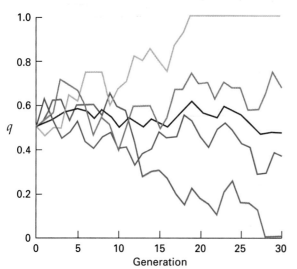

quencies in five populations over 30 generations. Although all populations begin with an allelic frequency equal to 0.50, the frequencies in each population change over time as a result of sampling error. In each generation, the gene frequency may increase or decrease, and over time the frequencies wander randomly or drift (hence the name "genetic drift"). Sometimes, just by chance, the gene frequency reaches a value of 0.0 or 1.0. At this point, one allele is lost from the population and the population is said to be *fixed* for the remaining allele. Once an allele has reached fixation, no further change in gene frequency can occur, unless the other allele is reintroduced through mutation or migration. The probability of fixation in a population increases with time, as shown in Figure 23.12. If the initial gene frequencies are equal, which allele becomes fixed is strictly random. If, on the other hand, initial gene frequencies are not equal, the rare allele is more likely to be lost. During this process of genetic drift and fixation, the number of heterozygotes in the population also decreases, and after fixation, the population heterozygosity is zero. As heterozygosity decreases and alleles become fixed, populations lose genetic variation; thus, the second effect of genetic drift is a reduction in genetic variation within populations.

Since genetic drift causes random change in gene frequency, the allelic frequencies in separate, individual populations will not change in the same direction. Therefore, populations diverge in their gene frequencies through genetic drift. This is illustrated in Figures 23.11 and 23.12; in both figures all the populations begin with p and q equal to 0.5. After a few generations, the gene frequencies of the populations diverge, and this divergence increases over generations. The maximum divergence in gene frequencies is reached when all populations are fixed for one or the other allele. If gene frequencies are initially equal to 0.5, approximately half of the populations will be fixed for one allele, and half will be fixed for the other.

Because genetic drift is greater in small populations and leads to genetic divergence, we expect more variance in gene frequency among small populations than

~ FIGURE 23.12

Fixation of alleles through genetic drift. The graphs illustrate the results of 400 experiments: in each experiment a population has 8 diploid individuals with an initial frequency of 0.5 for allele *A*. Gene frequencies in each population are allowed to drift randomly over time. The upper graph shows the distribution of *A* in the 400 populations after a single generation. The distributions of *A* after 4, 8, and 32 generations are shown in the lower graphs. (From Mettler, Gregg, and Shaffer, *Population Genetics and Evolution*, 2nd ed. Copyright © 1988 by Prentice-Hall, Inc. Reprinted by permission of Prentice-Hall, Inc., Englewood Cliffs, NJ.)

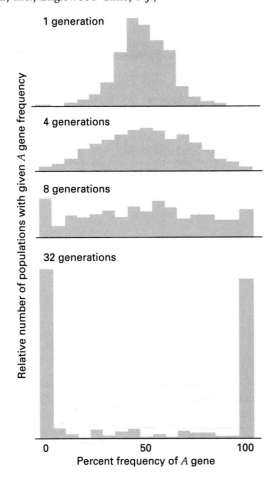

among large populations. Such a relationship has been observed in studies of natural populations. R. K. Selander, for example, studied genetic variation in populations of the house mouse inhabiting barns in Texas. Through systematic trapping, he was able to estimate population size, and using electrophoresis he examined the variance in gene frequency at two loci, a locus coding for the enzyme esterase (Est-3) and a locus coding for hemoglobin (Hbb). Selander found that the variance in gene frequency among small populations was several times larger than that among large populations (Table 23.7), an observation consistent with our understanding of how genetic drift leads to population divergence.

𝒦EYNOTE

Genetic drift, or chance changes in gene frequency due to sampling error, can be an important evolutionary force in small populations. Genetic drift leads to loss of genetic variation within populations, genetic divergence among populations, and random fluctuation in the gene frequencies of a population over time.

Migration

One of the assumptions of the Hardy-Weinberg law is that the population is closed and not influenced by outside evolutionary forces. Many populations are not completely isolated, however, and exchange genes with other populations of the same species. Individuals migrating into a population may introduce new alleles to the gene pool, altering the frequencies of existing alleles. Thus **migration** has the potential to disrupt Hardy-Weinberg equilibrium and may influence the evolution of gene frequencies within populations.

The term *migration* usually implies movement of organisms. In population genetics, however, we are

~ TABLE 23.7

Variance in Gene Frequency as a Function of Population Size

TYPE OF POPULATION	NUMBER OF POPULATIONS	MEAN ALLELIC FREQUENCY		VARIANCE IN ALLELIC FREQUENCY	
		Est-3[b]	Hbb[8]	Est-3[b]	Hbb[8]
Small ($N < 50$)	29	0.418	0.849	0.051	0.188
Large ($N > 50$)	13	0.372	0.843	0.013	0.008

Data from Selander, R. K. 1970. Behavior and genetic variation in natural populations *(Mus musculus)*. *Am Zoologist* 10:53–66.

interested in the movement of genes, which may or may not occur when organisms move. Movement of genes takes place only when organisms or gametes migrate, and contribute their genes to the gene pool of the recipient population. This process is also referred to as **gene flow.**

Gene flow has two major effects on a population. First, it introduces new alleles to the population. Because mutation is generally a rare event, a specific mutant allele may arise in one population and not in another. Gene flow spreads unique alleles to other populations and, like mutation, is a source of genetic variation for the population. Second, when the gene frequencies of migrants and the recipient population differ, gene flow changes the allelic frequencies within the recipient population. Through exchange of genes, populations remain similar, and thus, migration is a homogenizing force that tends to prevent populations from evolving genetic differences.

To illustrate the effect of migration on gene frequencies, we will consider a simple model in which gene flow occurs in only one direction, from population I to population II. Suppose that the frequency of allele A in population I (p_I) is 0.8 and the frequency of A in population II (p_{II}) is 0.5. Each generation some individuals migrate from population I to population II, and these migrants are a random sample of the genotypes in population I. After migration, population II actually consists of two groups of individuals: the migrants with $p_I = 0.8$, and the residents with $p_{II} = 0.5$. The migrants now make up a proportion of population II, which we will designate m. The frequency of A in population II after migration (p'_{II}) is

$$p'_{II} = mp_I + (1 - m)p_{II}$$

We see that the frequency of A after migration is determined by the proportion of A alleles in the two groups that now comprise population II. The first component, mp_I, represents the A alleles in the migrants—we multiply the proportion of the population that consists of migrants (m) by the gene frequency of the migrants (p_I). The second component represents the A alleles in the residents, and equals the proportion of the population consisting of residents ($1 - m$) multiplied by the gene frequency in the residents (p_{II}). Adding these two components together gives us the gene frequency of A in population II after migration. This model of gene flow is diagramed in Figure 23.13.

The change in gene frequency in population II as a result of migration (Δp) equals the original frequency of A subtracted from the frequency of A after migration:

$$\Delta p = p'_{II} - p_{II}$$

~ **FIGURE 23.13**

Theoretical model illustrating the effect of migration on the gene pool of a population. After migration, population II consists of two groups, the migrants with gene frequency of p_I and the original residents with gene frequency p_{II}.

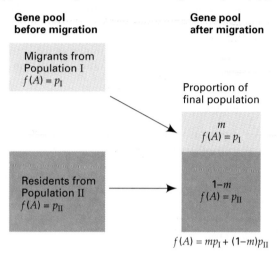

$$f(A) = mp_I + (1-m)p_{II}$$

In the previous equation, we found that p'_{II} equaled $mp_I + (1 - m)p_{II}$, so the change in gene frequency can be written as

$$\Delta p = mp_I + (1 - m)p_{II} - p_{II}$$

Multiplying $(1 - m)$ by p_{II} in the above equation, we obtain

$$\Delta p = mp_I + p_{II} - mp_{II} - p_{II}$$
$$\Delta p = mp_I - mp_{II}$$
$$\Delta p = m(p_I - p_{II})$$

This final equation indicates that the change in gene frequency from migration depends upon two factors: the proportion of the migrants in the final population and the difference in gene frequency between migrants and the residents. If no differences exist in the gene frequency of migrants and residents ($p_I - p_{II} = 0$), then we can see that the change in gene frequency is zero. Populations must differ in their gene frequencies in order for migration to affect the makeup of the gene pool. With continued migration, p_I and p_{II} become increasingly similar, and, as a result, the change in gene frequency due to migration decreases. Eventually, gene frequencies in the two populations will be equal, and no further change will occur. This is only true, however, when other factors besides migration do not influence gene frequencies.

~ FIGURE 23.14

Extensive gene flow occurs among populations of the monarch butterfly (a). The butterflies overwinter in Mexico (b) and then migrate north during spring and summer to breeding grounds as far away as northern Canada. Extensive gene flow occurs during the migration period and monarch populations display relatively little genetic divergence.

a)

b)

An additional point to be noted is that migration among populations tends to increase the effective size of the populations. As we have seen, small population size leads to genetic drift, and genetic drift causes populations to diverge. Migration, on the other hand, reduces divergence among populations, effectively increasing the size of the individual populations. Extensive gene flow has been shown to occur, for example, among populations of the monarch butterfly (Figure 23.14). Even a small amount of gene flow can reduce the effect of genetic drift. Calculations have shown that a single migrant moving between two populations each generation will prevent the two populations from becoming fixed for different alleles.

KEYNOTE

Migration of individuals into a population may alter the makeup of the population gene pool, if the genes carried by the migrants differ from the residents of the population. Migration, also termed gene flow, tends to reduce genetic divergence among populations and increases the effective size of the population. _____

Natural Selection

We have now examined three major evolutionary forces capable of changing gene frequencies and producing evolution—mutation, genetic drift, and migration. These forces alter the gene pool of a population, and they certainly influence the evolution of a species. However, mutation, migration, and genetic drift do not result in adaptation. Adaptation is the process by which traits evolve that make organisms more suited to their immediate environment; these traits increase the organism's chances of surviving and reproducing. Adaptation is responsible for the many extraordinary traits seen in nature—wings that enable a hummingbird to fly backward, leaves of the pitcher plant that capture and devour insects, brains that allow humans to speak, read, and love. These biological features and countless other exquisite traits are the product of adaptation (Figure 23.15; p. 738). Genetic drift, mutation, and migration all influence the pattern and process of adaptation, but adaptation arises chiefly from natural selection. Natural selection is the dominant force in the evolution of many traits and has shaped much of the phenotypic variation observed in nature.

Charles Darwin (Figure 23.16; p. 738) and Alfred Russel Wallace independently developed the concept of natural selection in the middle of the nineteenth century, although some earlier naturalists had similar ideas. In 1858, Darwin's and Wallace's theory was presented to the Linnaean Society of London and was enthusiastically received by other scientists. Darwin pursued the theory of evolution further than Wallace, amassing hundreds of observations to support it, and publishing his ideas in a book entitled *On the Origin of Species*. For his innumerable contributions to our understanding of natural selection, Darwin is frequently regarded as the "father" of evolutionary theory.

Natural selection can be defined as differential reproduction of genotypes. It simply means that individuals with certain genes produce more offspring than

~ FIGURE 23.15

Natural selection produces organisms that are finely adapted to their environment, such as this lizard with cryptic coloration, which allows it to blend in with its natural surroundings.

~ FIGURE 23.16

Charles Darwin was the first person to develop fully the theory of evolution by means of natural selection.

others; therefore, those genes increase in frequency in the next generation, as discussed in Chapter 22. Through natural selection, traits that contribute to survival and reproduction increase over time. In this way organisms adapt to their environment.

SELECTION IN NATURAL POPULATIONS. A classic example of selection in natural populations is the evolution of melanic (dark) forms of moths in association with industrial pollution, a phenomenon known as "industrial melanism." Melanic phenotypes have appeared in a number of different species of moths found in the industrial regions of Europe, North America, and England. One of the best studied cases involves the peppered moth, *Biston betularia*. The common phenotype of this species, termed the *typical* form, is a greyish white color with black mottling over the body and wings.

Prior to 1848, all peppered moths collected in England possessed this *typical* phenotype, but in 1848, a single black moth was collected near Manchester, England. This new phenotype, called *carbonaria*, presumably arose by mutation, and rapidly increased in frequency around Manchester and in other industrial regions. By 1900, the *carbonaria* phenotype had reached a frequency of more than 90 percent in several populations. High frequencies of *carbonaria* appeared to be associated with industrial regions, whereas the *typical* phenotype remained common in more rural districts. Laboratory studies by a number of investigators, including E. B. Ford and R. Goldschmidt, demonstrated that the *carbonaria* phenotype was dominant to the *typical* phenotype. A third phenotype was also discovered, which was somewhat intermediate to *typical*

and *carbonaria;* this phenotype, *insularia*, was produced by a dominant allele at a different locus.

H. B. D. Kettlewell investigated color polymorphism in the peppered moth, demonstrating that the increase in the *carbonaria* phenotype occurred as a result of strong selection against the *typical* form in polluted woods. Peppered moths are nocturnal; during the day they rest on the trunks of lichen-covered trees. Birds frequently prey upon the moths during the day, but because the lichens that cover the trees are naturally grey in color, the *typical* form of the peppered moth is well camouflaged against this background (Figure 23.17a). In industrial areas, however, extensive pollution beginning with the industrial revolution in the mid-nineteenth century had killed most of the lichens and covered the tree trunks with black soot. Against this black background, the *typical* phenotype was quite conspicuous and was readily consumed by birds. In contrast, the *carbonaria* form was well camouflaged against the blackened trees and had a higher rate of survival than the *typical* phenotype in polluted areas (Figure 23.17b). Because *carbonaria* survived better in polluted woods, more *carbonaria* genes were transmitted to the next generation; thus, the *carbonaria* phenotype increased in frequency in industrial areas. In rural areas, where pollution was absent, the *carbonaria* phenotype was conspicuous and the *typical* form was camouflaged; in these regions the frequency of the *typical* form remained high.

Kettlewell demonstrated that selection affected the frequencies of the two phenotypes by conducting a series of mark-and-recapture experiments involving dark and light moths in smoky, industrial Birmingham, England, and in nonindustrialized Dorset. As predicted,

~ FIGURE 23.17

Biston betularia, the peppered moth, and its dark form *carbonaria* (a) on the trunk of a white tree in the unpolluted countryside, and (b) on the trunk of a tree blackened by industrial pollution from Birmingham, England. On the white tree, the dark form is readily seen, whereas the light form is well camouflaged. On the polluted tree, the dark form is well camouflaged.

a)

b)

the *typical* phenotype was favored in Dorset, and *carbonaria* was favored in Birmingham.

FITNESS AND COEFFICIENT OF SELECTION. Darwin described natural selection primarily in terms of survival, and even today, many nonbiologists think of natural selection in terms of a struggle for survival. However, what is most important in the process of natural selection is the relative number of genes that are contributed to future generations. Certainly the ability to survive is important, but survival alone will not ensure that genes are passed on—reproduction must also occur. Therefore, we measure natural selection by assessing reproduction. Natural selection is measured with **fitness**, which is defined as the relative reproductive ability of a genotype.

Fitness is usually symbolized as *W*. Since fitness is a measure of the relative reproductive ability, population geneticists usually assign a fitness of 1 to a genotype that produces the most offspring. The fitnesses of the other genotypes are assigned relative to this. For example, suppose that the genotype G^1G^1 on the average produces 8 offspring, G^1G^2 produces an average of 4 offspring, and G^2G^2 produces an average of 2 offspring. The G^1G^1 genotype has the highest reproductive efficiency, so its fitness is 1 ($W_{11} = 1.0$). Genotype G^1G^2 produces on the average 4 offspring for the 8 produced by the most fit genotype, so the fitness of G^1G^2 (W_{12}) is 4/8 = 0.5. Similarly, G^2G^2 produces 2 offspring for the 8 produced by G^1G^1, so the fitness of G^2G^2 (W_{22}) is 2/8 = 0.25. Table 23.8 illustrates the calculation of fitness values.

~ TABLE 23.8

Computation of Fitness Values and Selection Coefficients of Three Genotypes

	GENOTYPES		
	G^1G^1	G^1G^2	G^2G^2
Number of breeding adults in one generation	16	10	20
Number of offspring produced by all adults of the genotype in the next generation	128	40	40
Average number of offspring produced per breeding adult	128/16 = 8	40/10 = 4	40/20 = 2
Fitness *W* (relative number of offspring produced)	8/8 = 1	4/8 = 0.5	2/8 = 0.25
Selection coefficient (1 − *W*)	1 − 1 = 0	1 − 0.5 = 0.5	1 − 0.25 = 0.75

Fitness tells us how well a genotype is doing in terms of natural selection. A related measure is the **selection coefficient,** which is a measure of the relative intensity of selection against a genotype. The selection coefficient is symbolized by s and equals $1 - W$. In our example, the selection coefficients for G^1G^1 are $s = 0$; for G^1G^2, $s = 0.5$; for G^2G^2, $s = 0.75$.

EFFECT OF SELECTION ON GENE FREQUENCIES. Natural selection produces a number of different effects. At times, natural selection eliminates genetic variation, and at other times it maintains variation; it can change gene frequencies or prevent gene frequencies from changing; it can produce genetic divergence among populations or maintain genetic uniformity. Which of these effects occurs depends primarily on the relative fitness of the genotypes and on the frequencies of the alleles in the population.

The change in gene frequency that results from natural selection can be calculated by constructing a table such as Table 23.9. This "table method" can be used for any type of single-locus trait, whether the trait is dominant, codominant, recessive, or overdominant. To use the table method we begin by listing the genotypes (A^1A^1, A^1A^2, and A^2A^2) and their initial frequencies. If random mating has just taken place, the genotypes are in Hardy-Weinberg proportions and the initial frequencies are p^2, $2pq$, and q^2. We then list the fitnesses for each of the genotypes, W_{11}, W_{12}, and W_{22}. Now, suppose that selection occurs and only some of the genotypes survive. The contribution of each genotype to the next generation will be equal to the initial frequency of the genotype multiplied by its fitness. For A^1A^1 this will be $p^2 \times W_{11}$. Notice that the contributions of the three genotypes do not add up to one. The sum of the contributions equals $p^2W_{11} + 2pqW_{12} + q^2W_{22} = \overline{W}$, which is called the *mean fitness of the population.* Next, we normalize the relative contributions by dividing each by the mean fitness of the population, which gives us the relative frequencies of the genotypes after selection. We then calculate the new gene frequency (p') from the genotypes after selection, using our familiar formula, $p' = $ (frequency of A^1A^1) + ($1/2 \times$ frequency of A^1A^2). Finally, the change in gene frequency resulting from selection equals $p' - p$. A sample calculation using some actual gene frequencies and fitness values is presented in Table 23.10.

SELECTION AGAINST A RECESSIVE TRAIT. Many important traits and most new mutations are recessive and have reduced fitness. When a trait is completely recessive, both the heterozygote and the dominant homozygote have a fitness of 1, while the recessive homozygote has reduced fitness, as shown below.

Genotype	Fitness
A A	1
A a	1
a a	$1 - s$

~ TABLE 23.9

General Method of Determining Change in Gene Frequency Due to Natural Selection

	GENOTYPES		
	A^1A^1	A^1A^2	A^2A^2
Initial genotypic frequencies	p^2	$2pq$	q^2
Fitness[a]	W_{11}	W_{12}	W_{22}
Frequency after selection	p^2W_{11}	$2pqW_{12}$	q^2W_{22}
Relative genotypic frequency after selection	$P' = \dfrac{p^2W_{11}}{\overline{W}^b}$	$H' = \dfrac{2pqW_{12}}{\overline{W}}$	$Q' = \dfrac{q^2W_{22}}{\overline{W}}$

Gene frequency after selection $= p' = P' + 1/2(H')$
$$q' = 1 - p'$$

Change in gene frequency due to selection $= \Delta p = p' - p$

[a]Note: For simplicity, fitness in this example is considered to be the probability of survival. Change in gene frequency due to differences in the number of offspring produced by the genotypes is calculated in the same manner.
[b]$\overline{W} = p^2W_{11} + 2pqW_{12} + q^2W_{22}$

~ TABLE 23.10

General Method of Determining Change in Gene Frequency Due to Natural Selection when Initial Gene Frequencies are p-0.6 and q-0.4

	GENOTYPES		
	A^1A^1	A^1A^2	A^2A^2
Initial genotypic frequencies	p^2 $(0.6)^2 = 0.36$	$2pq$ $2(0.6)(0.4) =$ 0.48	q^2 $(0.4)^2 = 0.16$
Fitness	$W_{11} = 0$	$W_{12} = 0.4$	$W_{22} = 1$
Frequency after selection	$p^2W_{11} =$ $(0.36)(0) = 0$	$2pqW_{12} =$ $(0.48)(0.4) =$ 0.19	$q^2W_{22} =$ $(0.16)(1)$ $= 0.16$
Relative genotypic frequency after selection	$P' = \dfrac{p^2W_{11}}{\overline{W}*}$ $P' = 0/0.35 = 0$	$H' = \dfrac{2pqW_{12}}{\overline{W}}$ $H' = 0.19/0.35$ $= 0.54$	$Q' = \dfrac{q^2W_{22}}{\overline{W}}$ $Q' = 0.16/0.35$ $= 0.46$

Gene frequency after selection $p' = P' + 1/2(H')$
$$p' = 0 + 1/2(0.54) = 0.27$$
$$q' = 1 - p' = 1 - 0.27 = 0.73$$

Change in gene frequency due to selection $= \Delta p = p' - p$
$$\Delta p = 0.27 - 0.6 = -0.33$$

$*\overline{W} = p^2W_{11} + 2pqW_{12} + q^2W_{22}$
$\overline{W} = 0 + 0.19 + 0.16$
$\overline{W} = 0.35$

If the genotypes are initially in Hardy-Weinberg proportions, the contribution of each genotype to the next generation will be the frequency times the fitness.

$A A$	$p^2 \times 1 = p^2$
$A a$	$2pq \times 1 = 2pq$
$a a$	$q^2 \times (1 - s) = q^2 - sq^2$

The mean fitness of the population is $p^2 + 2pq + q^2 - sq^2$. Since $p^2 + 2pq + q^2 = 1$, the mean fitness becomes $1 - sq^2$, and the normalized genotypic frequencies after selection are

$A A$	$\dfrac{p^2}{1 - sq^2}$
$A a$	$\dfrac{2pq}{1 - sq^2}$
$a a$	$\dfrac{q^2 - sq^2}{1 - sq^2}$

To obtain q', the frequency after selection, we add the frequency of the aa homozygote and half the frequency of the heterozygote:

$$q' = \frac{q^2 - sq^2}{1 - sq^2} + \frac{1}{2} \times \frac{2pq}{1 - sq^2}$$

$$= \frac{q^2 - sq^2}{1 - sq^2} + \frac{pq}{1 - sq^2}$$

$$= \frac{q^2 - sq^2 + pq}{1 - sq^2}$$

$$= \frac{q^2 + pq - sq^2}{1 - sq^2}$$

$$= \frac{q(q + p) - sq^2}{1 - sq^2}$$

Since $(q + p) = 1$,

$$q' = \frac{q - sq^2}{1 - sq^2}$$

Therefore, the change in the frequency of a after one generation of selection is

$$\Delta q = q' - q$$

$$= \frac{q - sq^2}{1 - sq^2} - q$$

$$= \frac{q - sq^2}{1 - sq^2} - \frac{q(1 - sq^2)}{(1 - sq^2)}$$

$$= \frac{q - sq^2 - q(1 - sq^2)}{1 - sq^2}$$

$$= \frac{q - sq^2 - q + sq^3}{1 - sq^2}$$

$$= \frac{-sq^2 + sq^3}{1 - sq^2}$$

$$= \frac{-sq^2(1 - q)}{1 - sq^2}$$

So $\Delta q = -spq^2/(1 - sq^2)$ because $1 - q = p$. When $\Delta q = 0$, no further change occurs in allelic frequencies. Notice that there is a negative sign in the equation to the left of spq^2; because the values of s, p, and q are always positive or zero, Δq is negative or zero. Thus, the value of q will decrease with selection.

Selection also depends on the actual frequencies of genes in the population. This is because the relative proportions of Aa and aa individuals at various frequencies of allele a influence how effectively selection can reduce a detrimental recessive trait. When the frequency of a recessive gene is relatively high, many homozygous recessive individuals are present in the population and will have low fitness, causing a large change in the gene frequency. When the gene frequency is low, however, the homozygous recessive genotype is rare, and little change in gene frequency occurs.

Table 23.11 shows the magnitude of change in gene frequency each generation in three populations with different initial gene frequencies. Population 1 begins with gene frequency q equal to 0.9, population 2 begins with q equal to 0.5, and population 3 begins with q equal to 0.1. In this example the homozygous recessive genotype (aa) has a fitness of 0 and the other two genotypes (AA and Aa) have a fitness of 1. When the frequency of q is high, as in population 1, the change in gene frequency is large; in the first generation, q drops from 0.9 to 0.47. However, when q is small, as in population 3, the change in q is much less; here q drops from 0.1 to 0.091 in the first generation. So, as q becomes smaller, the change in q becomes less. Because of this diminishing change in frequency, it is virtually impossible to eliminate entirely a recessive trait from the population. This result only applies, however, to completely recessive traits; if the fitness of the heterozygote is also reduced, the change in gene frequency will be more rapid, because now selection also acts against the heterozygote in addition to the homozygote. The effect of dominance on changes in gene frequency as a result of selection is illustrated in Figure 23.18.

We have gone through a lengthy discussion of the effects of selection on a recessive trait to illustrate how the formula for change in gene frequency can be derived from our general table method of gene frequency change under selection. Similar derivations can

~ **TABLE 23.11**

Effectiveness of Complete Selection Against a Recessive Phenotype at Different Initial Frequencies

| | GENOTYPE (q^2 aa) AND ALLELE (q a) FREQUENCIES | | | | | |
| | POPULATION 1 | | POPULATION 2 | | POPULATION 3 | |
GENERATION	q^2	q	q^2	q	q^2	q
0	0.810	0.900	0.250	0.500	0.010	0.100
1	0.224	0.474	0.112	0.333	0.008	0.091
2	0.103	0.321	0.063	0.250	0.007	0.083
3	0.059	0.243	0.040	0.200	0.006	0.077
4	0.038	0.195	0.028	0.167	0.005	0.071
5	0.027	0.163	0.020	0.143	0.004	0.066
.						
.						
.						
10			0.007	0.08		

~ TABLE 23.12

Formulas for Calculating Change in Gene Frequency After One Generation of Selection

TYPE OF SELECTION	FITNESSES OF GENOTYPES			CALCULATION OF CHANGE IN GENE FREQUENCY
	A^1A^1	A^1A^2	A^2A^2	
Selection against recessive homozygote	1	1	$1 - s$	$\Delta q = \dfrac{-spq^2}{1 - sq^2}$
Selection against a dominant allele	$1 - s$	$1 - s$	1	$\Delta q = \dfrac{-spq^2}{1 - s + sq^2}$
Selection with no dominance	1	$(1 - s/2)$	$1 - s$	$\Delta q = \dfrac{-spq/2}{1 - sq}$
Selection which favors the heterozygote (overdominance)	$1 - s$	1	$1 - t$	$\Delta q = \dfrac{pq(sp - tq)}{1 - sp^2 - tq^2}$
Selection against the heterozygote	1	$1 - s$	1	$\Delta q = \dfrac{spq(q - p)}{1 - 2spq}$
General	W_{11}	W_{12}	W_{22}	$\Delta q = \dfrac{pq[p(W_{12} - W_{11}) + q(W_{22} - W_{12})]}{\overline{W}^*}$

*Note: For calculation of $\overline{W}$ see Table 23.9.

~ FIGURE 23.18

Comparison of changes in frequency of an allele with different types of dominance, holding selection coefficient constant. A_1 is the favored allele.

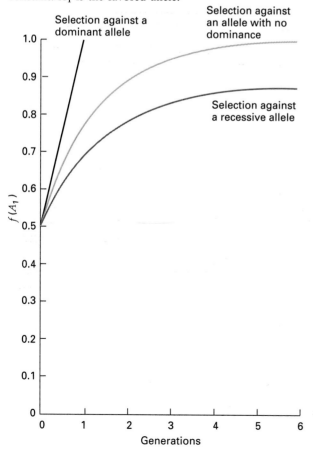

be carried out for dominant traits and codominant traits. We will not present those derivations here, but the appropriate formulas for calculating changes in gene frequency under different types of dominance are presented in Table 23.12. Remember, however, it is possible to calculate changes in gene frequency for any type of trait using the table method.

$\mathcal{K}$EYNOTE

Natural selection involves differential reproduction of genotypes and is measured in terms of fitness, the relative reproductive contribution of a genotype. The effects of selection depend not only on the fitness of the different genotypes, but also on the dominance, or lack of dominance, of the alleles and on the frequencies of the alleles in the population. _____

Mutation and Selection

As we have seen, natural selection reduces the frequency of a deleterious recessive allele, but as its frequency becomes low, the change in frequency each generation diminishes. When the allele is rare, the change in frequency is very slight. Opposing this reduction in the allele's frequency due to selection is mutation pressure, which will continually produce new alleles and tend to increase the frequency. Eventually a balance, or equilibrium, is reached, in which the input of new alleles by recurrent mutation is exactly counterbalanced by the loss of alleles through natural selection. When equilibrium is obtained, the allele's frequency remains stable, in spite of the fact that selection and mutation continue, unless perturbed by some other evolutionary force.

Consider a population in which selection occurs against a deleterious recessive allele, *a*. As we saw on p. 743, the amount *a* will change in one generation (Δq) as a result of selection is

$$\Delta q = \frac{-spq^2}{1 - sq^2}$$

For a rare recessive allele, q^2 will be near 0 and the denominator in this equation, $1 - sq^2$, will be approximately 1, so that the decrease in frequency caused by selection is given by

$$\Delta q = -spq^2$$

At the same time, the frequency of the *a* allele increases as a result of mutation from *A* to *a* (pp. 727–730). Provided the frequency of *a* is low, the reverse mutation of *a* to *A* essentially can be ignored. Equilibrium between selection and mutation occurs when the decrease in gene frequency produced by selection is the same as the increase produced by mutation:

$$spq^2 = up$$

We can predict what the frequency of *a* at equilibrium ($\hat{q}$) will be by rearranging this equation:

$$sq^2 = u$$
$$q^2 = u/s$$

and

$$\hat{q} = \sqrt{u/s}$$

If the recessive homozygote is lethal ($s = 1$), the equation becomes

$$\hat{q} = \sqrt{u}$$

As an example of the balance between mutation and selection, consider a recessive gene for which the mutation rate is 10^{-6} and *s* is 0.1. At equilibrium, the frequency of the gene will be $\hat{q} = \sqrt{10^{-6}/0.1} = 0.0032$. Most recessive deleterious traits remain within a population at low frequency because of equilibrium between mutation and selection.

For a dominant allele *A*, the frequency at equilibrium ($\hat{q}$) is

$$\hat{q} = u/s$$

If the mutation rate is 10^{-6} and *s* is 0.1, as in the above example, the frequency of the dominant gene at equilibrium will be $10^{-6}/0.1 = 0.00001$, which is considerably less than the equilibrium frequency for a recessive allele with the same fitness and mutation rate. This is because selection cannot act on a recessive allele in the heterozygote state, whereas both the homozygote and the heterozygote for a dominant allele have reduced fitness. For this reason, detrimental dominant alleles generally are less common than recessive ones.

Overdominance

An equilibrium of gene frequencies also arises when the heterozygote has higher fitness than either of the homozygotes. This situation is called **overdominance** or **heterozygote advantage.** In overdominance, both alleles are maintained in the population, because both are favored in the heterozygote genotype. Gene frequencies will change as a result of selection until the equilibrium point is reached and then will remain stable. The gene frequencies at which the population reaches equilibrium depend upon the relative fitnesses of the two homozygotes. If the selection coefficient of *AA* is *s* and the selection coefficient of *aa* is *t*, it can be shown algebraically that at equilibrium

$$\hat{p} = f(A) = t/(s + t)$$

and

$$\hat{q} = f(a) = s/(s + t)$$

An example of overdominance in humans is sickle-cell anemia. Sickle-cell anemia results from a mutation in the gene coding for beta hemoglobin. In some populations, there are three hemoglobin genotypes: *Hb-A/Hb-A*, *Hb-A/Hb-S*, and *Hb-S/Hb-S*. Individuals with the *Hb-A/Hb-A* genotype have completely normal red blood cells; *Hb-S/Hb-S* individuals have sickle-cell anemia; and *Hb-A/Hb-S* individuals have sickle-cell trait, a mild form of sickle-cell anemia. In an environment in which malaria is common, the heterozygotes are at a selective advantage over the two homozygotes. Apparently, the abnormal hemoglobin mixture in the heterozygotes provides an unfavorable environment for the growth or maintenance of the malarial parasite in the red cell. The heterozygotes therefore have greater resistance to malaria and thus higher fitness than *Hb-A/Hb-A* individuals. The *Hb-S/Hb-S* individuals are at a serious selective disadvantage because they have sickle-cell anemia. As a result, in malaria-infested areas

~ FIGURE 23.19

The distribution of malaria caused by the parasite *Plasmodium falciparum* coincides with distribution of the *Hb-S* allele for sickle-cell anemia. The frequency of *Hb-S* is high in areas where malaria is common, because *Hb-A/Hb-S* heterozygotes are resistant to malarial infection. (After A. C. Allison. 1961. Abnormal hemoglobin and erythrocyte enzyme-deficiency traits. In *Genetic Variation in Human Populations,* ed G. A. Harrison. Pergamon, NY: pp. 16–40.)

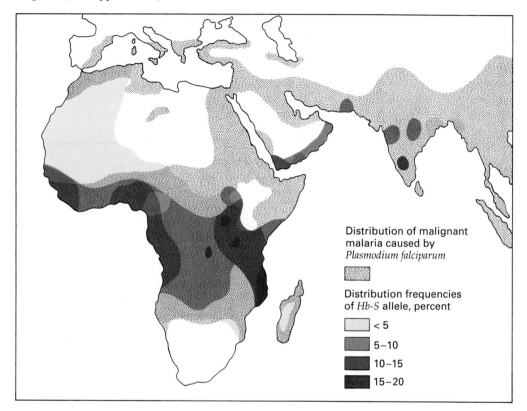

Distribution of malignant malaria caused by *Plasmodium falciparum*

Distribution frequencies of *Hb-S* allele, percent

< 5
5–10
10–15
15–20

in which the *Hb-S* gene is also found, an equilibrium state is established in which a significant number of *Hb-S* alleles are found in the heterozygotes because of the selective advantage of this genotype. The distributions of malaria and the *Hb-S* allele are illustrated in Figure 23.19.

NONRANDOM MATING

A fundamental assumption of the Hardy-Weinberg law is that members of the population mate randomly. But, many populations do not mate randomly for some traits, and when nonrandom mating occurs, the genotypes will not exist in the proportions predicted by the Hardy-Weinberg law.

One form of nonrandom mating is **positive assortative mating,** which occurs when individuals with similar phenotypes mate preferentially. Positive assortative mating is common in natural populations. For exam-

ple, humans mate assortatively for height; tall men and tall women marry more frequently and short men and short women marry more frequently than would be expected on a random basis. **Negative assortative mating** occurs when phenotypically dissimilar individuals mate more often than randomly chosen individuals. If humans exhibited negative assortative mating for height, tall men and short women would marry preferentially and short men and tall women would marry preferentially. Neither positive nor negative assortative mating affects the gene frequencies of a population, but they may influence the genotypic frequencies if the phenotypes for which assortative mating occurs are genetically determined.

Two other departures from random mating are **inbreeding** and **outbreeding.** As we learned in Chapter 22, inbreeding involves preferential mating between close relatives, and thus, inbreeding is really positive assortative mating for relatedness. Outbreeding is preferential mating between nonrelated individuals and is a

~ TABLE 23.13

Relative Genotype Distributions Resulting from Self-Fertilization over Several Generations Starting with an *Aa* Individual

GENERATION	GENOTYPES		
	AA	*Aa*	*aa*
0		1	
1	$\frac{1}{4}$	$\frac{1}{2}$	$\frac{1}{4}$
2	$\frac{1}{4} + \frac{1}{8} = \frac{3}{8}$	$\frac{1}{4}$	$\frac{1}{4} + \frac{1}{8} = \frac{3}{8}$
3	$\frac{3}{8} + \frac{1}{16} = \frac{7}{16}$	$\frac{1}{8}$	$\frac{3}{8} + \frac{1}{16} = \frac{7}{16}$
4	$\frac{7}{16} + \frac{1}{32} = \frac{15}{32}$	$\frac{1}{16}$	$\frac{7}{16} + \frac{1}{32} = \frac{15}{32}$
5	$\frac{15}{32} + \frac{1}{64} = \frac{31}{64}$	$\frac{1}{32}$	$\frac{15}{32} + \frac{1}{64} = \frac{31}{64}$
n	$[1 - (\frac{1}{2})^n]/2$	$(\frac{1}{2})^n$	$[1 - (\frac{1}{2})^n]/2$
∞	$\frac{1}{2}$	0	$\frac{1}{2}$

form of negative assortative mating. The most extreme case of inbreeding is self-fertilization, which occurs in many plants and a few animals, such as some snails.

The effects of self-fertilization are illustrated in Table 23.13. Assume that we begin with a population consisting entirely of *Aa* heterozygotes, and that all individuals in this population reproduce by self-fertilization. After one generation of self-fertilization, the progeny will consist of $\frac{1}{4}$ *AA*, $\frac{1}{2}$ *Aa*, and $\frac{1}{4}$ *aa*. Now only half of the population consists of heterozygotes. When this generation undergoes self-fertilization, the *AA* homozygotes will produce only *AA* progeny, and the *aa* homozygotes will produce only *aa* progeny. When the heterozygotes reproduce, however, only half of their progeny will be heterozygous like the parents, and the other half will be homozygous ($\frac{1}{4}$ *AA* and $\frac{1}{4}$ *aa*). This means that in each generation of self-fertilization, the percentage of heterozygotes decreases by 50 percent. After an infinite number of generations, there will be no heterozygotes and the population will be divided equally between the two homozygous genotypes. Note that the population was in Hardy-Weinberg proportions in the first generation after self-fertilization, but after further rounds the proportion of homozygotes is greater than that predicted by the Hardy-Weinberg law. The result of continued self-fertilization is to increase homozygosity at the expense of heterozygosity. The frequencies of alleles *A* and *a* remain constant, while the frequencies of the three genotypes change significantly. When less intensive inbreeding occurs, similar, but less pronounced, effects occur. On the other hand, outbreeding increases heterozygosity.

KEYNOTE

Inbreeding involves preferential mating between close relatives, and outbreeding is preferential mating between unrelated individuals. Continued inbreeding increases homozygosity within a population, whereas outbreeding tends to increase heterozygosity.

SUMMARY OF THE EFFECTS OF EVOLUTIONARY FORCES ON THE GENE POOL OF A POPULATION

Let us now review the major effects of the different evolutionary forces on (1) changes in gene frequency within a population, (2) genetic divergence among populations, and (3) increases and decreases in genetic variation within populations.

Mutation, migration, genetic drift, and selection all have the potential to change the gene frequencies of a population over time. Mutation, however, usually occurs at such a low rate that the change resulting from mutation pressure alone is frequently negligible. Genetic drift will produce substantial changes in gene frequency only when the population size is small. Furthermore, mutation, migration, and selection may lead to equilibrium, where evolutionary forces continue to act, but the gene frequencies no longer change. Nonrandom mating does not change gene frequencies, but it does affect the genotypic frequencies of a population: inbreeding leads to increases in homozygosity and outbreeding produces an excess of heterozygotes.

Several evolutionary forces lead to genetic divergence among populations. Because genetic drift is a random process, gene frequencies in different populations may drift in different directions, so genetic drift can produce genetic divergence among populations. Migration among populations has just the opposite effect, increasing effective population size and equalizing gene frequency differences among populations. If the population size is small, different mutations may arise in different populations, and, therefore, mutation may contribute to population differentiation. Natural selection can increase genetic differences among populations by favoring different alleles in different populations, or it can prevent divergence by keeping gene frequencies among populations uniform. Nonrandom mating will not, by itself, generate genetic differences among populations, although it may contribute to the effects of other forces by increasing or decreasing effective population size.

Gene flow and mutation tend to increase genetic variation within populations by introducing new alleles to the gene pool. Genetic drift produces the opposite effect, decreasing genetic variation within small populations through loss of alleles. Because inbreeding leads to increases in homozygosity, it also diminishes genetic variation within populations; outbreeding, on the other hand, increases genetic variation by increasing heterozygosity. Natural selection may increase or decrease genetic variation. If one particular allele is favored, other alleles decrease in frequency and may be eliminated from the population by selection. Alternatively, natural selection may increase genetic variation within populations through overdominance and other forms of balancing selection.

In practice, these evolutionary forces never act in isolation but combine and interact in complex ways. In most natural populations, the combined effects of these forces and their interaction determine the pattern of genetic variation observed in the gene pool over time.

MOLECULAR EVOLUTION

In recent years population geneticists have begun to apply molecular genetic techniques to studies of genetic variation within populations and to questions about the molecular basis of evolution. By using restriction mapping and DNA sequencing methods (see Chapter 15), biologists can now examine evolution at the most basic genetic level, the level of the DNA. These studies have not altered our basic principles of population

genetics, but they have provided a more complete and detailed picture of how the forces of nature produce evolutionary change.

DNA Sequence Variation

An important question in the study of evolution is how the patterns and rates of evolution differ among genes and among different parts of the same gene. Rates of evolution of given DNA sequences can be measured by comparing sequences in two different organisms that diverged from a common ancestor at some known time in the past. We assume that the common ancestor had a single DNA sequence. Then, after the two organisms diverged from this ancestor, their DNA sequences underwent independent evolutionary changes, producing the differences we see in their sequences today. For example, it is believed that most major groups of mammals diverged from a common ancestor some 65 million years ago. Suppose we examined a particular DNA sequence, perhaps the gene for growth hormone, in mice and humans; we might find that the sequences for this gene in mouse and human differ in 20 nucleotides. These 20 nucleotides must have changed during the past 65 million years since the split between the mouse and human evolutionary lines. To calculate the rate of evolutionary change in this gene, we first compute the number of nucleotide substitutions (changes) that occurred to produce the 20 nucleotide differences we observe today. There are a number of different mathematical methods for estimating this number, which is usually expressed as nucleotide substitutions per nucleotide site, so that the rate of evolution is independent of the length of the sequences compared. To obtain rates of change, we then divide our value of nucleotide substitutions per nucleotide site by the number of years of evolutionary time that separate the two organisms. The rate of change obtained is then expressed as substitutions per nucleotide site per year. In our example with the growth hormone gene, we might find that the rate of change was 4×10^{-9} substitutions per site per year.

Studies of nucleotide sequences in numerous genes have revealed that different parts of genes evolve at different rates. Recall from our discussion of molecular genetics that a typical eukaryotic gene is made up of some nucleotides that specify the amino acid sequence of a protein (coding sequences) and other nucleotides that do not code for amino acids in a protein (noncoding sequences). Noncoding sequences include introns, leader regions, and trailer regions which are transcribed but not translated, and 5′ and 3′ flanking sequences that are not transcribed. Additional noncoding sequences include pseudogenes, which are nucleotide sequences that no longer produce a functional gene product because of mutations. Even within the coding regions of a functional gene, not all nucleotide substitutions produce a corresponding change in the amino acid sequence of a protein. In particular, many mutations occurring at the third position of the codon have no effect on the amino acid sequence of the protein, because such mutations produce a synonymous codon—one that codes for the same amino acid (see Chapter 14).

Different regions of the gene are apparently subject to different evolutionary forces, and evolve at different rates. Table 23.14 shows relative rates of change in different parts of mammalian genes. Within functional genes, notice that the highest rate of change involves synonymous changes in the coding sequences. The rate of synonymous nucleotide change is about five times greater than the observed rate of nonsynonymous changes. Synonymous changes do not alter the amino acid sequence of the protein. Thus, the high rate of evolutionary change seen there is not unexpected, because these changes do not affect a protein's functioning. Synonymous and nonsynonymous mutations probably arise with equal frequency. However, nonsynonymous changes that arise within coding sequences are usually detrimental to fitness and are eliminated by natural selection, whereas synonymous mutations are rarely detrimental, so they are tolerated.

You will notice in Table 23.14 that high rates of evolutionary change also occur in the 3' flanking regions of functional genes. Like synonymous changes, sequences in the 3' flanking regions have no known

~ TABLE 23.14

Relative Rates of Evolutionary Change in DNA Sequences of Mammalian Genes

SEQUENCE	NUCLEOTIDE SUBSTITUTIONS PER SITE PER YEAR ($\times 10^{-9}$)
Functional Genes	
5' Flanking region	2.36
Leader	1.74
Coding sequence—synonymous	4.65
Coding sequence—nonsynonymous	0.88
Intron	3.70
Trailer	1.88
3' Flanking region	4.46
Pseudogenes	4.85

From Li, W., C. Luo, and C. Wu. 1985. Evolution of DNA sequences. In R. J. MacIntyre, ed. *Molecular Evolutionary Genetics.* New York, Plenum Press.

effect on the amino acid sequence and usually have little effect on gene expression; most mutations that occur here will be tolerated by natural selection. Rates of change in introns are also high, but not as high as the synonymous changes and those in the 3' flanking regions. Although the sequences in the introns do not code for proteins, the intron must be properly spliced out for the mRNA to be translated into a functional protein. A few sequences within the intron are critical for proper splicing; these include the consensus sequences at the 5' and 3' ends of the intron and the sequence at the branch point (see Chapter 13). As a result, not all changes in introns will be tolerated, so the overall rate of evolution is a bit lower than that seen in synonymous coding sequences and in the 3' flanking region.

Lower rates of evolutionary change are seen in the 5' flanking region. Although this region is neither transcribed nor translated, it does contain the promoter for the gene; thus, sequences in the 5' flanking region are important for gene expression. Important consensus sequences, such as the TATA box, are found here. Mutations in consensus sequences may prevent the gene from being transcribed and thus will have detrimental effects on the fitness of the organism; natural selection eliminates these mutations and keeps change in this region low.

Next in evolutionary rate are the leader and trailer regions (Table 23.14), which have somewhat lower rates than the 5' flanking region. Although leaders and trailers are not translated, they are transcribed, and they provide important signals for processing the mRNA and for attachment of the ribosome to the mRNA. Nucleotide substitutions in these regions are therefore limited. The lowest rate of evolution is seen in the nonsynonymous coding sequences. Alteration of these nucleotides changes the amino acid sequence of the protein, and most mutations that occur here are eliminated by natural selection.

One final thing to note in Table 23.14 is that the highest rate of evolution is seen in nonfunctional pseudogenes. Among human globin pseudogenes, for example, the rate of nucleotide change is approximately 10 times that observed in the coding sequences of functional globin genes. The high rate of evolution observed in these sequences occurs because pseudogenes no longer code for proteins; since changes in these genes do not affect an individual's fitness, changes are not eliminated by natural selection. In summary, we observe the highest rates of evolutionary change in those sequences that have the least effect on function.

An interesting observation from the study of nucleotide sequences is that although the rate of evolution for synonymous nucleotide changes is much higher than that of nonsynonymous changes, rates of synonymous changes are often not quite as high as those observed in pseudogenes. This observation suggests that synonymous mutations are not completely synonymous; natural selection may favor some synonymous codons over others. This hypothesis is reinforced by the finding that synonymous codons are not used equally throughout the coding sequences. For example, six different codons specify the amino acid leucine (UUA, UUG, CUU, CUC, CUA, and CUG), but 60 percent of the leucine codons in bacteria are CUG, and in yeast, 80 percent are UUG. Since the synonymous codons specify the same amino acid, selection must be favoring some synonymous codons over others. Remember that some synonymous codons pair with different tRNAs that carry the same amino acid. Therefore, a mutation to a synonymous codon does not change the amino acid, but it may change the tRNA. Studies of the different tRNAs reveal that within a cell, the amounts of the isoacceptor tRNAs (different tRNAs that accept the same amino acid) differ, and the most abundant tRNAs are those that pair with the most frequently used codons. Selection may favor one synonymous codon over another because the tRNA for the codon is more abundant, and translation of that codon is more efficient. Alternatively, the bonding energy between codon and anticodon of synonymous codons may differ because different bases are paired, and this may be subject to natural selection.

In addition to differences in evolutionary rate among parts of a gene, different functional genes evolve at different rates (Table 23.15). For example, the rate of nonsynonymous nucleotide substitution in the prolactin gene of mammals is over 300 times higher than that found in the histone H4 gene of mammals.

DNA Length Polymorphisms

In addition to evolution of nucleotide sequences through nucleotide substitution, variation frequently occurs in the number of nucleotides found within a gene. These variations are called DNA length polymorphisms, and they arise through deletions and insertions of relatively short stretches of nucleotides. For example, DNA length polymorphisms have been observed in the alcohol dehydrogenase gene of *Drosophila melanogaster*. In a study mentioned earlier, Martin Krietman sequenced 11 copies of this gene in fruit flies. In addition to extensive variation in nucleotide sequence, Krietman found six insertions and deletions in the 11 copies of the gene he examined. All of these were confined to introns and flanking regions of the DNA—none were found within exons. Insertions and deletions within exons usually alter the reading frame, so they will be selected against. As a result, insertions and deletions are most commonly found in noncoding regions of the DNA. However, some insertions and deletions have been found in the coding regions of certain genes. Another class of DNA length polymorphisms involves variation in the number of copies of a particular gene. For example, among individual fruit flies, the number of copies of ribosomal genes varies extensively. The number of copies of transposons also varies extensively among individuals and is responsible for some DNA length polymorphisms.

Evolution of Multigene Families Through Gene Duplication

In eukaryotic organisms, we frequently find multiple copies of genes, all having identical or similar sequences. A group of such genes is termed a **multigene family**, which is defined as a set of related genes that have evolved from some ancestral gene through the process of gene duplication. Members of a gene family may be clustered together or dispersed on different chromosomes.

An example of a multigene family is the globin gene family, which consists of the genes that code for the polypeptide chains making up the hemoglobin molecule. The organization of this multigene family in humans was discussed in Chapter 17. The globin multigene family in humans is comprised of seven α-like genes found on chromosome 16 and six β-like genes found on chromosome 11. Globin genes are also found in other animals, and globin-like genes are even found in plants, suggesting that this is a very ancient gene family. Almost all functional globin genes in animal species have the same general structure, consisting of three exons separated by two introns. However, the numbers of globin genes and their order varies among

~ TABLE 23.15

Relative Rates of Evolutionary Change in DNA Sequences of Different Mammalian Genes*

GENE	NONSYNONYMOUS RATE	SYNONYMOUS RATE
Histone H4	0.004	1.43
Insulin	0.16	5.41
Prolactin	1.29	5.59
α-globin	0.56	3.94
β-globin	0.87	2.96
Albumin	0.92	6.72
α-fetoprotein	1.21	4.90

* All rates are in nucleotide substitutions per site per year $\times 10^{-9}$. Data from Li, W. et al. 1985. *Mol. Bio. Evol.* 2:150–174.

~ FIGURE 23.20

Organization of the globin gene families in several mammalian species.

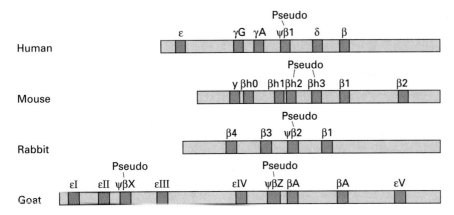

species, as is shown for the β-like genes in Figure 23.20. Because all globin genes have similarities in structure and sequence, it appears that an ancestral globin gene (perhaps related to the present day myoglobin gene) duplicated and diverged to produce an ancestral α-like gene and an ancestral β-like gene. These two genes then underwent repeated duplications, giving rise to the various α-like and β-like genes found in vertebrates today. Repeated gene duplication, such as that giving rise to the globin gene family, appears to be a frequent evolutionary occurrence. Indeed, the number of copies of globin genes varies in some human populations. For example, most humans have two α-globin genes on chromosome 16, as shown in Figure 17.20. However, some individuals have a single α-globin gene on chromosome 16; other individuals have three or even four copies of the α-globin gene on one of their chromosomes. These observations indicate that duplication and deletion of genes in a multigene family are constant, ongoing processes. Gene duplications and deletions often arise as a result of misalignment of sequences during crossing over, a process termed *unequal crossing over*. Duplications may also arise through transposition (see Chapter 19). Following gene duplication, the separate copies of a gene may undergo changes in sequence. In some cases, mutations arise that render a copy of the gene nonfunctional, creating a pseudogene. In other cases, the change of nucleotide sequence may lead to different functions for the protein product of a gene.

Evolution in Mitochondrial DNA Sequences

In Chapter 21 we discussed the mitochondrial genome. We saw that animal mitochondrial DNA (mtDNA) is comprised of approximately 15,000 base pairs and that this DNA encodes 2 rRNAs, 22–23 tRNAs, and 10–12 proteins. A number of recent studies have examined sequence variation in mtDNA. These studies reveal that mtDNA evolution differs from evolution typically observed in nuclear DNA. For example, nucleotide sequences within animal mtDNA evolve at a faster rate than coding sequences in nuclear genes. In fact, the mtDNA of mammals undergoes evolutionary change at a rate that is 5 to 10 times faster than that typically observed in mammalian nuclear DNA. However, the rapid rate of evolution observed in animal mtDNA is not universal for all eukaryotes. For unknown reasons, plant mtDNA appears to evolve at an even slower rate than plant nuclear DNA.

It is not entirely clear why animal mtDNA undergoes rapid evolutionary change, but the reason may be related to a higher mutation rate in mtDNA sequences. Higher mutation rates could occur in mtDNA because DNA repair mechanisms are lacking or because the DNA polymerase involved in mtDNA replication is more prone to errors. Another possible reason for the higher rate of evolution in animal mtDNA is that the selection pressure that normally eliminates many mutations in nuclear genes is relaxed in the mitochondria. In other words, changes in the proteins, tRNAs, and rRNAs encoded by the mtDNA sequences might be less detrimental to individual fitness than similar changes in the proteins, tRNAs, and rRNAs encoded by nuclear genes.

Whatever the cause of accelerated evolution of mtDNA, the fact that these sequences change rapidly makes them ideal for assessing evolutionary relationships among groups of closely related organisms. For example, analyses of mtDNA sequences have been used to investigate the relationships among humans, chimpanzees, gorillas, orangutans, and gibbons, a group whose precise evolutionary relationships have been controversial.

Mitochondrial DNA also differs from nuclear DNA in that all mtDNA is inherited clonally from the mother; mitochondria are located in the cytoplasm and only the egg cell contributes cytoplasm to a zygote. As a consequence, mtDNA does not undergo meiosis, and all offspring should be identical to the maternal genotype for mtDNA sequences (the offspring are clones for mtDNA genes). This pattern of inheritance allows matriarchal lineages (descendants from one female) to be traced, and provides a means for examining family structure in some populations. An example of geographic variation in mtDNA sequences is shown in Figure 23.21.

Concerted Evolution

One of the surprising findings from studies of DNA sequence variation is the observation that some molecular force, or forces, maintain uniformity of sequence in multiple copies of a gene. This phenomenon has been termed **concerted evolution** or **molecular drive**. As we discussed in previous chapters, some genes in eukaryotes exist in multiple copies. Ribosomal RNA genes in complex organisms, for example, typically exist in hundreds or thousands of copies. Undoubtedly, these multiple copies arose through duplication. Following duplication, individual copies of a gene might be expected to acquire mutations and diverge. Selection might limit mutations in coding regions, but if many

copies exist, we would expect some divergence to occur, especially in the noncoding sequences. Contrary to this expectation, numerous studies have revealed that nucleotide sequences in the different copies of a gene are frequently quite homogeneous. Furthermore, the noncoding sequences are also homogeneous, which suggests that purifying selection is not responsible. When the same genes are examined in a second, closely related species, that group's sequences are also homogeneous, but are frequently different from the homogeneous sequence found in the first species.

These observations have led to the conclusion that some molecular process continually enforces uniformity among multiple copies of the same sequence within a species. At the same time, the process allows for rapid differentiation among species. The mechanism of concerted evolution is not fully understood, but concerted evolution has important consequences for how genes evolve, and it represents an evolutionary force that was unknown before the application of modern molecular techniques to population genetics.

Evolutionary Relationships Revealed by RNA and DNA Sequences

Within the past few years, molecular genetics has provided powerful tools for deciphering the evolutionary history of life. Because evolution is defined as genetic change, genetic relationships are of primary importance in the construction of evolutionary trees. However, until recently, it was impossible to examine the genes directly. In the past, evolutionary biologists were forced to rely entirely on comparison of phenotypes to infer genetic similarities and differences. They assumed that if the phenotypes were similar, the genes coding for the phenotypes were similar; if the phenotypes were different, the genes were different. Thus, phenotypes were used for evolutionary studies. Originally the phenotypes examined consisted largely of gross anatomical features. Later, behavioral, ultrastructural, and biochemical characteristics were also studied. Comparisons of such traits were successfully used to construct evolutionary trees for many groups of plants and animals, and, indeed, are still the basis of most evolutionary studies today.

However, relying on the study of such traits has limitations. Sometimes, similar phenotypes can evolve in organisms that are distantly related. For example, if a naive biologist tried to construct an evolutionary tree on the basis of whether wings were present or absent, he or she might place birds, bats, and insects in the same evolutionary group, since all have wings. In this particular case, it is fairly obvious that these three organisms are not closely related—they differ in many features besides presence of wings, and the wings them-

~ FIGURE 23.21

The geographic distribution of mitochondrial DNA phenotypes in the pocket gopher, *Geomys pinetis*, in southeastern U.S., showing geographic separation of mitochondrial lineages. *Bam*HI M and *Bam*HI N are different restriction patterns revealed when the mitochondrial DNA is cleaved by the enzyme *Bam*HI.

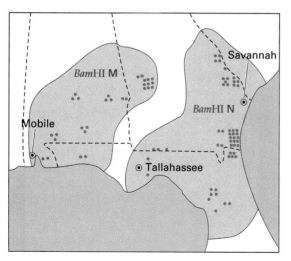

selves are very different. But this extreme example shows that phenotypes can sometimes be misleading about evolutionary relationships, and phenotypic similarities do not necessarily reflect genetic similarities.

Another problem with relying on a comparison of phenotypes is that not all organisms have a number of easily studied phenotypic features. For example, the study of evolutionary relationships among bacteria has always been problematic, because bacteria have few obvious traits that correlate with the degree of genetic relatedness. A third problem arises when we try to compare distantly related organisms. How do we compare, for example, bacteria and mammals, which have very few traits in common?

We have seen how DNA sequencing methods and analysis of restriction fragment length polymorphisms provide information about DNA sequences. DNA sequences provide the most accurate and reliable information upon which to infer evolutionary relationships. They allow direct comparison of the genetic differences among organisms, they are easily quantified, and all organisms have them (at least all organisms have some genes in common such as tRNA genes, rRNA genes, and genes for a few proteins). Because of these considerable advantages, many evolutionary biologists have turned to DNA sequences for assessing evolutionary relationships and for constructing evolutionary trees.

One case where sequence data has provided new information about evolutionary relationships is in our understanding of the primary divisions of life. Many years ago, biologists divided all of life into two major groups, the plants and the animals. As more organisms were discovered and their features examined in more detail, this simple dichotomy became unworkable. It was later recognized that organisms can be divided into prokaryotes and eukaryotes on the basis of cell structure. More recently, several primary divisions of life have been recognized, such as the five kingdoms (prokaryotes, protista, plants, fungi, and animals) proposed by Whittaker.

Within the past 10 years, RNA and DNA sequences have been used to uncover the primary lines of evolutionary history among all organisms. In one study, Norman Pace and his colleagues constructed an evolutionary tree of life based upon the sequences found in the 16S rRNA, which all organisms possess. As illustrated in Figure 23.22, their evolutionary tree revealed three major evolutionary groups: the eubacteria (the traditional prokaryotes), the eukaryotes, and the archaebacteria (a relatively little known group of

~ **FIGURE 23.22**

An evolutionary tree of life revealed by comparison of 16s rRNA sequences. From Pace, et. al., 1986. Cell 45:325–326.

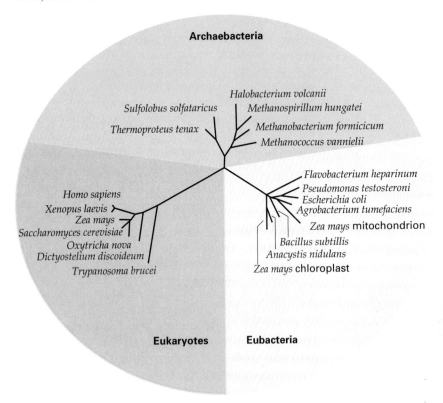

bacteria). The eubacteria and the archaebacteria, although both prokaryotic, were found to be as different genetically as were the eubacteria and eukaryotes. The deep evolutionary differences that separate the eubacteria and the archaebacteria were not obvious on the basis of phenotype; this became clear only after their nucleotide sequences were compared. Sequences of other genes, including 5S rRNAs, large rRNAs, and genes for some basic proteins, support the idea that three major evolutionary groups exist among living organisms.

Another field where DNA sequences are being used to study evolutionary relationships is in human evolution. In contrast to the extensive variation that is observed in size, body shape, facial features, skin color, etc., genetic differences among human populations are relatively small. For example, analysis of mtDNA sequences shows that the mean difference in sequence between two human populations is about one third of one percent. Other primates exhibit much larger differences. For example, the two subspecies of orangutan differ by as much as five percent. This indicates that all human groups are closely related. Nevertheless, some genetic differences do occur among different human groups. Surprisingly, the greatest differences are not found between populations located on different continents, but are seen among human populations residing in Africa. All other human populations show fewer differences than we find among the African populations. Many experts interpret these findings to mean that humans experienced their origin and early evolutionary divergence in Africa. After a number of genetically differentiated populations had evolved in Africa, it is hypothesized that a small group of humans may have migrated out of Africa and given rise to all other human populations. This hypothesis has been termed the "out-of-Africa" theory. Sequence data from both mitochondrial DNA and nuclear genes are consistent with this hypothesis. Although the out-of-Africa theory is not universally accepted, DNA sequence data is playing an increasingly important role in the study of human evolution and indeed in the study of the evolution of many groups.

SUMMARY

Population genetics is the study of patterns of genetic variation found within groups and how these patterns change or evolve over time. The gene pool of a population is the total of all genes within a Mendelian population, and it is described in terms of the gene and genotypic frequencies. The Hardy-Weinberg law describes what happens to gene and genotypic frequencies of a large, randomly mating population free from evolutionary forces; when these conditions are met, gene frequencies do not change, and genotypic frequencies stabilize after one generation in the proportions p^2, $2pq$, q^2, where p and q equal the gene frequencies of the population.

The classical, balance, and neutral mutation models are hypotheses to explain how much genetic variation should exist within natural populations and what forces are responsible for the variation observed. Protein electrophoresis showed that most populations of plants and animals contain large amounts of genetic variation, proving that the classical model was wrong. However, attempts to determine whether most genetic variation is maintained by natural selection (the balance model) or by the neutral forces of genetic drift and mutation (the neutral mutation model) have provided no clear answer.

Mutation, genetic drift, migration, and natural selection are forces that can alter gene frequencies of a population. Recurrent mutation changes the gene frequencies of a population; the relative rates of forward and reverse mutation will determine gene frequencies of a population in the absence of other forces. Genetic drift, chance change in gene frequencies due to small effective population size, leads to a loss of genetic variation within a population, genetic divergence among populations, and change of gene frequency within a population. Migration tends to reduce genetic divergence among populations and increases effective population size. Natural selection is differential reproduction of genotypes, and it is measured by fitness, the relative reproductive contribution of genotypes. The effects of natural selection depend upon the fitnesses of the genotypes, the degree of dominance, and the frequencies of the alleles in the population. Nonrandom mating affects only the genotypic frequencies of a population; the gene frequencies are unaffected. One type of nonrandom mating, inbreeding, leads to an increase in homozygosity.

New techniques of molecular genetics, including analysis of restriction fragment length polymorphisms and RNA and DNA sequences, have provided new insight into evolutionary processes. Different parts of a gene are found to evolve at different rates; those parts of the gene that have the least effect on fitness appear to evolve at the highest rates. In addition to changes in nucleotide sequences, molecular evolution involves variation in DNA length polymorphisms. Multigene families evolve through repeated duplication of genes, followed by genetic divergence of their sequences. Mitochondrial DNA of animals appears to evolve at a faster rate than the DNA of nuclear genes. RNA and DNA sequences can be used for inferring evolutionary rates and relationships among organisms.

ANALYTICAL APPROACHES FOR SOLVING GENETICS PROBLEMS

Q.1 In a population of 2000 gaboon vipers a genetic difference with respect to venom exists at a single locus. The alleles are incompletely dominant. The population shows 100 individuals homozygous for the *t* allele (genotype *tt*, nonpoisonous), 800 heterozygous (genotype *Tt*, mildly poisonous), and 1100 homozygous for the *T* allele (genotype *TT*, deadly poisonous).

a. What is the frequency of the *t* allele in the population?

b. Are the genotypes in Hardy-Weinberg equilibrium?

A.1 This question addresses the basics of calculating gene frequencies and relating them to the genotype frequencies expected of a population in Hardy-Weinberg equilibrium.

a. The *t* frequency can be calculated from the information given, since the trait is an incompletely dominant one. There are 2000 individuals in the population under study, meaning a total of 4000 alleles at the *T*/*t* locus. The number of *t* alleles is given by

$$(2 \times tt \text{ homozygotes}) + (1 \times Tt \text{ heterozygotes})$$
$$= (2 \times 100) + (1 \times 800) = 1000$$

This calculation is straightforward, since both alleles in the nonpoisonous snakes are *t*, while only one of the two alleles in the mildly poisonous snakes is *t*. Since the total number of alleles under study is 4000, the frequency of *t* alleles is 1000/4000 = 0.25. This system is a two-allele system, so the frequency of *T* must be 0.75.

b. For the genotypes to be in Hardy-Weinberg equilibrium, the distribution must be p^2 *TT* + 2*pq* *Tt* + q^2 *tt* genotypes, where *p* is the frequency of the *T* allele and *q* is the frequency of the *t* allele. In *part a* we established that the frequency of *T* is 0.75 and the frequency of *t* is 0.25. Therefore *p* = 0.75 and *q* = 0.25. Using these values, we can determine the genotype frequencies expected if this population is in Hardy-Weinberg equilibrium:

$$(0.75)^2 \ TT + 2(0.75)(0.25) \ Tt + (0.25)^2 \ tt$$

This expression gives 0.5625 *TT* + 0.3750 *Tt* + 0.0625 *tt*. Thus with 2000 individuals in the population we would expect 1125 *TT*, 750 *Tt*, and 125 *tt*. These values are close to the values given in the question, suggesting that the population is indeed in genetic equilibrium.

To check this result, we should perform a chi-square analysis (see Chapter 5, pp. 137–139), using the given numbers (not frequencies) of the three genotypes

as the observed numbers and the calculated numbers as the expected numbers. The chi-square analysis is as follows, where *d* = (observed − expected):

GENOTYPE	OBSERVED	EXPECTED	d	d^2	d^2/e
TT	1100	1125	−25	625	0.556
Tt	800	750	+50	2500	3.334
tt	100	125	−25	625	5.000
Totals	2000	2000	0		8.890

Thus the chi-square value (i.e., the sum of all the d^2/e values) is 8.89. For the reasons discussed in the text for a similar example, there is only one degree of freedom. Looking up the chi-square value in the chi-square table (Table 5.2, p. 139), we find a *P* value of approximately 0.0025. So about 25 times out of 10,000 we would expect chance deviations of the magnitude observed. In other words, our hypothesis that the population is in Hardy-Weinberg equilibrium is not substantiated. In this case our guess that it was in equilibrium was inaccurate. Nonetheless, the population is not greatly removed from an equilibrium state.

Q.2 About one normal allele in 30,000 mutates to the X-linked recessive allele for hemophilia in each human generation. Assume for the purposes of this problem that one *h* allele in 300,000 mutates to the normal alternative in each generation. (Note that in reality it is difficult to measure the reverse mutation of a human recessive allele that is essentially lethal, like the allele for hemophilia.) The mutation frequencies are indicated in the following equation:

$$h^+ \underset{v}{\overset{u}{\rightleftarrows}} h$$

where *u* = 10*v*. What gene frequencies would prevail at equilibrium under mutation pressures alone in these circumstances?

A.2 This question seeks to test our understanding of the effects of mutation on gene frequencies. In the chapter we discussed the consequences of mutation pressure. The conclusion was that if *A* mutates to *a* at *n* times the frequency that *a* mutates back to *A*, then at equilibrium the value of *q* will be $\hat{q} = u/(u + v)$ or $\hat{q} = nv/(n + 1)v$. Applying this general derivation to this particular problem, we simply use the values given. We are told that the forward mutation rate is 10 times the reverse mutation rate, or $u = 10v$. At equilibrium the value of q will be $\hat{q} = u/(u + v)$. Since $u = 10v$, this equation becomes $\hat{q} = 10v/11v$, so $\hat{q} = 10/11$, or 0.909. Therefore, at equilibrium brought about by mutation pressures the frequency of h (the hemophilia allele) will be 0.909, and the frequency of h^+ (the normal allele) will be $\hat{p}$, that is, $(1 - q) = (1 - 0.909) = 0.091$.

QUESTIONS AND PROBLEMS

*23.1 In the European land snail, *Cepaea nemoralis*, multiple alleles at a single locus determine shell color. The allele for brown (C^B) is dominant to the allele for pink (C^P) and to the allele for yellow (C^Y). Pink is recessive to brown, but is dominant to yellow, and yellow is recessive to pink and brown. Thus, the dominance hierarchy among these alleles is $C^B > C^P > C^Y$. In one population of *Cepaea*, the following color phenotypes were recorded:

Brown	236
Pink	231
Yellow	33
Total	500

Assuming that this population is in Hardy-Weinberg equilibrium (large, randomly mating, and free from evolutionary forces), calculate the frequencies of the C^B, C^Y, and C^P alleles.

23.2 Three alleles are found at a locus coding for malate dehydrogenase (MDH) in the spotted chorus frog. Chorus frogs were collected from a breeding pond, and each frog's genotype at the MDH locus was determined with electrophoresis. The following numbers of genotypes were found:

M^1M^1	8
M^1M^2	35
M^2M^2	20
M^1M^3	53
M^2M^3	76
M^3M^3	62
Total	254

a. Calculate the frequencies of the M^1, M^2, and M^3 alleles in this population.
b. Using a chi-square test, determine whether the MDH genotypes in this population are in Hardy-Weinberg proportions.

23.3 In a large interbreeding population 81 percent of the individuals are homozygous for a recessive character. In the absence of mutation or selection, what percentage of the next generation would be homozygous recessives? Homozygous dominants? Heterozygotes?

*23.4 Let A and a represent dominant and recessive alleles whose respective frequencies are p and q in a given interbreeding population at equilibrium (with $p + q = 1$).
a. If 16 percent of the individuals in the population have recessive phenotypes, what percentage of the total number of recessive genes exist in the heterozygous condition?
b. If 1.0 percent of the individuals were homozygous recessive, what percentage of the recessive genes would occur in heterozygotes?

*23.5 A population has eight times as many heterozygotes as homozygous recessives. What is the frequency of the recessive gene?

23.6 In a large population of range cattle the following ratios are observed: 49 percent red (RR), 42 percent roan (Rr), and 9 percent white (rr).
a. What percentage of the gametes that give rise to the next generation of cattle in this population will contain allele R?
b. In another cattle population only 1 percent of the animals are white and 99 percent are either red or roan. What is the percentage of r alleles in this case?

23.7 In a population gene pool the alleles A and a have initial frequencies of p and q, respectively. Prove that the gene frequencies and zygotic frequencies do not change from generation to generation as long as there is no selection, mutation, or migration, the population is large, and the individuals mate at random.

*23.8 The $S–s$ antigen system in humans is controlled by two codominant alleles, S and s. In a group of 3146 individuals the following genotypic frequencies were found: 188 SS, 717 Ss, and 2241 ss.
a. Calculate the frequency of the S and s alleles.
b. Determine whether the genotypic frequencies conform to the Hardy-Weinberg equilibrium by using the chi-square test.

23.9 Refer to Problem 23.8. A third allele is sometimes found at the S locus. This allele S^u is recessive to both the S and the s alleles and can only be detected in the homozygous state. If the frequencies of the alleles S, s, and S^u are p, q, and r, respectively, what would be the expected frequencies of the phenotypes $S–$, Ss, $s–$, and S^uS^u?

23.10 In a large interbreeding human population 60 percent of individuals belong to blood group O (genotype I^O/I^O). Assuming negligible mutation and no selective advantage of one blood type over another, what percentage of the grandchildren of the present population will be type O?

*23.11 A selectively neutral, recessive character appears in 0.40 of the males and in 0.16 of the females in a randomly interbreeding population. What is the gene's frequency? How many females are heterozygous for it? How many males are heterozygous for it?

23.12 Suppose you found two distinguishable types of individuals in wild populations of some organism in the following frequencies:

	TYPE 1	TYPE 2
Females	99%	1%
Males	90%	10%

The difference is known to be inherited. What is its genetic basis?

*23.13 Red-green color blindness is due to a sex-linked recessive gene. About 64 women out of 10,000 are color-blind. What proportion of men would be expected to show the trait if mating is random?

23.14 About 8 percent of the men in a population are red-green color-blind (owing to a sex-linked recessive gene). Answer the following questions, assuming random mating in the population, with respect to color blindness.
a. What percentage of women would be expected to be color-blind?
b. What percentage of women would be expected to be heterozygous?
c. What percentage of men would be expected to have normal vision two generations later?

23.15 List some of the basic differences in the classical, balance, and neutral-mutation models of genetic variation.

*23.16 Two alleles of a locus, A and a, an be interconverted by mutation:

$$A \xrightleftharpoons[v]{u} a$$

and u is a mutation rate of 6.0×10^{-7}, and v is a mutation rate of 6.0×10^{-8}. What will be the frequencies of A and a at mutational equilibrium, assuming no selective difference, no migration, and no random fluctuation caused by genetic drift?

*23.17 a. Calculate the effective population size (N_e) for a breeding population of 50 adult males and 50 adult females.
b. Calculate the effective population size (N_e) for a breeding population of 60 adult males and 40 adult females.

c. Calculate the effective population size (N_e) for a breeding population of 10 adult males and 90 adult females.
d. Calculate the effective population size (N_e) for a breeding population of 2 adult males and 98 adult females.

23.18 In a population of 40 adult males and 40 adult females, the frequency of allele A is 0.6 and the frequency of allele a is 0.4.
a. Calculate the 95 percent confidence limits of the gene frequency for A.
b. Another population with the same gene frequencies consists of only 4 adult males and 4 adult females. Calculate the 95 percent confidence limits of the gene frequency for A in this population.
c. What are the 95 percent confidence limits of A if the population consists of 76 females and 4 males?

23.19 The land snail Cepaea nemoralis is native to Europe but has been accidentally introduced into North America at several localities. These introductions occurred when a few snails were inadvertently transported on plants, building supplies, soil, or other cargo. The snails subsequently multiplied and established large, viable populations in North America.

Assume that today the average size of Cepaea populations found in North America is equal to the average size of Cepaea populations in Europe. What predictions can you make about the amounts of genetic variation present in European and North American populations of Cepaea? Explain your reasoning.

*23.20 A population of 80 adult squirrels resides on campus, and the frequency of the Est^1 allele among these squirrels is 0.70. Another population of squirrels is found in a nearby woods, and there, the frequency of Est^1 allele is 0.5. During a severe winter, 20 of the squirrels from the woods population migrate to campus in search of food and join the campus population. What will be the gene frequency of Est^1 in the campus population after migration?

23.21 Upon sampling three populations and determining genotypes, you find the following three genotype distributions. What would each of these distributions imply with regard to selective advantages of population structure?

POPULATION	AA	Aa	aa
1	0.04	0.32	0.64
2	0.12	0.87	0.01
3	0.45	0.10	0.45

23.22 The frequency of two adaptively neutral alleles in a large population is 70 percent A : 30 percent a. The population is wiped out by an epidemic, leaving only four individuals, who produce many offspring. What is the probability that the population several years later will be 100 percent *AA*? (Assume no mutations.)

23.23 A completely recessive gene, owing to changed environmental circumstances, becomes lethal in a certain population. It was previously neutral, and its frequency was 0.5.

a. What was the genotype distribution when the recessive genotype was not selected against?

b. What will be the gene frequency after one generation in the altered environment?

c. What will be the gene frequency after two generations?

*__**23.24**__ Human individuals homozygous for a certain recessive autosomal gene die before reaching reproductive age. In spite of this removal of all affected individuals, there is no indication that homozygotes occur less frequently in succeeding generations. To what might you attribute the constant rate of appearance of recessives?

23.25 A completely recessive gene (Q^1) has a frequency of 0.7 in a large population, and the Q^1Q^1 homozygote has a relative fitness of 0.6.

a. What will be the frequency of Q^1 after one generation of selection?

b. If there is no dominance at this locus (the fitness of the heterozygote is intermediate to the fitnesses of the homozygotes), what will the gene frequency be after one generation of selection?

c. If Q^1 is dominant, what will the gene frequency be after one generation of selection?

*__**23.26**__ As discussed earlier in this chapter, the gene for sickle-cell anemia exhibits overdominance. An individual who is an *Hb-A/Hb-S* heterozygote has increased resistance to malaria and therefore has greater fitness than the *Hb-A/Hb-A* homozygote, who is susceptible to malaria, and the *Hb-S/Hb-S* homozygote, who has sickle-cell anemia. Suppose that the fitness values of the genotypes in Africa are as presented below:

^{1} *Hb-A/Hb-A* = 0.88

^{2} *Hb-A/Hb-S* = 1.00

$^{1/2}$*Hb-S/Hb-S* = 0.14

Give the expected equilibrium frequencies of the sickle-cell gene (*Hb-S*).

*__**23.27**__ Achondroplasia, a type of dwarfism in humans, is caused by an autosomal dominant gene. The mutation rate for achondroplasia is about 5×10^{-5} and the fitness of achondroplastic dwarfs has been estimated to be about 0.2, compared with unaffected individuals. What is the equilibrium frequency of the achondroplasia gene based on this mutation rate and fitness value?

23.28 The frequencies of the L^M and L^N blood group alleles are the same in each of the populations I, II, and III, but the genotypes' frequencies are not the same, as shown below. Which of the populations is most likely to show each of the following characteristics: random mating, inbreeding, genetic drift. Explain your answers.

	$M-M$	$M-N$	$N-N$
I	.50	.40	.10
II	.49	.42	.09
III	.45	.50	.05

*__**23.29**__ DNA was collected from 100 people randomly sampled from a given human population and was digested with the restriction enzyme *Bam*HI, electrophoresed, and Southern blotted. The blots were probed with a particular cloned sequence. Three different patterns of hybridization were seen on the blots. Some DNA samples (56 of them) showed a single band of 6.3 kb, others (6) showed a single band at 4.1 kb, and yet others (38) showed both the 6.3 and the 4.1 kb bands.

a. Interpret these results in terms of *Bam*HI sites.

b. What are the frequencies of the restriction site alleles?

c. Does this population appear to be in Hardy-Weinberg equilibrium for the relevant restriction site(s)?

23.30 DNA was isolated from 10 nine-banded armadillos and cut with the restriction enzyme *Hin*dIII. *Hin*dIII recognizes the six-base sequence $\begin{array}{l}5' = \text{AAGCTT} = 3'\\3' = \text{TTCGAA} = 5'\end{array}$. The DNA fragments that resulted from the restriction reaction were separated with agarose electrophoresis and transferred to nitrocellulose using Southern blotting. A labeled probe for the β-hemoglobin gene was added, which resulted in the following set of restriction patterns. Note: +/+ indicates that the restriction site was present on both chromosomes of the individual, +/− indicates that the restriction site was present on one chromosome and absent on one chromosome of the individual, and −/− indicates that the restriction site was absent on both chromosomes of the individual:

+/+ −/− +/− +/− −/− +/− +/+ +/− −/− +/+

___   ___ ___   ___ ___ ___   ___

___   ___ ___   ___ ___ ___   ___

Calculate the expected heterozygosity in nucleotide sequence.

*__**23.31**__ Fifty tiger salamanders from one pond in west Texas were examined for genetic variation by using the technique of protein electrophoresis. Each salamander was genotyped for five loci (AmPep, ADH, PGM, MDH, and LDH-1). No variation was found at AmPep, ADH, and LDH-1; in other words, all individuals were homozygous for the same allele at these loci. The following numbers of genotypes were observed at the MDH and PGM loci.

MDH GENOTYPES	NUMBER OF INDIVIDUALS
AA	11
AB	35
BB	4

PGM GENOTYPES	NUMBER OF INDIVIDUALS
DD	35
DE	10
EE	5

Calculate the proportion of polymorphic loci and the heterozygosity for this population.

23.32 What factors cause genetic drift?

23.33 What are the primary effects of the following evolutionary forces on the gene and genotypic frequencies of a population?
a. mutation
b. migration
c. genetic drift
d. inbreeding

23.34 Explain how overdominance leads to an increased frequency of sickle-cell anemia in areas where malaria is widespread.

*23.35 Suppose we examine the rates of nucleotide substitution in two 300-nucleotide sequences of DNA isolated from humans. In the first sequence (sequence A), we find a nucleotide substitution rate of 4.88×10^{-9} substitutions per site per year. The substitution rate is the same for synonymous and nonsynonymous substitutions. In the second sequence (sequence B), we find a synonymous substitution rate of 4.66×10^{-9} substitutions per site per year and a nonsynonymous substitution rate of 0.70×10^{-9} substitutions per site per year. Referring to Table 23.14, what might you conclude about the possible functions of sequence A and sequence B?

23.36 What are some of the characteristics of mitochondrial DNA evolution in animals?

23.37 What is concerted evolution?

23.38 What are some of the advantages of using DNA sequences to infer evolutionary relationships?

GLOSSARY

SUGGESTED READINGS

SOLUTIONS TO
SELECTED PROBLEMS

CREDITS

INDEX

GLOSSARY

[Numbers in parentheses indicate the page(s) on which the term is defined]

acrocentric chromosome A chromosome with a centromere near the end, such that it has one long arm plus a stalk and a satellite. (13)

additive genetic variance (V_A) Genetic variance that arises from the cumulative effects of genes on the phenotype. (689)

adenine (A) A purine base found in RNA and DNA; in double-stranded DNA, adenine pairs with the pyrimidine thymine. (275)

allele One of two or more alternative forms of a single gene locus. Different alleles of a gene each have a unique nucleotide sequence, and their activities are all concerned with the same biochemical and developmental process, although their individual phenotypes may differ. (40; 43)

allelic frequencies (gene frequencies) The frequencies of alleles at a locus occurring among individuals in a population. (710)

allelomorph (allele) A term coined by William Bateson; literally means "alternative form"; later shortened by others to *allele*. (52)

allopolyploidy Polyploidy involving two or more genetically distinct sets of chromosomes. (602)

alternation of generations The two distinct reproductive phases of green plants; the stages alternate between the sporophyte (characterized by diploid cells) and the gametophyte (characterized by haploid cells). (29)

Ames test Developed by Bruce Ames in the early 1970s, this test investigates new or old environmental chemicals for carcinogenic effects. It uses the bacterium *Salmonella typhimurium* as a test organism for mutagenicity of compounds. (573)

amino acids The building blocks of polypeptides. There are 20 different amino acids. (251)

aminoacyl-tRNA A tRNA molecule covalently bound to an amino acid. This complex brings the amino acid to the ribosome so that is can be used in polypeptide synthesis. (416)

aminoacyl-tRNA synthetase An enzyme that catalyzes the addition of a specific amino acid to a tRNA molecule. Since there are 20 amino acids, there are also 20 synthetases. (415)

amniocentesis A procedure in which a sample of fluid is withdrawn from the amniotic sac of a developing fetus, and cells of the fluid are cultured and examined for chromosomal abnormalities. (249)

analysis of variance (ANOVA) A series of statistical procedures for examining differences in means and for partitioning variance. (683)

anaphase The stage in mitosis or meiosis during which the sister chromatids (mitosis) or homologous chromosomes (meiosis) separate and migrate toward the opposite poles of the cell. (18)

anaphase II The second stage of meiosis during which the centromeres (and therefore the chromatids) are pulled to the opposite poles of the spindle. The separated chromatids are now referred to as chromosomes in their own right. (25)

aneuploidy The abnormal condition in which one or more whole chromosomes of a normal set of chromosomes either are missing or are present in more than the usual number of copies. (72; 597)

antibody A protein molecule that recognizes and binds to a foreign substance introduced into the organism. (79; 539)

anticodon A three-nucleotide sequence that pairs with a codon in mRNA by complementary base pairing. (394)

antigen Any large molecule which stimulates the production of specific antibodies or which binds specifically to an antibody. (79; 539)

artificial selection Human manipulation of plant or animal mating by determining which individuals will survive and reproduce. If the selected traits have a genetic basis, they will change and evolve. (694)

asexual (vegetative) reproduction Reproduction in which a new individual develops from either a single cell or from a group of cells in the absence of any sexual process. (13)

attenuation In certain bacterial biosynthetic operons, a regulatory mechanism that controls gene expression by causing RNA polymerase to terminate transcription. (493)

autonomous development The development of a cell solely according to the genetic information it contains, with no influence by the environment. (236)

autonomously replicating sequences (ARS elements) Specific sequences (e.g., in baker's yeast, *Saccharomyces cerevisiae*) that, when included as part of an extrachromosomal, circular DNA molecule, confer upon that molecule the ability to replicate autonomously. (347)

autopolyploidy Polyploidy involving more than two chromosome sets of the same species. (602)

autosome A chromosome other than a sex chromosome. (13; 69)

auxotroph A mutant strain of a given organism that is unable to synthesize a molecule required for growth; therefore the molecule must be supplied in the growth medium in order for the organism to grow. (199)

auxotrophic mutation (nutritional, biochemical) A mutation that affects an organism's ability to make a particular molecule essential for growth. (573; 580)

bacteria Prokaryotic organisms, characterized as spherical, rod-shaped, or spiral-shaped, single-cellular or multi-cellular, filamentous. (3)

balance model A hypothesis of genetic variation proposing that selection maintains large amounts of genetic variation within populations. (719)

Barr body A highly condensed mass of chromatin found in the cell nuclei of normal females, but not in the cell nuclei of normal male cells. The mass represents a cytologically condensed and inactivated X chromosome. (78)

base analog A chemical with a molecular structure that is extremely similar to the bases normally found in DNA. (568)

base-pair substitution mutation A change in a gene such that one base pair is replaced by another base pair, e.g., an AT is replaced by a GC pair. (559)

bidirectional replication The DNA synthesis that takes place in both directions away from the origin of replication point. (335)

binary fission The cell reproduction process in which a parental cell divides into two daughter cells of approximately equal size. (5)

binomial distribution The theoretical frequency distribution of events that have two possible outcomes. (675)

biochemical mutation See auxotrophic mutation. (573; 580)

biparental inheritance Plant zygotes that show traits indicating chloroplast chromosomes from both parents are present and active. (663)

bivalent A pair of homologous, synapsed chromosomes during the first meiotic division. (21)

bottleneck effect A form of genetic drift that occurs when a population is drastically reduced in size. In the reduction some genes may be lost from the gene pool as a result of chance. (733)

branch-point sequence The consensus sequence in mammalian cells, YNCURAY (where Y is a pyrimidine, R is a purine, and N is any base), to which the free 5′ end of the intron loops and binds to the A nucleotide in the sequence during intron splicing. (391)

broad-sense heritability A quantity representing the proportion of phenotypic variance that consists of genetic variance. (688; 689)

C_0t plot A graph of the kinetics of renaturation of a sample of single-stranded DNA molecules. The graph is a plot of the fraction of remaining single-stranded DNA molecules versus the product of DNA concentration at time zero (C_0) and time (t)—hence the term C_0t. (318)

C value The amount of DNA found in the haploid set of chromosomes. (303)

CAAT element One of the eukaryotic promoter elements found approximately 80 base pairs upstream of the initiation site. The element can function at a number of other locations and in either orientation with respect to the start point. The consensus sequence is 5′-GGCCAATCT-3′. (373)

cancer Diseases characterized by the uncontrolled and abnormal division of eukaryotic cells and by the spread of the disease (metastasis) to disparate sites in the organism. (628)

5′-capping The addition of a methylated guanine nucleotide (a "cap") to the 5′ end of a premessenger RNA molecule; the cap is retained on the mature mRNA molecule. (383)

catabolite repression (glucose effect) The inactivation of an inducible bacterial operon when glucose is present, even though the operon's inducer is also present. (489)

cDNA DNA copies made from an RNA template that is catalyzed by the enzyme reverse transcriptase. (442)

cDNA cloning A process of constructing fragments to be cloned. The mRNA molecules are isolated from cells, and DNA copies of these RNAs are made and inserted into a cloning vector. The end result is a cDNA library. See cDNA; cDNA library. (444)

cDNA library The collection of molecular clones; the collection contains cDNA copies of the entire mRNA population of a cell. See cDNA. (442; 444)

cell The smallest building unit of a multicellular organism and, as a unit, in itself an elementary organism. (3)

cell cycle The cyclical process of cellular growth and reproduction in unicellular and multicellular eukaryotes. The cycle includes nuclear division, or mitosis, and cell division, or cytokinesis. (14)

cell division A process whereby one cell divides to produce two cells. (14)

cell-free, protein-synthesizing system A system, isolated from cells, which contains ribosomes, mRNA, tRNAs with amino acids attached, and all the necessary protein factors for the *in vitro* synthesis of polypeptides. (411)

cell theory The basic principles defining cells and their functions. Cell theory has three basic tenets: (1) The cell is the smallest building unit of a multicellular organism; (2) each cell in a multicellular organism has a specific role; (3) a cell can be produced from another cell only by cell division. (3)

cellular oncogene (c-onc) The genes, present in a functional state in cancerous cells, which are responsible for the cancerous state. (633)

centi-Morgan (cM) The map unit. It is sometimes called a centi-Morgan in honor of T. H. Morgan. (140)

centromere (kinetochore) A specialized chromosome region, seen as a constriction under the microscope. This region is important in chromosome activities during cellular division. (13)

chain-terminating codon One of three codons for which no normal tRNA molecule exists with an appropriate anticodon. Also called a *nonsense codon*, this codon in mRNA specifies the termination of polypeptide synthesis. (413)

character An observable phenotypic feature of the developing or fully developed organism; the result of gene action alone. (34)

charged tRNA The product of combining an amino acid and a tRNA. (416)

charging The act of adding the amino acid to the tRNA. (416)

chiasma A cross-shaped structure formed during crossing-over and visible during the diplonema stage of meiosis. (21)

chiasma interference (chromosomal interference) The physical interference caused by the breaking and rejoining chromatids that reduces the probability that more than one crossing-over event occurs near another in one part of the meiotic tetrad. (151)

chi-square test A statistical procedure that determines what constitutes a significant difference between observed results and the results expected on the basis of a particular hypothesis; a goodness-of-fit test. (137)

chloroplast The cellular organelle found only in green plants; the site of photosynthesis in the cells containing it. (651)

chorionic villus sampling A procedure in which a sample of chorionic villus tissue of a developing fetus is collected and examined for chromosomal abnormalities. (250)

chromatid One of the two visibly distinct longitudinal subunits of all replicated chromosomes; becomes visible between early prophase and metaphase of mitosis. (14)

chromatin The piece of DNA-protein complex that is studied and analyzed. Each chromatin fragment reflects the general features of chromosomes but not the specifics of any individual chromosome. (307)

chromosomal aberration (chromosomal mutation) A change in the structure or number of chromosomes. (558)

chromosomal interference See chiasma interference. (151)

chromosome The genetic material of the cell, complexed with protein and organized into a number of linear structures. Literally the term means "colored body," because the threadlike structures are visible under the microscope only after they are stained with dyes. (7)

chromosome aberration The variation from the wild-type condition in either chromosome number or chromosome structure. (588)

chromosome theory of heredity The theory that chromosomes are carriers of genes. The first clear formulation of the theory was made by both Sutton and Boveri, who independently recognized that the transmission of chromosomes from one generation to the next closely paralleled the pattern of gene transmission from one generation to the next. (62)

cis-dominance The control of a gene or a DNA sequence over only those genes that are on the same contiguous piece of DNA. (482)

cis-trans (complementation) test A test, developed by E. Lewis, for determining whether two different mutations are within the same cistron (gene). (222)

classical model A hypothesis of genetic variation, proposing that natural populations contain little genetic variation as a result of strong selection for one allele. (719)

clonal selection A process whereby cells with a specific type of antibody on their surfaces are stimulated to proliferate and secrete that antibody. (539)

cloning The generation of many copies of a DNA molecule (e.g., a recombinant DNA molecule) by replication in a suitable host. (432)

cloning vector (cloning vehicle) A double-stranded DNA molecule that is able to replicate autonomously in a host cell and with which a DNA fragment (or fragments) can be bonded to form a recombinant DNA molecule for cloning. (432; 436)

clustered repeated sequences DNA sequences that are repeated one after another in a row. (320)

coding sequence The part of an mRNA molecule that specifies the amino acid sequence of a polypeptide during translation. (383; 386)

codominance The situation in which the heterozygote exhibits the phenotypes of both homozygotes. (100)

codon A group of three adjacent nucleotides in an mRNA molecule that specifies either one amino acid in a polypeptide chain or the termination of polypeptide synthesis. (383; 410)

coefficient of coincidence A number that expresses the extent of chiasma interference throughout a genetic map; the ratio of *observed* double-crossover frequency to *expected* double-crossover frequency. Interference is equal to 1 minus the coefficient of coincidence. (151)

combinatorial gene regulation Transcriptional control (i.e. whether a gene is active or inactive) achieved by combining relatively few regulatory proteins (negative and positive) that bind to particular DNA sequences. (518)

complementary-base pairing The hydrogen bonding between a particular purine and a particular pyrimidine in double-stranded nucleic acid molecules (DNA-DNA, DNA-RNA, or RNA-RNA). The major specific pairings are guanine with cytosine and adenine with thymine or uracil. (268)

complementation test *See* cis-trans test. (222)

complementary DNA *See* cDNA. (442)

complete dominance The case in which one allele is dominant to the other so that, at the phenotypic level, the heterozygote is essentially indistinguishable from the homozygous dominant. (97)

complete recessiveness The situation in which an allele is phenotypically expressed only when it is homozygous. (97)

concerted evolution (molecular drive) A poorly understood evolutionary process that produces a uniform sequence in multiple copies of a gene. (751)

concordant A condition in twins indicating that both do or do not possess a trait. (119)

conditional mutant A mutant organism that is normal under one set of conditions but becomes seriously impaired or dies under other conditions. (332)

conjugation A process having a unidirectional transfer of genetic information through direct cellular contact between a donor ("male") or a recipient ("female") bacterial cell. (196; 200)

consensus sequence The most commonly encountered nucleotide sequence that is believed to provide the same function in different genes. Examples are promoter sequences for the initiation of RNA transcription or splicing sites for the excision of introns from premessenger RNA. (348; 367)

conservative model A DNA replication scheme in which the two parental strands of DNA remain together and serve as a template for the synthesis of a new daughter double helix. (326)

constitutive gene A gene whose products are essential to normal cell functioning, no matter what the life-supporting environmental conditions are. These genes are always active in growing cells. (477)

constitutive heterochromatin Condensed chromatin that is always genetically inactive, and is found at homologous sites on chromosome pairs. (307)

continuous trait *See* quantitative (continuous) traits. (672)

contributing allele An allele that contributes to the phenotype. (685)

controlling site A specific sequence of nucleotide pairs adjacent to a gene where transcription of that gene occurs in response to a particular molecular event. (478)

coordinate induction The simultaneous transcription and translation of two or more genes brought about by the presence of an inducer. (480)

core enzyme The portion of the *E. coli* RNA polymerase that is the active enzyme; can be written as $\alpha_2\beta\beta'\omega$. (365)

correlation coefficient A statistic that measures the strength of the association between two variables. (679)

cosmids Cloning vectors derived from plasmids; structures capable of cloning large fragments of DNA. In addition to the features of plasmid cloning vectors (i.e., origin of replication and selectable marker[s] for growth in bacteria), cosmids contain phage lambda *cos* sites, which permit the packaging of recombinant DNA molecules that are constructed into the lambda phage head *in vitro*. (438)

cotransduction The simultaneous transduction of two or more bacterial genes; a good indication that the genes are closely linked. (211)

cotransformational transport The simultaneous synthesis and movement of a protein into the ER. (424)

coupling An arrangement in which two wild-type alleles of a gene are on one homologous chromosome and two recessive mutant alleles are on the other. (140)

covariance A statistic used to calculate the correlation coefficient between two variables. The covariance is calculated by taking the sum of $(x-\bar{x})(y-\bar{y})$ over all pairs of values for the variables x and y, where $\bar{x}$ is the mean of the x values and $\bar{y}$ is the mean of all y values. (679)

crisscross inheritance A type of gene transmission in which a trait is passed from a male parent to a female child to a male grandchild. (69)

cross *See* cross-fertilization. (43)

cross-fertilization (cross) A term used for the fusion of male and female gametes from different individuals; the bringing together of genetic material from different individuals for the purpose of genetic recombination. (38)

crossing-over a term introduced by Morgan and E. Cattell, in 1912, to describe the process of reciprocal chromosomal interchange by which recombinants arise. (21; 131)

cytohet The genetic condition of plant zygotes that display biparental inheritance; the term is derived from "*cyto*plasmically *het*erozygous." (663)

cytokinesis A term that refers to the division of the cytoplasm. The process compartmentalizes the two new nuclei into separate daughter cells, completing the mitotic cell division process. (18)

cytological marker A cytologically distinguishable feature of a chromosome. (132)

cytoplasm The semisolid matrix of cell material in the surrounding cell body. The term was coined by E. Strasburger. (9; 61)

cytoskeleton A complex network of protein filaments that extend throughout the cytoplasm. (9)

cytosine (C) A pyrimidine base found in RNA and DNA. In double-stranded DNA, cytosine pairs with the purine guanine. (275)

dark repair *See* excision repair. (576)

daughter chromosome The chromatid after anaphase (of mitosis) or anaphase II (of meiosis). (14)

degeneracy A multiple coding; more than one codon per amino acid. (413)

degradation control Regulation of the RNA breakdown rate in the cytoplasm. (531)

deleted When a chromosome spontaneously breaks off and is lost. (184)

deletion (deficiency) A chromosome mutation resulting in the loss of a segment of the genetic material (and the genetic information the segment contains). (589)

deoxyribonuclease (DNase) An enzyme that catalyzes the degradation of DNA to nucleotides. (270)

deoxyribonucleic acid (DNA) A polymeric molecule consisting of deoxyribonucleotide building blocks; in a double-stranded, double-helical form, DNA is the genetic material of most organisms. (3)

deoxyribonucleotide The basic building block of DNA, consisting of deoxyribose (a sugar), a base, and a phosphate. (275)

deoxyribose The pentose (five-carbon) sugar found in DNA. (275)

development The process of regulated growth that results from the interaction of the genome with cytoplasm and the environment. The process involves a programmed sequence of phenotypic events that are typically irreversible. (532)

diakinesis The stage that follows diplonema and during which the four chromatids of each tetrad are most condensed and the chiasmata often terminalize. (21)

dicentric bridge *See* dicentric chromosome. (592)

dicentric chromosome A chromosome with two centromeres. For example, as a result of the crossover between genes B and C in the inversion loop, one recombinant chromatid becomes stretched across the cell as the two centromeres begin to migrate, forming a dicentric bridge. (592)

dideoxy nucleotide A modified nucleotide which has a 3'-H on the deoxyribose sugar rather than a 3'-OH. If a dideoxy nucleoside triphosphate (ddNTP) is used in a DNA synthesis reaction, the ddNTP can be incorporated into the growing chain. However, no further DNA synthesis can occur because no phosphodiester bond can be formed with an incoming DNA precursor. (461)

dideoxy (Sanger) sequencing A method of rapid sequencing of DNA molecules developed by Fred Sanger. This technique incorporates the use of dideoxy nucleotides in a DNA polymerase-catalyzed DNA synthesis reaction. (458)

differentiation An aspect of development that involves the formation of different types of cells, tissues, and organs from a zygote through the specific regulation of gene expression. (532)

dihybrid cross A cross between two dihybrids of the same type. Individuals that are heterozygous for two pairs of alleles at two different loci are called dihybrid. (47)

dioecious A term referring to plant species that have male sex organs on one type of individual and female sex organs on another type of individual. (80)

diploid A eukaryotic cell with two sets of chromosomes. (12)

diplonema The second stage of prophase I, in which the chromosomes begin to repel one another and tend to move apart. (21)

discontinuous DNA replication A DNA replication involving the synthesis of short DNA segments that are subsequently linked to form a long polynucleotide chain. (335)

discontinuous trait A heritable trait in which the mutant phenotype is sharply distinct from the alternative, wild-type phenotype. (671)

discordant A condition in twins, indicating that one possesses a trait and the other twin does not. (119)

disjunction The process in anaphase during which sister chromatid pairs undergo separation. (18)

dispersive model A DNA replication scheme in which the parental double helix is cleaved into double-stranded DNA segments which act as templates for the synthesis of new double-stranded DNA segments. Somehow, the segments reassemble into complete DNA double helices, with parental and progeny DNA segments interspersed. (326)

dizygotic twins *See* fraternal twins. (119)

DNA *See* deoxyribonucleic acid. (3)

DNA fingerprinting The use of restriction fragment length polymorphism (RFLP) DNA analysis to identify an individual. (467)

DNA helicase An enzyme that catalyzes the unwinding of the DNA double helix during replication in *E. coli*; product of the *rep* gene. (334)

DNA ligase (polynucleotide ligase) An enzyme that catalyzes the formation of a covalent bond between free single-stranded ends of DNA molecules during DNA replication and DNA repair. (335)

DNA polymerase An enzyme that catalyzes the synthesis of DNA. (330)

DNA polymerase I An *E. coli* enzyme that catalyzes DNA synthesis, originally called the Kornberg enzyme. (330)

DNA primase The enzyme in DNA replication that catalyzes the synthesis of a short nucleic acid primer. (334)

docking protein An integral protein membrane of the endoplasmic reticulum (ER) to which the nascent polypeptide-signal recognition particle (SRP)-ribosome complex binds to facilitate the binding of the polypeptide's signal sequence and associated ribosome to the ER. (424)

dominance variance (V_D) Genetic variance that arises from the dominance effects of genes. (689)

dominant An allele or phenotype that is expressed in either the homozygous or the heterozygous state. (44)

dominant lethal allele Allele that will exhibit a lethal phenotype when present in the heterozygous condition. (111)

dosage compensation A mechanism in mammals which compensates for X chromosomes in excess of the normal complement. *See* Barr body. (78)

double crossover Two crossovers occurring in a particular region of a chromosome during meiosis. (143)

double fertilization An event found only in the life cycle of flowering plants. In addition to the fertilization of the egg by one sperm, a second sperm fertilizes the two nuclei of the gametophyte's central cell to form the cell that will become the seed's endosperm. (28)

Down syndrome (trisomy-21) A human clinical condition characterized by various abnormalities. Typically, it is caused by the presence of an extra copy of chromosome 21. (597)

duplication A chromosome mutation that results in the doubling of a segment of a chromosome. (590)

effective population size The effective number of adults contributing gametes to the next generation. (731)

effector A small molecule involved in the control of expression of many regulated genes. (478)

endosperm A triploid (3N) tissue in flowering plants that surrounds and nourishes the embryo. It develops from the union of a sperm nucleus and the polar nuclei of the egg. (28)

enhancer sequence (enhancer element) In eukaryotes, a type of DNA sequence element having a strong, positive effect on transcription by RNA polymerase II. (373)

environmental variance (V_E) Any nongenetic source of phenotypic variation among individuals. (688)

episome An autonomously replicating plasmid (a circular, double-stranded DNA molecule) that is capable of integrating into the host cell's chromosome. (617)

epistasis A form of gene interaction in which one gene interferes with the phenotypic expression of another nonallelic gene; therefore, when both genes are present in the genotype, the phenotype is governed by the former gene and not the latter. (104)

erythroblasts Red blood cell precursors. (518)

essential genes Genes which, when they are mutated, can result in a lethal phenotype. (111)

euchromatin Chromatin that is condensed during division but becomes uncoiled during interphase. (307)

eukaryote A term that literally means "true nucleus." Eukaryotes are organisms that have cells in which the genetic material is located in a membrane-bound nucleus. Eukaryotes can be unicellular or multicellular. (5)

excision repair (dark repair) An enzyme-catalyzed, light-independent process of repair of ultraviolet-light-induced thymine dimers in DNA that involves removal of the dimers and synthesis of a new piece of DNA complementary to the undamaged strand. (576)

ex-conjugant A pair of bacterial cells, one male and one female, that have separated during or following conjugation. The female ex-conjugant contains a fragment of male cell DNA. *See* trans-conjugant. (205)

exon *See* coding sequence. (383; 386)

expressivity The degree to which a particular genotype is expressed in the phenotype. (54; 113)

F₁ generation The first filial generation produced by crossing two parental strains. (39)

F₂ generation The second filial generation produced by selfing the F₁. (39)

F-pili (sex pili) Hairlike cell surface components produced by cells containing the F factor, which allow the physical union of F^+ and F^- cells, or Hfr and F^- cells to take place. (200)

facultative heterochromatin Chromatin that may become condensed throughout the cell cycle and may contain genes that are inactivated when the chromatin becomes condensed. (307)

familial trait A trait shared by members of a family. (691)

fine-structure mapping A high-resolution mapping of allelic sites within a gene. (217)

first filial generation *See* F₁ generation; the offspring that result from the first experimental crossing of animals or plants. (39)

first law *See* principle of segregation. (42)

fitness The relative reproductive ability of a genotype. (739)

formylmethionine (fMet) A specially modified amino acid involving the addition of a formyl group to the amino group of methionine. It is the first amino acid incorporated into a polypeptide chain in prokaryotes and in eukaryotic cellular organelles. (416)

forward mutation A mutational change from a wild-type allele to a mutant allele. (560)

founder effect A phenomenon that occurs when the isolate effect is exhibited by a small breeding unit formed by migration of a small number of individuals from a large population. (732)

frameshift mutation A mutational addition or deletion of a base pair in a gene such that the normal reading frame of an mRNA, which is read in groups of three bases, is disrupted. (560)

fraternal (dizygotic) twins A result of the separate fertilization of two eggs released by the ovaries at the same time. The two eggs are likely to be very different genetically, since most humans are heterozygous for many of their genes. (119)

frequency distribution A means of summarizing the phenotypes of a continuous trait. The population is described in terms of the proportion of individuals that have each phenotype. (674)

gametes Mature reproductive cells that are specialized for sexual fusion. Each gamete is haploid, and fusion of two gametes produces a diploid zygote. (12)

gametogenesis The formation of male and female gametes by meiosis. (20)

gametophyte The haploid sexual generation in the life cycle of plants that produces the gametes. (27)

GC element A eukaryotic promoter element with the consensus sequence 5′-GGGCGGG-3′, found in either orientation upstream of the transcription initiation site. The GC elements appear to help the RNA polymerase bind near the transcription start point. (373)

gene (Mendelian factor) The determinant of a characteristic of an organism. Genetic information is coded in the DNA, which is responsible for species and individual variation. A gene's nucleotide sequence specifies a polypeptide or RNA and is subject to mutational alteration. (34)

gene flow The movement of genes that takes place when organisms migrate from one population to another and then reproduce, contributing their genes to the gene pool of the recipient population. (736)

gene frequency (allele frequency) The proportion of one particular type of allele to the total of all alleles at this genetic locus in a Mendelian breeding population. (710)

gene locus *See* locus. (43)

gene marker *See* genetic marker. (132)

gene mutation A heritable alteration of the genetic material, usually from one allelic form to another. (558)

gene pool The total genetic information encoded in the total genes of a breeding population at a specific point in time. (708)

gene redundancy A situation in which tRNA genes occur two or more times in the *E. coli* chromosome. (396)

gene regulatory elements *See* transcription-controlling sequence. (363)

gene segregation *See* principle of segregation (first law). (43)

generalized transduction A type of transduction in which any gene may be transferred between bacteria. (208)

genetic code The base-pair information that specifies the amino acid sequence of a polypeptide. (410)

genetic correlation An association between the genes that determine two traits. (697)

genetic counseling Evaluation and explanation of the risk that prospective parents may have a child who expresses a genetic disease. The genetic counselor typically makes predictions about the probabilities that particular traits (deleterious or not) may occur in a couple's child. (52; 248)

genetic drift Any change in gene frequency in a population due to chance. (713; 730)

genetic engineering The alteration of the genetic constitution of cells or individuals by directed and selective modification, insertion, or deletion of an individual gene or genes. In some cases, novel gene combinations are made by joining DNA fragments from different organisms. (432)

genetic map (linkage map) A representation of the genetic distance separating nonallelic gene loci in a linkage structure. (137)

genetic mapping The using of genetic crosses to locate genes, relative to one another, on chromosomes. (137)

genetic marker (gene marker) Any genetically controlled phenotypic difference used in genetic analysis, particularly in the detection of genetic recombination events. (132)

genetic recombination A process by which parents with different genetic characters give rise to progeny, so that genes in which the parents differed are associated in new combinations. For example, from *A B* and *a b* the recombinants *A b* and *a B* are produced. (21)

genetic variance (represented by V_G) Genetic sources of phenotypic variation among individuals of a population. Includes dominance genetic variance, additive genetic variance, and epistatic genetic variance. (688)

genetics The science of heredity, involving the structure and function of genes and the way genes are passed from one generation to the next. (2)

genome The total amount of genetic material in a cell; in eukaryotes the haploid set of chromosomes of an organism. (12)

genome mutation A mutation characterized by a change in the number of chromosomes. (558)

genomic library The collection of molecular clones that contain at least one copy of every DNA sequence in the genome. (442)

genotype The genetic constitution of an organism. (34; 43)

genotypic frequencies The frequencies or percentages of different genotypes found within a population. (709)

genotypic sex determination The process by which the sex chromosomes play a decisive role in the inheritance and determination of sex. (74)

germ-line mutations Mutations in the germ-line of sexually reproducing organisms may be transmitted by the gametes to the next generation, giving rise to an individual with the mutant state in both its somatic and germ-line cells. (558)

glucose effect *See* catabolite repression. (489)

Goldberg-Hogness box (TATA box, or TATA element) Found approximately at position − 30 from the transcription initiation site. The Goldberg-Hogness sequence is considered to be the likely eukaryotic promoter sequence. The consensus sequence for the Goldberg-Hogness box is TATAAAAA. (373)

group I-intron self-splicing *See* self-splicing. (405)

guanine (G) Purine base found in RNA and DNA. In double-stranded DNA, guanine pairs with the pyrimidine cytosine. (275)

haploid A cell or an individual with one copy of each nuclear chromosome. (12)

haploidization The formation of a haploid nucleus from a diploid heterokaryon. (177)

Hardy-Weinberg law (Hardy-Weinberg equilibrium, Hardy-Weinberg law of genetic equilibrium) An extension of Mendel's laws of inheritance; describes the expected relationship between gene frequencies in natural populations and the frequencies of individuals of various genotypes in the same populations. (709; 712; 714)

harlequin chromosomes 5-bromodeoxyuridine (5-BUdR), a thymidine analog, is incorporated into DNA during replication. When both DNA strands contain 5-BUdR, the chromatid stains less intensely than when only one DNA strand contains the analog. When cells are grown in the presence of 5-BUdR for two replication cycles, the two sister chromatids stained differentially are called harlequin chromosomes. (346)

helix-destabilizing proteins *See* single-strand DNA-binding proteins. (337)

hemizygous The condition of X-linked genes in males. Males that have an X chromosome with an allele for a particular gene but do not have another allele of that gene in their gene complement are hemizygous. (68)

hereditary trait A genetically controlled characteristic that is transmitted from one generation to another. (34)

heritability The proportion of phenotypic variation attributable to genetic factors in a population. (687)

hermaphroditic For animals (e.g. nematode), species in which each individual has both testes and ovaries. In plants, species that have both stamens and pistils on the same flower. (80)

heterochromatin Chromatin that remains condensed throughout the cell cycle and is genetically active. (307)

heterogametic sex The sex that has sex chromosomes of different types (e.g., XY); therefore each sex chromosome produces its own distinctive kinds of gametes. (64)

heterogeneous nuclear RNA (hnRNA) The RNA molecules of various sizes that exist in a large population in the nucleus. Some hnRNA molecules are precursors to mature mRNAs. (388)

heterokaryon A cell or collection of cells (as in a mycelium) possessing genetically different nuclei (regardless of their number) in a common cytoplasm. (177)

heterosis The phenomenon in which the heterozygous genotypes, because of one or more characteristics, are superior to the corresponding homozygous genotypes in terms of growth, survival, phenotypic expression, and fertility. (698)

heterozygosity The proportion of individuals heterozygous at a locus. (720)

heterozygote advantage *See* overdominance. (744)

heterozygous A term describing a diploid organism having different alleles of one or more genes. Therefore the organism produces gametes of different genotypes. (42)

Hfr (high-frequency recombination) cell A male cell in *E. coli*, with the *F* factor integrated into the bacterial chromosome. When the *F* factor promotes conjugation with a female (F^-) cell, bacterial genes are transferred to the female cell with high frequency. (202)

highly repetitive sequence A DNA sequence that is repeated between 10^5 and 10^7 times in the genome. (319)

histocompatibility The acceptance by one organism a graft of another organism's tissue. (79)

histone One of a class of basic proteins that are complexed with DNA in chromosomes and that play a major role in determining the structure of eukaryotic nuclear chromosomes. (307)

homeobox A 180-bp consensus sequence found in the protein-coding sequences of genes that regulate development. (548)

homeodomain The 60-amino-acid part of specific proteins that corresponds to the homeobox sequence of genes. All homeodomain-containing proteins appear to be located in the nucleus. (548)

homoeotic mutations Mutations that alter the identity of particular segments, transforming them into copies of other segments. (547)

homogametic sex The gender in the species, most often the female, that produces only the X sex chromosome. (64)

homolog Each individual member of a pair of homologous chromosomes. (12)

homologous chromosomes The members of a chromosome pair that are identical in the arrangement of genes they contain and in their visible structure. (12)

homozygous A term describing a diploid organism having the same alleles at one or more genes and therefore producing gametes of identical genotypes. (41; 43)

homozygous dominant A diploid organism that has the same dominant allele for a given gene locus on both members of a homologous pair of chromosomes. (43)

homozygous recessive A diploid organism that has the same recessive allele for a given gene locus on both members of homologous chromosomes. (43)

human genome project A project coordinated by James Watson to obtain the sequence of the complete 3 billion (3×10^9) nucleotide pairs of the human genome, and to map all of the estimated 50,000 to 100,000 human genes. (468)

H-Y antigen The product of the Y-linked H-Y locus (the histocompatibility locus). (79)

hybrid dysgenesis The appearance of a series of defects, including mutations, chromosomal aberrations, and sterility, when certain strains of *Drosophila melanogaster* are crossed. (627)

hypersensitive sites (hypersensitive regions) In the regions of DNA around transcriptionally active genes, sites that are highly sensitive to digestion by DNase I. (520)

identical (monozygotic) twins The result of the fertilization of a single egg and the subsequent separation of the two cells produced by the first division of the zygote. These two individuals are genetically identical because they come from the same fertilized egg. (119)

imaginal discs In the *Drosophila* blastoderm, undifferentiated cells that will develop into adult tissue and organs. (545)

immunoglobulins Specialized proteins (antibodies) secreted by B cells. These antibodies circulate in the blood and lymph, which are responsible for humoral immune responses. (539)

inbreeding Preferential mating between close relatives. (698; 745)

incomplete (partial) dominance The absence of dominance by either allele in a heterozygote. The heterozygote has a phenotype between that of individuals homozygous for either allele involved. An example of incomplete dominance is flower color in the snapdragon. (97)

induced mutation A mutation that results from deliberate treatment with mutagens. (558)

inducer For bacterial operons a chemical or environmental agent that brings about transcription of the operon. (478)

induction Synthesis of a gene product (or products) in response to the action of an inducer, i.e., a chemical or environmental agent. (478)

insertion sequence (IS) element The simplest transposable genetic element found in prokaryotes. This element is a mobile DNA segment that contains genes required for inserting the DNA segment into a chromosome and for mobilizing the element to different locations. (613)

interaction variance (V_I) Genetic variance that arises from epistatic interactions among genes. (689)

intergenic suppressor A mutation whose effect is to suppress the phenotypic consequences of another mutation in a gene distinct from the one in which the suppressor mutation is located. (561)

internal control region (ICR) Promoter sequence, recognized by RNA polymerase III, that is located within the gene sequence, e.g., in tRNA genes and 5S rRNA genes of eukaryotes. (375)

intervening sequence (ivs) *See* intron. (383; 385)

intragenic suppressors A mutation that suppresses the phenotypic consequences of another mutation within the same gene in which the suppressor mutation is located. (561)

intron A nucleotide sequence in eukaryotes that must be excised from a structural gene transcript in order to convert the transcript into a mature messenger RNA molecule containing only coding sequences that can be translated into the amino acid sequence of a polypeptide. (383; 385)

inversion A chromosome mutation that results when a segment of a chromosome is excised, then reintegrated in an 180° orientation from the original orientation. (592)

IS element *See* insertion sequence (IS) element. (613)

isogenic strain A group of individuals (e.g., produced by inbreeding or cloning) that have the same genotype. (79)

karyotype A complete set of all the metaphase chromatid pairs in a cell (literally, "nucleus type"). (76; 300)

kinetochore *See* centromere. (13)

Klinefelter syndrome A human clinical syndrome that results from disomy for the X chromosome in a male, producing a 47,XXY male. Many affected males are mentally deficient, have underdeveloped testes, and are taller than average. (76)

lagging strand In DNA replication, the DNA strand that is synthesized discontinuously in the 5′-to-3′ direction away from the replication fork. (335)

leader sequence One of three main parts of the mRNA molecule. The leader sequence is located at the 5′ end of the mRNA molecule and contains the coded information that tells the ribosome and special proteins where to begin polypeptide synthesis. (383)

leading strand In DNA replication, the DNA strand synthesized continuously in the 5′-to-3′ direction toward the replication fork. (335)

leptonema The stage during meiosis in prophase I at which the chromosomes have begun to coil and are visible. (21)

lethal allele An allele that results in the death of an organism. (110)

light repair *See* photo-reactivation. (573)

LINES (long interspersed repeated sequences) The dispersed families of repeated sequences in mammals that are several thousand base pairs in length and occur > 20,000 times in the genome. (320)

linkage A term describing genes located on the same chromosome. (128)

linkage map *See* genetic map. (137)

linked genes Genes that are located on the same chromosome. (128)

linker *See* restriction site linker. (446)

locus (*plural, loci*) The position of a gene on a genetic map; the specific place on a chromosome where a gene is located. (140)

lyonization A mechanism in mammals that allows them to compensate for X chromosomes in excess of the normal complement. The excess X chromosomes are cytologically condensed and inactivated; they do not play a role in much of the individual's development. The name derives from the discoverer of the phenomenon, Mary Lyon. (78)

lysogenic A term describing a bacterium that contains a temperate phage in the prophage state. The bacterium is said to be lysogenic for that phage. Upon induction, phage reproduction is initiated, progeny phages are produced, and the bacterial cell lyses. (208)

lysogenic pathway An alternative path to the lytic cycle for a temperate phage. The chromosome does not replicate; instead, it inserts itself physically into a specific region of the host cell's chromosome, much like *F* factor integration. (207)

lysogeny The insertion of a temperate phage chromosome into a bacterial chromosome. The phage chromosome then replicates when the bacterial chromosome replicates. In this state the phage genome is repressed and is said to be in the prophage state. (208)

lytic cycle A phage life cycle in which the phage takes over the bacterium and directs its growth and reproductive activities to express the phage's genes and to produce progeny phages. (207)

macromolecule A large molecule (such as DNA, RNA, or proteins) that has a molecular weight of at least a few thousand daltons. (275)

mapping functions Mathematical formulas used to correct the observed recombination values for the incidence of multiple crossovers. (146)

map unit (mu) A unit of measurement used for the distance between two pairs on a genetic map. A crossover frequency of 1 percent between two genes equals 1 map unit. *See* centi-Morgan. (140)

maternal effect The predetermination of gene-controlled traits by the maternal nuclear genotype, prior to the fertilization of the egg. (655)

maternal inheritance Expression of the mother's phenotype exclusively. (654)

mating types A genic system in which two sexes are morphologically indistinguishable but carry different alleles and will mate. (80)

Maxam-Gilbert sequencing A method of rapid sequencing of DNA molecules, developed by Allan Maxam and Walter Gilbert. The technique uses specific chemical reactions to break DNA at specific nucleotides. The DNA is first radiolabeled with ^{32}P at the 5′ or 3′ end of a chain. Secondly, the DNA is chemically modified and cleaved at various points along the backbone. Thirdly, the DNA fragments are analyzed with polyacrylamide gel electrophoresis. (458)

mean The average of a set of numbers, calculated by adding all the values represented and dividing by the number of values. (677)

megasporogenesis In flowering plants the formation of megaspores and the production of the embryo sac (the female gametophyte). (28)

meiosis Two successive nuclear divisions of a diploid nucleus to form haploid gametes or meiospores having one-half the genetic material of the original cell. (20)

meiosis I The first meiotic division, which reduces the number of chromosomes. This division consists of four stages: prophase I, metaphase I, anaphase I, and telophase I. (21)

meiosis II The second meiotic division, which separates the chromatids. (21)

Mendelian factor *See* gene. (43)

Mendelian population An interbreeding group of individuals sharing a common gene pool; the basic unit of study in population genetics. (708)

messenger RNA (mRNA) The RNA molecule that contains the coded information for the amino acid sequence of a protein. (363; 382)

metacentric chromosome A chromosome in which the centromere falls approximately half-way along the chromosome's length. (13)

metaphase A stage in mitosis or meiosis in which chromosomes become aligned along the equatorial plane of the spindle. (18)

metaphase II The second stage of meiosis, during which the centromeres line up on the equator of the second-division spindles (in each of two daughter cells formed from meiosis I). (25)

metaphase plate The plane where chromosomes become aligned during metaphase. (18)

microsporogenesis In flowering plants the formation of microspores in the anthers and the production of the male gametophyte (pollen), normally from diploid microsporocytes. (28)

migration Movement of organisms from one location to another. (735)

missense mutation A gene mutation in which a base-pair change in the DNA causes a change in an mRNA codon. As a result, an amino acid is inserted into the polypeptide that is different than the one specified by the wild-type codon. (559)

mitochondria Organelles found in the cytoplasm of all aerobic animal and plant cells; the principal sources of energy in the cell. (643)

mitosis The process of nuclear division in haploid or diploid cells. The result is daughter nuclei that contain chromosome complements that are genetically identical to one another and to the parent nucleus from which they arose. (14)

mitotic crossing-over (mitotic recombination) A genetic recombination that occurs following the rare pairing of homologs during mitosis of a diploid cell. (173)

moderately repetitive sequence A DNA sequence that is reiterated from a few to as many as 10^3 to 10^5 times in the genome. (319)

molecular cloning *See* cloning. (432)

molecular drive *See* concerted evolution. (751)

molecular genetics A subdivision of the science of genetics, involving how genetic information is encoded within the DNA and how the cell's biochemical processes translate the genetic information into the phenotype. (708)

monoecious A term referring to plants in which male and female gametes are produced in the same individual. (80)

monohybrid cross A cross between two individuals heterozygous for the same pair of alleles (e.g., $Aa \times Aa$). By extension, the term also refers to crosses involving the pure-breeding parents that differ with respect to the alleles of one locus (e.g., $AA \times aa$). (39)

monoploidy An aberrant, aneuploid state in a normally diploid cell or organism in which one chromosome is missing, leaving one chromosome with no homolog. (597; 601)

monosomy An aberrant, aneuploid state in a normally diploid cell or organism in which one chromosome is missing, leaving one chromosome with no homolog. (597)

monozygotic twins *See* identical twins. (119)

mRNA splicing A process whereby an intervening sequence between two coding sequences in an RNA molecule is excised and the coding sequences ligated (spliced) together. (390)

multifactorial trait A trait influenced by multiple genes and by environmental factors. (673)

multigene family A set of related genes that have evolved from an ancestoral gene through the process of gene duplication. (749)

multiple alleles Many alternative forms of a single gene. (93)

multiple cloning site *See* polylinker. (437)

multiple crossovers In meiosis, more than one crossover occurring in a particular region of a chromosome. (143)

mutagen Any physical or chemical agent that significantly increases the frequency of mutational events above the rate of spontaneous mutation. (558)

mutant allele Any alternative to the wild-type allele of a gene. Mutant alleles may be dominant or recessive to wild-type alleles. (64; 93)

mutation Any detectable and heritable change in the genetic material not caused by genetic recombination. (558)

mutation frequency The number of occurrences of a particular kind of mutation in a population of cells or individuals. (563)

mutation rate The probability of a particular kind of mutation as a function of time. (563)

narrow-sense heritability The proportion of the phenotypic variance that results from additive genetic variance. (688; 689)

natural selection Differential reproduction of genotypes. (737; 694)

negative assortative mating A mating that occurs between dissimilar individuals more often than between randomly chosen individuals. (745)

neutral mutation A base-pair change in the gene that changes a mRNA codon such that there is no change in the function of the protein translated from that message. (559)

neutral-mutation hypothesis A hypothesis that replaced the classical model by acknowledging the presence of extensive genetic variation in proteins, but proposing that this variation is neutral with regard to natural selection. (721)

nick translation A process for labeling a double-stranded DNA molecule radioactively, using DNase I and DNA polymerase. The result is a labeled probe. (450)

nitrogenous base A nitrogen-containing base that, along with a pentose sugar and a phosphate, is one of the three parts of a nucleotide, the building block of RNA and DNA. (275)

nonautonomous development (c.f. **autonomous development**) When the development of a cell's phenotype is affected by the environment in which it develops. (236)

noncontributing alleles Alleles that do not have any effect on the phenotype of the quantitative trait. (685)

nondisjunction (primary disjunction) A failure of homologous chromosomes or sister chromatids to separate at anaphase. (71)

nonhistone A type of acidic or neutral protein found in chromatin. (307)

nonhomologous chromosomes Chromosomes containing dissimilar genes that do not pair during meiosis. (12)

non-Mendelian inheritance (cytoplasmic inheritance) The inheritance of characters determined by genes located not on nuclear chromosomes but on mitochondrial or chloroplast chromosomes. Such genes show inheritance patterns distinctly different from those of nuclear genes. (654)

nonparental-ditype (NPD) One of three types of tetrads possible when two genes segregate in a cross. The NPD tetrad contains four nuclei, all of which are recombinant (nonparental) genotypes, i.e., two of each possible type. (166)

nonsense codon See chain-terminating codon. (413)

nonsense mutation A gene mutation in which a base-pair change in the DNA causes a change in an mRNA codon from an amino acid-coding codon to a chain-terminating (nonsense) codon. As a result, polypeptide chain synthesis is terminated prematurely and is therefore either nonfunctional or, at best, partially functional. (559)

nontranscribed spacer (NTS) sequence Sequences that are not transcribed, found between transcription units in rDNA. Important sequences that control transcription of rDNA are within the NTS. (377)

normal distribution A probability distribution in statistics, graphically displayed as a bell-shaped curve. (675)

northern blot analysis A similar technique to Southern blotting except that RNA rather than DNA is separated and transferred to a filter for hybridization with a probe. (456)

nuclease An enzyme that catalyzes the degradation of a nucleic acid by breaking phosphodiester bonds. Nucleases specific for DNA are termed deoxyribonucleases (DNases), and nucleases specific for RNA are termed ribonucleases (RNases). (269)

nucleofilament A fiber seen in chromatin. Each is approximately 10 nm in diameter and consists of DNA wrapped around nucleosome cores. (308)

nucleoplasm The semisolid matrix of cell material in the nucleus. (61)

nucleoside phosphate See nucleotide. (276)

nucleosome The basic structural unit of eukaryotic nuclear chromosomes. The unit conists of about 140 base pairs of DNA wound almost two times around an octamer of histones. Connecting the unit to adjacent nucleosomes are about 60 base pairs of DNA complexed with another histone. (308)

nucleosome core particle A 146-bp DNA fragment associated with two each of the core histones H2A, H2B, H3, and H4. (309)

nucleotide A monomeric molecule of RNA or DNA, consisting of three distinct parts: a pentose (ribose in RNA, deoxyribose in DNA), a nitrogenous base, and a phosphate group. (275)

nucleus A discrete structure within the cell, bound by a nuclear membrane. The nucleus contains most of the cell's genetic material. (45)

nullisomy The aberrant, aneuploid state in a normally diploid cell or organism in which there is a loss of one pair of homologous chromosomes. (597)

nutritional mutation See auxotrophic mutation. (573; 580)

Okazaki fragments In discontinuous DNA replication, the relatively short, single-stranded DNA fragments that are synthesized during DNA replication and that subsequently are covalently joined to make a continuous strand. (335)

oligomers ("oligo" = few) Short DNA molecules. (281)

oncogene A gene that transforms a normal cell to a tumorous (cancerous) state. (629)

oncogenesis Tumor (cancer) initiation in an organism. (628)

one gene-one enzyme hypothesis The hypothesis, based on Beadle and Tatum's studies in biochemical genetics, that each gene controls the synthesis of one enzyme. When it was found that enzymes could consist of more than one polypeptide, the one gene-one enzyme hypothesis was updated to the one gene-one polypeptide hypothesis. (241)

oogenesis The development of the female animal germ cell (egg cell) within the female gonad. (26)

open reading frame In a segment of DNA, a potential protein-coding sequence identified by an initiator codon in frame with a chain-terminating codon—that is, separated by a number of base pairs divisible by three. (462)

operator The controlling site, adjacent to a promoter, that controls the transcription of genes that are contiguous to the promoter. (481)

operon A cluster of genes whose expressions are regulated together by operator-regulator protein interactions, plus the operator region itself and the promoter. (482)

ordered tetrads A structure resulting from meiosis in which the order of the four meiotic products is an exact reflection of the orientation of the four chromatids at the metaphase plate in meiosis I. (135)

origin A specific site on the chromosome where the double helix denatures into single strands and continues to unwind as the replication fork(s) migrates. (200; 333)

origin of replication In prokaryotes, a specific DNA sequence required for the initiation of DNA replication. (333)

outbreeding Preferential mating between nonrelated individuals. (745)

overdominance (heterozygote advantage) Condition in which the heterozygote has greater fitness than either of the homozygotes. (744)

ovum A mature egg cell. In the second meiotic division the secondary oocyte produces two haploid cells; the larger cell rapidly matures into the ovum. (27)

P generation The parental generation, i.e., the immediate parents of an F_1. (39)

pachynema The stage in meiosis (mid-prophase I) during which the homologous pairs of chromosomes exchange chromosome regions. (21)

paracentric inversion An inversion in which the inverted segment occurs on one chromosome arm and does not include the centromere. (592)

parasexual system A system that achieves genetic recombination by means other than the regular alternation of meiosis and fertilization. (177)

parental-ditype (PD) One of three types of tetrads possible when two genes are segregating in a cross. The PD tetrad contains four nuclei, all of which are parental genotypes, with two of one parent and two of the other. (166)

parental genotypes (parental classes, parentals) Among progeny of crosses, individuals that have combinations of genetic markers like one or other of the parents in the parental generation. (129)

partial dominance *See* incomplete dominance. (97)

partial linkage Linkage that occurs when homologous chromosomes exchange corresponding parts during meiosis through crossing-over; a linkage between two alleles that is not complete. (128)

particulate factors Mendel's term for the structures that carried hereditary information and were transmitted from parents to progeny through the gametes. We now know these factors by the name *genes*. (40)

pedigree analysis A family tree investigation that involves the careful compilation of a family's phenotypic records over several generations. (52)

penetrance The frequency with which a dominant or homozygous recessive gene manifests itself in an individual's phenotype. (54; 112)

pentose sugar A 5-carbon sugar that, along with a nitrogenous base and a phosphate group, is one of the three parts of a nucleotide, the building block of RNA and DNA. (275)

peptide bond A covalent bond in a polypeptide chain, joining the α-carboxyl group of one amino acid to the α-amino group of the adjacent amino acid. (251)

peptidyl transferase The enzyme that catalyzes formation of the peptide bond in protein synthesis. (421)

pericentric inversion An inversion in which the inverted segment includes the parts of both chromosome arms and, therefore, the centromere. (592)

phage lysate The progeny phages released by lysis of phage-infected bacteria. (207)

phage vector A phage that carries pieces of bacterial DNA from one bacterial strain to another in the process of transduction. (208)

phenocopy An abnormal individual resulting from special environmental conditions. It mimics a similar phenotype caused by gene mutation. (118)

phenotype The physical manifestation of a genetic trait, resulting from a specific genotype and its interaction with the environment. (34; 43)

phenotypic correlation An association between two traits. (696)

phenotypic sex determination The process by which the environment plays a major role in determining the sex of an organism. (80)

phenotypic variance (V_P) A measure of a trait's variability. (688)

phosphate group A component, along with a pentose sugar and a nitrogenous base, of a nucleotide, the building block of RNA and DNA. Because phosphate groups are acidic, DNA and RNA are called nucleic acids. (275)

phosphodiester bond A covalent bond between a sugar and a phosphate in RNA and DNA. Phosphodiester bonds form the repeating sugar-phosphate array of the backbone of DNA and RNA. (277; 330)

photoreactivation (light repair) One way by which thymine dimers can be repaired. The dimers are reverted directly to the original form by exposure to visible light in the wavelength range 320–370 nm. (575)

pistil The female reproductive organ in a flowering plant; the structure typically consists of the stigma, the style, and the ovary. (27)

plaque A round, clear area in a lawn of bacteria on solid medium; the result of the lysis of cells by repeated cycles of phage lytic growth. (214)

plasma membrane Lipid bilayer which surrounds the cytoplasm of both animal and plant cells. (7)

plasmid An extrachromosomal genetic element consisting of double-stranded DNA that replicates autonomously from the host chromosome. (200; 436; 617)

point mutants Organisms whose phenotypes result from an alteration (mutation) of a single nucleotide pair. (221)

point mutation A mutation caused by a substitution of one base pair for another. (558)

polarity A term referring to a bacterial operon that codes for a polygenic mRNA. In polarity, certain nonsense mutations not only result in the loss of activity of the enzyme encoded by the gene in which they are located; there also is a significant reduction or absence of the synthesis of enzymes coded by structural genes on the operator-distal side of the mutation. The mutations are called polar mutations. (480)

poly(A) addition site The 3′ end of mRNA to which 50 to 250 adenine nucleotides are added as part of mRNA posttranscriptional modification. (385)

poly(A) polymerase The enzyme which catalyzes the production of the 3′ poly(A) tail. (385)

poly(A) tail A sequence of 50 to 250 adenine nucleotides added as a posttranscriptional modification at the 3′ ends of most eukaryotic mRNAs. (385)

polygene (multiple-gene) hypothesis for quantitative inheritance The hypothesis that quantitative traits are controlled by many genes. (684)

polygenic mRNA (polycistronic mRNA) In prokaryotic organisms, a single mRNA transcript of two or more adjacent structural genes specifying the amino acid sequences of the corresponding polypeptides. (480)

polygenic traits Traits encoded by many loci. (673)

polylinker (multiple cloning site) A region of clustered unique restriction sites in a cloning vector. (437)

polymerase chain reaction (PCR) A method used to selectively and repeatedly replicate defined DNA sequences from a DNA mixture. (462)

polynucleotide A linear sequence of nucleotides in DNA or RNA. (278)

polypeptide A polymeric, convalently bonded, linear arrangement of amino acids joined by peptide bonds. (251)

polyploidy A cell or organism that has more than its normal number of sets of chromosomes. (597; 601)

polyribosome (polysome) The complex between an mRNA molecule and all the ribosomes that are translating it simultaneously. (421)

polytene chromosome A special type of chromosome representing a bundle of numerous chromatids that are the result of repeated cycles of replication of single chromatids without nuclear division. This type of chromosome is characteristic of various tissues of Diptera. (536; 589)

population A group of interbreeding individuals that share a set of genes. (674)

population genetics A branch of genetics that describes in mathematical terms the consequences of Mendelian inheritance at the population level. (708)

positive assortative mating A mating that occurs more frequently between individuals who are phenotypically similar than between randomly chosen individuals. (745)

posttranslational transport Transport in which protein synthesis is completed before import into the organelle takes place. (425)

precursor mRNA (primary transcripts; pre-mRNA) The initial transcript of a gene that is modified and/or processed to produce a mature, functional mRNA molecule. In eukaryotes, for example, the transcript is modified at both the 5′ and the 3′ ends, and in a number of cases RNA sequences that do not code for amino acids are present and must be excised. (383)

precursor RNA molecule (primary transcripts; pre-RNA) The initial transcript whose processing may involve addition and/or removal of bases, chemical modification of some bases, or cleavage of sequences from the precursors. (371)

precursor rRNA (pre-rRNA) A primary transcript of adjacent rRNA genes (16S, 23S, and 5S rRNA genes in prokaryotes; 18S, 5.8S, and 28S rRNA genes in eukaryotes) plus flanking and spacer DNA. All must be processed to release mature rRNA molecules. (377; 400)

precursor tRNA (pre-tRNA) A primary transcript of a tRNA gene. The bases of the transcript must be extensively modified; it must be processed to remove extra RNA sequences in order to produce the mature tRNA molecule. In some cases the primary transcript may contain the sequences of two or more tRNA molecules. (394)

Pribnow box A part of the promoter sequence in prokaryotic genomes, located about 10 base pairs upstream from the transcription starting point. The consensus sequence for the Pribnow box is TATAAT. The Pribnow box is often referred to as the TATA box. (367)

primary nondisjunction (nondisjunction) A rare event in which sister chromatids (in mitosis) or chromosomes contained in pairing configurations (in meiosis) fail to be distributed to opposite poles. (71)

primary transcripts See precursor RNA molecules (pre-RNAs). (382)

primer In DNA replication a pre-existing polynucleotide chain to which new nucleotides can be added. (334)

primosome A complex of *E. coli* primase, helicase, and perhaps other polypeptides that together become functional in catalyzing the initiation of DNA synthesis. (334)

prophase I The first stage of meiosis, during which the chromosomes become shorter and thicker, crossing-over occurs, the spindle apparatus forms, and the nuclear membrane and nucleolus/nucleoi disappear. There are several stages of prophase I, including leptonema, zygonema, pachynema, diplonema, and diakinesis. (21)

prophase II The second stage of meiosis, during which there is chromosome contraction. (25)

proportion of polymorphic loci A ratio calculated by determining the number of polymorphic loci and dividing by the total number of loci examined. (720)

propositus (proband) In human genetics the affected person with whom the study of a character in a family begins. (52)

protein One of a group of high-molecular weight, nitrogen-containing organic compounds of complex shape and composition. (251)

protein degradation control Regulation of the protein degradation rate. (531)

proto-oncogenes A gene which, in normal cells, functions to control the normal proliferation of cells, and which, when mutated or changed in any other way, becomes an oncogene. (633)

prototroph A strain that is the wild type for all nutritional requirement genes and thus requires no supplements in its growth medium. (199)

Punnett square A matrix that describes all the possible gametic fusions that will give rise to the zygotes of the next generation. (42)

pure-breeding See true-breeding. (38)

purine A type of nitrogenous base. In DNA and RNA the purines are adenine and guanine. (275)

pyrimidine A type of nitrogenous base. Cytosine is a pyrimidine in DNA and RNA; thymine is a pyrimidine in DNA; and uracil is a pyrimidine in RNA. (275)

Q banding A staining technique in which metaphase chromosomes are stained with quinacrine mustard to produce temporary fluorescent Q bands on the chromosomes. (303)

quantitative genetics Study of the inheritance of quantitative traits. (672)

quantitative (continuous) traits Traits that show a continuous variation in phenotype over a range. (672)

random mating Matings between genotypes occurring in proportion to the frequencies of the genotypes in the population. (713)

rDNA repeat units The tandem arrays of rRNA genes (18S-5.8S-28S), repeated many times along the chromosome. (402)

recessive An allele or phenotype that is expressed only in the homozygous state. (41)

recessive lethal allele An allele that causes lethality when it is homozygous. (111)

reciprocal cross Crosses between two strains exhibiting alternative traits, with sexes reversed. In the garden pea, for example, reciprocal crosses for smooth and wrinkled seeds are smooth female × wrinkled male and wrinkled female × smooth male. (39)

recombinant chromosome A chromosome that emerges from meiosis with a combination of genes different from a parental combination of genes. (21)

recombinant DNA molecule A unique DNA sequence that has been constructed or engineered in the test tube from two or more distinct DNA sequences. (432)

recombinant DNA techology A collection of experimental procedures that allow molecular biologists to splice a DNA fragment from one organism into DNA from another organism, then clone the new recombinant DNA molecule. This technology includes development and application of particular molecular techniques, such as biotechnology or genetic engineering. The technology is important, for example, in the production of antibiotics, hormones, and other medical agents used in the diagnosis and treatment of certain genetic diseases. (432)

recombinants Individuals or cells that have nonparental combinations of genes as a result of the processes of genetic recombination. (129)

regression A statistical analysis that assesses the association between two variables. (681)

regression line A mathematically computed line that represents the best fit of a line to the points. (681)

regulated gene A gene whose activity is controlled in order to respond to the needs of a cell or organism. (477)

regulatory factors Proteins active in the activation or repression of gene transcription. (372)

release factors See termination factors. (422)

replica plating The procedure for transferring the pattern of colonies from a master plate to a new plate. In this procedure a velveteen pad on a cylinder is pressed lightly onto the surface of the master plate, thereby picking up a few cells from each colony to inoculate onto the new plate. (580)

replication bubble Opposing replication forks found with the local denaturing of DNA during replication. (333)

replication fork A Y-shaped structure formed when a double-stranded DNA molecule unwinds to expose the two single-stranded template strands for DNA replication. (334)

replication machine (replisome) The complex formed by close association of the key proteins used during DNA replication. (339)

replicon (replication unit) The stretch of DNA in eukaryotes from the origin of replication to the two termini of replication on each side of the origin. (347)

repressor *See* repressor gene. (481)

repressor gene A regulatory gene whose product is a protein that controls the transcriptional activity of a particular operon. (482)

repressor molecule The protein product of a repressor gene. (482)

repulsion An arrangement in which one homologous chromosome carries the wild-type allele of one gene and the other homologous chromosome carries a mutant allele of the wild type. (140)

restriction endonucleases (restriction enzymes) Enzymes with the ability to cleave double-stranded DNA molecules at specific nucleotide pair sequences. Important for analyzing DNA and for constructing recombinant DNA molecules. (432)

restriction fragment length polymorphisms (RFLPs) The different restriction maps that result from different distribution patterns of restriction sites. RFLPs are detected by the presence of restriction fragments of different lengths on gels. (465)

restriction map A genetic map of DNA showing the relative positions of restriction enzyme cleavage sites. (452)

restriction site linker (linker) A relatively short, double-stranded oligodeoxyribonucleotide about 8 to 12 nucleotide pairs long, synthesized by chemical means and containing within its sequence the cleavage site for a specific restriction enzyme. (446)

retroviruses Single-stranded RNA viruses that replicate via double-stranded DNA intermediates. The DNA integrates into the host's chromosome, where it can be transcribed. (626)

reverse genetics An experimental approach in which a research mutates a gene into a cell, then investigates the resulting phenotypic changes in order to understand the normal gene's function. (573)

reverse mutation (reversion) A mutational change from a mutant allele back to a wild-type allele. (560)

reverse transcriptase An enzyme (an RNA-dependent DNA polymerase) that makes a complementary DNA copy of an mRNA strand. (445)

ribonuclease (RNase) An enzyme that catalyzes the degradation of RNA to nucleotides. (270; 363)

ribonucleic acid (RNA) Usually a single-stranded polymeric molecule consisting of ribonucleotide building blocks. RNA is chemically very similar to DNA. Three major types of RNA in prokaryotic and eukaryotic cells—ribosomal RNA (rRNA), transfer RNA (tRNA), and messenger RNA (mRNA)—perform essential roles in protein synthesis (translation). Eukaryotic cells contain a fourth type of RNA, small nuclear RNA (snRNA). The several species of snRNA are involved in RNA processing. In some viruses, RNA is the genetic material. (3)

ribonucleotide The basic building block of RNA, consisting of a sugar (ribose), a base, and a phosphate. (275)

ribose The pentose (five-carbon) sugar component of the nucleotide building block of RNA. (275)

ribosomal DNA (rDNA) The regions of DNA that contain the genes for rRNAs in prokaryotes and eukaryotes. (400)

ribosomal proteins Proteins that, along with rRNA molecules, comprise the ribosomes of prokaryotes and eukaryotes. (399)

ribosomal RNA (rRNA) RNA molecules of discrete sizes that, along with ribosomal proteins, comprise ribosomes of prokaryotes and eukaryotes. (363; 399; 400)

ribosome A complex cellular particle, composed of ribosomal protein and rRNA molecules, that is the site of amino acid polymerization during protein synthesis. (382)

ribosome-binding site On an mRNA molecule, the nucleotide sequence where the ribosome becomes oriented in the correct reading frame to initiate protein synthesis. (416)

ribozyme (RNA enzyme) An RNA molecule that can function as an enzyme to cleave RNA molecules at specific sequences. (406)

R looping (R loops) A technique developed by M. Thomas, R. White, and R. Davis. When molecules of double-stranded DNA are incubated at temperatures below their denaturing temperature, short stretches of the DNA double helix open up so that single-stranded RNA molecules can begin to form DNA/RNA hybrids where the two are complementary. The DNA/RNA hybrid forms an R loop by displacing a single-stranded section of DNA. (388)

RNA *See* ribonucleic acid. (3)

RNA enzyme *See* ribozyme. (407)

RNA ligase An enzyme that splices together RNA pieces once an intervening sequence is removed from the pre-tRNA. (397)

RNA polymerase An enzyme that catalyzes the synthesis of RNA molecules from a DNA template in a process called transcription. (334; 365)

RNA polymerase I Found in the nucleolus of eukaryotes, this enzyme catalyzes the transcription of the 18S, 5.8S, and 28S rRNA genes. (371)

RNA polymerase II An enzyme in eukaryotes, found only in the nucleoplasm of the nucleus. It catalyzes the transcription of mRNA-coding genes. (372)

RNA polymerase III An enzyme found only in the nucleoplasm of eukaryotes. It catalyzes the transcription of the tRNA and 5S rRNA genes. (372)

RNA primer *See* primer. (334)

RNA processing control The second level of control of gene expression in eukaryotes. This level involves regulating the production of mature RNA molecules from precursor-RNA molecules. (528)

RNA synthesis *See* transcription. (410)

Robertsonian translocation A type of nonreciprocal translocation, in which the long arms of two nonhomologous acrocentric chromosomes become attached to a single centromere. (598)

sample The subset used to give information about a population. A sample must be of reasonable size, and it must be a random subset of the larger group in order to provide accurate information about the population. (674)

sampling error The phenomenon in which chance deviations from expected proportions arise in small samples. (731)

satellite DNA In an equilibrium density analysis the DNA that forms a band that is distinct from the band constituting the majority of the genomic DNA. The difference is the result of a different buoyant density. (321)

second law *See* principle of independent assortment. (47)

secondary nondisjunction A nondisjunction of the Xs in the progeny of females produced by primary nondisjunction. (72)

secondary oocyte A large cell produced by the primary oocyte. In the ovaries of female animals the diploid primary oocyte goes through meiosis I and unequal cytokinesis to produce two cells; the larger cell is called the secondary oocyte. (27)

second-site mutation *See* suppressor mutation. (560)

selection coefficient A measure of the relative intensity of selection against a genotype. (740)

selection differential In natural and artificial selection, the difference between the mean phenotype of the selected parents and the mean phenotype of the unselected population. (695)

selection response The amount that a phenotype changes in one generation when selection is applied to a group of individuals. (695)

self-assembly The ability of certain complex biological structures (e.g., bacterial ribosomes) to assemble into functional structures from component molecules. (401)

self-fertilization (selfing) The union of male and female gametes from the same individual. (38)

self-splicing The excision of introns from some precursor RNA molecules. This process is a protein-independent reaction in some organisms. (405)

semiconservative replication model A DNA replication scheme in which each daughter molecule includes one of the parental strands. (326)

semidiscontinuous Concerning DNA replication, when one new strand is synthesized continuously and the other discontinuously. *See* discontinuous DNA replication. (337)

sense codon In an mRNA molecule a sense codon (as opposed to a nonsense codon) specifies an amino acid in the corresponding polypeptide. (413)

sex chromosome In eukaryotic organisms a chromosome that is represented differently in the two sexes. In many organisms one sex possesses a pair of visibly different chromosomes. One is an X chromosome, and the other is a Y chromosome. Commonly, the XX sex is female and the XY sex is male. (12; 62)

sex-influenced traits The traits that appear in both sexes but either there is a difference in the frequency of occurrence in the two sexes, or there is a different relationship between genotype and phenotype. (115)

sex-limited trait A genetically-controlled character that is phenotypically exhibited in only one of the two sexes. (114)

sex-linked *See* X-linked. (69)

sexual reproduction Reproduction involving the fusion of haploid gametes produced by meiosis. (13)

shuttle vector A cloning vector that can replicate in two or more host organisms. Shuttle vectors are used for experiments in which recombinant DNA is introduced into organisms other than *E. coli*. (441)

signal hypothesis The hypothesis that a protein is secreted from a cell through the binding of a hydrophobic amino terminal extension to the membrane; subsequently the extension is removed and degraded in the cisternal space of the endoplasmic reticulum. (424)

signal peptidase In the cisternal space of the ER, an enzyme that catalyzes removal of the signal sequence from a polypeptide. (424)

signal recognition particle (SRP) In eukaryotes, a complex of a small RNA molecule with six proteins. The complex can temporarily halt protein synthesis by recognizing the signal sequence of a nascent polypeptide destined to be translocated through the ER, binding to the signal sequence, thereby blocking further translation of the mRNA. (424)

signal sequence The hydrophobic, amino terminal extension (terminus) found on proteins that are secreted from a cell. The terminus is removed and degraded in the cisternal space of the endoplasmic reticulum. (424)

silencer *See* silencer element. (604)

silencer element In eukaryotes, a transcriptional regulatory element that decreases RNA transcription rather than stimulating it, like other enhancer elements. (373)

silent mutation A mutational change resulting in a protein that still has a wild-type function because of an unchanged amino-acid sequence. (560)

simple telomeric sequences Simple, tandemly repeated DNA sequences at, or very close to, the extreme ends of the chromosomal DNA molecules. (316)

SINES (short interspersed repeated sequences) One class of interspersed and highly repeated sequences. This class consists of dispersed families with unit lengths of fewer than 500 base pairs and repeated for as many as hundreds of thousands of copies in the genome. (320)

single-strand DNA binding (SSB) proteins (helix-destabilizing proteins) Proteins which help the DNA unwinding process by stabilizing single-stranded DNA. (337)

sister chromatid A chromatid derived from replication of one chromosome during interphase of the cell cycle. (14)

slope (regression coefficient) The change in one variable (y) associated with a unit increase in another variable (x). (683)

small nuclear ribonucleoprotein particles (snRNPs) Complexes formed by small nuclear RNAs and proteins in which the processing of pre-mRNA molecules occurs. (392)

small nuclear RNA (snRNA) Found only in eukaryotes, one of four major classes of RNA molecules produced by transcription. snRNAs are used in the processing of pre-mRNA molecules. (363)

somatic cell hybridization The fusion of two genetically different somatic cells of the same or different species to generate a somatic hybrid for genetic analysis. (181)

somatic mutation A mutation in a cell, producing a mutant spot or area; the mutant characteristic is not passed on to the succeeding generation. (558)

Southern blot technique A technique (invented by E. M. Southern) used in analyzing genes and gene transcripts; DNA fragments are transferred from a gel to a nitrocellulose filter. (452)

spacer sequences Transcribed sequences found flanking, and between, coding RNA sequences. Spacer sequences are removed during processing of pre-rRNA and pre-tRNA to produce mature molecules. (377)

specialized transducing phage A temperate bacteriophage that can transduce only a certain section of the bacterial chromosome. (213)

specialized transduction A type of transduction in which only specific genes are transferred. (208)

spermatogenisis Development of the male animal germ cell within the male gonad. (26).

sperm cells (spermatozoa) The male gametes; the spermatozoa produced by the testes in male animals. (26)

spliceosomes The splicing complexes formed by the association of several snRNPs bound to the pre-mRNA. (392)

spontaneous mutations The mutations that occur without the use of chemical or physical mutagenic agents. (558; 562)

sporophyte The diploid, asexual generation in the life cycle of plants that produces haploid spores by meiosis. (27)

stamen In a flowering plant the male reproductive organ, which usually consists of a stalk, called a filament, bearing a pollen-producing anther. (27)

standard deviation The square root of the variance. It measures the extent to which each measurement in the data set differs from the mean value and is used as a measure of the extent of variability in a population. (678)

standard error of gene frequency A measure of the amount of variation among the gene frequencies of populations. It is the square root of the variance of gene frequency. (732)

steroid response element (REs) The DNA sequence to which steroid hormones will bind to activate a gene. (525)

stop codon *See* chain-terminating codon. (413)

stringent response (stringent control) A rapid shutdown of essential cellular activities. (498)

structural gene A gene that codes for an mRNA molecule, hence, for a polypeptide chain. (363)

submetacentric chromosome A chromosome that has the centromere nearer one end than the other. Such chromosomes appear J-shaped at anaphase. (13)

sum rule The rule that the probability of either one of two mutually exclusive events occurring is the sum of their individual probabilities. (44)

suppressor gene A gene that suppresses mutations in other genes. (561)

suppressor mutation A secondary mutation at a second site, totally or partially restoring a function lost because of a primary mutation at another site. (560)

Svedberg units (S values) The conversions of sedimentation rates in sucrose density centrifugation. Svedberg units are used as a rough indication of relative sizes of the components being analyzed. (387)

synapsis The specific pairing of homologous chromosomes during the zygonema stage of meiosis. (21)

synaptinemal complex A complex structure spanning the region between meiotically paired (synapsed) chromosomes. This complex is involved with crossing-over, rather than with chromosome pairing. (21)

synkaryon A fusion nucleus produced following the fusion of cells with genetically different nuclei. (182)

syntenic The genes that are localized to a particular chromosome by using an experimental approach (literally "together thread," i.e., another term for *linked*). (184)

TATA element *See* Goldberg-Hogness box. (373)

TATA factor A key transcription factor (TFIID), which binds to the TATA element. (374)

tautomeric shift The change in the chemical form of a DNA (or RNA) base. (564)

tautomers Alternate chemical forms of DNA (or RNA) bases. (563)

telocentric chromosome A chromosome that has the centromere more or less at one end. (13)

telomere-associated sequences Repeated, complex DNA sequences extending from the molecular gene of chromosomal DNA. Suspected to mediate many of the telomere-specific interactions. (316)

telophase A stage during which the migration of daughter chromosomes to the two poles is completed. (18)

telophase II The last stage of meiosis II, during which a nuclear membrane forms around each set of chromosomes and cytokinesis takes place. (25)

template strand The unwound single strand of DNA upon which new strands are made (following complementary base pairing rules). (333)

termination factors (release factors, RF) In polypeptide synthesis (translation), specific proteins that read the chain termination codons, then initiate a series of specific events to terminate polypeptide synthesis. (422)

terminator *See* transcription terminator sequence. (367)

testcross A cross of an individual of unknown genotype (usually expressing the dominant phenotype) with a homozygous recessive individual in order to determine the genotype of the individual. (45)

tetrad analysis Genetic analysis of all the products of a single meiotic event. Tetrad analysis is possible in those organisms in which the four products of a single nucleus that has undergone meiosis are grouped together in a single structure. (163)

tetrasomy The aberrant, aneuploid state in a normally diploid cell or organism. An extra chromosome pair results in the presence of four copies of one chromosome type and two copies of every other chromosome type. (597)

tetratype (T) One of the three possible types of tetrads when two genes are segregating in a cross. The T tetrad contains two parental and two recombinant nuclei, one of each parental type and one of each recombinant type. (166)

three-point testcross A test involving three genes within a relatively short section of the chromosome; used to map gene order in the chromosome and to determine the distance between them. (146)

thymine (T) A pyrimidine base found in DNA but not in RNA. In double-stranded DNA thymine pairs with adenine. (275)

topoisomerases A class of enzymes that catalyze DNA supercoiling. (291)

totipotency The capacity of a nucleus to direct events through all the stages in development and therefore produce a normal adult. (534)

trailer sequence A sequence of the mRNA molecule beginning at the end of the amino acid-coding sequence and ending at the 3′ end of the mRNA. The trailer sequence is not translated and varies in length from molecule to molecule. (383)

transconjugants In bacteria, the recipients inheriting donor DNA in the process of conjugation. (196)

transcription The transfer of information from a double-stranded DNA molecule to a single-stranded RNA molecule. It is also called RNA synthesis. (362; 410)

transcriptional control The first level of control of gene expression in eukaryotes. This level involves regulating whether or not a gene is to be transcribed and the rate at which transcripts are produced. (516)

transcription factors (TFs) Specific proteins required for the initiation of transcription by each of the three eukaryotic RNA polymerases. Each polymerase uses its own set of TFs. (374)

transcription terminator sequence (terminator) A transcription regulatory sequence, located at the distal end of a gene, that signals the termination of transcription. (367)

transdetermination During development, a process whereby an imaginal disc does not totally dedifferentiate but switches to another determined path. (547)

transdominant The phenomenon of a gene or DNA sequence controlling genes that are on a different piece (strand) of DNA. (482)

transducing phage The phage that is the vehicle by which genetic material is shuttled between bacteria. (210)

transducing retroviruses Retroviruses that have picked up an oncogene from the cellular genome. (631)

transductants In bacteria, the recipients inheriting donor DNA in the process of transduction. (197)

transduction A process by which bacteriophages mediate the transfer of bacterial genetic information from one bacterium (the donor) to another (the recipient); a process whereby pieces of bacterial DNA are carried between bacterial strains by a phage. (196)

transfer RNA (tRNA) One of the four classes of RNA molecules produced by transcription and involved in protein synthesis; molecules that bring amino acids to the ribosome, where they are matched to the transcribed message on the mRNA. (363; 382)

transformant The genetic recombinant generated by the transformation process. (196)

transformation a) A process in which genetic information is transferred by means of extracellular pieces of DNA in bacteria. (222) b) The failure of cells to remain constrained in their growth properties and give rise to tumors. (628)

transit peptidase The enzyme that removes transit sequences from proteins transported into organelles. (425)

transit sequences Extra sequences at the N-terminal ends of proteins, necessary and sufficient for posttranslation transport into organelles. (425)

transition mutation A specific type of base-pair substitution mutation, involving a change in the DNA from one purine-pyrimidine base pair to the other purine-pyrimidine base pair at a particular site (e.g., AT to GC). (559)

translation (protein synthesis) The conversion in the cell of the mRNA base sequence information into an amino acid sequence of a polypeptide. (362; 382; 410)

translational control The regulation of protein synthesis by ribosome synthesis among mRNAs. (531)

translocation (transposition) a) A chromosome mutation involving a change in position of a chromosome segment (or segments) and the gene sequences it contains. (184; 593) b) In polypeptide synthesis, translocation is the movement of the ribosome, one codon at a time, along the mRNA toward the 3′ end. (421)

transmission genetics (classical genetics) A subdivision of the science of genetics, primarily dealing with how genes are passed from one individual to another. (708)

transport control Regulating the number of transcripts that exit the nucleus to the cytoplasm. (530)

transposable genetic element (TGE) In both prokaryotic and eukaryotic chromosomes, a genetic element that has the capacity to mobilize itself and move from one location to another in the genome. (612)

transposase Encoded by the IS element of a tranposon, this enzyme catalyzes transposition activity of a transposable genetic element (TGE). (613)

transposon (Tn) A mobile DNA segment that contains genes for the insertion of the DNA segment into the chromosome and for mobilization of the element to other locations on the chromosomes. (614)

transversion mutation A specific type of base-pair substitution mutation that involves a change in the DNA from a purine-pyrimidine base pair to a pyrimidine-purine base pair at the same site (e.g., AT to TA or GC to TA). (559)

trihybrid cross A cross between individuals of the same type that are heterozygous for three pairs of alleles at three different loci. (50)

trisomy An aberrant, aneuploid state in a normally diploid cell or organism in which there are three copies of a particular chromosome instead of two. (597)

trisomy-21 See Down syndrome. (597)

true-breeding (pure-breeding) strain A strain allowed to self-fertilize for many generations to ensure that the traits to be studied are inherited and unchanging. (38)

true reversion A point mutation from mutant back to wild-type—the change codes for the original amino acid of the wild type. (560)

tumor viruses Viruses which induce cells to dedifferentiate and to divide to produce a tumor. (628)

Turner syndrome A human clinical syndrome resulting from monosomy for the X chromosome in the female, which gives a 45,X female. These females fail to develop secondary sexual characteristics, tend to be short, have weblike necks, have poorly developed breasts, are usually infertile, and exhibit mental deficiencies. (76)

twin spots Two adjacent cell groups that differ in genotype and phenotype. They result from mitotic crossing-over within the somatic cells of a heterozygous individual. (173)

uniparental inheritance A phenomenon, usually exhibited by extranuclear genes, in which all progeny have the phenotype of only one parent. (654)

unique (single-copy) sequence A class of DNA sequences that has one to a few copies per genome. (319)

upstream activator sequences (UAS) In yeast, elements that are functionally similar to enhancers in other eukaryotes. UASs can function in either orientation and at variable distances upstream of the promoter. (374)

uracil (U) A pyrimidine base found in RNA but not in DNA. (275)

variance A statistical measure of how values vary from the mean. (677)

variance of gene frequency The variance in the frequency of an allele among a group of populations. (731)

vegetative reproduction See asexual reproduction. (13)

viral oncogene A viral gene that transforms a cell it infects to a cancerous state. See oncogenesis; cellular oncogene. (631)

virulent phage A phage like T4, which always follows the lytic cycle when it infects bacteria. (207)

virus A noncellular organism that can reproduce only within a host cell. It contains genetic material within a membrane or protein coat. Once a virus is within a cell, its genetic material causes the cellular machinery to produce progeny viruses. (3)

visible mutation A mutation that affects the morphology, or physical appearance, of an organism. (580)

wild type A strain, organism, or gene of the type that is designated as the standard for the organism with respect to genotype and phenotype. (64)

wild-type allele The allele designated as the standard ("normal") for a strain of organism. (54; 93)

wobble hypothesis A theory proposed by Francis Crick which proposes that the base at the 5′ end of the anticodon (3′ end of the codon) is not as constrained as the other two bases. This feature allows for less exact base pairing so that the 5′ end of the anticodon can potentially pair with one of three different bases at the 3′ end of the codon. (414)

X-linked dominant trait A trait due to a dominant mutant gene carried on the X chromosome. (82)

X-linked recessive trait A trait due to a recessive mutant gene carried on the X chromosome. (82)

X chromosome A sex chromosome present in two copies in the homogametic sex and in one copy in the heterogametic sex. (63)

X chromosome-autosome balance system A genotypic sex determination system. The main factor in sex determination is the ratio between the numbers of X chromosomes and autosomes. Sex is determined at the time of fertilization, and sex differences are assumed to be due to the action during development of two sets of genes located in the X chromosomes and in the autosomes. (75)

X chromosome nondisjunction An event occurring when the two X chromosomes fail to separate in meiosis. The eggs that are produced have either two X chromosomes or no X chromosomes, instead of the usual one X chromosome. (71)

X-linked Referring to genes located on the X chromosome. (69)

Y chromosome A sex chromosome that, when present, is found in one copy in the heterogametic sex, along with an X chromosome, and is not present in the homogametic sex. Not all organisms with sex chromosomes have a Y chromosome. (63)

Y-intercept In a regression analysis, the value of y when x is zero. (682)

Y-linked (holandric ["wholly male"]) trait A trait due to a mutant gene carried on the Y chromosome but with no counterpart on the X. (84)

zygonema The stage during meiosis in prophase I at which homologous chromosomes begin to pair in a highly specific way (like a zipper). (21)

zygote The cell produced by the fusion of the male and female gametes. (12; 43)

SUGGESTED READING

CHAPTER 1: VIRUSES, CELLS, AND CELLULAR REPRODUCTION

Brachet, J., and Mirsky, A. E., eds. 1961. *The cell*, vol. 3, *Meiosis and mitosis*. New York: Academic Press.

Sturtevant, A. H. 1965. *A history of genetics*. New York: Harper & Row.

CHAPTER 2: MENDELIAN GENETICS

Bateson, W. 1909. *Mendel's principles of heredity*. Cambridge: Cambridge University Press.

Mendel, G. 1866. Experiments in plant hybridization (translation). In *Classic papers in genetics*, edited by J. A. Peters, 1959. Englewood Cliffs, NJ: Prentice-Hall.

Peters, J. A., ed. 1959. *Classic papers in genetics*. Englewood Cliffs, NJ: Prentice-Hall.

Sandler, I., and Sandler, L. 1985. A conceptual ambiguity that contributed to the neglect of Mendel's paper. *Hist. Phil. Life Sci.* 7:3–70.

Tschermak-Seysenegg, E. von. 1951. The rediscovery of Mendel's work. *J. Hered.* 42:163–171.

CHAPTER 3: CHROMOSOMAL BASIS OF INHERITANCE, SEX DETERMINATION, AND SEX LINKAGE

Baker, B., Nagoshi, R. N., and Burtin, K. C. 1987. Molecular genetic aspects of sex determination in *Drosophila*. *BioEssays* 6:66–70.

Bodmer, W. F., and Cavalli-Sforza, L. L. 1976. *Genetics, evolution, and man*. San Francisco: Freeman.

Boggs, R. T., Gregor, P., Idriss, S., Belote, J. M., and McKeown, M. 1987. Regulation of sexual differentiation in *Drosophila melanogaster* via alternative splicing of RNA from the transformer. *Cell* 50:739–47.

Bridges, C. B. 1916. Nondisjunction as a proof of the chromosome theory of heredity. *Genetics* 1:1–52, 107–163.

———. 1925. Sex in relation to chromosomes and genes. *Am. Naturalist* 59:127–137.

Dice, L. R. 1946. Symbols for human pedigree charts. *J. Hered.* 37:11–15.

Eicher, E. M., and Washburn, L. L. 1986. Genetic control of primary sex determination in mice. *Annu. Rev. Genet.* 20:327–360.

Farabee, W. C. 1905. Inheritance of digital malformations in man. *Papers Peabody Museum Amer. Arch. Ethnol. (Harvard Univ.)* 3:65–78.

Hodgkin, J. 1980. More sex-determination mutants of *Caenorhabditis elegans*. *Genetics* 96:649–664.

Hodgkin, J. 1985. Males, hermaphrodites, and females: sex determination in *Caenorhabditis elegans*. *Trends Genet.* 1:85–88.

Hodgkin, J. 1989. Drosophila sex determination: a cascade of regulated splicing. *Cell* 56:905–906.

Levitan, M., and Montagu, A. 1971. *Textbook of human genetics*. New York: Oxford University Press.

McClung, C. E. 1902. The accessory chromosome—sex determinant? *Biol. Bull.* 3:43–84.

McKusick, V. A. 1965. The royal hemophilia. *Sci. Am.* 213:88–95.

McLaren, A., Simpson, E., Tomonari, K., Chandler, P., and Hogg, H. 1984. Male sexual differentiation in mice lacking H-Y antigen. *Nature* 312:552–555.

Morgan, L. V. 1922. Non criss-cross inheritance in *Drosophila melanogaster*. *Biol. Bull.* 42:267–274.

Morgan, T. H. 1910. Sex-limited inheritance in *Drosophila*. *Science* 32:120–122.

———. 1911. An attempt to analyze the constitution of the chromosomes on the basis of sex-limited inheritance in *Drosophila*. *J. Exp. Zool.* 11:365–414.

Page, D. C. 1985. Sex-reversal: Deletion mapping of the male-determining function of the human Y chromosome. *Cold Spring Harbor Symp. Quant. Biol.* 51:229–235.

———, de la Chapelle, A., and Weissenbach, J. 1985. Chromosome Y-specific DNA in related human XX males. *Nature* 315:224–226.

———, Mosher, R., Simpson, E. M., Fisher, E. M. C., Mardon, G., Pollack, J., McGillivray, B., de la Chapelle, A., and Brown, L. G. 1987. The sex-determining region of the human Y chromosome encodes a finger protein. *Cell* 51:1091–1104.

Palmer, M. S., Sinclair, A. H., Berta, P., Ellis, N. A., Goodfellow, P. N., Abbas, N. E., and Fellous, M. 1990. Genetic evidence that ZFY is not the testis-determining factor. *Nature* 342:937–939.

Simpson, E., Chandler, P., Goulmy, E., Disteche, C. M., Ferguson-Smith, M. A., and Page, D. 1987. Separation of the genetic loci for the H-Y antigen and for testis determination on the human Y chromosome. *Nature* 326:876–878.

Stern, C., Centerwall, W. P., and Sarkar, Q. S. 1964. New data on the problem of Y-linkage of hairy pinnae. *Am. J. Hum. Genet.* 16:455–471.

Sutton, W. S. 1903. The chromosomes in heredity. *Biol. Bull.* 4:231–251.

Villeneuve, A. M., and Meyer, B. J. 1987. *sdc-1*: A link between sex determination and dosage compensation in *C. elegans*. *Cell* 48:25–37.

Wolfner, M. F. 1988. Sex-specific gene expression in somatic tissues of *Drosophila melanogaster*. *Trends Genet.* 4:333–337.

Wilson, E. B. 1905. The chromosomes in relation to the determination of sex in insects. *Science* 22:500–502.

CHAPTER 4: EXTENSIONS OF MENDELIAN GENETIC ANALYSIS

Ginsburg, V. 1972. Enzymatic basis for blood groups. *Methods Enzymol.* 36:131–149.

Landauer, W. 1948. Hereditary abnormalities and their chemically induced phenocopies. *Growth Symposium* 12:171–200.

R–1

Landsteiner, K., and Levine, P. 1927. Further observations on individual differences of human blood. *Proc. Soc. Exp. Biol. Med.* 24:941–942.

Reed, T. E., and Chandler, J. H. 1958. Huntington's chorea in Michigan. I. Demography and genetics. *Am. J. Hum. Genet.* 10:201–225.

CHAPTER 5: LINKAGE, CROSSING-OVER, AND GENE MAPPING IN EUKARYOTES

Bateson, W., Saunders, E. R., and Punnett, R. G. 1905. Experimental studies in the physiology of heredity. *Rep. Evol. Committee R. Soc.* II:1–55 and 80–99.

Blixt, S. 1975. Why didn't Mendel find linkage? *Nature* 256:206.

Creighton, H. S., and McClintock, B. 1931. A correlation of cytological and genetical crossing-over in *Zea mays. Proc. Natl. Acad. Sci. USA.* 17:492–497.

Morgan, T. H. 1910. Sex-limited inheritance in *Drosophila. Science* 32:120–122.

———. 1910. The method of inheritance of two sex-limited characters in the same animal. *Proc. Soc. Exp. Biol. Med.* 8:17.

———. 1911. An attempt to analyze the constitution of the chromosomes on the basis of sex-limited inheritance in *Drosophila. J. Exp. Zool.* 11:365–414.

———. 1911. Random segregation versus coupling in Mendelian inheritance. *Science* 34:384.

———, Sturtevant, A. H., Muller, H. J., and Bridges, C. B. 1915. *The mechanism of Mendelian heredity.* New York: Henry Holt.

Sturtevant, A. H. 1913. The linear arrangement of six sex-linked factors in *Drosophila,* as shown by their mode of association. *J. Exp. Zool.* 14:43–59.

Sutton, W. S. 1903. The chromosomes in heredity. *Biol. Bull.* 4:231–251.

CHAPTER 6: ADVANCED GENETIC MAPPING IN EUKARYOTES

Barratt, R. W., Newmeyer, D., Perkins, D. D., and Garnjobst, L. 1954. Map construction in *Neurospora crassa. Adv. Genet.* 6:1–93.

Chaleff, R. S., and Carlson, P. S. 1974. Somatic cell genetics of higher plants. *Annu. Rev. Genet.* 8:267–278.

Ephrussi, B., and Weiss, M. C. 1969. Hybrid somatic cells. *Sci. Am.* 220:26–35.

Fincham, J. R. S., Day, P. R., and Radford, A. 1979. *Fungal genetics.* 3rd ed. Oxford: Blackwell Scientific.

Kao, F., Jones, C., and Puck, T. T. 1976. Genetics of somatic mammalian cells: Genetic, immunologic, and biochemical analysis with Chinese hamster cell hybrids containing selected human chromosomes. *Proc. Natl. Acad. Sci. USA* 73:193–197.

McKusick, V. A. 1971. The mapping of human chromosomes. *Sci. Am.* 224:104–113.

Pontecorvo, G. 1956. The parasexual cycle in fungi. *Annu. Rev. Microbiol.* 20:151–168.

———, and Kafer, E. 1958. Genetic analysis based on mitotic recombination. *Adv. Genet.* 9:71–104.

Pritchard, R. H. 1955. The linear arrangement of a series of alleles of *Aspergillus nidulans. Heredity* 9:343–371.

Ruddle, F. H., and Kucherlapati, R. S. 1974. Hybrid cells and human genes. *Sci. Am.* 231:36–44.

Stern, C. 1936. Somatic crossing over and segregation in *Drosophila melanogaster. Genetics* 21:625–730.

CHAPTER 7: GENETIC RECOMBINATION IN BACTERIA AND BACTERIOPHAGES

Archer, L. J. 1973. *Bacterial transformation.* New York: Academic Press.

Benzer, S. 1959. On the topology of the genetic fine structure. *Proc. Natl. Acad. Sci. USA* 45:1607–1620.

———. 1961. On the topography of the genetic fine structure. *Proc. Natl. Acad. Sci. USA* 47:403–415.

———. 1962. The fine structure of the gene. *Sci. Am.* 206:70–84.

Campbell, A. 1969. *Episomes.* New York: Harper & Row.

Curtiss, R. 1969. Bacterial conjugation. *Annu. Rev. Microbiol.* 23:69–136.

Delbruck, M. 1940. The growth of bacteriophage and lysis of the host. *J. Gen. Physiol.* 23:643–660.

Ellis, E. L., and Delbruck, M. 1939. The growth of bacteriophage. *J. Gen. Physiol.* 22:365–384.

Fincham, J. 1966. *Genetic complementation.* Menlo Park, CA: Benjamin.

Hayes, W. 1968. *The genetics of bacteria and their viruses,* 2nd ed. New York: Wiley.

Hershey, A. D., and Rotman, R. 1949. Genetic recombination between host-range and plaque-type mutants of bacteriophage in single bacterial cells. *Genetics* 34:44–71.

Hotchkiss, R. D., and Gabor, M. 1970. Bacterial transformation with special reference to recombination processes. *Annu. Rev. Genet.* 4:193–224.

Jacob, F., and Wollman, E. L. 1951. *Sexuality and the genetics of bacteria.* New York: Academic Press.

Ravin, A. W. 1961. The genetics of transformation. *Adv. Genet.* 10:61–163.

Susman, M. 1970. General bacterial genetics. *Annu. Rev. Genet.* 4:135–176.

Vielmetter, W., Bonhoeffer, F., and Schutte, A. 1968. Genetic evidence for transfer of a single DNA strand during bacterial conjugation. *J. Mol. Biol.* 37:81–86.

Wollman, E. L., Jacob, F., and Hayes, W. 1962. Conjugation and genetic recombination in *E. coli K-12. Cold Spring Harbor Symp. Quant. Biol.* 21:141–162.

Zinder, N., and Lederberg, J. L. 1952. Genetic exchange in *Salmonella. J. Bacteriol.* 64:679–699.

CHAPTER 8: THE BEGINNINGS OF MOLECULAR GENETICS: GENE FUNCTION

Beadle, G. W., and Ephrussi, B. 1937. Development of eye colors in *Drosophila:* Diffusible substances and their interrelationships. *Genetics* 22:76–86.

———, and Tatum, E. L. 1942. Genetic control of biochemical reactions in *Neurospora. Proc. Natl. Acad. Sci. USA* 27:499–506.

Galjaard, H. 1986. Biochemical diagnosis of genetic diseases. *Experientia* 42:1075–1085.

Garrod, A. E. 1909. *Inborn errors of metabolism.* New York: Oxford University Press.

Gilbert, F., Kucherlapati, R., Creagan, R. P., Murnane, M. J., Darlington, G. J., and Ruddle, F. H. 1975. Tay-Sachs' and Sandhoff's diseases: The assignment of genes for hexosaminidase A and B to individual human chromosomes. *Proc. Natl. Acad. Sci. USA* 72:263–267.

Gusella, J. F., Wexler, N. S., Conneally, P. M., Naylor, S. L., Anderson, M. A., Tanzi, R. E., Watkins, P. C., Ottina, K., Wallace, M. R., Sakaguchi, A. Y., Young, A. B., Shoulson, I., Bonilla, E., and Martin, J. B. 1983. A polymorphic DNA marker genetically linked to Huntington's disease. *Nature* 306:234–238.

Guttler, F., and Woo, S. L. C. 1986. Molecular genetics of PKU. *J. Inherited Metab. Dis.* 9 Suppl. 1:58–68.

Harris, H. 1975. *The principles of human biochemical genetics.* Amsterdam: North-Holland.

Ingram, V. M. 1963. *The hemoglobins in genetics and evolution.* New York: Columbia University Press.

Kan, Y. W., and Dozy, A. M. 1978. Polymorphism of DNA sequence adjacent to human β-globin structural gene: Relationship to sickle mutation. *Proc. Natl. Acad. Sci. USA* 75:5631–5635.

Maniatis, T., Fritsch, E. F., Lauer, J., and Lawn, R. M. 1980. The molecular genetics of human hemoglobins. *Annu. Rev. Genet.* 14:145–178.

Milunsky, A. 1976. Prenatal diagnosis of genetic disorders. *N. Engl. J. Med.* 295:377–380.

Motulsky, A. G. 1964. Hereditary red cell traits and malaria. *Am. J. Trop. Med. Hyg.* 13:147–158.

———. 1973. Frequency of sickling disorders in U.S. blacks. *N. Engl. J. Med.* 288:31–33.

Neel, J. V. 1949. The inheritance of sickle-cell anemia. *Science* 110:64–66.

Pauling, L., Itano, H. A., Singer, S. J., and Wells, J. C. 1949. Sickle-cell anemia, a molecular disease. *Science* 110:543–548.

Scriver, C. R., and Clow, C. L. 1980. Phenylketonuria and other phenylalanine hydroxylation mutants in man. *Annu. Rev. Genet.* 14:179–202.

Srb, A. M., and Horowitz, N. H. 1944. The ornithine cycle in *Neurospora* and its genetic control. *J. Biol. Chem.* 154:129–139.

Woo, S. L. C., Lidsky, A. S., Guttler, F., Chandra, T., and Robson, K. J. H. 1983. Cloned human phenylalanine hydroxylase gene allows prenatal diagnosis and carrier detection of classical phenylketonuria. *Nature* 300:151–155.

CHAPTER 9: THE STRUCTURE OF GENETIC MATERIAL

Avery, O. T., MacLeod, C. M., and McCarty, M. 1944. Studies on the chemical nature of the substance inducing transformation of pneumococcal types. Induction of transformation by a deoxyribonucleic acid fraction isolated from pneumococcus type III. *J. Exp. Med.* 79:137–158.

Azorin, F., and Rich, A. 1985. Isolation of Z-DNA binding proteins from SV40 minichromosomes: Evidence for binding to the viral control region. *Cell* 41:365–374.

Chargaff, E. 1951. Structure and function of nucleic acids as cell constituents. *Fed. Proc.* 10:654–659.

Davidson, J. N. 1972. *The biochemistry of the nucleic acids*, 8th ed. New York: Academic Press.

Dickerson, R. E. 1983. The DNA helix and how it is read. *Sci. Am.* 249:(December):94–111.

Fraenkel-Conrat, H., and Singer, B. 1957. Virus reconstitution: Combination of protein and nucleic acid from different strains. *Biochim. Biophys. Acta* 24:540–548.

Franklin, R. E., and Gosling, R. 1953. Molecular configuration of sodium thymonucleate. *Nature* 171:740–741.

Geis, I. 1983. Visualizing the anatomy of A, B and Z-DNAs. *J. Biomol. Struct. Dynam.* 1:581–591.

Gierer, A., and Schramm, G. 1956. Infectivity of ribonucleic acid from tobacco mosaic virus. *Nature* 177:702–703.

Griffith, F. 1928. The significance of pneumococcal types. *J. Hyg. (Lond)* 27:113–159.

Hershey, A. D., and Chase, M. 1952. Independent functions of viral protein and nucleic acid in growth and bacteriophage. *J. Gen. Physiol.* 36:39–56.

Krishna, P., Kennedy, B. P., van de Sande, J. H., and McGhee, J. D. 1988. Yolk proteins from nematodes, chickens, and frogs bind strongly and preferentially to left-handed Z-DNA. *J. Biol. Chem.* 263:19066–19070.

Pauling, L., and Corey, R. B. 1956. Specific hydrogen-bond formation between pyrimidines and purines in deoxyribonucleic acids. *Arch. Biochem. Biophys.* 65:164–181.

Rich, A., Nordheim, A., and Wang, A. H.-J. 1984. The chemistry and biology of left-handed Z-DNA. *Annu. Rev. Biochem.* 53:791–846.

"Structures of DNA." 1982. *Cold Spring Harbor Symp. Quant. Biol.* 47. New York: Cold Spring Harbor Laboratory.

Wang, A. H.-J., Quigley, G. J., Kolpak, F. J., Crawford, J. L., van Boom, J. H., van der Marel, G., and Rich, A. 1979. Molecular structure of a left-handed double helical DNA fragment at atomic resolution. *Nature* 282:680–686.

Watson, J. D. 1968. *The double helix.* New York: Atheneum.

———, and Crick, F. H. C. 1953. Genetical implications of the structure of deoxyribonucleic acid. *Nature* 171:964–969.

———. 1953. Molecular structure of nucleic acids. A structure for deoxyribose nucleic acid. *Nature* 171:737–738.

Wilkins, M. H. F., Stokes, A. R., and Wilson, H. R. 1953. Molecular structure of deoxypentose nucleic acids. *Nature* 171:738–740.

Wing, R. M., Drew, H. R., Takano, T., Broka, C., Tanaka, S., Itakura, K., and Dickerson, R. E. 1980. Crystal structure analysis of a complete turn of B-DNA. *Nature* 287:755–758.

Wittig, B., Dorbic, T., and Rich, A. 1989. The level of Z-DNA in metabolically active, permeabilized mammalian cell nuclei is regulated by torsional strain. *J. Cell Biol.* 108:755–764.

Wu, H.-M., and Crothers, D. M. 1984. The locus of sequence-directed and protein-induced DNA bending. *Nature* 308:509–513.

CHAPTER 10: THE ORGANIZATION OF DNA IN CHROMOSOMES

Amati, B. B., and Gasser, S. M. 1988. Chromosomal ARS and CEN elements bind specifically to the yeast nuclear scaffold. *Cell* 54:967–978.

Blackburn, E. H. 1984. Telomeres: Do the ends justify the means? *Cell* 37:7–8.

———, and Szostak, J. W. 1984. The molecular structure of centromeres and telomeres. *Annu. Rev. Biochem.* 53:163–194.

Bloom, K. S., Amaya, E., Carbon, J., Clarke, L., Hill, A., and Yeh, E. 1984. Chromatin conformation of yeast centromeres. *J. Cell Biol.* 99:1559–1568.

Boy de la Tour, E., and Laemmli, U. K. 1988. The metaphase scaffold is helically folded: sister chromatids have predominantly opposite helical handedness. *Cell* 55:937–944.

Britten, R. J., and Kohne, D. E. 1968. Repeated sequences in DNA. *Science* 161:529–540.

Burlingame, R. W., Love, W. E., Wang, B.-C., Hamlin, R., Xuang, N.-H., and Moudranakis, E. N. 1985. Crystallographic structure of the octameric histone core of the nucleosome at a resolution of 3.3 Å. *Science* 228:546–553.

Cai, M., and Davis, R. W. 1990. Yeast centromere binding protein CBF1, of the helix-loop-helix protein family, is required for chromosome stability and methionine prototrophy. *Cell* 61:437–446.

Carbon, J. 1984. Yeast centromeres: structure and function. *Cell* 37:351–353.

Cold Spring Harbor Symposia on Quantitative Biology. 1978. *Chromatin*, vol. 42. New York: Cold Spring Harbor Laboratory.

Comings, D. 1978. Mechanisms of chromosome banding and implications for chromosome structure. *Annu. Rev. Genet.* 12:25–46.

D'Ambrosio, E., Waitzikin, S. D., Whitney, F. R., Salemme, A., and Furano, A. V. 1985. Structure of the highly repeated, long interspersed DNA family (LINE or L1Rn) of the rat. *Mol. Cell. Biol.* 6:411–424.

DuPraw, E. J. 1970. *DNA and chromosomes.* New York: Holt, Rinehart and Winston.

Eisenberg, J. C., Cartwright, I. L., Thomas, G. H., and Elgin, S. C. R. 1985. Selected topics in chromatin structure. *Annu. Rev. Genet.* 19:485–536.

Freifelder, D. 1978. *The DNA molecule. Structure and properties.* San Francisco: Freeman.

Gellert, M. 1981. DNA topoisomerases. *Annu. Rev. Biochem.* 50:879–910.

Jelinek, W. R., and Schmid, C. W. 1982. Repetitive sequences in eukaryotic DNA and their expression. *Annu. Rev. Biochem.* 51:813–844.

Korenberg, J. R., and Rykowski, M. C. 1988. Human genome organization: Alu, Lines, and the molecular structure of metaphase chromosome bands. *Cell* 53:391–400.

Kornberg, R. D. 1977. Structure of chromatin. *Annu. Rev. Biochem.* 46:931–954.

———, and Klug, A. 1981. The nucleosome. *Sci. Am.* 244 (2):52–64.

Lewin, B. 1980. *Gene expression*, 2nd ed., vol. 2, *Eucaryotic chromosomes*. New York: Wiley.

Long, E. O., and Dawid, I. B. 1980. Repeated genes in eukaryotes. *Annu. Rev. Biochem.* 49:727–764.

MacHattie, L. A., Ritchie, D. A., and Thomas, C. A. 1967. Terminal repetition in permuted T2 bacteriophage DNA molecules. *J. Mol. Biol.* 23:355–363.

Marmur, J., Rownd, R., and Schildkraut, C. L. 1963. Denaturation and renaturation of deoxyribonucleic acid. *Prog. Nucleic Acid Res. Mol. Biol.* 1:231–300.

Mirkovich, J., Mirault, M.-E., and Laemmli, U. K. 1984. Organization of the higher-order chromatin loop: specific DNA attachment sites on nuclear scaffold. *Cell* 39:223–232.

Morse, R. H., and Simpson, R. T. 1988. DNA in the nucleosome. *Cell* 54:285–287.

Moyzis, R. K., Buckingham, J. M., Cram, L. S., Dani, M., Deaven, L. L., Jones, M. D., Meyne, J., Ratliff, R. L., and Wu, J.-R. 1988. A highly conserved repetitive DNA sequence, (TTAGGG)$_n$, present at the telomeres of human chromosomes. *Proc. Natl. Acad. Sci. USA* 85:6622–6626.

Murray, A. W., and Szostak, J. W. 1983. Chromosome structure and behaviour. *Trends Biochem. Sci.* (Mar.):112–115.

Olins, A. L., Carlson, R. D., and Olins, D. E. 1975. Visualization of chromatin substructure: nu-bodies. *J. Cell Biol.* 64:528–537.

Richard, T. J., Finch, J. T., Rushton, B., Rhodes, D., and Klug, A. 1984. Structure of the nucleosome core particle at 7 Å resolution. *Nature* 311:532–537.

Richards, E. J., and Ausubel, F. M. 1988. Isolation of a higher eukaryotic telomere from *Arabidopsis thaliana*. *Cell* 53:127–136.

Singer, M. F. 1982. Highly repeated sequences in mammalian genomes. *Int. Rev. Cytol.* 76:67–112.

———. 1982. SINEs and LINEs: Highly repeated short and long interspersed sequences in mammalian genomes. *Cell* 28:133 134.

———, and Skowronski, J. 1985. Making sense out of LINES: Long interspersed repeat sequences in mammalian genomes. *Trends Biochem. Sci.* (Mar.): 119–121.

Sinsheimer, R. L. 1959. A single-stranded deoxyribonucleic acid from bacteriophage ΦX174. *J. Mol. Biol.* 1:43–53.

Streisinger, G., Edgar, R. S., and Denhardt, G. H. 1964. Chromosome structure in phage T4, I. Circularity of the linkage map. *Proc. Natl. Acad. Sci. USA* 5:775–779.

Thomas, C. A., and MacHattie, L. A. 1967. The anatomy of viral DNA molecules. *Annu. Rev. Biochem.* 36:485–518.

Walmsley, R. W., Chan, C. S. M., Tye, B.-K., and Petes, T. D. 1984. Unusual DNA sequences associated with the ends of yeast chromosomes. *Nature* 310:157–160.

Williamson, J. R., Raghuraman, M. K., and Cech, T. R. 1989. Monovalent cation-induced structure of telomeric DNA: the G-quartet model. *Cell* 59:871–880.

Woodcock, C. L. F., Frado, L.-L. Y., and Rattner, J. B. 1984. The higher-order structure of chromatin: Evidence for a helical ribbon arrangement. *J. Cell Biol.* 99:42–52.

Worcel, A. 1978. Molecular architecture of the chromatin fiber. *Cold Spring Harbor Symp. Quant. Biol.* 42:313–324.

———, and Benyajati, C. 1977. Higher order coiling of DNA in chromatin. *Cell* 12:83–100.

———, and Burgi, E. 1972. On the structure of the folded chromosome of *Escherichia coli*. *J. Mol. Biol.* 71:127–147.

Yokoyama, R., and Yao, M.-C. 1986. Sequence characterization of *Tetrahymena* macronuclear ends. *Nucleic Acids Res.* 14:2109–2122.

Zimmerman, S. B. 1982. The three-dimensional structure of DNA. *Annu. Rev. Biochem.* 51:395–427.

CHAPTER 11: DNA REPLICATION AND RECOMBINATION

Alberts, B. M. 1985. Protein machines mediate the basic genetic processes. *Trends Genet.* 1:26–30.

Biswas, S. B., and Biswas, E. E. 1990. ARS binding factor I of the yeast *Saccharomyces cerevisiae* binds to sequences in telomeric and nontelomeric autonomously replicating sequences. *Mol. Cell. Biol.* 10:810–815.

Bollum, F. J. 1975. Mammalian DNA polymerases. *Prog. Nucleic Acid Res. Mol. Biol.* 15:109–144.

Bramhill, D., and Kornberg, A. 1988. Duplex opening by *dnaA* protein at novel sequences in initiation of replication at the origin of the *E. coli* chromosome. *Cell* 52:743–755.

Campbell, J. L. 1986. Eucaryotic DNA replication. *Annu. Rev. Biochem.* 55:733–772.

Cold Spring Harbor Symposia on Quantitative Biology. 1968. *Replication of DNA in microorganisms.* Vol. 33. New York: Cold Spring Harbor Laboratory.

Cozzarelli, N. R. 1980. DNA gyrase and the supercoiling of DNA. *Science* 207:953–960.

DeLucia, P., and Cairns, J. 1969. Isolation of an *E. coli* strain with a mutation affecting DNA polymerase. *Nature* 224:1164–1166.

DePamphilis, M. S. 1988. Transcriptional elements as components of eukaryotic origins of DNA replication. *Cell* 52:635–638.

De Pamphilis, M. L., and Wassarman, P. M. 1980. Replication of eukaryotic chromosomes: A close-up of the replication fork. *Annu. Rev. Biochem.* 49:627–666.

Gefter, M. L. 1975. DNA replication. *Annu. Rev. Biochem.* 44:45–78.

———, Hirota, Y., Kornberg, T., Wechsler, J. A., and Barnoux, C. 1971. Analysis of DNA polymerase II and III in mutants of *E. coli* thermosensitive for DNA synthesis. *Proc. Natl. Acad. Sci. USA* 68:3150–3153.

Gilbert, W., and Dressler, D. 1968. DNA replication: The rolling circle model. *Cold Spring Harbor Symp. Quant. Biol.* 33:473–484.

Holliday, R. 1964. A mechanism for gene conversion in fungi. *Genet. Res.* 5:282–304.

Huberman, J. A. 1987. Eukaryotic DNA replication: A complex picture partially clarified. *Cell* 48:7–8.

———, and Riggs, A. D. 1968. On the mechanism of DNA replication in mammalian chromosomes. *J. Mol. Biol.* 32:327–341.

Klein, A., and Bonhoeffer, F. 1972. DNA replication. *Annu. Rev. Biochem.* 41:301–332.

Kornberg, A. 1960. Biologic synthesis of deoxyribonucleic acid. *Science* 131:1503–1508.

———. 1980. *DNA replication.* San Francisco: Freeman.

Kornberg, A. 1988. DNA replication. *J. Biol. Chem.* 263:1–4.

Masters, M., and Broda, P. 1971. Evidence for the bidirectional replication of the *E. coli* chromosome. *Nature New Biol.* 232:137–140.

Meselson, M., and Radding, C. M. 1975. A general model for genetic recombination. *Proc. Natl. Acad. Sci. USA* 72:358–361.

Meselson, M., and Stahl, F. W. 1958. The replication of DNA in *Escherichia coli. Proc. Natl. Acad. Sci. USA* 44:671–682.

Nossal, N. G. 1983. Prokaryotic DNA replication systems. *Annu. Rev. Biochem.* 58:581–615.

Ogawa, T., Baker, T. A., van der Ende, A., and Kornberg, A. 1985. Initiation of enzymatic replication at the origin of the *Escherichia coli* chromosome: Contributions of RNA polymerase and primase. *Proc. Natl. Acad. Sci. USA* 82:3562–3566.

———, and Okazaki, T. 1980. Discontinuous DNA replication. *Annu. Rev. Biochem.* 49:424–457.

Okazaki, R. T., Okazaki, K., Sakobe, K., Sugimoto, K., and Sugino, A. 1968. Mechanism of DNA chain growth. I. Possible discontinuity and unusual secondary structure of newly synthesized chains. *Proc. Natl. Acad. Sci. USA* 59:598–605.

Orr-Weaver, T. L., and Szostak, J. W. 1985. Fungal recombination. *Microbiol. Rev.* 49:33–58.

Pardee, A. B., Dubrow, R., Hamlin, J. L., and Kleitzien, R. F. 1978. Animal cell cycle. *Annu. Rev. Biochem.* 47:715–750.

Prescott, D. M. 1976. *Reproduction of eukaryotic cells.* New York: Academic Press.

Radding, C. 1982. Homologous pairing and strand exchange in genetic recombination. *Annu. Rev. Genet.* 16:405–437.

Recombination at the DNA level. 1984. Cold Spring Harbor Symposium on Quantitative Biology. Vol. 49. Cold Spring Harbor Laboratory, Cold Spring Harbor, NY.

Reynolds, A. E., McCarroll, E. M., Newlon, C. S., and Fangman, W. L. 1989. Time of replication of ARS elements along yeast chromosome III. *Mol. Cell. Biol.* 9:4488–4494.

Simchen, G. 1978. Cell cycle mutants. *Annu. Rev. Genet.* 12:161–191.

Szostak, J., Orr-Weaver, T., Rothstein, R., and Stahl, F. 1983. The double-strand break repear model for recombination. *Cell* 33:25–35.

Taylor, J. H. 1970. The structure and duplication of chromosomes. In *Genetic organization,* edited by E. Caspari and A. Ravin, vol. 1, pp. 163–221. New York: Academic Press.

Van der Ende, A., Baker, T. A., Ogawa, T., and Kornberg, A. 1985. Initiation of enzymatic replication at the origin of the *Escherichia coli* chromosome: Primase as the sole priming enzyme. *Proc. Natl. Acad. Sci. USA* 82:3954–3958.

Wang, J. C., and Liu, L. F. 1979. DNA topoisomerases; enzymes that catalyze the concerted breakage and rejoining of DNA backbone bonds. In *Molecular Genetics,* edited by J. H. Taylor, part 3, pp. 65–88. New York: Academic Press.

Weissbach, A. 1977. Eukaryotic DNA polymerases. *Annu. Rev. Biochem.* 46:25–47.

Wickner, S. H. 1978. DNA replication proteins of *Escherichia coli. Annu. Rev. Biochem.* 47:1163–1191.

Williamson, D. H. 1985. The yeast ARS element, six years on: A progress report. *Yeast* 1:1–14.

Zyskind, J. W., and Smith, D. W. 1986. The bacterial origin of replication, *oriC. Cell* 46:489–490.

CHAPTER 12: TRANSCRIPTION

Baker, S. M., and Platt, T. 1986. Pol I transcription: Which comes first, the end or the beginning? *Cell* 47:839–840.

Bell, S. P., Pikaard, C. P., Reeder, R. H., and Tjian, R. 1989. Molecular mechanisms governing species-specific transcription of ribosomal RNA. *Cell* 59:489–497.

Birnsteil, M. L., Busslinger, M., and Struhl, K. 1985. Transcription termination and 3′ processing: The end is in site! *Cell* 41:349–359.

Bogenhagen, D. F., Sakonju, S., and Brown, D. D. 1980. A control region in the center of the 5S RNA gene directs specific intiation of transcription II: The 3′ border of the region. *Cell* 19:27–35.

Brand, A. H., Breeden, L., Abraham, J., Sternglanz, R., and Nasmyth, K. 1987. Characterization of a "silencer" in yeast: A DNA sequence with properties opposite to those of a transcriptional enhancer. *Cell* 41:41–48.

Brennan, C. A., Dombroski, A. J., and Platt, T. 1987. Transcription termination factor rho is an RNA-DNA helicase. *Cell* 48:945–952.

Buratowski, S., Hahn, S., Guarente, L., and Sharp, P. A. 1989. Five intermediate complexes in transcription initiation by RNA polymerase II. *Cell* 58:549–561.

Corden, J., Wasylyk, B., Buchwalder, A., Sassone-Corsi, P., Kedinger, C., and Chambon, P. 1980. Promoter sequences of eukaryotic protein-coding genes. *Science* 209:1406–1414.

Dynan, W. S. 1989. Modularity in promoters and enhancers. *Cell* 58:1–4.

Geiduschek, E. P., and Tocchini-Valentini, G. P. 1988. Transcription by RNA polymerase III. *Annu. Rev. Biochem.* 57:873–914.

Gluzman, Y., ed. 1985. Eukaryotic transcription: The role of cis- and trans-acting elements in initiation. Cold Spring Harbor Laboratory, Cold Spring Harbor, NY.

Guarente, L. 1984. Yeast promoters: Positive and negative elements. *Cell* 36:799–800.

Guarente, L. 1988. UASs and enhancers: common mechanism of transcriptional activation in yeast and mammals. *Cell* 52:303–305.

Hawley, D. K., and McClure, W. R. 1983. Compilation and analysis of *Escherichia coli* promoter DNA sequences. *Nucleic Acids Res.* 11:2237–2255.

Jacob, S. T. 1986. Transcription of eukaryotic ribosomal RNA genes. *Mol. Cell. Biochem.* 70:11–20.

Kassavetis, G. A., Braun, B. R., Nguyen, L. H. and Geiduschek, E. P. 1990. *S. cerevisiae* TFIIIB is the transcription initiation factor proper of RNA polymerase III, while TFIIIA and TFIIIC are assembly factors. *Cell* 60:235–245.

Labhart, P., and Reeder, R. H. 1987. Ribosomal precursor 3′ end formation requires a conserved element upstream of the promoter. *Cell* 50: 51–57.

Losick, R., and Chamberlin, M., eds. 1976. *RNA polymerase.* New York: Cold Spring Harbor Laboratory.

Marmur, J., Greenspan, C. M., Palecek, E., Kahan, F. M., Levine, J., and Mandel, M. 1963. Specificity of the complementary RNA formed by *Bacillus subtilis* infected with bacteriophage SP8. *Cold Spring Harbor Symp. Quant. Biol.* 28:191–199.

McKnight, S. L., and Kingsbury, R. 1982. Transcriptional control signals of a eukaryotic protein-coding gene. *Science* 217:316–324.

McStay, B., and Reeder, R. H. 1986. A termination site for Xenopus RNA polymerase I also acts as an element of an adjacent promoter. *Cell* 47:913–920.

Platt, T., and Bear, D. 1983. The role of RNA polymerase, rho factor, and ribosomes in transcription termination, in *Gene Function in Prokaryotes,* edited by J. Beckwith, J. Davies, and J. Gallant, pp. 123–162. New York: Cold Spring Harbor Laboratory.

Pruss, G. J., and Drlica, K. 1989. DNA supercoiling and prokaryotic transcription. *Cell* 56:521–523.

Reznikoff, W. S., Siegele, D. A., Cowing, D. W., and Gross, C. A. 1985. The regulation of transcription initiation in bacteria. *Annu. Rev. Genet.* 19:355–387.

Rodriguez, R., and Chamberlin, M., eds. 1982. *Promoters: Structure and Function.* New York: Praeger.

Sollner-Webb, B. 1988. Surprises in polymerase III transcription. *Cell* 52:153–154.

Stewart, P. R., and Letham, D. S., eds. 1977. *The ribonucleic acids,* 2nd ed. New York: Springer-Verlag.

Struhl, K. 1987. Promoters, activator proteins, and the mechanism of transcriptional initiation in yeast. *Cell* 49:295–297.

Van der Sande, C. A. F. M., Kulkens, T., Kramer, A. B., de Wijs, I. J., van Heerikhuizen, H., Klootwijk, J., and Planta, R. J. 1989. Termination of transcription by yeast RNA polymerase I. *Nucleic Acids Res.* 17:9127–9146.

Voss, S. D., Schlokat, U., and Gruss, P. 1986. The role of enhancers in the regulation of cell-type-specific transcriptional control. *Trends Biochem. Sci.* (July): 287–289.

CHAPTER 13: RNA MOLECULES AND RNA PROCESSING

Abelson, J. N., Brody, E. N., Cheng, S.-C., Clark, M. W., Green, P. R., Dalbadie-McFarland, G., Lin, R.-J., Newman, A. J., Phizicky, E. M., and Vijayraghavan, U. 1986. RNA splicing in yeast. *Chemica Scripta* 26B:127–137.

Banerjee, A. K. 1980. 5'-terminal cap structure in eucaryotic messenger ribonucleic acids. *Microbiol. Rev.* 44:175–205.

Black, D. L., Chabot, B., and Steitz, J. A. 1985. U2 as well as U1 small nuclear ribonucleoproteins are involved in pre-messenger RNA splicing. *Cell* 42:737–750.

Breathnach, R., and Chambon, P. 1981. Organization and expression of eucaryotic split genes coding for proteins. *Annu. Rev. Biochem.* 50:349–383.

———, Mandel, J. L., and Chambon, P. 1977. Ovalbumin gene is split in chicken DNA. *Nature* 270:314–318.

Breitbart, R. E., Andreadis, A., and Nadal-Ginard, B. 1987. Alternative splicing: A ubiquitous mechanism for the generation of multiple protein isoforms from single genes. *Annu. Rev. Biochem.* 56:467–495.

Brimacombe, R., Stoffler, G., and Wittmann, H. G. 1978. Ribosome structure. *Annu. Rev. Biochem.* 47:217–249.

Brody, E., and Abelson, J. 1985. The "spliceosome": Yeast premessenger RNA associates with a 40S complex in a splicing dependent reaction. *Science* 228:963–967.

Cech, T. R. 1983. RNA splicing: Three themes with variations. *Cell* 34:713–716.

———. 1985. Self-splicing RNA: Implications for evolution. *Int. Rev. Cytol.* 93:3–22.

———. 1986. Ribosomal RNA gene expression in *Tetrahymena:* Transcription and RNA splicing. *Mol. biol. ciliated protozoa,* pp. 203–225. New York: Academic Press.

———. 1986. The generality of self-splicing RNA: Relationship to nuclear mRNA splicing. *Cell* 44:207–210.

———. 1988. Group I intron structure. *Gene* 73:259–271.

———. 1988. Ribozymes and their medical implications. *JAMA* 260: 3030–3034.

Chabot, B., Black, D. L., LeMaster, D. M., and Steitz, J. A. 1985. The 3' splice site of pre-messenger RNA is recognized by a small nuclear ribonucleoprotein. *Science* 230:1344–1349.

———, and Steitz, J. A. 1987. Multiple interactions between the splicing substrate and small nuclear ribonucleoproteins in spliceosomes. *Mol. Cell. Biol.* 7:281–293.

Chambliss, G., Craven, G. R., Davies, J., Davis, K., Kahan, L., and Nomura, M., eds. 1980. *Ribosomes. Structure, function, and genetics.* Baltimore: University Park Press.

Chambon, P. 1975. Eukaryotic nuclear RNA polymerases. *Annu. Rev. Biochem.* 44:613–638.

Cheng, S.-C., and Abelson, J. 1986. Fractionation and characterization of a yeast mRNA splicing extract. *Proc. Natl. Acad. Sci. USA* 83:2387–2391.

Choi, Y. D., Grabowski, P. J., Sharp, P. A., and Dreyfuss, G. 1986. Heterogeneous nuclear ribonucleoproteins: Role in RNA splicing. *Science* 231:1534–1539.

Chu, F. K., Maley, G. F., West, D. K., Belfort, M., and Maley, F. 1986. Characterization of the intron in the phage T4 thymidylate synthase gene and evidence for its self-excision from the primary transcript. *Cell* 45:157–166.

Crick, F. H. C. 1979. Split genes and RNA splicing. *Science* 204:264–271.

Furuichi, Y., Morgan, M., Shatkin, A. J., Jelinek, W., Salditt-Georgieff, M., and Darnell, J. E. 1975. Methylated, blocked 5' termini in HeLa cell mRNA. *Proc. Natl. Acad. Sci. USA* 72:1904–1908.

Gilbert, W. 1985. Genes-in-pieces revisited. *Science* 228:823–824.

Grabowski, P. J., Seiler, S. R., and Sharp, P. A. 1985. A multicomponent complex is involved in the splicing of messenger RNA precursors. *Cell* 42:355–367.

Green, M. R., 1986. Pre-mRNA splicing. *Annu. Rev. Genet.* 20:671–708.

———. 1989. Pre-mRNA processing and mRNA nuclear export. *Curr. Opinion Cell Biol.* 1:519–525.

Guthrie, C. 1986. Finding functions for small nuclear RNAs in yeast. *Trends Biochem. Sci.* (Oct.): pp. 430–434.

Jeffreys, A. J., and Flavell, R. A. 1977. The rabbit beta-globin gene contains a large insert in the coding sequence. *Cell* 12:1097–1108.

Konarska, M. M., and Sharp, P. A. 1987. Interactions between small nuclear ribonucleoprotein particles in the formation of spliceosomes. *Cell* 49:763–774.

Maniatis, T., and Reed, R. 1987. The role of small nuclear ribonucleoprotein particles in pre-mRNA splicing. *Nature* 325:673–678.

Moore, P. B. 1988. The ribosome returns. *Nature* 331:223–227.

Nomura, M. 1973. Assembly of bacterial ribosomes. *Science* 179:864–873.

———, Morgan, E. A., and Jaskunas, S. R. 1977. Genetics of bacterial ribosomes. *Annu. Rev. Genet.* 11:297–347.

Padgett, R. A., Grabowski, P. J., Konarska, M. M., and Sharp, P. A. 1985. Splicing messenger RNA precursors: Branch sites and lariat RNAs. *Trends Biochem. Sci.* (April):154–157.

Perry, R. P. 1976. Processing of RNA. *Annu. Rev. Biochem.* 45:605–629.

Reed, R., and Maniatis, T. 1985. Intron sequences involved in lariat formation during pre-mRNA splicing. *Cell* 41:95–105.

Reeder, R. H. 1989. Regulatory elements of the generic ribosomal gene. *Curr. Opinion Cell Biol.* 1:466–474.

Reich, C., Olsen, G. J., Pace, B., and Pace, N. R. 1988. Role of the protein moiety of ribonuclease P, a ribonucleoprotein enzyme. *Science* 239:178–181.

Schmidt, F. J. 1985. RNA splicing in prokaryotes: Bacteriophage T4 leads the way. *Cell* 41:339–340.

Sharp, P. A. 1985. On the origin of RNA splicing and introns. *Cell* 42:397–400.

Smith, C. W. J., Porro, E. B., Patton, J. G., and Nadal-Ginard, B. 1989. Scanning from an independently specified branch point defines the 3' splice site of mammalian introns. *Nature* 342:243–247.

Sollner-Webb, B., and Tower, J. 1986. Transcription of cloned eukaryotic ribosomal RNA genes. *Annu. Rev. Biochem.* 55:801–830.

Starzyk, R. M. 1986. Prokaryotic RNA processing. *Trends Biochem. Sci.* (Feb.):60.

Sussman, J. L., and Kim, S. H. 1976. Three-dimensional structure of a transfer RNA in two crystal forms. *Science* 192:853–858.

Sylvester, J. E., Whiteman, D. A., Podolsky, R., Pozsgay, J. M., Respess, J., and Schmickel, R. D. 1986. The human ribosomal RNA genes: Structure and organization of the complete repeating unit. *Hum. Genet.* 73:193–198.

Tilghman, S. M., Curis, P. J., Tiemeier, D. C., Leder, P., and Weissman, C. 1978. The intervening sequence of a mouse β-globin gene is transcribed within the 15S β-globin mRNA precursor. *Proc. Natl. Acad. Sci. USA* 75:1309–1313.

———, Tiemeier, D. C., Seidman, J. G., Peterlin, B. M., Sullivan, M., Maizel, J. V., and Leder, P. 1978. Intervening sequence of DNA identified in the structural portion of a mouse beta-globin gene. *Proc. Natl. Acad. Sci. USA* 78:725–729.

Weinstock, R., Sweet, R., Weiss, M., Cedar, H., and Axel, R. 1978. Intragenic DNA spacers interrupt the ovalbumin gene. *Proc. Natl. Acad. Sci. USA* 75:1299–1303.

Zaug, A. J., and Cech, T. R. 1986. The intervening sequence RNA of Tetrahymena is an enzyme. *Science* 231:470–475.

Chapter 14: The Genetic Code and the Translation of the Genetic Message

Bachmair, A., Finley, D., and Varshavsky, A. 1986. In vivo half-life of a protein is a function of its amino-terminal residue. *Science* 234:179–186.

Blobel, G., and Dobberstein, B. 1975. Transfer of proteins across membranes. I. Presence of proteolytically processed and unprocessed nascent immunoglobulin light chains on membrane-bound ribosomes of murine myeloma. *J. Cell Biol.* 67:835–851.

Brenner, S., Jacob, F., and Meselson, M. 1961. An unstable intermediate carrying information from genes to ribosomes for protein synthesis. *Nature* 190:576–581.

Burgess, T. L., and Kelly, R. B. 1987. Constitutive and regulated secretion of proteins. *Annu. Rev. Cell Biol.* 3:243–293.

Cold Spring Harbor Symposia for Quantitative Biology, Vol. 31, 1966: *The genetic code*. New York: Cold Spring Harbor Laboratory.

Colman, A., and Robinson, C. 1986. Protein import into organelles: hierarchical targeting signals. *Cell* 46:321–322.

Crick, F. H. C. 1966. Codon-anticodon pairing: The wobble hypothesis. *J. Mol. Biol.* 19:548–555.

———, Barnett, L., Brenner, S., and Watts-Tobin, R. J. 1961. General nature of the genetic code for proteins. *Nature* 192:1227–1232.

Dingwall, C. 1985. The accumulation of proteins in the nucleus. *Trends Biochem. Sci.* (Feb.):pp. 64–66.

———, and Laskay, R. A. 1986. Protein import into the cell nucleus. *Annu. Rev. Cell Biol.* 2:367–390.

Garen, A. 1968. Sense and nonsense in the genetic code. *Science* 160:149–159.

Griffiths, G., and Simons, K. 1986. The trans Golgi network: Sorting at the exit site of the Golgi complex. *Science* 234:438–442.

Horowitz, S., and Gorovsky, M. A. 1985. An unusual genetic code in nuclear genes of *Tetrahymena*. *Proc. Natl. Acad. Sci. USA* 82:2452–2455.

Khorana, H. G. 1966–67. Polynucleotide synthesis and the genetic code. *Harvey Lectures* 62:79–105.

Kozak, M. 1983. Comparison of initiation of protein synthesis in procaryotes, eucaryotes, and organelles. *Microbiol. Rev.* 47:1–45.

Maitra, U., Stringer, E. A., and Chaudhuri, A. 1982. Initiation factors in protein biosynthesis. *Annu. Rev. Biochem.* 51:869–900.

Meyer, D. I. 1982. The signal hypothesis—A working model. *Trends Biochem. Sci.* 7:320–321.

Moldave, K. 1985. Eukaryotic protein synthesis. *Annu. Rev. Biochem.* 54:1109–1150.

Morgan, A. R., Wells, R. D., and Khorana, H. G. 1966. Studies on polynucleotides. LIX. Further codon assignments from amino acid incorporation directed by ribopolynucleotides containing repeating trinucleotide sequences. *Proc. Natl. Acad. Sci. USA* 56:1899–1906.

Nirenberg, M., and Leder, P. 1964. RNA code words and protein synthesis. *Science* 145:1399–1407.

———, and Matthaei, J. H. 1961. The dependence of cell-free protein synthesis in *E. coli* upon naturally occurring or synthetic polyribonucleotides. *Proc. Natl. Acad. Sci. USA* 47:1588–1602.

Perara, E., Rothman, R. E., and Lingappa, V. R. 1986. Uncoupling translocation from translation: Implications for transport of proteins across membranes. *Science* 232:348–352.

Pfeffer, S. R., and Rothman, J. E. 1987. Biosynthetic protein transport and sorting by the endoplasmic reticulum and Golgi. *Annu. Rev. Biochem.* 56:829–852.

Rogers, S., Wells, R., and Rechsteiner, M. 1986. Amino acid sequences common to rapidly degraded proteins: The PEST hypothesis. *Science* 234:364–368.

Schekman, R. 1985. Protein localization and membrane traffic in yeast. *Annu. Rev. Cell Biol.* 1:115–143.

Schmidt, G. W., and Mishkind, M. L. 1986. The transport of proteins into chloroplasts. *Annu. Rev. Biochem.* 55: 879–912.

Shine, J., and Delgarno, L. 1974. The 3′-terminal sequence of *Escherichia coli* 16S ribosomal RNA: Complementarity to nonsense triplet and ribosome binding sites. *Proc. Natl. Acad. Sci. USA* 71:1342–1346.

Silhavy, T. J., Benson, S. A., and Emr, S. D. 1983. Mechanisms of protein localization. *Microbiol. Rev.* 47:313–344.

Verner, K., and Schatz, G. 1988. Protein translocation across membranes. *Science* 241:1307–1313.

Walter, P., Gilmore, R., and Blobel G. 1984. Protein translocation across the endoplasmic reticulum. *Cell* 38:5–8.

Walter, P., and Lingappa, V. 1986. Mechanism of protein translocation across the endoplasmic reticulum membrane. *Annu. Rev. Cell Biol.* 2:499–516.

Watson, J. D. 1963. The involvement of RNA in the synthesis of proteins. *Science* 140:17–26.

Chapter 15: Recombinant DNA Technology and the Manipulation of DNA

Aaij, C., and Borst, P. 1972. The gel electrophoresis of DNA. *Biochim. Biophys. Acta* 269:192–200.

Arber, W. 1965. Host-controlled modification of bacteriophage. *Annu. Rev. Microbiol.* 19:365–378.

———, and Dussoix, D. 1962. Host specificity of DNA produced by *Escherichia coli* I. Host controlled modification of bacteriophage lambda. *J. Mol. Biol.* 5:18–36.

Bobrow, M. 1988. Prenatal diagnosis. *J. Chem. Tech. Biotechnol.* 43:285–291.

Boyer, H. W. 1971. DNA restriction and modification mechanisms in bacteria. *Annu. Rev. Microbiol.* 25:153–176.

Danna, K., and Nathans, D. 1971. Specific cleavage of simian virus 40 DNA by restriction endonuclease of *Haemophilus influenzae*. *Proc. Natl. Acad. Sci. USA* 68:2913–2917.

Donis-Keller, H., Barker, D., Knowlton, R., Schumm, J., and Braman, J. 1986. Applications of RFLP probes to genetic mapping and clinical diagnosis in humans. In *Current communications in molecular biology: DNA probes— Applications in genetic and infectious disease and cancer*, edited by L. S. Lerman, pp. 73–81. Cold Spring Harbor Laboratory, Cold Spring Harbor, NY.

Drayna, D., and White, R. 1985. The genetic linkage map of the human X chromosome. *Science* 230:753–758.

Dussoix, D., and Arber, W. 1962. Host specificity of DNA produced by *Escherichia coli*. II. Control over acceptance of DNA from infecting phage lambda. *J. Mol. Biol.* 5:37–49.

Feinberg, A. P., and Vogelstein, B. 1983. A technique for radiolabeling DNA restriction endonuclease fragments to high specific activity. *Anal. Biochem.* 132:6–13.

Feinberg, A. P., and Vogelstein, B. 1984. Addendum: A technique for radiolabeling DNA restriction endonuclease fragments to high specific activity. *Anal. Biochem.* 137:266–267.

Freifelder, D. 1978. *Recombinant DNA: Readings from Scientific American*. San Francisco:W. H. Freeman.

Kessler, C., Neumaier, P. S., and Wolf, W. 1985. Recognition sequences of restriction endonucleases and methylases—A review. *Gene* 33:1–102.

Klee, H., Horsch, R., and Rogers, S. 1987. *Agrobacterium*-mediated plant transformation and its further applications to plant biology. *Annu. Rev. Plant Physiol.* 38:467–486.

Luria, S. E. 1953. Host induced modification of viruses. *Cold Spring Harbor Symp. Quant. Biol.* 18:237–244.

Maxam, A. M., and Gilbert, W. 1977. A new method for sequencing DNA. *Proc. Natl. Acad. Sci. USA* 74:560–564.

Moores, J. C. 1987. Current approaches to DNA sequencing. *Anal. Biochem.* 163:1–8.

Mullis, K. B. 1990. The unusual origin of the polymerase chain reaction. *Sci. Am.* (Apr): pp. 56–65.

Mullis, K. B., and Faloona, F. A. 1987. Specific synthesis of DNA *in vitro* via a polymerase-catalyzed chain reaction. *Meth. Enzymol.* 155:335–350.

Sambrook, J., Fritsch, E. F., and Maniatis, T. 1989. Molecular cloning: A laboratory manual. 2nd ed. Cold Spring Harbor Laboratory, Cold Spring Harbor, NY.

Sanger, F., and Coulson, A. R. 1975. A rapid method for determining sequences in DNA by primed synthesis with DNA polymerase. *J. Mol. Biol.* 94:441–448.

Schell, J., and Van Montagu, M. 1983. The Ti plasmids as natural and practical gene vectors for plants. *Bio/Technology* 1:175–180.

Southern, E. M. 1975. Detection of specific sequences among DNA fragments separated by gel electrophoresis. *J. Mol. Biol.* 98:503–517.

Struhl, K., Cameron, J. R., and Davis, R. W. 1976. Functional genetic expression of eukaryotic DNA in *Escherichia coli*. *Proc. Natl. Acad. Sci. USA* 73:1471–1475.

White, T. J., Arnheim, N., and Erlich, H. A. 1989. The polymerase chain reaction. *Trends Genet.* 5:185–188.

CHAPTER 16: REGULATION OF GENE EXPRESSION IN BACTERIA AND BACTERIOPHAGES

Aloni, Y., and Hay, N. 1985. Attenuation may regulate gene expression in animal viruses and cells. *CRC Crit. Rev. Biochem.* 18:327–383.

Beckwith, J. R., and Zipser, D. 1970. *The lactose operon*. Cold Spring Harbor Laboratory, Cold Spring Harbor, NY.

Bertrand, K., Korn, L., Lee, F., Platt, T., Squires, C. L., Squires, C., and Yanofsky, C. 1975. New features of the structure and regulation of the tryptophan operon of *Escherichia coli*. *Science* 189:22–26.

———, and Yanofsky, C. 1976. Regulation of transcription termination in the leader region of the tryptophan operon of *Escherichia coli* involves tryptophan as its metabolic product. *J. Mol. Biol.* 103:339–349.

Calos, M. P. 1978. DNA sequence for a low-level promoter of the *lac* repressor gene and an "up" promoter mutation. *Nature* 274:762–765.

Cashel, M. 1975. Regulation of bacterial ppGpp and pppGpp. *Annu. Rev. Microbiol.* 29:301–318.

Dickson, R. C., Abelson, J., Barnes, W. M., and Reznikoff, W. S. 1975. Genetic regulation: The *lac* control region. *Science* 187:27–35.

Fisher, R. F., Das, A., Kolter, R., Winkler, M. E., and Yanofsky, C. 1985. Analysis of the requirements for transcription pausing in the tryptophan operon. *J. Mol. Biol.* 182:397–409.

Gilbert, W., Maizels, N., and Maxam, A. 1974. Sequences of controlling regions of the lactose operon. *Cold Spring Harbor Symp. Quant. Biol.* 38:845–855.

———, and Muller-Hill, B. 1966. Isolation of the *lac* repressor. *Proc. Natl. Acad. Sci. USA* 56:1891–1898.

Ho, Y-S., Wulff, D. L., and Rosenberg, M. 1983. Bacteriophage λ protein *cII* binds promoters on the opposite face of the DNA helix from RNA polymerase. *Nature* 304:703–708.

Jacob, F., and Monod, J. 1961. Genetic regulatory mechanisms in the synthesis of proteins. *J. Mol. Biol.* 3:318–356.

———. 1965. Genetic mapping of the elements of the lactose region of *Escherichia coli*. *Biochem. Biophys. Res. Commun.* 18:693–701.

Lamond, A. I., and Travers, A. A. 1985. Stringent control of bacterial transcription. *Cell* 41:6–8.

Lee, F., and Yanofsky, C. 1977. Transcription termination at the *trp* operon attenuators of *Escherichia coli* and *Salmonella typhimurium*: RNA secondary structure and regulation of termination. *Proc. Natl. Acad. Sci. USA* 74:4365–4369.

Maizels, N. 1974. *E. coli* lactose operon ribosome binding site. *Nature (New Biol.)* 249:647–649.

Miller, J. H., and Reznikoff, W. S. 1978. *The operon*. Cold Spring Harbor Laboratory, Cold Spring Harbor, NY.

Pabo, C. O., Sauer, R. T., Sturtevant, J. M., and Ptashne, M. 1979. The λ repressor contains two domains. *Proc. Natl. Acad. Sci. USA* 76:1608–1612.

Platt, T. 1981. Termination of transcription and its regulation in the tryptophan operon of *E. coli*. *Cell* 24:10–23.

Ptashne, M. 1984. Repressors. *Trends Biochem. Sci.* 9:142–145.

———. 1986. *A genetic switch*. Oxford: Cell Press and Blackwell Scientific Publications.

———. 1967. Isolation of the λ phage repressor. *Proc. Natl. Acad. Sci. USA* 57:306–313.

———, and Gilbert, W. 1970. Genetic repressors. *Sci. Am.* 222:36–44.

Ullmann, A. 1985. Catabolite repression 1985. *Biochimie* 67:29–34.

Winkler, M. E., and Yanofsky, C. 1981. Pausing of RNA polymerase during in vitro transcription of the tryptophan operon leader region. *Biochemistry* 20:3738–3744.

Yanofsky, C. 1981. Attenuation in the control of expression of bacterial operons. *Nature* 289:751–758.

Yanofsky, C. 1987. Operon-specific control by transcription attenuation. *Trends Genet.* 3:356–360.

———, and Kolter, R. 1982. Attenuation in amino acid biosynthetic operons. *Annu. Rev. Genet.* 16:113–134.

CHAPTER 17: REGULATION OF GENE EXPRESSION AND DEVELOPMENT IN EUKARYOTES

Ashburner, M., Chihara, C., Meltzer, P., and Richards, G. 1974. Temporal control of puffing activity in polytene chromosomes. *Cold Spring Harbor Symp. Quant. Biol.* 38:655–662.

Baker, W. 1978. A genetic framework for *Drosophila* development. *Annu. Rev. Genet.* 12:451–470.

Beato, M. 1989. Gene regulation by steroid hormones. *Cell* 56:335–344.

Beerman, W., and Clever, U. 1964. Chromosome puffs. *Sci. Am.* 210:50–58.

Blackwell, T. K. and F. W. 1988. Immunoglobin genes, in *Molecular Immunology*, edited by B. D. Homes and D. M. Glover. Washington, D.C.: IRL Press, pp. 1–60.

Boggs, R. T., Gregor, P., Idriss, S., Belote, J. M., and McKeown, M. 1987. Regulation of sexual differentiation in D. *melanogaster* via alternative splicing of RNA from the *transformer* gene. *Cell* 50:739–747.

Bonner, J. J., and Pardue, M. L. 1976. Ecdysone-stimulated RNA synthesis in imaginal discs of *Drosophila melanogaster*. *Chromosoma* 58:87–99.

Brawerman, G. 1989. mRNA decay: finding the right targets. *Cell* 57:9–10.

Cedar, H. 1988. DNA methylation and gene activity. *Cell* 53:3–4.

Cooper, D. N. 1983. Eukaryotic DNA methylation. *Hum. Genet.* 64:315–333.

Davidson, E. H. 1976. *Genetic activity in early development,* 2nd ed. New York: Academic Press.

De Robertis, E. M., and Gurdon, J. B. 1977. Gene activation in somatic nuclei after injection into amphibian oocytes. *Proc. Natl. Acad. Sci. USA* 74:2470–2474.

Doerfler, W. 1983. DNA methylation and gene activity. *Annu. Rev. Biochem.* 52:93–124.

Efstratiadis, A., Posakony, J. W., Maniatis, T., Lawn, R. M., O'Connell, C., Spritz, R. A., DeRiel, J. K., Forget, B. G., Weissman, S. M., Slighton, J. L., Blechl, A. E., Smithies, O., Baralle, F. E., Shoulders, C. C., and Proudfoot, N. J. 1980. The structure and evolution of the human β-globin gene family. *Cell* 21:653–668.

Eissenberg, J. C., Cartwright, I. L., Thomas, G. H., and Elgin, S. C. R. 1985. Selected topics in chromatin structure. *Annu. Rev. Genet.* 19:485–536.

Gehring, W. 1979. Developmental genetics of *Drosophila*. *Annu. Rev. Genet.* 10:209–252.

Green, M. R. 1989. Pre-mRNA processing and mRNA nuclear export. *Curr. Opinion Cell Biol.* 1:519–525.

Gross, D. S., and Garrard, W. T. 1987. Poising chromatin for transcription. *Trends Biochem. Sci.* (Aug.), pp. 293–297.

———. 1988. Nuclease hypersensitive sites in chromatin. *Annu. Rev. Biochem.* 57:159–197.

Gurdon, J. B. 1968. Transplanted nuclei and cell differentiation. *Sci. Am.* 219:24–35.

Hadorn, E. 1968. Transdetermination in cells. *Sci. Am.* 219:110–120.

Hodgkin, J. 1989. *Drosophila* sex determination: a cascade of regulated splicing. *Cell* 56:905–906.

Karlsson, S., and Nienhuis, A. W. 1985. Development regulation of human globin genes. *Annu. Rev. Biochem.* 54:1071–1078.

Lucas, G. L., and Opitz, J. M. 1966. The nail-patella syndrome. *J. Pediatr.* 68:273–288.

McGinnis, W., Levine, M. S., Hafen, E., Kuroiwa, A., and Gehring, W. J. 1984. A conserved DNA sequence homeotic genes of the *Drosophila Antennapedia* and *bithorax* complexes. *Nature* 308:428–433.

O'Malley, B. W., and Schrader, W. T. 1976. The receptors of steroid hormones. *Sci. Am.* 234:32–43.

O'Malley, B., Towle, H., and Schwartz, R. 1977. Regulation of gene expression in eucaryotes. *Annu. Rev. Genet.* 11:239–275.

Peifer, M. F., Karch, F., and Bender, W. 1987. The bithorax complex: control of segmental identity. *Genes Dev.* 1:891–898.

Postlethwait, J. H., and Schneiderman, H. A. 1973. Developmental genetics of *Drosophila* imaginal discs. *Annu. Rev. Genet.* 7:381–433.

Reeves, R. 1984. Transcriptionally active chromatin. *Biochim. Biophys. Acta* 782:343–393.

Rogers, J. O., Early, H., Carter, C., Calame, K., Bond, M., Hood, L., and Wall, R. 1980. Two mRNAs with different 3' ends encode membrane-bound and secreted forms of immunoglobin chain. *Cell* 20:303–312.

Scott, M. P., and Carroll, S. B. 1987. The segmentation and homeotic gene network in early *Drosophila* development. *Cell* 51:689–698.

Scott, M. P., Tamkun, J. W., and Hartzell III, G. W. 1989. The structure and function of the homeodomain. *Biochim. Biophys. Acta* 989:25–48.

Scott, M. P., Weiner, A. J., Hazelrigg, T. I., Polisky, B. A., Pirotta, V., Scalenghe, F., and Kaufman, T. C. 1983. The molecular organization of the *Antennapedia* locus of *Drosophila*. *Cell* 35:763–776.

Spelsberg, T. C., Littlefield, B. A., Seelke, R., Dani, G. M., Toyoda, H., Boyd-Leinen, P., Thrall, C., and Kon, O. I. 1983. Role of specific chromosomal proteins and DNA sequences in the nuclear binding sites for steroid receptors. *Recent Prog. Horm. Res.* 39:425–517.

Vesell, E. S. 1965. Genetic control of isozyme patterns in human tissues. *Prog. Med. Genet.* 4:128–175.

Wang, T. Y., Kostraba, N. C., and Newman, R. S. 1976. Selective transcription of DNA mediated by nonhistone proteins. *Prog. Nucleic Acid Res. Mol. Biol.* 19:447–462.

Weatherall, D., and Clegg, J. 1979. Recent developments in the molecular genetics of human hemoglobins. *Cell* 18:467–479.

Yamamoto, K. R. 1985. Steroid receptor regulated transcription of specific genes and gene networks. *Annu. Rev. Genet.* 19:209–252.

Yamamoto, K. R., and Alberts, B. M. 1976. Steroid receptors: Elements for modulation of eukaryotic transcription. *Annu. Rev. Biochem.* 45:721–746.

CHAPTER 18: GENE MUTATIONS

Ames, B. N., Durston, W. E., Yamasaki, E., and Lee, F. D. 1973. Carcinogens are mutagens: A simple test system combining liver homogenates for activation and bacteria for detection. *Proc. Natl. Acad. Sci. USA* 70:2281–2285.

———, and Gold, L. S. 1990. Too many rodent carcinogens: mitogenesis increases mutagenesis. *Science* 249:970–971.

Auerbach, C. 1976. *Mutation Research*. London: Chapman and Hall.

Boyce, R. P., and Howard-Flanders, P. 1964. Release of ultraviolet light-induced thymine dimers from DNA in *E. coli* K12. *Proc. Natl. Acad. Sci. USA* 51:293–300.

Cairns, J., Overbaugh, J., and Miller, S. 1988. The origin of mutants. *Science* 335:142–145.

Collins, A., Johnson, R. T., and Boyle, J. M., eds. 1987. Molecular biology of DNA repair. Supplement 6, *J. Cell Sci.*

Devoret, R. 1979. Bacterial tests for potential carcinogens. *Sci. Am.* 241:40–49.

Drake, J. W. 1970. *The Molecular Basis of Mutation.* San Francisco: Holden-Day.

Haseltine, W. A. 1983. Ultraviolet light repair and mutagenesis revisited. *Cell* 33:13–17.

Lederberg, J., and Lederberg, E. M. 1952. Replica plating and indirect selection of bacterial mutants. *J. Bacteriol.* 63:399–406.

Moat, A. G., Peters, N., and Srb, A. M. 1959. Selection and isolation of auxotrophic yeast mutants with the aid of antibiotics. *J. Bacteriol.* 77:673–681.

Modrich, P. 1987. DNA mismatch correction. *Annu. Rev. Biochem.* 56:435–466.

Radman, M., and Wagner, R. 1986. Mismatch repair in *Escherichia coli. Annu. Rev. Genet.* 20:523–538.

Setlow, R. B., and Carrier, W. L. 1964. The disappearance of thymine dimers from DNA: An error-correcting mechanism. *Proc. Natl. Acad. Sci. USA* 51:226–231.

Walker, G. C., Marsh, L., and Dodson, L. A. 1985. Genetic analyses of DNA repair: inference and extrapolation. *Annu. Rev. Genet.* 19:103–126.

Woodward, V. W., DeZeeuw, J. R., and Srb, A. M. 1954. The separation and isolation of particular biochemical mutants of *Neurospora* by differential germination of conidia, followed by filtration and selective plating. *Proc. Natl. Acad. Sci. USA* 40:192–200.

CHAPTER 19: CHROMOSOMAL ABERRATIONS

Auerbach, C. 1976. *Mutation research.* London: Chapman and Hall.

Barr, M. L. and Bertram, E. G. 1949. A morphological distinction between neurones of the male and female, and the behavior of the nucleolar satellite during accelerated nucleoprotein synthesis. *Nature* 163:676–677.

Bloom, A. D. 1972. Induced chromosome aberrations in man. *Adv. Hum. Genet.* 3:99–153.

Borst, P., and Greaves, D. R. 1987. Programmed gene rearrangements altering gene expression. *Science* 235:658–667.

Dalla-Favera, R., Martinotti, S., Gallo, R., Erickson, J., and Croce, C. 1983. Translocation and rearrangements of the *c-myc* oncogene locus in human undifferentiated B-cell lymphomas. *Science* 219:963–997.

DeKlein, A., van Kessel, A. G., Grosveld, G., Bartram, C. R., Hagemeijer, A., Bootsma, D., Spurr, N. K., Heisterkamp, N., Groffen, J., and Stephenson, J. R. 1982. A cellular oncogene is translocated to the Philadelphia chromosome in chronic myelocytic leukemia. *Nature* 300:765–767.

Herskowitz, I. 1985. Master regulatory loci in yeast and lambda. *Cold Spring Harbor Symp. Quant. Biol.* 50:565–574.

Lyon, M. F. 1961. Gene action in the X-chromosomes of the mouse *(Mus. musculus L).* *Nature* 190:372–373.

Nasmyth, K. A. 1982. Molecular genetics of yeast mating type. *Annu. Rev. Genet.* 16:439–500.

Penrose, L. S., and Smith, G. F. 1966. *Down's anomaly.* Boston: Little, Brown.

Rowley, J. D. 1973. A new consistent chromosomal abnormality in chronic myelogenous leukemia identified by quinacrine fluorescence and Giemsa staining. *Nature* 243:290–293.

Shaw, M. W. 1962. Familial mongolism. *Cytogenetics* 1:141–179.

CHAPTER 20: TRANSPOSABLE GENETIC ELEMENTS, TUMOR VIRUSES, AND ONCOGENES

Adams, S. E., Mellor, J., Gull, K., Sim, R. B., Tuite, M. F., Kingman, S. M., and Kingsman, A. J. 1987. The functions and relationships of Ty-VLP proteins in yeast reflect those of mammalian retroviral proteins. *Cell* 49:111–119.

Astrin, S. M., and Rothberg, P. G. 1983. Oncogenes and cancer. *Cancer Investigation* 1:355–364.

Baltimore, D. 1985. Retroviruses and retrotransposons: The role of reverse transcription in shaping the eukaryotic genome. *Cell* 40:481–482.

Bishop, J. M. 1983. Cancer genes come of age. *Cell* 32:1018–1020.

———. 1983. Cellular oncogenes and retroviruses. *Annu. Rev. Biochem.* 52:301–354.

Boeke, J. D., Garfinkel, D. J., Styles, C. A., and Fink, G. R. 1985. Ty elements transpose through an RNA intermediate. *Cell* 40:491–500.

Bukhari, A. I., Shapiro, J. A., and Adhya, S. L., eds. 1977. *DNA insertion elements, plasmids and episomes.* Cold Spring Harbor Laboratory, Cold Spring Harbor, NY.

Busch, H. 1984. *Onc* genes and other new targets for cancer chemotherapy. *J. Cancer Res. Clin. Oncol.* 107:1–14.

Calos, M. P., and Miller, J. H. 1980. Transposable elements. *Cell* 20:579–595.

Cohen, S. N., and Shapiro, J. A. 1980. Transposable genetic elements. *Sci. Am.* 242:40–49.

Cold Spring Harbor Symposia for Quantitative Biology. 1980. Vol. 45. *Movable genetic elements.* Cold Spring Harbor Laboratory, Cold Spring Harbor, NY.

DeVita, V., Hellman, S., and Rosenberg, S., eds. 1986. *Important advances in oncology.* Philadelphia: Lippincott.

Engels, W. R. 1983. The P family of transposable elements in *Drosophila. Annu. Rev. Genet.* 17:315–344.

Federoff, N. V. 1989. About maize transposable elements and development. *Cell* 56:181–191.

———. 1983. Controlling elements in maize, in *Mobile genetic elements,* edited by J. Shapiro, pp. 1–63. New York: Academic Press.

Finnegan, D. J. 1985. Transposable elements in eukaryotes. *Int. Rev. Cytol.* 93:281–326.

Foster, T. J., Davis, M. A., Roberts, D. E., Takashita, K., and Kleckner, N. 1981. Genetic organization of transposon Tn10. *Cell* 23:201–213.

Garfinkel, D. J., Boeke, J. D., and Fink, G. R. 1985. Ty element transposition: Reverse transcriptase and virus-like particles. *Cell* 42:507–517.

Gordon, H. 1985. Oncogenes. *Mayo Clin. Proc.* 60:697–713.

Kingston, R. E., Baldwin, A. S., and Sharp, P. A. 1985. Transcription control by oncogenes. *Cell* 41:3–5.

Kleckner, N. 1981. Transposable elements in prokaryotes. *Annu. Rev. Genet.* 15:341–404.

Krontiris, T. G. 1983. The emerging genetics of human cancer. *N. Engl. J. Med.* 309:404–409.

McClintock, B. 1961. Some parallels between gene control systems in maize and in bacteria. *Am. Naturalist* 95:265–277.

———. 1965. The control of gene action in maize. *Brookhaven Symp. Biol.* 18:162 ff.

Mount, S. M., and Rubin, G. M. 1985. Complete nucleotide sequence of the *Drosophila* transposable element copia: Homology between copia and retroviral proteins. *Mol. Cell. Biol.* 5:1630–1638.

Ratner, L., Gallo, R. C., and Wong-Staal, F., 1985. Cloning of human oncogenes. In *Recombinant DNA research and virus,* edited by Y. Becker. Boston: Martinus Nijhoff.

———, Josephs, S. F., and Wong-Staal, F. 1985. Oncogenes: Their role in neoplastic transformation. *Annu. Rev. Microbiol.* 39:419–449.

Shapiro, J. A. ed. 1983. *Mobile genetic elements.* New York: Academic Press.

Slamon, D. J., deKernion, J. B., Verma, I. M., and Cline, M. J. 1984. Expression of cellular oncogenes in human malignancies. *Science* 224:256–262.

Spradling, A. C., and Rubin, G. M. 1981. *Drosophila* genome organization: Conserved and dynamic aspects. *Annu. Rev. Genet.* 15:219–264.

Varmus, H. E. 1982. Form and function of retroviral proviruses. *Science* 216:812–821.

Weiss, R. A., Teich, N., Varmus, H. E., and Coffin, J. M., eds. 1985. *Molecular Biology of Tumor Viruses: RNA Tumor Viruses,* 2nd ed. (2 vols.). Cold Spring Harbor Laboratory, Cold Spring Harbor, NY.

CHAPTER 21: ORGANIZATION AND GENETICS OF EXTRANUCLEAR GENOMES

Ashwell, M., and Work, T. S. 1970. The biogenesis of mitochondria. *Annu. Rev. Biochem.* 39:251–290.

Birky, C. W. 1978. Transmission genetics of mitochondria and chloroplasts. *Annu. Rev. Genet.* 12:471–512.

Boardman, N. K., Linnane, A. W., and Smillie, R. M., eds. 1971. *Autonomy and biogenesis of mitochondria and chloroplasts.* Amsterdam: North-Holland.

Borst, P., and Grivell, L. A. 1978. The mitochondrial genome of yeast. *Cell* 15:705–723.

Clayton, D. A. 1982. Replication of animal mitochondrial DNA. *Cell* 28:693–705.

Collins, R. A., and Bertrand, H. 1978. Nuclear suppressors of the *[poky]* cytoplasmic mutant in *Neurospora crassa. Mol. Gen. Genet.* 161:267–273.

Ephrussi, B. 1953. *Nucleo-cytoplasmic relations in microorganisms.* New York: Oxford University Press.

Gillham, N. W. 1978. *Organelle heredity.* New York: Raven Press.

Kirk, J. T. O. 1971. Chloroplast structure and biogenesis. *Annu. Rev. Biochem.* 40:161–196.

Kuroiwa, T., Suzuki, T., Ogawa, K., and Kawano, S. 1981. The chloroplast nucleus: Distribution, number, size, and shape, and a model for the multiplication of the chloroplast genome during chloroplast development. *Plant Cell Physiol.* 22:381–396.

Lambowitz, A. M., and Luck, D. J. L. 1976. Studies on the *poky* mutant of *Neurospora crassa. J. Biol. Chem.* 251:3081–3095.

Levings, C.S. 1983. The plant mitochondrial genome and its mutants. *Cell* 32:659–661.

Linnane, A. W., and Nagley, P. 1978. Mitochondrial genetics in perspective: The derivation of a genetic and physical map of the yeast mitochondrial genome. *Plasmid* 1:324–345.

Rochaix, J. D. 1978. Restriction endonuclease map of the chloroplast DNA of *Chlamydomonas reinhardi. J. Mol. Biol.* 126:597–617.

Rush, M. G., and Misra, R. 1985. Extrachromosomal DNA in eukaryotes. *Plasmid* 14:177–191.

Sager, R. 1972. *Cytoplasmic genes and organelles.* New York: Academic Press.

Schmidt, R. J., Richardson, C. B., Gillham, N. W., and Boynton, J. E. 1983. Sites of synthesis of chloroplast ribosomal proteins in *Chlamydomonas. J. Cell Biol.* 96:1451–1463.

Van Winkle-Swift, K. P., and Birky, C. W. 1978. The non-reciprocality of organelle gene recombination in *Chlamydomonas reinhardi* and *Saccharomyces cerevisiae. Mol. Gen. Genet.* 166:193–209.

CHAPTER 22: QUANTITATIVE GENETICS

Blumer, M. G. 1980. *The mathematical theory of quantitative genetics.* Oxford: Clarendon Press.

East, E. M. 1910. A Mendelian interpretation of variation that is apparently continuous. *Am. Naturalist* 44:65–82.

———. 1916. Studies on size inheritance in *Nicotiana. Genetics* 1:164–176.

———, and Jones, D. F. 1919. *Inbreeding and outbreeding.* Philadelphia: Lippincott.

Falconer, D. S. 1989. *Introduction to quantitative genetics.* New York: Wiley.

Hill, W. G., ed. 1984. *Quantitative genetics,* Parts I and II. New York: Van Nostrand Reinhold Company, Inc.

Lander, E.S., and Botstein, D. 1989. Mapping Mendelian factors underlying quantitative traits using RFLP linkage maps. *Genetics* 121:185–199.

Mather, K. 1943. Polygenic inheritance and natural selection. *Biol. Rev.* 18:32–64.

Nilsson-Ehle, H. 1909. Kreuzungsuntersuchungen an Hafer und Weizen. *Lunds Univ. Aarskr. N.F. Atd.,* Ser. 2, 5 (2):1–122.

Paterson, A. H., Lander, E. S., Hewitt, J. D., Person, S., Lincoln, S. E., and Tanksley, S. D. 1988. Resolution of quantitative traits into Mendelian factors by using a complete RFLP linkage map. *Nature* 335:721–726.

Selander, R. K., and Kaufman, D. W. 1975. Self fertilization and genetic population structure in a colonizing land snail. *Proc. Natl. Acad. Sci. USA* 70:1186–1190.

Sokal, R. R., and Rohlf, F. J. 1981. *Biometry,* 2nd ed. San Francisco: W. H. Freeman.

———, and Thoday, J. M., eds. 1979. *Quantitative genetic variation.* New York: Academic Press.

Thompson, J. N., Jr., and Thompson, J. M., eds. 1979. *Quantitative genetic variation.* New York:Academic Press.

Weir, B. S., Eisen, E. J., Goodman, M. M., and Namkoong, G., eds. 1988. *Proceedings of the second international conference on quantitative genetics.* Sunderland, MA: Sinauer.

CHAPTER 23: POPULATION GENETICS

Avise, J. C. 1986. Mitochondrial DNA and the evolutionary genetics of higher animals. *Phil. Trans. Roy. Soc. London,* Ser. B 321:325–342.

Avise, J., Giblin-Davidson, C. C., Laerm, J., Patton, J. C., and Lansman, R. A. 1979. Mitochondrial DNA clones and matriarchal phylogeny within and among geographic populations of the pocket gopher *Geomys pinetis. Proc. Natl. Acad. Sci. USA* 76:6694–6698.

Ayala, F. J., ed. 1976. *Molecular evolution.* Sunderland, MA:Sinauer.

Buri, P. 1956. Gene frequency in small populations of mutant *Drosophila. Evolution* 10:367–402.

Clarke, C. A., and Sheppard, P. M. 1966. A local survey of the distribution of industrial melanic forms in the moth *Biston betularia* and estimates of the selective values of these in an industrial environment. *Proc. R. Soc. Lond. [Biol.]* 165:424–439.

Crow, J. F. 1986. *Basic concepts in population, quantitative, and evolutionary genetics.* New York: W. H. Freeman.

Darwin, C. 1860. *On the origin of species by means of natural selection, or the preservation of favoured races in the struggle for life.* New York: Appleton.

Dobzhansky, T. 1951. *Genetics and the origin of species,* 3rd ed. New York: Columbia University Press.

Fisher, R. A. 1930. *The genetical theory of natural selection.* Oxford: Clarendon Press.

Ford, E. B. 1971. *Ecological genetics,* 3rd ed. London: Chapman and Hall.

Gray, M. W. 1989. Origin and evolution of mitochondrial DNA. *Annal. Rev. Cell Biol.* 5:25–50.

Hardy, G. H. 1908. Mendelian proportions in a mixed population. *Science* 28:49–50.

Hartl, D. L., and Clark, A. G. 1988. *Principles of population genetics,* 2nd ed. Sunderland, MA: Sinauer.

Hedrick, P. H. 1983. *Genetics of populations.* Boston: Science Books International.

Hillis, D. M. and Moritz, C. 1990. *Molecular systematics.* Sunderland, MA: Sinauer.

———, Moritz, C., Porter, C. A., and Baker, R. J. 1991. Evidence for biased gene conversion in concerted evolution of ribosomal DNA. *Science* 251:308–309.

Kettlewell, H. B. D. 1961. The phenomenon of industrial melanism in the Lepidoptera. *Annu. Rev. Entomol.* 6:245–262.

Koehn, R., and Hilbish, T. J. 1987. The adaptive importance of genetic variation. *Amer. Sci.* 75:134–141.

Kreitman, M. 1983. Nucleotide polymorphism at the alcohol dehydrogenase locus of *Drosophila melanogaster. Nature* 304:412–417.

Lewontin, R. C. 1974. *The genetic basis of evolutionary change.* New York: Columbia University Press.

————, Moore, J. A., Provine, W. B., and Wallace, B. 1981. *Dobzhansky's genetics of natural populations I-XLIII*. New York: Columbia University Press.

————. 1985. Population genetics. *Annu. Rev. Genet.* 19:81–102.

Li, W-H. 1991. *Fundamentals of molecular evolution*. Sunderland, MA: Sinauer.

————, Luo, C. C., and Wu, C. I. 1985. Evolution of DNA sequences, pp. 1–94. *Molecular evolutionary genetics*, edited by R. J. MacIntyre. New York: Plenum.

MacIntyre, R. J., ed. 1985. *Molecular evolutionary genetics*. New York: Plenum.

Maniatis, T., Fritsch, E. F., Lauer, L., and Lawn, R. M. 1980. The molecular genetics of human hemoglobin. *Annal. Rev. Genet.* 14:145–178.

Maynard Smith, J. 1989. Evolutionary genetics. Oxford: Oxford University Press.

Nei, M. 1987. *Molecular evolutionary genetics*. New York: Columbia University Press.

Nei, M., and Koehn, R. K. 1983. *Evolution of genes and proteins*. Sunderland, MA: Sinauer.

Pace, N. R., Olsen, G. J., and Woese, C. R. 1986. Ribosomal RNA phylogeny and the primary lines of evolutionary descent. *Cell* 45:325–326.

Speiss, E. 1977. *Genes in populations*. New York: Wiley.

Stringer, C. B. 1990. The emergence of modern humans. *Sci Am.* (December):98–104.

Stringer, C. B., and Andrews, P. 1988. Genetic and fossil evidence for the origin of modern humans. *Science* 239:1263–1268.

Wallace, B. 1981. *Basic population genetics*. New York: Columbia University Press.

Woese, C. R. 1981. Archaebacteria. *Sci. Am.* 244:98–122.

Solutions to Selected Questions and Problems

Chapter 1

1.1 c

1.3 c

1.5 a. Yes, if a sexual mating system exists in that species. In that case, two haploid cells can fuse to produce a diploid cell, which can then go through meiosis to produce haploid progeny. The fungi *Neurospora crassa* and *Saccharomyces cerevisiae* exemplify this positioning of meiosis in the life cycle.

  b. No, because a diploid cell cannot be formed and meiosis only occurs starting with a diploid cell.

1.7 c

1.9 a. metaphase

  b. anaphase

1.13 a. $\frac{1}{8}$. The probability of a gamete receiving A is $\frac{1}{2}$, because the choice is A or A'. Similarly, the probability of receiving B is $\frac{1}{2}$ and of receiving C is $\frac{1}{2}$. Therefore, the probability of receiving A, B, and C is $\frac{1}{2} \times \frac{1}{2} \times \frac{1}{2} = \frac{1}{8}$ (like the probability of tossing three heads in a row, the individual probabilities are multiplied together).

  b. $\frac{6}{8}$ ($= \frac{3}{4}$). The probability of gametes containing chromosomes from both maternal and paternal origin is 1 minus the probability that the gametes will contain either all maternal or all paternal chromosomes. The probability of a gamete containing all maternal chromosomes is $\frac{1}{8}$ (part a) and that of a gamete containing all paternal chromosomes is also $\frac{1}{8}$. Therefore the answer is $1 - \frac{1}{8} - \frac{1}{8} = \frac{6}{8} = \frac{3}{4}$.

1.14 One of the long chromosomes and the short chromosome might be members of a heteromorphic pair—i.e., X- and Y-chromosomes, respectively.

1.16 False. Owing to the randomness of independent assortment and to crossing-over, both of which characterize meiosis, the probability of any two sperm cells being genetically identical is extremely remote.

Chapter 2

2.1 a. red

  b. 3 red, 1 yellow

  c. all red

  d. $\frac{1}{2}$ red, $\frac{1}{2}$ yellow

2.4 The F_2 genotypic ratio (if C is colored, c is colorless) is $\frac{1}{4}$ *CC* : $\frac{1}{2}$ *Cc* : $\frac{1}{4}$ *cc*. If we consider just the colored plants, there is a 1:2 ratio of *CC* homozygotes to *Cc* heterozygotes. Therefore, if a colored plant is picked at random, the probability that it is *CC* is $\frac{1}{3}$ (i.e., $\frac{1}{3}$ of the *colored* plants are homozygous) and the probability that it is *Cc* is $\frac{2}{3}$. Only if a *Cc* plant is selfed will more than one phenotypic class be found among its progeny; therefore, the answer is $\frac{2}{3}$.

2.5 a. Parents are *Rr* (rough) and *rr* (smooth); F_1 are *Rr* (rough) and *rr* (smooth).

  b. $Rr \times Rr \rightarrow \frac{3}{4}$ rough, $\frac{1}{4}$ smooth

2.7 Progeny ratio approximates 3:1, and so the parent is heterozygous. Of the dominant progeny there is a 1:2 ratio of homozygous to heterozygous, so $\frac{1}{3}$ will breed true.

2.8 Black is dominant to brown. If *B* is the allele for black and *b* for brown, then female X is *Bb* and female Y is *BB*. The male is *bb*.

2.10 a. F_1 intercross. If he crosses F_1 black babbits, the following matings could occur:

$$BB \times BB$$
$$Bb \times BB$$
$$BB \times Bb$$
$$Bb \times Bb$$

The only mating that can produce white progeny is the last one. The probability that any given black babbit is a heterozygote is

$$\frac{2Bb}{BB + 2\,Bb} = \frac{2}{3}$$

The probability that both F_1 parents are heterozygotes is $2/3 \times 2/3$.

And the probability that this mating produces a white offspring = 1/4.

The probability that in all the random F_1 matings, a white offspring is produced is $2/3 \times 2/3 \times 1/4 = 4/36 = 1/9$.

So 1/9 of the $F_1 \times F_1$ progeny will be white.

  b. The only way that he will produce white offspring is if he picks a *Bb* F_1. The probability that an F_1 black male is *Bb* = 2/3. But we know that the parental female is *Bb*. Therefore the probability that $F_1 \times P$ mating will produce a white babbit is $2/3 \times 1 \times 1/4 = 1/6$.

  c. His best strategy would be to backcross the F_1 *bb* male to the P *Bb* females.

  $Bb \times bb$ cross $\longrightarrow$ 50% of progeny would be *Bb*
  50% of progeny would be white (*bb*)

2.13 a. *WW Dd $\times$ ww dd*

  b. *Ww dd $\times$ Ww dd*

  c. *ww DD $\times$ WW dd*

  d. *Ww Dd $\times$ ww dd*

  e. *Ww Dd $\times$ Ww dd*

2.14 The cross is *Aa Bb Cc $\times$ Aa Bb Cc*.

  a. Considering the A gene alone, the probability of an offspring showing the A trait from *Aa $\times$ Aa* is $\frac{3}{4}$. Similarly the probability of showing the B trait from *Bb $\times$ Bb* is $\frac{3}{4}$ and the C trait from *Cc $\times$ Cc* is $\frac{3}{4}$. Therefore, the probability of a given progeny being phenotypically ABC (using the product rule) is $\frac{3}{4} \times \frac{3}{4} \times \frac{3}{4} = \frac{27}{64}$.

  b. Considering the A gene, the probability of an *AA* offspring from *Aa $\times$ Aa* is $\frac{1}{4}$. The same probability is the case for a *BB* offspring and for a *CC* offspring. Therefore, the probability of an *AA BB CC* offspring (from the product rule) is $\frac{1}{4} \times \frac{1}{4} \times \frac{1}{4} = \frac{1}{64}$.

2.17 a. The F_1 has the genotype *Aa Bb* Cc^h and is all agouti, black. The F_2 is 27/64 agouti, black; 9/64 agouti, black, Himalayan; 9/64 agouti, brown; 9/64 black; 3/64 agouti, brown, Himalayan; 3/64 black, Himalayan; 3/64 brown; 1/64 brown, Himalayan.

b. 27/64 of the F_2 are agouti, black. From the $F_1 \times F_1$ cross *Aa Bb* Cc^h × *Aa Bb* Cc^h, the proportion of F_2 mice with the genotype *Aa BB* Cc^h = ½ × ¼ × ½ = 1/16 = 4/64. Therefore, the proportion of F_2 agouti, blacks that are *Aa BB* Cc^h = 4/27.

c. F_2 Himalayans = 9/64 (agouti, black, Himalayan) + 3/64 (agouti, brown, Himalayan) + 3/64 (black, Himalayan) + 1/64 (brown, Himalayan) = 16/64. There are 3/64 + 1/64 = 4/64 brown, Himalayans; thus, the proportion of F_2 Himalayans that are brown = 4/16 = ¼

d. F_2 agoutis = 27/64 (agouti, black) + 9/64 (agouti, black, Himalayan) + 9/64 (agouti, brown) + 3/64 (agouti, brown, Himalayan) = 48/64. There are 27/64 + 9/64 = 36/64 F_2 black, agoutis. Therefore, the proportion of F_2 agoutis that are black = 36/48 = ¾

CHAPTER 3

3.1 The probability of a given homolog going to one particular pole is ½. The probability of all five paternal chromosomes going to the same pole is $(½)^5$ = 1/32. The same answer applies for all maternal chromosomes going to one pole.

3.2

	Mother's Mother	Mother's Father	Father's Mother	Father's Father
X chromosome	Yes	Yes	No	No
Y chromosome	No	No	No	Yes

3.4 The woman is heterozygous for the recessive color blindness allele, let us say c^+/c. The man is not colorblind and thus has the genotype c^+/Y. (It does not matter what phenotype his mother and father have because males are hemizygous for sex-linked genes; thus the genotype can be assigned directly from the phenotype.) All the female offspring will have normal color vision; half the male offspring will be colorblind, and half will have normal color vision.

3.6 If *c* is the red-green color-blindness allele, and *a* is the albino allele, the parent's genotypes are c^+c^+ *aa* ♀ and *cY* a^+a^+ ♂. All children will be normal visioned (c^+c^+ ♀ and c^+Y ♂) and normally pigmented (a^+a, both sexes).

3.7 The parentals are *ww* vg^+ vg^+ ♀ and w^+Y *vg vg*♂.

a. The F_1 males are all *w*Y vg^+ *vg*, white eyes, long wings. The F_1 females are all w^+w vg^+ *vg*, red eyes, long wings.

b. The F_2 females are ⅜ red, long, females; ⅜ white, long, females; ⅛ red, vestigial, females; ⅛ white, vestigial, females; the same ratios of the respective phenotypes apply for the males.

c. The cross of F_1 male with the parental female is *w*Y vg^+ *vg* × *ww* vg^+ vg^+. All progeny have white eyes and long wings. The cross of an F_1 female with the parental male is w^+w vg^+ *vg* × w^+Y *vg vg*. Female progeny: All have red eyes, half have long wings and half have vestigial wings. Male progeny: ¼ red, long; ¼ red, vestigial; ¼ white, long; ¼ white, vestigial.

3.10 The simplest hypothesis is that brown-colored teeth are determined by a sex-linked dominant mutant allele. Man A was *B*Y and his wife was *bb*. All sons will be *b*Y normals and cannot pass on the trait. All the daughters receive the X chromosome from their father, and so they are *Bb* with brown enamel. Half their sons will have brown teeth because half their sons receive the X chromosome with the *B* mutant allele.

3.12 a. To have produced a child with cystic fibrosis, both parents must have been heterozygous for the autosomal recessive mutant gene. Therefore the probability of their next child having cystic fibrosis is ¼ (i.e., ¼ of the progeny from *Aa* × *Aa* will be *aa*).

b. Non-affected children can be either *AA* or *Aa*. From a cross *Aa* × *Aa*, there will be a progeny ratio of 1 *AA*:2 *Aa*:1 *aa*. Therefore, ⅔ of the non-affected children will be *Aa* heterozygotes.

3.15 The answer is given in the following Punnett squares.

ww × w^+Y

Eggs \ Sperm	w^+	Y
ww	www^+	*ww*Y
O	w^+O	YO

Survivors are *ww* Y (white ♀) and w^+O (sterile, red ♂)

Backcross *ww* Y × w^+ Y

Eggs		Sperm w^+	Sperm Y
Normal	*w* Y	w^+ *w*Y Red ♀	*w* YY White ♂
Normal	*w*	w^+ *w* Red ♀	*w* Y White ♂
Secondary nondisjunction	*ww*	w^+ *ww* Triplo-X Usually dies	*ww* Y White ♀
Secondary nondisjunction	Y	w^+ Y Red ♂	Y Y Dies

3.19

	PEDIGREE A	PEDIGREE B	PEDIGREE C
Autosomal recessive	Yes	Yes	Yes
Autosomal dominant	Yes	Yes	No
X-linked recessive	Yes	Yes	No
X-linked dominant	No	No	No

3.21 a. X-linked recessive can be excluded, for example, because individual I.2 would be expected to pass on the trait to *all* sons and that is not the case. Also, the II.1, II.2 pairing would be expected to produce offspring all of whom express the trait.

Autosomal recessive can also be excluded because the II.1, II.2 pairing would be expected to produce offspring all of whom express the trait.

b. The remaining two mechanisms of inheritance are X-linked dominant and autosomal dominant. Genotypes can be written to satisfy both mechanisms of inheritance. X-linked dominance is perhaps more likely, for the II.6, II.7 pairing shows exclusively father-daughter inheritance; that is, all the daughters and none of the sons exhibit the trait: this is a characteristic of the segregation of X-linked dominant alleles. If the trait were autosomal dominant, then half the sons and half the daughters would be expected to exhibit the trait.

3.24 a is false: the father passes on his X to his daughters so *none* of the sons will be affected.

b is false: since the disease is rare, the mother would be heterozygous and only 50% of her daughters would receive the X with the dominant mutant allele.

c is true: all daughters receive the father's X chromosome.

d is false: only 50% of her sons would receive the X with the dominant mutant allele.

3.26 a is true: that is, $aa \times aa$ produces only aa progeny.

b is false: because the disease is autosomal, the sex of the progeny makes no difference in the fact that all progeny will be affected if both parents are affected.

c is false—a child must receive one autosomal recessive allele from each parent to have the disease, but neither the parents nor the grandparents need have had the disease; all could theoretically be carriers.

d is false—if the unaffected parent is a carrier, then affected progeny can result.

3.28 Because hemophilia is an X-linked trait, the most likely explanation is that random inactivation of X chromosomes (lyonization, see p. 78) produces individuals with different proportions of cells with the normal allele. Thus, if some women had only 40% of their cells with an active h^+ allele, and 60% with the h (hemophilia) allele, but other women had 60% of their cells with an active h^+ allele, and 40% with the h allele, there would be a significant difference in the amount of clotting factor these two individuals would make.

3.29 a. Females are XY, and males are XX. (The data also fit the model that males are homozygous for a recessive sex-determining allele while females are heterozygous for that and for a dominant, female-determining allele.) In the mating of estrogen-raised females to normal males, half of the matings were the usual XY ♀ × XX ♂, and gave the expected 50% XX and 50% XY offspring. The other half of matings were XX sex reversed (♀) × normal XX (♂), and could produce only XX (♂) offspring. In the matings of androgen-raised males to normal females, half of the matings were the usual XX × XY, but the other half were XY sex-reversed (♂) × XY (♀). These yielded 25% XX (♂), 50% XY (♀) and 25% YY (lethal) = 1 XX ♂ : 2 XY ♀.

b. There would be two kinds. Two-thirds of the matings would be XX × XY and yield 1 female: 1 male. One-third of the matings would be XX × YY and would yield all XY (♀) offspring.

CHAPTER 4

4.2 6 possible genotypes: w/w, $w/w1$, $w/w2$, $w1/w1$, $w1/w2$, $w2/w2$.

4.6 The woman's genotype is I^A/I^B and the man's genotype is I^A/I^O.

a. $\frac{1}{2} \times \frac{1}{2} = \frac{1}{4}$.

b. Zero. A blood group O baby is not possible.

c. $\frac{1}{2}$ (probability of male) × $\frac{1}{4}$ (probability of AB) × $\frac{1}{2}$ (probability of male) × $\frac{1}{4}$ (probability of B) = $\frac{1}{64}$ probability that all four conditions will be fulfilled.

4.8 Blood type O, since genotype is I^O/I^O.

4.10 Half will be R/r and hence will resemble the parents.

4.12 a. The cross is F/F o/o × f/f O/O. The F_1 is F/f O/o, which is fuzzy with round leaf glands.

b. Interbreeding the F_1 gives $\frac{3}{16}$ fuzzy, oval-glanded; $\frac{6}{16}$ fuzzy, round-glanded; $\frac{3}{16}$ fuzzy, no-glanded; $\frac{1}{16}$ smooth, oval-glanded; $\frac{2}{16}$ smooth, round-glanded; $\frac{1}{16}$ smooth, no-glanded.

c. Cross is F/f O/o × f/f O/O. The progeny are $\frac{1}{4}$ fuzzy, oval-glanded; $\frac{1}{4}$ fuzzy, round-glanded; $\frac{1}{4}$ smooth, oval-glanded; $\frac{1}{4}$ smooth, round-glanded.

4.18 To show no segregation among the progeny, the chosen plant must be homozygous. The genotypes comprising the $\frac{9}{16}$ col-

ored plants are: 1 A/A B/B : 2 A/a B/B : 2 A/A B/b : 4 A/a B/b. Only one of these genotypes is homozygous; the answer is $\frac{1}{9}$.

4.20 a. If Y governs yellow and y governs agouti, then Y/Y are lethal, Y/y are yellow, and y/y are agouti. Let C determine colored coat and c determine albino. The parental genotypes, then, are Y/y C/c (yellow) and Y/y c/c (white).

b. The proportion is 2 yellow:1 agouti:1 albino. None of the yellows breed true because they are all heterozygous, with homozygous YY individuals being lethal.

4.22 a. Y/Y R/R (crimson) × y/y r/r (white) gives Y/y R/r F_1 plants, which have magenta-rose flowers. Selfing the F_1 gives an F_2 as follows: $\frac{1}{16}$ crimson (Y/Y R/R), $\frac{2}{16}$ orange-red (Y/Y R/r), $\frac{1}{16}$ yellow (Y/Y r/r), $\frac{2}{16}$ magenta (Y/y R/R), $\frac{4}{16}$ magenta-rose (Y/y R/r), $\frac{2}{16}$ pale yellow (Y/y r/r), and $\frac{4}{16}$ white (y/y R/R, y/y R/r, and y/y r/r). Progeny of the F_1 backcrossed to the crimson parent are $\frac{1}{4}$ crimson, $\frac{1}{4}$ orange-red, $\frac{1}{4}$ magenta, and $\frac{1}{4}$ magenta-rose.

b. Y/Y R/r × Y/y r/r gives $\frac{1}{4}$ orange-red (Y/Y R/r), $\frac{1}{4}$ yellow (Y/Y r/r), $\frac{1}{4}$ magenta-rose (Y/r R/r), $\frac{1}{4}$ pale yellow (Y/y r/r).

c. Y/Y r/r (yellow) × y/y R/r (white) gives $\frac{1}{2}$ magenta-rose (Y/y R/r) and $\frac{1}{2}$ pale yellow (Y/y r/r).

4.24 a. The simplest approach is to calculate the proportion of progeny that will be black and then subtract that answer from 1. The black progeny have the genotype $A/-$ $B/-$ $C/-$, and the proportion of these progeny is $(\frac{3}{4})^3$. Therefore the proportion of colorless progeny is $1 - (\frac{3}{4})^3 = 1 - \frac{27}{64} = \frac{37}{64}$.

b. Black is produced only when c/c $A/-$ $B/-$ results, and this offspring occurs with the frequency $(\frac{1}{4})(\frac{3}{4})(\frac{3}{4}) = \frac{9}{64}$ black, which gives $\frac{55}{64}$ colorless.

4.26 In males H/H and H/h are horned, and h/h is hornless; in females H/H is horned, and H/h and h/h are hornless. The cross is a H/H W/W male × h/h w/w female. The F_1 is H/h W/w, which gives horned white males and hornless white females. Interbreeding the F_1 gives the following F_2:

	MALE	FEMALE
$\frac{3}{16}$ H/H $W/-$	horned, white	horned, white
$\frac{6}{16}$ H/h $W/-$	horned, white	hornless, white
$\frac{3}{16}$ h/h $W/-$	hornless, white	hornless, white
$\frac{1}{16}$ H/H w/w	horned, black	horned, black
$\frac{2}{16}$ H/h w/w	horned, black	hornless, black
$\frac{1}{16}$ h/h w/w	hornless, black	hornless, black

In sum, the ratios are $\frac{9}{16}$ horned white: $\frac{3}{16}$ hornless white: $\frac{3}{16}$ horned black: $\frac{1}{16}$ hornless black males and $\frac{3}{16}$ horned white: $\frac{9}{16}$ hornless white: $\frac{1}{16}$ horned black: $\frac{3}{16}$ hornless black females.

4.28 Ewe A is H/h w/w; B is H/h W/w or h/h W/w; C is H/H w/w; D is H/h W/w; the ram is H/h W/w.

4.29 c

CHAPTER 5

5.3 From the χ^2 test, $\chi^2 = 16.10$; P is less than 0.01 at three degrees of freedom. This test reveals that the two genes do not fit a 1:1:1:1 ratio. It does not say why. Linkage might seem reasonable until it is realized that the minority classes are not reciprocal classes (both carry the aa phenotype). If the segregation at each locus is considered, however, the $B/-:b/b$ ratio is about 1:1 (203:197), but the $A/-:a/a$ ratio is not (240:160). The departure, then, specifically results from a deficiency of a/a individuals. This departure should be confirmed in other crosses that test the segregation at locus A. In corn, further evidence would be that the a/a deficiency might show up as a class of ungerminated seeds or of seedlings that die early.

5.6 There are 158 progeny rabbits. The recombinant classes are the English plus Angora and the non-English plus short-haired, and so the map distance between the genes is $(11 + 6)/158 \times 100\% = 10.8\% = 10.8$ mu.

5.8 You were lucky in that a double mutant hatched. Based on the presence of one double mutant, you can eliminate linkage between *vg* and *m*. This is because there is no crossing-over in the *Drosophila* male, and there is no way to produce a recombinant *vg m* gamete in the male F_1. The crosses performed here were

$$\text{P} \qquad \frac{vg^+}{vg^+}\ \frac{m}{m} \times \frac{vg}{vg}\ \frac{m}{m}$$

$$\text{All } F_1 \qquad \frac{vg^+}{vg}\ \frac{m}{m^+} \text{ (wild-type)}$$

The results of this cross told you that *m* is *not* X-linked.

Therefore the F_1 cross was a dihybrid cross and we expect a 9:3:3:1 F_2 ($vg^+ m^+$ (wild type): $vg m^+$: $vg^+ m$: $vg m$) ratio. In the small sample, by chance, a double mutant appeared. This tells us that *vg* and *m* are segregating independently. If *vg* and *m* had been on the same chromosome, the crosses performed would have been

$$\text{P} \qquad \frac{vg^+\ m}{vg^+\ m} \times \frac{vg^+\ m}{vg\ m^+}$$

$$F_1 \qquad \frac{vg^+\ m}{vg\ m^+} \text{ (All wild-type)}$$

$$F_1 \times F_1$$
$$\downarrow$$
$$(F_2 \text{ (phenotypes)})$$

		♂ gametes	
		$vg^+ m$	$vg\ m^+$
♀ gametes	$vg^+ m$	maroon	wild
	$vg\ m^+$	wild	vestigial
	$vg^+ m^+$	wild	wild
	$vg\ m$	maroon	vestigial

The phenotypic ratio would be 4:2:2:0 (wild:vestigial:maroon: vestigial-maroon).

There is no way to get a double mutant from this cross *regardless* of the distance between the two linked genes (0–100 percent recombination). So if the two genes were linked one would never find a double mutant progeny in the F_2. Since a double mutant did hatch, this was enough to allow the preliminary (but correct) conclusion that *m* was not on chromosome 2. While *m* could be on chromosome 3 or 4, this experiment does not distinguish between these two possibilities.

5.9

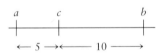

```
  c   e   a                              d         b
  |---|---|------------------------------|---------|
  ←7→←8→←------------ 38 --------------→←- 13 →
```

5.12 45% $a b^+$, 45% $a^+ b$, 5% $a^+ b^+$, 5% $a b$

5.13 a. $0.035 + 0.465 = 0.50$

b. All the daughters have $a^+ b^+$ phenotype.

5.16 a. The genotype of the F_1 is *A B*/*a b*. The gametes produced by the F_1 are 40% *A B*; 40% *a b*; 10% *A b*; 10% *a B*. From a testcross of the F_1 with *a b*/*a b* we get 40% *A B*/*a b*; 40% *a b*/*a b*; 10% *A b*/*a b*; 10% *a B*/*a b*.

b. The F_1 genotype is *A b*/*a B*. The gametes produced by the F_1 are 40% *A b*; 40% *a B*; 10% *A B*; 10% *a b*. From a testcross

of the F_1 with *a b*/*a b* we get 40% *A b*/*a b*; 40% *a B*/*a b*; 10% *A B*/*a b*; 10% *a b*/*a b*. This question illustrates that map distance is computed between sites in the chromosome, and it does not matter whether the genes are in coupling or in repulsion, the percentage of recombinants will be the same, even though the phenotypic classes constituting the recombinants differ.

5.18 Each chromosome pair segregates independently. We can compute the relative proportions of gametes produced for each homologous pair of chromosomes separately from the known map distances (P = parental; R = recombinant):

P	R	P	R	P	R
A B 0.4	A b 0.1	C D 0.45	C d 0.05	E F 0.35	E f 0.15
a b 0.4	a B 0.1	c d 0.45	c D 0.05	e f 0.35	e F 0.15

To answer the questions, simply multiply the probabilities of getting the particular gamete from the F_1 multiple heterozygote.

a. $A\,B\,C\,D\,E\,F = 0.4 \times 0.45 \times 0.35 = 0.063$ (6.3%)

b. $A\,B\,C\,d\,e\,f = 0.4 \times 0.05 \times 0.35 = 0.007$ (0.7%)

c. $A\,b\,c\,D\,E\,f = 0.1 \times 0.05 \times 0.15 = 0.00075$ (0.075%)

d. $a\,B\,C\,d\,e\,f = 0.1 \times 0.05 \times 0.35 = 0.00175$ (0.175%)

e. $a\,b\,c\,D\,e\,F = 0.4 \times 0.05 \times 0.15 = 0.003$ (0.3%)

5.19 a. 47.5% each of *DPh* and *dph*; 2.5% each of *Dph* and *dPh*.

b. 23.75% each of *DpH*, *Dph*, *dPH*, and *dPh*; 1.25% each of *DPH*, *DPh*, *dph*, and *dph*.

5.25 a. *FmW* and *fMw* will be the least frequent (because of double crossovers).

b. *MfW* and *mFw* will be the least frequent.

c. *MwF* and *mWf* will be the least frequent.

5.27 a. By doing a three-point mapping analysis as described in the chapter, we find that the order of genes in the chromosome is *dp-b-hk*, with 35.5 mu between *dp* and *b*, and 5.4 mu between *b* and *hk*.

b. (1) The frequency of observed double crossovers is 1.4%, and the frequency of expected double crossovers is 1.9%. The coefficient of coincidence, therefore, is $1.4/1.9 = 0.73$. (2) The interference value is given by $1 -$ coefficient of coincidence, 0.27.

5.31 a. $a^+ c\ b$/$a\ c^+\ b^+$

b. $a^+ c^+\ b^+$/Y

c.

```
      a            c                    b
      |------------|--------------------|
      ←-- 5 --→←------ 10 ------→
```

5.34 a. *a*, *b*, *c*, and *d* are linked on the same chromosome, since the percentage of recombinations is less than 50. *e* is on a separate chromosome because it segregates independently from the four other genes. The map is constructed by calculating the recombination frequency for all possible pairs of *a*, *b*, *c*, and *d*. The map distances allows an order to be determined. The map is shown below.

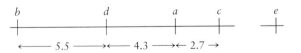

```
  b                 d            a      c           e
  |-----------------|------------|------|-----------|
  ←------ 5.5 ------→←-- 4.3 --→←2.7→
```

b. To get an $a^+ b^+ c^+ d^+ e^+$ fly, there must be a crossover between *b* and *d*, *d* and *a*, and *a* and *c*. ½ of the progeny from those crossovers are $a^+ b^+ c^+ d^+$, and the other ½ are *a b c d*. Then ½ are e^+. The answer is 1.6×10^{-5}, that is, $0.055 \times 0.043 \times 0.027 \times 0.5$ (for the half of the progeny produced by the triple crossover that have all the wild-type alleles) $\times 0.5$ (to give the proportion that are e^+).

CHAPTER 6

6.4

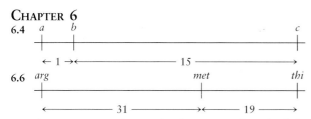

6.6

6.9 The two genes can be considered to assort independently because the number of parental ditype (20) is approximately equal to the number of nonparental ditype (18).

6.12 a. From applying the tetrad analysis formula,

$$\text{map distance} = \frac{\frac{1}{2}T + NPD}{Total} \times 100\%$$

the *a-b* distance is 19.6 map units, the *b-c* distance is 11 map units, and the *a-c* distance is 14 map units. Thus the gene order is *a-c-b*.

b. Centromere distances are calculated from the formula
(% second division segregation tetrads)/2

Since we have all gene-gene distances, all we need to know is whether *a* or *b* is nearer the centromere. *a* is 20.7 map units from the centromere, and *b* is 6 mu. Thus, the map looks like this:

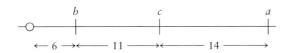

6.13 Map distance is ½ × % of second division segregation asci, which in this case = 5 map units.

6.16 There are three linkage groups, as determined by which blocks of genes always stay together during haploidization: *sm-phe*, *pu-w-ad*, and *y-bi*.

CHAPTER 7

7.1 Strain A is *thr leu*+ and B is *thr*+ *leu*. Transformed B should be *thr*+ *leu*+, and its presence should be detected on a medium containing neither threonine nor leucine.

7.3 The whole chromosome would have to be transferred in order for the recipient to become a donor in an *Hfr* × *F*− cross; that is, the *F* factor in the *Hfr* strain is transferred to the *F*− cell last. This transfer takes approximately 100 min, and usually the conjugal unions break apart before then.

7.4

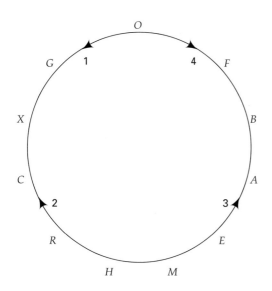

7.9 a. GT   f. ST
b. ST   g. ST
c. ST   h. B
d. GT   i. N
e. GT

7.11 The *trpB* marker cotransduced with the *cysB*+ selected marker more frequently than did the *trpE* marker; thus *trpB* is closer to *cysB*.

7.14 Working from the principle that a relatively high cotransduction frequency means that the genes are relatively close, the order of the genes is *supD-cheA-cheB-eda*.

7.15 Only the volume of phage suspension plated is relevant to the calculation. 0.1 ml of the diluted phage suspension produced 20 plaques, meaning there were 200 phages/ml of the diluted suspension. The dilution factor was 10^8, and so the concentration in the original suspension was 200×10^8 or 2×10^{10}

7.16 0.07 mu. The plaques produced on *K12* (λ) are wild-type *r*+ phages, and in the undiluted lysate there are 470×5 per milliliter (since 0.2 mL was plated), or 2350/mL. The *r*+ phages were generated by recombination between the two *rII* mutations. The other product of the recombination event is the double mutant, and it does not grow on *K12*(λ). Therefore the true number of recombinants in the population is actually equivalent to twice the number of *r*+ phages, since for every wild-type phage produced, there ought to be a doubly mutant recombinant produced. Thus there are 4700 recombinants/mL. The total number of phages in the lysate is $627 \times$ (dilution factor) × (1 mL divided by the sample size plated) per milliliter, or $672 \times 1000 \times 10 = 6,720,000$/mL. The map distance between the mutations is (4700/6,720,000) × 100 = 0.07%.

7.18 Two answers are compatible with the data:

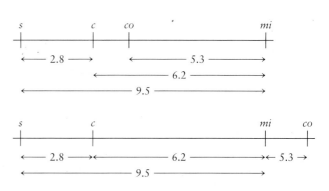

7.20 0.5% recombination (the numbers of plaques counted are so small that this value is a rather rough approximation).

7.23

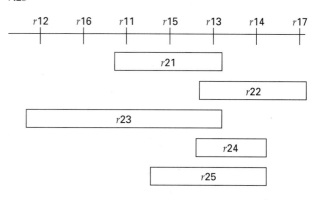

7.25

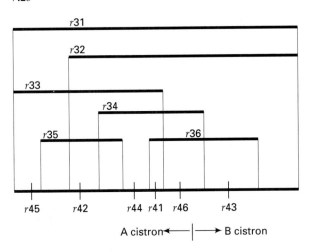

A cistron ◄───|───► B cistron

7.27 a. 3 genes

b. *A*, *D*, and *F* are in one; *B* and *G* are in the second; and *C* and *E* are in the third.

CHAPTER 8

8.2 The double homozygote should have PKU but not AKU. The PKU block should prevent most homogentisic acid from being formed, so that it could not accumulate to high levels.

8.4 a. The simplest approach is to calculate the proportion of the F_2 that are colored and to subtract that answer from one. In this case the noncolorless are the brown or black progeny. The proportion of progeny that make at least the brown pigment is given by the probability of having the following genotype: $a^+/- b^+/- c^+/- (d^+/d^+, d^+/d, $ or $d/d)$. The answer is $3/4 \times 3/4 \times 3/4 \times 1 = 27/64$. Therefore the proportion of colorless is $1 - 27/64 = 37/64$.

b. The brown progeny have the following genotype: $a^+/- b^+/- c^+/- d/d$. The probability of getting individuals with this genotype is $3/4 \times 3/4 \times 3/4 \times 1/4 = 27/256$.

8.6 a. The colorless F_2s are $a/a\ b/b\ c/c$ in genotype, the probability for which is $1/4 \times 1/4 \times 1/4 = 1/64$.

b. $67/256$.

8.8 a. Half have white eyes (the sons) and half have fire red eyes (the daughters).

b. All have fire red eyes.

c. All have brown eyes.

d. All are w^+; a fourth are $bw^+/- st^+/-$, red; a fourth are $bw^+/- st/st$, scarlet; a fourth are $bw/bw\ st^+/-$, brown; a fourth are $bw/bw\ st/st$, the color of 3-hydroxykynurenine plus the color of the precursor to biopterin, or colorless.

8.9 Wild-type T4 will produce progeny phages at all three temperatures. Let us suppose that model (1) is correct. If cells infected with the double mutant are first incubated at 17°C and then shifted to 42°C, then progeny phages will be produced and the cells will lyse. The explanation is as follows: The first step, *A* to *B*, is controlled by a gene whose product is heat-sensitive. At 17°C the enzyme works and *A* is converted to *B*, but *B* cannot then be converted to mature phages because that step is cold-sensitive. When the temperature is raised to 42°C, the *A*-to-*B* step is now blocked, but the accumulated *B* can be converted to mature phages for the enzyme involved with that step is cold-sensitive, and so the enzyme is functional at the high temperature. If model 2 is the correct pathway, then progeny phages should be produced in a 42°-to-17°C temperature shift, but not vice versa. In general, two gene product functions can be ordered by this method when-

ever one temperature shift allows phage production and the reciprocal shift does not, according to the following rules: (1) If a low-to-high temperature results in phages but a high-to-low temperature does not, then the *hs* step precedes the *cs* step (model 1); (2) If a high-to-low temperature results in phages but a low-to-high temperature does not, then the *cs* step precedes the *hs* step (model 2).

8.11 A mutant cell can take up a substance and grow on it only if that substance occurs in the metabolic pathway at a point after the mutant's own block. Since *trpE* can be fed by *C*, *D*, *F* and *A*, *E* is blocked earlier in the pathway than the others. *F* is also earlier than *B*, since *B* can feed *F*. Thus anthranilate synthetase is the first enzyme in the pathway. *F* can feed *C* and *C* can feed *D*. Since *A* and *B* both feed *F*, the order of the steps is *E*, *D*, *C*, *F*, [*AB*]. After anthranilate synthetase, the enzymes are, in order, PPA transferase, PRA isomerase, IGP synthetase and tryptophan synthetase.

8.13 a. $c\ d^+$ and $c^+\ d$

b. The genes are not linked because parental ditype (PD) and nonparental ditype (NPD) tetrads occur in equal frequencies.

c. The pathway is Y to X to Z, with *d* blocking the synthesis of Y and *c* blocking the synthesis of X and Y.

8.16 c linear, folded chains of amino acids

8.18 Baby Joan, whose blood type is O, must belong to the Smith family. Mrs. Jones, being of blood type AB, could not have an O baby. Therefore Baby Jane must be hers.

CHAPTER 9

9.3 d

9.4 (a), (b), and (c) C, N and H are all present both in proteins and in DNA, so label would be located both within and on the surface of the host cell.

9.6 3′ and 5′ carbons are connected by a phosphodiester bond.

9.11 The evidence for only two base pair combinations in DNA, that is A-T and G-C, comes from quantitative measurements of the four bases in double-stranded DNA isolated from a wide variety of organisms. In all cases, the amount of A was shown to equal the amount of T, and the amount of G was shown to equal the amount of C. Moreover, different DNAs exhibit different base ratios (stated as the %GC) so that while A = T and G = C, in most organisms (A + T) does not equal (G + C). The simplest hypothesis, therefore, is that there are two base pairs in DNA, A-T and G-C, and the proportion of the two base pairs varies from organism to organism.

9.13 a. 3′ T C A A T G G A C T A G C A T 5′

b. 3′ A A G A G T T C T T A A G G T 5′

9.14 The adenine-thymine base pair is held together by two hydrogen bonds but the guanine-cytosine base pair is held together by three hydrogen bonds. Thus, the guanine-cytosine base pair requires more energy to break it and is the harder pair to break apart.

9.16 b, c, and d

9.19 C = 17%, therefore G = 17% and the % GC = 34. Therefore, the % AT = 66 and the % A = 66 ÷ 2 = 33.

9.27 a. 200,000

b. 10,000

c. 3.4×10^4 nm

9.28 The probability any group of four bases would be GUUA is $(0.3)(0.25)(0.25)(0.2) = 0.00375$. A molecule 10^6 nucleotides long contains very nearly 10^6 groups of four bases. (The first group of four is bases 1, 2, 3 and 4, the second group of four is bases 2, 3, 4 and 5, etc.) Thus the number of occurrences of GUUA should be about $(0.00375)(10^6) = 3750$.

CHAPTER 10

10.1 a.

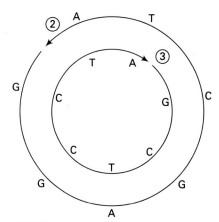

b. 3 $\dfrac{\text{GCTCCTA}}{\text{CGAGGAT}}$ 4

10.3 a. T4 chromosomes are circularly permuted, terminally redundant molecules. Upon denaturation and renaturation, there would be a few perfectly matched, double-stranded molecules, but most molecules would have single-stranded ends. Molecules with very long single-stranded ends might form complex partially circularized structures.

b. T7 chromosomes have a unique sequence that is the same for all molecules. Therefore, all molecules would be perfectly matched.

c. λ chromosomes are linear; each chromosome has the same sequence, and there are single-stranded "sticky ends" that permit the molecules to form circles. Therefore, all molecules would have single-stranded ends or be circularized (to nicked circles).

10.6 a. 9/24 = 0.375. Since S is 37.5% of the cycle, 37.5% of cells should be in S at any instant, and these are the ones that would incorporate label.

b. A little over 4 h. Such cells would have to take up some ^{3}H (a few minutes), pass through G_2 (4 h), and pass through mitotic prophase (several minutes).

c. Both. Each chromatid is a double-stranded DNA molecule containing one old and one newly synthesized DNA strand.

d. A little more than 13 h. Such cells would have to pass through nearly all of S, all of G_2, and through prophase.

10.10

DNA TYPE	CHROMATIN TYPE
Barr body (inactivated DNA)	Facultative heterochromatin
Centromere	Constitutive heterochromatin
Telomere	Constitutive heterochromatin
Most expressed genes	Euchromatin

10.14 See pp. 314–316. Each of the known *Saccharomyces cerevisiae* centromeres has a common core centromere region, which consists of three sequence domains. All three domains are important for centromere function. The centromeres of other eukaryotic organisms that have been characterized are quite different from the *Saccharomyces* centromeres.

10.16 Most protein-coding genes are found in unique-sequence DNA.

10.19 a. Assuming the virus and the C_ot 3×10^4 amphibian species both represent unique sequences, the amphibian genome is 5×10^9 bp. There is a 10^5 fold difference in C_ot between the two species, but the unique sequences are only half of the amphibian genome.

b. The two phases of the curve result from the renaturation of repeated (C_ot 3) and unique (C_ot 3×10^4) sequences.

c. This population has a C_ot 10^4 times lower than the unique sequences, so it must be 10^4 fold more abundant. Thus there are 10^4 copies per genome.

d. The 10^4 copies represent half of the 5×10^9 genome, or 2.5×10^9 bp. Each copy is thus $(2.5 \times 10^9)/10^4$ or 2.5×10^5 bp long.

CHAPTER 11

11.3 Key: ^{15}N – ^{15}N DNA = HH; ^{15}N – ^{14}N DNA = HL; ^{14}N – ^{14}N DNA = LL.

a. Generation 1: all HL; 2: ½ HL, ½ LL; 3: ¼ HL, ¾ LL; 4: ⅛ HL, ⅞ LL; 6: ³¹⁄₃₂ HL, ³¹⁄₃₂ LL; 8: ¹⁄₁₂₈ HL, ¹²⁷⁄₁₂₈ LL

b. Generation 1: ½ HH, ½ LL; 2: ¼ HH, ¾ LL; 3: ⅛ HH, ⅞ LL; 4: ¹⁄₁₆ HH, ¹⁵⁄₁₆ LL; 6: ¹⁄₆₄ HH, ⁶³⁄₆₄ LL; 8: ¹⁄₂₅₆ HH, ²⁵⁵⁄₂₅₆ LL.

11.7 See text, pp. 329–330. DNA polymerase; intact, high-molecular-weight DNA; dATP, dGTP, dTTP, and dCTP; and magnesium ions are needed.

11.8 Two principal lines of evidence show that the Kornberg enzyme is not the enzyme involved in *E. coli* chromosome duplication *in vivo*: First, a mutant that is deficient in the Kornberg enzyme nonetheless grows, replicates the DNA, and divides. Second, other mutants which are temperature-sensitive for DNA synthesis and which do not replicate the DNA or divide have Kornberg enzyme activity at the nonpermissive temperature.

11.11 A primer strand is a nucleic acid sequence that is extended by DNA synthesis activities. A template strand directs the base sequence of the DNA strand being made; for example, an A on the template causes a T to be inserted on the new chain, and so on.

11.12 a. One base pair is 0.34 nm, and the chromosome is 1100 μm. Thus the number of base pairs is (1100/0.34) × 1000 = 3.24×10^6.

b. There are 10 base pairs per turn in a normal DNA double helix; therefore it has a total of 3.24×10^5 turns.

c. 3.24×10^5 turns and 60 min for unidirectional synthesis; therefore $(3.24 \times 10^5/60)$ turns per minute = 5400 revolutions per minute.

d.

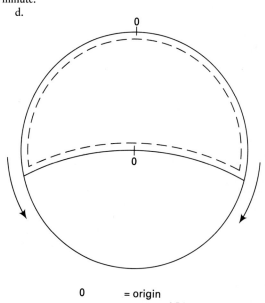

0 = origin

——— = parental DNA

- - - - = new DNA

11.17 See text, pp. 337–339.

11.22 See Tables 11.1 and 11.3.

11.24 Assuming these cells have the typical 4 h G_2 period, you would add ^{3}H thymidine to the medium, wait 4.5 h, and prepare a slide of metaphase chromosomes. Autoradiography would then be done on this chromosome preparation. Regions of chromosomes displaying silver grains are then the late-replicating regions. Cells which were at earlier stages in the S period when they began to take up ^{3}H will not have had sufficient time to reach metaphase.

11.26 See text, pp. 325. Hybrid DNA resulting from branch migration may contain mismatched sequences. Mismatched sequences are recognized by special enzymes that remove a short segment of one DNA strand and then fill in the gap. The mismatch repair does not recognize which is the correct sequence.

11.27 In each case there is evidence that the segregation of one of the alleles in the tetrad has resulted from gene conversion caused by mismatch repair of heteroduplex DNA.

For the $a1\ a2^+ \times a1^+\ a2$ cross the tetrad shows 2:2 segregation of $a1^+$:$a1$ but 3:1 segregation of $a2^+$:$a2$, indicating gene conversion of an $a2$ allele to its wild-type counterpart. Similarly, the $a1\ a3^+ \times a1^+\ a3$ cross shows 2:2 segregation of $a3^+$:$a3$ and 3:1 segregation of $a1^+$:$a1$ resulting from gene conversion of an $a1$ allele to $a1^+$. In the $a2\ a3^+ \times a2^+\ a3$ cross, the $a2$ allele segregates in a Mendelian fashion while the $a3$ allele segregates 3:1 $a3^+$:$a3$ as a result of gene conversion of one $a3$ allele to $a3^+$.

Chapter 12

12.1 The DNA contains deoxyribose and thymine, whereas RNA contains ribose and uracil, respectively. Also, DNA is usually double-stranded, but RNA is usually single-stranded.

12.5 See text, pp. 366–367, 371–375.

12.7 See text, pp. 370–371.

12.8 RNA polymerase I transcribes the major rRNA genes that code for 18S, 5.8S and 28S rRNAs; RNA polymerase II transcribes the protein-coding genes to produce mRNA molecules; and RNA polymerase III transcribes the 5S rRNA genes, the tRNA genes, and the genes for small nuclear RNA (snRNA) molecules.

In the cell the 18S, 5.8S, 28S, and 5S rRNAs are structural components of ribosomes. The mRNAs are translated to produce proteins, the tRNAs bring amino acids to the ribosome to donate to the growing polypeptide chain during protein synthesis, and at least some of the snRNAs are involved in RNA processing events.

12.10 An enhancer element is defined as a DNA sequence that somehow, without regard to its position relative to the gene or its orientation in the DNA, increases the amount of DNA synthesized from the gene it controls (discussed in the chapter on pp. 373–374).

Chapter 13

13.2 The three classes of RNA are mRNA (p. 383), tRNA, (pp. 394–399), and rRNA (pp. 399–403).

13.4 For mRNA: 5′ capping and 3′ polyadenylation in eukaryotes. For tRNA: modification of a number of the bases, removal of 5′ leader and 3′ trailer sequences (if present), and addition of a three-nucleotide sequence (5′-CCA-3′) at the 3′ end of the tRNA. For rRNA in prokaryotes a pre-rRNA molecule is processed to the mature 16S, 23S, and 5S rRNAs. In eukaryotes a precursor molecule is processed to the 18S, 5.8S, and 28S rRNA molecules; 5S rRNA is made elsewhere. The large rRNAs in both types of organisms are methylated. Additionally, the same rRNAs in eukaryotes are pseudouridylated.

13.8 a. This would prevent splicing out of any sequences corresponding to introns, and would result in abnormal sequences of most proteins. This should be lethal.

b. If splicing is not affected, this should have no phenotypic consequence.

c. This would prevent the splicing out of intron 2 material. Since the 5′ cut could still be made, it is likely this mutation would lead to the absence of functional mRNA and the absence of β-globin. (This should produce a phenotype similar to that seen when the β-globin gene is deleted. In that case there is anemia compensated by increased production of fetal hemoglobin. The condition is called β-thalassemia.)

13.10 a. This would be possible because the 5S genes occur in clusters apart from the other rRNA genes.

b. This would not be possible, because the 18S gene copies are interspersed among 5.8S and 28S genes.

c. This would be possible. These three occur in tandem arrays.

d. This would not be possible, because the 5S genes are located separately within the chromosomes.

Chapter 14

14.1 b mRNA

14.9 a. 4 A:6 C

$$AAA = \left(\frac{4}{10}\right)\left(\frac{4}{10}\right)\left(\frac{4}{10}\right) = 0.064, \quad \text{or} \quad 6.4\% \quad \text{Lys}$$

$$AAC = \left(\frac{4}{10}\right)\left(\frac{4}{10}\right)\left(\frac{6}{10}\right) = 0.096, \quad \text{or} \quad 9.6\% \quad \text{Asn}$$

$$ACA = \left(\frac{4}{10}\right)\left(\frac{6}{10}\right)\left(\frac{4}{10}\right) = 0.096, \quad \text{or} \quad 9.6\% \quad \text{Thr}$$

$$CAA = \left(\frac{6}{10}\right)\left(\frac{4}{10}\right)\left(\frac{4}{10}\right) = 0.096, \quad \text{or} \quad 9.6\% \quad \text{Gln}$$

$$CCC = \left(\frac{6}{10}\right)\left(\frac{6}{10}\right)\left(\frac{6}{10}\right) = 0.216, \quad \text{or} \quad 21.6\% \quad \text{Pro}$$

$$CCA = \left(\frac{6}{10}\right)\left(\frac{6}{10}\right)\left(\frac{4}{10}\right) = 0.144, \quad \text{or} \quad 14.4\% \quad \text{Pro}$$

$$CAC = \left(\frac{6}{10}\right)\left(\frac{4}{10}\right)\left(\frac{6}{10}\right) = 0.144, \quad \text{or} \quad 14.4\% \quad \text{His}$$

$$ACC = \left(\frac{4}{10}\right)\left(\frac{6}{10}\right)\left(\frac{6}{10}\right) = 0.144, \quad \text{or} \quad 14.4\% \quad \text{Thr}$$

In sum, 6.4% Lys, 9.6% Asn, 9.6% Gln, 36.0% Pro, 24.0% Thr, and 14.4% His.

b. 4 G:1 C

$$GGG = \left(\frac{4}{5}\right)\left(\frac{4}{5}\right)\left(\frac{4}{5}\right) = 0.512, \quad \text{or} \quad 51.2\% \quad \text{Gly}$$

$$GGC = \left(\frac{4}{5}\right)\left(\frac{4}{5}\right)\left(\frac{1}{5}\right) = 0.128, \quad \text{or} \quad 12.8\% \quad \text{Gly}$$

$$CCG = \left(\frac{4}{5}\right)\left(\frac{1}{5}\right)\left(\frac{4}{5}\right) = 0.128, \quad \text{or} \quad 12.8\% \quad \text{Ala}$$

$$CGG = \left(\frac{1}{5}\right)\left(\frac{4}{5}\right)\left(\frac{4}{5}\right) = 0.128, \quad \text{or} \quad 12.8\% \quad \text{Arg}$$

$$CCC = \left(\frac{1}{5}\right)\left(\frac{1}{5}\right)\left(\frac{1}{5}\right) = 0.008, \quad \text{or} \quad 0.8\% \quad \text{Pro}$$

$$CCG = \left(\frac{1}{5}\right)\left(\frac{1}{5}\right)\left(\frac{4}{5}\right) = 0.032, \quad \text{or} \quad 3.2\% \quad \text{Pro}$$

$$CGC = \left(\frac{1}{5}\right)\left(\frac{4}{5}\right)\left(\frac{1}{5}\right) = 0.032, \quad \text{or} \quad 3.2\% \quad \text{Arg}$$

$$GCC = \left(\frac{4}{5}\right)\left(\frac{1}{5}\right)\left(\frac{1}{5}\right) = 0.032, \quad \text{or} \quad 3.2\% \quad \text{Ala}$$

In sum, 64.0% Gly, 16.0% Ala, 16.0% Arg, and 4.0% Pro.

   c. 1 A:3 U:1 C; the same logic is followed here, using ⅕ as the fraction for A, ⅗ for U, and ⅕ for C.

AAA = 0.008, or 0.8% Lys
AAU = 0.024, or 2.4% Asn
AUA = 0.024, or 2.4% Ile
UAA = 0.024, or 2.4% Chain terminating
AUU = 0.072, or 7.2% Ile
UAU = 0.072, or 7.2% Tyr
UUA = 0.072, or 7.2% Leu
UUU = 0.216, or 21.6% Phe
AAC = 0.008, or 0.8% Asn
ACA = 0.008, or 0.8% Thr
CAA = 0.008, or 0.8% Gln
ACC = 0.008, or 0.8% Thr
CAC = 0.008, or 0.8% His
CCA = 0.008, or 0.8% Pro
CCC = 0.008, or 0.8% Pro
UUC = 0.072, or 7.2% Phe
UCU = 0.072, or 7.2% Ser
CUU = 0.072, or 7.2% Leu
UCC = 0.024, or 2.4% Ser
CUC = 0.024, or 2.4% Leu
CCU = 0.024, or 2.4% Pro
UCA = 0.024, or 2.4% Ser
UAC = 0.024, or 2.4% Tyr
CUA = 0.024, or 2.4% Leu
CAU = 0.024, or 2.4% His
AUC = 0.024, or 2.4% Ile
ACU = 0.024, or 2.4% Thr

In sum, 0.8% Lys, 3.2% Asn, 12.0% Ile, 2.4% Chain terminating, 9.6% Tyr, 19.2% Leu, 28.8% Phe, 4.0% Thr, 0.8% Gln, 3.2% His, 4.0% Pro, and 12.0% Ser. The likelihood is that the chain would not be long because of the chance of the chain-terminating codon.

   d. 1 A:1 U:1 G:1 C; all 64 codons will be generated. The probability of each codon is ¹⁄₆₄, so there is a ³⁄₆₄ chance of the codon being a chain-terminating codon. With those exceptions the relative proportion of amino acid incorporation is directly dependent on the codon degeneracy for each amino acid, and that can be determined by inspecting the code word dictionary.

14.10

	Word size	Number of combinations
a.	5	$2^5 = 32$
b.	3	$3^3 = 27$
c.	2	$5^2 = 25$

(The minimum word sizes must uniquely designate twenty amino acids.)

14.12 a. AAA ATA AAA ATA etc.
   b. TTT TAT TTT TAT etc.
   c. AAA for Phe and AUA for Tyr

14.14 No chain-terminating codons can be produced from only As and Gs. On the other hand, the stop codon UAA can be made from As and Us. Therefore the population A proteins will be longer than those from population B. Most of the population B proteins will be free in solution rather than attached to ribosomes.

14.16 Met-Val-Ser-Ser-Pro-Ile-Gly-Ala-Ala-Ile-Ser . . . (In fact either of the Ile residues might be replaced by Met in a particular molecule.) The normal tRNA recognizes the AUC codon, and therefore carries Ile. The mutant tRNA will recognize the codon AUG and insert Ile there (although the normal tRNA-Met will compete for these sites). The N terminal Met will not be replaced by Ile since it requires the special tRNA-fMet for initiation to occur.

14.18 Normal: Met-Phe-Ser-Asn-Tyr- . . . -Met-Gly-Trp-Val.
   Mutant a: Met-Phe-Ser-Asn
   Mutant b: Starts at later AUG to give: Met-Gly-Trp-Val.
   Mutant c: Met-Phe-Ser-Asn-Tyr- . . . -Met-Gly-Trp-Val.
   Mutant d: Met-Phe-Ser-Lys-Tyr- . . . -Met-Gly-Trp-Val.
   Mutant e: Met-Phe-Ser-Asn-Ser- . . . -Trp-Gly-Gly-Trp . . . (no stop codon; protein continues)
   Mutant f: Met-Phe-Ser-Asn-Tyr- . . . -Met-Gly-Trp-Val-Trp . . . (no stop codon; protein continues)

CHAPTER 15

15.1 At any one position in the DNA, there are four possibilities for the base pair: A-T, T-A, G-C, and C-G. Therefore the length of the base-pair sequence that the enzyme recognizes is given by the power to which four must be raised to equal (or approximately equal) the average size of the DNA fragment produced by enzyme digestion. The answer in this case is 6; that is, 4 to the power of 6 = 4096. If the enzyme instead recognized a four base-pair sequence, the average size of the DNA fragment would be 4 to the power of 4 = 256.

15.5 Genomic libraries were discussed on pp. 442–446, and cloning in a lambda vector by replacing a central section of the λ chromosome with foreign DNA (see p. 439). The steps to make a yeast genomic library are:

1. Isolate high-molecular-weight DNA from yeast nuclei;
2. Perform a partial digest of the DNA with a restriction enzyme that cuts frequently (e.g., *Sau*3A), and isolate DNA fragments of the correct size range for cloning in the λ vector by sucrose gradient centrifugation;
3. Remove the central section of an appropriate λ vector by digestion with *Bam*HI;
4. Ligate the left and right λ arms to the yeast DNA fragments—the *Sau*3A and *Bam*HI sticky ends are complementary (see p. 443);
5. Package the recombinant DNA molecules *in vitro* into λ phage particles;
6. Infect *E. coli* cells with the λ phage population, and collect progeny phages produced by cell lysis. These phages represent the yeast genomic library.

15.7 Use cDNA. Human genomic DNA contains introns. mRNA transcribed off genomic DNA needs to be processed, before it can be translated. Bacteria do not contain these processing enzymes. So in the bacterium, even if the human mRNA were translated, the protein it would code for would not be insulin. cDNA is the complementary copy of a functional mRNA molecule. So when this is the template, its mRNA transcript will be functional and when translated, human (pro-)insulin will be synthesized.

15.10 The restriction map is:

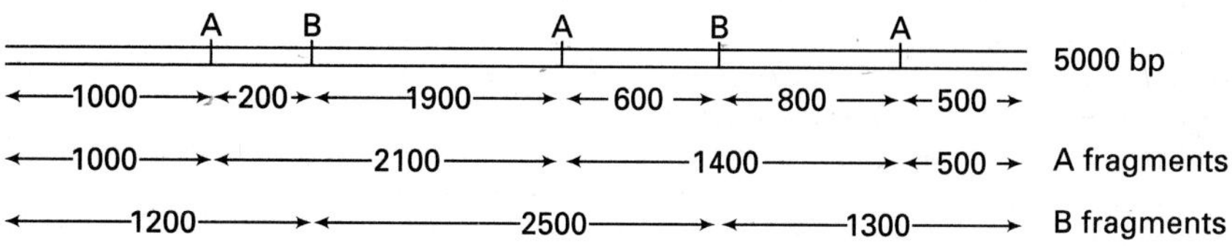

The map is built by considering the large fragments first. For example, the 2500 bp B fragment is cut with A to produce 1900-bp and 600-bp fragments. The 2100-bp A fragment is cut with B to produce the same 1900-bp fragment, and a 200-bp fragment. Thus the 200-bp and 600-bp fragments must be on opposite sides of the 1900-bp fragment. The map is extended in a step-by-step fashion, by next considering other cuts which produced 200-bp fragments, 600-bp fragments, and so on.

15.11 The sequencing gel banding pattern is shown in the following figure:

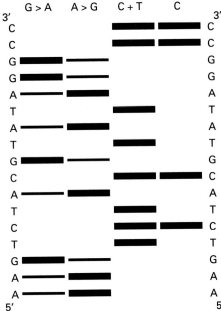

15.13 I-1 is heterozygous for the RFLP and the disease. Inspection of Figure 15.B (p. 474) shows three different "haplotypes" (DNA types) are segregating in this family. Let us name them A, B, and C in decreasing order of size. Let us use *d* for the normal allele of the disease gene, and *D* for the allele causing the disease. I-1 is then *Dd* and AB, but we don't know whether he is *D*A/*d*B or *D*B/*d*A, so we cannot tell which of his offspring are recombinant and which are nonrecombinant. We can tell that I-2 is *d*B/*d*B, so we know the genotypes of II-2 and II-5 are both *D*A/*d*B. II-2 has six offspring. All of these except III-6 have received either *D*A or *d*B, and are non-recombinants. III-6 has received *d* and A, and is a recombinant. Of the six offspring of II-5, III-12 and III-13 have received recombinant chromosomes (*d*A for III-12 and *D*B for III-13). The three offspring of III-2 all received non-recombinant chromosomes, as did three of the four offspring of III-13. However, IV-6 received a recombinant chromosome (*d*B). Overall there are 19 individuals whose recombinant *vs.* non-recombinant status can be ascertained. Of these, four are recombinant. 4/19 = 0.21, which is less than the 0.5 expected from independent assortment. But is it *significantly* less? To find out, we must do the χ^2 test. Independent assortment predicts 19/2 = 9.5 recombinants and the same number of non-recombinants. The value of χ^2 is thus 6.368, with one degree of freedom. The χ^2 table (Chapter 5, p. 139) indicates this difference is significant, so we can conclude the two loci are linked.

15.16 Let us use the symbol "A" to represent the heavier molecular weight band detected by probe A, and "a" to represent the lighter molecular weight band. Similarly, let us use "B" to symbolize the heavier molecular weight band detected by probe B, and "b" to symbolize the doublet of lighter bands. Then the genotype of strain J can be designated as Ab, and that of strain K as aB.

First, the data indicate the region homologous to probe A is linked to that homologous to probe B. This is reflected in the large excess of parental ditype asci (50) over non-parental ditypes (2). Therefore we need to map only one centromere.

The data show that the region homologous to probe A is on the opposite side of the centromere from the region homologous to probe B. This is reflected in the fact that the two single exchange classes (32 asci and 12 asci) show first division segregation for the A region only (12) or for the B region only (32), and you get second division segregation for both the A and B regions only in the double exchange asci (4, 2 and 2).

Considering the distance from region A to the centromere, we see that A shows second division segregation in 20 of the 100 asci. Gene-centromere distance is given by this formula:

$$\frac{\% \text{ second-division segregation asci}}{2}$$

Here, then, the gene-centromere distance is 20%/2 = 10 mu. To calculate this distance another way, the 20 asci contain 80 meiotic products, and half of these have had an exchange between A and the centromere. Thus 40 of a total of 400 meiotic products have experienced an exchange between A and the centromere. 40/400 = 0.1, so the A region is 10 mu from the centromere.

The B region shows second division segregation in 40 asci, out of 100 assayed giving a gene-centromere distance of 40%/2 = 20 mu. Calculating gene-centromere distance another way, the 40 asci contain 80 meiotic products that have inherited chromosomes which have crossed over between B and the centromere. 80/400 = 0.2, so the B region is 20 mu from the centromere.

Given that A is 10 mu from the centromere, and B is 20 mu from the centromere, the one remaining fact to determine is whether A and B are on the same side of a centromere, or on opposite sides. If they were on the same side, the prediction would be that they are about 10 mu apart. If they were on the opposite sides of the centromere, the prediction would be that they are about 30 mu apart. Map distance using tetrad analysis is given by the formula

$$\frac{\frac{1}{2}T + NPD}{\text{Total}} \times 100$$

$$= \frac{\frac{1}{2}(48) + 2}{100} \times 100$$

$$= 26 \text{ mu}$$

Thus our map is

A ←—10—→ mu ←————20————→ mu B

Chapter 16

16.3 A constitutive phenotype can be the result of either an *lacI*⁻ or an *lacO*ᶜ mutation.

16.4 a. A missense mutation results in partial or complete loss of β-galactosidase activity, but there would be no loss of permeases and transacetylase activities. (b) A nonsense mutation is likely to have polar effects unless the mutation is very close to the normal chain-terminating codon for β-galactosidase. Thus permease and transacetylase activities would be lost in addition to the loss of β-galactosidase activity if nonsense mutation occurred near the 5' end of the *lacZ* gene.

16.6 $lacI^+ lacO^c lacZ^+ lacY^-/lacI^+ lacO^+ lacZ^- lacY^+$
(It cannot be ruled out that one of the repressor genes is $lacI^-$)
16.8 Answer is in Table 16.A.
16.11 The CAP, in a complex with cAMP, is required to facilitate RNA polymerase binding to the *lac* promoter. The RNA polymerase binding occurs only in the absence of glucose and only if the operator is not occupied by a repressor (i.e., if lactose is absent). A mutation in the CAP gene, then, would render the *lac* operon incapable of expression because RNA polymerase would not be able to recognize the promoter.
16.14 See text, 498–499. Stringent control is the regulatory mechanism that operates in bacteria to stop essential cellular biochemical activities when the cell is in a harsh nutritional condition, as when a carbon source runs out or when an auxotroph is starved for the nutrient involved. Characteristically, the unusual nucleotides ppGpp and pppGpp accumulate rapidly when the stringent-control system is in effect.
16.17 The *cI* gene codes for repressor protein that functions to keep the lytic functions of the phage repressed when lambda is in the lysogenic state. A *cI* mutant strain would be unable to establish lysogeny, and thus it would also follow the lytic pathway.

CHAPTER 17
17.1 The final product of the rRNA genes is an rRNA molecule. Hence a large number of genes are required to produce the large number of rRNA molecules required for ribosome biosynthesis. Ribosomal proteins, in contrast, are the end products of the translation of mRNAs, which can be "read" over and over to produce the large number of ribosomal protein molecules required for ribosome biosynthesis.
17.2

GLOBIN GENE	TISSUE		
	EMBRYONIC YOLK SAC	SPLEEN	ADULT BONE MARROW
β	equal	less	equal
γ	equal	equal	less
δ	equal	less	equal
ζ	greater	less	less
ϵ	greater	less	less

17.6 The data show clearly that the synthesis of ovalbumin is dependent upon the presence of the hormone estrogen. The data do not indicate at what level estrogen works. Theoretically, it could act to increase transcription of the ovalbumin gene, to stabilize the ovalbumin mRNA, to stabilize the ovalbumin protein, or to stimulate transport of the ovalbumin mRNA out of the nu-

TABLE 16.A

GENOTYPE	INDUCER ABSENT		INDUCER PRESENT	
	β-GALACTO-SIDASE	PERMEASE	β-GALACTO-SIDASE	PERMEASE
a. $lacI^+ lacP^+ lacO^+ lacZ^+ lacY^+$	–	–	+	+
b. $lacI^+ lacP^+ lacO^+ lacZ^- lacY^+$	–	–	–	+
c. $lacI^+ lacP^+ lacO^+ lacZ^+ lacY^-$	–	–	+	–
d. $lacI^- lacP^+ lacO^+ lacZ^+ lacY^+$	+	+	+	+
e. $lacI^s lacP^+ lacO^+ lacZ^+ lacY^+$	–	–	–	–
f. $lacI^+ lacP^+ lacO^c lacZ^+ lacY^+$	+	+	+	+
g. $lacI^s lacP^+ lacO^c lacZ^+ lacY^+$	+	+	+	+
h. $lacI^+ lacP^+ lacO^c lacZ^+ lacY^-$	+	–	+	–
i. $lacI^{-d} lacP^+ lacO^+ lacZ^+ lacY^+$	+	+	+	+
j. $lacI^- lacP^+ lacO^+ lacZ^+ lacY^+/$ $lacI^+ lacP^+ lacO^+ lacZ^- lacY^-$	–	–	+	+
k. $lacI^- lacP^+ lacO^+ lacZ^+ lacY^-/$ $lacI^+ lacP^+ lacO^+ lacZ^- lacY^+$	–	–	+	+
l. $lacI^s lacP^+ lacO^+ lacZ^+ lacY^-/$ $lacI^+ lacP^+ lacO^+ lacZ^- lacY^+$	–	–	–	–
m. $lacI^+ lacP^+ lacO^c lacZ^- lacY^+/$ $lacI^+ lacP^+ lacO^+ lacZ^+ lacY^-$	–	+	+	+
n. $lacI^- lacP^+ lacO^c lacZ^+ lacY^-/$ $lacI^+ lacP^+ lacO^+ lacZ^- lacY^+$	+	–	+	+
o. $lacI^s lacP^+ lacO^+ lacZ^+ lacY^+/$ $lacI^+ lacP^+ lacO^c lacZ^+ lacY^+$	+	+	+	+
p. $lacI^{-d} lacP^+ lacO^+ lacZ^+ lacY^-/$ $lacI^+ lacP^+ lacO^+ lacZ^- lacY^+$	+	+	+	+
q. $lacI^+ lacP^- lacO^c lacZ^+ lacY^-/$ $lacI^+ lacP^+ lacO^+ lacZ^- lacY^+$	–	–	–	+
r. $lacI^+ lacP^- lacO^+ lacZ^+ lacY^-/$ $lacI^+ lacP^+ lacO^c lacZ^- lacY^+$	–	+	–	+
s. $lacI^- lacP^- lacO^+ lacZ^+ lacY^+/$ $lacI^+ lacP^+ lacO^+ lacZ^- lacY^-$	–	–	–	–
t. $lacI^- lacP^+ lacO^+ lacZ^+ lacY^-/$ $lacI^+ lacP^- lacO^+ lacZ^- lacY^+$	–	–	+	–

cleus. Experiments in which the levels of the ovalbumin mRNA were measured have shown that the production of ovalbumin is primarily regulated at the level of transcription.

17.11 four: A_3, A_2B, AB_2, B_3

17.16 Experiment A results in all the DNA becoming radioactively labeled, and so because DNA is a fundamental and major component of polytene chromosomes, radioactivity is evident throughout the chromosomes. Experiment B results in radioactive labeling of RNA molecules. Because radioactivity is first found only around puffs and later in the cytoplasm, we can hypothesize that the puffs are sites of transcriptional activity. Initially, radioactivity is found in RNA that is being synthesized, and the later appearance of radioactivity in the cytoplasm reflects the completed RNA molecules that have left the puffs and are being translated in the cytoplasm. Experiment C provides additional support for the hypothesis that transcriptional activity is associated with puffs. That is, the inhibition of RNA transcription by actinomycin D blocks the appearance of ^{3}H-uridine (which would be in RNA) at puffs. In fact, puffs are much smaller, indicating that the puffing process is intimately associated with the onset of transcriptional activity for the gene(s) in that region of the chromosomes.

17.17 Pre-existing mRNAs (stored in the oocyte) are recruited into polysomes as development begins following fertilization.

17.20

Mutant	Class
a	segmentation gene (segment polarity)
b	maternal gene (anterior-posterior gradient)
c	segmentation gene (gap)
d	homeotic (eye to wing transdetermination)
e	segmentation gene (gap)

17.21 Preexisting mRNA that was by the mother and deposited in the egg prior to fertilization is translated up until the gastrula stage. After gastrulation new mRNA synthesis is necessary for production of proteins needed for subsequent embryo development.

17.22 The tissue taken from the blastula/gastrula has not yet been committed to its final differentiated state in terms of its genetic programming; that is, it is not yet determined. Thus when it is transplanted into the host, the determined tissues surrounding the transplant in the host communicate with the transplanted tissue and cause it to be determined in the same way as they are; for example, tissue transplanted to a future head area will differentiate into head material, and so on. In contrast, tissues in the neurula stage are now determined as to their final tissue state once development is complete. Thus tissue transplanted from a neurula to an older embryo cannot be influenced by the determined, surrounding tissues and will develop into the tissue type for which it is determined, in this case an eye.

Chapter 18

18.1 b

18.3 e None of these (i.e., all are classes of mutations).

18.4 c induction of thymine dimers and their persistence or imperfect repair.

18.6 a. The normal anticodon was 5′-CAG-3′, while the mutant one is 5′-CAC-3′. The mutational event was a CG to GC transversion.

   b. Presumably Leu

   c. Val

   d. Leu

18.8 Acridine is an intercalating agent, and so can be expected to induce frame shift mutations. 5BU, on the other hand, is incorporated in place of T, but is relatively likely to be read as C by DNA polymerase because of keto to enol shift. Thus, 5BU induced mutations would be expected to be point mutations, usually TA to CG transitions. If these expectations are realized, *lac z-1* would probably contain a single amino acid difference from the normal β-galactosidase, although it could be a truncated normal protein due to a nonsense point mutation. *Lac z-2*, on the other hand, should have a completely altered amino acid sequence after some point, and might also be truncated.

18.9 a. Six.

   b. Three. The mutation results in replacement of the UGG codon in the mRNA by UAG, which is a chain termination codon.

18.11 a. UAG: CAG Gln; AAG Lys; GAG Glu; UUG Leu; UCG Ser; UGG Trp; UAU Tyr; UAC Tyr; UAA chain terminating.

   b. UAA: CAA Gln; AAA Lys; GAA Glu; UUA Leu; UCA Ser; UGA chain terminating; UAU Tyr; UAC Tyr; UAG chain terminating

   c. UGA: CGA Arg; AGA Arg; GGA Gly; UUA Leu; UCA Ser; UAA chain terminating; UGU Cys; UGC Cys; UGG Trp

18.13

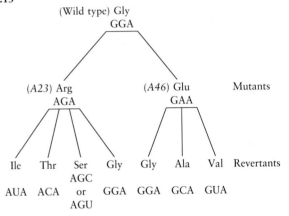

18.16 a. 5BU in its normal form is a T analog; in its rare form it resembles C. The mutation is a AT-to-GC transition.

$$A\text{-}T \rightarrow A\text{-}5BU \rightarrow G\text{-}5BU \rightarrow G\text{-}C$$

   b. Nitrous acid can deaminate C to U, and so the reversion is to the original AT pair.

18.18 The reaction stops when it needs dGTP; only dGMP is present.

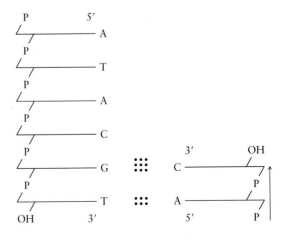

18.21 HNO$_2$ converts A to H, C to U, and G to X.

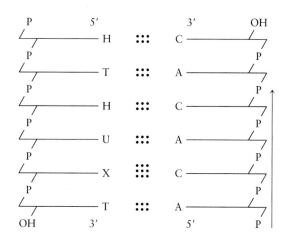

18.24 *ara*$^+$ to *ara*-1: This mutation is CG to AT for it is reverted by base analogs but not by HA or the frameshift mutagen. *ara*$^+$ to *ara*-2: This mutation is AT to GC because it is reverted by base analogs and HA but not by frameshift mutagen. *ara*$^+$ to *ara*-3: This mutation is AT to GC for the same reasons as given for the second mutant. Mutagen X causes transition mutations in both directions because mutants are revertible by base analogs, some are revertible by HA, and none (if this is a representative sample) by frameshift mutagens.

CHAPTER 19

19.1 a. pericentric inversion (D∘E F inverted)
b. nonreciprocal translocation (BC moved to other arm)
c. tandem duplication (EF duplicated)
d. reverse tandem duplication (EF duplicated)
e. deletion (C deleted)

19.2 See text, pp. 592–593.

19.5 a.

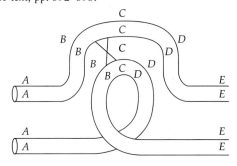

b. From part a the crossover between *B* and *C* is as follows:

```
 A  B  C  D  E

 A    B    C    D    A

 E  B  C  D  E

 A  D  C  B  E
```

c. Paracentric inversion, because the centromere is not included in the inverted DNA segment.

19.8 a. Mr. Lambert is heterozygous for a pericentric inversion of chromosome 6. One of the breakpoints is within the fourth light band up from the centromere, and the other is in the sixth dark band below the centromere. Mrs. Lambert's chromosomes are normal.

b. When Mr. Lambert's chromosome 6s paired, they formed an inversion loop which included the centromere. Crossing over occurred within the loop, and gave rise to the partially duplicated, partially deficient 6, which the child received.

c. The child's abnormalities stem from having three copies of some, and only one copy of other, chromosome 6 regions. The top part of the short arm is duplicated, and there is a deficiency of the distal part of the long arm in this case.

d. The inversion appears to cover more than half of the length of chromosome 6, so crossing over will occur in this region in the majority of meioses. In the minority of meioses where crossing-over has occurred outside the loop or where it has occurred within the loop but the child receives an uncrossed-over chromatid, the child can be normal. There is significant risk for abnormality, so monitoring of fetal chromosomes should be done.

19.9 a. He has a paracentric inversion within the long arm of one of his number 12 chromosomes.

b. Crossing-over within the inversion loop should produce dicentric chromatids, which would form anaphase bridges. These chromatid bridges would be visible where they join the two chromatid masses at anaphase I.

c. The inversion is large, so bridges will be formed in the majority of meioses. Cells that form a bridge do not complete meiosis or form sperm in mammals.

d. Nothing can be done to increase his sperm count.

19.10 a → c → e → d
                      ↘ b

That is, the following sequence occurred:

a. *A B C D E F G H I*
   ↓   └──────↓ Inversion

c. *A B F E D C G H I*
   ↓   └────↓ Inversion

e. *A B F E H G C D I*

d and b are derived from e by separate inversion events:

e. *A B F E H G C D I*
   ↓   └────↓ Inversion

d. *A B F C G H E D I*

and

e. *A B F E H G C D I*
   ↓   └────↓ Inversion

b. *H E F B A G C D I*

19.11

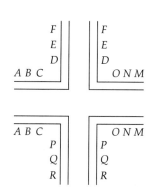

19.13 a. Mr. Denton has normal chromosomes. Mrs. Denton is heterozygous for a balanced reciprocal translocation between chromosomes 6 and 12. Most of the short arm of chromosome 6 has been reciprocally translocated onto the long arm of chromosome 12. The breakpoints appear to be in the thick dark band just above the centromere of 6 and in the third dark band below the centromere of 12.

b. The child received a normal 6 and a normal 12 from his father. In meiosis in Mrs. Denton, the 6s and 12s paired and formed a cross-like figure in prophase I, segregation of adjacent non-homologous centromeres to the same pole occurred, and the child received a gamete containing a normal 6 and one of the translocation chromosomes. (See Figure 19.12.)

c. The child received a normal chromosome 6 and a normal chromosome 12 from Mr. Denton. The child is not phenotypically normal because of partial trisomy (it has three copies of part of the short arm of chromosome 6), and partial monosomy (it has only one copy of most of the long arm of chromosome 12).

d. Given that segregation of adjacent homologous centromeres to the same pole will be relatively rare in this case, there should be about a 50% chance that a given conception will be chromosomally unbalanced. Of the 50% balanced ones, 50% would be translocation heterozygotes.

e. Prenatal monitoring of fetal chromosomes could be done, followed by therapeutic abortion of chromosomally unbalanced fetuses.

19.16 a. 45
b. 47
c. 23
d. 69
e. 48

19.17 b. An individual with three instead of two chromosomes is said to be *trisomic*.

19.18 a. Mother. The color-blind Turner syndrome child must have received her X chromosome from her father because the father carries the only color-blind allele. Therefore, the mother must have produced an egg with no X chromosomes.

b. Father. Because the Turner syndrome child does not have colorblindness, she must have received her X chromosome from the homozygous normal mother. Therefore, the father must have produced a sperm with no X or Y chromosome.

19.20 a. *AA aa*

b. If we label the four alleles *A*1, *A*2, *a*1, and *a*2, there are 6 possible gamete types: *A*1 *A*2, *A*1 *a*1, *A*1 *a*2, *A*2 *a*1, *A*2 *a*2, and *a*1 *a*2. i.e., 1/6 *AA*, 4/6 *Aa*, and 1/6 *aa*. The possible gamete pairings are as follows:

	$\frac{1}{6}$ *AA*	$\frac{4}{6}$ *Aa*	$\frac{1}{6}$ *aa*
$\frac{1}{6}$ *AA*	$\frac{1}{36}$ *AAAA*	$\frac{4}{36}$ *AAAa*	$\frac{1}{36}$ *AAaa*
$\frac{4}{6}$ *Aa*	$\frac{4}{36}$ *AAAa*	$\frac{16}{36}$ *AAaa*	$\frac{4}{36}$ *Aaaa*
$\frac{1}{6}$ *aa*	$\frac{1}{36}$ *AAaa*	$\frac{4}{36}$ *Aaaa*	$\frac{1}{36}$ *aaaa*

Phenotypically this gives 35/36 *A*:1/36 *a*

19.23 (7N + 10N) × 2 = 34 (4N)

CHAPTER 20

20.4 Renaturation is ordinarily dependent on concentration because two single-stranded molecules must collide in order for renaturation to begin. Obviously the probability of collision increases as concentration increases. If renaturation of a particular kind of DNA is independent of concentration, it means that collision is not required for renaturation to occur. This in turn means that the renaturing sequences are on the same single-stranded molecule, indicating closely spaced inverted repeats. Thus the results indicate that 2% of the DNA of this fungus contains closely spaced inverted repeats, some of which may represent transposable elements.

20.6 Since a transposition results in the loss of the intron, it may be hypothesized that transposition in A occurs via an RNA intermediate. That is, intron removal occurs only at the RNA level.

The lack of intron removal during B transposition suggests that in B there is a DNA-DNA transposition mechanism, without any RNA intermediate.

20.7 In engineering the retrovirus in the way he plans, the investigator would probably be creating a new cancer virus, in which the cloned nerve growth factor gene would be the oncogene. It is, of course, an advantage that the engineered virus would not be able to reproduce itself, but we know that many "wild" cancer viruses are also defective and reproduce with the help of other viruses. If the engineered virus were to infect cells carrying other viruses (for example wild-type versions of itself) that could supply the *env* and *gag* functions, the new virus could be reproduced and spread. Presumably, infection of normal nerve cells *in vivo* by the engineered retrovirus would sometimes result in abnormally high levels of nerve growth factor, and thus perhaps to the production of nervous system cancers.

In Avian myeloblastosis virus (AMV) the *pol* and *env* genes are partially deleted. Thus, like our investigator's virus, AMV needs a helper virus to reproduce. In AMV the *myb* oncogene has been inserted, which encodes a nuclear protein presumably involved in control of gene expression. In our new virus, the oncogene would be the cloned nerve growth factor gene.

CHAPTER 21

21.5 The two codes are different in codon designations. The mitochondrial code has more extensive wobble so that fewer tRNAs are needed to read all possible sense codons. As a consequence, fewer mitochondrial genes are necessary. The advantage is that fewer tRNAs are needed, and hence fewer tRNA genes need be present than for cytoplasmic tRNAs.

21.9 The features of extranuclear inheritance are differences in reciprocal-cross results (not related to sex), non-mappability to known nuclear chromosomes, Mendelian segregation not followed, and the indifference to nuclear substitution.

21.12 The parental snails were *D/d* female and *d/—* male. The F$_1$ snail is *d/d*. (Given the F$_1$ genotype, the male can either be homozygous *d* or heterozygous, but the determination cannot be made from the data given.)

21.14 The first possibility is that the results are the consequence of a sex-linked lethal gene. The females would be homozygous for a dominant gene *L* that is lethal in males but not in females. In this case mating the F$_1$ females of an *L/L* × +/Y cross to +/Y males should give a sex ratio of 2 females:1 male in the progeny flies.

The second possibility is that the trait is cytoplasmically transmitted via the egg and is lethal to males. In this case the same F$_1$ females should continue to have *only* female progeny when mated with +/Y males.

21.17 ½ petite, ½ wild-type *(grande)*

21.19 a. Parents: [*po*]F♀ × [N] + ♂; progeny: [*po*]F and [*po*]+ in equal numbers. Because standard *poky* [*po*]+ is found among the offspring, gene *F* must not effect a permanent alteration of *poky* cytoplasm. The 1:1 ratio of *poky* to *fast-poky* indicates that all progeny have the *poky* mitochondrial genotype (by maternal inheritance) and that the *F* gene must be a nuclear gene segregating according to Mendelian principles. Thus the *poky* progeny are [*po*]+, and the *fast-poky* are [*po*]F.

b. Parents: [N]+ × [po]F; progeny: [N]+ and [N]F in equal numbers. These two types are phenotypically indistinguishable.

21.21 a. If normal cytoplasm is [N] and male-sterile cytoplasm is [Ms], then the F_1 genotype is [Ms]Rf/rf, and the phenotype is male-fertile.

b. The cross is [Ms]Rf/rf ♀ × [N]rf/rf ♂ giving 50 percent [Ms]Rf/rf and 50 percent [Ms]rf/rf progeny. Thus, the phenotypes are 1/2 male-fertile and 1/2 male-sterile.

21.22 a. Carlos Mendoza will have inherited his mitochondrial DNA from his mother, and she will have inherited it from her mother. If Mrs. Mendoza and Mrs. Escobar have different mitochondrial RFLPs, then it can be determined which of them contributed mitochondria to Carlos.

b. None of the potential grandfathers need to be tested, since they will not have given any mitochondria to Carlos. In addition, there is no point in testing Mrs. Sanchez. She may have given mitochondria to Carlos' father, but the father did not pass them on to Carlos.

c. If Mrs. Mendoza and Mrs. Escobar do not differ in RFLP, the data will not be helpful. If they do differ, and if Carlos matches Mrs. Mendoza, the case should be dismissed. If Carlos matches Mrs. Escobar, then the Escobar and Sanchez couples are indeed the grandparents, and the Mendozas have claimed a stolen child.

21.24 There are various possibilities. You could isolate the minor peak and examine it using the electron microscope. If the molecules in this peak are circular, they are unlikely to be nuclear fragments. You could grow the yeast in the presence of an intercalating agent such as acridine, and see whether this treatment causes the minor peak species to disappear. If it does, the minor peak is organellar in origin. You could isolate the minor peak, label it (by nick translation, for example), and hybridize this labeled DNA to DNA within suitably prepared yeast cells. Then, using the electron microscope, you could determine whether the label is found over the nucleus or the mitochondria. Finally, you could isolate the minor peak DNA and study its homology to other yeast mitochondria.

Chapter 22

22.1 a. Head width: Mean = 3.15, Standard deviation = 0.49, Wing length: Mean = 35.21, Standard deviation = 7.68.

b. Correlation coefficient = $r = 0.973$

c. Conclusions: Head width and wing length display a strong positive association, meaning that ducks with wider heads tend to have longer wings.

22.3 252/1024

22.4 Each pure-breeding parent is homozygous for the genes (however many there are) controlling the size character, and hence each parent is homogeneous in type. A cross of two pure-breeding strains will generate an F_1 heterozygous for those loci controlling the size trait. Since the F_1 is genetically homogeneous (all heterozygotes), it shows no greater variability than the parents.

22.7 a. 8 cm

b. 1 (2 cm) :6 (4 cm) :15 (6 cm) :20 (8 cm) :15 (10 cm) :6 (12 cm) :1 (14 cm)

c. Proportion of F_2 of AA BB CC is $(1/4)^3$; proportion of F_2 of aa bb cc is $(1/4)^3$; proportion of F_2 with heights equal to one or other is $(1/4)^3 + (1/4)^3 = 2/64$.

d. The F_1 height of 8 cm can be produced only by maintaining heterozygosity at no less than one gene pair (e.g., AA Bb cc). Thus although 20 of 64 F_2 plants are expected to be 8 cm tall, none of these will breed true for this height.

22.10 It is a quantitative trait. Variation appears to be continuous over a range rather than falling into three discrete classes. Also, the pattern in which the F_1 mean falls between the parental means and in which the F_2 individuals show a continuous range of varia-

tion from one parental extreme to the other is typical of quantitative inheritance.

22.13 The cross is aa bb cc (3 lb) × AA BB CC (6 lb), which gives an F_1 that is Aa Bb Cc and which weighs 4.5 lb. The distribution of phenotypes in the F_2 is, 1 (3 lb) :6 (3.5 lb) :15 (4 lb) :20 (4.5 lb) :15 (5 lb) :6 (5.5 lb) :1 (6 lb).

22.15 a. The progeny are Aa Bb Cc Dd, which are 18 dm high.

b. (1) The minimum number of capital-letter alleles is one and the maximum number is four, giving a height range of 12 to 18 dm. (2) The minimum number is one, and the maximum number is four, giving a height range of 12 to 18 dm. (3) The minimum number is four, and the maximum number is six, giving a height range of 18 to 22 dm. (4) The minimum number is zero, and the maximum number is seven, giving a height range of 10 to 24 dm.

22.18 The selection differential is 0.14 g. The selection response is 0.08 g. $h^2 = 0.08/0.14 = 0.57$.

22.19 0, because there would be no genetic variability in the population to affect blood pressure. Blood pressure would vary only because of salt exposure.

22.20 Heritability is specific to a particular population and to a specific environment and cannot be used to draw conclusions about the basis of populational differences. Because the environments of the farms in Kansas and in Poland differ, and because the two wheat varieties differ in their genetic makeup, the heritability of yield calculated in Kansas cannot be applied to the wheat grown in Poland. Furthermore, the yield of TK138 would most likely be different in Poland, and might even be less than the yield of the Russian variety when grown in Poland.

22.22 Selection differential = 14.3 − 9.7 = 4.6

Selection response = 13 − 9.7 = 3.3

Narrow-sense heritability = selection response/selection differential = 3.3/4.6 = 0.72

22.24 a. Fathers: Mean = 71.9, variance = 11.6
Sons: Mean = 70, variance = 10.25

b. Correlation = 0.49

c. Slope = $b = 0.46$

Narrow-sense heritability = $2(b) = 2(0.46) = 0.92$

22.25 Selection response = narrow-sense heritability × selection differential.

Selection response = $0.60 \times 10 \text{ g} = 6 \text{ g}$

Chapter 23

23.1 Brown = $C^B C^B, C^B C^P, C^B C^Y = p^2 + 2pq + 2pr$
$$= 236/500 = 0.472$$

Pink = $C^P C^P, C^P C^Y = q^2 + 2qr = 231/500 = 0.462$
Yellow = $C^Y C^Y = r^2 = 33/500 = 0.066$
$f(C^Y C^Y) = r^2$
$\sqrt{f(C^Y C^Y)} = r^2$
$\sqrt{33/500} = r = \sqrt{0.066} = 0.26$
$f(C^P C^P + C^P C^Y + C^Y C^Y) = q^2 + 2qr + r^2$
$f(C^P C^P + C^P C^Y + C^Y C^Y) = (q + r)^2$
$$\frac{231 + 33}{500} = (q + r)^2$$
$$0.528 = (q + r)^2$$
$$\sqrt{0.528} = q + r$$
$$0.727 = q + r$$
$$r = 0.26$$
$$0.727 - r = q$$
$$q = 0.467$$
$$p = 1 - q - r$$
$$= 1 - 0.467 - 0.26 = 0.273$$
and so, $f(C^B) = p = 0.273$
$f(C^Y) = r = 0.26$
$f(C^P) = q = 0.467$

23.4 a. $\sqrt{0.16} = 0.4 = 40\%$ = frequency of recessive alleles; $1 - 0.4 = 0.6 = 60\%$ = frequency of dominant alleles; $2pq = (2)(0.4)(0.6) = 0.48$ = probability of heterozygous diploids. Then $(0.48)/[(2 \times 0.16) + 0.48] = 0.48/0.80 = 60\%$ of recessive alleles are heterozygotes.

b. If $q^2 = 1\% = 0.01$, then $q = 0.1$, $p = 0.9$, and $2pq = 0.18$ heterozygous diploids. Therefore $(0.18)/[(2 \times 0.01) + 0.18] = 0.18/0.20 = 90\%$ of recessive alleles in heterozygotes.

23.5 $2pq/q^2 = 8$, and so $2p = 8q$; then $2(1 - q) = 8q$, and $2 = 10q$, or $q = 0.2$

23.8 a. Let p = the frequency of S and q = the frequency of s. Then

$$p = \frac{2(188)\ SS + 717\ Ss}{2(3146)} = \frac{1093}{6292} = 0.1737$$

$$q = \frac{717\ Ss + 2(2241)\ ss}{2(3146)} = \frac{5199}{6292} = 0.8263$$

b.

CLASS	OBSERVED	EXPECTED	d	d^2/e
SS	188	94.9	+ 93.1	91.235
Ss	717	903.1	−186.1	38.361
ss	2241	2148.0	+ 93.0	4.032
	3146	3145.9	0	133.628

There is only one degree of freedom because the three genotypic classes are completely specified by two gene frequencies, namely, p and q. Thus the number of degrees of freedom = number of genes − 1. The χ^2 value for this example is 133.628, which, for one degree of freedom, gives a P value less than 0.0001. Therefore the distribution of genotypes differs significantly from the Hardy-Weinberg equilibrium.

23.11 Since the frequency of the trait is different in males and females, the character might be caused by a sex-linked recessive gene. If the frequency of this gene is q, females would occur with the character at a frequency of q^2, and males with the character would occur at a frequency of q. The frequency of males is $q = 0.4$, and thus we may predict that the frequency of females would be $(0.4)^2 = 0.16$ if this is a sex-linked gene. This result fits the observed data. Therefore the frequency of heterozygous females is $2pq = 2(0.6)(0.4) = 0.48$. For sex-linked genes no heterozygous males exist.

23.13 64/10,000 are color blind, i.e., $0.0064 = q^2$, and so $q = 0.08$ = probability of color-blind male.

23.16

$$q = \frac{u}{u + v} = \frac{6 \times 10^{-7}}{(6 \times 10^{-7}) + (6 \times 10^{-8})}$$

$$= \frac{6 \times 10^{-7}}{(6 \times 10^{-7}) + (0.6 + 10^{-7})} = \frac{6}{6.6} = 0.91$$

$$p = 1 - q = 1 - 0.91 = 0.09$$

Thus, the frequencies are 0.008 AA, 0.16 Aa, and 0.828 aa

23.17 a. 100
   b. 96
   c. 36
   d. 7.8

23.20 $p'_{II} = mp_I + (1 - m)p_{II}$
   $p'_{II} = (0.20)(0.5) + (1 - 0.20)(0.70) = 0.66$

23.24 mutation of A to a

23.26 $q = s/(s + t) = 0.12/(0.12 + 0.86) = 0.12$

23.27 $q = u/s = (5 \times 10^{-5})/0.8 = 0.0000625$

23.29 a. The data fit the idea that a single BamHI site varies. The probe is homologous to a region wholly within the 4.1-kb piece bounded on one end by the variable BamHI site, and on the other end by a constant site. When the variable site is present, the hybridized fragment is 4.1 kb. When the variable site is absent, the fragment extends to the next constant BamHI site, and is 6.7 kb long. People with only 4.1 or only 6.7 kb bands are homozygotes, people with both are heterozygotes.

b. The "+" allele of the variable site is present in 2(6) + 38 = 50 chromosomes, and the "−" allele is present in 2(56) + 38 = 150 chromosomes. Thus f(+) is 0.25 and f(−) is 0.75.

c. If the population is in Hardy-Weinberg equilibrium, we would expect $(0.25)^2$, or 0.0625 of the sample to show only the 4.1 kb band. This would be 6.25 individuals. We observed 6. We expect $(0.75)^2$ or 0.5625 to be homozygous for the 6.7 kb band, which is 56.25 individuals. We saw 56. Finally, we would expect 2(0.25)(0.75) or 0.375 to be heterozygotes, or 37.5 individuals. We observed 38. The observed numbers are so close to the expected that a χ^2 test is unnecessary.

23.31 To calculate the percent polymorphic loci we divide the number of loci with more than one allele (2) by the total number of loci examined (5).

$$\frac{2}{5} = 0.40$$

Heterozygosity is calculated by averaging the frequency of heterozygotes for each locus. The frequency of heterozygotes for the $AmPep$, ADH, and LDH-1 loci is zero. 35 out of 50 individuals are heterozygous at the MDH locus and 10 out of 50 individuals are heterozygous at the PGM locus. Thus,

$$\frac{0 + 0 + 0 + \dfrac{35}{50} + \dfrac{10}{50}}{5} = \frac{0.9}{5} = 0.18$$

23.33 a. Leads to change in gene frequencies within a population if no other forces are acting. Introduces new genetic variation. If population size is small, mutation may lead to genetic differentiation among populations.

b. Increases population size and increases genetic variation within populations. Equalizes gene frequencies among populations.

c. Reduces genetic variation within populations and leads to genetic change over time. Increases genetic differences among populations. Increases homozygosity within populations.

d. Increases homozygosity within populations. Decreases genetic variation.

23.35 Comparison with the rates of nucleotide substitution in Table 23.14 indicates that sequence A has as high a rate as that typically observed in mammalian pseudogenes. In addition, the rates of synonymous and nonsynonymous substitutions are the same. These observations suggest that sequence A is either a pseudogene or is a sequence that provides no function, since high rates of substitution are observed when sequences are functionless.

Sequence B has a relatively low rate of nonsynonymous substitution but a relatively high rate of synonymous substitution. This is the pattern we expect when a sequence codes for a protein; thus, sequence B probably encodes a protein.

CREDITS

TEXT CREDITS

Figure 1.8: From Bruce Alberts, et al., *Molecular biology of the cell,* p. 650. Copyright © 1989 by Bruce Alberts, Dennis Bray, Julian Lewis, Martin Raff, Keith Roberts, and James D. Watson. Reprinted by permission of Garland Publishing Inc.

Figure 1.11: From Curtis and Barnes, *Biology,* 5th ed., Worth Publishers, New York, 1989. Reprinted by permission.

Figure 1.28: Reproduced from William T. Keeton and James L. Gould with Carol Grant Gould, *Biological science,* 4th ed. Illustrated by Michael Reingold. Copyright © 1986, 1980, 1979, 1978, 1972, 1967 by W. W. Norton and Company, Inc. Reprinted by permission of W. W. Norton and Company.

Table 4.4: From A. E. Mourant, A. C. Kopec, and K. Domaniewska-Sobczak, *The distribution of the human blood groups,* 2nd ed. (London: Oxford University Press, 1976). Reprinted by permission.

Table 4.6: Reprinted with permission of Macmillan Publishing Company from George W. Burns, *The science of genetics,* 5th ed., p. 44. Copyright © 1983 by George W. Burns.

Figure 5.5: From Ursula Goodenough, *Genetics,* 2nd ed., Fig. 12-1. Copyright © 1978 by Holt, Rinehart and Winston. Reprinted by permission of the publisher.

Table 5.2: Taken from Table IV of Fisher and Yates, *Statistical tables for biological, agricultural, and medical research,* 6th ed., published by Longman Group Ltd. London, 1974 (previously published by Oliver and Boyd Ltd. Edinburgh), by permission of the authors and publishers.

Figure 6.8: Source: Robert H. Tamarin, *Principles of genetics.* Boston: Prindle, Weber and Schmidt Publishers, 1982.

Figure 6.17: From Bruce Alberts, et al., *Molecular biology of the cell.* Copyright © 1983 by Bruce Alberts, Dennis Bray, Julian Lewis, Martin Raff, Keith Roberts, and James D. Watson. Reprinted by permission of Garland Publishing Inc.

Figure 7.31: From W. H. Wood and H. R. Revel, The genome of bacteriophage T4, in *Bacteriological Review* 40(1976):847–68. Reprinted by permission.

Table 8.3: From A. Milunsky, Prenatal diagnosis of genetic disorders. Reprinted by permission of *The New England Journal of Medicine* 295(1976):337.

Figure 9.20: From R. E. Dickerson, The DNA helix and how it is read, in *Scientific American,* vol. 249, December 1983. Reprinted by permission.

Figure 10.6: From Benjamin Lewin, *Genes IV.* Published by Oxford University Press and Cell Press, 1990. Reprinted by permission of Benjamin Lewin.

Figure 10.16: Adapted from R. J. Britten and E. Davidson, Distribution of genome size in animals, in *Quarterly Review of Biology* 46(1971):111. Reprinted by permission.

Figure 10.24: From Bruce Alberts, et al., *Molecular biology of the cell,* p. 498. Copyright © 1989 by Bruce Alberts, Dennis Bray, Julian Lewis, Martin Raff, Keith Roberts, and James D. Watson. Reprinted by permission of Garland Publishing Inc.

Figure 10.29: From Bruce Alberts, et al., *Molecular biology of the cell,* p. 504. Copyright © 1989 by Bruce Alberts, Dennis Bray, Julian Lewis, Martin Raff,

Keith Roberts, and James D. Watson. Reprinted by permission of Garland Publishing Inc.

Figure 10.30: From Bruce Alberts, et al., *Molecular biology of the cell,* Fig. 8-24. © 1983 by Bruce Alberts, Dennis Bray, Julian Lewis, Martin Raff, Keith Roberts, and James D. Watson. Reprinted by permission of Garland Publishing Inc.

Figures 10.34, 10.35, 10.36: From R. J. Britten and D. E. Kohne, Repeated sequences in DNA, in *Science,* vol. 161, August 9, 1978, pp. 529–40. Copyright 1978 by the AAAS. Reprinted by permission.

Table 10.4: From S. C. R. Elgin and H. Weintraub, Chromosomal proteins and chromatin structure. Reproduced, with permission, from *Annual Review of Biochemistry,* vol. 44. Copyright © 1975 by Annual Reviews Inc.

Figure 11.5: From Bruce Alberts, et al., *Molecular biology of the cell,* p. 235. Copyright © 1989 by Bruce Alberts, Dennis Bray, Julian Lewis, Martin Raff, Keith Roberts, and James D. Watson. Reprinted by permission of Garland Publishing Inc.

Figure 11.7: From James E. Watson, et al., *Molecular Biology of the Gene,* 4th ed. Copyright 1965, 1970, 1976, 1987, by the Benjamin/Cummings Publishing Company, Inc. Reprinted by permission.

Figure 11.8: From Bruce Alberts, et al., *Molecular biology of the cell,* p. 232. Copyright © 1989 by Bruce Alberts, Dennis Bray, Julian Lewis, Martin Raff, Keith Roberts, and James D. Watson. Reprinted by permission of Garland Publishing Inc.

Figure 11.9: From Bruce Alberts, et al., *Molecular biology of the cell,* p. 231. Copyright © 1989 by Bruce Alberts,

Figure 23.1: From E. B. Ford, *Ecological genetics*. Reprinted by permission of Chapman and Hall Ltd.

Figure 23.3: From R. K. Koehn, et al. in *Evolution* 30(1976):6. Reprinted by permission of the Society for the Study of Evolution.

Figure 23.10: From P. Buri in *Evolution* 10 (1956):367. Reprinted by permission of the Society for the Study of Evolution.

Figure 23.11: From Philip W. Hedrick, *Genetics of populations*, 1983. Reprinted by permission of Jones and Bartlett Publishers, Inc.

Table 23.3: From R. K. Selander, Genetic variation in natural populations, in F. J. Ayala, *Molecular evolution*, 1976. Reprinted by permission of Sinauer Associates, Inc., Publishers.

Table 23.5: From Monroe W. Strickberger, *Genetics,* 3rd ed. Copyright © 1985 by Monroe W. Strickberger. Reprinted by permission of Macmillan Publishing Company.

PHOTO CREDITS

Unless otherwise acknowledged, all photographs are the property of ScottForesman. Page abbreviations are as follows: (T)top, (C)center, (B)bottom, (L)left, (R)right, (IN-S)inset.

CHAPTER OPENERS

Chapter One:
p. 1: EM of mitosis in root tip of Zea Mays, Dr. Jeremy Burgess/SS/Photo Researchers

Chapter Two:
p. 33: Pea plants with pods, Walter Chandoha

Chapter Three:
p. 60: EM of Drosophila Melanogaster (fruit fly), Dr. Jeremy Burgess/SPL/Photo Researchers

Chapter Four:
p. 92: Snapdragons, E. R. Degginger

Chapter Five:
p. 126: Rat Kangaroo cells in anaphase of mitosis, Mark S. Ladinsky, from *Mitosis* by J. R. McIntosh and Michael P. Koonce, *Science* © 1989, Vol. 246, p. 622, by copyright permission of the AAAS.

Chapter Six:
p. 162: SEM of yeast cells, Dr. Jeremy Burgess/SPL/Photo Researchers

Chapter Seven:
p. 194: EM of bacteriophage which infects E. Coli, Lee Simon/Photo Researchers

Chapter Eight:
p. 233: Computer graphic of hemoglobin molecule from human red blood cell, Laboratory of Molecular Biology/MRC/SPL/Photo Researchers

Chapter Nine:
p. 266: DNA computer graphic, Langride/Dan McCoy/Rainbow

Chapter Ten:
p. 288: False-color light micrograph of normal human chromosomes, CNRI/SPL/Photo Researchers

Chapter Eleven:
p. 325: E. Coli dividing, Tom Broker/Rainbow

Chapter Twelve:
p. 361: Immunofluorescence photo of polytene puffs on chromosomes, Dr. John T. Lis, Cornell University

Chapter Thirteen:
p. 381: Spliceosomes, Jack Griffith

Chapter Fourteen:
p. 409: Computer graphic of enzyme, t-RNA synthetase, Physics Dept./Imperial College, London/SPL/Photo Researchers

Chapter Fifteen:
p. 430: False-color EM of plasmids, Dr. Sopal Murti/SPL/Photo Researchers

Chapter Sixteen:
p. 476: Model of C1 Lambda repressor & operator, Wayne F. Anderson, Department of Biochemistry, Vanderbilt University, Nashville, TN.

Chapter Seventeen:
p. 514: Drosophila Melanogaster embryo, Courtesy Dr. Matthew Scott/Kim Schuske/Stanford University

Chapter Eighteen:
p. 556: Movie poster from "Them," The Kobal Collection/SuperStock International

Chapter Nineteen:
p. 587: DNA probe for prenatal diagnoses of Down Syndrome, Courtesy Oncor, Inc., Photo by H. F. Willard, Ph.D.

Chapter Twenty:
p. 611: False color EM of AIDS virus particles inside stricken cell, Prof. Luz Montagnier, Institut/Photo Researchers

Chapter Twenty-one:
p. 642: Four o'clock flowers, Eric L. Heyer/Grant Heilman Photography

Chapter Twenty-two:
p. 670: Milkweed bugs, Bruce S. Cushing/VU

Chapter Twenty-three:
p. 707: Monarch butterfly, Larry West/Bruce Coleman, Inc.

IN-CHAPTER CREDITS

4BC A. N. Eroars, B. J. Panessa, and J. F. Gennaro, Jr.

4TL Valentine, R. C. & Pereira, H. G., "Antigens and Structure of the Adenovirus," *J.Mol. Biol.*, 13,13–20. © 1965 by copyright permission of Academic Press, Inc., Orlando, Fl

4TR K. G. Murti/Visuals Unlimited

4CL Centers for Disease Control, Atlanta

4CR K. G. Murti/Visuals Unlimited

4BL K. G. Murti/Visuals Unlimited

4BR David M. Phillips/Visuals Unlimited

5TL David Scharf/Peter Arnold, Inc.

5CL Lederle Laboratories/Photo Researchers

5CR Omikron/Photo Researchers

5R J. J. Duda & J. M. Slack, West Virginia University/Biological Photo Service

5B T. J. Beveridge, University of Guelph/Biological Photo Service

6TC J. B. Woolsey Associates

6TL J. Forsdyke/Gene Cox/SPL/Photo Researchers

6TR Supplied by Carolina Biological Supply Company

6CL John Colwell/Grant Heilman Photography

6CR Larry Lefever/Grant Heilman Photography

6BL Grant Heilman Photography

6BC Courtesy John Sulston, Medical Research Council/Laboratory of Molecular Biology

7 John D. Cunningham/Visuals Unlimited

8T M. C. Ledbetter/Brookhaven National Laboratory and NYU Medical Center

8B R. Rodewald, University of Virginia/Biological Photo Service

9T D. D. Kunkel, University of Washington/Biological Photo Service

9B Courtesy of K. E. Carr, University of Glasgow, and P. G. Toner, Glasgow Royal Informary. From *Cell structure,* p. 169. © 1982 Longman Group Limited. Used by permission of Churchill Livingstone, London.

10T H. S. Pankratz, Michigan State University/Biological Photo Service

10B G. T. Cole, University of Texas, Austin/Biological Photo Service

11T P. W. Johnson & G. McN. Sieburth, University of Rhode Island/Biological Photo Service

549C F. R. Turner, Indiana University, Bloomington

549B From David Suzuki et al., *Introduction to genetic analysis.* Copyright 1986, W. H. Freeman and Company. Used with permission.

597 Mary Elene Trarter

653R Biophoto Associates/Photo Researchers

653L Supplied by Carolina Biological Supply Company

655 Kim Taylor/Bruce Coleman, Inc.

657 E. R. Degginger

721 © Chip Clark

730 Photo: Hildegard Adler

737T Kevin Byron/Bruce Coleman, Inc.

737B Hal Clason M./Tom Stack & Associates

738L C. B. & D. W. Frith/Bruce Coleman, Inc.

738R Down House/The Royal College of Surgeons of England

739ALL Breck P. Kent

INDEX

Page numbers in *italics* indicate material in figures and tables.